Experience Psychology

Fifth Edition

LAURA A. KING

University of Missouri, Columbia

McGraw Hill

EXPERIENCE PSYCHOLOGY

Published by McGraw Hill LLC, 1325 Avenue of the Americas, New York, NY 10019. Copyright ©2022 by McGraw Hill LLC. All rights reserved. Printed in the United States of America. No part of this publication may be reproduced or distributed in any form or by any means, or stored in a database or retrieval system, without the prior written consent of McGraw Hill LLC, including, but not limited to, in any network or other electronic storage or transmission, or broadcast for distance learning.

Some ancillaries, including electronic and print components, may not be available to customers outside the United States.

This book is printed on acid-free paper.

1 2 3 4 5 6 7 8 9 LWI 26 25 24 23 22 21

ISBN 978-1-266-13819-5
MHID 1-266-13819-6

Cover Image: *Dimitris66/Getty Images*

mheducation.com/highered

Courtesy Lisa Jensen

Laura A. King

Laura King did her undergraduate work at Kenyon College, where she began studying toward an English major. In the second semester of her junior year, she declared a second major in psychology. She completed her A.B. in English with high honors and distinction and in psychology with distinction in 1986. Laura then did graduate work at Michigan State University and the University of California, Davis, receiving her Ph.D. in personality psychology in 1991.

Laura began her career at Southern Methodist University (SMU) in Dallas, moving to the University of Missouri in 2001, where she is now a Curators' Distinguished Professor of Psychology. In addition to seminars in the development of character, social psychology, and personality psychology, she has taught undergraduate lecture courses in introductory psychology, introduction to personality psychology, social psychology, and the Art and Science of Living. She has also taught the Psychology of the Good Life for the community at large. At SMU, she received six different teaching awards, including the "M" award for "sustained excellence" in 1999. At the University of Missouri, she received the Chancellor's Award for Outstanding Research and Creative Activity in 2004.

Her research, which has been funded by the National Institute of Mental Health and the National Science Foundation, has focused on a variety of topics relevant to the question of what it is that makes for a good life. She has studied goals, life stories, happiness, well-being, and meaning in life. In general, her work reflects an enduring interest in understanding what is good and healthy in people.

She has published over 100 articles and chapters (typically with graduate and undergraduate student collaborators). She received the Carol and Ed Diener Award for Distinguished Contributions to Personality in 2011; the award for Distinguished Service to SPSP in 2015; and the 2018 Jack Block Award for Distinguished Research in Personality Psychology, the highest honor in her field. She was elected President of the Society for Personality and Social Psychology in 2021. Laura has served as editor or associate editor for a number of journals and was the first woman to edit the *Journal of Personality and Social Psychology: Personality Processes and Individual Differences,* the top outlet in personality and social psychology.

In "real life," Laura is an accomplished cook and enjoys listening to music (mostly jazz vocalists and singer-songwriters), running with her faithful dog John, and learning about advanced calculus from her son, Sam.

about the author

For Sam

brief contents

connect |PSYCHOLOGY McGraw Hill Education Psychology APA
Documentation Style Guide

contents

9 Motivation and Emotion   331

10 Personality   369

14   Health Psychology   527

connect
|PSYCHOLOGY   McGraw Hill Education Psychology APA
Documentation Style Guide

guide to diversity, equity, and inclusion

Throughout the text, numerous changes have been made to provide more inclusive language and more diverse examples when discussing human groups—including those based on gender, gender identity, sexuality, disability status, veteran status, immigrant status, and religious faith. New photos for the fifth edition were chosen to ensure representation of the broad range of people who are of interest to psychologists, including stigmatized groups such as people with overweight, people with disabilities, and people of gender and sexual minorities. In addition, we strived to include images that show people of different backgrounds interacting in nonstereotypical ways, such as white people being helpful to Black people. The fifth edition revision was completed with an emphasis on combating anti-Black racism and ensuring that all students—regardless of gender, gender identity, sexual orientation, religious faith, disability status, immigrant status, indigenous identities, socioeconomic status, military veteran status—would recognize themselves in *Experience Psychology*.

CHAPTER 1 THE SCIENCE OF PSYCHOLOGY

- Example of forgiveness drawn from the Amish was replaced with civil rights hero John Lewis, and it includes a description of the man who beat Lewis and his statements about his own racist views.

- *Challenge Your Thinking: Where Is Everybody?* confronts the lack of diversity in the history of psychology, as well as in contemporary science (Roberts & others, 2020). Scholars who are highlighted include Mary Whiton Caulkins, Henry Turner, and Inez Beverly Prosser. The feature box notes that to address the pressing problems facing psychology we need everyone included in the conversation—regardless of not only race and gender but disability status, gender identity, and sexual orientation.

- *Psychology in Our World: Big Data in Psychological Science* reviews the ways big data has been used to measure racism and health disparities and to estimate the number of sexual and gender minority people.

- A summary is provided on research (Biswas-Diener & others, 2005) into happiness in groups of people not generally included in psychological studies—Inughuits, Maasai, and Old Order Amish.

- The concept of WEIRD samples is introduced and defined.

- The section on values in psychological research focuses on the example of research on gay and lesbian parents (Farr & others, 2019, 2020).

CHAPTER 2 THE BRAIN AND BEHAVIOR

- The discussion of the links between heredity and environmental influences reviews research on how genetic essentialism impacts our judgments of others, including people with psychological disorders or with diverse sexual orientations (Lynch & others, 2019).

- A review of gender differences is provided related to the neurotransmitter oxytocin (Feldman & Bakermans-Kranenburg, 2017; Lieberz & others, 2020; von Dawans & others, 2019).

CHAPTER 3 SENSATION AND PERCEPTION

- Throughout the chapter, biases and shortcuts that characterize perception are pointed out to be problematic when applied to people.
- Expectations are discussed in terms of person perception, gender, race/ethnicity, wealth indicators, and other factors, tying perceptual processes to the diverse world. The impact of stereotypes on accurate perception is presented (Moradi & others, 2020; Petsko & Bodenhausen, 2020).
- Cultural influences on perception are discussed, highlighting research on differences between Japanese and American study participants (Masuda & Nisbett, 2001, 2006).
- Morgan Freeman and James Earl Jones are used as examples for the concept of timbre.
- The chapter reviews research on the undertreatment of pain in Black patients, including research on lower painkiller prescription rates for broken bones (Todd & others, 2000) and appendicitis in children (Goyal & others, 2015). Cultural (Feng & others, 2020; Onishi & others, 2017) and gender (Day & others, 2020; Nascimento & others, 2020) differences in reporting pain are also discussed.
- *Intersection: Sensation and Perception and Social Psychology: Do False Beliefs about Race Affect Pain Treatment?* reviews research on false beliefs about racial differences in pain perception among medical students and residents (Hoffman & others, 2016). The key factor in these false beliefs was ingroup empathy, suggesting the need for more diversity in the medical profession (Hagedorn & others, 2020).
- Cultural impacts on the sensation of taste are presented, including the example of the role of umami in Asian cuisine.

CHAPTER 4 STATES OF CONSCIOUSNESS

- The concept of dehumanization is introduced.
- *Intersection: Consciousness and Social Psychology: Could Robots Make Us Kinder to Each Other?* reviews research showing that reminders of robots that boost a sense of shared humanity reduce outgroup animosity (Castelo & others, 2020).
- The link between lovingkindness meditation and reductions in prejudice is reviewed (Kang & others, 2014; Price-Blackshear & others, 2017).
- The discussion of alcohol use disorder has been revised to remove objectifying labels.

CHAPTER 5 LEARNING

- The chapter-opening vignette focuses on service dogs trained to assist military veterans who have trauma-related physical disabilities or psychological disorders.
- *Intersection: Learning and Social Psychology: Can Observational Learning Lead to Bias?* highlights research by an outstanding scholar of color, Sylvia Perry, examining how observing someone react negatively to a target can lead observers to be biased toward that target (Skinner & Perry, 2020).
- The discussion of modeling includes consideration of the idea that if similarity to a model promotes learning, then we need more diverse models in the world. The section points to lack of diversity in STEM as well as the new role models provided by the elections of President Barack Obama in 2008 and Vice President Kamala Harris in 2020.
- Research is reviewed on how multicultural experiences enhance insight learning (Leung & others, 2008). The chapter also reviews the impact of cultural values on the application of learning processes and the content of learning.

CHAPTER 6 MEMORY

- Review of evidence for the role of racial prejudice and other forms of bias in faulty memory (Behrman & Davey, 2001; Brigham & others, 2007; Loftus, 1993).

- *Psychology in Our World: Using Psychological Research to Improve Police Lineups* describes strategies for reducing the effects of bias in eye-witness identifications (Brewer & Wells, 2011; Wells & others, 2020).

CHAPTER 7 THINKING, INTELLIGENCE, AND LANGUAGE

- The problem of using prototypes to make social judgments is reviewed, including examples of a woman candidate to be a CEO and a Latina girl wanting to be a doctor.

- *Challenge Your Thinking: Can Artificial Intelligence Be Racist?* reviews how AI systems are created and how although it might seem that computers would be less biased than people, they are not (Cossins, 2018; Heaven, 2020).

- Cultural differences in definitions of intelligence are reviewed.

- A critique is provided of Robert Sternberg's definition of intelligence as "the capacity to success in whatever context" as circular and biased and supporting the status quo. The discussion highlights that women, people of color, indigenous people, and people with disabilities may have extraordinarily high ability but still not succeed in an unfair system.

- The science of intelligence is acknowledged as having "a difficult, even shameful, often racist history." The link between intelligence and eugenics is more explicitly examined (Winston, 2020). Spearman's idea of requiring intelligence before allowing people to be parents is discussed/critiqued.

- Cultural bias in testing is discussed, including rural-urban differences and disadvantages for people of diverse backgrounds who speak no English or nonstandard English (Cathers-Shiffman & Thompson, 2007; Scarr, 1984). The difficulties in creating culture-fair intelligence tests are reviewed.

- The section "Genes, Environment, and Group Differences" reviews race as a social construction, the meaning of skin color and its general irrelevance to genes associated with intelligence, the influence of skin color on access to resources such as education, nutrition, healthcare, and related factors. The text also highlights racial disparities in wealth (Kraus & others, 2019) and the role of financial differences in enduring differences in intelligence (Weiss & Saklofske, 2020).

- The chapter dismantles the hereditarian argument about group differences in IQ by pushing students to recognize the role of the environment in characteristics regardless of their heritability. It emphasizes that because encountering complex environments leads to more insight learning, our diverse world is pulling us to higher levels of cognitive sophistication.

- A review is provided of prejudice and bias in testing for giftedness, suggesting that bias in who gets tested helps to explain the lack of diversity in gifted education, in terms of race/ethnicity, language differences, physical disabilities, and learning disabilities (Carman, 2011).

- Forms of classification of intellectual disability are reviewed (Giesbers & others, 2020; McNicholas & others, 2017), including the use of assessment of capacities in addition to IQ scores (American Association on Intellectual and Developmental Disabilities, 2010).

- The discussion of the bilingual advantage critiques the idea that for bilingualism to be a positive thing it must be linked to some external cognitive skills. The text emphasizes that regardless of its other correlates, knowing more than one language is good skill.

- A critique is provided of the "30-million word gap" as a way to think about socioeconomic differences in verbal ability and academic performance.

CHAPTER 8 HUMAN DEVELOPMENT

- Diverse examples are included throughout. Examples provided of people with optimal life themes include Martin Luther King Jr., Mother Teresa, Nelson Mandela, and Oprah Winfrey. Nelba Marquez-Greene is featured in a self-quiz question about life themes. The Japanese island of Okinawa is used as an example in the discussion of aging.

- The discussions of puberty, pregnancy, and gender feature gender-inclusive language and a diversity of examples of identities and experiences.

- Vygotsky's sociocultural approach is reviewed.

- Important women scholars are highlighted, including Elizabeth Spelke (infant cognitive development), Mary Ainsworth (infant socioemotional development), Carol Gilligan (moral development), and Elisabeth Kübler-Ross (death and dying).

- The cultural context of parenting is reviewed (Pinquart & Kauser, 2018).

- The importance of the value of diverse identities in adolescence is stressed with citations to research showing the importance of such identities to minoritized groups, including racial and ethnic minorities, sexual and gender minorities, disability-related groups, religious minorities, and immigrant groups (Pohjola, 2020; Quam & others, 2020; Raifman & others, 2020). The concept of biculturalism is discussed (Ferguson & others, 2020; Romero & others, 2020).

- The risk of bullying for sexual and gender minority youth is reviewed (Kaufman & others, 2020) along with the importance of supportive families and communities (Bouris & Hill, 2017; Hall, 2017). The section ultimately concludes, "Creating a context where all identities are celebrated allows youth from all different groups to find safe, nurturing, and positive context to be themselves."

- John Gottman's inclusion of same-gender couples in his long-running research on successful relationships is noted.

- The "Gender Development" section has been updated to reflect inclusive language and includes revised definitions for gender, gender identity, sexual orientation, gender expression, and gender roles.

- Research on differences in infant toy preferences as a function of assigned gender is revised and expanded to include recent evidence that calls large differences into question (Davis & Hines, 2020).

- *Psychology in Our World: Human Identities and the Changing Gender Landscape* reviews the ways that our understanding of gender has changed, rapidly. Elliot Page provides an example.

CHAPTER 9 MOTIVATION AND EMOTION

- *Intersection: Developmental Psychology and Social Psychology: Are There "Boy Foods" and "Girl Foods"?* highlights the gendered nature of eating (Graziani & others, 2020).

- The concept of affirmative consent is reviewed in the discussion of cognitive and sensory/perceptual factors in sexuality.

- Cultural factors in sexuality are reviewed, including classic studies by Messenger (1971) and Marshall (1971).

- The section "Sexual Behavior and Orientation" reviews definitions of sexual behavior that consider the diversity of sexual orientations. The section acknowledges in the discussion (and with a photo) the place of sexual activity in a fulfilling life for people with disabilities (McGrath & others, 2020).

- Gender differences in sexual attitudes and behavior are reviewed, including issues with study design that have called some earlier findings into question (Conley, 2011).

- The section on sexual orientation includes revised definitions of gay, lesbian, bisexual, pansexual, and asexual orientations. It also notes that any explanation of sexual orientation must explain *heterosexuality.*

- Research on gay and lesbian well-being and parenting is reviewed (Farr & Patterson, 2013; Golombok & others, 2014; Oakley & others, 2017; Patterson, 2013, 2014; Patterson & Farr, 2014; Sumontha & others, 2017).

- Maslow's hierarchy is critiqued for classism, and the text notes that Maslow's perspective appears to derive from the Blackfoot Nation, a First Nation of Canada.

- The influence of culture in emotional expression is reviewed, including sociocultural display rules (Hudson & Jacques, 2014; Zhu & others, 2013).

CHAPTER 10 PERSONALITY

- Freud's approach is noted as situated in the gender binary of his time period.

- Cross-cultural studies in research on personality traits are reviewed (De Raad & others, 2010; Lovik & others, 2017; Paunonen & others, 1992; Thalmayer & others, 2020; Zhou & others, 2009).

- In the discussion of Maslow, his list of self-actualized individuals is noted as biased, limited to people who were able to be successful in society in a particular historical context—mostly men, mostly white people, and not including people of diverse gender identities or sexual orientations or people with disabilities.

- *Intersection: Personality and Social Psychology: Do Personality Traits Predict Prejudice?* reviews recent research on traits and other individual differences as predictors of racism (Banton & others, 2020; Esses, 2020; Federico & Aguilera, 2019; Kocaturk & Bozdag, 2020; Lin & Alvarez, 2020; Marsden & Barnett, 2020; Metin-Orta & Metin-Camgöz, 2020; Stern & Crawford, 2020; Ziller & Berning, 2019).

- *Challenge Your Thinking: Does It Really Matter How Long a Child Waits for That Second Treat?* reviews the role of social class in delay of gratification among young children (Watts & others, 2018).

- The chapter notes issues with Hans Eysenck as a scholar who promoted ideas about genetic explanations for group differences in IQ.

CHAPTER 11 SOCIAL PSYCHOLOGY

- This chapter, of course, contains a plethora of topics and studies relevant to issues of prejudice, racism, stereotype threat, stereotyping, ethnocentrism, and other topics. The chapter narrative and photo program highlight the diversity of important scholars in social psychology.

- Stereotype threat (Steele, 1997, 2012; Steele & Aronson, 1995, 2004) is reviewed, including discussion of which groups are impacted (Cadaret & others, 2017; Gonzalez & others, 2020; Jordano & Touron, 2017; Lewis & Sekaquaptewa, 2016; Robinson & others, 2020; Wegmann, 2017; Weiss & Perry, 2020), when it is activated, what impact it can have (Bullock & others, 2020; Kalokerinos & others, 2017), and what interventions might be effective (Liu & others, 2020; Vallée & others, 2020).

- Sociocultural influences on aggression are discussed, including cultural norms about masculine pride and family honor and cultures of honor (Cohen, 2001; Cohen & others, 1996; Gul & others, 2020; Hadi, 2020; Nawata, 2020).

- Cross-cultural research on interpersonal attraction is reviewed.

- The lack of diverse samples in close relationship research is highlighted.

- Coverage of deindividuation (Levine & others, 2010) includes consideration of KKK as well as multiple perpetrator rape, highlighting the potential role of a leader whose voice matters to the group (Woodhams & others, 2020).

- The importance of social and ethnic identities is reviewed, including as a resource for coping with bias and injustice (Crocker & others, 1998; Marks & others, 2015).

- The discussion of prejudice includes expanded coverage of anti-Black prejudice, systemic racism (Liverpool, 2020), and the continuing influence of race on many aspects of U.S. life (Bertrand & Mullainathan, 2004; Copur-Gencturk & others, 2020; Eberhardt, 2020; Ge & others, 2020; Shin & others, 2016). Prejudice against people of Asian descent during the COVID-19 pandemic is also noted.

- Microaggressions are defined and discussed, with examples and impacts (Sue, 2010; Williams, 2020).

- The discussion of implicit racism highlights each person's personal responsibility for working on their own biases and not assuming that because such prejudice is automatic it is somehow beyond personal accountability.

- Research on factors that may improve intergroup relations is reviewed (Onyeador & others, 2020; Shelton & Richeson, 2014), including specific types of intergroup contact (Hässler & others, 2020; Marinucci & others, 2020; Mousa, 2020; Pettigrew & Tropp, 2006; Stott & others, 2012; Turner & others, 2020; White & others, 2020). The example of Christian refugees and Muslims playing on soccer teams together in Iraq is noted (Mousa, 2020).

- A final section on "Breaking the Prejudice Habit" describes research on the hard work of reducing prejudice (Devine & others, 2012; Paluck & others, 2021).

CHAPTER 12 PSYCHOLOGICAL DISORDERS

- The chapter-opening vignette focuses on a student who experienced a psychotic break, was diagnosed with bipolar disorder, and went on to develop a play about mental illness that aims to overcome stigma (Porter, 2019).

- Diverse examples are included throughout: for example, Lin Manuel Miranda, Naomi Osaka, and Patrick Mahomes represent atypical people who are not "abnormal"; choreographer Alvin Ailey is provided as an example of a person with bipolar disorder.

- The cultural context of disorders reviews how people struggling for social justice have often been labeled deviant or mentally ill (Potter, 2012) as well as the misuse of psychological diagnoses in immigrants who differ culturally (Alallawi & others, 2020; Carney, 2020).

- The idea of valuable neurodiversity is introduced (Bertilsdotter Rosqvist & others, 2019; Chapman, 2020; Ortiz, 2020) in the discussion of autism spectrum disorder.

- Patterns of associations with race, gender, and socioeconomic status in the diagnoses of neurodevelopmental disorders are reviewed (Maenner & others, 2020).

- The section on trauma and stress-related disorders includes coverage of sexual assault, sexual victimization on campus, and their links to PTSD (Coulter & Rankin, 2020; Gilmore & others, 2017; Hannan & others, 2020).

- *Psychology in Our World: Sexual Victimization on Campus* outlines statistics about sexual violence on campus among diverse groups (Cantor & others, 2019), Title IX protections for students, and suggested interventions.

- Sociocultural factors in depression are reviewed, including socioeconomic status and gender (Hodes & others, 2017; Joshi & others, 2017; Kisely & others, 2017; Linder & others, 2020; Lorant & others, 2007; Salk & others, 2017).

- Figure 3 reviews gender differences in depression rates in different cultures. Gender differences in rates of eating disorders are also discussed.

- Sociocultural factors in schizophrenia highlight the role of poverty and industrialization (Jablensky, 2000; Myers, 2010).

- Rates of suicide are reviewed for U.S. groups, with special attention to Native American/Native Alaskan adolescents (Hoffmann & others, 2020; Kerr & others, 2017). Worldwide rates of suicide are also discussed (World Population Review, 2018).

- The final section of the chapter addresses combating stigma. The concept of illusory correlation and prejudice against those with psychological disorders (stereotyped as "dangerous") is explicitly confronted.

- The Intersection feature *Clinical Psychology and Social Psychology: How Does the Stigma of Mental Illness Affect Social Interactions?* examines how stigma affects social interactions (Lucas & Phelan, 2019).

CHAPTER 13 THERAPIES

- The controversy over applied behavior analysis for the neurodiversity community is noted.
- *Challenge Your Thinking: Who Should Decide What Treatment Is Best for a Person?* addresses the importance of autonomy among people with psychological disorders in deciding on their treatment.
- Community mental health is reviewed, along with the important goal of empowerment.
- The sociocultural approach to therapy is reviewed along with the concepts of cultural competence (Tehee, 2020) and cultural humility (Davis & others, 2018; Jones & Branco, 2020).
- Ethnicity (Akhtar, 2006; Jackson & Greene, 2000), gender (Zerbe Enns & others, 2020), and gender identity and sexual orientation (Cronin & others, 2020; Huffman & others, 2020) are reviewed as important factors in therapy.
- A quiz question requires students to put themselves in the shoes of a young Asian American man seeking therapy and considering how to choose a therapist.

CHAPTER 14 HEALTH PSYCHOLOGY

- Research on military personnel and the importance of social support to PTSD and physical symptoms is reviewed (Luciano & McDevitt-Murphy, 2017).
- The role of church-based support for Blacks as a protective buffer against depression is discussed (Chatters & others, 2014).
- *Challenge Your Thinking: How Powerful Is the Power of Positive Thinking?* addresses victim blaming and stigmatization of those with physical illness.
- Type A research is critiqued for focusing on people unlikely to be at risk for coronary heart disease.
- A new section, "Stress and Prejudice," reviews research on the ways prejudice affects stress among sexual minority groups, people of diverse gender identities, and people who differ in immigrant status, veteran status, and race (e.g., Albuja & others, 2019; Feinstein & others, 2020; Hipes & Gemoets, 2019; Kassing & others, 2020; Zia & Ma, 2020). It reviews the meaning of health disparities and the differences in life expectancy for Black and white Americans (Xu & others, 2020) and emphasizes how COVID-19 likely exacerbated these (Wrigley-Field, 2020). The killing of George Floyd is discussed, and racism is described as a public health emergency.
- Prejudice against people with overweight and obesity is reviewed, including its impact on chronic stress (Tomiyama, 2019).
- *Intersection: Health Psychology and Social Psychology: Can Weight-Based Bias Affect the Health Consequences of Obesity?* reviews the health consequences of weight bias (Puhl & others, 2020) and a longitudinal study (Daly & others, 2019) showing that perceived discrimination independently predicts ill health among individuals with obesity even controlling for the health effects of obesity itself.
- The role of civic activism as a way to positively reappraise injustice among Black Americans is reviewed (Riley & others, 2020).
- *Psychology in Our World: Environments That Support Active Lifestyles* notes the role of systemic racism in access to environments that promote health (Bell & others, 2019).

preface

Some Students Take Psychology . . . Others Experience It!

Informed by student data, *Experience Psychology* helps students understand and appreciate psychology as an integrated whole. The personalized, adaptive learning program, thought-provoking examples, and interactive assessments help students see psychology in the world around them and experience it in everyday life. *Experience Psychology* is about, well, experience—our own behaviors; our relationships at home and in our communities, in school, and at work; and our interactions in different learning environments. Grounded in meaningful real-world contexts, *Experience Psychology*'s contemporary examples, personalized author notes, and applied exercises speak directly to students, allowing them to engage with psychology and to learn verbally, visually, and experientially—by reading, seeing, and doing. Function is introduced before dysfunction, building student understanding by looking first at typical, everyday behavior before delving into the less common—and likely less personally experienced—rare and abnormal behavior. *Experience Psychology* places the science of psychology, and the research that helps students see the academic foundations of the discipline, at the forefront of the course.

With *Experience Psychology*, students do not just "take" psychology but actively *experience* it. Paired with McGraw Hill Education Connect, a digital assignment and assessment platform that strengthens the link between faculty, students, and coursework, instructors and students accomplish more in less time. Connect Psychology is particularly useful for remote and hybrid courses, and includes assignable and assessable videos, quizzes, exercises, and interactivities, all associated with learning objectives. Interactive assignments and videos allow students to experience and apply their understanding of psychology to the world with fun and stimulating activities.

Experience a Personalized Approach

PERSONAL NOTES FROM THE AUTHOR THAT PROMOTE UNDERSTANDING

Experience Psychology emphasizes a personal approach, with an abundance of personal pedagogical "asides" communicated directly by author Laura King to students to guide their understanding and stimulate their interest as they read. Some of these notes highlight important terms and concepts; others prompt students to think critically about the complexities of the issues; and still others encourage students to apply what they have learned to their prior reading or to a new situation. These mini-conversations between the author and the reader help develop students' analytical skills for them to carry and apply well beyond their courses.

This is very "meta," but you're reading one right now. Think of it as a chance to nudge the reader's attention. And also to say thanks for taking on Intro Psych!

Students study more effectively with Connect and SmartBook 2.0.

How many students *think* they know everything about introductory psychology but struggle on the first exam?

- SmartBook 2.0 helps students study more efficiently by highlighting what to focus on in the chapter, asking review questions, and directing them to resources until they understand.

- Connect's assignments help students contextualize what they've learned through application, so they can better understand the material and think critically.

- SmartBook 2.0 personalizes learning to individual student needs, continually adapting to pinpoint knowledge gaps and focus learning on topics that need the most attention. Study time is more productive and, as a result, students are better prepared for class and coursework.

- Connect reports deliver information regarding performance, study behavior, and effort so instructors can quickly identify students who are having issues or focus on material that the class hasn't mastered.

With McGraw Hill's free **ReadAnywhere app,** students can read or study when it's convenient for them—anytime, anywhere. Available for iOS or Android smartphones or tablets, ReadAnywhere gives users access to McGraw Hill tools including the eBook and SmartBook 2.0 in Connect. Students can take notes, highlight, and complete assignments offline – their work will sync when they open the app with WiFi access.

Experience the Power of Data

Experience Psychology harnesses the power of data to improve the instructor and student experiences. Whether a class is face-to-face, hybrid, or entirely online, McGraw Hill Connect provides the tools needed to reduce the amount of time and energy instructors spend administering their courses. Easy-to-use course management tools allow instructors to spend less time administering and more time teaching, while reports allow students to monitor their progress and optimize their study time.

- The **At-Risk Student Report** provides instructors with one-click access to a dashboard that identifies students who are at risk of dropping out of the course due to low engagement levels.

- The **Category Analysis Report** details student performance relative to specific learning objectives and goals, including APA learning goals and outcomes and levels of Bloom's taxonomy.

- The **SmartBook Reports** allow instructors and students to easily monitor progress and pinpoint areas of weakness, giving each student a personalized study plan to achieve success.

Expand each category to see scores.

	Questions	Students submitted	Category score (*Best assignment attempt*)
Bloom's			
➕ Analyze	38	30/35	78%
➕ Apply	214	32/35	87%
➕ Create	8	29/35	86%
➕ Evaluate	24	31/35	92%
➕ Remember	257	35/35	93%
➕ Understand	238	34/35	89%

Expand each category to see scores.

	Questions	Students submitted	Category score (*Best assignment attempt*)
APA Outcome			
➕ 1.1: Describe key concepts, principles, and overarching themes in psychology	315	34/35	89.15%
➕ 1.2: Develop a working knowledge of psychology's content domains	459	33/35	88.75%
➕ 1.3: Describe applications of psychology	132	35/35	90.5%
➕ 2.1: Use scientific reasoning to interpret psychological phenomena	299	28/35	78.9%
2.2: Demonstrate psychology information literacy	304	34/35	83.5%
2.3: Engage in innovative and integrative thinking and problem solving	1	35/35	85.5%
: Interpret, design, and conduct basic psychological research	16	34/35	81.7%
Apply ethical standards to evaluate psychological science in practice	6	33/35	92.5%
➕ 5.1: Apply psychological content and skills to career goals	35	29/35	73.8%
➕ 5.2: Exhibit self-efficacy and self-regulation	24	33/35	81.6%

Experience an Emphasis on Critical Thinking

Experience Psychology stimulates critical reflection and analysis. The **Challenge Your Thinking** features involve students in debates relevant to findings from contemporary psychological research. Thought-provoking questions encourage examination of the evidence on both sides of a debate or issue. For example, the Challenge in the "Thinking, Intelligence, and Language" chapter considers whether artificial intelligence can be biased (it can be). The Challenge in the "Psychological Disorders" chapter considers whether birth month predicts an ADHD diagnosis (it does).

INTERSECTION

Cognition and Motivation: Why Do People Believe in Conspiracy Theories?

Why do people believe that important events are controlled by secret, powerful, and often evil groups? On the cognitive side, conspiracy beliefs are linked to system 1 processing—that aspect of thinking that functions rapidly and is often based on gut feelings (Tomljenovic & others, 2020). System 1 is also responsible for our capacity to recognize patterns and connections in the environment (important capacities for associative learning). Conspiracy theories often involve seeing patterns and making connections, even if these are very unusual. Engaging in rational reflection (that is, using system 2) can help to reduce conspiracy beliefs (Swami & others, 2014) and education predicts lower conspiracy theory belief (van Prooijen, 2017). Still, if you have ever talked to someone who believes in conspiracy theories, you probably already know that presenting facts and logic does not necessarily budge the person's beliefs. So, there must be something more to these beliefs than simply faulty reasoning, such as emotion and motivation—how we feel and what we want.

Conspiracy beliefs appear to be more emotionally than cognitively driven. Faced with facts and logic, a person might say that the simplest (factual) explanation just does not "feel right" (Tomljenovic & others, 2020). Emotional aspects of events and experiences may be more important than logic in conspiracy beliefs. Research has shown that feelings of vulnerability (Poon & others, 2020) and powerlessness (Biddlestone & others, 2020) predict higher conspiracy theory belief. Conspiracy beliefs are linked to our motivation to understand the world and to feel in control of events (Douglas & others, 2017).

The fact that conspiracy theories are commonplace suggests that they may be an outgrowth of adaptive processes that were useful long ago in our evolutionary history. According to one approach, accurately detecting potential hostile factions was important for our ancestors to survive (van Prooijen & van Vugt, 2018). The idea is that long ago our ancestors needed to be vigilant for actual malevolent factions that might do them harm. When contemporary people espouse conspiracy beliefs these same ancient mechanisms may be at work.

You might have noticed that most conspiracy believers do not appear happy about the explanations they have found for life events. Indeed, although conspiracy beliefs may be motivated by a desire to understand the world and their place in it—to find a way to "feel right" about the way things are—such beliefs do not actually help people. Conspiracy believers tend to feel *less* personal control and *less* autonomous than nonbelievers (Douglas & others, 2017). Combatting conspiracy beliefs is important because those who do not believe in the scientific facts that underlie vaccines or climate science can put themselves and others at risk through their behavior. Understanding these strange but commonplace beliefs is an important goal for psychology.

\\ How do you think conspiracy theories can be effectively addressed?

ibreakstock/Shutterstock

Experience Psychology's **Intersection** features are also designed to spark critical thought. Showcasing studies in different areas of psychological research that focus on the same topic, the Intersections shed light on the links between, and the reciprocal influences of, this exciting work, and they raise provocative questions for student reflection and class discussion. For example, the selection for the "Thinking, Intelligence, and Language" chapter brings together cognition and motivation to examine recent research on the development and persistence of conspiracy theories.

Each chapter of *Experience Psychology* includes a **Psychology in Our World** section that demonstrates the relevance of psychology in a variety of real-world contexts, including the workplace, the media, and current events. For example, in the "Therapies" chapter, the Psychology in Our World section focuses on apps and online options for therapy. In the "Science of Psychology" chapter, the Psychology in Our World feature reviews the use of big data in psychological research.

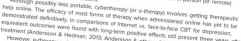

PSYCHOLOGY IN OUR WORLD

Seeking Therapy? There Is Probably an App for That

You have probably seen ads for smartphone apps that are meant to help improve psychological functioning. You may have even used one for yourself. While those apps target things like relaxation and mindfulness (Gál & others, 2021), psychotherapy apps, sometimes called "mHealth," meaning *mobile health-related interventions*, are more interactive—providing immediate feedback aimed at modifying thoughts and behaviors. Therapy apps are meant to be mobile extensions of evidence-based therapies. The idea of using apps to administer psychotherapy is exciting because it might meet an enormous need for psychological help to many people (Schueller & Torous, 2020). However, demonstrating the effectiveness of therapy apps, in general and across specific disorders, remains a goal for the future, rather than a current reality (Wright & Mishkind, 2020; Porras-Segovia & others, 2020). Still, apps may be used by psychotherapists as a complement to their in-person (or remote) interventions.

one photo/Shutterstock

Although possibly less portable, *cybertherapy* (or *e-therapy*) involves getting therapeutic help online. The efficacy of most forms of therapy when administered online has yet to be demonstrated definitively. In comparisons of Internet vs. face-to-face CBT for depression, equivalent outcomes were found with long-term positive effects still present three years after treatment (Andersson & Hedman, 2013; Andersson & others, 2013).

However, e-therapy websites and mHealth apps are controversial among mental health professionals (Emmelkamp, 2011; Lui & others, 2017). For one thing, many of these sites do not include the most basic information about the therapists' qualifications (Norcross & others, 2013). It is notable that, because cybertherapy occurs at a distance, these sites typically exclude people who are having thoughts of suicide. Further, confidentiality, a crucial aspect of the therapeutic relationship, cannot always be guaranteed on a website. On the plus side, though, people who are unwilling or unable to seek out face-to-face therapy may be more disposed to get help online, and studies indicate some success of Internet-based therapies (Andersson & Titov, 2014; Norcross & others, 2013). Certainly the COVID-19 pandemic nudged many psychotherapists and their clients into online rather than in-person formats (Feijt & others, 2020), which will likely increase what we know about their effectiveness.

In addition, the **Psychological Inquiry** features draw students into analyzing and interpreting figures and photos by embedding a range of critical thinking questions within the caption.

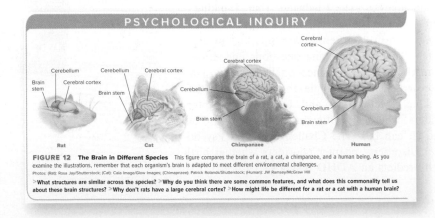

PSYCHOLOGICAL INQUIRY

Rat — Brain stem, Cerebellum, Cerebral cortex
Cat — Cerebellum, Cerebral cortex, Brain stem
Chimpanzee — Cerebral cortex, Cerebellum, Brain stem
Human — Cerebral cortex, Cerebellum, Brain stem

FIGURE 12   The Brain in Different Species   This figure compares the brain of a rat, a cat, a chimpanzee, and a human being. As you examine the illustrations, remember that each organism's brain is adapted to meet different environmental challenges.
Photos: (Rat): Rosa Jay/Shutterstock; (Cat): Caia Image/Glow Images; (Chimpanzee): Patrick Rolands/Shutterstock; (Human): JW Ramsey/McGraw Hill

>What structures are similar across the species? >Why do you think there are some common features, and what does this commonality tell us about these brain structures? >Why don't rats have a large cerebral cortex? >How might life be different for a rat or a cat with a human brain?

Experience Active Engagement

Experience Psychology offers several ways to actively engage with the course content. Through **Do It!**, a series of brief, recurring sidebar activities linked to the text reading, students get an opportunity to test their assumptions and learn through hands-on exploration and discovery. Reinforcing that the science of psychology requires active participation, Do It! selections include, for example, an exercise on conducting an informal survey to observe and classify behaviors in a public setting, as well as an activity guiding students on how to research a "happiness gene." Such exercises provide vibrant and involving experiences that get students thinking like psychologists.

Do It!

Go on a caffeine hunt. Check out the ingredient lists on your favorite beverages, snacks, and painkillers. Which of these contain caffeine? You might be surprised by how much caffeine you consume every day without even knowing it.

McGraw Hill Connect offers several ways to actively engage students. New for the fifth edition, the **Writing Assignment** tool delivers a learning experience to help students improve their written communication skills and conceptual understanding. As an instructor you can assign, monitor, grade, and provide feedback on writing more efficiently and effectively. Writing Assignment Premium promotes student's critical thinking with auto-scored writing prompts, helping to save you time providing feedback and grading. For students, Writing Assignment is tablet ready and provides time-saving tools with a just-in-time basic writing and originality checking. **Power of Process** guides students through the process of critical reading and analysis. Faculty can select or upload content, such as journal articles, and assign guiding questions to gain insight into students' understanding of the scientific method while helping them improve upon their information literacy skills.

Concept Clips help students comprehend some of the most difficult ideas in introductory psychology. Colorful graphics and stimulating animations describe core concepts in a step-by-step manner, engaging students and aiding in retention. Concept Clips can be used as a presentation tool in the classroom or for student assessment. Concept Clips are also embedded in the ebook to offer an alternative presentation of these challenging topics.

Interactivities engage students with content through experiential activities. Topics include Explicit and Implicit Bias; Cognitive Dissonance; Correlations; Neurons; The Stages of Sleep; Levels of Processing; Naturalistic Observation; Observational Learning; Defense Mechanisms; and Heuristics.

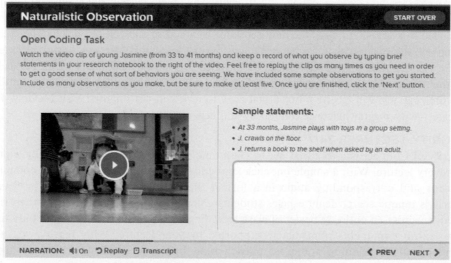

Through the connection of psychology to students' own lives, concepts become more relevant and understandable. Located in Connect, **NewsFlash** is a multi-media assignment tool that ties current news stories, TedTalks, blogs, and podcasts to key psychological principles and learning objectives. Students interact with relevant news stories and are assessed on their ability to connect the content to the research findings and course material. NewsFlash is updated twice a year and uses expert sources to cover a wide range of topics including emotion, personality, stress, drugs, COVID-19, ableism, disability, social justice, stigma, bias, inclusion, gender, LGBTQA+, and many more.

At the Apply level of Bloom's taxonomy, expanded **Application-Based Activities** provide a means for experiential learning. These are highly interactive, automatically graded, online learn-by-doing exercises that offer students a safe space to apply their knowledge and problem-solving skills to real-world scenarios. Each scenario addresses key concepts and skills that students must use to work through and solve course-specific problems, resulting in improved critical thinking.

At the Apply and Analyze levels of Bloom's taxonomy, **Scientific Reasoning Activities** found in Connect offer in-depth arguments to sharpen students' critical thinking skills and prepare them to be more discerning consumers of psychology in their everyday lives. For each chapter, there are multiple sets of arguments accompanied by auto-graded assessments requiring students to think critically about claims presented as facts. These exercises can also be used in Connect as group activities or for discussion.

Brain and nervous system "tours" are embedded in the ebooks to encourage learners to engage with key structures. These tours provide students with practice in grasping key biological structures and processes that are essential to an appreciation of the role of science in psychology and success in the course. The digital tours are also available in the instructor's materials to be used as presentation resources.

Psychology at Work videos, assignable and assessable within McGraw Hill Connect, highlight careers in which knowledge of psychology is beneficial in the workplace. Each video introduces a person at work, who specifies how knowledge gained from taking introductory psychology in college is applied to the work environment.

Experience the Course You Want to Teach

SUPPORTING INSTRUCTORS WITH TECHNOLOGY
With McGraw Hill Education, you can develop and tailor the course you want to teach.

CREATE: YOUR BOOK, YOUR WAY
McGraw Hill's Content Collections Powered by Create® is a self-service website that enables instructors to create custom course materials—print and eBooks—by drawing upon McGraw Hill's comprehensive, cross-disciplinary content. Choose what you want from our high-quality text-books, articles, and cases. Combine it with your own content quickly and easily, and tap into other rights-secured, third-party content such as readings, cases, and articles. Content can be arranged in a way that makes the most sense for your course and you can include the course name and information as well. Choose the best format for your course: color print, black-and-white print, or eBook. The eBook can be included in your Connect course and is available on the free ReadAnywhere app for smartphone or tablet access as well. When you are finished customizing, you will receive a free digital copy to review in just minutes! Visit McGraw Hill Create®–www.mcgrawhillcreate.com–today and begin building!

TEGRITY: LECTURES 24/7
Tegrity in Connect is a tool that makes class time available 24/7 by automatically capturing every lecture. With a simple one-click start-and-stop process, you capture all computer screens and corresponding audio in a format that is easy to search, frame by frame. Tegrity's unique search feature helps students efficiently find what they need, when they need it, across an entire semester of class recordings. Help turn your students' study time into learning moments immediately supported by your lecture. With Tegrity, you also increase intent listening and class participation by easing students' concerns about

note-taking. Using Tegrity in Connect will make it more likely you will see students' faces, not the tops of their heads.

PROCTORIO: REMOTE PROCTORING & BROWSER-LOCKING CAPABILITIES

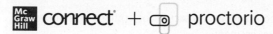

Remote proctoring and browser-locking capabilities, hosted by Proctorio within Connect, provide control of the assessment environment by enabling security options and verifying the identity of the student. Seamlessly integrated within Connect, these services allow instructors to control students' assessment experience by restricting browser activity, recording students' activity, and verifying students are doing their own work. Instant and detailed reporting gives instructors an at-a-glance view of potential academic integrity concerns, thereby avoiding personal bias and supporting evidence-based claims.

TRUSTED SERVICE AND SUPPORT

McGraw Hill Education's Connect offers comprehensive service, support, and training throughout every phase of your implementation. If you're looking for some guidance on how to use Connect, or want to learn tips and tricks from super users, you can find tutorials as you work. Our Digital Faculty Consultants and Student Ambassadors offer insight into how to achieve the results you want with Connect.

OLC-ALIGNED COURSES: IMPLEMENTING HIGH-QUALITY ONLINE INSTRUCTION AND ASSESSMENT THROUGH PRECONFIGURED COURSEWARE

In consultation with the Online Learning Consortium (OLC) and our certified Faculty Consultants, McGraw Hill has created pre-configured courseware using OLC's quality scorecard to align with best practices in online course delivery. This turnkey courseware contains a combination of formative assessments, summative assessments, homework, and application activities, and can easily be customized to meet an individual's needs and course outcomes. For more information, visit https://www.mheducation.com/highered/olc.

INTEGRATION WITH YOUR LEARNING MANAGEMENT SYSTEM

McGraw Hill integrates your digital products from McGraw Hill Education with your school LMS for quick and easy access to best-in-class content and learning tools. Build an effective digital course, enroll students with ease, and discover how powerful digital teaching can be.

Available with Connect, integration is a pairing between an institution's learning management system (LMS) and Connect at the assignment level. It shares assignment information, grades, and calendar items from Connect into the LMS automatically, creating an easy-to-manage course for instructors and simple navigation for students.

Instructor Supplements

Instructor's Manual The instructor's manual provides a wide variety of tools and resources for presenting the course, including learning objectives and ideas for lectures and discussions.

Test Bank and Test Builder Organized by chapter, questions in the bank are designed to test factual, conceptual, and applied understanding. Test Builder is a cloud-based tool that enables instructors to format tests that can be printed or administered within an LMS. Test Builder offers a modern, streamlined interface for easy content configuration that matches course needs, without requiring a download. Test Builder enables instructors to:

- access all test bank content from a particular title

- easily pinpoint the most relevant content through robust filtering options

- manipulate the order of questions or scramble questions and/or answers

- pin questions to a specific location within a test
- choose the layout and spacing
- add instructions and configure default settings

Test Builder provides a secure interface for better protection of content and allows for just-in-time updates to flow directly into assessments.

PowerPoint Presentations The PowerPoint presentations, available in both dynamic, lecture-ready and accessible, WCAG-compliant versions, highlight the key points of the chapter and include supporting visuals. All of the slides can be modified to meet individual needs.

Image Gallery The Image Gallery features the complete set of downloadable figures and tables from the text. These can be easily embedded by instructors into their own PowerPoint slides.

Chapter-by-Chapter Changes

All chapters are updated with current research and data, when available.

CHAPTER 1 THE SCIENCE OF PSYCHOLOGY

- New chapter-opening vignette, "The Art and Science of Waiting," referring to COVID-19 lock-downs and the concept of flow
- New example of counterintuitive results, based on a study of expectations about seeing *Star Wars: The Last Jedi*
- New example of forgiveness using events in the life of John Lewis
- New Challenge Your Thinking feature, "Where Is Everybody?" that looks at lack of diversity in the history of psychology
- New Psychology in Our World Feature, "Big Data in Psychological Science" that examines how big data may be used in psychological research
- New Intersection feature, "Cognitive Psychology and Health Psychology: How Can We Combat COVID-19 Misinformation Online?"
- Updated figure showing the settings in which psychologists are expected to be working in 2030
- New example of experimenter bias from research on oxytocin.
- Expanded coverage of issue of lack of diversity in research samples, with new definition of WEIRD samples
- Expanded coverage of replication

CHAPTER 2 THE BRAIN AND BEHAVIOR

- New chapter-opening vignette, "Seeing Strange Things," describing a case of a rare neurological condition (Riddoch phenomenon)
- New Intersection feature, "Neuroscience and Language: What Is a Word to a Dog?" that describes studies using fMRI in dogs
- New Challenge Your Thinking feature, "How Should We Think about Genes and Behavior?"

- Updated treatment of the human genome, traumatic brain injury, and issues with the reproducibility of genome-wide association studies
- Updated research on brain tissue implants in rats

CHAPTER 3 SENSATION AND PERCEPTION

- New chapter-opening vignette, "Is Everything Cake?" describing the popular 2020 meme
- New Psychology in Our World feature, "Are You Listening Safely?"
- New Intersection feature, "Sensation, Perception, and Social Psychology: Do False Beliefs about Race Affect Pain Treatment?"
- New Challenge Your Thinking feature, "Could You Be Fooled into Thinking You Have Six Fingers?"
- Updated discussion with new research on disparities in pain perception and on how stereotypes may affect perception
- New research on subliminal priming
- Updated treatment of person perception
- New discussion of proprioceptive drift

CHAPTER 4 STATES OF CONSCIOUSNESS

- New chapter-opening vignette, "A Star in His Own Imagination," describing a case of locked-in syndrome
- New Intersection feature, "Consciousness and Social Psychology: Could Robots Make Us Kinder to Each Other?"
- New Challenge Your Thinking feature, "Does Legalized Medical Marijuana Reduce Opioid Abuse and Overdoses?"
- Expanded treatment of theory of mind and the social nature of consciousness
- Updated and expanded treatment of sleep and memory and insomnia treatments
- New coverage of sleep problems associated with the COVID-19 pandemic
- New research on health problems associated with shift work and the impact of artificial light on sleep
- Updated statistics on drug and alcohol use and medical marijuana
- New discussion of vaping

CHAPTER 5 LEARNING

- Revised chapter-opening vignette, "Service Dogs: Helping Heroes Heal"
- New Challenge Your Thinking feature, "Do Machines Actually Learn?"
- New Intersection feature, "Learning and Social Psychology: Can Observational Learning Lead to Bias?"
- Expanded treatment of observational learning
- Updated treatment with new research on applied behavior analysis
- Updated discussion of embedded marketing

CHAPTER 6 MEMORY

- New chapter-opening vignette about the use of nostalgia to cope with COVID-19-related challenges

- Updated statistics on divorce
- Updated treatment of the neuroscience of conformity
- Expanded coverage of group influence

CHAPTER 12 PSYCHOLOGICAL DISORDERS

- New chapter-opening vignette featuring the example of a young man newly diagnosed with bipolar disorder, with a focus on stigma
- New Challenge Your Thinking feature, "Does Birth Month Predict ADHD Diagnosis?"
- New Intersection feature, "Clinical Psychology and Social Psychology: How Does the Stigma of Mental Illness Affect Social Interactions?"
- Updated statistics on psychological disorders and suicide
- New discussion of potential biomarkers for autism spectrum disorder
- Updated treatment of the potential origins of social phobia and OCD
- Updated discussion of biological factors in depressive disorders and risk factors for suicide

CHAPTER 13 THERAPIES

- New chapter-opening vignette focused on play therapy for children who have been exposed to trauma
- New Intersection feature, "Neuroscience and Psychotherapy: Does Oxytocin Reflect the Therapeutic Alliance?"
- Updated Psychology in Our World feature, "Seeking Therapy? There Is Probably an App for That"
- New Challenge Your Thinking feature, "Who Should Decide What Treatment Is Best for a Person?"
- Updated discussion of drug therapies, including experimental use of ketamine
- Updated treatment of ECT and transcranial magnetic stimulation, including side effects

CHAPTER 14 HEALTH PSYCHOLOGY

- New chapter-opening vignette focused on the COVID-19 pandemic, what it revealed about human psychology and how its related effects impacted physical health
- New Intersection feature, "Health Psychology and Social Psychology: Can Weight-Based Bias Affect the Health Consequences Obesity?"
- New section on stress and prejudice and their links to health disparities among the U.S. population
- Expanded discussion of public health
- Updated discussions of optimism and healthy eating
- Updated and expanded coverage of stress and its links to disease
- Updated treatment of STIs, including prevention strategies and PrEP for HIV

acknowledgments

The quality of *Experience Psychology* is a testament to the skills and abilities of so many people, and I am tremendously grateful to the following individuals for their insightful contributions during the project's development.

Alisa Beyer, *Chandler Gilbert Community College*

Gerald Braasch, *McHenry County College*

Cynthia Campbell, *Ivy Tech Community College of Indiana–Fort Wayne*

Sharon E. Chacon, *Northeast Wisconsin Technical College*

Amy Cianci, *Westchester Community College*

Marc Coutanche, *University of Pittsburgh*

Myra Cox, *Harold Washington College, City College of Chicago*

Ronald Fredin, *Delgado Community College*

Douglass Godwin, *Nashville State Community College*

Christopher Green, *Ivy Tech Community College of Indiana–Bloomington*

Cassidy Hawf, *Iowa Central Community College*

Brett Heintz, *Delgado Community College*

Melissa Ivey, *Northern Maine Community College*

Michael James, *Ivy Tech Community College of Indiana–Bloomington*

Margaret Keaton, *Ivy Tech Community College of Indiana–Indianapolis*

Patricia Kemerer, *Ivy Tech Community College of Indiana–Fort Wayne*

Linda Kieffer, *Delgado Community College*

Sarah Kirk, *Palomar College*

Douglas Lalama, *Ivy Tech Community College of Indiana–Lake County*

Tracy Litzinger, *Flagler College*

Lynda Mae, *Arizona State University*

Ed McGee, *West Georgia Technical College*

Bradley Mitchell, *Ivy Tech Community College of Indiana–Valparaiso*

Tonya Nascimento, *University of West Florida*

Blake Nielsen, *Columbia College*

Michael Poulakis, *Ivy Tech Community College of Indiana–Indianapolis*

Tramaine Presley, *Ivy Tech Community College of Indiana–Indianapolis*

Sandra Prince, *Delgado Community College*

Michael Rader, *Johnson County Community College*

Alexandria Reynolds, *The University of Virginia's College at Wise*

Cynthia Rickert, *Ivy Tech Community College of Indiana–Indianapolis*

Sheldon Rifkin, *Kennesaw State University*

Kelsi Rugo, *Salt Lake Community College*

Patricia Rousselo, *Siena Heights University*

Michael Skibo, *Westchester Community College*

Amy Skinner, *Shelton State Community College*

Deborah Stipp, *Ivy Tech Community College of Indiana–Lake County*

Melissa Sutherland, *San Antonio College*

Brad Thurmond, *Ivy Tech Community College of Indiana–Bloomington*

Kyle van Ittersum, *Angelo State University*

Jill Wallen, *Westchester Community College*

Carmon Weaver Hicks, *Ivy Tech Community College of Indiana–Indianapolis*

Jillian Whatley, *Georgia State University*

Personal Acknowledgments

Returning to *Experience Psychology* for this fifth edition in the midst of a global pandemic has been incredibly challenging. I very much appreciate all those at McGraw Hill Education who brought their characteristic levels of innovation, enthusiasm, and encouragement to bear. I wish to thank Kirstan Price for her editorial acumen and especially Sandy Wille who went above and beyond on this revision.

This revision, completed in the forced solitude of the pandemic, while teaching online for the first time, owes a great deal to the people who touched my life during a difficult time, even if only via Zoom. I want to thank, especially, Deb White and Korinne Cikanek from Normandale Community College. These brilliant women gave up an afternoon to talk with me about introductory psychology and their wisdom and insight was a true life line. Thanks as well to the psychology departments at Seminole State College and Des Moines Area Community College for taking the time to Zoom with me. Thanks to my wonderful 300 students in Introduction to Personality at Mizzou who inspired me with their engagement even in an online course. I owe special debts of gratitude to Megan Edwards, Hope Rose, Jake Womick, and Chris Sanders, my graduate students, who have patiently endured having an advisor who is often very busy while generating their own very exciting scholarship. Finally, a heartfelt thank-you goes to my family and friends who have supported and inspired me, especially Lisa and Sam. You are my life and I could not have made it through the last year without your delightful company.

1 The Science of Psychology

The Art (and Science) of Waiting

I n spring 2020, the world was rocked by a global pandemic. The novel coronavirus (COVID-19) placed many lives on hold. Important events were canceled or postponed. Efforts to quell the virus meant that many people spent a great deal more time just waiting than ever before. Waiting can be stressful, especially during uncertain times.

During the pandemic, many people coped with waiting by distracting themselves with jigsaw puzzles or baking bread. The psychology of waiting shows that some distractions are better than others. The best way to wait might be to experience *flow*. Flow refers to a state of mind that happens when we engage in activities that present an optimal challenge for our abilities (Csikszentmihalyi, 1990). When we experience flow, we lose track of time and experience ourselves as at one with what we are doing. A series of studies (Rankin & others, 2019) showed that for people stuck waiting—whether it was budding lawyers awaiting bar exam results or hopeful candidates on the job market—engaging in activities that promoted flow enhanced positive emotions and lowered negative feelings. Our capacity to experience the deep pleasure that comes from engagement with challenging tasks can make a difficult, uncertain time a little easier.

You might be surprised to learn that psychologists study the best way to wait. It turns out psychologists are interested in just about everything people do. Even ordinary human behavior can become extraordinary when viewed in the right light, with a close lens. Psychologists are scientists who look at the world with just such a lens. Right now, thousands of dedicated scientists are studying things about you that you might have never considered, like how your eyes adjust to a sunny day or how your brain registers the face of a friend. There is not a single thing people do that is not fascinating to some psychologist somewhere.

This chapter begins by defining psychology and reviewing the history of the field. Next we survey seven broad approaches that characterize psychological science today. Then, in sequence, we examine the elements of the scientific method, review the different kinds of research psychologists do, and consider the importance of conducting psychological research according to ethical guidelines. We conclude with a look at applications of psychology to daily life—a central focus of *Experience Psychology*.

1 Defining Psychology and Exploring Its Roots

Formally defined, **psychology** is the scientific study of behavior and mental processes. Let's consider the three key terms in this definition: *science, behavior,* and *mental processes.*

What is your definition of psychology? When you think of the word psychology, what first comes to mind?

As a **science**, psychology uses systematic methods to observe human behavior and draw conclusions. The goals of psychological science are to describe, predict, and explain behavior. In addition, psychologists are often interested in controlling or changing behavior, and they use scientific methods to examine interventions that might, for example, help to reduce violence or promote happiness.

Researchers might be interested in knowing whether individuals will help a stranger who has fallen down. The researchers could devise a study in which they observe people walking past a person who needs help. Through many observations, the researchers could come to *describe* helping behavior by counting how many times it occurs in particular circumstances. They might also try to *predict* who will help, and when, by examining characteristics of the people studied. Are happy people more likely to help? Does gender relate to helping? After the psychologists have analyzed their data, they also will want to *explain* why helping behavior occurred when it did. Finally, they might be interested in changing helping behavior, such as by devising strategies to increase helping.

Behavior is everything we do that can be directly observed—two people kissing, a baby crying, a college student knitting to wait out the pandemic. **Mental processes** are the thoughts, feelings, and motives that each of us experiences privately but that cannot be

psychology
The scientific study of behavior and mental processes.

science
The use of systematic methods to observe the natural world, including human behavior, and to draw conclusions.

mental processes
The thoughts, feelings, and motives that people experience privately but that cannot be observed directly.

behavior
Everything we do that can be directly observed.

Behavior includes the observable act *of two people kissing; mental processes include their* unobservable thoughts *about kissing.*
(Both): Nora Pelaez/Visual Ideas/Getty Images

observed directly. Although we cannot directly see thoughts and feelings, they are none-theless real. They include *thinking* about kissing someone, a baby's *feelings* when its mother leaves the room, and a student's *memory* of learning to knit.

The Psychological Frame of Mind

What makes for a good job, a good marriage, or a good life? Psychologists approach these big life questions as scientists. Psychology is a rigorous discipline that tests assumptions, bringing scientific data to bear on the questions of central interest to human beings (Brinkmann, 2020; Henriques & Michalski, 2020). Psychologists conduct research and rely on that research to provide evidence for their conclusions. They examine the available evidence about some aspect of mind and behavior, evaluate how strongly the data (information) support their hunches, analyze disconfirming evidence, and carefully consider whether they have explored all possible factors and explanations. At the core of this scientific approach are four attitudes: *critical thinking, curiosity, skepticism,* and *objectivity.*

critical thinking
The process of thinking deeply and actively, asking questions, and evaluating the evidence.

Like all scientists, psychologists are *critical thinkers.* **Critical thinking** is the process of thinking deeply and actively, asking questions, and evaluating the evidence (Sternberg & Halpern, 2020). Critical thinkers question and test what some people say are facts. They examine research to see how soundly it supports an idea. Critical thinking reduces the likelihood that conclusions will be based on unreliable personal beliefs, opinions, and emotions. Critical thinking also comes into play when scientists consider the conclusions they draw from research. As critical thinkers who are open to new information, scientists must tolerate uncertainty, knowing that even long-held views are subject to revision.

Critical thinking is very important as you read *Experience Psychology.* Some of what you read might fit with your beliefs, and some might challenge you to reconsider them. Actively engaging in critical thinking is vital to making the most of psychology. As you study the field, think about how what you are learning relates to your life experiences and your assumptions about human behavior.

Scientists are also *curious.* The scientist notices things in the world (a star in the sky, an insect, a happy person) and wants to know what it is and why it is that way. Science involves asking questions, even very big questions (e.g., Where did the earth come from? How does love between two people endure for 50 years?). Thinking like a psychologist means opening your mind and imagination to wondering why things are the way they are.

In addition, scientists are *skeptical.* Skeptical people require evidence before accepting something as true. Being skeptical can mean questioning what "everybody knows." There was a time when "everybody knew" that women were morally inferior to men, that race could influence a person's IQ, and that the earth was flat. Psychologists, like all scientists, look at assumptions in new and questioning ways. Psychology is different from common sense because psychologists are skeptical of commonsensical answers.

Psychological research often turns up the unexpected in human behavior. Such results are called *counterintuitive* because they contradict our intuitive impressions of how the world works. To get a sense of what counterintuitive means, consider the advice, "Don't get your hopes up." This saying suggests that if you have high hopes for an experience, you are more likely to experience great disappointment if things do not go well, or even that you might jinx the experience itself. Do you think this concern is well placed? Recent research addressed this question in the context of going to see the movie, *Star Wars: The Last Jedi* (Bonus & others, 2020). Before the movie was released, people were asked how much they thought seeing the movie would make them happy and nostalgic. (Nostalgia refers to wistful feelings for meaningful experiences from the past.) Three weeks after the film was released, the participants were asked if they saw the movie, to rate how it made them feel, and then evaluate the movie overall. Although experiencing the movie as producing less happiness than expected was

*You might be wondering about the names and dates in parentheses. They are **research citations** that identify the authors of particular studies and the year each study was published. If you see an especially interesting study, you might look it up in the References and check it out online or in your school's library.*

associated with negative evaluations of the film, those who expected it to be bad were least likely to enjoy it. So, high hopes may be less problematic than low ones. Knowing if what we think we know is true requires scientific evidence.

Last, practicing science also means being *objective*. Being objective involves trying to see things as they really are, not just as the observer would like them to be. Scientific knowledge ultimately is based on objective evidence. It also means that sometimes knowledge changes in the face of new evidence. Thinking like a scientist means being open to that new evidence even when it challenges our assumptions.

To gather objective evidence, scientists rely on empirical methods. The **empirical method** involves gaining knowledge by observing events, collecting data, and reasoning logically. For scientists, objectivity means waiting to see what the evidence tells them rather than going with their hunches. Does playing a new smartphone app reduce depression? A scientist would say, "That's an empirical question," meaning that hard evidence is required to answer it. An objective thinker insists on sound evidence before drawing conclusions. Like critical thinking, relying on evidence to provide the foundation for conclusions means being open to uncertainty. Empirical evidence provides the best answers to questions at any given moment.

This is why researchers often say that a study "supports" a particular prediction, but rarely if ever say that it "proves" anything.

empirical method Gaining knowledge through the observation of events, the collection of data, and logical reasoning.

Once you start to think like a psychologist, the world begins to look like a different place. Easy answers and simple assumptions will not do. As you can probably imagine, psychologists, as a group, are people with many different opinions about many different things. If a number of these critical thinkers were to gather around a table, it is a safe bet that they would have a lively conversation.

Indeed, as you will see throughout *Experience Psychology,* there are many things about which psychologists disagree, and psychology, like any science, is filled with debate and controversy. We will address many of these controversies in sections called "Challenge Your Thinking." Each of these boxes will give you a chance to think critically about a topic.

So, debate and controversy are a natural part of thinking like a psychologist. Psychology has advanced as a field because psychologists do not always agree with one another about why mind and behavior work the way they do. Psychologists have reached a more accurate understanding of human behavior because psychology fosters controversies and because psychologists think deeply and reflectively and examine the evidence on all sides. A good place to try out your critical thinking skills is by revisiting the definition of psychology.

Psychology as the Science of All Human Behavior

As you consider the general definition of psychology as the science of human behavior, you might be thinking, okay, where's the couch? Where's the mental illness? Psychology certainly includes the study of therapy and psychological disorders. Clinical psychologists are psychologists who specialize in studying and treating psychological disorders. By definition, though, psychology is a much more general science, practiced in several environments in addition to clinical settings (Figure 1). In fact, some of the common places to find psychologists are school settings. How did we end up with the idea that psychology is only about mental illness? Surely, psychological disorders are very interesting, and the media often portray psychologists as therapists. Yet the view of psychology as the science of what is wrong with people started long before television was even invented.

When they think of psychology, many people think of Sigmund Freud (1856–1939). Freud believed that most of human behavior is caused by dark, even horrific, unconscious impulses pressing for expression. For Freud, even the average person on the street is a mysterious well of unconscious desires. Certainly, Freud has had a lasting impact on psychology and on society. Consider, though, that Freud based his ideas about human nature on the patients he saw in his clinical practice—people who were struggling with psychological problems. His experiences with these patients, as well as his analysis of himself, colored his outlook on all of humanity. Freud

You have probably heard of a "Freudian slip." Freud's name has become part of our everyday language.

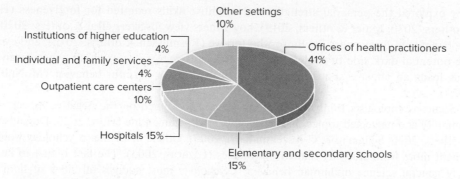

FIGURE 1 Settings in Which Psychologists Work—Projections for 2030
This figure shows the settings where individuals who have PhDs in psychology are projected to be working in 2030. As you can see, many are employed in education and medical contexts.

SOURCE: American Psychological Association. (2018). APA Fact Sheet Series on Psychologist Supply and Demand Projections 2015-2030: Demand by Employment Setting. Washington, DC: Author.

once wrote, "I have found little that is 'good' about human beings on the whole. In my experience most of them are trash" (1918/1996).

Freud's view of human nature has crept into general perceptions of what psychology is all about. Imagine, for example, that you are seated on a plane, having a pleasant conversation with the woman, a stranger, sitting next to you. At some point you ask your seatmate what she does for a living, and she informs you she is a psychologist. You might think to yourself, "Uh oh. What have I already told this person? What secrets does she know about me that I don't know about myself? Has she been analyzing me this whole time?" Would you be surprised to discover that this psychologist studies happiness? Or intelligence? Or the processes related to sight? The study of psychological disorders is a very important aspect of psychology, but it represents only one part of the science of psychology.

Psychology seeks to understand the truths of human life in *all* its dimensions, including people's best and worst experiences, and everything in between. Research on the human capacity for forgiveness demonstrates this point (Balliet & others, 2011; Costa & Neves, 2017; Harper & others, 2014; McCullough & others, 2011, 2013). Forgiveness is the act of letting go of anger and resentment toward someone who has done something harmful to us. Through forgiveness we cease seeking revenge or avoiding the person who did us harm, and we might even wish that person well (Lin & others, 2014; Tuck & Anderson, 2014).

In the summer of 2020, the United States lost a civil rights hero when U.S. Representative John Lewis died (Seelye, 2020). Lewis had served 17 terms as a representative of the Atlanta area. As a young man, Lewis was a member of the Freedom Riders, civil rights activists who risked their lives protesting to end racial segregation in the United States. On May 9, 1961, a man beat Lewis to the ground. The representative's only memory of the event was white fists, punching him to the ground. Nearly 50 years later, Elwin Wilson, the man who owned those fists, visited Lewis's office to offer not violence but a sincere apology. "It was wrong for people to be like I was," he said, "but I am not that man anymore" (Dys, 2009). He said he was sorry. Without hesitation, John Lewis replied, "I forgive you. I hold no grudge. Hate is too heavy a burden to bear" (Dys, 2009).

John Lewis was an amazing person. His willingness to forgive a horrible, racist crime is both remarkable and puzzling. Can we scientifically understand the human ability to forgive even what might seem to be unforgivable? A number of psychologists have taken up the topic of forgiveness in research on racism (Brooks & others, 2020) and even genocide (Ordóñez-Carabaño & others, 2020). For John Lewis, his deep religious faith was certainly part of his capacity to embrace forgiveness (Finchman, 2020). Researchers also

U.S. Representative John Lewis's capacity to forgive Elwin Wilson is one of the remarkable human characteristics that fascinate psychologists. The men made many appearances together, such as this one, to spread a message of hope and healing.
Andy Burris/The Herald/AP Images

have explored the personal strengths and cognitive skills required for forgiveness (Ho & others, 2020; Maier & others, 2019), how others view forgivers (Raj & others, 2020), the benefits of forgiveness for those who forgive (Rasmussen & others, 2019), and even the potential dark side of forgiveness, which might emerge, for example, when forgiveness leads an abusive spouse to feel free to continue a harmful behavior (McNulty, 2020).

Some psychologists think the field has focused too much on the negative aspects of humanity and neglected topics that reflect the best of human life (Hart, 2020; Donaldson & others, 2020; Seligman & Csikszentmihalyi, 2000). Others insist that psychology would benefit more from studying human weaknesses (Lazarus, 2003). The fact is that to be a truly general science of human behavior, psychology must address all sides of human experience. Surely, controversy is a part of any science. Healthy debate characterizes the field of psychology, and a new psychological perspective sometimes arises when one scientist questions the views of another. Such ongoing debate is a sign of a lively discipline. Indeed, the very birth of the field was marked by debate. Great minds do not always think alike, especially when they are thinking about psychology.

Psychology in Historical Perspective

Psychology seeks to answer questions that people have been asking for thousands of years—for example:

■ How do we learn?

■ What is memory?

■ Why does one person grow and flourish while another struggles?

The notion that such questions might be answered through scientific inquiry is relatively new. From the time human language included the word *why* and became rich enough to let people talk about the past, we have been creating myths to explain why things are the way they are. Ancient myths attributed most important events to the pleasure or displeasure of the gods: When a volcano erupted, the gods were angry; if two people fell in love, they had been struck by Cupid's arrows. Gradually, myths gave way to *philosophy*—the rational investigation of the underlying principles of being and knowledge. People attempted to explain events in terms of natural rather than supernatural causes.

Western philosophy came of age in ancient Greece in the fourth and fifth centuries B.C.E. Socrates, Plato, Aristotle, and others debated the nature of thought and behavior, including the possible link between the mind and the body. Later philosophers, especially René Descartes, argued that the mind and body were completely separate, and they focused their attention on the mind. Psychology grew out of this tradition of thinking about the mind and body. The influence of philosophy on contemporary psychology persists today, as researchers who study emotion still talk about Descartes, and scientists who study happiness often refer to Aristotle (Seaborn & others, 2020; Ward & King, 2016).

In addition to philosophy, psychology has roots in the natural sciences of biology and physiology (Wertheimer & Puente, 2020). Indeed, it was Wilhelm Wundt (1832–1920), a German philosopher-physician, who put together the pieces of the philosophy–natural science puzzle to create the academic discipline of psychology. Some historians like to say that modern psychology was born in December 1879 at the University of Leipzig, when Wundt and his students (most notably E. B. Titchener) performed an experiment to measure the time lag between the instant a person heard a sound and when that person pressed a telegraph key to signal that he had heard it.

What was so special about this experiment? Wundt's study was about the workings of the brain: He was trying to measure the time it took the human

Wilhelm Wundt (1832–1920)
Wundt founded the first psychology laboratory (with his two coworkers) in 1879 at the University of Leipzig in Germany.
bilwissedition/imageBROKER/REX/Shutterstock

brain and nervous system to translate information into action. At the heart of this experiment was the idea that mental processes could be measured. This focus ushered in the new science of psychology.

Wundt and his collaborators concentrated on discovering the basic elements, or "structures," of mental processes. Their approach was called **structuralism** because of its focus on identifying the elemental parts or structures of the human mind. The method they used in the study of mental structures was *introspection,* which literally means "looking inside." You have likely engaged in introspection when you have thought deeply about your feelings or have sought to quietly monitor your own responses to some event. For Wundt's introspection research, a person sat in a laboratory and was asked to think (to introspect) about what was going on mentally as various events took place. For example, the individual might be subjected to a sharp, repetitive clicking sound and then might be asked to report whatever conscious feelings the clicking produced. What made this method scientific was the systematic, detailed self-reports required of the person in the controlled laboratory setting.

Although Wundt is most often regarded as the founder of modern psychology, it was psychologist and philosopher William James (1842–1910), perhaps more than anyone else, who gave the field an American stamp. From his perspective, the key question for psychology is not so much what the mind is (i.e., its structures) as what it is for (its purpose or function). James's view was eventually named *functionalism.*

Structuralism emphasized the components of the mind. In contrast, **functionalism** probed the functions or purposes of the mind and behavior in the individual's adaptation to the environment. Whereas structuralists were looking inside the mind and searching for its structures, functionalists focused on human interactions with the outside world to understand the purpose of thoughts. If structuralism is about the "what" of the mind, functionalism is about the "why."

A central question in functionalism is, why is human thought *adaptive?* When we talk about whether a characteristic is adaptive, we are concerned with how it makes an organism better able to survive. So, the functionalist asks, Why are people better off because they can think than they would be otherwise? Unlike Wundt, James did not believe in the existence of rigid structures of the mind. Instead, James saw the mind as flexible and fluid, characterized by constant change in response to a continuous flow of information from the world. Fittingly, James called the natural flow of thought a "stream of consciousness."

Functionalism fit well with the theory of evolution through natural selection proposed by British naturalist Charles Darwin (1809–1882). In 1859, Darwin published his ideas in *On the Origin of Species.* He proposed the principle of **natural selection,** an evolutionary process in which organisms that are best adapted to their environment will survive and, importantly, produce offspring. Darwin noted that members of any species are often locked in competition for scarce resources such as food. Natural selection is the process by which the environment determines who wins that competition. Darwin said that organisms with biological features that led to survival and reproduction would be better represented in subsequent generations. Over many generations, organisms with these characteristics would constitute a larger percentage of the population. Eventually this process could change an entire species. If environmental conditions changed, however, other characteristics might become favored by natural selection, moving the process in a different direction.

If you are unfamiliar with Darwin's theory of evolution, it is helpful to consider the simple question, Why do giraffes have long necks? An early explanation might have been that giraffes live in places where the trees are very tall, and so the creatures must stretch their necks to get their food—leaves. Lots of stretching might lead to adult giraffes that have longer necks. This explanation does not tell us, though, why giraffes

structuralism
Wundt's approach to discovering the basic elements, or structures, of mental processes.

functionalism
James's approach to mental processes, emphasizing the functions and purposes of the mind and behavior in the individual's adaptation to the environment.

natural selection
Darwin's principle of an evolutionary process in which organisms that are best adapted to their environment will survive and produce offspring.

Introspection has its limits. Many behaviors are hard to explain using introspection. Think about talking, for example. You somehow know where you are heading even as the words are tumbling out of your mouth, but you cannot say where those words are coming from.

William James (1842–1910)
James's approach became known as functionalism.
Historia/REX/Shutterstock

It turns out that giraffes don't eat food from tall trees. Instead, they use their long necks to fight!

are *born* with long necks. A characteristic cannot be passed from one generation to the next unless it is recorded in the *genes,* those collections of molecules that are responsible for heredity.

According to evolutionary theory, species change through random genetic mutation. This means that, essentially by accident, some members of a species are born with genetic characteristics that make them different from other members (for instance, some lucky giraffes being born with unusually long necks). If these changes are adaptive (e.g., if they help those giraffes compete for food, survive, and reproduce), they become more common in members of the species. So, presumably long, long ago, some giraffes were genetically predisposed to have longer necks, and some giraffes were genetically predisposed to have shorter necks. Only those with the long necks survived to reproduce, giving us the giraffes we see today. The survival of the giraffes with long necks is a product of natural selection. Evolutionary theory implies that the way we are, at least partially, is the way that is best suited to survival in our environment.

Darwin's theory continues to influence psychologists today because it is strongly supported by observation. We can make such observations every day. Right now, for example, in your kitchen sink, various bacteria are locked in competition for scarce resources in the form of those tempting food particles from your last meal. When you use an antibacterial cleaner, you are playing a role in natural selection, because you are effectively killing off the bacteria that cannot survive the cleaning agents. However, you are also letting the bacteria that are genetically adapted to survive that cleaner take over the sink. The same principle applies to taking an antibiotic medication at the first sign of a sore throat or an earache. By killing off the bacteria that may be causing the illness, you are creating an environment where their competitors (so-called antibiotic-resistant bacteria) may flourish. These observations powerfully demonstrate Darwinian selection in action.

Wundt and James are recognized as the twin founders of psychological science. Everyone who holds a PhD in psychology today can trace their intellectual family tree back to one of these two men. These two founders of psychology are both similar in an important way—they are both white men. Psychology has long struggled to be a truly diverse science. To read about this issue, see the Challenge Your Thinking.

If structuralism won the battle to be the birthplace of psychology, functionalism won the war. To this day, psychologists continue to talk about the adaptive nature of human characteristics. Indeed, from these beginnings, psychologists have branched out to study more aspects of human behavior than Wundt or James might have imagined. We now examine various contemporary approaches to the science of psychology.

self-quiz

1. Which of the following statements is correct?
   A. There are many controversies in the field of psychology.
   B. Psychologists on the whole agree among themselves on most aspects of the field.
   C. Psychologists do not engage in critical thinking.
   D. There are few controversies in the field of psychology.

2. Of the following, the characteristic that is *not* at the heart of the scientific approach is
   A. skepticism.
   B. critical thinking.
   C. prejudging.
   D. curiosity.

3. Charles Darwin's work is relevant to psychology because
   A. Darwin's research demonstrated that there are few differences between humans and animals.
   B. Darwin's principle of natural selection suggests that human behavior is partially a result of efforts to survive.
   C. Darwin stated that humans descended from apes, a principle that allows psychologists to understand human behavior.
   D. Darwin created functionalism.

APPLY IT! 4. Two psychologists, Clayton and Sam, are interested in studying emotional expressions. Clayton wants to determine whether emotional expression is healthy and if it has an influence on well-being. Sam is interested in describing the types of emotions people express and building a catalog of all the emotions and emotional expressions that exist. In this example, Clayton is most like _____ and Sam is most like _____.
   A. Wilhelm Wundt; William James
   B. William James; Wilhelm Wundt
   C. Wilhelm Wundt; Sigmund Freud
   D. Sigmund Freud; Wilhelm Wundt

Challenge YOUR THINKING

Where Is Everybody?

Wundt, James, Darwin—those recognized as the founders of psychology—are not a diverse group. Where are the women? Where are the people of color? Women and people of color faced the barrier of discrimination in seeking to contribute to psychology. Yet, they did contribute. For instance, Mary Whiton Calkins (pictured below) studied with William James at Harvard. She completed all the requirements for a PhD. Harvard University refused to award her the degree because she was a woman. Still, Calkins contributed to the early science of psychology. She wrote four books and over a hundred scholarly articles on memory, dreams, and a person's sense of self. She was the first woman president of the American Psychological Association (APA, 2011).

Racism also prevented many talented people from contributing to psychology in its early days. Charles Henry Turner, who received a PhD in zoology in 1907, is often recognized as the first Black person to conduct psychological research. He was interested in insect behavior and learning, especially the perceptual capacities of honeybees. He published 70 scholarly articles (Abramson, 2009). Sadly, this brilliant scholar was never able to secure a faculty position in a research-oriented university.

It was not until 1933 that the first Black American woman received a PhD in psychology. Inez Beverly Prosser was forced to leave her home state of Texas to seek a PhD because of segregation (Benjamin & others, 2005). Although she died just a year after receiving her degree, her legacy remains in the

Wellesley College Archives, photographer Charles W. Hearn

students whose lives she improved and in her existence as a role model for those who came after her.

There has been progress, but even today psychology remains marred by a lack of diversity among the scientists creating knowledge (Eagly, 2020; Roberts & others, 2020), the ideas tested (that largely reflect Western and European conceptions of mind and behavior) (Brennan & Houde, 2017), and the people studied (Henrich & others, 2010). We know less about human behavior because we have not listened to diverse voices and benefited from their contribution.

Psychologists ask complex, difficult questions. That is why it is vital that everyone with something to contribute—regardless of gender, gender identity, race/ethnicity, disability status, or sexual orientation—has a place at the table. Creating a truly representative science of human behavior remains a continuing goal. Consider yourself invited to imagine how you, personally, might change our science. Who knows what the next William James might look like?

What Do You Think?

- When you think of a typical psychologist, what does that person look like?

- Do you think it is especially important for psychology to be a diverse field? Why or why not?

---

2 Contemporary Approaches to Psychology

In this section we survey seven different approaches—biological, behavioral, psychodynamic, humanistic, cognitive, evolutionary, and sociocultural—that represent the intellectual backdrop of psychological science.

The Biological Approach

biological approach
An approach to psychology focusing on the body, especially the brain and nervous system.

Some psychologists examine behavior and mental processes through the **biological approach,** which is a focus on the body, especially the brain and nervous system. For example, researchers might investigate the way your heart races when you are afraid or how your hands sweat when you tell a lie. Although a number of physiological systems may be involved in thoughts and feelings, perhaps the largest contribution to physiological psychology has come through the emergence of neuroscience.

Neuroscience is the scientific study of the structure, function, development, genetics, and biochemistry of the nervous system. Neuroscience emphasizes that the brain and

neuroscience
The scientific study of the structure, function, development, genetics, and biochemistry of the nervous system, emphasizing that the brain and nervous system are central to understanding behavior, thought, and emotion.

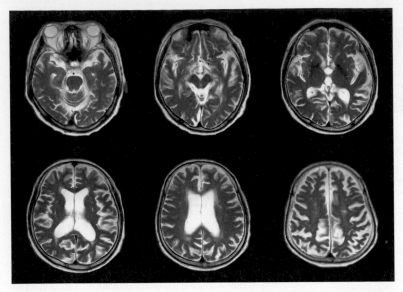

Magnetic resonance imaging allows scientists to get a view of different "slices" of the brain in a living person.
AkeSak/Shutterstock

nervous system are central to understanding behavior, thought, and emotion. Neuroscientists believe that thoughts and emotions have a physical basis in the brain. Electrical impulses zoom throughout the brain's cells, releasing chemical substances that enable us to think, feel, and behave. Our remarkable human capabilities would not be possible without the brain and nervous system, which constitute the most complex, intricate, and elegant system imaginable. Although biological approaches might sometimes seem to reduce complex human experience to simple physical structures, developments in neuroscience have allowed psychologists to understand the brain as an amazingly complex organ, perhaps just as complex as the psychological processes linked to its functioning.

The Behavioral Approach

The **behavioral approach** emphasizes the scientific study of observable behavioral responses and their environmental determinants. It focuses on an organism's visible behaviors, not thoughts or feelings. The psychologists who adopt this approach are called *behaviorists*. Under the intellectual leadership of John B. Watson (1878–1958) and B. F. Skinner (1904–1990), behaviorism dominated psychological research during the first half of the twentieth century.

Skinner (1938) emphasized that psychology should be about what people do—their actions and behaviors—and should not concern itself with things that cannot be seen, such as thoughts, feelings, and goals. He believed that rewards and punishments determine our behavior. For example, a child might behave in a well-mannered fashion because her parents have rewarded this behavior. We do the things we do, say behaviorists, because of the environmental conditions we have experienced and continue to experience.

Contemporary behaviorists still emphasize the importance of observing behavior to understand an individual, and they use rigorous methods advocated by Watson and Skinner. They also continue to stress the importance of environmental determinants of behavior. However, not every behaviorist today accepts the earlier behaviorists' rejection of thought processes, which are often called *cognition*.

behavioral approach
An approach to psychology emphasizing the scientific study of observable behavioral responses and their environmental determinants.

The Psychodynamic Approach

The **psychodynamic approach** emphasizes unconscious thought, the conflict between biological drives, such as the drive for sex, and society's demands, and early childhood family experiences. Practitioners of this approach believe that sexual and aggressive impulses buried deep within the unconscious mind influence the way people think, feel, and behave.

Sigmund Freud, the founder of the psychodynamic approach, theorized that early relationships with parents shape an individual's personality. Freud's theory (1917) was the basis for the therapeutic technique that he called *psychoanalysis,* which involves an analyst's unlocking a person's

Sigmund Freud (1856–1939) *Freud was the founder of the psychodynamic approach.*
Bettmann/Getty Images

psychodynamic approach
An approach to psychology emphasizing unconscious thought, the conflict between biological drives (such as the drive for sex) and society's demands, and early childhood family experiences.

unconscious conflicts by talking with the individual about child-hood memories, dreams, thoughts, and feelings. Certainly, Freud's views have been controversial, but they remain a part of contemporary psychology. Today's psychodynamic theories tend to place less emphasis on sexual drives and more on cultural or social experiences as determinants of behavior (Eagle, 2020; Hogan & Sherman, 2020; Lane, 2020).

The Humanistic Approach

According to humanistic psychologists, warm, supportive behavior toward others helps us to realize our tremendous capacity for self-understanding.
Blend Images/Alamy Stock Photo

humanistic approach
An approach to psychology emphasizing a person's positive qualities, the capacity for positive growth, and the freedom to choose any destiny.

The **humanistic approach** emphasizes a person's positive qualities, the capacity for positive growth, and the freedom to choose one's destiny. Humanistic psychologists stress that people have the ability to control their lives and are not simply controlled by the environment (Maslow, 1971; Rogers, 1961). They theorize that rather than being driven by unconscious impulses (as the psychodynamic approach dictates) or by external rewards (as the behavioral approach emphasizes), people can choose to live by higher human values such as *altruism* (an unselfish concern for another person's well-being) and free will. Many aspects of this optimistic approach appear in research on motivation, emotion, and personality psychology (Lynch & Sheldon, 2020; Pritchard & others, 2020; Sheldon & others, 2020).

The Cognitive Approach

cognitive approach
An approach to psychology emphasizing the mental processes involved in knowing: how we direct our attention, perceive, remember, think, and solve problems.

According to cognitive psychologists, your brain houses a "mind" whose mental processes allow you to remember, make decisions, plan, set goals, and be creative. The **cognitive approach,** then, emphasizes the mental processes involved in knowing: how we direct our attention, perceive, remember, think, and solve problems. For example, cognitive psychologists want to know how we solve math problems, why we remember some things for only a short time but others for a lifetime, and how we can use our imaginations to plan for the future.

Cognitive psychologists view the mind as an active and aware problem-solving system. This view contrasts with the behavioral outlook, which portrays behavior as controlled by external environmental forces. From the cognitive perspective, an individual's mental processes are in control of behavior through memories, perceptions, images, and thinking.

The Evolutionary Approach

evolutionary approach
An approach to psychology centered on evolutionary ideas such as adaptation, reproduction, and natural selection as the basis for explaining specific human behaviors.

Although arguably much of psychology emerges out of evolutionary theory, some psychologists emphasize an **evolutionary approach** that uses evolutionary ideas such as adaptation, reproduction, and natural selection as the basis for explaining specific human behaviors. Evolutionary psychologists argue that just as evolution molds our physical features, such as body shape, it also influences our decision making, level of aggressiveness, fears, and mating patterns (Buss & others, 2020). Thus, evolutionary psychologists say, the way we are is traceable to problems early humans faced in adapting to their environments (Cosmides, 2011).

Evolutionary psychologists believe that their approach provides an umbrella that unifies the diverse fields of psychology. Not all psychologists agree with this conclusion, however. For example, some critics stress that the evolutionary approach provides an inaccurate explanation of gender and social roles, and it does not adequately account for cultural diversity and experiences (Eagly & Wood, 2013). Yet keep in mind that even psychologists who disagree with the application of the evolutionary approach to psychological characteristics still agree with the general principles of evolutionary theory.

Human beings originally evolved long ago in a very different environment than we occupy today. The survivors were those who were most able to endure extremely difficult circumstances, struggling to find food, avoid predators, and create social groups. What do you think were the most adaptive traits for these early people? To what specific environments are humans adapting even now?

The Sociocultural Approach

The **sociocultural approach** examines the influences of social and cultural environments on behavior. Socioculturalists argue that understanding a person's behavior requires knowing about the cultural context in which the behavior occurs. (*Culture* refers to the shared knowledge, practices, and attitudes of groups of people and can include language, customs, and beliefs about what behavior is appropriate and inappropriate.) The sociocultural approach often includes *cross-cultural* research, meaning research that compares individuals in various cultures to see how they differ on important psychological attributes. Cross-cultural research is important for testing the assumption that findings for one culture also generalize to other cultural contexts, and as such it allows psychologists to test for the possibility that some characteristics are universal.

The sociocultural approach focuses not only on comparisons of behavior across countries but also on the behavior of individuals from different ethnic and cultural groups within a country. In light of rising cultural diversity in the United States in recent years, there has been increasing interest in understanding the behavior of African Americans, Latinos, and Asian Americans, especially in terms of the factors that have restricted or enhanced their ability to adapt and cope with living in a predominantly non-Latino white society.

sociocultural approach
An approach to psychology that examines the ways in which social and cultural environments influence behavior.

Summing Up the Seven Contemporary Approaches

These seven approaches to studying psychology provide different views of behavior, and therefore each may contribute uniquely valuable insights. Think about the simple experience of sleeping, something we all do. Scientists from the biological approach might probe the brain processes that help us fall asleep, stay asleep, and wake up. Behaviorists might be interested in how rewards and punishers in our environment promote sleep or wakefulness. Psychodynamic thinkers might be most interested in our dreams during sleep and what they reveal about unconscious desires. Humanistic psychologists might study the role of sleep in self-care and how we can create a world that allows people to get enough sleep. The cognitive approach might examine the role that sleep plays in memory consolidation. Evolutionary psychologists might seek to explain the function of sleep and why it is better for members of our species to sleep when and how they do. Finally, the sociocultural approach might probe cultural customs around sleep, such as whether a child sleeps alone or with others in a "family bed."

Although these approaches differ from each other, they all fit under the umbrella of the **biopsychosocial approach**. From this perspective, behavior is influenced by biological factors (such as genes), psychological factors (such as childhood experiences, learning histories, thoughts, and emotions), and sociocultural factors (such as gender, ethnicity, or socioeconomic status). Taking a biopsychosocial approach means acknowledging that all of these factors can combine and influence one another and behavior. From the biopsychosocial perspective, biological, psychological, and social factors are all significant ingredients in producing behavior.

biopsychosocial approach
A perspective on human behavior that asserts that biological, psychological, and social factors are all significant ingredients in producing behavior. All of these levels are important to understanding human behavior.

These broad approaches are reflected in the variety of specialties within which psychologists work (Figure 2). Many of these specialties are represented by chapters in *Experience Psychology*. As you read, keep in mind that psychology is a science in which psychologists work together collaboratively to examine a wide range of research questions. Indeed, many times scholars from different specialties within psychology join forces to understand some aspect of human behavior. It is the purpose of the "Intersection" feature to review research that represents a collaboration among scientists from different specialties to answer the same question.

Specialization	Focus of Specialists
Behavioral Neuroscience	Behavioral neuroscience focuses on biological processes, especially the brain's role in behavior.
Sensation and Perception	Sensation and perception researchers focus on the physical systems and psychological processes of vision, hearing, touch, and smell that allow us to experience the world.
Learning	Learning specialists study the complex process by which behavior changes to adapt to shifting circumstances.
Cognitive	Cognitive psychology examines attention, consciousness, information processing, and memory. Cognitive psychologists are also interested in cognitive skills and abilities such as problem solving, decision making, expertise, and intelligence.
Developmental	Developmental psychology examines how people become who they are, from conception to death, concentrating on biological and environmental factors.
Motivation and Emotion	Researchers from a variety of specializations are interested in these two aspects of experience. Motivation researchers examine questions such as how individuals attain difficult goals. Emotion researchers study the physiological and brain processes that underlie emotional experience, the role of emotional expression in health, and the possibility that emotions are universal.
Personality	Personality psychology focuses on the relatively enduring characteristics of individuals, including traits, goals, motives, genetics, and personality development.
Social	Social psychology studies how social contexts influence perceptions, social cognition, and attitudes. Social psychologists study how groups influence attitudes and behavior.
Clinical and Counseling	Clinical and counseling psychology, the most widely practiced specialization, involves diagnosing and treating people with psychological problems.
Health	Health psychology emphasizes psychological factors, lifestyle, and behavior that influence physical health.
Industrial and Organizational (I/O)	I/O psychology applies findings in all areas of psychology to the workplace.
Community	Community psychology is concerned with providing accessible care for people with psychological problems. Community-based mental health centers are one means of delivering such services as outreach programs.
School and Educational	School and educational psychology centrally concerns children's learning and adjustment in school. School psychologists in elementary and secondary school systems test children, make recommendations about educational placement, and work on educational planning teams.
Environmental	Environmental psychologists explore the effects of physical settings in most major areas of psychology, including perception, cognition, and learning, among others. An environmental psychologist might study how different room arrangements influence behavior or what strategies might be used to reduce human behavior that harms the environment.
Psychology of Women	Psychology of women stresses the importance of integrating information about women with current psychological knowledge and applying that information to society and its institutions.
Forensic	Forensic psychology applies psychology to the legal system. Forensic psychologists might help with jury selection or provide expert testimony in trials.
Sport	Sport psychology applies psychology to improving sport performance and enjoyment of sport participation.
Cross-Cultural	Cross-cultural psychology studies culture's role in understanding behavior, thought, and emotion, with a special interest in whether psychological phenomena are universal or culture specific.

FIGURE 2 **Areas of Specialization in Psychology** Psychology has many overlapping subfields.

1. The approach to psychology that is most interested in early childhood relationships is
   A. evolutionary psychology.
   B. cognitive psychology.
   C. psychodynamic psychology.
   D. behavioral psychology.

2. The approach to psychology that views psychological distress as a result of persistent negative thoughts is
   A. the humanistic approach.
   B. the behavioral approach.
   C. the sociocultural approach.
   D. the cognitive approach.

3. The approach to psychology that focuses on self-fulfillment, altruism, and personal growth is
   A. the cognitive approach.
   B. the behavioral approach.

C. the psychodynamic approach.
D. the humanistic approach.

APPLY IT! 4. In 2007 a father posted a video clip of his young sons on YouTube. Widely known as "Charlie Bit My Finger," the clip, which quickly went viral, shows a British baby laughing hysterically as he bites his crying brother's finger. The clip is still one of the most popular videos on YouTube. If you haven't seen it, take a look: www.youtube.com/watch?v=_OBlgSz8sSM. What explains the clip's enduring appeal? Each of the contemporary approaches we have reviewed might offer an explanation. Which of the following is most like what a *psychodynamic* thinker might say?

A. Human beings have been *rewarded* for watching children bite each other.
B. Adorable children are *universally* loved.
C. Human beings have an *unconscious* desire to harm their siblings, which is disguised by the humor of the clip.
D. This clip demonstrates that cuteness is an important *adaptation*. Cute kids are more likely to survive and reproduce.

3 Psychology's Scientific Method

Science is not defined by *what* it investigates, but by *how* it investigates. Whether you study photosynthesis, butterflies, Saturn's moons, or happiness, the *way* you study your question of interest determines whether your approach is scientific. The scientific method is how psychologists gain knowledge about mind and behavior. A key theme in the scientific method is that knowledge comes from empirical research.

It is the use of the scientific method that makes psychology a science. Indeed, most of the studies psychologists publish in research journals follow the scientific method, which may be summarized in these five steps (Figure 3):

1. Observing some phenomenon

2. Formulating hypotheses and predictions

3. Testing through empirical research

4. Drawing conclusions

5. Evaluating conclusions

1. OBSERVING SOME PHENOMENON   The first step in conducting a scientific inquiry involves observing some phenomenon in the world. The critical thinking, curious psychologist sees something and wants to know why or how it is the way it is. Inspiration for scientific inquiry can come from contemporary social problems, current events, personal experiences, and more. The phenomena that scientists study are called variables, a word related to the verb *to vary*. A **variable** is anything that can change.

variable
Anything that can change.

For example, one variable that interests psychologists is happiness. Some people seem to be happier than others. What might account for these differences? As scientists consider answers to such questions, they often develop theories. A **theory** is a broad idea or set of closely related ideas that attempts to explain observations. Theories seek to explain why certain things are as they are or why they have happened. Theories can be used to make predictions about future observations. For instance, some psychologists theorize that the most important human need is the need to belong to a social group (Leary, 2020; Hudson & others, 2020). This theory would seek to explain human

theory
A broad idea or set of closely related ideas that attempts to explain observations and to make predictions about future observations.

spend on yourself

spend on someone else

1
Observing Some Phenomenon

We feel good when we give someone a gift. However, do we genuinely feel better giving something away than we might feel if we could keep it? Elizabeth Dunn, Lara Aknin, and Michael Norton (2008) decided to test this question.

2
Formulating Hypotheses and Predictions

These researchers hypothesized that spending money on other people would lead to greater happiness than spending money on oneself.

3
Testing Through Empirical Research

In an experiment designed to examine this prediction, the researchers randomly assigned undergraduate participants to receive money ($5 or $20) that they had to spend either on themselves or on someone else by 5 P.M. that day. Those who spent the money on *someone else* reported greater happiness that night.

4
Drawing Conclusions

The experiment supported the hypothesis that spending money on others can be a strong predictor of happiness. Money might not buy happiness, the researchers concluded, but spending money in a particular way, that is, on other people, may enhance happiness.

5
Evaluating Conclusions

The experimental results were published in the prestigious journal *Science*. Now that the findings are public, other researchers might investigate related topics and questions inspired by this work, and their experiments might shed further light on the original conclusions.

FIGURE 3   Steps in the Scientific Method: Is It Better to Give Than to Receive? This figure shows how the steps in the scientific method were applied in a research experiment examining how spending money on ourselves or others can influence happiness (Dunn, Aknin, & Norton, 2008). The researchers theorized that although money does not typically buy happiness, the way we spend it might well predict happy feelings.

(Gift): Stockbyte/Getty Images; (Multiple faces): Flashpop/Getty Images; (Woman with money): JUPITERIMAGES/Brand X/Alamy Stock Photo

> **For each step in the process, what decisions did the researchers make, and how did those decisions influence the research?** > **Are the findings counterintuitive or not?** > **This study was inspired by the saying, "It is better to give than to receive." Does the study do a good job of evaluating that cliché? How else might a researcher have addressed this question?**

behaviors through the need to belong, and scientists might expect that when our need to belong is not met we should be especially distressed. For instance, a recent study showed that those whose belongingness needs were not met during the COVID-19 pandemic were at higher risk of suicide. Compared to concerns about job loss, thwarted belongingness was a stronger predictor of suicide risk (Gratz & others, 2020).

A key characteristic of a scientific theory is that it must be *falsifiable,* meaning that even a scientist who believes that a theory is true must be able to generate ideas about research that would prove the theory wrong and test those ideas. This is what separates scientific theories from beliefs and opinions.

A scientist must be able to anticipate being wrong and remain open to that possibility.

2. FORMULATING HYPOTHESES AND PREDICTIONS

hypothesis
A testable prediction that derives logically from a theory.

The second step in the scientific method is stating a hypothesis. A **hypothesis** is a testable prediction that derives logically from a theory. A theory can generate many hypotheses. If more and more hypotheses related to a theory turn out to be true, the theory gains in credibility. So, a researcher who believes that social belonging is the most important aspect of human functioning might predict that people who belong to social groups will be happier than

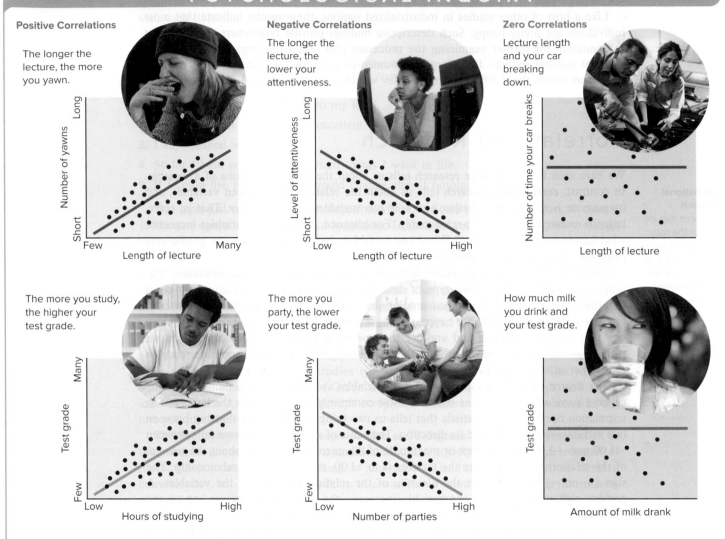

Positive Correlations

The longer the lecture, the more you yawn.

The more you study, the higher your test grade.

Negative Correlations

The longer the lecture, the lower your attentiveness.

The more you party, the lower your test grade.

Zero Correlations

Lecture length and your car breaking down.

How much milk you drink and your test grade.

FIGURE 4 **Scatter Plots Showing Positive and Negative Correlations** A positive correlation is a relationship in which two factors vary in the same direction, as shown in the two scatter plots on the left. A negative correlation is a relationship in which two factors vary in opposite directions, as shown in the two scatter plots on the right. Note that each dot on these graphs represents one person's (or one class's) score on the two variables of interest.

(Yawning): Doug Menuez/Getty Images; (Attentiveness): Ariel Skelley/Blend Images LLC; (Mechanics): ColorBlind Images/Getty Images; (Male): Andriy Popov/123RF; (Group): Stockbyte/Getty Images; (Female): JGI/Blend Images LLC

For each graph, consider these questions:
> How might one of these variables cause the other? Can you imagine a way that the causal direction could be reversed? > What is a possible third variable that might account for these relationships? > Identify two variables that you think are positively correlated and two that are negatively correlated. > What would the graphs look like if the two variables were not systematically related?

Crime, then, is probably also "linked" to air-conditioner sales and repair profits, as well as swimsuit sales.

To understand the third variable problem, consider the following example. A researcher measures two variables: the number of ice cream cones sold in a town and the number of violent crimes that occur in that town throughout the year. The researcher finds that ice cream cone sales and violent crimes are positively correlated, to the magnitude of +.50. This high positive correlation would indicate that as ice cream sales increase, so does violent crime. Would it be reasonable for the local paper to run the headline "Ice Cream Consumption Leads to Violence"? Should concerned citizens gather outside the local Frosty Freeze to stop the madness? Probably not. Perhaps you have already thought of the third variable that might explain this correlation—heat. Research has shown that crime goes up when the temperature rises (Miles-Novelo & Anderson, 2019), and, as any ice cream shop manager will tell you, ice cream sales are higher when it is warm outside. Given the potential problems with third variables, why do researchers conduct correlational studies? Three important reasons are as follows:

■ Some important questions can be investigated *only* by using a correlational design. Such questions may involve variables that can only be measured or observed, such as personality traits, genetic factors, and ethnic background.

■ Sometimes the variables of interest are real-world events that influence people's lives, such as the global COVID-19 pandemic.

■ Finally, for some research questions it would be unethical to do the research in any other way. For example, it would be unethical for an experimenter to direct expectant mothers to smoke varying numbers of cigarettes to see how cigarette smoke affects birth weight and fetal activity. Instead a correlational study might measure these variables.

Although we have focused on relationships between just two variables, researchers often measure many variables in their studies. This way, they can examine whether a relationship between two variables is explained by a third variable (or a fourth or fifth variable). An interesting research question that has been addressed in this fashion is, do happy people live longer? In one study, 2,000 Mexican Americans aged 65 and older were interviewed twice over the course of 2 years (Ostir & others, 2000). In the first assessment, participants completed measures of happiness but also reported about potential third variables such as diet, physical health, smoking, marital status, and distress. Two years later, the researchers contacted the participants again to see who was still alive. Even with these many potential third variables taken into account, happiness predicted who was still living 2 years later.

C Squared Studios/Photodisc/Getty Images

Correlational studies are useful, too, when researchers are interested in everyday experience. For example, some correlational researchers use the *experience sampling method (ESM)* to study people in their natural settings. This approach involves having people report on their daily experiences in a diary a few times a day or complete measures of their mood and behavior whenever they are beeped by an electronic organizer or smartphone.

LONGITUDINAL DESIGNS   One way that correlational researchers can deal with the issue of causation is to employ a special kind of systematic observation called a **longitudinal design**. Longitudinal research involves observing and measuring the same variables periodically over time. Longitudinal research can suggest potential causal relationships because if one variable is thought to cause changes in another, it should at least come before that variable in time (Grimm & others, 2017).

longitudinal design
A special kind of systematic observation, used by correlational researchers, that involves obtaining measures of the variables of interest in multiple waves over time.

One intriguing longitudinal study is the Nun Study, conducted by Snowdon and other scholars (Mortimer & others, 2009; SantaCruz & others, 2011; Snowdon, 2003; Weinstein & others, 2019). The study began in 1986 and has followed a sample of 678 School Sisters of Notre Dame ever since. The nuns ranged in age from 75 to 103 when the study began. These women completed a variety of psychological and physical measures annually. This sample is unique in many respects. However, some characteristics render the participants an excellent group for correlational research. For one thing, many potential extraneous third variables are relatively identical for all the women in the group. Their gender, living conditions, diet, activity levels, marital status, and religious participation are essentially held constant, providing little chance that differences in these variables can explain results.

In one study, researchers examined the relationship between happiness and longevity using this rich dataset. All of the nuns had been asked to write a spiritual autobiography when they entered the convent

Do It!

Pull out some old "sent messages" in your email from as far back as you can find. Count up your positive emotion words and negative emotion words. How do the themes in those messages reflect your life at that time and today? Does looking at your old emails change the way you might think about the results of the Nun Study? Explain.

Big Data in Psychological Science

Consider all the public records that accumulate in the course of a human life: birth certificates, drivers licenses, diplomas, marriage licenses (and divorce decrees), and death certificates. Such records tell us about important human behavior: who gets married at what age, how long marriages last, how long people live. Information about communities, such as how many restaurants, churches, hospitals, schools, childcare facilities, or prisons exist in a location, may reveal important characteristics of those communities. Records of how much money people spend on healthcare, childcare, or vacations can also be sources of information about people. All of these records are (or could be) data that provide insight into human life.

Big data refers to these and other large, naturally occurring sources of data. Increasingly, psychologists have used big data to address their research questions (Giusino & others, 2019; Oswald, 2020; Strauss & Grand, 2020).

Big data are not just public records. Most of what you do online is tracked by someone, somewhere. Carrying your phone with you wherever you go produces enormous amounts of data. Psychologists use big data in many ways, such as tracking language used on Twitter (Kovacs & Kleinbaum, 2020) or coding images posted on Instagram (Yu & others, 2020). Big data have been used to predict risk for psychological disorders (Thorstad & Wolff, 2019), to measure racism and health disparities (Chae & others, 2015), and to estimate the number of LGBT people in the United States (Stephens-Davidowitz, 2015), among other things.

Big data are valuable to psychologists because they often indicate behavior (Boyd & others, 2020). People may be inaccurate in their reports of what they do for a host of reasons, including forgetting and embarrassment (Webster & others, 2020). Using naturally occurring data provides a direct look at human life. Still, when public records and social media activity become scientific data, issues of ethics and privacy come to the fore (Favaretto & others, 2020). It is always important to read user agreements to know how your data might be used. In addition, being cautious online is important. Keep in mind that public posts to social media come with no assurance of confidentiality and may be seen by potential employers and others.

Rafal Olechowski/Shutterstock

(for some, as many as 80 years before). Danner and colleagues (2001) were given access to these documents and used them as indicators of happiness earlier in life by counting the number of positive emotions expressed in the autobiographies (note that here we have yet another operational definition of happiness). Women who included positive emotion in their autobiographies when they were in their early 20s were 2.5 times more likely to survive when they were in their 80s and 90s.

Longitudinal designs provide ways by which correlational researchers may attempt to demonstrate causal relations among variables. Still, it is important to be aware that even in longitudinal studies, causal relationships are not completely clear. For example, the nuns who wrote happier autobiographies may have had happier childhood experiences that might be influencing their longevity, or a particular genetic factor might explain both their happiness and their survival. As you read about numerous correlational research studies throughout *Experience Psychology,* do so critically, and with some skepticism, and consider that even the brightest scientist may not have thought of all of the potential third variables that might have explained the results. Psychologists who conduct correlational research often have a sense of the causal order they assume exists for the variables they study (Grosz & others, 2020). Nevertheless, keep in mind how easy it is to assume causality when two events or characteristics are merely correlated. Think about those innocent ice cream cones and critically evaluate conclusions that may be drawn from simple observation.

Experimental Research

experiment
A carefully regulated procedure in which the researcher manipulates one or more variables that are believed to influence some other variable.

To determine whether a causal relationship exists between variables, researchers must use experimental methods. An **experiment** is a carefully regulated procedure in which the researcher manipulates one or more variables that are believed to influence some other variable. Imagine that a researcher notices that people who listen to classical music seem to be of above-average intelligence. A correlational study on this question would not tell us if listening to classical music *causes* increases in intelligence. To demonstrate causation, the researcher would manipulate whether or not people listen to classical music. The researcher might create two groups: one that listens to classical music and one that listens to pop music. To test for differences in intelligence, the researcher would then measure intelligence.

If that manipulation led to differences between the two groups on intelligence, we could say that the manipulated variable *caused* those differences. In other words, the experiment has demonstrated cause and effect. This notion that experiments can demonstrate causation is based on the idea that if participants are randomly assigned to groups, the only systematic difference between them must be the manipulated variable. **Random assignment** means that researchers assign participants to groups by chance. This technique reduces the likelihood that the experiment's results will be due to any preexisting differences between groups. The logic of random assignment is that if only chance determines which participants are assigned to each group in an experiment, the potential differences on other characteristics will cancel out over the long run. So, for instance, in the example of classical music and intelligence, we might wonder if it is possible that the groups differed on intelligence to begin with. Because participants were randomly assigned, we assume that intelligence is spread across the groups evenly.

random assignment
Researchers' assignment of participants to groups by chance, to reduce the likelihood that an experiment's results will be due to preexisting differences between groups.

Random assignment does not always work. One way to improve its effectiveness is to start with a relatively large pool of people. Let's say that you decide to do that study—examining whether listening to classical music (as compared to no music) prior to taking an intelligence test leads to higher scores on the test. Although you wisely use random assignment, you begin with just 10 people. Unbeknownst to you, there are two geniuses (people with extraordinarily high IQs) in that small pool of participants. Each person has a 50-50 chance of ending up in either group, so there is a 25 percent chance that both geniuses will end up in the same group, meaning there is a 25 percent chance that your groups will differ systematically in intelligence before you even start the study. In contrast, if your study had begun with, say, 100 people, intelligence scores will likely be more evenly spread in the overall pool. When these individuals are randomly assigned, differences in intelligence would be much more likely to cancel out across the two groups. That is why it is important that random assignment is allowed to work its magic on a larger pool of people.

Fuse/Getty Images

To get a sense of what experimental studies, as compared to correlational studies, can tell us, consider the following example. Psychologists have long assumed that experiencing one's life as meaningful is an important aspect of psychological well-being (Frankl, 1946/1984; Steger & Frazier, 2005). Because surveys that measure meaning in life and well-being correlate positively (i.e., the more meaningful your life, the happier you are), the assumption has been that experiencing meaning in life causes greater happiness. Because the studies involved in exploring this relationship have been correlational, however, the causal pathway is unknown. Meaning in life may lead people to be happier, but the reverse might also be true: Happiness might make people feel that their lives are more meaningful.

To address this issue, King and colleagues (2006; King & Hicks, 2020) conducted a series of laboratory experiments. The researchers had some people listen to happy music and other participants listen to neutral music. Participants who listened to happy music rated their lives as more meaningful than did individuals who listened to neutral music. Note that participants were randomly assigned to one of two conditions, happy music or neutral music, and then rated their meaning in life using a questionnaire. In this case happiness was operationally defined by the type of music participants listened to, and meaning in life was operationally defined by ratings on a questionnaire. Because participants were randomly assigned to conditions, we can assume that the only systematic difference between the two groups was the type of music they heard. As a result, we can say that the happy music caused people to rate their lives as more meaningful.

INDEPENDENT AND DEPENDENT VARIABLES   Experiments have two types of variables: independent and dependent. An **independent variable** is a manipulated experimental factor. The independent variable is the variable that the experimenter changes to see what its effects are; it is a potential cause. Any experiment may include several independent variables or factors that are manipulated to determine their effect on some outcome. In the study of positive mood and meaning in life, the independent variable is mood (positive versus neutral), operationally defined by the type of music participants listened to.

Sometimes the independent variable is the social context. Social psychologists often manipulate the social context with the help of a **confederate**, a person who is given a role to play in a study so that the social context can be manipulated. For example, if researchers are interested in reactions to being treated rudely, they might have a confederate treat participants rudely (or not).

> 🖊 *The independent variable and dependent variable are two of the **most important concepts** in psychological research. You should **memorize their meanings now**. Remember that the independent variable is the cause, and the dependent variable is the effect.*

A **dependent variable** is the outcome—the factor that can change in an experiment in response to changes in the independent variable. As researchers manipulate the independent variable, they measure the dependent variable for any resulting effect. In the study of mood and meaning in life, meaning in life was the dependent variable. The similarity in their names has plagued students of psychology forever, but independent and dependent variables are very different things. Independent variables are manipulated. Dependent variables are measured.

EXPERIMENTAL AND CONTROL GROUPS   Experiments can involve one or more experimental groups and one or more control groups. In an experiment, the researcher manipulates the independent variable to create these groups. An **experimental group** consists of the participants in an experiment who receive the treatment that is of interest to the researcher, or a particular drug under study—that is, the participants who are exposed to the change that the independent variable represents. A **control group** in an experiment is as much like the experimental group as possible and is treated in every way like the experimental group except for that change. The control group provides a comparison against which the researcher can test the effects of the independent variable. In the study of meaning in life described earlier, participants who listened to happy music were the experimental group, and those who heard neutral music were the control group.

> 🖊 *Coming up with a good control group can be a challenge. Say you want to do a study on the influence of smiling on social behaviors. Your experimental group will interact with a confederate who smiles a lot during the experiment. What will you have happen in your control group? Will the confederate keep a blank expression? Frown? Smile just a little?*

Even very subtle experimental manipulations can have impact on how people think, feel, and behave. To read about research using experimental research to understand the sharing of misinformation online, see the Intersection.

Within-Participant Designs   One way to ensure that a control group and an experimental group are as similar as possible is to use a *within-participant design,* in which the participants serve as their own control group. In such a design, rather than relying on

independent variable
A manipulated experimental factor; the variable that the experimenter changes to see what its effects are.

confederate
A person who is given a role to play in a study so that the social context can be manipulated.

dependent variable
The outcome; the factor that can change in an experiment in response to changes in the independent variable.

experimental group
The participants in an experiment who receive the drug or other treatment under study—that is, those who are exposed to the change that the independent variable represents.

control group
The participants in an experiment who are as much like the experimental group as possible and who are treated in every way like the experimental group except for a manipulated factor, the independent variable.

Cognitive Psychology and Health Psychology: How Can We Combat COVID-19 Misinformation Online?

Generally, people are motivated to share information on social media that is true (not false) (Pennycook & others, 2020). Yet, when engaging with social media, people may be distracted by aspects of a message—such as whether it fits with their preexisting attitudes or their political views—and share it without considering whether it is true. Indeed, a key obstacle in battling the COVID-19 pandemic was misinformation about the virus, spread on social media. On Facebook and Twitter, people shared information that was objectively false—for instance, that coconut oil or silver could cure the virus. People also shared false information about the effectiveness of preventive measures, such as wearing masks and social distancing. This false information likely harmed public health efforts to control the virus in the United States. How can we combat "fake news" online? Two recent studies sought to answer this question and see if it might be possible to nudge people subtly to be more discerning in their social media posts.

In a first study, the researchers (Pennycook & others, 2020) asked participants to read 15 true and 15 false headlines about COVID-19. They rated each headline in two ways: whether they thought it was true or false and whether they would share the headline on social media. This descriptive study showed that people were actually pretty good at telling what was true and what was false. However, the truth of a headline did not consistently lead people to be more likely to share it.

Fokusiert/Getty Images

The second study was an experiment. Participants were assigned randomly to one of two groups. In the experimental condition, participants were first asked to judge whether a headline that was unrelated to COVID-19 was true or false. The control group did not complete this very brief activity. Then all participants completed the dependent variable—they rated 15 true and 15 false headlines about COVID-19, for whether they thought each headline was true and whether they would share it online.

The results showed that truth mattered more to participants who first considered the truth or falsehood of an unrelated headline. Just thinking for a moment about whether an unrelated headline was true made them more likely to share true COVID-19 information and less likely to share false COVID-19 information (Pennycook & others, 2020).

This simple study shows how the experimental method can be used to address very important questions. The results tell us that when the stakes are life and death, even simple interventions can help clear up misinformation.

\\ Did you encounter fake news about COVID-19 online? Did you share it? Why or why not?

random assignment to produce equivalent groups, a researcher has the same group of participants experience the various conditions in the study. Let's say that a researcher predicts that the presence of other people (the independent variable) decreases math performance (the dependent variable). They might have participants complete math problems first while alone and then with others, and compare performance in the two conditions to test that prediction. The advantages to a within-participant design include requiring only half the number of participants and knowing that the groups are the very same people in each condition. The disadvantages include concerns about whether the two math tests are really equivalent and how the order of the conditions might influence their effects on performance.

Quasi-Experimental Designs Another approach to experimental research is to use a *quasi-experimental design*. As the prefix *quasi-* ("as if") suggests, this type of design is similar to an experiment, but it is not quite

When the participants who are in the experimental and control groups are different people, the design is called a *between-participant design*.

the same thing. The key difference is that a quasi-experimental design does not include random assignment of participants to a condition, because such assignment is either impossible or unethical.

Quasi-experimental designs might be used for studies that examine the differences between groups of people who have had different experiences—say, soldiers who have seen combat versus those who have not, or children whose school was destroyed by a tornado versus those in a neighboring town whose school was not affected. In a quasi-experimental design, a researcher examines participants in varying groups, but their assignment to groups is not determined by random assignment.

For example, a researcher interested in the influence of using online learning tools on performance in introductory psychology classes might compare students from two different sections of a class—one that uses online tools and one that does not. Of course, students typically choose what section of a course they take, so the experimenter cannot randomly assign them to sections. Examining differences between the groups might provide information about the merits of online learning tools. However, there might be confounds (e.g., whether students are morning people or not) that account for differences between the groups. Although quasi-experimental designs are common, it is important to keep in mind that they do not allow for the strong causal conclusions that can be drawn from experiments.

> The late Donald Campbell, an eminent psychologist, referred to quasi-experimental designs as "queasy experimental designs" because of the way they often make hard-core experimentalists feel.

> Quasi-experimental designs tend to be high on external validity.

SOME CAUTIONS ABOUT EXPERIMENTAL RESEARCH

Validity refers to the soundness of the conclusions that a researcher draws from an experiment. Two broad types of validity matter to experimental designs. The first is **external validity**, which refers to the degree to which an experimental design really reflects the real-world issues it is supposed to address. That is, external validity is concerned with the question, do the experimental methods and the results generalize—do they apply—to the real world?

Imagine, for example, that a researcher is interested in the influence of stress (the independent variable) on creative problem solving (the dependent variable). The researcher randomly assigns individuals to be blasted with loud noises at random times during the session (the high-stress or experimental group) or to complete the task in relative quiet (the control group). As the task, the researcher gives all participants a chance to be creative by asking them to list all of the uses they can think of for a cardboard box. Counting up the number of uses that people list, the researcher discovers that those in the high-stress group generated fewer uses of the box. This finding might indicate that stress reduces creativity. In considering the external validity of this study, however, we might ask some questions: How similar are the blasts of loud, random noises to the stresses individuals experience every day? Is listing uses for a cardboard box really an indicator of creativity? We are asking, in other words, if these operational definitions do a good job of reflecting the real-world processes they are supposed to represent.

The second type of validity is **internal validity**, which refers to the degree to which changes in the dependent variable are due to the manipulation of the independent variable (Ejelöv & Luke, 2020). In the case of internal validity, we want to know whether the experimental methods are free from biases and logical errors that may render the results suspect. Although experimental research is a powerful tool, it requires safeguards. Expectations and biases can, and sometimes do, tarnish results, as we next consider.

Experimenter Bias

Demand characteristics are any aspects of a study that communicate to the participants how the experimenter wants them to behave. One such influence is experimenter bias. **Experimenter bias** occurs when the experimenter's expectations influence the outcome of the research. No one designs an experiment without wanting meaningful results. Consequently, experimenters can sometimes subtly communicate to participants what they want the participants to do. For example, consider research on the effects of oxytocin, a neurotransmitter that is sometimes considered a "love drug" in interpersonal trust. In one study the researchers showed that those randomly assigned to receive oxytocin in a nasal spray (the experimental group) were more trusting than a group

validity
The soundness of the conclusions that a researcher draws from an experiment. In the realm of testing, the extent to which a test measures what it is intended to measure.

external validity
The degree to which an experimental design actually reflects the real-world issues it is supposed to address.

internal validity
The degree to which changes in the dependent variable are due to the manipulation of the independent variable.

demand characteristics
Any aspects of a study that communicate to the participants how the experimenter wants them to behave.

experimenter bias
The influence of the experimenter's expectations on the outcome of research.

who received a control substance (the control group) (Mikolajczak & others, 2010). However, when these investigators tried to replicate this study, they failed to get the same results (Lane & others, 2015, 2016; Mierop & others, 2020). The reason? The research team realized that, in the first study, the experimenter who interacted with participants was aware of the condition assignment the participants had received. Even without intending it, those who interact with participants can influence participant behavior.

In a classic study, Rosenthal (1966) turned college students into experimenters. He randomly assigned the participants rats from the same litter. Half of the students were told that their rats were "maze bright," whereas the other half were told that their rats were "maze dull." The students then conducted experiments to test their rats' ability to navigate mazes. The results were stunning. The so-called maze-bright rats were more successful than the maze-dull rats at running the mazes. The only explanation for the results is that the college students' expectations affected the rats' performance.

Often the participants in psychological studies are not rats but people. Imagine that you are an experimenter and you know that a participant is going to be exposed to disgusting pictures in a study. Is it possible that you might treat the person differently than you would if you were about to show them photos of cute kittens? The reason experimenter bias is important is that it introduces systematic differences between the experimental and control groups, so that we cannot know if those who looked at disgusting pictures were more, say, upset because of the pictures or because of different treatment by the experimenter. Like third variables in correlational research, these systematic biases are called *confounds*. In experimental research, confounds are factors that "ride along" with the experimental manipulation, systematically and undesirably influencing the dependent variable. Experimenter bias, demand characteristics, and confounds may all lead to differences between groups on the dependent variable that threatens the validity of experimental results.

Research Participant Bias and the Placebo Effect

Like experimenters, research participants may have expectations about what they are supposed to do and how they should behave, and these expectations may affect the results of experiments. **Research participant bias** occurs when the behavior of research participants during the experiment is influenced by how they think they are supposed to behave or their expectations about what is happening to them.

One example of the power of participant expectations is the placebo effect. The **placebo effect** occurs when participants' expectations, rather than the experimental treatment, produce an outcome. Participants in a drug study might be assigned to an experimental group that receives a pill containing an actual painkiller or to a control group that receives a placebo pill. A **placebo** is a harmless substance that has no physiological effect. This placebo is given to participants in a control group so that they are treated identically to the experimental group except for the active agent—in this case, the painkiller. Giving individuals in the control group a placebo pill allows researchers to determine whether changes in the experimental group are due to the active drug agent and not simply to participants' expectations.

Another way to ensure that neither the experimenter's nor the participants' expectations affect the outcome is to design a **double-blind experiment**. In this design, neither the experimenter administering the treatment nor the participants are aware of which participants are in the experimental group and which are in the control group until the results are calculated. This setup ensures that the experimenter cannot, for example, make subtle gestures signaling who is receiving a drug and who is not. A double-blind study allows researchers to identify the specific effects of the independent variable from the possible effects of the experimenter's and the participants' expectations about it.

Replication

Recall that a key aspect of the scientific method is evaluating conclusions that have been reached in a research study. The process of replication, defined earlier, has a critical role to play in this process. Recently, psychologists have become especially concerned about whether

research participant bias
In an experiment, the influence of participants' expectations, and of their thoughts on how they should behave, on their behavior.

placebo effect
The situation where participants' expectations, rather than the experimental treatment, produce an experimental outcome.

placebo
In a drug study, a harmless substance that has no physiological effect, given to participants in a control group so that they are treated identically to the experimental group except for the active agent.

double-blind experiment
An experimental design in which neither the experimenter nor the participants are aware of which participants are in the experimental group and which are in the control group until the results are calculated.

Volunteering for a double-blind drug study might seem risky. Would you do it? How might differences in willingness to volunteer to participate in this research influence its external validity?

If you are a data nerd you can actually access these data and run analyses on your own.

results of published studies are, in fact, replicable. When other investigators have tried to reproduce results from many studies they have been unable to produce the same results (Open Science Collaboration, 2015; Shrout & Rodgers, 2018). Studies conducted in fields that often rely on experimental methods have been particularly called into question (Baucal & others, 2020; Klein & others, 2014). There have been a number of responses to this issue (Simmons & Simonsohn, 2017; Wingen & others, 2020). Two important principles have emerged to guide researchers. First, scientists should use larger sample sizes to ensure equivalence between groups. They should consider pooling data across labs when a single lab can only collect a small number of participants, for example, in infant research (Gennetian & others, 2020) and neuroimaging studies (Szucs & Ioannidis, 2020). Second, researchers should be transparent and thorough in the report of methods to allow work to be replicated correctly. Currently, researchers are more likely to provide public access to their original data so that others can repeat the statistical analyses used.

Applications of the Three Types of Research

All three types of research that we have considered—descriptive, correlational, and experimental—can be used to address the same research topic (Figure 5). For instance, researchers have been interested in examining the role of intensely positive experiences in human functioning. Maslow (1971) envisioned the healthiest, happiest people in the world as capable of having intense moments of awe, and he used the descriptive case study approach to examine the role of such "peak experiences" in the lives of individuals who seemed to exemplify the best in human life. Through correlational research, McAdams (2001) probed individuals' descriptions of the most intensely positive experiences of their lives. He found that people who were motivated toward warm interpersonal experiences tended to mention such experiences as the

	Descriptive Research			Correlational Research	Experimental Research
Goal	To determine the basic dimensions of a phenomenon.			To determine how variables change together.	To determine whether a causal relationship exists between two variables.
Sample Research Questions	**Observation** How much time are individuals spending each day on Facebook and similar sites? How many people use different kinds of social media?	**Interviews/Surveys** How do people describe their use of social media? Are they accurate? Do people view the social media they use positively or negatively?	**Case Studies** Does the social media page of a particular person (e.g., someone who has committed suicide) reveal important information about the individual?	What is the relationship between the number of hours spent on social media and face-to-face interactions? How does personal page content relate to personality characteristics?	How does going "cold turkey" on social media influence stress levels? If we randomly assign a smiling picture to a personal page, does it produce more friend requests than the identical page with a frowning picture?
Strengths and Weaknesses	The findings would lay the groundwork for future research by establishing the types of questions that ought to be addressed. This work would not tell us, however, about the processes involved or provide generalizable conclusions.			This research would give us information about how variables change together. However, it would not allow us to make causal conclusions.	This research would permit causal conclusions. The potential artificiality of the manipulations might raise concerns about external validity.

FIGURE 5 **Psychology's Research Methods Applied to Studying Social Media Use** Psychologists can apply different methods to study the same phenomenon. The popularity of social media has opened up a host of new research questions for psychologists.

best memories of their lives. Experimental researchers have also investigated this question. In their work, people who were randomly assigned to write about their most intensely positive experiences for a few minutes each day for 2 or 3 days experienced enhanced positive mood as well as fewer physical illnesses 2 months later, compared to individuals in control groups who wrote about unemotional topics (Burton & King, 2004, 2008).

self-quiz

1. A correlation of −.67 indicates
   A. a strong positive relationship.
   B. a strong negative relationship.
   C. a weak positive relationship.
   D. a weak negative relationship.

2. A study on obesity had four groups, each with a different assignment. One group of participants read a brochure about diet and nutrition; another group had a 30-minute nutrition counseling session; a third group read the newspaper; a fourth group watched a video about exercise and fitness. The control group is
   A. the group that had a counseling session.

   B. the group that read the newspaper.
   C. the group that read the brochure.
   D. the group that watched the video.

3. Which of the following statements is *correct?*
   A. Only correlational research allows researchers to determine causality.
   B. Only experimental research allows researchers to determine causality.
   C. Both correlational and experimental research allow researchers to determine causality.
   D. Neither correlational nor experimental research allows researchers to determine causality.

APPLY IT! 4. Jacob wants to study the relationship between falling in love and a person's academic performance. He asks students to fill out a questionnaire in which they answer "true" or "false" to the question, "Did you fall in love this semester?" Then he asks them for their GPA for the semester. Jacob's study is _____ study.
   A. a correlational
   B. an experimental
   C. a sociological
   D. a longitudinal

5 Research Samples and Settings

Among the important decisions to be made about collecting data are whom to choose as the participants and where to conduct the research. Will the participants be people or nonhuman animals? Will they be children, adults, or both? Where will the research take place—in a lab or in a natural setting?

The Research Sample

When psychologists conduct a study, they usually want to be able to draw conclusions that will apply to a larger group of people than the participants they actually study. The entire group about which the investigator wants to draw conclusions is the **population**. The subset of the population chosen by the investigator for study is a **sample**. The researcher might be interested only in a particular group, such as all children who are gifted and talented, all young women who embark on science and math careers, or all gay men. The key is that the sample studied must be representative of the population to which the investigator wants to generalize the results. That is, the researcher might study only 100 gifted adolescents but want to apply these results to all gifted and talented adolescents.

To mirror the population as closely as possible, the researcher uses a **random sample**, a sample that gives every member of the population an equal chance of being selected. A representative sample would reflect population factors such as age, socioeconomic status, ethnic origin, marital status, geographic location, and religion. A random sample provides much better grounds for generalizing the results to a population than a nonrandom sample, because random selection improves the chances that the sample is representative of the population. A random sample is still a smaller subset of the population but it is more likely to reflect the characteristics of the population.

In selecting a sample, researchers must strive to minimize bias, including gender bias. Because psychology is the scientific study of human behavior, it should pertain to *all*

population
The entire group about which the researcher wants to draw conclusions.

sample
The subset of the population chosen by the investigator for study.

random sample
A sample that gives every member of the population an equal chance of being selected.

The research sample might include a particular group, such as environmental activists or women runners.
(Rally): M. Stan Reaves/REX/Shutterstock; (Runners): Grant Falvey/LNP/REX/Shutterstock

humans, and so the participants in psychological studies ought to be representative of humanity as a whole. For a long time, the human experience studied by psychologists was primarily the white male experience.

There is a growing realization that psychological research needs to include more people from diverse groups (Roberts & others, 2020). Because a great deal of psychological research involves college student participants, individuals from groups that have not had as many educational opportunities have not been well represented in that research. Given the fact that individuals from diverse ethnic groups were excluded from psychological research for so long, we might reasonably conclude that people's real lives are more varied than past research data have indicated.

These issues are important because scientists want to be able to predict *human* behavior, not just non-Latino white male college student behavior. Contemporary psychological research continues to be plagued by a lack of true diversity in the samples used. Indeed, even research that is focused on the issue of race has relied on mostly white samples and is conducted by mostly white researchers (Roberts & others, 2020). Much of the research we will review has relied on samples that have been described as W.E.I.R.D., meaning, Western, Educated, Industrialized, Rich, Democratic societies (Henrich & others, 2010). As you learn about key findings in psychology, ask yourself whether you think these findings apply to all people or just those from this privileged group.

The Research Setting

All three types of research we examined in the preceding section can take place in different settings. Common settings include the research laboratory and natural settings.

Because psychological researchers often want to control as many aspects of the situation as possible, they conduct much of their research in a laboratory, a controlled setting with many of the complex factors of the real world, including potential confounds, removed. Although laboratory research provides a great deal of control, doing research in the laboratory has drawbacks:

■ It is almost impossible to conduct research in the lab without the participants knowing they are being studied.

■ The laboratory setting is not the real world and therefore can cause the participants to behave unnaturally.

■ People who are willing to go to a university laboratory may not be representative of groups from diverse cultural backgrounds.

■ Some aspects of the mind and behavior are difficult if not impossible to examine in the laboratory.

Natural settings and laboratories are common locales for psychological studies. (first) Jane Goodall, who specializes in animal behavior, has carried out extensive research on chimpanzees in natural settings. Her work has contributed a great deal to our understanding of these intelligent primates. (second) Barbara L. Fredrickson, a psychologist at the University of North Carolina, Chapel Hill, whose work investigates topics such as positive emotions and human flourishing, conducts a laboratory study.

(Goodall): Michael Nichols/National Geographic/Getty Images; (Frederickson): Courtesy of Barbara Fredrickson, University of North Carolina

naturalistic observation
The observation of behavior in a real-world setting.

Research can also take place in a natural setting. **Naturalistic observation** is observing behavior in a real-world setting. Psychologists conduct naturalistic observations at sporting events, childcare centers, work settings, shopping malls, and other places where people frequent. If you wanted to study the level of civility on your campus for a research project, most likely you would include naturalistic observation of how people treat one another in such gathering places as the cafeteria and the library reading room.

In another example of a natural setting, researchers who use survey methods are increasingly relying on web-based assessments that allow participants to complete the measures online. These studies might include volunteers who stumble upon a website and decide to take a survey. They might also include individuals who make money by completing surveys online. Researchers also conduct experiments online. Although collecting data online comes with problems, online studies allow researchers to reach a more diverse array of people than they might otherwise.

The type of research a psychologist conducts, the operational definitions of the variables of interest, and the choice of sample and setting are decisions that ideally are guided by the research question itself. However, sometimes these decisions represent a compromise between the psychologist's key objective (e.g., to study a representative sample of Americans) and the available resources (e.g., a sample of 100 college students).

self-quiz

1. The *entire group* of people about whom a researcher wants to draw conclusions is the
   A. sample.
   B. random sample.
   C. population.
   D. field.

2. When a researcher decides to study a particular group, such as Latino factory workers, the researcher is specifically determining the study's
   A. population.
   B. sample.
   C. research setting.
   D. scope.

3. A drawback of laboratory research is that

   A. it is hard to conduct without the participants' knowledge that they are being studied.
   B. people who participate in lab studies may not be representative of the diverse world in which we live.
   C. the lab setting is unnatural and may thus cause participants to behave unnaturally.
   D. all of the above

APPLY IT! 4. Emily, a committed environmentalist, reads a report that among a nationally representative sample, 60 percent of people polled support drilling for oil off the shores of California. The poll includes 1,000 people. Emily scoffs at the results, noting that all of the

people she knows do not support offshore drilling. The poll must be flawed, she insists. How do you evaluate Emily's statement?
   A. Emily is likely to be wrong. A representative sample is more likely to reflect the general population than the small sample of Emily's friends.
   B. Emily is likely to be wrong because 1,000 people is a high number.
   C. Emily is probably right because, as an environmentalist, she is probably more in tune with these issues than any polling organization.
   D. Emily is probably right because even representative polls are usually biased.

6 Conducting Ethical Research

Ethics is an important consideration for all science. This fact came to the fore in the aftermath of World War II, for example, when it became apparent that Nazi doctors had used concentration camp prisoners as unwilling participants in experiments. These atrocities spurred scientists to develop a code of appropriate behavior—a set of principles about the treatment that participants in research have a right to expect. In general, ethical principles of research focus on balancing the rights of the participants with the rights of scientists to ask important research questions.

The issue of ethics in psychological research may affect you personally if at some point you serve as a participant in a study. In that event, you need to know your rights as a participant and the researchers' responsibilities in ensuring that these rights are safeguarded. Experiences in research can have unforeseen effects on people's lives.

Consider, for instance, that many researchers interested in close relationships might ask romantic couples to keep diaries tracking the quality of their interactions (Michalowski & others, 2020). Might just paying close attention to this variable influence it? Other researchers have couples come into the lab and discuss a topic that is a source of conflict in that relationship (Campbell & others, 2013). Such procedures are important to measuring how couples handle conflicts. But, could such interactions influence the couple long after the study is over? Researchers have a responsibility to anticipate the personal problems their study might cause and, at least, to inform the participants of the possible fallout.

Leo Patrizi/Getty Images

Ethics comes into play in every psychological study. Even smart, conscientious students sometimes think that members of their church, athletes in the Special Olympics, or residents of the local nursing home where they volunteer present great samples for psychological research. Without proper permission, though, the most well-meaning, kind, and considerate researchers still violate the rights of the participants.

Ethics Guidelines

Various guidelines have been developed to ensure that research is conducted ethically. At the base of all of these guidelines is the notion that people participating in psychological research should be no worse off coming out of the study than they were on the way in.

Today colleges and universities have a review board, typically called the *institutional review board,* or *IRB,* which evaluates the ethical nature of research conducted at their institutions. Proposed research plans must pass the scrutiny of a research ethics committee before the research can be initiated. In addition, the American Psychological Association (APA) has developed ethics guidelines for its members. The APA code of ethics instructs psychologists to protect their participants from mental and physical harm. The participants' best interests need to be kept foremost in the researcher's mind. APA's guidelines address four important issues:

- *Informed consent:* All participants must know what their participation will involve and what risks might develop. For example, participants in a study on dating should be told beforehand that a questionnaire might stimulate thoughts about issues in their relationships that they have not considered. Participants also should be informed that in some instances a discussion of the issues might improve their relationships but that in others it might worsen the relationships and possibly end them. Even after informed consent is given, participants must retain the right to withdraw from the study at any time and for any reason.

- *Confidentiality:* Researchers are responsible for keeping all of the data they gather on individuals completely confidential and, when possible, completely anonymous. Confidential data are not the same as anonymous. When data are confidential, it is possible to link a participant's identity to their data.

- *Debriefing:* After the study has been completed, the researchers should inform the participants of its purpose and the methods they used. In most cases, the experimenters also can inform participants in a general manner beforehand about the purpose of the research without leading the participants to behave in a way that they think that the experimenters are expecting. When preliminary information about the study is likely to affect the results, participants must be debriefed after the study's completion.

- *Deception:* This is an ethical issue that psychologists debate extensively. In some circumstances, telling the participants beforehand what the research study is about substantially alters the participants' behavior and invalidates the researcher's data. For example, suppose a psychologist wants to know whether bystanders will report a theft. A mock theft is staged, and the psychologist observes which bystanders report it. Had the psychologist informed the participants beforehand that the study intended to discover the percentage of bystanders who will report a theft, the whole study would have been ruined. Thus, the researcher deceives participants about the purpose of the study, perhaps leading them to believe that it has some other purpose. In all cases of deception, however, the psychologist must ensure that the deception will not harm the participants and that the participants will be told the true nature of the study (will be debriefed) as soon as possible after the study is completed. Note that in studies using deception, participants cannot be fully informed prior to giving consent. This is why participants in studies involving deception should have the option of withdrawing consent after they find out what the study is actually about. As a student of psychology, you will encounter many studies that employ deception. Researchers who employ deception in their studies must be able to justify lying to participants, because doing so is vital to the scientific merit of their work. Psychological researchers take deception seriously and employ it only when no other options would allow them to ask the questions they seek to answer.

The federal government also takes a role in ensuring that research involving human participants is conducted ethically. The Federal Office for Protection from Research Risks is devoted to safeguarding the well-being of participants in research studies. Over the years, the office has dealt with many challenging and controversial issues—among them, informed consent rules for research on psychological disorders, regulations governing research on pregnant women and fetuses, and ethical issues regarding AIDS vaccine research.

Of considerable concern is the ways that the principles of ethical research apply when studies involve *vulnerable populations,* which include children, individuals with psychological disorders, incarcerated individuals, and others who may be especially susceptible to coercion. For example, when children participate in research, parental consent is essential. Children may be asked to agree to participate as well. When prisoners or individuals who are on parole participate in research, it must be clear that their treatment and decisions about their future release or parole will not be influenced by their willingness to participate.

Students who receive extra credit for participating in research might also be considered vulnerable to coercion. They must have an option to receive that same extra credit without agreeing to participate in research.

The Ethics of Research with Animals

For generations, psychologists have used animals in some research. Animal studies have provided a better understanding of and solutions for many human problems (Fischer, 2019). Miller (1985) listed the following areas in which research employing animals has benefited humans:

- Psychotherapy techniques and behavioral medicine
- Rehabilitation of neuromuscular disorders

- Alleviation of the effects of stress and pain
- Drugs to treat anxiety and severe mental illness
- Methods for avoiding drug addiction and relapse
- Treatments to help premature infants gain weight so they can leave the hospital sooner
- Methods used to alleviate memory deficits in old age

Rosa Jay/Shutterstock

Only about 5 percent of APA members use animals in their research. Rats and mice account for 90 percent of all psychological research with animals. It is true that researchers sometimes use procedures with animals that would be unethical with humans, but they are guided by a set of standards for housing, feeding, and maintaining the psychological and physical well-being of their animal subjects. Researchers are required to weigh potential benefits of the research against possible harm to the animal and to avoid inflicting unnecessary pain. Animal abuse is not as common as animal activist groups charge. In short, researchers must follow stringent ethical guidelines, whether animals or humans are the subjects in their studies.

The Place of Values in Psychological Research

Questions are asked not only about the ethics of psychology but also about its values and its standards for judging what is worthwhile and desirable. Some psychologists argue that psychology should be value free and morally neutral. From their perspective, the psychologist's role as a scientist is to present facts as objectively as possible. Others believe that because psychologists are human, they cannot possibly be value free. Indeed, some people go so far as to argue that psychologists should take stands on issues about which they have expertise. For example, psychological research shows that children reared by gay and lesbian parents are no more likely to be gay than other children and that the children of gay and lesbian partners tend to demonstrate levels of psychological health that are equal to or higher than those of children reared by heterosexual parents (Farr & others, 2019, 2020). When people argue against the rights of gay and lesbian individuals to adopt children or to retain custody of their biological children, psychologists may have a role to play in the debate about these issues.

self-quiz

1. Providing research participants with information about the purpose of a study at the study's conclusion is called
   A. informed consent.
   B. deception.
   C. debriefing.
   D. confidentiality.

2. The organization that provides ethical guidelines for psychologists is the
   A. American Psychiatric Association.
   B. Institutional Review Board.
   C. American Medical Association.
   D. American Psychological Association.

3. A study could possibly put participants at risk of harm, but the participants are not told about that risk. The ethical standard that has been violated is
   A. debriefing.
   B. informed consent.
   C. deception.
   D. confidentiality.

APPLY IT! 4. Amanda is participating in a psychological study as part of her Introductory Psychology course. While filling out items on a questionnaire, Amanda finds that some of them embarrass her, and she decides to skip them. As she leaves the study, the experimenter notices these blank questions and asks Amanda to complete them because the research will be ruined without complete data from all participants. Which of the following accurately assesses the ethics of this situation?
   A. Amanda should really complete those questions. What's the big deal?
   B. Amanda is within her rights to leave any question blank if she chooses, and the experimenter has definitely "crossed a line."
   C. Amanda is ethically wrong because she agreed to be in the study, and so she must see it through.
   D. If Amanda read and signed the consent form, she is obligated to do as the experimenter says.

7 Learning About Psychology Means Learning About You

Throughout your life you have been exposed to a good deal of information about psychological research. In *Experience Psychology* and your introductory psychology class, you will also learn about a multitude of research findings. In this last section, we consider the ways that learning about psychological studies can help you learn about yourself. We start by looking at some guidelines for evaluating psychological research findings that you might encounter in your everyday life.

Encountering Psychology in Everyday Life

Not all psychological information that is presented for public consumption comes from professionals with excellent credentials and reputations at colleges or universities or in applied mental health settings. Journalists, television reporters, and other media personnel are not usually trained in psychological research. As a result, they often have trouble sorting through the widely varying material they find and making sound decisions about the best information to present to the public. In addition, the media often focus on sensationalistic and dramatic psychological findings to capture public attention. They tend to go beyond what actual research articles and clinical findings really say.

Even when the media present the results of excellent research, they have trouble accurately informing people about the findings and their implications for people's lives (Lewis & Wai, 2020). *Experience Psychology* is dedicated to carefully introducing, defining, and elaborating on key concepts and issues, research, and clinical findings. The media, however, do not have the luxury of so much time and space to detail and specify the limitations and qualifications of research. In the end, you have to take responsibility for evaluating media reports on psychological research. To put it another way, you have to consume psychological information critically and wisely. Five guidelines follow.

AVOID OVERGENERALIZING BASED ON LITTLE INFORMATION
Media reports of psychological information often leave out details about the nature of the sample used in a given study. Without information about sample characteristics—such as the number of participants, their gender and ethnic representation—it is wise to take research results with a grain of salt. Perhaps most importantly, try to find information about the sample size for any given study. Remember that in an experiment, a random assignment will only produce equal groups if the sample was large.

DISTINGUISH BETWEEN GROUP RESULTS AND INDIVIDUAL NEEDS
Just as we cannot generalize from a small group to all people, we also cannot apply conclusions from a group to an individual. When you learn about psychological research through the media, you might be disposed to apply the results to your life. It is important to keep in mind that statistics about a group do not necessarily represent each individual in the group equally well. Imagine, for example, taking a test in a class and being told that the class average was 75 percent, but you got 98 percent. It is unlikely that you would want the instructor to apply the group average to your score.

Sometimes consumers of psychological research can get the wrong idea about whether their experience is "normal" if it does not match group statistics. New parents face this issue all the time. They read about developmental milestones that supposedly characterize an entire age group of children; one such milestone might be that most 2-year-olds are conversing with their parents. However, this group information does not necessarily characterize all children who are developing normally. Albert Einstein did not start talking until he was at the ripe old age of 3.

Ollyy/Shutterstock

LOOK FOR ANSWERS BEYOND A SINGLE STUDY   The media might identify an interesting piece of research and claim that it is something phenomenal with far-reaching implications. Although such pivotal studies do occur, they are rare. It is safer to assume that no single study will provide conclusive answers to an important question, especially answers that apply to all people. In fact, in most psychological domains that prompt many investigations, conflicting results are common. Answers to questions in research usually emerge after many scientists have conducted similar investigations that yield similar conclusions. Remember that you should not take one research study as the absolute, final answer to a problem, no matter how compelling the findings.

AVOID ATTRIBUTING CAUSES WHERE NONE HAVE BEEN FOUND
Drawing causal conclusions from correlational studies is one of the most common mistakes the media make. When a true experiment has not been conducted—that is, when participants have not been randomly assigned to treatments or experiences—two variables might have only a noncausal relationship to each other. Remember from the discussion of correlation earlier in the chapter that causal interpretations cannot be made when two or more factors are simply correlated. We cannot say that one causes the other. When you hear about correlational studies, be skeptical of words indicating causation until you know more about the particular research.

CONSIDER THE SOURCE OF PSYCHOLOGICAL INFORMATION
Studies conducted by psychologists are not automatically accepted by the rest of the research community. The researchers usually must submit their findings to a journal for review by their colleagues, who make a decision about whether to publish the paper, depending on its scientific merit. This process, called *peer review,* means that research that is published in scholarly journals has survived the scrutiny of experts in a particular area. Although the quality of research and findings is not uniform among all psychology journals, in most cases journals submit the findings to far greater scrutiny than the popular media do.

Within the media, though, you can usually draw a distinction. The reports of psychological research in respected newspapers such as the *New York Times* and the *Washington Post,* as well as in credible magazines such as *Time* and *Newsweek,* are more trustworthy than reports in tabloids such as the *National Enquirer* and *Star.*

Finally, it is not unusual to read about psychological research online, where bloggers and others might comment on the validity of researchers' findings. When you encounter research-related information on the web, see if you can find the actual study and read it. As helpful as a blog or Wikipedia entry might be, reading the actual science is crucial to evaluating the conclusions drawn. Many articles are available via a search on Google Scholar. Whatever the source—serious publication, tabloid, blog, online news outlet, or even academic journal—you are responsible for reading the details behind the reported findings and for analyzing the study's credibility.

> ### *Do It!*
>
> In the next few days, look through several newspapers, magazines, and your favorite online news sources for reports about psychological research. Apply the guidelines for being a wise consumer of information about psychology to these media reports.

Appreciating Psychology as the Science of You

In taking introductory psychology, you have an amazing opportunity. You will learn a great deal about human beings, especially one particular human being: you. Whether the psychological research presented is about emotions and motivation or the structures of the nervous system, it is still essentially about you.

When you think of psychology, you might think first and foremost about the mind and its complex mental processes such as those involved in love, gratitude, hate, and anger.

However, psychology has come to recognize more and more that the mind and its operations are intricately connected to the body. As you will see when we examine neuroscience, observations of the brain at work reveal that when mental processes change, so do physical processes. This mind–body link has fascinated philosophers for centuries. Psychology occupies the very spot where the mind and the body meet.

It might be helpful to think concretely about the ways the mind and body can relate to each other even as they are united in the physical reality of a person. Let's say you experience a mental event such as watching a documentary about the dangers of obesity. You decide to embark on a workout regimen for six-pack abdominals. Dedication, goal setting, and self-discipline will be the kinds of mental processes necessary to transform your body. The mind can work on the body, prompting changes to its shape and size.

Similarly, the body can dramatically influence the mind. Consider how fuzzy your thinking is after you stay out too late and how much easier it is to solve life's problems after a good night's sleep. Also, think about your outlook on the first day of true recovery from a nagging cold: Everything seems better. Your mood and your work improve. Clearly, physical states such as illness and health influence how we think.

The relationship between the body and mind is illustrated in a major dilemma that countless psychologists have faced: the impact of nature versus nurture. Essentially, psychologists have wondered and debated which of the two is more important to a person—nature (i.e., genetic heritage) or nurture (i.e., social experiences). The influence of genetics on a variety of psychological characteristics, and the ways that genetic influence can itself be altered by experience, will be addressed in many of the main topics in *Experience Psychology,* from development to personality traits to psychological disorders. You will see that at every turn, your physical and mental self are intertwined in ways you may have never considered.

Throughout *Experience Psychology*, we investigate the ways that all of the various approaches to psychology matter to your life. Psychology is crucially about you, essential to your understanding of your life, your goals, and the ways that you can use the insights of thousands of scientists to make your life healthier and happier.

Andrew Rich/The Agency Collection/Getty Images

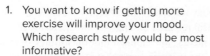
self-quiz

1. You want to know if getting more exercise will improve your mood. Which research study would be most informative?
   A. A correlational study with a sample of 40 adults.
   B. An experiment that involves randomly assigning 200 students similar to yourself to either a physical activity condition or a control group, and then measuring mood.
   C. A case study of one psychology student's experience with exercise.
   D. A longitudinal study tracking exercise and mood over 2 years in a sample of 100 grade school children.

2. In a blog post, a scientist describes a recently completed study in which she found that people who were assigned to an exercise group reported 10 times more happiness than those in a control group. She and her colleagues plan to submit the findings for publication. What would a wise reader of science think about this post?
   A. The results are clearly important and represent a major contribution to science.
   B. The results are likely worthless because you cannot believe anything you read in a blog post.
   C. The results are not surprising and are therefore probably true.
   D. The jury is still out on the findings because the research has not gone through peer review.

3. A study reports that height is surprisingly correlated with lifetime income. Specifically, tall people tend to make more money than short people. April is the shortest person in her class and when she reads about the study she becomes upset. What advice would you have for April?
   A. She can still lead a successful life even if she cannot ever make as much money as a tall person.
   B. Correlation does not imply causation and group statistics cannot be assumed to apply to every single case.
   C. Scientists base their conclusions on data. The data show that she is simply very likely to be poor.
   D. Being a critical thinker means accepting whatever the data say.

APPLY IT! 4. Sarah's mother forwards her a chain email warning of an association between drinking orange juice and developing arthritis. Her mom tells her that three of her closest friends are big orange juice fans and they recently developed arthritis. Sarah's mother tells her to stop drinking orange juice and to tell her friends to do the same. Rather than forwarding the email to everyone she knows, Sarah should
   A. gently explain to her mother the importance of considering the source of scientific information.
   B. respectfully tell her mother that even if an association between two variables is found, it does not imply a causal relationship.
   C. kindly ask her mother if the study was published in a scientific journal and if so, which one.
   D. All of the above.

1 Defining Psychology and Exploring Its Roots

Psychology is the scientific study of human behavior and mental processes. Psychologists approach human behavior as scientists who think critically and who are curious, skeptical, and objective. Psychology emerged as a science from philosophy and physiology. Two founders of the science of psychology are Wilhelm Wundt and William James.

2 Contemporary Approaches to Psychology

Approaches to psychology include biological, behavioral, psychodynamic, humanistic, cognitive, evolutionary, and sociocultural. All of these perspectives consider important questions about human behavior from different but complementary vantage points. The biopsychosocial perspective stresses that behavior is explained by processes at these many levels.

3 Psychology's Scientific Method

Psychologists use the scientific method to address research questions. This method involves starting with a theory and then making observations, formulating hypotheses, testing these through empirical research, drawing conclusions, and evaluating these conclusions. The science of psychology is an ongoing conversation among scholars.

4 Types of Psychological Research

Three types of research commonly used in psychology are descriptive research (finding out about the basic dimensions of some variable), correlational research (finding out if and how two variables change together), and experimental research (determining the causal relationship between variables). In an experiment, the independent variable is manipulated to see if it produces changes in the dependent variable. Experiments involve comparing two groups: the experimental group (the one that receives the treatment or manipulation of the independent variable) and the control group (the comparison group or baseline that is equal to the experimental group in every way except for the independent variable). Experimental research relies on random assignment to ensure that the groups are roughly equivalent before the manipulation of the independent variable.

5 Research Samples and Settings

Two important decisions that must be made for psychological research are whom to study and where to study them. A sample is the group that participates in a study; the population is the group to which the researcher wishes to generalize the results. A random sample is the best way of ensuring that the sample reflects the population. Research settings include the laboratory as well as real-world, naturalistic contexts. The laboratory allows a great deal of control, but naturalistic settings may give a truer sense of natural behavior.

6 Conducting Ethical Research

For all kinds of research, the ethical treatment of participants is very important. When participants leave a psychological study, they should be no worse off than they were when they entered. Guiding principles for ethical research in psychology include informed consent, confidentiality, debriefing (participants should be fully informed about the purpose of a study once it is over), and explaining fully the use of deception in a study.

7 Learning About Psychology Means Learning About You

In your everyday life and in introductory psychology, you will be exposed to psychological research findings. In encountering psychological research in the media, you should adopt the attitude of a scientist and critically evaluate the research as presented. In introductory psychology, you should make the most of the experience by applying it to your life. Psychology is, after all, the scientific study of you—your behavior, thoughts, goals, and well-being.

KEY TERMS

behavior
behavioral approach
biological approach
biopsychosocial approach
case history
case study
cognitive approach
confederate
control group
correlational research
critical thinking
demand characteristics

dependent variable
double-blind experiment
empirical method
evolutionary approach
experiment
experimental group
experimenter bias
external validity
functionalism
humanistic approach
hypothesis
independent variable

internal validity
longitudinal design
mental processes
natural selection
naturalistic observation
neuroscience
operational definition
placebo
placebo effect
population
psychodynamic approach
psychology

random assignment
random sample
replication
research participant bias
sample
science
sociocultural approach
structuralism
theory
third variable problem
validity
variable

Section 1:   1. A; 2. C; 3. B; 4. B
Section 2:   1. C; 2. D; 3. D; 4. C
Section 3:   1. D; 2. D; 3. B; 4. B

Section 4:   1. B; 2. B; 3. B; 4. A
Section 5:   1. C; 2. B; 3. D; 4. A
Section 6:   1. C; 2. D; 3. B; 4. B

Section 7:   1. B; 2. D; 3. B; 4. D

2 The Brain and Behavior

Seeing Strange Things

Nineteen years ago, when Milena Canning was 30 years old, she suffered an infection that led to several major strokes. After 8 weeks in a coma, she emerged completely blind. Months after the coma, she started noticing strange images, like sparkles bubbling into sight. Eventually, she found that she could see some objects, sometimes. She could see the steam rising out of her coffee (but not the cup of coffee itself). She could see her daughter's ponytail bobbing behind her as she walked (but not her daughter's face). She could see and catch a ball thrown to her (but not see the ball sitting still). She could see many things, *but only when they were moving* (University of Western Ontario, 2018). Recently, neuroscientists were able to find out what might explain Milena's very rare condition, known as *Riddoch phenomenon* (the ability to see things only when they are moving) (Arcaro & others, 2019). Using brain imaging, the researchers found that Milena's brain was missing an apple-sized portion at the back—the area responsible for processing visual experience. But the damage wrought by the strokes was not the end of the road for Milena's brain. Over time, her brain developed a "workaround" using what remained of her visual system, including the area responsible for seeing motion. Essentially, Milena's brain rewired itself to make the most of what she had, allowing her to develop a surprisingly rich visual life. She no longer sees as she did before her illness. She now sees a world of motion.

This example illuminates the brain's role in precious human experience, its amazing capacity to adapt itself to change, and its sometimes mysterious nature. Imagine: This intricate 3-pound structure that you are reading about is the engine that is doing the work of reading itself. The brain is at once the object of study and the reason we are able to study it.

In this chapter, our focus is the nervous system and its command center—the brain. We will study the essentials of what the brain has come to know about itself, including the biological foundations of human behavior and the brain's extraordinary capacity for adaptation and repair. The chapter concludes with a look at how genes influence behavior.

1 The Nervous System

nervous system
The body's electrochemical communication circuitry.

The **nervous system** is the body's electrochemical communication circuitry. The field that studies the nervous system is called *neuroscience,* and the people who study it are *neuroscientists.*

The human nervous system is made up of billions of communicating cells, and it is likely the most intricately organized aggregate of matter on the planet. A single cubic centimeter of the human brain consists of well over 50 million nerve cells, each of which communicates with many other nerve cells in information-processing networks that make the most elaborate computer seem primitive.

One cubic centimeter of brain = 50 million nerve cells. That's about the size of a snack cube of cheese.

Characteristics of the Nervous System

The brain and nervous system guide our interactions with the world, moving the body and directing our adaptation to our environment. Several extraordinary characteristics allow the nervous system to direct our behavior: complexity, integration, adaptability, and electrochemical transmission.

COMPLEXITY   The human brain and nervous system are enormously complex. The orchestration of all the billions of nerve cells in the brain—to allow you to sing, dance, write, talk, and think—is an awe-inspiring achievement. Right now, your brain is carrying out a multitude of tasks, including seeing, reading, learning, and breathing. Extensive assemblies of nerve cells participate in each of these activities, all at once.

INTEGRATION   The brain is the "great integrator" (Hyman, 2001), meaning that the brain does a wonderful job of pulling information together. Sounds, sights, touch, taste, smells—the brain integrates all of these as we function in the world.

The brain and the nervous system have different levels and many different parts. Brain activity is integrated across these levels through countless interconnections of brain cells and extensive pathways that link different parts of the brain. Each nerve cell can make connections with thousands of others, making an astronomical number of connections. The evidence for these connections is observable, for example, when a loved one takes your hand. How does your brain know, and tell you, what has happened? Bundles of interconnected nerve cells relay information about the sensation in your hand through the nervous system in an orderly fashion, all the way to the areas of the brain involved in recognizing that someone you love is holding your hand. Then the brain might send a reply back and prompt your hand to give that person a little squeeze.

ADAPTABILITY   The world around us is constantly changing. To survive, we must change, or adapt to new conditions. Our brain and nervous system together serve as our agent in adapting to the world. Although nerve cells reside

Adaptation, adaptability, and adapt: Psychologists use these terms when referring to the ability to function in a changing world.

in certain brain regions, they are not fixed, unchanging structures. They have a hereditary, biological foundation, but they are constantly adapting to changes in the body and the environment.

The term **plasticity** denotes the brain's special capacity for change. You might believe that thinking is a mental process, not a physical one. Yet thinking *is* a physical event, because your every thought is reflected in physical activity in the brain. Moreover, the brain can be changed by experience. London cab drivers who have developed a familiarity with the city show increases in the size of the brain area thought to be responsible for reading maps (Maguire & others, 2000). Think about that: When you change the way you think, you are literally changing the brain's physical processes and even its shape. Experiences throughout life can have long-lasting impact on the brain. For example, people who have managerial experience, especially those who supervised a large number of employees, are likely to show less age-related decreases in the size of the hippocampus later in life (Suo & others, 2017). As we will see later, the hippocampus is a brain structure that plays an important role in memory. Your experiences contribute to the wiring or rewiring of your brain (Mercure & others, 2020; Pliatsikas, 2020), just as the experiences of those London cab drivers did.

plasticity
The brain's special capacity for change.

ELECTROCHEMICAL TRANSMISSION   The brain and the rest of the nervous system work as an information-processing system, powered by electrical impulses and chemical messengers. When an impulse travels down a nerve cell, or *neuron,* it does so electrically. When that impulse gets to the end of the line, it communicates with the next neuron using chemicals, as we will consider in detail later in this chapter.

Pathways in the Nervous System

As we interact with and adapt to the world, the brain and the nervous system receive and transmit sensory input (e.g., sounds, smells, and flavors), integrate the information received from the environment, and direct the body's motor activities. Information flows into the brain through input from our senses, and the brain makes sense of that information, pulling it together and giving it meaning. In turn, information moves out of the brain to the rest of the body, directing all of the physical things we do.

Afferent and efferent are hard to keep straight. It might be helpful to remember that afferent nerves arrive at the brain and spinal cord, while efferent nerves exit the brain and spinal cord—A for afferent and arrive, and E for efferent and exit.

The nervous system possesses specialized pathways that are adapted for different functions. These pathways are made up of afferent nerves, efferent nerves, and neural networks. Afferent and efferent nerves are like two sets of "one-way" streets that carry information to and from the nervous system. **Afferent nerves,** or *sensory nerves*, carry information to the brain and spinal cord. These sensory pathways communicate information about the external environment (e.g., seeing a sunrise) and internal body processes (e.g., feeling tired or hungry) from sensory receptors to the brain and spinal cord. **Efferent nerves,** or *motor nerves*, carry information out of the brain and spinal cord—that is, they carry the nervous system's output. These motor pathways communicate information from the brain and spinal cord to other areas of the body, including muscles and glands, telling them to get busy. Efferent neurons are called motor neurons because they tell our muscles what to do as we move.

Most information processing occurs through **neural networks.** These are interconnected groups of nerve cells that integrate sensory input and motor output (Taherkhani & others, 2020). For example, as you read your class notes, the input from your eyes is transmitted to your brain and

afferent nerves
Also called sensory nerves; nerves that carry information about the external environment to the brain and spinal cord via sensory receptors.

efferent nerves
Also called motor nerves; nerves that carry information out of the brain and spinal cord to other areas of the body.

neural networks
Networks of nerve cells that integrate sensory input and motor output.

Sharing our gaze with another, electrical charges and chemical messages pulse through our brain, knitting the cells together into pathways and networks for processing the information.
Eric Herchaft/SuperStock

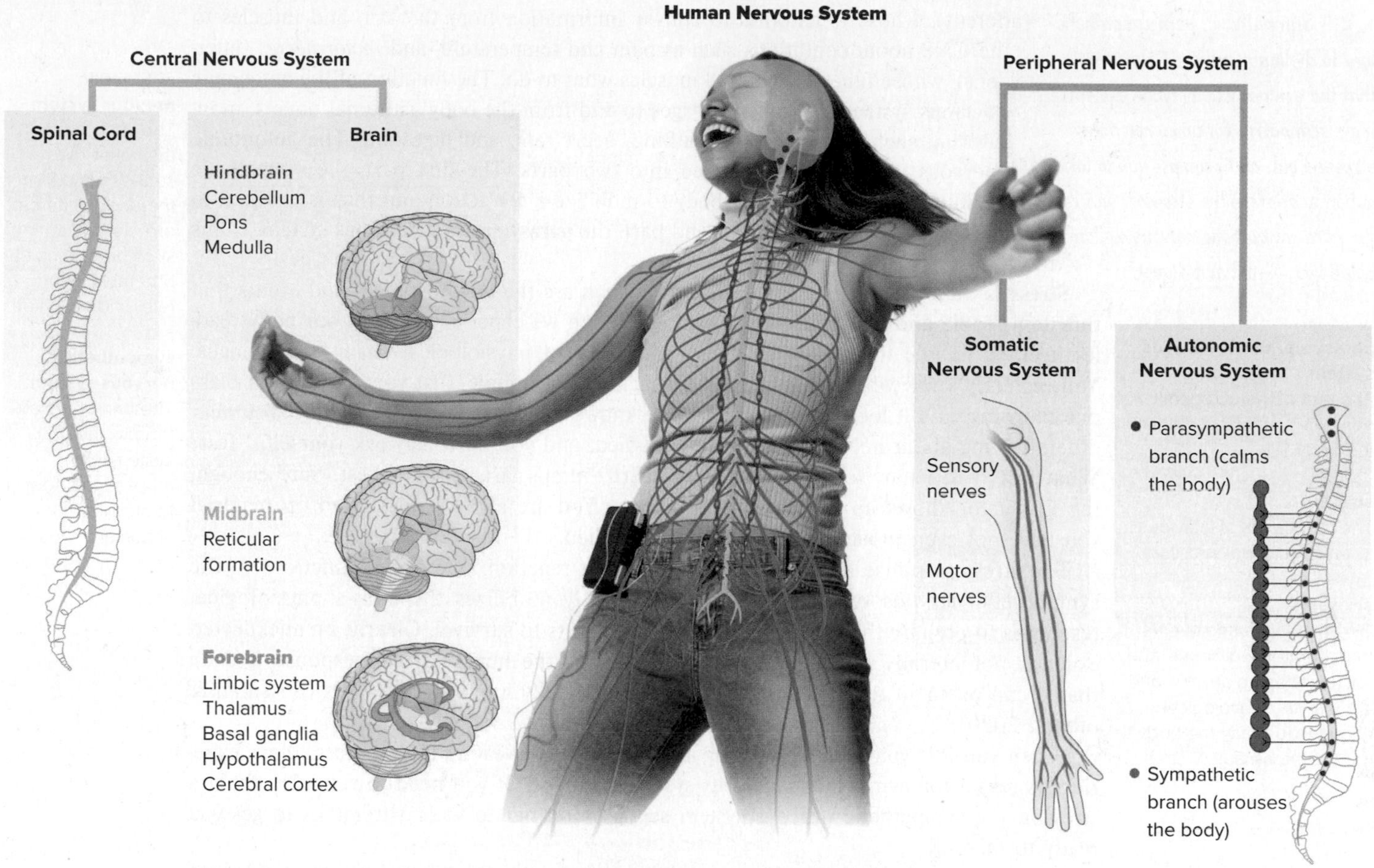

FIGURE 1 **Major Divisions of the Human Nervous System** The nervous system has two main divisions. One is the *central nervous system* (*left*), which comprises the brain and the spinal cord. The nervous system's other main division is the *peripheral nervous system* (*right*), which itself has two parts—the *somatic nervous system,* which controls sensory and motor neurons, and the *autonomic nervous system,* which monitors processes such as breathing, heart rate, and digestion. These complex systems work together to help us successfully navigate the world.

(Photo): Photoplay/Media Bakery

then passed through many neural networks, which translate the characters on the page into neural codes for letters, words, associations, and meanings. Some of the information is stored in the neural networks, and if you read aloud, some is passed on as messages to your lips and tongue. Neural networks make up most of the brain. Working in networks amplifies the brain's computing power.

Divisions of the Nervous System

This truly elegant system is highly ordered and organized for effective function. Figure 1 shows the two primary divisions of the human nervous system: the central nervous system and the peripheral nervous system.

The **central nervous system (CNS)** is the brain and spinal cord. More than 99 percent of all our nerve cells are located in the CNS. The **peripheral nervous system (PNS)** is the network of nerves that connects the brain and spinal cord to other parts of the body. The functions of the peripheral nervous system are to bring information to and from the brain and spinal cord and to carry out the commands of the CNS to execute various muscular and glandular activities.

The peripheral nervous system has two major divisions: the somatic nervous system and the autonomic nervous system. The **somatic nervous system** consists of *sensory nerves*

somatic nervous system
The body system consisting of the sensory nerves, whose function is to convey information from the skin and muscles to the central nervous system about conditions such as pain and temperature, and the motor nerves, whose function is to tell muscles what to do.

central nervous system (CNS)
The brain and spinal cord.

peripheral nervous system (PNS)
The network of nerves that connects the brain and spinal cord to other parts of the body.

parasympathetic nervous system
The part of the autonomic nervous system that calms the body.

stress
The responses of individuals to environmental stressors.

stressors
Circumstances and events that threaten individuals and tax their coping abilities and that cause physiological changes to ready the body to handle the assault of stress.

(afferent), whose function is to convey information from the skin and muscles to the CNS about conditions such as pain and temperature, and *motor nerves* (efferent), whose function is to tell muscles what to do. The function of the **autonomic nervous system** is to take messages to and from the body's internal organs, monitoring such processes as breathing, heart rate, and digestion. The autonomic nervous system also is divided into two parts. The first part, the **sympathetic nervous system,** arouses the body to mobilize it for action and thus is involved in the experience of stress; the second part, the **parasympathetic nervous system,** calms the body.

Stress is the body's response to **stressors,** which are the circumstances and events that threaten people and tax their coping abilities. When we experience stress, our body readies itself to handle the assault of stress; a number of physiological changes take place. You certainly know what stress feels like. Imagine, for example, that you show up for class one morning, and it looks as if everyone else knows that there is a test that day. You hear others talking about how much they have studied, and you nervously ask yourself: "Test? What test?" You start to sweat, and your heart thumps fast in your chest. Sure enough, the instructor shows up with a stack of exams. You are about to be tested on material you have not even thought about, much less studied.

The stress response begins with a fight-or-flight reaction, one of the functions of the sympathetic nervous system. This reaction quickly mobilizes the body's physiological resources to prepare the organism to deal with threats to survival. Clearly, an unexpected exam is not literally a threat to your survival, but the human stress response is such that it can occur in reaction to any threat to personally important motives (Sussman & others, 2020).

When you feel your heart pounding and your hands sweating under stress, those experiences reveal the sympathetic nervous system in action. If you need to run away from a stressor, the sympathetic nervous system sends blood out to your extremities to get you ready to take off.

When we undergo stress, we also experience the release of *corticosteroids,* which are powerful stress hormones. Corticosteroids in the brain allow us to focus our attention on what needs to be done *now.* For example, in an emergency, people sometimes report feeling strangely calm and doing just what has to be done, whether it is calling 911 or applying pressure to a serious cut. Such experiences reveal the benefits of corticosteroids for humans in times of emergency (Langer & others, 2020). *Acute stress* is the momentary stress that occurs in response to life experiences. When the stressful situation ends, so does acute stress.

However, we are not in a live-or-die situation most of the time when we experience stress. Indeed, we can even "stress ourselves out" just by thinking. *Chronic stress*—that is, stress that goes on continuously—may lead to persistent autonomic nervous system arousal. While the sympathetic nervous system is working to meet the demands of whatever is stressing us out, the parasympathetic nervous system is not getting a chance to do its job of maintenance and repair, of digesting food, or of keeping our organs in good working order. Thus, over time, chronic autonomic nervous system activity can break down the immune system (Kempuraj & others, 2020; Lafuse & others, 2020). Chronic stress is clearly best avoided, although this objective is sometimes easier said than done.

Yet the brain, an organ that is itself powerfully affected by chronic stress, can be our ally in preventing such continuous stress. Consider that when you face a challenging situation, you can use the brain's abilities and interpret the experience in a less stressful way. For example, you might approach an upcoming exam or an audition for a play not so much as a stressor but as an opportunity to shine. When we reinterpret potentially stressful situations as challenges rather than threats, we can avoid the experience of stress.

Many cognitive therapists believe that changing the way people think about their life opportunities and experiences can help them live less stressfully (Pearlstein & others, 2020). By changing the way, we think about potentially stressful events, we can reduce stress. Coming up with ways to consider potential stressors as challenges (not threats) can help to reduce the body's stress response.

autonomic nervous system
The body system that takes messages to and from the body's internal organs, monitoring such processes as breathing, heart rate, and digestion.

sympathetic nervous system
The part of the autonomic nervous system that arouses the body to mobilize it for action and thus is involved in the experience of stress.

self-quiz

1. The characteristics that allow the nervous system to direct behavior are its complexity, integration, electrochemical transmission, and
   A. constancy.
   B. adaptability.
   C. sensitivity.
   D. fight-or-flight response.

2. Neural networks are networks of nerve cells that integrate sensory input and
   A. the fight-or-flight response.
   B. electrochemical transmission.
   C. bodily processes such as heart rate and digestion.
   D. motor output.

3. When you are in danger, the part of the nervous system that is responsible for an increase in your heart rate is the
   A. central nervous system.
   B. peripheral nervous system.
   C. sympathetic nervous system.
   D. parasympathetic nervous system.

APPLY IT! 4. Shannon and Terrell are two college students. Shannon is in a constant state of low-level stress. She spends a lot of time worrying about what might happen, and she gets herself worked up about imagined catastrophes. Terrell is more easygoing, but on his way to class one day he is in a near-miss traffic accident—at the moment he sees the truck coming at him, his body tenses up, his heart races, and he experiences extreme panic. Which answer most accurately identifies the individual who is most likely to catch the cold that is going around the dorm this semester?
   A. Shannon, who is experiencing chronic stress
   B. Terrell, who is experiencing acute stress
   C. Shannon, who is experiencing acute stress
   D. Terrell, who is experiencing chronic stress

2 Neurons

Within each division of the nervous system, much is happening at the cellular level. Nerve cells, chemicals, and electrical impulses work together to transmit information at speeds of up to 330 miles per hour. As a result, information can travel from your brain to your hands (or vice versa) in just milliseconds.

There are two types of cells in the nervous system: neurons and glial cells. **Neurons** are the nerve cells that handle the information-processing function. The human brain contains about 100 billion neurons. The average neuron is a complex structure with as many as 10,000 physical connections with other cells. Researchers have been especially interested in a special type of neuron called *mirror neurons*. Mirror neurons seem to play a role in imitation and are activated (in primates and humans) when we perform an action but also when we watch someone else perform that same task (Schmidt & others, 2020). In addition to imitation, these neurons may play a role in empathy and in our understanding of others (Rehman & others, 2020).

Glial cells (or glia) provide support, nutritional benefits, and other functions in the nervous system (Hirbec & others, 2020). Glial cells are the most common cells in the nervous system. In fact, for every neuron there are about 10 glial cells. They have many functions. Glial cells keep neurons running smoothly. These cells are not specialized to process information in the same way as neurons.

neurons
One of two types of cells in the nervous system; neurons are the nerve cells that handle the information-processing function.

glial cells
The second of two types of cells in the nervous system; glial cells (also called glia) provide support, nutritional benefits, and other functions and keep neurons running smoothly.

 330 mph is fast! Most of us will never experience driving a car that fast. The supersonic rocket car that holds the world record can drive over 700 miles per hour. Its British developers are shooting for over 1,000 miles per hour. Now, the wisdom of driving a car faster than we can think is another story...

You might think of glial cells as the pit crew of the nervous system.

Specialized Cell Structure

Not all neurons are alike, as they are specialized to handle different information-processing functions. However, all neurons do have some common characteristics. Most neurons are created very early in life, but their shape, size, and connections can change throughout the life span. The way neurons function reflects the major characteristic of the nervous system described at the beginning of the chapter: plasticity. Neurons can and do change.

Every neuron has a cell body, dendrites, and an axon (Figure 2). The **cell body** contains the nucleus, which directs the manufacture of substances that the neuron needs for growth and maintenance. **Dendrites,** treelike fibers projecting from a neuron, receive information and orient it toward the neuron's cell body. Most nerve cells have numerous dendrites, which increase their surface area, allowing each neuron to receive input from many other

cell body
The part of the neuron that contains the nucleus, which directs the manufacture of substances that the neuron needs for growth and maintenance.

dendrites
Treelike fibers projecting from a neuron, which receive information and orient it toward the neuron's cell body.

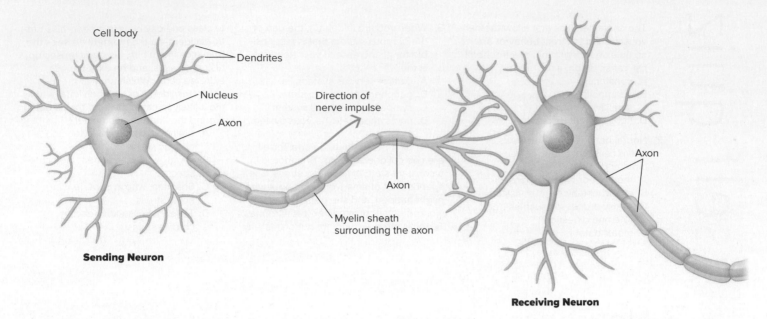

Cell body

Dendrites

Nucleus

Direction of nerve impulse

Axon

Axon

Axon

Myelin sheath surrounding the axon

Sending Neuron

Receiving Neuron

FIGURE 2 **The Neuron** The drawing shows the parts of a neuron and the connection between one neuron and another. Note the cell body, the branching of dendrites, and the axon with a myelin sheath.

neurons. The **axon** is the part of the neuron that carries information away from the cell body toward other cells. Although extremely thin (1/10,000 inch—a human hair by comparison is 1/1,000 inch), axons can be very long, with many branches. Some extend more than 3 feet—from the top of the brain to the base of the spinal cord.

Covering all surfaces of neurons, including the dendrites and axons, are very thin cellular membranes that are much like the surface of a balloon. The neuronal membranes are semipermeable, meaning that they contain tiny holes, or channels, that allow only certain substances to pass into and out of the neurons.

A **myelin sheath,** consisting of a layer of cells containing fat, encases and insulates most axons. By insulating axons, myelin sheaths speed up transmission of nerve impulses. Glial cells provide myelination (Abati & others, 2020). Numerous disorders are associated with problems in either the creation or the maintenance of this vital insulation. One of them is multiple sclerosis (MS), a degenerative disease of the nervous system in which myelin hardens, disrupting the flow of information through the neurons. Symptoms of MS include blurry and double vision, tingling sensations throughout the body, and general weakness.

The myelin sheath developed as the nervous system evolved. As brain size increased, it became necessary for information to travel over longer distances in the nervous system. Axons without myelin sheaths are not very good conductors of electricity. With the insulation of myelin sheaths, axons transmit electrical impulses and convey information much more rapidly (Raasakka & Kursula, 2020). We can compare the myelin sheath's development to the evolution of interstate highways as cities grew. Highways keep fast-moving, long-distance traffic from getting snarled by slow local traffic.

axon
The part of the neuron that carries information away from the cell body toward other cells.

myelin sheath
A layer of fat cells that encases and insulates most axons.

BG015/Bauer-Griffin/GC Images/Getty Images

The Neural Impulse

To transmit information to other neurons, a neuron sends brief electrical impulses (let's call them "blips") through its axon to the next neuron. As you navigate down this page while you read, hundreds of such impulses will stream down the axons in your arm to tell your muscles when to flex and how quickly. By changing the rate of the signals, or blips, the neuron can vary its message. Those impulses traveling down the axon are electrical. How does a neuron—a living cell—generate electricity? To answer this question, let's examine the axon.

The axon is a tube encased in a membrane. The membrane has hundreds and thousands of tiny gates in it. These gates are generally closed, but they can open. We call this membrane *semipermeable* because fluids can sometimes flow in and out of the gates. Indeed, there are fluids both inside and outside the axon. Floating in those fluids are electrically charged particles called *ions.*

Some of these ions, notably sodium and potassium, carry positive charges. Negatively charged ions of chlorine and other elements also are present. The membrane surrounding the axon prevents negative and positive ions from randomly flowing into or out of the cell. The neuron creates electrical signals by moving positive and negative ions back and forth through its outer membrane. How does the movement of ions across the membrane occur? Those tiny gates mentioned earlier, called *ion channels,* open and close to let the ions pass into and out of the cell. Normally when the neuron is *resting,* or not transmitting information, the ion channels are closed, and a slight negative charge is present along the inside of the cell membrane. On the outside of the cell membrane, the charge is positive. Because of the difference in charge, the membrane of the resting neuron is said to be *polarized,* with most negatively charged ions on the inside of the cell and most positively charged ions on the outside. This polarization creates a voltage between the inside and the outside of the axon wall (Figure 3). That voltage, called the neuron's **resting potential,** is between −60 and −75 millivolts. (A millivolt is 1/1,000 of a volt.)

For ions like magnets, it is true that opposites attract. The negatively charged ions inside the membrane and the positively charged ions outside the membrane will rush to each other if given the chance. Impulses that travel down the neuron do so by opening and closing ion channels, allowing the ions to flow in and out.

A neuron becomes activated when an incoming impulse—a reaction to, say, a pinprick or the sight of someone's face—raises the neuron's voltage, and the sodium gates at the

The rate of the blips determines the intensity of the impulse. So, if you are dying of suspense while reading about neural impulses (and who isn't?), the blips are happening faster as you rush down each screen or turn each page.

resting potential
The stable, negative charge of an inactive neuron.

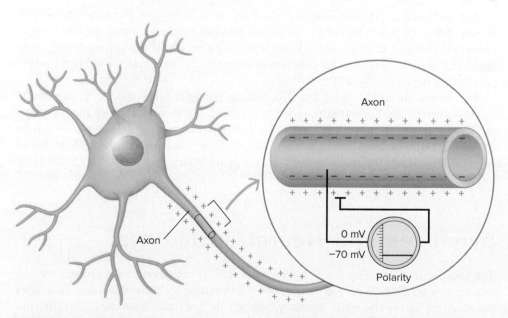

FIGURE 3 **The Resting Potential** An oscilloscope measures the difference in electrical potential between two electrodes. When one electrode is placed inside an axon at rest and one is placed outside, the electrical potential inside the cell is −70 millivolts (mV) relative to the outside. This potential difference is due to the separation of positive (+) and negative (−) charges along the membrane.

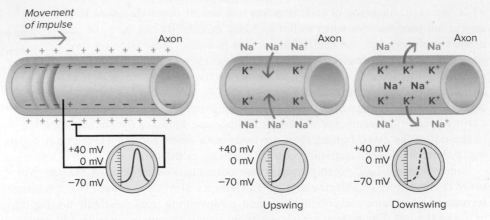

(a) Action potential generated by an impulse within a neuron

(b) Movement of sodium (Na⁺) and potassium (K⁺) ions responsible for the action potential

FIGURE 4 **The Action Potential** An action potential is a brief wave of positive electrical charge that sweeps down the axon as the sodium channels in the axon membrane open and close. (*a*) The action potential causes a change in electrical potential as it moves along the axon. (*b*) The movements of sodium ions (Na⁺) and potassium ions (K⁺) into and out of the axon cause the electrical changes.

base of the axon open briefly. This action allows positively charged sodium ions to flow into the neuron, creating a more positively charged neuron and *depolarizing* the membrane by decreasing the charge difference between the fluids inside and outside the neuron. Then potassium channels open, and positively charged potassium ions move out through the neuron's semipermeable membrane. This outflow returns the neuron to a negative charge. Then the same process occurs as the next group of channels flips open briefly. So it goes all the way down the axon, like a long row of cabinet doors opening and closing in sequence. It is hard to imagine, but this simple system of opening and closing tiny doors is responsible for the fluid movements of a ballet dancer and the flying fingers of a pianist playing a concerto.

action potential
The brief wave of positive electrical charge that sweeps down the axon.

The term **action potential** describes the brief wave of positive electrical charge that sweeps down the axon (Figure 4). An action potential lasts only about 1/1,000 second, because the sodium channels can stay open for only a very brief time. They quickly close again and become reset for the next action potential. When a neuron sends an action potential, it is commonly said to be "firing."

all-or-nothing principle
The principle that once the electrical impulse reaches a certain level of intensity (its threshold), it fires and moves all the way down the axon without losing any intensity.

The action potential follows the **all-or-nothing principle:** Once the electrical impulse reaches a certain level of intensity, called its *threshold,* it fires and moves all the way down the axon without losing any of its intensity. The impulse traveling down an axon can be compared to the burning fuse of a firecracker. Whether you use a match or blowtorch to light the fuse, once the fuse has been lit, the spark travels quickly and with the same intensity down the fuse.

Synapses and Neurotransmitters

The movement of an impulse down an axon may be compared to a crowd's "wave" motion in a stadium. With the wave, there is a problem, however—the aisles. How does the wave get across the aisle? Similarly, neurons do not touch each other directly, and electricity cannot travel over the space between them. Yet somehow neurons manage to communicate. This is where the chemical part of electrochemical transmission comes in. Neurons communicate with each other through chemicals that carry messages across the space. This connection between one neuron and another is one of the most

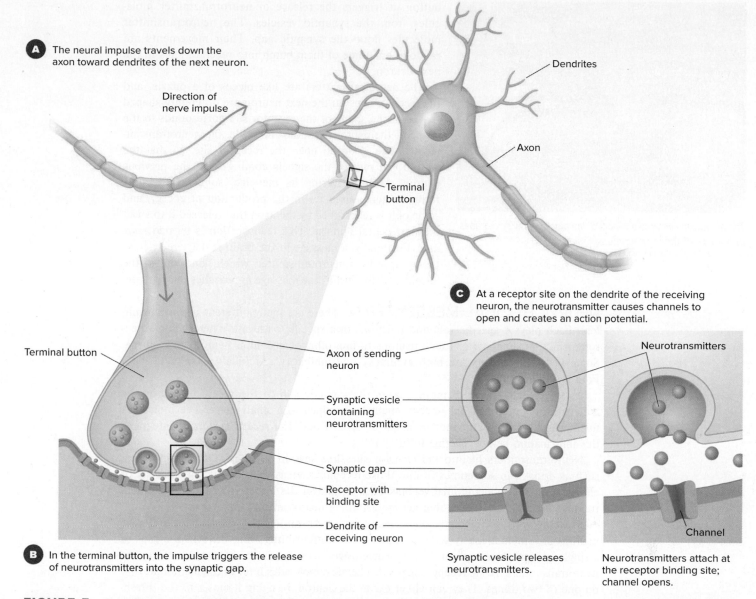

A The neural impulse travels down the axon toward dendrites of the next neuron.

Dendrites

Direction of nerve impulse

Axon

Terminal button

C At a receptor site on the dendrite of the receiving neuron, the neurotransmitter causes channels to open and creates an action potential.

Terminal button

Axon of sending neuron

Synaptic vesicle containing neurotransmitters

Synaptic gap

Receptor with binding site

Dendrite of receiving neuron

Neurotransmitters

B In the terminal button, the impulse triggers the release of neurotransmitters into the synaptic gap.

Synaptic vesicle releases neurotransmitters.

Neurotransmitters attach at the receptor binding site; channel opens.

Channel

FIGURE 5 **How Synapses and Neurotransmitters Work** (*A*) The axon of the *presynaptic* (sending) neuron meets dendrites of the *postsynaptic* (receiving) neuron. (*B*) This is an enlargement of one synapse, showing the synaptic gap between the two neurons, the terminal button, and the synaptic vesicles containing a neurotransmitter. (*C*) This is an enlargement of the receptor site. Note how the neurotransmitter opens the channel on the receptor site, triggering the neuron to fire.

intriguing and highly researched areas of contemporary neuroscience (Girault, 2020; Zhang & Bramham, 2020). Figure 5 gives an overview of how this connection between neurons takes place.

neurotransmitters
Chemical substances that are stored in very tiny sacs within the neuron's terminal buttons and involved in transmitting information across a synaptic gap to the next neuron.

SYNAPTIC TRANSMISSION **Synapses** are tiny spaces between neurons; the gap between neurons is referred to as a *synaptic gap*. Most synapses lie between the axon of one neuron and the dendrites or cell body of another neuron. Before an impulse can cross the synaptic gap, it must be converted into a chemical signal.

Each axon branches out into numerous fibers that end in structures called *terminal buttons*. Stored in very tiny synaptic vesicles (sacs) within the terminal buttons are chemical substances called **neurotransmitters**. As their name suggests, neurotransmitters transmit, or carry, information across the synaptic gap to the next neuron. When a nerve impulse reaches the terminal

synapses
Tiny spaces between neurons; the gaps between neurons are referred to as synaptic gaps.

Having synapses between neurons makes the nervous system more plastic.

The neurotransmitter-like venom of the black widow spider does its harm by disturbing neurotransmission.
Centers for Disease Control and Prevention

button, it triggers the release of neurotransmitter molecules from the synaptic vesicles. The neurotransmitter molecules flood the synaptic gap. Their movements are random, but some of them bump into receptor sites in the next neuron.

The neurotransmitters are like pieces of a puzzle, and the receptor sites on the next neuron are differently shaped spaces. If the shape of the receptor site corresponds to the shape of the neurotransmitter molecule, the neurotransmitter acts like a key to open the receptor site, so that the neuron can receive the signals coming from the previous neuron. After delivering its message, some of the neurotransmitter is used up in the production of energy, and some of it is reabsorbed by the axon that released it to await the next neural impulse. This reabsorption is termed *reuptake*. Essentially, a message in the brain is delivered across the synapse by a neurotransmitter, which pours out of the terminal button just as the message approaches the synapse.

NEUROCHEMICAL MESSENGERS There are many different neurotransmitters. Each plays a specific role and functions in a specific pathway. Whereas some neurotransmitters stimulate or excite neurons to fire, others can inhibit neurons from firing. Some neurotransmitters are both excitatory *and* inhibitory (Archibald & others, 2020; Ford & others, 2020).

As the neurotransmitter moves across the synaptic gap to the receiving neuron, its molecules might spread out or they might be confined to a small space. The molecules might come in rapid sequence or might be spaced out. The receiving neuron integrates this information before reacting to it.

Neurotransmitters fit into the receptor sites like keys in keyholes. Other substances, such as drugs, can sometimes fit into those receptor sites as well, producing a variety of effects. Similarly, many animal venoms, such as that of the black widow spider, are neurotransmitter-like substances that act by disturbing neurotransmission.

Most neurons secrete only one type of neurotransmitter, but often many different neurons are simultaneously secreting different neurotransmitters into the synaptic gaps of a single neuron. At any given time, a neuron is receiving a mixture of messages from the neurotransmitters. At its receptor sites, the chemical molecules bind to the membrane and do one of two things: They can either excite the neuron, bringing it closer to the threshold at which it will fire, or they can inhibit the neuron from firing. Usually the binding of an excitatory neurotransmitter from one neuron will not be enough to trigger an action potential in the receiving neuron. Triggering an action potential often takes a number of neurons sending excitatory messages simultaneously or fewer neurons sending rapid-fire excitatory messages.

The precise number of neurotransmitters that exist is unknown. Scientists have identified more than 100 neurotransmitters in the brain alone, each with a unique chemical makeup. In organisms ranging from snails to whales, neuroscientists have found the same neurotransmitter molecules that our own brains use. To get a better sense of what neurotransmitters do, let's consider eight of these chemicals.

Acetylcholine *Acetylcholine (ACh)* usually stimulates the firing of neurons and is involved in the action of muscles, learning, and memory (Crans & others, 2020; Inayat & others, 2020). ACh is found throughout the central and peripheral nervous systems. The venom of the black widow spider causes ACh to gush out of the synapses between the spinal cord and skeletal muscles, producing violent spasms.

People with Alzheimer disease, a degenerative brain disorder that involves a decline in memory, have an acetylcholine deficiency (Coughlin & others, 2020;

Botox injections contain botulin, a poison that, by destroying ACh, blocks the recipient's facial muscles from moving. Wrinkles, as well as many genuine facial expressions, are thereby prevented.

Thompson & Tobin, 2020). Some medications used to treat the symptoms of Alzheimer disease do so by compensating for the loss of the brain's supply of acetylcholine (Ellermann & others, 2020; Potasiewicz & others, 2020).

GABA GABA (gamma-aminobutyric acid) is found throughout the central nervous system. It is believed to be the neurotransmitter in as many as one-third of the brain's synapses. GABA is important in the brain because it keeps many neurons from firing (Andersen & others, 2020; Spurny & others, 2020). In this way, it helps to control the precision of the signal being carried from one neuron to the next. Low levels of GABA are linked with anxiety (Green & others, 2020). Antianxiety drugs increase the inhibiting effects of GABA.

You can think of GABA as the brain's brake pedal.

Glutamate Glutamate has a key role in exciting many neurons to fire and is especially involved in learning and memory (Marcondes & others, 2020). Too much glutamate can overstimulate the brain and trigger migraine headaches or even seizures. Researchers have recently proposed that glutamate also is a factor in anxiety, depression, schizophrenia, Alzheimer disease, and Parkinson disease (Jia & others, 2020; McCutcheon & others, 2020; Olajide & others, 2020). Because of the widespread expression of glutamate in the brain, glutamate receptors have increasingly become the targets of drug treatment for a number of neurological and psychological disorders (Cans & others, 2020).

Norepinephrine Norepinephrine inhibits the firing of neurons in the central nervous system, but it excites the heart muscle, intestines, and urogenital tract. Stress stimulates the release of norepinephrine (Giustino & others, 2020). This neurotransmitter also helps to control alertness. Too little norepinephrine is associated with depression, and too much triggers agitated, manic states. For example, amphetamines and cocaine cause hyperactive, manic states of behavior by rapidly increasing brain levels of norepinephrine (Underhill & others, 2020).

Recall from the beginning of the chapter that one of the most important characteristics of the brain and nervous system is integration. In the case of neurotransmitters, they may work in teams of two or more. For example, norepinephrine works with acetylcholine to regulate states of sleep and wakefulness.

Dopamine Dopamine helps to control voluntary movement and affects sleep, mood, attention, learning, and the ability to recognize rewards and other important signals in the environment (Bek & others, 2020; Katthagen & others, 2020; Tang & others, 2020; Wang & others, 2020). Dopamine is related to the personality trait of extraversion (being outgoing and gregarious), as we will see when we look at personality. Stimulant drugs such as cocaine and amphetamines produce excitement, alertness, elevated mood, decreased fatigue, and sometimes increased motor activity mainly by activating dopamine receptors (Wise & Robble, 2020).

Low levels of dopamine are associated with Parkinson disease, in which physical movements deteriorate (Bek & others, 2020; Williams & others, 2020). Problems regulating dopamine are associated with schizophrenia (Katthagen & others, 2020), a psychological disorder that we will examine when we look at psychological disorders.

Serotonin Serotonin is involved in the regulation of sleep, mood, attention, and learning. In regulating states of sleep and wakefulness, it teams with acetylcholine and norepinephrine. Serotonin is also a key to maintaining the brain's neuroplasticity (i.e., allowing the brain to change with experience) (Pawluski & others, 2020). Lowered levels of serotonin are associated with depression (Tiger & others, 2020). The antidepressant drug Prozac is thought to work by slowing the reuptake of serotonin into terminal buttons, thereby increasing brain levels of serotonin (Muller & Cunningham, 2020). Figure 6 shows the

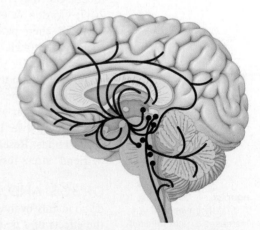

FIGURE 6 **Serotonin Pathways** Each of the neurotransmitters in the brain has specific pathways in which it functions. Shown here are the pathways for serotonin.

brain pathways for serotonin. There are 15 known types of serotonin receptors in the brain (Hoyer & others, 2002; Martin & others, 2020; Murphy & others, 2020) and each type of antidepressant drug has its effects on different receptors.

Endorphins *Endorphins* are natural opiates that mainly stimulate the firing of neurons. Endorphins shield the body from pain and elevate feelings of pleasure. A long-distance runner, a woman giving birth, and a person in shock after a car wreck all have elevated levels of endorphins (Goldfarb & others, 2020; MacGillavry & Ullrich, 2020; Werner & others, 2020).

As early as the fourth century B.C.E., the Greeks used wild poppies to induce euphoria. More than 2,000 years later, the magical formula behind opium's addictive action was finally discovered. In the early 1970s, scientists found that opium plugs into a sophisticated system of natural opiates that lie deep within the brain's pathways (Pert, 1999; Pert & Snyder, 1973). Morphine (the most important narcotic of opium) mimics the action of endorphins by stimulating receptors in the brain involved with pleasure and pain. This means that morphine, like natural endorphins in the brain, reduces the experience of pain.

Research has linked the hormone oxytocin to bonding between parents and their newborn.
SelectStock/Getty Images

Oxytocin *Oxytocin* is a hormone and neurotransmitter that plays an important role in the experience of love and social bonding (Zhang & others, 2020). A powerful surge of oxytocin is released in mothers who have just given birth, and oxytocin is related to the onset of lactation and breast feeding (Gust & others, 2020). Oxytocin, however, is not only involved in a mother's ability to provide nourishment for her baby. It is also a factor in the experience of parents who find themselves "in love at first sight" with their newborn (Eldred, 2017; Lecompte & others, 2020). Oxytocin is released as part of the sexual orgasm and is thought to play a role in the human tendency to feel pleasure during orgasm and to form emotional bonds with romantic partners (Alley & Diamond, 2020).

Oxytocin release is stimulated by birth and lactation in mothers. Might it also be released in fathers? In one study, oxytocin levels were checked in fathers when their babies were 6 weeks old and when they were 6 months old; when fathers engaged in more stimulating contact with the babies, encouraged the babies' exploration, and directed the babies' attention to objects, the fathers' oxytocin levels increased (Feldman & Bakermans-Kranenburg, 2017; Gordon & others, 2010). One study found that when fathers were administered oxytocin, their parenting behavior improved as evidenced in increased positive affect, social gaze, touch, and vocal synchrony when interacting with their infants (Weisman & others, 2014).

In addition to its role in social bonding, oxytocin may have a role to play in responding to stress, especially for women. Taylor (2011) proposed "tend and befriend theory," which states that when they experience stress women do not experience the classic fight-or-flight response. Instead women experience a surge of oxytocin that leads them to seek social bonds. Research supports the idea that oxytocin is associated with this "tend and befriend" stress response in women (Lieberz & others, 2020; von Dawans & others, 2019).

DRUGS AND NEUROTRANSMITTERS

agonist
A drug that mimics or increases a neurotransmitter's effects.

antagonist
A drug that blocks a neurotransmitter's effects.

Most drugs that influence behavior do so mainly by interfering with the work of neurotransmitters. Drugs can mimic or increase the effects of a neurotransmitter, or they can block those effects. An **agonist** is a drug that mimics or increases a neurotransmitter's effects. For example, the drug morphine mimics the actions of endorphins by stimulating receptors in the brain and spinal cord associated with pleasure and pain. An **antagonist** is a drug that blocks a neurotransmitter's effects. For example, drugs used to treat schizophrenia interfere with the activity of dopamine.

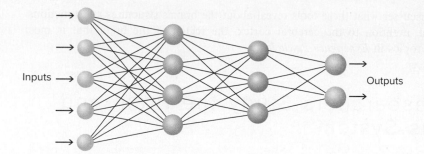

FIGURE 7   An Example of a Neural Network Inputs (information from the environment and from sensory receptors, such as the details of a person's face) become embedded in extensive connections between neurons in the brain. This embedding process leads to outputs such as remembering the person's face.

Inputs → Outputs

Neural Networks

So far, we have focused mainly on how a single neuron functions and on how a nerve impulse travels from one neuron to another. Now let's look at how large numbers of neurons work together to integrate incoming information and coordinate outgoing information. Figure 7 shows a simplified drawing of a neural network, or pathway. This diagram gives you an idea of how the activity of one neuron is linked with that of many others.

Some neurons have short axons and communicate with other, nearby neurons. Other neurons have long axons and communicate with circuits of neurons some distance away. Neural networks are not static. They can be altered through changes in the strength of synaptic connections. Any piece of information, such as a name, might be embedded in hundreds or even thousands of connections between neurons. In this way, human activities such as being attentive, memorizing, and thinking are distributed over a wide range of connected neurons. The strength of these connected neurons determines how well you remember the information.

There's a big hint here for how to study successfully. When your goal is to remember something, the best way is to build a neural network. That means making connections between the material and other things in your life—experiences, family, everyday habits. Actively engaging with the material will create neural networks to help you remember.

3 Structures of the Brain and Their Functions

Of course, the human body's extensive networks of neurons are not visible to the naked eye. Fortunately technology is available to help neuroscientists form pictures of the structure and organization of neurons and the larger structures they make up without harming the organism being studied. This section explores techniques that scientists use in brain research and discusses what these tools reveal about the brain's structures and functions. We pay special attention to the cerebral cortex, the region of the brain that is most relevant to the topics in *Experience Psychology*.

How Researchers Study the Brain and Nervous System

Early knowledge of the human brain came mostly from studies of individuals who had suffered brain damage from injury or disease or who had brain surgery to relieve another condition. Modern discoveries have relied largely on technology that enables researchers to "look inside" the brain while it is at work. Let's examine some of these innovative techniques.

BRAIN LESIONING   *Brain lesioning* is an abnormal disruption in the tissue of the brain resulting from injury or disease. In a lab setting, neuroscientists produce lesions in nonhuman laboratory animals to determine the effects on the animal's behavior (Vaidya & others, 2019; Ziess, 2020). They create the lesions by surgically removing brain tissue, destroying tissue with a laser, or eliminating tissue by injecting it with a drug (Buhidma & others, 2020; Medina & others, 2020). Examining the person or animal that has the lesion gives the researchers a sense of the function of the part of the brain that has been damaged.

Do you know anyone who has experienced a stroke or brain-damaging head injury? These experiences create lesioned areas in the brain.

FIGURE 8   An EEG Recording The electroencephalograph (EEG) is widely used in psychological research. It involves measuring electrical impulses that sweep across the surface the of the brain.

annedde/E+/Getty Images

ELECTRICAL RECORDING   The *electroencephalograph (EEG)* records the brain's electrical activity. Electrodes placed on the scalp detect brain-wave activity, which is recorded on a chart known as an *electroencephalogram* (Figure 8). This device can assess brain damage, epilepsy (a condition that produces seizures, caused by abnormal electrical surges in the brain), and other problems. In research, it can be used to assess brain activity associated with specific stimuli (Schneider & others, 2020; Travis, 2020). An advantage of EEG recordings is the electrical activity measured is very rapid and can be measured precisely so that it is clear how various stimuli affect the brain.

Ekman and colleagues (1990) measured EEG activity during emotional experiences provoked by film clips. Individuals in this study watched amusing film clips (such as a puppy playing with flowers, and monkeys taking a bath) as well as clips likely to provoke fear or disgust (a leg amputation and an operation on an eye). How does the brain respond to such stimuli? The researchers found that while watching the amusing clips, people tended to exhibit more left than right prefrontal activity, as shown in EEGs. In contrast, when the participants viewed the fear-provoking films, the right prefrontal area was generally more active than the left.

Do these differences generalize to overall differences in feelings of happiness? They just might. Urry and colleagues (2004) found that individuals who have relatively more left than right prefrontal activity (what is called *prefrontal asymmetry*) tend to rate themselves higher on a number of measures of well-being, including self-acceptance, positive relations with others, purpose in life, and life satisfaction. Such asymmetry has been found to predict positive outcomes in other groups as well (Kong & others, 2020).

Not every recording of brain activity is made with surface electrodes that are attached to the scalp. In *single-unit recording,* which provides information about a single neuron's electrical activity, a thin probe is inserted in or near an individual neuron (Hermiz & others, 2020). The probe transmits that neuron's electrical activity to an amplifier so that researchers can "see" the activity.

BRAIN IMAGING   For years, medical practitioners have used X-rays to reveal damage inside and outside the body, both in the brain and in other locations. A single X-ray of the brain is hard to interpret, however, because it shows only a two-dimensional image of the three-dimensional interior of the brain. A better technique, *computer axial tomography* (CAT scan or CT scan), produces a three-dimensional image obtained from X-rays of the head that are assembled into a composite image by a computer. The CT scan provides valuable information about the location and extent of damage involving stroke, language disorder, or loss of memory (Madhok & others, 2020). The capacity to provide a three-dimensional image of the brain is what makes CT scans superior to single X-rays.

Positron-emission tomography (PET scan) is based on metabolic changes in the brain related to activity (Nettis & others, 2020). PET measures the amount of glucose in various areas of the brain and then sends this information to a computer for analysis. Neurons use glucose for energy, so glucose levels vary with the levels of activity throughout the brain. Tracing the amounts of glucose generates a picture of activity levels throughout the brain.

An interesting application of the PET technique is the work of Kosslyn and colleagues (1996) on mental imagery, the brain's ability to create perceptual states in the absence of external stimuli. For instance, if you were to think of your favorite song right now, you could "hear" it in your mind's ear; or if you reflected on a loved one's face, you could probably "see" it in your mind's eye. Research using PET scans has shown that often the same area of the brain—a location called Area 17—is activated when we think of seeing something as when we are actually seeing it. However, Area 17 is not always activated for all of us when we imagine a visual image. Kosslyn and colleagues asked their participants to visualize a letter in the alphabet and then asked those individuals to answer some yes/no questions about the letter. For instance, a person might be thinking of the letter *C* and have to answer the question, "Does it have curvy lines?" The answer would be yes. If the person was thinking of *F*, the answer would be no. The fascinating result of this work was that people who showed brain activation on the PET scan in Area 17 while engaged in the visualization task answered the questions faster than those who were not using Area 17.

So, although human brains are similar to one another in some ways, in other ways all brains are unique.

Another technique, *magnetic resonance imaging (MRI),* involves creating a magnetic field around a person's body and using radio waves to construct images of the person's tissues and biochemical activities. The magnetic field of the magnet used to create an MRI image is extremely powerful, many times more powerful than the earth's magnetic field. MRI generates very clear pictures of the brain's interior, does not require injecting the brain with a substance, and (unlike X-rays) does not pose a problem of radiation overexposure. Getting an MRI scan involves lying still in a large metal barrel-like tunnel. MRI scans provide an excellent picture of the architecture of the brain and allow us to see if and how experience affects brain structure. For example, a number of studies have shown how training motor skills can affect brain structure (Hamano & others, 2020). Consider, for instance, how learning to play the piano or to dance can involve forging new connections in the brain (Barrett & others, 2020; Yue & others, 2020; Zhao & others, 2020). In one of the first MRI studies on this topic, Amunts and colleagues (1997) documented a link between the number of years a person has practiced musical skills (e.g., playing the piano) and the size of the brain region that is responsible for controlling hand movements.

Although MRI reveals considerable information about brain structure, it cannot portray brain function. Other imaging techniques, however, can serve as a window on the brain in action. One such method, *functional magnetic resonance imaging (fMRI)* allows scientists literally to see what is happening in the brain while it is working (Figure 9). Like the PET scan, fMRI rests on the idea that mental activity is associated with changes in the brain. While PET is about the use of glucose as fuel for thinking, fMRI exploits changes in

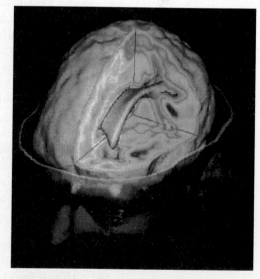

FIGURE 9   Functional Magnetic Resonance Imaging (fMRI) Through fMRI, scientists can literally see what areas of the brain are active during a task by monitoring oxygenated blood levels.

blood oxygen that occur in association with brain activity. When part of the brain is working, oxygenated blood rushes into the area. This oxygen, however, is more than what is needed. In a sense, fMRI is based on the fact that thinking is like running sprints. When you run the 100-yard dash, blood rushes to the muscles in your legs, carrying oxygen. Right after you stop, you might feel a tightness in your leg, because the oxygen has not all been used. Similarly, if an area of the brain is hard at work (e.g., solving a math problem), the increased activity leads to a surplus of oxygenated blood. This "extra" oxygen allows the brain activity to be imaged.

Getting an fMRI involves reclining in the same large metal barrel as for an MRI, but in the case of fMRI, the person can be doing something—listening to audio signals sent by the researcher through headphones or watching visual images that are presented on a screen mounted overhead. Pictures of the brain are taken, both while the brain is at rest and while it is engaging in an activity such as listening to music, looking at a picture, or making a decision. By comparing the at-rest picture to the active picture, scientists can identify what specific brain activity is associated with the mental experience being studied. fMRI technology is one of the most exciting methodological advances to hit psychology in a long time.

Note that saying that fMRI tells us about the brain activity *associated* with a mental experience is a *correlational* statement. Correlations point to the association between variables, not to the potential causal link between them. Although, for example, identifying a picture as a cat may relate to activation in a particular brain area, we do not know if recognizing the cat *caused* the brain activity.

Functional MRI is especially useful for examining cognitive activity in organisms that cannot tell us what they are thinking or feeling. For example, fMRI can be used with infants (Azhari & others, 2020) who are too young to tell us about their experiences. Similarly, fMRI can be used to reveal what is going on in the heads of nonhuman animals. The Intersection reviews interesting recent research using fMRI with dogs.

An additional method for studying brain functioning, and one that *does* allow for causal inferences, is *transcranial magnetic stimulation (TMS)* (Catricalà & others, 2020). TMS is often combined with brain-imaging techniques to establish causal links between brain activity and behavior, to examine neuronal functioning following brain-injuring events such as accidents and strokes, and even to treat some neurological and psychological disorders.

In the TMS procedure, magnetic coils are placed over the person's head and directed at a particular brain area. TMS uses a rapidly changing magnetic field to induce brief electric current pulses in the brain, and these pulses trigger action potentials in neurons. Immediately following this burst of action potentials, activity in the targeted brain area is inhibited, causing what is known as a *virtual lesion*. Completely painless, this technique, when used with brain imaging, allows scientists to examine the role of various brain regions. If a brain region is *associated* with a behavior, as demonstrated using fMRI or PET, then the temporary disruption of processing in that area should disrupt that behavior as well. So, for instance, if researchers were doing a study involving the cat recognition example described earlier, they might use TMS to disrupt the brain area that was associated with cat recognition and see whether the study's participants are temporarily unable to identify a picture of the feline.

fMRI is also used to study the brain AT REST! It can tell us what the brain is doing, even when it is resting. Such research tells us about what is called the "default network."

Sorry, lefties! Most fMRI studies include only right-handed people. As we will see later, handedness can influence brain structure.

It sounds kinda scary, huh? But it's not. TMS is also used to treat some psychological disorders.

How the Brain Is Organized

As a human embryo develops inside its mother's womb, the nervous system begins forming as a long, hollow tube on the embryo's back. At 3 weeks or so after conception, cells making up the tube differentiate into a mass of neurons, most of which then develop into three major regions of the brain: the hindbrain, which is adjacent to the top part of the spinal cord; the midbrain, which rises above the hindbrain; and the forebrain, which is the uppermost region of the brain (Figure 10).

Neuroscience and Language: What Is a Word to a Dog?

Dogs learn to respond to many different words. Ask a family dog if it wants a treat, a ride, or a walk and you will likely see a great deal of enthusiasm. But does a dog "know" that the words "treat," "ride," or "walk" refer to different things? If an owner said "dance" with the same intonation would the dog be just as eager? To find out, researchers (Prichard & others, 2018) used fMRI with dogs.

The participants were 12 pet dogs (with names like Ninja, Eddie, and Truffles) of a variety of sizes and breeds. The dogs had been trained previously to lie still in an fMRI machine. This is important because imaging requires the subject to be very still while being tested (Karl & others, 2020). Also, the scanner makes loud noises, so it was important that the dogs were used to those noises. In the weeks prior to the study, the dogs were trained to associate two words with two different toys. Owners played fetch and tug-of-war with the objects and their dogs, always saying the object names during the training. For example, Ninja's two objects were a block (called "block") and a toy monkey (called "monkey"). Ninja was trained to fetch each of the objects when its name was called by her owner. Before the scanning session, the dogs had to show that they were able to fetch each object in response to its name.

Lenka_N/Shutterstock

The dogs laid on their bellies in the scanner, with heads toward the opening, where their owners stood. While their brains were scanned, the dogs experienced different types of trials. On some trials, owners said one of the two object names five times and then showed the dog that object. On other trials, the owner said a completely fake word that was new to the dog (such as "doba" or "bobbu") and showed the dog a new object. Finally, to keep the dogs engaged, some trials just involved the dogs getting treats for being such good dogs and remaining still for so long. Brain activity was compared to see how the dogs' brains responded to the known versus unknown words.

The results? The dogs' brains showed stronger activation to the unknown, fake words that preceded the new objects. This result differs from what would be expected in humans, who are relatively more responsive to real (versus fake) words (Prichard & others, 2018). The researchers surmised that the dogs were engaged in learning the association between the new word and the new object that followed it. Dogs, it seems, are especially motivated to learn new things. Interestingly, in half the dogs, the area activated by the new words was analogous to the area in the human brain that processes word meanings. Overall, the results show that dogs *do* have neural representations of words.

Words are especially important to humans. Dogs are more attuned to smells (Horowitz & Franks, 2020). This study shows that, even if words do not convey as much to dogs as smells do, a dog's capacity to learn words goes beyond simply being rewarded for their responses. Comedian Groucho Marx once quipped, "Outside of a dog, a book is a man's best friend. Inside of a dog, it's too dark to read." This study sheds some light on what goes on inside of a dog.

\\ **What other questions might be addressed using fMRI with nonhuman animals?**

HINDBRAIN   The **hindbrain,** located at the skull's rear, is the lowest portion of the brain. The three main parts of the hindbrain are the medulla, cerebellum, and pons. Figure 11 locates these structures.

The *medulla* begins where the spinal cord enters the skull. This structure controls many vital functions, such as breathing and heart rate. It also regulates our reflexes.

The *cerebellum* extends from the rear of the hindbrain, just above the medulla. It consists of two rounded structures thought to play important roles in motor coordination (Singh & others, 2020). Leg and arm movements are coordinated by the cerebellum, for example. When we play golf, practice the piano, or learn a new dance, the cerebellum is hard at work. If another portion of the brain commands us to write the number 7, it is the cerebellum that integrates the muscular activities required to do so. Damage to the cerebellum impairs the performance of coordinated movements. When this damage occurs, people's movements become awkward and jerky. Extensive damage to the cerebellum

hindbrain
Located at the skull's rear, the lowest portion of the brain, consisting of the medulla, cerebellum, and pons.

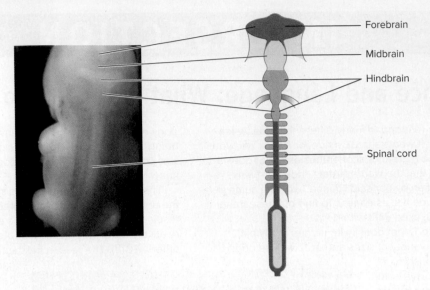

Forebrain

Midbrain

Hindbrain

Spinal cord

FIGURE 10 **Embryological Development of the Nervous System** The photograph shows the primitive tubular appearance of the nervous system at 6 weeks in the human embryo. The drawing shows the major brain regions and spinal cord as they appear early in the development of a human embryo.

(Photo): Petit Format/Science Source

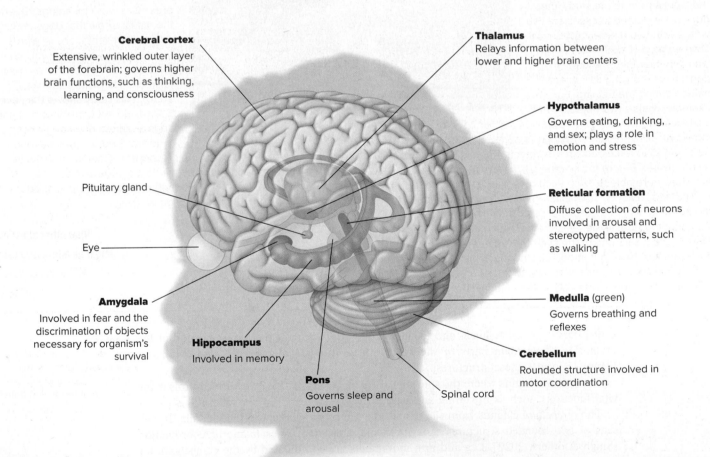

Cerebral cortex
Extensive, wrinkled outer layer of the forebrain; governs higher brain functions, such as thinking, learning, and consciousness

Thalamus
Relays information between lower and higher brain centers

Hypothalamus
Governs eating, drinking, and sex; plays a role in emotion and stress

Pituitary gland

Reticular formation
Diffuse collection of neurons involved in arousal and stereotyped patterns, such as walking

Eye

Amygdala
Involved in fear and the discrimination of objects necessary for organism's survival

Hippocampus
Involved in memory

Medulla (green)
Governs breathing and reflexes

Cerebellum
Rounded structure involved in motor coordination

Pons
Governs sleep and arousal

Spinal cord

FIGURE 11 **Structure and Regions in the Human Brain** To get a feel for where these structures are in your own brain, use the eye (pictured on the left of the figure) as a landmark. Note that structures such as the thalamus, hypothalamus, amygdala, pituitary gland, pons, and reticular formation reside deep within the brain.

makes it impossible even to stand up. Although the cerebellum is recognized generally for its role in motor behavior, it has a role in social thinking and social behavior such as mimicry (Van Overwalle & others, 2020).

The *pons* is a bridge in the hindbrain that connects the cerebellum and the brain stem. It contains several clusters of fibers involved in sleep and arousal (Kashiwagi & others, 2020).

A region called the **brain stem** includes much of the hindbrain (it does not include the cerebellum) and the midbrain (which we examine later) and gets its name because it looks like a stem. Embedded deep within the brain, the brain stem connects at its lower end with the spinal cord and then extends upward to encase the reticular formation in the midbrain. The most ancient part of the brain, the brain stem evolved more than 500 million years ago (Manger, 2020). Clumps of cells in the brain stem determine alertness and regulate basic survival functions such as breathing, heartbeat, and blood pressure (Park & others, 2020).

brain stem
The stemlike brain area that includes much of the hindbrain (excluding the cerebellum) and the midbrain; connects with the spinal cord at its lower end and then extends upward to encase the reticular formation in the midbrain.

MIDBRAIN
The **midbrain,** located between the hindbrain and forebrain, is an area in which many nerve-fiber systems ascend and descend to connect the higher and lower portions of the brain (Blaess & others, 2020). In particular, the midbrain relays information between the brain and the eyes and ears. The ability to attend to an object visually, for example, is linked to one bundle of neurons in the midbrain. Parkinson disease, a deterioration of movement that produces rigidity and tremors, damages a section near the bottom of the midbrain.

Two systems in the midbrain are of special interest. One is the **reticular formation** (see Figure 11), a diffuse collection of neurons involved in stereotyped patterns of behavior such as walking, sleeping, and turning to attend to a sudden noise. The other system consists of small groups of neurons that use the neurotransmitters serotonin, dopamine, and norepinephrine. Although these groups contain relatively few cells, they send their axons to a remarkable variety of brain regions, an operation that perhaps explains their involvement in complex, integrative functions.

midbrain
Located between the hindbrain and forebrain, an area in which many nerve-fiber systems ascend and descend to connect the higher and lower portions of the brain; in particular, the midbrain relays information between the brain and the eyes and ears.

reticular formation
A system in the midbrain comprising a diffuse collection of neurons involved in stereotyped patterns of behavior such as walking, sleeping, and turning to attend to a sudden noise.

FOREBRAIN
You try to understand what all of these terms and parts of the brain mean. You talk with friends and plan a party for this weekend. You remember that it has been 6 months since you went to the dentist. You are confident you will do well on the next exam in this course. All of these experiences and millions more would not be possible without the **forebrain,** the brain's largest division and its most forward part.

Before we explore the structures and function of the forebrain, let's stop for a moment and examine how the brain evolved. The brains of the earliest vertebrates (that is, animals with a backbone or spinal column) were smaller and simpler than those of later animals. Genetic changes during the evolutionary process were responsible for the development of more complex brains with more parts and more interconnections (Knudsen, 2020; Trevino & others, 2020). Figure 12 compares the brains of a rat, cat, chimpanzee, and human. In both the chimpanzee's brain and (especially) the human's brain, the hindbrain and midbrain structures are covered by a forebrain structure called the *cerebral cortex.* The human hindbrain and midbrain are similar to those of other animals, so it is the relative size of the forebrain that mainly differentiates the human brain from the brains of animals such as rats, cats, and chimps. The human forebrain's most important structures are listed next, with brief descriptions of their locations.

forebrain
The brain's largest division and its most forward part.

- *The limbic system*: A network of structures under the cerebral cortex.
- *The thalamus*: A structure that sits on top of the brain stem, in the central core of the brain.
- *The basal ganglia:* A cluster of neurons that sits below the cerebral cortex and atop the thalamus.
- *The hypothalamus*: A small structure just below the thalamus.
- *The cerebral cortex:* The outer layer of the brain.

We will take a look at these first four structures before we embark on an exploration of the last. Remember that these are all parts of the forebrain.

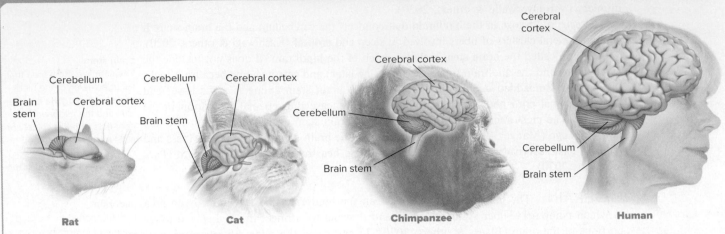

FIGURE 12 The Brain in Different Species This figure compares the brain of a rat, a cat, a chimpanzee, and a human being. As you examine the illustrations, remember that each organism's brain is adapted to meet different environmental challenges.

Photos: (Rat): Rosa Jay/Shutterstock; (Cat): Caia Image/Glow Images; (Chimapnzee): Patrick Rolands/Shutterstock; (Human): JW Ramsey/McGraw Hill

> What structures are similar across the species? > Why do you think there are some common features, and what does this commonality tell us about these brain structures? > Why don't rats have a large cerebral cortex? > How might life be different for a rat or a cat with a human brain?

limbic system
A set of subcortical brain structures central to emotion, memory, and reward processing.

Our amygdalae respond automatically to stimuli–to cute puppies, scary dogs, and attractive potential romantic partners–without our ever noticing.

hippocampus
The structure in the limbic system that has a special role in the storage of memories.

thalamus
The forebrain structure that sits at the top of the brain stem in the brain's central core and serves as an important relay station.

basal ganglia
Large neuron clusters located above the thalamus and under the cerebral cortex that work with the cerebellum and the cerebral cortex to control and coordinate voluntary movements.

Limbic System The **limbic system,** a loosely connected network of structures under the cerebral cortex, is important in both memory and emotion (Kamali & others, 2020; Talami & others, 2020). Its two principal structures are the amygdala and the hippocampus (see Figure 11).

The **amygdala** is an almond-shaped structure located inside the brain toward the base. In fact, there is an amygdala on each side of the brain. The amygdala is involved in the discrimination of objects that are necessary for the organism's survival, such as appropriate food, mates, and social rivals. Neurons in the amygdala often fire selectively at the sight of such stimuli, and lesions in the amygdala can cause animals to engage in inappropriate behavior such as attempting to eat, fight, or even mate with an object like a chair. In *Experience Psychology,* you will encounter the amygdala when we investigate intense emotions such as fear and rage. The amygdala also is involved in emotional awareness and expression through its many connections with a variety of brain areas.

amygdala
An almond-shaped structure within the base of the temporal lobe that is involved in the discrimination of objects that are necessary for the organism's survival, such as appropriate food, mates, and social rivals.

The **hippocampus** has a special role in the storage of memories (Urgolites & others, 2020). Individuals who suffer extensive hippocampal damage cannot retain any new conscious memories after the damage. It is fairly certain, though, that memories are not stored "in" the limbic system. Instead, the limbic system seems to determine what parts of the information passing through the cortex should be "printed" into durable, lasting neural traces in the cortex.

Thalamus The **thalamus** is a forebrain structure that sits at the top of the brain stem in the central core of the brain (see Figure 11). It serves as an essential relay station, functioning much like a server in a computer network. That is, an important function of the thalamus is to sort information and send it to the appropriate places in the forebrain for further integration and interpretation. For example, one area of the thalamus receives information from the cerebellum and projects it to the motor area of the cerebral cortex. Indeed, most neural input to the cerebral cortex goes through the thalamus. Whereas one area of the thalamus works to orient information from the sense receptors (hearing, seeing, etc.), another region seems to be involved in sleep and wakefulness, having ties with the reticular formation.

Basal Ganglia Above the thalamus and under the cerebral cortex lie large clusters, or *ganglia,* of neurons called **basal ganglia.** The basal ganglia work with the cerebellum and the cerebral cortex to control and coordinate voluntary movements. Basal ganglia enable people to engage in habitual behaviors such as riding a bicycle and typing a text message. People with damage to basal ganglia suffer from either unwanted movement, such as constant writhing or jerking of limbs, or too little movement, as in the slow and deliberate movements of those with Parkinson disease (Modreanu & others, 2020).

Hypothalamus

The **hypothalamus,** a small forebrain structure just below the thalamus, monitors three pleasurable activities—eating, drinking, and sexual behavior—as well as emotion, stress, and reward (see Figure 11 for the location of the hypothalamus). As we will see later, the hypothalamus also helps direct the endocrine system.

hypothalamus
A small forebrain structure, located just below the thalamus, that monitors three pleasurable activities—eating, drinking, and sexual behavior—as well as emotion, stress, and reward.

Perhaps the best way to describe the function of the hypothalamus is as a regulator of the body's internal state. It is sensitive to changes in the blood and neural input, and it responds by influencing the secretion of hormones and neural outputs. For example, if the temperature of circulating blood near the hypothalamus is increased by just 1 or 2 degrees, certain cells in the hypothalamus start increasing their rate of firing. As a result, a chain of events is set in motion. Increased circulation through the skin and sweat glands occurs immediately to release this heat from the body. The cooled blood circulating to the hypothalamus slows the activity of some of the neurons there, stopping the process when the temperature is just right—37.1 degrees Celsius (98.6 degrees Fahrenheit). These temperature-sensitive neurons function like a finely tuned thermostat in maintaining the body in a balanced state.

Pleasure center receptors can become inactive after the use of drugs such as Ecstasy and methamphetamine. The damaging effects of these drugs on the brain's reward system are what seduce individuals into a hopeless pursuit of the same feelings they had during their first high—but they will never feel that high again.

The hypothalamus also is involved in emotional states, playing an important role as an integrative location for handling stress. Much of this integration is accomplished through the hypothalamus's action on the pituitary gland, an important endocrine gland located just below it.

If certain areas of the hypothalamus are electrically stimulated, a feeling of pleasure results. In a classic experiment, Olds and Milner (1954) implanted an electrode in the hypothalamus of a rat's brain. When the rat ran to a corner of an enclosed area, a mild electric current was delivered to its hypothalamus. The researchers thought the electric current would cause the rat to avoid the corner. Much to their surprise, the rat kept returning to the corner. Olds and Milner believed they had discovered a pleasure center in the hypothalamus. Olds (1958) conducted further experiments and found that rats would press bars until they dropped over from exhaustion just to continue to receive a mild electric shock to their hypothalamus. One rat pressed a bar more than 2,000 times an hour for a period of 24 hours to receive the stimulation to its hypothalamus (Figure 13). Today researchers agree that the hypothalamus is involved in pleasurable feelings but that other areas of the brain, such as the limbic system and a bundle of fibers in the forebrain, are also important in the link between the brain and pleasure.

The Olds studies have implications for drug addiction. The rat pressed the bar mainly because this action produced a positive, rewarding effect (pleasure), not because it wanted to avoid or escape a negative effect (pain). Cocaine users talk about the drug's ability to heighten pleasure in food, sex, and a variety of activities, highlighting the reward aspects of the drug. We will look into the effects of drugs on the brain's reward centers when we talk about consciousness and learning.

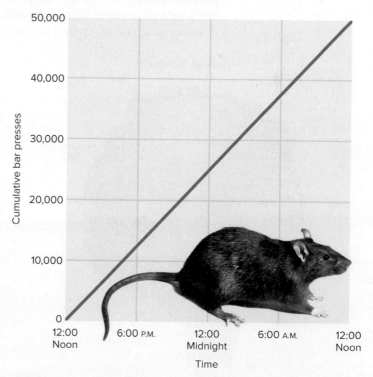

FIGURE 13 **Results of the Experiment on the Role of the Hypothalamus in Pleasure** The graphed results for one rat show that it pressed the bar more than 2,000 times an hour for a period of 24 hours to receive stimulation to its hypothalamus.
(Photo): Digital Vision Ltd./SuperStock

The Cerebral Cortex

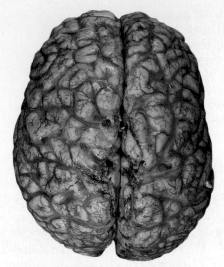

FIGURE 14   The Human Brain's Hemispheres The two halves (hemispheres) of the human brain can be seen clearly in this photograph.

Arthur Glauberman/Science Source

The **cerebral cortex** is part of the forebrain and is the most recently developed part of the brain in the evolutionary scheme. The word *cortex* means "bark" (like the bark of a tree) in Latin, and the cerebral cortex is the outer layer of the brain. It is in the cerebral cortex that the most complex mental functions, such as thinking and planning, take place.

The **neocortex** (or "new bark") is the outermost part of the cerebral cortex. In humans, this area makes up 80 percent of the cortex (compared with just 30 to 40 percent in most other mammals). The size of the neocortex in mammals is strongly related to the size of the social group in which the organisms live. Some scientists theorize that this part of the human brain, which is responsible for complex thinking, evolved so that we could understand one another (Bzdok & Dunbar, 2020; Filley, 2020).

The neural tissue that makes up the cerebral cortex covers the lower portions of the brain like a sheet that is laid over the brain's surface. In humans the cerebral cortex is greatly convoluted with lots of grooves and bulges, and these considerably enlarge its surface area (compared with a brain with a smooth surface). The cerebral cortex is highly connected with other parts of the brain. Millions of axons connect the neurons of the cerebral cortex with those located elsewhere in the brain.

cerebral cortex
Part of the forebrain, the outer layer of the brain, responsible for the most complex mental functions, such as thinking and planning.

neocortex
The outermost part of the cerebral cortex, making up 80 percent of the human brain's cortex.

LOBES The wrinkled surface of the cerebral cortex is divided into two halves called *hemispheres* (Figure 14). Each hemisphere is subdivided into four regions, or *lobes*—occipital, temporal, frontal, and parietal (Figure 15).

The **occipital lobes,** located at the back of the brain, respond to visual stimuli. Connections among various areas of the occipital lobes allow for the processing of information about such aspects of visual stimuli as their color, shape, and motion. A person can have perfectly functioning eyes, but the eyes only detect and transport information. That information must be interpreted in the occipital lobes for the viewer to "see it." A stroke or a wound in an occipital lobe can cause blindness or wipe out a portion of the person's visual field.

occipital lobes
Structures located at the back of the head that respond to visual stimuli.

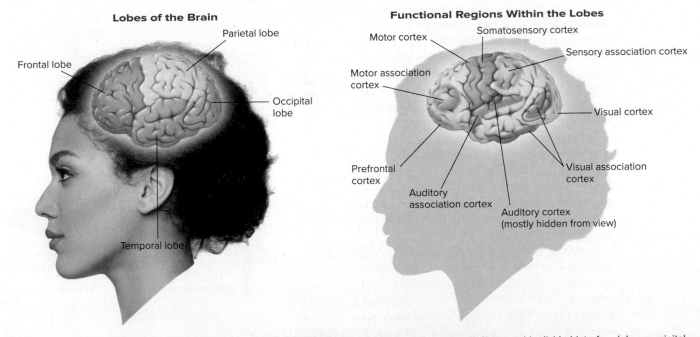

FIGURE 15   The Cerebral Cortex's Lobes and Association Areas The cerebral cortex (*left*) is roughly divided into four lobes: occipital, temporal, frontal, and parietal. The cerebral cortex (*right*) also consists of the motor cortex and somatosensory cortex. Further, the cerebral cortex includes association areas, such as the visual association cortex, auditory association cortex, and sensory association cortex.

(Photo): Paffy69/Getty Images

temporal lobes
Structures in the cerebral cortex that are located just above the ears and are involved in hearing, language processing, and memory.

The **temporal lobes,** the part of the cerebral cortex just above the ears, are involved in hearing, language processing, and memory. The temporal lobes have a number of connections to the limbic system. For this reason, people with damage to the temporal lobes cannot file experiences into long-term memory.

frontal lobes
The portion of the cerebral cortex behind the forehead, involved in personality, intelligence, and the control of voluntary muscles.

The **frontal lobes,** the portion of the cerebral cortex behind the forehead, are involved in personality, intelligence, and the control of voluntary muscles. A fascinating case study illustrates how damage to the frontal lobes can significantly alter personality. Phineas T. Gage, a 25-year-old foreman who worked for the Rutland and Burlington Railroad, was the victim of a terrible accident in 1848. Phineas and several coworkers were using blasting powder to construct a roadbed. The crew drilled holes in the rock and gravel, poured in the blasting powder, and then tamped down the powder with an iron rod. While Phineas was still tamping it down, the powder exploded, driving the iron rod up through the left side of his face and out through the top of his head. Although the wound in his skull healed in a matter of weeks, Phineas had become a different person. Previously he had been a mild-mannered, hardworking, emotionally calm individual, well liked by all who knew him. Afterward he was obstinate, moody, irresponsible, selfish, and incapable of participating in any planned activities. Damage to the frontal lobe area of his brain had dramatically altered Phineas' personality.

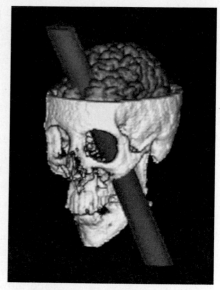

A computerized reconstruction of Phineas T. Gage's accident, based on measurements taken of his skull.
Patrick Landmann/Science Source

Without intact frontal lobes, humans are emotionally shallow, distractible, listless, and insensitive to social contexts. People with frontal lobe damage become so distracted by irrelevant stimuli that they often cannot carry out basic directions. In one such case, an individual, when asked to light a candle, struck a match correctly, but instead of lighting the candle, he put it in his mouth and acted as if he were smoking it (Luria, 1973).

The frontal lobes of humans are especially large when compared with those of other animals. For example, the frontal cortex of rats barely exists; in cats, it occupies just 3.5 percent of the cerebral cortex; in chimpanzees, 17 percent; and in humans, approximately 30 percent.

An important part of the frontal lobes is the **prefrontal cortex,** which is at the front of the motor cortex (see Figure 15) and is involved in higher cognitive functions such as planning, reasoning, and self-control. The prefrontal cortex functions as an executive control system, monitoring and organizing thinking.

prefrontal cortex
An important part of the frontal lobes that is involved in higher cognitive functions such as planning, reasoning, and self-control.

The **parietal lobes,** located at the top and toward the rear of the head, are involved in registering spatial location, attention, and motor control (Yi & Kim, 2020). Thus, the parietal lobes are at work when you are judging how far you have to throw a ball to get it to someone else, when you shift your attention from one activity to another (turn your attention away from the television to a noise outside), and as you continue to read. The brilliant physicist Albert Einstein said that his reasoning often was best when he imagined objects in space. It turns out that his parietal lobes were 15 percent larger than average (Witelsonet & others, 1999; Shi & others, 2017).

parietal lobes
Structures at the top and toward the rear of the head that are involved in registering spatial location, attention, and motor control.

A word of caution is in order about going too far in localizing function within a particular lobe or brain region. Although this discussion has attributed specific functions to a particular lobe (such as spatial location, attention, and motor control in the parietal lobe), integration and connection are extensive between any two or more lobes and between lobes and other parts of the brain. In addition, no two brains are exactly alike (Becht & Mills, 2020). We generalize about the brain's structures and functions, but it is important to bear in mind that people's brains are different, and a brain area might not serve the function that is typically associated with it in all people.

SOMATOSENSORY CORTEX AND MOTOR CORTEX

Two other important regions of the cerebral cortex are the somatosensory cortex and the motor cortex (see Figure 15). The **somatosensory cortex** processes information about body sensations.

somatosensory cortex
A region in the cerebral cortex that processes information about body sensations, located at the front of the parietal lobes.

It is located at the front of the parietal lobes. The **motor cortex,** at the rear of the frontal lobes, processes information about voluntary movement.

The map in Figure 16 shows which parts of the somatosensory and motor cortexes are associated with different parts of the body. It is based on research done by Penfield (1947), a neurosurgeon at the Montreal Neurological Institute. Penfield worked with patients with severe epilepsy, and he often performed surgery to remove portions of the epileptic patients' brains. However, he was concerned that removing a portion of the brain might impair some of the individuals' functions. Penfield's solution was to map the cortex during surgery by stimulating different cortical areas and observing the responses of the patients, who were given a local anesthetic so they would remain awake during the operation. He found that when he stimulated certain somatosensory and motor areas of the brain, patients reported feeling different sensations, or different parts of a patient's body moved. For both somatosensory and motor areas, there is a point-to-point relation between a part of the body and a location on the cerebral cortex. In Figure 16, the face and hands are given proportionately more space than other body parts because the face and hands are capable of finer perceptions and movements than are other body areas and therefore need more cerebral cortex representation.

motor cortex A region in the cerebral cortex, located just behind the frontal lobes, that processes information about voluntary movement.

Penfield's technique has influence to this day. Specifically, during brain surgery, patients are often awake. The brain cannot feel pain, so keeping patients awake allows surgeons to ask about what they are feeling, hearing, and seeing, to be sure that the surgery does not damage brain areas that are crucial for consciousness, speech, and other important functions.

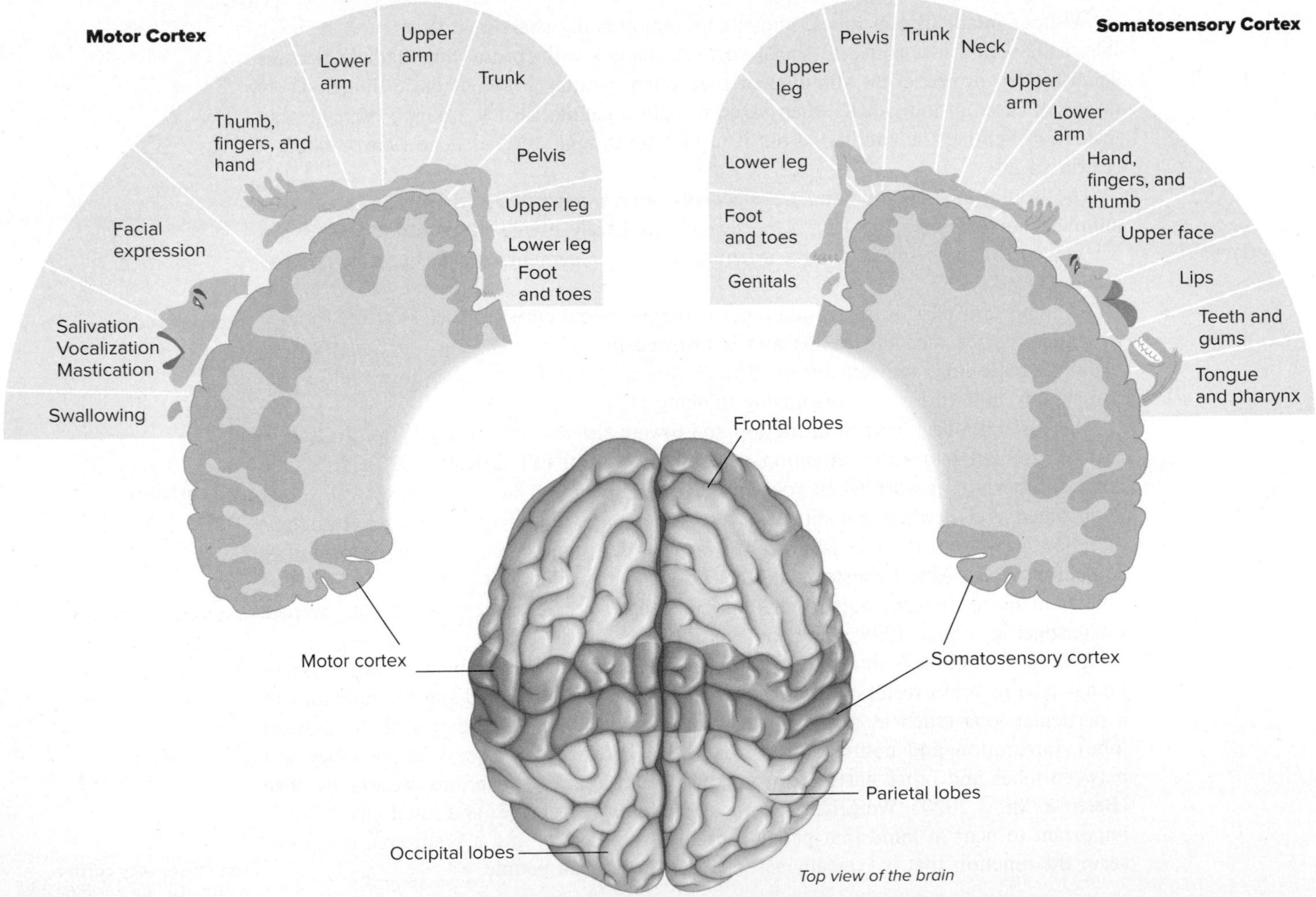

FIGURE 16 **Disproportionate Representation of Body Parts in the Motor and Somatosensory Areas of the Cortex** The amount of cortex allotted to a body part is not proportionate to the body part's size. Instead, the brain has more space for body parts that require precision and control. Thus, the thumb, fingers, and hand require more brain tissue than does the arm.

The point-to-point mapping of somatosensory fields onto the cortex's surface is the basis of our orderly and accurate perception of the world (Chéreau & others, 2020). When something touches your lip, for example, your brain knows what body part has been touched because the nerve pathways from your lip are the only pathways that project to the lip region of the somatosensory cortex.

ASSOCIATION CORTEX

association cortex Sometimes called association areas, the region of the cerebral cortex that is the site of the highest intellectual functions, such as thinking and problem solving.

Embedded in the brain's lobes, association cortex makes up 75 percent of the cerebral cortex (see Figure 15). **Association cortex** (sometimes called *association areas*) refers to regions of the cerebral cortex that integrate sensory and motor information. The highest intellectual functions, such as thinking and problem solving, occur in association cortex. There are association areas throughout the brain, and each sensory system has its own association area in the cerebral cortex.

Interestingly, damage to a specific part of the association cortex often does not result in a specific loss of function. With the exception of language areas (which are localized), loss of function seems to depend more on the extent of damage to the association cortex than on the specific location of the damage. By observing brain-damaged individuals and using a mapping technique, scientists have found that the association cortex is involved in linguistic and perceptual functioning.

The term association cortex applies to cortical material that is not somatosensory or motor cortex, but is NOT filler space!

The largest portion of association cortex is located in the frontal lobes, directly behind the forehead. Damage to this area does not lead to somatosensory or motor loss but rather to problems in planning and problem solving or what are called *executive functions* (Zelazo, 2020). Personality also may be linked to the frontal lobes. Recall the misfortune of Phineas Gage, whose personality radically changed after he experienced frontal lobe damage.

The Cerebral Hemispheres and Split-Brain Research

Recall that the cerebral cortex is divided into two halves—left and right (see Figure 14). Do these hemispheres have different functions? In 1861, French surgeon Paul Broca saw a patient who had received an injury to the left side of his brain about 30 years earlier. The patient became known as Tan because *tan* was the only word he could speak. Tan suffered from *aphasia,* a language disorder associated with brain damage. Tan died several days after Broca evaluated him, and an autopsy revealed that the injury was to a precise area of the left hemisphere. Today we refer to this area of the brain as *Broca's area,* and we know that it plays an important role in the production of speech. Another area of the brain's left hemisphere that significantly figures in language is *Wernicke's area,* which, if damaged, causes problems in comprehending language. Figure 17 locates Broca's area and Wernicke's area. Research continues to probe the degree to which the brain's left hemisphere or right hemisphere is involved in various aspects of thinking, feeling, and behavior (Gerrits & others, 2020).

corpus callosum The large bundle of axons that connects the brain's two hemispheres, responsible for relaying information between the two sides.

THE ROLE OF THE CORPUS CALLOSUM

For many years, scientists speculated that the **corpus callosum,** the large bundle of axons connecting the brain's two hemispheres, has something to do with relaying information between the two

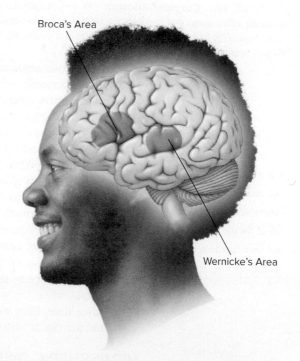

FIGURE 17 Broca's Area and Wernicke's Area
Broca's area is located in the brain's left hemisphere, and it is involved in the control of speech. Individuals with damage to Broca's area have problems saying words correctly. Also shown is Wernicke's area, the portion of the left hemisphere that is involved in understanding language. Individuals with damage to this area cannot comprehend words; they hear the words but do not know what they mean.
(Photo): Ranta Images/Getty Images

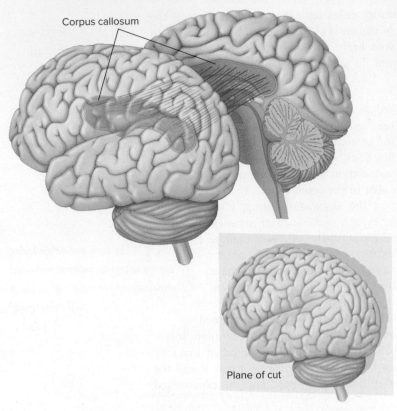

FIGURE 18 **Corpus Callosum** The corpus callosum is a thick band of about 80 million axons that connects the brain cells in one hemisphere to those in the other. In healthy brains, the two sides engage in a continuous flow of information via this neural bridge.

Corpus callosum

Plane of cut

sides (Figure 18). Sperry (1974) confirmed this in an experiment in which he cut the corpus callosum in cats. He also severed certain nerves leading from the eyes to the brain. After the operation, Sperry trained the cats to solve a series of visual problems with one eye blindfolded. After a cat learned the task—say, with only its left eye uncovered—its other eye was blindfolded, and the animal was tested again. The "split-brain" cat behaved as if it had never learned the task. It seems that the memory was stored only in the left hemisphere, which could no longer directly communicate with the right hemisphere.

Further evidence of the corpus callosum's function has come from studies of patients with severe, even life-threatening, forms of epilepsy. Epilepsy is caused by electrical "brainstorms" that flash uncontrollably across the corpus callosum. In one famous case, neurosurgeons severed the corpus callosum of a patient with epilepsy, now known as W. J., in a final attempt to reduce his unbearable seizures. Sperry (1968) examined W. J. and found that the corpus callosum functions the same in humans as in animals—cutting the corpus callosum seemed to leave the patient with "two separate minds" that learned and operated independently. By having W. J. focus on a specific point (as shown in Figure 19), the researchers could send information to each hemisphere independently. W. J. was able to point with his left hand to what only the right hemisphere had seen. He could say, outloud, what only the left hemisphere saw.

Figure 19 is tricky. Notice that, unlike the hand, each eye is actually split in half, so that half of the information that each eye sees goes to a different hemisphere. The information on the same side of the eye as the nose crosses over, and the information that is on the outside of each eye stays put.

As it turns out, the right hemisphere receives information only from the left side of the body, and the left hemisphere receives information only from the right side of the body. When you hold an object in your left hand, for example, only the right hemisphere of your brain detects the object. When you hold an object in your right hand, only the left hemisphere of the brain detects it (Figure 19). In individuals with a normally functioning corpus callosum, both hemispheres receive this information eventually, as it travels between the hemispheres through the corpus callosum. In fact, although we might have two minds, we usually use them in tandem.

You can appreciate how well and how rapidly the corpus callosum integrates your experience by thinking about the challenge of doing two things at once. Recall, for example, when you were a kid and you tried to tap your head and rub your stomach at the same time. Even with two separate hands controlled by two separate hemispheres, such dual activity is very difficult.

HEMISPHERIC DIFFERENCES IN FUNCTIONING In people with intact brains, specialization of function, or what is sometimes called *lateralization,* occurs in some areas. Researchers have uncovered evidence for hemispheric differences in function by sending different information to each ear. Remember, the left hemisphere gets its information (first) from the right ear, and the right hemisphere hears what is going on (first) in the left ear. This research has shown that the brain tends to divide its functioning into one hemisphere or the other as follows:

■ *Left hemisphere:* The most extensive research on the brain's two hemispheres has focused on language. Speech and grammar are localized to the left hemisphere (Ishkhanyan

FIGURE 19 Information Pathways from the Eyes to the Brain Each of our eyes receives sensory input from both our left and our right field of vision. Information from the left half of our visual field goes to the brain's right hemisphere (which is responsible for simple comprehension), and information from the right half of our visual field goes to the brain's left hemisphere (the brain's main language center, which controls speech and writing). The input received in either hemisphere passes quickly to the other hemisphere across the corpus callosum. When the corpus callosum is severed, however, this transmission of information cannot occur.

(Photos): Nukul Chanada/Shutterstock

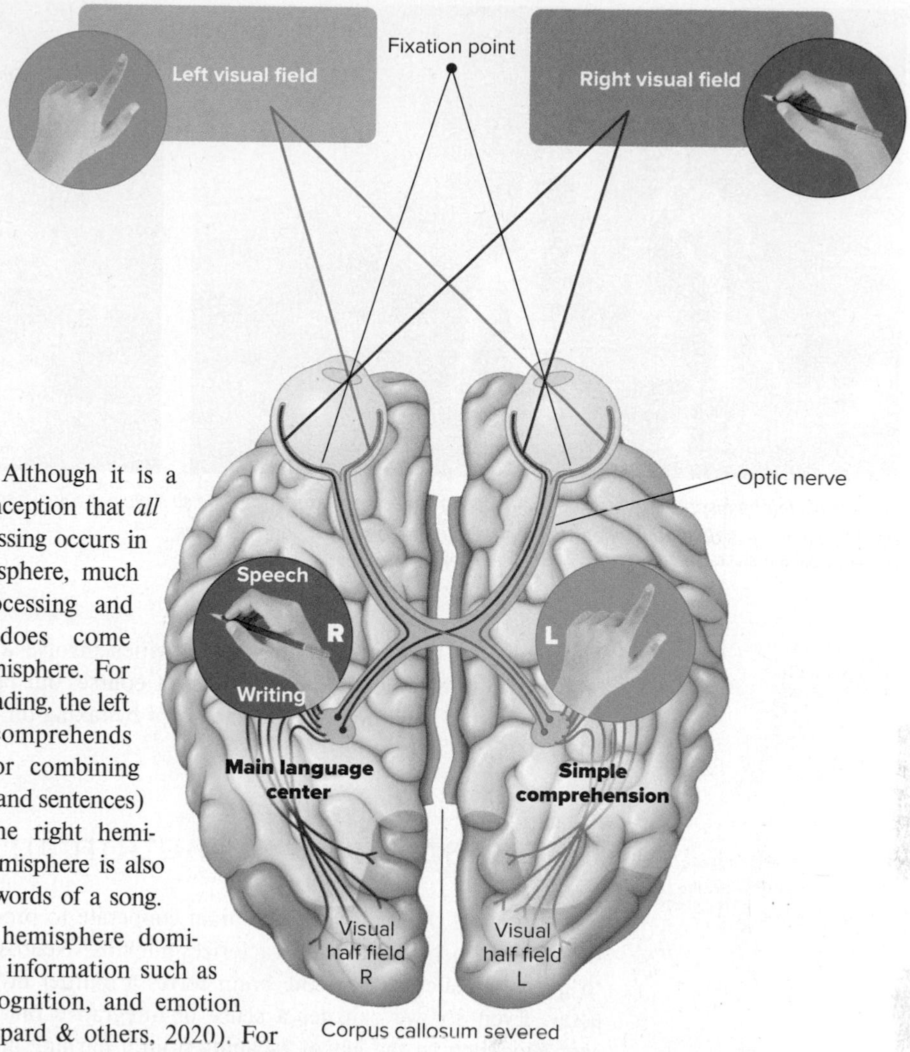

The right hemisphere is expert at recognizing faces. Researchers have asked people to watch images on a computer screen and to press a button when they see a face. Even right-handed people are much faster at this task when they use their left hand because the information goes directly from the right hemisphere to the hand that hemisphere controls.

& others, 2020). Although it is a common misconception that *all* language processing occurs in the left hemisphere, much language processing and production does come from this hemisphere. For example, in reading, the left hemisphere comprehends syntax (rules for combining words into phrases and sentences) and grammar, but the right hemisphere does not. The left hemisphere is also keenly involved in singing the words of a song.

■ *Right hemisphere:* The right hemisphere dominates in processing nonverbal information such as spatial perception, visual recognition, and emotion (Morris & others, 2020; Sheppard & others, 2020). For example, one part of the right hemisphere, the fusiform face area in the right hemisphere, is mainly at work when we process information about people's faces (Kanwisher, 2006; Wu & others, 2020).

■ The right hemisphere also may be more involved than the left hemisphere in processing information about emotions, both when we express emotions ourselves and when we recognize others' emotions (Sheppard & others, 2020). People are more likely to remember emotion words if they hear them in the left ear. Much of our sense of humor resides in the right hemisphere (Kovarski & others, 2017; Marinkovic & others, 2011). In fact, if you want to be sure that someone laughs at your joke, tell it to the person's left ear.

■ The right hemisphere is also adept at interpreting story meanings and voice intonations (Seydell-Greenwald & others, 2020). Further, the right hemisphere excels at picking up a song melody. Importantly, though, it is difficult to learn exactly what the right hemisphere can do, because it cannot just tell us. We have to come up with a way for the right hemisphere to communicate what it knows. The right hemisphere certainly has some verbal abilities, for instance, because people with split brains can draw (with their left hand) pictures of words that have been spoken to them (in the left ear).

Could this be why women, even right-handed women (but not men), automatically carry a baby in the left hand?

Because differences in the functioning of the brain's two hemispheres are known to exist, people commonly use the phrases *left-brained* (meaning logical and rational) and *right-brained* (meaning creative or artistic) as a way of categorizing themselves and others. Such generalizations have no scientific basis, and that is a good thing. We have both hemispheres because we use them both. Regardless of how much fun it might be to label ourselves "right-brained" or "left-brained," we are fortunate to be whole-brained, period.

The brain's left hemisphere is intricately involved in speech and language, and so it plays a role when we recall song lyrics. The fusiform face area of the brain's right hemisphere is dominant when we process information about people's faces.
(Singer) Liudmila P. Sundikova/Shutterstock; (friends) Hola Images/age fotostock

The reality is that most day-to-day activities involve a complex interplay between the brain's two hemispheres. Furthermore, of course, our brains can adapt to changes, such as injuries, and each hemisphere can adapt by taking on other functions (Gertel & others, 2020; Hope & others, 2017).

Integration of Function in the Brain

How do all of the regions of the brain cooperate to produce the wondrous complexity of thought and behavior that characterizes humans? Neuroscience still does not have answers to questions such as how the brain solves a murder mystery or composes a poem or an essay. Even so, we can get a sense of integrative brain function by using a real-world scenario, such as the act of escaping from a burning building.

Imagine that you are sitting at your computer writing an email when a fire breaks out behind you. The sound of crackling flames is relayed from your ear through the thalamus, to the auditory cortex, and on to the auditory association cortex. At each stage, the stimulus is processed to extract information, and at some stage, probably at the association cortex level, the sounds are finally matched with something like a neural memory representing sounds of fires you have heard previously. The association "fire" sets new machinery in motion. Your attention (guided in part by the reticular formation) shifts to the auditory signal being held in your association cortex and on to your auditory association cortex, and simultaneously (again guided by reticular systems) your head turns toward the noise. Now your visual association cortex reports in: "Objects matching flames are present." In other regions of the association cortex, the visual and auditory reports are synthesized ("We have things that look and sound like fire"), and neural associations representing potential actions ("flee") are activated. However, firing the neurons that code the plan to flee will not get you out of the chair. The basal ganglia must become engaged, and from there the commands will arise to set the brain stem, motor cortex, and cerebellum to the task of transporting you out of the room. All of this happens in mere seconds.

Which part of your brain did you use to escape? Virtually all systems had a role. By the way, you would probably remember this event because your limbic circuitry would likely have started memory formation when the association "fire" was triggered. The next time the sounds of crackling flames reach your auditory association cortex, the associations triggered would include this most recent escape. In sum, considerable integration of function takes place in the brain (Crowell & others, 2020). All of the parts of the nervous system work together as a team to keep you safe and sound.

1. Four ways that researchers study the brain and the nervous system are electrical recording, imaging, staining, and
   A. biopsy.
   B. lesioning.
   C. lobotomy.
   D. neurosurgery.

2. The brain's three major regions are the hindbrain, the midbrain, and the
   A. brain stem.
   B. reticular formation.

   C. forebrain.
   D. temporal lobes.

3. The most recently developed level of the human brain is the
   A. midbrain.
   B. forebrain.
   C. reticular formation.
   D. brain stem.

APPLY IT! 4. Because Miles suffers from extreme seizures, a surgeon severs his corpus callosum. Using a special technique, researchers present a picture of a flower to Miles's right brain and a picture of a bumblebee to Miles's left brain. When Miles is asked to say aloud what he sees, he is likely to answer,
   A. "A flower."
   B. "I don't know."
   C. "A bee."
   D. There is no way to know.

4 The Endocrine System

endocrine system
The body system consisting of a set of glands that regulate the activities of certain organs by releasing their chemical products into the bloodstream.

glands
Organs or tissues in the body that create chemicals that control many bodily functions.

The **endocrine system** consists of a set of glands that regulate the activities of certain organs by releasing their chemical products into the bloodstream. **Glands** are organs or tissues in the body that create chemicals that control many bodily functions. Neuroscientists have discovered that the nervous system and endocrine system are intricately interconnected. They know that the brain's hypothalamus connects the nervous system and the endocrine system and that the two systems work together to control the body's activities. Yet the endocrine system differs significantly from the nervous system in a variety of ways. For one thing, the parts of the endocrine system are not all connected in the same way as the parts of the nervous system. For another thing, the endocrine system works more slowly than the nervous system, because the chemicals released by the endocrine glands are transported through the circulatory system, in the blood. The heart does a mind-boggling job of pumping blood through the body, but blood moves far more slowly than neural impulses do.

The chemical messengers produced by the endocrine glands are called **hormones.** The bloodstream carries hormones to all parts of the body, and the membrane of every cell has receptors for one or more hormones.

The endocrine glands consist of the pituitary gland, the thyroid and parathyroid glands, the adrenal glands, the pancreas, the ovaries, and the testes (Figure 20). In much the same way as the brain's control of muscular activity is constantly monitored and altered to suit the

hormones
Chemical messengers that are produced by the endocrine glands and carried by the bloodstream to all parts of the body.

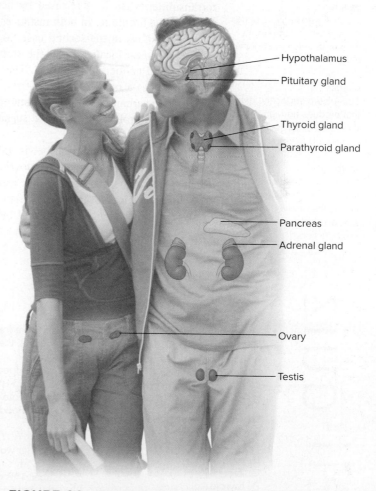

- Hypothalamus
- Pituitary gland
- Thyroid gland
- Parathyroid gland
- Pancreas
- Adrenal gland
- Ovary
- Testis

FIGURE 20 **The Major Endocrine Glands** The pituitary gland releases hormones that regulate the hormone secretions of the other glands. The pituitary gland is regulated by the hypothalamus.
(Photo): Laurence Mouton/PhotoAlto/Getty Images

FIGURE 21

The Pituitary Gland The pituitary gland, which hangs by a short stalk from the hypothalamus, regulates the hormone production of many of the body's endocrine glands. Here it is enlarged 30 times.

MedicalRF.com

pancreas
A dual-purpose gland under the stomach that performs both digestive and endocrine functions.

ovaries
Sex-related endocrine glands that produce hormones involved in sexual development and reproduction.

testes
Sex-related endocrine glands in the scrotum that produce hormones involved in sexual development and reproduction.

information received by the nervous system, the action of the endocrine glands is continuously monitored and changed by nervous, hormonal, and chemical signals. Recall from earlier in the chapter that the autonomic nervous system regulates processes such as respiration, heart rate, and digestion. The autonomic nervous system acts on the endocrine glands to produce a number of important physiological reactions to strong emotions, such as rage and fear.

The **pituitary gland,** a pea-sized gland just beneath the hypothalamus, controls growth and regulates other glands (Figure 21). The anterior (front) part of the pituitary is known as the *master gland* because almost all of its hormones direct the activity of target glands elsewhere. In turn, the anterior pituitary gland is controlled by the hypothalamus.

The **adrenal glands,** located at the top of each kidney, regulate mood, energy level, and the ability to cope with stress. Each adrenal gland secretes epinephrine (also called *adrenaline*) and norepinephrine (also called *noradrenaline*). Unlike most hormones, epinephrine and norepinephrine act quickly. Epinephrine helps a person get ready for an emergency by acting on smooth muscles, the heart, stomach, intestines, and sweat glands. In addition, epinephrine stimulates the reticular formation, which in turn arouses the sympathetic nervous system, and this system subsequently excites the adrenal glands to produce more epinephrine. Norepinephrine also alerts the individual to emergency situations by interacting with the pituitary gland and the liver. You may remember that norepinephrine functions as a neurotransmitter when it is released by neurons. In the adrenal glands, norepinephrine is released as a hormone. In both instances, norepinephrine conveys information—in the first case, to neurons; in the second case, to glands.

The **pancreas,** located under the stomach, is a dual-purpose gland that performs both digestive and endocrine functions. The part of the pancreas that serves endocrine functions produces a number of hormones, including insulin. This part of the pancreas, the islets of Langerhans, turns out hormones like a little factory. Insulin is an essential hormone that controls glucose (blood sugar) levels in the body and is related to metabolism, body weight, and obesity.

The **ovaries,** located in the pelvis on either sides of the uterus, and **testes,** located in the scrotum, are the sex-related endocrine glands that produce hormones related to sexual development and reproduction. These glands and the hormones they produce play important roles in developing sexual characteristics such as breasts and facial hair. They are also involved in other characteristics and behaviors, as you will see as you learn more about psychology.

pituitary gland
A pea-sized gland just beneath the hypothalamus that controls growth and regulates other glands.

adrenal glands
Glands at the top of each kidney that are responsible for regulating moods, energy level, and the ability to cope with stress.

self-quiz

1. The endocrine glands produce chemicals called
   A. hormones.
   B. neurotransmitters.
   C. endocrine secretions.
   D. bile.

2. The endocrine glands include all of the following except the
   A. pituitary.
   B. pancreas.
   C. liver.
   D. thyroid.

3. The adrenal glands regulate energy level, the ability to deal with stress, and
   A. appetite.
   B. digestion.
   C. motor coordination.
   D. mood.

APPLY IT! 4. Diabetes, a common disorder worldwide, involves problems in the body's regulation of glucose, or blood sugar. This disorder is often treated by diet, but sometimes individuals with

diabetes must inject themselves with insulin. The endocrine system gland that is involved in diabetes is the
   A. pituitary.
   B. ovaries.
   C. pancreas.
   D. adrenal.

5 Brain Damage, Plasticity, and Repair

Recall from earlier in this chapter that plasticity is an example of the brain's remarkable adaptability. Neuroscientists have studied plasticity, especially following brain damage, and have charted the brain's ability to repair itself. Brain damage can produce horrific effects, including paralysis, sensory loss, memory loss, and personality deterioration. When such damage occurs, can the brain recover some or all of its functions? Recovery from brain damage varies considerably, depending on the age of the individual, the extent of the damage, the area of the brain affected, and the characteristics of the person (Beuriat & others, 2020; Sivandzade & others, 2020).

The Brain's Plasticity and Capacity for Repair

For much of the twentieth century, it was generally concluded that the younger children are, the better their recovery will be from a traumatic brain injury. However, age alone is often not a good indicator of the brain's ability to recover from a traumatic injury. Although the young child's brain has more plasticity than an older child's, because of its immaturity it also is more vulnerable to insults (Anderson & others, 2009; Beuriat & others, 2020). Thus, assessing outcomes of brain insults based on age alone can be misleading. A research review concluded that children's outcomes following injury to the brain depend on factors related to the injury (nature, severity, and timing of insult), physical factors (age, cognitive capacity, and genetic makeup), and environmental influences (intervention and quality of rehabilitation, family functioning, and socioeconomic status) (Gomez-Nicola & Perry, 2014). Regarding age, in general, a more severe insult is more damaging to the young child's brain than the older child's brain (Anderson & others, 2005).

A significant factor in recovery is whether some or all of the neurons in an affected area are just damaged or are completely destroyed (Huang & Chang, 2009). If the neurons have not been destroyed, brain function often becomes restored over time as the brain repairs itself.

There are three ways that such repair might take place:

1. *Collateral sprouting,* in which the axons of some healthy neurons adjacent to damaged cells grow new branches (Freeman & others, 2020; Lemaitre & others, 2020).

2. *Substitution of function,* in which the damaged region's function is taken over by another area or areas of the brain.

3. *Neurogenesis,* the process by which new neurons are generated. Researchers have found that neurogenesis occurs in mammals such as mice. It is now accepted that neurogenesis can occur in humans (Dillen & others, 2020; Durante & others, 2020). However, to date, the presence of new neurons has been documented only in the hippocampus, which is involved in memory (Berger & others, 2020b). Neurogenesis in other mammals has been found in the olfactory bulb (involved in the sense of smell), but few if any of such cells are newly formed in adult humans (Berger & others, 2020a). Researchers are exploring how the grafting of neural stem cells to various regions of the brain, such as the hypothalamus, might increase neurogenesis (Mauceri & others, 2020; Shahror & others, 2020). If researchers can discover how new neurons are generated, possibly the information can be used to fight degenerative diseases of the brain such as Alzheimer disease and Parkinson disease.

Protecting the Athlete's Brain

In his first two games as a professional American football player, Jahvid Best scored a remarkable five touchdowns for the Detroit Lions. His NFL career seemed to be on a fast track to greatness. However, he suffered a concussion during a preseason game the following year. This injury was only one of multiple serious head injuries leading to concussions he had experienced as a football player in high school and college (Elman, 2020). And this one ended his football career at the age of just 24. Best was puzzled by the gravity of the situation: How could his life be upended by experiences he could not even remember happening?

A concussion or mild traumatic brain injury (MTBI) is a head injury that leads to temporary loss of brain function (Moore & others, 2020). Symptoms include headache, dizziness, nausea, unequal pupil size, and lack of memory for events surrounding the trauma. Importantly, these symptoms may not be immediately apparent, and contrary to common belief concussions do not necessarily lead to a lack of consciousness.

John Mersits/CSM/REX/Shutterstock

MTBIs are an increasing concern among health professionals who work with athletes at every level. Among high school athletes, MTBIs are thought to make up at least 15 percent of sports injuries (Hollis & others, 2014; Wallace & others, 2017). Although concussions can occur in any sport, head injuries are a special concern for youth football players because of the risk of tackles. A review of deaths among high school football players, spanning 2010 to 2014, found that 24 deaths occurred due to catastrophic brain and spinal cord injuries (Kucera & others, 2017).

Treatment for concussion involves rest and careful monitoring (Wallace & others, 2017). The brain requires time to recover from injury. When the brain has not had time to recover from a concussion, a second injury can cause rapid swelling of the brain, leading to brain stem failure and eventual coma or death (Cobb & Battin, 2004).

Of course, for a concussion to be treated properly, someone must first notice that it has occurred and then must respond appropriately. This responsibility ought not to fall on the injured athlete. Indeed, a person suffering from a concussion may not recognize their own injury, because the trauma has damaged the very organ that would provide the person with insight on the injury.

Psychologists know that motivation can influence perception. Our motives can shape what we see and how we interpret that evidence. Teams, players, and fans want to win, and having star players on the field is seen as crucial to victory. In this context, coaches might reason that injured players are well enough to return to the field because they "seem" normal. And players, even if suffering from a powerful headache, might not complain because they are so motivated to win, to show team loyalty, and to avoid appearing weak.

The dangers of not accurately diagnosing and treating concussions are great. TBI among young athletes is of special concern because such injuries may require special educational services (Nagele & others, 2019). Long-term effects are also troubling. In one study, researchers examined 111 brains that were donated by former NFL players (Mez & others, 2017). Among these people, all but one showed signs of a condition called chronic traumatic encephalopathy (CTE). CTE is a degenerative condition of the brain that is caused by repeated injuries. Symptoms of CTE include difficulty thinking, memory loss, depression, and suicidal thoughts.

Fortunately for Jahvid Best, the end of his football career was not the end of his life as an athlete. He ran in the 2016 Olympics as a member of the San Lucia national track team, becoming the first NFL player to compete in an Olympics (Elman, 2020), although his performance placed him far from the medal stand.

Brain Tissue Implants

The brain naturally recovers some functions that are lost following damage, but not all. Scientists have been exploring the possibility that full recovery may be made possible through *brain grafts*—implants of healthy tissue into damaged brains. The potential success of brain grafts is much better when brain tissue from the fetal stage (an early stage in prenatal development) is used (Wahlberg, 2020). The neurons of the fetus are still growing and have a much higher probability of making connections with other neurons than do the neurons of adults. In a number of studies, researchers have damaged part of an adult rat's brain, waited until the animal recovered as much as possible by itself, and assessed its behavioral deficits. Then they take the corresponding area of a fetal rat's brain and transplant it into the damaged brain of the adult rat. In these studies, the rats that receive the brain transplants demonstrate considerable behavioral recovery (Xu & others, 2020).

The human body contains more than 220 different types of cells, but **stem cells** are unique because they are primitive cells that have the capacity to develop into most types of human cells. Because of their amazing plasticity, stem cells might potentially replace damaged cells in the human body, including cells involved in spinal cord injury and brain damage. Recent research shows that implanted human stem cells can generate connections within the adult cerebral cortex (Grønning Hansen & others, 2020). One promising direction for research is the treatment of Parkinson disease. Parkinson disease impairs coordinated movement to the point that just walking across a room can be a major ordeal. Fetal dopamine cells are transplanted into the basal ganglia of the person with Parkinson disease in an effort to improve motor performance. Results for such studies have been limited but encouraging (Chen & others, 2020; Doi & others, 2020).

stem cells
Unique primitive cells that have the capacity to develop into most types of human cells.

Typically, researchers harvest the stem cells from frozen embryos left over from in vitro fertilization procedures. In these procedures, a number of eggs, or ova, are collected from a woman's ovaries to be fertilized in a lab (rather than in the body). The ova are brought together with sperm, producing human embryos. Because the procedure is difficult and delicate, doctors typically fertilize many eggs with the hope that some will survive when implanted in the uterus. In the typical procedure, there are leftover embryos, in the *blastocyst* stage, which occurs 5 days after conception. At this stage the embryo has not attached to the uterus. The blastocyst has no brain, no central nervous system, and no mouth—it is an undifferentiated ball of cells.

Do you support or oppose stem cell research? Why?

Supporters of stem cell technology emphasize that using these cells for research and treatment might relieve a great deal of human suffering. Opponents of abortion disapprove of the use of stem cells in research or treatment on the grounds that the embryos die when the stem cells are removed. (Leftover embryos are likely to be destroyed in any case.) In the United States, whether federal funds may be used for stem cell research (as they were under President Barack Obama) or not (as was the case under President Donald J. Trump) has often depended on politics.

self-quiz

1. Repair of the damaged brain might take place by all of the following *except*
   A. substitution of function.
   B. psychotherapy.
   C. collateral sprouting.
   D. neurogenesis.

2. The process by which the axons of healthy neurons adjacent to damaged cells grow new branches is called
   A. substitution of function.
   B. neurogenesis.
   C. collateral sprouting.
   D. dendritic branching.

3. The primitive cells that have the capacity to develop into most types of human cells are called
   A. stem cells.
   B. blastocysts.
   C. collateral cells.
   D. neurogenetic cells.

APPLY IT! 4. Taylor is injured in a serious car accident, suffering head injuries. After the accident, Taylor, who used to be talkative, seems to be unable to speak. Which of the following would best predict that Taylor is likely to regain the ability to talk?
   A. Taylor is male.
   B. Taylor is under 5 years old.
   C. Taylor is over the age of 21.
   D. Taylor is female.

6 Genetics and Behavior

In addition to the brain and nervous system, other aspects of our physiology also have consequences for psychological processes. Genes are one important contributor to these processes. The particular influences of nature (genetic endowment) and of nurture (experience) on psychological characteristics have long fascinated psychologists. Here we begin by examining the central agent of nature: our genetic heritage.

Chromosomes, Genes, and DNA

chromosomes
In the human cell, threadlike structures that come in 23 pairs, one member of each pair originating from each parent, and that contain DNA.

deoxyribonucleic acid (DNA)
A complex molecule in the cell's chromosomes that carries genetic information.

genes
The units of hereditary information, consisting of short segments of chromosomes composed of DNA.

Within the human body are literally trillions of cells. The nucleus of each human cell contains 46 **chromosomes,** threadlike structures that come in 23 pairs, one member of each pair originating from each parent. Chromosomes contain the remarkable substance **deoxyribonucleic acid (DNA),** a complex molecule that carries genetic information. **Genes,** the units of hereditary information, are short segments of chromosomes composed of DNA. What do genes do? Their main function is to manufacture the proteins that are necessary for maintaining life.

It is not the case that each gene is translated into one and only one protein; moreover, a gene does not act independently. Indeed, rather than being a group of independent genes, the human genome (*genome* refers to an organism's complete genetic material) consists of many genes that collaborate both with one another and with nongenetic factors inside and outside the body. The cellular machinery mixes, matches, and links small pieces of DNA to reproduce the genes, and that machinery is influenced by what is going on around it. Figure 22 illustrates the relationship among cells, chromosomes, genes, and DNA.

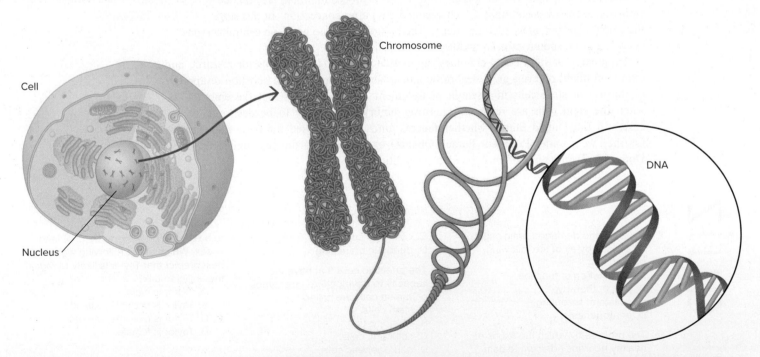

Cell

Nucleus

Chromosome

DNA

FIGURE 22   Cells, Chromosomes, Genes, and DNA (*Left*) The body contains trillions of cells, which are the basic structural units of life. Each cell contains a central structure, the nucleus. (*Middle*) Chromosomes and genes are located in the nucleus of the cell. Chromosomes are made up of threadlike structures composed mainly of DNA molecules. (*Right*) A gene is a segment of DNA that contains the hereditary code. The structure of DNA resembles a spiral ladder.

Genome refers to an organism's complete genetic material. The exact number of genes on the human genome continues to be debated (Willyard, 2018). Current estimates suggest there are 19,000 to 30,000 genes, each producing three proteins (NIH, 2020). Although scientists are still a long way from unraveling all the mysteries about the way genes work, some aspects of this process are well understood, starting with the fact that multiple genes interact to give rise to observable characteristics.

The Study of Genetics

Historically speaking, genetics is a relatively young science. Its origins go back to the mid-nineteenth century, when an Austrian monk named Gregor Mendel studied heredity in generations of pea plants. By cross-breeding plants with different characteristics and noting the characteristics of the offspring, Mendel discovered predictable patterns of heredity and laid the foundation for modern genetics.

Danny Smythe/Shutterstock

Mendel noticed that some genes seem to be more likely than others to show up in the physical characteristics of an organism. In some gene pairs, one gene is dominant over the other. If one gene of a pair is dominant and one is recessive, according to the **dominant-recessive genes principle,** the dominant gene overrides the recessive gene. A recessive gene exerts its influence only if both genes of a pair are recessive. If you inherit a recessive gene from only one biological parent, you may never know you carry the gene. In the world of dominant-recessive genes, brown eyes, farsightedness, and dimples rule over blue eyes, nearsightedness, and freckles. If you inherit a recessive gene for a trait from both of your biological parents, you will show the trait. That is why two brown-eyed parents can have a blue-eyed child: Each biological parent would have a dominant gene for brown eyes and a recessive gene for blue eyes. Because dominant genes override recessive genes, the parents have brown eyes. However, the child can inherit a recessive gene for blue eyes from each biological parent. With no dominant gene to override them, the recessive genes make the child's eyes blue.

dominant-recessive genes principle
The principle that if one gene of a pair is dominant and one is recessive, the dominant gene overrides the recessive gene. A recessive gene exerts its influence only if both genes of a pair are recessive.

Do It!

Search the web for information about a happiness gene. How would you evaluate research on such a gene given what you have read so far in *Experience Psychology*? What (if anything) would the existence of such a gene mean for your ability to find happiness in your life?

Unlike eye color, complex human characteristics such as personality and intelligence are likely influenced by many different genes. Scientists use the term *polygenic inheritance* to describe the influences of multiple genes on behavior. The term *gene-gene interaction* is used to describe the ways two or more genes combine to influence characteristics, behavior, diseases, and development (Biton & others, 2020; Serdarevic & others, 2020).

Today researchers continue to apply Mendel's methods, as well as modern technology, in their quest to expand knowledge about genetics. This section discusses four ways to study genetics: molecular genetics, selective breeding, and behavior genetics.

MOLECULAR GENETICS The field of *molecular genetics* involves the manipulation of genes using technology to determine their effect on behavior. Researchers who study molecular genetics are trying to discover the specific locations on genes that determine an individual's susceptibility to many diseases and other aspects of health and well-being (Euesden & others, 2020; Liyanarachchi & others, 2020).

SELECTIVE BREEDING *Selective breeding* is a genetic method in which organisms are chosen for reproduction based on how much of a particular trait they display. Mendel developed this technique in his studies of pea plants. An example involving behavior is the classic selective breeding study conducted by Tryon (1940), who chose to study maze-running ability in rats. After training a large number of rats to run a complex maze, Tryon then mated the rats that were best at maze running ("maze bright") with each other and the ones that were the worst ("maze dull") with each other. He continued this process with 21 generations of rats. After several generations, the maze-bright rats significantly outperformed the maze-dull rats.

Selective breeding studies have demonstrated that genes are an important influence on behavior, but that does not mean that experience is unimportant. For example, in another study, maze-bright and maze-dull rats were reared in one of two environments: (1) an impoverished environment that consisted of a barren wire-mesh group cage or (2) an enriched environment that contained tunnels, ramps, visual displays, and other stimulating objects (Cooper & Zubek, 1958). When they reached maturity, only the maze-dull rats that had been reared in an impoverished environment made more maze-learning errors than the maze-bright rats.

Psychologists have learned a lot from studying rats in mazes.

GENOME-WIDE ASSOCIATION METHOD   The *genome-wide association method* means that researchers take all of the information they can from people's genes and probe for genetic variations linked to a particular disease, such as cancer, cardiovascular disease, psychological disorders, or Alzheimer disease (Andrews & others, 2020; Galesloot & others, 2020; Li & others, 2020; Niarchou & others, 2020). To conduct a genome-wide association study, researchers obtain DNA from individuals who have the disease under study and from those who do not. Then, each participant's complete set of DNA, or genome, is scanned to determine markers of genetic variation. If certain genetic variations occur more frequently in people who have the disease, the variations point to the region in the human genome where the disease-causing problem exists.

Genes that are close to one another in our DNA are more likely to be inherited together. This link between genes is used in what is called *linkage analysis*. This analysis may help identify the location of certain genes by referring to those genes whose position is already known. This strategy is often used to search for genes associated with risk for a disease. Gene linkage studies are now being conducted on a wide variety of disorders and health issues. A key challenge with genome-wide association and linkage studies is that results from these do not consistently replicate, which means that one study might produce a set of conclusions that other studies do not find. As such, it is important for researchers to combine information from many of these studies to ensure conclusions are reproducible (Andrews & others, 2020; Montag & others, 2020).

BEHAVIOR GENETICS   *Behavior genetics* is the study of the degree and nature of heredity's influence on behavior. Behavior genetics is less invasive than molecular genetics and selective breeding. Using methods such as the *twin study,* behavior geneticists examine the extent to which individuals are shaped by their heredity and their environmental experiences (Abramson & others, 2020). In the most common type of twin study, researchers compare the behavioral similarity of identical twins with the behavioral similarity of fraternal twins. Identical twins develop from a single fertilized egg that splits into two genetically identical embryos, each of which becomes a person. Fraternal twins develop from separate eggs and separate sperm, and so they are genetically no more similar than nontwin siblings. They may even be of different sexes.

By comparing groups of identical and fraternal twins, behavior geneticists capitalize on the fact that identical twins are more similar genetically than are fraternal twins. In a recent study, researchers were interested in examining the role of genes in affectionate communication (i.e., the tendency to share feelings of warmth and intimacy with a romantic partner) (Floyd & others, 2020). They compared 229 identical twin pairs to 235 fraternal twin pairs who all completed measures of affectionate communication. The results showed that the identical twins were more similar than the fraternal twins, suggesting genes matter to this important behavior.

One problem with twin studies is that adults might stress the similarities of identical twin children more than those of fraternal twins, and identical twins might perceive themselves as a "set" and play together more than fraternal twins do. If so, observed similarities in identical twins might be strongly influenced by environmental (not just genetic) factors.

In another type of twin study, researchers evaluate identical twins who have been reared in separate environments. If their behavior is similar, the assumption is that heredity has played an important role in shaping their behavior. This strategy is the basis for the

Minnesota Study of Twins Reared Apart, directed by Bouchard and his colleagues (1996). They bring identical twins who have been reared apart to Minneapolis from all over the world to study their behavior. They ask thousands of questions about their family, childhood, interests, and values. Detailed medical histories are obtained, including information about diet, smoking, and exercise habits.

One pair of twins in the Minnesota study, Jim Springer and Jim Lewis, were separated at 4 weeks of age and did not see each other again until they were 39 years old. They had an uncanny number of similarities, even though they had lived apart for decades. For example, they both worked as part-time deputy sheriffs, had vacationed in Florida, had owned Chevrolets, had dogs named Toy, and had married and divorced women named Betty. Both liked math but not spelling. Both were good at mechanical drawing. Both put on 10 pounds at about the same time in their lives, and both started suffering headaches at 18 years of age. They did have a few differences. For example, one expressed himself better orally, and the other was more proficient at writing. One parted his hair over his forehead; the other wore his hair slicked back with sideburns.

Critics argue that some of the separated twins in the Minnesota study had been together several months prior to their adoption, that some had been reunited prior to testing (in certain cases, for a number of years), that adoption agencies often put identical twins in similar homes, and that even strangers are likely to have some coincidental similarities (Joseph, 2006). To think critically about this issue, see the Psychological Inquiry.

Identical Twins *We've all heard stories about identical twins who were separated at birth. When these twins meet up in adulthood, people often find the similarities between them to be uncanny. Are these similarities evidence of the extraordinary power of genes? Let's take a closer look.*
Kenneth Sponsler/Getty Images

>Imagine that you did not see the photo of twins provided and were simply asked how similar two people of the same gender, ethnicity, appearance, and age might be. >How might even very different environments respond to these individuals similarly? >How many people of this same gender, age, and ethnicity might enjoy similar hobbies? Have similar jobs? >What does this Psychological Inquiry tell you about the power of vivid and unusual cases in the conclusions we reach?

Genes and the Environment

So far, we have talked a lot about genes, and you are probably getting the picture that genes are a powerful force in an organism. The role of genetics in some characteristics may seem obvious; for instance, how tall you are depends to a large degree on how tall your parents are. However, imagine a person growing up in a severely impoverished environment—with poor nutrition, inadequate shelter, little or no medical care, and a mother who had received no prenatal care. This person may have the genetic potential for the height of an NBA or a WNBA center, but without environmental support for this genetic capacity, may never reach that genetically programmed height. Thus, the relationship between an individual's genes and the actual person we see before us is not a perfect one-to-one correspondence. Even for a characteristic such as height, genes do not fully determine where a person will stand on this variable. We need to account for the role of nurture, or environmental factors, in the characteristics we see in the fully grown person.

If the environment matters for an apparently simple characteristic such as height, imagine the role it might play in complex characteristics such as being outgoing or intelligent. For these psychological characteristics, genes are, again, not directly reflected in the characteristics of the person. Indeed, genes cannot tell us exactly what a person will be like. Genes are simply related to some of the characteristics we see in a person.

Our height depends significantly on the genes we inherit. However, even if we have genes that call for the stature of a basketball center, we may not reach that genetically programmed height if we lack good nutrition, adequate shelter, and medical care.
Ron Waite/CSM/REX/Shutterstock

To account for this gap between genes and actual observable characteristics, scientists distinguish between a genotype and a phenotype. A **genotype** is a person's genetic heritage, the actual genetic material. A **phenotype** is the person's observable characteristics. The relationship between a genotype and a phenotype is not always obvious. Recall that some genetic characteristics are dominant and others are recessive. Seeing that a person has brown eyes (the phenotype) tells us nothing about whether the person might also have a gene for blue eyes (the genotype) hiding out as well. The phenotype is influenced by the genotype but also by environmental factors.

The term *phenotype* applies to both physical and psychological characteristics. Consider a trait such as extraversion, the tendency to be outgoing and sociable. Even if we knew the exact genetic recipe for extraversion, we still could not perfectly predict a person's level of (phenotypic) extraversion from their genes because at least some of this trait comes from the person's experience. We will revisit the concepts of genotype and phenotype throughout *Experience Psychology*. How does the process from genotype to phenotype work? It's very complex, but at a very basic level in a cell, DNA information is transcribed to RNA (ribonucleic acid), which in turn is translated into amino acids that will become proteins. Once assembled, proteins become capable of producing characteristics in the phenotype.

Whether a gene is "turned on"—working to assemble proteins—is also a matter of collaboration. The activity of genes *(genetic expression)* is affected by their environment. For example, hormones that circulate in the blood make their way into the cell where they can turn genes on and off. The flow of hormones, too, can be affected by environmental conditions, such as light, day length, nutrition, and behavior. Factors such as stress, radiation, and temperature can influence gene expression (Biton & others, 2020).

Studies are exploring how interactions between heredity and environment influence development, including interactions that involve specific DNA sequences (Burman & others, 2020; Chubar & others, 2020; Davies & Cicchetti, 2014). An important lesson from this work is that people with the same genetic characteristics can, because of environmental factors, differ from one another in important ways. For instance, research has found that variations in dopamine-related genes interact with supportive or unsupportive family environments to influence children's aggressive behavior (Golds & others, 2020). Children with a particular genetic characteristic were more likely to behave aggressively if their parents were harsh and insensitive. This type of research concerns **gene × environment (G × E) interaction**—the interaction of a specific measured variation in DNA and a specific measured aspect of the environment. G × E interactions can be tricky to understand. They simply tell us that just because two people share genetic similarities, they are still likely to be different. Even when a gene has a strong relationship to a particular phenotypic characteristic, that characteristic may not show itself in a person if the person's experiences do not lead the gene to express itself (Williams, 2020).

You may know someone with a family history of a particular condition, such as alcoholism. A family history might suggest genetics is involved. Yet, that person may or may not develop the condition if they never drink or do not encounter experiences that promote drinking.

genotype
An individual's genetic heritage; their actual genetic material.

phenotype
An individual's observable characteristics.

✎ Environmental influences can affect things like hair color. Just because Mom has dark brown hair doesn't mean she's genetically a brunette! Only her hairdresser knows for sure.

gene × environment (g × e) interaction
The interaction of a specific measured variation in DNA and a specific measured aspect of the environment.

How Should We Think About Genes and Behavior?

Imagine two young classical violinists, Jake and Chris. Both are outstanding at their craft. They play in competitions, and garner awards and acclaim. Now, imagine a scientist discovers a gene that is associated with musical genius. Jake and Chris are both tested for the gene. Chris has it but Jake does not. How does this genetic information affect the way you think about Jake and Chris? Is one more talented than the other? Is one more likely to be successful? Now, consider if Jake and Chris are not musicians but teenagers who have engaged in criminal behavior. If Chris has a gene that is linked to criminality and Jake does not, how does that affect how you think about their futures or how likely they are to continue engaging in criminal acts as adults?

Finding out that a characteristic "is genetic" can affect how we think about that characteristic. *Genetic essentialism* refers to biased thinking about genes. Genetic essentialism means thinking that genes directly determine characteristics and that those characteristics cannot change (Lynch & others, 2019). This way of thinking about genetics is mistaken, of course. Nothing we have learned about genes suggests that they possess a magical ability to determine human behavior. Although genetic mutations can have profound effects on people, it is important to be mindful that genes combine with each other and with environmental factors to predict important outcomes.

Nevertheless, people seem to attribute a great deal of importance to genes when they judge others. Genetic explanations can lead to more negative views of those with psychological disorders but more positive views of those from diverse sexual orientations (Lynch & others, 2019). Genetic explanations can affect whether we hold someone morally accountable for their actions. Is Chris as guilty as Jake if Chris's genes led him to a life of crime?

Hill Street Studios/Getty Images

The far-reaching implications of genetic essentialism suggest that it is important to think critically whenever you encounter such information.

One way to think about the relationship between genes and human behavior is the concept of risk factor. **Risk factors** are characteristics, experiences, or exposures that increase the likelihood of a person developing a disorder or disease. Genetic characteristics can serve as such risk factors. Although typically applied to disorders and diseases, the concept of risk factor provides a way to think about the link between genes and positive and negative outcomes. Genes are correlated with these outcomes. Of course, correlation does not imply causation. A person can have a particular gene that is related to a psychological characteristic but never develop that characteristic.

Many of the biases that characterize the way we think about genes are also at play when we learn about other biological aspects of people, especially the brain (Haslam, 2011). Biological explanations for behavior may feel like they carry an extra helping of scientific fact. But these explanations must be met with the same scrutiny as any other explanations.

What Do You Think?

- If you received information that you are genetically at risk for a disease, how would you feel and how would it change your behavior?
- Do you think a person is less blameworthy for their behaviors if those behaviors are linked to genes? Why or why not?

Genetic and DNA testing are part of contemporary life. Long cold criminal cases are solved by DNA testing. Scientists have discovered links between specific genes and thousands of diseases, including various forms of cancer, diabetes, and neuromuscular diseases. Like over 20 million others worldwide, you may have already collected genetic information about yourself from companies like 23andMe or AncestryDNA. From a saliva sample these companies document a person's genotype and may provide information about the links between genes and many important life outcomes, including psychological and physical disease risk. It can be tricky to understand the influence genes might have on psychological characteristics and behaviors. To think critically about this issue, see Challenge Your Thinking.

The biological foundations of psychology are in evidence across the entire nervous system, including the brain, the intricately working neurotransmitters, the endocrine system, and the genes. These physical realities of the body work in amazing concert to produce behaviors, thoughts, and feelings. The activities you do each day, from large to small, are all signs of the spectacular success of this physical system. Your mastery of the material in this chapter is but one reflection of the extraordinary capabilities of this biological feat.

risk factor
Characteristics, experiences, or exposures that increase the likelihood that a person will develop a psychological disorder.

self-quiz

1. The threadlike structures that are present in the cell nucleus and contain genes and DNA are called
   A. genomes.
   B. polygenic markers.
   C. chromosomes.
   D. stem cells.

2. Researchers study genetics through all of the following methods *except*
   A. twin studies.
   B. selective breeding.
   C. environmental impact studies.
   D. molecular genetics.

3. The individual's *observable* characteristics, influenced by both genetic

and environmental factors, are called the
   A. genome.
   B. genotype.
   C. phenotype.
   D. prototype.

APPLY IT! 4. Sarah and Jack both have brown hair. When their son Trent is born, he has bright red hair. Family and friends start making jokes about any male friends of Sarah's who have red hair. Should Jack be worried that he is not Trent's biological father?
   A. Jack should not be worried because brown hair color is part of

the phenotype, not necessarily Sarah's and Jack's genotypes.
   B. Jack should not be worried because Trent's hair color is part of his genotype, not necessarily his phenotype.
   C. Jack should be worried because there is no way for two brunettes to have a baby with red hair.
   D. Jack should be worried because Trent's phenotype should match his parents' types exactly.

SUMMARY

① The Nervous System

The nervous system is the body's electrochemical communication circuitry. Four important characteristics of the brain and nervous system are complexity, integration, adaptability, and electrochemical transmission. The brain's special ability to adapt and change is called plasticity.

Decision making in the nervous system occurs in specialized pathways of nerve cells. Three of these pathways involve sensory input, motor output, and neural networks.

The nervous system is divided into two main parts: central (CNS) and peripheral (PNS). The CNS consists of the brain and spinal cord. The PNS has two major divisions: somatic and autonomic. The autonomic nervous system consists of two main divisions: sympathetic and parasympathetic. In particular, the sympathetic nervous system is involved in the experience of stress.

② Neurons

Neurons are cells that specialize in processing information. They make up the communication network of the nervous system. The three main parts of the neuron are the cell body, dendrite (receiving part), and axon (sending part). A myelin sheath encases and insulates most axons and speeds up transmission of neural impulses.

A neuron sends information along its axon in the form of brief electric impulses. Resting potential is the stable, slightly negative charge of an inactive neuron. The brief wave of electrical charge that sweeps down the axon, called the action potential, is an all-or-nothing response. The synapse is the space between neurons. At the synapse, neurotransmitters are released from the sending neuron, and some of these attach to receptor sites on the receiving neuron, where they stimulate another electrical impulse. Neurotransmitters include acetylcholine, GABA, glutamate, norepinephrine, dopamine, serotonin, oxytocin, and endorphins. Neural networks are clusters of neurons that are interconnected to process information.

③ Structures of the Brain and Their Functions

Techniques used to study the brain include brain lesioning, electrical recording, and brain imaging. These methods have revealed much about the brain's major divisions—hindbrain, midbrain, and forebrain.

The cerebral cortex makes up most of the outer layer of the brain, and it is here that higher mental functions such as thinking and planning take place. The wrinkled surface of the cerebral cortex is divided into hemispheres, each with four lobes: occipital, temporal, frontal, and parietal. There is considerable integration and connection among the brain's lobes.

The brain has two hemispheres. Two areas in the left hemisphere that involve specific language functions are Broca's area (speech) and Wernicke's area (language comprehension). The corpus callosum is a large bundle of fibers that connects the two hemispheres. Research suggests that the left brain is more dominant in processing verbal information (such as language) and the right brain in processing nonverbal information (such as spatial perception, visual recognition, faces, and emotion). Nonetheless, in a person whose corpus callosum is intact, both hemispheres of the cerebral cortex are involved in most complex human functioning.

④ The Endocrine System

The endocrine glands release hormones directly into the bloodstream for distribution throughout the body. The pituitary gland is the master endocrine gland. The adrenal glands play important roles in mood, energy level, and ability to cope with stress. Other parts of the endocrine system include the pancreas, which produces insulin, and the ovaries and testes, which produce sex hormones.

Brain Damage, Plasticity, and Repair

The human brain has considerable plasticity, although this plasticity is greater in young children than later in development. Three ways in which a damaged brain might repair itself are collateral sprouting, substitution of function, and neurogenesis. Brain grafts are implants of healthy tissue into damaged brains. Brain grafts are more successful when fetal tissue is used. Stem cell research may allow for new treatments for damaged nervous systems.

 Genetics and Behavior

Chromosomes are threadlike structures that occur in 23 pairs, with one member of each pair coming from each parent. Chromosomes contain the genetic substance deoxyribonucleic acid (DNA). Genes, the units of hereditary information, are short segments of chromosomes composed of DNA. According to the dominant-recessive genes principle, if one gene of a pair is dominant and one is recessive, the dominant gene overrides the recessive gene.

Two important concepts in the study of genetics are the genotype and phenotype. The genotype is an individual's actual genetic material. The phenotype is the observable characteristics—both physical and psychological—of the person.

Four methods of studying heredity's influence are molecular genetics, selective breeding, genome-wide association studies, and behavior genetics. Two methods used by behavior geneticists are twin studies and adoption studies. Both genes and environment play a role in determining the phenotype of an individual. Even for characteristics in which genes play a large role (such as height and eye color), the environment also is a factor. Gene × environment (G × E) interactions mean that people with the same gene may still differ from each other because of experience.

KEY TERMS

action potential
adrenal glands
afferent nerves
agonist
all-or-nothing principle
amygdala
antagonist
association cortex
autonomic nervous system
axon
basal ganglia
brain stem
cell body
central nervous system (CNS)
cerebral cortex
chromosomes
corpus callosum

dendrites
deoxyribonucleic acid (DNA)
dominant-recessive genes principle
efferent nerves
endocrine system
forebrain
frontal lobes
gene × environment (G × E) interaction
genes
genotype
glands
glial cells
hindbrain
hippocampus
hormones
hypothalamus

limbic system
midbrain
motor cortex
myelin sheath
neocortex
nervous system
neural networks
neurons
neurotransmitters
occipital lobes
ovaries
pancreas
parasympathetic nervous system
parietal lobes
peripheral nervous system (PNS)
phenotype
pituitary gland

plasticity
prefrontal cortex
risk factor
resting potential
reticular formation
somatic nervous system
somatosensory cortex
stem cells
stress
stressors
sympathetic nervous system
synapses
temporal lobes
testes
thalamus

ANSWERS TO SELF-QUIZZES

Section 1: 1. B; 2. D; 3. C; 4. A
Section 2: 1. D; 2. D; 3. A; 4. B

Section 3: 1. B; 2. C; 3. B; 4. C
Section 4: 1. A; 2. C; 3. D; 4. C

Section 5: 1. B; 2. C; 3. A; 4. B
Section 6: 1. C; 2. C; 3. C; 4. A

3 Sensation and Perception

Is Everything Cake?

In the summer of 2020, the #EverthingIsCake meme invaded social media (Lorenz, 2020). People shared videos of objects—pickles, a bar of soap, an eggplant, a croc shoe, a salad, even a human hand—that all turned out to be cake. The videos attracted millions of views as people watched the ultimate reveal: a knife slice through the extremely realistic objects to reveal an expertly shaped and frosted cake. For many, those cake videos were a silly distraction during a difficult time. But these videos also present a fascinating glimpse of the interplay of our senses. Any child knows the essentials of sensation. We see with our eyes, hear with our ears, smell with our noses, and taste with our tongues. What could be simpler? Expert chefs often warn, "We taste with our eyes first." What about a cake in the shape of an onion or lemons? Natalie Sideserf, who baked those strange concoctions, assured viewers that the cakes were delicious: "I would never put this amount of time and effort into something that doesn't taste as good as it looks" (Lorenz, 2020). Do you think your favorite cake would taste just as good if shaped like a croc?

What we see, hear, smell, and taste is all that we know of reality. Our reality is the product of our senses combining in fascinating ways—even when everything just might be cake.

This chapter explores sensation and perception, the processes by which we engage with the external world. We first examine vision and then probe hearing, the skin senses, taste, smell, and the kinesthetic and vestibular senses. Without the senses, we would be isolated from the world around us; we would live in dark silence—and in a tasteless, colorless, feelingless void.

1 How We Sense and Perceive the World

Sensation and perception researchers represent a broad range of specialties, including *ophthalmology,* the study of the eye's structure, function, and diseases; *audiology,* the science concerned with hearing; *neurology,* the scientific study of the nervous system; and many others. Understanding sensation and perception requires comprehending the physical properties of the objects of our perception—light, sound, the texture of material things, and so on. The psychological approach to these processes involves understanding the physical structures and functions of the sense organs, as well as the brain's conversion of the information from these organs into experience.

The Processes and Purposes of Sensation and Perception

Our world is alive with stimuli—all the objects and events that surround us. A stimulus can be anything that we detect in the world. Sensation and perception are the processes that allow us to detect and understand these various stimuli. We do not actually experience these stimuli directly; rather, our senses allow us to get information about aspects of our environment, and we then take that information and form a perception of the world. **Sensation** is the process of receiving stimulus energies from the external environment and transforming those energies into neural energy. Physical energy such as light, sound, and heat is detected by specialized receptor cells in the sense organs—eyes, ears, skin, nose, and tongue. When the receptor cells register a stimulus, the energy is converted into an electrochemical impulse or action potential that relays information about the stimulus through the nervous system to the brain (Baldwin & Ko, 2020). Recall that an action potential is the brief wave of electrical charge that sweeps down the axon of a neuron for possible transmission to another neuron. When it reaches the brain, the information travels to the appropriate area of the cerebral cortex (Fan & others, 2020).

The brain gives meaning to sensation through perception. **Perception** is the process of organizing and interpreting sensory information so that it makes sense. Receptor cells in our eyes record—that is, sense—a sleek silver object in the sky, but they do not "see" a jet plane. Sensation is about the biological

sensation
The process of receiving stimulus energies from the external environment and transforming those energies into neural energy.

perception
The process of organizing and interpreting sensory information so that it makes sense.

If you've ever begged someone to try your favorite food only to have the person give you a shrug and "meh," that's the difference between sensation and perception. Both tongues had the same experience. But perception is subjective.

Through sensation we take in information from the world; through perception we identify meaningful patterns in that information. Sensation and perception work hand in hand when we see a beautiful face and enjoy the sweet fragrance of a flower.
Westend61/Getty Images

processing that occurs between our sensory systems and the environment. Perception is our experience of those processes in action. Sensation is the process of the person encountering stimuli. Perception is the brain providing meaning to those encounters.

BOTTOM-UP AND TOP-DOWN PROCESSING

Psychologists distinguish between bottom-up and top-down processing in sensation and perception (Barker & Hicks, 2020; Sakai, 2020). In **bottom-up processing**, sensory receptors register information about the external environment and send it to the brain for interpretation. Bottom-up processing means taking in information and trying to make sense of it. Bottom-up processing begins with the external world. In contrast, **top-down processing** starts with cognitive processing in the brain. In top-down processing we begin with some sense of what is happening and apply that framework to incoming information from the world. Top-down processing is reflected when our expectations affect what we see, hear, or feel.

One way to understand the difference between bottom-up and top-down processing is to think about how you experience a song you have never heard before versus that same song when you have heard it many times (Figure 1). The first time you hear the song, you listen carefully to get a "feel" for it. That is bottom-up processing: taking the incoming information of the music and relying on that external experience. Now, once you have a good feel for the song, you listen to it in a different way. You have expectations and know what comes next. That is top-down processing. You might even sing along with the song when it comes on the radio or at a club. Be careful, though, because top-down expectations are not always accurate, and you might find yourself singing to a remix that differs from what you are used to.

Top-down processing can happen even in the absence of any stimulus at all. You can experience top-down processing by "listening" to your favorite song in your head. As you "hear" the song in your mind's ear, you are engaged in a perceptual experience produced by top-down processing.

Eating strawberries, cherries, and red popsicles, little kids often get the idea that red = sweet . . . then they get a taste of red beets! It's always a fun moment when bottom-up experience collides with top-down expectations.

Bottom-up and top-down processing work together in sensation and perception to allow us to function accurately and efficiently (Gordon & others, 2019; Riddle & others, 2019). By themselves our ears provide only incoming information about sound in the environment. Only when we consider both what the ears hear (bottom-up processing) and what the brain interprets (top-down processing) can we fully understand how we perceive sounds in our world. In everyday life, the two processes of sensation and perception are essentially inseparable. For this reason, most psychologists consider sensation and perception as a unified information-processing system.

bottom-up processing
The operation in sensation and perception in which sensory receptors register information about the external environment and send it up to the brain for interpretation.

top-down processing
The operation in sensation and perception, launched by cognitive processing at the brain's higher levels, that allows the organism to sense what is happening and to apply that framework to information from the world.

Top-Down Processing

...that the brain interprets as music

Thinking about the music...

...creates a perceptual experience in the mind's ear.

Taking in the sounds...

Bottom-Up Processing

FIGURE 1   Top-Down and Bottom-Up Processes in Perception When you listen to a song for the first time, bottom-up processing allows you to get a feel for the tune. Once you know the song well, you can create a perceptual experience in your mind's ear, by "playing" it in your head. That's top-down processing.

Most predatory animals have eyes at the front of the face; most animals that are prey have eyes on the sides of their head. Through these adaptations, predators perceive their prey accurately, and prey gain a measure of safety from their panoramic view of their environment.
(Owl): Michael Cummings/Moment/Getty Images; (Rabbit): Andrew Howe/Vetta/Getty Images

Top-down expectations can influence what we think we see or hear (Brooks & Freeman, 2019; Eckstein & others, 2019). For example, in a series of studies, researchers asked people to find a toothbrush in a picture of a bathroom. Participants had no trouble picking out a (regular sized) toothbrush on the counter in front of the sink. However, what they did not seem to notice was a gigantic toothbrush, nearly as big as the counter itself, just behind the sink (Eckstein & others, 2017). In fact, people missed giant objects 13 percent more often than much smaller, but regularly sized objects. The explanation for this difference is that we look for objects based on our expectations for what they should look like, including size.

When the objects of our perception are other people, such expectations can include stereotypes, or our expectations for a person's behavior because of their membership in a particular group. If you hold a stereotype of professors as helpful you might be more likely to see a professor's neutral face as smiling. Similarly, stereotypes we hold can influence our ability to accurately perceive a broad range of other people, sometimes with important consequences (Moradi & others, 2020). When we perceive a person, we see many aspects of identity all at once. We might see the person's gender, race or ethnicity, cues to the person's level of wealth, and so forth. These various cues can affect how we treat the person in an automatic way (Petsko & Bodenhausen, 2020). Recognizing how our expectations affect perception is important not only because of how they influence our perceptions of toothbrushes, music, or tea, but because they may color our interactions with people in our diverse world.

THE PURPOSES OF SENSATION AND PERCEPTION Why do we perceive the world? From an evolutionary perspective, the purpose of sensation and perception is adaptation that improves a species' chances for survival (DeCasien & Higham, 2019; Iwaniuk & Wylie, 2020). An organism must be able to sense and respond quickly and accurately to events in the immediate environment, such as the approach of a predator, the presence of prey, and the appearance of a potential mate. Not surprisingly, therefore, most animals—from goldfish to gorillas to humans—have eyes and ears, as well as sensitivities to touch and chemicals (smell and taste). Furthermore, a close comparison of sensory systems in animals reveals that each species is adapted exquisitely to the habitat in which it evolved. Animals that are primarily predators generally have their eyes at the front of their face so that they can perceive their prey accurately. In contrast, animals that are more likely to be someone else's lunch have their eyes on either side of their head, giving them a wide view of their surroundings at all times.

Dogs can smell better than humans. But that's because dogs need to and humans don't.

Sensory Receptors and the Brain

sensory receptors
Specialized cells that detect stimulus information and transmit it to sensory (afferent) nerves and the brain.

All sensation begins with sensory receptors. **Sensory receptors** are specialized cells that detect stimulus information and transmit it to sensory (afferent) nerves and the brain. Sensory receptors are the openings through which the brain and nervous system experience the world. Figure 2 shows the human sensory receptors for vision, hearing, touch, smell, and taste.

*There is that word again, **afferent**. Remember that **afferent** = **arrives** at the brain.*

FIGURE 2 **Human Senses: Organs, Energy Stimuli, and Sensory Receptors** The receptor cells for each sense are specialized to receive particular types of energy stimuli.
(Eye): Barbara Penoyar/Photodisc/Getty Images; (Ear): Triocean/iStock/Getty Images; (Skin): Jill Braaten/McGraw Hill; (Nose): Sigrid Olsson/ZenShui/Getty Images; (Tongue): Glow Images/Superstock

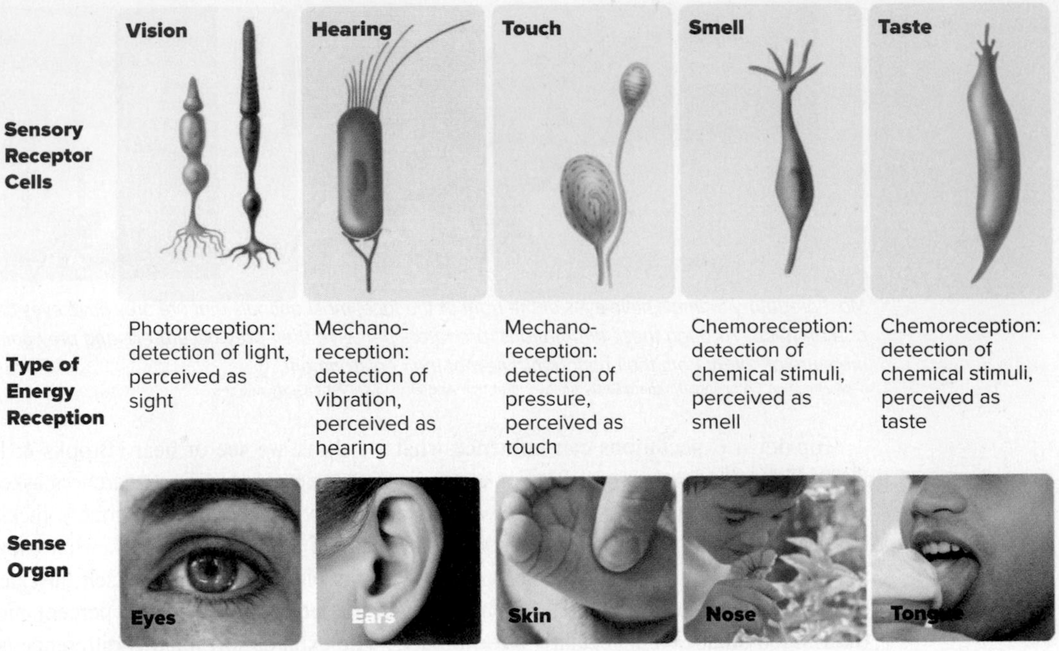

	Vision	Hearing	Touch	Smell	Taste
Sensory Receptor Cells					
Type of Energy Reception	Photoreception: detection of light, perceived as sight	Mechano-reception: detection of vibration, perceived as hearing	Mechano-reception: detection of pressure, perceived as touch	Chemoreception: detection of chemical stimuli, perceived as smell	Chemoreception: detection of chemical stimuli, perceived as taste
Sense Organ	Eyes	Ears	Skin	Nose	Tongue

Figure 3 depicts the flow of information from the environment to the brain. Sensory receptors take in information from the environment, creating local electrical currents. These currents are graded; that means they are sensitive to the intensity of stimulation, such as the difference between a dim and a bright light. These receptors trigger action potentials in sensory neurons, which carry that information to the central nervous system. Because sensory neurons (like all neurons) follow the all-or-nothing principle, the intensity of the stimulus cannot be communicated to the brain by changing the strength of the action potential. Instead, the receptor varies the *frequency* of action potentials sent to the brain. So, if a stimulus is very intense, like the bright sun on a hot day, the neuron will fire more frequently (but with the same strength) to let the brain know that the light is indeed very, very bright.

*Remember the **frequency** of firing communicates intensity for all of our sensory neurons. A loud noise leads to more frequent pulses, too. And a very painful jab from a needle leads to more frequent pulses communicating, OUCH!*

Other than frequency, the action potentials of all sensory nerves are alike. This sameness raises an intriguing question: How can an animal distinguish among sight, sound, odor, taste, and touch? The answer is that sensory receptors are selective and have different neural pathways. They are specialized to absorb a particular type of energy—light energy, sound vibrations, or chemical energy, for example—and convert it into an action potential.

Sensation involves detecting and transmitting information about different kinds of energy. The sense organs and sensory receptors fall into several main classes based on the type of energy that is transmitted. The functions of these classes include the following:

■ *Photoreception:* detection of light, perceived as sight

■ *Mechanoreception:* detection of pressure, vibration, and movement, perceived as touch, hearing, and equilibrium (or balance)

■ *Chemoreception:* detection of chemical stimuli, perceived as smell and taste

Each of these processes belongs to a particular class of receptors and brain processes. There are rare cases, however, in which the senses can become confused. The term *synaesthesia* describes an experience in which one sense (say, sight) induces an experience in another sense (say, hearing) (Arend & others, 2020; Ward & Filiz, 2020). A person might "see" music or "taste" a color, for example. One woman was able to taste sounds, so that a piece of music might taste like tuna fish (Beeli & others, 2005). Neuroscientists are exploring the neurological bases of synaesthesia, especially in the connections between

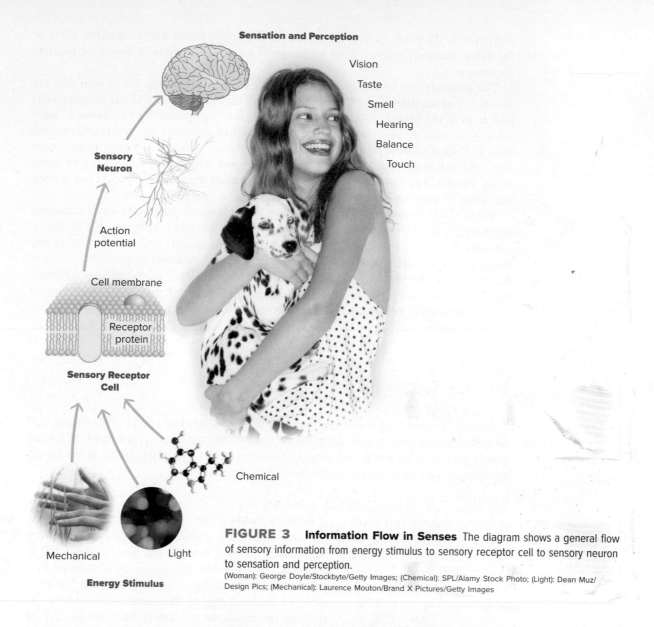

Sensation and Perception

Vision
Taste
Smell
Hearing
Balance
Touch

Sensory Neuron

Action potential

Cell membrane

Receptor protein

Sensory Receptor Cell

Chemical

Mechanical   Light

Energy Stimulus

FIGURE 3   Information Flow in Senses The diagram shows a general flow of sensory information from energy stimulus to sensory receptor cell to sensory neuron to sensation and perception.
(Woman): George Doyle/Stockbyte/Getty Images; (Chemical): SPL/Alamy Stock Photo; (Light): Dean Muz/Design Pics; (Mechanical): Laurence Mouton/Brand X Pictures/Getty Images

the various sensory regions of the cerebral cortex (Arend & others, 2020). For example, one fMRI study identified the parietal cortex, an area of the brain associated with integrating sensory experiences, as the key auditory-visual location in the brain for individuals with synaesthesia who experienced music as color (Neufeld & others, 2012). Another study found, in a blind participant, that "seeing" sounds as color involved activity in the occipital lobe, the brain area responsible for vision (Yong & others, 2017).

In the brain, nearly all sensory signals pass through the thalamus. Remember, the thalamus is the brain's relay station. It receives signals from sensory neurons and then routes those signals to the sensory areas of the cerebral cortex, where they are modified and spread throughout a vast network of neurons.

Certain areas of the cerebral cortex are specialized to handle different sensory functions. Visual information is processed mainly in the occipital lobes; hearing, in the temporal lobes; and pain, touch, and temperature, in the parietal lobes. Keep in mind, however, that the interactions and pathways of sensory information are complex, and the brain often must coordinate extensive information and interpret it.

Occipital = the back of the brain; temporal = on the sides; parietal = on the top. (You're welcome!)

An important part of perception is interpreting sensory messages. Many top-down factors determine this meaning, including signals from different parts of the brain, prior learning, the person's goals, and the degree of alertness. Moving in the opposite direction, bottom-up signals from a sensory area may help other parts of the brain maintain arousal, form an image of where the body is in space, or regulate movement.

The principles we have surveyed so far apply to all of the senses. We've seen that the senses are about detecting different energies and that all have specialized receptor cells and areas of the brain that serve their functions. You have probably heard about a "sixth sense"—*extrasensory perception (ESP)*. ESP means that a person can detect information from the world without receiving concrete sensory input. Examples of ESP include *telepathy* (the ability to read another person's mind) and *precognition* (the ability to sense future events). The vast majority of psychologists reject the idea that people can foretell the future or read each other's minds (Cameron, 2016; Reber & Alcock, 2020).

Think about something like precognition in the ways we have considered sensation and perception. Sensation involves detecting energy from the environment. If ESP exists, consider this: Which afferent neurons send psychic messages from the future to the brain, and what sort of energy conveys these messages? Believers in ESP have not produced answers to these important questions. The success of gambling casinos, daily experiences with surprise, and scientific evidence (Barušs & Rabier, 2014) converge to tell us that human beings cannot feel the future. Perhaps we are fine with just the five senses we have.

Thresholds

Any sensory system must be able to detect varying degrees of energy. This energy can take the form of light, sound, chemical, or mechanical stimulation. How much of a stimulus is necessary for you to see, hear, taste, smell, or feel something? What is the lowest possible amount of stimulation that will still be detected?

ABSOLUTE THRESHOLD   One way to think about the lowest limits of perception is to assume that there is an **absolute threshold**, or minimum amount of stimulus energy that a person can detect. In other words, the absolute threshold is the dimmest light, the faintest sound, or the softest touch a person can still see, hear, or feel (Heil & Matysiak, 2020). When the energy of a stimulus falls below this absolute threshold, we cannot detect its presence; when the energy of the stimulus rises above the absolute threshold, we can detect the stimulus. As an example, find a clock that ticks; put it on a table and walk far enough away that you no longer hear it. Then gradually move toward the clock. At some point, you will begin to hear it ticking. Hold your position and notice that occasionally the ticking fades, and you may have to move forward to reach the threshold; at other times, it may become loud, and you can move backward.

In this experiment, if you measure your absolute threshold several times, you likely will record several different distances for detecting the stimulus. For example, the first time you try it, you might hear the ticking at 25 feet from the clock. However, you probably will not hear it every time at 25 feet. Maybe you hear it only 38 percent of the time at this distance, but you hear it 50 percent of the time at 20 feet away and 65 percent of the time at 15 feet. People have different thresholds. Some have better hearing than others, and some have better vision. Figure 4 shows one person's measured absolute threshold for detecting a clock's ticking sound.

absolute threshold The minimum amount of stimulus energy that a person can detect.

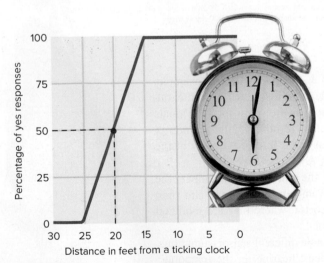

FIGURE 4   **Measuring Absolute Threshold** Absolute threshold is the minimum amount of energy we can detect. To measure absolute threshold, psychologists have arbitrarily decided to use the criterion of detecting the stimulus 50 percent of the time. In this graph, the person's absolute threshold for detecting the ticking clock is at a distance of 20 feet.
(Photo): Wavebreakmedia Ltd/Wavebreak Media/Getty Images

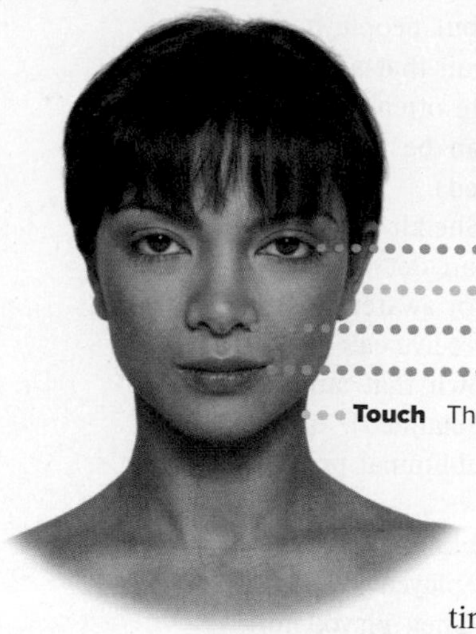

FIGURE 5 **Approximate Absolute Thresholds for Five Senses** These thresholds show the amazing power of our senses to detect even very slight variations in the environment.

Vision A candle flame at 30 miles on a dark, clear night

Hearing A ticking clock at 20 feet under quiet conditions

Smell One drop of perfume diffused throughout three rooms

Taste A teaspoon of sugar in 2 gallons of water

Touch The wing of a fly falling on your neck from a distance of 1 centimeter

Psychologists have arbitrarily decided that absolute threshold is the point at which the individual detects the stimulus 50 percent of the time—in this case, 20 feet away. Using the same clock, another person might have a measured absolute threshold of 26 feet, and yet another, 18 feet. Figure 5 lists the approximate absolute thresholds of five senses.

Under ideal circumstances, our senses have very low absolute thresholds, so we can be remarkably good at detecting small amounts of stimulus energy. You might be surprised to learn that the human eye can see a candle flame at 30 miles on a dark, clear night. However, our environment seldom gives us ideal conditions with which to detect stimuli. If the night were cloudy or the air smoky, for example, you would have to be much closer to see the candle flame. In addition, other lights on the horizon—car or house lights—would hinder your ability to detect the candle's flicker. **Noise** is the term given to irrelevant and competing stimuli—not just sounds but any distracting stimuli for our senses (Nesti & others, 2014; Zhao & others, 2017). When we are in a noisy environment, it is difficult to detect accurately specific stimuli.

noise
Irrelevant and competing stimuli—not only sounds but also any distracting stimuli for the senses.

DIFFERENCE THRESHOLD In addition to studying how much energy is required for a stimulus to be detected, psychologists investigate the degree of difference that must exist between two stimuli before the difference is detected. This is the **difference threshold**, or just noticeable difference (Festa & others, 2020; Nilsson, 2020). An artist might detect the difference between two similar shades of color. A fashion designer might notice a difference in the texture of two fabrics. How different must the colors and textures be for someone to say, "These are different"? Like the absolute threshold, the difference threshold is the smallest difference in stimulation required to discriminate one stimulus from another 50 percent of the time.

difference threshold
The degree of difference that must exist between two stimuli before the difference is detected.

Difference thresholds increase as a stimulus becomes stronger. That means that at very low levels of stimulation, small changes can be detected, but at very high levels, small changes are less noticeable. When music is playing softly, you may notice when your roommate increases the volume by even a small amount. If, however, they turn the volume up an equal amount when the music is playing very loudly, you may not notice. **Weber's law**, discovered by German physiologist E. H. Weber more than 150 years ago, is the principle that two stimuli must differ by a constant proportion to be perceived as different. For example, we add 1 candle to 20 candles and notice a difference in the brightness of the candles; we add 1 candle to 120 candles and do not notice a difference, but we would notice the difference if we added 6 candles to 120 candles. Weber's law holds true in many contexts (Ozana & Ganel, 2019; Rinaldi & Marelli, 2020).

Note that 1:20 = 6:120.

Weber's law
The principle that two stimuli must differ by a constant minimum percentage (rather than a constant amount) to be perceived as different.

SUBLIMINAL PERCEPTION Can sensations that occur below our absolute threshold affect us without our being aware of them? **Subliminal perception** refers to the detection of information below the level of conscious awareness. In 1957, James Vicary, an advertising executive, announced that he was able to increase popcorn and soft drink sales by secretly flashing the words "EAT POPCORN" and "DRINK COKE" on a movie

subliminal perception
The detection of information below the level of conscious awareness.

Burke/Triolo/Brand X Pictures/Getty Images

screen in a local theater (Weir, 1984). Vicary's claims were a hoax, but people have continued to wonder whether behavior can be influenced by stimuli that are presented so quickly that we cannot perceive them. These stimuli are often referred to as "primes"—subtle reminders of ideas or activities that can be presented outside of awareness (e.g., too fast to be consciously recognized).

Studies have shown that the brain (Detloff & others, 2020; Sheikh & others, 2019) and cardiovascular system (Pottratz & others, 2020; van der Ploeg & others, 2020) respond to information that is presented outside of awareness.

Whether and how information that is presented too fast to perceive can affect behavior is controversial. Although some experiments have shown that stimuli presented outside of awareness can affect behavior (as Vicary claimed), other studies have not replicated these effects. Overall, the effects of subliminal primes on behavior are likely to be weak (Weingarten & others, 2016).

To get a sense of these experiments, consider an example of subliminal priming that appears to affect behavior. In a series of studies, participants played the card game Black Jack against a computer (Payne & others, 2016). They played for points and were instructed to win as many points as possible. After the participant and the computer both were dealt two cards, players were asked whether they wanted to bet (i.e., risk more points) or pass (and not risk any points). Unbeknownst to the participants, in the moments before they made their decisions, they were exposed to words presented so fast they could not be seen. In this within-person experiment, participants served as their own control group. The words presented were related either to betting (such as *gamble, wager,* or *bet*) or to passing (such as *pass, fold,* or *stay*). These words were presented for less than 0.3 second (that is, very fast). To participants, they likely looked like random flashes. Did these words affect decisions? They did! Participants were more likely to bet after being primed with words related to betting and more likely to pass when primed with words related to passing. Importantly, though, behavior did not simply follow the primes in an automatic way. Rather, the primes affected behavior most when the value of the participant's hand was moderate and did not suggest an obvious choice to bet (if the hand was very high) or pass (if it was low) (Payne & others, 2016).

Although such results can give priming a magical quality, keep in mind that subliminally presented information is registering on some level. Moreover, some have argued that even though primes are presented rapidly, some people may still detect them, consciously (Sand & Nilsson, 2016). Perhaps even more importantly, presenting information so that it can be consciously processed can have stronger effects on behavior than subliminal presentation (Aoyama & others, 2017).

The notion that stimuli we do not consciously perceive can influence our behavior challenges the usefulness of the idea of thresholds. Can you see why? If stimuli that fall below the threshold can still influence us, what do thresholds really tell us? Further, you might have noticed that the definition of absolute threshold is not very absolute. It refers to the intensity of stimulation detected *50 percent of the time.* How can something absolute change from one trial to the next?

If, for example, you tried the ticking clock experiment described earlier, you might have found yourself making judgment calls. Sometimes you felt very sure you could hear the clock, but other times you were uncertain and probably took a guess. Sometimes you guessed right, and other times you were mistaken. Now, imagine that someone offered to pay you $50 for every correct answer you gave—would the presence of that incentive change your judgments? Alternatively, what if you were charged $50 for every time you said you heard the clock and it was not ticking? In fact, perception is often about making such judgment calls.

An alternative approach to the question of whether a stimulus is detected would emphasize that saying (or not saying), "Yes, I hear that ticking" is actually a *decision.* This approach is called **signal detection theory**, which focuses on decision making about stimuli under conditions of uncertainty. In signal detection theory, detection of sensory stimuli depends on a variety of factors besides the physical intensity of the stimulus and the sensory abilities of the observer (DeCarlo, 2020; Jayakumar & Simpson, 2020; Wixted, 2020; Ye & others, 2020). These factors include individual and contextual

signal detection theory
An approach to perception that focuses on decision making about stimuli in the presence of uncertainty.

variations, such as fatigue, expectations, and the urgency of the moment. Figure 6 shows how signal detection works.

Perceiving Sensory Stimuli

As we just saw, the perception of stimuli is influenced by more than the characteristics of the environmental stimuli themselves. Two important factors in perceiving sensory stimuli are attention and perceptual set.

	Observer's Response	
	"Yes, I see the signal."	*"No, I don't see the signal."*
Signal Present	Hit (correct)	Miss (mistake)
Signal Absent	False alarm (mistake)	Correct rejection (correct)

FIGURE 6 **Four Outcomes in Signal Detection** Signal detection research helps to explain when and how perceptual judgments are correct or mistaken.

ATTENTION

The world holds a lot of information to perceive. At this moment you are perceiving the letters and words that make up this sentence. Now gaze around you and fix your eyes on something other than what you are reading. Afterward, curl up the toes on your right foot. In each of these circumstances, you engaged in **selective attention**, which involves focusing on a specific aspect of experience while ignoring others. A familiar example of selective attention is the ability to focus on one voice among many in a crowded airline terminal or noisy restaurant. Psychologists call this common occurrence the *cocktail party effect* (Kuyper, 1972; Bosker & others, 2020).

Not only is attention selective, but it also is *shiftable*. For example, you might be paying close attention to your instructor's lecture, but if the person next to you starts texting a friend, you might look to see what is going on over there. The fact that we can attend selectively to one stimulus and shift readily to another indicates that we must be monitoring many things at once.

Certain features of stimuli cause people to attend to them. *Novel stimuli* (those that are new, different, or unusual) often attract our attention. If a Ferrari convertible whizzes by, you are more likely to notice it than you would a sedan. Size, color, and movement also influence attention. Objects that are large, vividly colored, or moving are more likely to grab our attention than objects that are small, dull colored, or stationary.

Sometimes even very interesting stimuli can be missed if our attention is otherwise occupied. *Inattentional blindness* refers to the failure to detect unexpected events when attention is engaged by a task (Wood & Simons, 2019). When we are working intently on something, such as finding a seat in a packed movie theater, we might not even see an unusual stimulus, such as a friend waving to us in the crowd. Research by Simons and Chabris (1999) provides a remarkable example of inattentional blindness. In that study, participants were asked to watch a video of two teams playing basketball. The participants were instructed to closely count the number of passes thrown by each team. During the video, a small woman dressed in a gorilla suit walked through the action, clearly visible for 5 seconds. Surprisingly, over half of the participants (who were apparently deeply engaged in the counting task) never noticed the gorilla. Inattentional blindness is more likely to occur when a task is difficult (Greene & others, 2020) and when the distracting stimulus is very different from stimuli that are relevant to the task at hand (White & Aimola Davies, 2008; Wiemer & others, 2012). Interestingly, when people are primed with the goal to detect and the monitoring task is only moderately difficult, they are more likely to notice the gorilla (Légal & others, 2017).

Research on inattentional blindness highlights the dangers of multitasking when one of the tasks is driving. Engaging in a task such as talking on a phone, reading social media, or texting

jinga80/Getty Images

selective attention The act of focusing on a specific aspect of experience while ignoring others.

can so occupy attention that little is left over for the important task of piloting a motor vehicle. In 2018 alone, 2,814 people lost their lives due to distracted drivers (National Highway Traffic Safety Administration, 2020). Research shows that individuals who text while they drive face 23 times the risk of a crash or near-crash compared to non-distracted drivers (Blanco & others, 2009; Hanowski & others, 2009). In this research, cameras continuously observed drivers for more than 6 million miles of driving. Texting drew the drivers' eyes away from the road long enough for the vehicle to travel the length of a football field at 55 miles an hour. The statistics are sobering. In the United States, more than eight people are killed every day due to distracted drivers, and at any given daylight moment an estimated 660,000 drivers are using cellphones or other electronic devices (FCC, 2017).

Culture influences which stimuli we attend to as we perceive the world (Senzaki & others, 2014). People from Western cultures are more likely to attend to objects in the foreground of scenes (or *focal objects*), while people from East Asia looking at the same scenes are more likely to notice aspects of the context. For example, in one study (Masuda & Nisbett, 2001), participants from the U.S. and Japan were shown video clips of underwater scenes. When asked to describe what they had seen, the Americans were more likely to talk about the colorful fish swimming around, and Japanese participants were more likely to talk about the locations of objects and aspects of the setting. Such differences have led psychologists to conclude that Westerners take a more analytical orientation, while East Asians are more likely to see the big picture. Culture also influences the kinds of stimuli that are missed in inattentional blindness. Research on *change blindness* (the tendency to miss changes that have occurred in a scene) shows that when objects in the foreground change, Americans are more likely to notice, while Japanese people are more likely to notice when changes occur in the context (Masuda & Nisbett, 2006).

What might explain these cultural differences in attention? One possibility may be differences in the environments that people in these cultures typically encounter (Senzaki & others, 2014). Comparing photographs, Miyamoto and colleagues (2006) found that Japanese hotels and schools had more detail and were more complex than American hotels and schools. People in Japan, then, may develop the tendency to look at the whole picture because navigating their world requires such attention. Interestingly, in another study, the same researchers had American and Japanese participants watch brief video clips of American or Japanese scenes (Miyamoto & others, 2006). They found that American and Japanese participants alike noticed changes in focal objects in American scenes but noticed changes in the context in Japanese scenes. This research suggests that while the mechanics of sensation are the same for all human beings, the experience of perception can be shaped by the physical environment in which each person lives.

PERCEPTUAL SET   Cover the playing cards on the right in the illustration and look at the playing cards on the left. As quickly as you can, count how many aces of spades you see. Then cover the cards on the left and count the number of aces of spades among the cards on the right.

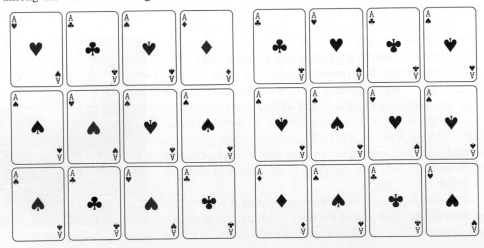

Most people report that they see two or three aces of spades in the set of cards on the left. However, if you look closely, you will see that there are five. Two of the aces of spades are black and three are red. When people look at the set of cards on the right, they are more likely to count five aces of spades. Why do we perceive the two sets of cards differently? We expect the ace of spades to be black because it is always black in a regular deck of cards. We do not expect red spades, so we skip right over the red ones: Expectations influence perceptions.

Psychologists refer to a predisposition or readiness to perceive something in a particular way as a **perceptual set**. Perceptual sets act as "psychological" filters in processing information about the environment (Matthews & others, 2020). Perceptual sets reflect top-down influences on perception. Interestingly, young children are more accurate at the task involving the ace of spades than adults are. Why? Because they have not built up the perceptual set that the ace of spades is black.

The processes of sensation and perception take on special significance when applied to people. Social psychologists study the phenomenon of "person perception" as a special case of coming to understand the most complex and important stimuli we encounter—*each other*. Certainly, person perception occurs using the same sensory receptors we have reviewed so far. When we perceive other people we are also susceptible to same forces at work when we perceive other stimuli in the environment. Top-down expectations, inattentional blindness, and perceptual sets mean that when we encounter people who do not fit with what we expect, we may make errors in perception or fail to notice important aspects of those people.

> **perceptual set**
> A predisposition or readiness to perceive something in a particular way.

Sensory Adaptation

Turning out the lights in your bedroom at night, you stumble across the room to your bed, blind to the objects around you. Gradually the objects reappear and become clearer. The ability of the visual system to adjust to a darkened room is an example of **sensory adaptation**—a change in the responsiveness of the sensory system based on the average level of surrounding stimulation (Liu & Engel, 2020).

You have experienced sensory adaptation countless times in your life. You adjust to the water in an initially "freezing" swimming pool. You turn on your windshield wipers while driving, and shortly you are unaware of their rhythmic sweeping back and forth. When you first enter a room, you might be bothered by the hum of the air conditioner, but after a while you get used to it. All of these experiences represent sensory adaptation.

In the example of adapting to the dark, when you turn out the lights, everything is black. Conversely, when you step out into the bright sunshine after spending time in a dark basement, light floods your eyes and everything appears light. These momentary blips in sensation arise because adaptation takes time.

> **sensory adaptation**
> A change in the responsiveness of the sensory system based on the average level of surrounding stimulation.

self-quiz

1. Every day, you see, hear, smell, taste, and feel stimuli from the outside world. Collecting data about that world is the function of _____, and interpreting the data collected is the function of _____.
   A. the brain; the spinal cord
   B. the spinal cord; the brain
   C. sensation; perception
   D. perception; sensation

2. The main classes into which the sense organs and sensory receptors fall include all of the following *except*
   A. chemoreception.
   B. electroreception.
   C. photoreception.
   D. mechanoreception.

3. An architect is designing apartments and wants them to be soundproof. She asks a psychologist what the smallest amount of sound is that can be heard. Her question is most related to
   A. the absolute threshold.
   B. the difference threshold.
   C. Weber's law.
   D. the sensory receptors.

APPLY IT! 4. Trina, a first-year college student, goes home at Thanksgiving break after being away from home (for the first time) for 3 months. She feels as if she has changed a lot, but her parents still treat her like a kid in high school. At Thanksgiving dinner she confronts them, bursting out, "Stop top-down processing me!" Her parents think Trina has lost her mind. Which of the following explains her outburst?
   A. Trina feels that her parents are judging her sophisticated college ways too harshly.
   B. Trina probably ate too much turkey.
   C. Trina feels that her parents have spent too much time analyzing her behavior.
   D. Trina believes that her parents are letting their preconceived ideas of who she is prevent them from seeing her as the person she has become.

2 The Visual System

When Michael May of Davis, California, was 3 years old, an accident left him visually impaired, with only the ability to perceive the difference between night and day. He lived a rich, full life, marrying and having children, founding a successful company, and becoming an expert skier. Twenty-five years passed before doctors transplanted stem cells into May's right eye, a procedure that gave him partial sight (Bach, 2015; Kurson, 2007). May can now see; his right eye is functional and allows him to detect color and negotiate the world without the use of a cane or reliance on his service dog. His visual experience is unusual, however: He sees the world as if it is an abstract painting. He can catch a ball thrown to him by his sons, but he cannot recognize his wife's face. His brain has to work at interpreting the new information that his right eye is providing. May's experience highlights the intimate connection between the brain and our sense organs in producing perception. Vision is a remarkable process that involves the brain's interpretation of the visual information sent from the eyes. We now explore the physical foundations of the visual system and the processes involved in the perception of visual stimuli.

The Visual Stimulus and the Eye

Our ability to detect visual stimuli depends on the sensitivity of our eyes to differences in light.

LIGHT   *Light* is a form of electromagnetic energy that can be described in terms of wavelengths. Light travels through space in waves. The *wavelength* of light is the distance from the peak of one wave to the peak of the next. Wavelengths of visible light range from about 400 to 700 nanometers (nm). (A nanometer is 1 billionth of a meter.) The wavelength of light that is reflected from a stimulus determines its hue or color.

Outside the range of visible light are longer radio and infrared radiation waves and shorter ultraviolet and X-rays (Figure 7). These other forms of electromagnetic energy continually bombard us, but we do not see them.

We can also describe waves of light in terms of their height, or *amplitude,* which determines the brightness of the stimulus. Finally, the *purity* of the wavelengths— whether they are all the same or a mix of waves—determines the perceived *saturation,* or richness, of a visual stimulus (Figure 8). The color tree shown in Figure 9 can help you understand saturation. White light is a combination of color wavelengths that is perceived as colorless, like sunlight. Very pure colors have no white light in them. They are located on the outside of the color tree. Notice how, the closer we get to the center of the color tree, the more white light has been added to the single wavelength of a particular color. In other words, the deep colors at the edge fade into pastel colors toward the center.

THE STRUCTURE OF THE EYE   The eye, like a camera, is constructed to get the best possible picture of the world. An accurate picture is in focus, is not too dark or too light, and has good contrast between the dark and light parts. Each of several structures in the eye plays an important role in this process.

If you look closely at your eyes in the mirror, you will notice three parts—the sclera, iris, and pupil (Figure 10). The *sclera* is the white, outer part of the eye that helps to maintain the shape of the eye and to protect it from injury. The *iris* is the colored part of the eye, which might be light blue in one individual and dark brown in another. The *pupil,* which appears black, is the opening in the center of the iris. The iris contains

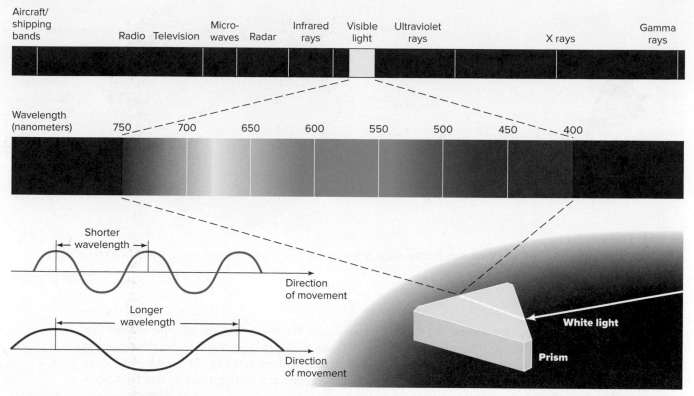

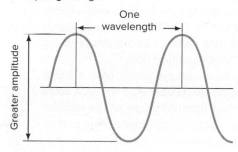

FIGURE 7 **The Electromagnetic Spectrum and Visible Light** (*top*) Visible light is only a narrow band in the electromagnetic spectrum. Visible light wavelengths range from about 400 to 700 nanometers. X-rays are much shorter, radio waves much longer. (*bottom*) The two graphs show how waves vary in length between successive peaks. Shorter wavelengths are higher in frequency, as reflected in blue colors; longer wavelengths are lower in frequency, as reflected in red colors.

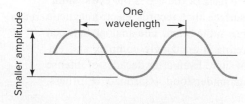

FIGURE 8 **Light Waves of Varying Amplitude** The top graph might suggest a spotlight on a concert stage; the bottom, a candlelit dinner.

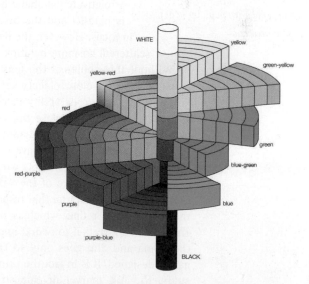

FIGURE 9 **A Color Tree Showing Color's Three Dimensions: Hue, Saturation, and Brightness** Hue is represented around the color tree—saturation horizontally and brightness vertically.
Universal Images Group/Getty Images

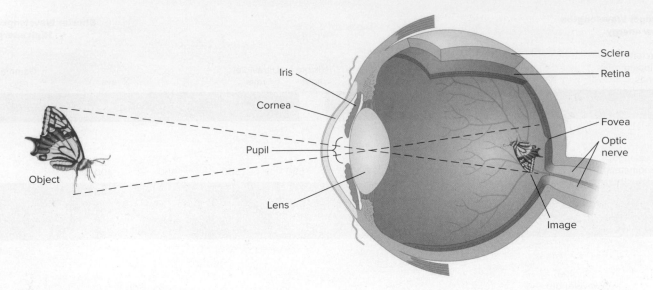

Labels: Iris, Cornea, Pupil, Object, Lens, Sclera, Retina, Fovea, Optic nerve, Image

FIGURE 10 **Parts of the Eye** Note that the image of the butterfly on the retina is upside down. The brain allows us to see the image right side up.

muscles that control the size of the pupil, and hence regulates the amount of light that enters the eye. To get a good picture of the world, the eye needs to be able to adjust the amount of light that enters. In this sense, the pupil acts like the aperture of a camera, opening to let in more light when it is needed and closing to let in less light when there is too much.

Two structures bring the image into focus, the cornea and the lens. The *cornea* is the first structure to encounter the world. It is a clear membrane at the very front of the eye. The *lens* is a transparent and somewhat flexible, disk-like structure filled with a gelatin-like material. Light hits the cornea and then the lens. The function of both of these structures is to bend the light falling on the surface of the eye just enough to focus it at the back. The curved surface of the cornea does most of this bending, while the lens fine-tunes things. When you are looking at faraway objects, the lens has a relatively flat shape because the light reaching the eye from faraway objects is parallel and the bending power of the cornea is sufficient to keep things in focus. However, the light reaching the eye from objects that are close is more scattered, so more bending of the light is required to achieve focus.

If you have ever had your pupils dilated for an eye exam, your doctor has basically opened the pupil up to get a good look at your eye!

Without this ability of the lens to change its curvature, the eye would have a tough time focusing on close objects such as reading material. As people age, the lens loses flexibility and hence its ability to change from its normal flattened shape to the rounder shape needed to bring close objects into focus. That is why many people with normal vision throughout their young adult lives require reading glasses later in life.

The parts of the eye we have considered so far work together to give us the sharpest picture of the world. This effort would be useless, however, without a vehicle for recording the images the eyes take of the world—in essence, the film of the camera. Photographic film is made of a material that responds to light. At the back of the eye is the eye's "film," the multilayered **retina**, which is the light-sensitive surface that records electromagnetic energy and converts it to neural impulses for processing in the brain. The analogy between the retina and film goes only so far, however. The retina is amazingly complex and elegantly designed. It is in fact the primary mechanism of sight. Even after decades of intense study, the full marvel of this structure is far from understood (Cardozo & others, 2020; Reichenbach & Bringmann, 2020).

The human retina has approximately 126 million receptor cells. They turn the electromagnetic energy of light into a form of energy that the nervous system can process. There are two kinds of visual receptor cells: rods and cones. Rods and cones differ both in how

retina
The multilayered light-sensitive surface in the eye that records electromagnetic energy and converts it to neural impulses for processing in the brain.

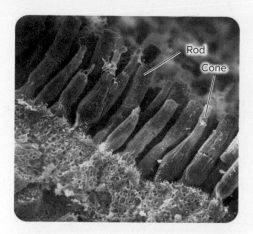

FIGURE 11 Rods and Cones In real life, rods and cones look somewhat like stumps and corncobs. To get a sense of how well the cones in the fovea work, try reading out of the corner of your eye. It is difficult because the fovea doesn't get to do the reading for you. The visual information in the retina that is closest to the nose crosses over, and the visual information on the outer side of the retina stays on that side of the brain.
Omikron/Science Source

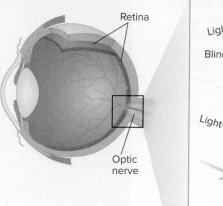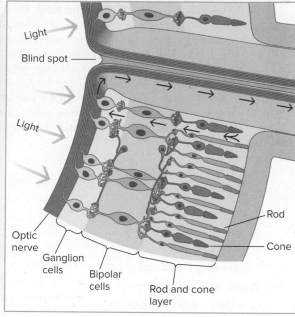

FIGURE 12 Direction of Light in the Retina After light passes through the cornea, pupil, and lens, it falls on the retina. Three layers of specialized cells in the retina convert the image into a neural signal that can be transmitted to the brain. First, light triggers a reaction in the rods and cones at the back of the retina, transducing light energy into electrochemical neural impulses. The neural impulses activate the bipolar cells, which in turn activate the ganglion cells. Then light information is transmitted to the optic nerve, which conveys it to the brain. The arrows indicate the sequence in which light information moves in the retina.

rods
The receptor cells in the retina that are sensitive to light but not very useful for color vision.

cones
The receptor cells in the retina that allow for color perception.

optic nerve
The structure at the back of the eye, made up of axons of the ganglion cells, that carries visual information to the brain for further processing.

they respond to light and in their patterns of distribution on the surface of the retina. **Rods** are the receptors in the retina that are sensitive to light, but they are not very useful for color vision. Rods function well when there is little light; they are hard at work at night. Humans have about 120 million rods. **Cones** are the receptors that we use for color perception. Like rods, cones are light sensitive; however, they require a larger amount of light to respond than the rods do, so they operate best in daylight or under high illumination. There are about 6 million cone cells in human eyes. Figure 11 shows what rods and cones look like.

The most important part of the retina is the *fovea,* a tiny area in the center of the retina at which vision is at its best (see Figure 10). The fovea contains only cones and is vital to many visual tasks. Rods are found almost everywhere on the retina except in the fovea.

Figure 12 shows how the rods and cones at the back of the retina convert light into electrochemical impulses. The signal is transmitted to the *bipolar cells* and then moves on to another layer of specialized cells called *ganglion cells* (Steketee & others, 2014). The axons of the ganglion cells make up the **optic nerve**, which carries the visual information to the brain for further processing. Figure 13 summarizes the characteristics of rods and cones.

If you want to see a very faint star, you should gaze slightly away from it, to allow your rods to do their work.

Characteristics	Rods	Cones
Type of vision	Black and white	Color
Responses to light conditions	Dimly lit	Well lit
Shape	Thin and long	Short and fat
Distribution	Not on fovea	On fovea and scattered outside of fovea

FIGURE 13 Characteristics of Rods and Cones Rods and cones differ in shape, location, and function.

FIGURE 14 **The Eye's Blind Spot** There is a normal blind spot in your eye, a small area where the optic nerve leads to the brain. To find your blind spot, hold this image at arm's length, cover your left eye, and stare at the red pepper on the left with your right eye. Move the image slowly toward you until the yellow pepper disappears. To find the blind spot in your left eye, cover your right eye, stare at the yellow pepper, and adjust the distance until the red pepper disappears.
David Tietz/Editorial Image LLC

Interestingly, squid (and other cephalopods) do not have blind spots because unlike us, their optic nerves do not block the visual field.

Do It!

If you'd like to explore your blind spot more, search for "blind spot demonstrations" online.

One place on the retina contains neither rods nor cones. This area, the *blind spot,* is the area on the retina where the optic nerve leaves the eye on its way to the brain (see Figure 12). We cannot see anything that reaches only this part of the retina. To prove to yourself that you have a blind spot, look at Figure 14. Once you have seen the yellow pepper disappear, you have probably noticed it took a while to succeed at this task. Now shut one eye and look around. You see a perfectly continuous picture of the world around you; there is no blind spot. This is a great example of top-down processing and a demonstration of the constructive aspect of perception. Your brain fills in the gap for you (the one that ought to be left by your blind spot) with some pretty good guesses about what must be in that spot, like a creative artist painting in the blind spot.

Visual Processing in the Brain

The eyes are just the beginning of visual perception. The next step occurs when neural impulses generated in the retina are dispatched to the brain for analysis and integration (Rasmussen & Yonehara, 2020).

What's on the nose side of the retina crosses over.

The optic nerve leaves the eye, carrying information about light toward the brain. Light travels in a straight line; therefore, stimuli in the left visual field are registered in the right half of the retina in both eyes, and stimuli in the right visual field are registered in the left half of the retina in both eyes (Figure 15). In the brain, at a point called the *optic chiasm,* the optic nerve fibers divide, and approximately half of the nerve fibers cross over the midline of the brain. As a result, the visual information originating in the right halves of the two retinas is transmitted to the right side of the occipital lobe in the cerebral cortex, and the visual information coming from the left halves of the retinas is transmitted to the left side of the occipital lobe. These crossings mean that what we see in the left side of our visual field is registered in the right side of the brain, and what we see in the right visual field is registered in the left side of the brain (see Figure 15). Then this information is processed and combined into a recognizable object or scene in the visual cortex.

THE VISUAL CORTEX

The **visual cortex**, located in the occipital lobe at the back of the brain, is the part of the cerebral cortex involved in vision. Most visual information travels to the primary visual cortex, where it is processed, before moving to other visual areas for further analysis.

An important aspect of visual information processing is the specialization of neurons. Like the cells in the retina, many cells in the primary visual cortex are highly specialized

visual cortex
Located in the occipital lobe, the part of the cerebral cortex involved in vision.

feature detectors Neurons in the brain's visual system that respond to particular features of a stimulus.

(Murgas & others, 2020). **Feature detectors** are neurons in the brain's visual system that respond to particular features of a stimulus. These detectors pick up the edges, shapes, colors, and contours of stimuli. Hubel and Wiesel (1963) won a Nobel Prize for their research on feature detectors. By recording the activity of a single neuron in a cat while it looked at patterns that varied in size, shape, color, and movement, the researchers found that the visual cortex has neurons that are individually sensitive to different types of lines and angles. One neuron might show a sudden burst of activity when stimulated by lines of a particular angle; another neuron might fire only when moving stimuli appear; yet another neuron might be stimulated when the object in the visual field has a combination of certain angles, sizes, and shapes.

Hubel and Wiesel (1963) also noted that when deprived of certain types of visual stimulation early on, kittens lost the ability to perceive these patterns. This finding suggested that there might be a critical period in visual development and that the brain requires stimulation in its efforts to delegate its resources to different perceptual tasks. The brain "learns" to perceive through experience. This explains Michael May's unusual experience, described at the beginning of our examination of the visual system. Once deprived of stimulation, the brain will redistribute its resources to other tasks.

parallel processing The simultaneous distribution of information across different neural pathways.

PARALLEL PROCESSING Sensory information travels quickly through the brain because of **parallel processing**, the simultaneous distribution of information across different neural pathways (Li & others, 2020; Munn & others, 2020). A sensory system designed to process information about sensory qualities one at a time (such as processing first the shapes of images, then their colors, then their movements, and finally their locations) would be too slow to keep us current with a rapidly changing world. To function, we need to "see" all of these characteristics at once, which is parallel processing. There is some evidence suggesting that parallel processing also occurs for sensations of touch and hearing (Spence & Frings, 2020).

BINDING Some neurons respond to color, others to shape, and still others to movement; but note that all of these neurons are involved in responding to a given stimulus (e.g., a toddler running toward you). How does the brain know that these physical features, communicated by different neurons, all belong to the same object of perception? The answer is, binding.

binding In the sense of vision, the bringing together and integration of what is processed by different neural pathways or cells.

One of the most exciting topics in visual perception, **binding** is the bringing together and integration of what is processed by different neural pathways or cells (Hitch & others, 2020). When you see a toddler running toward you, you see the whole cute little person. You perceive a running child, not just bits and pieces. Through binding, you can integrate information about the toddler's body shape, smile, and movement into a complete image in the cerebral cortex. How binding occurs is a puzzle that fascinates

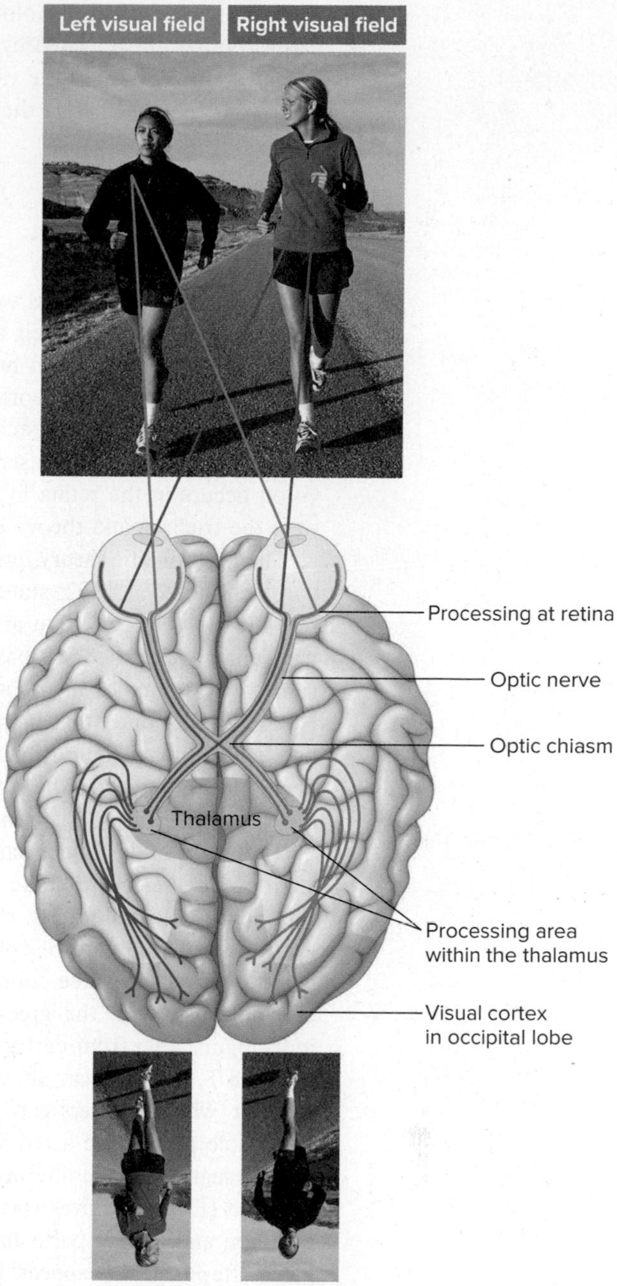

FIGURE 15 Visual Pathways to and through the Brain Light from each side of the visual field falls on the opposite side of each eye's retina. Visual information then travels along the optic nerve to the optic chiasm, where most of the visual information crosses over to the other side of the brain. From there visual information goes to the occipital lobe at the rear of the brain. All these crossings mean that what we see in the left side of our visual field (here, the shorter, dark-haired woman) is registered in the right side of our brain, and what we see in the right visual field (the taller, blonde woman) is registered in the left side of our brain.
(Photos): RubberBall Productions/Getty Images

neuroscientists (Dowd & Golomb, 2020; Laub & Frings, 2020). All the neurons throughout pathways that are activated by a visual object pulse together at the same frequency (Engel & Singer, 2001). Within the vast network of cells in the cerebral cortex, this set of neurons appears to *bind* together all the features of the objects into a unified perception (Velik, 2012).

Color Vision

Imagine how dull the world would be without color. Art museums are filled with paintings that are remarkable for their use of color, and flowers would lose much of their beauty if we could not see their rich hues. The process of color perception starts in the retina, the eyes' film. Interestingly, theories about how the retina processes color were developed long before methods existed to study the anatomical and neurophysiological bases of color perception. Instead, psychologists made some extraordinarily accurate guesses about how color vision occurs in the retina by observing how people see. The two main theories proposed were the trichromatic theory and opponent-process theory. Both turned out to be correct.

The **trichromatic theory**, proposed by Thomas Young in 1802 and extended by Hermann von Helmholtz in 1852, states that color perception is produced by three types of cone receptors in the retina that are particularly sensitive to different but overlapping ranges of wavelengths. The theory is based on experiments showing that a person with normal vision can match any color in the spectrum by combining three other wavelengths. Young and Helmholtz reasoned that if the combination of any three wavelengths of different intensities is indistinguishable from any single pure wavelength, the visual system must base its perception of color on the relative responses of three receptor systems—cones sensitive to red, blue, and green.

The study of dysfunctional color vision, or *color blindness* (Figure 16), provides further support for the trichromatic theory. Complete color blindness is rare; most people who experience color blindness, most of whom are biologically male, can see some colors but not others. The nature of color blindness depends on which of the three kinds of cones is inoperative. The three cone systems are green, red, and blue. In the most common form of color blindness, the green cone system malfunctions in some way, rendering green indistinguishable from certain combinations of blue and red.

In 1878, the German physiologist Ewald Hering observed that some colors cannot exist together, whereas others can. For example, it is easy to imagine a greenish blue but nearly impossible to imagine a reddish green. Hering also noticed that trichromatic theory could not adequately explain afterimages. *Afterimages* are sensations that remain after a stimulus is removed (Figure 17 gives you a chance to experience an afterimage). Color afterimages are common and involve particular pairs of colors. If you look at red long enough, eventually a green afterimage will appear. If you look at yellow long enough, eventually a blue afterimage will appear. These associations between viewing a particular color (yellow) and then experiencing an afterimage of another color (blue) cannot be explained by trichromatic theory.

trichromatic theory Theory stating that color perception is produced by three types of cone receptors in the retina that are particularly sensitive to different, but overlapping, ranges of wavelengths.

FIGURE 16 **Examples of Stimuli Used to Test for Color Blindness** People with normal vision see the number 8 in the left circle and the number 5 in the right circle. People with red-green color blindness may see just the 8, just the 5, or neither. A complete color-blindness assessment involves the use of 15 stimuli.
Alexander Kaludov/123RF

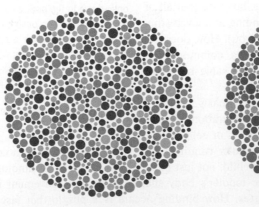
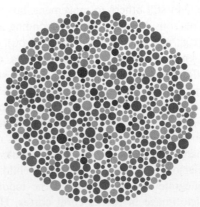

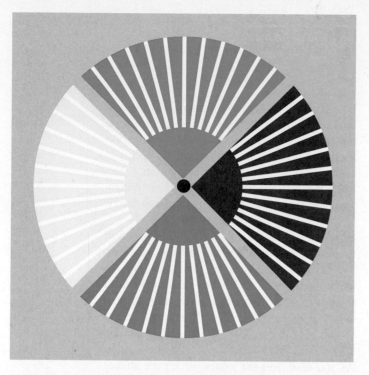

FIGURE 17 **Negative Afterimage—Complementary Colors** If you gaze steadily at the dot in the colored panel on the left for a few moments, then shift your gaze to the gray box on the right, you will see the original hues' complementary colors. The blue appears as yellow, the red as green, the green as red, and the yellow as blue. This pairing of colors has to do with the fact that color receptors in the eye are apparently sensitive as pairs: When one color is turned off (when you stop staring at the panel), the other color in the receptor is briefly turned on. The afterimage effect is especially noticeable with bright colors.

opponent-process theory
Theory stating that cells in the visual system respond to complementary pairs of red-green and blue-yellow colors; a given cell might be excited by red and inhibited by green, whereas another cell might be excited by yellow and inhibited by blue.

Hering's observations led him to propose that there were not three types of color receptor cones (as proposed by trichromatic theory) but four, organized into complementary pairs: red-green and blue-yellow. Hering's view, **opponent-process theory**, states that cells in the visual system respond to red-green and blue-yellow colors; a given cell might be excited by red and inhibited by green, whereas another cell might be excited by yellow and inhibited by blue. Opponent process theory does indeed explain afterimages. If you stare at red, for instance, your red-green system seems to "tire," and when you look away, it rebounds and gives you a green afterimage. Try it out and you will see that the opponent-process theory does an excellent job of explaining afterimages.

If the trichromatic theory of color perception is valid, and we do have three kinds of cone receptors like those predicted by Young and Helmholtz, then how can the opponent-process theory also be accurate? The answer is that the red, blue, and green cones in the retina are connected to retinal ganglion cells in such a way that the three-color code is immediately translated into the opponent-process code (Figure 18). For example, a green cone might inhibit and a red cone might excite a particular ganglion cell. Thus, both the trichromatic and opponent-process theories are correct—the eye and the brain use both methods to code colors.

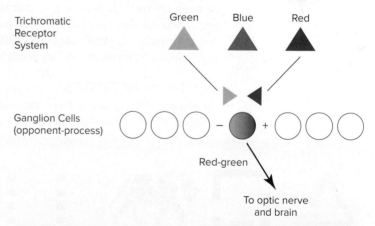

FIGURE 18 **Trichromatic and Opponent-Process Theories: Transmission of Color Information in the Retina** Cones responsive to green, blue, or red light form a trichromatic receptor system in the retina. As information is transmitted to the retina's ganglion cells, opponent-process cells are activated. As shown here, a retinal ganglion cell is inhibited by a green cone (−) and excited by a red cone (+), producing red-green color information.

Perceiving Shape, Depth, Motion, and Constancy

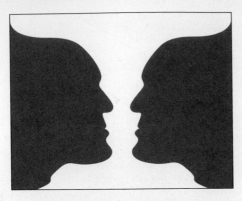

FIGURE 19   Reversible Figure-Ground Pattern Do you see the silhouette of a goblet or a pair of faces in profile? Use this figure to think again about bottom-up and top-down processes.
> **What processes did you use the first moment you looked at the picture—top-down or bottom-up?** > **Now try to see the opposite image (if you saw a goblet, look for the faces; if you saw the faces, look for the goblet). Is this top-down or bottom-up processing?** > **Ask some friends what they see first in this image. What do they report?** > **What do you think accounts for the differences?**

Perceiving visual stimuli means organizing and interpreting the fragments of information that the eye sends to the visual cortex. Information about the dimensions of what we are seeing is critical to this process and critical to our capacity to navigate the world. Among these dimensions are shape, depth, motion, and constancy.

SHAPE   Think about the visible world and its shapes—buildings against the sky, boats on the horizon, the letters on this page. We see these shapes because they are marked off from the rest of what we see by *contour,* a location at which a sudden change of brightness occurs (Baker & others, 2020; Phillips & Fleming, 2020). Now think about the letters on this page. As you read, you see letters, which are shapes or figures, in a white field or background. The **figure-ground relationship** is the principle by which we organize the perceptual field into stimuli that stand out (*figure*) and those that are left over (*background,* or *ground*). Generally this principle works well for us, but some figure-ground relationships are highly ambiguous, and it may be difficult to tell what is figure and what is ground. Figure 19 shows a well-known ambiguous figure-ground relationship. As you look at this illustration, your perception is likely to shift from seeing two faces to seeing a single goblet.

The figure-ground relationship is a gestalt principle (Figure 20 shows others). *Gestalt* is German for "configuration" or "form," and **gestalt psychology** is a school of thought that probes how people naturally organize their perceptions according to certain patterns. One of gestalt psychology's main principles is that the whole is different from the sum of its parts. For example, when you watch a movie, the motion you see in the film cannot be found in the film itself; if you examine the film, you see only separate frames. When you watch the film, the frames move past a light source at a rate of many per second, and you perceive a whole that is very different from the separate frames that are the film's parts. Similarly, thousands of tiny pixels make up an image (whole) on a computer screen.

DEPTH PERCEPTION   Images appear on our retinas in two-dimensional form, yet remarkably we see a three-dimensional world. **Depth perception** is the ability to perceive objects three dimensionally. Look around you. You do not see your surroundings

figure-ground relationship
The principle by which we organize the perceptual field into stimuli that stand out (figure) and those that are left over (ground).

gestalt psychology
A school of thought interested in how people naturally organize their perceptions according to certain patterns.

depth perception
The ability to perceive objects three dimensionally.

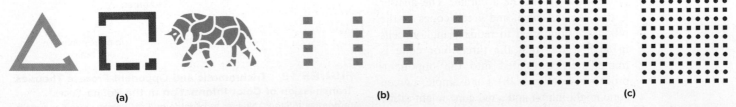

(a) (b) (c)

FIGURE 20   Gestalt Principles of Closure, Proximity, and Similarity (*a*) *Closure:* When we see disconnected or incomplete figures, we fill in the spaces and see them as complete figures. (*b*) *Proximity:* When we see objects that are near each other, they tend to be seen as a unit. You are likely to perceive the grouping as four columns of four squares, not one set of 16 squares. (*c*) *Similarity:* When we see objects that are similar to each other, they tend to be seen as a unit. Here, you are likely to see vertical columns of circles and squares in the left box but horizontal rows of circles and squares in the right box.

as flat. You see some objects farther away, some closer. Some objects overlap each other. The scene and objects that you are looking at have depth. How do you see depth? To perceive a world of depth, we use two kinds of information, or cues—binocular and monocular.

Because we have two eyes, we get two views of the world, one from each eye. **Binocular cues** are depth cues that depend on the combination of the images in the left and right eyes and on the way the two eyes work together. The pictures are slightly different because the eyes are in slightly different positions. Try holding your hand about 10 inches from your face. Alternately close and open your left and right eyes so that only one eye is open at a time. The image of your hand will appear to jump back and forth, because the image is in a slightly different place on the left and right retinas. The *disparity*, or difference, between the images in the two eyes is the binocular cue the brain uses to determine the depth, or distance, of an object. The combination of the two images in the brain, and the disparity between them in the eyes, give us information about the three-dimensionality of the world (Aida & others, 2020). Thus, the disparity between the images presented by our two eyes contributes to our capacity to see a world in three dimensions. Glasses that are used to see films in three dimensions (3D glasses) capitalize on the fact that our brain will do a great job of "making sense" of disparate images.

Convergence is another binocular cue to depth and distance. When we use our two eyes to look at something, they are focused on the same object. If the object is near us, our eyes converge, or move together, almost crossing. If the object is farther away, we can focus on it without pulling our eyes together. The muscle movements involved in convergence provide information about how far away or how deep something is.

In addition to using binocular cues to get an idea of objects' depth, we rely on a number of **monocular cues**, available from the image in one eye, either right or left. Monocular cues are powerful, and under normal circumstances they can provide a compelling impression of depth. Try closing one eye—your perception of the world still retains many of its three-dimensional qualities. Examples of monocular cues are listed next.

1. *Familiar size and relative size:* This cue to the depth and distance of objects is based on what we have learned from experience about the standard sizes of objects. We know how large oranges tend to be, so we can tell something about how far away an orange is likely to be by the size of its image on the retina. We also use relative sizes to give us information about objects in the world. We know that dogs are smaller than cars, and we can get a sense for how big a dog is if it is standing in front of a car.

2. *Height in the field of view:* All other things being equal, objects positioned higher in a picture are seen as farther away.

3. *Linear perspective:* Objects that are farther away take up less space on the retina. So, things that appear smaller are perceived to be farther away. As Figure 21 shows, as an object recedes into the distance, parallel lines in the scene appear to converge. If you were drawing a picture of a hill covered with trees, you would likely draw the trees at the front of the picture taller than those farther away, high on the hill. Objects in the horizon appear smaller than those that are closer.

4. *Overlap:* We perceive an object that partially conceals or overlaps another object as closer.

5. *Shading:* This cue involves changes in perception due to the position of the light and the position of the viewer. Consider an egg under a desk lamp. If you walk around the desk, you will see different shading patterns on the egg.

6. *Texture gradient:* Texture becomes denser and finer the farther away it is from the viewer (Figure 22).

Depth perception is a remarkably complex adaptation. People with only one functioning eye cannot see depth in the way that those with two eyes can. Other disorders of the eye can also lead to a lack of depth perception. The late Oliver Sacks (2006) described the case of Susan Barry, who had been born with crossed eyes. The operation to correct her eyes left her cosmetically typical, but she was unable to perceive depth throughout

FIGURE 21 **An Artist's Use of the Monocular Cue of Linear Perspective**
Image courtesy National Gallery of Art

FIGURE 22 **Texture Gradient** The gradients of texture create an impression of depth on a flat surface.

her life. As an adult, she became determined to see depth. With a doctor's aid, she found special glasses and performed eye muscle exercises to improve her chances of perceiving in three dimensions. It was a difficult and long process, but one day she noticed things starting to "stick out" at her—as you might when watching a film in 3D. Although Barry had successfully adapted to life in a flat visual world, she had come to realize that relying on monocular cues was not the same as experiencing the rich visual world of binocular vision. She described flowers as suddenly appearing "inflated." She noted how "ordinary things looked extraordinary" as she saw the leaves of a tree, an empty chair, and her office door projecting out from the background. For the first time, she had a sense of being inside the world she was viewing.

MOTION PERCEPTION Motion perception plays an important role in the lives of many species (Carter & others, 2020; Héjja-Brichard & others, 2020). Indeed, for some animals, motion perception is critical for survival. Both predators and their prey depend on being able to detect motion quickly (Hirai & Senju, 2020). Frogs and some other simple vertebrates may not even see an object unless it is moving. For example, if a dead fly is dangled motionlessly in front of a frog, the frog cannot sense its winged meal. The bug-detecting cells in the frog's retinas are wired only to sense movement.

Whereas the retinas of frogs can detect movement, the retinas of humans and other primates cannot. In humans and other primates, the brain takes over the job of analyzing motion through highly specialized pathways.

How do humans perceive movement? First, we have neurons that are specialized to detect motion. Second, feedback from our body tells us whether we are moving or whether someone or some object is moving (e.g., you move your eye muscles as you watch a ball coming toward you). Third, the environment we see is rich in cues that give us information about movement. For example, when we run, our surroundings appear to be moving (Shayman & others, 2020).

Psychologists are interested in both real movement and **apparent movement**, which occurs when we perceive a stationary object as moving (Sperling & others, 2020). You

apparent movement
The perception that a stationary object is moving.

can experience apparent movement at IMAX movie theaters. In watching a film of a climb of Mount Everest, you may find yourself feeling breathless as your visual field floods with startling images. In theaters without seats, viewers of these films are often warned to hold the handrail because perceived movement is so realistic that they might fall.

PERCEPTUAL CONSTANCY

Retinal images change constantly. Yet even though the stimuli that fall on our retinas change as we move closer to or farther away from objects, or as we look at objects from different orientations and in light or dark settings, our perception of them remains stable. **Perceptual constancy** is the recognition that objects are constant and unchanging even though sensory input about them is changing.

We experience three types of perceptual constancy—size constancy, shape constancy, and color constancy—as follows:

- *Size constancy* is the recognition that an object remains the same size even though the retinal image of the object changes (Figure 23). Experience is important to size perception. No matter how far away you are from your car, you know how large it is.

- *Shape constancy* is the recognition that an object retains the same shape even though its orientation to you changes. Look around. You probably see objects of various shapes (e.g., chairs and tables). If you walk around the room, you will view these objects from different sides and angles. Even though the retinal image of the object changes as you walk, you still perceive the objects as having the same shape (Figure 24).

perceptual constancy
The recognition that objects are constant and unchanging even though sensory input about them is changing.

FIGURE 23 **Size Constancy** Even though our retinal images of the hot air balloons vary, we still realize the balloons are approximately the same size. This illustrates the principle of size constancy.
Steve Allen/Getty Images

FIGURE 24 **Shape Constancy** The various projected images from an opening door are quite different, yet you perceive a rectangular door.

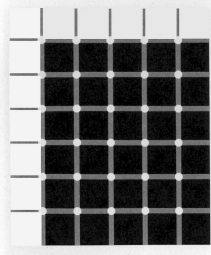

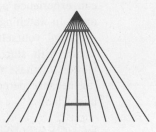

Ponzo Illusion
The top line looks much longer than the bottom, but they are the same length.

Blinking Effect Illusion
Stare at the white circles and notice the intermittent blinking effect. Your eyes make the static figure seem dynamic, attempting to fill in the white circle intersections with the black of the background.

Rotational Illusion
The two rings appear to rotate in different directions when we approach or move away from this figure while fixing our eyes on the center.

Pattern Recognition
Although the diagram contains no actual triangles, your brain "sees" two overlapping triangles. The explanation is that the notched circles and angled lines merely suggest gaps in which complete objects should be. The brain fills in the missing information.

Induction Illusion
The yellow patches are identical, but they look different and seem to take on the characteristics of their surroundings when they appear against different-color backgrounds.

FIGURE 25 **Perceptual Illusions** These illusions show how adaptive perceptual cues can lead to errors when taken out of context. They are definitely fun, but keep in mind that these illusions are based on processes that are quite adaptive in real life. Remember, not everyone sees these illusions. In cultures where exposure to two-dimensional representations is not common, individuals are less fooled by geometric illusions.

■ *Color constancy* is the recognition that an object retains the same color even though different amounts of light fall on it. For example, if you are reaching for a green Granny Smith apple, it looks green to you whether you are having it for lunch in the bright noon sun or as an evening snack in the pale pink of sunset.

Perceptual constancy tells us about the crucial role of interpretation in perception: We interpret sensation. That is, we perceive objects as having particular characteristics regardless of the retinal image detected by our eyes. Images may flow across the retina, but experiences are made sensible through perception. The many cues we use to visually perceive the real world can lead to *optical illusions* when they are taken out of that real-world context, as you can experience for yourself in Figure 25. Optical illusions involve "tricking" the eye by using cues that work very well in real-life vision to create perceptions that do not reflect reality.

As you look over these illusions, consider that culture can influence the extent to which people experience these illusions. In cultures where two-dimensional images, such as drawings on a piece of paper, are not typically used, geometrical illusions are less likely to lead to errors (Cohn, 2020; Segall & others, 1966).

1. When we refer to the hue of a light wave, we are referring to what we perceive as
   A. intensity.
   B. radiation.
   C. brightness.
   D. color.

2. To read this question, you are looking at it. After the light passes into your eyes, the incoming light waves are recorded by receptor cells located in the
   A. retina.
   B. cornea.
   C. blind spot.
   D. optic chiasm.

3. If you are in a well-lighted room, your rods are being used _____ and cones are being used _____.
   A. infrequently; frequently
   B. infrequently; infrequently

   C. frequently; infrequently
   D. frequently; frequently

APPLY IT! 4. Sondra was driving in the country one afternoon. There was not much traffic on the long, straight road, though Sondra noticed a man walking along the roadside some distance away. Suddenly, as she approached the person, he drifted toward the middle of the road, and Sondra, with screeching brakes, was shocked to realize she had nearly hit a child. Fortunately, the child was not harmed. It had become clear to Sondra that what had seemed like a man some distance away was actually a child who was much closer than she realized. What explains this situation?
   A. Sondra's occipital lobe must be damaged.

   B. Because objects that are smaller on the retina are typically farther away, Sondra was fooled by relative size.
   C. Because objects in the mirror are closer than they appear, Sondra was not able to detect the just-noticeable difference.
   D. Because objects that are smaller on the retina are typically closer than they appear, Sondra was fooled by shape constancy.

3 The Auditory System

Just as light provides us with information about the environment, so does sound. Sounds tell us about the presence of a person behind us, the approach of an oncoming car, the force of the wind, and the mischief of a 2-year-old. Perhaps most important, sounds allow us to communicate through language and song.

The Nature of Sound and How We Experience It

At a fireworks display, you may feel the loud boom of the explosion in your chest. At a concert, you might have sensed that the air around you was vibrating. Bass instruments are especially effective at creating mechanical pulsations, even causing the floor to vibrate. When the bass is played loudly, we can sense air molecules being pushed forward in waves from the speaker. How does sound generate these sensations?

Sound waves are vibrations in the air that are processed by the *auditory* (hearing) system. Remember that light waves are much like the waves in the ocean moving toward the beach. Sound waves are similar. Sound waves also vary in length. Wavelength determines the sound wave's *frequency,* that is, the number of cycles (full wavelengths) that pass through a point in a given time interval. *Pitch* is the perceptual experience of the frequency of a sound, whether it is high like a whistle or low like a bass horn. We perceive high-frequency sounds as having a high pitch, and low-frequency sounds as having a low pitch. A soprano voice sounds high pitched; a bass voice sounds low pitched. As with the wavelengths of light, human sensitivity is limited to a range of sound frequencies. It is common knowledge that dogs, for example, can hear higher frequencies than humans can.

Sound waves vary not only in frequency but also, like light waves, in amplitude. A sound wave's *amplitude,* measured in decibels (dB), is the amount of pressure the sound wave produces relative to a standard. The typical standard, 0 dB, is the weakest sound the human ear can detect. *Loudness* is the perception of the sound wave's amplitude. In general,

courtyardpix/Shutterstock

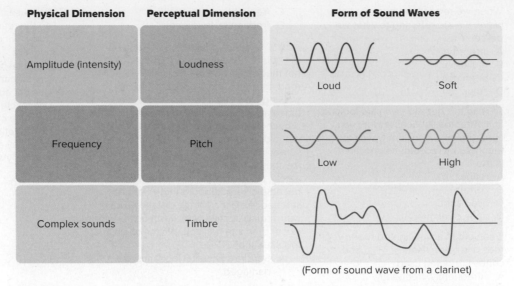

Physical Dimension	Perceptual Dimension	Form of Sound Waves
Amplitude (intensity)	Loudness	Loud Soft
Frequency	Pitch	Low High
Complex sounds	Timbre	(Form of sound wave from a clarinet)

FIGURE 26 **Physical Difference in Sound Waves and the Qualities of Sound They Produce** Here we can see how the input of sound stimuli requires our ears and brain to attend to varying characteristics of the rich sensory information that is sound.

the higher the amplitude of the sound wave, or the higher the decibel level, the louder we perceive the sound to be. Thus, in terms of amplitude, the air is pressing more forcibly against you and your ears during loud sounds and more gently during quiet sounds.

So far we have been describing a single sound wave with just one frequency. A single sound wave is similar to the single wavelength of pure colored light, discussed in the context of color matching. Most sounds, including those of speech and music, are complex sounds, those in which numerous frequencies of sound blend together. *Timbre* is the tone saturation, or the perceptual quality, of a sound. Timbre is responsible for the perceptual difference between a trumpet and a trombone playing the same note and for the quality differences we hear in human voices. Figure 26 illustrates the physical differences in sound waves that produce the various qualities of sounds.

Morgan Freeman and James Earl Jones do a lot of voice-over work because their voices have marvelously rich timbres!

Structures and Functions of the Ear

What happens to sound waves once they reach your ear? How do various structures of the ear transform sound waves into signals that the brain will recognize as sound? Functionally the ear is analogous to the eye. The ear serves the purpose of transmitting a high-fidelity version of sounds in the world to the brain for analysis and interpretation. Just as an image needs to be in focus and sufficiently bright for the brain to interpret it, a sound needs to be transmitted in a way that preserves information about its location, its frequency (which helps us distinguish the voice of a child from that of an adult), and its timbre (which allows us to identify the voice of a friend on the phone). The ear is divided into three parts: outer ear, middle ear, and inner ear (Figure 27).

OUTER EAR   The **outer ear** consists of the pinna and the external auditory canal. The funnel-shaped *pinna* (plural, *pinnae*) is the outer, visible part of the ear. (Elephants have very large pinnae.) The pinna collects sounds and channels them into the interior of the ear. The pinnae of many animals, such as cats, are movable and serve a more important role in sound localization than do the pinnae of humans. Cats turn their ears in the direction of a faint and interesting sound.

outer ear
The outermost part of the ear, consisting of the pinna and the external auditory canal.

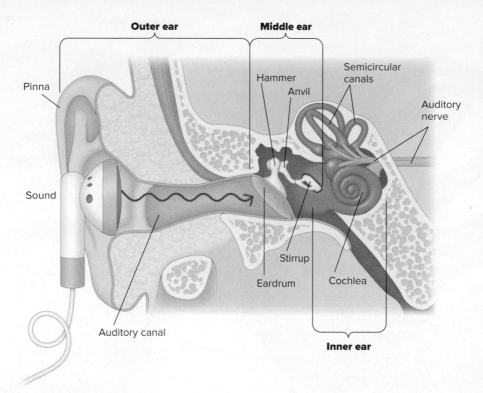

Outer ear
Middle ear

Pinna

Hammer
Anvil

Semicircular
canals

Auditory
nerve

Sound

Stirrup

Eardrum

Cochlea

Auditory canal

Inner ear

FIGURE 27 The Outer, Middle, and Inner Ear On entering the outer ear, sound waves travel through the auditory canal, where they generate vibrations in the eardrum. These vibrations are transferred via the hammer, anvil, and stirrup to the fluid-filled cochlea in the inner ear. There the mechanical vibrations are converted into an electrochemical signal that the brain will recognize as sound.

middle ear
The part of the ear that channels and amplifies sound through the eardrum, hammer, anvil, and stirrup to the inner ear.

MIDDLE EAR   After passing the pinna, sound waves move through the auditory canal to the middle ear. The **middle ear** channels and amplifies the sound through the eardrum, hammer, anvil, and stirrup to the inner ear. The *eardrum,* or *tympanic membrane,* separates the outer ear from the middle ear and vibrates in response to sound. The eardrum is the first structure that sound touches in the middle ear. The *hammer, anvil,* and *stirrup* are an intricately connected chain of very small bones. When they vibrate, they transmit sound waves to the fluid-filled inner ear (Oxenham, 2018). The muscles that operate these tiny bones take the vibration of the eardrum and transmit it to the *oval window,* the opening of the inner ear.

If you have gone swimming, you know that sound travels far more easily in air than in water. Sound waves entering the ear travel in air until they reach the inner ear. At the border between the middle and the inner ear—which, as we will see later, is a border between air and fluid—sound meets the same kind of resistance as do shouts directed at an underwater swimmer when the shouts hit the surface of the water. To compensate, the muscles of the middle ear can maneuver the hammer, anvil, and stirrup to amplify the sound waves. Importantly, these muscles, if necessary, can also work to decrease the intensity of sound waves, to protect the inner ear.

The hammer, anvil, and stirrup are also called the **ossicles.** *These are the tiniest bones in the human body.*

inner ear
The part of the ear that includes the oval window, cochlea, and basilar membrane and whose function is to convert sound waves into neural impulses and send them to the brain.

INNER EAR   The function of the **inner ear**, which includes the oval window, cochlea, and basilar membrane, is to convert sound waves into neural impulses and send them on to the brain (Raufer & others, 2020). The stirrup is connected to the *oval window.* The oval window is a membrane-covered opening that leads from the middle ear to the inner ear. The oval window transmits sound waves to the cochlea. The *cochlea* is a tubular, fluid-filled structure that is coiled up like a snail (Figure 28). The *basilar membrane* lines the inner wall of the cochlea and runs its entire length. It is narrow and rigid at the base of the cochlea but widens and becomes more flexible at the top. The variation in width and flexibility allows different areas of the basilar membrane to vibrate more intensely when exposed to different sound frequencies (Oxenham, 2018). For example, the high-pitched tinkle of a little bell stimulates the narrow region of the basilar membrane at the base of the cochlea, whereas the low-pitched tones of a tugboat whistle stimulate the wide end.

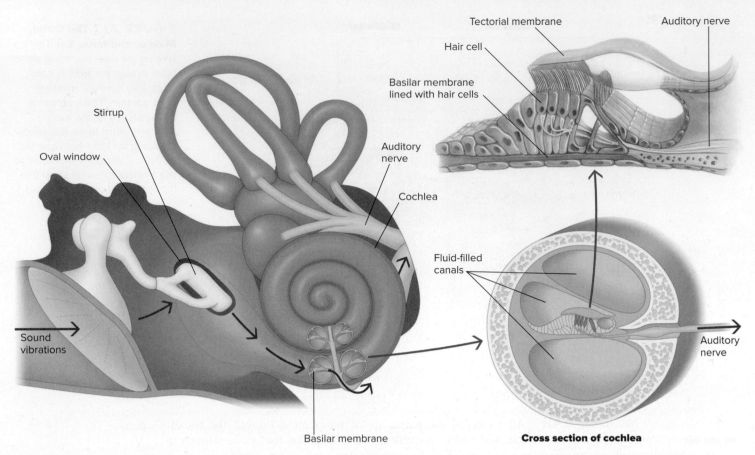

Stirrup

Oval window

Sound vibrations

Basilar membrane

Auditory nerve

Cochlea

Tectorial membrane

Hair cell

Basilar membrane lined with hair cells

Auditory nerve

Fluid-filled canals

Auditory nerve

Cross section of cochlea

FIGURE 28   **The Cochlea** The cochlea is a spiral structure consisting of fluid-filled canals. When the stirrup vibrates against the oval window, the fluid in the canals vibrates. Vibrations along portions of the basilar membrane correspond to different sound frequencies. The vibrations exert pressure on the hair cells (between the basilar and tectorial membranes); the hair cells in turn push against the tectorial membrane, and this pressure bends the hairs. This triggers an action potential in the auditory nerve.

In humans and other mammals, hair cells line the basilar membrane (see Figure 28). These *hair cells* are the sensory receptors of the ear. They are called hair cells because of the tufts of fine bristles, or *cilia,* that sprout from the top of them. The movement of the hair cells against the *tectorial membrane,* a jellylike flap above them, generates resulting impulses that the brain interprets as sound (Han & others, 2020; Sellon & others, 2017). Hair cells are so delicate that exposure to loud noise can destroy them, leading to deafness or difficulties in hearing. Once lost, hair cells cannot regenerate. However, researchers are seeking to develop ways to regenerate these crucial aspects of hearing including through the use of stem cells (Nie & Hashino, 2020).

Cochlear implants are devices that were specifically developed to replace damaged hair cells. A *cochlear implant*—a small electronic device that is surgically implanted in the ear and head—allows deaf or profoundly hard-of-hearing individuals to detect sound. Cochlear implants are not hearing aids. A hearing aid amplifies sound so a person can hear it. A cochlear implant does not amplify sound. Rather, it works by using electronic impulses to directly stimulate whatever working auditory nerves the recipient has in their cochlea. In the United States, approximately 58,000 adults and 38,000 children have had cochlear implants (NIDCD, 2017).

Theories of Hearing

One of the auditory system's mysteries is how the inner ear registers the frequency of sound. Two theories aim to explain this mystery: place theory and frequency theory.

Are You Listening Safely?

Whether it is for listening to music or podcasts, gaming, or participating in online classes, people increasingly wear earphones (including earbuds or headsets). Wearing earphones can help to block out distractions as you gear up your best "concentration music" to study. Can the common practice of using earphones harm our hearing?

Noise-induced hearing loss (NIHL) refers to the effects of exposure to loud sounds on the human ability to hear. You may have noticed your ears ringing or sounds being muffled after attending a loud concert or a fireworks display. The effects of these experiences are often temporary. But very loud noises can rapture the ear drum. Loud noises can also damage the bones of the ear and damage or kill the delicate hair cells of the cochlea. Hair cell damage may occur without us even knowing it. Between 30 and 50 percent of hair cells can be damaged before a person notices a loss of hearing (CDC, 2018). Early signs of NIHL include a buzzing or roaring sound in the ears and a feeling that sounds are muffled. Perhaps in part due to the increased use of headphones, nearly 1 in 5 teenagers experiences some form of hearing loss (AOA 2020; NICHD, 2019). Young men may be particularly at risk (Alessio & others, 2020).

Hill Street Studios/Getty Images

NIHL can be prevented. Because damage to the ears depends on both the level of the sound and the length of time of exposure, experts suggest that 60/60 rule—meaning setting your volume to no more than 60 percent of the maximum and listening for no more than 60 minutes (AOA, 2020).

If you use earbuds to block out distractions, noise-canceling headphones are a good alternative. These devices may help prevent volume creep—the temptation to turn the volume up to drown out outside noise.

The very ubiquity of headphones tells us that, for many people, hearing is vital to creating life's soundtrack. Taking care of our hearing means using earphones wisely.

---

place theory
Theory on how the inner ear registers the frequency of sound, stating that each frequency produces vibrations at a particular spot on the basilar membrane.

Place theory states that each frequency produces vibrations at a particular place on the basilar membrane. Von Békésy (1960) studied the effects of vibration applied at the oval window on the basilar membrane of human cadavers. Through a microscope, he saw that this stimulation produced a traveling wave on the basilar membrane. A traveling wave is like the ripples that appear in a pond when you throw in a stone. However, because the cochlea is a long tube, the ripples can travel in only one direction, from the oval window at one end of the cochlea to the far tip of the cochlea. High-frequency vibrations create traveling waves that maximally displace, or move, the area of the basilar membrane next to the oval window; low-frequency vibrations maximally displace areas of the membrane closer to the tip of the cochlea.

Von Békésy won a Nobel Prize in 1961 for his research on the basilar membrane.

Place theory adequately explains high-frequency sounds but not low-frequency sounds. A high-frequency sound, like the screech of a referee's whistle or the piercing high note of an opera diva, stimulates a precise area on the basilar membrane, just as the theory suggests. However, a low-frequency sound, like the tone of a tuba or the croak of a bullfrog, causes a large part of the basilar membrane to be displaced, making it hard to identify an exact location that is associated with hearing this kind of sound. Looking only at the movement of the basilar membrane, you would get the impression that humans are probably not very good at hearing low-frequency sounds, and yet we are. Therefore, some other factors must be at play in low-frequency hearing.

Frequency theory gets at these other influences by stating that the perception of a sound's frequency depends on how often the auditory nerve fires. Higher frequency sounds cause the auditory nerve to fire more often than do lower frequency sounds. One limitation of frequency theory, however, is that a single neuron has a maximum firing rate of about 1,000 times per second. Therefore, frequency theory does not apply to tones with frequencies that would require a neuron to fire more rapidly.

To deal with this limitation of frequency theory, researchers developed the **volley principle**, which states that a cluster of nerve cells can fire neural impulses in rapid succession, producing a volley of impulses. Individual neurons cannot fire faster than 1,000 times per second, but if the neurons team up and alternate their neural firing, they can attain a combined frequency above that rate. To get a sense for how the volley principle works, imagine a troop of soldiers who are all armed with guns that can fire only one round at a time and that take time to reload. If all the soldiers fire at the same time, the frequency of firing is limited and cannot go any faster than it takes to reload those guns. If, however, the soldiers are coordinated as a group and fire at different times, some of them can fire while others are reloading, leading to a greater frequency of firing. Frequency theory better explains the perception of sounds below 1,000 times per second, whereas a combination of frequency theory and place theory is needed to account for sounds above 1,000 times per second.

Phew! Two theories and a principle just to explain how we hear.

As we considered in the discussion of the visual system, once our receptors pick up energy from the environment, that energy must be transmitted to the brain for processing and interpretation. We saw that in the retina, the responses of the rod and cone receptors feed into ganglion cells and leave the eye via the optic nerve. In the auditory system, information about sound moves from the hair cells of the inner ear to the **auditory nerve**, which carries neural impulses to the brain's auditory areas. Remember that it is the movement of the hair cells that transforms the physical stimulation of sound waves into the action potential of neural impulses (Oxenham, 2018).

Auditory Processing in the Brain

Auditory information moves up the auditory pathway via electrochemical transmission in a more complex manner than does visual information in the visual pathway. Many synapses occur in the ascending auditory pathway, with most fibers crossing over the midline between the hemispheres of the cerebral cortex, although some proceed directly to the hemisphere on the same side as the ear of reception (Lee & others, 2020). This means that most of the auditory information from the left ear goes to the right side of the brain, but some also goes to the left side of the brain. The auditory nerve extends from the cochlea to the brain stem, with some fibers crossing over the midline. The cortical destination of most of these fibers is the temporal lobes of the brain, beneath the temples of the head. Like visual information, researchers have found that features are extracted from auditory information and transmitted along parallel pathways in the brain (Lesicko & Llano, 2020).

Do It!

Imagine hearing impossible sounds, nonexistent words, or three voices where only two exist. Welcome to the world of auditory illusions. In Figure 25 you tried out some visual illusions. Did you know that there are auditory illusions as well? Search the web for "auditory illusions" and try some out. They can be truly amazing and baffling! Keep in mind, just as you did when looking at the optical illusions, that these illusions emerge as a function of capacities that work very well in the "real world."

Localizing Sound

When we hear a fire engine's siren or a dog's bark, how do we know where the sound is coming from? The basilar membrane gives us information about the frequency, pitch, and complexity of a sound, but it does not tell us where a sound is located.

frequency theory
Theory on how the inner ear registers the frequency of sound, stating that the perception of a sound's frequency depends on how often the auditory nerve fires.

volley principle
Principle addressing limitations of the frequency theory of hearing, stating that a cluster of nerve cells can fire neural impulses in rapid succession, producing a volley of impulses.

auditory nerve
The nerve structure that receives information about sound from the hair cells of the inner ear and carries these neural impulses to the brain's auditory areas.

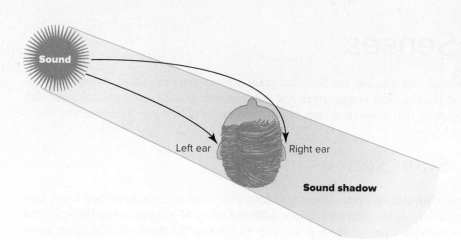

FIGURE 29 The Sound Shadow The sound shadow is caused by the listener's head, which forms a barrier that reduces the sound's intensity. Here the sound is to the person's left, so the sound shadow will reduce the intensity of the sound that reaches the right ear.

Earlier in the chapter we saw that because our two eyes see slightly different images, we can determine how near or far away an object is. Similarly, having two ears helps us to localize a sound because each receives somewhat different stimuli from the sound source. A sound coming from the left has to travel different distances to the two ears, so if a barking dog is to your left, your left ear receives the sound sooner than your right ear. Also, your left ear will receive a slightly more intense sound than your right ear in this case. The sound reaching one ear is more intense than the sound reaching the other ear for two reasons: (1) It has traveled less distance, and (2) the other ear is in what is called the *sound shadow* of the listener's head, which provides a barrier that reduces the sound's intensity (Figure 29). People who are blind use the sound shadow to orient themselves. Compared to other animals, the human capacity to locate sounds is characterized by high levels of precision (Ege & others, 2019).

So, differences in both the *timing* of the sound and the *intensity* of the sound help you to localize a sound. You often have difficulty localizing a sound that is coming from a source directly in front of you because it reaches both ears simultaneously. The same is true for sounds directly above your head or directly behind you, because the disparities that provide information about location are not present. So, we are able to localize sounds because we have two ears on the opposite sides of our head. This anatomical setup allows us to use time and intensity to know where a sound is coming from.

self-quiz

1. Your mother's and sister's voices have the same pitch and loudness, but you can tell them apart on the telephone. This is due to the perceptual quality, or _____, of their voices.
   A. timbre
   B. wavelength
   C. frequency
   D. amplitude

2. The major function of the hammer, anvil, and stirrup of the middle ear is
   A. to soften the tone of incoming stimuli for appropriate processing.
   B. to stir cochlear fluid so that bone conduction hearing can occur.
   C. to amplify vibrations and pass them on to the inner ear.

   D. to clean the external auditory canal of any potential wax buildup.

3. The bones of the middle ear are set into motion by vibrations of the
   A. cochlea.
   B. eardrum.
   C. saccule.
   D. basilar membrane.

APPLY IT! 4. The late conservative radio personality Rush Limbaugh experienced sudden hearing loss in 2001, after which he received a cochlear implant. He described his ability to listen to music as dependent on what he heard before becoming deaf. If he had heard a song prior to becoming deaf, he could hear it, but if it was a new song, he could not

make sense of it. Which of the following explains Limbaugh's experience?
   A. He was no longer able to listen to music from a top-down perspective.
   B. He was able to engage in top-down listening, but not bottom-up listening.
   C. He was likely to have experienced damage to the temporal lobes.
   D. He was not able to experience any auditory sensation.

4 Other Senses

Beyond vision and hearing, the body has other sensory systems. These include the skin senses and the chemical senses (smell and taste), as well as the kinesthetic and vestibular senses (systems that allow us to stay upright and to coordinate our movements).

The Skin Senses

You know when a friend has a fever by putting your hand to their head; you know how to find your way to the light switch in a darkened room by groping along the wall; and you know whether a pair of shoes is too tight by the way the shoes touch different parts of your feet when you walk. Many of us think of our skin as a canvas rather than a sense. We color it with cosmetics, dyes, and tattoos. In fact, the skin is our largest sensory system, draped over the body with receptors for touch, temperature, and pain. These three kinds of receptors form the *cutaneous senses*.

TOUCH   Touch is one of the senses that we most often take for granted, yet our ability to respond to touch is astounding. What do we detect when we feel "touch"? What kind of energy does our sense of touch pick up from our external environment? In vision we detect light energy. In hearing we detect the vibrations of air or sound waves pressing against our eardrums. In touch we detect mechanical energy, or pressure against the skin. The lifting of a single hair causes pressure on the skin around the hair shaft. This tiny bit of mechanical pressure at the base of the hair is sufficient for us to feel the touch of a pencil point. More commonly we detect the mechanical energy of the pressure of a car seat against our buttocks or of a pencil in our hand. Is this energy so different from the kind of energy we detect in vision or hearing? Sometimes the only difference is one of intensity—the sound of a rock band playing softly is an auditory stimulus, but at the high volumes that make a concert hall reverberate, this auditory stimulus is also felt as mechanical energy pressing against our skin.

Your ability to find a nickel in your pocket without looking is a truly amazing feat of touch. Not even the most sophisticated robot can do it.

How does information about touch travel from the skin through the nervous system? Sensory fibers arising from receptors in the skin enter the spinal cord. From there the information travels to the brain stem, where most fibers from each side of the body cross over to the opposite side of the brain. Next the information about touch moves on to the thalamus, which serves as a relay station. The thalamus then projects the map of the body's surface onto the somatosensory areas of the parietal lobes in the cerebral cortex (Kumar & others, 2020; Sereno, 2017).

Just as the visual system is more sensitive to images on the fovea than to images in the peripheral retina, our sensitivity to touch is not equally good across all areas of the skin. Human toolmakers need excellent touch discrimination in their hands, but they require much less touch discrimination in other parts of the body, such as the torso and legs. The brain devotes more space to analyzing touch signals coming from the hands than from the legs.

TEMPERATURE   We not only can feel the warmth of a comforting hand on our hand, we also can feel the warmth or coolness of a room. We must be able to detect temperature in order to maintain our body temperature. **Thermoreceptors**, sensory nerve endings under the skin, respond to temperature changes at or near the skin and provide input to keep the body's temperature at 98.6 degrees Fahrenheit. There are two types of thermoreceptors: warm and cold. Warm thermoreceptors respond to the warming of the skin, and cold thermoreceptors respond to the cooling of the skin. When warm and cold receptors that are close to each other in the skin are stimulated simultaneously, we experience the sensation of hotness. Figure 30 illustrates this "hot" experience.

Warm water   Cold water

FIGURE 30 A "Hot" Experience When two pipes, one containing cold water and the other warm water, are braided together, a person touching the pipes feels a sensation of "hot." The perceived heat coming from the pipes is so intense that the individual cannot touch them for longer than a couple of seconds.

thermoreceptors
Sensory nerve endings under the skin that respond to changes in temperature at or near the skin and provide input to keep the body's temperature at 98.6 degrees Fahrenheit.

pain
The sensation that warns an individual of damage to the body.

PAIN **Pain** is the sensation that warns us of damage to the body. When contact with the skin takes the form of a sharp pinch, our sensation of mechanical pressure changes from touch to pain. When a pot handle is so hot that it burns our hand, our sensation of temperature becomes one of pain. Intense stimulation of any one of the senses can produce pain—too much light, very loud sounds, or too many habanero peppers, for example. Our ability to sense pain is vital for our survival as a species. It functions as a quick-acting messenger that tells the brain's motor systems that they must act fast to minimize or eliminate damage.

Pain receptors are dispersed widely throughout the body—in the skin, in the sheath tissue surrounding muscles, in internal organs, and in the membranes around bone. Although all pain receptors are anatomically similar, they differ in the type of physical stimuli to which they most readily react, with some responding to pressure, others to heat, and others to both. Many pain receptors are chemically sensitive and respond to a range of pain-producing substances.

Pain receptors have a much higher threshold for firing than receptors for temperature and touch (Hill & Bautista, 2020; Park & others, 2019). Pain receptors react mainly to physical stimuli that distort them or to chemical stimuli that irritate them into action. Inflamed joints or sore, torn muscles produce *prostaglandins,* fatty acids that stimulate the receptors and cause the experience of pain. Drugs such as aspirin likely reduce the feeling of pain by reducing prostaglandin production.

Two different neural pathways transmit pain messages to the brain: a fast pathway and a slow pathway (Boddice, 2017; Ciotu & others, 2019). In the *fast pathway,* fibers connect directly with the thalamus and then to the motor and sensory areas of the cerebral cortex. This pathway transmits information about sharp, localized pain, as when you cut your skin. The fast pathway may serve as a warning system, providing immediate information about an injury—it takes less than a second for the information in this pathway to reach the cerebral cortex. In the *slow pathway,* pain information travels through the limbic system, a detour that delays the arrival of information at the cerebral cortex by seconds. The unpleasant, nagging pain that characterizes the slow pathway may function to remind the brain that an injury has occurred and that we need to restrict normal activity and monitor the pain.

Many neuroscientists believe that the brain actually generates the experience of pain. There is evidence that turning pain signals on and off is a chemical process that probably involves *endorphins.* Recall that endorphins are neurotransmitters that function as natural opiates in producing pleasure and pain. Endorphins are believed to be released mainly in the synapses of the slow pathway.

Perception of pain is complex and often varies from one person to the next (Letzen & others, 2020). Some people rarely feel pain; others seem to be in great pain if they experience a minor bump or bruise. When people experience the same potentially pain-causing injury, some may find it to be relatively mild while others are made quite miserable. A person's pain thresholds can also change with age (González-Roldán & others, 2020). To some degree, these individual variations may be physiological. A person who experiences considerable pain even with a minor injury may have a neurotransmitter system that is deficient in endorphin production. However, perception of pain goes beyond physiology. Although it is true that all sensations are affected by factors such as motivation, expectation, and other related decision factors, the perception of pain is especially susceptible to these factors. For example, the experience of stress and trauma can alter pain thresholds (López-López & others, 2020; Lunde & Sieberg, 2020; Tsur & others, 2020).

Our understanding of pain has been hindered because historically pain research on animals has used only male animals (Mogil, 2020). In humans, women experience more clinical pain and suffer more pain-related distress than men (Fuensalida-Novo & others, 2020; Sylwander & others, 2020). However, this difference is not consistently found, and men and women are very similar in pain perception, intensity, and types of pain experienced (Racine & others, 2020). In laboratory studies, women report lower pain thresholds than men (Day & others, 2020; McHugh & others, 2020). An important factor to consider

Do It!

People with mirror-touch synaesthesia feel what other people are physically feeling. To learn more about mirror-touch synaesthesia, check out the Entanglement episode of the *Invisibilia* Podcast. www.npr.org/podcasts/510307/Invisibilia

in the realm of gender differences in pain is the role of cultural expectations (Berke & others, 2017). Men may be more likely than women to downplay their experience of pain and to tolerate higher levels of pain to live up to standards of masculinity (Nascimento & others, 2020). In addition, compared to women, men are more likely to view pain in the laboratory as a challenge rather than a threat (Day & others, 2020).

Cultural and ethnic contexts influence the degree to which people experience or report pain. Compared to people in the United States and Western Europe, people in Japan consider expressing physical pain to be inappropriate and report lower levels of pain (Feng & others, 2020; Onishi & others, 2017). Of course, pain can mean different things to different people. For example, one pain researcher described a ritual performed in India in which a chosen person travels from town to town delivering blessings to the children and the crops while suspended from metal hooks embedded in his back (Melzack, 1973). The person apparently reports no sensation of pain and appears to be in ecstasy.

People who experience pain often seek medical attention, not only to cure a physical problem but also to reduce their pain. Research consistently shows racial disparities in pain treatment. This means that the equality of medical care provided to people depends, unfairly, on their skin color. For example, one study found that Black patients were less likely than white patients to receive painkillers for *broken bones*, despite reporting the same amount of pain (Todd & others, 2000). Another study of nearly 1 million children with appendicitis found that Black children were less likely to receive pain medication than white children (Goyal & others, 2015). Many studies have shown that medical professionals do not properly treat pain in Black patients (Anastas & others, 2020; Aronowitz & others, 2020; Dreyer, 2020). Why do people with medical training show racial bias in pain treatment? To read about research addressing this question, see the Intersection.

The Chemical Senses

The information processed through our senses comes in many diverse forms: electromagnetic energy in vision, sound waves in hearing, and mechanical pressure and temperature in the skin senses. The two senses we now consider, smell and taste, are responsible for processing chemicals in our environment. Through smell, we detect airborne chemicals, and through taste we detect chemicals that have been dissolved in saliva. Smell and taste are frequently stimulated simultaneously. We notice the strong links between the two senses when a nasty cold with lots of nasal congestion takes the pleasure out of eating. Our favorite foods become "tasteless" without their characteristic smells. Despite this link, taste and smell are two distinct systems.

norikko/Shutterstock

TASTE Think of your favorite food. Why do you like it? Imagine that same food without its flavor. The thought of giving up a favorite taste, such as chocolate, can be depressing. Indeed, eating food we love is a major source of pleasure.

How does taste happen? To get at this question, try this. Take a drink of milk and allow it to coat your tongue. Then go to a mirror, stick out your tongue, and look carefully at its surface. You should be able to see rounded bumps above the surface. Those bumps, called **papillae**, contain *taste buds,* the receptors for taste. Your tongue houses about 10,000 taste buds, which are replaced about every 2 weeks. As people age, however, this

papillae
Rounded bumps above the tongue's surface that contain the taste buds, the receptors for taste.

Sensation, Perception, and Social Psychology: Do False Beliefs about Race Affect Pain Treatment?

In pain research, Black participants report higher pain sensitivity and lower pain thresholds than white participants (Kim & others, 2019; Klonoff, 2009; Losin & others, 2020; Trawalter & Hoffman, 2015). Yet, many people believe the exact opposite—that Black people are less susceptible to pain than white people are. Could such a false belief about race and pain help to explain the racial disparity in pain treatment?

A study of medical students and medical residents (i.e., individuals who have completed medical school and are doctors in training) measured false beliefs about race and how they influence medical decision making (Hoffman & others, 2016). The false beliefs included that compared to white people, Black people's nerve endings are less

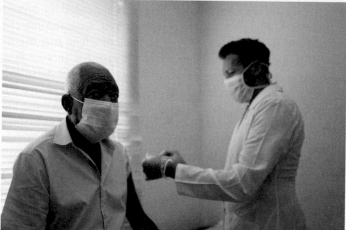

FG Trade/Getty Images

sensitive, their blood coagulates faster, their skin is thicker, and they age more slowly. *These beliefs are all completely false.* Yet, half of the medical students and residents rated at least one of the false statements as probably or definitively true. Those who endorsed more false beliefs about race also rated painful experiences as less severe for a Black person than for themselves. In a later decision-making task, participants were asked how they would treat a hypothetical Black or white patient. Those who endorsed these false beliefs were less likely to recommend the appropriate treatment for pain for Black patients compared to white patients. Those who did not endorse these false beliefs showed no difference in recommendation.

Probing the data, the researchers found that medical students and residents were motivated by higher levels of empathy for a white person in pain. Such results highlight the need for greater diversity and inclusion in the medical profession (Hagedorn & others, 2020). They also demonstrate the damaging effect of racial prejudice in contemporary life.

\\ **Where do you think these false beliefs about pain come from?**

replacement process is not quite as efficient, and an older individual may have just 5,000 working taste buds at any given moment. As with all of the other sensory systems we have studied, the information picked up by these taste receptors is transmitted to the brain for analysis and, when necessary, for a response (e.g., spitting something out).

The taste fibers leading from a taste bud to the brain often respond strongly to a range of chemicals spanning multiple taste elements, such as salty and sour. The brain processes these somewhat ambiguous incoming signals and integrates them into a perception of taste (Iannilli & others, 2012). So, although people often categorize taste sensations along the four dimensions of sweet, bitter, salty, and sour, our tasting ability goes far beyond these. Researchers and chefs have been exploring a taste called *umami* (Wang & others, 2020). *Umami* is the Japanese word for "delicious" or "yummy." The taste of umami, one that Asian cooks have long recognized, is the flavor of L-glutamate. What is that taste? Umami is a savory flavor that is present in many seafoods as well as soy sauce, parmesan and mozzarella cheeses, anchovies, mushrooms, and hearty meat broths.

Culture certainly influences the experience of taste. European Americans watching the Japanese version of the TV series *Iron Chef* quickly notice that some people enjoy the flavor of sea urchin, while others just do not get the appeal. In some cultures, food that

W Some people have more papillae than others. A "supertaster" is someone who has as many as 40 papillae in a 6-mm area of the tongue (about the size of a pencil eraser). Supertasters can taste substances that others cannot and thus are valuable to the food and wine industries.

is so spicy as to be practically inedible for the outsider may be viewed as quite delicious or umami. The culture in which we live can influence the foods we are exposed to as well as our sense of what tastes good. In some cultures, very spicy food is introduced slowly into children's diets so that they can learn what is delicious at an early age.

SMELL   Why do we have a sense of smell? One way to appreciate the importance of smell is to think about animals with a more sophisticated sense of smell than our own. A dog, for example, can use smell to find its way back from a long stroll, to distinguish friend from foe, and even (with practice) to detect illegal drugs concealed in a suitcase. In fact, dogs can detect odors in concentrations 100 times lower than those detectable by humans. Given the nasal feats of the average dog, we might be tempted to believe that the sense of smell has outlived its usefulness in humans.

What do humans use smell for? For one thing, humans need the sense of smell to decide what to eat. We can distinguish rotten food from fresh food and remember (all too well) which foods have made us ill in the past. The smell of a food that has previously made us sick is often by itself enough to make us feel nauseated. For another thing, although tracking is a function of smell that we often associate only with nonhuman animals, humans are competent odor trackers. We can follow the odor of gas to a leak, the smell of smoke to a fire, and the aroma of a hot apple pie to a windowsill. In 2020, the loss of the sense of smell was recognized as an early sign of infection with COVID-19 (Malnic & Glezer, 2020).

What physical equipment do we use to process odor information? Just as the eyes scan the visual field for objects of interest, the nose is an active instrument. We actively sniff when we are trying to track down the source of a fire or an unfamiliar chemical odor. The **olfactory epithelium** lining the roof of the nasal cavity contains a sheet of receptor cells for smell (Figure 31), so sniffing maximizes the chances of detecting an odor. The receptor cells are covered with millions of minute, hairlike antennae that project through the mucus in the top of the nasal cavity and make contact with air on its way to the throat and lungs (Mullol & others, 2020). Interestingly, unlike the neurons of most sensory systems, the neurons in the olfactory epithelium tend to replace themselves after injury (Berger & others, 2020).

olfactory epithelium
The lining of the roof of the nasal cavity, containing a sheet of receptor cells for smell.

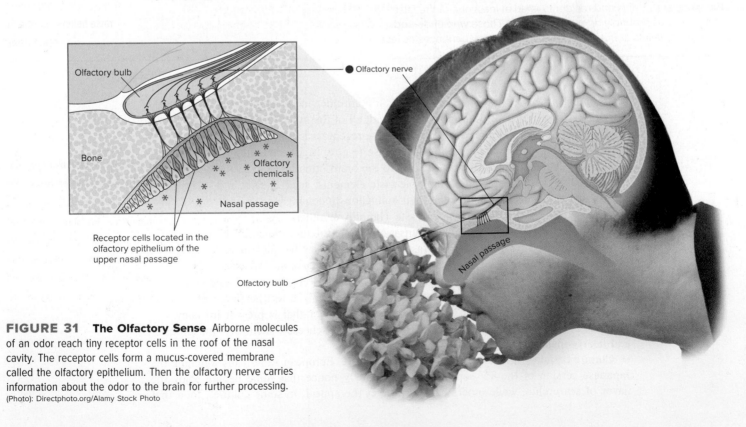

Olfactory bulb

Olfactory nerve

Bone

Olfactory chemicals

Nasal passage

Receptor cells located in the olfactory epithelium of the upper nasal passage

Olfactory bulb

Nasal passage

FIGURE 31 **The Olfactory Sense** Airborne molecules of an odor reach tiny receptor cells in the roof of the nasal cavity. The receptor cells form a mucus-covered membrane called the olfactory epithelium. Then the olfactory nerve carries information about the odor to the brain for further processing.
(Photo): Directphoto.org/Alamy Stock Photo

What is the neural pathway for information about smell? The answer to this question reveals an important difference between the sense of smell and other senses. All other sensory pathways pass through the thalamus but the pathway for smell does not. In smell, the neural pathway goes straight to the olfactory areas in the temporal lobes. Then it projects to various brain regions, especially the limbic system, which is involved in emotion and memory (Kazour & others, 2020). Unlike the other senses, smells take a superhighway to emotion and memory, a phenomenon we will consider in more detail when those topics are discussed.

Smell might have a role to play in the chemistry of interpersonal attraction (Oren & Shamay-Tsoory, 2019). From an evolutionary perspective, the goal of human mating is to find someone with whom to produce the healthiest offspring. Mates with differing sets of genes (known as the *major histocompatibility complex [MHC]*) produce healthier offspring with the broadest immune systems (Mueller, 2010). How do we find these people, short of taking a blood test? One way may be by sense of smell (Mahmut & Croy, 2019). Haselton (2006) has conducted studies on interpersonal attraction using the "smelly T-shirt" paradigm. In this research, heterosexual men are asked to wear a T-shirt to bed every day for a week without washing it. After they have been thoroughly imbued with a person's personal scent, the T-shirts are presented to heterosexual women to smell and rate for attractiveness. The women rated men whose MHCs differed from their own as more attractive, on the basis of the aroma of the T-shirts. Thus, although the eyes may be the window to the soul, the nose might be the gateway to love.

The Kinesthetic and Vestibular Senses

You know the difference between walking and running and between lying down and sitting up. To perform even the simplest act of motor coordination, such as reaching out to take a book off a shelf or getting up out of a chair, the brain must constantly receive and coordinate information from every part of the body. Your body has two kinds of senses that give you information about your movement and orientation in space, as well as help you to maintain balance. The **kinesthetic senses** provide information about movement, posture, and orientation. The **vestibular sense** provides information about balance and movement.

No specific organ contains the kinesthetic senses. Instead, they are embedded in muscle fibers and joints. As we stretch and move, these receptors signal the state of the muscle. *Kinesthesia,* the name for the kinesthetic senses, is a sense that you often do not even notice until it is gone. Try walking when your leg is "asleep" or smiling (never mind talking) when you have just come from a dentist's office and are still under the effects of Novocain.

We can appreciate the sophistication of kinesthesis when we think about it in terms of memory. Even a mediocre typist can bang out 20 words per minute—but how many of us could write down the order of the letters on a keyboard without looking? Typing on a keyboard is a skill that relies on very coordinated sensitivity to the orientation, position, and movements of our fingers. We say that our fingers "remember" the positions of the keys. Likewise, the complicated movements a pitcher uses to throw a baseball cannot be

kinesthetic senses
Senses that provide information about movement, posture, and orientation.

vestibular sense
Sense that provides information about balance and movement.

The kinesthetic and vestibular senses play an essential role in a baseball pitcher's windup and delivery.
Flying Colours Ltd/Digital Vision/Getty Images

written down or communicated easily using language. They involve nearly every muscle and joint in the body. Most information about the kinesthetic senses is transmitted from the joints and muscles along the same pathways to the brain as information about touch.

The vestibular sense tells us whether our head (and hence usually our body) is tilted, moving, slowing down, or speeding up. It works in concert with the kinesthetic senses to coordinate our *proprioceptive feedback,* which is information about the position of our limbs and body parts in relation to other body parts. Consider the combination of sensory abilities involved in the motion of a hockey player skating down the ice, cradling the puck, and pushing it forward with the hockey stick. The hockey player is responding simultaneously to a multitude of sensations, including those produced by the slickness of the ice, the position of the puck, the speed and momentum of the forward progression, and the requirements of the play to turn and to track the other players on the ice.

The **semicircular canals** of the inner ear contain the sensory receptors that detect head motion caused when we tilt or move our head and/or body (Figure 32). These canals consist of three fluid-filled, circular tubes that lie in the three planes of the body—right-left, front-back, and up-down. We can picture these as three intersecting hula hoops. As you move your head, the fluid of the semicircular canals flows in different directions and at different speeds (depending on the force of the head movement). Our perception of head movement and position is determined by the movements of these receptor cells (Dakin & others, 2020; Hag & others, 2020). This ingenious system of using the motion of fluid in tubes to sense head position is similar to the auditory system of the inner ear. However, the fluid movement in the cochlea results from the pressure sound exerts on the oval window, whereas the movements in the semicircular canals reflect physical movements of the head and body. Vestibular sacs in the semicircular canals contain hair cells embedded in a gelatin-like mass. Just as the hair cells in the cochlea trigger hearing impulses in the brain, the hair cells in the semicircular canals transmit information about balance and movement.

The brain pathways for the vestibular sense show the interesting nature of this sense and the importance of structures used for hearing for our sense of balance. Impulses from the vestibular sense begin in the auditory nerve, which contains both the cochlear nerve (with information about sound) and the vestibular nerve (which has information about balance and movement). Most of the axons of the vestibular nerve connect with the medulla, although some go directly to the cerebellum. There also appear to be vestibular projections to the temporal cortex, but research has not fully charted their specific pathways.

You have had the same body your entire life. Our sense of our own body as constant provides the context of our experience of ourselves as a stable whole person. We certainly have top-down expectations for our body—that we have two hands, with five fingers on each hand. Yet, our perceptions of our body are surprisingly flexible. *Proprioceptive drift* refers to the perception that something other than our actual body "belongs" to us and is part of our body. Studies show that we can come to experience proprioceptive ownership of body parts that defy reality. For example, using virtual reality people can come to own and use extra arms, a different face, and even a tail (Steptoe & others, 2013). The illusion of bodily changes does not require sophisticated technology. When we cannot

semicircular canals
Three fluid-filled circular tubes in the inner ear containing the sensory receptors that detect head motion caused when an individual tilts or moves the head and/or the body.

FIGURE 32 **The Semicircular Canals and the Vestibular Sense** The semicircular canals provide feedback to the gymnast's brain as her head and body tilt in different directions. Any angle of head rotation is registered by hair cells in one or more semicircular canals in both ears. (*inset*) The semicircular canals.
(Gymnast): Dominique Douieb/PhotoAlto/Getty Images; (Inner ear) MedicalRF.com/Getty Images

Could You Be Fooled into Thinking You Have Six Fingers?

You have likely had five fingers on each hand your entire life. Yet, a simple procedure has been shown to lead to the illusion of a sixth finger on one hand. These studies use a device called a mirror box. A mirror box is a rectangular box with a mirror along one side (facing outwards). The box can accommodate a person's body part, such as a leg, arm, or hand. When one limb is placed alongside the box (and the other is inside the box, unseen) the reflection in the mirror looks like it is the opposing (hidden) limb. The ends of the box are open, so an experimenter can touch the limb inside it.

So, imagine yourself seated at a table in front of a mirror box. Your left hand is inside the box and your right hand is on the table next to the mirror. The reflection of your right hand would look like it is your left hand, sitting next to it. You are told to look at the reflection, not at your actual visible (right) hand.

To create the illusion of a sixth finger on your left hand, the experimenter would stroke the top of each finger of both hands from the knuckles to the tip, simultaneously, starting at the thumb and then moving along through to the ring finger. Each time, both corresponding fingers on each hand are stroked at the same time.

At the pinky, things change. The experimenter strokes the side of the pinky of the hand inside the box while stroking the top of the pinky of the visible hand. In the mirror, the reflection shows the pinky being stroked along the top. Then, as seen in the mirror, the experimenter strokes the *empty space* next to the pinky while actually stroking the top of the pinky of the hand inside the box (Cadete & Longo, 2020). This simple action can produce the sensation of having a sixth finger—an extra pinky! The illusion usually last only a few moments

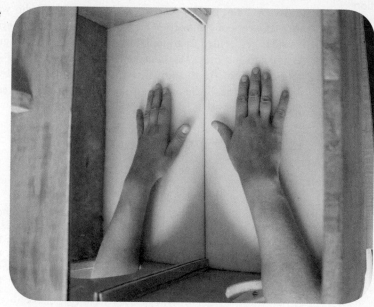

Media for Medical SARL/Alamy Stock Photo

but it can continue for longer with a slightly different procedure, stroking the fingers repeatedly, up and down and doing so 20 times. This procedure produces a sensation lasting several seconds that the empty space is actually part of the person's body—a sixth finger (Cadete & Longo, 2020).

These and other studies suggest that our perceptions of our body can defy even a lifetime of experience with our body. Interestingly, the illusion that body parts are getting bigger or longer is easier to induce than the illusion that they are shrinking (Byrne & Preston, 2019). Why the difference? One reason may be that we have seen our body (and its parts) grow throughout our lives—so it is not as surprising to perceive our body growing. As we have seen, the brain represents each of our appendages in specialized areas. Nevertheless, remarkably, the brain is able to accommodate even very strange information that suggests despite a lifetime of living with five fingers that we, suddenly, appear to have six.

What Do You Think?

- Visual illusions are adaptive in the natural world. Why might it be adaptive for the brain to be flexible in its perception of our own body?

- Might some people be more susceptible to proprioceptive illusions? Who and why?

see our hands, for example, we can come to perceive our fingers as stretching out longer than they really are (Byrne & Preston, 2019). We can also be led to perceive ourselves as having a sixth finger on one hand. To read about this fascinating work, see Challenge Your Thinking.

Information from the sense of vision supplements the combination of kinesthetic and vestibular senses. This principle causes a motorist to slam on the brakes in his tiny sports car when the big truck next to him starts to move forward. When everything in our visual field appears to be moving, it is generally because we are moving.

Do It!

If you have a few minutes and a strong stomach, give your vestibular system a workout. Spin around quickly and repeatedly for 1 minute. You can do it in a swivel chair or standing in the center of a room (be careful of sharp edges). When you stop, you will feel dizzy. Here's what's happening. The fluid in the semicircular canals moves rather slowly and changes direction very slowly. When we spin for a while, the fluid eventually catches up with our rate of motion and starts moving in the same direction. When we stop moving, however, the slow-moving fluid keeps on moving. It tells the hair cells in the vestibular canals (which in turn tell the brain) "We are still spinning"—and we feel as if we are.

Throughout this chapter, we have viewed sensation and perception as our connections to the world. Sensation builds a bridge between the objects in our environment and the creative interpreter that is our brain. Through perception, our sensations become meaningful and our mental life engages with the environment. Sensation and perception allow us to survive in that environment but also to experience the world in all its vibrancy. Sue Barry, who achieved the ability to perceive depth only after a long, arduous effort, described her encounter with nature on a snowy day. "I felt myself within the snow fall, among the snow-flakes . . . . I was overcome with a sense of joy. A snow fall can be quite beautiful—especially when you see it for the first time" (quoted in Sacks, 2006, p. 73). Also, recall Michael May, who was able to see after 25 years of blindness. One night, with his seeing-eye dog Josh at his side, he decided to go look at the sky. Lying on the grass in a field, he opened his eyes. He thought he was "seeing stars"—in the metaphorical sense. He thought that the thousands of white lights in the sky could not really be real, but they were. As he remarked in his vision diary: "How sweet it is" (May, 2003; Stein, 2003).

self-quiz

1. Taste buds are bunched together in
   A. taste cells.
   B. the papillae.
   C. salivary glands.
   D. the olfactory epithelium.

2. The _____ is/are involved in the sense of smell.
   A. papillae
   B. olfactory epithelium
   C. thalamus
   D. pinnae

3. The inner-ear structures that contain the sensory receptors that detect head motion, as when we move our head and/or body, are the
   A. stirrups.
   B. semicircular canals.
   C. hammer.
   D. cochlea.

APPLY IT! 4. Sean loves anchovy, mushroom, and double-cheese pizza on a whole-wheat crust from his hometown pizzeria. He brings a pie back from home to give his roommate Danny a chance to taste it. Sean is stunned by Danny's reaction to the pizza: "Dude! Epic fail!" (in other words, he hates it). What does this example demonstrate?
   A. Danny may not have the taste receptors for umami.
   B. Although Sean and Danny have similar tongue anatomy, perception is still a subjective process. Sean is apparently a big umami fan, but Danny is not.
   C. Danny may have a disorder of the olfactory epithelium.
   D. Danny is engaged in top-down processing.

SUMMARY

1 How We Sense and Perceive the World

Sensation is the process of receiving stimulus energies from the environment. Perception is the process of organizing and interpreting sensory information to give it meaning. Perceiving the world involves both bottom-up and top-down processing. All sensation begins with sensory receptors, specialized cells that detect and transmit information about a stimulus to sensory neurons and the brain. Sensory receptors are selective and have different neural pathways.

Psychologists have explored the limits of our abilities to detect stimuli. Absolute threshold refers to the minimum amount of energy that people can detect. The difference threshold, or just noticeable difference, is the smallest difference in stimulation required to discriminate one stimulus from another 50 percent of the time.

Perception is influenced by attention, beliefs, and expectations. Sensory adaptation is a change in the responsiveness of the sensory system based on the average level of surrounding stimulation, essentially the ways that our senses start to ignore a particular stimulus once it is around long enough.

2 The Visual System

Light is the stimulus that is sensed by the visual system. Light can be described in terms of wavelengths. Three characteristics of light waves determine our experience: wavelength (hue), amplitude (brightness), and purity (saturation).

In sensation, light passes through the cornea and lens to the retina, the light-sensitive surface in the back of the eye that houses light receptors called rods (which function in low illumination) and cones (which react to color). The fovea of the retina contains only cones and sharpens detail in an image. The optic nerve transmits neural impulses to the brain. There it diverges at the optic chiasm, so that what we see in the left visual field is registered in the right side of the brain and vice versa. In the occipital lobes of the cerebral cortex, the information is integrated.

The trichromatic theory of color perception holds that three types of color receptors in the retina allow us to perceive three colors (green, red, and blue). The opponent-process theory states that cells in the visual system respond to red-green and blue-yellow colors. Both theories are probably correct—the eye and the brain use both methods to code colors.

Shape perception is the ability to distinguish objects from their background. Depth perception is the ability to perceive objects three dimensionally and depends on binocular (two eyes) cues and monocular (one eye) cues. Motion perception by humans depends on specialized neurons, feedback from the body, and environmental cues. Perceptual constancy is the recognition that objects are stable despite changes in the way we see them.

 ## The Auditory System

Sounds, or sound waves, are vibrations in the air that are processed by the auditory system. These waves vary in important ways that influence what we hear. Pitch (how high or low in tone a sound is) is the perceptual interpretation of wavelength frequency. Amplitude of wavelengths, measured in decibels, is perceived as loudness. Complex sounds involve a blending of frequencies. Timbre is the tone saturation, or perceptual quality, of a sound.

The outer ear consists of the pinna and external auditory canal and acts to funnel sound to the middle ear. In the middle ear, the eardrum, hammer, anvil, and stirrup vibrate in response to sound and transfer the vibrations to the inner ear. Important parts of the fluid-filled inner ear are the oval window, cochlea, and basilar membrane. The movement of hair cells between the basilar membrane and the tectorial membrane generates nerve impulses.

Place theory states that each frequency produces vibrations at a particular spot on the basilar membrane. Place theory adequately explains high-frequency sounds but not low-frequency sounds. Frequency theory holds that the perception of a sound's frequency depends on how often the auditory nerve fires. The volley principle states that a cluster of neurons can fire impulses in rapid succession, producing a volley of impulses.

Information about sound moves from the hair cells to the auditory nerve, which carries information to the brain's auditory areas. The cortical destination of most fibers is the temporal lobes of the cerebral cortex. Localizing sound involves both the timing of the sound and the intensity of the sound arriving at each ear.

 ## Other Senses

The skin senses include touch, temperature, and pain. Touch is the detection of mechanical energy, or pressure, against the skin. Touch information travels through the spinal cord, brain stem, and thalamus and on to the somatosensory areas of the parietal lobes. Thermoreceptors under the skin respond to increases and decreases in temperature. Pain is the sensation that warns us about damage to our bodies.

The chemical senses of taste and smell enable us to detect and process chemicals in the environment. Papillae are bumps on the tongue that contain taste buds, the receptors for taste. The olfactory epithelium contains a sheet of receptor cells for smell in the roof of the nose.

The kinesthetic senses provide information about movement, posture, and orientation. The vestibular sense gives us information about balance and movement. Receptors for the kinesthetic senses are embedded in muscle fibers and joints. The semicircular canals in the inner ear contain the sensory receptors that detect head motion.

KEY TERMS

absolute threshold	frequency theory	papillae	sensory receptors
apparent movement	gestalt psychology	parallel processing	signal detection theory
auditory nerve	inner ear	perception	subliminal perception
binding	kinesthetic senses	perceptual constancy	thermoreceptors
binocular cues	middle ear	perceptual set	top-down processing
bottom-up processing	monocular cues	place theory	trichromatic theory
cones	noise	retina	vestibular sense
convergence	olfactory epithelium	rods	visual cortex
depth perception	opponent-process theory	selective attention	volley principle
difference threshold	optic nerve	semicircular canals	Weber's law
feature detectors	outer ear	sensation	
figure-ground relationship	pain	sensory adaptation	

ANSWERS TO SELF-QUIZZES

Section 1:   1. C; 2. B; 3. A; 4. D
Section 2:   1. D; 2. A; 3. A; 4. B

Section 3:   1. A; 2. C; 3. B; 4. B

Section 4:   1. B; 2. B; 3. B; 4. B

Design elements: (Preview icon): Jiang Hongyan/Shutterstock; (Marginal notes): Shutterstock/Vadarshop

4 States of Consciousness

A Star in His Own Imagination

P aul Allen was a retired project manager and fervent amateur actor and singer who appeared in community the-aters productions when, at the age of 56, he suffered a stroke that left him with *locked-in syndrome*. A person with locked-in syndrome is immobile but has typical cognitive and emotional function. Paul cannot move or speak. He can feel an itch on his face but cannot reach up to scratch it. He is able to blink his left eye and that limited movement is how he communicates with his wife and caregivers. It is also how he wrote a book, *A Star in His Own Imagination* (*London News*, 2020). In his book, Paul describes his experiences and the things he misses most. He also expresses his sense of wonder at those who have stayed by his side: "It has shown me how hard all the wonderful people all work for me" (*London News*, 2020). To care for Paul, to love and support him, takes a great deal of effort. But his wife and caregivers take on the extraordinary challenge because, despite the fact that he cannot do even the smallest things, they know that inside his immobile body he is still Paul.

You might be surprised to learn that patients with locked-in syndrome feel like "themselves" and report high quality of life (Iyer & others, 2018; Vidal, 2018). To some, living as a mind literally locked inside an immobile body might sound like a nightmare. Yet patients with locked-in syndrome and those dedicated to helping them connect with the world demonstrate a simple truth: A conscious mind, even if it is locked in a body that cannot move, is still a person very much worth reaching. Such is the power of consciousness.

Portra/Getty Images

1 The Nature of Consciousness

Consciousness is a crucial part of human experience. Our conscious awareness represents that private inner mind where we think, feel, plan, wish, pray, imagine, and quietly relive experiences. Consider that if we did not have private thoughts and feelings, we could not tell a lie.

In the late nineteenth and early twentieth centuries, psychology pioneer William James (1950) described the mind as a **stream of consciousness,** a continuous flow of changing sensations, images, thoughts, and feelings. The content of our awareness changes from moment to moment. Information moves rapidly in and out of consciousness. Our minds can race from one topic to the next—from the person approaching us to our physical state today to the café where we will have lunch to our strategy for the test tomorrow.

William James thought of just about everything psychologists currently study. Who knows what he might have accomplished with the neuroscience tools we have today?

stream of consciousness
Term used by William James to describe the mind as a continuous flow of changing sensations, images, thoughts, and feelings.

In his description of the stream of consciousness, James included aspects of our awareness that he described as on the "fringe" of the stream of consciousness. This fringe includes all of the thoughts and feelings that we have *about* our thoughts. We are aware not only of those things that take center stage in our mental life, those shiny fish in the stream of consciousness, but also of all the thoughts and feelings that surround those fish.

Today, psychologists use the term *metacognition* to describe the processes by which we think about thinking (Norman & others, 2019). This term includes our awareness of the fringe elements of the conscious stream. When we read a text, for instance, the difficulty or ease with which we comprehend what is written can influence how we feel about what we read. When written text is easy to read, we are more likely to think that what we are reading is true and accurate (Brashier & Marsh, 2020).

The metacognitive experience of ease can impact our thought processes in surprising ways. Consider the following two items, taken from a questionnaire that measures the experience of meaning in life (Steger & others, 2006). The top one is printed in a difficult-to-read font, the bottom one in a clear, easy-to-read font:

I HAvE found *a really significant meaning in* my life.

I have found a really significant meaning in my life.

One study found that participants rated their meaning in life to be lower when the scale used the difficult-to-read font (Trent & others, 2013). Other research has shown that when young people experienced metacognitive difficulty in thinking about their life goals, they were less likely to believe they could reach those goals (Fisher & Oyserman, 2017; Walter & others, 2020). The logic behind such results is that while thinking about their life goals, the person might reason, "If it is this hard for me to even imagine myself pursuing these goals, they must not be very possible."

During much of the twentieth century, psychologists focused less on the study of mental processes and more on the study of observable behavior. More recently, the study of consciousness has regained widespread respectability in psychology. Indeed, scientists from many different fields are interested in consciousness.

Defining Consciousness

We can define consciousness in terms of its two parts: awareness and arousal. **Consciousness** is a person's awareness of external events and internal sensations under a condition of arousal.

Awareness includes awareness of the self and thoughts about one's experiences. Consider that on an autumn afternoon, when you see a beautiful tree, vibrant with color, you are not simply perceiving the colors; you are also *aware* that you are seeing them. Having awareness means that our sensory experiences are more than the sum of their parts (Adams & Browning, 2020; Dennett, 2020).

Arousal, the second part of consciousness, is the physiological state of being engaged with the environment. Thus, a sleeping person is not conscious in the same way that they would be while awake.

Consciousness and the Brain

There has been a dramatic increase in theoretical and research interest in determining more specifically how the brain functions to produce consciousness (Thomas, 2019). The two aspects of consciousness—awareness and arousal—are associated with different parts of the brain.

*Notice the word **associated**. Keep in mind that studies showing links between the brain and neural activity are generally correlational studies.*

The awareness aspect of consciousness refers to a feeling that we have about our experiences. This feeling is associated with many parts of the brain working in synchrony. It appears that the subjective feeling of awareness occurs in what has been termed a *global brain workspace* that involves various brain areas working in parallel (Mashour & others, 2020; Panagiotaropoulos & others, 2020). These locations include the front-most part of the brain—the prefrontal cortex—as well as the *anterior cingulate* (an area associated with acts of will) and the *association areas* (Mashour & Hudetz, 2017). This wide-reaching brain workspace is an assembly of neurons that are thought to work in cooperation to produce the subjective sense of consciousness. These neurons work together when it makes sense to broadcast sensory information throughout the brain. Areas of the prefrontal cortex appear to be especially involved in the ways that awareness goes beyond the input of sensory information. For instance, these areas of the brain are active when we taste complex flavors and track the subjective pleasure that accompanies rewarding experiences (Londerée & Wagner, 2020). Scientists continue to probe the precise neural correlates of consciousness. A key challenge is accounting for that somewhat mysterious feeling we have of our minds and their workings (Black, 2020; Brown & LeDoux, 2020; Graziano & others, 2020).

Although awareness is linked to many different brain areas working together, arousal is more specific. In the brain, arousal is regulated by the **reticular activating system,** a network of structures including the brain stem, medulla, and thalamus. Even with a perfectly functioning frontal cortex, a person with damage to these brain areas can remain unconscious (Kwak & Chang, 2020). Arousal refers to the ways that awareness is regulated: If we are in danger, we might need to be on "high alert," but if we are in a safe environment with no immediate demands, we can relax, and our arousal may be low.

Theory of Mind

You might think of consciousness as the mind, that part of yourself that contains your private thoughts and feelings. It might seem obvious that other people have private thoughts and feelings as well, but the human ability to recognize the subjective experience of another is a true developmental accomplishment.

Creatas/Getty Images Plus/Getty Images

Developmental psychologists who study children's ideas about mental states use the phrase **theory of mind** to refer to individuals' understanding that they and others think, feel, perceive, and have private experiences (Goffin & others, 2020; Graziano & others, 2020; Richardson & Saxe, 2020).

Developmental psychologists have used a procedure called the *false belief task* to examine children's theory of mind (Quesque & Rossetti, 2020). In one version of the false belief task, the child is asked to consider the following situation (Wellman & Woolley, 1990). Anna is a little girl who has some chocolate that she decides to save for later. She puts it in a blue cupboard and goes outside to play. While Anna is gone, her mother moves the chocolate to the red cupboard. When Anna comes back in, where will she look for her chocolate? Three-year-olds give the wrong answer—they assume that Anna will look in the red cupboard because they know (even though Anna does not) that Anna's mom moved the chocolate to the red one. Four-year-olds answer correctly—they recognize that Anna does not know everything they do and that she will believe the chocolate is where she left it (Wellman & Woolley, 1990). Success at the false belief task is associated with enhanced cognitive development in young children (Osterhaus & others, 2020), academic skill development (Cavadel & Frye, 2017; Ebert, 2020), and social competence (Lecce & others, 2017). Children who perform the false belief task well are better liked by their peers, in part because they are more likely to behave kindly (Ball & others, 2017; Spenser & others, 2020).

Theory of mind is essential to valuable social capacities, such as empathy and sympathy (Spenser & others, 2020). One approach to **autism spectrum disorder,** a neurodevelopmental disorder that involves difficulties in social interactions and communication, suggests that people on the autism spectrum have a deficit in theory of mind (Baron-Cohen, 2011), which would account for their social difficulties (Andreou & Skrimpa, 2020; Stewart & others, 2020). We will explore this and other explanations for autism spectrum disorder later.

The idea of theory of mind highlights the social nature of consciousness. We come to understand the idea of having private experiences not only by observing them inside ourselves but by observing other people (Graziano & others, 2020). It has even been suggested that one good reason we feel ourselves as possessing consciousness is that we will confer that same experience to others (Humphrey, 2008). If we know other people have a private experience that contains the kinds of rich thoughts and feelings we experience, we might be expected to treat people with kindness. Yet, we often do not. We often treat people who are different from us in ways that are profoundly unkind—with unfairness and bias. *Dehumanization* refers to a process by which we deprive other people of positive human qualities, such as having a mind. Dehumanization means robbing others of their humanity. Recent research examined how the presence of nonhumans who lack consciousness (robots) might actually help to remind us that we are all, after all, humans. To read about this work, see the Intersection.

Levels of Awareness

The flow of sensations, images, thoughts, and feelings that William James spoke of can occur at different levels of awareness. Although we might think of consciousness as either present or not, there are in fact shades of awareness, observed in comatose patients as well as in everyday life. Here we consider five levels of awareness: higher-level consciousness, lower-level consciousness, altered states of consciousness, subconscious awareness, and no awareness (Figure 1).

HIGHER-LEVEL CONSCIOUSNESS

In **controlled processes,** the most alert states of human consciousness, people actively focus their efforts toward a goal (Boag & others, 2019). For example, observe a classmate as they struggle to master the unfamiliar buttons on a new smartphone. They do not hear you humming or notice the intriguing

Consciousness and Social Psychology: Could Robots Make Us Kinder to Each Other?

The fact that we all possess consciousness could be the binding force that allows all humans to live peacefully together. Yet, too often we seem to forget that other people, no matter their differences from us, are, actually, people. Prejudice refers to unjustified negative attitudes about a person based on group membership. People are treated unfairly because of their race, religion, gender, or sexuality because of prejudice. How might we use our shared humanity to help people see that each of us warrants respect and fair treatment? One surprising way to do so may be to remind people of robots.

The increasing automation of the American workforce means that many people have become concerned that they might be replaced by robots. The very existence of robots highlights those things that make us human and may help people recognize that we are all members of one large group—humanity. Inspired by this possibility, researchers (Castelo & others, 2020) sought to test the idea that robots might remind us that humans are not that different from each other after all.

In one study, participants were assigned randomly to one of two conditions. In the experimental condition, they read a description of the increasing use of robots in the workplace ("Robots: Here to Stay?"). In the control condition, they read a description suggesting

Olena Yakobchuk/Shutterstock

robots are not on the rise ("Robots: Just a Fad?"). Then, participants completed measures of prejudice (negative attitudes about people who differ from themselves in terms of race, religion, immigrant status, etc.), the dependent variable. The results showed that participants who read about robots taking over the American workforce were less prejudiced (toward other humans) than those in the control condition (Jackson & others, 2020). In other studies, participants who were reminded of the existence of robots were fairer in their distribution of resources to people of different races than in a control condition in which robots were not mentioned.

What explains the differences? The researchers found that reminding people of robots, potentially threatening American workers, led participants to see people from many different walks of life as part of humanity (Jackson & others, 2020).

Think about it. Reminders of the existence of beings that differ from humanity can spur us to enlarge the group of people we care about so that it is not only our family, our friends, or people who are very much like us. Reminders of robots can help people recognize the common thread of humanity that unites us.

\\ **What other factors might help unite humanity?**

shadow on the wall. Their state of focused awareness illustrates the idea of controlled processes. Controlled processes require selective attention: the ability to concentrate on a specific aspect of experience while ignoring others (Parry & le Roux, 2019). A key aspect of controlled processing is executive function. **Executive function** refers to higher-order, complex cognitive processes, including thinking, planning, and problem solving. These cognitive processes are linked to the functioning of the brain's prefrontal cortex (Baum & others, 2020; Zelazo, 2020). Executive function is the person's capacity to harness consciousness, to focus in on specific thoughts while ignoring others. Because controlled processes demand attention and effort, they are slower than automatic processes. Often, after we have practiced an activity a great deal, we no longer have to think about it while doing it. It becomes automatic and faster.

executive function
Higher-order, complex cognitive processes, including thinking, planning, and problem solving.

LOWER-LEVEL CONSCIOUSNESS   Beneath the level of controlled processes are other levels of conscious awareness. Lower levels of awareness include automatic processes and daydreaming.

Automatic Processes   A few weeks after acquiring that smartphone, your classmate sends a text message in the middle of a conversation with you. They do not have to concentrate

Level of Awareness	Description	Examples
Higher-Level Consciousness	Involves controlled processing, in which individuals actively focus their efforts on attaining a goal; the most alert state of consciousness	Doing a math or science problem; preparing for a debate; taking an at-bat in a baseball game
Lower-Level Consciousness	Includes automatic processing that requires little attention, as well as daydreaming	Punching in a number on a cell phone; typing on a keyboard when one is an expert; gazing at a sunset
Altered States of Consciousness	Can be produced by drugs, trauma, fatigue, possibly hypnosis, and sensory deprivation	Feeling the effects of having taken alcohol or psychedelic drugs; undergoing hypnosis to quit smoking or lose weight
Subconscious Awareness	Can occur when people are awake, as well as when they are sleeping and dreaming	Sleeping and dreaming
No Awareness	Freud's belief that some unconscious thoughts are too laden with anxiety and other negative emotions for consciousness to admit them	Having unconscious thoughts; being knocked out by a blow or anesthetized

FIGURE 1

Levels of Awareness Each level of awareness has its time and place in human life.

on the keys and hardly seem aware of the device in their hand as they continue to talk to you while finishing lunch. Using that phone has reached the point of automatic processing.

Automatic processes are states of consciousness that require little attention and do not interfere with other ongoing activities. Automatic processes require less conscious effort than controlled processes (Greenwald & Lai, 2020). When we are awake, our automatic behaviors occur at a lower level of awareness than controlled processes, but they are still conscious behaviors. Your classmate pushed the right buttons, so at some level they apparently were aware of what they were doing. This kind of automatic behavior suggests that we can be aware of stimuli on some level without paying attention to them.

automatic processes
States of consciousness that require little attention and do not interfere with other ongoing activities.

Daydreaming   Another state of consciousness that involves a low level of conscious effort is *daydreaming,* which lies between active consciousness and dreaming while asleep. It is a little like dreaming while we are awake (Domhoff, 2017; Eeles & others, 2020). Daydreams usually begin spontaneously when we are doing something that requires less than our full attention.

Mind wandering is probably the most obvious type of daydreaming (Preiss & others, 2020). We regularly take brief side trips into our own private kingdoms of imagery and memory while reading, listening, or working. When we daydream, we drift into a world of fantasy. We perhaps imagine ourselves on a date, at a party, on television, in a faraway place, or at another time in our life. Sometimes our daydreams are about everyday events such as paying the rent, walking to class, and meeting with someone at school or work. Daydreams can seem like a waste of time but they can also spark creativity (Baer & others, 2020; Zedelius & others, 2020).

One of the most interesting things about mind wandering is that we often don't realize when it happens. We just fall into it. **Hey! Are you paying attention?**

In the brain, mind wandering is related to the activity of a broad set of regions in the brain, called the *default mode network.* This array of connections works together to process information that is not about receiving information from the outside world (Philippi & others, 2020).

Peter Cade/Photodisc/Getty Images

ALTERED STATES OF CONSCIOUSNESS *Altered states of consciousness* or *awareness* are mental states that are noticeably different from normal awareness. Altered states of consciousness can range from losing one's sense of self-consciousness to hallucinating. Such states can be produced by drugs, trauma, fever, fatigue, sensory deprivation, meditation, and possibly hypnosis. Drug use can also induce altered states of consciousness, as we will consider later.

SUBCONSCIOUS AWARENESS A great deal of brain activity occurs beneath the level of conscious awareness. Some psychologists are interested in the subconscious processing of information, which can take place while we are awake or asleep (Hurme & others, 2017; Railo & others, 2014).

Waking Subconscious Awareness

When we are awake, processes are going on just below the surface of our awareness. For example, while we are grappling with a problem, the solution may pop into our head. Such insights can occur when a subconscious connection between ideas is so strong that it rises into awareness.

Incubation refers to the subconscious processing that leads to a solution to a problem after a break from conscious thought about the problem (Shin & Grant, 2020). During incubation, we can be working on a problem without really knowing we are doing it. Clearly, during incubation, information is being processed even if we are unaware of that processing. Interestingly, successful incubation requires that we first expend effort thinking carefully about the problem beforehand (Pachai & others, 2016; Gonzalez-Vallejo & others, 2008).

Evidence that we are not always aware of our brain's processing of information comes from studies of people with certain neurological disorders. In one case, a woman who suffered neurological damage was unable to describe or report the shape or size of objects in her visual field, although she was capable of describing other physical perceptions that she had (Milner & Goodale, 1995). Nonetheless, when she reached for an object, she could accurately adjust the size of her grip to allow her to grasp the object. Thus, she did possess some subconscious knowledge of the size and shape of objects, even though she had no awareness of this knowledge.

Subconscious information processing can occur simultaneously in a distributed manner along many parallel tracks. (Recall the previous discussion of parallel processing of visual information.) For example, when you look at a dog running down the street, you are consciously aware of the event but not of the subconscious processing of the object's identity (a dog), its color (black), and its movement (fast). In contrast, conscious processing occurs in sequence and is slower than much subconscious processing. Note that the various levels of awareness often work together. You might rely on controlled processing when memorizing material for class, but later, the answers on a test just pop into your head as a result of automatic or subconscious processing.

Subconscious Awareness During Sleep and Dreams

When we sleep and dream, our level of awareness is lower than when we daydream, but sleep and dreams are not best regarded as the absence of consciousness (Domhoff, 2017). Rather, they are low levels of consciousness.

When people are asleep, they remain aware of external stimuli to some degree. In sleep laboratories, when people are clearly asleep (as determined by physiological monitoring devices), they are able to respond to faint tones by pressing a handheld button (Ogilvie & Wilkinson, 1988). In one study, the presentation of pure auditory tones to sleeping individuals activated auditory processing regions of the brain, whereas participants' names activated language areas, the amygdala, and the prefrontal cortex (Stickgold, 2001). We return to the topics of sleep and dreams in the next section.

NO AWARENESS The term *unconscious* generally applies to someone who has been knocked out by a blow or anesthetized, or who has fallen into a deep, prolonged unconscious state. However, Sigmund Freud (1917) used the term *unconscious* in a very different way: **Unconscious thought,** said Freud, is a reservoir of unacceptable wishes, feelings, and thoughts that are beyond conscious awareness. In other words, Freud's interpretation viewed the unconscious as a storehouse for vile thoughts and impulses. He believed that some aspects of our experience remain unconscious for good reason, as if we are better off not knowing about them. For example, from Freud's perspective, the human mind is full of disturbing impulses such as a desire to have sex with our parents.

Although Freud's interpretation remains controversial, psychologists now widely accept that unconscious processes do exist. Contemporary researchers acknowledge that many mental processes (thoughts, emotions, goals, and perceptions) can occur outside of awareness (di Giannantonio & others, 2020; King & Mendoza, 2020; Mikkelsen & others, 2020; Wang & others, 2020).

unconscious thought
According to Freud, a reservoir of unacceptable wishes, feelings, and thoughts that are beyond conscious awareness.

self-quiz

1. The term *metacognition* refers to
   A. the study of cognitive research.
   B. thinking about thought processes.
   C. cognition that is subliminal.
   D. stream of consciousness thinking.

2. All of the following are examples of automatic processes *except*
   A. a student solving a math problem.
   B. a runner jogging.
   C. a person talking to a friend.
   D. a taxi driver operating a car.

3. Sleep and dreams are most accurately viewed as
   A. the absence of consciousness.
   B. intermittent consciousness.
   C. low levels of consciousness.
   D. high levels of consciousness.

APPLY IT! 4. Xavier is a student in Intro Psych. He has attended every class and taken detailed notes. Before the first exam, he studies and takes practice tests. On the morning of the test, however, he feels suddenly blank. Xavier takes a deep breath and starts answering each question one by one, taking guesses on many questions. The next week, Xavier is thrilled to learn that he got a 100 percent on the exam. He later tells a friend, "I just took guesses and still got a perfect score. Next time I should not even bother studying!" Which of the following is true of Xavier's plan?
   A. It is a good idea, since a lot of psychology is just common sense.

   B. It is a bad idea. Xavier's grade shows that controlled and subconscious processes work together. Controlled processes helped him learn the material so that he was able to take good guesses.
   C. Xavier's plan is not a good idea, because his subconscious helped him on this test, but such processing cannot be counted on to work all the time.
   D. Xavier's plan is a good one. His experience demonstrates that controlled processes are generally not helpful in academic work.

2 Sleep and Dreams

sleep
A natural state of rest for the body and mind that involves the reversible loss of consciousness.

You already know that sleep involves a decrease in body movement and (typically) closed eyes. What is sleep, more precisely? We can define **sleep** as a natural state of rest for the body and mind that involves the reversible loss of consciousness. Sleep must be important, because it takes up about one-third of our lifetime, more than anything else we do. *Why* is sleep so crucial? Before tackling this question, let's first see how sleep is linked to our internal biological rhythms.

How, where, and with whom we sleep are all influenced by culture. In the United States, infants typically sleep alone, and parents are discouraged from sleeping with their infants. In other countries, such cosleeping is an accepted practice. Some cultures encourage a family bed where everyone sleeps together.

Biological Rhythms and Sleep

biological rhythms
Periodic physiological fluctuations in the body, such as the rise and fall of hormones and accelerated/decelerated cycles of brain activity, that can influence behavior.

Biological rhythms are periodic physiological fluctuations in the body. We are unaware of most biological rhythms, such as the rise and fall of hormones and accelerated and decelerated cycles of brain activity, but they can influence our behavior. These rhythms are controlled by biological clocks, which include annual or seasonal cycles such as the migration of birds and the hibernation of bears, and 24-hour cycles such as the sleep/wake cycle and temperature changes in the human body. Let's further explore the body's 24-hour cycles.

Cerebral cortex

Hypothalamus

Suprachiasmatic
nucleus (SCN)

Reticular
formation

FIGURE 2
**Suprachiasmatic
Nucleus** The
suprachiasmatic nucleus (SCN)
plays an important role in
keeping our biological clock
running on time. The SCN is
located in the hypothalamus.
It receives information from
the retina about light, which
is the external stimulus that
synchronizes the SCN. Output
from the SCN is distributed to
the rest of the hypothalamus
and to the reticular formation.
(Photo): Image Source/Alamy Stock
Photo

CIRCADIAN RHYTHMS

Circadian rhythms are daily behavioral or physiological cycles. Daily circadian rhythms involve the sleep/wake cycle, body temperature, blood pressure, and blood sugar level (Qian & others, 2020; Sorensen & others, 2020). For example, body temperature fluctuates about 3 degrees Fahrenheit in a 24-hour day, peaking in the afternoon and reaching its lowest point between 2 A.M. and 5 A.M.

The body monitors the change from day to night by means of the **suprachiasmatic nucleus (SCN),** a small brain structure that uses input from the retina to synchronize its own rhythm with the daily cycle of light and dark (Harvey & others, 2020; Reyes-Mendez & others, 2020). Output from the SCN allows the hypothalamus to regulate daily rhythms such as temperature and hunger and the reticular formation to regulate daily rhythms of sleep and wakefulness (Figure 2). Although a number of biological clocks seem to be involved in regulating circadian rhythms, the SCN is the most important (Hastings & others, 2020).

Many people who are totally blind experience lifelong sleeping problems because their retinas cannot detect light. These people have a kind of permanent jet lag and periodic insomnia because their circadian rhythms often do not follow a 24-hour cycle (Emens, 2020). It is also possible for sighted individuals to experience this disorder.

circadian rhythms
Daily behavioral
or physiological
cycles that involve
the sleep/wake
cycle, body tem-
perature, blood
pressure, and
blood sugar level.

**suprachiasmatic
nucleus (SCN)**
A small brain struc-
ture that uses input
from the retina to
synchronize its
own rhythm with
the daily cycle of
light and dark; the
body's way of
monitoring the
change from day
to night.

Circadian means conforming to a 24-hour cycle. It means DAILY.

DESYNCHRONIZING THE BIOLOGICAL CLOCK

Biological clocks can become *desynchronized* or thrown off their regular schedules. Among the circumstances of life that can introduce irregularities into our sleep are jet travel, changing work shifts, and *insomnia* (the inability to sleep). What effects might such irregularities have on circadian rhythms?

If you fly from Los Angeles to New York and then go to bed at 11 P.M. eastern time, you may have trouble falling asleep because your body is still on west coast time. Even if you sleep for 8 hours that night, you may have a hard time waking up at 7 A.M. eastern time, because your body thinks it is 4 A.M. If you stay in New York for several days, your body will adjust to this new schedule.

The jet lag you experience when you fly from Los Angeles to New York occurs because your body time is out of phase, or synchronization, with clock time (van Rensburg & others, 2020). Jet lag is the result of two or more body rhythms being out of sync. You usually go to bed when your body temperature begins to drop, but in your new location, you might be trying to go to sleep when it is rising. In the morning, your adrenal glands release large doses of the hormone cortisol to help you wake up. In your new geographic time zone, the glands may be releasing this chemical just as you are getting ready for bed at night.

Changing to a night-shift job can desynchronize our biological clocks and affect our circadian rhythms and performance.
FG Trade/Getty Images

Circadian rhythms may also become desynchronized when shift workers change their work hours (Blum & Zeitzer, 2020). Shift-work problems most often affect night-shift workers who never fully adjust to sleeping in the daytime after work. Such workers may fall asleep at work and are at risk for cardiovascular disease (Kervezee & others, 2020), pregnancy complications (Rada & others, 2020), weight gain and diabetes (Gao & others, 2020), as well as depression, anxiety, and sleep disturbance (Brown & others, 2020).

RESETTING THE BIOLOGICAL CLOCK If your biological clock for sleeping and waking becomes desynchronized, how can you reset it? With regard to jet lag, if you take a transoceanic flight and arrive at your destination during the day, it is a good idea to spend as much time outside in the daylight as possible. Bright light during the day, especially in the morning, increases wakefulness, whereas bright light at night delays sleep. A recent U.S. national study showed that adolescents growing up in areas with high levels of artificial light (as measured by satellites) tended to go to bed later while those with the lowest artificial light slept for the most hours on weeknights (Parksarian & others, 2020). High levels of artificial light were also linked with anxiety and depression.

Why would melatonin be particularly helpful for eastward but not westward travel?

Researchers are studying melatonin, a hormone that increases at night in humans, for its possible effects in reducing jet lag and other sleep problems (Mazur & others, 2020; Wei & others, 2020). Studies have shown that a small dosage of melatonin can reduce jet lag by advancing the circadian clock—an effect that makes it useful for eastward but not westward jet lag (Pena-Orbea & others, 2020).

Why Do We Need Sleep?

All animals require sleep. Furthermore, the human body regulates sleep, as it does eating and drinking, and this fact suggests that sleep may be just as essential for survival. Yet why we need sleep remains a bit of a mystery.

THEORIES ON THE NEED FOR SLEEP A variety of theories have been proposed for the need for sleep (Freiberg, 2020; Nicholas & others, 2017). First, from an evolutionary perspective, sleep may have developed because animals needed to protect themselves at night. The idea is that it makes sense for animals to be inactive when it is dark, because nocturnal inactivity helps them to avoid both becoming other animals' prey and injuring themselves due to poor visibility.

A second possibility is that sleep is a way to conserve energy. Spending a large chunk of any day sleeping allows animals to conserve their calories, especially when food is scarce (Siegel, 2005). For some animals, moreover, the search for food and water is easier and safer when the sun is up. When it is dark, it is adaptive for these animals to save their energy. Animals that are likely to serve as someone else's food sleep the least of all. Figure 3 illustrates the average amount of sleep per day of various animals.

A third explanation for the need for sleep is that sleep is restorative (Fattinger & others, 2017). Scientists have proposed that sleep restores, replenishes, and rebuilds the brain and body, which the day's waking activities can wear out. This idea fits with the feeling of being tired before we go to sleep and restored when we wake up. In support of the theory of a restorative function of sleep, many of the body's cells show increased production and reduced breakdown of proteins during deep sleep (Picchioni & others, 2014). Protein molecules are the building blocks needed for cell growth and for repair of damage from factors

Hours of Sleep per 24-Hour Period

Bat	19.9
Armadillo	18.5
Cat	14.5
Fox	9.8
Rhesus monkey	9.6
Rabbit	8.4
Human	8.0
Cow	3.9
Sheep	3.8
Horse	2.9

FIGURE 3

From Bats to Horses: The Wide Range of Sleep in Animals We might feel envious of bats, which sleep nearly 20 hours a day, and more than a little in awe of horses, still running on just under 3 hours of rest.

(Bat): Ewen Charlton/Getty Images; (Horse): Fuse/Getty Images such as stress. Short sleep duration is associated with stress and lowered immune function (Eissa & others, 2020). In a recent study, healthy young adults kept a sleep diary 3 days before and then 10 days after receiving a flu vaccine. The results showed that shorter sleep duration predicted a less effective response to the vaccine (Prather & others, 2020).

A final explanation for the need for sleep centers on the role of sleep in brain plasticity (Dissel, 2020; Sun & others, 2020). Recall that the plasticity of the brain refers to its capacity to change in response to experience. Sleep has been recognized as playing an important role in the ways that experiences influence the brain. For example, sleep following experiences enhances synaptic connections between neurons (Sun & others, 2020). The lack of sleep leads to weaker connections in the brain (Poh & Chee, 2017), so the brain possesses a less clear representation of new learning. Findings such as these suggest an important role for sleep in the consolidation of memories (Batterink & Paller, 2017; Friedrich & others, 2020). Sleep is vital to the consolidation of memory, whether memory for specific information, for skills, or for emotional experiences (Latchoumane & others, 2017; Muehlroth & others, 2020). Lost sleep often results in lost memories. As we will see, during sleep the hippocampus is busy and memories are consolidated throughout a night's rest.

So, if you are thinking about studying all night for your next test, you might want to think again. Sleep can enhance your memory. In one study, participants who had studied word lists the day before performed better in a recall test for those words if they had a good night's sleep before the test (Racsmány & others, 2010). Participants who were tested prior to sleep did not perform as well. Not only can sleep improve your memory, but just losing a few hours of sleep can have negative effects on many cognitive processes, including attention, language, reasoning, and decision making.

THE EFFECTS OF CHRONIC SLEEP DEPRIVATION Experts recommend that adults get 7 to 9 hours of sleep a night (National Sleep Foundation, 2020a). Lack of sleep is stressful and has an impact on the body and the brain. When deprived of sleep, people have trouble paying attention to tasks and solving problems (Hudson & others, 2020). Sleep deprivation decreases brain activity in the thalamus, the prefrontal cortex, and the brain's reward centers (Javaheipour & others, 2019; Krause & others, 2017; Nechifor & others, 2020) and leads to reduced complexity of brain activity (Grant & others, 2018). The tired brain must compensate by using different pathways or alternative neural networks when thinking (Koenis & others, 2011; Peng & others, 2020).

Although sleep is unquestionably key to optimal physical and mental performance, we have only to look around us (or into a mirror) to notice that many people do not get sufficient sleep. In a national survey of more than 1,500 U.S. adults conducted by the National Sleep Foundation (2011), 43 percent of 19- to 64-year-olds reported that they rarely or ever get a good night's sleep on weeknights. Sixty percent said that they experience a sleep problem every night or almost every night, such as waking during the night, waking too early, or feeling tired when they wake up in the morning. A majority stated that they get slightly less than 7 hours of sleep a night on weeknights, and 15 percent indicated that they sleep less than 6 hours a night.

Why do Americans get too little sleep? Pressures at work and school, family responsibilities, and social obligations often lead to long hours of wakefulness and irregular sleep/wake schedules. Not having enough hours to do all that we want or need to do in a day, we cheat on our sleep. As a result we may suffer from a "sleep debt," an accumulated level of exhaustion.

For many people, the COVID-19 pandemic presented sleep problems. Worries about the virus and changes in schedules and routines interfered with sleep (Altena & others, 2020; Fu &

others, 2020; Kokou-Kpolou & others, 2020). Good sleep practices are important at all times but especially in the context of a pandemic. As we have seen, sleep promotes immune function and psychological resources. It is not easy to cope with serious difficulties while sleep deprived. Some steps that can improve life during a pandemic include maintaining regular hours, avoiding taking too many naps while stuck at home, spending time outside in the sunshine, and reserving one's bed for sleeping (National Sleep Foundation, 2020b).

Stages of Wakefulness and Sleep

Have you ever been awakened from your sleep and been totally disoriented? Have you ever awakened in the middle of a dream and suddenly gone right back into the dream as if it were a movie running just under the surface of your consciousness? These two circumstances reflect two distinct stages in the sleep cycle.

Stages of sleep correspond to massive electrophysiological changes that occur throughout the brain as the fast, irregular, and low-amplitude electrical activity of wakefulness is replaced by the slow, regular, high-amplitude waves of deep sleep. Using the electroencephalography (EEG) to monitor the brain's electrical activity, scientists have identified four stages of sleep.

The following stages (of wakefulness and sleep) are defined by both the brain's activity and muscle tone. The stages are named by letters and numbers that represent what is going on at that stage, including whether the person is awake or asleep and experiencing rapid eye movement (REM). Non-REM sleep is characterized by a lack of rapid eye movement and little dreaming, as described next.

STAGE W   The "W" here stands for "wake." During this stage, when people are awake, their EEG patterns exhibit two types of waves: beta and alpha. *Beta waves* reflect concentration and alertness. These waves are the highest in frequency and lowest in amplitude. This means that the waves go up and down a great deal, but they do not have very high peaks or very low ebbs. They also are more *desynchronous* than other waves, meaning they do not form a very consistent pattern. Inconsistent patterning makes sense given the extensive variation in sensory input and activities we experience when we are awake.

When we are relaxed but still awake, our brain waves slow down, increase in amplitude, and become more *synchronous,* or regular. These waves, associated with relaxation or drowsiness, are called *alpha waves.*

STAGE N1 (NON-REM1) SLEEP   When people are just falling asleep, they enter the first stage of non-REM sleep. The "N" stands for "non-REM" meaning that rapid eye movements do not occur during these stages. Stage N1 is characterized by drowsy sleep. In this stage, the person may experience sudden muscle movements called *myoclonic jerks*. If you watch someone in your class fighting to stay awake, you might notice their head jerking upward. This reaction demonstrates that this first stage of sleep often involves the feeling of falling. EEGs of people in stage N1 sleep are characterized by *theta* waves, which are even slower in frequency and greater in amplitude than alpha waves. The difference between being relaxed and being in stage N1 sleep is gradual. Figure 4 shows the EEG pattern of stage N1 sleep, along with the EEG patterns for the other stages of wakefulness and sleep.

Watch people in your classes fight to stay awake-you'll see their heads jerk up. This first stage of sleep often involves the feeling of falling.

STAGE N2 (NON-REM2) SLEEP   In stage N2 sleep, muscle activity decreases, and the person is no longer consciously aware of the environment. Theta waves continue but are interspersed with a defining characteristic of stage N2 sleep: sleep spindles. Sleep spindles are brief high-frequency bursts of neurons firing simultaneously (Blume & others, 2017). Sleep spindles are a definitive characteristic of stage N2 sleep. Sleep spindles are

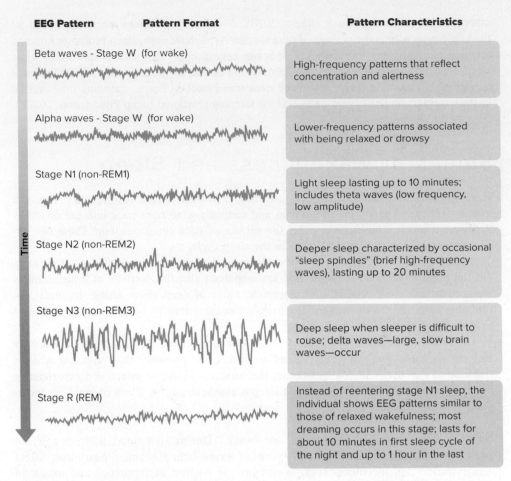

EEG Pattern	Pattern Format	Pattern Characteristics
Beta waves - Stage W (for wake)		High-frequency patterns that reflect concentration and alertness
Alpha waves - Stage W (for wake)		Lower-frequency patterns associated with being relaxed or drowsy
Stage N1 (non-REM1)		Light sleep lasting up to 10 minutes; includes theta waves (low frequency, low amplitude)
Stage N2 (non-REM2)		Deeper sleep characterized by occasional "sleep spindles" (brief high-frequency waves), lasting up to 20 minutes
Stage N3 (non-REM3)		Deep sleep when sleeper is difficult to rouse; delta waves—large, slow brain waves—occur
Stage R (REM)		Instead of reentering stage N1 sleep, the individual shows EEG patterns similar to those of relaxed wakefulness; most dreaming occurs in this stage; lasts for about 10 minutes in first sleep cycle of the night and up to 1 hour in the last

FIGURE 4 **Characteristics and Formats of EEG Recordings During Stages of Wakefulness and Sleep** Even while you are sleeping, your brain is busy. No wonder you sometimes wake up feeling tired.

important to memory consolidation and play a role in the communication between the hippocampus and the neocortex (Friedrich & others, 2020). Stages N1 and N2 are both relatively light stages of sleep, and if people awaken during one of these stages, they often report not having been asleep at all.

STAGE N3 (NON-REM3) SLEEP   Stage N3 sleep is characterized by delta waves, the slowest and highest-amplitude brain waves during sleep. *Delta sleep* is our deepest sleep, the time when our brain waves are least like our brain waves while we are awake. Delta sleep is also called *slow-wave sleep*. This is also the stage when bedwetting (in children), sleepwalking, and sleep talking occur. When awakened during this stage, people usually are confused and disoriented.

STAGE R (REM) SLEEP   After going through stages N1 to N3, sleepers drift up through the sleep stages toward wakefulness. Instead of reentering stage N1, however, they enter stage R, a different form of sleep called REM (rapid eye movement) sleep. **REM sleep** is an active stage of sleep during which the most vivid dreaming occurs (Reinoso-Suárez & others, 2020). The EEG pattern for REM sleep shows fast waves similar to those of relaxed wakefulness, and the sleeper's eyeballs move up and down and from left to right (Figure 5). Interestingly, REM sleep is characterized by theta waves in the hippocampus, which are also present during waking activity (Lugaresi & others, 2020).

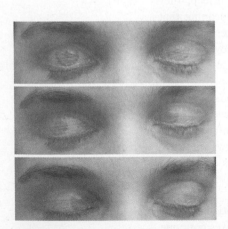

FIGURE 5   **REM Sleep** During REM sleep, your eyes move rapidly, as if following the images moving in your dreams.

Allan Hobson/Science Source

REM sleep
A stage of sleep characterized by rapid eye movement when most vivid dreams occur.

A person who is awakened during REM sleep is more likely to report having dreamed than when awakened at any other stage. Even people who claim they rarely dream frequently report dreaming when they are awakened during REM sleep. The longer the period of REM sleep, the more likely the person will report dreaming. Dreams also occur during slow-wave or non-REM sleep, but the frequency of dreams in these stages is relatively low and we are less likely to remember these dreams. Reports of dreaming by individuals awakened from REM sleep are typically longer, more vivid, more physically active, more emotionally charged, and less related to waking life than reports by those awakened from non-REM sleep.

We can't know for sure what they experience, but all mammals show REM sleep and might all be dreaming!

We have reviewed the important role of sleep in memory consolidation. Sleep spindles during stage N2 are involved in memory consolidation. REM sleep is also important in memory (Scullin & others, 2019). Neurons in the hippocampus, the brain structure deeply involved in memory, are activated during REM (Peyron & Rampon, 2020).

SLEEP CYCLING THROUGH THE NIGHT The five stages of sleep we have considered make up a normal cycle of sleep. As shown in Figure 6, one of these cycles lasts about 90 to 100 minutes and recurs several times during the night. The amount of deep sleep (stage N3) is much greater in the first half of a night's sleep than in the second half. Most stage R sleep takes place toward the end of a night's sleep, when the REM stage becomes progressively longer. The night's first REM stage might last for only 10 minutes, but the final REM stage might continue for as long as an hour. During a normal night of sleep, individuals will spend about 60 percent of sleep in light sleep (stages N1 and N2), 20 percent in delta or deep sleep (stage N3), and 20 percent in stage R (Webb, 2000). So, you can think of your night's sleep starting with deep sleep and ending with the big show of the night's REM.

SLEEP AND THE BRAIN The five sleep stages are associated with distinct patterns of neurotransmitter activity initiated in the reticular formation, the core of the brainstem. In all vertebrates, the reticular formation plays a crucial role in sleep and arousal (Jones, 2020) (see Figure 2). As previously noted, damage to the reticular formation can result in coma and death.

Three important neurotransmitters involved in sleep are serotonin, norepinephrine, and acetylcholine. As sleep begins, the levels of neurotransmitters sent to the forebrain from the reticular formation start dropping, and they continue to fall until they reach their lowest levels during the deepest sleep stage—stage N3. Stage R sleep is initiated by a rise in acetylcholine, which activates the cerebral cortex while the rest of the brain remains relatively inactive. REM sleep ends when there is a rise in serotonin and norepinephrine, which increases the level of forebrain activity nearly to the awakened state (El-Sheikh, 2013; Lushington & others, 2014; Miller & O'Callaghan, 2006). You are most likely to wake up just after an REM period. If you do not wake up then, the level of the neurotransmitters falls again, and you enter another sleep cycle.

In addition to the three neurotransmitters reviewed earlier, GABA, the brain's break pedal, also plays a role in sleep (Cissé & others, 2020). In particular, GABA receptors appear to help regulate the process of sleep, moving the person from one stage to the next.

Remember these three neurotransmitters? Serotonin plays a role in regulating our emotions; norepinephrine regulates arousal; and acetylcholine is involved in muscle movements. AND they help you sleep!

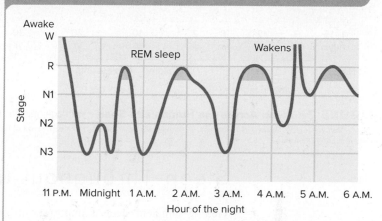

PSYCHOLOGICAL INQUIRY

FIGURE 6 Cycling Through a Night's Sleep During a night's sleep, we go through several cycles. Depth of sleep decreases, and REM sleep (shown in green) increases as the night progresses. In this graph, the person is depicted as awakening at about 5 A.M. and then going back to sleep for another hour.

> How many sleep cycles are presented? > How many times does the sleeper wake up? > Trace the rise and fall of the neurotransmitters acetylcholine, serotonin, and norepinephrine in the sleep cycle depicted. Has this sleeper achieved a good night's rest? Why or why not?

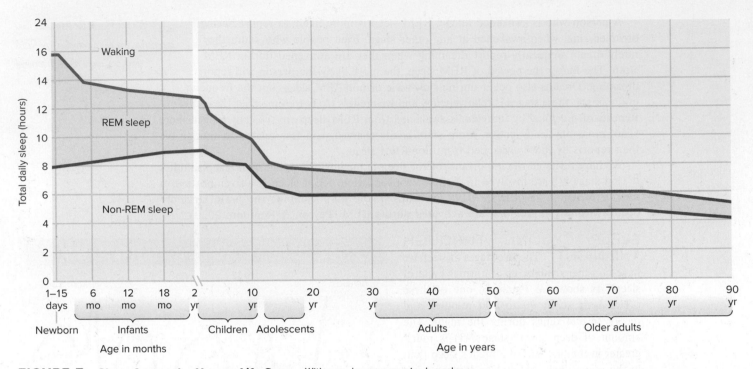

FIGURE 7 **Sleep Across the Human Life Span** With age, humans require less sleep.

SOURCE: Roff Warg, H. P., Muzio, J. N., & Dement, W. C. (1966, 29 April). Ontogenetic development of human dream-sleep cycle. *Science* 152, 604–619 (Figure 1, p. 608). http://www.sciencemag.org/content/152/3722/604.extract

Sleep Throughout the Life Span

Getting sufficient sleep is important at every period in human life. Figure 7 shows how total sleep time and time spent in each type of sleep vary over the human life span.

Sleep may benefit physical growth and brain development in infants and children. For example, deep sleep coincides with the release of growth hormone in children. Children are more likely to sleep well when they avoid caffeine, experience a regular bedtime routine, are read to before going to bed, and do not have a television in their bedroom.

As children age, their sleep patterns change. Many adolescents stay up later at night and sleep longer in the morning than they did when they were children, and these shifting sleep patterns may influence their academic work. During adolescence, the brain, especially the cerebral cortex, is continuing to develop, and the adolescent's need for sleep may be linked to this brain development (Shimizu & others, 2020).

Studies of adolescent sleep patterns show that, if given the opportunity, adolescents will sleep about 9.5 hours a night (Tarokh & Carskadon, 2010). Most, however, get considerably less than 9 hours of sleep, especially during the week. This shortfall creates a sleep debt that adolescents often attempt to make up on the weekend.

The reason for this adolescent sleep pattern appears to be a delay in the nightly release of the sleep-inducing hormone melatonin. Melatonin is secreted at about 9:30 P.M. in younger adolescents and approximately an hour later in older adolescents. Experts have begun to call for more attention to the unique sleep needs of adolescents (Carskadon, 2020; Carskadon & Barker, 2020). Delaying school start times appears to be a way to ensure that high school students get more and higher quality sleep (Cheng & Carroll, 2020; Widome & others, 2020). A number of communities have recognized that adolescents require more sleep and have instituted later start times for school, to good effect (Alfonsi & others, 2020). When students in middle school and high school start their days a bit later, health and grades benefit.

Do sleep patterns change in emerging adulthood (18–25 years of age)? Research indicates that they do (May & others, 2020). In one study, the weekday bedtimes and rise times of first-year college students were approximately 1 hour 15 minutes later than those of high school seniors (Lund & others, 2010). However, the first-year college students had later bedtimes and rise times than third- and fourth-year college students, indicating that at about 20 to 22 years of age, a reverse in the timing of bedtimes and rise times occurs.

Sleep patterns also change as people age through the middle-adult (40s and 50s) and late-adult (60s and older) years (Jonasdottir & others, 2020). Many adults in these age spans go to bed earlier at night and wake up earlier in the morning than they did in their younger years. As well, beginning in the 40s, people report that they are less likely to sleep through the entire night than when they were younger. Middle-aged adults also spend less time in deep sleep than they did before their middle years.

One study found that changes in sleep duration across five years in middle age were linked to cognitive abilities such as problem solving and memory (Ferrie & others, 2011). In this study, a decrease from 6, 7, or 8 hours of sleep and an increase from 7 or 8 hours were related to lower scores on most cognitive assessments. In late adulthood, approximately 50 percent of older adults complain of having difficulty sleeping (Neikrug & Ancoli-Israel, 2010). Better sleep patterns relate to lower risk of cardiovascular disease, even taking into account genetic risk for the disease (Fan & others, 2020).

Sleep and Disease

Sleep plays a role in a large number of health problems, diseases, and disorders. For example, stroke and asthma attacks are more common during the night and in the early morning, probably because of changes in hormones, heart rate, and other characteristics associated with sleep (Teodorescu & others, 2006). Sleeplessness is also associated with obesity, diabetes, psychological disorders, and heart disease (Liew & Aung, 2020; Tubbs & others, 2020).

Neurons that control sleep interact closely with the immune system (Imeri & Opp, 2009). As anyone who has had the flu knows, infectious diseases make us sleepy. The probable reason is that chemicals called cytokines, produced by the body's cells while we are fighting an infection, are powerfully sleep-inducing (Besedovsky & others, 2012; Davis & Raizen, 2017; Krueger, 2020). Sleep may help the body conserve energy and other resources it needs to overcome infection. Sleeplessness is also associated with less effective immune system functioning (Silva & others, 2020; Suzuki & others, 2017).

Brian Hagiwara/Getty Images

Sleep problems afflict most people who have psychological disorders, including those with depression (Tubbs & others, 2020). Individuals with depression often awaken in the early hours of the morning and cannot get back to sleep, and they often spend less time in delta wave or deep sleep than do individuals who are not depressed.

Sleep problems are common in many other disorders as well, including Alzheimer disease, stroke, cancer, among others. In some cases, however, these problems may be due not to the disease itself but to the drugs used to treat the disease.

Sleep Disorders

Many people suffer from undiagnosed and untreated sleep disorders that leave them struggling through the day, feeling unmotivated and exhausted. Some of the major sleep problems are insomnia, sleepwalking and sleep talking, nightmares and night terrors, narcolepsy, and sleep apnea.

INSOMNIA A common sleep problem is *insomnia,* the inability to sleep. Insomnia can involve a problem in falling asleep, waking up during the night, or waking up too early. Between three and five adults suffer from insomnia (Li & others, 2017). Insomnia is more common among women and older adults, as well as individuals who are thin, stressed, or depressed (Sateia & others, 2017).

For short-term insomnia, most physicians prescribe sleeping pills. However, most sleeping pills stop working after several weeks of nightly use, and their long-term use can interfere with good sleep.

In many cases, mild insomnia often can be reduced by simply practicing good sleep habits. Here are a few tips:

- Always go to bed at the same time, even on weekends.
- Sleep in a dark, quiet place.
- Use your bed only for sleeping, not other tasks.
- Take a hot bath before bed.

Behavioral changes (such as avoiding naps and setting an alarm in the morning) can help insomniacs increase their sleep time and awaken less frequently in the night. In addition, meditation, which we will discuss later in this chapter, may be reduce insomnia (Thimmapuram & others, 2020). Psychotherapy (particularly cognitive-behavioral theory, which we will review later) is a first-line treatment for chronic insomnia (Baglioni & others, 2020; Crönlein & others, 2020). A recent research review supported the effectiveness of melatonin but concluded that other treatments, including light therapy and exercise, show promise but await strong evidence (Baglioni & others, 2020).

SLEEPWALKING AND SLEEP TALKING *Somnambulism* is the formal term for sleepwalking, which occurs during the deepest stages of sleep (Gedam & others, 2018). Somnambulism takes place during stage N3, when a person is unlikely to be dreaming. So, it is not the case that people who are sleepwalking are acting out their dreams.

The specific causes of sleepwalking have not been identified, but it is more likely to occur when people are sleep deprived or have been drinking alcohol. Sleepwalking is also a side effect of some medications. There is nothing abnormal about sleepwalking, and despite superstition, it is safe to awaken sleepwalkers. In fact, they probably should be awakened, as they may harm themselves wandering around in the dark.

Another quirky night behavior is sleep talking, or *somniloquy.* If you interrogate sleep talkers, can you find out what they did, for instance, last Thursday night? Probably not. Although sleep talkers will converse with you and make fairly coherent statements, they are soundly asleep. Thus, even if a sleep talker mumbles a response to your question, do not count on its accuracy.

There is some evidence of an even rarer sleep behavior—sleep eating. Ambien is a widely prescribed sleep medication for insomnia. Some Ambien users began to notice odd things upon waking up from a much-needed good night's sleep, such as candy wrappers strewn around the room, crumbs in the bed, and food missing from the refrigerator. One woman gained 100 pounds without changing her awake eating or exercise habits. How could this be? Sleep eating is a noted side effect of this sleep aid (Wong & others, 2017).

The phenomenon of sleep eating illustrates that even when we feel fast asleep, we may be "half-awake"—and capable of putting together some unusual late-night snacks, including buttered cigarettes, salt sandwiches, and raw bacon. The maker of Ambien has noted this unusual side effect on the label of the drug. Even more alarming than sleep eating is sleep driving. Sleep experts agree that sleep driving while taking Ambien is rare and extreme but plausible.

For people who are battling persistent insomnia, a drug that provides a good night's rest may be worth the risk of these unusual side effects. Furthermore, no one should abruptly stop taking any medication without consulting a physician.

NIGHTMARES AND NIGHT TERRORS

A *nightmare* is a frightening dream that often awakens a dreamer from REM sleep. The nightmare's content invariably involves danger—the dreamer is chased, robbed, or thrown off a cliff. Nightmares are common. Most of us have had them, especially as young children. Nightmares peak at 3 to 6 years of age and then decline, although the average college student experiences four to eight nightmares a year (Hartmann, 1993). Reported increases in nightmares or worsening nightmares are often associated with an increase in life stressors. Perhaps not surprisingly, front-line healthcare workers reported poorer sleep quality and more nightmares during the COVID-19 pandemic (San Martin & others, 2020). People with a history of stress-related disorders also reported more nightmares during this time (Gupta, 2020).

A *night terror* features sudden arousal from sleep and intense fear. Night terrors are accompanied by a number of physiological reactions, such as rapid heart rate and breathing, loud screams, heavy perspiration, and movement (Hunter, 2020). Night terrors, which peak at 5 to 7 years of age, are less common than nightmares, and unlike nightmares, they occur during slow-wave, non-REM sleep.

NARCOLEPSY

The disorder *narcolepsy* involves the sudden, overpowering urge to sleep (Zee & others, 2020). The urge is so uncontrollable that the person may fall asleep while talking or standing up. People with narcolepsy immediately enter REM sleep rather than progressing through the typical sleep stages. People with narcolepsy are often very tired during the day. Narcolepsy can be triggered by extreme emotional reactions, such as surprise, laughter, excitement, and anger. The disorder appears to involve problems with the hypothalamus and amygdala (Adamantidis & others, 2020).

SLEEP APNEA

Sleep apnea is a sleep disorder in which individuals stop breathing because the windpipe fails to open or because brain processes involved in respiration fail to work properly. People with sleep apnea experience numerous brief awakenings during the night so that they can breathe better, although they usually are not aware of their awakened state. During the day, these people may feel sleepy because they were deprived of sleep at night. A common sign of sleep apnea is loud snoring punctuated by silence (the apnea).

Sleep apnea affects between 9 and 38 percent of the population and is more common in men than in women (Senaratna & others, 2017). The disorder is most common among infants and adults over the age of 65. Sleep apnea also occurs more frequently among people with obesity, men, and people with large necks and recessed chins. Untreated sleep apnea can cause high blood pressure and stroke (Bouloukaki & others, 2020) as well as heart attack (Ottaviani & Buja, 2020). In addition, the daytime sleepiness caused by sleep apnea can result in accidents, lost productivity, and relationship problems (Devita & others, 2017; Hartenbaum & others, 2006). Sleep apnea is commonly treated by weight-loss programs, side sleeping, propping the head on a pillow, or wearing a device (called a CPAP, for *continuous positive airway pressure*) that sends pressurized air through a mask that prevents the airway from collapsing.

Sleep apnea may also be a factor in *sudden infant death syndrome (SIDS),* the unexpected sleep-related death of an infant less than 1 year old (Ottaviani & Buja, 2020). SIDS is typically confirmed with an autopsy that reveals no specific cause of death (Chowdhury & others, 2017; Shapiro-Mendoza & others,

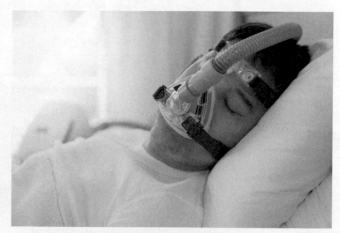

Continuous Positive Airway Pressure (CPAP) *Sleep apnea is often treated with the use of a CPAP.*
sbw18/Shutterstock

Do It!

Keep a sleep journal for several nights. Compare your sleep patterns with those described in the text. Do you have a sleep debt? If so, which stages of sleep are you most likely missing? Does a good night's sleep affect your behavior? Keep a record of your mood and energy levels after a short night's sleep and then after you have had at least 8 hours of sleep in one night. What changes do you notice, and how do they compare with the changes predicted by research on sleep deprivation described in the chapter?

2014). It is common for infants to have short pauses in their breathing during sleep, but for some infants frequent sleep apnea may be a sign of problems in regulating arousal (Rognum & others, 2014). There is evidence that infants who die of SIDS in fact experience multiple episodes of sleep apnea in the days before the fatal event (Kahn & others, 1992). Why are these infants prone to sleep apnea? Possible explanations include genetics (Köffer & others, 2020) and problems in regulating the neurotransmitters (Erickson, 2020).

Dreams

Have you ever dreamed that you left your long-term romantic partner for a former lover? If so, did you tell your partner about that dream? Probably not. However, you would have likely wondered about the dream's meaning, and if so you would not be alone. Since the dawn of language, human beings have attributed great meaning to dreams. As early as 5000 B.C.E., Babylonians recorded and interpreted their dreams on clay tablets. Egyptians built temples in honor of Serapis, the god of dreams. Dreams are described at length in more than 70 passages in the Bible. Psychologists have also examined this fascinating topic.

Sigmund Freud put great stock in dreams as a key to our unconscious minds. He believed that dreams (even nightmares) symbolize unconscious wishes and that analysis of dream symbols could uncover our hidden desires. Freud distinguished between a dream's manifest content and its latent content. **Manifest content** is the dream's surface content, which contains dream symbols that disguise the dream's true meaning; **latent content** is the dream's hidden content, its unconscious—and true—meaning. For example, if a person had a dream about riding on a train and talking with a friend, the train ride would be the dream's manifest content. Freud thought that this manifest content expresses a wish in disguised form. To get to the latent or true meaning of the dream, the person would have to analyze the dream images. In our example, the dreamer would be asked to think of all the things that come to mind when they think of a train, the friend, and so forth. By following these associations to the objects in the manifest content, the latent content of the dream could be brought to light. Artists have sometimes incorporated the symbolic world of dreaming in their work (Figure 8).

More recently, psychologists have approached dreams not as expressions of unconscious wishes but as mental events that come from various sources. Research has revealed a great deal about the nature of dreams (Domhoff, 2017). A common misconception is that dreams are typically bizarre or strange, but many studies of thousands of dreams, collected from individuals in sleep labs and sleeping at home, have shown that dreams generally are not especially strange. Instead, research shows that dreams are often very similar to waking life (Domhoff, 2017; Schwartz, 2010).

Although some aspects of dreams are unusual, dreams often are no more bizarre than a typical fairy tale, TV show episode, or movie plot. Dreams do generally contain more negative emotion than everyday life; and certainly some unlikely characters, including dead people, sometimes show up in dreams.

There is also no evidence that dreams provide opportunities for problem solving or advice on handling life's difficulties. We may dream about a problem we are dealing with, but we typically find the solution while we are awake and thinking about the dream, not during the dream itself (Domhoff, 2017). There is also no

manifest content
According to Freud, the surface content of a dream, containing dream symbols that disguise the dream's true meaning.

latent content
According to Freud, a dream's hidden content; its unconscious and true meaning.

FIGURE 8 **Artist's Portrayal of a Dream**
Timothy Banks

evidence that people who remember their dreams are better off than those who do not (Blagrove & Akehurst, 2000).

So, if the typical dream involves doing ordinary things, what are dreams? The most prominent theories that attempt to explain dreams are cognitive theory and activation-synthesis theory.

Why might we believe dreams are stranger than they really are? Which type of dreams are we more likely to remember? Why?

COGNITIVE THEORY OF DREAMING

The **cognitive theory of dreaming** proposes that we can understand dreaming by applying the same cognitive concepts we use in studying the waking mind. The theory rests on the idea that dreams are essentially subconscious cognitive processing. Dreaming involves information processing and memory. Indeed, thinking during dreams appears to be very similar to thinking in waking life (Domhoff, 2017).

In the cognitive theory of dreaming, there is little or no search for the hidden, symbolic content of dreams that Freud sought. Instead, dreams are viewed as dramatizations of general life concerns that are similar to relaxed daydreams. Even very unusual aspects of dreams—such as odd activities, strange images, and sudden scene shifts—can be understood as metaphorically related to a person's preoccupations while awake (Domhoff, 2017; Domhoff & Schneider, 2020). The cognitive theory also ties the brain activity that occurs during dreams to the activity that occurs during waking life. Earlier we discussed the *default mode network* that refers to a collection of neurons that are active during mind wandering and daydreaming, essentially whenever we are not focused on a task. Research suggests that dreaming during sleep may also emerge from the activity of this network (Gott & others, 2020; Perogamvros & others, 2017).

The cognitive theory of dreaming strongly argues that dreams should be viewed as a kind of mental simulation that is very similar in content to our everyday waking thoughts. The same themes that occupy us in our waking life occupy our dreams. That is why this perspective has sometimes been called the *continuity hypothesis* because it emphasizes the link between dreaming and waking thought. This perspective on dreams contrasts with activation-synthesis theory of dreams.

cognitive theory of dreaming
Theory proposing that one can understand dreaming by applying the same cognitive concepts used in studying the waking mind.

ACTIVATION-SYNTHESIS THEORY

According to **activation-synthesis theory,** dreaming occurs when the cerebral cortex synthesizes neural signals generated from activity in the lower part of the brain. Dreams result from the brain's attempts to find logic in random brain activity that occurs during sleep (Hobson, 1999; Hobson & Voss, 2011).

When we are awake and alert, our conscious experience tends to be driven by *external* stimuli—all those things we see, hear, and respond to. During sleep, according to activation-synthesis theory, conscious experience is driven by internally generated stimuli that have no behavioral consequence. A key source of such internal stimulation is spontaneous neural activity in the brainstem (Hobson, 2000). Some of the neural activity that produces dreams comes from external sensory experiences. If a fire truck with sirens blaring drives past your house, you might find yourself dreaming about an emergency. Many of us have had the experience of incorporating the sound of our alarm clock going off in an early morning dream.

Supporters of activation-synthesis theory have suggested that neural networks in other areas of the forebrain play a significant role in dreaming (Hobson & others, 2000). Specifically, they believe that the same regions of the forebrain that are involved in certain waking behaviors also function in particular aspects of dreaming (Lu & others, 2006). As levels of neurotransmitters rise and fall during the stages of sleep, some neural networks are activated and others shut down. Random neural firing in various areas of the brain leads to dreams that are the brain's attempts to

activation-synthesis theory
Theory that dreaming occurs when the cerebral cortex synthesizes neural signals generated from activity in the lower part of the brain and that dreams result from the brain's attempts to find logic in random brain activity that occurs during sleep.

You may have noticed how internal states influence your dreams if you have ever been very thirsty while sleeping, and you dream that you get a glass of water.

Rubberball/Alamy Stock Photo

make sense of the activity. So, firing in the primary motor and sensory areas of the forebrain might be reflected in a dream of running and feeling wind on your face. From the activation-synthesis perspective, our nervous system is cycling through various activities, and our consciousness is simply along for the ride (Hobson, 2000, 2004). Dreams are merely a flashy sideshow, not the main event (Hooper & Teresi, 1993). Indeed, one activation-synthesis theorist has referred to dreams as so much "cognitive trash" (Hobson, 2002, p. 23).

Like all dream theories, activation-synthesis theory has its critics. A key criticism is that damage to the brainstem does not necessarily reduce dreaming, suggesting that this area of the brain is not the only starting point for dreaming. Furthermore, life experiences stimulate and shape dreaming more than activation-synthesis theory acknowledges (Domhoff, 2017; Malcolm-Smith & others, 2008).

self-quiz

1. The brain structure that is responsible for the synchronization of circadian rhythm is the
   A. cerebral cortex.
   B. hypothalamus.
   C. reticular formation.
   D. suprachiasmatic nucleus.

2. Immediately entering REM sleep is a symptom of
   A. sleep apnea.
   B. narcolepsy.
   C. night terrors.
   D. somnambulism.

3. The brain waves that are active when we are awake and focused are
   A. alpha waves.
   B. beta waves.

   C. delta waves.
   D. theta waves.

APPLY IT! 4. Bobby and Jill have a friendly competition going in their psychology class. Both have spent several hours studying for the final exam over the last few weeks of school. The night before the final, Bobby declares that he is going to pull an all-nighter, adding 12 full hours to his study time compared to Jill. Altogether, Jill studies 23 hours for the exam, while Bobby studies 35 hours. All other things being equal, who is likely to do better on the exam and why?
   A. Bobby will do better because he studied much more than Jill did.

   B. Jill will do better because she studied a great deal and has the benefit of a good night's sleep, allowing her memory for course material to consolidate.
   C. Bobby will do better because even though he missed some hours of sleep, his memories will be fresher than Jill's.
   D. Jill will do better because Bobby is probably overprepared—35 hours is too long to study anything.

3 Psychoactive Drugs

One way that people seek to alter their own consciousness is through the use of psychoactive drugs. In fact, illicit drug use is a global problem. According to the United Nations Office on Drugs and Crime (UNDOC, 2020), nearly 270 million people worldwide used drugs in 2018, a 30 percent increase from 2009. Drug consumption among youth is a special concern because of its links to problems such as unsafe sex, sexually transmitted infections, unplanned pregnancy, depression, school-related difficulties, and death. The use of drugs among U.S. secondary school students declined in the 1980s but began to increase in the early 1990s (Johnston & others, 2020). Since the late 1990s, the proportion of secondary school students reporting the use of any illicit drug declined or leveled off at a low level. The one exception being vaping, which we will discuss later.

Drug use by U.S. high school seniors since 1975 and by U.S. eighth- and tenth-graders since 1991 has been tracked in a national survey called *Monitoring the Future* (Johnston & others, 2020). Figure 9 shows the trends for these groups from 2008 to 2019. Although recent declines in drug use are encouraging, the United States still has one of the highest rates of adolescent drug use of any industrialized nation.

Uses of Psychoactive Drugs

psychoactive drugs
Drugs that act on the nervous system to alter consciousness, modify perception, and change moods.

Psychoactive drugs act on the nervous system to alter consciousness, modify perceptions, and change moods. Some people use psychoactive drugs as a way to deal with life's

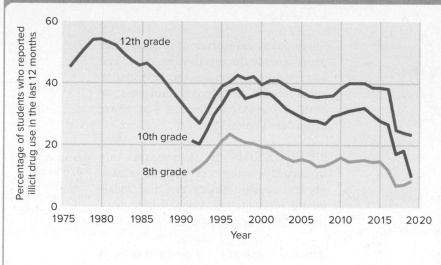

FIGURE 9 **Trends in Drug Use by U.S. Eighth-, Tenth-, and Twelfth-Grade Students** This graph shows the percentage of U.S. eighth-, tenth-, and twelfth-grade students who reported having taken an illicit drug in the last 12 months from 1991 to 2019 (for eighth- and tenth-graders) and from 1975 to 2019 (for twelfth-graders) (Johnston & others, 2020).

SOURCE: Johnston, L. D., O'Malley, P. M., Bachman, J. G., & Schulenberg, J. E. "MONITORING THE FUTURE NATIONAL SURVEY RESULTS ON DRUG USE, 1975-2019." Institute for Social Research The University of Michigan. https://cdn.ymaws.com/www.fadaa.org/resource/resmgr/files/resource_center/mtf-overview2019.pdf.

> Note that data were not collected from eighth- and tenth-graders until 1991. Why do you think these groups were added? > After the mid-1990s, all age groups show a similar pattern of decline in drug use. Why might this pattern have occurred in all three groups? > What are the implications of using self-reports from children and adolescents in this research? Do you think each group would be similarly likely to be honest, to overreport, or to underreport their drug use?

tolerance
The need to take increasing amounts of a drug to get the same effect.

physical dependence
The physiological need for a drug that causes unpleasant withdrawal symptoms such as physical pain and a craving for the drug when it is discontinued.

psychological dependence
The strong desire to repeat the use of a drug for emotional reasons, such as a feeling of well-being and reduction of stress.

addiction
A physical or a psychological dependence, or both, on a drug.

difficulties. Drinking, smoking, and taking drugs reduce tension, relieve boredom and fatigue, and help people to escape from the harsh realities of life. Some people use drugs because they are curious about their effects.

The use of psychoactive drugs, whether it is to cope with problems or just for fun, can carry a high price tag. These include losing track of one's responsibilities, problems in the workplace and in relationships, drug dependence, and increased risk for serious, sometimes fatal diseases. For example, drinking alcohol may initially help people relax and forget about their worries. If, however, they turn more and more to alcohol to escape reality, they may develop a dependence that can destroy relationships, careers, and their bodies.

Continued use of psychoactive drugs leads to **tolerance,** the need to take increasing amounts of a drug to get the same effect. For example, the first time someone takes 5 milligrams of the tranquilizer Valium, the person feels very relaxed. However, after taking the pill every day for six months, they may need to consume twice as much to achieve the same calming effect.

Continuing drug use can also result in **physical dependence,** the physiological need for a drug that causes unpleasant withdrawal symptoms such as physical pain and a craving for the drug when it is discontinued. **Psychological dependence** is the strong desire to repeat the use of a drug for emotional reasons, such as a feeling of well-being and reduction of stress. **Addiction** refers to a physical or psychological dependence, or both, on the drug. Addiction means that the person's body requires a drug to avoid withdrawal symptoms. **Substance use disorder** is a psychological disorder in which a person's use of a psychoactive drug affects their health and abilities to work and engage in social relationships.

How does the brain become addicted? Psychoactive drugs increase dopamine levels in the brain's reward pathways (Volkow & others, 2019). This reward pathway is located in the ventral tegmental area (VTA) and nucleus accumbens (NAc) (Figure 10). Only the limbic and prefrontal areas of the brain are directly activated by dopamine, which comes from the VTA. Although different drugs have different mechanisms of action, each drug increases the activity of the reward pathway by increasing dopamine transmission. The neurotransmitter dopamine plays a vital role in the experience of rewards.

substance use disorder
A psychological disorder in which a person's use of psychoactive drugs (such as alcohol or opiates) affects their health, ability to work, and engage in social relationships.

Ventral tegmental area and nucleus accumbens are mouthfuls, but these areas of the brain are vital to the experience of pleasure. Remember these structures; they will come up again and again.

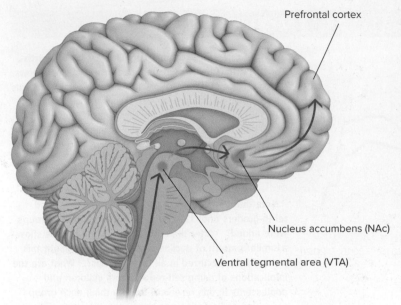

Prefrontal cortex

Nucleus accumbens (NAc)

Ventral tegmental area (VTA)

FIGURE 10 **The Brain's Reward Pathway for Psychoactive Drugs**
The ventral tegmental area (VTA) and nucleus accumbens (NAc) are important locations in the reward pathway for psychoactive drugs. Information travels from the VTA to the NAc and then up to the prefrontal cortex. The VTA is located in the midbrain just above the pons, and the NAc is located in the forebrain just beneath the prefrontal cortex.

SOURCE: National Institute of Drug Abuse. (2001). *Teaching packet for psychoactive drugs* (slide 9). National Institute of Drug Abuse.

Types of Psychoactive Drugs

Three main categories of psychoactive drugs are depressants, stimulants, and hallucinogens. All have the potential to cause health or behavior problems, or both. To evaluate whether you abuse drugs, see Figure 11.

DEPRESSANTS **Depressants** are psychoactive drugs that slow mental and physical activity. Among the most widely used depressants are alcohol, barbiturates, tranquilizers, and opiates.

depressants
Psychoactive drugs that slow down mental and physical activity.

Alcohol Alcohol is a powerful drug. It acts on the body primarily as a depressant and slows the brain's activities. This effect might seem surprising, as people who tend to be inhibited may begin to talk, dance, and socialize after a few drinks. However, people "loosen up" after a few drinks because the brain areas involved in inhibition and judgment slow down. As people drink more, their inhibitions decrease even further, and their judgment becomes increasingly impaired. Activities that require intellectual functioning and motor skills, such as driving, become harder to perform.

Respond "yes" or "no" to the following statements:

Yes	No	
☐	☐	I have gotten into problems because of using drugs.
☐	☐	Using alcohol or other drugs has made my college life unhappy at times.
☐	☐	Drinking alcohol or taking other drugs has been a factor in my losing a job.
☐	☐	Drinking alcohol or taking other drugs has interfered with my studying for exams.
☐	☐	Drinking alcohol or taking drugs has jeopardized my academic performance.
☐	☐	My ambition is not as strong since I've been drinking a lot or taking drugs.
☐	☐	Drinking or taking drugs has caused me to have difficulty sleeping.
☐	☐	I have felt remorse after drinking or taking drugs.
☐	☐	I crave a drink or other drugs at a definite time of the day.
☐	☐	I want a drink or other drug in the morning.
☐	☐	I have had a complete or partial loss of memory as a result of drinking or using other drugs.
☐	☐	Drinking or using other drugs is affecting my reputation.
☐	☐	I have been in the hospital or another institution because of my drinking or taking drugs.

College students who responded yes to items similar to these on the Rutgers Collegiate Abuse Screening Test were more likely to be substance abusers than those who answered no. If you responded yes to just 1 of the 13 items on this screening test, consider going to your college health or counseling center for further screening.

FIGURE 11 **Do You Abuse Drugs?** Take this short quiz to see if your use of drugs and alcohol might be a cause for concern.

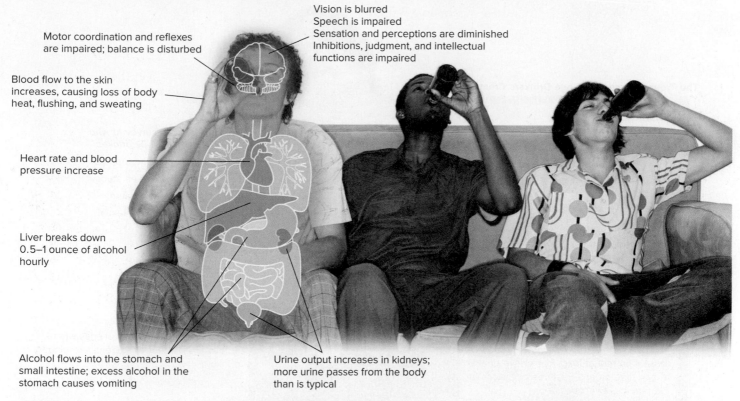

Motor coordination and reflexes are impaired; balance is disturbed

Vision is blurred
Speech is impaired
Sensation and perceptions are diminished
Inhibitions, judgment, and intellectual functions are impaired

Blood flow to the skin increases, causing loss of body heat, flushing, and sweating

Heart rate and blood pressure increase

Liver breaks down 0.5–1 ounce of alcohol hourly

Alcohol flows into the stomach and small intestine; excess alcohol in the stomach causes vomiting

Urine output increases in kidneys; more urine passes from the body than is typical

FIGURE 12 **The Physiological and Behavioral Effects of Alcohol** Alcohol has a powerful impact throughout the body. Its effects touch everything from the operation of the nervous, circulatory, and digestive systems to sensation, perception, motor coordination, and intellectual functioning. (Photo): George Doyle/Stockbyte/Getty Images

Eventually the drinker falls asleep. With extreme intoxication, the person may lapse into a coma and die. Figure 12 illustrates alcohol's main effects on the body.

The effects of alcohol vary from person to person. Factors in this variation are body weight, the amount of alcohol consumed, individual differences in the way the body metabolizes alcohol, and the presence or absence of tolerance. Because of differences in body fat as well as stomach enzymes, women are likely to be more strongly affected by alcohol than men.

Sometimes friends think someone who is dangerously drunk just needs to "sleep it off." Drinking to the point of passing out is a symptom of alcohol poisoning. Call 911.

How does alcohol affect the brain? Like other psychoactive drugs, alcohol goes to the VTA and the NAc (Ji & others, 2017). Alcohol also increases the concentration of the neurotransmitter GABA, which is widely distributed in many brain areas, including the cerebral cortex, cerebellum, hippocampus, amygdala, and nucleus accumbens (Prisciandaro & others, 2019). Alcohol consumption affects areas of the frontal cortex involved in judgment and impulse control (Gerchen & others, 2019).

After caffeine, alcohol is the most widely used drug in the United States. Over 86 percent of U.S. adults report having drunk an alcoholic beverage at some point in their lives (SAMHSA, 2018). The common use of alcohol is related to other serious problems, including death and injury from driving while drinking. Research has also found a link between alcohol and violence and aggression (Littleton, 2014). More than 60 percent of homicides involve alcohol use by the offender or the victim, and 65 percent of acts of sexual violence against women are associated with alcohol consumption by the offender.

Alcohol consumption among U.S. secondary school and college students has long been a concern. However, in the Monitoring the Future survey, the percentage of U.S. high

Henryk Sadura/Alamy Stock Photo

This explains how getting the next drink can become more important than anything else in the person's life.

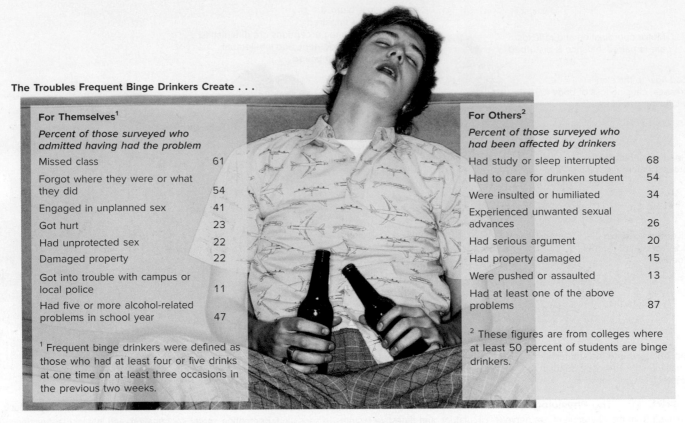

The Troubles Frequent Binge Drinkers Create . . .

For Themselves[1]		For Others[2]	
Percent of those surveyed who admitted having had the problem		*Percent of those surveyed who had been affected by drinkers*	
Missed class	61	Had study or sleep interrupted	68
Forgot where they were or what they did	54	Had to care for drunken student	54
Engaged in unplanned sex	41	Were insulted or humiliated	34
Got hurt	23	Experienced unwanted sexual advances	26
Had unprotected sex	22	Had serious argument	20
Damaged property	22	Had property damaged	15
Got into trouble with campus or local police	11	Were pushed or assaulted	13
Had five or more alcohol-related problems in school year	47	Had at least one of the above problems	87

[1] Frequent binge drinkers were defined as those who had at least four or five drinks at one time on at least three occasions in the previous two weeks.

[2] These figures are from colleges where at least 50 percent of students are binge drinkers.

FIGURE 13 **Consequences of Binge Drinking** Binge drinking has wide-ranging negative consequences.

(Photo): George Doyle/Stockbyte/Getty Images: (Text): Source: "Health and Behavioral Consequences of Binge Drinking in College: A National Survey of Students at 140 Campuses." Journal of the American Medical Association, vol. 272. 7 Dec, 1994. https://www.csus.edu/faculty/M/fred.molitor/docs/Binge%20Drinking%20and%20Consequences.pdf.

school seniors who reported consuming alcohol in the last 30 days dropped from 72 percent in 1980 to 29.3 percent in 2019 (Johnston & others, 2020). This large change is very good news.

Binge drinking often increases during the first two years of college and peaks at 21 to 22 years of age (Johnston & others, 2020). As Figure 13 shows, it can take its toll on students. In a national survey of drinking patterns on college campuses, almost half of the binge drinkers reported problems such as missed classes, injuries, trouble with police, and unprotected sex (Wechsler & others, 2000, 2002). Binge-drinking college students were 11 times more likely to fall behind in school, 10 times more likely to drive after drinking, and twice as likely to have unprotected sex as college students who did not binge drink. Many emerging adults, however, decrease their alcohol use as they assume adult responsibilities such as a permanent job, marriage or cohabitation, and parenthood (Chen & Jacobson, 2012).

alcoholism
Disorder that involves long-term, repeated, uncontrolled, compulsive, and excessive use of alcoholic beverages and that impairs the drinker's health and social relationships.

Alcoholism is a disorder that involves long-term, repeated, uncontrolled, compulsive, and excessive use of alcoholic beverages and that impairs the drinker's health and social relationships. Approximately 18 million people in the United States suffer from alcoholism (MedlinePlus, 2014). A longitudinal study linked early onset of drinking to later alcohol problems (Hingson & others, 2006). People who began drinking alcohol before 14 years of age were more likely to become alcohol dependent than their counterparts who began drinking alcohol at 21 years of age or older. Among youth, aggression increases after they first begin to drink alcohol (Staff & others, 2019).

One in nine individuals who drink continues down the path to alcoholism. Family studies consistently find a high frequency of alcoholism in the close biological relatives of people with alcoholism (Holla & others, 2019). A possible explanation is that the brains of people genetically predisposed to alcoholism may be unable to produce adequate dopamine, the neurotransmitter that can make us feel pleasure (Edenberg & others, 2019). For these individuals, alcohol may increase dopamine concentration and resulting pleasure to the point where it leads to addiction (Hansson & others, 2019).

Like other psychological characteristics, though, alcoholism is not all about nature: Nurture matters, too. Indeed, research shows that experience plays a role in alcoholism (Knobloch & others, 2020). Many people with alcoholism do not have close relatives with the disorder (Duncan & others, 2006), a finding that points to environmental influences.

What does it take to stop alcoholism? About one-third of people who are addicted to alcohol recover whether they are in a treatment program or not. This finding came from a long-term study of 700 people (Vaillant, 2003). The researcher, George Vaillant, followed the participants for over 60 years, and formulated the so-called one-third rule for alcoholism: By age 65, one-third will be terribly ill or even die; one-third are still trying to recover from their addiction; and one-third are abstinent or drinking only socially. Vaillant found that recovery from alcoholism was predicted by (1) having a strong negative experience with drinking, such as a serious medical emergency; (2) finding a substitute dependency, such as meditation, exercise, or overeating (which has its own adverse health effects); (3) developing new, positive relationships; and (4) joining a support group such as Alcoholics Anonymous.

barbiturates
Depressant drugs, such as Nembutal and Seconal, that decrease central nervous system activity.

Barbiturates **Barbiturates**, such as Nembutal and Seconal, are depressant drugs that decrease central nervous system activity. These types of drugs are likely to make a person feel groggy. Physicians once widely prescribed barbiturates as sleep aids. In heavy dosages, they can lead to impaired memory and decision making. When combined with alcohol (e.g., sleeping pills taken after a night of binge drinking), barbiturates can be lethal. Heavy doses of barbiturates by themselves can cause death. For this reason, barbiturates are the drug most often used in suicide attempts. Abrupt withdrawal can produce seizures. Because of the addictive potential and relative ease of toxic overdose, barbiturates have largely been replaced by tranquilizers in the treatment of insomnia.

Handy Survival Tips: Never share prescription drugs. Never take someone else's prescription drugs. Never mix prescription drugs with alcohol.

Tranquilizers **Tranquilizers**, such as Valium and Xanax, are depressant drugs that reduce anxiety and induce relaxation. In small doses tranquilizers can induce a feeling of calm; higher doses can lead to drowsiness and confusion. Tolerance for tranquilizers can develop within a few weeks of usage, and these drugs are addictive. Widely prescribed in the United States to treat anxiety, tranquilizers can produce withdrawal symptoms when use is stopped. Prescription tranquilizers were part of the lethal cocktail of drugs that, in 2008, ended the life of actor Heath Ledger, who won an Academy Award for his portrayal of the Joker in the Batman film *The Dark Knight*.

tranquilizers
Depressant drugs, such as Valium and Xanax, that reduce anxiety and induce relaxation.

Opioids **Opioids** are a class of drugs that act on the brain's endorphin receptors. These include opium and its natural derivatives (sometimes called *opiates*) as well as chemicals that do not occur naturally but that have been created to mimic the activity of opium. These drugs (also called narcotics) depress activity in the central nervous system.

Opioids are powerful painkillers. The most common drugs derived from opium—morphine and heroin—affect synapses in the brain that use endorphins as their neurotransmitter. When they bind to the brain's opioid receptors, they drive up the production of dopamine, producing feelings of pleasure (Lefevre & others, 2020). When these drugs leave the brain, the affected synapses become under-stimulated. For several hours after taking an opiate, the person feels euphoric and pain-free and has an increased appetite for food and sexual activity. Opioids are highly addictive, and users experience craving and painful withdrawal when the drug becomes unavailable.

The risk of death from opioid overdose is very high. Remember, these drugs work by attaching to receptors for endorphins. Such receptors are present in the brainstem, the

opioids
A class of drugs that act on the brain's endorphin receptors. These include opium and its natural derivatives (sometimes called opiates) as well as chemicals that do not occur naturally but that have been created to mimic the activity of opium. These drugs (also called narcotics) depress activity in the central nervous system and eliminate pain.

region that controls breathing. So, if someone takes too much of an opioid (or an opioid mixed with another drug such as alcohol) they may stop breathing (NIDA, 2020a). Naloxone, a drug used to treat opioid overdose, can restore normal breathing. Currently, opioid overdoses in the United States represent a national crisis. In 2018, an estimated 128 people died from opioid overdose every day (NIDA, 2020a). A person does not have to be addicted to opioids to die of an overdose of these powerful drugs. Opioids should never be mixed with other substances.

Many people who abuse opioids first encountered them as (legal) prescription pain killers. *Fentanyl* is a fast-acting opioid prescription pain killer that has been implicated in numerous overdoses, including the death of music legend Prince in 2016. Fentanyl is estimated to be 50 to 100 times more powerful than morphine (NIDA, 2020a).

In 2016, prescription opioids accounted for more deaths than heroin and cocaine combined (NIDA, 2020a). The availability of prescription opioids is a public health emergency. Because these drugs are prescribed, people may not realize how powerful and dangerous they are. One healthcare expert cautioned, "Prescription pain pills are similar to having heroin in the medicine cabinet" (Hazelden Betty Ford Foundation, 2015). Even children who live in households where adults have prescription drugs are at risk for overdose (Turkewitz, 2017). Importantly, a dose of nonopioid medications (such as Tylenol or Motrin) can be just as effective as a dose of opioids in treating acute pain (Chang & others, 2017).

Often when a person experiences an overdose, it falls to family and friends to respond. The symptoms of an opioid overdose are as follows (WHO, 2020):

- Pinpoint pupils
- Unconsciousness
- Slowed breathing

If you think someone has experienced an overdose, immediately call 911 to get help. Bringing pill bottles to the hospital can help professionals know what to do.

STIMULANTS

Stimulants are psychoactive drugs that increase the central nervous system's activity. The most widely used stimulants are caffeine, nicotine, amphetamines, and cocaine.

stimulants
Psychoactive drugs, including caffeine, nicotine, amphetamines, and cocaine, that increase the central nervous system's activity.

Caffeine Often overlooked as a drug, caffeine is the world's most widely used psychoactive drug. Caffeine is a stimulant and a natural component of the plants that are the sources of coffee, tea, and cola drinks. Caffeine also is present in chocolate, in many nonprescription medications, and in energy drinks such as Red Bull. People often perceive the stimulating effects of caffeine as beneficial for boosting energy and alertness, but some experience unpleasant side effects.

Caffeinism refers to overindulgence in caffeine. It is characterized by mood changes, anxiety, and sleep disruption. Caffeinism often develops in people who drink five or more cups of coffee (at least 500 milligrams) each day. Common symptoms are insomnia, irritability, headaches, ringing ears, dry mouth, increased blood pressure, and digestive problems (Uddin & others, 2017).

Caffeine affects the brain's pleasure centers, so it is not surprising that it is difficult to kick the caffeine habit. When people who regularly consume caffeinated beverages remove caffeine from their diet, they typically experience headaches, lethargy, apathy, and concentration difficulties. These symptoms of withdrawal are usually mild and subside after several days.

Nicotine Nicotine is the main psychoactive ingredient in all forms of smoking and smokeless tobacco. Even with all the publicity given to the enormous health risks posed by tobacco, we sometimes overlook the highly addictive nature of nicotine. Nicotine stimulates the brain's reward centers by raising dopamine levels. Behavioral effects of nicotine include improved

PSYCHOLOGY IN OUR WORLD

Responding to the Opioid Crisis

There is not a group in the United States that has not been affected by the opioid crisis. The number of adults aged 24 or younger who died from opioid overdose doubled from 2005 to 2015 (Spencer, 2017). In 2018, drug overdoses were the leading cause of death in the United States (CDC, 2020a), with two-thirds of those deaths being attributed to opioids. College students are not immune. A survey found that a third of young adults reported opioids to be easy to get and 16 percent reported using prescription drugs not prescribed for them at some point in their lives (Hazelden Betty Ford Foundation, 2015).

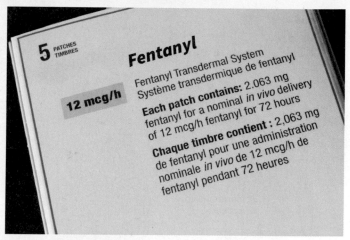

RobertDupuis/Getty Images

Student athletes may be particularly at risk as they are likely to receive opioids to treat injuries. Communities are taking action to reduce the use and abuse of opioids.

Colleges and universities have begun to develop living groups around issues related to drug addiction. These include sober, drug-free housing and specialized living groups for students in recovery (Spencer, 2017). In these living groups, students can pursue the difficult path of recovery surrounded by others on the same path who understand how difficult it can be.

Increasingly, society at large has begun to recognize that addiction is a disease not a crime. Policy makers have begun to devise innovative ways to treat those addicted to opioids to prevent overdose deaths. Judge Craig Hannah of Buffalo, New York (himself a person in recovery from addiction), instituted a special Opiate Crisis Intervention Court (Helmore, 2017). Offenders who report themselves as opioid users are offered the chance to enter a treatment program. Those who accept can look forward to a personal relationship with the judge, as he serves as support on the long road to recovery. It is Judge Hannah's goal to reduce overdose deaths by breaking addiction's hold on people's lives.

attention and alertness, reduced anger and anxiety, and pain relief. Figure 14 shows the main effects of nicotine on the body.

Tolerance develops for nicotine both in the long run and on a daily basis, so that cigarettes smoked later in the day have less effect than those smoked earlier. Withdrawal from nicotine often quickly produces strong, unpleasant symptoms such as irritability, craving, inability to focus, sleep disturbance, and increased appetite. Withdrawal symptoms can persist for months or longer.

Tobacco poses a grave threat to public health. Tobacco is involved in nearly one in every five deaths in the United States, more than the total number killed by AIDS, alcohol, motor vehicles, homicide, illegal drugs, and suicide combined (CDC, 2020b). Smoking is related to cancer not only of the lungs but also of many different bodily organs.

Do It!

Go on a caffeine hunt. Check out the ingredient lists on your favorite beverages, snacks, and painkillers. Which of these contain caffeine? You might be surprised by how much caffeine you consume every day without even knowing it.

Attention and alertness improve

At high levels, muscles become more relaxed and anxiety and anger may be reduced; pleasant feelings induce the smoker to smoke more

Circulation to extremities decreases

Heart rate and blood pressure increase

During pregnancy, nicotine freely passes through the placenta wall into amniotic fluid

Smoker loses appetite for carbohydrates

FIGURE 14 **The Physiological and Behavioral Effects of Nicotine** Smoking has many physiological and behavioral effects. Highly addictive, nicotine delivers pleasant feelings that make the smoker smoke more, but tobacco consumption poses very serious health risks to the individual.

Cigarette smoking is decreasing among both adolescents and college students. Figure 15 shows this decline as detected in the Monitoring the Future survey (Johnston & others, 2020). Cigarette smoking peaked in 1996 and 1997 and then decreased dramatically from 1998 to 2019 (see Figure 15).

The drop in cigarette use by U.S. youth may have several sources, including higher cigarette prices, less tobacco advertising reaching adolescents, more antismoking advertisements, and more negative publicity about the tobacco industry than before. Increasingly, adolescents report perceiving cigarette smoking as dangerous, disapprove of it, are less accepting of being around smokers, and prefer to date nonsmokers (Johnston & others, 2020). Among college students and young adults, smoking has shown a smaller decline than adolescent and adult smoking.

Although these patterns for cigarette smoking are encouraging, these gains in public health are threatened by increases in vaping. *Vaping* involves inhaling aerosols using a battery-powered e-cigarette. These products can be used to inhale substances containing nicotine, marijuana, or just flavorings. Vaping has increased dramatically among U.S. youth–doubling or tripling over the last three years across the three age groups included in the Monitoring the Future survey (Miech & others, 2020). These increases are the largest ever recorded for substance use by American youth. Nearly a third of high school seniors reported having engaged in vaping in the last 30 days (Johnston

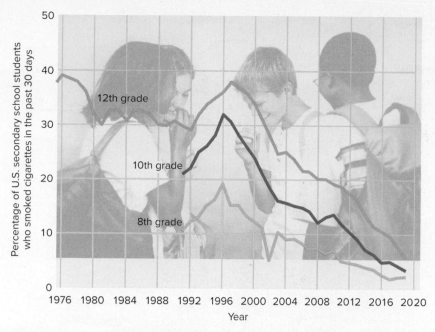

FIGURE 15 **Trends in Cigarette Smoking by U.S. Secondary School Students** Fortunately, cigarette smoking by U.S. high school students is on the decline.
(Photo): Flying Colours Ltd/Getty Images

& others, 2020). Nicotine is highly addictive, and vaping involves exposing oneself to that highly addictive substance. Exposure among youth is particularly concerning because of the effects on the developing brain. Even vaping just flavors can lead to permanent lung damage (Butt & others, 2019). Vaping can also damage the gums and teeth (Froum & Neymark, 2019). E-cigarettes contain a number of chemicals such as formaldehyde that are harmful to the human body. The bottom line is that there is no such thing as safe smoking.

Amphetamines *Amphetamines,* or "uppers," are stimulant drugs that people use to boost energy, stay awake, or lose weight. Often prescribed in the form of diet pills, these drugs increase the release of dopamine, which enhances the user's activity level and pleasurable feelings.

Perhaps the most insidious illicit drug for contemporary society is crystal methamphetamine, or crystal meth. Smoked, injected, or swallowed, crystal meth (also called "crank" or "tina") is a synthetic stimulant that causes a powerful feeling of euphoria, particularly the first time it is ingested. Meth is made using household products such as battery acid, cold medicine, drain cleaner, and kitty litter, and its effects have been devastating, notably in rural areas of the United States.

Crystal meth releases enormous amounts of dopamine in the brain, producing intense feelings of pleasure. The drug is highly addictive. The extreme high of crystal meth leads to a severe "come down" experience that is associated with strong cravings. Crystal meth also damages dopamine receptors, so that a person can be chasing a high their brain can no longer produce. Because a person's very first experience with crystal meth can lead to ruinous consequences, the Drug Enforcement Agency has started a website, designed by and targeted at teenagers, www.justthinktwice.com, to share the hard facts of the horrific effects of this and other illicit substances.

Seriously, don't try it—not even once.

Cocaine Cocaine is an illegal drug that comes from the coca plant, native to Bolivia and Peru. Cocaine is either snorted or injected in the form of crystals or powder. Used

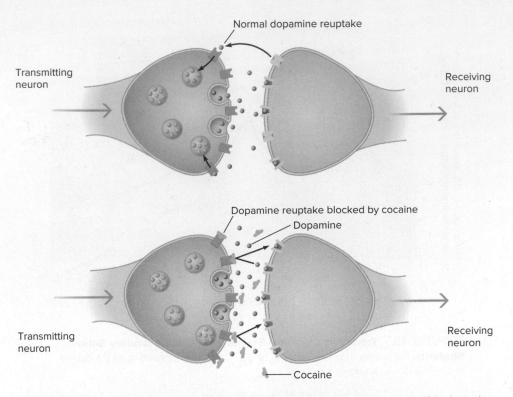

FIGURE 16 **Cocaine and Neurotransmitters** Cocaine concentrates in areas of the brain that are rich in dopamine synapses such as the ventral tegmental area (VTA) and the nucleus accumbens (NAc). (*Top*) What happens in normal reuptake. The transmitting neuron releases dopamine, which stimulates the receiving neuron by binding to its receptor sites. After binding occurs, dopamine is carried back into the transmitting neuron for later release. (*Bottom*) What happens when cocaine is present in the synapse. Cocaine binds to the uptake pumps and prevents them from removing dopamine from the synapse. The result is that more dopamine collects in the synapse, and more dopamine receptors are activated.

this way, cocaine floods the bloodstream rapidly, producing a rush of euphoric feelings that lasts for about 15 to 30 minutes. Because the rush depletes the brain's supply of the neurotransmitters dopamine, serotonin, and norepinephrine, an agitated, depressed mood usually follows as the drug's effects decline (Isenschmid, 2020). Figure 16 shows how cocaine affects dopamine levels in the brain.

Crack is a potent form of cocaine, consisting of chips of pure cocaine that are usually smoked. Scientists believe that crack is one of the most addictive substances known. Treatment of cocaine addiction is difficult. Cocaine's addictive properties are so strong that, six months after treatment, more than 50 percent of abusers return to the drug, a statistic that highlights the importance of prevention.

MDMA (Ecstasy)

MDMA (Ecstasy) MDMA—called Ecstasy, X, XTC, Molly—is an illegal synthetic drug with both stimulant and hallucinogenic properties. MDMA leads to feelings of warmth, pleasure, and alertness. People have called Ecstasy an "empathogen" because under its influence, users tend to feel warm bonds with others. MDMA produces its effects by releasing serotonin, dopamine, and norepinephrine. The effects of the drug on serotonin are particularly problematic. MDMA depletes the brain of this important neurotransmitter, producing lingering feelings of listlessness that often continue for days after use (NIDA, 2020b).

MDMA impairs memory and cognitive processing. Heavy users of Ecstasy show cognitive deficits that persist even two years after they begin to abstain (Rogers & others, 2009). Because MDMA destroys axons that release serotonin, repeated use might lead to susceptibility to depression (Cadoni & others, 2017; Cowan & others, 2008). Other researchers have examined whether MDMA might be useful to help individuals with psychological disorders (Gorman & others, 2020; Wagner & others, 2017).

HALLUCINOGENS **Hallucinogens** are psychoactive drugs that modify a person's perceptual experiences and produce visual images that are not real. Hallucinogens are also called *psychedelic* (from the Greek meaning "mind-revealing") drugs. Marijuana has a mild hallucinogenic effect; LSD, a stronger one.

Brand X Pictures/Getty Images

Marijuana   Marijuana is the dried leaves and flowers of the hemp plant *Cannabis sativa,* which originated in Central Asia but is now grown in most parts of the world. The plant's dried resin is known as *hashish.* The active ingredient in marijuana is THC (delta-9-tetrahydrocannabinol). Unlike other psychoactive drugs, THC does not affect a specific neurotransmitter. Rather, marijuana disrupts the membranes of neurons and affects the functioning of a variety of neurotransmitters and hormones. Here we classify marijuana as a mild hallucinogen, although it has diverse effects. Marijuana was the most used illicit substance worldwide in 2018 (UNDOC, 2020).

The physical effects of marijuana include increased pulse rate and blood pressure, reddening of the eyes, and dry mouth. Psychological effects include a mixture of excitatory, depressive, and mildly hallucinatory characteristics. Marijuana can trigger spontaneous unrelated ideas and laughter; distorted perceptions of time and place; increased sensitivity to sounds, tastes, smells, and colors; and erratic verbal behavior. The drug can also impair attention and memory. Further, when used daily in large amounts, marijuana can alter sperm count and change hormonal cycles (Carroll & others, 2020). Marijuana use should be avoided during pregnancy. Research indicates a number of negative birth outcomes associated with prenatal exposure to marijuana (Volpe, 2020), including birth defects and later problems in cognitive ability.

Marijuana is the illegal drug most widely used by high school students (Johnston & others, 2020). One concern about adolescents' use of marijuana is that the drug might be a gateway to the use of other more serious illicit substances. Although there is a correlational relationship between using marijuana and using other illicit drugs, evidence for the notion that marijuana use leads to the use of other drugs is mixed (Arnold & Sade, 2020).

Currently, the use of marijuana for medical purposes is legal in 33 U.S. states, as well as the District of Columbia. In addition, adults over the age of 21 can purchase and use marijuana legally for recreational purposes in Alaska, California, Colorado, Illinois, Maine, Massachusetts, Michigan, Nevada, Oregon, Vermont, and Washington. However, U.S. federal law continues to treat marijuana as an illegal substance. Recently, large-scale correlational studies have shown that states with legal medical marijuana have lower incidents of opioid use and abuse. Marijuana has less damaging and life-threatening effects than opioids (Lake & others, 2020). Might medical marijuana present a legal alternative to opioids and help in resolving the opioid crisis? To answer this question, let's take a deep dive into the data, in Challenge Your Thinking.

A large number of states in the United States have legalized medical marijuana, and some have legalized it for recreational use among adults. What do you think about these trends?

LSD   LSD (lysergic acid diethylamide) is a hallucinogen that even in low doses produces striking perceptual changes. Objects change their shapes and glow. Colors become kaleidoscopic and astonishing images unfold. LSD-induced images are sometimes pleasurable and sometimes grotesque. LSD can also influence a user's sense of time so that brief glances at objects are experienced as deep, penetrating, and lengthy examinations, and minutes turn into hours or even days. A bad LSD trip can trigger extreme anxiety, paranoia, and suicidal or homicidal impulses.

LSD's effects on the body can include dizziness, nausea, and tremors. LSD acts primarily on the neurotransmitter serotonin in the brain, though it also can affect dopamine (Halberstadt, 2017; Preller & others, 2017). Emotional and cognitive effects may include rapid mood swings and impaired attention and memory. The use of LSD peaked in the 1960s and 1970s, and its consumption has been decreasing in the twenty-first century (Johnston & others, 2020). The potential for LSD to be used for therapeutic purposes has been examined as well. It might be useful in small doses and in controlled

Does Legalized Medical Marijuana Reduce Opioid Abuse and Overdoses?

Ever since certain U.S. states began legalizing medical marijuana, scientists have been interested in gauging the effects of legal medical marijuana on opioid overdoses. A 2014 study comparing the 13 states that had legalized medical marijuana to the 37 states that had not (Bachhuber & others, 2014) showed that states with legal medical marijuana had nearly 25 percent fewer deaths from opioid overdose. States that legalized medical marijuana also showed 33 percent drops in opioid mortality rates. Subsequent studies showed a negative correlation between legalization of marijuana opioid overdose (Powell & others, 2018). However, more recent research has called those initial findings into question, showing that over a longer period of time, the relationship between legalized marijuana and opioid overdose did not remain consistently negative (Shover & others, 2019).

Many people first encounter opioids as legally prescribed painkillers. Do medical professionals prescribe fewer opioids when they can turn to marijuana instead? Population-based studies examining opioid prescriptions have shown that these drop by 5 to 7 percent after laws legalizing medical and recreational marijuana go into effect (Wen & Hockenberry, 2018). The change in prescription rates for all opioids amounts to over 3.7 million fewer daily doses per year following the legalization of medical marijuana (Bradford & others, 2018). Even more recently, large-scale analyses have shown that legalization of marijuana is associated with decreases in opioid prescriptions (Kropp Lopez & others, 2020; McMichael & others, 2020).

Do these findings indicate that a move to more liberalized marijuana laws is a way to solve the opioid epidemic? Answering this question is more complicated than it might seem. Remember that correlation does not imply causation, even when the correlation might seem to offer an answer to a difficult societal problem. A multitude of third variables might account for the negative association between legalized medical marijuana and opioid deaths (Hall & others, 2018; Shover & others, 2019). For example, programs

Medical marijuana is now legal in about half the United States.
Wiktor Szymanowicz/REX/Shutterstock

that have been in place to combat opioid abuse (which happened at the same time as changes to marijuana laws) may be having an effect. Perhaps growing awareness of opioid addiction has led physicians to reduce their prescriptions or patients to be more careful with their medications. High-profile overdose deaths of celebrities might also have reduced opioid use. Large-scale studies cannot tell us about what is going on in specific cases. We might assume that these findings mean people are being prescribed marijuana instead of opioids or that those who would have used powerfully addictive opioids are substituting marijuana, but the current data cannot tell us whether this is what is happening. The data are not relevant to those experiences "on the ground." The results of a large-scale study of the link between medical marijuana on opioid death rates demonstrates this complexity (Powell & others, 2018). The availability of medical marijuana was associated with reduced rates of opioid-related deaths; however, this effect was only true for states that maintained liberal policies about the distribution of marijuana. When states were more stringent in their regulation of marijuana dispensaries, the negative relation to opioid deaths was weaker. The researchers concluded that legalized marijuana does seem to facilitate substitution of marijuana for more seriously addictive opioids, but this benefit is only apparent when states allow access in a fairly liberal fashion. Such nuances may be important as more and more states legalize medical marijuana.

What Do You Think?

- Do you think marijuana should be legalized broadly? Why or why not?
- How might we test for a causal role of medical marijuana in opioid use and abuse?

environments for pain relief (Ramaekers & others, 2020) and to help those dealing with terminal illness (Fuentes & others, 2019).

Indeed, some of the psychoactive drugs we have surveyed (as well as drugs not reviewed here) have been considered useful for medical purposes. Figure 17 summarizes the effects of a variety of psychoactive drugs and some of their potential medical uses.

Drug Classification	Potential Medical Uses	Short-Term Effects	Overdose Effects	Health Risks	Risk of Physical/ Psychological Dependence
Depressants					
Alcohol	Pain relief	Relaxation, depressed brain activity, slowed behavior, reduced inhibitions	Disorientation, loss of consciousness, even death at high blood-alcohol levels	Accidents, brain damage, liver disease, heart disease, ulcers, birth defects	Physical: moderate Psychological: moderate
Barbiturates	Sleeping pill	Relaxation, sleep	Breathing difficulty, coma, possible death	Accidents, coma, possible death	Physical and psychological: moderate to high
Tranquilizers	Anxiety reduction	Relaxation, slowed behavior	Breathing difficulty, coma, possible death	Accidents, coma, possible death	Physical: low to moderate Psychological: moderate to high
Opiates (narcotics)	Pain relief	Euphoric feelings, drowsiness, nausea	Convulsions, coma, possible death	Accidents, infectious diseases such as AIDS	Physical: high Psychological: moderate to high
Stimulants					
Amphetamines	Weight control	Increased alertness, excitability; decreased fatigue, irritability	Extreme irritability, feelings of persecution, convulsions	Insomnia, hypertension, malnutrition, possible death	Physical: possible Psychological: moderate to high
Cocaine	Local anesthetic	Increased alertness, excitability, euphoric feelings; decreased fatigue, irritability	Extreme irritability, feelings of persecution, convulsions, cardiac arrest, possible death	Insomnia, hypertension, malnutrition, possible death	Physical: possible Psychological: moderate (oral) to very high (injected or smoked)
MDMA (Ecstasy)	Psychotherapy and anxiety reduction	Mild amphetamine and hallucinogenic effects; high body temperature and dehydration; sense of well-being and social connectedness	Brain damage, especially memory and thinking	Cardiovascular problems; death	Physical: possible Psychological: moderate
Caffeine	None	Alertness and sense of well-being followed by fatigue	Nervousness, anxiety, disturbed sleep	Possible cardiovascular problems	Physical: moderate Psychological: moderate
Nicotine	None	Stimulation, stress reduction, followed by fatigue, anger	Nervousness, disturbed sleep	Cancer and cardiovascular disease	Physical: high Psychological: high
Hallucinogens					
LSD	Psychotherapy for terminally ill and others	Strong hallucinations, distorted time perception	Severe mental disturbance, loss of contact with reality	Accidents	Physical: none Psychological: low
Marijuana*	Treatment of the eye disorder glaucoma; appetite and well-being	Euphoric feelings, relaxation, mild hallucinations, time distortion, attention and memory impairment	Fatigue, disoriented behavior	Accidents, respiratory disease	Physical: very low Psychological: moderate

*Classifying marijuana is difficult because of its diverse effects.

FIGURE 17 **Categories of Psychoactive Drugs: Depressants, Stimulants, and Hallucinogens** Note that these various drugs have different effects and negative consequences. Marijuana has diverse effects but is typically considered a mild hallucinogen.

1. The most widely consumed drug in the world as a whole is
   A. nicotine.
   B. marijuana.
   C. cocaine.
   D. caffeine.

2. Roger used to feel the effect of one or two alcoholic drinks; he now needs four or five to feel the same effect. Roger is experiencing
   A. physical dependence.
   B. psychological dependence.
   C. withdrawal symptoms.
   D. tolerance.

3. Of the following drugs, the one that is *not* a depressant is
   A. nicotine.
   B. alcohol.
   C. barbiturates.
   D. opiates.

APPLY IT! 4. In high school, Kareem was a star student, and drinking alcohol did not fit with his academic ambitions. In college, he started drinking and eventually drank heavily every weekend. He has a lot of trouble making it to his Monday classes, and his grades have dropped from mostly *A*s to mainly *C*s, but otherwise he feels he is doing pretty well. A friend asks him about his drinking, and Kareem declares, "I will definitely stop drinking once I am out of college. Besides, no one in my family has ever been an alcoholic, so I am not at risk." Which of the statements is an accurate assessment of Kareem's belief?
   A. Kareem is probably not at risk for alcoholism, but he might be putting himself at risk for alcohol-related problems, such as drunk driving.
   B. Kareem is at risk for alcohol-related problems only if he also drives while drinking.
   C. Kareem is clearly already suffering from alcohol-related problems. Despite his lack of family history, his alcohol use could put him at risk for dependence. Furthermore, if Kareem is binge drinking, he is risking death by alcohol poisoning.
   D. Kareem appears to have his alcohol use under control and accurately states his lack of risk—he has no family history of alcoholism, and he did not start drinking until he was college-age.

4 Hypnosis

Fifty-three-year-old Shelley Thomas entered a London hospital for a 30-minute pelvic surgery. Before the operation, with her hypnotherapist guiding her, Shelley counted backward from 100 and entered a hypnotic trance. Her surgery was performed with no anesthesia (Song, 2006); rather, Shelley relied on hypnosis to harness her mind's powers to overcome pain.

Maybe you have seen a hypnotist on television or in a nightclub, putting a person into a trance and then perhaps making them act like a chicken or pretend to be a contestant on *The Voice* or enact some similarly strange behavior. When we observe someone in such a trance, we might be convinced that hypnosis involves a powerful manipulation of another person's consciousness. What is hypnosis, really? The answer to this question is itself the source of some debate.

Some psychologists think of hypnosis as an altered state of consciousness, while others believe that it is simply a product of more mundane processes such as focused attention and expectations (Kihlstrom, 2020). In fact, both views are reasonable, and we may define **hypnosis** as an altered state of consciousness or as a psychological state of altered attention and expectation in which the individual is unusually receptive to suggestions. People have used basic hypnotic techniques since the beginning of recorded history, in association with religious ceremonies, magic, and the supernatural (Klocek, 2017).

Today, psychology and medicine recognize hypnosis as a legitimate process, although researchers still have much to learn about how it works. In addition, there is continuing debate about whether hypnosis truly is an altered state of consciousness.

hypnosis
An altered state of consciousness or a psychological state of altered attention and expectation in which the individual is unusually receptive to suggestions.

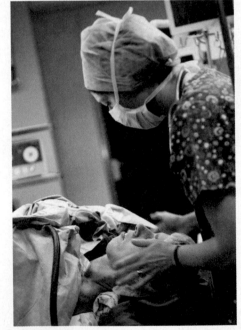
Hypnosis has made its way into some operating rooms as an alternative to traditional anesthesia.
Virginia Mayo/AP Images

The Nature of Hypnosis

When Shelley Thomas was in a hypnotic trance, what exactly was happening in her brain and mind? Patterns of brain activity during the hypnotic state suggest that hypnosis produces a state of consciousness similar to other states of consciousness. For example, individuals in a hypnotic state, when monitored by an EEG, display a predominance of alpha and beta waves, characteristic of persons in a relaxed waking state (Williams & Gruzelier, 2001). In a brain-imaging study, widespread areas of the cerebral cortex—including the occipital lobes, parietal lobes, sensorimotor cortex, and prefrontal cortex—were activated when individuals were in a hypnotic state (Faymonville & others, 2006; Parris, 2017). A similar activation pattern is found in individuals in a nonhypnotic waking state who are engaging in mental imagery. How does the hypnotist lead people into this state of relaxation and imagery?

THE FOUR STEPS IN HYPNOSIS Hypnosis involves four steps. The hypnotist does the following:

1. Minimizes distractions and makes the person to be hypnotized comfortable.

2. Tells the person to concentrate on something specific, such as an imagined scene or the ticking of a watch.

3. Informs the person what to expect in the hypnotic state, such as relaxation or a pleasant floating sensation.

4. Suggests certain events or feelings the person knows will occur or observes occurring, such as "Your eyes are getting tired." When the suggested effects occur, the person interprets them as being caused by the hypnotist's suggestions and accepts them as an indication that something is happening. This increase in the person's expectations that the hypnotist will make things happen in the future makes the person even more suggestible.

INDIVIDUAL VARIATIONS IN HYPNOSIS Some people are more easily hypnotized than others, and some are more strongly influenced by hypnotic suggestions. *Hypnotizability* refers to the extent to which a person's responses *are changed* when they are hypnotized (DeSouza & others, 2020). There is no easy way to know if a person is hypnotizable without first trying to hypnotize the individual. If you have the capacity to immerse yourself deeply in an imaginative activity—listening to a favorite piece of music or reading a novel, for example—you might be a likely candidate (Spiegel, 2010). Still, the relationship between the ability to become completely absorbed in an experience and hypnotizability is weak (Nash, 2001).

Explaining Hypnosis

How does hypnosis have its effects? Contemporary theorists disagree as to whether hypnosis is a divided state of consciousness or simply a learned social behavior.

A DIVIDED STATE OF CONSCIOUSNESS Hilgard (1977, 1992), in his **divided consciousness view of hypnosis,** proposed that hypnosis involves a special state of consciousness in which consciousness is split into separate components. One component follows the hypnotist's commands, while another component acts as a "hidden observer." Hilgard thought that the "hidden observer" explained why people who are hypnotized will not commit acts that would violate their code of ethics.

Hilgard placed one hand of hypnotized people in a bucket of ice-cold water and told them that they would not feel pain but that a part of their mind—a hidden part that would be aware of what was going on—could signal any true pain by pressing a key with the hand that was not submerged (Figure 18). The individuals under hypnosis reported afterward that they had not experienced any pain; yet while their hand had

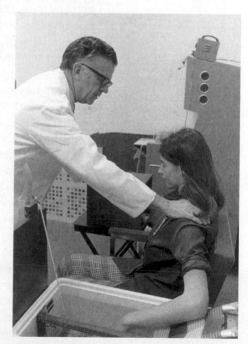

FIGURE 18 Hilgard's Divided Consciousness Experiment Ernest Hilgard tests a participant in the study in which he had individuals place one arm in ice-cold water. Stanford News Service

divided consciousness view of hypnosis Hilgard's view that hypnosis involves a splitting of consciousness into two separate components, one following the hypnotist's commands and the other acting as a "hidden observer."

been submerged in the ice-cold water, they had pressed the key with their nonsubmerged hand, and they had pressed it more frequently the longer their hand was in the cold water. Hilgard thus concluded that in hypnosis, consciousness has a hidden part that stays in contact with reality and feels pain while another part of consciousness feels no pain.

Critics of Hilgard's view suggest that the hidden observer simply demonstrates that the hypnotized person is not in an altered state of consciousness at all. From this perspective, the hidden observer is simply the person themselves, having been given permission to admit to the pain that they were always feeling (Green & others, 2005). This argument is part of the social cognitive behavior view of hypnosis.

social cognitive behavior view of hypnosis

The perspective that hypnosis is a normal state in which the hypnotized person behaves the way they believe that a hypnotized person should behave.

SOCIAL COGNITIVE BEHAVIOR

Some experts are skeptical that hypnosis is an altered state of consciousness (Chaves, 2000; Lynn & Green, 2011). In the **social cognitive behavior view of hypnosis,** hypnosis is a normal state in which the hypnotized person behaves the way they believe a hypnotized person should behave. The social cognitive perspective frames the important questions about hypnosis around cognitive factors—the attitudes, expectations, and beliefs of good hypnotic participants—and around the powerful social context in which hypnosis occurs (Accardi & others, 2013; Lynn & others, 2017). People being hypnotized surrender their responsibility to the hypnotist and follow the hypnotist's suggestions; and they have expectations about what hypnosis is supposed to be like.

Experts have continued to debate whether hypnosis is indeed an altered state of consciousness (Kihlstrom, 2020) or simply a reaction to a special social situation (Lynn & Green, 2011). Although there may be no consensus about what hypnosis is, health professionals have begun to apply this powerful technique to a number of problems.

Uses of Hypnosis

As psychologists' interest in studying consciousness has grown, hypnosis has emerged as a useful tool (Beebe, 2014; Flor, 2014). Some researchers employ hypnosis in a way similar to transcranial magnetic stimulation, to dampen brain processes experimentally (Cox & Bryant, 2008). Combining hypnosis with brain imaging allows researchers to understand both the effects of hypnosis itself and the brain's functioning (Oakley & Halligan, 2011).

Beyond its role in basic research, hypnosis has been applied to a variety of problems. In the United States, practitioners of hypnosis use the technique to treat alcoholism, somnambulism, depression, suicidal tendencies, posttraumatic stress disorder, migraines, overeating, diabetes, and smoking. Whether hypnosis actually works for these diverse problems remains debatable (Brown, 2007). People in hypnosis-based treatment programs rarely achieve dramatic results unless they are already motivated to change. Hypnosis is most effective when combined with psychotherapy.

A long history of research and practice has clearly demonstrated that hypnosis can reduce the experience of pain (Jensen & Patterson, 2014; Syrjala & others, 2014). A fascinating study examined the pain perceptions of hypnotized individuals, with the goal of changing their pain threshold. In this study, the brain of each participant was monitored while each received painful electrical shocks (rated 8 or higher on a 1 to 10 pain scale) (Schulz-Stübner & others, 2004). Those who were hypnotized to find the shocks less painful did rate them as lower in pain (giving them a 3 or less). Furthermore, the brain-scanning results were most interesting: The subcortical brain areas (the brainstem and midbrain) of the hypnotized patients responded the same as those of the patients who were not hypnotized, a finding suggesting that these brain structures recognized the painful stimulation. However, the sensory cortex was not activated in the hypnotized patients, an indication that although they sensed pain on some level, they were never conscious of it. In essence, the "ouch" signal never made it to awareness. If a patient is allergic to anesthesia or it is ill advised for other reasons (the patient has a history of addiction), hypnosis may be a viable alternative. Research shows that it can be effective in pain management for a variety of contexts (Aravena & others, 2020; Garcia & others, 2020; Montenegro & others, 2017).

In summary, although the nature of hypnosis remains a mystery, evidence is increasing that hypnosis can play a role in a variety of health contexts, and it can influence the brain in fascinating ways. For psychologists, part of the ambiguity about the definition of hypnosis arises from the fact that it has been studied in specific social contexts, involving a hypnotist. It is also possible, however, to experience altered states of consciousness without these special circumstances, as we next consider.

self-quiz

1. The type of brain waves that hypnotized people display include
   A. alpha waves.
   B. delta waves.
   C. gamma waves.
   D. theta waves.

2. The divided consciousness theory of hypnosis receives support from evidence that
   A. hypnosis can block sensory input.
   B. hypnosis can affect voluntary, but not involuntary, behaviors.
   C. hypnotized people often seem to play the role of "good hypnotic subjects."
   D. hypnotized people can be aware of pain sensation without experiencing emotional distress.

3. Hypnosis treatments tend to work best when they are accompanied by

A. daily meditation.
B. physical exercise.
C. yoga.
D. psychotherapy.

APPLY IT! 4. Ryan and his friends attend a show by the Great Chorizo, a hypnotist. Chorizo asks for volunteers to be hypnotized, and he picks the first five people who raise their hands. He puts the five people into a trance, and within minutes he has them lying on stage sizzling like slices of bacon in a frying pan. When it is all over, one of Ryan's friends remarks that Chorizo must have amazing powers: "That guy could make a person do anything!" Ryan, who has been working on his critical thinking and the psychology of hypnosis, wisely notes which of the following about Chorizo's act?

A. As long as Chorizo followed the steps of hypnosis described in this text, he probably does have amazing powers of suggestion.
B. Ryan would need to see Chorizo's training and qualifications prior to rendering judgment.
C. Chorizo selected the first five volunteers, and these individuals might have been especially motivated, suggestible, and likely to believe in the effects of hypnosis. There is no way to gauge whether Chorizo could have influence over anyone else.
D. Hypnotizability is similar for all people, so if Chorizo was able to get five people to act like frying bacon, he could probably do just about anything.

5 Meditation

Hypnosis involves a powerful social context, but harnessing the power of consciousness is also possible without the aid of a hypnotist—through meditation. **Meditation** involves attaining a peaceful state of mind in which thoughts are not occupied by worry; the meditator is mindfully present to their thoughts and feelings but is not consumed by them. Let's look at how meditation can enhance well-being and examine more closely what it is.

There are many types of meditative practice (Newcombe & O'Brien-Kop, 2020). They share at least two characteristics: focused attention and open monitoring. *Focused attention* means bringing one's awareness to one's inner life and attending to one's thoughts. It means being psychologically present as one thinks. *Open monitoring* refers to the capacity to observe one's thoughts as they happen without getting preoccupied by them; that is, through open monitoring, the person is able to reflect without becoming attached to a particular thought or idea. Research shows that experienced meditators show a pattern of brain activity that is different from both effortful engagement in thought and sleep or drowsiness (Hinterberger & others, 2014). Meditation (or contemplative practice) is related to developing cognitive skills such as attentional control and executive function (Cásedas & others, 2020; Izzetoglu & others, 2020).

meditation
The attainment of a peaceful state of mind in which thoughts are not occupied by worry; the meditator is mindfully present to their thoughts and feelings but is not consumed by them.

Mindfulness Meditation

Among those practicing meditation are Zen monks who explore the Buddha-nature at the center of their being.
Erin Koran/McGraw Hill

Melissa Munroe, a Canadian woman diagnosed with Hodgkin lymphoma (a cancer of the immune system), was tormented by excruciating pain. Seeking ways to cope with the agony, Munroe enrolled in a meditation program. She was skeptical at first. "What I didn't realize," she said, "is that if people have ever found themselves taking a walk in the countryside or in the forest or on a nice pleasant autumn day . . . and find themselves in a contemplative state, that's a form of meditation." Munroe worked hard to use meditation to control her pain. Interestingly, the way she harnessed the power of her mind to overcome pain was by concentrating her thoughts on the pain—not trying to avoid it.

Using *mindfulness meditation,* a technique practiced by yoga enthusiasts and Buddhist monks, Munroe focused on her pain. By doing so, she was able to isolate the pain from her emotional response to it and to her cancer diagnosis. She grew to see her physical discomfort as bearable. Munroe's success shows that contrary to what a nonmeditator might think, meditation is not about avoiding one's thoughts. Indeed, the effort involved in avoidance steers the person away from the contemplative state. Munroe described her thoughts as like people striding by her on the street, walking in the other direction; she explained, "They come closer and closer, then they pass you by." Her sentiment reflects the open monitoring that is common to many forms of meditation.

Jon Kabat-Zinn (2006, 2009) pioneered using meditation techniques in medical settings. Research by Kabat-Zinn and colleagues has demonstrated the beneficial effects of mindfulness meditation for a variety of conditions, including depression, panic attacks, and anxiety (Miller & others, 1995), chronic pain (Kabat-Zinn & others, 1985), and stress and the skin condition psoriasis (Kabat-Zinn & others, 1998). Many of these effects have also been shown to be long-lasting (Crane & others, 2017).

Richard Davidson and colleagues (including Kabat-Zinn) have studied the brain and immune system changes that might underlie the health and wellness effects of meditation (Davidson & others, 2003; Ferrarelli & others, 2013; Kabat-Zinn & Davidson, 2012). They performed magnetic resonance imaging (MRI) on the brains of individuals who were in a standard eight-week meditation-training program. After the training program and as compared to a control group, those in the meditation program reported reduced anxiety and fewer negative emotions. Furthermore, brain scans revealed that these individuals showed increased activation in the left hemisphere. Such activation is associated with happiness. In addition, the meditators had a better immune system response to a flu vaccine (Davidson & others, 2003). These results suggest that our conscious minds may have a role to play in enhancing our psychological and physical health (Kabat-Zinn, 2019).

Lovingkindness Meditation

Another form of meditation is called *lovingkindness* meditation. The goal of this meditative practice is the development of loving acceptance of oneself and others. In lovingkindness meditation, the meditator begins by developing warm, accepting feelings toward oneself. Then, the person moves to meditate about a very close other, such as a family member the person loves and respects. Over time, lovingkindness meditation widens to include an ever-broadening circle of people. Lovingkindness fosters feelings of warmth, friendliness, compassion, and appreciative joy. At its highest level, the person experiences a sense of equanimity, or a feeling of calm acceptance, an openness to their thoughts and feelings without becoming preoccupied with them (Kok & others, 2013; Koopmann-Holm & others, 2020).

Lovingkindness meditation leads to heightened feelings of social connection, positive emotions (Fredrickson & others, 2008), better coping with stress (Fredrickson & others, 2003), greater optimism (Koopmann-Holm & others, 2020), and slower biological aging (Le Nguyen & others, 2019). Given its focus on compassion toward others, might this type of contemplative practice also possess payoffs for the social world? Some research suggests that this form of meditation may help to combat prejudice. Research has begun to support the idea that mindfulness practices can help reduce prejudice (Kang & others, 2014; Price-Blackshear & others, 2017).

The Meditative State of Mind

What actually is the meditative state of mind? As a physiological state, meditation shows qualities of sleep and wakefulness yet is distinct from both. You may have experienced a state called *hypnagogic reverie*—an overwhelming feeling of wellness right before you fall asleep, the sense that everything is going to work out. Meditation has been compared to this relaxed sense that all is well (Friedman & others, 1998).

In a study of Zen meditators, researchers examined what happens when people switch from their normal waking state to a meditative state (Ritskes & others, 2003). Using functional MRI (fMRI), the experimenters got images of the brain before and after the participants entered the meditative state. They found that the switch to meditation involved initial increases in activation in the basal ganglia and prefrontal cortex (the now familiar area that is often activated during consciousness). However, and interestingly, they also found that these initial activations led to decreases in the anterior cingulate, a brain area that is thought to be associated with acts of will. These results provide a picture of the physical events of the brain that are connected with the somewhat paradoxical state of meditation—controlling one's thoughts in order to let go of the need to control.

Getting Started with Meditation

Would you like to experience the meditative state? If so, you can probably reach that state by following some simple instructions:

- Find a quiet place and a comfortable chair.
- Sit upright in the chair, rest your chin comfortably on your chest, and place your arms in your lap. Close your eyes.
- Now focus on your breathing. Every time you inhale and every time you exhale, pay attention to the sensations of air flowing through your body, the feeling of your lungs filling and emptying.
- After you have focused on several breaths, begin to repeat silently to yourself a single word every time you breathe out. You can make a word up, use the word *one*, or try a word associated with an emotion you want to produce, such as *trust, love, patience, happy*, or *lovingkindness*. Experiment with several different words to see which one works for you.
- If you find that thoughts are intruding and you are no longer attending to your breathing, refocus on your breathing and say your chosen word each time you exhale.

After you have practiced this exercise for 10 to 15 minutes, twice a day, every day for two weeks, you will be ready for a shortened version. If you notice that you are experiencing stressful thoughts or circumstances, simply meditate, on the spot, for several minutes. If you are in public, you do not have to close your eyes; just fix your gaze on a nearby object, attend to your breathing, and say your word silently every time you exhale.

Meditation is an age-old practice. Without explicitly mentioning meditation, some religions advocate related practices such as daily prayer and peaceful introspection. Whether the practice involves praying over rosary beads, chanting before a Buddhist shrine, or taking a moment to commune with nature, a contemplative state clearly has broad appeal and conveys many benefits. Research on the contemplative state suggests that there are good reasons why human beings have been harnessing its beneficial powers for centuries.

self-quiz

1. A presleep state of calmness and wellness, with an accompanying feeling of optimism, is called
   A. a trance.
   B. a hypnagogic reverie.
   C. meditation.
   D. mindfulness.

2. Which of the following is *true* about mindfulness meditation?
   A. It involves clearing the mind of all thoughts.
   B. It has not been demonstrated to be effective in pain management.
   C. It increases activation of the right hemisphere of the brain.
   D. It focuses thoughts on specific bodily sensations.

3. In terms of physiology, meditation shows characteristics of
   A. sleep and wakefulness.
   B. sleep and hypnosis.
   C. sleep and hallucinations.
   D. sleep and daydreaming.

APPLY IT! 4. Patricia enjoys walking by a river next to her apartment building nearly every day. As she walks, she loses track of time and just thinks quietly about her life and experiences, letting her thoughts come and go without concern. Sometimes she stops to gaze at the river and enjoys its quiet glistening. After her walks, she always feels refreshed and ready for life's next challenges. One day a friend sees her while walking and says, "Oops, I didn't mean to interrupt your meditation." Patricia says, "Oh I'm not meditating, I'm just taking a walk." Is Patricia right?
   A. Patricia is right. If she does not think she is meditating, then she cannot be meditating.
   B. Patricia is right because she is not sitting down, and she is not repeating a word over and over.
   C. Patricia is not correct. She may not know it, but she is engaged in a contemplative state.
   D. Patricia is not correct. She is meditating because she is in a hypnotic state.

SUMMARY

① The Nature of Consciousness

Consciousness is the awareness of external events and internal sensations, including awareness of the self and thoughts about experiences. Most experts agree that consciousness is likely distributed across the brain. The association areas and prefrontal lobes are believed to play important roles in consciousness.

William James described the mind as a stream of consciousness. Consciousness occurs at different levels of awareness that include higher-level awareness (controlled processes and selective attention), lower-level awareness (automatic processes and daydreaming), altered states of consciousness (produced by drugs, trauma, fatigue, and other factors), subconscious awareness (waking subconscious awareness, sleep, and dreams), and no awareness (unconscious thought).

② Sleep and Dreams

Sleep is a natural state of rest for the body and mind that involves the reversible loss of consciousness. The biological rhythm that regulates the daily sleep/wake cycle is the circadian rhythm. The part of the brain that keeps our biological clocks synchronized is the suprachiasmatic nucleus, a small structure in the hypothalamus that registers light. Such things as jet travel and work shifts can desynchronize biological clocks. Some strategies are available for resetting the biological clock.

We need sleep for physical restoration, adaptation, growth, and memory. Research studies increasingly reveal that people do not function optimally when they are sleep deprived.

Stages of sleep correspond to massive electrophysiological changes that occur in the brain and that can be assessed by an EEG. The human sleep cycle is defined by stages. In Stage W, the person is awake. In the non-REM stages (stages N1 to N3), the person does not experience rapid eye movement but moves from light sleep to deep sleep. Stage N3 is the deepest sleep. Most dreaming occurs during stage R or REM sleep. A sleep cycle of five stages lasts about 90 to 100 minutes and recurs several times during the night. The REM stage lasts longer toward the end of a night's sleep.

The sleep stages are associated with distinct patterns of neurotransmitter activity. Levels of the neurotransmitters serotonin, norepinephrine, and acetylcholine decrease as the sleep cycle progresses from stage N1 to N3. Stage R, REM sleep, begins when the reticular formation raises the level of acetylcholine.

Sleep plays a role in a large number of diseases and disorders. Neurons that control sleep interact closely with the immune system, and when our bodies are fighting infection our cells produce a substance that makes us sleepy. Individuals with depression often have sleep problems.

Many people in the United States suffer from chronic, long-term sleep disorders that can impair normal daily functioning. These include insomnia, sleepwalking and sleep talking, nightmares and night terrors, narcolepsy, and sleep apnea.

Contrary to popular belief, most dreams are not bizarre or strange. Freud thought that dreams express unconscious wishes in disguise. The cognitive theory of dreaming attempts to explain dreaming in terms of the same cognitive concepts that are used in studying the waking mind. According to activation-synthesis theory, dreaming occurs when the cerebral cortex synthesizes neural signals emanating from activity in the lower part of the brain. In this view, the rising level of acetylcholine during REM sleep plays a role in neural activity in the brainstem that the cerebral cortex tries to make sense of.

Psychoactive Drugs

Psychoactive drugs act on the nervous system to alter states of consciousness, modify perceptions, and change moods. Humans are attracted to these types of drugs because they ease adaptation to change.

Addictive drugs activate the brain's reward system by increasing dopamine concentration. The reward pathway involves the ventral tegmental area (VTA) and nucleus accumbens (NAc). The abuse of psychoactive drugs can lead to tolerance, psychological and physical dependence, and addiction—a pattern of behavior characterized by a preoccupation with using a drug and securing its supply.

Depressants slow down mental and physical activity. Among the most widely used depressants are alcohol, barbiturates, tranquilizers, and opioids.

After caffeine, alcohol is the most widely used drug in the United States. The high rate of alcohol abuse by high school and college students is especially alarming. Alcoholism is a disorder that involves long-term, repeated, uncontrolled, compulsive, and excessive use of alcoholic beverages that impairs the drinker's health and work and social relationships.

Stimulants increase the central nervous system's activity and include caffeine, nicotine, amphetamines, cocaine, and MDMA (Ecstasy). Hallucinogens modify a person's perceptual experiences and produce visual images that are not real. Marijuana has a mild hallucinogenic effect; LSD has a strong one.

 ## Hypnosis

Hypnosis is a psychological state or possibly altered attention and awareness in which the individual is unusually receptive to suggestions. The hypnotic state is different from a sleep state, as confirmed by EEG recordings. Inducing hypnosis involves four basic steps, beginning with minimizing distractions and making the person feel comfortable and ending with the hypnotist's suggesting certain events or feelings that the person knows will occur or observes occurring.

There are substantial individual variations in people's susceptibility to hypnosis. People in a hypnotic state are unlikely to do anything that violates their morals or that involves a real danger.

Two theories have been proposed to explain hypnosis. In Hilgard's divided consciousness view, hypnosis involves a divided state of consciousness, a splitting of consciousness into separate components. One component follows the hypnotist's commands; the other acts as a hidden observer. In the social cognitive behavior view, hypnotized individuals behave the way they believe hypnotized individuals are expected to behave.

 ## Meditation

Meditation refers to a state of quiet reflection. Meditation has benefits for a wide range of psychological and physical illnesses. Meditation can also benefit the body's immune system. Research using fMRI suggests that meditation allows an individual to control their thoughts in order to "let go" of the need to control.

Mindfulness meditation is a powerful tool for managing life's problems. How we think about our lives and experiences plays a role in determining whether we feel stressed and worried or challenged and excited about life. Seeking times of quiet contemplation can have a positive impact on our abilities to cope with life's ups and downs. The goal of lovingkindness meditation is the development of loving acceptance of oneself and others. This type of meditation fosters feelings of warmth, friendliness, compassion, and appreciative joy.

KEY TERMS

activation-synthesis theory
addiction
alcoholism
autism spectrum disorder
automatic processes
barbiturates
biological rhythms
circadian rhythms
cognitive theory of dreaming
consciousness

controlled processes
depressants
divided consciousness view of hypnosis
executive function
hallucinogens
hypnosis
latent content
manifest content
meditation

opioids
physical dependence
psychoactive drugs
psychological dependence
REM sleep
reticular activating system
sleep
social cognitive behavior view of hypnosis
stimulants

stream of consciousness
substance use disorder
suprachiasmatic nucleus (SCN)
theory of mind
tolerance
tranquilizers
unconscious thought

ANSWERS TO SELF-QUIZZES

Section 1: 1. B; 2. A; 3. C; 4. B
Section 2: 1. D; 2. B; 3. B; 4. B

Section 3: 1. D; 2. D; 3. A; 4. C
Section 4: 1. A; 2. D; 3. D; 4. C

Section 5: 1. B; 2. D; 3. A; 4. C

5 Learning

Service Dogs: Helping Heroes Heal

Returning home after serving in combat can be stressful and lonely for many military veterans. Civilian life may include coping with the wounds of war: physical injury and psychological trauma. Across the United States, nonprofit groups have paired returning military veterans with rescue dogs trained to help them make the transition from military to civilian life. Some dogs provide the simple comfort of a loving pet and help with many tasks of daily living. For example, Pups4Patriots trains rescued dogs as service animals, especially to assist former military members with post-traumatic stress disorder (a trauma-related psychological disorder) and traumatic brain injury (Cahn, 2020). Canine Companions for Independence trains dogs to assist their owners by fetching medications, helping with household tasks, and so forth. Such services rely on training that turns rescue dogs into trained professionals. The hundreds of thousands of service dogs working in the United States are trained to aid people with a variety of disabilities. Their skills are amazing. They provide sound discrimination for people who are hard of hearing, assist those with limited mobility, and retrieve items that are out of reach; they locate people, bathrooms, elevators, and lost cell phones. They open and close doors, help people dress and undress, flush toilets, and even put clothes in a washer and dryer. Truly, service dogs are highly skilled professionals. Service dogs are trained to perform these complex acts using the principles that psychologists have uncovered studying the processes that underlie learning, the focus of this chapter.

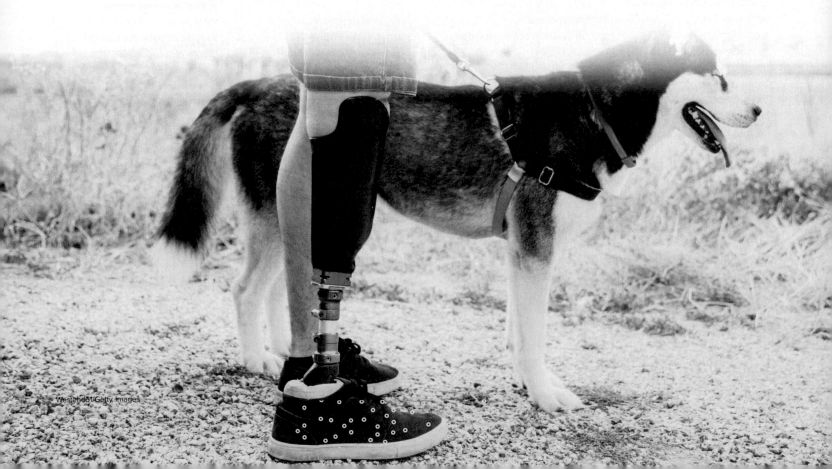

This chapter begins by defining learning and sketching out its main types—associative learning and observational learning. We then turn to two types of associative learning—classical conditioning and operant conditioning—followed by a close look at observational learning. We next explore the role of cognitive processes in learning, before finally considering biological, cultural, and psychological constraints on learning. As you read, ask yourself about your own beliefs concerning learning. If a dog can learn to perform household chores, then surely the human potential for learning has barely been tapped.

1 Types of Learning

Learning anything new involves change. Once you learned the alphabet, it did not leave you; it became part of a "new you" who had been changed through the process of learning. Similarly, once you learn how to drive a car, you do not have to go through the process again at a later time. If you decide to take up a new hobby, like snowboarding, you may break a few bones along the way, but at some point you probably will learn a trick or two through the experience, changing from a novice to an enthusiast who can at least stay on top of a snowboard.

By way of experience, too, you may have learned that you have to study to do well on a test, that there usually is an opening act at a music concert, and that a field goal in U.S. football adds 3 points to the score. Putting these pieces together, we arrive at a definition of **learning**: a systematic, relatively permanent change in behavior that occurs through experience.

If someone were to ask you what you learned in class today, you might mention new ideas you heard about, lists you memorized, or concepts you mastered. However, how would you define learning if you could not refer to unobservable mental processes? You might follow the lead of behavioral psychologists. **Behaviorism** is a theory of learning that focuses on observable behaviors. From the behaviorist perspective, understanding the causes of behavior requires looking at the environmental factors that produce them. Behaviorists view internal states like thinking, wishing, and hoping as behaviors that are caused by external factors as well. Psychologists who examine learning from a behavioral perspective define learning as relatively stable, observable changes in behavior. The behavioral approach has emphasized general laws that guide behavior change and make sense of some of the puzzling aspects of human life.

Behaviorism maintains that the principles of learning are the same whether we are talking about nonhuman animals or people. Because of the influence of behaviorism, psychologists' understanding of learning started with studies of rats, cats, pigeons, and even raccoons. A century of research on learning in animals and in humans suggests that many of the principles generated initially in research on animals also apply to humans (Domjan, 2015).

In this chapter we look at two types of learning: associative learning and observational learning. **Associative learning** occurs when we make a connection, or an association, between two events. *Conditioning* is the process of learning these associations. There are two types of conditioning: classical and operant, both of which have been studied by behaviorists.

In *classical conditioning,* organisms learn the association between two stimuli. As a result of this association, organisms learn to anticipate events. For example, lightning is associated with thunder and regularly precedes it. Thus, when we see lightning, we anticipate that we will hear thunder soon afterward. Fans of horror films know the power of classical conditioning. Watching a

learning
A systematic, relatively permanent change in behavior that occurs through experience.

behaviorism
A theory of learning that focuses solely on observable behaviors, discounting the importance of such mental activity as thinking, wishing, and hoping.

associative learning
Learning that occurs when an organism makes a connection, or an association, between two events.

Learning is RELATIVELY permanent-sometimes we forget what we've learned. Also, learning involves EXPERIENCE. Changes in behavior that result from physical maturation would not be considered learning.

This is going to sound very abstract right now. Hang on-once we get to the details, it will make sense.

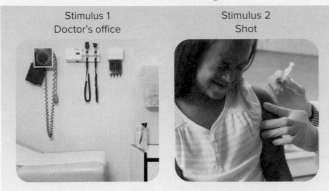

Stimulus 1
Doctor's office

Stimulus 2
Shot

Behavior

Consequences

FIGURE 1 Associative Learning: Comparing Classical and Operant Conditioning (*left*) In this example of classical conditioning, a child associates a doctor's office (stimulus 1) with getting a painful injection (stimulus 2). (*right*) In this example of operant conditioning, performing well in a swimming competition (behavior) becomes associated with getting awards (consequences).

(doctor office): Image Source/Jupiterimages; (shot): Amble Design/Shutterstock; (swimmer): Ryan McVay/Getty Images; (medalist): technotr/Getty Images

terrifying film, we know that when the camera focuses on a character, but leaves the frame open to show what is behind that person, something terrible is about to happen.

In *operant conditioning,* organisms learn the association between a behavior and a consequence, such as a reward. As a result of this association, organisms learn to increase behaviors that are followed by rewards and to decrease behaviors that are followed by punishment. For example, children are likely to repeat their good manners if their parents reward them with candy after they have shown good manners. Also, if children's bad manners are followed by scolding words and harsh glances by parents, the children are less likely to repeat the bad manners. Figure 1 compares classical and operant conditioning.

Much of learning, however, is not a result of direct consequences but rather of exposure to models performing a behavior or skill. For instance, as you watch someone shoot baskets, you get a sense of how it is done. The learning that takes place when a person observes and imitates another's behavior is called **observational learning**. Observational learning is a common way that people learn in educational and other settings. Observational learning is different from the associative learning described by behaviorism because it relies on mental processes: The learner has to pay attention, remember, and reproduce what the model did. Observational learning is especially important to human beings. In fact, watching other people is one way that human infants acquire skills (Bohn & others, 2020).

observational learning
Learning that occurs through observing and imitating another's behavior.

Have you ever noticed that humans' eyes differ from other animals' eyes because the 'whites' can be seen? It might be that this characteristic allows humans to model one another closely—because we can see what the model is looking at.

Human infants differ from baby monkeys in their strong reliance on imitation (Marshall & Meltzoff, 2014). After watching an adult model perform a task, a baby monkey will figure out its own way to do it, but a human infant will do exactly what the model did. Imitation may be the human baby's way to solve the huge problem it faces: to learn the vast amount of cultural knowledge that is part of human life. Many of our behaviors are rather arbitrary. Why do we clap to show approval or wave "hello" or "bye-bye"? The human infant has a lot to learn and may be well served to follow the old adage, "When in Rome, do as the Romans do."

Learning applies to many areas of acquiring new behaviors, skills, and knowledge (Woolcott, 2020).

Interestingly, the human capacity to learn has inspired computer scientists and engineers who work in the area of artificial intelligence. Artificial intelligence involves creating machines capable of performing activities that require intelligence when people do them. Machine learning is a branch of artificial intelligence that focuses on creating machines that can change their behavior in response to data without a human being stepping in. Do machines actually learn? To explore this provocative question, see the Challenge Your Thinking.

Do Machines Actually *Learn*?

Entering a word in a text box on a smartphone, you have likely noticed the predictive text that appears just above the keyboard, offering suggestions to finish your sentence. You type in "what" and it suggests, "time" or "happened." This is just one example of machine learning, that branch of artificial intelligence in which computers use data to reach conclusions, make decisions, offer suggestions, and so forth. That predictive text is a result of a machine learning about human language and your typical texting behavior. Machine learning occurs when a computer program is able to use data to direct its own actions without the intervention of a human programmer. The first example of machine learning was a program for a checkers game that learned from its successes and failures, invented by Arthur Samuel in 1959. Today, across a wide range of contexts, computer scientists "train" computers to rely on ever expanding databases to direct their actions.

Phonlamai Photo/Shutterstock

The question is, are computers actually learning? As you consider this question you might recognize that the answer depends very much on what is meant by learning (Burgos, 2018). Consider the definition offered in the main text earlier: "a systematic, relatively permanent change in behavior that occurs through experience." Nowhere in this definition do the words "living organism" appear. Does that mean that nonliving entities, such as computers, can learn?

Let's apply our definition to an example of machine learning, an email spam filter. Spam emails are messages often involving sales or even fraud. For example, someone tells you that they will deposit $1 million into your bank account if you give them your account information. Emails like this often appear with the warning "SUSPECTED SPAM." This warning is the product of machine learning. The spam filter relies on a huge database of all words and phrases that might reveal whether a message is likely being sent with malicious intent.

The program searches all incoming messages for these indicators, calculating the probability that any given message is spam. The program monitors its own success or failure in identifying spam and then tinkers with the database—perhaps adding a phrase or deleting another. Over time, it becomes more and more precise at detecting spam emails. Has that program then, learned? What separates what it has done from what a dog, a cat, or human has done when learning occurs?

In computer science, deep learning refers to modeling computer programs after the human brain, sensory processes, and behavior. We might ask, If computers simply model human behavior, can we say they are really learning (Burgos, 2018)? Consider that when this type of program starts processing, it starts changing itself and its behavior based on experience.

Media depictions of artificial intelligence often imply that humans will be in grave danger if intelligent machines begin to think for themselves. Are we all at risk of being enslaved by robot overlords? Probably not, but the feeling that machines might replace us may explain our reluctance to call their relatively permanent change in behavior based on experience true learning. If a computer program alters its own behavior based on experience, might that be considered learning?

What Do You Think?

- Other than predictive text and spam filters, what is one example of machine learning you have encountered in your daily life?

- Do you think computers truly learn? Why or why not?

self-quiz

1. Any situation that involves learning
   A. requires some relatively permanent change to occur.
   B. requires a great deal of effort.
   C. involves conscious determination.
   D. is relatively automatic.

2. A cat that associates the sound of a can opener with being fed has learned through
   A. behaviorism.
   B. operant conditioning.
   C. classical conditioning.
   D. observational learning.

3. Which one of the following statements is *true* about learning?
   A. Learning can be accomplished only by higher-level species, such as mammals.
   B. Learning is never permanent.
   C. Learning occurs through experience.
   D. Learning processes in humans are distinct from learning processes in animals.

APPLY IT! 4. After seeing dogs catching Frisbees in the park, Lionel decides that he wants to teach his dog Ivan to do it, too. He takes Ivan to the park and sits with him, making sure that he watches the other dogs successfully catching Frisbees. What technique is Lionel using on Ivan, and what are the chances for success?
   A. He is using associative learning, and his chances for success are very good, because dogs and humans both learn this way.
   B. He is using operant conditioning, and his chances for success are very good, because dogs and humans both learn this way.
   C. He is using observational learning, and his chances for success are pretty bad, because dogs are not as likely as people to learn in this way.
   D. He is using classical conditioning, and his chances for success are pretty bad, because dogs are much less likely than people to learn in this way.

2 Classical Conditioning

Early one morning, Bob is in the shower. While he showers, his spouse enters the bathroom and flushes the toilet. Scalding hot water bursts down on Bob, causing him to yell in pain. The next day, Bob is back for his morning shower, and once again his spouse enters the bathroom and flushes the toilet. Panicked by the sound of the toilet flushing, Bob yelps in fear and jumps out of the shower stream. Bob's panic at the sound of the toilet illustrates the learning process of **classical conditioning**, in which a neutral stimulus (the sound of a toilet flushing) becomes associated with a meaningful stimulus (the pain of scalding hot water) and acquires the capacity to elicit a similar response (panic).

classical conditioning Learning process in which a neutral stimulus becomes associated with an innately meaningful stimulus and acquires the capacity to elicit a similar response.

Pavlov's Studies

Even before beginning this course, you might have heard about Pavlov's dogs. The Russian physiologist Ivan Pavlov's work is very well known. Still, it is easy to take its true significance for granted. Importantly, Pavlov demonstrated that neutral aspects of the environment can attain the capacity to evoke responses through pairing with other stimuli and that bodily processes can be influenced by environmental cues.

In the early 1900s, Pavlov was interested in the way the body digests food. In his experiments, he routinely placed meat powder in a dog's mouth, causing the dog to salivate. By accident, Pavlov noticed that the meat powder was not the only stimulus that caused the dog to salivate. The dog salivated in response to a number of stimuli associated with the food, such as the sight of the food dish, the sight of the person who brought the food into the room, and the sound of the door closing when the food arrived. Pavlov recognized that the dog's association of these sights and sounds with the food was an important type of learning, which came to be called *classical conditioning*.

Pavlov wanted to know why the dog salivated in reaction to various sights and sounds before eating the meat powder. He observed that the dog's behavior included both unlearned and learned components. The unlearned part of classical conditioning is based on the fact that some stimuli automatically produce certain responses apart from any prior

Pavlov (the white-bearded man in the center) is shown demonstrating the nature of classical conditioning to students at the Military Medical Academy in Russia.
Sovfoto/Universal Images Group/Getty Images

learning; in other words, they are innate (inborn). *Reflexes* are such automatic stimulus-response connections. They include salivation in response to food, nausea in response to spoiled food, shivering in response to low temperature, coughing in response to throat congestion, pupil constriction in response to light, and withdrawal in response to pain. An **unconditioned stimulus (US)** is a stimulus that produces a response without prior learning; food was the US in Pavlov's experiments. An **unconditioned response (UR)** is an unlearned reaction that is automatically elicited by the US. Unconditioned responses are involuntary; they happen in response to a stimulus without conscious effort. In Pavlov's experiment, salivating in response to food was the UR.

Now, what about the learned part? In classical conditioning, learning is demonstrated in a response to a stimulus that does not evoke an automatic response. In classical conditioning, a **conditioned stimulus (CS)** is a previously neutral stimulus that eventually elicits a conditioned response after being paired with the unconditioned stimulus. The **conditioned response (CR)** is the learned response to the conditioned stimulus that occurs after CS–US pairing (Pavlov, 1927). Sometimes conditioned responses are quite similar to unconditioned responses, but typically they are not as strong.

In studying a dog's response to various stimuli associated with meat powder, Pavlov rang a bell before giving meat powder to the dog. Until then, ringing the bell did not have a particular effect on the dog, except perhaps to wake the dog from a nap. The bell was a neutral stimulus. However, the dog began to associate the sound of the bell with the food and salivated when it heard the bell. The bell had become a conditioned (learned) stimulus (CS), and salivation was now a conditioned response (CR). In the case of Bob's interrupted shower, the sound of the toilet flushing was the CS, and panicking was the CR after the scalding water (US) and the flushing sound (CS) were paired. Figure 2 summarizes how classical conditioning works, identifying each step in Pavlov's procedure.

unconditioned stimulus (US)
A stimulus that produces a response without prior learning.

conditioned response (CR)
The learned response to the conditioned stimulus that occurs after a conditioned stimulus–unconditioned stimulus pairing.

unconditioned response (UR)
An unlearned reaction that is automatically elicited by the unconditioned stimulus.

conditioned stimulus (CS)
A previously neutral stimulus that eventually elicits a conditioned response after being paired with the unconditioned stimulus.

Note that the association between food and salivating is natural (unlearned), while the association between a bell and salivating is learned.

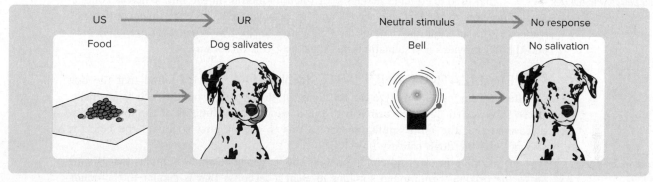

Before Conditioning

US → UR
Food → Dog salivates

Neutral stimulus → No response
Bell → No salivation

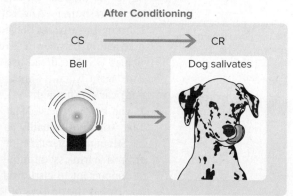

Conditioning

Neutral stimulus + US → UR
Bell + Food → Dog salivates

After Conditioning

CS → CR
Bell → Dog salivates

FIGURE 2 Pavlov's Classical Conditioning In one experiment, Pavlov presented a neutral stimulus (bell) just before an unconditioned stimulus (food). The neutral stimulus became a conditioned stimulus by being paired with the unconditioned stimulus. Subsequently, the conditioned stimulus (bell) by itself was able to elicit the dog's salivation.

Research has shown that salivation can be used as a conditioned response not only in dogs and humans but also in, of all things, cockroaches. In one study, researchers paired the smell of peppermint (the CS, which was applied to the cockroaches' antennae) with sugary water (the US) (Watanabe & Mizunami, 2007). Cockroaches naturally salivate (the UR) in response to sugary foods, and after repeated pairings between the peppermint smell and sugary water, the cockroaches salivated in response to the peppermint scent (the CR). Collecting and measuring the cockroach saliva, the researchers found that the cockroaches had slobbered over that scent for two minutes.

W Awesome addition to any résumé: Cockroach Saliva Technician.

ACQUISITION

Whether it is human beings, dogs, or cockroaches, the first part of classical conditioning is called acquisition. **Acquisition** is the initial learning of the connection between the CS and US when these two stimuli are paired (as with the peppermint scent and the sugary water). During acquisition, the CS is repeatedly presented followed by the US. Eventually, the CS will produce a response. Note that classical conditioning is a type of learning that occurs without awareness or effort, based on the presentation of two stimuli together. For this pairing to work, however, two important factors must be present: contiguity and contingency.

Contiguity simply means that the CS and US are presented very close together in time—even a mere fraction of a second (Austen & Sanderson, 2020). In Pavlov's work, if the bell had rung 20 minutes before the presentation of the food, the dog probably would not have associated the bell with the food. However, pairing the CS and US close together in time is not all that is needed for conditioning to occur.

Contingency means that the CS must not only precede the US closely in time; it must also serve as a reliable indicator that the US is on its way (Mertens & Engelhard, 2020; Rescorla, 1966, 1988, 2009). To get a sense of the importance of contingency, imagine that the dog in Pavlov's experiment is exposed to a ringing bell at random times all day long. Whenever the dog receives food, the delivery of the food always immediately follows a bell ring. However, in this situation, the dog will not associate the bell with the food, because the bell is not a reliable signal that food is coming: It rings a lot when no food is on the way. Whereas contiguity refers to the fact that the CS and US occur close together in time, contingency refers to the information value of the CS relative to the US. When contingency is present, the CS provides a systematic signal that the US is on its way.

W So, contiguity means the CS and US are close in time. Contingency means the CS is reliably and specifically paired with the US.

acquisition
The initial learning of the connection between the unconditioned stimulus and the conditioned stimulus when these two stimuli are paired.

GENERALIZATION AND DISCRIMINATION

Pavlov found that the dog salivated in response not only to the bell tone but also to other sounds, such as a whistle. These sounds had not been paired with the unconditioned stimulus of the food. Pavlov discovered that the more similar the noise was to the original sound of the bell, the stronger was the dog's salivary flow.

Generalization in classical conditioning is the tendency of a new stimulus that is similar to the original conditioned stimulus to elicit a response that is similar to the conditioned response (Andreatta & others, 2020). Generalization has value in preventing learning from being tied too much to a specific stimuli. For example, once you learn the association between a given CS (say, flashing police lights behind your car) and a particular US (the dread associated with being pulled over), you do not have to learn it all over again when a similar stimulus presents itself (a never-before seen police car behind your car on the highway). In classical conditioning, this is called stimulus generalization. If you become nervous at the sight of a car that looks like a police car (but is not one), that would also qualify as stimulus generalization.

Stimulus generalization is not always beneficial. For example, the cat that generalizes from a harmless minnow to a dangerous piranha has a major problem; therefore, it is important to also discriminate among stimuli. **Discrimination** in classical conditioning is the process of learning to respond to certain stimuli and not others (Bergstrom, 2020). To produce discrimination, Pavlov gave food to the dog only after ringing the bell and not after other sounds. In this way, the dog learned to distinguish between the bell and other sounds.

generalization (classical conditioning)
The tendency of a new stimulus that is similar to the original conditioned stimulus to elicit a response that is similar to the conditioned response.

W Note that 'discrimination' here simply means that the organism responds differently to different stimuli.

discrimination (classical conditioning)
The process of learning to respond to certain stimuli and not others.

EXTINCTION AND SPONTANEOUS RECOVERY

extinction (classical conditioning)
The weakening of the conditioned response when the unconditioned stimulus is absent.

After conditioning the dog to salivate at the sound of a bell, Pavlov rang the bell repeatedly in a single session and did not give the dog any food. Eventually the dog stopped salivating. This result is **extinction,** which in classical conditioning is the weakening of the conditioned response when the unconditioned stimulus is absent (Konrad & others, 2020). Without continued association with the US, the CS loses its power to produce the CR. Research using brain imagery has shown that extinction of rewarding associations is reflected in enhanced brain activity in the amygdalae and nucleus accumbens (part of the brain's reward center) (Kruse & others, 2020).

spontaneous recovery
The process in classical conditioning by which a conditioned response can recur after a time delay, without further conditioning.

Extinction is not always the end of a conditioned response (Gallistel & Papachristos, 2020). The day after Pavlov extinguished the conditioned salivation to the sound of a bell, he took the dog to the laboratory and rang the bell but still did not give the dog any meat powder. The dog salivated, indicating that an extinguished response can spontaneously recur. **Spontaneous recovery** is the process in classical conditioning by which a conditioned response can recur after a time delay, without further conditioning. Consider an example of spontaneous recovery you may have experienced: You thought that you had forgotten about (extinguished) a former romantic partner, but then you found yourself in a particular context (perhaps the restaurant where you always dined together), and you suddenly got a mental image of your ex, accompanied by an emotional reaction to this person from the past (spontaneous recovery).

Figure 3 shows the sequence of acquisition, extinction, and spontaneous recovery. Spontaneous recovery can occur several times, but as long as the conditioned stimulus is presented alone (that is, without the unconditioned stimulus), spontaneous recovery becomes weaker and eventually ceases.

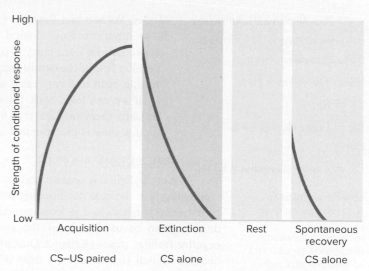

FIGURE 3   The Strength of a Classically Conditioned Response During Acquisition, Extinction, and Spontaneous Recovery During acquisition, the conditioned stimulus and unconditioned stimulus become associated. As the graph shows, when this association occurs, the strength of the conditioned response increases. During extinction, the conditioned stimulus is presented alone, and as can be seen, the result is a decrease in the conditioned response. After a rest period, spontaneous recovery appears, although the strength of the conditioned response is not nearly as great at this point as it was after a number of CS–US pairings. When the CS is presented alone again, after spontaneous recovery, the response is extinguished rapidly.

Classical Conditioning in Humans

Classical conditioning has a great deal of survival value for human beings. Here we review examples of classical conditioning at work in human life.

EXPLAINING FEARS   Classical conditioning provides an explanation of fears (Andreatta & others, 2020; Bergstrom 2020). John B. Watson (who coined the term *behaviorism*) and Rosalie Rayner (1920) demonstrated classical conditioning's role in the development of fears with an infant named Albert.

First, they showed Albert a white laboratory rat to see if he was afraid of it. He was not (so the rat was a neutral stimulus or CS). As Albert played with the rat, the researchers sounded a loud noise behind his head (the noise was then the US). The noise caused little Albert to cry (the UR). After only seven pairings of the loud noise with the white rat, Albert began to fear the rat even when the noise was not sounded (the CR). Albert's fear was generalized to a rabbit, a dog, and a sealskin coat.

Watson and Rayner conditioned 11-month-old Albert to fear a white rat by pairing the rat with a loud noise. When little Albert was later presented with other stimuli similar to the rat, such as the rabbit shown here with Albert, he was afraid of them, too. This study illustrates stimulus generalization in classical conditioning, as well as many ethical issues.
Professor Benjamin Harris, University of New Hampshire

counterconditioning
A classical conditioning procedure for changing the relationship between a conditioned stimulus and its conditioned response.

Today, Watson and Rayner's (1920) study would violate the ethical guidelines of the American Psychological Association. In any case, Watson correctly concluded that we learn many of our fears through classical conditioning. We might develop fear of the dentist because of a painful experience, fear of driving after having been in a car crash, and fear of dogs after having been bitten by one.

If we can learn fears through classical conditioning, we also can possibly unlearn them through that process (Lipp & others, 2020). In fact, classical conditioning is incorporated into treatments for phobias.

BREAKING HABITS   Psychologists have applied classical conditioning to helping individuals unlearn certain feelings and behaviors. For example, **counterconditioning** is a classical conditioning procedure for changing the relationship between a conditioned stimulus and its conditioned response (Keller & others, 2020). Counterconditioning can be used to break the association between certain stimuli (spiders) and negative feelings (fear) (Keller & Dunsmoor, 2020). It can also be used to break the link between stimuli (for instance, a drug of abuse) with positive feelings (the pleasure produced by the drug) (Keller & others, 2020).

Aversive conditioning is a form of treatment that involves repeated pairings of a stimulus with a very unpleasant stimulus. Electric shocks, loud noises, and nausea-inducing substances are examples of noxious stimuli that can be used in aversive conditioning. In a treatment to reduce drinking, for example, every time a person drinks an alcoholic beverage, they also consume a mixture that induces nausea. In classical conditioning terminology, the alcoholic beverage is the conditioned stimulus, and the nausea-inducing agent is the unconditioned stimulus. Through a repeated pairing of alcohol with the nausea-inducing agent, alcohol becomes the conditioned stimulus that elicits nausea, the conditioned response. As a consequence, alcohol no longer is associated with something pleasant but rather something highly unpleasant. Antabuse (disulfiram), a drug treatment for alcoholism since the late 1940s, is based on this association (Ullman, 1952). When someone takes this drug, ingesting even the smallest amount of alcohol will make the person quite ill, even if the exposure to the alcohol is through mouthwash or cologne. Antabuse continues to be used in the treatment of alcoholism (Brewer, 2020).

aversive conditioning
A form of treatment that consists of repeated pairings of a stimulus with a very unpleasant stimulus.

Classical conditioning is likely to be at work whenever we engage in mindless, habitual behavior (Bouton & others, 2020). Cues in the environment serve as conditioned stimuli, evoking feelings and behaviors without thought. These associations become implicit "if-then" connections: If you are sitting in front of your laptop, then you check your email. These automatic associations can function for good (for instance, you get up every morning and go for a run without even thinking) or ill (you walk into the kitchen and open the fridge for a snack without even thinking).

Let's consider an imaginary Pavlov's dog, called Bill. After years of service in Pavlov's lab, Bill is taken in by a caring family. Now, Bill is not kept in a state of hunger but is allowed to eat whenever he wants. But Bill's family might notice that he runs to his dish, salivating, and seems to want to eat whenever the doorbell rings—even if he is not hungry. Why? Because Bill has acquired an association in which ringing bells evoke food-related behaviors. Bill, our imaginary dog, eats when he is not hungry because of a learned association. Could such a link help explain overeating and obesity in humans? It might. Indeed, the principles of classical conditioning have been used to combat overweight and obesity in humans by targeting such links (Goldschmidt & others, 2017).

CLASSICAL CONDITIONING AND THE PLACEBO EFFECT   The *placebo effect* is the effect of a substance (such as a pill taken orally) or procedure (such as using a syringe to inject fluid) that researchers use as a control to identify the actual effects of a treatment. Placebo effects are observable changes (such as a drop in pain) that cannot be explained by the effects of an actual treatment. The principles of classical conditioning help to explain some of these effects (Kaptchuk & others, 2020; Skvortsova & others, 2020). In this case, the pill or syringe serves as a CS, and the actual drug is the US. After the experience of pain relief following the consumption of a drug, for instance, the pill or syringe might lead to a CR of lowered pain even in the absence of actual painkiller. The strongest evidence for the role of classical conditioning on placebo effects comes from research on the immune system and the endocrine system.

CLASSICAL CONDITIONING AND THE IMMUNE AND ENDOCRINE SYSTEMS

Even the human body's internal organ systems can be classically conditioned. The immune system is the body's natural defense against disease. Robert Ader and Nicholas Cohen conducted a number of studies that reveal that classical conditioning can produce *immunosuppression,* a decrease in the production of antibodies, which can lower a person's ability to fight disease (Ader, 2000; Ader & Cohen, 1975, 2000).

The initial discovery of the link between classical conditioning and immunosuppression came as a surprise. In studying classical conditioning, Ader (1974) was examining how long a conditioned response would last in some laboratory rats. He paired a conditioned stimulus (saccharin solution) with an unconditioned stimulus, a drug called Cytoxan, which induces nausea. Afterward, while giving the rats saccharin-laced water without the accompanying Cytoxan, Ader watched to see how long it would take the rats to forget the association between the two.

Unexpectedly, in the second month of the study, the rats developed a disease and began to die off. In analyzing this unforeseen result, Ader looked into the properties of the nausea-inducing drug he had used. He discovered that one of its side effects was immunosuppression. It turned out that the rats had been classically conditioned to associate sweet water not only with nausea but also with the shutdown of the immune system. The sweet water apparently had become a conditioned stimulus for immunosuppression. So, Ader found that immunosuppression can occur as a result of classical conditioning.

Conditioned immune responses also occur in humans (Schiller & others, 2020). For example, in one study, patients with multiple sclerosis were given a flavored drink prior to receiving a drug that suppressed the immune system. After this pairing, the flavored drink by itself lowered immune functioning similarly to the drug (Giang & others, 1996).

This is pretty wild. Your body is learning things without your even noticing it.

Similar results have been found for the endocrine system (Hadamitzky & others, 2020; Skvortsova & others, 2020). The endocrine system is a loosely organized set of glands that produces and circulates hormones. Placebo pills can influence the secretion of hormones if patients had previous experiences with pills containing actual drugs that affected hormone secretion (Benedetti & others, 2003). The sympathetic nervous system (the part of the autonomic nervous systems that responds to stress) plays an important role in the learned associations between conditioned stimuli and immune and endocrine functioning (Geuter & others, 2017).

TASTE AVERSION LEARNING

Consider this scenario: Mike goes out for sushi with some friends and eats tekka maki (tuna roll), his favorite dish. He then proceeds to a jazz concert. Several hours later, he becomes very ill with stomach pains and nausea. A few weeks later, he tries to eat tekka maki again but cannot stand it. Importantly, Mike does not experience an aversion to jazz, even though he attended the jazz concert that night before getting sick. Mike's experience exemplifies *taste aversion:* a special kind of classical conditioning involving the learned association between a particular taste and nausea (Garcia & Koelling, 1966; Nakajima, 2020; Pebsworth & Radhakrishna, 2020).

Taste aversion is special because it typically requires only one pairing of a neutral stimulus (a taste) with the unconditioned response of nausea to seal that connection, often

The U.S. Fish and Wildlife Service is trying out taste aversion as a tool to prevent Mexican gray wolves from preying on cattle. To instill taste aversion for beef, the agency is deploying bait made of beef and cowhide but that also contains odorless and flavorless substances that induce nausea (Bryan, 2012). The hope is that wolves that are sickened by the bait will no longer prey on cattle and might even rear their pups to enjoy alternative meals.
Nagel Photography/Shutterstock

for a very long time. As we consider later, it is highly adaptive to learn taste aversion in only one trial. Consider what would happen if an animal required multiple pairings of a taste with poison. It would likely not survive the acquisition phase. It is notable, though, that taste aversion can occur even if the taste experience had nothing to do with getting sick—perhaps, in Mike's case, he was simply coming down with a stomach bug. Taste aversion can even occur when a person has been sickened by a completely separate event, such as being spun around in a chair (Klosterhalfen & others, 2000). Although taste aversion is often considered an exception to the rules of learning, Domjan (2015) has suggested that this form of learning demonstrates how classical conditioning works in the natural world, where associations matter to survival.

Across species, aversions like taste aversion can be learned on the basis of the way animals select their food. As humans, we pick our food based on its flavor. So, it makes sense that we would learn aversions to flavors. In other animals, aversions might be learned based on the scent, color, or other aspects of food (Ko & others, 2020).

Remember, in taste aversion, the taste or flavor is the CS; the agent that made the person sick (it could be a roller-coaster ride or salmonella, for example) is the US; nausea or vomiting is the UR; and taste aversion is the CR.

Taste aversion learning is especially important in the context of the traditional treatment of some cancers. Radiation and chemotherapy for cancer can produce nausea in patients, with the result that people sometimes develop strong aversions to foods they ingest prior to treatment (Holmes, 1993; Jacobsen & others, 1993). Consequently, they may experience a general tendency to be turned off by food, a situation that can lead to nutritional deficits (Mahmoud & others, 2011).

These results show discrimination in classical conditioning—the kids developed aversions only to the specific scapegoat flavors.

Classical conditioning principles can be used to combat these taste aversions, especially in children, for whom antinausea medication is often ineffective (Frumkin, 2020; Skolin & others, 2006) and for whom aversion to protein-rich food is particularly problematic (Ikeda & others, 2006). Early studies demonstrated that giving children a "scapegoat" conditioned stimulus prior to chemotherapy would help limit the taste aversion to only one flavor (Broberg & Bernstein, 1987). For example, children might be given a particular flavor of Lifesaver candy or ice cream before receiving treatment. For these children, the nausea would be more strongly associated with the Lifesaver or the ice cream flavor than with the foods they needed to eat for good nutrition. Fortunately, contemporary cancer treatments may be less likely to cause nausea, and better treatments for nausea itself have been developed.

DRUG HABITUATION   Over time, a person who has taken a psychoactive drug (for example as a treatment for, say, anxiety) might develop a tolerance for that drug and need a higher and higher dose of the substance to get the same effect. Classical conditioning helps to explain this process of **habituation**, which refers to the decreased responsiveness to a stimulus after repeated presentations. In this case, a mind-altering drug is an unconditioned stimulus (US): It naturally produces a response in the person's body. This unconditioned stimulus is often paired systematically with a previously neutral stimulus (CS). For instance, the physical appearance of the drug in a pill or syringe, and the room where the person takes the drugs, are all conditioned stimuli that are paired with the US of the drug. These repeated pairings should produce a conditioned response, and they do—but it is different from those we have considered so far.

habituation
Decreased responsiveness to a stimulus after repeated presentations.

The conditioned response to a drug can be the body's way of *preparing* for the effects of a drug (Vogel & others, 2020). In this case, the body braces itself for the drug effects with a conditioned response (CR) that is the opposite of the unconditioned response (UR). For instance, if the drug (the US) leads to an increase in heart rate (the UR), the CR might be a drop in heart rate. The CS—the previously neutral stimulus—serves as a warning that the drug is coming, and the CR in this case is the body's compensation for the drug's effects (Figure 4). In this situation the conditioned response works to decrease the effects of the unconditioned stimulus, making the drug experience less intense. Some drug users try to prevent habituation by varying the physical location of where they take the drug.

The role of classical conditioning in habituation can play a role in deaths caused by drug overdoses. How might classical conditioning be involved? A person typically takes a

US

CS

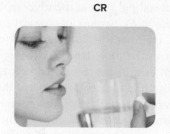

CR

The psychoactive drug is an unconditioned stimulus (US) because it naturally produces a response in a person's body.

Appearance of the drug tablets and the room where the person takes the drug are conditioned stimuli (CS) that are paired with the drug (US).

The body prepares to receive the drug in the room. Repeated pairings of the US and CS have produced a conditioned response (CR).

FIGURE 4 **Drug Habituation** Classical conditioning is involved in drug habituation. As a result of conditioning, the drug user needs to take more of the drug to get the same effect as before the conditioning. Moreover, if the user takes the drug without the usual conditioned stimulus or stimuli—represented in the middle panel by the bathroom and the drug tablets—overdosing is likely.

(pills): Jupiterimages/Thinkstock/Getty Images; (glass): Olga Miltsova/Shutterstock; (woman): Rick Gomez/Corbis

PSYCHOLOGY IN OUR WORLD

Marketing Between the Lines

Classical conditioning is the foundation for many of the commercials bombarding us daily. (Appropriately, when John Watson left the field of psychology, he went on to advertising.) Think about it: Advertising involves creating an association between a product and pleasant feelings (buy that Caffè Misto grande and be happy).

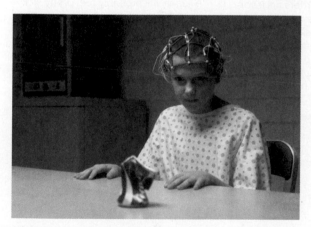

PictureLux/The Hollywood Archive/Alamy Stock Photo

Watching television shows and movies, you can see how advertisers cunningly apply classical conditioning principles to consumers by showing ads that pair something pleasant (a US) with a product (a CS) in hopes that you, the viewer, will experience those positive feelings toward the product (CR). Even on streaming services that do not have commercials, advertisers exploit classical conditioning principles—for instance, through the technique of product placement, or what is known as *embedded marketing*.

Here is how embedded marketing works. Viewing a TV show or movie, you notice that a character is drinking a particular brand of soft drink or eating a particular type of cereal. By placing their products in the context of a show or movie you like, advertisers are hoping that your positive feelings about the show, movie plot, or a character (the UR) carry over to their product (the CS). Sure, it may seem like a long shot—but all they need to do is enhance the chances that, say, navigating through a car dealership or a grocery store, you will feel attracted to their product. Consider the toys in the *Toy Story* movies or the Subway sandwiches and drinks in *To All the Boys I've Loved Before.* And if you catch a rerun of the syndicated comedy *The Office*, you might recognize that Jim classically conditioned Dwight Schrute with breath mints, modeling Pavlov's work, as you can check out on YouTube (search for *The Office*, Pavlov's Theory). This pop culture moment explicitly demonstrated classical conditioning while also using classical conditioning in product placement for those curiously strong mints, Altoids.

drug in a particular setting, such as a bathroom, and acquires a conditioned response to this location (Siegel, 2016). Because of classical conditioning, as soon as the person walks into the bathroom, their body begins to prepare for and anticipate the drug dose in order to lessen the effect of the drug. However, if they take the drug in a location other than the usual one, such as at a concert, the drug's effect is greater because no conditioned responses have built up in the new setting, and therefore the body is not prepared for the drug. In cases in which heroin causes death, researchers often have found that the individuals took the drug under unusual circumstances, at a different time or in a different place, relative to the context in which they usually took the drug. In these cases, with no CS signal, the body is unprepared for (and tragically overwhelmed by) the drug's effects. This process can explain why at times, people can die of an "overdose" when, in fact, the dose of the drug they have ingested would not have typically led to death (Siegel, 2016).

3 Operant Conditioning

Recall from early in the chapter that classical conditioning and operant conditioning are forms of associative learning, which involves learning that two events are connected. In classical conditioning, organisms learn the association between two stimuli (US and CS). Classical conditioning involves a form of *respondent behavior,* behavior that occurs in automatic response to a stimulus such as a nausea-producing drug, and later to a conditioned stimulus such as sweet water that was paired with the drug.

Classical conditioning explains how neutral stimuli become associated with unlearned, *involuntary* responses. Classical conditioning is not as effective, however, in explaining *voluntary* behaviors such as a student's studying hard for a test, a gambler's playing slot machines in Las Vegas, or a dog's searching for and finding his owner's lost cell phone. Operant conditioning is usually much better than classical conditioning at explaining such voluntary behaviors.

Defining Operant Conditioning

Operant conditioning (or **instrumental conditioning**) is a form of associative learning in which the consequences of a behavior change the probability of the behavior's occurrence. American psychologist B. F. Skinner (1938) developed the concept of operant conditioning. Skinner chose the term *operant* to describe the behavior of the organism. According to Skinner, an operant behavior occurs spontaneously, and the consequences that follow such a behavior determine whether it will be repeated.

Adam Sternin/Cavan Images/Image Source

Imagine, for example, that you spontaneously decide to take a different route while driving to campus one day. You are more likely to repeat that route on another day if you have a pleasant experience—for instance, arriving at school faster or finding a great new coffee place to try—than if you have a lousy experience such as getting stuck in traffic. In either case, the consequences of your spontaneous act influence whether that behavior happens again.

Recall that *contingency* is an important aspect of classical conditioning in which the occurrence of one stimulus can be predicted from the presence of another one. Contingency also plays a key role in operant conditioning. For example, when a rat pushes a lever (behavior) that delivers food, the delivery of food (consequence) is contingent on that behavior. The principle of contingency helps explain why passersby should never praise, pet, or feed a service dog while he is working (at least without asking first). Providing rewards during such times might interfere with the dog's training.

> People sometimes confuse "operant" and "classical" conditioning. Remember that operant conditioning is about what comes AFTER a behavior. Classical conditioning is really about what comes BEFORE an unconditioned stimulus.

Thorndike's Law of Effect

Although Skinner emerged as the primary figure in operant conditioning, the experiments of E. L. Thorndike (1898) first established the power of consequences in determining voluntary behavior. At about the same time that Pavlov was conducting classical conditioning experiments with salivating dogs, Thorndike, an American psychologist, was studying cats in puzzle boxes. Thorndike put a hungry cat inside a box and placed a piece of fish outside. To escape from the box and obtain the food, the cat had to learn to open the latch inside the box. At first the cat made a number of ineffective responses. It clawed or bit at the bars and thrust its paw through the openings. Eventually the cat accidentally stepped on the lever that released the door bolt. When the cat returned to the box, it went through the same random activity until it stepped on the lever once more. On subsequent trials, the cat made fewer and fewer random movements until finally it immediately stepped on the lever to open the door (Figure 5). Thorndike's resulting **law of effect** states that behaviors followed by satisfying outcomes are strengthened and that behaviors followed by frustrating outcomes are weakened (Domjan, 2016).

The law of effect is important because it presents the basic idea that the consequences of a behavior influence the likelihood of that behavior's recurrence. Quite simply, a behavior can be followed by something good or something bad, and the probability of a behavior's being repeated depends on these outcomes. As we now explore, Skinner's operant conditioning model expands on this idea.

> The law of effect lays the foundation for operant conditioning. What happens AFTER a given behavior determines whether the behavior will be repeated. In 1898? You go, Thorndike!

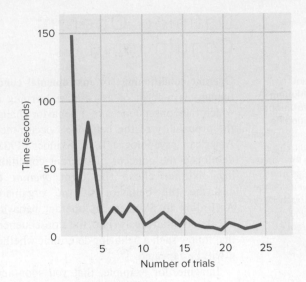

FIGURE 5 **Thorndike's Puzzle Box and the Law of Effect** (*left*) A box typical of the puzzle boxes Thorndike used in his experiments with cats to study the law of effect. Stepping on the treadle released the door bolt; a weight attached to the door then pulled the door open and allowed the cat to escape. After accidentally pressing the treadle as it tried to get to the food, the cat learned to press the treadle when it wanted to escape the box. (*right*) One cat's learning curve over 24 separate trials. Notice that the cat escaped much more quickly after about five trials. It had learned from the consequences of its behavior.

Skinner's Approach to Operant Conditioning

Skinner believed that the mechanisms of learning are the same for all species. This conviction led him to study animals in the hope that he could discover the components of learning with organisms simpler than humans, including pigeons. During World War II, Skinner trained pigeons to pilot missiles. Although top navy officials just could not accept pigeons piloting their missiles in a war, Skinner congratulated himself on the degree of control he was able to exercise over the pigeons (Figure 6).

Skinner and other behaviorists made every effort to study organisms under precisely controlled conditions so that they could examine the connection between the operant behavior and the specific consequences in minute detail. One of Skinner's creations in the 1930s to control experimental conditions was the operant conditioning chamber, sometimes called a Skinner box (Figure 7). A device in the box delivered food pellets into a tray at random. After a rat became accustomed to the box, Skinner installed a lever and observed the rat's behavior. As the hungry rat explored the box, it occasionally pressed the lever, and a food pellet was dispensed. Soon the rat learned that the consequences of pressing the lever were positive: It would be fed. Skinner achieved further control by soundproofing the box to ensure that the experimenter was the only influence on the organism. In many of the experiments, the responses were mechanically recorded, and the food (the consequence) was dispensed automatically. These precautions aimed to prevent human error (like cheering for the little rat's performance).

FIGURE 6 **Skinner's Pigeon-Guided Missile** Skinner wanted to help the military during World War II by using pigeons' tracking behavior. A gold electrode covered the tip of the pigeons' beaks. Contact with the screen on which the image of the target was projected sent a signal informing the missile's control mechanism of the target's location. A few grains of food occasionally given to the pigeons maintained their tracking behavior.

Shaping

Imagine trying to teach even a really smart dog how to do the laundry. The challenge might seem insurmountable, as it is quite unlikely that a dog will spontaneously start putting the clothes in the washing machine. You could wait a very long time for such a feat to happen. It is possible, however, to train a dog or another animal to perform highly complex tasks through the process of shaping.

Shaping refers to rewarding successive approximations of a desired behavior (Fernandez, 2020). For example, shaping can be used to train a rat to press a bar to obtain food. When a rat is first placed in the conditioning box, it rarely presses the bar. Thus, the experimenter may start off by giving the rat a food pellet if it is in the same half of the cage as the bar. Then the experimenter might reward the rat's behavior only when it is within 2 inches of the bar, then only when it touches the bar, and finally only when it presses the bar.

Returning to the service dog, rather than waiting for the dog spontaneously to put the clothes in the washing machine, we might reward the dog for fetching the cloths and eventually carrying the clothes to the laundry room. We would then only reward the dog for bringing the clothes nearer and nearer the washing machine. Finally, we might reward the dog only when it puts the clothes inside the washer. Indeed, trainers use this type of shaping technique extensively in teaching animals to perform tricks. A dolphin that jumps through a hoop held high above the water has been trained to perform this behavior through shaping.

Operant conditioning relies on the notion that a behavior is likely to be repeated if it is followed by a reward. A reasonable question is, what makes a reinforcer rewarding? Research examining this question tracks how positive events following a behavior are related to activation of dopamine receptors in the brain (Dabney & others, 2020).

shaping
Rewarding successive approximations of a desired behavior.

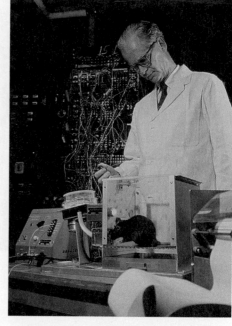

FIGURE 7 **Skinner's Operant Conditioning Chamber** B. F. Skinner conducts an operant conditioning study in his behavioral laboratory. The rat being studied is in an operant conditioning chamber, sometimes referred to as a Skinner box.
Nina Leen/The LIFE Picture Collection/Getty Images

Principles of Reinforcement

We noted earlier that a behavior can be followed by something good or something bad. *Reinforcement* refers to those good things that follow a behavior. **Reinforcement** is the process by which a stimulus or event (a *reinforcer*) following a particular behavior increases the probability that the behavior will happen again. Such consequences of a behavior fall into two types, called *positive reinforcement* and *negative reinforcement*. Both types of consequences increase the frequency of a behavior.

reinforcement
The process by which a stimulus or an event (a reinforcer) following a particular behavior increases the probability that the behavior will happen again.

Although Thorndike talked about 'satisfying' outcomes strengthening behaviors, Skinner took the need for satisfying states out of the equation. For Skinner, if a stimulus increased a behavior, it was reinforcing—no need to talk about how the animal feels.

POSITIVE AND NEGATIVE REINFORCEMENT
In **positive reinforcement,** the frequency of a behavior increases because it is followed by the presentation of something that increases the likelihood the behavior will be repeated. For example, if someone you meet smiles at you after you say, "Hello, how are you?" and you keep talking, the smile has reinforced your talking. The same principle of positive reinforcement is at work when you teach a dog to "shake hands" by giving it a treat when it lifts its paw.

positive reinforcement
The presentation of a stimulus following a given behavior in order to increase the frequency of that behavior.

In contrast, in **negative reinforcement** the frequency of a behavior increases because it is followed by the *removal* of something. For example, if your parent nagged you to clean out the garage and kept nagging until you cleaned out the garage, your response (cleaning out the garage) removed the unpleasant stimulus (your parent's nagging). Taking an aspirin when you have a headache works the same way: A reduction of pain reinforces the act of taking an aspirin. Similarly, if your laptop is making an irritating buzzing sound, you might give it a good smack on the side, and if the buzzing stops, you are more likely to smack it again

negative reinforcement
The removal of a stimulus following a given behavior in order to increase the frequency of that behavior.

Through operant conditioning, animal trainers can coax some amazing behaviors from their star performers.
FilipSinger/EPA-EFE/REX/Shutterstock

if the buzzing resumes. Ending the buzzing sound rewards the laptop smacking. Even though it is labeled "negative," this kind of reinforcement is about increasing the likelihood that a behavior will be repeated.

Notice that both positive and negative reinforcement involve following a behavior with reward—but they do so in different ways. Positive reinforcement means following a behavior with the addition of something, and negative reinforcement means following a behavior with the removal of something. Remember that in this case, "positive" and "negative" have nothing to do with "good" and "bad." Rather, they refer to processes in which something is given (positive reinforcement) or removed (negative reinforcement). Whether it is positive or negative, reinforcement is about increasing a behavior. Figure 8 provides further examples to illustrate the distinction between positive and negative reinforcement.

Positive Reinforcement

Behavior	Rewarding Stimulus Provided	Future Behavior
You turn in homework on time.	Teacher praises your performance.	You increasingly turn in homework on time.
You wax your skis.	The skis go faster.	You wax your skis the next time you go skiing.
You randomly press a button on the dashboard of a friend's car.	Great music begins to play.	You deliberately press the button again the next time you get into the car.

Negative Reinforcement

Behavior	Stimulus Removed	Future Behavior
You turn in homework on time.	Teacher stops criticizing late homework.	You increasingly turn in homework on time.
You wax your skis.	People stop zooming by you on the slope.	You wax your skis the next time you go skiing.
You randomly press a button on the dashboard of a friend's car.	An annoying song shuts off.	You deliberately press the button again the next time the annoying song is on.

FIGURE 8 Positive and Negative Reinforcement Positive reinforcers involve adding something (generally something pleasant). Negative reinforcers involve taking away something (generally something aversive).

avoidance learning
An organism's learning that it can altogether avoid a negative stimulus by making a particular response.

A special kind of response to negative reinforcement is avoidance learning. **Avoidance learning** occurs when the organism learns that by making a particular response, a negative stimulus can be altogether avoided. For instance, a student who receives one bad grade might thereafter always study hard in order to avoid the negative outcome of bad grades in the future. Even when they never get another bad grade, the pattern of behavior sticks. Avoidance learning is very powerful in the sense that the behavior is maintained even in the absence of any aversive stimulus. For example, animals that have been trained to avoid a negative stimulus, such as an electrical shock, by jumping into a safe area may always thereafter gravitate toward the safe area, even when the risk of shock is no longer present.

Do It!

Positive reinforcement and negative reinforcement can be difficult concepts to grasp. The real-world examples and accompanying practice exercises on the following website should help to clarify the distinction for you: http://psych.athabascau.ca/html/prtut/

learned helplessness
Through experience with unavoidable aversive stimuli, an organism learns that it has no control over negative outcomes.

Experience with unavoidable negative stimuli can lead to a particular deficit in avoidance learning called **learned helplessness**, in which the organism, exposed to uncontrollable aversive stimuli, learns that it has no control over negative outcomes (Reznik & others, 2017). Learned helplessness was first identified by Martin Seligman and his colleagues (Altenor & others, 1979; Hannum & others, 1976), who found that dogs that were first exposed to inescapable shocks were later unable to learn to avoid those shocks, even when they could avoid them (Seligman & Maier, 1967). This inability to learn to escape was persistent: The dogs would suffer painful shocks hours, days, and even weeks later and never attempt to escape. Exposure to unavoidable negative circumstances may also set the stage for humans' inability to learn avoidance, such as with the experience of depression and despair (Trindade & others, 2020). Learned helplessness has aided psychologists in understanding a variety of perplexing issues, such as why some people in abusive relationships fail to flee their terrible situation and why some students respond to failure at school by giving up trying.

Yes, dog lovers, many have questioned the ethics of this research. What do you think?

TYPES OF REINFORCERS Psychologists classify positive reinforcers as primary or secondary based on whether the rewarding quality of the consequence is innate or learned. A **primary reinforcer** is innately satisfying; that is, a primary reinforcer does not require any learning on the organism's part to make it pleasurable. Food, water, and sexual satisfaction are primary reinforcers. A **secondary reinforcer**, on the other hand, acquires its positive value through an organism's experience; a secondary reinforcer is a learned or conditioned reinforcer. We encounter hundreds of secondary reinforcers in our lives, such as getting an *A* on a test and a paycheck for a job. Although we might think of these as quite positive outcomes, they are not innately positive. We learn through experience that *A*'s and paychecks are good. Secondary reinforcers can be used in a system called a token economy. In a *token economy,* behaviors are rewarded with tokens (such as poker chips or stars on a chart) that can be exchanged later for desired rewards (such as candy or money).

primary reinforcer
A reinforcer that is innately satisfying; one that does not take any learning on the organism's part to make it pleasurable.

secondary reinforcer
A reinforcer that acquires its positive value through an organism's experience; a secondary reinforcer is a learned or conditioned reinforcer.

Parents who are potty-training toddlers often use token economies.

GENERALIZATION, DISCRIMINATION, AND EXTINCTION Generalization, discrimination, and extinction are important not only in classical conditioning. They also are key principles in operant conditioning.

generalization (operant conditioning)
Performing a reinforced behavior in a different situation.

Generalization In operant conditioning, **generalization** means performing a reinforced behavior in a different situation. For example, in one study pigeons were reinforced for pecking at a disk of a particular color (Guttman & Kalish, 1956). To assess stimulus generalization, researchers presented the pigeons with disks of varying colors. As Figure 9 shows, the pigeons were most likely to peck at disks closest in color to the original. When a student who gets excellent grades in a calculus class by studying the course material every night starts to study psychology and history every night as well, generalization is at work. So, you can think of generalization as exhibiting learning that occurred in one setting in a variety of other similar settings. For instance, a service dog may be trained by a professional or volunteer trainer but then must generalize what it has learned to an owner who needs help.

discrimination

An unjustified negative or harmful action toward a member of a group simply because the person belongs to that group.

Discrimination   In operant conditioning, **discrimination** means responding appropriately to stimuli that signal that a behavior will or will not be reinforced. For example, you go to a restaurant that has a "University Student Discount" sign in the front window, and you enthusiastically flash your student ID with the expectation of getting the reward of a reduced-price meal. Without the sign, showing your ID might get you only a puzzled look, not cheap food.

The principle of discrimination helps to explain how service dogs "know" when they are working. Typically, the dogs wear training harnesses while on duty but not at other times. Thus, when a service dog is wearing a harness, it is important to treat the dog like a working professional. Similarly, an important aspect of the training of service dogs is the need for selective disobedience. Selective disobedience means that in addition to obeying commands from their human partners, service dogs must at times override such commands if the context provides cues that obedience is not the appropriate response. So, if a guide dog is standing at the corner with their visually impaired human, and the human commands the dog to move forward, the dog might refuse if the "Don't Walk" sign is seen flashing. Stimuli in the environment serve as cues, informing the organism if a particular reinforcement contingency is in effect.

If you are accustomed to using your fingers to stretch out text or an image on your smartphone or tablet, you might find yourself trying to do the same thing with a computer monitor and looking foolish. That's a lack of discrimination.

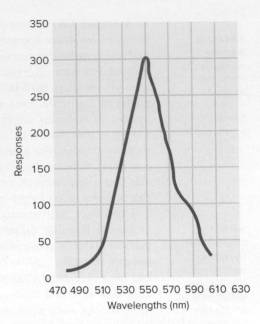

FIGURE 9   Stimulus Generalization
In the experiment by Norman Guttman and Harry Kalish (1956), pigeons initially pecked a disk of a particular color (in this graph, a color with a wavelength of 550 nm) after they had been reinforced for this wavelength. Subsequently, when the pigeons were presented disks of colors with varying wavelengths, they were likelier to peck those that were similar to the original disk.
SOURCE: Guttman, N., & Kalish, H. I. (1956). Discriminability and stimulus generalization. *Journal of Experimental Psychology, 51*(1), 79–88.

Extinction   In operant conditioning, **extinction** occurs when a behavior is no longer reinforced and decreases in frequency. If, for example, a soda machine that you frequently use starts "eating" your coins without dispensing soda, you quickly stop inserting more coins. Several weeks later, you might try to use the machine again, hoping that it has been fixed. Such behavior illustrates spontaneous recovery in operant conditioning.

extinction (operant conditioning)
Decreases in the frequency of a behavior when the behavior is no longer reinforced.

SCHEDULES OF REINFORCEMENT   Most of the examples of reinforcement we have considered so far involve *continuous reinforcement,* in which a behavior is reinforced every time it occurs. When continuous reinforcement takes place, organisms learn rapidly. However, when reinforcement stops, extinction also takes place quickly. A variety of conditioning procedures have been developed that are more resistant to extinction. These involve *partial reinforcement,* in which a reinforcer follows a behavior only a portion of the time. Partial reinforcement characterizes most life experiences. For instance, a track star does not win every race they enter; a chess whiz does not win every match they play; a student does not receive praise each time they solve a problem.

Schedules of reinforcement are specific patterns that determine when a behavior will be reinforced. There are four main schedules of partial reinforcement: fixed ratio, variable ratio, fixed interval, and variable interval. *Ratio schedules* rely on the number of behaviors that must be performed prior to reward. *Interval schedules* are based on the amount of time that must pass before a behavior is rewarded. In a *fixed schedule,* the number of behaviors or the amount of time is always the same. In a *variable schedule,* the required number of behaviors or the amount of time that must pass changes and is unpredictable from the perspective of the learner. Let's look concretely at how each of these schedules of reinforcement influences behavior.

schedules of reinforcement
Specific patterns that determine when a behavior will be reinforced.

A *fixed-ratio schedule* reinforces a behavior after a set number of behaviors. For example, a factory might require a line worker to produce 100 items in order to get paid a particular amount.

Fixed ratio schedules are easy for a learner to figure out. Typically, a factory worker is told what goal to shoot for. Now think about a more mysterious situation, playing a slot machine at a casino. If the reward schedule for a slot machine were easy to figure out, casinos would not be successful. What makes gambling so tantalizing is the unpredictability of wins (and losses). Unlike factory lines, slot machines are on a *variable-ratio schedule,* a system in which behaviors are rewarded an average number of times but on an unpredictable basis. For example, a slot machine might pay off at an average of every 20th time, but the gambler

Slot machines are on a variable-ratio schedule of reinforcement.
Juan Silva/Getty Images

does not know when this payoff will be. The slot machine might pay off twice in a row and then not again until after 58 coins have been inserted. This averages out to a reward for every 20 behavioral acts, but precisely which behavior will lead to the reward is unpredictable. Variable-ratio schedules produce high, steady rates of behavior that are more resistant to extinction than the other reinforcement schedules. Note that to the player, the payoff might seem to depend on the time that has elapsed, but time is not the determining factor. Rather, on ratio schedules, the factor that determines reinforcement is the number of responses emitted. On variable-ratio schedules, that number keeps changing.

Whereas ratio schedules of reinforcement are based on the *number of behaviors* that occur, interval reinforcement schedules are determined by the *time elapsed* since the last behavior was rewarded. A *fixed-interval schedule* reinforces the first behavior after a fixed amount of time has passed. If you take a class that has four scheduled exams, you might procrastinate most of the semester and cram just before each test. Classes with scheduled exams are on fixed-interval schedules. Fixed-interval schedules of reinforcement are also responsible for the fact that pets seem to be able to "tell time," eagerly sidling up to their food dish at 5 P.M. in anticipation of dinner. On a fixed-interval schedule, the rate of a behavior increases rapidly as the time approaches when the behavior likely will be reinforced. For example, a government official who is running for reelection may intensify their campaign activities as Election Day draws near.

A *variable-interval* schedule is a timetable in which a behavior is reinforced after a variable amount of time has elapsed. Pop quizzes occur on a variable-interval schedule. You have to keep studying (the behavior that is reinforced) throughout the course of a class because you never know when you might be quizzed. Fishing is also a type of variable-interval schedule. You do not know if the fish will bite in the next minute, in a half hour, in an hour, or ever. Because it is difficult to predict when a reward will come, behavior is slow and consistent on a variable-interval schedule.

Pop quizzes lead to more consistent levels of studying compared to the cramming that might be seen with scheduled exams because they are on variable interval schedules.

To sharpen your sense of the differences between fixed- and variable-interval schedules, consider the following example. Latisha and Maria both design slot machines for their sorority's charity casino night. Latisha puts her slot machine on a variable-interval schedule of reinforcement; Maria puts hers on a fixed-interval schedule of reinforcement. On average, both machines will deliver a reward every 20 minutes. Whose slot machine is likely to make the most money for the sorority charity? Maria's machine is likely to lead to long lines just before the 20-minute mark, but people will be unlikely to play on it at other times. In contrast, Latisha's is more likely to entice continuous play, because the players never know when they might hit a jackpot. The magic of variable schedules of reinforcement is that the learner can never be sure exactly when the reward is coming. Figure 10 shows how the different schedules of reinforcement result in different rates of responding.

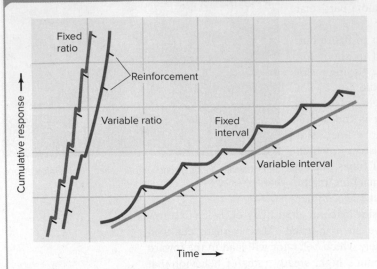

FIGURE 10 **Schedules of Reinforcement and Different Patterns of Responding** In this figure, each hash mark indicates the delivery of reinforcement. Notice on the fixed-ratio schedule the dropoff in responding after each response; on the variable-ratio schedule the high, steady rate of responding; on the fixed-interval schedule the immediate dropoff in responding after reinforcement, and the increase in responding just before reinforcement (resulting in a scalloped curve); and on the variable-interval schedule the slow, steady rate of responding.

> **Which schedule of reinforcement represents the "most bang for the buck"? That is, which is associated with the most responses for the least amount of reinforcement?** > **Which schedule would be best if you have very little time for training?** > **Which schedule of reinforcement is most common in your life?**

PUNISHMENT We began this section by noting that behaviors can be followed by something good or something bad. So far, we have explored only the good things—reinforcers that are meant to increase behaviors. Sometimes, however, the goal is to decrease a behavior, and in such cases the behavior might be followed by something unpleasant. **Punishment** is a consequence that *decreases* the likelihood that a behavior will occur. For instance, a child plays with a matchbox and gets burned when they light one of the matches; the child consequently is less likely to play with matches in the future. As another example, a parent makes fish tacos for their kids but the kids hate them, complain, and refuse to eat them. This consequence—the children's rejection—makes the parent less likely to serve fish tacos in the future. In punishment, a behavior decreases because of its unpleasant consequences. The key difference between a punisher and a reinforcer is that punishing consequences reduce a behavior; reinforcing consequences increase a behavior.

The terms *positive* and *negative* apply to punishment in the same way they apply to reinforcement. Positive means "adding something," and negative means "taking something away." So, in **positive punishment**, a behavior *decreases* when it is followed by the presentation of a stimulus. An example of positive punishment is scolding a spouse who forgot to call when they were running late at work. Another example is a coach who makes their team run wind sprints after a lackadaisical practice. Whenever a behavior is followed by the addition of something unwanted, that is positive punishment and should reduce the behavior.

In **negative punishment** a behavior decreases when a stimulus is removed. Negative punishment means taking away something pleasant to reduce a behavior. Time-out is a form of negative punishment in which a child is removed from a positive reinforcer, such as their toys. Getting grounded and having one's phone taken away for missing curfew are examples of negative punishment. They both involve taking a teenager away from the fun things in life. Figure 11 compares positive reinforcement, negative reinforcement, positive punishment, and negative punishment. These concepts can be tricky, so be sure to take a moment to review Figure 11.

punishment
A consequence that decreases the likelihood that a behavior will occur.

positive punishment
The presentation of a stimulus following a given behavior in order to decrease the frequency of that behavior.

negative punishment
The removal of a stimulus following a given behavior in order to decrease the frequency of that behavior.

*Punishment is sometimes confused with negative reinforcement. Any kind of reinforcement **increases** the behavior it follows. Punishment **decreases** the behavior it follows.*

TIMING AND THE CONSEQUENCES OF BEHAVIOR
In operant conditioning, learning is more efficient when the interval between a behavior and its consequence is a few seconds rather than minutes or hours, especially in nonhuman (or primate) animals. For a rat, for example, if a food reward is delayed for more than 30 seconds after the rat presses a bar, the food is virtually ineffective as reinforcement. Similarly, if one wants to train a cat to stop scratching the furniture, immediately squirting the cat with water when it starts scratching will be more effective than waiting until the cat has scratched for few minutes. Unlike rats and cats, though, humans have the ability to respond to delayed consequences. Let's take a look at how humans respond to immediate versus delayed reinforcement and punishment.

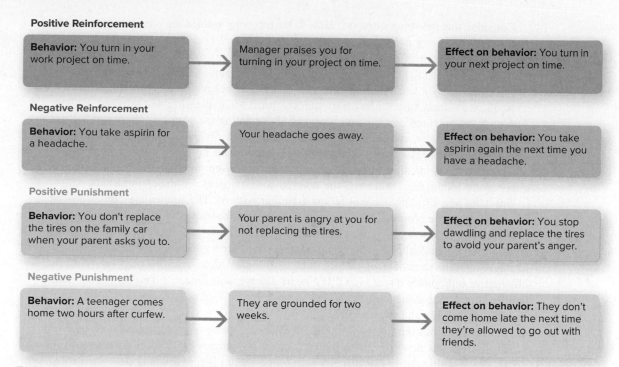

Positive Reinforcement

| **Behavior:** You turn in your work project on time. | → | Manager praises you for turning in your project on time. | → | **Effect on behavior:** You turn in your next project on time. |

Negative Reinforcement

| **Behavior:** You take aspirin for a headache. | → | Your headache goes away. | → | **Effect on behavior:** You take aspirin again the next time you have a headache. |

Positive Punishment

| **Behavior:** You don't replace the tires on the family car when your parent asks you to. | → | Your parent is angry at you for not replacing the tires. | → | **Effect on behavior:** You stop dawdling and replace the tires to avoid your parent's anger. |

Negative Punishment

| **Behavior:** A teenager comes home two hours after curfew. | → | They are grounded for two weeks. | → | **Effect on behavior:** They don't come home late the next time they're allowed to go out with friends. |

FIGURE 11 **Positive Reinforcement, Negative Reinforcement, Positive Punishment, and Negative Punishment** The fine distinctions here can sometimes be confusing. With respect to reinforcement, note that both types of reinforcement are intended to increase behavior, either by presenting a stimulus (in positive reinforcement) or by taking away a stimulus (in negative reinforcement). Punishment is meant to decrease a behavior either by presenting something (in positive punishment) or by taking away something (in negative punishment). The words *positive* and *negative* mean the same things in both cases.

Immediate versus Delayed Reinforcement and Punishment Sometimes important life decisions involve whether to seek and enjoy a small, immediate reinforcer or to wait for a delayed but more highly valued reinforcer. **Delay of gratification** means putting off the pleasure of an immediate reward to gain a more valuable reward later. For example, let's say you have a job that pays you weekly. Each week, you could take the pleasure of your paycheck and spend it on clothes, concert tickets, or eating out. Doing so, you might spend all the money earned in one week. Alternatively, you might decide to delay the pleasure of that paycheck, save your money up so that months down the line you can enjoy a bigger reward: a car. In life, we often have the choice between immediate pleasures (for example, going out with friends on a Friday night) and delayed rewards (for example, getting into professional school or getting a great job based on your academic accomplishments).

Timing can also affect human responses to punishing consequences. Not studying at the beginning of a semester can lead to poor grades much later. Fortunately, humans have the capacity to notice that this early behavior contributed to the negative outcome. We can often avoid negative outcomes by imagining them in advance and behaving accordingly.

Many daily behaviors revolve around both rewards and punishments that are immediate and delayed. We might put off going to the dentist to avoid a small punisher (such as the discomfort that comes with getting a cavity filled). However, this procrastination might contribute to greater pain later (such as the pain of having a tooth pulled). Sometimes life is about enduring a little pain now to avoid a lot of pain later.

delay of gratification Putting off the pleasure of an immediate reward in order to gain a larger, later reward.

DO NOT SKIP FIGURE 11! GO BACK!!! These terms are easy to confuse. It is worth your time to review these terms. Trust me!

Jacob Wackerhausen/Getty Images

Consider the following question: How does receiving immediate small reinforcement versus delayed strong punishment affect human behavior? For instance, should Dave eat the cupcake in front of him (an immediate small pleasure) or risk the possible negative consequences that might occur down the line (delayed punishers)? One reason that obesity is such a major health problem is that eating is a behavior with immediate positive consequences—food tastes great and quickly provides a pleasurable, satisfied feeling. Although the potential delayed consequences of overeating are negative (obesity and other possible health risks), the immediate consequences are difficult to override. When the delayed consequences of behavior are punishing and the immediate consequences are reinforcing, the immediate consequences usually win, even when the immediate consequences are minor reinforcers and the delayed consequences are major punishers.

Smoking and drinking alcohol follow a similar pattern. The immediate consequences of smoking are reinforcing for most people who smoke—the powerful combination of positive reinforcement (enhanced attention, energy boost) and negative reinforcement (tension relief, removal of craving). The primarily long-term effects of smoking are punishing and include shortness of breath, a chronic sore throat and/or coughing, chronic obstructive pulmonary disease (COPD), heart disease, and cancer. Likewise, the immediate pleasurable consequences of drinking override the delayed consequences of a hangover or even alcoholism and liver disease.

Now think about the following situations. Why are some of us so reluctant to take up a new sport, try a new dance step, run for office on campus or in local government, or do almost anything out of our comfort zone? One reason is that learning new skills often involves minor punishing consequences, such as initially looking and feeling awkward, not knowing what to do, and maybe even putting up with sarcastic comments from others. In these circumstances, reinforcing consequences are often delayed. For example, it may take a long time to become a good enough basketball player or a good enough dancer to enjoy these activities, but persevering through the rough patches just might be worth it.

Applied Behavior Analysis

Although behavioral approaches have been criticized for ignoring mental processes and focusing only on observable behavior, these approaches provide an optimistic perspective for people interested in changing their behaviors. That is, rather than concentrating on factors such as the type of person you are, behavioral approaches imply that you can modify even long-standing habits by changing the reward contingencies that maintain those habits (Schlinger, 2017).

One real-world application of operant conditioning to promote better functioning is applied behavior analysis. **Applied behavior analysis** (also called **behavior modification**) is the use of operant conditioning principles to change human behavior. In applied behavior analysis, the rewards and punishers that exist in a particular setting are carefully analyzed and manipulated to change behaviors. Applied behavior analysis seeks to identify the rewards that might be maintaining unwanted behaviors and to enhance the rewards of more appropriate behaviors. From this perspective, we can understand all human behavior as being influenced by the consequences of behavior. If we can figure out what rewards and punishers are controlling a person's behavior, we can change them—and eventually change the behavior itself.

applied behavior analysis or behavior modification The use of operant conditioning principles to change human behavior.

A manager who rewards staff members with a half day off if they meet a particular work goal is employing applied behavior analysis. So are a therapist and a client when they establish clear consequences of the client's behavior in order to reinforce more adaptive actions and discourage less adaptive ones. A teacher who notices that a troublesome student seems to enjoy the attention they receive—even when that attention is scolding—might use applied behavior analysis by changing responses to the child's behavior, ignoring it instead, that is, using negative punishment. These examples show how attending to the consequences of behavior can be used to improve performance in settings such as a workplace and a classroom. Advocates of applied behavior analysis believe that many emotional and behavioral problems stem from inadequate or inappropriate consequences (Wilder & others, 2020).

fizkes/Shutterstock

Applied behavior analysis has been effective in a wide range of situations. Practitioners have used it, for example, with people on the autism spectrum (Alves & others, 2020; Stanislaw & others, 2020; Yu & others, 2020) and residents of mental health facilities (Phillips & Mudford, 2008); to create better classroom behavior in children (DeJager & others, 2020); instruct individuals in effective parenting (Ahmann, 2014); to enhance environmentally conscious behaviors such as recycling and not littering (Geller, 2002); to get people to wear seatbelts (Streff & Geller, 1986); and to promote workplace safety (Geller, 2006). Applied behavior analysis can also be used to increase effective handwashing and mask wearing during a pandemic (Jess & Dozier, 2020). Applied behavior analysis can help people improve their self-control in many aspects of mental and physical health.

self-quiz

1. A parent takes away their child's favorite toy when the child misbehaves. This action is an example of
   A. positive reinforcement.
   B. negative reinforcement.
   C. positive punishment.
   D. negative punishment.

2. The schedule of reinforcement that results in the greatest increase in behavior is
   A. fixed ratio.
   B. variable ratio.
   C. fixed interval.
   D. variable interval.

3. Kelley is scolded each time she teases her little brother. Her mother notices that the frequency of teasing has decreased. Scolding Kelley is an effective
   A. negative reinforcer.
   B. negative punisher.
   C. conditioner.
   D. positive punisher.

APPLY IT! 4. Kevin's girlfriend is very moody, and he never knows what to expect from her. When she is in a good mood, he feels as if he is in heaven, but when she is in a bad mood, she makes him incredibly unhappy. His friends all think that he should dump her, but Kevin finds that he just cannot break it off. Kevin's girlfriend has him on a _____ schedule of reinforcement.
   A. variable
   B. fixed
   C. continuous
   D. nonexistent

4 Observational Learning

Would it make sense to teach a 15-year-old how to drive with either classical conditioning or operant conditioning procedures? Driving a car is a voluntary behavior, so classical conditioning would not apply. In terms of operant conditioning, we could ask the teen to try to drive down the road and then reward positive behaviors. Not many of us would want to be on the road, though, during the training session. Albert Bandura (2012) pointed out that if we learned only in such a trial-and-error fashion, learning would be exceedingly tedious and at times hazardous. Instead, Bandura says, many complex behaviors are the result of exposure to competent models. By observing other people, we can acquire knowledge, skills, rules, strategies, beliefs, and attitudes.

Observational learning occurs when a person observes and imitates someone else's behavior.
monkeybusinessimages/Getty Images

Bandura's approach is called social learning theory and it incorporates not only the principles of associative learning but observational learning as well. *Observational learning,* also called *imitation* or *modeling,* is learning that occurs when a person observes and imitates behavior. The capacity to learn by observation eliminates trial-and-error learning. Often observational learning takes less time than operant conditioning. A famous example of observational learning is the Bobo doll study (Bandura & others, 1961), in which children who had watched an aggressive adult model punch the doll were more likely to behave aggressively when left alone than were children who had observed a nonaggressive model.

Learning from someone else's actions is different from the types of learning we have discussed so far. If you want to become a great basketball player, watching a WNBA game might be a good start. But you have to watch it in a particular way, paying close attention and seeing what sorts of behaviors lead to the best outcomes. Indeed, there are four main processes that are crucial to observational learning: attention, retention, motor reproduction, and reinforcement (Bandura, 1986). Let's review each of these.

In observational learning, the first process that must occur is *attention*. To reproduce a model's actions, you must attend to what the model is saying or doing. In an online class, sitting in front of your computer, you might miss important parts of your instructor's analysis of a problem if you are distracted by your dog. A teen learning to drive might have to pay attention to the person driving a car, rather than looking at their phone. As a further example, imagine that you decide to take a class to improve your drawing skills. To succeed, you need to attend to the instructor's words and hand movements. Characteristics of the model can influence attention to the model. Warm, powerful, atypical people, for example, command more attention than do cold, weak, typical people.

Retention is the second process required for observational learning. To reproduce a model's actions, you must remember the information and keep it in memory so that you can retrieve it later. A simple verbal description, or a vivid image of what the model did, assists retention. In the example of taking a class to sharpen your drawing ability, you will need to remember what the instructor said and did in modeling good drawing skills.

Motor reproduction, a third element of observational learning, is the process of imitating the model's actions. People might pay attention to a model and encode what they have seen, but limitations in motor ability or development might make it difficult for them to reproduce the model's action. Thirteen-year-olds might see a professional basketball player do a reverse two-handed dunk but be unable to reproduce the pro's play. Similarly, in your drawing class, if you lack fine motor skills, you might be unable to reproduce the instructor's example.

Reinforcement is a final component of observational learning. In this case, the question is whether the model's behavior is followed by a consequence. Seeing a model attain a reward for an activity increases the chances that an observer will repeat the behavior—a process called *vicarious reinforcement*. So, if a basketball player is more likely to make the basket when their feet are appropriately planted, then that is a behavior a budding player is more likely to seek to repeat. On the other hand, seeing the model punished makes the observer less likely to repeat the behavior—a process called *vicarious punishment*. If an observer sees that shooting the basket off balance leads to missed shots, that behavior is less likely to be repeated.

In sum, when an observer attends to a model, retains that information in memory, is able to reproduce the model's actions, and sees the model receive reinforcement, that observer has a good chance of learning to do what the model does.

To appreciate the steps of observational learning, consider the experience of a 10-year-old Michigan boy, Jacob O'Connor. By modeling behavior, Jacob saved his two-year-old brother, Dylan, from drowning (Shapiro, 2017). Jacob happened upon Dylan floating face down in their grandmother's swimming pool. Jacob remembered seeing Dwayne "The Rock" Johnson in the movie *San Andreas*. In the film, Johnson saves his daughter from drowning by performing chest compressions. With just that as his training, Jacob modeled the behavior and saved his brother's life. Note that Jacob paid attention to the disaster film to learn this skill. He *remembered it* when the time was right. He also *saw the behavior lead to a positive outcome*: The Rock saved his daughter's life in the movie. And, of course, he was able to *reproduce the action*s that led to the positive outcome.

Observational learning can occur in subtle ways. Imagine it's your first day of work at a new tech company. The lunch room is luxurious, with comfortable chairs and an espresso bar. Entering the room, you notice that most tables have three or four people eating and chatting. There is one person seated alone at a table and people seem to be making a point of not sitting with the person. Would their behavior affect your attitude toward that person, even if you two had never met? When we observe others responding negatively to another person, we may develop positive or negative attitudes about that person. To read about research on these issues, see the Intersection.

INTERSECTION

Learning and Social Psychology: Can Observational Learning Lead to Bias?

Interpersonal biases, based on factors such as gender, race, religion, disability status, or sexuality, are a key problem in contemporary society. In a diverse world, it is important that people interact with each other without preexisting biases. Yet, bias all too often seems to take on a life of its own. How do people learn who to dislike or mistrust? According to social learning theory, observational learning is the key way attitudes are conveyed and learned (Bandura & Rosenthal, 1966).

A recent series of studies demonstrated how observing others respond to another person in even subtle negative ways can influence attitudes toward that person (Skinner & Perry, 2020). In the studies, participants were shown video clips. The clips were taken from the TV show *Ally McBeal*—you have probably not heard of it and none of the participants had either. In the brief clips, a character was shown making either positive

sanjeri/Getty Images

facial expressions (such as smiling) or negative facial expressions (like frowning) toward another character (the target). After watching the clips, participants rated their own feelings about the target. Across a number of studies, results consistently showed that seeing an unfamiliar person express negative feelings toward an unknown target led participants to express more negative attitudes about that target, as well.

Now, these studies might not seem surprising to you. But consider their implications for human interactions and bias. The negative attitudes developed by participants occurred without any information about the target of the negative expressions. Indeed, they occurred without any information about the person making the expressions. Seeing a complete stranger direct negative facial expressions to another complete stranger led to negative attitudes about the target. Simply observing how a person is treated can influence our views of that person. When we rely on others' treatment of someone we rob that person of the opportunity to get to know us.

\\ Why do we rely on how others behave toward people instead of getting to know them for ourselves?

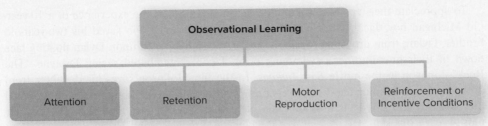

FIGURE 12 **Bandura's Model of Observational Learning** In terms of Bandura's model, if you are learning to ski, you need to attend to the instructor's words and demonstrations. You need to remember what the instructor did and the instructor's tips for avoiding disasters. You also need the motor abilities to reproduce what the instructor has shown you. Praise from the instructor after you have completed a few moves on the slopes should improve your motivation to continue skiing.

Observational learning has been studied in a variety of contexts. Researchers have explored observational learning, for example, as a means by which nonhuman primates learn from one another (Whiten, 2020; Zhang & others, 2020). They have also studied it as a process by which people learn whether stimuli are likely to be painful (Yakunchikov & others, 2017), what food to eat (Edenbrandt & others, 2020), and as a tool people use to make economic decisions (Wang & Yu, 2017). Researchers are also interested in comparing learning from experience, learning through observation, and direct instruction (Glogger-Frey & others, 2017; Lanter & Singer-Dudek, 2020).

Observational learning can be an important factor in the functioning of role models in inspiring people and changing their perceptions. Whether a model is similar to us can influence that model's effectiveness in modifying our behavior. The shortage of role models for women and people of color in science and engineering has often been suggested as a reason for the lack of diversity in these fields. After the election of Barack Obama as president of the United States, many commentators noted that, for the first time, Black children could see concretely that they might also attain the nation's highest office someday. You may have seen the photo of five-year-old Jacob Philadelphia feeling President Obama's hair, to see if it was just like his (Calmes, 2012). The election of Vice President Kamala Harris in 2020 provided yet another role model in whom children might see themselves.

Figure 12 summarizes Bandura's model of observational learning.

> ## Do It!
>
> Having positive role models and mentors you can observe can be a significant factor in your learning and success. Make a list of your most important role models and mentors. Next to each, briefly describe how they have influenced you. What would your *ideal* role model or mentor be like?

self-quiz

1. Another name for observational learning is
   A. replication.
   B. modeling.
   C. trial-and-error learning.
   D. visualization.

2. According to Bandura, _____ occurs first in observational learning.
   A. motor reproduction
   B. retention
   C. attention
   D. reinforcement

3. A friend shows you how to do a card trick. However, you forget the second step in the trick and are thus unable to replicate the card trick. There has been a failure in
   A. motor reproduction.
   B. retention.
   C. attention.
   D. reinforcement.

APPLY IT! 4. Shawna is a 15-year-old high school girl whose mother is a highly paid accountant. Shawna's mom works long hours, often complains about her workplace and how much she hates her boss, and seems tired most of the time. When she is asked what she might do when she grows up, Shawna says she does not think she wants to pursue a career in accounting. Her mother is shocked and cannot understand why Shawna would not want to follow in her footsteps. Which of the following is the most likely explanation for this situation?
   A. Shawna has not observed her mother being reinforced for her behavior. She has only experienced vicarious punishment.
   B. Shawna is not aware that her mother is an accountant.
   C. Shawna is too different from her mother for her mother to be an effective role model.
   D. Shawna has not been paying attention to her mother.

5 Cognitive Factors in Learning

In learning about learning, we have looked at cognitive processes only as they apply in observational learning. Skinner's operant conditioning and Pavlov's classical conditioning focus on the environment and observable behavior, not what is going on in the head of the learner. Many contemporary psychologists, including some behaviorists, recognize the importance of cognition and believe that learning involves more than environment-behavior connections (Bandura, 2012; Bjork & others, 2013). A good starting place for considering cognitive influences in learning is the work of E. C. Tolman.

Purposive Behavior

Tolman (1932) emphasized the *purposiveness* of behavior—the idea that much of behavior is goal-directed. Tolman believed that it is necessary to study entire behavioral sequences in order to understand why people engage in particular actions. For example, high school students whose goal is to attend college or university study hard in their classes. If we focused only on their studying, we would miss the purpose of their behavior. Students do not always study hard just because they have been reinforced for studying in the past. Rather, studying is a means to intermediate goals (learning, high grades) that in turn improve their likelihood of getting into the college or university of their choice.

We can see Tolman's legacy today in the extensive interest in the role of goal setting in human behavior (van Lent & Souverijn, 2020). Researchers are especially curious about how people self-regulate and self-monitor their behavior to reach a goal (Valenzuela & others, 2020).

EXPECTANCY LEARNING AND INFORMATION

In studying the purposiveness of behavior, Tolman went beyond the stimuli and responses of Pavlov and Skinner to focus on cognitive mechanisms. Tolman said that when classical conditioning and operant conditioning occur, the organism acquires certain expectations. In classical conditioning, the young boy fears the rabbit because he expects it will hurt him. In operant conditioning, a person works hard all week because they expect a paycheck on Friday. So, according to this line of reasoning, when we learn through conditioning, we are not only acquiring associations but expectations about the world. Expectancies are acquired from people's experiences with their environment. Expectancies influence a variety of human experiences. We set the goals we do because we believe that we can reach them.

Research supports an important role for expectancies in classical conditioning (Pfeuffer & others, 2020; Taylor & others, 2017), including in the placebo effect. Many painkillers have been shown to be more effective in reducing pain when patients can see the intravenous injection sites than when they cannot (Price & others, 2008). If patients can see that they are getting a drug, they can harness their own expectations for pain reduction.

Tolman (1932) emphasized that the information value of the conditioned stimulus is important as a signal or an expectation that an unconditioned stimulus will follow. Anticipating contemporary thinking, Tolman believed that the information the CS provides is the key to understanding classical conditioning. One contemporary view of classical conditioning describes an organism as an information seeker, using logical and perceptual relations among events, along with preconceptions, to form a representation of the world (Rescorla, 2003, 2005, 2009).

A classic experiment conducted by Leon Kamin (1968) illustrates the importance of an organism's history and the information provided by a conditioned stimulus in classical conditioning. Kamin conditioned a rat by repeatedly pairing a tone (CS) and a shock (US) until the tone alone produced fear (conditioned response). Then he continued to pair the tone with the shock, but he turned on a light (a second CS) each

Vasiliy Koval/Shutterstock

time the tone sounded. Even though he repeatedly paired the light (CS) and the shock (US), the rat showed no conditioning to the light (the light by itself produced no CR). Conditioning to the light was blocked, almost as if the rat had not paid attention. The rat apparently used the tone as a signal to predict that a shock would be coming; information about the light's pairing with the shock was redundant with the information already learned about the tone's pairing with the shock. In this experiment, conditioning was governed not by the contiguity of the CS and US but instead by the rat's history and the information it received. The light provided nothing new and was therefore not relevant to learning. Contemporary classical conditioning researchers are further exploring the role of information in an organism's learning (Hermans & others, 2014; Rescorla & Wagner, 2009).

LATENT LEARNING   Experiments on latent learning provide other evidence to support the role of cognition in learning. **Latent learning** (or **implicit learning**) is unreinforced learning that is not immediately reflected in behavior. Latent learning is sometimes called incidental learning because it "just happens" as a result of experience. It happens even without reinforcement. In one study, researchers put two groups of hungry rats in a maze and required them to find their way from a starting point to an end point (Tolman & Honzik, 1930). The first group found food (a reinforcer) at the end point; the second group found nothing there. In the operant conditioning view, the first group should learn the maze better than the second group, which is exactly what happened. However, when the researchers subsequently took some of the rats from the nonreinforced group and gave them food at the end point of the maze, they quickly began to run the maze as effectively as the reinforced group. The nonreinforced rats apparently had learned a great deal about the maze as they roamed around and explored it. However, their learning was latent (or hidden), stored cognitively in their memories but not yet expressed behaviorally. When these rats were given a good reason (reinforcement with food) to run the maze speedily, they called on their latent learning to help them reach the end of the maze more quickly.

Outside a laboratory, latent learning is evident when you walk around a new setting to get "the lay of the land." The first time you visited your college campus, you may have wandered about without a specific destination in mind. Exploring the environment made you better prepared when the time came to find that 8 A.M. class.

latent learning or implicit learning
Unreinforced learning that is not immediately reflected in behavior.

Insight Learning

Like E. C. Tolman, the German gestalt psychologist Wolfgang Köhler believed that cognitive factors play a significant role in learning. Köhler spent four months in the Canary Islands during World War I observing the behavior of apes. There he conducted two fascinating experiments—the stick problem and the box problem. Although these two experiments are basically the same, the solutions to the problems are different. In both situations, the ape discovers that it cannot reach an alluring piece of fruit, either because the fruit is too high or because it is outside of the ape's cage and beyond reach. To solve the stick problem, the ape has to insert a small stick inside a larger stick to reach the fruit. To master the box problem, the ape must stack several boxes to reach the fruit (Figure 13).

According to Köhler (1925), solving these problems does not involve trial and error or simple connections between stimuli and responses. Rather, when the ape realizes that its customary actions are not going to help it get the fruit, it often sits for a period of time and appears to ponder how to solve the problem. Then it quickly rises, as if it has had a flash of insight, piles the boxes on top of one another, and gets the fruit. **Insight learning** is a form of problem solving in which the organism develops a sudden insight into or understanding of a problem's solution.

The idea that insight learning is essentially different from learning through trial and error or through conditioning has always been controversial (Spence, 1938). Insight learning appears to entail both gradual and sudden processes, and understanding how these lead to problem

insight learning
A form of problem solving in which the organism develops a sudden insight into or understanding of a problem's solution.

FIGURE 13 **Insight Learning** Sultan, one of Wolfgang Köhler's brightest chimps, is faced with the problem of reaching a cluster of bananas overhead. He solves the problem by stacking boxes on top of one another to reach the bananas. Köhler called this type of problem solving "insight learning."
(all): 3LH/SuperStock

solving continues to fascinate psychologists. In one study, researchers observed orangutans trying to figure out a way to get a tempting peanut out of a clear plastic tube (Mendes & others, 2007). The primates wandered about their enclosures, experimenting with various strategies. Typically, they paused for a moment before finally landing on a solution: Little by little they filled the tube with water that they transferred by mouth from their water dishes to the tube. Once the peanut floated to the top, the clever orangutans had their snack. Chimps can solve the floating peanut task through observational learning (Tennie & others, 2010).

Insight learning requires thinking "outside the box," setting aside previous expectations and assumptions. One way that insight learning can be enhanced in human beings is through multicultural experiences (Leung & others, 2008). Correlational studies have shown that time spent living abroad is associated with higher insight learning performance among MBA students (Maddux & Galinsky, 2007). Furthermore, experimental studies have demonstrated that exposure to other cultures can influence insight learning. In one study, U.S. college students were randomly assigned to view one of two slide shows—one about Chinese and U.S. cultures and the other about a control topic. Those who saw the multicultural slide show scored higher on measures of creativity and insight, and these changes persisted for a week (Leung & others, 2008). Being exposed to other cultures and other ways of thinking can be a key way to enhance insight and creativity, and a person does not have to travel to enjoy the learning benefits of multicultural experience.

What makes insight learning unique is that 'Aha!' moment—but that moment often comes after some trial and error during which many of the 'wrong' answers have been thoroughly dismissed.

self-quiz

1. E. C. Tolman emphasized the *purposiveness* of behavior—the idea that much of behavior is oriented toward the achievement of
   A. immortality.
   B. altruism.
   C. goals.
   D. self-esteem.

2. When the answer to a problem just pops into your head, you have experienced
   A. latent learning.
   B. insight learning.
   C. implicit learning.
   D. expectancy learning.

3. A type of learning that does *not* involve trial and error is
   A. insight learning.
   B. shaping.
   C. expectancy learning.
   D. implicit learning.

APPLY IT! 4. Derek is rehearsing his lines and songs for an upcoming production of *Grease*. He is playing the lead role of Danny Zucco. His friend Maria helps him practice his lines and learn the words to his songs. Maria is not in the play and wouldn't even think of appearing onstage. On opening night, Maria is in the audience, and halfway through "Summer Lovin'" people sitting around her are complaining because she is singing along. She also has been saying all of Danny's lines under her breath. What is the explanation?
   A. Maria is demonstrating the power of latent learning.
   B. Maria is demonstrating insight learning.
   C. Maria is showing purposive behavior.
   D. Maria secretly dreams of playing Danny Zucco in an all-female version of *Grease* someday.

6 Biological, Cultural, and Psychological Factors in Learning

Albert Einstein had many special talents. He combined enormous creativity with keen analytic ability to develop some of the twentieth century's most important insights into the nature of matter and the universe. Genes obviously endowed Einstein with extraordinary intellectual skills that enabled him to think and reason at a very high level, but cultural factors also contributed to his genius. Einstein received an excellent, rigorous education, and he experienced the freedom and support believed to be important in creative exploration. Would Einstein have been able to develop his skills fully and to make such brilliant insights if he had grown up in a less advantageous environment? It is unlikely. Clearly, both biological and cultural factors contribute to learning.

Biological Constraints

Humans cannot breathe under water, fish cannot roller skate, and cows cannot solve math problems. The structure of an organism's body permits certain kinds of learning and inhibits others. For example, chimpanzees cannot learn to speak English because they lack the necessary vocal equipment.

Dogs who are trained for various tasks can sometimes illustrate the limits of learning principles. Lulu, a black Labrador dog, made headlines when the U.S. Drug Enforcement Agency tweeted her photo to the world along with the news that she had flunked out of "sniffer dog" training (Chorney, 2017). Lulu simply was not interested in sniffing out illegal drugs on clothes or in luggage or other containers. Fortunately, Lulu was adopted by one of her trainers and currently lives in a happy home where tasks are more to her liking. This case shows that even in the same breed of dogs, some may be able to quickly learn to complete tasks while others may not.

INSTINCTIVE DRIFT   Keller and Marion Breland (1961), students of B. F. Skinner, used operant conditioning to train animals to perform at fairs and conventions and in television advertisements. They applied Skinner's techniques to teach pigs to cart large wooden nickels to a piggy bank and deposit them. They also trained raccoons to pick up a coin and place it in a metal tray. Although the pigs and raccoons, as well as chickens and other animals, performed most of the tasks well (raccoons became adept basketball players, for example—see Figure 14), some of the animals began acting strangely. Instead of picking up the large wooden nickels and carrying them to the piggy bank, the pigs dropped the nickels on the ground, shoved them with their snouts, tossed them in the air, and then repeated these actions. The raccoons began to hold on to their coins rather than dropping them into the metal tray. When two coins were introduced, the raccoons rubbed them together in a miserly fashion. Somehow these behaviors overwhelmed the strength of the reinforcement. This example of biological influences on learning illustrates **instinctive drift**, the tendency of animals to revert to instinctive behavior that interferes with learning.

Why were the pigs and the raccoons misbehaving? The pigs were rooting, an instinct that is used to

FIGURE 14   **Instinctive Drift** This raccoon's skill in using its hands made it an excellent basketball player, but because of instinctive drift, the raccoon had a much more difficult time dropping coins in a tray.

Keystone-France/Hulton Archive/Getty Images

instinctive drift
The tendency of animals to revert to instinctive behavior that interferes with learning.

uncover edible roots. The raccoons were engaging in an instinctive food-washing response. Their instinctive drift interfered with learning.

PREPAREDNESS   Some animals learn readily in one situation but have difficulty learning in slightly different circumstances (Garcia & Koelling, 1966, 2009). The difficulty might result not from some aspect of the learning situation but from the organism's biological predisposition (Seligman, 1970). **Preparedness** is the species-specific biological predisposition to learn in certain ways but not others.

preparedness
The species-specific biological predisposition to learn in certain ways but not others.

Evidence for preparedness comes from research on taste aversion (Garcia, 1989; Garcia & Koelling, 2009). Recall that taste aversion involves a single trial of learning the association between a particular taste and nausea. Animals that choose their food based on taste are prepared to learn the association between flavors and illness. If you have ever gotten sick on a food, you might never eat it again for the rest of your life. The speed and long-term nature of such effects cannot be accounted for by classical conditioning, which would argue that a single pairing of the conditioned and unconditioned stimuli would not last that long. Thus such aversions are an example of preparedness.

Another example of preparedness comes from research on conditioning humans and monkeys to associate snakes with fear. Susan Mineka and Arne Ohman (2002; Ohman & Mineka, 2003) have investigated the fascinating natural power of snakes to evoke fear in many mammals. Many monkeys and humans fear snakes, and both monkeys and humans are very quick to learn the association between snakes and fear. In classical conditioning studies, when pictures of snakes (CS) are paired with electrical shocks (US), the snakes are likely to quickly and strongly evoke fear (the CR). Interestingly, pairing pictures of, say, flowers (CS) with electrical shocks produces much weaker associations (Mineka & Ohman, 2002; Ohman & Soares, 1998). Even more significantly, pictures of snakes can serve as conditioned stimuli for fearful responses, even when the pictures are presented so rapidly that they cannot be consciously perceived (Ohman & Mineka, 2001).

The link between snakes and fear has been demonstrated outside of classical conditioning paradigms. Monkeys that have been raised in the lab and that have never seen a snake rapidly learn to fear snakes, even entirely by observational learning. Lab monkeys that see a video of a monkey expressing fear toward a snake learn to be afraid of snakes faster than monkeys seeing the same fear video spliced so that the feared object is a rabbit, a flower, or a mushroom (Ohman & Mineka, 2003).

Mineka and Ohman (2002) suggest that these results demonstrate preparedness among mammals to associate snakes with fear and aversive stimuli. They argue that this association is related to the amygdala (the part of the limbic system that is linked to emotion) and is difficult to modify. These researchers theorize that this preparedness for fear of snakes has emerged out of the threat that reptiles likely posed to our evolutionary ancestors.

PSYCHOLOGICAL INQUIRY

On the Indonesian island of Bali, young children learn traditional dances, whereas in Norway children commonly learn to ski early in life. As cultures vary, so does the content of learning.
(dancers): J.W.Alker/Shutterstock; (skiers): Lasse Bolstad/Alamy Stock Photo

> **Think of some activities you learned at an early age; how are they related to the culture in which you grew up?** > **What sorts of basic knowledge did these experiences allow you to acquire?** > **How might such early experiences relate to your current learning?** > **How did the adults in your life respond to success or failure, and how has that shaped your experience of learning?**

Cultural Influences

Traditionally, interest in the cultural context of human learning has been limited, partly because the organisms in those contexts typically were animals. The question arises, how might culture influence human learning? Most psychologists agree that the principles of classical conditioning, operant conditioning, and observational learning are universal and are powerful learning processes in every culture. However, culture can influence the degree to which these learning processes are used. The values of a culture will surely affect which actions are reinforced and which are punished.

In addition, culture can determine the content of learning. We cannot learn about something we do not experience. The four-year-old who grows up among the Bushmen of the Kalahari Desert is unlikely to learn about taking baths and eating with a knife and fork. Similarly, a child growing up in Chicago is unlikely to be skilled at tracking animals and finding water-bearing roots in the desert. Learning often requires practice, and certain behaviors are practiced more often in some cultures than in others. In Bali, many children are skilled dancers by the age of six, whereas Norwegian children are much more likely to be good skiers and skaters by that age.

Psychological Constraints

Are there psychological constraints on learning? For animals, the answer is probably no. For humans, the answer may well be yes. This section opened with the claim that fish cannot roller skate. The truth of this statement is clear. Biological circumstances make it impossible. If we put biological considerations aside, we might ask ourselves about times in our lives when we feel like a fish trying to roller skate—when we feel that we just do not have what it takes to learn a skill or master a task. Some people believe that humans have particular learning styles that make it easier for them to learn in some ways but not others.

Learning styles refers to the idea that people differ in terms of the method of instruction that will be most effective for them. You may have heard, for example, that someone can be a *visual learner* (they learn by seeing), an *aural learner* (they learn by listening), or a *kinesthetic learner* (they learn through hands-on experience). These ideas are popular but there is no evidence that learning styles actually matter to learning (An & Carr, 2017). Although children and adults report consistent preferences for particular learning styles, there is *no evidence* that tailoring instructional methods to "visual," "aural," or "kinesthetic" learners produces better learning (Pashler & others, 2008). Interestingly, learning styles predict the kind of materials participants prefer and attend to; but the match between a person's learning style and the mode of instruction is unrelated to learning (Koć-Januchta & others, 2017; Massa & Mayer, 2006).

FatCamera/E+/Getty Images

Research shows that when material is presented in a way that matches people's learning style, they are likely to *feel like* they have learned more (Knoll & others, 2017). However, performance on an actual test shows those subjective feelings are misleading: No differences emerged in actual learning. There is no evidence that learning styles matter to learning.

Carol Dweck (2012, 2017; Burnette & others, 2020) uses the term *mindset* to describe the way our beliefs about ability dictate what goals we set for ourselves, what we think we *can* learn, and ultimately what we *do* learn. Dweck identifies two mindsets: a *fixed mindset,* in which they believe that their qualities are carved in stone and cannot change; or a *growth mindset,* in which they believe their qualities can change and improve through their effort. These two mindsets have implications for the meaning of failure. From a fixed mindset, failure means lack of ability. From a growth mindset, however, failure tells the person what they still need to learn. Your mindset influences whether you will be optimistic or pessimistic, what your goals will be and how hard you will strive to reach those goals.

Dweck (2017) studied first-year pre-med majors taking their first chemistry class in college. Students with a growth mindset got higher grades than those with a fixed mindset. Even when they did not do well on a test, the growth-mindset students bounced back on the next test. Fixed-mindset students typically read and reread the text and class notes or tried to memorize everything verbatim. The fixed-mindset students who did poorly on tests concluded that chemistry and maybe pre-med were not for them. By contrast, growth-mindset students took charge of their motivation and learning, searching for themes and principles in the course and going over mistakes until they understood why they made them. According to Dweck, the key to their success was that they were studying to learn, not just to get a good grade.

Following are some effective strategies for developing a growth mindset (Dweck, 2017):

- *Understand that your intelligence and thinking skills are not fixed but can change.* Even if you are extremely bright, with effort you can increase your intelligence.

- *Become passionate about learning and stretch your mind in challenging situations.* It is easy to withdraw into a fixed mindset when the going gets tough; but as you bump up against obstacles, keep growing, work harder, stay the course, and improve your strategies; you will become a more successful person.

- *Think about the growth mindsets of people you admire.* Possibly you have a hero, someone who has achieved something extraordinary. You may have thought their accomplishments came easy because the person is so talented. However, find out more about this person and how they work and think. You likely will discover that much hard work and effort over a long period of time were responsible for this individual's achievements.

- *Begin now.* If you have a fixed mindset, commit to changing now. Think about when, where, and how you will begin using your new growth mindset.

Dweck's work challenges us to consider the limits we place on our own learning. When we think of the lack of diversity math and science professions, we might consider the messages these groups have received about whether they have what it takes to succeed in these domains. Our beliefs about ability profoundly influence what we try to learn. You never know what you can do until you try.

self-quiz

1. When a pig's rooting behavior interferes with its learning, the phenomenon is an example of
   A. preparedness.
   B. learned helplessness.
   C. a taste aversion.
   D. instinctive drift.

2. Mineka and Ohman suggest that humans' preparedness for fear of snakes emerged because of
   A. cultural myths.
   B. the religious symbolism of snakes.
   C. the danger that snakes and other reptiles posed to earlier humans.
   D. the limitations of human learning.

3. Believing that hard work is part of learning is an example of
   A. a growth insight.
   B. a growth mindset.
   C. preparedness.
   D. a fixed mindset.

APPLY IT! 4. Frances is a dog person who has just adopted her first-ever cat. Given her experience in housebreaking her pet dogs, Frances is shocked that her new kitty, Tolman, uses the litter box the very first day and never has an accident in the house. Frances thinks that Tolman must be a genius cat. Tolman's amazing ability demonstrates
   A. psychological constraints on learning.
   B. biological preparedness.
   C. cultural constraints on learning.
   D. that dog people are not very bright.

Types of Learning

Learning is a systematic, relatively permanent change in behavior that occurs through experience. Associative learning involves learning by making a connection between two events. Observational learning is learning by watching what other people do.

Conditioning is the process by which associative learning occurs. In classical conditioning, organisms learn the association between two stimuli. In operant conditioning, they learn the association between behavior and a consequence.

 Classical Conditioning

Classical conditioning occurs when a neutral stimulus becomes associated with a meaningful stimulus and comes to elicit a similar response. Pavlov discovered that an organism learns the association between an unconditioned stimulus (US) and a conditioned stimulus (CS). The US automatically produces the unconditioned response (UR). After conditioning (CS–US pairing), the CS elicits the conditioned response (CR) by itself. Acquisition in classical conditioning is the initial linking of stimuli and responses, which involves a neutral stimulus being associated with the US so that the CS comes to elicit the CR. Two important aspects of acquisition are contiguity and contingency.

Generalization in classical conditioning is the tendency of a new stimulus that is similar to the original conditioned stimulus to elicit a response that is similar to the conditioned response. Discrimination is the process of learning to respond to certain stimuli and not to others. Extinction is the weakening of the CR in the absence of the US. Spontaneous recovery is the recurrence of a CR after a time delay without further conditioning.

In humans, classical conditioning has been applied to explaining and eliminating fears, breaking habits, combating taste aversion, and understanding such different experiences as pleasant emotions and drug habituation.

 Operant Conditioning

Operant conditioning is a form of learning in which the consequences of behavior produce changes in the probability of the behavior's occurrence. Skinner described the behavior of the organism as operant: The behavior operates on the environment, and the environment in turn operates on the organism. Whereas classical conditioning involves respondent behavior, operant conditioning involves operant behavior. In most instances, operant conditioning is better at explaining voluntary behavior than is classical conditioning.

Thorndike's law of effect states that behaviors followed by positive outcomes are strengthened, whereas behaviors followed by negative outcomes are weakened. Skinner built on this idea to develop the notion of operant conditioning.

Shaping is the process of rewarding approximations of desired behavior in order to shorten the learning process. Principles of reinforcement include the distinction between positive reinforcement (the frequency of a behavior increases because it is followed by a rewarding stimulus) and negative reinforcement (the frequency of behavior increases because it is followed by the removal of an aversive stimulus). Positive reinforcement can be classified as primary reinforcement (using reinforcers that are innately satisfying) and secondary reinforcement (using reinforcers that

acquire positive value through experience). Reinforcement can also be continuous (a behavior is reinforced every time) or partial (a behavior is reinforced only a portion of the time). Schedules of reinforcement—fixed ratio, variable ratio, fixed interval, and variable interval—determine when a behavior will be reinforced.

Operant conditioning involves generalization (giving the same response to similar stimuli), discrimination (responding to stimuli that signal that a behavior will or will not be reinforced), and extinction (a decreasing tendency to perform a previously reinforced behavior when reinforcement is stopped).

Punishment is a consequence that decreases the likelihood that a behavior will occur. In positive punishment, a behavior decreases when it is followed by a (typically unpleasant) stimulus. In negative punishment, a behavior decreases when a positive stimulus is removed from it.

Applied behavior analysis involves the application of operant conditioning principles to a variety of real-life behaviors.

 Observational Learning

Observational learning occurs when a person observes and imitates someone else's behavior. Bandura identified four main processes in observational learning: attention, retention, motor reproduction, and reinforcement.

 Cognitive Factors in Learning

Tolman emphasized the purposiveness of behavior. Purposiveness refers to Tolman's belief that much of behavior is goal-directed. In studying purposiveness, Tolman went beyond stimuli and responses to discuss cognitive mechanisms. Tolman believed that expectancies, acquired through experiences with the environment, are an important cognitive mechanism in learning.

Köhler developed the concept of insight learning, a form of problem solving in which the organism develops a sudden insight into or understanding of a problem's solution.

 Biological, Cultural, and Psychological Factors in Learning

Biological constraints restrict what an organism can learn from experience. These constraints include instinctive drift (the tendency of animals to revert to instinctive behavior that interferes with learned behavior), preparedness (the species-specific biological predisposition to learn in certain ways but not in others), and taste aversion (the biological predisposition to avoid foods that have caused sickness in the past).

Although most psychologists agree that the principles of classical conditioning, operant conditioning, and observational learning are universal, cultural customs can influence the degree to which these learning processes are used. Culture also often determines the content of learning.

In addition, there is no evidence that learning styles actually affect learning. However, what we learn is determined in part by what we believe we can learn. Dweck emphasizes that individuals benefit enormously from having a growth mindset rather than a fixed mindset.

KEY TERMS

acquisition

applied behavior analysis or behavior modification

associative learning

aversive conditioning

avoidance learning

behaviorism

classical conditioning

conditioned response (CR)

conditioned stimulus (CS)

counterconditioning

delay of gratification

discrimination (classical conditioning)

discrimination (operant conditioning)

extinction (classical conditioning)

extinction (operant conditioning)

generalization (classical conditioning)

generalization (operant conditioning)

habituation

insight learning

instinctive drift

latent learning or implicit learning

law of effect

learned helplessness

learning

negative punishment

negative reinforcement

observational learning

operant conditioning or instrumental conditioning

positive punishment

positive reinforcement

preparedness

primary reinforcer

punishment

reinforcement

schedules of reinforcement

secondary reinforcer

shaping

spontaneous recovery

unconditioned response (UR)

unconditioned stimulus (US)

ANSWERS TO SELF-QUIZZES

Section 1: 1. A; 2. C; 3. C; 4. C
Section 2: 1. B; 2. A; 3. D; 4. B

Section 3: 1. D; 2. B; 3. D; 4. A
Section 4: 1. B; 2. C; 3. B; 4. A

Section 5: 1. C; 2. B; 3. A; 4. A
Section 6: 1. D; 2. C; 3. B; 4. B

6 Memory

Traveling (mentally) through time

The COVID-19 pandemic meant that many people all over the world were thrust into a present moment that included the threat of a potentially deadly virus, working and going to school in unusual ways, and missing opportunities to gather with loved ones. One way people handled this challenging present moment was engaging in nostalgia. Nostalgia, a sentimental longing for the past, was a source of joy and a way to cope (Gammon & Ramshaw, 2020). At home, people watched old movies, streamed TV shows from their childhood, and watched replays of sporting events from the past. One analysis showed that of the nearly 17 trillion songs accessed on Spotify during the first few months of the COVID-19 lockdown, an unusually high proportion was old songs—suggesting that people were using music to engage in nostalgia and distance themselves from the present (Yeung, 2020). When the here and now is especially difficult, the past may provide solace. Our memories allow us to access a storehouse of experiences that promise a better future.

Certainly, memory provides crucial support for many mundane activities—for example, it allows us to know what we were looking for when we opened the fridge, where we left our running shoes, and when we need to mail Aunt Lucy's birthday card. But our memories are also precious because they represent a lasting imprint of our experiences, moments from the past that give our lives meaning.

LeoPatrizi/Getty Images

There are few moments when your life is not steeped in memory. Memory is at work with each step you take, each thought you think, and each word you speak. Through memory, you weave the past into the present and plan for the future. In this chapter, we explore the key processes of memory, including how information gets into memory and how it is stored, retrieved, and sometimes forgotten. We also probe what the science of memory reveals about the best way to study and retain course material, as well as how memory processes can enrich our lives.

1 The Nature of Memory

The stars are shining and the moon is full. A beautiful evening is coming to a close. You look at your significant other and think, "I'll never forget this night." How is it possible that in fact you never will forget it? Years from now, you might tell your children about that one special night so many years ago, even if you had not thought about it in the years since. How does one perfect night become a part of your enduring life memories?

memory
The retention of information or experience over time as the result of three key processes: encoding, storage, and retrieval.

Psychologists define **memory** as the retention of information or experience over time. Memory occurs through three important processes: encoding, storage, and retrieval. Memory requires taking in information (encoding the sights and sounds of that night), storing it or representing it (retaining it in some mental storehouse), and then retrieving it for a later purpose (recalling it when someone asks, "So, how did you two end up together?"). In the next three sections, we focus on these phases of memory: encoding, storage, and retrieval (Figure 1).

Except for the annoying moments when your memory fails or the upsetting situation where someone you know experiences memory loss, you most likely do not consider how much everything you do and say depends on the smooth operation of your memory systems. Think about asking for someone's phone number when you have no pencil, paper, or cell phone handy. You must attend to what the person tells you and rehearse the digits in your head until you can store them someplace permanently. Then, when the time comes to record the numbers, say, in your phone, you have to retrieve the identity of the person and the reason you got that phone number to begin with. Was it to ask the person out or to borrow notes for your psychology class? Human memory systems are truly remarkable considering how much information we put into memory and how much we must retrieve to perform life's activities.

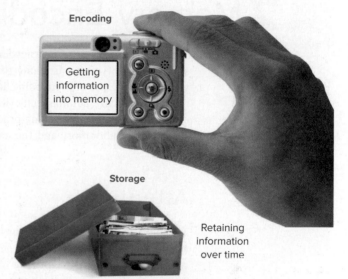

Encoding

Getting information into memory

Storage

Retaining information over time

Retrieval

Taking information out of storage

FIGURE 1 **Processing Information in Memory** As you read about the many aspects of memory in this chapter, think about the organization of memory in terms of these three main activities.

(camera): Garret Bautista/iStock/Getty Images; (box): Grata Victoria/iStock/Getty Images; (woman): Gary He/McGraw Hill

1. Memory is the _____ of information or experience over a period of time.
   A. rehearsal
   B. intake
   C. association
   D. retention

2. When we take in information in the course of daily life, such as the words and diagrams presented during a lecture, we are using the memory process of
   A. retention.
   B. encoding.
   C. retrieval.
   D. fixation.

3. The three processes of memory are encoding, _____, and retrieval.
   A. storage
   B. rehearsal
   C. recollection
   D. fixation

APPLY IT! 4. James and Adam are very good friends and often sit next to each other in Intro Psych. James spends a lot of time in class working on homework for his biology lab, while Adam listens to the lecture and takes lots of notes. Before the first exam, James asks to borrow Adam's notebook from Intro and studies those notes very carefully. In fact, both James and Adam study for 10 hours for the test. After the exam, James finds out he got a C, while Adam got an A. James cannot understand how they could have studied the same notes yet gotten such different grades. The most likely, most accurate explanation is that
   A. James and Adam encoded the information differently.
   B. Adam simply has a better memory than James.
   C. James is taking too many hard courses and could not retrieve the information as well as Adam because of stress.
   D. Adam probably gave James fake notes to torpedo his work.

2 Memory Encoding

The first step in memory is **encoding**, the process by which information gets into memory storage. When you are listening to a lecture, watching a play, reading a book, or talking with a friend, you are encoding information into memory. Some information gets into memory virtually automatically, whereas encoding other information takes effort. Let's examine some of the encoding processes that require effort. These include attention, deep processing, elaboration, and the use of mental imagery.

encoding
The first step in memory; the process by which information gets into memory storage.

Attention

To begin the process of memory encoding, we have to pay attention to information (Urgolites & others, 2020). *Selective attention* involves focusing on a specific aspect of experience while ignoring others. Attention is selective because the brain's resources are limited. We cannot attend to everything. These limitations mean that we must attend selectively to some things in our environment and ignore others (Dunne & Opitz, 2020). So, on that special night with your romantic partner, you never noticed the bus that roared by or the people whom you passed as you strolled along the street. Those details did not make it into your enduring memory.

In addition to selective attention, psychologists have described two other ways that attention may be allocated: divided attention and sustained attention. **Divided attention** involves concentrating on more than one activity at the same time. If you are listening to music or the television while reading this chapter, you are dividing your attention. **Sustained attention** (also called *vigilance*) is the ability to maintain attention to a selected stimulus for a prolonged period of time (Unsworth & Robison, 2020). For example, paying close attention to your notes while studying for an exam is a good application of sustained attention. **Executive attention** involves action planning, allocating attention to goals, error detection and compensation, monitoring progress on tasks, and dealing with novel or difficult circumstances. Thus, executive attention involves directing attention to engage in higher-level cognitive functioning.

Not paying attention can be especially detrimental to encoding. *Multitasking,* which in some cases involves dividing attention not just between two activities but among three

divided attention
Concentrating on more than one activity at the same time.

sustained attention
The ability to maintain attention to a selected stimulus for a prolonged period of time.

executive attention
The ability to plan action, allocate attention to goals, detect errors and compensate for them, monitor progress on tasks, and deal with novel or difficult circumstances.

How many times a day do you find yourself multitasking like this individual?
eva_blanco/Shutterstock

or more, may be the ultimate in divided attention (Tassone & others, 2020). It is not unusual for high school and college students to divide their attention among homework, instant messaging, TikTok videos, and looking at an iTunes playlist, all at once. Multitaskers are often very confident in their multitasking skills (Pattillo, 2010). However, research shows that multitasking during a lecture is related to lower scores on a memory test for the material (Jamet & others, 2020). Moreover, experimental evidence points to the causal impact of multitasking (especially using media) on poorer academic outcomes (Liu & Gu, 2020). Trying to listen to a lecture in class while texting or playing a game on your phone (or checking out social media on your laptop) is likely to impede your ability to pay adequate attention to the lecture. Divided attention decreases encoding. We remember information better when we devote our attention to it.

Remember that fact the next time you sit down to study in front of the TV or a computer.

Levels of Processing

Another factor that influences memory is whether we engage with information superficially or really get into it. Fergus Craik and Robert Lockhart (1972) first suggested that encoding can be influenced by levels of processing. The term **levels of processing** refers to a continuum from shallow to intermediate to deep, with deeper processing producing better memory.

Imagine that you are asked to memorize a list of words, including the word *mom.* Shallow processing includes noting the physical features of a stimulus, such as the shapes of the letters in the word *mom.* Intermediate processing involves giving the stimulus a label, as in reading the word *mom.* The deepest level of processing entails thinking about the meaning of a stimulus—for instance, thinking about the meaning of the word *mom* and about your own mother, your mother's face, and her special qualities.

The more deeply we process, the better the memory (Howes, 2006; Rose & Craik, 2012). For example, researchers have found that if we encode something meaningful about a face and make associations with it, we are more likely to remember the face (Harris & Kay, 1995). A restaurant server who strives to remember the face of the customer and to imagine them eating the food they have ordered is using deep processing (Figure 2).

levels of processing
A continuum of memory processing from shallow to intermediate to deep, with deeper processing producing better memory.

Level of Processing	Process	Examples
Shallow	Physical and perceptual features are analyzed.	The lines, angles, and contour that make up the physical appearance of an object, such as a car, are detected.
Intermediate	Stimulus is recognized and labeled.	The object is recognized as a car.
Deep	Semantic, meaningful, symbolic characteristics are used.	Associations connected with car are brought to mind—you think about the Porsche or Ferrari you hope to buy or the fun you and friends had on spring break when you drove a car to the beach.

FIGURE 2
Depth of Processing
According to the levels of processing principle, deeper processing of stimuli produces better memory of them.

Elaboration

Effective encoding of a memory depends on more than just depth of processing. Within deep processing, the more extensive the processing, the better the memory (Dudukovic & Kuhl, 2017). **Elaboration** refers to the formation of a number of different connections around a stimulus at any given level of memory encoding. Elaboration is like creating a huge spider web of links between some new information and everything one already knows, and it can occur at any level of processing. In the case of the word *mom,* a person can elaborate on mom even at a shallow level—for example, by thinking of the shapes of the letters and how they relate to the shapes of other letters, say, how an *m* looks like two *n*'s. At a deep level of processing, a person might focus on what a mother is or might think about various mothers they know, images of mothers in art, and portrayals of mothers on television and in film. Generally speaking, the more elaborate the processing, the better memory will be. Deep, elaborate processing is a powerful way to remember.

When you engage in elaborate, deep processing, you are setting the stage for later remembering by leaving lots of paths to follow when you need to recall the material. Who knows which of these cues will help you remember, but you'll have lots of possibilities.

For example, rather than trying to memorize the definition of memory, you would do better to weave a complex spider web around the concept of memory by coming up with a real-world example of how information enters your mind, how it is stored, and how you can retrieve it. Thinking of concrete examples of a concept is a good way to understand it. *Self-reference*—relating material to your own experience—is another effective way to elaborate on information, drawing mental links between aspects of your own life and new information (Daley & others, 2020; Yin & others, 2019) (Figure 3).

The process of elaboration is evident in the physical activity of the brain. Neuroscience research has shown a link between elaboration during encoding and brain activity. In one study, researchers placed individuals in magnetic resonance imaging (MRI) machines and flashed one word every 2 seconds on a screen inside (Wagner & others, 1998). Initially, the participants simply noted whether the words were in uppercase or lowercase letters. As the study progressed, they were asked to determine whether each word meant something concrete, such as *chair* or *book,* or abstract, such as *love* or *democracy.* The participants showed more neural activity in the left frontal lobe of the brain during the concrete/abstract task than they did when they were asked merely to state whether the words were in uppercase or lowercase letters. Further, they demonstrated better memory in the concrete/abstract task. The researchers concluded that greater elaboration of information is linked with neural activity, especially in the brain's left frontal lobe, and with improved memory.

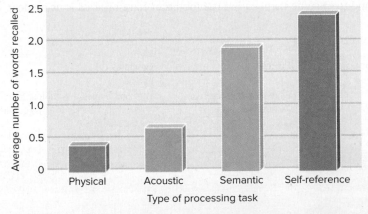

FIGURE 3 **Memory Improves When Self-Reference Is Used**
In one study, researchers (Hunt & Ellis, 2014) asked participants to remember lists of words according to the words' physical, acoustic (sound), semantic (meaning), or self-referent characteristics. As the figure illustrates, when individuals generated self-references for the words, they remembered them better.

Memorization of...	Record Holder	Country	Year	Record
Written numbers in 1 minute, no errors	Simon Reinhard	Germany	2016	104 numbers
Random words in 15 minutes*	Katie Kermode	Great Britain	2016	318 words
Speed to recall a single deck of 52 shuffled playing cards, no errors	Shijir-Erdene Bat-Enkh	Mongolia	2018	12.74 seconds
Historic dates in 5 minutes	Prateek Yadav	India	2019	148 dates
Abstract images in 15 minutes	Hu Jiabao	China	2018	804 images

FIGURE 4
World Champions of Memory For memorization wizards such as these world record holders, imagery is a powerful encoding tool.
SOURCE: The International World Record Breakers' Club

*Participants view random words in columns of 25 words. Scoring is tabulated by column: 1 point for each word. One mistake reduces the score for that column by half, and the second mistake reduces the score for that column to zero.

Imagery

One of the most powerful ways to make memories distinctive is to use mental imagery. Images are remembered better than verbal labels (Ensor & others, 2019). Mental imagery entails visualizing material that we want to remember in ways that create a lasting portrait. Imagery functions as a powerful encoding tool for all of us, certainly including the world champions of memory listed in Figure 4. Consider, for instance, Jonas von Essen, who, in 2020, recited the digits of pi to the first 100,000 decimal places. Think about memorizing a list of 100,000 digits with no pattern. How would you go about it? One way would be to use mental imagery to create a kind of visual mental walkthrough of the digits. To memorize the first eight digits of pi (3.1415926), one might say, "3 is a chubby fellow who walks with a cane (1), up to a takeout window (4), and orders 15 hamburgers. The cook (9), who has very large biceps (2), slips on his way to deliver the burgers (6)."

Mental imagery comes in handy in everyday life. Think about a restaurant server. After reciting your rather complicated order to a server, you notice that they are not writing anything down. Waiting patiently through your friends' orders, you wonder, "How can they possibly remember all this?" When the meal arrives, however, everything is exactly right. Servers seem to commit remarkable acts of memory routinely. How do they do it? Asked to share his secrets, a college student who moonlights in food service explained: "I always try to remember the person's face and imagine them eating the food they ordered."

If you are a graphic novel fan, you may resonate with Paivio's theory. Having both words and images can increase a person's involvement with media!

Classic studies by Allan Paivio (1971, 1986, 2007) have documented how imagery can improve memory. Paivio argues that memory is stored in one of two ways: as a verbal code (a word or a label) or an image code. Paivio thinks that the image code, which is highly detailed and distinctive, produces better memory than the verbal code. His *dual-code theory* claims that memory for pictures is better than memory for words because pictures—at least those that can be named—are stored as both image codes and verbal codes (Paivio & Sadoski, 2011; Welcome & others, 2011). Thus, when we use imagery to remember, we have two potential avenues by which we can retrieve information. We

Easy, right? Remembering a picture means remembering two things: the image AND the word that goes with it.

JohnnyGreig/Getty Images

encode stimuli based on the modality in which they are presented. So, memory of verbal information can be boosted by linking it to imagery. Recent research shows that even when stimuli are not visual (for instance, a spoken word), we encode both acoustic and visual information (Guitard & Cowan, 2020).

3 Memory Storage

storage
The retention of information over time and how this information is represented in memory.

Atkinson-Shiffrin theory
Theory stating that memory storage involves three separate systems: sensory memory, short-term memory, and long-term memory.

The quality of encoding is not the only thing that determines the quality of memory. A memory also must be stored properly after it is encoded. **Storage** encompasses how information is retained over time and how it is represented in memory.

We remember some information for less than a second, some for half a minute, and some for minutes, hours, years, or even a lifetime. Richard Atkinson and Richard Shiffrin (1968) formulated an early influential theory of memory that acknowledged the varying time span of memories (Figure 5). The **Atkinson-Shiffrin theory** separates memory storage into three systems:

- Sensory memory: time frames of a fraction of a second to several seconds
- Short-term memory: time frames up to 30 seconds
- Long-term memory: time frames up to a lifetime

As you read about these three memory storage systems, you will find that time frame is not the only thing that makes them different from one another. Each type of memory also operates in a distinctive way and has a special purpose.

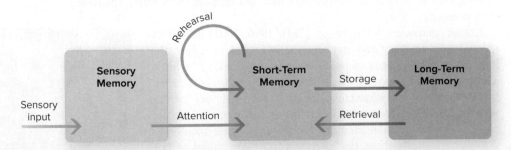

FIGURE 5   Atkinson and Shiffrin's Theory of Memory In this model, sensory input goes into sensory memory. Through the process of attention, information moves into short-term memory, where it remains for 30 seconds or less unless it is rehearsed. When the information goes into long-term memory storage, it can be retrieved over a lifetime.

Sensory Memory

sensory memory
Memory system that involves holding information from the world in its original sensory form for only an instant, not much longer than the brief time it is exposed to the visual, auditory, and other senses.

Sensory memory holds information from the world in its original sensory form for only an instant, not much longer than the brief time it is exposed to the visual, auditory, and other senses. Sensory memory is very rich and detailed, but we lose the information in it quickly unless we use certain strategies that transfer it into short-term or long-term memory.

Think about the sights and sounds you might encounter on a walk around your neighborhood. Literally thousands of stimuli come into your field of vision and hearing—cracks in the sidewalk, chirping birds, a noisy motorcycle, the blue sky, faces and voices of hundreds of people. You do not process all of these stimuli, but you do process a number of them. In general, you process many more stimuli at the sensory level than you consciously notice. Sensory memory retains this information from your senses, including a large portion of what you think you ignore. However, sensory memory does not retain the information very long.

Echoic memory (from the word *echo*) refers to auditory sensory memory, which is retained for up to several seconds. Imagine standing in an elevator with a friend who suddenly asks, "What was that song?" about the piped-in tune that just ended. If your friend asks this question quickly enough, you just might have a trace of the song left in your sensory registers.

Iconic memory (from the word *icon,* which means "image") refers to visual sensory memory, which is retained only for about 0.25 second (Figure 6). Visual sensory memory is responsible for our ability to "write" in the air using a sparkler on the Fourth of July—the residual iconic memory is what makes a moving point of light appear to be a line. The sensory memory for other senses, such as smell and touch, has received little attention in research studies.

The first scientific research on sensory memory focused on iconic memory. Remember that iconic memory is that part of sensory memory that includes images. In George Sperling's (1960) classic study, participants viewed patterns of stimuli such as those in Figure 7. As you look at the letters, you have no trouble recognizing them. However, Sperling flashed the letters on a screen for very brief intervals, about 0.05 second. Afterward, the participants could report only four or five letters. With such a short exposure, reporting all nine letters was impossible.

Some participants in Sperling's study reported feeling that for an instant, they could see all nine letters within a briefly flashed pattern. They ran into trouble when they tried to name all the letters they had initially seen. One hypothesis to explain this experience is that all nine letters *were* initially processed at the iconic sensory memory level. This is why all nine letters were seen. However, forgetting from iconic memory occurred so

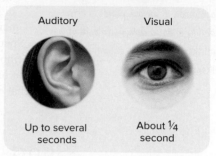

Type of Sensory Register

Auditory | Visual

Up to several seconds | About ¼ second

FIGURE 6 **Auditory and Visual Sensory Memory** If you hear this bird's call while walking through the woods, your auditory sensory memory holds the information for several seconds. If you see the bird, your visual sensory memory holds the information for only about 0.25 second.

(bird): steve_byland/123RF; (eye): ColorBlind Images/Blend Images LLC; (ear): Geoff du Feu/Alamy Stock Photo

L H V
R F Z
D T C

FIGURE 7
Sperling's Sensory Memory Experiment This array of stimuli is similar to those flashed for about 0.05 second to the participants in Sperling's study.

simonkr/Getty Images

Sperling's solution is truly remarkable. He realized that by giving the participants the signal, he could help them to scan their mental image quickly so that they could find specific pieces of the information that it contained in various places. Their ability to do so demonstrates that all the material was actually there. Fantastic!

rapidly that the participants did not have time to transfer all the letters to short-term memory, where they could be named.

Sperling reasoned that if all nine letters are actually processed in sensory memory, they should all be available for a brief time. To test this possibility, Sperling sounded a low, medium, or high tone just after a block of letters was shown. This tone signaled to participants which row they would be asked to report—only the letters from the bottom, middle, or top row. Under these conditions, the participants performed much better. This outcome suggests a brief memory for most or all of the letters in the display did exist for participants. Sperling showed that an entire array of information is briefly present in iconic memory. To experience this phenomenon, glance at this page for just a second. All of the letters are present in your sensory memory for an instant, creating a mental image that momentarily exists in its entirety.

Short-Term Memory

short-term memory
Limited-capacity memory system in which information is usually retained for only as long as 30 seconds unless strategies are used to retain it longer.

Much information goes no further than the stage of auditory and visual sensory memory. We retain this information for only a brief instant. However, some information, especially that to which we pay attention, proceeds into short-term memory. **Short-term memory** is a limited-capacity memory system in which information is usually retained for only as long as 30 seconds unless we use strategies to retain it longer. Compared with sensory memory, short-term memory is limited in capacity, but it can store information for a longer time.

George Miller (1956) examined the limited capacity of short-term memory in the classic paper "The Magical Number Seven, Plus or Minus Two." Miller pointed out that on many tasks people are limited in how much information they can keep track of without external aids. Usually the limit is in the range of 7 ± 2 items. The most widely cited example of this phenomenon involves *memory span,* the number of digits an individual can report back in order after a single presentation of them. Most college students can remember eight or nine digits without making errors (think about how easy it is to recall a phone number). Longer lists pose problems because they exceed short-term memory capacity. If you rely on simple short-term memory to retain longer lists, you probably will make errors.

What are some important numbers in your life? Do they follow the 7 ± 2 rule?

CHUNKING AND REHEARSAL

Two ways to improve short-term memory are chunking and rehearsal. *Chunking* involves grouping or "packing" information that exceeds the 7 ± 2 memory span into higher-order units that can be remembered as single units. Chunking works by making large amounts of information more manageable (Isbilen & others, 2020; Norris & others, 2020; Thalmann & others, 2019).

To get a sense of chunking, consider this word list: *hot, city, book, forget, tomorrow,* and *smile.* Hold these words in memory for a moment and then write them down. If you recalled the words, you succeeded in holding 30 letters, grouped into six chunks, in memory. Now hold the following list in memory and then write it down:

O LDH ARO LDAN DYO UNGB EN

How did you do? Do not feel bad if you did poorly. This string of letters is very difficult to remember, even though it is arranged in chunks. The problem is that the chunks lack meaning. If you re-chunk the letters to form the meaningful words "Old Harold and Young Ben," they become much easier to remember.

Another way to improve short-term memory involves *rehearsal*: the conscious repetition of information (Himmer & others, 2019). Information stored in short-term memory lasts half a minute or less without rehearsal. However, if rehearsal is not interrupted, information can be retained indefinitely. Rehearsal is often verbal, giving the impression of an inner voice, but it can also be visual or spatial, giving the impression of a private inner eye.

Rehearsal works best when we must briefly remember a list of numbers or items such as entrées from a dinner menu. When we need to remember information for longer periods of time, as when we are studying for a test coming up next week or even an hour from now, other strategies usually work better. A main reason rehearsal does not work well for retaining information over the long term is that rehearsal often involves just mechanically repeating information, without imparting meaning to it. The fact that, over the long term, we remember information best when we add meaning to it demonstrates the importance of deep, elaborate processing.

You probably recognize rehearsal as the first thing you would do if given, say, a list of numbers to remember. By repeating the list over and over in your head, you can keep the numbers until you have time to write them down or put them in your phone. Interestingly, research on the actual causal role of rehearsal in maintaining information in memory is not as strong as is often assumed (Oberauer, 2019). Research on this topic continues to probe both what people naturally do to maintain information in memory and whether those strategies work.

Though useful, Atkinson and Shiffrin's theory of the three time-linked memory systems fails to capture the dynamic way short-term memory functions (Baddeley, 2008, 2012). Short-term memory is not just about storing information; it is about attending to information, manipulating it, and using it to solve problems (Cowan & others, 2011a, 2011b; Murphy & others, 2020). How can we understand these processes? The concept of working memory is one way psychologists have addressed this question.

working memory
A combination of components, including short-term memory and attention, that allow individuals to hold information temporarily as they perform cognitive tasks; a kind of mental workbench on which the brain manipulates and assembles information to guide understanding, decision making, and problem solving.

WORKING MEMORY   **Working memory** refers to a combination of components, including short-term memory and attention, that allow us to hold information temporarily as we perform cognitive tasks (Cowan, 2008, 2017; Cowan & others, 2012). Working memory is not the same as short-term memory. For instance, a person can hold a list of words in short-term memory by rehearsing them over and over. A measure of short-term memory capacity, then, might simply involve counting how many words in the list the person can remember. Working memory capacity is separable from short-term memory capacity. We cannot be rehearsing information while we are working on solving a problem. This may be why measures of short-term memory capacity are not strongly related to cognitive aptitudes, such as intelligence, while working memory capacity is (Cowan, 2008). Because short-term memory capacity can rely on rehearsal, 7 ± 2 chunks are generally manageable. However, in working memory, if the chunks are relatively complex, most young adults can remember only 4 ± 1, that is, three to five chunks (Cowan, 2010). Working memory is not a passive storehouse with shelves to store information until it moves to long-term memory, but rather it is an active memory system.

Working memory can be thought of as a mental blackboard, a place where we can imagine and visualize. In this sense, working memory is the context for conscious thought (Gordon & others, 2020; Honig & others, 2020). Anthropologists, archaeologists, and psychologists have been interested in understanding how working memory evolved. For example, prehistoric tools (Haidle, 2010) and works of art (Wynn & others, 2009) reveal how early humans thought. Consider the *Lion Man,* an ivory sculpture archaeologists found in a cave in Germany. The 28-cm figurine, with the head of a lion and the body of a man, is believed to have been created 32,000 years ago (Balter, 2010). This ancient work of art must have been the product of a person who had the capacity to see two things and, in working memory, ask something like, "What would they look like if I combined them?" Some commentators have suggested that working memory lays the foundation for creative culture (Haidle, 2010). Working memory has been proposed as a key difference between Neanderthals and *Homo sapiens* (Wynn & Coolidge, 2010).

Working memory has served as a helpful framework for addressing practical problems outside the laboratory. For example, advances in understanding working memory have allowed researchers to identify students at risk for academic underachievement (Gonthier & Roulin, 2020; Luo & Zhou, 2020). Training programs aimed at improving working memory in healthy children have not, however, been found to have general benefits (Sala & Gobet, 2020). However, working memory

Think about it! Working memory is, in some ways, the very essence of consciousness—your ability to quietly think to yourself in the private little workshop in your head.

training may be helpful for older individuals and those with neurological disorders such as Parkinson disease or stroke (Giehl & others 2020; Peers & others, 2020).

Working memory is a kind of mental workbench on which the brain manipulates and assembles information to help us understand, make decisions, and solve problems. If, say, all of the information on the hard drive of your computer is like long-term memory, then working memory is comparable to what you have open and active at any given moment. Working memory has a limited capacity, and, to take the computer metaphor further, the capacity of the working memory is like RAM.

How does working memory work? British psychologist Alan Baddeley (1993, 1998, 2003, 2008, 2012, 2013) has proposed an influential model of working memory featuring a three-part system that allows us to hold information temporarily as we perform cognitive tasks.

Figure 8 shows Baddeley's view of the three components of working memory. The three components are the *central executive,* the *phonological loop,* and the *visuo-spatial sketchpad.* Think of them as a boss (the central executive) who has two assistants (the phonological loop and the visuo-spatial sketchpad) to help do the work. Be sure to examine Figure 8 carefully.

1. The *phonological loop* is specialized to briefly store speech-based information about the sounds of language (Baddeley & Hitch, 2019). The phonological loop contains two separate components: an acoustic code (the sounds we heard), which decays in a few seconds, and rehearsal, which allows us to repeat the words in the phonological store.

2. The *visuo-spatial sketchpad* stores visual and spatial information, including visual imagery (Baddeley & others, 2019). As in the case of the phonological loop, the capacity of the visuo-spatial sketchpad is limited. If we try to put too many items in the visuo-spatial sketchpad, we cannot represent them accurately enough to retrieve them successfully. The phonological loop and the visuo-spatial sketchpad function independently. We can rehearse numbers in the phonological loop while making spatial arrangements of letters in the visuo-spatial sketchpad.

3. The *central executive* integrates information not only from the phonological loop and the visuo-spatial sketchpad but also from long-term memory. In Baddeley's view, the central executive plays important roles in attention, planning, and organizing. The central executive acts like a supervisor who monitors which information deserves our attention and which we should ignore. It also selects which strategies to use to process information and solve problems (Hitch & others, 2020; Hu & others, 2014). Like the phonological loop and the visuo-spatial sketchpad, the central executive has a limited capacity. If working memory is like the files you have open on your computer, the central executive is you. You pull up information you need, close out other things, and so forth.

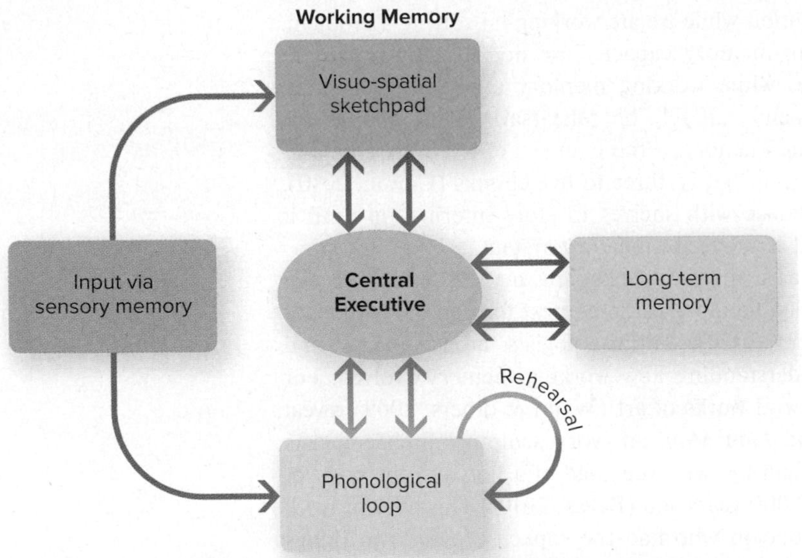

> *If the connection to RAM is not working for you, think about the last time your phone started acting 'janky' and you realized you had 47 apps all open at the same time.*

Working Memory

FIGURE 8 **Baddeley's View of Working Memory** In Baddeley's working memory model, working memory consists of three main components: the phonological loop, the visuo-spatial sketchpad, and the central executive. The phonological loop and the visuo-spatial sketchpad serve as assistants, helping the central executive do its work. Input from sensory memory goes to the phonological loop, where information about speech is stored and rehearsal takes place, and to the visuo-spatial sketchpad, where visual and spatial information, including imagery, is stored. Working memory is a limited capacity system, and information is stored there only briefly. Working memory interacts with long-term memory, drawing information from long-term memory and transmitting information to long-term memory for longer storage.

Though it is compelling, Baddeley's notion of working memory is simply a conceptual model describing processes in memory. Neuroscientists continue to search for brain areas and activity that might be responsible for these processes (Murphy & others, 2020).

Long-Term Memory

long-term memory
A relatively permanent type of memory that stores huge amounts of information for a long time.

Long-term memory is a relatively permanent type of memory that stores huge amounts of information for a long time. The capacity of long-term memory is staggering. John von Neumann (1958), a distinguished mathematician, put the size at 2.8×10^{20} (280 quintillion) bits, which in practical terms means that our storage capacity is virtually unlimited. Von Neumann assumed that we never forget anything; but even considering that we do forget things, we can hold several billion times more information than a large computer.

An interesting question is how the availability of information on the Internet has influenced memory. If we know we can look something up on the web, why bother storing it in our head? A series of studies by Betsy Sparrow and her colleagues (Sparrow & others, 2011) demonstrated that in the face of difficult memory tasks, people are likely to think immediately of looking to the computer for the answer rather than doing the hard work of remembering.

Long-term memory capacity is amazing. When you get a new memory, you don't have to pitch an old one out. It's like a gigantic closet with unlimited space.

COMPONENTS OF LONG-TERM MEMORY   Long-term memory is complex, as Figure 9 shows. At the top level, it is divided into substructures of explicit memory and implicit memory (Heyselaar & others, 2017). Explicit memory can be further subdivided into episodic and semantic memory. Implicit memory includes the systems involved in procedural memory, classical conditioning, and priming.

In simple terms, explicit memory has to do with remembering who, what, where, when, and why; implicit memory has to do with remembering how. To explore the distinction, let's look at the case of a person known as H. M. Afflicted with severe epilepsy, H. M. underwent surgery in 1953, when he was 27 years old, that involved removing the hippocampus and a portion of the temporal lobes of both hemispheres in his brain (Carey, 2009). H. M.'s epilepsy improved, but something devastating happened to his memory. Most dramatically, he was unable to form new memories that could outlive working

H. M. spent most of his life in the present moment. His legacy to cognitive science is unforgettable. When he died, many of the psychologists who had studied his case felt as if they had lost a friend–even if a friend who had never been able to remember them

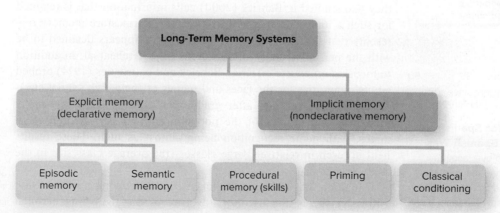

FIGURE 9   Systems of Long-Term Memory Long-term memory stores huge amounts of information for long periods of time, much like a computer's hard drive. The hierarchy in the figure shows the division of long-term memory at the top level into explicit memory and implicit memory. Explicit memory can be further divided into episodic and semantic memory; implicit memory includes procedural memory, priming, and classical conditioning.

memory. H. M.'s memory time frame was only a few minutes at most, so he lived, until his death in 2008, in a perpetual present and could not remember past events (explicit memory). In contrast, his memory of how to do things (implicit memory) was less affected. For example, he could learn new physical tasks, even though he had no memory of how or when he learned them.

H. M.'s situation demonstrates a distinction between explicit memory, which was dramatically impaired in his case, and implicit memory, which in his case was less influenced by his surgery. Let's explore the subsystems of explicit and implicit memory more thoroughly.

Explicit Memory **Explicit memory (declarative memory)** is the conscious recollection of information, such as specific facts and events and, at least in humans, information that can be verbally communicated (Tulving, 1989, 2000). Examples of using explicit, or declarative, memory include recounting the events in a movie you have seen and recalling which politicians are in the president's cabinet.

How long does explicit memory last? Explicit memory includes things you are learning in your classes even now. Will this information stay with you? Research by Harry Bahrick has examined this question. Ohio Wesleyan University, where Bahrick is a professor of psychology, is a small (about 1,800 students) liberal arts school that boasts very loyal alumni who faithfully return to campus for reunions and other events. Bahrick (1984) took advantage of this situation to conduct an ingenious study on the retention of course material over time. He gave vocabulary tests to people who had taken Spanish in college as well as to a control group of college students who had not taken Spanish. The individuals chosen for the study had used Spanish very little since their college courses. Some people were tested at the end of an academic year (just after having taken their Spanish courses), but others were tested years after graduation—as many as 50 years later. When Bahrick assessed how much the participants had forgotten, he found a striking pattern (Figure 10): Forgetting tended to occur in the first three years after taking the classes and then leveled off, so that adults maintained considerable knowledge of Spanish vocabulary words up to 50 years later.

Bahrick (1984) assessed not only how long ago the adults had studied Spanish but also how well they did in Spanish during college. Those who had earned an *A* in their courses 50 years earlier remembered more Spanish than adults who had gotten a *C* when taking Spanish only one year earlier. Thus, how well students initially learned the material was even more important than how long ago they had studied it. Bahrick (2000) calls information that is retained for such a long time "permastore" content. Permastore memory represents that portion of original learning that appears destined to be with the person virtually forever, even without rehearsal. In addition to focusing on course material, Bahrick and colleagues (1974) probed adults' memories for the faces and names of their high school classmates. Thirty-five years after graduation, the participants visually recognized 90 percent of the portraits of their high school classmates, with name recognition being almost as high. These results held up even in relatively large classes (the average class size in the study was 294).

Canadian cognitive psychologist Endel Tulving (1972, 1989, 2000) has been the foremost advocate of distinguishing between two subtypes of explicit memory: episodic and semantic. **Episodic memory** is the retention of information about the where, when, and what of life's happenings—how we remember life's episodes. Episodic memory is autobiographical. For example, episodic memory includes the details of where you were when your younger

> *What course material do you think you will never forget?*

explicit memory or declarative memory The conscious recollection of information, such as specific facts or events and, at least in humans, information that can be verbally communicated.

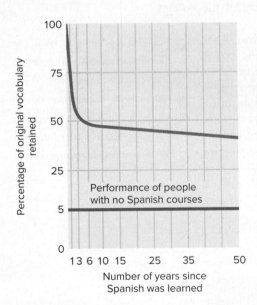

FIGURE 10 **Memory for Spanish as a Function of Age Since Spanish Was Learned** An initial steep drop over about a three-year period in remembering the vocabulary learned in Spanish classes occurred. However, there was little dropoff in memory for Spanish vocabulary from three years after taking Spanish classes to 50 years after taking them. Even 50 years after taking Spanish classes, individuals still remembered almost 50 percent of the vocabulary.

SOURCE: Bahrick, Harry P. Journal of Experimental Psychology: General, 113(1), Mar 1984, 1–29.

episodic memory The retention of information about the where, when, and what of life's happenings—that is, how individuals remember life's episodes.

sibling was born, what happened on your first date, and what you ate for breakfast this morning.

Semantic memory is a person's knowledge about the world. It includes one's areas of expertise, general knowledge of the sort learned in school, and everyday knowledge about the meanings of words, famous individuals, important places, and common things. For example, semantic memory is involved in a person's knowledge of chess, of geometry, and of who the Dalai Lama, Serena Williams, and Lady Gaga are. An important aspect of semantic memory is that it appears to be independent of an individual's personal identity with the past. You can access a fact—such as the detail that Lima is the capital of Peru—and not have the foggiest notion of when and where you learned it.

The difference between episodic and semantic memory is also demonstrated in certain cases of amnesia (memory loss). A person with amnesia might forget entirely who they are—their name, family, career, and all other vital information about themselves—yet still be able to talk, know what words mean, and have general knowledge about the world, such as what day it is or who currently holds the office of U.S. president (Rosenbaum & others, 2005). In such cases, episodic memory is impaired, but semantic memory is functioning.

Your recollection of your first day on campus is an episodic memory. If you take a history class, your memory of the information you need to know for a test involves semantic memory.

Learning a new language, you learn the words and general grammatical structure (semantic memory) and you might also remember how you learned them (episodic memory). Language is vital to memory. We use words (part of semantic memory) to tell the stories of our lives (episodic memory). Think about the distinction between episodic and semantic memory. It is remarkable that these two memory systems do not collide. Both semantic and episodic memory depend on the hippocampus. The balance that the brain strikes is pretty remarkable. How does the brain manage to encode both types of information without getting them mixed up? Babies present an excellent context for probing this question. To read about research demonstrating the brain's remarkable capacity to foster both kinds of memories, see the Intersection.

Figure 11 summarizes some aspects of the episodic/semantic distinction. The differences that are listed are controversial. One criticism is that many cases of explicit, or declarative, memory arc neither purely episodic nor purely semantic but fall in a gray area in between. Consider your memory for what you studied last night. You probably added knowledge to your semantic memory—which was, after all, the reason you were studying. You probably remember where you were studying, as well as about when you started and when you stopped. You probably also can remember some minor occurrences, such as a burst of loud laughter from the room next door or the coffee you spilled on the desk. Is episodic or semantic memory involved here? Tulving (1983, 2000) argues that semantic and episodic systems often work together in forming new memories. In such cases, the memory that ultimately forms might consist of an autobiographical episode and semantic information.

Implicit (Nondeclarative) Memory In addition to explicit memory, there is a type of long-term memory that is related to non-consciously remembering skills and sensory perceptions rather than con-sciously remembering facts. **Implicit memory (nondeclarative memory)** is memory in which behavior is affected by prior experience without a conscious recollection of that experience. Implicit memory comes into play, for example, in the skills of playing tennis and snowboarding, as well as in the physical act of text messaging. Another example of implicit memory is the repetition in your mind of a song you heard playing in the super-market, even though you had not noticed that song playing.

Characteristic	Episodic Memory	Semantic Memory
Units	Events, episodes	Facts, ideas, concept
Organization	Time	Concepts
Emotion	More important	Less important
Retrieval process	Deliberate (effortful)	Automatic
Retrieval report	"I remember"	"I know"
Education	Irrelevant	Relevant
Intelligence	Irrelevant	Relevant
Legal testimony	Admissible in court	Inadmissible in court

FIGURE 11 **Some Differences Between Episodic and Semantic Memory** These characteristics have been proposed as the main ways to differentiate episodic from semantic memory.

Sleep and Developmental Psychology: How Do Naps Allow Babies to Separate Episodic and Semantic Memories?

Human infants are born with a lot to accomplish in the realm of memory. They must acquire a huge amount of semantic information and yet also collect up some episodic memories. For the developing person, maintaining the separation between semantic and episodic memory is important. Imagine what it would be like—learning a constant stream of new words that are themselves embedded in experiences that are also important memories—without the capacity to separate semantic and episodic content. How does the infant brain keep those streams straight? A recent study showed that an important answer to this question is simple—naps!

The study (Friedrich & others, 2020) involved infants, ranging in age from 14 to 17 months. In the first part of the study, the encoding phase, infants were shown pictures of common objects, like dog, ball, and car. The name of each object was said aloud slowly as the object was shown. Next, the infants were assigned randomly to either take a nap or stay awake (the independent variable). After napping (or staying awake) infants were once again presented word-object pairs. This time, some of the pairs were the same ones from the encoding phase and some were new. This last phase of the study was the memory phase.

G-Stock Studio/Shutterstock

Now, infants cannot just tell us what they remember and whether it is semantic or episodic. So, the researchers used electroencephalography (or EEG). Recall that EEG involves attaching electrodes to the scalp to measure electrical activity across the brain. Scientists have already established EEG patterns that reflect semantic processing and patterns that reflect episodic processing. In the encoding phase, all of the babies responded to the word-object pairs as semantic information, suggesting that they were processing the labels for the objects (part of semantic memory).

The dependent variable was the brain's response to the old pairs in the memory phase. Did infants respond to the pairs they had seen before as semantic or episodic memory? The babies who had stayed awake continued to process the old pairs as semantic memory; they still treated the verbal labels only as words they were learning. However, the babies who napped processed the old pairs as episodic memory. The brain protected the new memories of these infants so that they might have a memory of learning, not just the knowledge that resulted from that learning (Friedrich & others, 2020). The brain seems to have temporarily disabled semantic processing for these specific word-object pairs that were, for these infants, *experiences*, not just labels. (Note that for the new word-object pairs, all infants showed semantic processing, so it is not that the infants who napped were unable to engage in semantic processing.)

Recall from our discussion of sleep that during stage N2 sleep, sleep spindles (quick and simultaneous firing of neurons), which allow the neocortex to communicate with the hippocampus, play an important role in memory consolidation. Aspects of sleep were also measured while babies in the nap condition slept. The researchers found that in the napping group, greater spindle activity predicted brain responses indicating episodic memory for the old word-object pairs (Friedrich & others, 2020).

This fascinating study reveals how the infant brain opens a space for episodic memory in human life. Yes, we need to learn the facts of our lives. But we also need a groundwork for the emerging story of that life.

\\ Why would protecting infant episodic memory from semantic processing be important?

Three subsystems of implicit memory are procedural memory, classical conditioning, and priming. All of these subsystems refer to memories that you are not aware of but that influence behavior.

This is also why you might find yourself knowing the words to a song you hate. You've heard it so many times that you've memorized it without even wanting to.

Procedural memory is a type of implicit memory process that involves memory for skills. For example, assuming that you are an expert typist, when you type a paper you are not conscious of where the keys are for the various letters; somehow, your well-learned, nonconscious skill of typing allows you to hit the right keys. Similarly, once you have learned to drive a car, you remember how to go about it: You do not have to remember consciously how to drive the car as you put the key in the ignition, turn the steering wheel, depress the gas pedal, and step on the brake pedal.

procedural memory
Memory for skills.

Thanks to implicit memory, we can perform a learned skill, such as playing tennis, without thinking about it consciously.
Karl Weatherly/Getty Images

priming
The activation of information that people already have in storage to help them remember new information better and faster.

Another type of implicit memory involves classical conditioning, a form of learning. Recall that classical conditioning involves the automatic learning of associations between stimuli, so that one stimulus comes to evoke the same response as the other. Classically conditioned associations involve nonconscious, implicit memory. So, without realizing it, you might start to like the person who sits next to you in your favorite class because they are around while you are feeling good.

Priming is the activation of information that people already have in storage to help them remember new information better and faster. In a common demonstration of priming, individuals study a list of words (such as *hope, walk,* and *cake*). Then they are given a standard recognition task to assess explicit memory. They must select all of the words that appeared in the list—for example, "Did you see the word *hope?* Did you see the word *form?*" Then participants perform a stem-completion task, which assesses implicit memory. In this task, they view a list of incomplete words (for example, *ho__, wa__, ca__*), called word stems, and must fill in the blanks with whatever word comes to mind. The results show that individuals more often fill in the blanks with the previously studied words than would be expected if they were filling in the blanks randomly. For example, they are more likely to complete the stem *ho__* with *hope* than with *hole*. This result occurs even when individuals do not recognize the words on the earlier recognition task. Because priming takes place even when explicit memory for previous information is not required, it is assumed to be an involuntary and unconscious process. Priming occurs when something in the environment evokes a response in memory, such as the activation of a particular concept.

To grasp the difference between explicit and procedural memory, imagine trying to describe in words how to tie a shoe—a procedure you can easily perform—without having a shoe around

Yeah baby! Classical conditioning is back! Yes, this automatic associative learning involves memory

HOW MEMORY IS ORGANIZED

Explaining the forms of long-term memory does not address the question of how the different types of memory are organized for storage. The word *organized* is important: Memories are not haphazardly stored but instead are carefully sorted.

Here is a demonstration. Recall the 12 months of the year as quickly as you can. How long did it take you? What was the order of your recall? Chances are, you listed them within a few seconds in chronological order (January, February, March, and so on). Now try to remember the months in alphabetical order. How long did it take you? Did you make any errors? Your memory for the months of the year is organized in a particular way. Indeed, one of memory's most distinctive features is its organization.

If people are encouraged to organize material simply, their memories of the material improve even if they receive no warning that their memories will be tested (Mandler, 1980). Psychologists have developed a variety of theories of how long-term memory is organized. Let's consider two of these more closely: schemas and connectionist networks.

schema
A preexisting mental concept or framework that helps people to organize and interpret information. Schemas from prior encounters with the environment influence the way individuals encode, make inferences about, and retrieve information.

Schemas You and a friend have taken a long drive to a new town where neither of you has ever been before. You stop at the local diner, have a seat, and look over the menu. You have never been in this diner before, but you know exactly what is going to happen. Why? Because you have a schema for what happens in a restaurant. When we store information in memory, we often fit it into the collection of information that already exists, as you do even in a new experience with a diner. A **schema** is a preexisting mental concept or framework that helps people to organize and interpret information (Guo & Yang, 2020). Schemas from prior encounters with the environment influence the way we handle information—how we encode it, what inferences we make about it, and how we retrieve it.

monkeybusinessimages/Getty Images

Schemas can also be at work when we recall information. Schema theory holds that long-term memory is not very exact. We seldom find precisely the memory that we want, or at least not all of what we want; hence, we have to reconstruct the rest. Our schemas support the reconstruction process, helping us fill in gaps between our fragmented memories.

We have schemas for lots of situations and experiences—for scenes and spatial layouts (a beach, a bathroom), as well as for common events (playing football, writing a term paper). A **script** is a schema for an event (Schank & Abelson, 1977). Scripts often have information about physical features, people, and typical occurrences. This kind of information is helpful when people need to figure out what is happening around them. For example, if you are enjoying your after-dinner coffee in an upscale restaurant and someone in a tuxedo comes over and puts a piece of paper on the table, your script tells you that is probably a waiter who has just given you the check. Scripts help to organize our storage of memories about events.

script
A schema for an event, often containing information about physical features, people, and typical occurrences.

Connectionist Networks   Schema theory has little or nothing to say about the role of the brain in memory. Thus, a new theory based on brain research has generated excitement among psychologists. **Connectionism,** or **parallel distributed processing (PDP),** is the theory that memory is stored throughout the brain in connections among neurons, several of which may work together to process a single memory (McClelland, 2011; Thibodeau & others, 2020). Recall the concept of neural networks and the idea of parallel sensory processing pathways. These concepts also apply to memory.

In the connectionist view, memories are not large knowledge structures (as in schema theories). Instead, memories are more like electrical impulses, organized only to the extent that neurons, the connections among them, and their activity are organized (Josselyn & Tonegawa, 2020; Kerzel & Andres, 2020; Xiao & others, 2017). Any piece of knowledge—such as your dog's name—is embedded in the strengths of hundreds or thousands of connections among neurons and is not limited to a single location.

connectionism (parallel distributed processing [PDP])
The theory that memory is stored throughout the brain in connections among neurons, several of which may work together to process a single memory.

This is why elaboration in encoding is so helpful. It is laying down all these pathways in your brain that light up when you try to remember.

How does the connectionist process work? A neural activity involving memory, such as remembering your dog's name, is spread across a number of areas of the cerebral cortex. The locations of neural activity, called *nodes,* are interconnected. When a node reaches a critical level of activation, it can affect another node across synapses. We know that the human cerebral cortex contains millions of neurons that are richly interconnected through hundreds of millions of synapses. Because of these synaptic connections, the activity of one neuron can be influenced by many other neurons. Owing to these simple reactions, the connectionist view argues that changes in the strength of synaptic connections are the fundamental bases of memory (McClelland & others, 2010). From the connectionist network perspective, memories are organized sets of neurons that are routinely activated together. The process by which these interconnected networks are formed is called *consolidation.* When neurons continually fire together in these interconnected ways, a memory forms; even though a memory may feel like "one thing," such as one fact or one story, it is represented in multiple connected paths in the brain.

Part of the appeal of the connectionist view is that it is consistent with what we know about brain function and allows psychologists to simulate human memory studies using computers (Marcus, 2001). Connectionist approaches also help to explain how priming a concept (for instance, with a photo of someone winning a race) can influence behavior (vigorously pursuing achievement goals; Chen & others, 2020). Furthermore, insights from this connectionist view support brain research undertaken to determine where memories are stored in the brain (Josselyn & Tonegawa, 2020; McClelland, 2011; McClelland & Rumelhart, 2009).

Indeed, so far we have examined the many ways cognitive psychologists think about *how* information is stored. The question remains, however, *where*? Although memory may seem to be a mysterious phenomenon, it, like all psychological processes, must occur in a physical place: the brain.

HOW MEMORIES ARE STORED
Karl Lashley (1950) spent a lifetime looking for a location in the brain in which memories are stored. He trained rats to discover the correct pathway in a maze and then cut out various portions of the animals' brains and retested their memory of the maze pathway. Experiments with thousands of rats showed that the loss of various cortical areas did not affect rats' ability to remember the pathway, leading Lashley to conclude that memories are not stored in a specific location in the brain. Other researchers, continuing Lashley's quest, agreed that memory storage is diffuse, but they developed additional insights. Canadian psychologist Donald Hebb (1949, 1980) suggested that assemblies of cells, distributed over large areas of the cerebral cortex, work together to represent information, just as the connectionist network perspective would predict.

Neurons and Memory
Modern neuroscience supports the idea that memories are composed of specific sets or circuits of neurons. Memories are not "spots" in the brain but collections of connected neurons working together. Within these connected networks, single neurons may respond to particular stimuli or features of those stimuli (Braun & others, 2012; Squire, 2007). Researchers who measure the electrical activity of single cells have found that some respond to faces and others to eye or hair color, for example. Still, in order for you to recognize your Uncle Max, individual neurons that provide information about hair color, size, and other characteristics must act together.

Yes, it is kind of weird, but although you may experience a memory as one thing—a story or a fact—it is actually represented not in one place in the brain but in many interconnected processes.

Neurotransmitters may be the ink with which memories are written. Recall that neurotransmitters are the chemicals that allow neurons to communicate across the synapse. These chemicals play a crucial role in forging the connections that represent memory.

Ironically, some of the answers to complex questions about the neural mechanics of memory come from studies on a very simple experimental animal—the inelegant sea slug. Eric Kandel and James Schwartz (1982) chose this large snail-without-a-shell because of the simple architecture of its nervous system, which consists of only about 10,000 neurons. (You might recall that the human brain has about 100 billion neurons.)

The sea slug is hardly a quick learner or an animal with a good memory, but it is equipped with a reliable reflex. When anything touches the gill on its back, it quickly withdraws it. So, first the researchers accustomed the sea slug to having its gill prodded. After a while, the animal ignored the prod and stopped withdrawing its gill. Next the researchers applied an electric shock to its tail when they touched the gill. After many rounds of the shock-accompanied prod, the sea slug violently withdrew its gill at the slightest touch. The researchers found that the sea slug remembered this message for hours or even weeks. They also determined that shocking the sea slug's gill releases the neurotransmitter serotonin at the synapses of its nervous system, and this chemical release basically provides a reminder that the gill was shocked. This "memory" informs the nerve cell to send out chemical commands to retract the gill the next time it is touched. If nature builds complexity out of simplicity, then the mechanism used by the sea slug may work in the human brain as well.

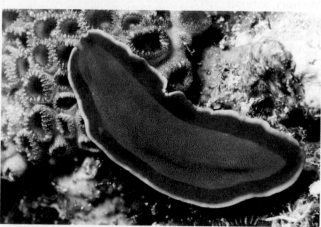
C.K.Ma/Shutterstock

The concept of *long-term potentiation* explains how memory functions at the neuron level. In line with connectionist theory, this concept means that if two neurons are activated at the same time, the connection between them—and thus the memory—may be strengthened (Dong & others, 2020). Long-term potentiation has been demonstrated experimentally by administering a drug that increases the flow of information from one neuron to another across the synapse, raising the possibility of someday improving memory through drugs that increase neural connections (Sumner & others, 2020)

In sum then, memory is *not* a place in the brain. It is best understood as a series of connections. These connections become well-worn paths in our brains but they are active processes, not locations. In fact, research shows that similar neurons are activated in the brain when we encode *and* when we retrieve a memory (Thakral & others, 2020). So, when the brain remembers, it recreates or reinstates the neural activity that encoded the information.

Brain Structures and Memory Functions No specific brain structure is "where" memories are. However, structures do play a role in the processes of memory. Many different parts of the brain and nervous system are involved in the rich, complex process that is memory. Although there is no one memory center in the brain, researchers have demonstrated that specific brain structures are involved in particular aspects of memory.

Figure 12 shows the location of brain structures active in different types of long-term memory. Note that implicit memory and explicit memory appear to involve different locations in the brain.

- *Explicit memory:* Neuroscientists have found that the hippocampus, the temporal lobes in the cerebral cortex, and other areas of the limbic system play a role in explicit memory. In many aspects of explicit memory, information is transmitted from the hippocampus to the frontal lobes, which are involved in both *retrospective* (remembering things from the past) and *prospective* (remembering things we need to do in the future) memory (Lee & others, 2020; Scullin & others, 2020). The left frontal lobe is especially active when we encode new information into memory; the right frontal lobe is more active when we subsequently retrieve it (Johansson & others, 2020). In addition, the amygdala, which is part of the limbic system, is involved in emotional memories (Dahlgren & others, 2020).

- *Implicit memory:* The cerebellum (the structure at the back and toward the bottom of the brain) is active in the implicit memory required to perform skills (Peter & others, 2020). Various areas of the cerebral cortex, such as the temporal lobes and hippocampus, function in priming (Kim, 2011; Roelke & Hofmann, 2020).

FIGURE 12

Structures of the Brain Involved in Different Aspects of Long-Term Memory Note that explicit memory and implicit memory appear to involve different locations in the brain.

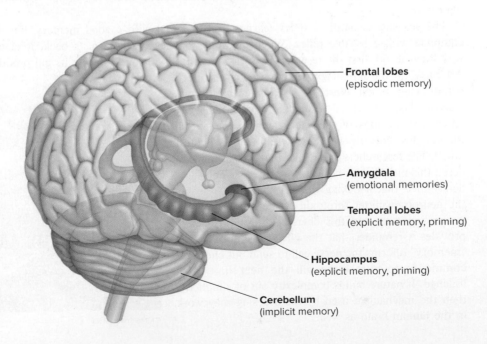

Frontal lobes (episodic memory)

Amygdala (emotional memories)

Temporal lobes (explicit memory, priming)

Hippocampus (explicit memory, priming)

Cerebellum (implicit memory)

Neuroscientists studying memory have benefited greatly from the use of MRI scans, which allow the tracking of neural activity during cognitive tasks (Luo & others, 2014). In one study, participants viewed color photographs of indoor and outdoor scenes while in an MRI machine (Brewer & others, 1998). The experimenters told them that they would receive a memory test about the scenes. After the MRI scans, the participants were asked which pictures they remembered well, vaguely, or not at all. The researchers compared their memories with the brain scans and found that the greater the activation in both prefrontal lobes and a particular region of the hippocampus during viewing, the better the participants remembered the scenes.

4 Memory Retrieval

Remember that unforgettable night of shining stars with your romantic partner? Let's say the evening has indeed been encoded deeply and elaborately in your memory. Through the years you have thought about the night a great deal and told your best friends about it. The story of that night has become part of the longer story of your life with your significant other. Fifty years later, your grandchild asks, "How did you two end up together?" You share that story you have been saving for just such a question. What are the retrieval processes that allow you to do so?

retrieval
The memory process that occurs when information that was retained in memory comes out of storage.

Retrieval takes place when information that was retained in memory comes out of storage. You might think of long-term memory as a library. You retrieve information in a fashion similar to the process you use to locate and check out a book in an actual library. To retrieve something from your mental data bank, you search your store of memory to find the relevant information. Retrieval means you take information out of storage.

The efficiency with which you retrieve information from memory is impressive. It usually takes only a moment to search through a vast storehouse to find the information you want. When were you born? What was the name of your first date? Who developed the first psychology laboratory? You can, of course, answer all of these questions instantly. Retrieval of memory is a complex and sometimes imperfect process.

Before examining ways that retrieval may fall short, let's look at some basic concepts and variables that are known to affect the likelihood that information

The first psychology lab was developed by Wundt...in case you couldn't retrieve that information...

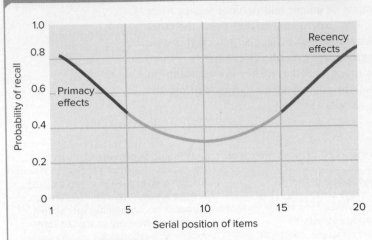

1.0
0.8
0.6 Primacy
effects
0.4
0.2
0
Probability of recall
Recency
effects
1   5   10   15   20
Serial position of items

FIGURE 13 **The Serial Position Effect** When a person is asked to memorize a list of words, the words memorized last usually are recalled best, those at the beginning next best, and those in the middle least efficiently. SOURCE: Murdock Jr., B. B. (1974). *Human Memory: Theory and Data.* Lawrence Erlbaum Associates.

>What is the probability that the item presented in the 15th position will be remembered? >Which is stronger—primacy or recency? Why? >Using the information in the graph, if it is the end of a semester and you are studying for finals, which information would it be best to brush up on, and why?

will be accurately encoded, stored, and ultimately retrieved. As we will see, retrieval depends heavily on the circumstances under which a memory was encoded and the way it was retained.

Serial Position Effect

The **serial position effect** is the tendency to recall the items at the beginning and end of a list more readily than those in the middle. If you are a reality TV fan, you might notice that you always seem to remember the first person to get voted off and the last few survivors. All those people in the middle, however, are a blur. The *primacy effect* refers to better recall for items at the beginning of a list; the *recency effect* refers to better recall for items at the end. Together with the relatively low recall of items from the middle of the list, this pattern makes up the serial position effect (Laming, 2010). See Figure 13 for a typical serial position effect that shows a weaker primacy effect and a stronger recency effect.

serial position effect The tendency to recall the items at the beginning and end of a list more readily than those in the middle.

Psychologists explain these effects using principles of encoding. With respect to the primacy effect, the first few items in the list are easily remembered because they are rehearsed more or because they receive more elaborative processing than do words later in the list (Atkinson & Shiffrin, 1968; Craik & Tulving, 1975). Working memory is relatively empty when the items enter, so there is little competition for rehearsal time. Moreover, because the items get more rehearsal, they stay in working memory longer and are more likely to be encoded successfully into long-term memory. In contrast, many items from the middle of the list drop out of working memory before being encoded into long-term memory.

As for the recency effect, the last several items are remembered for different reasons. First, when these items are recalled, they might still be in working memory. Second, even if these items are not in working memory, the fact that they were just encountered makes them easier to recall. Interestingly, both primacy and recency can influence how we feel about stimuli as well. In one study, wine tasters were more likely to prefer the first wine they sipped, an outcome demonstrating primacy (Mantonakis & others, 2009). In another study, participants felt that the best was saved for last when they evaluated paintings and *American Idol* audition tapes, an outcome demonstrating recency (Li & Epley, 2009).

Ever tried speed dating? How did the serial position effect influence the person you liked the best?

Retrieval Cues and the Retrieval Task

Two other factors are involved in retrieval: the nature of the cues that can prompt your memory and the retrieval task that you set for yourself. We consider each in turn.

Cues in the environment can help us retrieve memories. So, we might be more likely to retrieve information on a test if we are sitting in the same seat where we learned that information. If effective cues for what you are trying to remember do not seem to be available, you need to create them—a process that takes place in working memory. For example, if you have a block about remembering a new friend's name, you might go through the alphabet, generating names that begin with each letter. If you manage to stumble across the right name, you will probably recognize it.

We can learn to generate retrieval cues. One good strategy is to use different subcategories as retrieval cues. For example, write down the names of as many of your classmates from middle or junior high school as you can remember. When you run out of names, think about the activities you were involved in during those school years, such as math class, student council, lunch, drill team, and so on. Does this set of cues help you to remember more of your classmates?

Although cues help, your success in retrieving information also depends on the retrieval task you set for yourself. For instance, if you are simply trying to decide whether something seems familiar, retrieval is probably a snap. Let's say that you see a short, dark-haired woman walking toward you. You quickly decide that she is someone who shops at the same supermarket as you do. However, remembering her name or a precise detail, such as when you met her, can be harder. Such distinctions have implications for police investigations: A witness might be certain they have previously seen a face, yet they might have a hard time deciding whether it was at the scene of the crime or in a mug shot.

RECALL AND RECOGNITION   The presence or absence of good cues and the retrieval task required are factors in an important memory distinction: recall versus recognition. *Recall* is a memory task in which the person must retrieve previously learned information, as on essay tests. *Recognition* is a memory task in which the person only has to identify (recognize) learned items, as on multiple-choice tests. Recall tests such as essay tests have poor retrieval cues. For an essay in a history class you might be asked to recall certain material ("Discuss the factors that caused World War I"). In recognition tests such as multiple-choice tests, you merely judge whether a stimulus is familiar (such as that Archduke Franz Ferdinand was assassinated in 1914).

You probably have heard some people say that they never forget a face. However, recognizing a face is far simpler than recalling a face "from scratch," as law enforcement officers know. In some cases, police bring in an artist to draw a suspect's face from witnesses' descriptions (Figure 14). Recalling faces is difficult, and artists' sketches of suspects are frequently not detailed or accurate enough to result in apprehension.

Some people say they are better at essay tests, and some prefer multiple choice. Which type of person are you? From a memory perspective, multiple-choice tests should be easier—they rely on recognition.

 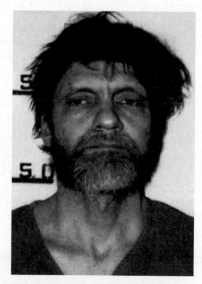

FIGURE 14   Remembering Faces (*first*) The FBI artist's sketch of Ted Kaczynski. Kaczynski, also known as the Unabomber, conducted a sequence of mail bombings targeting universities and airlines beginning in the late 1970s. (*second*) A photograph of Kaczynski. The FBI widely circulated the artist's sketch, which was based on bits and pieces of observations people had made of the infamous Unabomber, in the hope that someone would recognize him. Would you have been able to recognize Kaczynski from the artist's sketch? Probably not. Although most people say they are good at remembering faces, they usually are not as good as they think they are.

(sketch): Allan Tannenbaum/The LIFE Images Collection/Getty Images; (photo): Bureau of Prisons/Donaldson Collection/Archive Photos/Getty Images

Purestock/SuperStock

ENCODING SPECIFICITY Another consideration in understanding retrieval is the *encoding specificity principle,* which states that information present at the time of encoding or learning tends to be effective as a retrieval cue (Greene & Naveh-Benjamin, 2020). For example, you know your instructors when they are in the classroom setting—you see them there all the time. If, however, you run into one of them in an unexpected setting and in more casual attire, such as at the gym in workout clothes, the person's name might escape you. Your memory might fail because the cues you encoded are not available for use.

CONTEXT AT ENCODING AND RETRIEVAL

An important consequence of encoding specificity is that a change in context between encoding and retrieval can cause memory to fail (Ostendorf & others, 2020). In many instances, people remember better when they attempt to recall information in the same context in which they learned it—a process referred to as *context-dependent memory.* This better recollection is believed to occur because people have encoded features of the context in which they learned the information along with the actual information. Those features can then act as retrieval cues.

In one strange but classic study, scuba divers learned information on land and under water (Godden & Baddeley, 1975). Later they were asked to recall the information when they were either on land or under water. The divers' recall was much better when the encoding and retrieval contexts were the same (both on land or both under water).

Special Cases of Retrieval

We began this discussion by likening memory retrieval to looking for and finding a book in the library. However, the process of retrieving information from long-term memory is not as precise as the library analogy suggests. When we search through our long-term memory storehouse, we do not always find the exact "book" we want—or we might find the book but discover that several pages are missing. We have to fill in these gaps somehow.

Our memories are affected by a number of things, including the pattern of facts we remember, schemas and scripts, the situations we associate with memories, and the personal or emotional context. Certainly, everyone has had the experience of remembering a shared situation with a particular person, only to have the person remind us, "Oh, that wasn't me!" Such moments provide convincing evidence that memory may well be best understood as "reconstructive." This subjective quality of memory certainly has implications for important day-to-day procedures such as eyewitness testimony.

While the factors that we have discussed so far relate to the retrieval of generic information, various kinds of special memory retrieval also have generated a great deal of research. These memories have special significance because of their relevance to the self, to their emotional or traumatic character, or because they show unusually high levels of apparent accuracy. Cognitive psychologists have debated whether these memories rely on processes that are different from those already described or are simply extreme cases of typical memory processes (Romano & others, 2020). We now turn to these special cases of memory.

RETRIEVAL OF AUTOBIOGRAPHICAL MEMORIES **Autobiographical memory**, a special form of episodic memory, is a person's recollections of their own life experiences (Gülgöz & Sahin-Acar, 2020; Ross & others, 2020; Speer & Delgado, 2020).

autobiographical memory
A special form of episodic memory, consisting of a person's recollections of their life experiences.

An intriguing discovery about autobiographical memory is the *reminiscence bump*, which refers to the findings adults remember more events from the second and third decades of life than from other decades (Wolf & Zimprich, 2020). This reminiscence bump may occur because these are the times in our life when we have many novel experiences or because it is during our teens and 20s that we are forging a sense of identity (Rathbone & others, 2017). Generally, the very first few years of life are not characterized by many autobiographical memories. There is no evidence that memories from before we begin to talk are actually "real" (Akhtar & others, 2018).

Level	Label	Description
Level 1	Lifetime periods	Long segments of time measured in years and even decades
Level 2	General events	Extended composite episodes measured in days, weeks, or months
Level 3	Event-specific knowledge	Individual episodes measured in seconds, minutes, or hours

FIGURE 15   The Three-Level Hierarchical Structure of Autobiographical Memory When people relate their life stories, all three levels of information are typically present and intertwined.

Autobiographical memories are complex and seem to contain unending strings of stories and snapshots, but they can be categorized (Marsh & Roediger, 2013). For example, based on their research, Martin Conway and David Rubin (1993) sketched a structure of autobiographical memory that has three levels (Figure 15). The most abstract level consists of *life time periods;* for example, you might remember something about your life in high school. The middle level in the hierarchy is made up of *general events,* such as a trip you took with your friends after you graduated from high school. The most concrete level in the hierarchy is composed of *event-specific knowledge;* for example, from your postgraduation trip, you might remember the exhilarating experience you had the first time you jet-skied. When people tell their life stories, all three levels of information are usually present and intertwined.

Most autobiographical memories include some reality and some myth. Personality psychologist Dan McAdams (2001, 2018) argues that autobiographical memories are less about facts and more about meanings. Such memories provide a reconstructed, embellished telling of the past that connects the past to the present (Jiang & others, 2020).

Do It!

Write down a memory that you feel has been especially important in making you who you are. What are some characteristics of this self-defining memory? What do you think the memory says about you? How does it relate to your current goals and aspirations? Do you think of the memory often? You might find that this part of your life story can be inspiring when things are going poorly or when you are feeling down.

RETRIEVAL OF EMOTIONAL MEMORIES

When we remember our life experiences, the memories are often wrapped in emotion. Emotion affects the encoding and storage of memories and thus shapes the details that are retrieved. The role that emotion plays in memory is of considerable interest to contemporary researchers and has echoes in public life.

flashbulb memory
The memory of emotionally significant events that people often recall with more accuracy and vivid imagery than everyday events.

Flashbulb memory is the memory of emotionally significant events that people often recall with more accuracy and vivid imagery than everyday events (Muzzulini & others, 2020). Many adults can remember, for example, where they were when they first heard about important world events, such as the terrorist attacks on the United States on September 11, 2001. An intriguing dimension of flashbulb memories is that several decades later, people often remember where they were and what was going on in their lives at the time of such an emotionally charged event. These memories seem to be part of an adaptive system that fixes in memory the details that accompany important events so that they can be interpreted at a later time.

Most people express confidence about the accuracy of their flashbulb memories. However, flashbulb memories probably are not as accurately etched in our brain as we think. One way to gauge the accuracy of flashbulb memories is to probe how consistent the details of these memories remain over time. Research on memories of the 9/11 terrorist attacks shows that the accuracy of memory was predicted by

Many people have flashbulb memories of where they were and what they were doing when terrorists attacked the World Trade Center towers in New York City on September 11, 2001.
Robert J Fisch/Moment Open/Getty Images

people's physical proximity to the event; for instance, the memories of individuals in New York were more accurate than the recollections of those in Hawaii (Pezdek, 2003). In another study, Canadian students' memories for 9/11 were tested one week following the tragedy; they were tested again six months after the attacks. At six months after the events, students were better at remembering details about their own experience of hearing about the event than they were at recalling details of the event itself (Smith & others, 2003). These findings reflect our subjective experience of flashbulb memories. We might say, "I will never forget where I was when I heard about [some important event]." Such a statement fits with research showing that we may be more likely to remember our personal experiences of an event rather than the details of the event itself.

Still, on the whole, flashbulb memories do seem more durable and accurate than memories of day-to-day happenings (Demiray & Freund, 2014). One possible explanation is that flashbulb memories are quite likely to be rehearsed in the days following the event. However, it is not just the discussion and rehearsal of information that make flashbulb memories so long-lasting. The emotions triggered by flashbulb events also figure in their durability. Although we have focused on negative news events as typical of flashbulb memories, such memories can also occur for positive events. An individual's wedding day and the birth of a child are events that may become milestones in personal history and are always remembered.

MEMORY FOR TRAUMATIC EVENTS   In 1890, the American psychologist and philosopher William James said that an experience can be so emotionally arousing that it almost leaves a scar on the brain. Personal traumas are candidates for such emotionally stirring experiences.

Some psychologists argue that memories of emotionally traumatic events are accurately retained, possibly forever, in considerable detail (Langer, 1991). There is good evidence that memory for traumatic events is usually more accurate than memory for ordinary events (Kensinger & Ford, 2020). Consider the traumatic experience of some children who were kidnapped at gunpoint on a school bus in Chowchilla, California, in 1976 and then buried underground for 16 hours before escaping. The children had the classic signs of traumatic memory: detailed and vivid recollections.

However, when a child psychiatrist interviewed the children four to five years after the chilling episode, she noted striking errors and distortions in the memories of half of them (Terr, 1988). How can a traumatic memory be so vivid and detailed yet at the same time have inaccuracies? A number of factors can be involved. Some children might have made perceptual errors while encoding information because the episode was so shocking. Others might have distorted the information and recalled the episode as being less traumatic than it was in order to reduce their anxiety about it. Other children, in discussing the terrifying event with others, might have incorporated bits and pieces of these people's recollections of what happened.

A long-held belief is that memories of real-life traumas are more accurate and longer-lasting than memories of everyday events. Stress-related hormones likely play a role in memories that involve personal trauma. The release of stress-related hormones, signaled by the amygdala and regulated by the hippocampus (see Figure 12), likely accounts for some of the extraordinary durability and vividness of traumatic memories (Roozendaal & Mirone, 2020). However, research has not consistently supported the idea that memory for traumatic events is highly accurate or even long-lasting. One view is that in traumatic circumstances—such as combat trauma—our brains develop "gist" memories that allow us to avoid future dangers but may not include the necessary detail and precision required by the judicial system (Bahtiyar & others, 2020; Morgan & Southwick, 2014).

REPRESSED MEMORIES   Can an individual forget, and later recover, memories of traumatic events? A great deal of debate surrounds this question (Bruck & Ceci, 2012, 2013; Lamb & others, 2015). *Repression* is a defense mechanism by which a person is so traumatized by an event that the individual forgets it and then forgets the act of forgetting. According to psychodynamic theory, repression's main function is to protect the individual from threatening information.

The prevalence of repression is hotly contested (Brewin, 2020; Otgaar & others, 2020). Most studies of traumatic memory indicate that a traumatic life event such as childhood sexual abuse is very likely to be remembered. However, there is at least some evidence that childhood sexual abuse may not be remembered. Linda Williams and her colleagues conducted a number of investigations of memories of childhood abuse (Banyard & Williams, 2007; Liang & others, 2006; Williams, 2003, 2004). One study involved 129 women for whom hospital emergency room records indicated a childhood abuse experience (Williams, 1995). Seventeen years after the abuse incident, the researchers contacted the women and asked (among other things) whether they had ever been the victim of childhood sexual abuse. Of the 129 women, most reported remembering and never having forgotten the experience. Ten percent of the participants reported having forgotten about the abuse at least for some portion of their lives.

motivated forgetting
Forgetting that occurs when something is so painful or anxiety-laden that remembering it is intolerable.

If it does exist, repression can be considered a special case of **motivated forgetting**, which occurs when individuals forget something because it is so painful or anxiety laden that remembering it is intolerable (Haghighi & others, 2020). This type of forgetting may be a consequence of the emotional trauma experienced by victims of sexual violence or physical abuse, war veterans, and survivors of earthquakes, plane crashes, and other terrifying events. These emotional traumas may haunt people for many years unless they can put the details out of mind. Even when people have not experienced trauma, they may use motivated forgetting to protect themselves from memories of painful, stressful, or otherwise unpleasant circumstances (Anderson & Hanslmayr, 2014; Shu & others, 2011).

Cognitive psychologist Jonathan Schooler suggested that recovered memories are better termed *discovered memories* because, regardless of their accuracy, individuals do experience them as real (Geraerts & others, 2009; Schooler, 2002). Schooler and his colleagues (1997) investigated a number of cases of discovered memories of abuse, in which they sought independent corroboration by others. They were able to identify actual cases in which the perpetrator or some third party could verify a discovered memory. For example, Frank Fitzpatrick forgot abuse at the hands of a Catholic priest, but his report of the abuse, years later, was corroborated by witnesses who had also been abused (*Commonwealth of Massachusetts v. Porter,* 1993). The existence of such cases suggests that it is inappropriate to reject all claims by adults that they were victims of long-forgotten childhood sexual abuse.

How do psychologists consider these cases? Generally, there is consensus on a few key issues (Knapp & VandeCreek, 2000). First, all agree that child sexual abuse is an important and egregious problem that historically has not been acknowledged. Second, psychologists widely believe that most individuals who were sexually abused as children remember all or part of what happened to them and that these continuous memories are likely to be accurate. Third, there is broad agreement that it is possible for someone who was abused to forget those memories for a long time, and it is also possible to construct memories that are false but that feel very real to an individual. Finally, it is highly difficult to separate accurate from inaccurate memories, especially if methods such as hypnosis have been used in the "recovery" of memories.

Do It!

Cognitive psychologist Sam Sommers blogged about his experiences as an expert witness in an armed robbery case. He used his expertise on eyewitness memory in the case. Check out his account by googling "Sam Sommers eyewitness."

Do It!

To test your ability to be a good eyewitness, visit one of the following websites:

www.pbs.org/wgbh/pages/frontline/shows/dna/

https://public.psych.iastate.edu/glwells/theeyewitnesstest.html

Did this exercise change your opinion of the accuracy of eyewitness testimony? Explain.

EYEWITNESS TESTIMONY By now, you should realize that memory is not a perfect reflection of reality. Understanding the distortions of memory is particularly important when people are called on to report what they saw or heard in relation to a crime. Eyewitness testimonies, like other sorts of memories, may contain errors, and faulty memory in criminal matters has especially serious consequences. When eyewitness testimony is inaccurate, the wrong person might go to jail or even be put to death, or the perpetrator of the crime might not be prosecuted (Brewer & others, 2020). It is important to note, however, that witnessing a crime is often traumatic for the individual, and so this type of memory typically fits in the larger category of memory for highly emotional events.

Much of the interest in eyewitness testimony focuses on distortion, bias, and inaccuracy in memory (Frenda & others, 2011). One reason for distortion is that memory fades. In one study, people were able to identify pictures with 100 percent accuracy after a 2-hour time lapse. However, four months later they achieved an accuracy of only 57 percent; chance alone accounts for 50 percent accuracy (Shepard, 1967).

Unlike a video, memory can be altered by new information (Simons & Chabris, 2011). In one study, researchers showed students a film of an automobile accident and then asked them how fast the white sports car was going when it passed a barn (Loftus, 1975). Although there was no barn in the film, 17 percent of the students mentioned the barn in their answer.

Bias, especially racial prejudice, is also a factor in faulty memory (Brigham & others, 2007). People of one ethnic group are less likely to recognize individual differences among people of another ethnic group (Behrman & Davey, 2001). Latino eyewitnesses, for example, may have trouble distinguishing among several Asian suspects. In one experiment, a mugging was shown on a television news program (Loftus, 1993). Immediately after, a lineup of six suspects was broadcast, and viewers were asked to phone in and identify which one of the six individuals they thought had committed the robbery. Of the 2,000 callers, more than 1,800 identified the wrong person, and even though the robber was a white male, one-third of the viewers identified an African American or a Latino suspect as the criminal. Importantly, too, witness confidence is not necessarily a good indicator of accuracy of eyewitness accounts. One study showed that women's accounts were more accurate than men's but that men expressed greater confidence in their memories (Areh, 2011).

Using Psychological Research to Improve Police Lineups

In his book *Convicting the Innocent,* law professor Brandon Garrett (2011) traced the first 250 cases in the United States in which a convicted individual was exonerated—proved to be not guilty—by DNA evidence. Of those cases, 190 (or *76 percent*) involved mistaken eyewitness identification of the accused. Every year, more than 75,000 individuals are asked to identify suspects, and experts estimate that these identifications are wrong one-third of the time (The Week Staff, 2011).

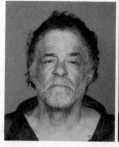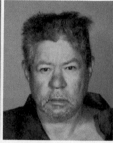

Cleveland PD/REX/Shutterstock

To reduce the chances that innocent individuals will be accused of crimes, law enforcement officials are applying psychological research findings to improve the way they conduct criminal lineups (Wells & others, 2020). In such lineups, a witness views a group of individuals in real time or looks at photos of individuals, one of whom is the suspect. The witness is asked to identify the perpetrator of the crime if the person is among those shown. Psychological research has influenced these procedures in two ways.

First, based on the suspicion that even very subtle bias can influence witness judgments, *double-blind* procedures have been adopted by many jurisdictions (Brewer & Wells, 2011). Recall that in a double-blind study, neither the participants nor the experimenter knows what condition the participants are in, to reduce the effects of bias on the results. The city of Dallas, for example, uses a double-blind approach to police lineups. A specific unit conducts all lineups, and no one involved in administering the lineup has any knowledge of the case or of which individual is suspected of the crime (Goode & Schwartz, 2011).

Another procedure supported by research involves sequential rather than simultaneous presentation of individuals or photographs in a lineup. Showing individuals or their photos in a lineup one at a time (sequentially) rather than all at once (simultaneously) is less likely to lead to false identifications (Steblay & others, 2011). The reason for the difference is that when presented with a set of suspects simultaneously, victims tend to choose the person who looks the most like their memory of the perpetrator. Studies show that when the actual perpetrator in a simulated crime is removed from a simultaneous lineup, people will pick the "filler person"—an individual who is not suspected of the crime and is simply filling up the lineup—who most closely resembles the perpetrator (Wells, 1993).

A large-scale field study, headed by Gary Wells (Wells & others, 2011), examined the effectiveness of double-blind sequential and simultaneous lineups in four different jurisdictions in the United States in actual criminal investigations. The results showed that sequential presentation of photo lineups led to fewer mistaken identifications than did simultaneous presentation. Wells and colleagues concluded that double-blind sequential lineups are a promising avenue for improving the accuracy of eyewitness identification.

TV crime dramas might give the impression that DNA evidence is widely available to protect innocent people from false accusations. However, researchers estimate that fewer than 5 percent of legal cases involving eyewitness identifications include biological evidence to potentially exonerate mistakenly identified convicts (Wells & others, 2011). So, for many crimes, eyewitness identification remains an important piece of evidence—and improving the validity of these identifications is thus a crucial goal.

Hundreds of people have been harmed by witnesses who have made a mistake (Frenda & others, 2011). One estimate indicates that each year approximately 7,500 people in the United States are arrested for and wrongly convicted of serious crimes (Huff, 2002).

Faulty memory is not just about accusing the wrong person. Understandably, given the circumstances, witnesses can be mistaken about many aspects of a crime. For many mass shootings, early reports from eyewitnesses indicate multiple shooters were involved in an attack.

Yet, in the end, the killings are attributable to a single gunmen, perhaps because the frightened witnesses could not fathom a single person being responsible for so many gunshots.

More recently, researchers have suggested that eyewitness accounts can be more accurate if the witnesses close their eyes. Why might that help?

Faulty memories complicated the search for the perpetrators in the sniper attacks that killed 10 people in the Washington, DC, area in 2002. Police released photos of the type of white truck or van that witnesses said they saw fleeing some of the crime scenes (bottom). In the end, however, the suspects were driving a blue car when law enforcement officials apprehended them (above).
AP Images

Marie P. Marzi/AP Images

self-quiz

1. The tendency to remember the items at the beginning and end of a list more easily than the items in the middle is the
   A. bookends effect.
   B. serial cues effect.
   C. serial position effect.
   D. endpoints effect.

2. Carrie prides herself on "never forgetting a face," although she frequently cannot put the correct name with a specific face. Carrie is really saying that she
   A. is better at recognition than at recall.
   B. is better at recall than at recognition.
   C. is better at memory retrieval than at memory reconstruction.

D. is better at memory reconstruction than at memory recall.

3. Faulty memory can occur due to
   A. bias.
   B. receipt of new information.
   C. distortion.
   D. all of the above.

APPLY IT! 4. Andrew is getting ready for a group interview for a job he really wants. The group session will take place at the beginning of the day, followed by individual interviews. When the manager who is conducting the interviews calls Andrew, he tells him that because he has not talked to any of the other candidates yet, Andrew can decide when he would like his individual interview to be.

There are five candidates. Which position should Andrew take?
   A. Andrew should go third because that way he will be right in the middle, and the interviewer will not be too nervous or too tired.
   B. Andrew should go either first or last, to be the candidate most likely to be remembered.
   C. Andrew should probably go second so that he will not be sitting around feeling nervous for too long—and besides, asking to go first might seem pushy.
   D. It will not matter, so Andrew should just pick a spot randomly.

5 Forgetting

Human memory has its imperfections, as we have all experienced. It is not unusual for two people to argue about whether something did or did not happen, each supremely confident that their memory is accurate and the other person's is faulty. We all have had the frustrating experience of trying to remember the name of some person or some place but not quite being able to retrieve it. Missed appointments, misplaced keys, the failure to recall the name of a familiar face, and inability to recall your password for Internet access are everyday examples of forgetting. Why do we forget?

One of psychology's pioneers, Hermann Ebbinghaus (1850–1909), was the first person to conduct scientific research on forgetting. In 1885, he made up and memorized a list of 13 nonsense syllables and then assessed how many of them he could remember as time passed. (Nonsense syllables are meaningless combinations of letters that are unlikely to have been learned already, such as *zeq, xid, lek,* and *riy.*) Even just an hour later, Ebbinghaus could recall only a few of the nonsense syllables he had memorized. Figure 16 shows Ebbinghaus's learning curve for nonsense syllables. Based on his research, Ebbinghaus concluded that most forgetting takes place soon after we learn something.

If we forget so quickly, why put effort into learning something? Fortunately, researchers have demonstrated that forgetting is not as extensive as Ebbinghaus envisioned (Hsieh & others, 2009). Ebbinghaus studied meaningless nonsense syllables. When we memorize more meaningful material—such as poetry, history, or the content of this text—forgetting is neither so rapid nor so extensive. Following are some of the factors that influence how well we can retrieve information from long-term memory.

Never forget: Ebbinghaus was the first to study forgetting.

Hermann Ebbinghaus (1850–1909)
Ebbinghaus was the first psychologist to conduct scientific research on forgetting.
Bettmann/Getty Images

Encoding Failure

Sometimes when people say they have forgotten something, they have not really forgotten it; rather, they never encoded the information in the first place. Encoding failure occurs when the information was never entered into long-term memory.

PSYCHOLOGICAL INQUIRY

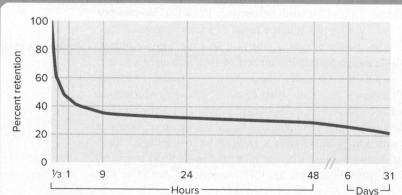

FIGURE 16 **Ebbinghaus's Forgetting Curve** This figure illustrates Ebbinghaus's conclusion about forgetting.

> When is information most likely to be forgotten? > What might explain differences in the slope of this curve for different individuals and different material? > Based on this graph, when is the best time to study new material to prevent it from being forgotten?

FIGURE 17

Which Is a Real U.S. Penny? In the original experiment, participants viewed 15 versions of pennies; only one version was an actual U.S. penny. This figure shows only 7 of the 15 versions, and as you likely can tell, the task is still very difficult. Why?

David A. Tietz/Editorial Image, LLC

(a) (b) (c) (d)

(e) (f) (g)

As an example of encoding failure, think about what the U.S. penny looks like. In one study, researchers showed 15 versions of the penny to participants and asked them which one was correct (Nickerson & Adams, 1979). Look at the pennies in Figure 17 and see whether you can tell which is the real penny. Most people do not do well on this task. Unless you are a coin collector, you probably have not encoded a lot of specific details about pennies. You may have encoded just enough information to distinguish them from other coins (pennies are copper-colored, dimes and nickels are silver-colored; pennies fall between the sizes of dimes and quarters). The correct answer? Penny c is the right one.

The penny exercise illustrates that we encode and enter into long-term memory only a small portion of our life experiences. In a sense, then, encoding failures really are not cases of forgetting; they are cases of not remembering.

Retrieval Failure

Problems in retrieving information from memory are clearly examples of forgetting. Psychologists have theorized that the causes of retrieval failure include problems with the information in storage, the effects of time, personal reasons for remembering or forgetting, and the condition of the brain (Fawcett & Hulbert, 2020).

INTERFERENCE   *Interference* is one reason that people forget (Maxcey & others, 2019). According to **interference theory**, people forget not because memories are lost from storage but because other information gets in the way of what they want to remember.

There are two kinds of interference: proactive and retroactive. **Proactive interference** occurs when material that was learned earlier disrupts the recall of material learned later (Corbett & Duarte, 2020). Remember that *pro-* means "forward in time." For example, suppose you had a good friend 10 years ago named Prudence and that last night you met someone named Patience. You might find yourself calling your new friend Prudence because the old information (Prudence) interferes with retrieval of new information (Patience). **Retroactive interference** occurs when material learned later disrupts the retrieval of information learned earlier. Remember that *retro-* means "backward in time." Suppose you have lately become friends with Ralph. In sending a note to your old friend Raul, you might mistakenly address it to Ralph because the new information (Ralph) interferes with the old information (Raul). Figure 18 depicts another example of proactive and retroactive interference.

Proactive and retroactive interference might both be explained as problems with retrieval cues. The reason the name Prudence interferes with the name Patience and the

interference theory
The theory that people forget not because memories are lost from storage but because other information gets in the way of what they want to remember.

proactive interference
Situation in which material that was learned earlier disrupts the recall of material that was learned later.

retroactive interference
Situation in which material that was learned later disrupts the retrieval of information that was learned earlier.

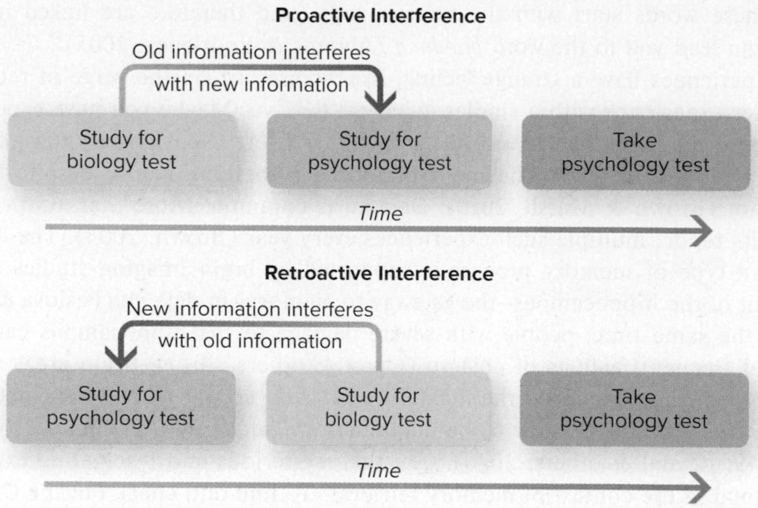

FIGURE 18

Proactive Interference

Old information interferes
with new information

| Study for biology test | Study for psychology test | Take psychology test |

Time

Retroactive Interference

New information interferes
with old information

| Study for psychology test | Study for biology test | Take psychology test |

Time

FIGURE 18
Proactive and Retroactive Interference *Pro-* means "forward"; in proactive interference, old information has a forward influence by getting in the way of new material learned. *Retro-* means "backward"; in retroactive interference, new information has a backward influence by getting in the way of material learned earlier.

name Ralph interferes with the name Raul might be that the cue you are using to remember the one name does not distinguish between the two memories. For example, if the cue you are using is "my good friend," it might evoke both names. The result might be retrieval of the wrong name or a kind of blocking in which each name interferes with the other and neither comes to mind. Retrieval cues (such as "friend" in our example) can become overloaded, and when that happens we are likely to forget or to retrieve incorrectly.

These are tricky. PRO-active means what you learned BEFORE interferes with later material; RETRO-active means what you learn now interferes with earlier material.

decay theory
Theory stating that when an individual learns something new, a neurochemical memory trace forms, but over time this trace disintegrates; suggests that the passage of time always increases forgetting.

DECAY Another possible reason for forgetting is the passage of time. According to **decay theory**, when we learn something new, a neurochemical memory trace forms, but over time this trace disintegrates. Decay theory suggests that the passage of time always increases forgetting.

Memories often do fade with the passage of time, but decay alone cannot explain forgetting. For example, under the right retrieval conditions, we can recover memories that we seem to have forgotten. You might have forgotten the face or name of someone in your high school class, for instance, but when you return to the setting where you knew the person, you might remember. Similarly, you may not have thought about someone from your past for a very long time, but when the person friends you on Facebook, you may remember a lot of prior experiences with them.

Many things we think we have forgotten are just waiting for the right cues to pop into our lives.

TIP-OF-THE-TONGUE PHENOMENON We are all familiar with the retrieval glitch called **tip-of-the-tongue (TOT) phenomenon**—a type of "effortful retrieval" that occurs when we are confident that we know something but cannot quite pull it out of memory. In a TOT state we usually can successfully retrieve characteristics of the word, such as the first letter and the number of syllables, but not the word itself. The TOT phenomenon arises when we can retrieve some of the desired information but not all of it (Cleary, 2019).

tip-of-the-tongue (TOT) phenomenon
A type of effortful retrieval associated with a person's feeling that they know something (say, a word or a name) but cannot quite pull it out of memory.

The TOT phenomenon demonstrates that we do not store all of the information about a particular topic or experience in one way. If you have ever struggled to think of a specific word, you probably came up with various words that mean the same thing as the word you were looking for, but you still had a nagging feeling that none was quite right. Sometimes you might find the solution in an unexpected way. For example, imagine that you are doing a crossword puzzle with the clue "colorful scarf" for a seven-letter word. You have a feeling you know this word. If you have not thought of the answer yet, say the following word aloud: *banana*. If you were experiencing the TOT phenomenon when doing the crossword, thinking of *banana* might have helped you come up with the correct answer, *bandana*. Although the meaning of *banana* is unrelated to that of *bandana,* the fact that these words start with the same sounds (and therefore are linked in verbal memory) can lead you to the word *bandana* (Abrams & Rodriguez, 2005).

TOT experiences have a strange feeling, like we are just on the verge of remembering. Another experience with a similar quality is déjà vu. Maybe you have experienced it—a strange feeling that you have done something before even if you know you never did. Déjà vu is the feeling of having experienced something before despite knowing you have not (Brown & Marsh, 2010). Déjà vu is common. More than 65 percent of young adults report multiple such experiences every year (Brown, 2004). The idea that déjà vu is a type of memory process is supported by brain imaging studies showing involvement of the hippocampus—the gateway to memory—in déjà vu (Pešlová & others, 2018). At the same time, people with severe damage to the hippocampus can report intense and frequent feelings of déjà vu (Curot & others, 2019). Brain areas that surround the hippocampus, called the *familiarity network*, appear to be implicated in this experience that things we know to be unfamiliar are suddenly very familiar (Curot & others, 2019; Revdal & others, 2020). Can this mysterious and fascinating experience be understood in the context of memory retrieval? To find out, check out the Challenge Your Thinking.

PROSPECTIVE MEMORY: REMEMBERING (OR FORGETTING) WHEN TO DO SOMETHING

The main focus of this chapter has been on **retrospective memory**, which is remembering the past. **Prospective memory** involves remembering information about doing something in the future; it includes memory for intentions (Strickland & others, 2019). Prospective memory includes both timing (when we have to do something) and content (what we have to do).

We can make a distinction between time-based and event-based prospective memory. *Time-based prospective memory* is our intention to engage in a given behavior after a specified amount of time has gone by, such as an intention to make a phone call to someone in one hour. In *event-based prospective memory,* we engage in the intended behavior when some external event or cue elicits it, as when we give a message to a roommate upon seeing her. The cues available in event-based prospective memory make it more effective than time-based prospective memory.

Some failures in prospective memory are referred to as *absentmindedness*. We are more absentminded when we become preoccupied with something else, are distracted by something, or are under a lot of time pressure. Absentmindedness often involves a breakdown between attention and memory storage (Schacter, 2001). Fortunately, our goals are encoded into memory along with the features of situations that would allow us to pursue them. Our memories, then, prepare us to recognize when a situation presents an opportunity to achieve those goals.

Researchers have found that older adults perform worse on prospective memory tasks than younger adults do, but typically these findings are true only for artificial lab tasks (Smith & others, 2012; Zollig & others, 2012). In real life, older adults generally perform as well as younger adults in terms of prospective memory (Kinsella & others, 2020; Rendell & Craik, 2000). Generally, prospective memory failure (forgetting to do something) occurs when retrieval is a conscious, effortful (rather than automatic) process (Henry & others, 2004).

Amnesia Recall the case of H. M. in the discussion of explicit and implicit memory. In H. M.'s surgery, the part of his brain that was responsible for laying down new memories was damaged beyond repair. The result was **amnesia**, the loss of memory. H. M. suffered from **anterograde amnesia**, a memory disorder that affects the retention of new information and events (*antero-* indicates amnesia that moves forward in time) (Levine & others, 2009). What he learned before the surgery (and thus before the onset of amnesia) was not affected. For example, H. M. could identify his friends,

retrospective memory
Remembering information from the past.

prospective memory
Remembering information about doing something in the future; includes memory for intentions.

Siede Preis/Getty Images

anterograde amnesia
A memory disorder that affects the retention of new information and events.

amnesia
The loss of memory.

How Can We Understand the Experience of Déjà vu?

Déjà vu involves two conflicting feelings about memory: a strong feeling of familiarity clashing with a strong feeling of knowing that one did not actually experience a particular event (Urquhart & O'Connor, 2014; Urquhart & others, 2018). No doubt, déjà vu has an uncanny, goose-bump inducing quality. Not only do you feel like you have "been here, done this," but you might feel that you know what is about to happen next. Déjà vu is unpredictable and often fades quickly—how could it be studied scientifically? Let's review research to see, first, how déjà vu can be created in the lab and, second, whether that feeling of knowing what is going to happen next has any validity. (Spoiler alert: It doesn't!)

In a number of studies (Cleary & Claxton, 2018; Cleary & others, 2019), researchers have induced déjà vu. First, they had people watch and encode 32 brief videos. The videos showed, from a first-person perspective, tours of various scenes. For instance, participants might see themselves walking through a manicured garden as the video led them through the setting. Next, participants were exposed to another set of 32 videos. For half of these the layout of the objects was identical to the layout in some of the videos seen during encoding, but the specific setting and objects differed. So, a junkyard might be identical in physical layout to the manicured garden. Instead of shrubs, the yard included piles of discarded wood. Participants were more likely to report experiencing déjà vu when watching these videos than when watching the others. The similarity in layout led to that feeling of "having been there before"—even though participants "knew" they had not been there before. They did not recognize the similarity of layouts. These results suggest that

Julietphotography/Shutterstock

déjà vu involves *a failure of retrieval*. Participants did have a memory of the layout, but they did not recall that fact (Cleary & Claxton, 2018).

The studies also probed the feeling of knowing what might happen next. The researchers stopped the videos prior to a crucial turn and asked participants which way they should go—right or left? The results showed that participants who had experienced déjà vu felt more sure that they knew which way to go. However, this strong feeling of knowing was not related to accuracy (Cleary & Claxton, 2018). In addition to a feeling of knowing what is to come, déjà vu is associated with another error in memory, the feeling that we knew all along how things would work out (Cleary & others, 2019).

These studies tell us that, surprisingly, when we experience déjà vu it is not that we are wrong about feeling we have been there before. Rather, we *have been* in a very similar situation before, but we don't recall that fact.

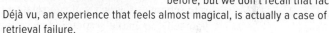

Déjà vu, an experience that feels almost magical, is actually a case of retrieval failure.

What Do You Think?

- Why do some people experience déjà vu more often than others?
- Do you think experiences like déjà vu have any special significance? Why or why not?

---

recall their names, and even tell stories about them—*if* he had known them before the surgery. People who met H. M. after the surgery remained strangers, even if they spent thousands of hours with him. H. M.'s postsurgical experiences were rarely encoded in his long-term memory.

retrograde amnesia
Memory loss for a segment of the past but not for new events.

Amnesia also occurs in a form known as **retrograde amnesia**, which involves memory loss for a segment of the past but not for new events (*retro-* indicates amnesia that moves back in time). Retrograde amnesia is much more common than anterograde amnesia and frequently occurs when the brain is assaulted by an electrical shock or a physical blow such as a head injury. In contrast to anterograde amnesia, in retrograde amnesia the forgotten information is old—it occurred prior to the event that caused the amnesia—and the ability to acquire new memories is not affected. Sometimes individuals have both anterograde and retrograde amnesia.

1. The term for the failure of information to enter into long-term memory is
A. rehearsal failure.
B. encoding failure.
C. proactive interference.
D. retroactive interference.

2. Marco, a college student, is sure that he knows the name of his old Little League coach, but he cannot quite pull the name out of his memory. Marco is experiencing
A. anterograde amnesia.
B. retrograde amnesia.
C. associative interference.
D. tip-of-the-tongue phenomenon.

3. Retrospective memory involves remembering information about doing something in the _____, and prospective memory involves remembering information about doing something in the _____.
A. past; present
B. past; future
C. present; future
D. present; past

APPLY IT! 4. Carl, who is fluent in Spanish, is planning a trip to Paris and decides to take a course in French. On Carl's first French vocabulary quiz, his instructor notes that he has mixed up a lot of Spanish words with French words. Carl cannot understand why, because he has always excelled at learning languages. What is happening to Carl?

A. Carl is experiencing retroactive interference. His knowledge of Spanish is interfering with learning the new French words.
B. Carl is experiencing proactive interference. His knowledge of Spanish is interfering with learning the new French words.
C. Carl is experiencing anterograde amnesia. He seems unable to learn new words.
D. Carl has hit his limit on languages. He should stick to Spanish and visit Barcelona instead.

6 Tips from the Science of Memory—for Studying and for Life

How can you apply your new knowledge of memory processes to improving your academic performance—and your life? No matter what model of memory you use, you can sharpen your memory by thinking deeply about the "material" of life and connecting the information to other things you know. Perhaps the one most well-connected node or most elaborate schema to which you can relate something is the self—what you know and think about yourself. To make something meaningful and to secure its place in memory, you must make it matter to yourself.

If you think about memory as a physical event in the brain, you can see that memorizing material is like training a muscle. Repeated recruitment of sets of neurons creates the connection you want to have available not only at exam time but throughout life.

Organizing, Encoding, Rehearsing, and Retrieving Course Content

ORGANIZE Before you engage the powerful process of memory, the first step in improving your academic performance is to make sure that the information you are studying is accurate and well organized. Just organizing material can lead to better memory.

Tips for Organizing

- Review your course notes routinely and catch potential errors and ambiguities early. There is no sense in memorizing inaccurate or incomplete information.

- Organize the material in a way that will allow you to commit it to memory effectively. Arrange information, rework material, and give it a structure that will help you to remember it.

- Experiment with different organizational techniques. One approach is to use a hierarchy such as an outline. You might create analogies (such as the earlier comparison of retrieval from long-term memory to finding a book in the library) that take advantage of your preexisting schemas.

ENCODE   Once you ensure that the material to be remembered is accurate and well organized, it is time to memorize. Although some types of information are encoded automatically, academic learning usually requires considerable effort.

Tips for Encoding

- Pay attention. Remember that staying focused on one thing is crucial. In other words, avoid divided attention.

- Process information at an appropriate level. Think about the material meaningfully and process it deeply.

- Elaborate on the points to be remembered. Make associations to your life and to other aspects of the material you want to remember.

- Use imagery. Devising images to help you remember (such as the mental picture of a computer screen to help you recall the concept of working memory) allows you to "double-encode" the information.

- Understand that encoding is not simply something that you should do before a test. Rather, encode early and often. During class, while reading, or in discussing issues, take advantage of opportunities to create associations to your course material. Once you have your notes for a lecture, rehearse that information right away.

REHEARSE   While learning material initially, relate it to your life and attend to examples that help you do so. After class, rehearse the course material over time to solidify it in memory.

Tips for Rehearsing

- Rewrite, type, or retype your notes. Some students find this exercise a good form of rehearsal.

- Talk to people about what you have learned and how it is important to real life in order to reinforce memory. You are more likely to remember information over the long term if you understand it and relate it to your world than if you mechanically rehearse and memorize it. Rehearsal works well for information in short-term memory, but when you need to encode, store, and then retrieve information from long-term memory, it is much less efficient. Thus, for most information, understand it, give it meaning, elaborate on it, and personalize it.

Notice that self-relevance is a great way to remember. Yet another reason why diversity and representation are important!

Caiaimage/Glow Images

- Test yourself. It is not enough to look at your notes and think, "Oh, yes, I know this." Sometimes recognition instills a false sense of knowing. If you look at a definition, and it seems so familiar that you are certain you know it, challenge yourself. What happens when you close the book and try to reconstruct the definition? Check your personal definition with the technical one in the book. How did you do?

- While reading and studying, ask yourself questions such as "What is the meaning of what I just read?" "Why is this important?" and "What is an example of the concept I just read about?" When you make a concerted effort to ask yourself questions about what you have read or about an activity in class, you expand the number of associations you make with the information you will need to retrieve later.

- Treat your brain kindly. If you are genuinely seeking to improve your memory performance, keep in mind that the brain is a physical organ. Perhaps the best way to promote effective memory storage is to make sure that your brain is able to function at maximum capacity. That means resting it, nourishing it by eating well, and keeping it free of mind-altering substances.

RETRIEVE   So, you have studied not just hard but deeply, elaborating on important concepts and committing lists to memory. You have slept well and eaten a nutritious breakfast, and now it is exam time. How can you best retrieve the essential information?

Tips for Retrieving

- Use retrieval cues. Sit in the same seat where you learned the material. Remember that the exam is full of questions about topics that you have thoughtfully encoded. Some of the questions on the test might help jog your memory for the answers to others.

- Sit comfortably, take a deep breath, and stay calm. Bolster your confidence by recalling that research on long-term memory has shown that material that has been committed to memory is there for a very long time—even among those who may experience a moment of panic when the test is handed out.

Memory is clearly crucial for learning and academic success, but memory serves far more functions. It is the fundamental ingredient of our life stories, as we next consider.

Autobiographical Memory and the Life Story

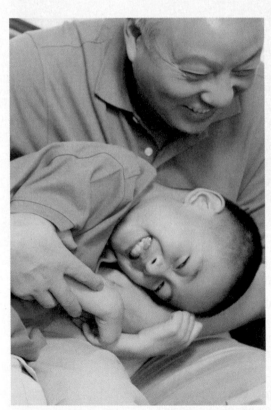

Our memories are an intimate way to share a part of ourselves with others, as a grandparent does with a grandchild.
Glow Asia/AP Images

Autobiographical memory may be one of the most important aspects of human life. For instance, one of the many functions that autobiographical memory serves is to allow us to learn from our experiences (Pillemer, 1998). In autobiographical memory, we store the lessons we have learned from life. These memories become a resource to which we can turn when faced with life's difficulties.

Autobiographical memory also allows us to understand ourselves and provides us with a source of identity (Jiang & others, 2020; McAdams, 2018). In studies of self-defining autobiographical memories, Jefferson Singer and his colleagues maintain that these internalized stories of personal experience serve as signs of the meaning we have created out of our life events and give our lives coherence (Baddeley & Singer, 2010; Singer & Blagov, 2004; Singer & Conway, 2008, 2011). According to Dan McAdams (2006, 2009, 2011a, 2011b), autobiographical memories form the core of our personal identity. A number of studies have shown that the stories we tell about our lives have important implications. For instance, McAdams and his colleagues have demonstrated that individuals who describe important life experiences that go from bad to better (*redemptive stories*) are more *generative*—that is, they are the kind of people who make a contribution to future generations, people who leave a legacy that will outlive them (Bauer & others, 2005). These individuals are also better adjusted than those whose self-defining memories go from good to bad (labeled *contamination stories*). Clearly, the construction and reconstruction of autobiographical memory may reveal important aspects of how people function, grow, and discover meaning in their lives (King & Hicks, 2006).

A final function of autobiographical memory, and perhaps its most vital aspect, is its role in social bonding (Speer & Delgado, 2020). Our memories are a valuable way to share a part of ourselves with others. Sharing personal experience is a way to foster intimacy, create bonds, and deepen existing ties. When we know a person's most cherished autobiographical memory, we know that person is no longer just an acquaintance but clearly a friend. To the extent that social bonds are necessary for survival, it makes sense that human beings can remember and share those memories with one another and that sharing our memories is a key pathway for sharing ourselves.

Do It!

Earlier in this chapter, you were invited to write about a self-defining memory. Pull out your description and see how it reflects the concepts of contamination and redemption. If the self-defining memory ends unhappily, think about stretching the story out further. Can you rewrite the memory so that it ends on a different note—not in a time of struggle and stress but now, as the person you are today?

Keeping Memory Sharp

As a process rooted in the brain, memory is also an indicator of brain functioning. Preserving memory is of vital importance as we age. A strong message from research on aging and memory is that, as for many things in life, the phrase "Use it or lose it" applies to memory.

Consider the case of Richard Wetherill, a retired lecturer and an uncommonly good chess player (Melton, 2005). Wetherill was so skilled that he was able to think eight moves ahead in a chess match. At one point, he noticed that he was having trouble playing chess—he could think only five moves ahead. He was sure that something was seriously wrong with him, despite his wife's assurances that she noticed no changes. A battery of cognitive tests revealed no abnormalities, and a brain scan was similarly reassuring. Two years later, Wetherill was dead, and the autopsy showed a brain ravaged by Alzheimer disease, a progressive, irreversible brain disorder that is characterized by gradual deterioration of memory, reasoning, language, and eventually physical functioning. Brain damage of this sort should indicate a person who was incapable of coherent thought. Wetherill's symptoms, however, had been limited to a small decline in his skill at playing chess.

Wetherill's case is surprising but also typical. People who lead active intellectual lives seem to be protected against the mental decline generally associated with age. Indeed, research has shown that individuals who are educated, have high IQs, and remain mentally engaged in complex tasks tend to cope better with a variety of assaults to the brain, including Alzheimer disease, stroke, head injury, and even poisoning with neurotoxins (Melton, 2005). Some research has suggested that an active mental life leads to the accumulation of a "cognitive store"—an emergency stash of mental capacity that allows individuals to avoid the negative effects of harm to the brain.

Yaakov Stern found that among a group of people with Alzheimer disease who appeared to be equal in terms of their outward symptoms, those who were

Mike Kemp/Rubberball/Getty Images

more educated were suffering from much worse brain damage, yet they were functioning at a level similar to others with relatively less damage (Stern & others, 1992). Stern and his colleagues (2004) also showed that intellectual pursuits such as playing chess and reading reduce the severity of Alzheimer symptoms. More recent research showed that while maintaining an active mental life can reduce the speed of cognitive decline associated with Alzheimer disease, it can also be related to more rapid decline once a person is diagnosed (Wilson & others, 2010).

Apparently, a lifetime of mental activity and engagement produces this cognitive reserve that allows the brain to maintain its ability to recruit new neural networks that compensate for damage. These brains are better able to adopt a backup plan to maintain the individual's level of functioning (Andel & others, 2005). The clear message from these studies is the importance of building up a cognitive reserve by staying mentally active throughout life. Remaining physically and cognitively active also seems to play a role in maintaining a sharp mind (Erickson & others, 2012; Park & Bischof, 2011). Modern conveniences can sometimes work against the goal of keeping the brain active. Individuals who rely on a GPS device to get around, for instance, may fail to exercise the hippocampus, a very important area for combating age-related cognitive decline (Bohbot & others, 2011).

Before we leave the science of memory, let's consider the role of memory in shaping meaningful experiences in daily life. Think of the most meaningful event of your life. Clearly, that event is one that you remember, among all the things you have experienced in your life. We all have certain particularly vivid autobiographical memories that stand out as indicators of meaning, such as those studied by Jefferson Singer that we reviewed earlier. Yet, everyday life is filled with potentially remarkable moments—a beautiful sunrise, a delicious meal prepared just for you, an unexpected telephone call from an old friend. Experiencing everyday life in its richness requires us to be attentive and engaged. Sometimes the daily chores and hassles of life lead us to feel that we are just going through the motions. This sort of mindless living may be a way to survive, but it is unlikely to be a way to thrive.

The processes of attention and encoding that we have explored in this chapter suggest that actively engaging in life—investing ourselves in the events of the day (Cantor & Sanderson, 1999)—is the way we can be assured that our life stories are rich and nuanced. That way, when someone asks, "So, tell me about yourself," we have a story to tell.

self-quiz

1. To improve your performance in your courses, you must make sure the information you are studying is
   A. well organized.
   B. accurate.
   C. complete.
   D. all of the above.

2. One way to improve the accuracy and efficiency of retrieval is through the use of
   A. retrieval cues.
   B. encoding cues.
   C. storage cues.
   D. none of the above.

3. Factors associated with keeping a sharp mind as we age include all of the following *except*

   A. educational attainment.
   B. being physically active.
   C. engaging in comfortable, easy tasks.
   D. having a high IQ.

 APPLY IT! 4. Albert is a retiree in his mid-60s. His wife, who is the same age, likes to read the latest novels and attend poetry readings. She also works the daily crossword puzzle in *The New York Times* and is taking a language class. Albert thinks that she is wasting her time and that retirement should be a time of kicking back. They both worked hard in school and had great careers, so what more does she want, Albert protests. Which of the

following might be good advice for Albert's wife?
   A. She should tell Albert that all of this mental activity is likely to keep her mind sharp and that he should join her in some of her pursuits.
   B. She should probably relax more. After all, at her age, how much more can she really learn?
   C. She should leave Albert alone; he is taking retirement seriously.
   D. She should continue in her activities, but they will not have any impact on her later memory ability.

SUMMARY

1 The Nature of Memory

Memory is the retention of information over time. The three processes involved in memory are encoding (getting information into storage), storage (retaining information over time), and retrieval (taking information out of storage).

2 Memory Encoding

Encoding requires attention, but the attention must be selective. Memory is negatively influenced by divided attention.

According to the theory of levels of processing, information is processed on a continuum from shallow (sensory or physical features are encoded) to intermediate (labels are attached to stimuli) to deep (the meanings of stimuli and their associations with other stimuli are processed). Deeper processing produces better memory. Elaboration, the extensiveness of processing at any given level of memory encoding, improves memory. Using imagery, or mental pictures, as a context for information can improve memory.

 ## Memory Storage

The Atkinson-Shiffrin theory describes memory as a three-stage process: sensory memory, short-term memory, and long-term memory.

Sensory memory holds perceptions of the world for only an instant. Visual sensory memory (iconic memory) retains information for about 0.25 second; auditory sensory memory (echoic memory) preserves information for several seconds.

Short-term memory is a limited-capacity memory system in which information is usually retained for as long as 30 seconds. Short-term memory's limitation is 7 ± 2 bits of information. Chunking and rehearsal can benefit short-term memory. Working memory involves a combination of short-term memory and attention that allow us to hold information temporarily as we perform cognitive tasks. Baddeley's model of working memory has three components: a central executive and two assistants (phonological loop and visuo-spatial sketchpad).

Long-term memory is a relatively permanent type of memory that holds huge amounts of information for a long time. Long-term memory has two main subtypes: explicit and implicit memory. Explicit memory is the conscious recollection of information, such as specific facts or events. Implicit memory affects behavior through prior experiences that are not consciously recollected. Explicit memory has two dimensions. One dimension includes episodic memory and semantic memory. The other dimension includes retrospective memory and prospective memory. Implicit memory is multidimensional, too, and includes systems for procedural memory, priming, and classical conditioning.

 ## Memory Retrieval

The serial position effect is the tendency to recall items at the beginning and the end of a list better than the middle items. The primacy effect is the tendency to recall items at the beginning of the list better than the middle items. The recency effect is the tendency to remember the items at the end of a list better than the middle items.

Retrieval is easier when effective cues are present. Another factor in effective retrieval is the nature of the retrieval task. Simple recognition of previously remembered information in the presence of cues is generally easier than recall of the information. According to the encoding specificity principle, information present at the time of encoding or learning tends to be effective as a retrieval cue, a process referred to as context-dependent memory.

Retrieval also benefits from priming, which activates particular connections or associations in memory. The tip-of-the-tongue phenomenon occurs when we cannot quite pull something out of memory. Five special cases of retrieval are autobiographical memory, emotional memory, memory for trauma, repressed memory, and eyewitness testimony.

Autobiographical memory is a person's recollections of his or her life experiences. Autobiographical memory has three levels: life time periods,

general events, and event-specific knowledge. Biographies of the self connect the past and the present to form our identity. Emotional memories may be especially vivid and enduring. Particularly significant emotional memories, or flashbulb memories, capture emotionally profound events that people often recall accurately and vividly.

Memory for personal trauma also is usually more accurate than memory for ordinary events, but it too is subject to distortion and inaccuracy. People tend to remember the core information about a personal trauma but might distort some of the details. Personal trauma can cause individuals to repress emotionally laden information so that it is not accessible to consciousness.

Repression means forgetting a particularly troubling experience because it would be too upsetting to remember it. Eyewitness testimony may contain errors due to memory decay or bias.

 ## Forgetting

Encoding failure is forgetting information that was never entered into long-term memory. Retrieval failure can occur for at least four reasons.

First, interference theory stresses that we forget not because memories are lost from storage but because other information gets in the way of what we want to remember. Interference can be proactive (as occurs when material learned earlier disrupts the recall of material learned later) or retroactive (as occurs when material learned later disrupts the retrieval of information learned earlier).

Second, decay theory states that when we learn something new, a neurochemical memory trace forms, but over time this chemical trail disintegrates.

The third reason for retrieval failure is that motivated forgetting occurs when we want to forget something. It is common when a memory becomes painful or anxiety-laden, as in the case of emotional traumas such as rape and physical abuse.

Finally, amnesia, the physiologically based loss of memory, can be anterograde, affecting the retention of new information or events; retrograde, affecting memories of the past but not memories of new events; or a combination of both.

⑥ Tips from the Science of Memory— for Studying and for Life

Effective encoding strategies when studying include paying attention and minimizing distraction, understanding the material rather than relying on rote memorization, asking yourself questions, and taking good notes. Research on memory suggests that the best way to remember course material is to relate it to many different aspects of your life.

Autobiographical memories, particularly self-defining memories, play a significant role in identity and social relationships. Our self-defining memories provide a unique source of identity, and sharing those memories with others plays a role in social bonding.

Taking on challenging cognitive tasks throughout life can stave off the effects of age on memory and lessen the effects of Alzheimer disease.

Engaging in everyday life means living memorably. Mindfulness to life events provides a rich reservoir of experiences upon which to build a storehouse of autobiographical memory.

KEY TERMS

amnesia
anterograde amnesia
Atkinson-Shiffrin theory
autobiographical memory
connectionism or parallel
 distributed processing (PDP)
decay theory
divided attention
elaboration
encoding
episodic memory

executive attention
explicit memory or declarative
 memory
flashbulb memory
implicit memory or nondeclarative
 memory
interference theory
levels of processing
long-term memory
memory
motivated forgetting

priming
proactive interference
procedural memory
prospective memory
retrieval
retroactive interference
retrograde amnesia
retrospective memory
schema
script
semantic memory

sensory memory
serial position effect
short-term memory
storage
sustained attention
tip-of-the-tongue (TOT)
 phenomenon
working memory

ANSWERS TO SELF-QUIZZES

Section 1: 1. D; 2. B; 3. A; 4. A
Section 2: 1. C; 2. D; 3. C; 4. B

Section 3: 1. D; 2. A; 3. B; 4. B
Section 4: 1. C; 2. A; 3. D; 4. B

Section 5: 1. B; 2. D; 3. B; 4. B
Section 6: 1. D; 2. A; 3. C; 4. A

Design elements: (Preview icon): Jiang Hongyan/Shutterstock; (Marginal notes): Shutterstock/Vadarshop

7

Thinking, Intelligence, and Language

Solving the Pandemic Dilemma

From the moment the novel coronavirus was identified, the world awaited a vaccine and scientists worked feverishly to develop a way to harness the body's immune system to combat the virus. Most vaccines involve exposing people to a weakened or dead form of a virus. The first vaccines developed for COVID-19 instead included an RNA molecule that sends a message to the immune system to learn to attack the spikes that cover the coronavirus (Corum & Zimmer, 2020). The development of such vaccines has been years in the making, and researchers were ready and waiting for a chance to apply their knowledge. In fact, the Moderna vaccine was designed by mid-January 2020, within days of the release of the DNA sequence for the virus (Wallace-Wells, 2020). That innovative design then underwent careful testing to see if it was both safe and effective. In all, the COVID-19 vaccines were produced faster than any vaccines in history.

The creation of these vaccines illustrates the harnessing of the human capacities for creativity, intelligence, and great effort to solve problems. The development of the Moderna vaccine also shows how sometimes support for great ideas can come from unexpected places. A close look at the report of the successful Moderna trial revealed that the vaccine was developed with the help of a $1 million donation from Dolly Parton (Andrew, 2020). When problems seem insurmountable, help from a country music superstar can't hurt.

Surely, COVID-19 challenged medical and scientific professionals to their maximum. The fact that this remarkable innovation is now being implemented shows that solving a problem is more than just reacting to the moment. In fact, the seeds of that idea were nurtured and perfected over the years, awaiting the great problem they would solve.

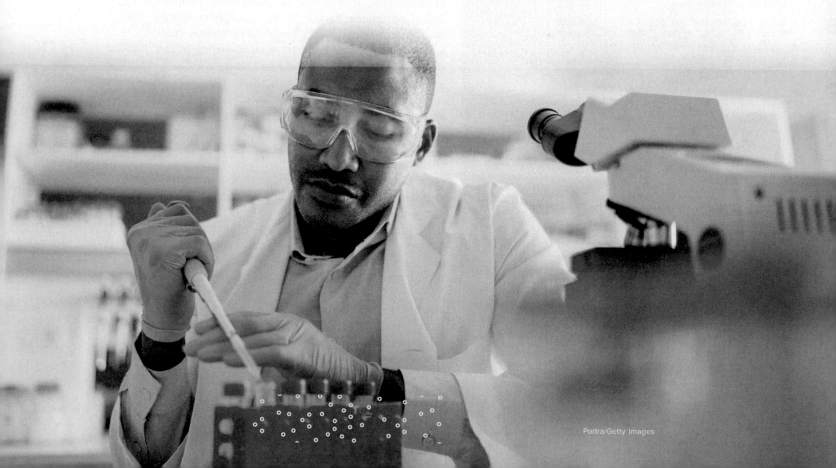

Cognitive psychology is the study of mental processes. This chapter investigates the basic cognitive processes of thinking, problem solving, reasoning, and decision making. We first define cognition and look at the cognitive revolution that led to new thinking about the workings of the human mind. We then review capacities associated with superior problem solving: critical thinking, creativity, and (perhaps most important) intelligence. We conclude by surveying the unique contributions of language to mental processes.

1 The Cognitive Revolution in Psychology

Cognitive psychologists study **cognition**—how information is processed and manipulated in remembering, thinking, and knowing. Cognitive psychology is a relatively young field, just over a half a century old. Let's begin by tracing its history.

Early in the twentieth century, behaviorism dominated the thinking of experimental psychologists. Behaviorists such as B. F. Skinner argued that the human mind is a "black box" best left to philosophers, and they considered observable behavior to be psychologists' proper focus. Behaviorism had little use for the mental processes occurring in that dark place between your ears.

In the 1950s psychologists' views began to change. Oddly enough, computer science was a key motivator in the birth of the study of human information processing. Because we could "see" what computers were doing internally, psychologists started thinking it might be acceptable to use those observations to study *human* mental processes. The first modern computer, developed by mathematician John von Neumann in the late 1940s, showed that machines could perform logical operations. In the 1950s, researchers speculated that computers might model some mental operations, and they believed that such modeling might shed light on how the human mind works (Núñez & others, 2019; Posner & Shulman, 2019).

Cognitive psychologists often use the computer as an analogy to help explain the relationship between cognition and the brain (Presti, 2016). They describe the physical brain as the computer's hardware and cognition as its software. Herbert Simon (1969) was among the pioneers in comparing the human mind to computer processing systems. In this analogy, the sensory and perceptual systems provide an "input channel," similar to the way data are entered into the computer (Figure 1). As input (information) comes into the mind, mental processes, or operations, act on it, just as the computer's software acts on the data. The transformed input generates information that remains in memory much in the way a computer stores what it has worked on. Finally, the information is retrieved from memory and "printed out" or "displayed" (so to speak) as an observable response.

cognition
The way in which information is processed and manipulated in remembering, thinking, and knowing.

Mathematician John von Neumann (1903–1957) pioneered in the early development of computers. The fact that his computer could perform logical operations led researchers to imagine that computers might model some mental processes and that such modeling might shed light on how the human mind functions.
Source: U.S. Department of Energy

Computers provide a logical and concrete, but vastly oversimplified, model of the mind's processing of information. Inanimate computers and human brains function quite differently in some respects. For example, most computers receive information from a human who has already coded the information and removed much of its ambiguity. In contrast, each brain neuron can respond to ambiguous information transmitted through sensory receptors such as the eyes and ears.

Computers can do some things better than humans. Think about the times you turn to a machine for help. A calculator can perform complex numerical calculations much faster and more accurately than humans could ever hope to. Computers can also apply and follow rules more consistently and with fewer errors than humans and can represent complex mathematical patterns better than humans. Research has shown that computers might be better than humans at cooperating to win a game (Snow, 2017).

Of course, the human mind is aware of itself; the computer is not. No computer is likely to approach the richness of human consciousness. In 2009, scientists in Europe created The Human Brain Project to simulate human information processing in a computer (Markram, 2012). A decade later, they still had not done so (Yong, 2019).

Still, the computer's role in cognitive psychology continues to increase. An entire scientific field called **artificial intelligence (AI)** focuses on creating machines capable of performing activities that require intelligence when they are done by people. AI can be especially useful for tasks requiring speed, persistence, and a vast memory. AI systems have been used in a broad range of contexts. An AI system called *Deepstack* defeated numerous professional poker players (Moravčík & others, 2017). NASA allowed the Mars lander computer to take on a role previously reserved for humans: choosing which materials should be collected for close examination with great success (Francis & others, 2017). During the global COVID-19 pandemic, AI was used in a multitude of ways (Raza, 2020), including improving diagnoses (Li & others, 2020), probing potential treatments (Zhou & others, 2020), and projecting the spread of the virus in different countries (da Silva & others, 2020).

Cognitive robotics is a field of study that seeks to endow a robot with intelligent behavior so it can function

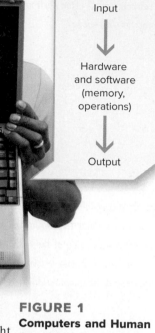

Human

Input

↓

Brain, mind, cognition (memory, problem solving, reasoning, consciousness)

↓

Output

Computers

Input

↓

Hardware and software (memory, operations)

↓

Output

In 2017, Facebook shut down an AI engine after it was discovered that two 'chatbots' were talking to each other in a language they created that was unknown to humans!

artificial intelligence (AI)
A scientific field that focuses on creating machines capable of performing activities that require intelligence when they are done by people.

Artificial intelligence (AI) researchers are exploring frontiers that were once the context for sci-fi movie plots. Cog is a human-form robot built by the Humanoid Robotics Group at the Massachusetts Institute of Technology. The sensors and structures in Cog's AI system model human sensory and motor activity as well as perception. Cog's creators have sought to achieve humanlike functioning and interactions with people—both of which, they hope, will lead to new, humanlike learning experiences for the robot. Think about it: How might research findings from experiments such as Cog be applied to real-world situations?
John B. Carnett/Popular Science/Getty Images

Have you noticed that intelligent computers on television and in the movies almost always turn out to be evil? Why do you think fictional treatments of AI often portray smart computers as scary?

in the world much like humans do. In this area, scientists strive to provide robots with processing equipment that will let them perceive, attend, memorize, learn, reason, and make decisions (Broz & others, 2014; Janecka, 2019; Paglieri & others, 2014). Such humanlike robots might, for instance, become helpers who would attend to the needs of humans with disabilities (Staffa & others, 2020). Cognitive robotics involves an array of scientists, including cognitive psychologists, neuroscientists, computer scientists, and engineers. A fascinating challenge faced by those seeking to model robots after humans is the *uncanny valley* (Mathur & others, 2020; Siebert &

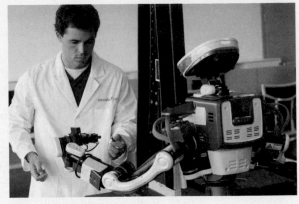

Javier Larrea/Alamy Stock Photo

others, 2020). The uncanny valley refers to the fact that robots that represent very close facsimiles of humans sometimes evoke revulsion in people. Something about these robots feels not quite right to human observers.

If you were creeped out by the "dead eyes" of animated characters in the movie Polar Express or the just-not-right quality of the humans in Monsters vs. Aliens, you've experienced the uncanny valley.

By the late 1950s the cognitive revolution was in full swing. The term *cognitive psychology* became a label for approaches that sought to explain observable behavior by investigating mental processes and structures that could not be directly observed. We now build on concepts introduced in the memory chapter by exploring thinking, problem solving, and decision making.

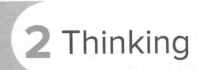

self-quiz

1. Behaviorists thought that psychology should properly focus on
   A. mental processes.
   B. the subconscious mind.
   C. private behavior.
   D. observable behavior.

2. The name for the scientific field that is concerned with making machines that mimic human information processing is
   A. cognitive science.
   B. cognitive neuroscience.
   C. artificial intelligence.
   D. computer science.

3. Cognition involves
   A. manipulating information.
   B. processing information.
   C. thinking.
   D. all of the above.

APPLY IT! 4. When Demarre plays chess against his friend, he almost always wins, but when he plays against his computer, he typically loses. What is the *most likely* explanation for Demarre's experience?
   A. Computers are smarter than human beings.

   B. When playing against his friend, Demarre always cheats.
   C. When playing against his friend, Demarre is able to use his human cognitive skills to "read" his opponent's facial expressions and predict what the latter will do. These cues are missing when he plays against his computer.
   D. Demarre's computer is better than his friend at picking up on Demarre's cues and predicting what he will do next.

2 Thinking

thinking
The process of manipulating information mentally by forming concepts, solving problems, making decisions, and reflecting critically or creatively.

When you save a computer file, you might hear a sound from inside or see a processing icon, and you know the computer is processing the work you have just done. Unlike a computer, the brain does not make noise to let us know it is working. Rather, the brain's processing is the silent operation of thinking. **Thinking** involves manipulating information mentally by forming concepts, solving problems, making decisions, and reflecting in a critical or creative manner. Let's explore the nature of concepts—the components of thinking—and investigate the cognitive processes of problem solving, reasoning, and decision making.

Concepts

concept
A mental category that is used to group objects, events, and characteristics.

A fundamental aspect of thinking is the notion of concepts. **Concepts** are mental categories that are used to group objects, events, and characteristics. Creating categories is one way we make sense of information in the world. We know that apples and oranges are both fruits. We know that poodles and collies are both dogs and that ants and ladybugs are both insects. These items differ from one another in various ways, and yet we recognize that they belong together because we have concepts for fruits, dogs, and insects. These points may seem pretty obvious but the existence of concepts is important for four reasons:

- First, **concepts allow us to generalize**. If we did not have concepts, each object and event in our world would be unique and brand new to us each time we encountered it.

- Second, **concepts allow us to associate experiences and objects**. Basketball, ice hockey, and track are sports. The concept *sport* gives us a way to compare these activities.

- Third, **concepts aid memory, making it more efficient so that we do not have to reinvent the wheel each time we come across a piece of information**. Imagine having to think about how to sit in a chair every time we find ourselves in front of one.

- Fourth, **concepts provide clues about how to react to a particular object or experience**. Perhaps you have had the experience of trying an exotic new cuisine and feeling puzzled as you consider the contents of your plate. If a friend tells you reassuringly, "That's food!" you know that given the concept *food,* it is okay to dig in.

Although it has a ducklike bill and lays eggs, the platypus is nevertheless a mammal like the tiger, as platypus females produce milk to feed their young. The prototypical birdlike characteristics of the platypus can lead us to think mistakenly that the platypus is a bird. Its atypical properties place the platypus on the extreme of the concept of mammal.
(platypus): JohnCarnemolla/Getty Images; (tiger): John Giustina/Getty Images

prototype model
A model emphasizing that when people evaluate whether a given item reflects a certain concept, they compare the item with the most typical item(s) in that category and look for a "family resemblance" with that item's properties.

What exactly determines whether something fits into a particular concept? One way that psychologists answer this question is the prototype model (Douven, 2019). The **prototype model** emphasizes that when people evaluate whether a given item reflects a certain concept, they compare the item with the most typical item(s) in that category and look for a "family resemblance" with that item's properties. Birds generally fly and sing, so we know that robins and sparrows are both birds. We recognize exceptions to these properties, however—we know that a penguin is still a bird even though it does not fly or sing. The prototype model maintains that people use characteristic properties to create a representation of the average or ideal member—the prototype—for each concept. Comparing individual cases to our mental prototypes may be a good way to decide quickly whether something fits a particular category.

Prototypes can be misleading when applied to people, as we will see.

Problem Solving

Concepts tell us *what* we think about but not *why* we think. *Why* do we bother to engage in the mental effort of thinking? Consider Levi Hutchins, an ambitious young clockmaker who, in 1787, invented the alarm clock. *Why* did he go to the trouble? He had a specific goal—he wanted to get up before sunrise every morning—yet he faced a problem in accomplishing that goal. A problem exists when a person has a goal with no clear way to achieve it. **Problem solving** means finding an appropriate way to attain a goal when the goal is not readily available (Vallée-Tourangeau & Vallée-Tourangeau, 2020). Problem solving entails following several steps, overcoming mental obstacles and developing expertise.

problem solving
The mental process of finding an appropriate way to attain a goal when the goal is not readily available.

FOLLOWING THE STEPS IN PROBLEM SOLVING   Psychological research points to four steps in the problem-solving process.

1. Find and Frame Problems
Recognizing a problem is the first step toward a solution (Vernon & others, 2016). Finding and framing problems involves asking questions in creative ways and "seeing" what others do not.

The ability to recognize and frame a problem is difficult to learn but it can be learned. For example, brainstorming in a group may be a great way to generate and define problems. Many real-world problems are ill defined or vague and have no clear-cut solutions. Inventors are visionaries who see problems that everyone else is content to live with. Recognizing problems involves being aware of and open to experiences (two mental habits we will examine later). It also means listening carefully to that voice in your head that occasionally sighs, "There must be a better way."

2. Develop Good Problem-Solving Strategies
Once we find a problem and clearly define it, we need to develop strategies for solving it. Among the effective strategies are subgoals, algorithms, and heuristics.

subgoals
Intermediate goals or problems to solve that put one in a better position for reaching a final goal or solution.

Subgoals are intermediate goals or intermediate problems to solve that put us in a better position for reaching the final goal or solution. Imagine that you are writing a paper for a psychology class. What are some subgoaling strategies for approaching this task? One might be locating the right books and research journals on your chosen topic. At the same time that you are searching for the right publications, you will likely benefit from establishing some subgoals within your time frame for completing the project. If the paper is due in two months, you might set a subgoal of a first draft of the paper two weeks before it is due, another subgoal of completing your reading for the paper one month before it is due, and still another subgoal of starting your library research tomorrow. Notice that in establishing the subgoals for meeting the deadline, we worked backward. Working backward in establishing subgoals is a good strategy. You first create the subgoal that is closest to the final goal and then work backward to the subgoal that is closest to the beginning of the problem-solving effort.

Algorithms are strategies that guarantee a solution to a problem. Algorithms come in different forms, such as formulas, instructions, and the testing of all possible solutions. We use algorithms in cooking (by following a recipe) and driving (by following directions to an address). Compared to humans, computers more efficiently apply algorithms (Jin & others, 2017). For example, a computer can quickly compare all possible combinations of, say, a string of letters to identify all possible words that can be formed from those letters.

algorithms
Strategies—including formulas, instructions, and the testing of all possible solutions—that guarantee a solution to a problem.

Allrightimages/age fotostock

For a human, an algorithmic strategy might take a long time to solve a simple problem. Staring at a rack of letters during a game of Scrabble, for example, you might find yourself moving the tiles around and trying all possible combinations to make a high-scoring word. Instead of using an algorithm to solve your Scrabble problem, however, you might rely on some rules of thumb about words and language.

heuristics
Shortcut strategies or guidelines that suggest a solution to a problem but do not guarantee an answer.

Heuristics are shortcut strategies or guidelines that suggest a solution to a problem but do not guarantee an answer (Banks & others, 2020; Štukelj, 2020). In your Scrabble game, you know that if you have a Q, you are going to need a U. If you have an X and a T, the T is probably not going to come right before the X. In this situation, heuristics allow you to be more efficient than algorithms would. In the real world, we are more likely to solve the types of problems we face by heuristics than by algorithms. Heuristics help us to narrow down the possible solutions and to find one that works.

3. Evaluate Solutions

Once we think we have solved a problem, we will not know how effective our solution is until we find out if it works. It helps to have in mind a clear criterion for the effectiveness of the solution. For example, what will your criterion be for judging the effectiveness of your solution to the psychology assignment, your psychology paper? Will you judge your solution to be effective if you simply complete the paper? If you get an *A*? If the instructor says that it is one of the best papers a student ever turned in on the topic?

4. Rethink and Redefine Problems and Solutions over Time

An important final step in problem solving is to rethink and redefine problems continually (Vernon & others, 2016). Good problem solvers tend to be more motivated than the average person to improve on their past performances and to make original contributions. Can we make the computer faster and more powerful? Can we make the iPhone even more distracting?

AN OBSTACLE TO PROBLEM SOLVING: BECOMING FIXATED   A key ingredient of being a good problem solver is to acknowledge that one does not know everything—that one's strategies and conclusions are always open to revision. Optimal problem solving may require a certain amount of humility, or the ability to admit that one is not perfect and that there may be better ways than one's tried and true methods to solve life's problems. Sometimes to solve a problem, we must forget what we think we know. It is easy to fall into the trap of becoming fixated on a particular strategy for solving a problem.

fixation
Using a prior strategy and failing to look at a problem from a fresh new perspective.

functional fixedness
Failing to solve a problem as a result of fixation on a thing's usual functions.

Fixation involves using a prior strategy and failing to look at a problem from a fresh, new perspective. **Functional fixedness** occurs when individuals fail to solve a problem because they are fixated on a thing's usual functions. If you have ever used a shoe to hammer a nail, you have overcome functional fixedness to solve a problem.

An example of a problem that requires overcoming functional fixedness is the Maier string problem, depicted in Figure 2 (Maier, 1931). The problem is to figure out how to tie two strings together when you must stand in one spot and cannot reach both at the same time. It seems as though you are stuck. However, there is a pair of pliers on a table. Can you solve the problem?

The solution is to use the pliers as a weight, tying them to the end of one string (Figure 3). Swing this string back and forth like a pendulum and grasp the stationary string. Your past experience with pliers and your fixation on their usual function makes this a difficult problem to solve. To do so, you need to find an unusual use for the pliers—in this case, as a weight to create a pendulum.

Effective problem solving often necessitates trying something new, or *thinking outside the box*—that is, exploring novel ways of approaching tasks and challenges and finding solutions. This might require admitting that one's past strategies were not ideal or do not readily translate to a particular situation. Students who are used to succeeding in high school by cramming for tests and relying on parental pressure to get homework done may find that in college these strategies are no longer viable ways to succeed. To explore how fixation might play a role in your own problem solving, see Figure 4.

Reasoning and Decision Making

In addition to forming concepts and solving problems, thinking includes the higher-order mental processes of reasoning and decision making. These activities require rich connections

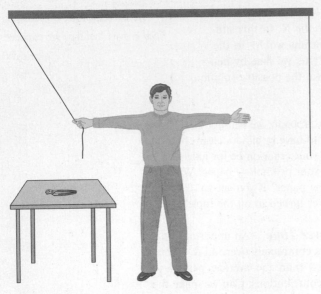

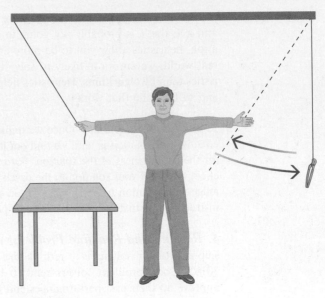

FIGURE 2
Maier String Problem How can you tie the two strings together if you cannot reach them both at the same time?

FIGURE 3
Solution to the Maier String Problem Use the pliers as a weight to create a pendulum motion that brings the second string closer.

The Candle Problem
How would you mount a candle on a wall so that it won't drip wax on a table or a floor while it is burning?

The Nine-Dot Problem
Take out a piece of paper and copy the arrangement of dots shown below. Without lifting your pencil, connect the dots using only four straight lines.

The Six-Matchstick Problem
Arrange six matchsticks of equal length to make four equilateral triangles, the sides of which are one matchstick long.

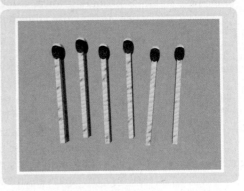

FIGURE 4   Examples of How Fixation Impedes Problem Solving These tasks help psychologists measure creative problem solving. *Solutions to the problems are presented at the end of the chapter.*

among neurons and the ability to apply judgment. The end result of this type of thinking is an evaluation, a conclusion, or a decision.

reasoning
The mental activity of transforming information to reach conclusions.

REASONING   **Reasoning** is the mental activity of transforming information to reach conclusions. Reasoning is involved in problem solving and decision making. It is also a skill closely tied to critical thinking (Fasko & Fair, 2020). Reasoning can be either inductive or deductive (Figure 5).

Inductive reasoning involves reasoning from specific observations to make generalizations (Lafraire & others, 2020; Stephens & others, 2020). Inductive reasoning is an important way that we form beliefs about the world. For instance, having turned on your smartphone many times without having it explode, you have every reason to believe that it will not explode the next time you turn it on. From your prior experiences with the phone, you form the general belief that

Sour milk notwithstanding, strong inductive reasoning is generally based on many observations. A woman who decides that all men are rats after two bad boyfriends is overgeneralizing.

inductive reasoning
Reasoning from specific observations to make generalizations.

it is not likely to become a dangerous object. Or, imagine taking a sip of milk from a container and finding that it tastes sour. Inductive reasoning is at work when you throw out the whole container even though you haven't tasted every drop.

Inductive reasoning is responsible for a great deal of scientific knowledge. Psychological research is often inductive. Whenever psychologists study a sample in order to make generalizations about a population from which the sample is drawn, that is inductive reasoning. Note that this involves going from a few observations to generalizing about a larger group. Conclusions drawn from inductive reasoning may be mistaken. Note that inductive reasoning starts with observations and then makes its way to general conclusions.

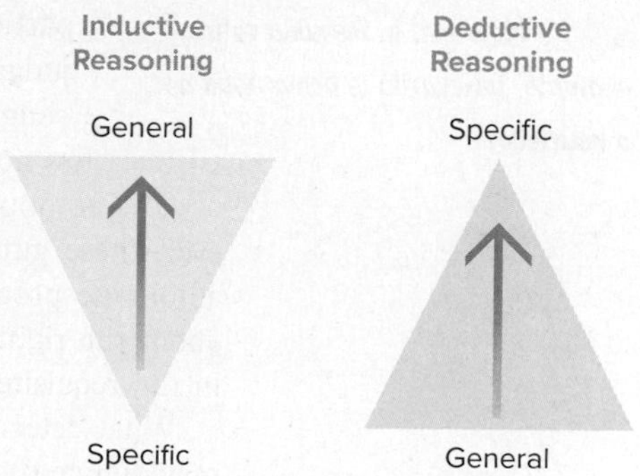

In contrast, **deductive reasoning** is reasoning from a general case that we know to be true to a specific instance. This kind of reasoning is sometimes called top-down logic. Using deductive reasoning, we draw conclusions based on logic and facts. Deductive reasoning is used, for instance, in mathematical proofs. Deductive reasoning rests on knowledge. For example, we know that all organisms require oxygen to survive. Squirrels are organisms; therefore, squirrels require oxygen to survive.

When psychologists and other scientists use theories to make predictions and then evaluate their predictions by making further observations, deductive reasoning is at work. When psychologists develop a hypothesis from a theory, they are using a form of deductive reasoning, because the hypothesis is a specific, logical extension of the general theory. If the theory is true, then the hypothesis will be true as well.

deductive reasoning Reasoning from a general case that is known to be true to a specific instance.

FIGURE 5 **Inductive and Deductive Reasoning** (*left*) The triangle represents inductive reasoning—going from specific to general. (*right*) The triangle represents deductive reasoning—going from general to specific.

DECISION MAKING

Think of all the decisions, large and small, that you have to make in life. Should you major in biology, psychology, or business? Should you go to graduate school right after college or get a job first? Should you establish yourself in a career before settling down to have a family? Do you want fries with that? **Decision making** involves evaluating alternatives and choosing among them.

Reasoning uses established rules to draw conclusions. In contrast, decision making is less clear cut and more uncertain. There are no established rules and we may not know the consequences of the decisions. Some of the information might be missing, and we might not trust all of the information we have. Making decisions means weighing information and coming to some conclusion that we feel will be best for us: Yes, we will have time to watch one more TikTok; no, we will not run that red light to get to class on time.

decision making The mental activity of evaluating alternatives and choosing among them.

TWO TYPES OF REASONING AND DECISION MAKING

Many psychologists divide the cognitive processes involved in reasoning and decision making into two levels—one that is automatic (often referred to as *Type 1* or *system 1*) and one that is controlled (often called *Type 2* or *system 2*) (Diederich & Trueblood, 2018). The automatic system involves processing that is rapid, heuristic, and intuitive; it entails following one's hunches or gut feelings about a particular decision or problem (Armstrong & others, 2020; Grayot, 2020). Intuitive judgment means knowing that something feels right even if the reason why is unknown. In contrast, the controlled system is slower, effortful, and analytical. It involves conscious reflection about an issue. This is the kind of thinking that might be required to solve a difficult math problem, for example.

Although conscious effortful thinking is invaluable for solving many problems, research has shown that intuitive processing also have an important role to play in decision making. Sometimes, compared to effortful reflection, intuitive decision making can be less biased and more efficient in decision making. For example, in one study (Halberstadt & Catty, 2008) participants were told that their job was to rate the popularity of particular songs. Participants listened to brief snippets of the songs. Half of the participants were asked to first reflect on the reasons a specific song might be popular. The other half simply rendered their judgments intuitively, without much thought. Those who intuitively judged the songs were actually more accurate in their popularity ratings. Why? These

Note that in the song rating example, familiarity is being used as a heuristic!

participants based their ratings on *how familiar* the song was to them. Basing judgments on the feeling of familiarity was a good idea in this case, as popular songs are likely to be familiar to everyone. Those who analyzed their reasons were less accurate because thinking about reasons disrupted this natural association.

The popular media sometimes portray intuitive hunches as sort of magical. However, these gut feelings do not emerge out of thin air. Rather, they are the product of automatic processes, of learned associations, and of implicit memory. Your gut feelings about the right answer on a test are certainly more likely to be accurate if you have put in the requisite hours of conscious effortful study.

What determines which system of reasoning we use? The type of problem to be solved may dictate the type of processing used. If a problem is difficult, we might harness system 2 processing. For less taxing decisions, we might go with system 1, following our immediate gut feeling. One factor that influences processing style is mood. A negative mood (feeling sad or worried) fosters more analytical, effortful cognitive processing (Shen & others, 2019). In contrast, a positive mood (feeling happy or cheerful) facilitates more rapid processing, promoting the use of heuristics and viewing the big picture. A positive mood may provide a signal that all is well and that we can safely go with our intuitive hunches for the task at hand (Maldei & others, 2020).

It is important to keep in mind that system 1 processes are as rapid as they are because they often rely on heuristics. The use of heuristics can lead to mistakes when these rules of thumb are applied inappropriately, as we now consider.

Biases and Heuristics

In many cases, our decision-making strategies are well adapted to deal with a variety of problems (Banks & others, 2020; Luan & others, 2019). Heuristics, for example, are intuitive and efficient ways of solving problems and making decisions; they are often at work when we make a decision by following a gut feeling. However, heuristics and gut feelings can lead to mistakes. Here we look at a few biases and heuristic errors, summarized in Figure 6.

One of the most powerful biases in human decision making is loss aversion. **Loss aversion** refers to the tendency to strongly prefer to avoid losses compared to acquiring gains. We dislike the prospect of losing something we have more than we enjoy the prospect of gaining something new, even when the prospect of a gain outweighs the loss (Kahneman & Tversky, 1984). Imagine that you have a *B+* in a class with an optional, but likely very challenging, final exam. If you do well on the final, you could nudge your grade to an *A−*.

loss aversion
The tendency to strongly prefer to avoid losses compared to attempting to acquire gains.

Loss Aversion	Confirmation Bias	Base Rate Neglect	Hindsight Bias	Representativeness Heuristic	Availability Heuristic
Description Tendency to weigh potential losses more heavily than potential gains **Example:** An investor decides not to buy stock in a new company even though the chances of financial gain outweigh the chances of financial loss.	**Description** Tendency to search for and use information that supports rather than refutes one's ideas **Example:** A politician accepts news that supports his views and dismisses evidence that runs counter to these views.	**Description** Tendency to ignore information about general principles in favor of very specific but vivid information **Example:** You read a favorable expert report on a television you are intending to buy, but you decide not to buy it when a friend tells you about a bad experience with that model.	**Description** Tendency to report falsely, after the fact, that one accurately predicted an outcome **Example:** You read about the results of a particular psychological study and say, "I always knew that," though in fact you have little knowledge about the issues examined in the study.	**Description** Tendency to make judgments about group membership based on physical appearances or one's stereotype of a group rather than available base rate information **Example:** The victim of a holdup, you view police photos of possible perpetrators. The suspects look very similar to you, but you choose the individual whose hair and clothing look dirtiest and most disheveled.	**Description** Prediction about the probability of an event based on the ease of recalling or imagining similar events **Example:** A teen from an extended family in which no family member ever attended college tells their parents that they want to be a doctor. The parents cannot imagine their child in such a career and suggest becoming a home healthcare aide.

FIGURE 6 **Decision-Making Problems: Biases and Heuristics** Biases and heuristics (rules of thumb) affect the quality of many of the decisions we make.

But if you do poorly, your *B+* could plummet to a *C*. Would you risk what you already have for the chance to get a better grade?

Loss aversion has been demonstrated in many different contexts and samples (Mrkva & others 2020; Sharma & others, 2020; Sokol-Hessner & Rutledge, 2019; Zhao & others, 2020). Indeed, so much research has been done concerning the way that human beings weigh losses more than gains that scholars have been able to estimate that losses appear to be two times more powerful than gains (Vohs & Luce, 2010).

Loss aversion helps to explain a variety of phenomena in psychology and economics. For example, the *endowment effect* means that people ascribe greater value to things they already own, compared to objects owned by someone else. In one study (Kahneman & others, 1990), some participants were shown a mug and were asked how much they would be willing to pay for it. Other participants were actually given the mug to keep and then were asked how much they would be willing to sell it for. In both groups the mugs were identical—simple coffee mugs with the university insignia on them. However, those who owned the mug believed it was worth $3 more than those who just looked it over. Somehow, just by owning it, it became more valuable.

> Can you see how loss aversion explains the endowment effect? The idea is that once we own something it becomes one of the many things we would hate to lose!

Loss aversion also explains why sometimes it is so hard to cut our losses when we are in a losing battle. The *sunk cost fallacy* refers to the fact that people are reluctant to give up on a venture because of past investment. If we were perfectly rational, we would make decisions based only on current circumstances, maximizing the benefits and minimizing costs. However, past investment biases our judgments. Economically, sunk cost fallacy means "throwing good money after bad."

Imagine that you have just suffered through two years of training to become an accountant. What you have discovered is that you neither enjoy accounting nor are you good at it. The courses have been a struggle and you have just managed to pass. Should you stick with it? If you were making a decision rationally based on current circumstances alone, you might decide to change course and switch majors. However, doing so would also mean recognizing that the last two years have been "wasted." You have sunk a lot of resources into something you no longer even want. But sunk costs may spur you to stick it out, or even try harder. Sunk costs reflect loss aversion in that we dread the thought of losing the effort, time, and money we have already put into a venture if we give up.

Like the endowment effect, the sunk cost fallacy derives from loss aversion. There are a number of other biases that have been identified in human decision making. Let's take a look at some of those.

confirmation bias
The tendency to search for and use information that supports one's ideas rather than refutes them.

Confirmation bias is the tendency to search for and use information that supports our ideas rather than refutes them. Confirmation bias is also sometimes called "Myside bias" because it is all about collecting up evidence that our own views are correct. Our decisions can become further biased because we tend to seek out and listen to people whose views confirm our own while we avoid those with dissenting views. It is easy to detect the confirmation bias in the way that many people think. Consider politicians. They often accept news that supports their views and dismiss evidence that runs counter to those views. Avoiding confirmation bias means applying the same rigorous analysis to *both* sides of an argument.

> Confirmation bias is one way to understand the polarization in U.S. politics. People get their news from different sources that agree with their previous views. On social media they limit themselves to information that agrees with them—and then algorithms operated by social media selectively expose people to information that supports their views.

hindsight bias
The tendency to report falsely, after the fact, that one has accurately predicted an outcome.

Hindsight bias is our tendency to report falsely, after the fact, that we accurately predicted an outcome. It is sometimes referred to as the "I knew it all along effect." With this type of bias, people tend to view events that have happened as more predictable than they were, and to represent themselves as being more accurate in their predictions than they actually were. Hindsight bias might sound self-serving in the sense that it means remembering ourselves as having known more than we really did, but cognitive psychologists recognize that this bias may be based on new learning and on updating our knowledge about the world (Nestler & others, 2010; Pezzo, 2011). One reason for hindsight bias is that actual events are more vivid in our minds than all those things that failed to happen, an effect called the availability heuristic.

availability heuristic
A prediction about the probability of an event based on the ease of recalling or imagining similar events.

The **availability heuristic** refers to a prediction about the probability of an event based on the ease of recalling or imagining similar events. For example, have you ever experienced a sudden fear of flying after hearing about an airplane crash? Shocking events such as plane crashes stick in our minds, making such disasters seem common. The chance of dying in a plane crash in a given year, however, is tiny (1 in 400,000) compared to the chance of dying in a car accident (1 in 6,500). Because car accidents are less newsworthy, they are less likely to catch our attention and remain in our awareness. The availability heuristic can reinforce generalizations about others in daily life (Park, 2020). Imagine, for instance, that Elvedina, a Latina girl, tells her mother that she wants to be a doctor. Her mother, who has never met a Latina doctor, finds it hard to conceive of her daughter's pursuing such a career and might suggest that she try a different career.

Also reflective of the impact of vivid cases on decision making is base rate neglect. **Base rate neglect** is the tendency to ignore information about general principles in favor of very specific but vivid information (Pilditch & others, 2020). Let's say that as a prospective car buyer, you read *Consumer Reports* and find that a panel of experts rates a particular vehicle exceptionally well. You might still be swayed in your purchasing decision, however, if a friend tells you about bad experiences with that car. Similarly, imagine being told that the average exam score for a test in your psychology class was 75 percent. If you were asked to guess a random student's score, 75 percent would be a good answer— the mean tells us the central tendency of any distribution. Yet if the student provided just a little bit of information, such as how many hours they studied, you might give too much weight to that specific information, losing sight of the valuable base rate information you have in the class mean.

base rate neglect
The tendency to ignore information about general principles in favor of very specific but vivid information.

To experience another heuristic in action, consider the following example. Your psychology professor tells you they have assembled 100 men in the hallway outside your classroom. The group consists of 5 librarians and 95 professional kick boxers. The professor will randomly select one man to enter the room, and you can win $100 if you accurately guess whether he is a librarian or a kick boxer. The man stands before you. He is in his 40s, with short graying hair, and he wears thick glasses, a button-down white shirt, a bow tie, neatly pressed slacks, and loafers. Is he a librarian or a kick boxer? If you guessed librarian, you have fallen victim to the representativeness heuristic.

representativeness heuristic
The tendency to make judgments about group membership based on physical appearances or the match between a person and one's stereotype of a group rather than on available base rate information.

The **representativeness heuristic** is the tendency to make judgments about group membership based on physical appearances or the match between a person and one's stereotype of a group rather than on available base rate information. A stereotype is the use of concepts to make generalizations about a group of people. In the example just described, the base rate information tells you that 95 times out of 100, the man in your class is likely to be a kick boxer. The best approach to winning the $100 might be simply to shut your eyes and guess kick boxer, no matter what the man looks like.

*Gut instincts can be powerful. Imagine ignoring your hunch that the man was a librarian, and yet it turned out that he **was** a librarian (despite the odds). How hard would you kick yourself then?*

The representativeness heuristic can be particularly damaging in the context of social judgments. Consider a scenario where a particular engineering corporation seeks to hire a new chief executive officer (CEO). Lori, a top-notch candidate with an undergraduate engineering degree and an MBA from an outstanding business school, applies. If there are few women in upper management at the firm, the company's board of directors might inaccurately view Lori as not fitting their view of the prototypical CEO—and miss the chance to hire an exceptional candidate.

Have you noticed that many of the biases and heuristics work pretty well for some things but are less effective when applied to decisions about humans? Considering the various biases that affect human decision making, it might seem like putting computers in charge of important decisions would be a good idea. After all, computers use algorithms that we might assume would be less prone to biases. However, even computers can be affected by bias when applied to people. To read about why this is the case, check out the Challenge Your Thinking.

*Note that the company's directors are using a **prototype model** for the category CEO—and making a mistake.*

Can Artificial Intelligence Be Racist?

Artificial intelligence (or AI) is useful in domains in which human problem solving faces insurmountable obstacles. For example, recently, an AI network developed by DeepMind (an offshoot of Google) produced an accurate three-dimensional model of a protein from its amino acid sequence. This feat had long evaded human scientists and promises to revolutionize biology (Callaway, 2020). If AI can solve complex problems better than humans can, should we rely on computers to make important decisions about people? Such decisions are among the most complex we ever make, including who to hire or if a person is likely to commit a crime. Could AI systems serve as perfectly fair, objective decision-makers in such circumstances? Unfortunately, AI systems can be biased. Let's take a look at why.

A first important consideration is something called the *training dataset*. An AI system does not bring years of experience to its decisions. Instead, such systems are trained on enormous datasets that include all available information about a topic. Remember, computers are especially good at problems that involve very large amounts of information and using algorithms that consider every possible combination of factors to get to the right answer.

Imagine we want to create an AI system that could safely land a plane in any circumstance. We would gather a vast amount of information, perhaps about all the plane landings ever recorded. The data would include relevant details of weather, time of day, terrain, type of plane, and so forth, along with information about success or failure to safely land the plane. That information would become the training dataset for our plane landing AI. It would "learn" from that training data and, if successful, dictate exactly what to do in any circumstance to safely land a plane. Landing a plane is a good example of how AI might be quite useful, right?

However, this process—using a vast amount of existing information to train AI to make decisions—is where bias creeps in. What an AI system

Phonlamai Photo/Shutterstock

"knows" depends on what is in the training dataset. Research has shown that biases in training datasets can lead to biased decisions even by computers (Heaven, 2020).

Especially concerning are predictive AIs used in the criminal justice system. These programs are used to guide decisions, such as whether a person who has been arrested will reoffend if released prior to standing trial. AIs make predictions based on the volumes of data that have been collected about people who have been arrested. In the United States, it is against the law for algorithms to use race as a predictor variable, but other variables related to race are used, such as socioeconomic status, education, and where a person lives (Heaven, 2020). Research has shown that such predictive AIs overestimate the likelihood that Black people will reoffend and underestimate the likelihood that white people will reoffend (Cossins, 2018).

Another important concern is that, generally, creators of AI systems do not allow public access to their training datasets so it is difficult to know with certainty the degree to which these systems are biased (Heaven, 2020). You may have heard the adage, "Garbage in, garbage out." AI systems can only be as fair as the data on which they are trained. The unfortunate conclusion is that if the datasets on which these systems are trained are biased by racism (or sexism or other prejudices) then the AI will make biased decisions.

What Do You Think?

- Is it better to use imperfect AI systems or people to make decisions about people?
- How might training datasets be improved to remove bias?

Heuristics help us make decisions rapidly. To solve problems accurately and make the best decisions, however, we must sometimes override these shortcuts and think more deeply, critically, and creatively.

Thinking Critically and Creatively

Problem solving and decision making are basic cognitive processes that we use multiple times each day. Certain strategies lead to better solutions and choices than others, and some people are especially good at these cognitive exercises. In this section we examine two skills associated with superior problem solving: critical thinking and creativity.

CRITICAL THINKING   Critical thinking means thinking reflectively and productively and evaluating the evidence. Scientists are critical thinkers. Critical thinkers grasp the deeper meaning of ideas, question assumptions, and decide for themselves what to believe or do. Critical thinking requires maintaining a sense of humility about what we know (and what we do not know). It means being motivated to see beyond the obvious.

Critical thinking is vital to effective problem solving. Colleges and universities are increasingly striving to instill critical thinking in students (Calma & Davies, 2020). A problem in elementary and high schools is that high stakes testing may move teachers to concentrate on getting students to give a single correct answer in an imitative way rather than on encouraging new ideas (Foote, 2012). Further, many people are inclined naturally to stay on the surface of problems rather than to stretch their mind. The cultivation of two mental habits is essential to critical thinking: mindfulness and open-mindedness.

Mindfulness means being alert and mentally present for one's everyday activities. The mindful person maintains an active awareness of the circumstances of their life. According to Ellen Langer (2000), mindfulness is a key to critical thinking. Langer distinguishes mindful behavior from mindless behaviors—automatic activities we perform without thought.

In a classic study, Langer found that people (as many as 90 percent) would mindlessly give up their place in line for a copy machine when someone asked, "Can I go first? I need to make copies" as compared to when the same person simply said, "Can I go first?" (just 60 percent) (Langer & others, 1978). For the mindless persons in the study, even a completely meaningless justification—after all, everyone in line was there to make copies—was reason enough to step aside. A mindless person engages in automatic behavior without careful thinking. In contrast, a mindful person is engaged with the environment, responding in a thoughtful way to various experiences.

Open-mindedness means being receptive to other ways of looking at things. People often do not even know that there is another side to an issue or evidence contrary to what they believe. Simple openness to other viewpoints can help to keep individuals from jumping to conclusions. As Socrates once said, knowing what it is you do not know is the first step to true wisdom.

Being mindful and maintaining an open mind may be more difficult than the alternative of going through life on automatic pilot. Critical thinking is valuable, however, because it allows us to make better predictions about the future, to evaluate situations objectively, and to effect appropriate changes. In some sense, critical thinking requires courage. When we expose ourselves to a broad range of perspectives, we risk finding out that our assumptions might be wrong. When we engage our critical mind, we may discover problems, but we are also more likely to have opportunities to make positive changes.

Critical thinking is a valuable skill. Its absence is especially problematic when it leads people to adopt beliefs that are false and potentially harmful to themselves or others. Conspiracy theories are explanations for events that involve secret plots by powerful, often evil groups who control world events (Butter & Knight, 2020; Douglas & others, 2017). Conspiracy theories include that global climate change is a hoax (Douglas & others, 2017), that the COVID-19 virus was developed as a bioweapon (Biddlestone & others, 2020; Sutton & Douglas, 2020), or that vaccines do more harm than good (Tomljenovic & others, 2020). Recently, conspiracy theories have become part of the conversation surrounding the 2020 U.S. presidential election (Beer, 2020). Conspiracy ideas spread rapidly on social media, but conspiracy theories predate these technologies and have been observed across the world, not just the United States (La France, 2020; van Prooijen & van Vugt, 2018). Conspiracy theories tend to be complex and hard to follow, and they defy many of the principles that guide scientists, such as the notion that the simplest explanation for a phenomenon is usually the right one. Moreover, conspiracy theories can be recognized as false with just a bit of research, logical reasoning, and critical thinking. Yet, these ideas continue to appeal to many people. How do scientists understand the power of conspiracy theories? To find out, check out the Intersection.

mindfulness
The state of being alert and mentally present for one's everyday activities.

open-mindedness
The state of being receptive to other ways of looking at things.

Cognition and Motivation: Why Do People Believe in Conspiracy Theories?

Why do people believe that important events are controlled by secret, powerful, and often evil groups? On the cognitive side, conspiracy beliefs are linked to system 1 processing—that aspect of thinking that functions rapidly and is often based on gut feelings (Tomljenovic & others, 2020). System 1 is also responsible for our capacity to recognize patterns and connections in the environment (important capacities for associative learning). Conspiracy theories often involve seeing patterns and making connections, even if these are very unusual. Engaging in rational reflection (that is, using system 2) can help to reduce conspiracy beliefs (Swami & others, 2014) and education predicts lower conspiracy theory belief (van Prooijen, 2017). Still, if you have ever talked to someone who believes in conspiracy theories, you probably already know that presenting facts and logic does not necessarily budge the person's beliefs. So, there must be something more to these beliefs than simply faulty reasoning, such as emotion and motivation— how we feel and what we want.

ibreakstock/Shutterstock

Conspiracy beliefs appear to be more emotionally than cognitively driven. Faced with facts and logic, a person might say that the simplest (factual) explanation just does not "feel right" (Tomljenovic & others, 2020). Emotional aspects of events and experiences may be more important than logic in conspiracy beliefs. Research has shown that feelings of vulnerability (Poon & others, 2020) and powerlessness (Biddlestone & others, 2020) predict higher conspiracy theory belief. Conspiracy beliefs are linked to our motivation to understand the world and to feel in control of events (Douglas & others, 2017).

The fact that conspiracy theories are commonplace suggests that they may be an outgrowth of adaptive processes that were useful long ago in our evolutionary history. According to one approach, accurately detecting potential hostile factions was important for our ancestors to survive (van Prooijen & van Vugt, 2018). The idea is that long ago our ancestors needed to be vigilant for actual malevolent factions that might do them harm. When contemporary people espouse conspiracy beliefs these same ancient mechanisms may be at work.

You might have noticed that most conspiracy believers do not appear happy about the explanations they have found for life events. Indeed, although conspiracy beliefs may be motivated by a desire to understand the world and their place in it—to find a way to "feel right" about the way things are—such beliefs do not actually help people. Conspiracy believers tend to feel *less* personal control and *less* autonomous than nonbelievers (Douglas & others, 2017). Combatting conspiracy beliefs is important because those who do not believe in the scientific facts that underlie vaccines or climate science can put themselves and others at risk through their behavior. Understanding these strange but commonplace beliefs is an important goal for psychology.

\\ **How do you think conspiracy theories can be effectively addressed?**

CREATIVE THINKING   In addition to thinking critically, coming up with the best solution to a problem may involve thinking creatively. The word *creative* can apply to an activity or a person, and creativity as a process may be open even to people who do not think of themselves as creative. When we talk about **creativity** as a characteristic of a person, we are referring to the ability to think about something in novel and unusual ways and to devise unconventional solutions to problems.

We can look at the thinking of creative people in terms of divergent and convergent thinking. **Divergent thinking** produces many solutions to the same problem. **Convergent thinking** produces the single best solution to a problem. Creative thinkers do both types of thinking (Sternberg, 2020). Divergent thinking occurs during *brainstorming,* which occurs when a group of people openly throw out a range of possible solutions to a problem, even some that might seem crazy. Having a lot of possible solutions, however, still

creativity
The ability to think about something in novel and unusual ways and to devise unconventional solutions to problems.

divergent thinking
Thinking that produces many solutions to the same problem.

convergent thinking
Thinking that produces the single best solution to a problem.

Do It!

To get a sense of the roles of divergent and convergent thinking in creativity, try the following exercise. First take 10 minutes and jot down all of the uses that you can think of for a cardboard box. Don't hold back—include every possibility that comes to mind. That list represents divergent thinking. Now look the list over. Which of the possible uses are most unusual or most likely to be worthwhile? That's convergent thinking.

requires that they come up with the solution that is best. That is where convergent thinking comes in. Convergent thinking means taking all of those possibilities and finding the right one for the job. Convergent thinking is best when a problem has only one right answer.

People who think creatively also show the following characteristics (Perkins, 1984).

■ *Flexibility and playful thinking:* Creative thinkers are flexible and play with problems. This trait gives rise to the paradox that although creativity takes hard work, the work goes more smoothly if it is taken lightly. In a way, humor greases the wheels of creativity (Goleman & others, 1993). When you are joking around, you are more likely to consider any possibility and to ignore the inner censor who can condemn your ideas as off base.

■ *Inner motivation:* Creative people often are motivated by the joy of creating. They tend to be less inspired than less creative people by grades, money, or favorable feedback from others. Thus, creative people are motivated more internally than externally (Hennessey, 2011).

■ *Willingness to face risk:* Creative people make more mistakes than their less imaginative counterparts because they come up with more ideas and more possibilities. They win some; they lose some. Creative thinkers know that being wrong is not a failure—it simply means that they have discovered that one possible solution does not work.

■ *Objective evaluation of work:* Most creative thinkers strive to evaluate their work objectively. They may use established criteria to make judgments or rely on the judgments of respected, trusted others. In this manner, they can determine whether further creative thinking will improve their work.

self-quiz

1. An example of a concept is
   A. a basketball.
   B. a daisy.
   C. a vegetable.
   D. an eagle.

2. Deductive reasoning starts at _____ and goes to _____.
   A. the general; the specific
   B. the specific; the general
   C. fixation; function
   D. function; fixation

3. All of the following are characteristic of creative thinkers *except*
   A. inner motivation.
   B. functional fixedness.
   C. objectivity.
   D. risk taking.

APPLY IT! 4. The students in an architecture class are given the assignment to design a new student center for their campus. They have one week to submit their first drafts. Jenny and David, two students in the class, spend the first day very differently. Jenny quickly decides on what her building will look like and starts designing it. David spends the first day doodling and devises 20 different styles he might use, including a Gothic version, a spaceship design, a blueprint with a garden growing on top, and another plan that looks like a giant elephant (the school's mascot).

When Jenny sees David's sketches, she scoffs, "You're wasting your time. We have only a week for the first draft!" On the second day, David selects his best effort and works on that one until he finishes it. Which of the following is true of these strategies?
   A. Jenny is effectively using divergent thinking in her strategy.
   B. David should be using heuristics given the limited time for the project.
   C. Jenny is criticizing David for engaging in deductive reasoning.
   D. David is using divergent thinking first and will get to convergent thinking later, while Jenny has engaged only in convergent thinking.

3 Intelligence

If you were asked, say, on an intelligence test to define the word "intelligence," how would you answer? If you think of the smartest person you know, would that person have the abilities that your definition includes? Many people in the United States think of intelligence in terms of reasoning and thinking skills but not all cultures view intelligence in this way. An intelligent person in Uganda is someone who knows what to do and follows through with appropriate action. Intelligence to the Iatmul people of Papua New Guinea involves the ability to remember the names of 10,000 to 20,000 clans. The residents of the widely dispersed Caroline Islands incorporate the talent of navigating by the stars into their definition of intelligence (Figure 7). Some languages do not even have a word that

FIGURE 7 **Iatmul and Caroline Island Intelligence** The intelligence of the Iatmul people of Papua New Guinea involves the ability to remember the names of many clans. On the 680 Caroline Islands in the Pacific Ocean east of the Philippines, the intelligence of the inhabitants includes the ability to navigate by the stars.
(Iatmul): National Geographic Creative/Alamy Stock Photo; (Caroline): Angelo Giampiccolo/Shutterstock

refers to intelligence as most Westerners think of it. For instance, Mandarin Chinese has words for specific abilities, such as wisdom, but not one single word that means, essentially, being smart (Yang & Sternberg, 1997a, 1997b).

In the United States, we generally define **intelligence** as an all-purpose ability to do well on cognitive tasks, to solve problems, and to learn from experience. The idea that intelligence captures a common general ability that is reflected in performance on various cognitive tests was introduced by Charles Spearman (1904). Spearman noted that schoolchildren who did well in math also did well in reading, and he came up with the idea that intelligence is a general ability, which he called *g*. This view of intelligence suggests that general intelligence underlies performance across a broad range of tasks. Spearman's *g* assumes that the intelligent person is a jack-of-all-cognitive trades. Many intelligence researchers today endorse this view (Rindermann & others, 2020).

Certainly, the way we define intelligence is important (Matias-Garcia & Cubero-Pérez, 2020). Robert Sternberg (2019) has asserted that we should think about intelligence as the capacity to succeed in whatever context a person finds oneself. Such a perspective means that the definition of intelligence would depend centrally on culture and the kinds of activities a culture values. However, note that this definition is circular: Rather than thinking of intelligence as a predictor of success it is equated with success. History is full of people, including women, indigenous people, people of color, people with disabilities, of remarkable talent who, because of societal constraints on their access to opportunities (like education), never enjoyed great success. Not all highly capable individuals are recognized as such during their lifetimes. Many highly capable people may never have earned the label "successful" due to the limitations of the place and time in which they lived.

The science of intelligence has a difficult, even shameful, often racist history (Winston, 2020). Early thinkers believed intelligence to be strongly genetically determined and therefore fixed or unchangeable. They believed that the way to increase intelligence was through, essentially, *selective breeding*, the process by which animals are mated to maximize characteristics such as speed in race horses. **Eugenics** is the belief in the possibility of improving the human species by discouraging reproduction among those with less desirable characteristics and enhancing reproduction among those with more desirable characteristics (such as high intelligence). Today we find such reasoning to be downright offensive, but a number of early intelligence scholars, including many whose work we will review later, embraced eugenic thinking to some degree. In 1921, for instance, Spearman wrote that it would be a good idea to require some minimal level of *g* for a person to vote or even to have children.

Eugenics was endorsed by the Nazis in World War II but it has had regrettable influence outside that context. For instance, in the United States prior to World War II, several

intelligence
All-purpose ability to do well on cognitive tasks, to solve problems, and to learn from experience.

eugenics
The belief in the possibility of improving the human species by discouraging reproduction among those with less desirable characteristics and enhancing reproduction among those with more desirable characteristics, such as high intelligence.

states had laws permitting the sterilization of individuals judged to be mentally deficient. Tragically, IQ tests were a tool to identify individuals for such procedures (Gould, 1981).

This disturbing history reminds us of the stakes involved when we draw conclusions about people based solely on their cognitive ability as measured by an intelligence test. Clearly, a person's abilities to contribute to the world may go far beyond a score on a single test. Moreover, even if we had tests that were perfect measures of cognitive ability (which we do not), the eugenic argument reduces a person's value to their cognitive ability, an argument that is plainly repugnant. Unfortunately, traces of eugenics are still present today in arguments about group differences in intelligence and the potential roles of genetics in these differences (Winston, 2020). We will review these issues later, but for now, with this context set, let's review what psychologists know about intelligence.

Measuring Intelligence

validity
The soundness of the conclusions that a researcher draws from an experiment. In the realm of testing, the extent to which a test measures what it is intended to measure.

Test validity means that a scale measures what it says it measures. In experiments, validity refers to the soundness of the conclusions that a researcher draws from an experiment.

Psychologists measure intelligence using tests that produce a score known as the person's *intelligence quotient (IQ)*. To understand how IQ is derived and what it means, let's first examine the criteria for a good intelligence test: validity, reliability, and standardization.

In the realm of testing, **validity** refers to the extent to which a test measures what it is intended to measure. If a test is supposed to measure intelligence, then it should measure intelligence, not some other characteristic, such as anxiety. One way to demonstrate validity is to show that test scores predict some outcome or *criterion,* thought to be related to the construct the test measures. For example, a psychologist might validate an intelligence test by asking employers of the people who took the test how intelligent they are at work. The employers' perceptions would be a criterion for measuring intelligence. When the scores on a measure relate to important outcomes (such as employers' evaluations), we say the test has high *criterion validity.*

Reliability is the extent to which a test gives a consistent, reproducible measure of performance. That is, a reliable test is one that produces the same score over time and repeated testing. Reliability and validity are related. If a test is valid, then it must be reliable, but a reliable test need not be valid. People can respond consistently on a test, but the test might not be measuring what it purports to measure.

reliability
The extent to which a test yields a consistent, reproducible measure of performance.

standardization
The development of uniform procedures for administering and scoring a test, and the creation of norms (performance standards) for the test.

Validity

Does the test measure what it purports to measure?

Reliability

Is test performance consistent?

Standardization

Are uniform procedures for administering and scoring the test used?

FIGURE 8

Test Construction and Evaluation Tests are a tool for measuring important abilities such as intelligence. Good tests show high reliability and validity and are standardized so that people's scores can be compared.

Good intelligence tests are not only reliable and valid but also standardized (Wright, 2020). **Standardization** involves developing uniform procedures for administering and scoring a test, as well as creating norms, or performance standards, for the test. Increasingly such testing is done on computers. Uniform testing procedures require that the testing environment be as similar as possible for all people. Norms are created by giving the test to a large group of individuals representative of the population for whom the test is intended. Norms tell us which scores are considered high, low, or average. Many tests of intelligence are designed for individuals from diverse groups. So that the tests are applicable to such different groups, many of them have norms for individuals of different ages, socioeconomic statuses, and ethnic groups (Urbina, 2011). Figure 8 summarizes the criteria for test construction and evaluation.

IQ TESTS In 1904, the French Ministry of Education asked psychologist Alfred Binet to devise a method that would determine which students did not learn effectively from regular classroom instruction. School officials wanted to reduce overcrowding by placing such students in special schools. Binet and his student Theophile Simon developed an intelligence test to meet this request. The test consisted of 30 items ranging from the ability to touch one's nose or

Alfred Binet (1857–1911) *Binet constructed the first intelligence test after being asked to create a measure to determine which children would benefit from instruction in France's schools.*
Source: National Library of Medicine/NIH

ear on command to the ability to draw designs from memory and to define abstract concepts. Binet's test is now known as the Stanford-Binet, and a version of the test is still used today.

To come up with a number reflecting intelligence, Binet developed the idea of comparing a person's mental abilities to the mental abilities that are typical for a particular age group. He developed the concept of **mental age (MA),** which is a person's level of mental development relative to that of others. Binet reasoned that a child of very low mental ability would perform like a typical child of a younger age. To think about a person's level of intelligence, then, we might compare the person's MA to their *chronological age (CA),* or age from birth. Within Binet's scheme, high intelligence would be reflected in an MA considerably above CA; low intelligence would be reflected in an MA below CA.

Using these ideas, the German psychologist William Stern devised the term **intelligence quotient (IQ)** in 1912. IQ consists of an individual's mental age divided by chronological age multiplied by 100:

$$IQ = (MA/CA) \times 100$$

This equation tells us that if a person's mental age is the same as chronological age, then the individual's IQ is 100, and 100 is considered average for IQ. If mental age is above chronological age, the IQ is more than 100 (above average). If mental age is below chronological age, the IQ is less than 100 (below average). For example, a six-year-old child with a mental age of eight has an IQ of 133, whereas a six-year-old child with a mental age of five has an IQ of 83.

In childhood, mental age increases as children age, but once they reach about age 16, the concept of mental age loses its meaning. That is why many experts today prefer to examine IQ scores in terms of how unusual a person's score is when compared to the scores of other adults. For this purpose, researchers and testers use standardized norms that they have identified in the many people who have been tested.

Another measure of intelligence is the Wechsler scales, developed by David Wechsler (1939). There are three versions of the scale. For those ages 16 and older, the Wechsler Adult Intelligence Scale (the WAIS) includes items such as vocabulary, working memory capacity, math problems, and the ability to complete jigsaw puzzles. For children between the ages of 6 and 16, the Wechsler Intelligence Scale for Children (the WISC) includes vocabulary and comprehension but also has tasks such as putting together blocks to fit a particular pattern. Finally, a version developed for children as young as 30 months is the Wechsler Pre-School and Primary Scale of Intelligence (the WPPSI, pronounced "whipsy"). On this measure, children are asked, for instance, to point to a picture that depicts a word the examiner says, to complete a block design, and to answer basic knowledge questions. If you have taken an IQ test, chances are it was one of the Wechsler scales. It is the most popular measure of intelligence (Beaujean & Woodhouse, 2020). In addition to summary scores for general IQ, the Wechsler scales include scores for areas such as verbal comprehension, perceptual reasoning, working memory, and processing speed.

Both the Stanford-Binet and the Wechsler scales provide measures of Spearman's *g.* Both of these measures of intelligence have a long history of research demonstrating their reliability, and they both feature standardized administration procedures and norms. Finally, they predict results that we would expect, including academic performance and a variety of life outcomes, including lower risk for psychological disorders (Wraw & others, 2016) and economic and career success (Bergman & others, 2014; Spengler & others, 2015).

Over the years, IQ tests like the Stanford-Binet and Wechsler scales have been given to thousands of children and adults of different ages selected at random from different parts of the United States. When the scores for many people are examined, these approximate a normal distribution (Figure 9). A **normal distribution** is a symmetrical, bell-shaped curve, with a majority of the scores falling in the middle of the possible range and few scores appearing toward the extremes of the range.

BIAS IN TESTING   Many early intelligence tests were culturally biased, favoring people who were from urban rather than rural environments, of middle rather than low

mental age (MA)
An individual's level of mental development relative to that of others.

intelligence quotient (IQ)
An individual's mental age divided by chronological age multiplied by 100.

normal distribution
A symmetrical, bell-shaped curve, with a majority of test scores (or other data) falling in the middle of the possible range and few scores (or other data points) appearing toward the extremes.

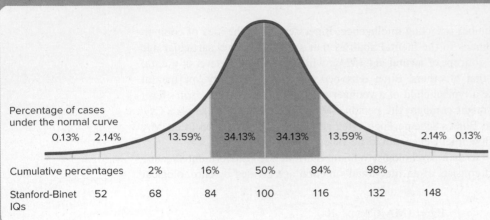

Percentage of cases
under the normal curve

| 0.13% | 2.14% | 13.59% | 34.13% | 34.13% | 13.59% | 2.14% | 0.13% |

Cumulative percentages		2%	16%	50%	84%	98%	
Stanford-Binet IQs	52	68	84	100	116	132	148

FIGURE 9 The Normal Curve and Stanford-Binet IQ Scores The distribution of IQ scores approximates a normal curve. Remember that the area under the curve represents the number of people who obtain a given score. >**Does most of the population fall in the low, middle, or high range of scores? How do you know?** >**If someone scored 132 on the test, how many people scored below that person's score?** >**Intelligence follows a normal distribution, meaning it produces this bell curve. What other human characteristics might have the same distribution?**

socioeconomic status, non-Latino white rather than Black, and men rather than women (Provenzo, 2002; Pezzuti & others, 2020). For example, a question on an early test asked what one should do if one finds a three-year-old child in the street. The correct answer was "call the police." However, children from inner-city families who perceive the police as scary are unlikely to choose this answer. Similarly, children from rural areas might not choose this answer if there is no police force nearby. Such questions clearly do not measure the knowledge necessary to adapt to one's environment or to be "intelligent" in an inner-city or a rural neighborhood (Scarr, 1984). In addition, people of diverse backgrounds may not speak English or may speak nonstandard English. Consequently, they may be at a disadvantage in trying to understand verbal questions that are framed in standard English, even if the content of the test is appropriate (Cathers-Shiffman & Thompson, 2007).

Researchers have sought to develop tests that accurately reflect a person's intelligence, regardless of cultural background. **Culture-fair tests** are intelligence tests that are intended to be culturally unbiased. One type of culture-fair test includes questions that are familiar to people from all socioeconomic and ethnic backgrounds. A second type contains no verbal questions. Figure 10 shows a question similar to ones found in the Raven Progressive Matrices Test. Even though tests such as the Raven Progressive Matrices are designed to be culture-fair, people with more education still score higher than do those with less education.

culture-fair tests
Intelligence tests that are intended to be culturally unbiased.

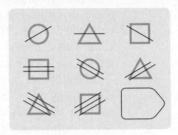

FIGURE 10 Sample Item from the Raven Progressive Matrices Test For this item, the respondent must choose which of the numbered figures would come next in the order. Why is number 6 the right answer?
Source: Raven's Progressive Matrices (Advanced Progressive Matrices), 1998.

Why is it so hard to create culture-fair tests? Just as the definition of intelligence may vary by culture, most tests of intelligence reflect what is important to the dominant culture. If tests have time limits, the test will be biased against groups not concerned with time. If languages differ, the same words might have different meanings for different language groups. Even pictures can produce bias, because some cultures have less experience with drawings and photographs (Anastasi & Urbina, 1996). Because of such difficulties, it may be best to think of tests as biased-reduced rather than bias-free. Any test that purports to measure cognitive ability should measure that ability and not factors such as experience with particular objects (such as furnace) or activities (such as riding on a bus).

Do It!

Try out a culture-fair IQ test for yourself. Go to http://psychologytoday. tests.psychtests.com/take_test. php?idRegTest=3202. Once you have checked it out, do a web search for intelligence tests and see if you get the same results when you take different tests. Do the websites provide information about the reliability and validity of the tests?

Genetic and Environmental Influences on Intelligence

There is no doubt that genes play a role in intelligence (Plomin, 2018).

Some researchers use a statistic called heritability to describe the extent to which the observable differences among people in a group can be explained by the genetic differences of the group's members. **Heritability** is the proportion of observable differences in a group that can be explained by differences in the genes of the group's members. For intelligence, this means that heritability tells us how much of the differences we observe in intelligence is attributable to differences in genes. Because heritability is a proportion, the highest degree of heritability is 100 percent. Twin studies have found the heritability of intelligence to be about 50 percent (Plomin, 2018), although some studies report lower and others higher estimates (Pesta & others, 2020; Trzaskowski & others, 2014).

PeopleImages/E+/Getty Images

heritability
The proportion of observable differences in a group that can be explained by differences in the genes of the group's members.

At this point, you might find yourself reflecting on some of the less than brilliant things your parents have done over the years, and you may be feeling discouraged. If IQ is heritable, is there any hope? There definitely is. Heritability, remember, is a statistic that provides information about a group, not a single person. This means that finding out that heritability for intelligence is 50 percent tells us nothing at all about the source of an individual person's intelligence. We cannot dissect your intelligence and determine that you got 50 percent of it from genes and 50 percent from environmental experiences. Heritability cannot be applied to a single case.

Also, heritability estimates can change over time and across different groups. If a group of people lives in the same advantageous setting (with good nutrition, supportive parents, excellent schools, stable neighborhoods, and plenty of opportunities), heritability estimates for intelligence might be quite high, as this optimal environment allows genetic characteristics to flourish to their highest potential. However, if a group of people lives in a highly variable environment (with some individuals experiencing rich, nurturing environments full of opportunity and others experiencing less supportive contexts), genetic characteristics may be less predictive of differences in intelligence in that group, relative to environmental factors.

Intelligence is a *polygenic trait,* meaning that a large number of genetic characteristics are involved in intelligence. It has been very difficult for scientists using molecular genetics to identify the precise genes that are involved in intelligence. Molecular genetic studies have identified some genes that work together to produce intelligence, but even these fail to explain all of the 50 percent that would account for heritability estimates (Plomin & von Stumm, 2018). One research review concluded that there may be more than 1,000 genes that affect intelligence, each possibly having a small influence on an individual's intelligence (Davies & others, 2011).

Even if the heritability of a characteristic is high, the environment still matters. Take height, for example. Heritability estimates suggest that more than 90 percent of the variation in height is explained by genetic variation. Generally speaking, humans continue to get taller and taller, however, and this trend demonstrates that environmental factors such as nutrition have an impact. Similarly, in the case of intelligence, environmental interventions can change IQ scores considerably (Protzko, 2017).

Indeed, research provides strong support for the conclusion that childhood experiences can profoundly affect IQ. Recent meta-analyses have summarized research from many studies examining the effects of various childhood interventions and experiences on IQ

(Protzko, 2017; Protzko & others, 2013). The results support the idea that specific experiences in childhood can influence IQ:

- **Dietary supplements** One type of dietary supplement that has been found to positively influence childhood IQ is long-chain polyunsaturated fatty acids, commonly referred to as omega-3 fatty acids. These acids are found in breast milk, fish oil, salmon, walnuts, spinach, and avocados. Researchers found that when pregnant women, nursing mothers, and infants received 1,000-milligram supplements of omega-3 fatty acids, the supplements led to a more than 3.5-point increase in IQ points.

- **Educational interventions** Research shows that early childhood education can improve the IQ of economically disadvantaged young children: Early educational interventions, especially those that involved training on complex tasks, led to an increase in IQ of more than 4 points.

- **Interactive reading** Interactive reading means that parents ask open-ended questions, encourage the child to read, and engage with the child actively about what they are reading together. Interactive reading raised a child's IQ by over 6 points, and this was especially true if the interventions occurred at younger ages.

- **Preschool** Sending a child to preschool increased IQ by more than 4 points (Protzko & others, 2013). Socioeconomic status played an important role in these results. For economically disadvantaged children, those whose parents might not have been able to afford preschool except for the studies in which they were enrolled, attending preschool raised IQs by as much as 7 points. Preschool curricula including a language-development component were especially effective. The researchers note that the effects of preschool may not be maintained if children are not continually exposed to complex cognitive challenges in their environments as they move to grade school.

One effect of the environment on intelligence is evident in increasing IQ test scores around the world, a phenomenon called the *Flynn effect,* named for James Flynn who died in 2020 (Flynn, 1999, 2006, 2011, 2013). Scores on these tests have been rising so fast that a high percentage of people regarded as having average intelligence at the turn of the twentieth century would be regarded as having below average intelligence today (Figure 11). Flynn (2013)

PSYCHOLOGICAL INQUIRY

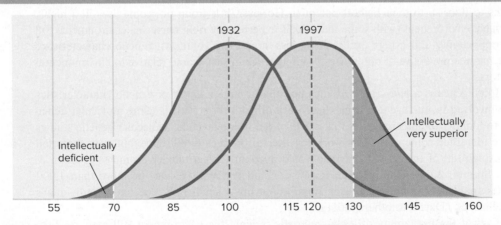

FIGURE 11 **The Increase in IQ Scores from 1932 to 1997** As measured by the Stanford-Binet intelligence test, American children seem to be getting smarter. Scores of a group tested in 1932 fell along a bell-shaped curve, with half below 100 and half above. Note that if children's scores from 1997 were plotted on this same scale, the average would be 120.

SOURCE: Winerman, L. (2013, March). Smarter than ever? *Monitor, 44*(1), 30.

> How would average children from 1932 compare to children from the more recent sample? Would they still be "average"? > Note the far ends (or tails) of the 1932 curve. How would these children be considered by the standards of the more recent scores? > It was reported that the average IQ rose by nearly 30 points from 1932 to 2012. Can you draw the curve for 2012 on this graph?

suggested that the world has undergone a change in valuing (and requiring) certain "habits of the mind" involving abstract and complex thought. More and more, the modern world requires complex ways of thinking, and Flynn believed that these changes help to explain the rapid increase in IQ. According to Flynn, when we compare one generation to the next, we are essentially comparing similar groups of people who differ primarily in terms of the richness and complexity of the cognitive demands of their environments (Flynn, 2009).

Genes, Environment, and Group Differences in IQ

One of the most controversial questions in psychology is explaining group differences in intelligence, focusing on comparisons of groups who differ in ethnicity or race (essentially skin color). Debates about the role of genetic differences in explaining IQ score differences between people of different races erupted almost as soon as IQ testing became widespread (Hunt & Carlson, 2007). The debate has continued for years, punctuated by notable (and sometimes notorious) publications (Herrnstein & Murray, 1994).

The center of this controversy is that, in large samples, there are mean level (or average) differences between different groups in IQ scores, with Asians (and Asian Americans) scoring higher than Europeans or European Americans, who score higher than Africans or African Americans. The question that has sparked debate is whether these differences are due to genetic differences between the groups. Those who advocate for the strong genetic argument (that is, that differences between racial groups are explained by genetic differences) assert that because IQ is largely genetic, it makes no sense to intervene to improve IQ in low-performing groups (Herrnstein & Murray, 1994).

Without trudging through all of these varied issues, we can use a brief consideration of race as well as what we have reviewed about intelligence to reject these ideas. Foremost, race is simply not a good indicator of genetics (Helms & others, 2005). Historically, race has been used as a means of classifying people based primarily on the color of their skin and other physical features (for instance, the shape of their eyes). Although it might have once seemed like there were, in fact, categorical differences between people based on these distinctions, we know that these features do not represent different kinds of people. Rather, race, now, is recognized as a *social category*, one that people may apply to themselves (for instance, identifying oneself as "white" or "Black" on a questionnaire) or others. There is no one biological characteristic that uniquely defines a race (Hunt & Carlson, 2007). Certainly, skin color is a very poor indicator of a person's genetic heritage (Nisbett, 2005). Considering how race continues to influence the lives of Americans today and all the ways that skin color can influence a person's experience, access to healthcare, nutrition, education, and so forth, it seems odd, indeed, to focus on potential genetic differences among individuals whose skin color differs. You might note the issue of cultural bias in testing that we have already considered.

Still, let's imagine for a moment that race is an excellent measure of genes, and IQ tests are unbiased measures of cognitive ability. Neither of these assertions is true, of course, but we can give advocates of the genetic argument both of these points and still their conclusions are unjustified. Can you see why? As you (now) know, even very high heritability does not limit what the environment might do for individuals. Even if genes play a strong role in intelligence, they do not set intelligence in stone. Genes can be strongly related to many outcomes (such as height) and still environment matters.

Certainly, the research reviewed earlier about the profound impact of childhood experience on IQ shows that environment matters very much for intelligence. Indeed, research over the years has shown that

One reason a genetic explanation for group differences in IQ has been controversial is that it is often closely associated with eugenics. Contemporary researchers who argue for a strong genetic explanation for the differences among groups in intelligence do not advocate for eugenics, but their reasoning, at times, comes perilously close.

Do It!

Many different intelligence tests are available online, such as at www.iqtest.com/. Give this one a try and then do a web search for intelligence tests and see if you get the same results when you take a different test. Do the websites tell you how reliable the tests are? Do they provide information on standardization or validity? If your scores on the two tests are very different, what might account for this variation?

IQ differences between groups continue to shrink (Flynn, 2009; Nisbett, 2005, 2009; Rindermann & Thompson, 2013). They may not be shrinking as fast as we would like, but they are shrinking.

Weighing in on this controversy, James Flynn (2013) noted that the Flynn effect shows how each new generation benefits, cognitively, from a changing world. He asserted that "heritability can be as high as you please without robbing the environment of its potency to create huge IQ gains over time" (Flynn, 2013 p. 364). Thus over time, each new generation moves toward its maximum potential, disproving the notion that categories such as "race" limit our abilities.

In fact, a key difference between Black and white Americans is wealth (Kraus & others, 2019). Such socioeconomic differences help to explain differences in IQ scores (Dolean & Călugăr, 2020). A recent meta-analysis found that accounting for socioeconomic differences between groups greatly reduced the IQ gap between Black and white Americans and completely erased the difference between Latinx and white Americans (Weiss & Saklofske, 2020). Clearly, differences in performance on IQ tests do not indicate genetic differences but rather differences in experiences, such as large differences in educational opportunities.

Intelligence is especially enhanced by experiences with complexity. One of the things that makes our social world complex is its diversity. We have the normal curve to thank for the fact that we will always encounter people who differ from us in various ways. That complexity, in turn, requires and allows us to sharpen our minds. Perhaps we can all appreciate, then, that the increasing diversity of our social world might just be making us all a little smarter.

Extremes of Intelligence

Intelligence, then, appears to emerge from a combination of genetic heritage and environmental factors. As we have seen, scores on IQ tests generally conform to the bell-shaped normal curve. We now examine the implications of falling on either tail of that curve.

GIFTEDNESS   There are people whose abilities and accomplishments outshine those of others—the *A+* student, the star athlete, the natural musician. People who are **gifted** have high intelligence (an IQ of 130 or higher) and/or superior talent in a particular area. People who are highly gifted are typically not gifted in many domains, and research on giftedness is increasingly focused on domain-specific developmental trajectories (Kell & Lubinski, 2014; Sternberg & Bridges, 2014; Thagard, 2014).

gifted
Possessing high intelligence (an IQ of 130 or higher) and/or superior talent in a particular area.

Lewis Terman (1925) conducted a study of 1,500 children whose Stanford-Binet IQs averaged 150, a score that placed them in the top 1 percent. A popular myth is that gifted children are maladjusted, but Terman found that his participants (the "Termites") were not only academically gifted but also socially well adjusted. Many of them later became successful doctors, lawyers, professors, and scientists. Do gifted children grow into gifted and highly creative adults? In Terman's research, gifted children typically did become experts in a well-established domain, such as medicine, law, or business, but the Termites did not become major creators or innovators (Winner, 2000, 2006).

In light of the sweeping social and economic changes of the digital age, are today's gifted children perhaps better able than the Termites to use their gifts in innovative and important ways in adulthood? The results from a longitudinal study of profoundly gifted children begun by Julian Stanley at Johns Hopkins University in 1971 seem to indicate just that. The Study of Mathematically Precocious Youth includes 320 participants whom researchers recruited before age 13 based

Olympic speed-skating gold medalist Joey Cheek well illustrates giftedness. Beyond his accomplishments on the ice, he studied economics and Chinese at Princeton University and is cofounder of Team Darfur, an organization dedicated to raising awareness of human rights abuses in Sudan.
Kalle Parkkinen/REX/Shutterstock

on IQ scores, with the group's average IQ estimated at 180. This group is said to represent the top 1 in 10,000 IQ scores (Lubinski & others, 2001). Following up on these individuals in their 20s, David Lubinski and his colleagues (2006) found that these strikingly gifted young people were doing remarkable things. At age 23, they were pursuing doctoral degrees at a rate 50 times higher than the average. Some reported achievements such as receiving creative writing awards, creating original art and music, publishing in scholarly journals, and developing commercially viable software and video games. Thus, unlike the Termites, this group has been extraordinarily creative and innovative (Wai & others, 2005).

Like intelligence itself, giftedness is likely a product of both heredity and environment. Gifted individuals often recall showing signs of high ability in a particular area at a very young age, prior to or at the beginning of formal training (Howe & others, 1995). This result suggests the importance of innate ability in giftedness. However, researchers also have found that the individuals who enjoy world-class status in the arts, mathematics, science, and sports all report strong family support and years of training and practice (Bloom, 1985). Deliberate practice is an important characteristic of individuals who become experts in a particular domain (Grigorenko & others, 2009).

Meeting the needs of gifted students in the classroom can be challenging. When students are not sufficiently challenged in their regular classes alternatives might include attending advanced classes in their domain of exceptional ability, as did Bill Gates, who took college math classes at 13, and famed cellist Yo-Yo Ma, who graduated from high school at 15 and then attended the Juilliard School of Music.

One important factor in gifted education is identifying children as gifted. Although many standardized tests are used to screen all children for IQ, parents or teachers may also nominate students as likely candidates for gifted instruction. Once these children are tested and identified as gifted, they are invited to enroll into such programs. This means that, for some students, qualifying for gifted instruction involves someone else's *social perception*. Being recognized as possibly gifted depends on an adult noticing that a child might be highly intelligent. Social perception can be influenced by stereotypes. Stereotypes are generalizations about individuals based on their group membership. Stereotypes do not take into account all the variability that group members likely possess. The notion that stereotypes about gifted children exist and that these stereotypes affect who is included in gifted programs is suggested by the fact that certain groups—such as Black Americans and Latinx Americans, those for whom English is a second language, and those with physical and learning disabilities—are underrepresented in U.S. gifted programs (Carman, 2011). Importantly, contrary to stereotypes that the gifted are poorly adjusted (Preckel & others, 2015), gifted children are not terribly different from their average-ability classmates in terms of personality and social characteristics (Wilson, 2015). Gifted students are likely to be more emotionally stable and less worrying than their average-ability counterparts (Ziedner & Shani-Zinovoch, 2011).

A serious U.S. public policy question is how to allocate resources: Should we be enhancing the experience of gifted students or bringing everyone up to a standard level? What's your opinion? How much money would you be willing to pay in taxes to accomplish both goals?

INTELLECTUAL DISABILITY

Just as some individuals are at the high extreme of intelligence, others are at the lower end. **Intellectual disability** is a condition of limited mental ability in which an individual has low cognitive abilities and has difficulty adapting to everyday life. In the United States, about 5 million people fit this definition of intellectual disability. Although in the past specific IQ score cutoffs might be given for the diagnosis of a person as intellectually disabled, current standards are more flexible. A person with an IQ of around 70 who has difficulty managing the challenges of life would likely be described as intellectually disabled (McNicholas & others, 2017).

It is a relatively new idea to think about intellectual disability as involving functional impairment—that is, having

intellectual disability
A condition of limited mental ability in which an individual has a low IQ, usually below 70 on a traditional intelligence test, and has difficulty adapting to everyday life.

Individuals with Down syndrome may excel in sensitivity toward others. The possibility that other strengths or intelligences coexist with cognitive ability (or disability) has led some psychologists to propose the need for expanding the concept of intelligence.
kali9/Getty Images

difficulty managing the responsibilities of life–rather than a low IQ test score. This change represents progress in understanding the role of adaptive behaviors in determining who requires additional supports and who does not. Note that a low IQ score in the context of a person who, say, holds down a job and has warm relationships with other people, would not indicate adaptive problems. It is not unusual to find clear *functional* differences between two people who have the same low IQ. For example, looking at two individuals with a similarly low IQ, we might find that one of them is married, employed, and involved in the community while the other requires constant supervision in an institution. Such differences in social competence have led psychologists to include deficits in adaptive behavior in their definition of intellectual disability (McNicholas & others, 2017).

Intellectual disability may have an organic cause, or it may be cultural and social in origin (McNicholas & others, 2017). *Organic intellectual disability* is caused by a genetic disorder or brain damage; *organic* refers to the tissues or organs of the body, so there is some physical damage in organic retardation. Down syndrome, one form of organic intellectual disability, occurs when an extra chromosome is present in the individual's genetic makeup. Most people who suffer from organic intellectual disability have an IQ between 0 and 50.

Cultural-familial intellectual disability is a cognitive disability with no evidence of organic brain damage. Individuals with this type of disability have an IQ between 55 and 70. Such deficits may result, at least in part, from growing up in an impoverished intellectual environment (Giesbers & others, 2020). As adults these individuals may not be terribly different from everyone else in many ways. It may be that the intelligence of such individuals increases as they move toward adulthood.

In the United States, Rosa's Law, passed in 2010, removed the offensive terms 'mentally retarded' and 'mental retardation' from all federal communications.

There are several classifications of intellectual disability (Hodapp & others, 2011). In one classification system, disability ranges from mild, to moderate, to severe or profound, according to the person's IQ (Heward, 2013). The large majority of individuals diagnosed with intellectual disability fall in the mild category. Many U.S. states and school systems still use this type of system, but categories based on IQ ranges are not perfect predictors of functioning. This is one reason why psychologists have been interested in incorporating social competence, not just IQ scores, in their definition of intellectual disability.

The American Association on Intellectual and Developmental Disabilities (2010) has developed an assessment that examines a person's level of adaptive behavior in three life domains:

- *Conceptual skills:* For example, literacy and understanding of numbers, money, and time.

- *Social skills:* For example, interpersonal skills, responsibility, self-esteem, and ability to follow rules and obey.

- *Practical skills:* For example, activities of daily living such as personal care, occupational skills, healthcare, travel/transportation, and use of the telephone.

Assessment of capacities in these areas can be used to determine the amount of care the person requires for daily living, not as a function of IQ but of the person's ability to negotiate life's challenges.

Individuals with Down syndrome may never accomplish the academic feats of those who are gifted. However, they may be capable of building close, warm relations with others; inspiring loved ones; and bringing smiles into an otherwise gloomy day (Van Riper, 2007). People with Down syndrome moreover might possess different kinds of intelligence, even if they are low on general cognitive ability. The possibility that other intelligences exist alongside cognitive ability (or disability) has inspired some psychologists to suggest that we need more than one concept of intelligence.

Theories of Multiple Intelligences

Is it more appropriate to think of an individual's intelligence as a general ability or, rather, as a number of specific abilities? Traditionally, most psychologists have viewed intelligence as a general, all-purpose problem-solving ability that, as we have seen, is sometimes referred to as *g*. Others have proposed that we think about different kinds of intelligence, such as *emotional intelligence*, the ability to perceive emotions in ourselves and others accurately. Robert Sternberg and Howard Gardner have developed theories suggesting there are *multiple intelligences*.

STERNBERG'S TRIARCHIC THEORY AND GARDNER'S MULTIPLE INTELLIGENCES
Robert Sternberg (1988) proposed the **triarchic theory of intelligence,** which says that intelligence comes in multiple (specifically, three) forms:

triarchic theory of intelligence
Sternberg's theory that intelligence comes in three forms: analytical, creative, and practical.

Analytical intelligence: The ability to analyze, judge, evaluate, compare, and contrast.

Creative intelligence: The ability to create, design, invent, originate, and imagine.

Practical intelligence: The ability to use, apply, implement, and put ideas into practice.

Howard Gardner (1983, 1993, 2002) suggests there are nine types of intelligence, or "frames of mind." These are described here, with examples of the types of vocations in which they are reflected as strengths (Campbell & others, 2004):

LightField Studios/Shutterstock

Verbal: The ability to think in words and use language to express meaning. Occupations: author, journalist, speaker.

Mathematical: The ability to carry out mathematical operations. Occupations: scientist, engineer, accountant.

Spatial: The ability to think three-dimensionally. Occupations: architect, artist, sailor.

Bodily-kinesthetic: The ability to manipulate objects and to be physically adept. Occupations: surgeon, craftsperson, dancer, athlete.

Musical: The ability to be sensitive to pitch, melody, rhythm, and tone. Occupations: composer, musician.

Interpersonal: The ability to understand and interact effectively with others. Occupations: teacher, mental health professional.

Intrapersonal: The ability to understand oneself. Occupations: theologian, psychologist.

Naturalist: The ability to observe patterns in nature and understand natural and human-made systems. Occupations: farmer, botanist, ecologist, landscaper.

Existentialist: The ability to grapple with the big questions of human existence, such as the meaning of life and death, with special sensitivity to issues of spirituality. Gardner has not identified an occupation for existential intelligence, but one career path would likely be philosopher.

According to Gardner, everyone has all of these intelligences to varying degrees. As a result, we prefer to learn and process information in different ways. We learn best when we can do so in a way that uses our stronger intelligences.

Do you know people some might call 'book smart' and others who are 'people smart'? What kinds of intelligence do they show?

EVALUATING THE APPROACHES OF MULTIPLE INTELLIGENCES
Sternberg's and Gardner's approaches have stimulated teachers to think broadly about what makes up children's competencies. They have motivated educators to develop programs that instruct students in multiple domains. These theories have also contributed to

interest in assessing intelligence and classroom learning in innovative ways, such as by evaluating student portfolios (Woolfolk, 2015).

Doubts about multiple intelligences persist, however. Some critics argue that a research base to support the three intelligences of Sternberg or the nine intelligences of Gardner has not yet emerged. People who excel at one type of intellectual task are likely to excel in others (Brody, 2007), which sounds very much like Spearman's *g*. People who do well at memorizing lists of digits are also likely to be good at solving verbal problems and spatial layout problems. Other critics ask, if musical skill, for example, reflects a distinct type of intelligence, why not also label the skills of outstanding chess players, prizefighters, painters, and poets as types of intelligence? In sum, controversy still characterizes whether it is more accurate to conceptualize intelligence as a general ability, specific abilities, or both (Gardner, 2014; Irwing & others, 2012; Trzaskowski & others, 2014).

Our examination of cognitive abilities has highlighted how individuals differ in the quality of their thinking and how thoughts may differ from one another. Some thoughts reflect critical thinking, creativity, or intelligence. Other thoughts are perhaps less inspired. One thing thoughts have in common is that they often involve language. Even when we talk to ourselves, we do so with words. The central role of language in cognitive activity is the topic to which we now turn.

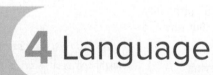
4 Language

language
A form of communication—whether spoken, written, or signed—that is based on a system of symbols.

Language is a form of communication, whether spoken, written, or signed, that is based on a system of symbols. We need language to communicate with others, listen to others, read, and write. Language is not just how we speak to others but how we talk to ourselves. Consider an occasion, for example, when you have experienced the feeling of a guilty conscience, of having done something you should not have. The little voice in your head that clamors, "You shouldn't have done that! Why did you do it?" speaks to you in your native tongue. In this section we first examine the fundamental characteristics of language and then trace the links between language and cognition.

PSYCHOLOGY IN OUR WORLD

What Do the Words We Use on Social Media Reveal about Us?

blvdone/Shutterstock.com

Every election season, it seems, social media become overrun with political debates and conversation. It is clear that, whether it is posted on Instagram or Twitter, people differ in the ways they communicate. Could those differences tell us anything about the people posting? Do the words people use when they post on social media relate to important characteristics? A recent large-scale study examined the words used by over 25,000 Twitter users to see if the language in their tweets related in systematic ways to the political views of the people tweeting (Sterling & others, 2020). The results showed that liberals and conservatives differed in the language they used in their tweets. Compared to liberals, conservatives used more negative emotion words and words related to power (such as "power," "manage," and "control") and tradition. Compared to conservatives, liberals were more likely to use words expressing benevolence (such as "improve," "care," and "nurture"). In addition, those in the middle of the political spectrum differed from those on the extremes. Those who were extreme in their views (regardless of whether they were liberal or conservative) expressed more certainty in their views.

This study is notable because it occurred "in the wild." The natural expression of language simply happened as people expressed themselves online. The results, then, are unlikely to be due to people managing the impressions they were making or reacting to an artificial laboratory situation. The current political climate is often described as conflictful, fractious, and polarized. These findings tell us that when they are talking anonymously online, partisans may be speaking different languages.

The Basic Properties of Language

All human languages have **infinite generativity**, the ability to produce an endless number of meaningful sentences. This superb flexibility comes from five basic rule systems:

- **Phonology:** a language's sound system. Language is made up of basic sounds, or *phonemes*. Phonological rules ensure that certain sound sequences occur (for example, in English, *sp, ba,* and *ar*) and others do not (for example, *zx* and *qp*) (Volenec & Reiss, 2020). A good example of a phoneme in the English language is /k/, the sound represented by the letter *k* in the word *ski* and by the letter *c* in the word *cat*. Although the /k/ sound is slightly different in these two words, the /k/ sound is described as a single phoneme in English.

- **Morphology:** a language's rules for word formation. Every word in the English language is made up of one or more morphemes. A *morpheme* is the smallest unit of language

infinite generativity
The ability of language to produce an endless number of meaningful sentences.

phonology
A language's sound system.

morphology
A language's rules for word formation.

that carries meaning. Some words consist of a single morpheme—for example, *help*. Others are made up of more than one; for example, *helper* has two morphemes, *help* + *er*. The morpheme *-er* means "one who"—in this case, "one who helps." As you can see, not all morphemes are words; for example, *pre-, -tion,* and *-ing* are morphemes. Just as the rules that govern phonemes ensure that certain sound sequences occur, the rules that govern morphemes ensure that certain strings of sounds occur in particular sequences (Gardani, 2020).

■ **Syntax:** a language's rules for combining words to form acceptable phrases and sentences (Freidin, 2020). If someone says, "John kissed Emily" or "Emily was kissed by John," you know who did the kissing and who was kissed in each case because you share that person's understanding of sentence structure. You also understand that the sentence "You didn't stay, did you?" is a grammatical sentence but that "You didn't stay, didn't you?" is unacceptable.

■ **Semantics:** the meaning of words and sentences in a particular language. Every word has a unique set of semantic features (Boleda, 2020). *Child* and *adult,* for example, share many semantic features (for instance, both signify human beings), but they differ semantically in regard to age. Words have semantic restrictions on how they can be used in sentences. The sentence, "The bicycle talked the child into buying a candy bar," is syntactically correct but semantically incorrect. The sentence violates our semantic knowledge that bicycles do not talk.

■ **Pragmatics:** the useful character of language and the ability of language to communicate even more meaning than is said (Rano & others, 2020). The pragmatic aspect of language allows us to use words to get the things we want. If you ever find yourself in a country in which you know only a little of the language, you will certainly take advantage of pragmatics. Wandering the streets of, say, Madrid, you might approach a stranger and ask, simply, "Autobus?" (the Spanish word for *bus*). You know that given your inflection and perhaps your desperate facial expression, the person will understand that you are looking for the bus stop.

With this basic understanding of language in place, we can examine the connections between language and cognition.

Language and Cognition

Language is a vast system of symbols capable of expressing most thoughts. Language is the vehicle for communicating most of our thoughts to one another. Although we do not always think in words, our thinking would be greatly impoverished without words. Moreover, the development of thinking and talking is strongly intertwined in children, adolescents, and young adults (Ebert, 2020; Ünal & Papafragou, 2020).

The connection between language and thought has been of considerable interest to psychologists. Some have even argued that we cannot think without language. This proposition has produced heated controversy. Is thought dependent on language, or is language dependent on thought?

THE ROLE OF LANGUAGE IN COGNITION   Recall that memory is stored in multiple ways–including sounds, images, and words. Language helps us think, make inferences, tackle difficult decisions, and solve problems. Language is a tool for representing ideas.

Today, most psychologists would accept these points. However, linguist Benjamin Whorf (1956) went a step further. He argued that language determines the way we think, a view that has been called the *linguistic relativity hypothesis*. Whorf and his mentor Edward Sapir were specialists in Native American languages, and they were fascinated by the possibility that people might perceive the world differently as the result of the different languages they speak. The Inuit in Alaska, for instance, have a dozen or more words to describe the various textures, colors, and physical states of snow. In contrast, English

syntax
A language's rules for combining words to form acceptable phrases and sentences.

semantics
The meaning of words and sentences in a particular language.

pragmatics
The useful character of language and the ability of language to communicate even more meaning than is verbalized.

has relatively few words to describe snow, and thus, according to Whorf's view, English speakers cannot see the different kinds of snow *because* they have no words for them.

Think of all the words we have for coffee drinks. What might this say about our society?

These bold claims appealed to many scholars. Some even tried to apply Whorf's view to gender differences in color perception. Asked to describe the colors of two sweaters, a woman might say, "One is mauve and the other is magenta," while a man might say, "They're both pink." Whorf's view of the influence of language on perceptual ability might suggest that women are able to see more colors than men simply because they have a richer color vocabulary (Hepting & Solle, 1973). It turns out, however, that men can learn to tell the differences among the various hues that women identify, and this outcome suggests that Whorf's view is not accurate (Stanlaw 2020).

Indeed, critics of Whorf's ideas say that words merely reflect, rather than cause, the way we think. The Inuits' adaptability and livelihood in Alaska depend on their capacity to recognize various conditions of snow and ice. A skier or snowboarder who is not Inuit might also know numerous words for snow, far more than the average person, and a person who does not know the words for the different types of snow might still be able to perceive these differences. Recent research shows that people with wider color vocabularies show better memory for color but not better perception of color (Hasantash & Afraz, 2020). These findings indicate that having words to go along with images enhances memories but not that such words alter perception.

Although the strongest form of Whorf's hypothesis—that language determines perception—seems doubtful, research has continued to demonstrate the influence of language on how we think, even about something as fundamental as our own personalities (Chen & others, 2014). For example, in a series of studies, researchers interviewed bilingual individuals (that is, people who fluently speak two languages, in this case Spanish and English) (Ramirez-Esparza & others, 2006). Each person rated their own personality characteristics, once in Spanish and once in English. Across all studies, and regardless of whether the individuals lived in a Spanish-speaking or an English-speaking country, respondents reported themselves as more outgoing, nicer, and more responsible when responding to the survey in English. Interestingly, a study of bilingual individuals fluent in both German and Spanish showed that people rated themselves

Whorf's view is that cultural experiences with a particular concept shape a catalog of names that can be either rich and complex or short and simple. Consider how rich your mental library of names for a camel might be if you had extensive experience with camels in a desert world, and how poor your mental library of names for snow might be if you lived in a tropical world of palm trees and parrots. Despite its intriguing appeal, psychologists do not believe Whorf's view of the pivotal role of language in shaping thought.
(Berber): eAlisa/Shutterstock; (Inuit): bikeriderlondon/Shutterstock

as more outgoing and more emotional when rating in Spanish (compared to German) (Veltkamp & others, 2013).

A final area of research demonstrating the link between language and thinking concerns the possible positive association between being fluent in more than one language and cognitive abilities, sometimes called the "bilingual advantage." Are people who are bilingual or multilingual better off cognitively than others? The research on this question is mixed, with some studies finding a positive association and others not (Bailey & others, 2020; D'Souza & others, 2020; Gunnerud & others, 2020; Moreno-Stokoe & Damian, 2020; Ware & others, 2020). Consider the complexity of the question itself. The link between language abilities and cognitive skills may depend on the specific languages and the specific cognitive skills considered. Keep in mind, too, that even if knowing a second (or third or fourth) language is not related to other cognitive abilities, it is, itself, a skill of considerable value.

Knowing more than one language means being able to talk to lots of different people including customers, coworkers, and collaborators.

THE ROLE OF COGNITION IN LANGUAGE

Clearly, language can influence cognition. Researchers also study the possibility that cognition is an important foundation for language.

One feature of human language that separates it from animal communication is the capacity to talk about objects that are not currently present (Hockett, 1960). A study comparing 12-month-old infants (who had not yet begun to talk) to chimpanzees suggests that this cognitive skill may underlie eventual language (Liszkowski & others, 2009). In this study, infants were more likely to communicate their desire for a toy by pointing to the place where the toy *used to be*. For many infants, this was the first thing they did to get their point across to another person who was present. In contrast, chimpanzees rarely pointed to where their desired object (food) had been, except as they desperately started pointing all over the place. So, even before they can talk, humans are communicating with others about what they want. Sometimes that communication demonstrates an appreciation of shared knowledge even about objects that are no longer present.

Consider how this research connects to theory of mind. Why would infants point to a place where an object used to be if they didn't know that something was going on in the head of a person with them? Why would we talk to one another at all if we didn't know that other people have a subjective awareness?

If language is a reflection of cognition in general, we would expect to find a close link between language ability and general intellectual ability. We would anticipate, for example, that general intellectual disability is accompanied by lowered language abilities. It is often but not always the case that individuals with intellectual disability have reduced language proficiency. For instance, individuals with Williams syndrome, a genetic disorder that affects about 1 in 20,000 births, tend to show extraordinary verbal, social, and musical abilities while having an extremely low IQ and difficulty with motor tasks and numbers. Williams syndrome demonstrates that intellectual disability is not always accompanied by poor language skills (Niego & Benítez-Burraco, 2019).

Consider as well individuals who have learning disabilities related to reading. **Dyslexia** is a learning difference characterized by problems in learning to read fluently and with accurate comprehension, despite typical intelligence (Snowling & others, 2020). People with dyslexia may have difficulty learning to decipher the meaning of letters on a page and may take longer to read than those who do not have the disorder. Dyslexia can occur in people with a range of IQs, and research shows that it is indeed independent of cognitive ability more generally (Moojen & others, 2020). Many extraordinarily gifted individuals have had dyslexia, including the late Steve Jobs, founder of Apple. Research on dyslexia may be a key to understanding the differences between language and cognition.

dyslexia
A learning disability characterized by difficulty with learning to read fluently and with accurate comprehension, despite normal intelligence.

Many people believe that language is what separates humans from other animals. But note that research on animal communication is conducted by humans—not other animals (Prat, 2019). Think about what that might mean...

In summary, although thought influences language and language influences thought, there is increasing evidence that language and thought are not part of a single system. Instead, they seem to have evolved as separate but related components of the mind.

Biological and Environmental Influences on Language

Everyone who uses language in some way "knows" its rules and has the ability to create an infinite number of words and sentences. Is this knowledge the product of biology, or is language learned and influenced by experiences in the environment?

BIOLOGICAL INFLUENCES Scientists believe that humans acquired language about 100,000 years ago. In evolutionary time, then, language is a very recent human ability. However, a number of experts believe that biological evolution that occurred long before language emerged undeniably shaped humans into linguistic creatures (Chomsky, 1975). The brain, nervous system, and vocal apparatus of our predecessors changed over hundreds of thousands of years. Consider just the differences that had to occur for the mouth, lips, and tongue to allow us to speak (Putt, 2020). Physically equipped to do so, *Homo sapiens* went beyond grunting and shrieking to develop abstract speech. This sophisticated language ability gave humans an enormous edge over other animals and increased their chances of survival (Pinker, 1994).

Language Universals American linguist Noam Chomsky (1975) has argued that humans come into the world biologically prewired to learn language at a certain time and in a certain way. According to Chomsky and many other language experts, the strongest evidence for language's biological basis is the fact that children all over the world reach language milestones at about the same time and in about the same order, despite vast variations in the language input they receive from their environments. For example, in some cultures adults never talk to infants under 1 year of age, yet these infants still acquire language.

Think about it—in school children are often taught the grammar of the language they already speak very well.

In Chomsky's view, children cannot possibly learn the full rules and structure of languages by only imitating what they hear. Rather, nature must provide children with a biological, prewired, universal grammar, allowing them to understand the basic rules of all languages and to apply these rules to the speech they hear. They learn language without an awareness of its underlying logic. Think about it: The terms we used earlier to define the characteristics of language—*phonology, morphology, semantics,* and so forth—may be new to you, but on some level you have mastered these principles. This mastery is demonstrated by your reading of this book, writing a paper for class, and texting a friend. Like all other humans, you are engaged in the use of a rule-based language system even without knowing that you know those rules.

Language and the Brain Strong evidence supports the idea that language has a biological basis. Neuroscience research has shown that the brain contains particular regions that are predisposed to language use (Jäncke & others, 2020; Versace & others, 2019; Wakita, 2020). Language processing, such as speech and grammar, mainly occurs in the brain's left hemisphere. Recall the importance of Broca's area, which contributes to speech production, and Wernicke's area, which is involved in language comprehension. Using brain-imaging techniques, researchers have found that when an infant is about 9 months old, the part of the brain that stores and indexes many kinds of memory becomes fully functional (Bauer, 2013). This is also the time when infants appear to be able to attach meaning to words, for instance to look at the ball if someone says "ball"—a development suggesting links among language, cognition, and the development of the brain.

ENVIRONMENTAL INFLUENCES Decades ago, behaviorists opposed Chomsky's hypothesis and argued that language represents nothing

Noam Chomsky (b. 1928) *MIT linguist Noam Chomsky was one of the early architects of the view that children's language development cannot be explained by environmental input. In Chomsky's view, language has strong biological underpinnings, with children biologically prewired to learn language at a certain time and in a certain way.*
Tracey Nearmy/Epa/REX/Shutterstock

more than chains of responses acquired through reinforcement (Skinner, 1957). A baby happens to babble "ma-ma"; mama rewards the baby with hugs and smiles; the baby says "mama" more and more. Bit by bit, said the behaviorists, the baby's language is built up. According to behaviorists, language is a complex learned skill, much like playing the piano or dancing.

Such a view of language development is simply not tenable, however, given the rapid way children learn language, as well as the lack of evidence that social environments carefully reinforce language skills (Brown, 1973). Still, the environment has a role in language development. A child's experiences, including the particular language to be learned and the context in which learning takes place, can strongly influence language acquisition.

Think about how often parents "reinforce" children for saying the WRONG thing: They laugh and smile when toddlers mispronounce words or use the wrong word altogether. Like calling boots "boops!" Too cute. But kids grow out of these habits eventually.

Cases of children who have lacked exposure to language provide evidence for the important role of the environment in language development. Such children rarely speak typically. However, in such cases the children are often exposed to abuse and so it is unclear whether their difficulties reflect that abuse or lack of exposure to language, specifically (Rymer, 1993).

A case of a woman who was exposed to language for the first time at age 32 provides an excellent case in point. The woman, called "Chelsea," grew up in a rural community in a large and loving family (Curtiss, 2014). Likely due to prenatal exposure to a virus, she was born with hearing disabilities. She did not receive an accurate diagnosis of partial deafness until the age of 32 when she was fitted with her first hearing aids. With those hearing aids, Chelsea's hearing was in the normal range. Scans of Chelsea's brain reveal that it is in the normal range and her IQ falls between 77 and 89.

After 32 years in silence, Chelsea was exposed, for the first time, to the sounds of language. For three decades, Chelsea has received training for both spoken and sign languages. Through the years, Chelsea has acquired and learned to use a range of words. Her vocabulary and understanding of the meaning of words is remarkable, though her performance on standardized tests does not always reflect the abilities she shows in everyday life. Yet, she has never acquired a sense of grammar. Verbal acts you likely never think about, such as putting a subject before a verb in a sentence, have escaped her. For example, in response to a picture of a boy riding a bike, Chelsea said, "Riding ride bike ride boy ride" (Curtiss, 2014, p. 124).

A remarkable aspect of this case is that in the context of these language deficits, Chelsea's math abilities appear unaffected. She has a sense of number meanings and is able to keep a checkbook and do her own shopping. She can use her watch to tell time. Thus, this case offers two important conclusions. First, Chelsea's language deficits support the idea of a "critical period" for language development, a special time in a child's life (usually the preschool years) during which language must develop or it never will. Second, the fact that Chelsea's mathematical ability appears to have been spared such deficits supports the idea (reviewed earlier) that, though connected, language and cognition are separable (Curtiss, 2014).

Whether or not the case of Chelsea confirms the idea of a critical period, her experiences and those of others who lack exposure to language in early childhood certainly support the idea that the environment is crucial for the development of language.

Most humans do not learn language in a social vacuum. From infancy, most children are exposed to a great deal of language. And infants' attention to the conversations going on with them and around them predicts not only language development (Abney & others, 2020). Parental conversation directed at infants predicts motor skills as well (Schreiner & others, 2020). The support and involvement of caregivers and teachers greatly facilitate a child's language learning (Hirsh-Pasek & Golinkoff, 2014). For example, one study showed that when mothers immediately smiled and touched their 8-month-old infants after they had babbled, the infants subsequently made more complex speechlike sounds than when mothers responded to their infants in a random manner (Goldstein & others, 2003) (Figure 12).

Environmental influences on language learning demonstrate how complex the foundations of language are. Language does not appear to be exclusively biologically programmed or completely social driven. We have to look at how biology and environment interact when children learn language. Children are biologically prepared to learn language but benefit enormously from being exposed to a competent language environment from an early age.

One environmental factor that predicts language development is socioeconomic status. Compared to children from affluent families, children from poorer economic backgrounds show lower levels of language skills, including vocabulary and understanding of syntax and grammar (Noble & Giebler, 2020; Perkins & others, 2013).

The language disparity between children in wealthier families and those from financially poorer families was highlighted in a now famous study conducted in the 1980s. The researchers compared families classified as upper class/affluent, middle class, lower class, and financially poor (Hart & Risley, 1995; Risley & Hart, 2006). Researchers visited each family for one hour a month from the time infants were about eight months until they were

FIGURE 12 **The Power of Smile and Touch** Research has shown that when mothers immediately smiled and touched their eight-month-old infants after they had babbled, the infants subsequently made more complex speechlike sounds than when mothers responded randomly to their infants.
Camille Tokerud/Getty Images

three years old. During the visit, the researcher sat quietly and taped the interactions among parents and babies, producing over 1,300 hours of taped conversation (Singhal, 2017).

What did the researchers find? First, by three years of age, all of the children developed typically, learned to talk, and acquired the basic rules of English and a fundamental vocabulary. The "big news" from the study was the large difference in the sheer amount of language to which the children were exposed depending on family income. In a typical hour, the affluent parents spent much more time talking to their children and exposing them to varied vocabulary in the process. Indeed, extrapolating from their findings, the researchers estimated that by four years of age, the average child from the poorer family group would be exposed to *30 million* fewer words than the average child from the affluent family group (Hart & Risley, 1995; Risley & Hart, 2006). This big number became famous as the "30 million word gap." Following some of the children into third grade, the researchers found that their language deficits at three years continued into grade school (Hart & Risley, 1995). This study gave rise to the idea that finding ways to fill the 30 million word gap might improve the lives of children from poorer families (Hindman & others, 2016; Singhal, 2017).

There are many reasons to take conclusions drawn from this study with a grain of salt, but two are especially clear. First, and most importantly, the study included just 42 families total, and just 6 families in the financially poor group. The wealthy group included 13 families. All of the financially poor families were Black, but only one of the wealthy families was Black. We cannot draw strong conclusions about all people from such small and ethnically imbalanced samples. Second, as you know, correlation does not imply causation. Because the number of words included in conversations was related to children's language development does not imply that word-exposure causes development. The fact that language deficits at age three correlated with grade school performance does not imply that these deficits cause that performance. Both of these associations might be due to the host of differences between families and especially the wide-ranging effects of poverty, including access to healthcare, education, nutrition, and experiences with discrimination.

How Do Nature and Nurture Influence Development?

Developmental psychologists are interested in understanding how nature and nurture contribute to development. **Nature** refers to a person's biological inheritance, especially genes. **Nurture** refers to the person's environmental and social experiences. Understanding development requires that we take into account the contributions of both genes (nature) and the environment (nurture).

What factors in your childhood environment influenced your expression of your gifts and abilities?

nature
An individual's biological inheritance, especially genes.

nurture
An individual's environmental and social experiences.

Recall that a *genotype* is the individual's genetic heritage—the actual genetic material. We also examined the idea of a *phenotype* (the person's observable characteristics). The phenotype shows the contributions of both nature (genetic heritage) and nurture (environment). Whether and how the genotype is expressed in the phenotype may depend on the environment. For example, a person might be born with the genes to be the next LeBron James, but without the necessary environmental factors such as good nutrition, sound medical care, access to a basketball court, and superb coaching, that potential might never be reached.

Consider the genetic condition called phenylketonuria (PKU). Caused by two recessive genes, PKU results in an inability to metabolize the amino acid phenylalanine. Phenylalanine is present in many foods (such as sweet potatoes) and in many artificial sweeteners. Decades ago, it was thought that the genotype for PKU led to a specific phenotype, namely, irreversible brain damage, intellectual disability, and seizures. However, experts now know that if people with the genotype for PKU stick to a diet that is very low in phenylalanine, these characteristics in the phenotype can be avoided (Cannett & others, 2020; Clocksin & others, 2020). These environmental precautions can change the phenotype associated with this genotype.

The PKU example tells us that a person's observable characteristics (phenotype) might not reflect their genetic heritage (genotype) very precisely because of the particular experiences the person has had. Instead, for each genotype, a *range* of phenotypes may be expressed, depending on environmental experiences. The person whom we see before us emerges out of an interplay of genetic and environmental experiences. Development is the product of nature, nurture, and the complex interaction of the two (Meaney, 2017).

A flock of birds flying in formation illustrates an emergent property. The birds may appear to be following a leader, but they aren't. Instead, each individual bird is following its own local rules. What you see as a flock of birds is in fact a collection of individual birds, each one "doing its own thing" but creating the formation (the emergent property) you recognize as a flock.

Although it might be easy to think of genes as the blueprint for a person, development is not a process that follows a genetic master plan (Turkheimer, 2011). In fact, it is difficult to tell a simple story about how development occurs. One way that scientists and philosophers think about complex processes such as development is through the concept of emergent properties. An *emergent property* is a big entity (like a person) that is a consequence of the interaction of multiple lower-level factors (Fesce, 2020; Gottlieb, 2007; Nalepka & others, 2017). Development is about the complex interactions of genes and experience that build the whole person.

Do Early Experiences Rule Us for Life?

The PKU example suggests the power of early experience (nurture) in human development. A key question in developmental psychology is the extent to which childhood experiences determine aspects of later life. Some research shows that unless infants experience warm, nurturing caregiving in the first year or so of life, they will not develop to their full potential (Boldt & others, 2020; Mares & McMahon, 2020). Other studies demonstrate the power of later experience in influencing development in adulthood (Asselmann & Specht, 2020; Woods & others, 2020). Life-span developmentalists stress that experiences throughout life contribute to development. Both early and later experience make significant contributions to development, so no one is doomed to be a prisoner of their childhood.

A key concept in understanding the role of negative early experiences in later development is resilience. **Resilience** refers to a person's ability to recover from or adapt to difficult times. Resilience means that despite encountering adversity, a person shows signs of positive functioning (Ernst & others, 2019; Masten & Motti-Stefanidi, 2020; Skinner & others, 2020). Resilience can involve factors that compensate for difficulties, buffering the person from the effects of these hardships. Moderate difficulties early in life can be strengthening experiences that lay the groundwork for effective future coping (Frankenhuis & Nettle, 2020; Lind & others, 2019; Seery & others, 2010). Although often studied as an aspect of childhood and adolescence, resilience can also characterize development in adulthood and old age (Infurna, 2020).

resilience
A person's ability to recover from or adapt to difficult times.

Nature, Nurture, and You

Because you cannot pick your genes or your parents, it might seem like you are stuck with the genes and environment you got at birth. However, the truth is that you have a vital role to play throughout development. As an active developer, you take the raw ingredients of nature and nurture and make them into the person you are (Turkheimer, 2011). Active developers, then, *interact* with their own genes and environment to produce the persons they become. In this sense, developers themselves can be a powerful force in their own development.

Indeed, some psychologists believe that people can develop beyond what their genetic inheritance and environment give them. They argue that a key aspect of development involves seeking optimal experiences in life (Halfon & Forrest, 2018). History is filled with examples of people who go beyond what life has given them to achieve extraordinary things. Such individuals author a unique developmental path, sometimes transforming apparent weaknesses into real strengths.

In individuals' efforts to experience life in optimal ways, they develop *life themes* that involve activities, social relationships, and life goals (Jeffers & others, 2020;

Human development is complex because it is the product of several processes. The hormonal changes of puberty, a baby's playing with blocks, and an older couple's embrace reflect physical, cognitive, and socioemotional processes, respectively.
(Teen): Arina Habich/Alamy Stock Photo; (Baby): JGI/Tom Grill/Getty Images; (Couple): Mike Watson Images/Getty Images

A person does not have to be famous to take an active role in shaping their development.

Pratt & others, 2020; Schofield & others, 2017). Some people are more successful at constructing optimal life experiences than others. Examples of public figures who have succeeded are Martin Luther King Jr., Mother Teresa, Nelson Mandela, and Oprah Winfrey. These individuals looked for and found meaningful life themes as they developed. Their lives were not restricted to biological survival or to settling for their particular life situations.

Three Domains of Development

The pattern of development is complex because it is the product of several processes in three domains of life—physical, cognitive, and socioemotional:

1. *Physical processes* involve changes in an individual's biological nature. Genes inherited from parents, the hormonal changes of puberty and menopause, and changes throughout life in the brain, height and weight, and motor skills all reflect the developmental role of biological processes. Physical development is often termed *maturation* because it involves an unfolding of biological processes.

2. *Cognitive processes* involve changes in a person's thought, intelligence, and language. It refers to a host of skills, such as learning to read, engaging in problem solving, and mastering calculus. Observing a colorful mobile as it swings above a crib, constructing a sentence about the future, imagining oneself as a contestant on *The Voice* or as president of the United States, memorizing a new telephone number—these activities reflect the role of cognitive processes in development.

3. *Socioemotional processes* involve changes in a person's relationships with other people, changes in emotions, and changes in personality. An infant's smile in response to a parent's touch, a child's development of assertiveness, an adolescent's joy at the senior prom, a young adult's ambition at work, and an older couple's affection for each other all reflect the role of socioemotional processes.

Throughout this chapter, we will trace these developmental processes along the broad periods of the life span, namely:

- *Childhood,* the period from infancy (birth to 24 months) through childhood (up to about age 10).

- *Adolescence,* the period beginning around ages 10 to 12 and spanning the transition from childhood to adulthood.

- *Adulthood,* the period that is generally separated into early (the 20s and 30s), middle (the 40s through the mid-60s), and late (mid-60s and beyond).

After tracing changes in physical, cognitive, and socioemotional development across the life span, we will turn to topics that demonstrate how strongly intertwined these processes are, including the development of gender and morality as well as death and dying. Think of Hannah, an infant whose parents place a teddy bear in her crib. As an infant she might simply look at the teddy bear when her parents jiggle it in front of her. Over time, she not only can see the teddy bear but also can reach for it. She might even remember that the teddy bear exists and might cry for it when it is not with her. As a toddler, when she carries it around, she is demonstrating her physical abilities to do so, as well as her capacity to use the teddy bear as a source of comfort. As an adolescent, Hannah may no longer sleep with her teddy, but she may give him a place of honor on a shelf. As you read this chapter's separate sections on physical, cognitive, and socioemotional development, remember that you are studying the development of an integrated human being in whom body, mind, emotion, and social relationships are interdependent.

Fancy Collection/SuperStock

1. Nelba Marquez-Greene lost her six-year-old daughter, Ana Grace, to the school shooting at Sandy Hook in 2012. She has dedicated her life to the issue of gun control and school safety. Her work illustrates
   A. a life theme.
   B. a genotype.
   C. a physical process.
   D. a phenotype.

2. Development can be best described as
   A. due entirely to nature.
   B. due entirely to nurture.
   C. the product of the interaction of nature and nurture.
   D. none of the above.

3. An example of an optimal experience is
   A. cooking food to eat.
   B. getting a great buy on those boots you wanted.
   C. volunteering time to teach adults to read.
   D. competing with others.

APPLY IT! 4. Sonja and Pete are engineers who met during college. They share a love of mathematics and science and have successful engineering careers. When their daughter Gabriella is born, they decorate her room with numbers and spend a great deal of time counting objects and talking about math with her. In her school years, Gabriella is particularly gifted in mathematics. Gabriella does not become an engineer, but she does have a career as a terrific math teacher. Which statement is most accurate in describing Gabriella?
   A. Her math ability is a direct result of her genetic heritage.
   B. Her math ability is a direct result of her environmental experiences.
   C. The fact that she became a teacher instead of an engineer shows that neither genetics nor environment matters that much to development.
   D. Gabriella's development shows the influence of genetics, environment, their interaction, and Gabriella's capacity to forge a life theme that is meaningful to her.

2 Physical Development

In this section, we trace the ways people grow and change *physically,* starting with *prenatal* ("before birth") physical development. Then we consider the physical changes associated with childhood, adolescence, and the phases of adulthood. Throughout, we consider the ways the brain changes in each developmental period.

Prenatal Physical Development

Prenatal development is a time of astonishing change, beginning with conception. Conception occurs when a single sperm cell merges with the ovum (egg) to produce a zygote, a single cell with an array of 46 chromosomes, 23 from each biological parent.

THE COURSE OF PRENATAL DEVELOPMENT Development from zygote to fetus is divided into three periods:

1. *Germinal period—weeks 1 and 2:* The germinal period begins with conception. In this period, the fertilized egg is called a zygote. After one week and many cell divisions, the zygote is made up of 100 to 150 cells. By the end of two weeks, the mass of cells has attached to the uterine wall.

2. *Embryonic period—weeks 3 through 8:* At this point, the zygote has become an embryo. The rate of cell differentiation intensifies, support systems for the cells develop, and the beginnings of organs appear (Figure 1a). In the third week, the neural tube, which eventually becomes the spinal cord, starts to take shape. Within the first 28 days after conception, the neural tube is formed and closes, encased inside the embryo. Problems in neural tube development can lead to conditions such as spina bifida, in which the spinal cord is not completely enclosed by the spinal column, or severe underdevelopment of the brain. Folic acid, a B vitamin found in orange juice and leafy green vegetables, greatly reduces the chances of problems in neural tube development. By the end of the embryonic period, the heart begins to beat, the arms and legs become more differentiated, the face starts to form, and the intestinal tract appears (see Figure 1b).

Note that 28 days is before most people even know they are pregnant. That's why doctors recommend that anyone who might become pregnant take folic acid supplements (about 400 micrograms a day) to promote neural tube development.

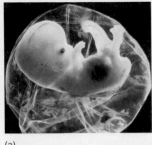
(a)

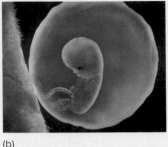
(b)

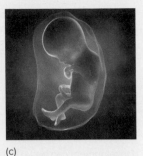
(c)

FIGURE 1 **From Embryo to Fetus** (*a*) At about four weeks, an embryo is about 0.2 inch (less than 1 centimeter) long. The head, eyes, and ears begin to show; the head and neck are half the length of the body. (*b*) At eight weeks, the developing embryo is about 1.6 inches (4 centimeters) long and has reached the end of its embryonic phase. It has become a fetus. The fetal stage is a period of growth and perfection of detail. The heart has been beating for a month, and the muscles have just begun their first exercises. (*c*) At four and a half months, the fetus is just over 7 inches (about 18 centimeters) long. If the thumb comes close to the mouth, the head may turn, and the lips and tongue begin their sucking motions—a reflex for survival.

(a): Science History Images/Alamy Stock Photo; (b): Stocktrek Images/Getty Images; (c): SCIEPRO/Science Photo Library/Getty Images

3. *Fetal period—months 2 through 9:* At two months, the fetus is the size of a kidney bean and has begun to move around. At four months, the fetus is 5 inches long and weighs about 5 ounces (see Figure 1c). At six months, the fetus has grown to 1.5 pounds. The last three months of pregnancy are the time when organ functioning increases, and the fetus puts on considerable weight and size, adding baby fat. The average newborn is about 19 inches long and weighs about 7 pounds.

Until about 80 years ago, people were unaware of the role that the diet and behavior of a pregnant person might play for the developing fetus. Although it floats in a comfortable, well-protected womb, the fetus is not immune to the wider environment. Indeed, sometimes prenatal development is disrupted by environmental insults.

THREATS TO THE FETUS A *teratogen* is any agent that causes a problem in prenatal development. Teratogens include chemical substances ingested by a pregnant person (such as nicotine or alcohol) and certain illnesses (such as rubella or German measles). Substances that are ingested by a pregnant person can lead to serious problems (Poels & others, 2020; Romeo & Običan, 2020). Heroin is an example of a teratogen. Prenatal heroin exposure increases risk for many problems, including premature birth, low birth weight, physical problems, breathing problems, and death.

Fetal alcohol spectrum disorders (FASD) are a cluster of problems that appear in babies due to prenatal alcohol exposure (Abel, 2020). Signs of FASD include a small head; facial characteristics such as wide-spaced eyes, a flattened nose, and an underdeveloped upper lip; limbs and heart abnormalities. For anyone who is pregnant or thinking of becoming pregnant, the best advice is to avoid alcohol.

A reality check for people who smoke during pregnancy: Imagine a baby puffing on a lit cigarette. Anyone who smokes during pregnancy is smoking for two.

The effects of teratogens on development depend on the timing of exposure. The body part or organ system that is developing when the fetus encounters the teratogen is most vulnerable. Genetic characteristics may buffer or worsen the effects of a teratogen. Perhaps most importantly, the environment children experience *after birth* can influence the ultimate effects of prenatal insults (Cheong & others, 2020).

Preterm infants (those born prior to 37 weeks after conception) and low birth-weight infants may be at risk for developmental difficulties. Many preterm and low birth-weight infants are healthy, but as a group they have more health and developmental problems than do typical birth-weight infants (Carlier & Harmony, 2020; Durankus & others, 2020; Steinbauer & others, 2020). The number and severity of these problems increase when infants are born very early and as their birth weight decreases. Survival rates for infants who are born very early and very small have risen, but with this improved survival rate has come increased rates of severe neurological problems (McNicholas & others, 2014; Pisani & others, 2020).

Postnatal experience plays a crucial role in determining the ultimate effects of preterm birth. For example, infant massage and kangaroo care, in which the newborn is held upright against the parent's bare chest, much as a baby kangaroo is carried by its mother, can improve developmental outcomes for preterm and low birth-weight infants and may have benefits for parents as well (Hardin & others, 2020; Lu & others, 2020).

Physical Development in Infancy and Childhood

Human infants are among the world's most helpless neonates. One reason for that helplessness is that we are born not quite finished. From an evolutionary perspective, what sets humans apart from other animals is our enormous brain. Getting that big brain out of the relatively small birth canal is a challenge that nature has met by sending human babies out of the womb before the brain has fully developed. The first months and years of life allow the developing human (and the environment) to put the finishing touches on that important organ.

REFLEXES Newborns come into the world equipped with several genetically wired reflexes that are crucial for survival. Babies are born with the ability to suck and swallow. If they are dropped in water, they will naturally hold their breath, contract their throats to keep water out, and move their arms and legs to stay afloat at least briefly. Some reflexes persist throughout life—coughing, blinking, and yawning, for example. Others, such as automatically grasping something that touches the fingers, disappear in the months following birth as infants develop voluntary control over many behaviors. Figure 2 shows some examples of infant reflexes.

MOTOR AND PERCEPTUAL SKILLS Relative to the rest of the body, a newborn's head is gigantic, and it flops around uncontrollably. Within 12 months, the infant becomes capable of sitting upright, standing, stooping, climbing, and often walking. During the second year, growth slows down, but rapid gains occur in such activities as running and climbing.

Motor skills and perceptual skills are coupled and depend on each other. To reach for something, the infant must be able to see it. Babies are continually coordinating their

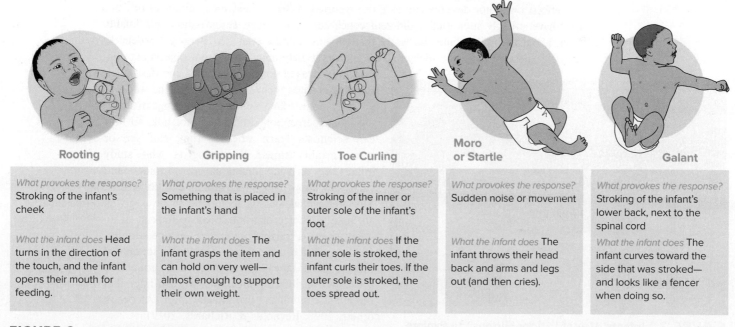

Rooting	Gripping	Toe Curling	Moro or Startle	Galant
What provokes the response? Stroking of the infant's cheek	*What provokes the response?* Something that is placed in the infant's hand	*What provokes the response?* Stroking of the inner or outer sole of the infant's foot	*What provokes the response?* Sudden noise or movement	*What provokes the response?* Stroking of the infant's lower back, next to the spinal cord
What the infant does Head turns in the direction of the touch, and the infant opens their mouth for feeding.	*What the infant does* The infant grasps the item and can hold on very well—almost enough to support their own weight.	*What the infant does* If the inner sole is stroked, the infant curls their toes. If the outer sole is stroked, the toes spread out.	*What the infant does* The infant throws their head back and arms and legs out (and then cries).	*What the infant does* The infant curves toward the side that was stroked—and looks like a fencer when doing so.

FIGURE 2 **Some Infant Reflexes** Infants are born with a number of reflexes to get them through life, and they are incredibly cute when they perform them. These reflexes disappear as infants mature.

movements with information they perceive through their senses to learn how to maintain their balance, reach for objects in space, and move across various surfaces and terrains (Adolph & Hoch, 2020; Newell, 2020). Consider what happens when a baby sees a fun toy across the room. Because they can see it, they are motivated to get it. They must perceive the current state of their body and learn how to use their limbs to get to their goal, the toy. In turn, action educates perception. For example, watching an object while holding and touching it helps infants to learn about its texture, size, and hardness. Moving from place to place in the environment teaches babies how objects and people look from different perspectives and whether surfaces will support their weight. Demonstrating the profound connections across levels of development, motor skills in infancy predict intelligence measured at four years of age (Heineman & others, 2017).

Infants are energetic developers. When infants are motivated to do something, they may create a new motor behavior, such as reaching out to grab a new toy or a caregiver's earrings. That new behavior is the result of many converging factors: the developing nervous system, the body's physical properties and its movement possibilities, the goal the infant is motivated to reach, and environmental support for the skill (Dinkel & Snyder, 2020; Valadi & Gabbard, 2020). The infant brain is very much tuned in to the world of action in which each child is embedded. Research shows that the infant brain responds differently to seeing other people moving if the babies have engaged in reflex walking while being held (Reid & others, 2019). The sensorimotor areas (that would be used in clapping) of the infant brain activate when they hear action sounds (like clapping) (Quadrelli & others, 2019). Researchers once thought that motor milestones (such as sitting up, crawling, and walking) unfolded as part of a genetic plan. Psychologists now recognize that motor development is not the consequence of nature or nurture alone.

C Squared Studios/Getty Images

Environmental experiences play a role in motor development. In one study, three-month-old infants participated in play sessions wearing "sticky mittens"—mittens with palms that stick to the edges of toys and allow the infants to pick up the toys (Needham & others, 2002, p. 279) (Figure 3). Infants who participated in sessions with the mittens grasped and manipulated objects earlier in their development than a control group of infants who did not receive the "mitten" experience. The experienced infants looked at the objects longer, swatted at them, and were more likely to put the objects in their mouth. The sticky mittens exercise is especially effective if the parents encourage the child while they use the mittens (Needham & others, 2017).

Infants' motor skills are limited by gravity. Infants can sometimes do a lot more when weightless. Holding a baby up in a kiddie pool, we can see how the infant can stand and even take some steps quite deftly, without body weight getting in the way.

Interestingly, studies have not consistently replicated (that is, reproduced) the same effects on motor development as those reported in the original sticky mittens study but they have strongly supported a different conclusion. Specifically, studies show that infants learn about themselves and the world through this training (van den Berg & Gredebäck, 2020). Experience with the sticky mittens appears to help infants grasp not objects but concepts, the ideas of goals and actions, and this understanding changes their expectations for the world.

Psychologists face a daunting challenge in studying infant perception. Infants cannot talk, so how can scientists learn whether they can see or hear certain things? Psychologists who study infants rely on what infants can do to understand what they know. One thing infants can do is look. The **preferential looking** technique involves giving an infant a choice of what object to look at. If an infant shows a reliable preference for one stimulus (say, a picture of a face) over another (a scrambled picture of a face) when these are repeatedly presented in differing locations, we can infer that the infant can tell the two images apart (Alcock & others, 2020; Reynolds & Richards, 2019).

preferential looking
A research technique that involves giving an infant a choice of what object to look at.

Using this technique with babies even within hours after they are born, researchers have shown that infants prefer to look at faces rather than other

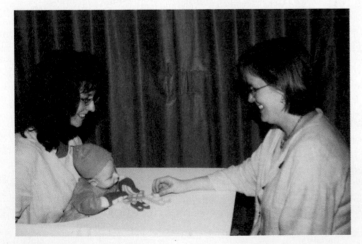

FIGURE 3 **Infants' Use of "Sticky Mittens" to Explore Objects** Amy Needham and her colleagues (2002) found that "sticky mittens" enhance young infants' object exploration skills.
Courtesy of Amy Needham, Duke University

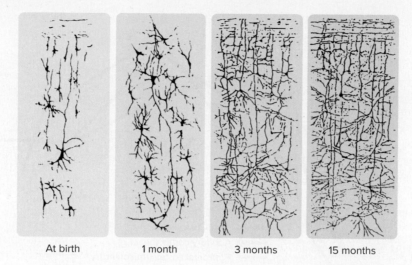

FIGURE 4
Dendritic Spreading
Note the increase in connections among neurons over the course of the first two years of life.

Leisman, Gerry, "Intentionality and 'Free-Will' From a Neurodevelopmental Perspective." *Frontiers in Integrative Science,* June 27, 2012, Figure 4. Copyright ©2012 by Gerry Leisman. All rights reserved. Used with permission.

At birth　　　1 month　　　3 months　　　15 months

objects and to look at attractive faces more than at unattractive ones (Johnson & Hannon, 2015; Lee & others, 2013). Also, as early as seven days old, infants are already engaged in organized perception of faces and are able to put together sights and sounds. If presented with two faces with mouths moving, infants will watch the face whose mouth matches the sounds they are hearing (Lewkowicz, 2010; Lewkowicz & Hansen-Tift, 2012; Pascalls & Kelly, 2008). At three months, infants prefer real faces to scrambled faces and their mother's face to a stranger's (Barrera & Maurer, 1981). Researchers also use eye-tracking technology that monitors what a child is looking at to track infant gaze (Manzi & others, 2020). Research using brain imaging suggests that infants may know more than even this clever strategy can tell us (Ellis & others, 2020).

THE BRAIN　　As an infant plays, crawls, shakes a rattle, smiles, and frowns, the brain is changing dramatically. At birth and in early infancy, the brain's 100 billion neurons have only minimal connections. The infant brain literally is ready and waiting for the experiences that will create these connections. During the first two years of life, the dendrites of the neurons branch out, and the neurons become far more interconnected (Figure 4); infancy is a time when the brain is all about making connections. Myelination, the process of encasing axons with fat cells (the myelin sheath), begins prenatally and continues after birth well into adolescence and adulthood (Kwon & others, 2020).

During childhood, *synaptic connections* increase dramatically. Recall that a synapse is a gap between neurons that is bridged by chemical neurotransmitters. Nearly twice as many synapses are available as will ever be used (Huttenlocher, 1999; Sakai, 2020). The connections that are made and used become stronger and will survive; the unused ones will be replaced by other neural pathways or disappear. In the language of neuroscience, these unused connections will be "pruned." Figure 5 illustrates the steep growth and later pruning of synapses during infancy in specific areas of the brain. As you can see, the degree of pruning that occurs is extensive; however, it is a precise process. Problems in the pruning process are related to later psychological disorders (Neniskyte & Gross, 2017; Sakai, 2020). Note as well that pruning is at least partly based on experience—unused connections are pruned away, demonstrating again the way the brain adapts to experience.

Brain imaging studies show that children's brains undergo amazing anatomical changes over time. Repeated brain scans of the same children for up to four years show that the amount of brain material in some areas can nearly double within as little as a year, followed by a drastic loss of tissue as unneeded cells are purged and the brain continues to reorganize itself. The overall size of the brain does not change very much, but local patterns within the brain change tremendously. From three to six years of age, the most rapid growth takes place in the frontal lobe areas, which are involved in planning and organizing new actions and in maintaining attention to tasks. These brain changes are not simply the result of nature; new experiences in the world also promote brain development (Humphreys & others, 2020; King & others, 2020). Thus, as in other areas of development, nature and nurture operate together.

The brain's plasticity makes it different from any other bodily organ. It comes into the world ready for whatever world it might encounter, and the features of that world influence its very structure. Other organs physically grow as we do. But the brain's very essence is attached to the world in which it lives.

FIGURE 5

Synaptic Density in the Human Brain from Infancy to Adulthood
The graph shows the dramatic increase and then pruning in synaptic density in three regions of the brain: visual cortex, auditory cortex, and prefrontal cortex. Synaptic density is believed to be an important indication of the extent of connectivity between neurons.

SOURCE: Huttenlocher, P. R., & A. S. Dabholkar. (1997). Regional differences in the synaptogenesis in the human cerebral cortex. *Journal of Comparative Neurology, 387*(2), 167–168.

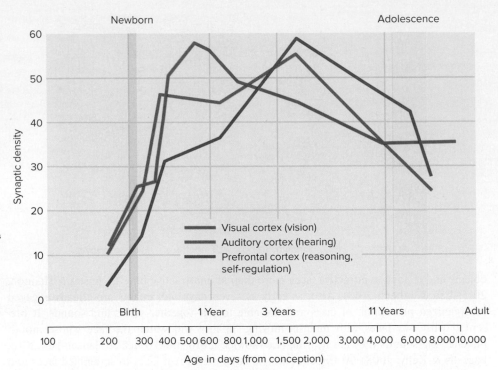

Visual cortex (vision)
Auditory cortex (hearing)
Prefrontal cortex (reasoning, self-regulation)

Physical Development in Adolescence

Adolescence refers to the developmental period spanning the transition from childhood to adulthood, beginning around 10 to 12 years of age and ending at 18 to 21 years of age. Dramatic physical changes characterize adolescence, especially early adolescence. Among the major physical changes of adolescence are those involving puberty and the brain.

PUBERTY   The signature physical change in adolescence is **puberty**, a period of rapid skeletal and sexual maturation that occurs mainly in early adolescence. We know when an individual is going through puberty in a general ways, but genetic sex may render it difficult to pinpoint its beginning. We will discuss the genetic contribution to sex in more detail later in this chapter. For now, it is important to know that the genes that affect the development of the genitals prenatally have impact as well on development of physical characteristics during puberty. Assigned sex refers to the categories (male, female, intersex) that are applied to infants, often based on the appearance of their genitals at birth. Intersex means that a person's chromosomes, genitals, or internal reproductive organs are not typical. For those who are assigned female at birth, *menarche* (the first menstrual cycle) is a clear marker of puberty. Girls who live in poorer countries (Leone & Brown, 2020), in more food-insecure households (Burris & others, 2020), and who experience more stressful events (Holdsworth & Appleton, 2020) are likely to start their periods earlier in life. For those who are assigned male at birth, there is no single marker that defines puberty the way menarche does. The first whisker or first nocturnal ejaculation (or wet dream) might mark the beginning of puberty but both may go unnoticed.

puberty
A period of rapid skeletal and sexual maturation that occurs mainly in early adolescence.

The jump in height and weight that characterizes pubertal change differs by assigned gender (Figure 6). In the United States today, the average beginning of the growth spurt is 9 years of age for those assigned female at birth and 11 years for children assigned male at birth.

Hormonal changes lie at the core of pubertal development. The concentrations of certain hormones increase dramatically during puberty

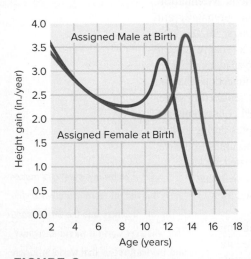

FIGURE 6

Pubertal Growth Spurt On average, the pubertal growth spurt begins and peaks about two years earlier for children assigned female at birth (starts at 9, peaks at 11.5) than for children assigned male at birth (starts at 11.5, peaks at 13.5).

SOURCE: Tanner, J. M., et al. (1965). Standards from birth to maturity for height, weight, height velocity: British children. *Archives of Diseases in Childhood, 41,* 613–635.

and are different based on biological sex. In children assigned male at birth, *testosterone,* an androgen, is associated with the development of genitals, an increase in height, and voice change. *Estradiol,* an estrogen, is associated with breast, uterine, and skeletal development among those assigned female at birth. Developmental psychologists believe that hormonal changes account for at least some of the emotional ups and downs of adolescence, but hormones are not alone responsible for adolescent behavior (Cameron & others, 2017; Deane & others, 2020; Jamieson & others, 2020; Vosberg & others, 2020).

From our discussion earlier in this chapter, recall that physical and socioemotional development are intertwined. This link is demonstrated in the implications of timing of puberty for socioemotional outcomes (Hoyt & others, 2020). Among those assigned male at birth, earlier maturation relative to peers predicts more positive socioemotional outcomes, such as being popular with their peers and having higher self-esteem (Cooper & Bidaisee, 2017; de Guzman & Nishina, 2014). In one study, male adolescents who matured early experienced more occupational success and were less likely to drink alcohol or smoke cigarettes than late-maturing male adolescents some 39 years later (Taga & others, 2006). In contrast, early maturation predicts female adolescents' vulnerability to a number of problems (Chen & others, 2017; Hamilton & others, 2014). Early-maturing female adolescents are at risk for smoking, drinking alcohol, depression, and disordered eating (Chen & others, 2017; Graber, 2013; Okada & others, 2020). A key concern is that adults may behave inappropriately toward these adolescents.

Recently, experts have noted that children seem to be entering puberty earlier than in previous times. What factors might explain this change?

THE ADOLESCENT BRAIN Based on experience and hormonal changes, in adolescence, brain changes first occur at the subcortical level in the **limbic system**, which is the seat of emotions and where rewards are experienced. The limbic system matures much earlier than the prefrontal cortex and is almost completely developed by early adolescence (Frere & others, 2020; Herting & Sowell, 2017). The limbic system structures that are especially involved in emotion are the amygdalae, the two almond-shaped structures deep inside the brain. Figure 7 shows the locations of the limbic system, amygdala, and prefrontal cortex. These changes in the brain may help to explain why adolescents often display very strong emotions, seek immediate rewards, and have trouble regulating their passions (Frere & others, 2020). Because of the relatively slow development of the prefrontal cortex, which continues to mature into early adulthood, adolescents may lack the cognitive skills to control their impulses effectively. This developmental disjunction may account for increased risk taking and other problems in adolescence (McIlvain & others, 2020).

One big question is, which comes first in development—biological changes in the brain or experiences that stimulate these changes? Consider a study in which the brain reactivity

limbic system
A set of subcortical brain structures central to emotion, memory, and reward processing.

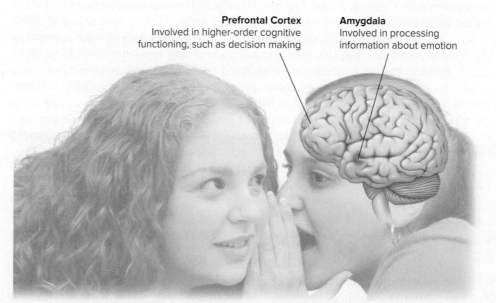

Prefrontal Cortex
Involved in higher-order cognitive functioning, such as decision making

Amygdala
Involved in processing information about emotion

FIGURE 7
Developmental Changes in the Adolescent's Brain
The amygdala, which is responsible for processing information about emotion, matures earlier than the prefrontal cortex, which is responsible for making decisions and other higher-order cognitive functions. The amygdala is located deep inside the outer layer where it is shown.
(Photo): Brand X Pictures/PunchStock

of children and adolescents was measured as they were shown images of facial expressions. The results showed that adolescents had stronger brain responses to changes in emotional facial expressions than children (Rosen & others, 2017). These results fit with the view that adolescents are more preoccupied with social standing and other people's feelings. Yet, were the results due to biology or to experience? Is it possible that adolescents' brains were so responsive because of exposure to the middle and high school social scene? This correlational study cannot answer that question, and once again we encounter the fascinating truth about the brain: It shapes and is shaped by experience.

Physical Development in Adulthood

As in other developmental periods, our bodies change during adulthood. Most of the changes that occur after adolescence involve declines in physical and perceptual abilities, as we now consider.

PHYSICAL CHANGES IN EARLY ADULTHOOD   Most adults reach their peak physical development during their 20s and are the healthiest then. Early adulthood, however, is also the time when many physical skills begin to decline. The downward trend in strength and speed often is noticeable in the 30s. Another realm in which physical changes occur with age is in the ability to perceive the world. Although it may not be noticeable, perceptual decline begins in early adulthood. For instance, starting as early as age 18, hearing begins a gradual decline.

PHYSICAL CHANGES IN MIDDLE AND LATE ADULTHOOD   Many physical changes in the 40s or 50s involve changes in appearance. The skin has begun to wrinkle and sag because of the loss of fat and collagen in underlying tissues. Small, localized areas of skin pigmentation produce age spots, especially in areas exposed to sunlight such as the hands and face. Hair becomes thinner and grayer due to a lower replacement rate and a decline in melanin production. People lose height in middle age as a result of bone loss in the vertebrae, and many gain weight. Once individuals reach their 40s, age-related changes to their vision usually become apparent, especially difficulty in seeing things up close. The sense of taste can also be affected by age, as taste buds are less likely to be replaced.

Biological sex plays a role in physical development in adulthood. Usually in the late 40s or early 50s, female adults' menstrual periods cease. With menopause comes a dramatic drop in the ovaries' production of estrogen. Estrogen decline produces uncomfortable symptoms for some, such as *hot flashes* (sudden, brief flushing of the skin and a feeling of elevated body temperature), nausea, fatigue, and rapid heartbeat. However, menopause overall is not the negative experience for most people it was once thought to be (Thomas & Daley, 2020).

With age, a variety of bodily systems are likely to show the effects of wear and tear as the body becomes less and less able to repair damage and regenerate itself (Chaudhry, 2020). Physical strength declines and motor speed slows; bones may become more brittle (especially for women). Nearly every bodily system changes with age. Significantly, however, even as age is associated with some inevitable decline, important aspects of successful aging are within the person's control. For instance, a healthy diet and regular exercise can help to prevent or slow these effects. Regular physical activity can have wide-reaching benefits not only for physical health but for cognitive functioning as well (de Souto Barreto & others, 2018; Gothe, 2020; Morrison & Mayer, 2017; Stenling & others, 2020; Zhu & others, 2016). Physical exercise interventions lead to increases in the size of the hippocampus and improved memory (Best & others, 2017; Lin, Kuo, & others, 2020; Rosano & others, 2017).

One way older adults navigate the physical changes associated with age is by changing their goals and developing new ways to engage in desired activities. Psychologists refer to this process as *selective optimization with compensation,* which means that older adults match their goals with their current abilities and compensate for declines by finding other ways to do the things they enjoy (Nimrod, 2020). A 75-year-old who can no longer drive because of vision changes might become an expert on their city's train and bus system, for example.

How is your current behavior building a foundation for a healthy late adulthood? If your future self could time-travel to visit you, would that self say "Thanks!"—or instead "What were you thinking?"

On the island of Okinawa (part of Japan), people live longer than anywhere else in the world, and Okinawa has the world's highest prevalence of *centenarians—* people who live to 100 years or beyond. Examination of Okinawans' lives provides insights into their longevity. Specific factors are diet (they eat nutritious foods such as grains, fish, and vegetables); lifestyle (they are easygoing and experience low stress); community (Okinawans look out for one another and do not isolate or ignore older adults); activity (they lead active lifestyles, and many older adults continue to work); and spirituality (they find a sense of purpose in spiritual matters) (Willcox & others, 2014; Willcox & Willcox, 2014). Just as physical changes are interwoven with socioemotional processes in childhood and adolescence, so they are as human beings enter the later stages of life.

Biological Theories of Aging

Why do people age? Of the many proposed biological theories of aging, three especially merit attention: cellular-clock theory, free-radical theory, and hormonal stress theory.

The *cellular-clock theory* is Leonard Hayflick's (1977) view that cells can divide a maximum of about 100 times and that, as we age, our cells become less capable of dividing. Hayflick found that the total number of cell divisions is roughly related to a person's age. Based on the way cells divide, Hayflick placed the upper limit of the human life span at about 120 years.

Why do cells lose their ability to divide? The answer may lie at the tips of chromosomes. Each time a cell divides, the *telomeres* protecting the ends of chromosomes shorten (Figure 8). After about 100 replications, the telomeres are dramatically reduced, and the cell no longer can divide (Shalev & others, 2013). Telomerase is an enzyme that helps extend telomeres. It might be possible to one day use genetic manipulation of chemical telomerase activators to maintain the length of telomeres. One factor that has been demonstrated to reduce telomere length is stress (Etzel & Shalev, 2021). The effect of stress on telomeres can be seen throughout development. For example, a recent study examined how prenatal stress and parenting predicted telomere length in a sample of children from age 6 to 10 (Beijers & others, 2020). The results showed that parenting only predicted telomere length among children who had experienced high prenatal stress. Among

People often say life's too short. But imagine living to be 120 years old. At age 70 you would still have 50 years to fill with life.

Do It!

Have you ever thought about how long you are likely to live? Might you be able to live to be 100 years old? Type in the following on your Internet browser: The Living to 100 Life Expectancy Calculator. That will give you access to Dr. Thomas Perls's website (www.livingto100.com), where in only about 10 minutes you can answer questions about different aspects of your life that will provide a number indicating how long you are likely to live. A special benefit is that you also will get feedback about how to improve your number. Dr. Perls is currently conducting one of the largest studies of *centenarians* (those who live to be 100), and his life expectancy calculator is based on research that has been conducted on the factors that predict longevity.

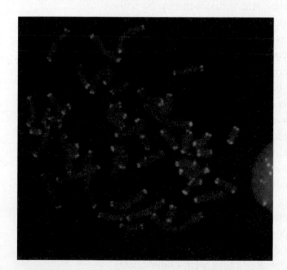

FIGURE 8 Telomeres and Aging The photograph shows telomeres lighting up the tips of chromosomes. The figure illustrates how the telomeres shorten every time a cell divides. Eventually, after about 100 divisions, the telomeres are greatly reduced in length. As a result, the cell can no longer reproduce and it dies.
(Photo): Los Alamos National Laboratory/Getty Images

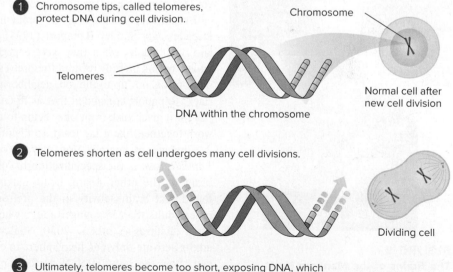

1. Chromosome tips, called telomeres, protect DNA during cell division.

Chromosome

Telomeres

DNA within the chromosome

Normal cell after new cell division

2. Telomeres shorten as cell undergoes many cell divisions.

Dividing cell

3. Ultimately, telomeres become too short, exposing DNA, which becomes damaged, and the cell dies. This is the normal life and death cycle of a cell.

Cell death

those children, having insensitive parents was linked with greater telomere erosion but having sensitive parents was linked with less telomere erosion. Another study showed that adolescents who experienced chaotic family environments with low levels of parental support felt more negative emotion, which in turn predicted decreases in telomere length (Brody & others, 2017).

If stress erodes telomeres, stress-reducing activities might, then, be expected to enhance telomeres, or levels of telomerase. One study showed that individuals who participated in a three-month meditation retreat (which included meditating for six hours a day) showed greater telomerase activity relative to a control group (Jacobs & others, 2011).

Another approach to aging, the *free-radical theory* of aging, states that people age because unstable oxygen molecules known as *free radicals* are produced inside their cells. These molecules damage DNA and other cellular structures (Matkarimov & Saparbaev, 2017; Yegorov, 2020). The damage done by free radicals may lead to a range of disorders, including cancer and arthritis (Gašparović, 2020).

Finally, *hormonal stress theory* argues that aging in the body's hormonal system can lower resistance to stress and increase the likelihood of disease. As people age, stress hormones stay in the bloodstream longer. Prolonged, elevated levels of stress hormones are linked to increased risks for many diseases, including cardiovascular disease, cancer, and diabetes (McEwan, 2017). The hormonal stress theory of aging has focused on the role of chronic stress in diminishing the functioning of the immune system (Guardino & others, 2017; Mate & others, 2014). As we have seen previously, prolonged stress can leave a person vulnerable to disease as the body is not given an opportunity to rest and repair.

Aging and the Brain For decades, scientists believed that no new brain cells are generated past early childhood. However, some new neurons are produced throughout life, in the hippocampus and the olfactory bulb (Brus & others, 2013; Huart & others, 2013). Scientists continue to study factors that might promote or inhibit *neurogenesis* (the term used to describe the generation of new neurons), such as drugs, exercise, or stress (Cutler & Kokovay, 2020; Marchetti & others, 2020).

Some brain areas shrink more than others with aging. The prefrontal cortex is one area that shrinks, and research has linked this shrinkage to a decrease in working memory and other cognitive activities in older adults (Wruck & Adjaye, 2020). The sensory regions of the brain—such as the primary visual cortex, primary motor cortex, and somatosensory cortex—are less vulnerable to the aging process (Rodrigue & Kennedy, 2011).

Even in late adulthood, the brain has remarkable repair capability. Decades ago, Stanley Rapaport (1994) compared the brains of younger and older adults when they were engaged in the same tasks. The older adults' brains literally rewired themselves to compensate for losses. If one neuron was not up to the job, neighboring neurons helped to pick up the slack. Rapaport concluded that as brains age, they can shift responsibilities for a given task from one region to another. Areas of the aging brain work together, like a tag team, to maintain a person's functioning.

Changes in lateralization may be one way the aging brain adapts. *Lateralization* is the specialization of function in one hemisphere of the brain or the other. Using neuroimaging techniques, researchers have found that brain activity in the prefrontal cortex is lateralized less in older adults than in younger adults when they are engaging in cognitive tasks (Esteves & others, 2020; Shalev & others, 2020). So, younger adults activate only one hemisphere to complete a task well. Older adults can show the same level of performance but reach that goal by using both hemispheres. The decrease in lateralization in older adults might play a compensatory role in the aging brain. That is, using both hemispheres may improve the cognitive functioning of older adults.

Results from the Nun Study support the role of experience in maintaining brain function. Recall that this study involves nearly 700 nuns in a convent in Mankato, Minnesota (Snowdon, 2003, 2007) (Figure 9).

FIGURE 9

The Brains of the Mankato Nuns
Nun Study participant Sister Nicolette Welter remained an active, contributing member of her community until her death in 2014, at age 102. (*inset*) A neuroscientist holds a brain donated by one of the Mankato Nun Study participants.
(Sister Nicolette): Scott Takushi/KRT/Newscom; (Brain): Steve Liss/Time Life Pictures/Getty Images

By examining the nuns' donated brains as well as others, neuroscientists have documented the aging brain's ability to grow and change (Pakhomov & Hemmy, 2014). Even the oldest Mankato nuns lead intellectually challenging lives, and neuroscientists believe that stimulating mental activities increase dendritic branching. Keeping the brain actively engaged in challenging activities can help to slow the effects of age.

In sum, in the physical domain across the life span, we see a dramatic pattern of growth and change in infancy, childhood, and adolescence followed by leveling off and decline in adulthood. How might this progression influence (and be influenced by) the way the developing person thinks about oneself and the world? To examine this question, we turn to the cognitive domain of human development.

In older adults, the brain hemispheres may work together more often—becoming jacks-of-all-trades to do what each used to do separately.

self-quiz

1. The first two weeks after conception are referred to as the
   A. fetal period.
   B. germinal period.
   C. embryonic period.
   D. zygotic period.

2. Puberty is generally characterized by all of the following *except*
   A. a decrease in concentrations of certain hormones.
   B. a dramatic increase in height and weight.
   C. a pause in physical growth.
   D. the development of thought.

3. The hormone associated with breast, uterine, and skeletal development during puberty is
   A. testosterone.
   B. estradiol.
   C. androgen.
   D. norepinephrine.

APPLY IT! 4. Gabriel's grandfather has always been a fitness enthusiast. Even at age 83, he continues to work out, lifting weights, swimming, and taking pride in his physique. After his grandfather beats him in a swimming race, Gabriel rolls his eyes and comments that all this work makes no sense at this point in his grandfather's life. Aside from calling him on his sour grapes, what would you say about Gabriel's viewpoint?
   A. Gabriel is right. All that working out is probably inappropriate and even unhealthy for someone in his 80s.
   B. Gabriel is wrong, because if his grandfather stays active, he can completely avoid all of the physical changes associated with age.
   C. Gabriel is wrong. His grandfather's hard work might be paying off in a variety of ways, including his physical health and his cognitive and emotional functioning generally.
   D. Gabriel is right, and his grandfather seems to be in denial about the effects of age on physical development.

3 Cognitive Development

Cognitive development refers to how thought, intelligence, and language processes change as people mature. *Cognition* refers to the operation of thinking and also to our cognitive skills and abilities. In this section, you will encounter one of the biggest names in all of psychology, Jean Piaget, who presented a theory of cognitive development that has had lasting impact. We examine Piaget's contributions and consider more recent research on the relationship between cognitive abilities and age.

Cognitive Development from Childhood into Adulthood

The Swiss developmental psychologist Jean Piaget (1896–1980) traced cognitive development through childhood into adulthood. Let's begin by reviewing Piaget's approach.

FIGURE 12 **Piaget's Conservation Task** The beaker test determines whether a child can think operationally—that is, can mentally reverse action and understand conservation of the substance. (*a*) Two identical beakers are presented to the child, each containing the same amount of liquid. As the child watches, the experimenter pours the liquid from B into C, which is taller and thinner than A and B. (*b*) The experimenter then asks the child whether beakers A and C have the same amount of liquid. The preoperational child says no. When asked to point to the beaker that has more liquid, the child points to the tall, thin one.
Marmaduke St. John/Alamy Stock Photo

symbolic than sensorimotor thought. In preschool years, children begin to represent their world with words, images, and drawings. Thus, their thoughts begin to exceed simple connections of sensorimotor information and physical action.

The type of symbolic thinking that children are able to accomplish during this stage is limited. The stage is called "preoperational" because they still cannot perform what Piaget called *operations,* by which he meant mental representations that are "reversible." Preoperational children have difficulty understanding that reversing an action may restore the original conditions from which the action began.

A well-known test of whether a child can think operationally is to present a child with two identical beakers, A and B, filled with liquid to the same height (Figure 12). Next to them is a third beaker (C). Beaker C is tall and thin, whereas beakers A and B are short and wide. The liquid is poured from B into C, and the child is asked whether the amounts in A and C are the same. The four-year-old child invariably says that the amount of liquid in the tall, thin beaker (C) is greater than that in the short, wide beaker (A). The eight-year-old child consistently says the amounts are the same. The four-year-old child, a preoperational thinker, cannot mentally reverse the pouring action; that is, she cannot imagine the liquid going back from container C to container B. Piaget said that such a child has not grasped the concept of *conservation,* a belief in the permanence of certain attributes of objects despite superficial changes. A child who has mastered conservation understands the principle of reversibility: That while superficially different, some aspects of objects can be restored when an operation is reversed.

The child's thought in the preoperational stage is egocentric. This does not mean that the child is self-centered or arrogant but that preoperational children cannot put themselves in someone else's shoes. They cannot take another's perspective.

Preschoolers aren't trying to frustrate their parents. They may say remarkably cruel things, but they are not able to gauge how their words might make someone else feel.

Preoperational thinking is also intuitive. This means that preoperational children make judgments based on gut feelings rather than logic. In reaching a basic level of operational understanding, the child progresses to the third of Piaget's cognitive stages.

Concrete Operational Stage

Piaget's **concrete operational stage** (7 to 11 years of age) involves using operations and replacing intuitive reasoning with logical reasoning in concrete situations. Children in the concrete operational stage can successfully complete the beaker task described earlier. They are able to mentally imagine the operation of reversing the pouring of the liquid back into the wide beaker. Many of the concrete operations identified by Piaget are related to the properties of objects. For instance, when playing with Play-doh, the child in the concrete operational stage realizes that *the amount* of Play-doh is not changed by changing its shape. One important skill at this stage of reasoning is the ability to classify or divide things into different sets or subsets and to consider their interrelations. (You might remember the childhood song that goes, "One of these things is not like the others," that aimed to coax you into concrete operations.)

Note that the movement from preoperational to concrete operational thought involves greater sophistication in comparing objects and understanding what characteristics are constant about the objects we see in the world. Concrete operational thought involves operational thinking, classification skills, and logical reasoning in concrete contexts. However, at this stage children are not thinking abstractly or hypothetically. According to Piaget, this kind of abstract, logical reasoning occurs in the fourth, and final, cognitive stage.

Consider that this review of cognitive development in childhood has covered life periods that you, yourself, have lived through. Yet, just thinking back on those times to

concrete operational stage Piaget's third stage of cognitive development, lasting from about 7 to 11 years of age, during which the individual uses operations and replaces intuitive reasoning with logical reasoning in concrete situations.

When Is Your First Memory?

What is the earliest memory you can recall? How old were you when the event happened? Most people report a "first memory" occurring when they were between three and four years old (Akhtar & others, 2018). This "time stamp" makes sense because that is when children are able to form and share autobiographical memories (Fivush, 2011).

There is little evidence that children can form verbal autobiographical memories of events that happened prior to being able to speak. Consider a clever and cute experiment in which developmental psychologists staged a very interesting event for children, some of whom were younger than three years. The event involved a "Magic Shrinking Machine." The machine was a large square box with a light on top, an on-and-off lever, and a crank on the side. The researchers brought the box and a suitcase full of toys to each child's

Owen D. Phillips/Shutterstock

home. The children were told that the machine could shrink things magically. They turned on the machine, placed a stuffed toy in the hole on the top, and cranked the crank. Then they opened a door on the bottom. Lo and behold, out popped a toy that was an exact replica but much smaller. They "shrank" a number of other toys during the session. As you can imagine, for the kids, this was a notable event. Yet, one year later, when the children were old enough to talk, none of them could describe their memory of the magic machine (Simcock & Hayne, 2002). This lack of memories for events early in life is sometimes called *infantile amnesia,* meaning that infants do not have the capacity to encode memories that can be shared later when they have language.

Now, interestingly, you might have answered our opening question with a memory of something that happened before you were three years old—like being in a crib or welcoming a new sibling home when you were just two years or even younger. In studies probing first memories, there are always some who claim to have a memory that occurred before three years of age. In a study of over 6,000 adults, 40 percent reported a first memory that occurred prior to three years of age (Akhtar & others, 2018), including some from the first year of life. The researchers concluded that such memories are likely *fictional*—emerging not out of experiences but from family stories, videos, and photographs. Much of our life story is created with and by the people with whom we share our lives. Even if fictional, first memories are not "real" memories; they reflect who we were, even before we had a sense of what that meant. Translating a preverbal experience into an autobiographical memory is much less likely than the alternative of creating a story around an event, told so many times it feels like a memory.

What Do You Think?

- Do you think first memories from infancy are fictional? Why or why not?
- What characteristics, other than age, might determine when a person is able to encode and share autobiographical memories?

---

remember how you thought seems impossible. Understanding the development of cognition in childhood might be a lot easier if we could all remember how we thought back then. Childhood memories have interested psychologists for a long time. If you think back to your first memory, when did it occur? Do you think that memory is real or might it be influenced by family stories or pictures you have seen? To read about first memories, see the Challenge Your Thinking.

Formal Operational Stage People enter the **formal operational stage** of cognitive development at 11 to 15 years of age. This stage continues through the adult years. Formal operational thought is more abstract and logical than concrete operational thought. Most importantly, formal operational thinking includes thinking about things that are not concrete, making predictions, and using logic to come up with hypotheses about the future. Formal operational thought can be directed at thought itself, as the reasoner comes to think about how they think.

Unlike elementary school children, adolescents can conceive of hypothetical, purely abstract possibilities. This type of thinking is called *idealistic* because it involves comparing how things are to how they might be. Adolescents also think more logically. They

formal operational stage
Piaget's fourth stage of cognitive development, which begins at 11 to 15 years of age and continues through the adult years; it features thinking about things that are not concrete, making predictions, and using logic to come up with hypotheses about the future.

begin to think more as a scientist thinks, devising plans to solve problems and systematically testing solutions. Piaget called this type of problem solving *hypothetical-deductive reasoning*. That mouthful of a phrase denotes adolescents' ability to develop hypotheses, or best hunches, about ways to solve a problem such as an algebraic equation. It also denotes their ability to systematically deduce, or come to a conclusion about, the best path for solving the problem. In adolescents and throughout the rest of life, Piaget believed that mature reasoning would be characterized by this capacity to hypothesize, reflect, and use logic to reach conclusions. In contrast, before adolescence, children are more likely to solve problems by trial and error.

In summary, over the course of Piaget's four developmental stages, a person progresses from sensorimotor cognition to abstract, idealistic, and logical thought. Piaget based his stages on careful observation of children's behavior, but there is always room to evaluate theory and research. Let's consider the current thinking about Piaget's theory of cognitive development.

EVALUATING AND EXPANDING ON PIAGET'S THEORY   Piaget opened up a new way of looking at how children's minds develop. We owe him for a long list of concepts that have enduring power and fascination, including schemas, assimilation, accommodation, cognitive stages, object permanence, egocentrism, and conservation. We owe Piaget for the currently accepted view of children as active, constructive thinkers who play a role in their own development. Nevertheless, just as other psychological theories have been criticized and amended, so have Piaget's.

The key challenge of the nature versus nurture debate is understanding what infants bring with them into the world (nature) and what they learn from experience (nurture). Piaget's approach emphasized the experiences that infants have with the world and how those experiences shape infant cognitive development. However, in the past several decades, many studies on infant development using sophisticated experimental techniques have challenged Piaget's view of sensorimotor development, as we now consider.

An Alternative View: The Nativist Approach   Nativist approaches to infant cognitive development suggest that infants bring a great deal more with them into the world than Piaget realized. For example, object permanence develops earlier than Piaget thought.

Research by Renée Baillargeon and her colleagues (Baillargeon & others, 2016; Lin, Stavans, & others, 2020; Stavans & others, 2019) shows that infants as young as three to four months expect objects to be *substantial* (in the sense that other objects cannot move through them) and *permanent* (in the sense that they continue to exist even when they are hidden).

Another example of a nativist theorist is Elizabeth Spelke (Huang & Spelke, 2014; Liu & others, 2019). She endorses a **core knowledge approach**, which states that infants are born with domain-specific knowledge systems. Among these knowledge systems are those involving space, number sense, object permanence, and language. Strongly influenced by evolution, the core knowledge domains are theorized to be "prewired" to allow infants to make sense of their world (Coubart & others, 2014). After all, Spelke asks, how could infants possibly grasp the complex world in which they live if they did not begin life equipped with core sets of knowledge? In this approach, core knowledge forms a foundation around which more mature cognitive functioning and learning develop. The core knowledge perspective argues that Piaget greatly underestimated the cognitive abilities of young infants (Liu & Spelke, 2017).

core knowledge approach
A perspective on infant cognitive development that holds that babies are born with domain-specific knowledge systems.

A major criticism of the nativist approach is that these scholars focus too much on what happens inside the infant's head apart from the rich social environments in which babies live (Nelson, 2013). Critics point out that even young infants have already accumulated hundreds, and in some cases thousands, of hours of experience in grasping what the world

Deborah Jaffe/Corbis

is about, which gives considerable room for the environment's role in the development of infant cognition (Highfield, 2008).

Whether infants bring this knowledge with them into the world, as the nativists believe, or it is acquired through early experience, many researchers conclude that young infants are far more competent than Piaget thought (Baillargeon & others, 2016; Johnson & Hannon, 2015). Such findings have led many developmental psychologists to assert that we must expand our appreciation for the perceptual and cognitive tools that are available to infants with very little experience (Baillargeon & others, 2016; Scott & Baillargeon, 2013; Spelke & others, 2013).

Vygotsky: Cognitive Development in Cultural Context

Piaget did not think that culture and education play important roles in cognitive development. For Piaget, the child's active interaction with the physical world was all that was needed for cognitive development. Russian psychologist Lev Vygotsky (1962) took a different approach, recognizing that cognitive development is an interpersonal process that happens in a social and cultural context. Vygotsky emphasized not only the importance of social interaction for learning and developing but also the wider cultural context. For him, talking about development outside that context makes no sense. Vygotsky did not embrace the idea of stages of development and was most interested in how the process of learning pushed development forward.

Vygotsky thought of children as apprentice thinkers who develop as they interact in dialogue with more knowledgeable others, such as parents and teachers. Vygotsky believed that these expert thinkers spur cognitive development by interacting with a child in a way that is just above the level of sophistication the child has mastered. In effect, these interactions provide *scaffolding* that allows the child's cognitive abilities to be built higher and higher.

Teachers and parents, in other words, provide a framework for thinking that is always just at a level the child can strive to attain. Furthermore, in Vygotsky's view, the goal of cognitive development is to learn the skills that will allow the person to be competent in their particular culture. These expert thinkers are not simply guiding a child into a level of cognitive sophistication, but also, along the way, sharing with the child important aspects of culture, such as language and customs. For Vygotsky, a child is not simply learning to think about the world—a child is learning to think about *their own world*.

Scaffolding is a way of learning that we use throughout life. Whenever we interact with someone who has more expertise, we have a new scaffold to climb.

Information-Processing Approach

The information-processing approach to development focuses on basic cognitive processes such as attention, memory, and problem solving (Barrouillet, 2015). For instance, from this perspective, researchers examine how the development of working memory (the mental workspace used for problem solving) relates to other aspects of children's development (Blankenship & others, 2019; Köster & others, 2020). Children who have better working memory are more advanced in reading comprehension, math skills, and problem solving than their counterparts with less effective working memory (Friedman & others, 2017; Peng & others, 2016). One study found that students with low working memory were at higher risk of dropping out of high school, even accounting for socioeconomic status and IQ (Fitzpatrick & others, 2015).

Another important aspect of cognitive development is **executive function**. Executive function refers to complex cognitive processes, including thinking, planning, and problem solving (Kenny & others, 2019). For children to be successful in school, they must be able to sit still, wait in line, and so forth. These simple tasks require self-control and the capacity to inhibit one's automatic responses. It is not surprising, then, that executive function during the preschool years is linked to school readiness (Pellicano & others, 2017). Executive function also predicts the development of social cognitive abilities, including theory of mind: the understanding that other people experience private knowledge and mental states (Devine & others, 2016).

Parents and teachers play important roles in the development of executive function. Parents who model executive function and self-control can serve as scaffolds for these skills (Mazursky-Horowitz & others, 2017). A variety of activities increase children's

executive function
Higher-order, complex cognitive processes, including thinking, planning, and problem solving.

executive function, such as training to improve working memory (Kirk & others, 2017; Sasser & others, 2017), aerobic exercise (Ishihara & others, 2017), and mindfulness training (Mak & others, 2017). Cognitive activities that require children to stretch the way they think can influence executive function. For example, five-year-olds who were instructed to complete an executive function measure as if they were someone else (for example, Batman) performed better than children without these special instructions (White & others, 2017). It is important to keep in mind that most interventions have not shown long-term benefits (Takacs & Kassai, 2019).

The information-processing approach differs from Piaget's classic perspective in that it does not look for qualitative differences across the lifespan. Instead, it examines how people think, using common processes, at every phase of life. From this perspective, we can ask whether children think like adults do for many different topics.

Revisionist Views of Adolescent and Adult Cognition   Researchers have also expanded on Piaget's view of adolescent cognition. In addition to advancing into Piaget's stage of formal operational thinking, another characteristic of adolescent thinking, especially in early adolescence, is *egocentrism*. Adolescent egocentrism involves the belief that others are as preoccupied with the adolescent as they are, that they are unique, and invincible (meaning unable to be harmed). Egocentrism at this developmental period is revealed, for example, when an adolescent perceives others to be noticing and watching them more than is the case. Think of the teenager who cannot be seen at the mall unless their hair is "just so." For that teen, everyone in the world will be watching when they make their entrance Posting pictures on social media that keep one's followers abreast of the minutiae of one's life might also express adolescent egocentrism. The sense of invincibility is the most dangerous aspect of adolescent egocentrism. This belief may lead to risky behaviors such as drag racing, drug use, and unsafe sex (Aalsma & others, 2006).

Piaget underestimated infant cognition but he may have overestimated adolescents' and adults' cognitive achievements. Formal operational thought does not emerge as consistently and universally in early adolescence as Piaget envisioned (Kuhn, 2008), and many adolescents and adults do not reason as logically as Piaget proposed. It may be that even in adulthood, we do not use logical reasoning to make decisions but rather to justify decisions that are the product of the intuitive hunches that characterize cognition in childhood (Mercier & Sperber, 2011).

At every age we put our own special stamp on egocentrism. Have you ever noticed that each of your instructors seems to think that their class is the only one you're taking?

Finally, developmental psychologists interested in cognition have noted that cognitive changes can occur *after* Piaget's formal operations stage. For Piaget, formal operational thought is the highest level of thinking, and he argued that no new qualitative changes in cognition take place in adulthood. Developmental psychologists, however, in expanding their focus to the entire life span, have tracked the ways that cognitive skills might change throughout adulthood.

Generally, approaches to cognitive development share the idea that cognition goes from simpler to more sophisticated—from more intuitive to more rational—over the course of life. Yet, there may be times when age is associated with being more biased. Sometimes, children might reason more logically than adults. To read about this fascinating topic, see the Intersection.

Cognitive Processes in Adulthood

Recall that for Piaget, each stage of cognitive development entails a way of thinking that is *qualitatively different* from the stage before. From Piaget's perspective, meaningful cognitive development ceases after the individual reaches the formal operational stage. Subsequent research has examined not qualitative differences in thinking over time, but the ebb and flow of cognitive abilities as a function of age. What kind of cognitive changes occur in adults?

Developmental Psychology and Cognitive Psychology: Can Young Children Be More Rational Than Adults?

Imagine being told there are 100 people in a park and only 5 of them are engineers. If a randomly selected person in the park is described as enjoying working with numbers and wears glasses, adults are likely to guess (mistakenly) that that person is an engineer. As we have already reviewed, the **representativeness heuristic** leads to mistakes when people ignore valuable numeric information and instead rely on vivid description or what psychologists call "individuating information." In the example above, the information about working with numbers and wearing glasses is the individuating information. It conflicts with the numbers given and leads many adults to less than optimal decisions. Would children make the same mistake? Researchers have probed this very question.

In this work, kids were told about robots of different colors in a park. A first study confirmed that children are able to use both types of information accurately, when each type is presented separately (Gualtieri & Denison, 2018). For instance, when they were told there were more blue than green robots in a park (numeric information), children correctly guessed that a robot wearing a white coat, so that its color could not be seen, was likely to be blue. Also, when told that green robots are naughty and blue robots are nice (individuating information), kids were able to guess that a naughty robot wearing a white coat was likely to be green.

What happens if the two types of information suggest different answers? To find out, the researchers asked children (aged four to six years) to make judgments when the numeric and individuating information conflicted. For example, knowing that there are more blue than green robots but that green robots are naughty, would children mistakenly guess that a robot who behaves in a naughty

Terry Vine/Getty Images

fashion is green? The answer is that younger children used base-rate information to make accurate judgments. Younger kids relied on the numbers! However, older children fell prey to the representativeness heuristic. Interestingly, the researchers tried out this task on college students and found that young adults were less likely than four-year-olds to use base-rate information when conflicting individuating information was provided (Gualtieri & Denison, 2018). That's right, four-year-olds rendered more rational judgments than college students.

These results challenge us to think about how, through development, people become less likely to trust the numbers and more likely to rely on social beliefs and stereotypes when rendering social judgments. Think about how these results might apply to real life. During the COVID-19 pandemic, people heard about numbers, including infection rates and deaths. Yet many failed to take precautions because they did not personally know anyone who was affected.

\\ **Why do you think people become less reliant on numbers and more reliant on individuating information as they age?**

representativeness heuristic
The tendency to make judgments about group membership based on physical appearances or the match between a person and one's stereotype of a group rather than on available base rate information.

COGNITION IN EARLY ADULTHOOD   Some experts on cognitive development argue that the typical idealism of Piaget's formal operational stage is replaced in young adulthood by more realistic, pragmatic thinking (Labouvie-Vief, 1986). Gisela Labouvie-Vief (2006) proposed that the increasing complexity of cultures in the past century has generated a greater need for reflective, more complex thinking that takes into account the changing nature of knowledge and the kinds of challenges contemporary thinkers face. She emphasizes that key aspects of cognitive development for young adults include deciding on a particular worldview or philosophy of life, recognizing that the worldview is subjective, and understanding that diverse worldviews should be acknowledged. In her perspective, only some individuals attain the highest level of thinking.

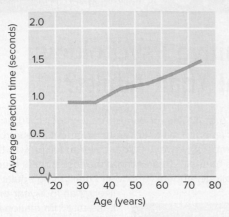

FIGURE 13   The Relationship between Age and Reaction Time In one study, the average reaction time slowed in the 40s, and this decline accelerated in the 60s and 70s (Salthouse, 1994). The task used to assess reaction time required individuals to match numbers with symbols on a computer screen.

> When is processing fastest? Slowest? > When does speed of processing start to slow down? > This was a cross-sectional study performed in 1994 using a computerized task. How might cohort effects explain the differences observed? > How would a longitudinal study examine this same question?

COGNITION IN MIDDLE ADULTHOOD

What happens to cognitive skills in middle adulthood? Although some cross-sectional studies indicate that middle adulthood is a time of cognitive decline, longitudinal evidence presents a different picture. K. Warner Schaie is conducting an extensive longitudinal study measuring a host of different intellectual abilities in adults repeatedly, starting in 1956 (Schaie, 1994, 2007, 2010, 2012). The highest level of functioning for four of the six intellectual abilities occurred in middle adulthood (Schaie, 2010, 2012). Only two of the six abilities declined in middle age. Based on the longitudinal data he has collected, Schaie concluded that middle adulthood is the period when many people reach their peak for a range of intellectual skills.

COGNITION IN LATE ADULTHOOD

Many contemporary psychologists conclude that some dimensions of intelligence decline in late adulthood whereas others are maintained or may even increase (Anderson & Craik, 2017; Dixon & others, 2013). One of the most consistent findings is that when the speed of processing information is involved, older adults do not perform as well as their younger counterparts (Figure 13). This decline in speed of processing is apparent in middle-aged adults and becomes more pronounced in older adults (Salthouse, 2012, 2017).

In general, older adults perform more poorly than younger adults in most, but not all, aspects of memory (Cansino & others, 2020; Ebaid & Crewther, 2020). In the area of memory involving knowledge of the world (for instance, the capital of Peru), older adults usually take longer than younger adults to remember the information, but they often are able to retrieve it (Kuo & others, 2014). In working memory, the important area of memory in which individuals manipulate and assemble information to solve problems and make decisions, decline occurs in older adults as well (Forsberg & others, 2020).

Some aspects of cognition might improve with age. One such area is **wisdom**, expert knowledge about the practical aspects of life (Barber & others, 2020; Ferrari & Westrate, 2013; Staudinger & Gluck, 2011). Wisdom may increase with age because of the buildup of life experiences, but individual variations characterize people throughout their lives (Grossman & others, 2010). Thus, not every older person has wisdom, and some young people are wise beyond their years.

wisdom
Expert knowledge about the practical aspects of life.

In younger individuals wisdom may come at a cost. They might gain wisdom from living through difficult life experiences.

Some factors can lessen the decline in cognitive ability among older adults. Education is consistently related to slower cognitive decline with age (Clouston & others, 2020; Seblova & others, 2020). Physical fitness is a powerful tool against cognitive decline (Leist & others, 2020). Maintaining sexual activity is also associated with less cognitive decline (Smith & others, 2020).

When older adults continue to increase their engagement in cognitive and physical activities, they are better able to maintain their cognitive functioning in late adulthood (Park & others, 2014; Rebok & others, 2014). Some software-based cognitive training games have been found to improve older adults' cognitive function (Hertzog & others, 2009; Nouchi & others, 2013). However, and importantly, training does not consistently generalize to other tasks or cognitive outcomes not targeted by the intervention (Guye & von Bastian, 2017; Souders & others, 2017). Indeed, many experts conclude that older adults are less able to adapt than younger adults and thus are limited in how much they can improve their cognitive skills (Finch, 2009; Salthouse, 2012).

self-quiz

1. Of the following activities, which are healthy older adults most likely to perform more poorly compared to younger adults?
   A. Reciting facts about the Civil War
   B. Reacting as fast as possible to a word presented on a computer screen
   C. Remembering how to perform tasks like tying a shoe
   D. Remembering the lyrics to a song they know well

2. Most cognitive skills reach their peak during _____, and most show decline during _____.
   A. early adulthood; middle adulthood
   B. middle adulthood; late adulthood
   C. adolescence; early adulthood
   D. adolescence; middle adulthood

3. Vygotsky stressed that cognitive development is an interpersonal process that happens in a _____ context.
   A. historical
   B. physical
   C. cultural
   D. positive

APPLY IT! 4. Tyrone is babysitting his younger cousins, who are three, four, and nine years of age. For lunch, each child will be drinking apple juice, which they all love. Tyrone has only three serving cups—one that is short and wide, and two that are tall and thin. Although Tyrone pours the same amount of juice into all the cups, the younger kids fuss and fight over who gets stuck with the short, wide cup. The nine-year-old shrugs and takes the wide cup. Tyrone later proclaims, "Those other two kids are really spoiled brats! Thank goodness the oldest is not so selfish." Which of the following best applies to Tyrone's conclusion?
   A. Tyrone is right—young kids are more likely to be spoiled and whiny.
   B. Tyrone does not understand that the younger kids do not recognize that the amount of juice in the cups is the same. The oldest child is not being unselfish; the nine-year-old knows the amounts are the same.
   C. The nine-year-old probably does not understand that the wider cup contains the same amount of juice as the other two.
   D. Tyrone probably got an *A* in Developmental Psychology.

4 Socioemotional Development

How do our social and emotional lives change through childhood and adulthood? Recall that *socioemotional processes* involve changes in person's social relationships, emotional life, and personality.

Socioemotional Development in Infancy

When we observe the newborns behind the window of a hospital nursery, one thing is clear: Humans differ from one another in terms of their emotional demeanor from the very beginnings of life. Some are easygoing, and some are prone to distress. Furthermore, in the earliest days of life, infants encounter a social network that will play an important role as they develop a sense of self and world. To begin exploring the socioemotional aspects of development, we focus first on the ingredients of emotional and social processes that are present very early in life—infant temperament and attachment (Groh & others, 2017).

TEMPERAMENT   **Temperament** refers to an individual's behavioral style and characteristic ways of responding. For infants, temperament centers on their emotionality and ways of reacting to stimuli in the environment. Does the infant look with interest at a new toy, or shy away? Is the infant easily soothed after an upset? Answers to these questions provide information about infant temperament.

temperament
An individual's behavioral style and characteristic way of responding.

There are a number of ways to think about infant temperament. For example, psychiatrists Stella Chess and Alexander Thomas (1977) identified three basic types of temperament in children:

1. The *easy child* generally is in a positive mood, quickly establishes regular routines in infancy, and easily adapts to new experiences.

2. The *difficult child* tends to be fussy and to cry frequently and engages in irregular daily routines.

3. The *slow-to-warm-up child* has a low activity level, tends to withdraw from new situations, and is very cautious in the face of new experiences.

Other researchers have suggested that we should think about infants as being high or low on different dimensions, such as *effortful control* or *self-regulation* (controlling arousal and not being easily agitated), *inhibition* (being shy and showing distress in an unfamiliar situation), and *negative affectivity* (tending to be frustrated or sad) (Kagan, 2013). Regardless of what the precise aspects of temperament might be, the emotional characteristics that a child brings into the world serve as a foundation for later personality (Tang & others, 2020), and the child's earliest social bonds might set the stage for later social relationships (Girme & others, 2020).

ATTACHMENT   Infants need warm social interaction to survive and develop. A classic (and controversial) study by Harry Harlow (1958) demonstrated the essential importance of warm contact. Harlow separated infant monkeys from their mothers at birth and placed them in cages in which they had access to two artificial "mothers." One of the mothers was a physically cold wire mother; the other was a warm, fuzzy cloth mother (the "contact comfort" mother). Each mother could be outfitted with a feeding mechanism. Half of the infant monkeys were fed by the wire mother, half by the cloth mother. The infant monkeys nestled close to the cloth mother and spent little time on the wire one, even if it was the wire mother that gave them milk (Figure 14). When afraid, the infant monkeys ran to the comfy mom. This study clearly demonstrates that what the researchers described as "contact comfort," not feeding, is crucial to the attachment of an infant to its caregiver.

Infant attachment is the close emotional bond between infant and caregiver. British psychiatrist John Bowlby (1969, 1989) theorized that the infant and the mother instinctively form an attachment, just like other animals. In Bowlby's view, the newborn comes into the world equipped to stimulate the caregiver to respond; it cries, clings, smiles, and coos. Bowlby thought that our early

infant attachment The close emotional bond between an infant and its caregiver.

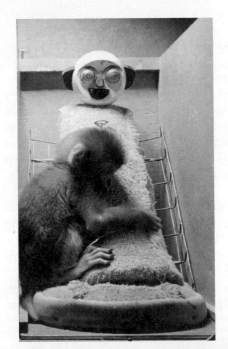

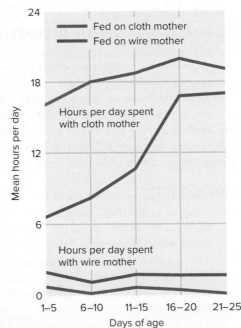

FIGURE 14   **Contact Time with Wire and Cloth Surrogate Mothers** Regardless of whether the infant monkeys were fed by a wire or a cloth mother, they overwhelmingly preferred to spend contact time with the cloth mother.
(Photo): Science Source

relationships with our caregivers are internalized and serve as our schemas for our sense of self and the social world. From Bowlby's perspective, infant experiences set the stage for future relationships, including our romantic relationships as adults. Many developmental psychologists concur that attachment during the first year provides an important foundation for later development (Groh & Narayan, 2019; Sroufe, 2020).

Mary Ainsworth devised a way to measure children's attachment, called the *strange situation test* (Ainsworth, 1979; Ainsworth & others, 1978). In this procedure, caregivers leave infants alone with a stranger and then return. Responses of children to this situation are used to classify their attachment style. Ainsworth used the term **secure attachment** to describe how infants use the caregiver, usually the mother, as a secure base from which to explore the environment. In the strange situation, the secure infant is upset when the mother leaves but calms down and appears happy to see her when she returns. Infants who are securely attached are more likely to have mothers who are consistent, responsive, and accepting, and who express affection toward them than are infants who are insecurely attached (Leerkes & others, 2017). The securely attached infant moves freely away from the mother but also keeps tabs on her by periodically glancing at her. An insecurely attached infant, in contrast, avoids the mother or is ambivalent toward her. In the strange situation, such an infant might not even notice the mother has gone (sometimes called *avoidant* or *dismissive attachment style*) or conversely might respond with intense distress, only to rage at the mother when she returns (sometimes called *anxious* or *preoccupied attachment style*) (Brownell & others, 2015). Most infants show a secure attachment style.

One criticism of attachment theory is that it does not adequately account for cultural variations (Mistry & others, 2013). For example, in some cultures infants show strong attachment to many people, not just to their primary caregiver (Rothbaum & others, 2000, 2007). Infants in agricultural societies tend to form attachments to older siblings who are assigned a major responsibility for younger siblings' care.

Another critique of attachment theory is that it may not account for temperamental differences among infants, which may color the attachment relationship (Kagan, 2013). A research review demonstrated that although attachment and temperament are linked they are also separable phenomena (Groh & others, 2017). Caregivers and infants likely share genetic characteristics, and it might be that the attachment relationship is really a product of these shared genes. Despite such criticisms there is ample evidence that secure attachment is important to development (Powell & others, 2014; Thompson, 2015a, 2015b; Thompson & others, 2014).

Equipped with the key basic ingredients of temperament and attachment, how does a human being develop in the socioemotional domain? This question was addressed by Erik Erikson, who devised a theory of what he called psychosocial development. Erikson's theory has powerfully guided thinking about how human beings' social and emotional capacities develop throughout the entire life span.

secure attachment
The ways that infants use their caregiver, usually their mother, as a secure base from which to explore the environment.

Erikson's Theory of Socioemotional Development

The life-span development theory of the influential psychologist Erik Erikson (1902–1994), who trained as a psychoanalyst under Sigmund Freud, proposed eight psychosocial stages of development from infancy through late adulthood. In calling the stages *psychosocial,* Erikson meant to emphasize how a person's psychological life is embedded in and shaped by social relationships and challenges faced by the developing person. Figure 15 illustrates all of Erikson's eight stages.

Take a thorough look at Figure 15. The details of Erikson's perspective are important. Think about your own life—what were you doing during these stages?

Trust versus Mistrust

Developmental period: Infancy (birth to 1.5 years)

Characteristics: A sense of trust requires a feeling of physical comfort and minimal amount of fear about the future. Infants' basic needs are met by responsive, sensitive caregivers.

Autonomy versus Shame and Doubt

Developmental period: Toddlerhood (1.5 to 3 years)

Characteristics: After gaining trust in their caregivers, infants start to discover that they have a will of their own. They assert their sense of autonomy, or independence. They realize their will. If infants are restrained too much or punished too harshly, they are likely to develop a sense of shame and doubt.

Initiative versus Guilt

Developmental period: Early childhood (preschool years, ages 3 to 5)

Characteristics: As preschool children encounter a widening social world, they are challenged more and need to develop more purposeful behavior to cope with these challenges. Children are now asked to assume more responsibility. Uncomfortable guilt feelings may arise, though, if the children are irresponsible and are made to feel too anxious.

Industry versus Inferiority

Developmental period: Middle and late childhood (elementary school years, age 6 to puberty)

Characteristics: At no other time are children more enthusiastic than at the end of early childhood's period of expansive imagination. As children move into the elementary school years, they direct their energy toward mastering knowledge and intellectual skills. The danger at this stage involves feeling incompetent and unproductive.

Identity versus Identity Confusion

Developmental period: Adolescence (10 to 20 years)

Characteristics: Individuals are faced with finding out who they are, what they are all about, and where they are going in life. An important dimension is the exploration of alternative solutions to roles. Career exploration is important.

Intimacy versus Isolation

Developmental period: Early adulthood (20s, 30s)

Characteristics: Individuals face the developmental task of forming intimate relationships with others. Erikson described intimacy as finding oneself yet losing oneself in another person.

Generativity versus Stagnation

Developmental period: Middle adulthood (40s, 50s)

Characteristics: A chief concern is to assist the younger generation in developing and leading useful lives.

Integrity versus Despair

Developmental period: Late adulthood (60s–)

Characteristics: Individuals look back and evaluate what they have done with their lives. The retrospective glances can be either positive (integrity) or negative (despair).

FIGURE 15 **Erikson's Eight Stages of Psychosocial Development** Erikson changed the way psychologists think about development by tracing the process of growth over the entire life span.

(First): Tari Faris/Getty Images; (Second): Stephan Hoeck/Getty Images; (Third): Ariel Skelley/Getty Images; (Fourth): Ariel Skelley/Blend Images LLC; (Fifth): fstop123/Getty Images; (Sixth): Blue Moon Stock/Alamy Stock Photo; (Seventh): Ken Karp/McGraw Hill; (Eighth): Ryan McVay/Getty Images

From Erikson's (1968) perspective, each stage represents the developmental task that the person must master at a particular place in the life span. According to Erikson, these developmental tasks are represented by two possible outcomes, such as trust versus mistrust (Erikson's first stage). If an infant's physical and emotional needs are well taken care of, the person will experience an enduring sense of trust in others. If, however, these needs are frustrated, the person might carry concerns about trust throughout life, with bits of this unfinished business being reflected in the rest of the stages. Erikson described each stage as a turning point with two opposing possible outcomes—one, greater personal competence; the other, greater weakness and vulnerability. Using Erikson's eight stages as a guide, let's consider the ways humans develop in terms of their capacities for interpersonal relationships and emotional well-being.

Erik Erikson (1902–1994) *Erikson generated one of the most important developmental theories of the twentieth century.*
Ted Streshinsky Photographic Archive/Getty Images

SOCIOEMOTIONAL DEVELOPMENT IN CHILDHOOD: FROM TRUST TO INDUSTRY

The first four of Erikson's eight stages apply to childhood. Each stage is linked with the activities associated with human life at each age:

- *Trust versus mistrust:* Infancy (birth to 1.5 years) is concerned with establishing trust in the social world. At this stage, the helpless infant depends on caregivers to establish a sense that the world is a predictable and friendly place. Without a sense of trust, the child lacks a firm foundation for social relationships.

- *Autonomy versus shame and guilt:* During toddlerhood (1.5 to 3 years), children, many of whom are going through toilet training, experience the beginnings of self-control. When these young children have the opportunity to experience control over their own behaviors, they develop the capacity for independence and confidence. This is the time when young children begin to say "no!" and assert themselves in interactions with others.

- *Initiative versus guilt:* In early childhood (3 to 5 years), preschoolers experience what it is like to forge their own interests and friendships and to take on responsibilities. If you have ever spent time with a three-year-old, you know how often the child wants to help with whatever an adult is doing. When they experience a sense of taking on responsibility, preschoolers develop initiative. Otherwise, according to Erikson, they may feel guilty or anxious.

- *Industry versus inferiority:* During middle and late childhood (6 years to puberty), children enter school and gain competence in academic skills. Just as the label *industry* would suggest, children find that this is the time to get to work, learn, achieve, and learn to enjoy learning.

From Erikson's perspective, then, children should grow toward greater levels of autonomy and self-confidence as they progress from infancy to school age and beyond. Is there a particular parenting style that is most likely to lead to these sorts of outcomes? Let's find out.

Parenting and Childhood Socioemotional Development

Researchers have tried to identify styles of parenting associated with positive developmental outcomes. Diana Baumrind (1991, 1993) described four basic styles of interaction between parents and their children:

1. **Authoritarian parenting** is a strict punitive style. The authoritarian parent firmly limits and controls the child with little verbal exchange. In a difference of opinion about how to do something, for example, the authoritarian parent might say, "You do it my way or else." Children of authoritarian parents sometimes lack social skills, show poor initiative, and compare themselves with others.

2. **Authoritative parenting** encourages the child to be independent but still places limits and controls on behavior. This parenting style is more

authoritarian parenting
A restrictive, punitive parenting style in which the parent exhorts the child to follow the parent's directions and to value hard work and effort.

authoritative parenting
A parenting style that encourages the child to be independent but that still places limits and controls on behavior.

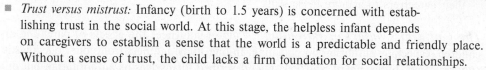 This is tricky: Authoritarian and authoritative sound very similar, but they are very different!

collaborative. Extensive verbal give-and-take is allowed, and parents are warm and nurturing toward the child. An authoritative parent might put his arm around the child in a comforting way and say, "You know you should not have done that; let's talk about how you can handle the situation better next time." Children of authoritative parents tend to be socially competent, self-reliant, and socially responsible.

3. **Neglectful parenting** is distinguished by a lack of parental involvement in the child's life. Children of neglectful parents might develop a sense that other aspects of the parents' lives are more important than they are. Children whose parents are neglectful tend to be less competent socially, to handle independence poorly, and (especially) to show poor self-control.

4. **Permissive parenting** places few limits on the child's behavior. Permissive parents let children do what they want. Some parents deliberately rear their children this way because they believe that the combination of warm involvement and few limits will produce a creative, confident child. However, children with very permissive parents typically rate poorly in social competence. They often fail to learn respect for others, expect to get their own way, and have difficulty controlling their behavior. Recall that socioemotional development means becoming increasingly adept at controlling and regulating one's emotions and behaviors. Children may require structure from their caregivers to acquire these skills.

The Cultural Context of Parenting   Culture influences the effects of parenting on children, especially authoritarian parenting. In one study, mothers from four collectivistic cultures (Iran, India, Egypt, and Pakistan) described themselves as authoritarian but did not express negative attitudes about their children, and the children did not show negative outcomes (Rudy & Grusec, 2006). A large-scale, cross-cultural meta-analysis found that across cultures and nations, some aspects of authoritative parenting are associated with positive child outcomes and some aspects of authoritarian parenting are associated with negative outcomes (Pinquart & Kauser, 2018). The researchers concluded that, across the globe, parents should be encouraged to adopt the authoritative style.

SOCIOEOMOTIONAL DEVELOPMENT IN ADOLESCENCE: THE EMERGENCE OF IDENTITY

Erikson (1968) viewed the key challenge of adolescence as identity versus identity confusion. In seeking an *identity,* adolescents must find out who they are, what they are all about, and where they are going in life. Adolescents are confronted with many new roles and adult statuses—from the vocational to the romantic. If they do not adequately explore their identities during this stage, they emerge confused about who they are. Adolescents who spend this time in their lives exploring alternatives can reach some resolution of the identity crisis and emerge with a new sense of self. Those who do not successfully resolve the crisis become confused, suffering what Erikson calls *identity confusion*. This confusion is expressed in one of two ways: Either individuals withdraw, isolating themselves from peers and family, or they lose themselves in the crowd.

Marcia's Theory of Identity Status   Building on Erikson's ideas, James Marcia proposed the concept of *identity status* to describe an adolescent's position in the development of an identity (Kroger & others, 2010; Marcia, 1980, 2002). Marcia described two dimensions, exploration and commitment. *Exploration* refers to a person's investigating various options for a career and for personal values. *Commitment* involves deciding which identity path to follow and personally investing in attaining that identity. Various combinations of exploration and commitment give rise to one of four identity statuses, according to Marcia (Figure 16):

- *Identity diffusion:* The adolescent has neither explored nor committed to an identity. Adolescents experiencing identity diffusion have not confronted the crisis of identity or are so overwhelmed by the challenge of answering the question "Who am I?" that they have withdrawn. Eventually, as they engage in life, these adolescents will begin the process of thinking about what matters (and what does not) and enter into the stage of exploration called *moratorium.*

- *Identity moratorium:* The adolescent is actively exploring and trying on new roles but has not committed to a particular identity. For example, consider an adolescent who passionately tries out a variety of internship opportunities to see what different jobs might be like.

- *Identity foreclosure:* The adolescent has committed to a particular identity but has done so without actually exploring their options. An example is an adolescent who, without much consideration, decides to pursue accounting as a major in college because everyone in their family is an accountant.

	Has the person made a commitment?	
	Yes	**No**
Has the person explored meaningful alternatives regarding some identity question? **Yes**	Identity achievement	Identity moratorium
No	Identity foreclosure	Identity diffusion

FIGURE 16

Marcia's Four Statuses of Identity Who are you? When you think of how you have come to identify yourself, which of these four statuses does your answer best represent?

- *Identity achievement:* After exploring the options and committing to an identity, the adolescent emerges with a sense of their own values and principles, a sense of the kind of person they wish to be, and goal commitments that provide life with a feeling of purpose.

Let's consider an example to illustrate Marcia's theory. Four teenagers, Mary, Tyron, Kris, and Riley, attend a summer camp for future engineers. Because they are all at the same camp, we might think they all identify with the idea of becoming an engineer. But they may differ in how much they have explored their identities or committed to this identity. For example, last summer, Mary attended a camp for future medical professionals; the following school year she interned at a law firm. Mary is still exploring her options. She has not committed to being an engineer and is therefore in *identity moratorium.* In contrast, Tyron has also deeply considered many different careers, imagining himself as a news reporter, a surgeon, and an engineer. He has shadowed different professionals in their jobs. Ultimately, he is excited about an engineering career. Tyron would be in *identity achievement.* The key difference between moratorium and achievement is that in moratorium the person is still exploring. Now, neither Kris nor Riley has explored their options. Kris wants to be an engineer because their parents are engineers. They never really considered alternative paths. Kris occupies *identity foreclosure.* Finally, Riley is attending the camp because his parents sent him, hoping it sparks an interest to start thinking about the future. Riley has neither explored nor committed to an identity and is therefore in *identity diffusion.* Note that Kris and Riley are similarly lacking in exploration but Kris is highly committed to the engineer identity.

Identity within a Larger Culture The adolescent search for identity occurs in a larger culture and is affected by the cultural messages conveyed by others. Developing an identity in adolescence can be especially challenging for people with marginalized identities, including minoritized racial or ethnic, sexual, and gender identity groups, Native American youth, religious minorities, immigrant, or disability-related groups (Pohjola, 2020; Quam & others, 2020; Raifman & others, 2020). As they mature, many adolescents become acutely aware of the evaluation of their groups by the majority culture. Ethnic minority youth may also face the challenge of *biculturalism*—identifying in some ways with their ethnic group and in other ways with the majority culture (Ferguson & others, 2020; Romero & others, 2020).

Adolescence can be a time when gender and sexual identities come to the fore (Eisenberg & others, 2017). Sexual and gender minority youth may face high levels of bullying and other social challenges (Earnshaw & others, 2017; Kaufman & others, 2020; Raifman & others, 2020). Like all adolescents, for these people, developing a positive view of sexual or gender minority identity also predicts better outcomes (Hall, 2017) as does an accepting, supportive family, school, and community environment (Bouris & Hill, 2017; Romijnders & others, 2017).

For ethnic minority youth, feeling both a positive attachment to their own group and an attachment to the larger culture is related to more positive academic and emotional outcomes (Rivas-Drake & others, 2014; Williams & others, 2014). Although it might seem that being a member of marginalized group would make life more stressful, having a strong positive identification with one's group can buffer adolescents from the effects of discrimination (Sladek & others, 2020). For both minority and majority adolescents, developing a positive identity is an important life theme (Chung & others, 2020; Palmqvist & others, 2020; Scroggs & Vennum, 2020). Living in a diverse society means that people will be different from each other in many ways. Creating a context where all identities are celebrated allows youth from all different groups to find a safe, nurturing, and positive context to be themselves.

Parents and Peers Parents and peers can help the adolescent answer the central questions of identity: "Who am I, and who do I hope to become?" To help adolescents reach their full potential, a key parental role is to be an effective manager—one who locates information, makes contacts, helps to structure adolescents' choices, and provides guidance. By assuming this managerial role, parents help adolescents to avoid pitfalls and to work their way through the decisions they face (Campione-Barr & others, 2020; Van Petegem & others, 2017).

For parents to play an active role in the development of their adolescent kids, they have to know what is going on in their adolescent's life. Research on adolescents' management of their parents' access to information, especially disclosing or concealing strategies about their activities, shows that adolescents are more willing to disclose information to parents when parents ask teenagers questions and when the parent–child relationship is characterized by a high level of trust and acceptance (McElvaney & others, 2014; Van Zalk & Van Zalk, 2020).

During adolescence, people spend more time with peers than they did in childhood. Peer influences can be positive or negative. A significant aspect of positive peer relations is having one or more close friends. Adolescents can learn to be skilled and sensitive partners in intimate relationships by forging close friendships with peers (Allen & others, 2020). However, some friends can negatively impact adolescents' development (Ray & others, 2017). Hanging out with delinquent peers in adolescence can be a strong predictor of substance abuse, delinquent behavior, and depression (Memmott-Elison & others, 2020; Widdowson & others, 2020).

Contemporary youth are often affected by their peers online as well as in person.

For Erikson, once the issues of identity have been resolved, the young adult turns to the important domain of intimate relationships. However, contemporary scholars note that many young people seem to be putting off the kinds of commitments to marriage, family, and career that we associate with adult life. Jeffrey Arnett (2006, 2007, 2010, 2012) introduced the concept of emerging adulthood to describe this transitional period that is partly an extended adolescence and partly a phase of experimenting with adult roles. If you are a traditional-age college student, you are at this point in the life span. Let's briefly examine socioemotional development in this life stage.

SOCIOEMOTIONAL DEVELOPMENT IN EMERGING ADULTHOOD

emerging adulthood
The transitional period from adolescence to adulthood, spanning approximately 18 to 25 years of age.

Emerging adulthood is the transitional period from adolescence to adulthood (Arnett, 2006, 2007, 2012, 2014; Arnett & Fischel, 2013). The age range for emerging adulthood is approximately 18 to 25 years. Experimentation and exploration characterize the emerging adult. Many emerging adults are still investigating their career path, their identity, and the kinds of close relationships they will have.

Jeffrey Arnett (2006, 2010) identified five main features of emerging adulthood:

- *Identity exploration, especially in love and work:* Emerging adulthood is the time of significant changes in identity for many people.
- *Instability:* Residential changes peak during emerging adulthood, a time during which there also is often instability in love, work, and education.

Some people enter adulthood earlier than others. Some enter into committed romantic relationships, start a family, and take up full-time employment right out of high school. Do you think these individuals experience "emerging adulthood"?

- *Self-focus:* Emerging adults are self-focused in that they may have little in the way of social obligations, little in the way of duties and commitments to others, which leaves them with a great deal of autonomy in running their own lives" (Arnett, 2006, p. 10).

- *Feeling "in between":* Many emerging adults consider themselves neither adolescents nor full-fledged adults.

- *The age of possibilities, a time when individuals have an opportunity to transform their lives:* Arnett (2006) describes two ways in which emerging adulthood is the age of possibilities: (1) Many emerging adults are optimistic about their future, and (2) for emerging adults who have experienced difficult times while growing up, emerging adulthood presents an opportunity to guide their lives in a more positive direction.

Emerging adults have more choices—and more control over those choices—in their daily life. The choices they make with regard to their life goals have implications for their later well-being (Hill & others, 2011), suggesting that this time of life sets the stage for later development. Like an extended adolescence, emerging adulthood implies continual identity exploration (Eriksson & others, 2020). Eventually, though perhaps later than previous generations, emerging adults adopt the mantle of adulthood. According to Erikson, the early part of adulthood is occupied with the experience of loving, intimate relationships.

How do you think economic conditions affect the existence of "emerging adulthood"?

SOCIOEMOTIONAL DEVELOPMENT IN EARLY ADULTHOOD: INTIMACY VERSUS ISOLATION

The sixth of Erikson's (1968) eight stages, *intimacy versus isolation,* refers to the challenge of forming intimate relationships with others or becoming socially isolated. Erikson describes intimacy as both finding oneself and losing oneself in another. If the young adult develops healthy friendships and an intimate relationship with a partner, intimacy will likely be achieved. One key way that young adults achieve intimacy is through long-term relationships with a romantic partner, often including marriage.

Marriage and Families

A striking change in social attitudes in recent decades has been the decreased stigma attached to people who do not maintain what were long-considered conventional families. Adults today choose many different ways of living and form many types of families. They live alone, live with a partner, marry or divorce. Contemporary families are sometimes blends of families from previous relationships. Sometimes the couples that head households are the same gender. Sometimes families include members who are chosen rather than a product of biological relationships.

Cohabitation refers to living together in a sexual relationship without being married, and there has been a dramatic increase in the number of cohabiting U.S. couples since 1970. According to the U.S. Census Bureau, 17 million adults in the United States are cohabiting, about 7 percent of the population (Gurrentz, 2019). The group of couples who are living together is more diverse than ever in terms of age, ethnicity, income, and education. Interestingly, cohabitation appears to be especially on the rise among divorced Americans aged 50 and older, who saw a 75 percent increase in cohabitation rates.

With regard to marriage, contemporary women are waiting much longer to marry. For example, in 1980, the median age for a first marriage in the United States was 24.7 for men and 22.0 for women. In 2020, the median age for a first marriage for men is estimated to be 30.5 and for women, 28.1 (U.S. Census Bureau, 2020).

Purestock/SuperStock

What makes a successful marriage? John Gottman has been studying married couples' lives since the early 1970s. He interviews couples, films them interacting with each other, and even measures their heart rate and blood pressure during their interactions (Gottman & others, 2002; Madhyastha & others, 2011). He also checks back with the couples every year to see how their marriages are faring. He and his colleagues continue to follow married heterosexual couples, as well as

Remember the median is the halfway point in a distribution. So the 50th percentile.

same-gender couples, to try to understand what makes relationships successful. Gottman (2006) has found these four principles at work in successful marriages:

1. *Nurturing fondness and admiration:* Partners sing each other's praises. When couples put a positive spin on their talk with and about each other, the marriage tends to work.

2. *Turning toward each other as friends:* Partners see each other as friends and turn toward each other for support in times of stress and difficulty.

3. *Giving up some power:* Bad marriages often involve one partner who is a powermonger.

4. *Solving conflicts together:* Couples work to solve problems, regulate their emotion during times of conflict, and compromise to accommodate each other.

SOCIOEMOTIONAL DEVELOPMENT IN MIDDLE ADULTHOOD: GENERATIVITY VERSUS STAGNATION

Erikson's seventh stage, *generativity versus stagnation,* occurs in middle adulthood. Generativity means creating something of value that will benefit future generations. Generative people believe they have left a lasting legacy and helped the next generation in important ways. This feeling that one has made a lasting and memorable contribution to the world is related to higher levels of psychological well-being (Homan & others, 2020; Moieni & others, 2020). The feeling of having done nothing of value for future generations is stagnation. Although Erikson did not think that parenting alone was a guarantee of generativity, he did believe that parenting could be a way to experience this important developmental accomplishment. Midlife adults can also experience generativity through mentoring, volunteering, and political activism.

SOCIOEMOTIONAL DEVELOPMENT AND AGING: INTEGRITY VERSUS DESPAIR

From Erikson's perspective, the person who has entered life's later years is engaged in looking back—evaluating their life, seeking meaning, and coming to terms with death. Erikson called this eighth and final stage *integrity versus despair.* If the person experiences life as a meaningful and coherent whole, they can face the later years with a strong sense of meaning and low fear of death. In the absence of integrity, the older adult is filled with despair and fear (Westerhof & others, 2017). Current research on socioemotional development and aging reveals that Erikson was correct in his view that meaning is a central concern for older adults. However, he may have overlooked that this meaning derives not necessarily from the past but also from the present.

In terms of social relationships, older adults become more selective about their social networks. At the same time, older adults report higher levels of happiness than their younger counterparts (Carstensen & others, 2011; Stanley & Isaacowitz, 2011). Laura Carstensen developed *socioemotional selectivity theory* to explain the narrowing of social contacts and the increase in positive emotion that occur with age (Carstensen, 2006; Carstensen & others, 2011; English & Carstensen, 2014). The theory states that because they recognize their time on earth is limited, older adults tend to be selective in their social interactions, striving to maximize positive, meaningful experiences. Although younger adults may gain a sense of meaning in life from long-term goals, older adults derive meaning from satisfying relationships and activities in the *present.* Unlike younger adults, who may be preoccupied with the future, older adults, and those experiencing limited time, embrace the present moment with increasing vitality (Hicks & others, 2012; Kotter-Grühn & Smith, 2011; Smith & Bryant, 2019). Socioemotional selectivity theory posits that it is not age itself but rather limited time that leads adults to focus on the present. Young adults who are asked to imagine having limited time show the same pattern of maximizing positive meaning in the present.

If you were about to take a yearlong trip, how, and with whom, would you choose to spend your last two weeks at home?

Just as physical changes can influence socioemotional experience, socioemotional factors can influence physical health. Older adults who remain connected to others, experience a high level of self-acceptance, and positive attitudes toward aging tend to live longer (Levy & others, 2002; Ng & others, 2020). Of course, even the most positive people will eventually die, and coming to terms with the truth of our mortality is part of human life as well, as we will consider later.

EVALUATING ERIKSON'S THEORY   Using Erikson's framework, we have followed the active developer from the early accomplishment of trust in the social world through to their capacity to experience a strong sense of meaning in a life well lived. Erikson's perspective has a number of strengths. He was the first to emphasize that development is a lifelong process and does not stop with childhood. He also developed important ideas about the central themes that often occupy human life at various ages. However, Erikson's conclusions have had their critics. First, Erikson mainly practiced case study research and based his ideas on the lives of single individuals not large-scale studies of representative samples. Second, because Erikson tried to capture each stage with a single theme (such as intimacy vs. isolation), Erikson left out other important developmental tasks that might occur in various stages. For example, Erikson said that the main task for young adults is to resolve the conflict between intimacy and isolation. However, another significant developmental task in early adulthood involves careers and work. In addition, many people live satisfying lives as singles, without marrying. Erikson's approach did not anticipate the kind of identity exploration that currently continues to occur in emerging adulthood. Despite these criticisms, Erikson is a giant in developmental psychology.

self-quiz

1. All of the following are examples of generativity *except*
   A. teaching the English language to recent U.S. immigrants.
   B. successfully investing one's earnings and enjoying a satisfying retirement.
   C. volunteering at a homeless shelter.
   D. acting as a scout leader.

2. Recent studies on socioemotional development and aging indicate that
   A. older adults put a great deal of stock in life's meaning.
   B. older adults may be more selective in their social interactions than younger adults.
   C. older adults are happier than their younger counterparts.
   D. all of the above are true.

3. With regard to ethnic identity, psychological research indicates that
   A. adolescents who downplay the importance of their ethnic identity do better in school.
   B. adolescents use their ethnic identity to avoid developmental tasks.
   C. adolescents who have a strong ethnic identity are better able to cope with stress.
   D. adolescents from different racial groups face the same challenges to their ethnic identity.

APPLY IT!   4. Rosemary's Grandpa Jack is 80 years old and wears a hearing aid and glasses, and he has always been very active. He and Rosemary's grandmother spend a great deal of time with their children and grandchildren, and both laugh a lot and seem genuinely to enjoy life. Grandpa Jack recently resigned his position as a member of the local senior citizen's council. He tells Rosemary that the council had become a hassle. Rosemary wonders whether this decision is normal and reasonable or whether her granddad is feeling depressed. Based on your reading of this example and this chapter, should Rosemary be worried?
   A. Rosemary should be worried because it appears that Grandpa Jack is withdrawing from opportunities for generativity.
   B. Rosemary should be worried because Grandpa Jack should stay involved in community activities given that such involvement is related to being happier and healthier.
   C. Rosemary should probably not be worried. Grandpa Jack just sounds as if he is going through what most people do when they are faced with limited time: maximizing meaningful social contacts.
   D. Rosemary should not be worried because Grandpa Jack has completed Erikson's stages, and it makes sense that he is probably preparing for death by withdrawing from the world.

5 Gender Development

So far, we have explored development across the human life span in the physical, cognitive, and socioemotional domains. In this section, we examine an important aspect of the human experience—gender development—that illustrates the links across the various domains of development. Understanding the development of gender requires attention to all three life domains—physical, cognitive, and socioemotional.

Gender refers to the social and psychological aspects of one's experience as male, female, both, or neither. The development of gender is a complex process that includes biological, cognitive, and socioemotional factors. Gender is not simply biological sex. Gender is people's understanding of the meaning of gender in their life, their gender identity, and gender expression. **Gender identity** is a person's inner concept of self in relation to the ideas of being male, female, both, or neither (Krieger, 2017). Gender identity consistent with or different from the sex one is assigned at birth and can be thought of as a continuum. Gender identity is distinct from **sexual orientation** (which we will review in Chapter 9). It is also separable from *gender expression,* the ways people present themselves through their

gender
The social and psychological aspects of being male, female, both, or neither.

gender identity
A person's inner concept of themselves in relation to the ideas of being male, female, both, or neither.

sexual orientation
The direction of an individual's erotic interests, today viewed as a continuum from exclusive male–female relations to exclusive same-gender relations.

appearance, including hairstyle, clothing choices, voice, and behavior. Finally, **gender role** refers to the types of behaviors society expects of people of different genders. To explore the varieties of gender identities, see the Psychology in Our World, below.

gender roles
Roles that reflect the society's expectations for how people of different genders should think, act, and feel.

Biology and Gender Development

Humans typically have 46 chromosomes arranged in pairs. The twenty-third pair may have an X shape (made by two X chromosomes) or it may look like an upside-down Y (one X and one Y chromosome). The XX pair produces a genetic female and the XY pair produces a genetic male.

In the first few weeks after conception, XX and XY embryos look alike. When the Y chromosome triggers the secretion of *androgens,* the main class of male sex hormones, male sex organs start to differentiate from female sex organs. Low androgen levels in a genetic female embryo allow for the typical development of female sex organs.

This is why those who are assigned male at birth are more likely to show characteristics like color blindness that are linked to the X chromosome. People assigned female at birth have a backup X chromosome.

You may have heard about studies showing that even as young infants, those assigned male at birth prefer "boy toys" (like trucks) and those assigned female at birth prefer "girl toys" (like dolls). Research does support the existence of such differences, especially if children are given choices specifically between trucks and dolls (Davis & Hines, 2020). In addition, similar differences have been reported in nonhuman primates (Lonsdord, 2017). However, the types of differences uncovered may depend on many factors, including the age of the infants, the types of toys considered, and parental attitudes about gender (Liu & others, 2020). Biological differences do not tell the whole story of gender development. Indeed, the biological organ we have emphasized in understanding human behavior, the brain, does not show large or consistent differences between people assigned male at birth and people assigned female at birth (Hyde & others, 2018; Joel, 2011; Joel & others, 2015).

Cognitive Aspects of Gender Development

Recall from Piaget's theory that a schema is a mental framework that organizes and guides an individual's thoughts. A *gender schema* is a mental framework for understanding what gender means in one's culture. Through experience, children learn that gender is an important organizing principle in social life, and they come to recognize that people who differ on gender are different in ways that matter. This gender schema then serves as a cognitive framework by which children interpret further experiences related to gender.

Because of where they are in cognitive development, very young children may believe that a person's gender can change depending on superficial features. They may reason that girls have long hair, and so if that hair is cut short, a girl becomes a boy.

How do gender schemas develop? Theorists suggest that children acquire these schemas through learning in the social world (Bandura & Bussey, 2004). Such learning occurs through processes, including reinforcement and punishment, observational learning, and modeling (Bandura & Bussey, 2004).

Children learn about gender through experience—by observing the activities and behaviors of others.
LightField Studios/Shutterstock

In subtle ways, children may be rewarded (or punished) for behavior depending on whether the behavior conforms to expectations for their gender. The child's social environment responds to behaviors in various ways, coloring their perception of their appropriateness. A girl might learn that pretending to be a professional football player is not a way to please her parents. A boy might pick up on his mother's subtle frown when he announces that he wants to try on her high-heeled shoes. Interestingly research has shown that the small gender difference in toy preferences grows with age and is most stark when children are being observed in a lab (rather than playing at home) (Todd & others, 2018). Such results suggest the pressure children feel as they age to conform to gender norms especially when they feel they are being watched.

PSYCHOLOGY IN OUR WORLD

Human Identities and the Changing Gender Landscape

Contemporary society includes many gender identities. Many of these identities have a long history in human experience but have only recently entered the mainstream conversation. In some ways, the science of gender has scarcely been able to keep up with the gender statuses people now occupy—*Facebook* currently offers more than 50 gender options for its user profiles. Let's review two of these, transgender and nonbinary identities.

Transgender refers to experiencing one's psychological gender (and gender expression) as different from one's assigned sex. Actor Elliot Page recently came out as trans and adopted the pronouns he/him and they/them. Cisgender refers to individuals who experience a match among their experienced, expressed, and assigned gender. It is difficult to know with any certainty how many transgender people live in the United States. There are more young people who identify as transgender, perhaps because they feel more free to be open about their identities (Hegarty & others, 2018; Krieger, 2017). For example, by the age of four years, Ellie Ford (assigned male at birth) demonstrated to her parents in many ways that she was a girl. At a princess-themed birthday party, she told her mother, "Mom, I'm not a

Gareth Cattermole/Contour/Getty Images

boy; I'm a girl in my heart and mind" (Peachman, 2017). Youth who identify as transgender have more options than in the past. Medications to delay puberty can provide a "time-out" from sexual development, preventing the permanent bodily changes spurred by puberty (Giordano & Holm, 2020). Gender-affirming surgery is a treatment pursued by some transgender adults, but many transgender people do not desire surgery. Some opt for hormone treatment only or no treatment. For some individuals, being transgender means living according to the belief that a person can be a man who happens to have a vagina or a woman who has a penis—as a person who occupies a different but valid gender territory.

Nonbinary gender identities include identification as a blend of male and female or as neither male nor female. People who identify as nonbinary may refer to themselves with various labels, including *genderqueer, genderfluid, transmasculine or transfeminine*, or *third gender* (Diamond, 2020; Krieger, 2017). It is not uncommon for nonbinary individuals to choose gender-neutral names and to choose, explicitly, the pronouns by which they prefer to be called ("they/them"). These changes can be difficult for friends and family to accommodate but they are expressions of the person's innermost self. To be *misgendered* can be quite hurtful (Dolan & others, 2020).

The changing gender landscape has inspired many different reactions, ranging from warm acceptance to outrage. However, controversies over restrooms should not cloud the fact that people who are different from typical are no less valuable. Learning new ways to understand the experience of others can be challenging but it is important keep in mind that for many people, labels like "transgender" or "nonbinary" reflect life-changing discoveries and important landmarks of identity.

According to Albert Bandura (1986), modeling is an especially potent mechanism for transmitting values. Children gain information about gender from the adults models they encounter. Who goes to work every day? Who does the housework? When children see their parents and other adults engaging in gender-related behavior (and as they observe whether and how these behaviors are reinforced), they learn about how gender matters in their world.

Socioemotional Experience and Gender Development

Social experiences clearly influence gender development. After all, behaviors are rewarded or punished by other people in the child's world (Liben & others, 2014). In addition to parents, peers also play an important role in gender development. Especially between the ages of 6 and 11, peer groups often segregate into boy groups and girl groups (Maccoby, 2002). Peers are stricter than most parents in rewarding behavior that conforms to gender rules and in punishing behavior that violates those rules (MacMullin & others, 2020). Children can be especially punitive toward children assigned male at birth who engage in behaviors that are not typically boyish (Wallien & others, 2010). As children grow up, they adopt gender roles, representing expectations of about how people of different genders ought to be behave, feel, and think (Eagly, 2009, 2012).

Again we see how cognitive development matters. At this age, kids are in Piaget's concrete operations stage, associated with putting things (and people) in categories.

Culture plays a vital role in the content of gender roles. Some cultures emphasize that children should be reared to adopt traditional gender roles. Parents in these cultures bring up children assigned male at birth to be powerful, aggressive, and independent. In contrast, those assigned female at birth are reared to be sensitive to others, concerned with social relationships, and less assertive. Other cultures emphasize rearing children to be assertive and self-reliant as well as sensitive and caring toward others, regardless of gender. The United States and many European countries have moved toward more egalitarian views of gender (Koenig & others, 2011; Mistry & others, 2013).

The gender differences we see in terms of pursuing math and science careers are not about differences in ability. Rather, they appear to be about the kinds of beliefs, expectations, and role models people have in these careers.

While acknowledging biologically influenced gender differences, many psychologists believe that social and cultural factors have a much stronger influence on eventual gender identity (Bandura & Bussey, 2004; Eagly & Wood, 2011, 2012). From this perspective, gender differences in career and life choices can be explained by differences in the availability of role models and beliefs about self-efficacy and personal control in these roles. Janet Shibley Hyde (2005, 2007, 2014), an expert on the psychology of gender, has periodically reviewed studies of gender differences in a wide range of characteristics. She has found that where differences do emerge, they are quite small. Overall, Hyde has found strong support for what she calls the **gender similarities hypothesis**—the idea that people of different genders are much more similar than they are different (Hyde, 2005, 2007, 2014).

gender similarities hypothesis Hyde's proposition that people of different genders are much more similar than they are different.

Nature and Nurture Revisited: The John/Joan Case

Although it seems clear that gender emerges as a function of a variety of intermingling processes, some have debated whether one of these processes is more important than the others. John Money, a well-known sex researcher, believed strongly that socialization was the main determinant of gender. In the 1960s, a case presented itself that gave Money the opportunity to test this theory. In 1965 assigned male twins were born, and a few months after birth, one twin's penis was destroyed during circumcision. Money persuaded the child's parents to allow him to surgically transform the injured male genitals into female genitals and to agree to treat the child as a girl. That child was then reared as a girl and, according to Money, essentially became a girl (Money & Tucker, 1975). The John/Joan case was a famous example of nurture's triumph over nature. For many years, this case was used as evidence for the amazing flexibility of gender.

Milton Diamond, a biologist and strong critic of Money's theory, followed up on Money's most famous case (Diamond & Sigmundson, 1997). Diamond found that over time, "Joan" became less and less interested in a female gender identity and eventually refused to continue the process of feminization that Money had devised. We now know that "Joan" was really David Reimer, whose biography, *As Nature Made Him,* written by John Colapinto (2000), revealed the difficulties of his life. David struggled with traumatic gender-related life experiences and depression, eventually dying by suicide in 2004. The John/Joan case of gender development illuminates the complex ways in which the physical, cognitive, and socioemotional domains of development intersect and influence one another.

Though David's story seems to indicate that biological factors powerfully guide gender development, remember that it is difficult to make generalizations based on a single case study. Other similar cases have had more positive outcomes.

1. Among other factors, the development of gender involves
   A. personality factors.
   B. cognitive factors.
   C. sensory factors.
   D. all of the above.

2. The term for a cognitive framework, developed through observational learning, for understanding what gender means is gender
   A. role.
   B. identity.
   C. diversity.
   D. schema.

3. Childhood peer groups
   A. are not concerned with rewarding and punishing gender-related behavior in members of the group.
   B. play a key role in a person's gender development.

   C. typically are mixed, in terms of gender.
   D. are less important than the media in gender development.

APPLY IT! 4. When Jerry announces he's planning to major in mechanical engineering, his proud parents are not surprised. They tell their friends, "We *knew* he was destined to be interested in mechanical things. Even as a tiny infant he loved mechanical toys and never gave his stuffed animals a second look! He's always been our little engineer." What does research on the psychology of gender development have to say about Jerry's parents' conclusion?
   A. Jerry may have always been their little engineer, but their comments to friends suggest that Jerry was not very different from most baby boys.

   B. Jerry's parents' observations are supported by research showing that very early interest in mechanical toys predicts later career decisions.
   C. Jerry's parents' observations cannot be true, because very young infants cannot tell the difference between mechanical toys and stuffed animals.
   D. Jerry's parents are probably biased by cultural beliefs, and so they remember his behavior in infancy fitting their preexisting beliefs about gender roles.

6 Moral Development

Moral development involves changes that occur with age in people's thoughts, feelings, and behaviors regarding the principles and values that guide them as they interact with others (Walker, 2014a, 2014b). Much research on moral reasoning and thinking has revolved around Lawrence Kohlberg's theory of moral development.

Kohlberg's Theory

Kohlberg (1958) began his study of moral thinking by asking children, adolescents, and adults questions about a series of stories. One of the stories goes something like this: A man, Heinz, whose wife is dying of cancer, knows about a drug that might save her life. He approaches the pharmacist who has the drug, but the pharmacist refuses to give it to him without being paid a very high price. Heinz is unable to scrape together the money and decides to steal the drug.

After reading the story, the interviewee was asked questions about the moral dilemma. Should Heinz have stolen the drug? Kohlberg was less interested in the answer to this question than he was to the response to the next one: Why? Based on the reasons people gave for their answers, Kohlberg (1986) evaluated their level of moral development. Kohlberg's stages of moral development consisted of three broad levels, which include the following:

How would you answer Kohlberg's question?

Lawrence Kohlberg (1927–1987)
Kohlberg created a provocative theory of moral development. In his view, "Moral development consists of a sequence of qualitative changes in the way an individual thinks."
Lee Lockwood/The LIFE Images Collection/Getty Images

1. *Preconventional:* The personl's moral reasoning is based primarily on the consequences of a behavior and on punishments and rewards from the external world. Reasoning is guided by not wanting Heinz to go to jail or by concern for the pharmacist's profits. A child who reasons that he should not steal the drug because someone might get mad would occupy this stage.

2. *Conventional:* The person abides by standards learned from parents or society's laws. At this level a person might reason that Heinz should follow the law no matter what.

3. *Postconventional:* The person recognizes alternative moral courses, explores the options, and then develops an increasingly personal moral code. At this level, the person might reason that Heinz's wife's life is more important than a law. When a person is able to think deeply and reason beyond consequences or external codes (such as laws), that person is reasoning at the postconventional level. At this level, the person might even feel guilty for *not* breaking a law that they viewed as unfair or unjust.

Kohlberg studied with Piaget, and his approach to moral reasoning emphasized the ability to think in sophisticated ways. For Kohlberg, a sense of justice was at the heart of moral reasoning, which he believed laid the foundation for moral behavior.

Critics of Kohlberg

Carol Gilligan (b. 1936) *Gilligan argues that Kohlberg's approach does not give adequate attention to relationships. In Gilligan's view, Kohlberg failed to listen to the ways women, especially, talked about moral dilemmas.*
Frazer Harrison/Getty Images

A criticism of Kohlberg is that his view does not adequately reflect concern for other people and social bonds (Ball & others, 2017; Killen & Smetana, 2014). Kohlberg's theory is called a *justice perspective* because it emphasizes the rights of the individual as the key to sound moral reasoning. In contrast, the *care perspective*, which is at the core of Carol Gilligan's (1982) approach to moral development, views people in terms of their connectedness with others and stresses interpersonal communication, relationships, and concern for others. From Gilligan's perspective this weakness in Kohlberg's approach explains why, using his measures, women generally scored lower than men on moral development.

Similarly, Kohlberg neglected the influence of culture on morality. Because of its central role in how we relate to each other, culture influences the ways individuals navigate moral dilemmas (Gibbs, 2014). Contemporary research on moral judgment shows that moral decisions are influenced by many different cultural values; it is not limited to care versus justice (Graham & others, 2011; Kivikangas & others, 2020). For instance, modern moral psychologists include not only care and justice but liberty, loyalty, and purity as important dimensions of moral reasoning.

You might have noticed a particular limitation of Kohlberg's emphasis on cognitive reasoning—that a person can know the right thing to do and yet might act in an immoral way. Contemporary research on moral development focuses on behavior that reflects moral goodness and on the socioemotional factors associated with this behavior.

Moral Development in a Socioemotional Context

Researchers interested in moral development have increasingly focused on the factors that predict **prosocial behavior**, behavior that is intended to benefit other people (Dovidio & others, 2017). Research has examined how attachment (Gross & others, 2017), cognitive abilities (Longobardi & others 2020; O'Toole & others, 2017), and brain reactivity (Decety & others, 2017) predict acts of kindness in children. Warm, empathic parenting is associated with children's prosocial behavior (Michalik & others, 2007), while rejecting and harsh parenting is linked to lower prosocial behavior and moral disengagement (Hyde & others, 2010). Childhood characteristics are important because longitudinal research shows that kind, moral children are more likely to be kind, moral adults (Eisenberg & Spinrad, 2014).

prosocial behavior
Behavior that is intended to benefit other people.

Other research on moral development has focused on when a child first shows signs of possessing a conscience (Scrimgeour & others, 2017). Having a conscience means hearing that voice in our head that tells us that something is morally good or bad. What sorts of experiences give rise to this inner voice? Researchers have examined the conversations toddlers have with their mothers to identify features of those interactions associated with moral development. Clear and elaborate parent–child interactions that are rich with emotional content and that include shared positive emotion foster conscience development (Laible & Karahuta, 2014).

You have likely heard the phrase, "Children can be cruel." This statement is often made after children have been unkind to a classmate or have said something inappropriate and hurtful (without, perhaps, being aware of its hurtfulness). Yet, anyone who has spent time with a three-year-old knows that "Can I help?" is often the first thing out of a child's mouth when a project is underway, whether it is baking cookies or cleaning the house (Pettygrove & others, 2013). Helping emerges early in human life. In one study, 18- and 30-month-olds were given two chances to help an adult in need. The majority of 18-month-olds and nearly all of the 30-month-olds helped on at least one task (Waugh & Brownell, 2017). A large body of evidence supports the idea that in the first two years of life, children experience empathy for people in need, begin to understand the principle of fairness, and show the ability to behave prosocially (Hammond & others, 2017; Newton & others, 2016; Sommerville & others, 2013; Warneken & Tomasello, 2007).

self-quiz

1. In Kohlberg's preconventional stage, moral reasoning centers on
   A. parental standards and established laws.
   B. personal moral standards.
   C. a behavior's consequences, as well as rewards and punishments from the external world.
   D. loyalty to other people.

2. Moral behavior is influenced by all of the following *except*
   A. early attachments.
   B. culture.
   C. parenting.
   D. the senses.

3. Efforts that are geared to helping others are referred to as
   A. the care perspective.
   B. prosocial behavior.
   C. the justice perspective.
   D. socioemotional behavior.

APPLY IT! 4. Chris and Amanda are parents to three-year-old Emma, who is always getting into trouble in preschool. She says mean things, snatches toys from other children, and engages in a lot of mischief. Her parents want to instill a strong sense of conscience in their little troublemaker. Which of the following strategies is best supported by the research you read about in this section?
   A. The parents should punish Emma for her mischief swiftly and without a word of discussion. Trying to reason with a three-year-old is a mistake.
   B. The parents should reward Emma consistently for good behavior and ignore her misdeeds. She is probably just trying to get attention.
   C. The parents should engage Emma in conversations about her behaviors, praising her for her good deeds and talking about the feelings that others might have when she gets in trouble. They should try to be clear and not focus only on the bad things she does.
   D. The parents should do nothing. The age of three years is too young to worry about these issues, and Emma will not understand what they are talking about until she is much older.

7 Death, Dying, and Grieving

Everything that lives will eventually die. This is, and always has been, the reality for living organisms. Because humans possess the capacity for consciousness, we are aware of our own mortality. Our ability to imagine ourselves in the future makes the reality of our own eventual demise inescapable. Researchers from a variety of perspectives have examined the ways that the awareness of death influences our thoughts, feelings, and behaviors.

Fireworks displays are thrilling rituals that make us feel like valued and invested participants in the larger culture.
Young Yun/Getty Images

Terror Management Theory: A Cultural Shield against Mortality

In the 1970s anthropologist Ernest Becker (1971) drew together theory and research from the social sciences to devise a grand theory of human life and culture. According to Becker, the human awareness that we will someday die creates the potential for overwhelming terror. Yet somehow we go about life without being preoccupied by the terrifying reality of death. Why is this so?

Becker proposed that as our capacity for self-awareness evolved, so did the human ability to create and invest in culture. Culture provides the customary beliefs, practices, religious rules, and social order for humans living together. Becker asserted that being part of a larger culture shields us from the terror of our own mortality. He maintained that by investing in a cultural worldview (our beliefs, routine practices, and standards for conduct), we are able to enjoy a sense of immortality. Culture provides both religious ideas about life after death and a context for accomplishments that will outlive us. As long as we feel that we are valued members of a culture, this status will buffer us against our fears of personal death.

Social psychologists Jeffrey Greenberg, Sheldon Solomon, and Tom Pyszczynski applied Becker's ideas in empirical research. They named their approach *terror management theory (TMT)* (Solomon & others, 1991; Pyszczynski & others 2020). A multitude of studies have explored TMT (Arndt & Goldenberg, 2017; Burke & others, 2010). This research shows that when reminded of their own death, people will endorse cultural beliefs more strongly (Arrowood & others, 2017), reject individuals who violate those beliefs (Leippe & others, 2017), and seek to maintain or enhance their self-esteem (Pyszczynski & others, 2004). Terror management studies support Becker's idea that cultural beliefs act as a buffer against the ultimate reality of our inevitable demise. Studies on the theory suggest that reminders of death can serve not only to promote defensiveness but also to spur people's creativity and motivation toward personal growth (Vail & others, 2012). Interestingly, research shows that compared to younger adults, older adults are less likely to strongly defend their cultural worldview after contemplating their own death (Maxfield & others, 2012), perhaps because they have come to accept that death is a fact of life.

Although terror management research tells us a great deal about how people cope with the reality of death in the abstract, it has yet to explore the process of dying itself. Even if one invests in a culture and in so doing establishes a sense of symbolic immortality, the reality of death remains. For aged or gravely ill persons, that reality can be experienced with particular vividness. How do people react when their own death is near?

Kübler-Ross's Stages of Dying

Elisabeth Kübler-Ross pioneered psychology's interest in the process of dying in her seminal book *On Death and Dying* (1969). Focusing on terminally ill people, Kübler-Ross identified five progressive stages of coping with death:

1. *Denial:* The person rejects the reality of impending death. They may express thoughts such as "I feel fine—this cannot be true."

2. *Anger:* Once the reality of death sets in, the person feels angry and asks, "Why me?" The unfairness of the situation may be especially upsetting.

3. *Bargaining:* The person may bargain with God, with doctors, or within their own head: "If I can just have a little more time, I'll do anything."

4. *Depression:* The person feels profound sadness and may begin to give up on life: "What's the point of doing anything?"

5. *Acceptance:* The person comes to terms with the difficult reality of their own death. A realization sets in that they will die and "it will be okay."

Jonathan Drake/Reuters/Alamy Stock Photo

Kübler-Ross eventually applied her stages of dying to a variety of life experiences, such as grieving for a loved one or mourning the loss of a job. Her book was important because it called on psychologists to confront the difficult problem of death. Still, her work has been criticized, and recent research has called into question a number of her ideas, such as the notion that everyone goes through these stages in the order she envisioned. One danger in the presentation of stages of dying is that it seems to represent a "right" way to die or the "best" way to grieve. Clearly, just as infants differ from one another from birth, so do individuals who face the end of life.

Bonanno's Theory of Grieving

Losing a loved one is certainly likely to be stressful. Research by George Bonanno and his colleagues has tracked individuals who have experienced bereavement, such as the loss of a spouse, over time (Bonanno, 2004, 2005; Chen & others, 2020; Maccallum & others, 2017). That research has identified four different patterns of grief:

1. *Resilience:* Resilient individuals experience immediate grief over their loss but only for a brief time and return quickly to their previous levels of functioning. Such individuals do not experience a profound disruption of life, despite having gone through a staggering loss.

2. *Recovery:* In the recovery pattern, the person experiences profound sadness and grief that dissipates more slowly. The individual will ultimately return to previous levels of functioning, but as a much slower and more gradual unfolding over time.

3. *Chronic dysfunction:* In this case, a traumatic grief experience leads to a long-term disruption of functioning in important life domains. Chronically dysfunctional people may ultimately be at risk for psychological disorders such as depression.

You might think that resilient individuals represent the lucky few. But Bonanno has found that resilience is actually the most common pattern.

4. *Delayed grief or trauma:* Some people do not experience the sadness or distress evoked by a loss immediately following that loss. Instead, these intense feelings may come over the person weeks or even months later.

Carving Meaning Out of the Reality of Death

Death is a difficult concept to contemplate. Consider, though, that the reality of death is one of the things that makes life so precious. Reminders of death can increase individuals' perception that life is meaningful and satisfying (King & Hicks, 2020). Furthermore, knowing that our loved ones will eventually die makes our commitments to our social relationships all the more heroic.

Indeed, the reality of death makes our commitment to life more meaningful. In the next section we will revisit that commitment, an important factor in development itself.

self-quiz

1. Studies on terror management theory indicate that when reminded of their own death, people
   A. lose interest in maintaining or building their self-esteem.
   B. endorse cultural beliefs more strongly.
   C. become less creative.
   D. reject the mainstream thinking of their culture.

2. Kübler-Ross's proposed stages of coping with death include all of the following *except*
   A. feeling angry.
   B. making amends to others.
   C. denying that death will come.
   D. accepting death.

3. In Bonanno's theory of grieving, the pattern in which the individual grieves for only a brief time and quickly resumes normal functioning is called
   A. recovery.
   B. rehabilitation.
   C. resilience.
   D. resolution.

APPLY IT! 4. When Lionel, Tawnya's beloved husband of 20 years, dies of a sudden heart attack, Tawnya is grief stricken. But just a few weeks later, she returns to work and is back to her routine, caring for her teenage sons. Considering her loss, she seems to be doing surprisingly well. Tawnya starts to wonder if there is something wrong with her. She loved Lionel deeply and remembers how, when her own father died, her mother seemed to be devastated for a long time. Tawnya confesses her feelings to a friend. According to research on grieving, which of the following would be the friend's best response?
   A. "Just work through your grief, and maybe join a support group so you can open up more. You need to get in touch with your feelings!"
   B. "So maybe your relationship with Lionel wasn't so great after all."
   C. "There are lots of ways to grieve. Your experience isn't unusual. Maybe it's your way of bouncing back. Talk about your loss if you wish, but don't assume there's something wrong."
   D. "You're holding in your emotions. You're going to get an ulcer or cancer if you keep this up. Let your feelings out!"

8 Active Development as a Lifelong Process

When you think about developmental psychology, child development may still be the first thing that pops into mind. However, development is in fact a lifelong process: You remain on this very day in the midst of your development, wherever you are in the adult life span. Further, as an active developer, you blaze your own developmental trail. Where will you carve that path as you go through life? How might you "grow yourself"?

One way that adults develop is through coping with life's difficulties. To understand how, let's revisit Piaget's ideas of assimilation and accommodation in childhood cognitive development and see how they apply to adult development (Block, 1982).

Recall that in assimilation, existing cognitive structures are used to make sense out of the current environment. Assimilation allows a person to enjoy a feeling of meaning because current experiences fit into their preexisting schemas (King & Hicks, 2007). However, life does not always conform to one's expectations. When experience conflicts with existing schemas, it is necessary to modify current ways of thinking. Accommodation is the process whereby existing schemas are modified or new structures are developed. Accommodation helps individuals to change so that they can make sense of life's previously incomprehensible events. For example, when people encounter a negative life circumstance such as an illness or a loss, they have the

All of us have experiences that we might regret. But even these experiences can have meaning to the extent that they contribute to who we are now.

opportunity to change—to develop and to mature (Blackie & others, 2017; Jayawickreme & Blackie, 2016; Jayawickreme & others, 2017; LoSavio & others, 2011). Indeed, research suggests that those who are faced with difficulties in life are more likely to come to a rich, complex view of themselves and the world when they are able to acknowledge, with mindfulness, the ways these experiences have changed them (King & Hicks, 2007).

Earlier, you had the opportunity to complete the Satisfaction with Life Scale. An item on that scale was, "If I could live my life over I would change almost nothing." One way to think about maturity is to revisit that item and consider the experiences you have had in your trip in the time machine of development. The mature person, dedicated to life while also acknowledging potentially negative experiences, may be the person who is able to give that item a high rating, valuing their life "warts and all" (King & Hicks, 2007).

Petar Paunchev/Shutterstock

Development involves change, so development in adulthood may mean maintaining an openness to changing and to being changed by experience. Consider someone who has spent much of their early adult life pursuing wealth and career success but then turns to more selfless pursuits in middle age. To contribute to the well-being of the next generation, the individual devotes energy and resources to helping others—say, by volunteering for a charity or working with young people.

Development, then, is an ongoing process we experience throughout life as we encounter opportunities to grow, change, and make a mark in the world, however grandly or humbly. Each period of life sets the stage for the next. But at each of these we might revise the script, rewriting the next chapters of our life.

SUMMARY

Exploring Human Development

Development is the pattern of change in human capabilities that begins at birth and continues throughout the life span. Both nature (biological inheritance) and nurture (environmental experience) influence development extensively. However, people are not at the mercy of either their genes or their environment when they actively construct optimal experiences. Resilience refers to the capacity of individuals to thrive during difficulties at every stage of development. Development is characterized along three interrelated levels—physical, cognitive, and socioemotional aspects of experience.

Physical Development

Prenatal development progresses through the germinal, embryonic, and fetal periods. Certain drugs, such as alcohol, can have an adverse effect on the fetus. Preterm birth is another potential problem, but its effects may depend on experiences after birth. The infant's physical development is dramatic in the first year, and a number of motor milestones are reached in infancy. Extensive changes in the brain, including denser connections between synapses, take place in infancy and childhood.

Puberty is a period of rapid skeletal and sexual maturation that occurs mainly in early adolescence. Hormonal changes lie at the core of

pubertal development. Most adults reach their peak physical performance during their 20s and are healthiest then. However, physical skills begin to decline during the 30s. Even in late adulthood, the brain has remarkable repair capacity and plasticity.

Cognitive Development

Jean Piaget introduced a theory of cognitive development. From his view, children use schemas to actively construct their world, either assimilating new information into existing schemas or adjusting schemas to accommodate it. Piaget identified four stages of cognitive development: the sensorimotor stage, the preoperational stage, the concrete operational stage, and the formal operational stage. According to Piaget, cognitive development in adolescence is characterized by the appearance of formal operational thought, the final stage in his theory. This stage involves abstract, idealistic, and logical thought.

Piaget argued that no new cognitive changes occur in adulthood. However, some psychologists have proposed that the idealistic thinking of adolescents is replaced by the more realistic, pragmatic thinking of young adults. Longitudinal research on intelligence shows that many cognitive skills peak in middle age. Overall, older adults do not do as well on memory and other cognitive tasks and are slower to process information than younger adults. However, older adults may have greater wisdom than younger adults.

Socioemotional Development

In infancy, among the key basic ingredients of socioemotional development are temperament—the child's overall emotional demeanor—and attachment. Erik Erikson proposed an influential theory of eight psychosocial stages that characterize socioemotional development from infancy to late adulthood. In each stage, the individual seeks to resolve a particular socioemotional conflict. The childhood stages involve a movement from trust to industry. Adolescents experience Erikson's fifth stage, identity versus identity confusion.

Marcia proposed four statuses of identity based on crisis and commitment. Psychologists refer to the period between adolescence and adulthood as emerging adulthood. This period is characterized by the exploration of identity through work and relationships, instability, and self-focus.

Erikson's three stages of socioemotional development in adulthood are intimacy versus isolation (early adulthood), generativity versus stagnation (middle adulthood), and integrity versus despair (late adulthood). Researchers have found that remaining active increases the likelihood that older adults will be happier and healthier. They also have found that older adults often reduce their general social affiliations. Older adults are motivated to spend more time with close friends and family members.

 ## Gender Development

Gender development involves physical (biological), cognitive, and socioemotional processes. Biological factors in gender development include sex chromosomes and hormones. Cognitive factors include gender schemas and gender roles, with the latter being strongly influenced by culture. Socioemotional aspects of gender include parental and peer responses to gender-related behavior. With regard to gender differences, research indicates that people of different genders are more similar than different.

 ## Moral Development

Moral development encompasses both cognitive and socioemotional processes. Kohlberg proposed a cognitive-developmental theory of moral development with three levels (preconventional, conventional, and postconventional). More recent research has focused on the development of prosocial behavior and the influence of socioemotional factors in putting moral reasoning into action.

 ## Death, Dying, and Grieving

Understanding the physical reality of death requires consideration of its socioemotional features. Terror management theory research has shown that awareness of death leads to investment in cultural worldviews. Elisabeth Kübler-Ross proposed a stage model of confronting death, in which the dying individual progresses from denial to acceptance. George Bonanno has shown that grief unfolds in four patterns—resilience, recovery, chronic dysfunction, and delayed grief. Awareness of death can render life more meaningful and satisfying.

Active Development as a Lifelong Process

Development is an ongoing process that is influenced by the active developer at every point in the life span. In adulthood, development can be especially shaped by the ways individuals confront significant life events. Coping with stressful experiences and being open to change can be catalysts for adult development.

KEY TERMS

accommodation
assimilation
authoritarian parenting
authoritative parenting
concrete operational stage
core knowledge approach
cross-sectional design
development
emerging adulthood

executive function
formal operational stage
gender
gender identity
gender roles
gender similarities hypothesis
infant attachment
limbic system
longitudinal study

nature
neglectful parenting
nurture
object permanence
permissive parenting
preferential looking
preoperational stage
prosocial behavior
puberty

representativeness heuristic
resilience
secure attachment
sensorimotor stage
sexual orientation
temperament
wisdom

ANSWERS TO SELF-QUIZZES

Section 1: 1. A; 2. C; 3. C; 4. D
Section 2: 1. B; 2. A; 3. B; 4. C
Section 3: 1. B; 2. B; 3. C; 4. B

Section 4: 1. B; 2. D; 3. C; 4. C
Section 5: 1. B; 2. D; 3. B; 4. A
Section 6: 1. C; 2. D; 3. B; 4. C

Section 7: 1. B; 2. B; 3. C; 4. C

Design elements: (Preview icon): Jiang Hongyan/Shutterstock; (Marginal notes): Shutterstock/Vadarshop

9 Motivation and Emotion

Racing to Virus Hot Spots

Thanksgiving 2020 was a different holiday for many. Because of the COVID-19 pandemic, many people spent the day avoiding gatherings with family and friends as a precaution against the virus. But for many travel nurses, the holiday was spent away from home and family as they raced to virus hot spots, helping out at hospitals all over the country. At least 25,000 people work as travel nurses in the United States (Bosman, 2020). These nurses use their skills wherever they are needed. Throughout 2020, as hospitals filled with the sick and dying, travel nurses arrived to help.

Of course, medical professionals emerged as true heroes during the deadly pandemic. They put their lives at risk every day, working long hours to try their best to save the many afflicted. Across the world, medical staff faced long days of striving, struggle, frustration, and, often, grief and hopelessness. As difficult as their work was, imagine working those same long hours but far from home, living in a hotel or Airbnb, serving not your own community but strangers in towns and cities you have never even visited. Why would people rush to places strapped for resources in the middle of a pandemic? For some, the answer was surely money as travel nurses are paid more than regular nurses (Bosman, 2020). For others, though, the call to help those in dire need was about more than these financial motivations—it was about enacting the mission of service to human welfare that guides the nursing profession.

The terms *motivation* and *emotion* come from the Latin word *movere*, which means "to move." Motivation and emotion are the "go" of human life, providing the steam that propels us to overcome obstacles and to accomplish the great and little things we do every day. Our emotions often define for us what we really want: We feel joy or sorrow depending on how events influence our most cherished life dreams.

Fly View Productions/Getty Images

1 Theories of Motivation

motivation
The force that moves people to behave, think, and feel the way they do.

Motivation is the force that moves people to behave, think, and feel the way they do. Motivated behavior is energized, directed, and sustained. Psychologists have proposed a variety of theories about why organisms are motivated to do what they do. In this section we explore some of the main theoretical approaches to motivation.

The Evolutionary Approach

In the early history of psychology, the evolutionary approach emphasized the role of instincts in motivation. *Ethology*—the study of animal behavior—also has described motivation from an evolutionary perspective. Studying the natural behavior of animals allows scientists to identify instincts.

instinct
An innate (unlearned) biological pattern of behavior that is assumed to be universal throughout a species.

An **instinct** is an innate (unlearned) biological pattern of behavior that is assumed to be universal throughout a species. Generally, an instinct is set in motion by a "sign stimulus"—something in the environment that turns on a fixed pattern of behavior. Instincts may explain a great deal of nonhuman animal behavior. In addition, some human behavior is instinctive. For example, let's consider infant reflexes. Babies do not have to learn to suck; they instinctively do it when something is placed in their mouth. So, for infants, an object touching the lips is a sign stimulus. After infancy, though, it is hard to think of specific behaviors that all human beings engage in when presented with a particular stimulus.

 You might think that some stimuli provide a strong case for instinctive responses. Doesn't everyone respond to chocolate with the behavior of eating? Not so fast—some people don't even like chocolate!

According to evolutionary psychologists, the motivations for sex, aggression, achievement, and other behaviors are rooted in our evolutionary past (Buss, 2020), and we can understand similarities among members of the human species through these shared evolutionary roots. Because evolutionary approaches emphasize the passing on of one's genes, these theories focus on domains of life that are especially relevant to reproduction. Evolutionary approaches focus on how members of a species are the same as each other and seek to establish principles that apply to everyone in the same way, similar to instincts in animals (Buss & Foley, 2020). However, in general, human beings are so different from each other and human behavior is so complex that explaining human motivation on the basis of species-specific instinct is inappropriate.

Human newborns display behavioral reflexes such as holding on to a rope so that they can be lifted. In our evolutionary past, this gripping reflex appeared in primates, allowing an infant to cling to their caregiver.
Bettmann/Getty Images

Indeed, it would hardly seem adaptive for humans to have a fixed action pattern that is invariably set in motion by a particular signal in the environment. To understand human behavior, psychologists have developed a variety of other approaches, as we now consider.

Drive Reduction Theory

Another way to think about motivation is through the constructs of need and drive. A **need** is a physical or biological deprivation that energizes the drive to eliminate or reduce the deprivation. A **drive** is an aroused state that occurs because of a physiological need. You can think of a drive as a psychological itch that requires scratching. Generally, psychologists think of needs as underlying our drives. You may have a need for water; the drive that accompanies that need is your feeling of being thirsty. Usually but not always, needs and drives are closely associated. For example, when your body needs food, your hunger drive will probably be aroused. An hour after you have eaten a hamburger, your body might still need essential nutrients (thus you need food), but your hunger drive might have subsided.

The following example should reinforce that drive pertains to a psychological state whereas need involves a physiological state, and that drives do not always follow from needs. If you are deprived of oxygen because of a gas leak, you may feel lightheaded but may not realize that your condition is the result of a gas leak that is creating a need for air. Your need for air fails to create a drive for oxygen that might lead you to open a window. In addition, drives sometimes seem to come out of nowhere. Imagine, for instance, having eaten a fine meal and feeling full to the point of not wanting another single bite—until the waiter wheels over the dessert cart. Suddenly you feel ready to tackle the double chocolate oblivion, despite your lack of hunger.

Drive reduction theory explains that as a drive becomes stronger, we are motivated to reduce it. The goal of drive reduction is **homeostasis,** the body's tendency to maintain an equilibrium, or steady state. Literally hundreds of biological states in the body must be maintained within a certain range: temperature, blood sugar level, potassium and sodium levels, oxygenation, and so on. When you dive into an icy swimming pool, your body uses energy to maintain its normal temperature. When you walk out of an air-conditioned room into the heat of a summer day, your body releases excess heat by sweating. These physiological changes occur automatically to keep your body in an optimal state of functioning.

Drive reduction theory does not provide a comprehensive framework for understanding motivation. Think about all your own favorite courses or leisure activities. Many of these probably involve hard work. In fact, many of them increase rather than reduce drives. Many things that people choose to do involve increasing (not decreasing) tensions—for example, taking a challenging course in school, raising a family, and working at a difficult job. Human motivation cannot be entirely explained by drive reduction because we seem motivated to increase (not decrease) many drives.

A great example of behavior that involves increasing drives is leadership. Whenever groups face difficulties, having a good leader is important. Being a leader involves shouldering responsibility, caring about group members, and sacrificing one's personal interest for the greater good. Why do people seek leadership? How does motivation explain whether a person actually becomes a leader? Surely, there are many reasons why people are motivated to lead. A strong predictor of becoming a leader and succeeding as a leader is the extent to which a person holds affective-identity motives to lead (Badura & others, 2020). This means that a person enjoys leading and self-identifies as a leader.

Optimum Arousal Theory

When psychologists talk about arousal, they are generally referring to a person's feelings of being alert and engaged. When we are very excited, our arousal levels are high. When we are bored, they are low. You have probably noticed that motivation influences arousal

This is why scents are added to gas—so that it can be smelled—and also why we need carbon monoxide detectors, because we cannot tell when we need oxygen!

need
A deprivation that energizes the drive to eliminate or reduce the deprivation.

drive
An aroused state that occurs because of a physiological need.

homeostasis
The body's tendency to maintain an equilibrium, or steady state.

levels. Sometimes you can want something (for example, to do well on a test) so much that you feel "overmotivated" and anxious. On the other hand, you might be so unmotivated for a task (such as doing dishes) that you can hardly force yourself to complete it.

Early in the twentieth century, two psychologists described how arousal can influence performance. According to their formulation, now known as the **Yerkes-Dodson law,** performance is best under conditions of moderate arousal rather than either low or high arousal (see the Psychological Inquiry below). At the low end of arousal, you may be too lethargic to perform tasks well; at the high end, you may not be able to concentrate. Think about how aroused you were the last time you took a test. If your arousal was too high, you might have felt too nervous to concentrate, and your performance likely suffered. If it was too low, you may not have worked fast enough to finish the test. Also think about performance in sports. Being too aroused usually harms athletes' performance; a thumping heart and rapid breathing have accompanied many golfers' missed putts. However, if athletes' arousal is too low, they may not concentrate well on the task at hand.

Yerkes-Dodson law
The psychological principle stating that performance is best under conditions of moderate arousal rather than either low or high arousal.

Overlearning is a crucial part of the training regimen by which the elite Navy SEALS prepare for missions, such as the 2011 raid on Osama bin Laden's compound in Pakistan and the al Qaeda leader's eventual death.
Source: U.S. Navy, photo by Kyle Gahlau

PSYCHOLOGICAL INQUIRY

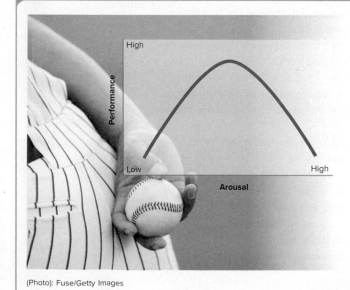
(Photo): Fuse/Getty Images

Obeying the (Yerkes-Dodson) Law

The graph displays the relationship between arousal (shown on the X, or horizontal, axis) and performance (shown on the Y, or vertical, axis). Note that the curve is an inverted U.

Using the figure as a reference, answer the following questions:

1. What was your arousal level the last time you took an exam? If you were very nervous, your arousal level would be considered high. If you were excited and engaged but not too worried, your level would be in the medium range. If you were feeling sluggish, your arousal level would be low.

2. How did you actually do on that test? Plot your performance on the graph. Does it fit with the Yerkes-Dodson prediction?

   A. Now think about performance in sports or the arts. Imagine your favorite athlete, musician, or actor. How might that person feel when they are on the spot, trying to sink a winning free throw, strike out the last batter, or impress an audience? How might arousal influence performance in those cases?

3. In many professions, individuals are forced to perform under conditions of very high arousal. These include EMTs, lifeguards, and emergency room staff. (Name some others.) How might such individuals train themselves to perform even under conditions of extreme arousal?

The relationship between arousal and performance is one reason individuals who have to perform well under stressful conditions (such as EMTs, lifeguards, and marines) are trained to *overlearn* important procedures so they do not require much thought. With this extra learning, when these individuals are under conditions of high arousal, they can rely on automatic pilot to do what needs to be done.

1. The force that moves people to behave, think, and feel the way they do is
   A. emotion.
   B. instinct.
   C. need.
   D. motivation.

2. Kai will be taking an exam today. According to the Yerkes-Dodson law, the condition that will allow Kai to score highest on the exam is
   A. no arousal.
   B. moderate arousal.
   C. high arousal.
   D. high relaxation.

3. Which of the following statements is correct?
   A. Instincts have little to do with animal behavior.
   B. Instincts are learned patterns of behavior.
   C. Instincts direct most aspects of human behavior.
   D. Instincts are innate and biological.

APPLY IT! 4. Jodi is a star player on her school's basketball team. In a crucial game, the score is tied with just a few seconds left on the clock, and Jodi finds herself at the free-throw line preparing to shoot the winning baskets. The opponents' coach calls a time-out to "ice" Jodi's nerves. Finally, as Jodi steps up to the line, the rival team's students scream and jump around in an attempt to psych Jodi out. Her heart racing, Jodi sinks both baskets, and her team wins. Which of the following is most likely true of this situation?
   A. The opposing team's coach and student fans know that low arousal leads to poor performance.
   B. Jodi has practiced free throws so many times that she can land them even when she is highly aroused.
   C. Jodi is showing the effects of very high arousal on performance.
   D. Jodi is generally a sluggish person, and her performance is helped by her low levels of arousal.

2 Hunger and Sex

Some of the influence of motivation in our lives is tied to physiological needs and adaptation. Two behaviors that are central to the survival of our species are eating and sex. In this section we examine the motivational processes underlying these behaviors.

The Biology of Hunger

You know you are hungry when your stomach growls and you feel those familiar hunger pangs. What role do such signals play in hunger?

GASTRIC SIGNALS In 1912, Walter Cannon and A. L. Washburn conducted an experiment that revealed a close association between stomach contractions and hunger (Figure 1). As part of the procedure, a partially inflated balloon was passed through a tube inserted in Washburn's mouth and pushed down into his stomach. A machine that measures air pressure was connected to the balloon to monitor Washburn's stomach contractions. Every time Washburn reported hunger pangs, his

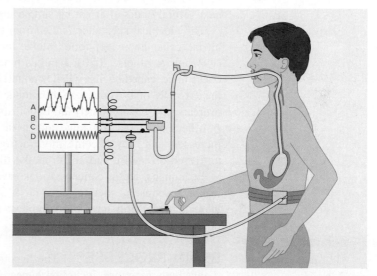

FIGURE 1 **Cannon and Washburn's Classic Experiment on Hunger** In this experiment, the researchers demonstrated that stomach contractions, which were detected by the stomach balloon, accompany a person's hunger feelings, which were indicated by pressing the key. Line A in the chart records increases and decreases in the volume of the balloon in the participant's stomach. Line B records the passage of time. Line C records the participant's manual signals of feelings of hunger. Line D records a reading from the belt around the participant's waist to detect movements of the abdominal wall and ensure that such movements are not the cause of changes in stomach volume.

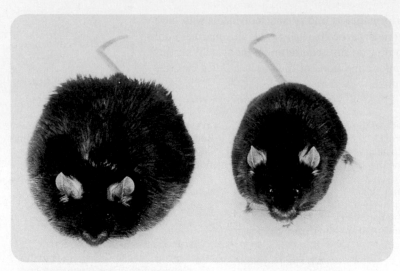

FIGURE 2 **Leptin and Obesity** The obese mouse on the left is untreated; the one on the right has been given injections of leptin.
The Rockefeller University/AP Images

stomach was also contracting. Sure enough, a growling stomach needs food. The stomach tells the brain not only how full it is but also how much nutrient is present, which is why rich food stops hunger faster than the same amount of water. The hormone cholecystokinin (CCK) helps start the digestion of food, travels to the brain through the bloodstream, and signals us to stop eating (Matsuda & others, 2020; Yeomans, 2020). Hunger involves a lot more than an empty stomach, however.

BLOOD CHEMISTRY Three key chemical substances play a role in hunger, eating, and satiety (the state of feeling full): glucose, insulin, and leptin.

Glucose (blood sugar) is an important factor in hunger, probably because the brain critically depends on sugar for energy. One set of sugar receptors, located in the brain itself, triggers hunger when sugar levels fall too low (Roh & others, 2016). Another set of sugar receptors is in the liver, which stores excess sugar and releases it into the blood when needed. The sugar receptors in the liver signal the brain when its sugar supply falls, and this signal also can make you feel hungry.

The hormone *insulin* plays a role in glucose control (Ferrannini & Mari, 2020). When we eat complex carbohydrates such as bread and pasta, insulin levels go up and fall off gradually. When we consume simple sugars such as candy, insulin levels rise and then fall sharply. So, we can feel quite energized by a chocolate bar but then experience a sudden drop in energy. Because of their effects on insulin, we are more likely to eat within the next several hours after eating simple sugars than after eating complex carbohydrates.

The chemical substance *leptin* (from the Greek word *leptos,* meaning "thin"), released by fat cells, decreases food intake and increases energy expenditure or metabolism (Koizumi & others, 2020). Leptin's functions were discovered in a strain of genetically obese mice, called *ob mice* (Pelleymounter & others, 1995). Because of a genetic mutation, the fat cells of ob mice cannot produce leptin. The ob mouse has a low metabolism, overeats, and gets extremely fat. Leptin appears to act as an antiobesity hormone (Schepers & others, 2020). If ob mice are given daily injections of leptin, their metabolic rate increases, and they become more active, eat less, and lose weight. Figure 2 shows an untreated ob mouse and an ob mouse that has received injections of leptin.

In humans, leptin concentrations have been linked with weight, body fat, and weight loss in response to dieting (Poetsch & others, 2020; Rebello & others, 2020). Problems in regulating leptin have been implicated in both obesity and depression (Simmons & others, 2020).

BRAIN PROCESSES The hypothalamus plays a central role in regulating important body functions, including hunger. The hypothalamus is located just above the midbrain and just under the thalamus (the brain's relay station). Activity in two areas of the hypothalamus regulates hunger. The *lateral hypothalamus* (*lateral* means "on the outer parts") is involved in stimulating eating. When an electrical current is passed through this area in a well-fed animal, the animal begins to eat. If this part of the hypothalamus is destroyed, even a starving animal will show no interest in food. The *ventromedial hypothalamus* (*ventromedial* means "on the inner locations") is involved in reducing hunger and restricting eating. When this area of an animal's brain is stimulated, the animal stops eating. When the area is destroyed, the animal eats profusely and quickly becomes obese.

This part can be tricky. These locations play opposite roles in hunger. Remember LATERAL means on the outside (like going out to eat when you are hungry). VENTROMEDIAL means on the inside (like staying in because you are not hungry).

Although the lateral and ventromedial areas of the hypothalamus both influence hunger, there is much more to the brain's role in determining hunger than these on/off centers in the hypothalamus. Neurotransmitters (the chemical messengers that convey information from neuron to neuron) and neural circuits (clusters of neurons that often involve different parts of the brain) also function in hunger. The neurotransmitter serotonin is partly responsible for the satiating effect of CCK, and serotonin agonists have been used to treat obesity in humans (Kapoor & others, 2020; Singh & Singh, 2020).

Obesity

Given that the brain and body are wired to regulate eating behavior, why do so many people in the United States overeat and suffer the effects of overeating? Around the world, about 39 percent of adults experience overweight or obesity (WHO, 2020). Among developed nations, more than half of adults and one in six children have overweight or obesity (OECD, 2017). Adult obesity rates are highest in the United States and Mexico and are lowest in Japan and South Korea (OECD, 2017). Overweight or obesity increase risk for a variety of health problems, including cardiovascular disease, type 2 diabetes, some cancers, and sleep problems, including sleep apnea (when the person stops breathing during sleep) (WHO, 2020). Overweight and obesity are global health problems.

*The World Health Organization coined the term **globesity** to refer to the worldwide problem of obesity.*

Why so many people overeat to the point of obesity is a motivational puzzle, because it involves eating when one is not in need of nutrition. As is the case with much behavior, in eating, biological, cognitive, and sociocultural factors interact in diverse ways in different people, making it difficult to point to a specific cause (Monnereau & others, 2017).

THE BIOLOGY OF OBESITY   Obesity clearly has a genetic component. After the discovery of the ob gene in mice, researchers found a similar gene in humans. Genes play a role in the human tendency to become overweight or obese. It is estimated that up to 70 percent of the differences we see in obesity can be explained by genetic differences (Golden & Kessler, 2020; Smoczek & others, 2020). Childhood obesity may be explained by different genes than those related to adult obesity (Greenhill, 2020; Littleton & others, 2020).

set point
The weight maintained when the individual makes no effort to gain or lose weight.

Another factor in weight is **set point,** the weight maintained when a person makes no effort to gain or lose weight (Geary, 2020). Set point is determined in part by the amount of stored fat in the body. Fat is stored in *adipose cells* or fat cells. When these cells are filled, you do not get hungry. When people gain weight their fat cells grow larger. In some people, fat cells create more fat cells. Fat cells are all about storing energy so those with more fat cells are more likely to feel hungry and may need to eat more to feel satisfied.

Comstock/Getty Images

PSYCHOLOGICAL FACTORS IN HUNGER AND OBESITY   Considering that food is an important part of every human life, it is surprising that psychologists have not studied eating as much as many other behaviors (Rozin, 2020). Eating is not just something we do to survive, it is part of many of our most memorable moments, including family meals, religious celebrations, and holiday feasts (Brumberg-Kraus, 2020). Think about all the many special occasions in your life. Chances are, food and eating with family and friends were a big part of those events. Food is more than physical nourishment, as the foods we eat implicate culture, ethnicity, and social class (Valentin, 2020).

Certainly, social context can affect when and how much people eat. Research has shown that women may select smaller portions and eat less (especially when eating with men) to appear more feminine, while men may eat larger portions to appear more masculine (Cavazza & others, 2017). Gender stereotypes can also affect what people eat. These stereotypes can vary by culture. For example, it appears to be an unusual feature

Developmental Psychology and Social Psychology: Are There "Boy Foods" and "Girl Foods"?

To probe children's gender stereotypes about food, researchers studied preschoolers (ages ranged from four to six years) (Graziani & others, 2020). Children's stereotypes were measured in two ways: implicitly and explicitly. First, to test for *implicit* (meaning automatic and nonconscious) stereotypes, the children were asked to respond as quickly as possible to pairs of pictures presented on a computer. Their job was to push a button to indicate whether two pictures go together or not. The pictures were female or male faces paired with pictures of food (for instance, a steak or a salad). The time it took to make the judgments was recorded. The idea behind this task is that unconscious or automatic stereotypes can be revealed by quicker responses to pairs that "go together." So if a child is quick to click a button saying that pictures of men go with pictures of steak, then that is evidence of an implicit stereotype.

To measure explicit stereotypes (that is, stereotypes of which the person is aware), the children played "the waiter game." In this game, children pretended they were a waiter at a special restaurant where the waiters pick the food for the customers. The children were shown pictures of a customer (either a woman or a man) and then possible foods to select for the person. The dishes included stereotypically feminine foods (such as a salad or grilled vegetables) or stereotypically masculine foods (such as steak or a hamburger). The number of times the

Ariel Skelley/Blend Images/Getty Images

child selected stereotypical foods for male or female customers was recorded.

Results showed that boys (children assigned male at birth), but not girls (children assigned female at birth), showed implicit gender stereotypes about food. They were quicker to associate men with masculine foods and women with feminine foods. In contrast, no evidence emerged for explicit gender stereotypes for food. Children were just as likely to give salads to the male customers as the female customers in the waiter game.

Interestingly, for their own favorite foods, girls, again, showed no gender stereotypes, but boys did. Boys expressed greater liking for stereotypically masculine foods.

These results show that although adults tend to show evidence of gender stereotypes about food, children generally do not. However, boys showed both greater implicit gender stereotypes about food and more liking for stereotypically masculine foods. Recall that boys may be more harshly treated when their behaviors conflict with gender-typical expectations. This research suggests that the gender socialization of boys may generalize as well to the realm of food. It is also interesting to consider how girls' relationships with food may change over their life course.

\\ **Where do boys' gender stereotypes about food come from?**

of the United States that craving chocolate is considered a feminine characteristic and many more American women than men report craving chocolate (Valentin, 2020). In many cultures, eating meat is associated with masculinity, while eating lighter, healthier foods is linked to femininity (Campos & others, 2020; Gal & Wilkie, 2010; Zhu & others, 2015).

During the pandemic, many people found themselves gathered around family dinner tables (Nordberg, 2020). Sitting at the dinner table is a time for families to eat but also to talk, argue, and bond (Bova, 2019; Hausken & others, 2019). Family meals are also an opportunity for children to learn an array of rich information about people, family, and culture (Walton & others, 2020). Sitting with parents and siblings at the table, children can learn about food and about who eats what. Although it might not often be noticed as such, food can be a part of gender roles, those expectations about what is proper for people of different genders to do. Research has shown that young children become more gender-stereotypical in their toy preferences as they age (Liu & others, 2020). Is it possible that they also demonstrate gender stereotypes for food and food preferences? To read about research with children probing stereotypes about gender and food preferences, see the Intersection.

From an evolutionary view, human taste preferences developed at a time when reliable food sources were scarce. Our earliest ancestors needed a lot of calories to survive in challenging circumstances. They likely developed a preference for sweet and fatty foods. Today many people still have a taste for such foods, though today's sweet and fatty foods provide far more calories than we need with far less nutritional value. Learned associations of food with a particular time and place are characteristic of many organisms, including humans (Maxwell & others, 2017). If it is noon, we eat lunch; in a movie theater, we eat popcorn.

Beyond these factors, subtle cues can influence how much people consume. Growing portion sizes have been implicated in the alarming rates of obesity in the United States (Zuraikat & others, 2020). Not surprisingly, a key predictor of portion size is liking of the food (Spence & others, 2016). If we like pizza, we may see a typical portion as three slices, not just one (Labbe & others, 2017).

It is puzzling, though, that portion sizes should matter. After all, people do not have to eat all of the food presented to them. Why do people eat an entire portion of food when just a bit of it might have led to the same feeling of fullness? Of course, parents often teach their children to "finish what's on your plate" (Potter & others, 2017). Many people rely not on a feeling of fullness but rather the sight of an empty plate or bowl as the cue to stop eating (Hinton & others, 2013). Increasingly, cleaning one's plate has involved eating more and more food. Since 1900, the diameter of the average plate has increased from about 9½ inches to nearly a foot (Marteau, 2015).

Relying on something like a clean plate to indicate that one should stop eating is a sign of eating mindlessly, "on automatic pilot." Mindless eating can involve eating until one has finished a portion (no matter how large) or not paying attention to the sheer amount one has consumed (Kristeller & Epel, 2014; Steenhuis & Poelman, 2017). One factor in mindless eating is the appearance of food (Reinders & others, 2017). A series of studies examined how people perceive the size of food portions when these are presented either stacked vertically or spread out (Szocs & Lefebvre, 2017). It appears that diners use the surface area of food on a plate as a guide to how much food the plate contains. This means that looking at equal amounts of, say, mashed potatoes on two plates, the one that is shaped like a tower may look like less food than the one presented in a flat mound (Szocs & Lefebvre, 2017). In these studies, participants appeared to be using the visual surface area of food as a *heuristic* to estimate portion size.

How might we reduce mindless eating? One way might be to adopt more mindful eating practices (Kabat-Zinn, 1990). Recall from Chapter 4 that being mindful means being present and open to one's thoughts without becoming preoccupied. When we are mindful, thoughts come and go. Eating mindfully means being mentally present while eating and being aware of all aspects of the food one is eating, its appearance, texture, smell, and taste. In one study, participants were randomly assigned to taste chocolate while listening to mindfulness instructions or to a news report (the control condition) (Mantzios & others, 2019). Afterward, they were told the study was over and they could snack on chocolate pieces while they waited for the experimenter to return in a few minutes. A key dependent measure was how much chocolate participants ate during this waiting period. Results showed that those who tasted the chocolate mindfully liked it just as much as those in the control condition, but they ate less of it during the waiting period. So, mindfulness might be a way to combat mindless eating. Being present even while eating chocolate might help people notice what they have eaten, how much, and to attend to bodily cues of being full.

One step that has been taken by policymakers and restaurants is to provide information about the calories and fat in various dishes on menus. To read about the effects of such information on eating, see Psychology in Our World.

Dieting is a continuing obsession in the United States. However, even if we are trying to lose weight, we must eat to survive. For our species to survive, we also have to have sex. Like hunger, sex has a strong physiological basis, as well as cognitive and sociocultural components.

What's on Menus and Labels and Why It Matters

Looking at a menu, it can be hard to know exactly how many calories each item contains. Some foods may look and sound healthy but be high in calories. On its own, lettuce is very low in calories. But a dinner-sized Caesar salad with chicken, dressing, cheese, and croutons can contain nearly 600 calories, which is about the same as two McDonald's cheeseburgers. In 2015, the U.S. Food and Drug Administration implemented a rule requiring chain restaurants to include caloric information on their menus. Can such information change diners' preferences? Research shows that it can (Birch & others, 2015; Cioffi & others, 2015; Haws & Liu, 2016). When diners are given information about the caloric content of food, they are more likely to think about it, remember it, and choose food with fewer calories (Hammond & others, 2015).

Olga Nayashkova/Shutterstock

Packaged foods also contain nutritional information on their labels. If you have ever taken a close look at that information on the side of a snack you might notice that the information is based on how many servings are included in a pack. So, even though it seems obvious that a whole pack of cookies is a serving, the information tells you there are actually two or three servings there. Could this kind of information affect how much people eat? Does it matter, for instance, if 11 tortilla chips are labeled as "one serving" versus "11 tortilla chips"? A series of studies (Lewis & Earl, 2018) addressed this question.

Participants all thought they were taking part in a taste test. They were presented with snack foods (in one study tortilla chips and in others gummy bears or mini rice cakes). All participants got the same amount of food in a plastic baggie. The only difference between them was the independent variable in the studies: how the food was presented. Participants were randomly assigned to be told that the baggie contained "one serving" of the snack or they were told the exact number of chips, gummy bears, or mini-rice cakes that were in the baggie. After tasting the snacks, participants rated how full they felt and how much of the snack they intended to eat in the future. The results? People felt more full and said they would need to eat fewer snacks in the future to feel full if they were told the actual amount of the snack they received (rather than being told it was one serving). These results are surprising—after all, everyone received the same amount of food. Yet, being given the exact count of the food presented led people to feel like it was more food. And, to the extent they thought it was more food, they actually ate less of it (Lewis & Earl, 2018). Imagine looking at that pack of cookies that seemed to hold one serving and being told that it actually contains 18 cookies. Perhaps small changes in the presentation of food could have cascading effects on the amount people eat and eventually contribute to healthier lifestyles.

The Biology of Sex

Given the central role of sexual behavior in reproduction, understanding the processes that underlie sexual behavior in all animals, including humans, is a vital goal for science. What brain areas are involved in sex? What role do hormones play in sexual motivation?

What is the nature of the human sexual response pattern? This section answers these central questions about the biology of sex.

THE HYPOTHALAMUS, CEREBRAL CORTEX, AND LIMBIC SYSTEM

Motivation for sexual behavior is centered in the hypothalamus (Kammel & Correa, 2020). However, like many other areas of motivation, brain functioning related to sex radiates outward to connect with a wide range of other brain areas in both the limbic system and the cerebral cortex (Grodd & others, 2020; Jennings & de Lecea, 2020).

Researchers have shown the importance of the hypothalamus in sexual activity by electrically stimulating or surgically removing it. Electrical stimulation of certain hypothalamic areas increases sexual behavior, while surgical removal of some hypothalamic areas inhibits sexual behavior. Electrical stimulation of the hypothalamus in a male can lead to as many as 20 ejaculations in an hour. The limbic system, which runs through the hypothalamus, also seems to be involved in sexual behavior. Its electrical stimulation can produce sexual arousal and orgasm.

Temporal lobe damage in male cats has been shown to impair the animals' ability to select an appropriate partner. The tomcats with temporal lobe damage try to copulate with everything in sight, including teddy bears, chairs—and even researchers! In humans, the temporal lobes of the neocortex (located on the sides of the brain) play an important role in moderating sexual arousal and directing it to an appropriate goal object (Carroll, 2013).

The brain tissues that produce sexual feelings and behaviors are activated by various neurotransmitters in conjunction with sex hormones. Like scratching an itch, sexual motivation also is characterized by a basic urge-reward-relief cycle. That means we become sexually aroused, feel a strong urge to engage in sexual behavior, engage in that behavior, and then experience a rewarding sensation, followed by feelings of calm relief. The motivation for sex is generated by excitatory neurotransmitters (Hull & Dominguez, 2006). The intense reward of orgasm is caused by a massive rush of dopamine, and the deep feeling of relaxation that follows is linked with the hormone oxytocin (Krüger & others, 2020).

estrogens
The class of sex hormones that predominates in females, produced mainly by the ovaries.

androgens
The class of sex hormones that predominates in males, produced by the testes in males and by the adrenal glands in all people.

SEX HORMONES   The endocrine system is deeply involved in sex. The two main classes of sex hormones are estrogens and androgens. **Estrogens,** are produced mainly by the ovaries. **Androgens,** such as testosterone, are produced by the testes in male children and adults and by the adrenal glands in all people. In men, higher androgen levels are associated with sexual motivation, orgasm frequency, and heightened well-being (Shea & others, 2014; Zitzmann, 2020). Research suggests that increasing testosterone in female adults increases sex drive and the frequency of satisfying sexual experiences (Braunstein, 2007; Davis, 2013), but it carries increased risk for ovarian and breast cancer (Ruth & others, 2020).

One substance that is known to decrease testosterone is black licorice. Some researchers have participants eat black licorice as a way to manipulate testosterone levels.

THE HUMAN SEXUAL RESPONSE PATTERN   What physiological changes do humans experience during sexual activity? To answer this question, William Masters and Virginia Johnson (1966) carefully observed and measured the physiological responses of 382 female and 312 male volunteers as they masturbated or had sexual intercourse. Masters and Johnson identified a **human sexual response pattern** consisting of four phases: excitement, plateau, orgasm, and resolution.

The *excitement phase* begins the process of erotic responsiveness. It lasts from several minutes to several hours, depending on the nature of the sexual activity involved. Engorgement of blood vessels and increased blood flow in genital areas and muscle tension characterize the excitement phase. The most obvious signs of response in this phase are the beginnings of lubrication of the vagina and partial erection of the penis.

The second phase of the human sexual response, the *plateau phase,* is a continuation and heightening of the arousal begun in the excitement phase. The increases in breathing, pulse rate, and blood pressure that occurred during the excitement phase become more intense, penile erection and vaginal lubrication are more complete, and orgasm is closer.

human sexual response pattern
According to Masters and Johnson, the characteristic sequence of physiological changes that humans experience during sexual activity, consisting of four phases: excitement, plateau, orgasm, and resolution.

Eric Audras/ONOKYO/Getty Images

The third phase of the human sexual response cycle is *orgasm,* which involves an explosive discharge of neuromuscular tension and an intensely pleasurable feeling. How long does orgasm last? Although it might feel like time is standing still, orgasm lasts for only about 3 to 15 seconds.

Following orgasm, the person enters the *resolution phase,* in which blood vessels return to their baseline state. At this point, women may be stimulated to orgasm again without delay. Men enter a refractory period during which they cannot have another orgasm.

Cognitive and Sensory/ Perceptual Factors in Sexuality

From experience, we know that our cognitive world plays an important role in sexuality. We have the cognitive capacity to think about the importance of respecting our partners and of being certain that they consent to sexual activity. We also have the cognitive capacity to generate sexual images—to become sexually aroused just by thinking about erotic images or our romantic partners (Stanton & others, 2014).

Sexual motivation is influenced by *sexual scripts,* stereotyped patterns of expectancies for how people should behave sexually (Ruvalcaba & Eaton, 2020; Sakaluk & others, 2014). We carry these mental scripts with us. Gender may influence the content of these scripts and they may be influenced by media portrayals (Seabrook & others, 2017). Some sexual scripts involve a double standard, such as, for example, judging that it is okay for male but not female adolescents to have sex. Sexual scripts have also been implicated when young women feel that they are being coerced into sexual behavior (French & Neville, 2017).

Cognitive interpretation of sexual activity also involves our perceptions of the person with whom we are having sex and that person's perceptions of us (Alvarez & Garcia-Marques, 2011; Cornelius & Kershaw, 2017). Amid the wash of hormones in sexual activity is the cognitive ability to control, reason about, and make sense of the activity.

Sexual scripts may be missing crucial information about the important role of consent in sexuality. The importance of personal autonomy and lack of coercion in sexual experiences is reflected in the movement toward affirmative consent at many universities (Shumlich & Fisher, 2020; Silver & Hovick, 2018). Affirmative consent means that, rather than simply assuming both partners are comfortable with sexual activity, each partner provides, by words or actions, a knowing and voluntary indication that they are willing to engage in sexual activity. Affirmative consent means that partners have provided unambiguous permission with regard to their willingness to engage in the sexual activity. Affirmative consent is a continuous process. People always have the option of changing their minds. Just because someone consented to a particular sexual behavior at one time does not mean that they have given up the right to say "no" later.

Coercion has no place in sexual intimacy.

People who are interested in being intimate with each other also have to be able to talk to each other.

Cultural Factors in Sexuality

The influence of culture on sexuality was demonstrated dramatically in a classic analysis by John Messenger (1971) of the people living on the small island of Inis Beag off the coast of Ireland. They knew nothing about tongue kissing or hand stimulation of the penis, and they detested nudity. For everyone, premarital sex was out of the question. Men avoided most sexual experiences because they believed that sexual intercourse reduced their energy level and was bad for their health. Under these repressive conditions, sexual intercourse occurred only at night, taking place as quickly as possible. As you might suspect, female orgasm was rare in this culture (Messenger, 1971).

In contrast, around the same time that Messenger was studying the people of Inis Beag, Donald Marshall (1971) was studying the Mangaian culture in the South Pacific. In Mangaia, young boys were taught about masturbation and encouraged to engage in it as much as they liked. At age 13, the boys underwent a ritual, initiating them into sexual manhood. First, their elders instructed them about sexual strategies, including how to aid their female partner in having orgasms. Two weeks later, the boy had intercourse with an experienced woman who helped him hold back from ejaculation until she experienced orgasm with him. By the end of adolescence, Mangaians had sex nearly every day. Mangaian women reported a high frequency of orgasm.

One way that societies teach youth about sex and sexuality is through formal education. Although many topics associated with sex and sexuality spur controversy, most people concerned with sex education share two relatively uncontroversial goals: to encourage the very young to delay sexual activity and to reduce teen pregnancy and sexually transmitted infections. When adolescents delay having penile-vaginal intercourse until after middle school they are more likely to use condoms and are likely to have fewer different sex partners, reducing the risk for unintended pregnancies and sexual transmitted infections (Lerner & Hawkins, 2016). As such, devising ways to encourage youth to delay sexual behavior is an important goal for all who care about their welfare. As we will see, though, achieving this uncontroversial goal can be controversial.

What are some aspects of your own culture that influence your sexual behavior?

SEX EDUCATION
Generally, sex education programs can be divided into two types. First, *comprehensive* sex education involves providing students with comprehensive knowledge about sexual behavior, birth control, and the use of condoms in protecting against sexually transmitted infections, while encouraging them to delay sexual activity and practice abstinence. Another form of sex education is the *abstinence-only* approach (or abstinence only until marriage). Abstinence-only educational programs claim that sexual behavior outside of marriage is harmful to people of any age. Instructors can present contraceptives and condoms only in terms of their failure rates. Abstinence-only sex education promotes the notion that abstinence is the only effective way to avoid pregnancy and sexually transmitted infections. Be aware that the first claim of abstinence-only sex education, that sexual behavior outside of marriage is harmful, has not been supported by research (Santelli & others, 2017).

Which approach to sex education most effectively delays sexual activity and prevents teen pregnancy? There is no doubt about the answer to this question: Comprehensive sex education programs do a better job than abstinence-only programs. Countless studies have demonstrated the validity of this conclusion (Beh, 2015; Goldfarb & Lieberman, 2020; Jaramillo & others, 2017; Kohler & others, 2008). Students exposed to abstinence-only programs are more likely to engage sexual behavior earlier in life, are unlikely to use protection when they do have sex, and are at risk for sexually transmitted infections and unplanned pregnancy (Santelli & others, 2017). Providing youth with comprehensive sex education does not make them more likely to have sex (Kirby, 2008). In 2010, the U.S. Congress cut funding for some abstinence-only programs and funded comprehensive sex education programs. However, states within the United States vary in terms of the ways that sex education is implemented and what information is shared (Santelli & others, 2017).

Teen births in the United States have declined every year since 2009, setting new record lows each year. In 2014, the number of teen births in the United States was 24.2 per 1,000 female teenagers. In 2018, this number dropped to 17.4 (Martin & others, 2019). The reasons for these declines are

Americans still have difficulty talking openly about sexual behavior, leading some to worry that online pornography has replaced serious sex education and open communication.

difficult to identify. Evidence indicates that American teens are delaying first intercourse and using birth control (Linbergh & others, 2016; Sedgh & others, 2015). Indeed, in a survey of U.S. teens from 2011 to 2015, nearly all of the young female respondents who reported being sexually active also reported using some type of birth control (Abma & Martinez, 2017). Even in the context of these improvements in teen pregnancy rates, it is important to bear in mind that the U.S. teen pregnancy rate is still much higher than those in other developed nations.

Sexual Behavior and Orientation

In the United States, varying sexual behaviors and attitudes reflect the country's diverse multicultural population, and Americans collectively fall somewhere in the middle of a continuum going from repressive to liberal. We are more conservative in our sexual habits than was once thought. However, we are quite a bit more open minded regarding sexual orientation than a century ago.

In this discussion, we first define sexual behavior and examine the frequencies of different sexual practices while noting the difficulties associated with doing research in these areas. We then turn to the factors involved in a person's sexual orientation.

DEFINING SEXUAL BEHAVIOR
What constitutes sexual behavior—what we commonly refer to as "sex"? Some people might answer that question with "penile-vaginal intercourse," but what about other sexual behaviors, such as anal sex and oral sex? If someone has engaged in these practices, are they still a "virgin"? If your significant other had recently engaged in oral sex with another person, would you consider that sexual infidelity? What about spending an hour sexting an attractive friend? These are the kinds of questions that come up in trying to define sexual behavior (McBride & others, 2017; Sewell & others, 2017).

One possibility is to define sex as activities that are involved in reproduction. By this interpretation, many sexually active gay men, lesbian women, and many others who engage exclusively in, say, oral sex or mutual masturbation have never had sex. Further, from this point of view, too, masturbation would not be a sexual behavior.

Another approach is to define sexual behavior by the arousal and sexual responses that occur when the behavior is performed. Though broader, this definition still might leave out people who themselves might say that they are engaged in sexual behavior. For instance, if a person is unable to experience sexual arousal but performs oral sex on a partner, has that person engaged in sexual behavior? Alternatively, we might broaden the definition a great deal and define sexual behaviors to include behaviors that are specific to each individual and that are pleasurable in a particular way—one that is unusually intimate and personal.

Confusion over what counts as sex can lead to potentially risky behavior. For example, oral sex has become relatively common during the teen years but many adolescents do not have much information about the practice (Bauer & others, 2020). They may view it as a safe alternative to intercourse. However, it is important to keep in mind that oral sex requires the use of protection to prevent sexually transmitted infections and that oral sex, like any sexual behavior, requires affirmative consent.

RESEARCH ON SEXUAL BEHAVIOR
When people in the United States engage in sexual behavior, what do they do, and how often? Alfred Kinsey and his colleagues conducted the earliest research on this topic in 1948. Kinsey is widely recognized as a pioneer who brought scientific attention to sexual behavior. He was interested in studying sex objectively, without concern about guilt or shame. He collected data wherever he could find it, interviewing anyone willing to discuss the intimate details of their sex life.

The *Kinsey Reports,* published in two volumes, presented his findings for men (Kinsey & others, 1948) and women (Kinsey & others, 1953). Among the data that shocked his readers were Kinsey's estimates of the frequency of bisexuality in men (nearly 12 percent) and women (7 percent) and his estimate that at least 50 percent of married men had been sexually unfaithful. Although acknowledged for initiating the scientific study of sexual behavior, Kinsey's work was limited by the lack of representative samples.

Subsequent surveys have provided a more nuanced picture of sexual behavior. For example, a national study of sexual behavior in the United States among adults 25 to 44 years of age found that 98 percent of the women and 97 percent of the men said that they had ever engaged in penile-vaginal intercourse (Chandra & others, 2011). Also in this study, 89 percent of the women and 90 percent of the men reported that they ever had oral sex with an opposite-gender partner, and 36 percent of the women and 44 percent of the men stated that they ever had anal sex with an opposite-gender partner.

How often do people have sex? A study of a sample of 278 newlyweds found that within a 2-week period, couples had sex an average of three to four times (Meltzer & McNulty, 2016). Additional interesting information about the frequency of sexual behavior comes from a survey given to large samples of adults in the United States. Starting in 1989, the survey asked participants about how many times they had had sex in the past year. Results spanning the years 1989 until 2014, including over 26,000 people, showed that overall sexual frequency declined over the years (Twenge & others, 2017). For instance, between 1989 and 1994, respondents estimated that they had had sex approximately 60 times in the past year. In comparison, from 2010 to 2014, that number had dropped to about 54 times. Interestingly, this drop appears to exist because younger generations are especially having less sex. Holding age constant, those born in the 1930s reported having sex about six times more often per year than those born in the 1990s. Another interesting trend from these data is a difference that has emerged in the link between marriage and sexual frequency. Prior to 2003, married (or partnered) adults reported having more sex than the unmarried (or unpartnered). However, that difference disappeared in 2003 and by 2014, unmarried (or unpartnered) individuals reported having more sex than married (or partnered) people (Twenge & others, 2017).

Does sexual frequency matter to couples and individuals? For couples in a relationship, it is important to keep in mind that sexual intimacy exists in the context of a larger relationship, but sexual satisfaction does matter to relationship satisfaction (Schoenfeld & others, 2017). Sexual intimacy can create a deep bond between partners (McNulty & others, 2016). A study of "sexual afterglow" in newlywed couples showed that having sex led to enhanced sexual satisfaction for 48 hours afterward (Meltzer & others, 2017). The stronger the couples' sexual afterglow, the more satisfied they were in their marriages over time. Sexual quality of life has been shown to decline with age, but older couples tend to focus more on the quality rather than quantity of sex with their partners (Forbes & others, 2017). Sexual activity is also an important component of a fulfilling life for people with disabilities (McGrath & others, 2020).

Gender Differences in Sexual Attitudes and Behavior Gender differences in sexual attitudes and sexual behaviors are areas on which a good deal of research has focused. Note that most of this research was conducted when gender was considered only as a binary concept. As such, what we know about gender and sexuality is limited and awaits a great deal more research. Still, let's review what we know about gender and sexuality in this admittedly limited context. However, many studies have relied on self-reports of sexual behaviors, so the observed differences may have more to do with contrasting societal expectations for people of different genders than with actual differences in behavior. A meta-analysis revealed that for the following factors, stronger gender differences in sexuality were found: Men engaged in more masturbation, viewed more pornography, reported engaging in more casual sex, and had more permissive attitudes about casual sex than did women (Petersen & Hyde, 2010).

Andrejs Pidjass/Alamy Stock Photo

Remember the importance of having a representative sample. Kinsey's research included only those who were willing to talk about their sex lives—a biased sample.

Research has called into question many gender differences in the domain of sexual behavior. For example, in a classic study, Russell Clark and Elaine Hatfield (1989) sent five men and five women experimenters to a college campus with a mission. They were to approach members of the opposite gender whom they found quite attractive and say, "I have been noticing you around campus. I find you very attractive." Then they were to ask one of three questions:

- "Would you like to go out with me?"
- "Would you like to go to my apartment with me?"
- "Would you like to go to bed with me?"

The independent variables were the gender of the person approached and the type of question asked. The dependent variable was whether that person said yes or no to the question. There was no gender difference in answers to the "going out" question—about half of the men and half of the women said yes. However, large differences emerged for the other two questions. Nearly 70 percent of men said yes to the "apartment" question, while most women said no. For the "bed" question, 75 percent of the men said yes, but *none* of the women did. For many years, this study was recognized as supporting the predictions that men are more interested in casual sex than are women and that women are choosier than men.

Can you identify a confound in the study design? Recall that a confound is a variable, other than the independent variable, that systematically differs across groups and that might explain differences between them on the dependent measure. Think about it—women were always approached *by men*, and men were always approached *by women*. Isn't it possible that the gender of the person doing the asking might have influenced whether those who were approached said yes to the proposal?

In a series of studies, Terri Conley (2011) showed that the proposer's characteristics influence whether the approached person accepts or rejects a proposal for casual sex. For instance, Conley found that both men and women rated a male stranger as potentially more dangerous than a female stranger. Would women be so choosy if they were approached by a familiar person rather than a stranger? Conley showed that women were more likely to report that they would say yes to casual sex if it was offered by a familiar person, such as an attractive friend or a celebrity.

Other established gender differences in sexuality have also fallen by the wayside (Conley & others, 2011; Peterson & Hyde, 2011). For instance, men often report having more sex partners than women do. In a study using a fake lie detector test (called the *bogus pipeline*), however, this difference disappeared when men and women thought that the researchers could tell if they were lying (Alexander & Fisher, 2003).

Do men at least *think about* having sex more often than women do? In one study, college students kept tallies of how many times they thought about sex, food, and sleep for a week (Fisher & others, 2012). Men did report thinking about sex more than women did. However, men also thought about food and sleep more than their female counterparts did. The researchers concluded that men may be more focused than women on their own physical needs.

Compared to men, women tend to show more changes in their sexual patterns and sexual desires over their lifetime (Baumeister, 2000; Diamond & others, 2017; Kanazawa, 2017). Women are more likely than men, for instance, to have had sexual experiences with same- and opposite-gender partners, even if they identify themselves strongly as heterosexual or lesbian (Santtila & others, 2008). In contrast, male sexual interest may be more limited to particular targets of attraction (Chivers, 2017; Chivers & Brotto, 2017). One study compared the sexual arousal of heterosexual women, lesbian women, heterosexual men, and gay men while they watched erotic films of various sexual acts featuring male and female actors or bonobo apes. The films included scenes of sexual activity between same- and opposite-gender human partners and between opposite-sex bonobos, and scenes of men and women masturbating alone or engaging in aerobic exercise while naked. Sexual arousal was measured physiologically by monitoring the genitals of men and women for indicators of arousal. Both heterosexual and lesbian women were aroused by all of the films showing sexual activity. However, gay men were aroused only by the films that included men, and heterosexual men were aroused only by the films that included women (Chivers & others, 2007).

SEXUAL ORIENTATION

Individuals' **sexual orientation** is a multifaceted, complex concept that includes the direction of their erotic interests but also their behaviors and identity. A person who identifies as heterosexual is generally sexually attracted to members of the other gender. Someone who identifies as gay or lesbian is generally sexually attracted to members of the same gender.

A person who experiences sexual interest in women and men might self-identify as *bisexual*. Some people identify as pansexual (Sprott & Benoit Hadcock, 2018). **Pansexual** orientation means that the person's sexual attractions do not depend on the biological sex, gender, or gender identity of others. Some people identify as **asexual**. Asexuality means the person experiences a lack of sexual attraction to others and may feel no sexual orientation (Bogaert & others, 2018).

Despite their widespread use, some researchers argue that labels such as "heterosexual," "gay," "lesbian," "bisexual," "pansexual," or "asexual" are misleading. Because a person's erotic attractions may be fluid, these commentators say, references to a construct such as a fixed sexual orientation ignore the potential flexibility of human sexual attraction and behavior (Diamond & Savin-Williams, 2015; Kanazawa, 2017).

It is difficult to know precisely how many gays, lesbians, and bisexuals there are in the world, partly because fears of discrimination may prevent people from answering honestly on surveys. Estimates of the percentage of the population who are gay and lesbian people range from 2 percent to 10 percent and are typically higher for gay men than for lesbian women (Zietsch & others, 2008). A Gallup poll found that 4.1 percent of Americans identified as LGBT (that is, lesbian, gay, bisexual, or transgender) (Gates, 2017). That percentage would indicate that there are 10 million people in the U.S. who identify as gay, lesbian, or bisexual, just less than the population of the state of Michigan. The number of millennials (those born between 1980 and 1998) identifying as lesbian, gay, bisexual, or transgender was 7.3 percent (Gates, 2017). Interestingly, the largest sexual minority group appears to be those who identify as bisexual (Brown, 2017). Generally, the newer the poll the larger the percentage of people identifying as lesbian, gay, bisexual, or transgender.

Gay and lesbian people are similar to their heterosexual counterparts in many ways (Diamond, 2013a, 2013b; Savin-Williams, 2015). Regardless of their sexual orientation, all people have similar physiological responses during sexual arousal and seem to be aroused by the same types of tactile stimulation. Investigators typically find no differences among lesbian, gay, bisexual, and heterosexual people in a wide range of attitudes, behaviors, and psychological adjustment (Allen & Diamond, 2012). An important aspect of sexual minority well-being is coping with prejudice and discrimination (Riggle & others, 2017).

What explains a person's sexual orientation? Speculation about this question has been extensive. Scientists have identified factors that *do not* predict sexual orientation. First, being reared by a gay or lesbian parent does not increase the chances of being gay (Farr & Patterson, 2013; Patterson & Farr, 2014). In fact, the vast majority of gay people have heterosexual parents. Nor does a particular parenting style relate to the emergence of homosexuality (Bell & others, 1981).

Think about it for a moment: Any theory of sexual orientation must be able to explain not only why some people are gay, lesbian, or bisexual but also why most people are heterosexual (Breedlove, 2017). Why is it that the majority of people feel sexually attracted to the opposite gender and never question that feeling? Given the many different ways in which parents interact with their children and the fact that the vast majority of people are heterosexual, it is extremely unlikely that the emergence of heterosexuality is explained by particular parenting strategies. The fact that most lesbian, gay, and bisexual people have heterosexual parents argues against the influence of observational learning or modeling in the development of sexual

sexual orientation
The direction of an individual's erotic interests, today viewed as a continuum from exclusive male–female relations to exclusive same-gender relations.

pansexual
A person's sexual attractions do not depend on the biological sex, gender, or gender identity of others.

asexual
A person experiences a lack of sexual attraction to others and may feel no sexual orientation.

An individual's sexual orientation is most likely determined by a combination of genetic, hormonal, and perhaps psychological factors.
Jakob Helbig/Getty Images

Homosexual behavior has been observed in nearly 1,500 species of animals, including rats, nonhuman primates, ostriches, goats, guppies, dolphins, and fruit flies (Bagemihl, 1999; Sommer & Vasey, 2006).

orientation. Further, same-gender sexual experience or experimentation in childhood does not predict eventual adult gay, lesbian, or bisexual sexual orientation (Bailey, 2003; Bogaert, 2000).

Researchers have examined a number of biological factors that might contribute to sexual attraction to the same gender. For example, twin studies have been used to estimate the heritability of sexual orientation. Heritability is a statistic that indicates the extent to which observed differences in a given characteristic can be explained based on differences in genes. A study of nearly 4,000 twins in Sweden demonstrated that the heritability of same-gender sexual behavior was about 35 percent in men and 19 percent in women (Langstrom & others, 2010). These heritability estimates suggest that although genes play a role in sexual orientation, genes are not as strong an influence as they are for other characteristics, such as intelligence. In addition to genes, researchers have found that, for women in particular, same-gender attraction may be rooted, at least in part, in prenatal exposure to androgens (Balthazart & Court, 2017; Breedlove, 2017).

Genetic explanations for same-gender attraction present a puzzle for evolutionary psychologists. How can a characteristic that decreases a person's likelihood of reproducing be passed down genetically (Chaladze, 2016)? One possibility is that some of the same genes that contribute to gay or lesbian sexual orientation may also lead to reproductive success for heterosexual individuals who possess them (Iemmola & Ciani, 2009). Research has examined twin pairs in which one twin is heterosexual and the other gay or lesbian (Zietsch & others, 2008). Heterosexual twins of gay men and lesbian women are likely to possess attractive qualities (such as, for men, being caring and gentle; and for women, being assertive and sexually open) and to have more sex partners than heterosexual individuals with heterosexual twins (Zietsch & others, 2008).

Clearly, much remains to be explained about the determination of sexual orientation. Similar to many other psychological characteristics, a person's sexual orientation most likely depends on a combination of genetic, hormonal, and perhaps psychological factors. No one factor alone causes sexual orientation, and the relative weight of each factor can vary from one person to the next (Balthazart & Court, 2017).

Whether heterosexual, gay, lesbian, bisexual, pansexual, or asexual, a person cannot be talked out of sexual orientation. Diverse sexual orientations are present in all cultures, regardless of whether a culture is accepting or intolerant. Research tells us that sexual orientation is not a choice but an integral part of the functioning human being and that person's sense of self (Breedlove, 2017).

Available evidence suggests that gay and lesbian households exist in 99 percent of counties throughout the United States, and approximately one in four of these households includes children (O'Barr, 2006). Children reared by gay men and lesbian women tend to be as well adjusted as those from heterosexual households, are no more likely to be gay themselves, and are no less likely to be accepted by their peers (Farr & Patterson, 2013; Oakley & others, 2017; Patterson, 2013, 2014; Patterson & Farr, 2014; Sumontha & others, 2017).

Gay and lesbian parents are likely to prioritize diversity in their parenting and emphasize with children the importance of respecting all kinds of families (Oakley & others, 2017). One study compared the incidence of co-parenting (the support that parents give each other in raising a child) in adoptive heterosexual, lesbian, and gay couples with preschool-aged children (Farr & Patterson, 2013). Both self-reports and observations indicated that lesbian and gay couples shared child care more than heterosexual couples did, with lesbian couples being the most supportive of each other in parenting. Another study revealed more positive parenting in adoptive gay father families and fewer child externalizing problems in these families than in heterosexual families (Golombok & others, 2014).

In the United States, marriage equality and gay and lesbian parenting have generated controversy, especially in political election years. The current legal status of

Support for marriage equality has risen, reaching 67 percent in the United States in 2020.
Ievgen Chabanov/Alamy Stock Photo

same-gender marriage fits with the scientific consensus that marriage equality for same-gender couples supports their well-being and the well-being of their children (American Psychological Association, 2004).

self-quiz

1. Obesity
   A. does not have a genetic component.
   B. is linked to good health.
   C. is associated with the body's set point.
   D. has most recently decreased in the United States.

2. The brain structure(s) *primarily* involved in motivation for sexual behavior is(are) the
   A. hypothalamus.
   B. temporal lobes.
   C. hippocampus.
   D. medulla.

3. Research indicates that one factor that can predict a person's sexual orientation is

A. the parenting style with which they grew up.
B. having a gay or lesbian parent.
C. having a heterosexual parent.
D. genetic background.

APPLY IT! 4. A small town's school board is considering what type of sex education program to adopt for the high school. A number of individuals have expressed concern that giving students information about contraception will send the message that it is okay to engage in sexual activity. Which of the following reflects the research relevant to this issue?
   A. Students who are given information about contraception generally

have sex earlier and more frequently than individuals who are not given this information.
B. Students who are given abstinence-only education are least likely to engage in sex at all.
C. Sex education in schools has shown no relationship to adolescent sexual activity.
D. Students who are given comprehensive information about contraception are less likely to become pregnant during adolescence and are not more likely to engage in sexual activity.

3 Beyond Hunger and Sex: Motivation in Everyday Life

Food and sex are crucial to human survival. Surviving is not all we do, of course. Think about the wide range of human actions and achievements reported in the news—everything from someone donating a kidney to a stranger to a person who grew up in poverty rising to be the CEO of a major corporation. Such behavior is not easily explained by motivational approaches that focus on physiological needs. Psychologists appreciate the role of the goals that people set for themselves in motivation. In this section, we explore the ways that psychologists explain the processes underlying everyday human behavior.

Maslow's Hierarchy of Human Needs

Humanistic theorist Abraham Maslow (1954, 1971) proposed a **hierarchy of needs** (Figure 3) that must be satisfied in the following sequence: physiological needs, safety, love and belongingness, esteem, and self-actualization. The strongest needs are at the base of the hierarchy (physiological), and the weakest are at the top (self-actualization). According to this hierarchy, people are motivated to satisfy their need for food first and to satisfy their need for safety before their need for love. If we think of our needs as calls for action, hunger and safety needs shout loudly, while the need for self-actualization beckons with a whisper. Maslow asserted that each lower need in the hierarchy comes from a deficiency, such as being hungry, afraid, or lonely, and that we see the higher-level needs in a person who is relatively sated in these basic needs. Such an individual can turn their attention to the fulfillment of a higher calling, achieving a sense of meaning by contributing something of lasting value to the world.

Self-actualization, the highest and most elusive of Maslow's needs, is the motivation to develop one's full potential as a human being. According to Maslow, self-actualization is possible only after the other needs in the hierarchy are met. Maslow cautions that most people stop moving up the hierarchy after they have developed a high level of esteem and thus do not become self-actualized.

hierarchy of needs
Maslow's theory that human needs must be satisfied in the following sequence: physiological needs, safety, love and belongingness, esteem, and self-actualization.

self-actualization
The motivation to develop one's full potential as a human being—the highest and most elusive of Maslow's proposed needs.

FIGURE 3
Maslow's Hierarchy of Needs Abraham Maslow developed the hierarchy of human needs to show that we have to satisfy basic physiological needs before we can satisfy other, higher needs.
(mountain) Photogl/iStock/Getty Images; (clapping) Colin Anderson/ Getty Images; (father and daughter) Digital Vision/Getty Images; (security) PNC/Getty Images; (eating) Brooke Fasani/Corbis

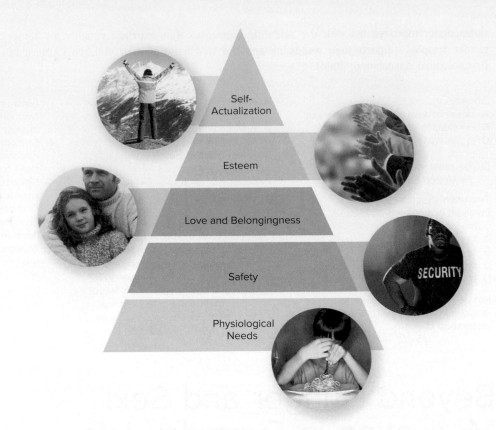

The idea that human motives are hierarchically arranged is appealing; however, Maslow's ordering of the needs is debatable. Some people, for example, might seek greatness in a career to achieve self-esteem while putting on hold their needs for love and belongingness. Certainly history is full of examples of individuals who, in the most difficult circumstances, were still able to perform acts of kindness that seem to come from higher-level needs. Research demonstrates that poor individuals are more likely than wealthy individuals to give generously to others (Kraus & others, 2012; Piff & others, 2010).

Maslow later added self-transcendence as a need even higher than self-actualization. Self-transcendence involves a level of experience that is beyond the self, including spirituality, compassion, and morality. What other needs do you think Maslow left out of his original hierarchy?

Another controversy surrounding Maslow's hierarchy of needs is the extent to which it originates with his own thinking. Maslow's scheme appears to derive from a hierarchy that is part of the tradition of the Blackfoot Nation, a First Nation (or indigenous tribe) of Canada. From the Blackfoot perspective, the triangle that you are studying was actually a *tipi*. The Blackfoot perspective on life includes the notions of self-actualization (though it provided the foundational bottom of the hierarchy). The possibility that Maslow borrowed these ideas without acknowledging their origin is supported by the fact that he spent time on a Blackfoot reservation prior to presenting his hierarchy (Lokensgard, 2014).

Perhaps Maslow's greatest contribution to our understanding of motivation is that he asked the key question about motivation for modern people: How can we explain what humans do, once their bellies are full? That is, how do we explain the "why" of human behavior when survival is not the most pressing need? This is the kind of questioning that inspired self-determination theory.

self-determination theory
Deci and Ryan's theory asserting that all humans have three basic, innate organismic needs: competence, relatedness, and autonomy.

Self-Determination Theory

Building from Maslow's humanistic approach, Richard Ryan and Edward Deci (2017) developed a theory of motivation that emphasizes particular kinds of needs as factors in optimal human functioning. Their **self-determination theory** asserts that there are three basic organismic needs: competence, relatedness, and autonomy. These psychological needs are innate and exist in every person. They are basic to human growth and functioning, just as water, soil, and sunshine are necessary for plant growth. This metaphor is especially apt, because once we plant a seed, all it requires to thrive and grow is a supportive environment. Similarly, self-determination theory holds that all of us have the capacity for growth and fulfillment in us, ready to emerge if our basic needs are met.

From the perspective of self-determination theory, these organismic needs do not arise from deficits. Self-determination theory is not a drive reduction theory. Like Maslow, Deci and Ryan argue that these needs concern personal growth, not the filling of deficiencies (Ryan & Deci, 2017). Let's examine each of these needs in depth.

The first organismic need described by self-determination theory, *competence,* is met when we feel that we are able to bring about desired outcomes. Competence motivation involves *self-efficacy* (the belief that you have the ability to accomplish a given goal or task) and *mastery* (the sense that you can gain skills and overcome obstacles). Competence is also related to expectancies for success. One domain in which competence needs may be met is in the realm of achievement. Some people spend considerable effort striving to excel.

Paralympics competitors demonstrate competence. Paralympic sprinting and long-jump medalist Kelly Cartwright says, "To advance in life you need to believe in yourself and you need to set goals for yourself. Push yourself, because you can achieve anything in life if you put your mind to it" (Paralympic Movement, 2012).
A.RICARDO/Shutterstock

The second organismic need described by self-determination theory is *relatedness*—the need to engage in warm relations with other people. The need for relatedness is reflected in the importance of caregivers' nurturing children's development, the intimate moments of sharing private thoughts in friendship, the uncomfortable feelings we have when we are lonely, and the powerful attraction we have for someone else when we are in love.

The third need proposed by self-determination theory is *autonomy*—the sense that we are in control of our own life. Autonomy is not just being independent and self-reliant; it is a key aspect of feeling that one's behavior is self-motivated and emerging from genuine interest (Sheldon & others, 2017). Of course, many of the behaviors we engage in may feel like things we are forced to do, but a sense of autonomy is strongly related to well-being (Sheldon & others, 2005). Teachers, bosses, parents, and caregivers can increase the well-being of students, employees, offspring, and partners when they support the person's sense of autonomy (Hobson & Maxwell, 2017; Nie & others, 2015; Niemiec & Coulson, 2017; Uysal & others, 2017; van der Kaap-Deeder & others, 2017).

Research on the role of motivation in well-being supports the idea that progress on goals that serve the three organismic needs is strongly related to well-being (Koletzko & others, 2015; Sheldon & Elliot, 1998). Further, valuing extrinsic qualities—such as money, prestige, and physical appearance—over organismic concerns is associated with lowered well-being, lowered self-actualization, and physical illness (Kasser & Ryan, 1996; Kasser & others, 2004). Self-determination theory maintains that when we are engaged in intrinsically motivated behavior that meets our needs for competence, relatedness, and autonomy, we are likely to experience personal growth and optimal well-being (Ryan & Deci, 2017).

Like any theory, self-determination theory has ignited some controversies. One important issue is the extent to which the three needs are universal. Cultures vary in how strongly they promote the needs for competence, relatedness, and autonomy. Many Western cultures—among them, the United States, Canada, and western European countries—are termed *individualist* because they emphasize individual achievement, independence, and self-reliance. In contrast, many Eastern cultures—such as China, Japan, and Korea—are called *collectivist* because they stress harmony, cooperation, and interdependence (Triandis, 2000). However, cross-cultural evidence suggests that the needs emphasized by self-determination theory are valued in both Western and Eastern cultures and similarly associated with well-being (Nalipay & others, 2020; Yu & others, 2017).

Self-determination theory maintains that one of the most important aspects of healthy motivation is the sense that we do the things we do because we have freely chosen to do them. When our behaviors follow from the needs for competence, autonomy, and relatedness, we experience intrinsic motivation (Ryan & Deci, 2020). When our behavior serves needs for other values, such as prestige, money, or approval, our behavior is extrinsically motivated. We examine this important distinction between intrinsic and extrinsic motivation next.

Intrinsic versus Extrinsic Motivation

One way psychologists understand the "why" of our goals is by distinguishing between intrinsic and extrinsic motivation. **Intrinsic motivation** is based on internal factors such as organismic needs (competence, relatedness, and autonomy), as well as curiosity, challenge, and fun. When we are intrinsically motivated, we explore, investigate, and master because doing so is enjoyable. **Extrinsic motivation** involves external incentives such as rewards and punishments. When we are extrinsically motivated, we engage in a behavior for some external payoff or to avoid an external punishment. Some students study hard because they are internally motivated to put forth considerable effort and achieve high quality in their work (intrinsic motivation). Other students study hard because they want to make good grades or avoid parental disapproval (extrinsic motivation).

If someone is producing shoddy work, seems bored, or has a negative attitude, offering an external incentive may improve motivation. There are times, though, when external rewards can diminish intrinsic motivation. The problem with using a reward as an incentive is that people may perceive that the reward rather than their own motivation caused their achievement behavior. Indeed, research comparisons often reveal that people whose motivation is intrinsic show more interest, excitement, and confidence in what they are doing than those whose motivation is extrinsic. Intrinsic motivation often results in improved performance, persistence, creativity, and self-esteem (Ryan & Deci, 2020). One study tested the associations between intrinsic and extrinsic motivation and supervisor ratings, work commitment, burnout, and work-family conflict, among employees in various industries (Kuvaas & others, 2017). The results showed that intrinsic motivation predicted higher evaluations and commitment and lower burnout and work-family conflict. In contrast, extrinsic motivation was negatively related (or unrelated) to positive work outcomes.

Of course, many very successful people are both intrinsically motivated (they have high personal standards of achievement and emphasize personal effort) and extrinsically motivated (they are strongly competitive). Elite athletes such as Olympic team members, as well as those who are highly successful in the business world, may be motivated by both intrinsic and extrinsic rewards. Indeed, many of us might think of the ideal occupation as one in which we get paid well (an extrinsic reward) for doing the very thing we love to do (intrinsic motivation).

Self-Regulation: The Successful Pursuit of Goals

Today many psychologists approach motivation by asking about goals and values and seeking to understand how these motivational forces shape behavior. Psychologists have referred to goals by various names, including *personal projects, best possible selves, life tasks,* and *personal strivings* (King, 2008). All of these terms reflect the goals a person is trying to accomplish in everyday life. Self-generated goals can range from trivial matters (such as letting a bad haircut grow out) to life tasks (such as becoming a good parent).

intrinsic motivation
Motivation based on internal factors such as organismic needs (competence, relatedness, and autonomy), as well as curiosity, challenge, and fun.

extrinsic motivation
Motivation that involves external incentives such as rewards and punishments.

Goal approaches to motivation include **self-regulation,** the process by which an individual effortfully controls behavior to pursue important objectives (McDaniel & Einstein, 2020). A key aspect of self-regulation is getting feedback about how we are doing in our goal pursuits (Lee, 2016). Our daily mood has been proposed as a way that we may receive this feedback—that is, we feel good or bad depending on how we are doing in the areas of life we value. Note that the role of mood in self-regulation means that we cannot be happy all the time. To pursue our goals effectively, we have to be open to the bad news that might occasionally come our way.

Putting our personal goals into action is a potentially complex process that involves setting goals, planning for their implementation, and monitoring our progress. Individuals' success improves when they set goals that are specific and moderately challenging. A fuzzy, nonspecific goal is, "I want to be successful." A concrete, specific goal is "I want to have a 3.5 average at the end of the semester." You can set both long-term and short-term goals. When you set long-term goals, such as "I want to be a clinical psychologist," make sure that you also create short-term goals as steps along the way, such as "I want to get an *A* on my next psychology test." Make commitments in manageable chunks. Planning how to reach a goal and monitoring progress toward the goal are critical aspects of achievement. Monitoring goal progress leads to higher achievement (Harkin & others, 2016).

Even as we keep our nose to the grindstone in pursuing short-term goals, it is also important to have a sense of the big picture. Dedication to a long-term dream or personal mission can enhance the experience of purpose in life. Research shows that thinking about the big picture of goal pursuit can also facilitate the accomplishment of smaller goals. One way to keep the big picture in mind is to think about having completed the goal. In a set of studies, participants were asked to imagine that they had completed a goal and then to plan backwards, from attainment to the subgoals themselves. This strategy led to better goal success (Jooyoung & others, 2017).

self-quiz

1. Rank-order the following needs according to Maslow's hierarchy: hunger, self-esteem, social relationships, safety.
   A. Social relationships must be fulfilled first, followed by hunger, safety, and finally self-esteem.
   B. Self-esteem must be fulfilled first, followed by social relationships, safety, and hunger.
   C. Hunger must be fulfilled first, followed by safety, social relationships, and finally self-esteem.
   D. Safety must be fulfilled first, followed by hunger, social relationships, and finally self-esteem.

2. Self-efficacy is most related to which need from self-determination theory?
   A. Autonomy
   B. Relatedness
   C. Competence
   D. Self-actualization

3. Of the following, the individual who will likely perform best is someone with
   A. high extrinsic motivation.
   B. low extrinsic motivation.
   C. high intrinsic motivation.
   D. low intrinsic motivation.

APPLY IT! 4. Kim cannot decide on a college major. She chose to major in biology because she likes it. She is not sure that is a good reason to pursue a goal. Based on the reading, what is your advice?
   A. Kim should definitely think about how much money she might be able to make as a biology major.
   B. Kim should should consider finding a life goal that promises to please her parents, rather than herself.
   C. Kim should not underestimate the importance of enjoying what she does. She might consider looking for ways to apply her passions in jobs using the skills she is gaining.
   D. Kim should not worry so much about the future. There is no reason to make long term plans

4 Emotion

As the concept of self-regulation implies, motivation and emotion are closely linked. We can feel happy or sad depending on how events influence the likelihood of our getting the things we want in life. Sometimes our emotions take us by surprise and give us a reality check about what we really want. We might think, for example, that we have lost interest in our romantic partner until that person initiates a breakup. Suddenly, we realize how

Regina King expressed a range of emotions, from surprise to joy to gratitude when she won the Academy Award for her performance in If Beale Street Could Talk, in 2019.
PA Images/Alamy Stock Photo

much they really mean to us. Anyone who has watched an awards show on television surely knows the link between motivation and emotion. Strolling in on the red carpet, the celebrities stress how honored they are to be nominated, but behind the Hollywood smiles is the longing to win. When the announcement is made, "And the Oscar goes to . . .," the cameras zoom in to catch a glimpse of real emotion: the winner's face lighting up with joy and, of course, the moment of disappointment for the others.

Emotions are certainly complex. The body, the mind, and the face play key roles in emotion, and psychologists debate which of these components is most significant in emotion and how they mix to produce emotional experiences (Kayser, 2017; Khan & others, 2017). For our purposes, **emotion** is feeling, or *affect,* that can involve physiological arousal (such as a fast heartbeat), conscious experience (thinking about being in love with someone), and behavioral expression (a smile or grimace).

emotion
Feeling, or affect, that can involve physiological arousal (such as a fast heartbeat), conscious experience (thinking about being in love with someone), and behavioral expression (a smile or grimace).

Biological Factors in Emotion

A friend whom you have been counseling about a life problem calls you to say, "We need to talk." As the time of your friend's visit approaches, you get nervous. What could be going on? You feel burdened—you have a lot of work to do, and you do not have time for a talk session. You also worry that they might be angry or disappointed about something you have done. When they arrive with a gift-wrapped package and a big smile, your nerves give way to relief. Your friend announces, "I wanted to give you this present to say thanks for all your help over the last few weeks." Your heart warms, and you feel a strong sense of your enduring bond with your friend. As you moved through the emotions of worry, relief, and joy, your body changed. Indeed, the body is a crucial part of our emotional experience.

AROUSAL Recall that the *autonomic nervous system (ANS)* takes messages to and from the body's internal organs, monitoring such processes as breathing, heart rate, and digestion. The ANS is divided into the sympathetic and the parasympathetic nervous systems (Figure 4). The *sympathetic nervous system (SNS)* is involved in the body's arousal; it is responsible for a rapid reaction to a stressor, sometimes referred to as the fight-or-flight response. The SNS immediately causes an increase in blood pressure, a faster heart rate, more rapid breathing for greater oxygen intake, and more efficient blood flow to the brain and major muscle groups. All of these changes prepare us for action. At the same time, the body stops digesting food, because it is not necessary for immediate action (which could explain why just before an exam, students usually are not hungry).

Remember:

Sympathetic = "fight or flight"

Parasympathetic = "rest and digest"

The *parasympathetic nervous system (PNS)* calms the body. Whereas the sympathetic nervous system prepares the individual for fighting or running away, the parasympathetic nervous system promotes relaxation and healing. When the PNS is activated, heart rate and blood pressure drop, stomach activity and food digestion increase, and breathing slows.

The sympathetic and parasympathetic nervous systems evolved to improve the human species' likelihood for survival, but it does not take a life-threatening situation to activate them. Emotions such as anger and fear are associated with elevated SNS activity as exemplified in heightened blood pressure and heart rate. States of happiness and contentment also activate the SNS to a lesser extent.

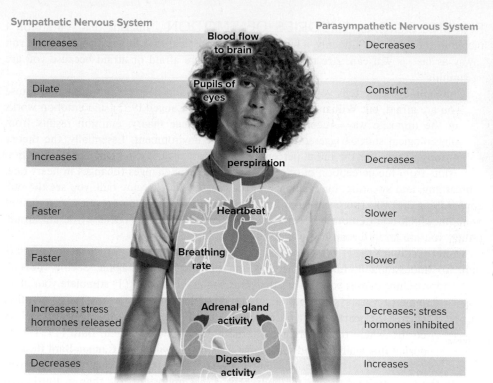

Sympathetic Nervous System		Parasympathetic Nervous System
Increases	Blood flow to brain	Decreases
Dilate	Pupils of eyes	Constrict
Increases	Skin perspiration	Decreases
Faster	Heartbeat	Slower
Faster	Breathing rate	Slower
Increases; stress hormones released	Adrenal gland activity	Decreases; stress hormones inhibited
Decreases	Digestive activity	Increases

FIGURE 4

The Autonomic Nervous System and Its Role in Arousing and Calming the Body The two parts of the autonomic nervous system work in different ways. The sympathetic nervous system arouses the body in reaction to a stressor, evoking the fight-or-flight response. In contrast, the parasympathetic nervous system calms the body, promoting relaxation and healing. Remember, the latter system functions to "rest and digest."
(Photo): James Woodson/Getty Images

MEASURING AROUSAL Because arousal includes a physiological response, researchers have been intrigued by how to measure it accurately. One aspect of emotional arousal is *skin conductance level (SCL)* response, a rise in the skin's electrical conductivity when sweat gland activity increases. A sweaty palm conducts electricity better than a dry palm, and this difference provides the basis for SCL, which produces an index of arousal that has been used in many studies of emotion.

Another measure of arousal is the **polygraph** or lie detector, a machine examiners use to try to determine whether someone is lying. The polygraph monitors changes in the body—heart rate, breathing, and SCL—thought to be influenced by emotional states.

In a typical polygraph test, the examiner asks the individual a number of neutral questions and several key, less neutral questions. If the individual's heart rate, breathing, and SCL responses increase substantially when the key questions are asked, the individual is assumed to be lying (Lerman, 2020; Palmatier & Rovner, 2014). How accurate is the lie detector? Although it measures the degree of arousal to a series of questions, no one has found a unique physiological response to telling lies (Lykken, 1987, 2001; Seymour & others, 2000). Heart rate and breathing can increase for reasons other than lying, and this effect can make it difficult to interpret the physiological indicators of arousal. Accurately identifying truth or deception is linked with the skill of both the examiner and the individual being examined. Body movements and the presence of certain drugs in the person's system can interfere with the polygraph's accuracy. Sometimes the mere presence of the polygraph and the person's belief that it is accurate in detecting deception trigger a confession of guilt. Police may use the polygraph in this way to get a suspect to confess (Gannon & others, 2014). However, in too many instances it has been misused and misrepresented. Experts argue that the polygraph errs just under 50 percent of the time, especially as it cannot distinguish between such feelings as anxiety and guilt (Iacono & Lykken, 1997).

The Employee Polygraph Protection Act of 1988 restricts polygraph testing outside government agencies, and most courts do not accept the results of lie detectors. The majority of psychologists argue against the polygraph's use because of its inability to tell who is lying and who is not (Iacono & Lykken, 1997; Lykken, 1998; Saxe & Ben-Shakhur, 1999; Steinbrook, 1992).

polygraph
A machine, commonly called a lie detector, that monitors changes in the body, and is used to try to determine whether someone is lying.

If a test confuses anxiety and guilt, how can it distinguish between a nervous person and a liar?

PHYSIOLOGICAL THEORIES OF EMOTION

Imagine that you are on a picnic in the country. Suddenly, a bull runs across the field toward you. Terrified, you run away as fast as you can. Are running because you are afraid or afraid because you are running?

Common sense tells you that you are trembling and fleeing from the bull because you are afraid, but William James (1950) and Carl Lange (1922) said emotion works in the opposite way. According to the **James-Lange theory,** emotion results from physiological states triggered by stimuli in the environment. Essentially, the theory proposes that after the initial perception of a stimulus, the experience of the emotion results from the perception of one's own physiological changes (changes in heart rate, breathing, and sweating, for example). In the case of the charging bull, you see the bull approaching and you run away. Your aroused body then sends sensory messages to your brain, at which point emotion is perceived. You do not run away because you are afraid; rather, you are afraid because you are running away.

James-Lange theory
The theory that emotion results from physiological states triggered by stimuli in the environment.

Walter Cannon (1927) presented an alternative physiologically based theory of emotion. To understand it, imagine the bull and the picnic once again. Seeing the bull approaching causes your brain's thalamus simultaneously to (1) stimulate your autonomic nervous system to produce the physiological changes involved in emotion (increased heart rate, rapid breathing) and (2) send messages to your cerebral cortex, where the experience of emotion is perceived. Philip Bard (1934) supported this analysis, and the theory became known as the **Cannon-Bard theory—** the proposition that emotion and physiological reactions occur simultaneously.

In the Cannon-Bard theory, the body plays a less important role than in the James-Lange theory.

> Remember for James-Lange: Bodily responses comes first, then we feel an emotion; for Cannon-Bard, these two things happen at the same time.

Cannon-Bard theory
The proposition that emotion and physiological reactions occur simultaneously.

NEURAL CIRCUITS AND NEUROTRANSMITTERS

Contemporary researchers are keenly interested in identifying the neurotransmitters and neural circuitry involved in emotions. For example, dopamine and endorphins are linked to positive emotions (Bowen, 2020). Norepinephrine functions in regulating arousal and anxiety (Jones & others, 2016). Low levels of serotonin are linked to anger and aggression (Alia-Klein & others, 2020; da Cunha-Bang & others, 2017) and the persistence of negative moods (van Roekel & others, 2017).

With regard to brain structures that play a role in emotion, a great deal of research concerns the amygdalae, the almond-shaped structures in the limbic system. The amygdalae house circuits that are activated when we encounter important, survival-relevant stimuli in the environment. The limbic system, including the amygdalae, is involved in the experience of positive emotions (Vrticka & others, 2014). However, most research has focused on the important role of the amygdalae in the experience of negative emotion, particularly fear.

Research by Joseph LeDoux and his colleagues (LeDoux, 1996, 2012, 2013, 2014; LeDoux & Pine, 2016) has documented the central role of the amygdalae in fear. When the amygdalae determine that danger is present, they shift into high gear, marshaling the brain's resources in an effort to protect the organism from harm. This fear system evolved to detect and respond to predators and other types of natural dangers that threaten survival or territory.

> The amygdalae's ability to respond quickly to threatening stimuli is adaptive. Think about early humans, facing a world filled with threatening predators. Members of our species who had to encounter a hungry tiger more than once before learning to avoid it probably didn't survive to reproduce. So, we have our great-great-great-great-great-great-great grandparents to thank for our amygdalae.

The brain circuitry that involves the emotion of fear can follow two pathways: a direct pathway from the thalamus to the amygdalae or an indirect pathway from the thalamus through the sensory cortex to the amygdalae (Figure 5). The direct pathway does not convey detailed information about the stimulus, but it has the advantage of speed—and speed clearly is an important characteristic of information for an organism facing a threat to its survival—like the bull that spoiled your picnic. The indirect pathway carries nerve impulses from the sensory organs (eyes and ears, for example) to the thalamus (recall that the thalamus is a relay station for incoming sensory stimuli); from the thalamus, the nerve impulses travel to the sensory cortex, which then sends appropriate signals to the amygdalae.

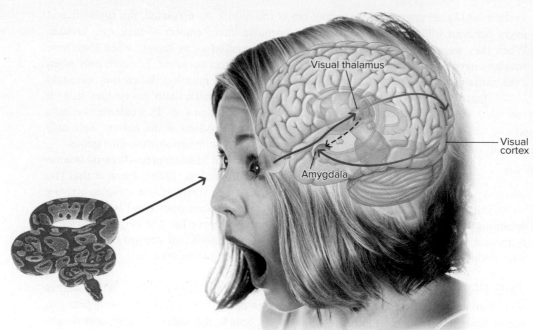

FIGURE 5
Direct and Indirect Brain Pathways in the Emotion of Fear
Information about fear can follow two pathways in the brain when an individual sees a snake. The direct pathway (*broken arrow*) conveys information rapidly from the thalamus to the amygdala. The indirect pathway (*solid arrows*) transmits information more slowly from the thalamus to the sensory cortex (here, the visual cortex) and then to the amygdala.
(snake) Simon Murrell/Getty Images; (woman) pathdoc/Shutterstock

Recall that the amygdalae are linked with emotional memories. LeDoux says that the amygdalae hardly ever forget (LeDoux, 2001). This quality is useful, because once we learn that something is dangerous, we do not have to relearn it. However, we pay a penalty for this ability. Many people carry fears and anxieties around with them that they would like to get rid of but cannot seem to shake. We will look at such fears when we explore phobias. Part of the reason fears are so difficult to change is that the amygdalae are well connected to the cerebral cortex, in which thinking and decision making primarily occur (Pina & Cunningham, 2017). Each amygdala is in a much better position to influence the cerebral cortex than the other way around, because they send more connections to the cerebral cortex than they get back. This may explain why it is so hard to control our emotions, and why, once fear is learned, it is so hard to erase.

Cognitive Factors in Emotion

Does emotion depend on the tides of the mind? Are we happy only when we think we are happy? Cognitive theories of emotion center on the premise that emotion always has a cognitive component (Frijda, 2007; Langeslag & Surti, 2017; Tong, 2015; Tong & Jia 2017). Thinking is said to be responsible for feelings of love and hate, joy and sadness. While cognitive theorists do recognize the role of the brain and body in emotion, they give cognitive processes the main credit for these responses.

THE TWO-FACTOR THEORY OF EMOTION

In the **two-factor theory of emotion** developed by Stanley Schachter and Jerome Singer (1962), emotion is determined by two factors: physiological arousal and cognitive labeling. When we feel aroused, Schachter and Singer argued, we look to the external world for an explanation of why we we feel that way. We interpret external cues and then label the emotion. So, from this view, we first have a state of arousal (the first factor) and then attribute that arousal to an emotion depending on external cues (the second factor).

To test their theory, Schachter and Singer (1962) injected volunteer participants with epinephrine, a drug that produces high arousal. After participants received the drug, they observed someone else behave in either a euphoric way (shooting balled-up papers at a

two-factor theory of emotion
Schachter and Singer's theory that emotion is determined by two factors: physiological arousal and cognitive labeling.

wastebasket) or an angry way (stomping out of the room). As predicted, the euphoric and angry behavior influenced the participants' cognitive interpretation of their own arousal. When they were with a happy person, they rated themselves as happy; when they were with an angry person, they said they were angry. This effect occurred, however, only when the participants were not told about the true effects of the injection. When they were told that the drug would increase their heart rate and make them jittery, they had no reason to attribute their own arousal to the other person. This pattern of results supports the two-factor theory because the emotions of the partner were only used as an explanation when the effects of the injection were ambiguous.

The labels we put on arousal are important. Remember the next time you are nervous about an important opportunity, you are not "nervous" you are "excited"!

In general, research supports the belief that misinterpreted arousal intensifies emotional experiences (Leventhal & Tomarken, 1986). Imagine that you are late for class on an important exam day. You sprint across campus, arriving just in time for the test. As you look over the questions, your heart is racing, your breathing is fast, and you feel sweaty. Are you nervous about the test or just recovering from your run to the classroom? The two-factor theory suggests that you just might mistake your bodily sensations as indications that you are scared of the test.

THE PRIMACY DEBATE: COGNITION OR EMOTION?   Which comes first, thinking or feeling? Fans of vintage episodes of TV's *Star Trek* may recognize this theme from the frequent arguments between Mr. Spock, the logical Vulcan, and Bones, the emotional doctor on the *Enterprise*. In the 1980s and 1990s, two eminent psychologists, Richard Lazarus (1922–2002) and Robert Zajonc (1923–2008) (whose name sounds like the word *science*), debated the question of which is central, cognition or emotion.

Lazarus (1991) argued for the primacy of thinking—he believed cognitive activity to be a precondition for emotion. Lazarus said that we cognitively appraise ourselves and our social circumstances. Our emotions depend on these appraisals. How we think determines our emotions. For examples, we feel unhappy if we fail to achieve a goal, because of how we think about that failure. Zajonc (1984) disagreed with Lazarus. Emotions are primary, he said, and our thoughts are a result of them. Zajonc famously argued that "preferences need no inferences," meaning that the way we feel about something on a "gut level" requires no thought.

Which of the two psychologists was right? Both were likely correct. Lazarus talked mainly about a cluster of related events that occur over a period of time, whereas Zajonc described single events or a simple preference for one stimulus over another. Lazarus was concerned with love over the course of months and years, a sense of value to the community, and plans for retirement; Zajonc spoke about a car accident, an encounter with a snake, and a preference for ice cream rather than spinach.

Some of our emotional reactions are virtually instantaneous and probably do not involve cognitive appraisal, such as shrieking upon detecting a snake. Other emotional circumstances, especially long-term feelings such as a depressed mood or anger toward a friend, are more likely to involve cognitive appraisal. Indeed, the direct and indirect brain pathways described earlier support the idea that some of our emotional reactions do not involve deliberate thinking, whereas others do (LeDoux, 2001).

Behavioral Factors in Emotion

Our definition of emotion includes not only physiological and cognitive components but also a behavioral component. The behavioral component can be verbal or nonverbal. Verbally, a person might show love for someone by professing it in words or might display anger by saying nasty things. Nonverbally, a person might smile, frown, show a fearful expression, look down, or slouch.

The most interest in the behavioral dimension of emotion has focused on the nonverbal behavior of facial expressions. Emotion researchers have long been intrigued by people's ability to detect emotion from a person's facial expression (Matsumoto & Hwang,

FIGURE 6 **Recognizing Emotions in Facial Expressions** Look at the six photographs and determine the emotion reflected in each of the faces. (first) Ioannis Pantzi/Shutterstock; (second) Monkey Business Images/Getty Images; (third) Marcos Mesa Sam Wordley/Shutterstock; (fourth) Mix and Match Studio/Shutterstock; (fifth) Ranta Images/Shutterstock; (sixth) rubberball/Getty Images

> First, without reading ahead, label each picture with the emotion you think it shows. > Second, match the pictures to each of the following emotions: anger, happiness, surprise, disgust, sadness, and fear. > Okay, the right answers for that second exercise are (top) happiness, anger, sadness; (bottom) surprise, disgust, fear. How does this analysis change your views of the universal quality of facial expressions of emotion?

2014; Nelson & Russell, 2014). In a typical research study, participants, when shown photographs like those in Figure 6, are usually able to identify six emotions: happiness, anger, sadness, surprise, disgust, and fear (Ekman & O'Sullivan, 1991).

Might our facial expressions not only reflect our emotions but also influence them? According to the **facial feedback hypothesis,** facial expressions can influence emotions as well as reflect them (Dzokoto & others, 2014). In this view, facial muscles send signals to the brain that help us to recognize the emotion we are experiencing (Lee & others, 2013). For example, we feel happier when we smile and sadder when we frown.

Support for the facial feedback hypothesis comes from an experiment by Paul Ekman and his colleagues (1983). In this study, professional actors moved their facial muscles in very precise ways, such as raising their eyebrows and pulling them together, raising their upper eyelids, and stretching their lips horizontally back to their ears (you might want to try this yourself). They were asked to hold their expression for 10 seconds, during which time the researchers measured their heart rate and body temperature. When the actors moved facial muscles in the ways described, they showed a rise in heart rate and a steady body temperature—physiological reactions that characterize fear. When they made an angry facial expression (with a penetrating stare, brows drawn together and downward, and lips pressed together or opened and pushed forward), their heart rate and body temperature both increased. The facial feedback hypothesis provides support for the James-Lange theory of emotion discussed earlier—namely, that emotional experiences can be generated by changes in and awareness of our own bodily states.

facial feedback hypothesis
The idea that facial expressions can influence emotions and reflect them.

✍ This description fits with Stanislavski's "method acting," which suggests that to feel a particular emotion, an actor should imitate the behavior of someone feeling that emotion.

Sociocultural Factors in Emotion

Are the facial expressions that are associated with different emotions largely innate, or do they vary across cultures? Are there gender variations in emotion? Answering these questions requires a look at research findings on sociocultural influences in emotions.

CULTURE AND THE EXPRESSION OF EMOTION   In *The Expression of the Emotions in Man and Animals,* Charles Darwin stated that the facial expressions of human beings are innate, not learned; are the same in all cultures around the world; and have evolved from the emotions of animals (Darwin, 1872/1965). Today psychologists still believe that emotions, including facial expressions of emotion, have strong biological ties. For example, children who are blind from birth and have never observed the smile or frown on another person's face smile or frown in the same way that children with normal vision do. If emotions and facial expressions that go with them are unlearned, then they should be the same the world over. Is that, in fact, the case?

Extensive research has examined the universality of facial expressions and the ability of people from different cultures accurately to label the emotion that lies behind facial expressions. Paul Ekman's careful observations reveal that the many faces of emotion do not differ significantly from one culture to another (Ekman, 1980, 1996, 2003). For example, Ekman and his colleague (Ekman & Friesen, 1969) photographed people expressing emotions such as happiness, fear, surprise, disgust, and grief. When they showed the photographs to people from the United States, Chile, Japan, Brazil, and Borneo (an Indonesian island in the western Pacific), the participants, across the various cultures, recognized the emotions the faces were meant to show (Ekman & Friesen, 1969). Another study focused on the way the Fore tribe, an isolated Stone Age culture in New Guinea, matched descriptions of emotions with facial expressions (Ekman & Friesen, 1971). Before Ekman's visit, most of the Fore had never seen a Caucasian face. Ekman's team showed them photographs of people's faces expressing emotions such as fear, happiness, anger, and surprise. Then they read stories about people in emotional situations and asked the Fore to pick out the face that matched the story. The Fore were able to match the descriptions of emotions with the facial expressions in the photographs. Figure 7 shows the similarity of facial expressions of emotions by many different people.

display rules
Sociocultural standards that determine when, where, and how emotions should be expressed.

Whereas facial expressions of basic emotions appear to be universal, display rules for emotion vary (Fischer, 2006; Fok & others, 2008). **Display rules** are sociocultural standards that determine when, where, and how emotions should be expressed (Hudson & Jacques, 2014; Zhu & others, 2013). For example, although happiness is a universally expressed emotion, when, where, and how people display it may vary from one culture to another. The same is true for other emotions, such as fear, sadness, and anger. Members of the Utku culture in Alaska, for example, discourage anger by cultivating acceptance and by dissociating themselves from any display of anger. If an unexpected snowstorm hampers a trip, the Utku do not express frustration but accept the storm and build an igloo. The importance of display rules is especially evident when we evaluate the emotional expression of another. Does that grieving man on a morning talk show seem appropriately distraught over his spouse's murder? Or might he be a suspect?

Display rules can also be seen as you watch sporting events. Think about the difference between the subdued crowd at Wimbledon versus the raucous U.S. Open.

Like facial expressions, some other nonverbal signals appear to be universal indicators of certain emotions. For example, regardless of where they live, when people are depressed, their emotional state shows not only in their sad facial expressions but also in their slow body movements, downturned heads, and slumped posture. Many nonverbal signals of emotion, though, vary from one culture to another. For example, male-to-male kissing is commonplace in Yemen but uncommon in the United States. The "thumbs up" sign, which in most cultures means either that everything is okay or that one wants to hitch a ride, is an insult in Greece, similar to a raised third finger in the United States—a cultural difference to keep in mind if you find yourself backpacking through Greece.

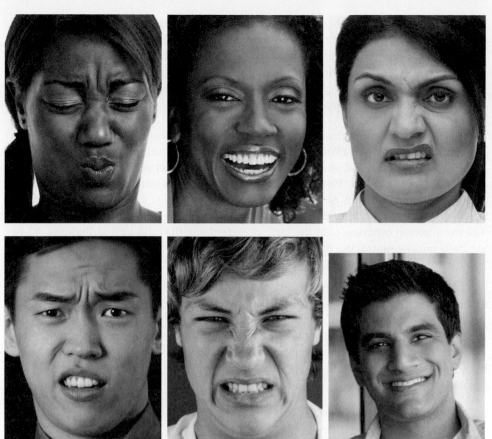

FIGURE 7
Disgust or Happiness?
Look at these pictures and as quickly as possible identify whether the faces indicate disgust or joy. Even though superficially similar (look at those mouths), across genders and ethnicities you are able to rapidly categorize these expressions. Notice the similarity within expressions of each emotion. Psychologists believe that many facial expressions of emotion are virtually the same, across people and in all cultures.
(first) DRB Images, LLC/E+/Getty Images; (second) digitalskillet/E+/Getty Images; (third) SIBSA Digital Pvt. Ltd./Alamy Stock Photo; (fourth) Vladimir Wrangel/Alamy Stock Photo; (fifth) themacx/Getty Images; (sixth) Jupiterimages/Getty Images

GENDER INFLUENCES   You probably know the stereotype about gender and emotion: Women are emotional; men are not. This stereotype, grounded, again, in the gender binary, is a powerful and pervasive image across cultures (Brescoll, 2016). However, researchers have found that people of different genders are often more alike in the way they experience emotion than the stereotype would lead us to believe (Carothers & Reis, 2013).

People of different genders often use the same facial expressions, adopt the same language, and describe their emotional experiences similarly when they keep diaries about their experiences. For many emotional experiences, researchers do not find gender differences (Hyde, 2014). Where differences do emerge, they suggest that women report more feelings of sadness and anxiety than men do, and men report more anger and irritability than women do (Schirmer, 2013).

To understand these differences, we must consider the contexts in which emotions are experienced and expressed (Brannon, 1999; Brody, 1999; Shields, 1991). Women may be judged harshly for expressions of anger, and men might be evaluated as "weak" if they express worry or sadness. Certainly, women and men are aware of the gendered stereotypes for their emotional behavior (Blakemore & others, 2009; Schirmer, 2013). Gender differences in emotion are much more tied to social context than to biological sex (Derntl & others, 2012).

Classifying Emotions

There are more than 200 words for emotions in the English language, indicating their complexity and variety. Not surprisingly, psychologists have created ways to classify emotions—to summarize these many emotions along various dimensions, including their valence and arousal.

VALENCE   The *valence* of an emotion refers to whether it feels pleasant or unpleasant. You probably are not surprised to know that happiness, joy, pleasure, and contentment are positively valenced emotions. In contrast, sadness, anger, worry, and feeling upset are negatively valenced emotions. Research has shown that emotions tend to go together based on their valence, so that if someone is sad, they are also likely to be angry or worried, and if a person is happy, they are also likely to be feeling confident, joyful, and content (Watson, 2001).

We can classify many emotional states on the basis of valence. Indeed, according to some experts in emotion (Watson, 2001), there are two broad dimensions of emotional experience: negative affect and positive affect. **Negative affect** refers to emotions such as anger, guilt, and sadness. **Positive affect** refers to emotions such as joy, happiness, and interest. Although it seems essential to consider the valence of emotions as a way to classify them, valence does not fully capture all that we need to know about emotional states. The joy a person experiences at the birth of a child and the mild high at finding a $5 bill are both positive states, but they clearly differ in important ways.

negative affect
Negative emotions such as anger, guilt, and sadness.

positive affect
Pleasant emotions such as joy, happiness, and interest.

AROUSAL LEVEL   The *arousal level* of an emotion is the degree to which the emotion is reflected in a person's being active, engaged, or excited versus being more passive, relatively disengaged, or calm. Positive and negative emotions can be high or low in arousal. Ecstasy and excitement are examples of high-arousal positive emotions, whereas contentment and tranquility are low-arousal positive emotions. Examples of high-arousal negative emotions are rage, fury, and panic, while irritation and boredom represent low-arousal negative emotions.

Valence and arousal level are independent dimensions that together describe a vast number of emotional states. Using these dimensions, we can effectively create a wheel of mood states (Figure 8). The illustration shows what psychologists call a *circumplex model* of mood (Posner & others, 2005). A circumplex is a type of graph that creates a circle from two independent dimensions. Using the two dimensions of valence and arousal level, we can arrange emotional states in an organized fashion (Valenza & others, 2014).

PSYCHOLOGICAL INQUIRY

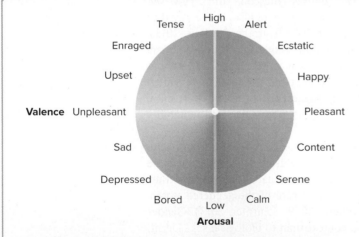

FIGURE 8   A Circumplex Model of Mood Using the dimensions of valence and arousal, this wheel-like figure shows a variety of emotional states. >Find "upset" and "sad" on the circumplex. According to the circumplex, these emotions differ primarily on the dimension of arousal. Which is higher on arousal? Do you agree with the placement of these emotions? Explain. >According to the circumplex, which emotion is the exact opposite of "serene"? >Where would you place the following emotions: embarrassed, proud, worried, angry?

Adaptive Functions of Emotions

In considering the functions of emotions, it is fairly easy to come up with a good reason for us to have emotions such as fear and anger. Negative emotions carry direct and immediate adaptive benefits in situations that threaten survival. Negative emotions indicate clearly that something is wrong and that we must take action. Positive emotions do not signal a problem. So, what is the adaptive function of positive emotions?

Confronting this question, Barbara Fredrickson proposed the **broaden-and-build model** of positive emotion (Fredrickson, 1998, 2013; Fredrickson & Siegel, 2017). She argues that the function of positive emotions lies in their effects on our attention and our ability to build resources. The broaden-and-build model begins

broaden-and-build model
Fredrickson's model of positive emotion, stating that the function of positive emotions lies in their effects on an individual's attention and ability to build resources.

Jacob Lund/Shutterstock

with the influence of positive emotions on cognitive processing.

Positive moods, such as joy and happiness, have been shown to broaden our attentional focus; they allow us to see the forest for the trees. As a result, when in a good mood, we may be more disposed to think outside the box—to see unusual possibilities that escaped us before. In addition, a good mood, Fredrickson says, gives us a chance to build resources—to make friends, to exercise to promote our health, to branch out in new ways. These activities allow us to build up strengths that we can use when we encounter life's difficulties (Kok & others, 2008). For example, joy broadens people by creating the urge to play, push the limits, and be creative. Interest broadens people by creating the motivation to explore, absorb new information and experiences, and expand the self. Positive emotions facilitate "approach" behavior (Otake & others, 2006; Watson, 2001), meaning that when we are feeling good, we are more likely to go after the rewards we want and to face our problems head on.

Of course, happiness is not the only positive emotion we experience. We also feel gratitude, awe, pride, and love. How can we understand these many different shades of positive feelings? Scholars have proposed a *family tree model* of positive emotions (Shiota & others, 2017). This model seeks to account for the many different positive feelings human being have. The root of the tree is the brain's reward center and the action of dopamine there. This makes sense, as all positive feelings likely have their root in this brain area that responds when something good happens. The trunk of the tree is enthusiasm, which can be considered the approach motivational aspect of positive feelings. The branches are the neurotransmitters that underlie each emotional state, which comprise the leaves of the tree. For example, oxytocin leads to nurturant love and opioids leads to pleasure and liking. The emergence of this model demonstrates the current movement of emotion science toward focusing on more differentiated emotional states and linking these to brain processes (Pressman & Cross, 2018; Shiota & others, 2017).

Positive emotions might play an important role in the ability of resilient individuals to cope successfully with life's challenges. Resilience refers to the ability to bounce back from negative experiences, to be flexible and adaptable when things are not going well. Resilient individuals might be thought of as tall trees that have the ability to bend but do not break in response to strong winds. In contrast, people who lack resilience might be characterized as more brittle—more likely to snap or break in the face of adversity (Block & Kremen, 1996).

Michelle Tugade, Barbara Fredrickson, and Lisa Feldman Barrett (2004) found that the superior coping of resilient individuals came from their ability to use positive emotions to spring back from negative emotional experiences. Using measures of cardiovascular activity, the researchers discovered that resilient individuals were better able to regulate their responses to stressful situations (for instance, being told they were about to give an important speech) by strategically experiencing positive emotion.

Resilient individuals seem to show a kind of emotional wisdom; they capitalize on the power of positive emotions to reverse the stress of negative feelings.

Do It!

Recall that some psychologists believe that the ability to identify and regulate one's emotions is a kind of intelligence. Emotionally intelligent people are also thought to be better at reading the emotional expressions of others. Do a web search for *emotional intelligence tests* and take some online quizzes, or just try this one at http://greatergood. berkeley.edu/ei_quiz/. Do you think you are emotionally intelligent? Does your performance on the test seem to reflect your actual experience? What is your opinion of the test you tried? Is there information on the site for its validity and reliability?

1. The James-Lange theory of emotion states that
   A. emotion happens first, followed by physiological reactions.
   B. physiological reactions happen first, followed by emotion.
   C. physiological reactions and emotion happen simultaneously.
   D. the body plays a minimal role in emotion.

2. In the case of fearful stimuli, *indirect* neural pathways go first to the thalamus and
   A. then to the hypothalamus, followed by the amygdala.
   B. then to the sensory cortex, followed by the amygdala.
   C. then to the hippocampus.

   D. then to the hypothalamus, followed by the sensory cortex, and finally the hippocampus.

3. The facial feedback hypothesis is consistent with the theory of emotion known as
   A. the James-Lange theory.
   B. the Cannon-Bard theory.
   C. direct theory.
   D. indirect theory.

APPLY IT! 4. Seymour is talking to his friend about his sadness over his recent breakup with his girlfriend. His girlfriend cheated on him, but Seymour was willing to forgive her. She was not interested, though and broke things off. As Seymour talks, his friend notices that

Seymour is clenching his teeth, making fists, and generally getting angry. The friend says, "You know, you sound more angry than sad." Why might Seymour have confused anger and sadness?
   A. Sadness and anger are similar in terms of their arousal level.
   B. Seymour's friend is probably wrong given that sadness produces the same facial expression as anger.
   C. Because he is a man, Seymour probably does not understand emotion very well.
   D. Sadness and anger have the same valence, so someone who is feeling sad is likely to also feel angry. Seymour is probably feeling both of these negative emotions.

5 Motivation and Emotion: The Pursuit of Happiness

Motivation is about what people want, and a quick scan of the best-seller list or the self-help section of any bookstore or on Amazon indicates that one thing people want very much is to be happy or happier. Can people get happier? Let's consider the evidence.

Biological Factors in Happiness

As we have seen, the brain is certainly at work in the experience of positive emotions. Genes also play a role. For instance, research on the heritability of well-being shows that a substantial proportion of well-being differences among people can be explained by genetic differences. The heritability estimates for happiness range from 50 to 80 percent (Lykken, 1999). Remember that heritability is a statistic that describes characteristics of a group, that heritability estimates can vary across groups and over time, and that even highly heritable characteristics can be influenced by experience. Thus, a person is not necessarily doomed to an unhappy life, even if the person has particularly miserable parents.

Recall the concept of *set point* in our discussion of weight. There may also be a happiness set point, a person's general level of happiness when the individual is not trying to increase happiness (Sheldon & Lyubomirsky, 2007, 2012). Like our weight, our happiness levels may fluctuate around this set point (Dijkhuizen & colleagues, 2017). In trying to increase happiness, we must consider the role of this powerful starting spot that is likely the result of genetic factors and personal dispositions.

Given these potential biological limitations, other factors also complicate the pursuit of happiness, including the hedonic treadmill and the dangers of striving for happiness itself.

Obstacles in the Pursuit of Happiness

The first key challenge individuals encounter in trying to increase their happiness is the hedonic (meaning "related to pleasure") treadmill (Brickman & Campbell, 1971; Fredrick & Loewenstein, 1999). The term *hedonic treadmill* captures the idea that any aspect of life that enhances one's positive feelings is likely to do so for only a short time, because

How Does Money Relate to Happiness?

How does money relate to happiness? The answer to this question, like so many in psychology, is "it depends." First, research shows that income does share a small positive correlation with well-being (Aknin & others, 2018; Ng & Diener, 2018). In addition, people's happiness shows a small but positive increase as their income grows (Diener & others, 2013). Finally, in general, the association between income and well-being is characterized by diminishing returns. So, the positive relationship between income and happiness primarily applies to low and medium levels of income, with the relationship leveling off. Once a person makes about $75,000 per year, more money no longer predicts greater happiness (Diener & Biswas-Diener, 2002; Ward & King, 2019).

Importantly, however, most studies have included very few people who might be considered extremely wealthy. So, the typical study of the association between money and happiness levels may be missing action at the very, very high end on the income scale. Two studies (Donnelly & others, 2018) included over 4,000 millionaires to see if the very wealthy (those with net worth over $8 million) might differ from the extremely wealthy (those with net worth over $10 million). The extremely wealthy were a bit happier than their very wealthy counterparts. Interestingly, this difference depended on the source of the money. Among the extremely wealthy, those who became that way via inheritance were not happier than the very wealthy. In contrast, those who *earned* the money were happier than the very wealthy (Donnelly & others, 2018).

everst/Shutterstock

Of course, money does not contribute directly to well-being. Rather, how that money is spent matters most. For example, spending money on experiences (like travel or concerts) rather than on things leads to higher boosts in well-being (Kumar & others, 2020). However, and importantly, research shows that experiences typically cost more than objects (Lee & others, 2018) and those who spend money on experiences often have more money to begin with. In addition, spending money on others (or donating to charity) instead of spending on oneself is associated with higher well-being (Aknin & others, 2020; Lok & others, 2020).

A key source of the money we have is the work we do. Research in this area reveals one of the ways work influences well-being: by providing financial resources. Where these resources come from and how we use them may affect our happiness, profoundly.

What Do You Think?

- Why do you think inherited wealth did not predict higher happiness, compared with earned wealth?
- Do you believe that money can buy happiness? Why or why not?

---

individuals generally adapt rapidly to any life change that would presumably influence their happiness. Winning the lottery, moving into a dream home, or falling in love may lead to temporary gains in the experience of joy, but eventually people go back to their baseline (Schkade & Kahneman, 1998). That is, what is first experienced as a life-changing improvement fades to a routine. How can people increase their happiness if such pleasure enhancers quickly lose their power? Clearly, happiness is not about shopping at the right stores, because new possessions will likely lead to only a momentary burst of pleasure, gradually giving way to the set point.

A second obstacle in the goal of enhancing happiness is that pursuing happiness for its own sake is rarely a good way to get happy or happier. When happiness is the goal, the pursuit may backfire (Schooler & others, 2003). Indeed, those who explicitly link the pursuit of their everyday goals to happiness fare quite poorly (McIntosh & others, 1995).

Now, thinking about what would make you happier, you might be thinking about money. The relationship between financial resources and happiness is complex. To read about how money relates to happiness, see the Challenge Your Thinking.

In light of this difficult path, how can we enhance our happiness without having any new capacity for joy become ho-hum? How might we achieve happiness without pursuing it in and of itself? A number of activities have been shown to lead to increases in happiness over time (Carr & others, 2020; Waters, 2020).

Happiness Activities and Goal Striving

Diane Collins and Jordan Hollender/Digital Vision/Getty Images

Sonja Lyubomirsky and her colleagues have proposed a promising approach to enhancing happiness (Lyubomirsky, 2008, 2011, 2013; Sheldon & Lyubomirsky, 2007; Sin & Lyubomirsky, 2009). They suggest that intentional activities like being physically active, expressing kindness, showing gratitude, being optimistic, dwelling less on negative experiences, and engaging in positive self-reflection all enhance positive affect (Kruse & others, 2014; Lyubomirsky & others, 2011a, 2011b; Sheldon & Lyubomirsky, 2007). Behaving altruistically—habitually helping others, especially through acts of service—is another powerful happiness booster, according to Lyubomirsky (2008, 2013).

One technique for practicing positive self-reflection is to keep a gratitude journal. Studies by Robert Emmons and Michael McCullough (2004) demonstrated that being grateful can enhance happiness and psychological well-being. In one study, participants kept a diary in which they counted their blessings every day. Those who did so were better off than others on various measures of well-being. Although some individuals seem to be naturally more grateful than others, experimental evidence indicates that even people who are not naturally grateful can benefit from counting their blessings (Emmons & McCullough, 2003; McCullough & others, 2002).

Another potentially useful approach to amplifying happiness is to commit to the pursuit of personally meaningful goals. Pause and write down the things you are trying to accomplish in your everyday behavior. You might identify goals such as "to get better grades" and "to be a good friend (or partner or parent)." Working toward such everyday goals relates strongly to subjective well-being (Brunstein, 1993; Sheldon, 2002). Goal pursuit provides the glue that meaningfully relates a chain of life events, endowing life with beginnings, middles, and ends (King, 2008).

The scientific literature on goal investment offers a variety of ideas about the types of goals that are likely to enhance happiness. To optimize the happiness payoffs of goal pursuit, one ought to set goals that are important and personally valuable and that reflect the intrinsic needs of relatedness, competence, and autonomy (Sheldon, 2002). These goals also should be moderately challenging and should share an instrumental relationship with each other so that the pursuit of one goal facilitates the accomplishment of another (Emmons & King, 1988).

With regard to the hedonic treadmill, goal pursuit has a tremendous advantage over many other ways of trying to enhance happiness. Goals change and are changed by life experience. As a result, goal pursuit may be less susceptible to the hedonic treadmill over time. Goals accentuate the positive but do not necessarily eliminate the negative. When we fail to reach our goals, we may experience momentary increases in unhappiness, which can be a very good thing. Because goals can make us happy and unhappy, they keep life emotionally interesting, and their influence on happiness does not wear off over time.

Overall, goal pursuit may lead to a happier life. Goals keep the positive possible and interesting. The conclusion to be drawn from the evidence, assuming that you want to enhance your happiness, is to strive mightily for the goals that you value. You may fail now and then, but missing the mark will only make your successes all the sweeter. Even in the pursuit of happiness, it is important to keep in mind that positive and negative emotions are both adaptive and that the best life is one that is emotionally rich.

SW Productions/Photodisc/Getty Images

1. Studies have demonstrated that variance in well-being is
   A. heritable.
   B. surprisingly unpredictable.
   C. extreme.
   D. dependent on the cultural context.

2. Similarly to our body weight, our personal happiness levels may fluctuate around
   A. our popularity.
   B. our stress levels.
   C. a set point.
   D. the seasons of the year.

3. Researchers have discovered that one way for individuals to engage in positive self-reflection and experience meaning in life is to

A. write down what they are thankful for in a diary.
B. pursue the hedonic treadmill.
C. practice transience.
D. refresh their coping skills.

APPLY IT! 4. Bonita works at a small advertising agency. She is committed to her work goals and always gives her all when she has a task to perform. She is deeply disappointed when a potential client decides to go with another firm after Bonita put a whole week into her presentation. A coworker notices Bonita's distress and says, "You know, you only feel so bad because you care too much. You should be like me. I don't care about anything, and I'm never disappointed."

What does the psychology of happiness tell us about this situation?
A. Although Bonita feels disappointed now, her overall approach will likely lead to greater happiness in the long term.
B. Bonita's colleague is right on. Bonita should disengage from her goals, and then she will never be disappointed.
C. Bonita's colleague will probably be happier than Bonita in the long term and will likely have a greater sense of purpose in life.
D. Bonita's happiness depends more on her genetic makeup than on any particular life experience.

SUMMARY

Theories of Motivation

Motivated behavior is energized, directed, and sustained. Early evolutionary theorists considered motivation to be based on instinct—the innate biological pattern of behavior.

A drive is an aroused state that occurs because of a physiological need or deprivation. Drive reduction theory was proposed as an explanation of motivation, with the goal of drive reduction being homeostasis: the body's tendency to maintain equilibrium.

Optimum arousal theory focuses on the Yerkes-Dodson law, which states that performance is best under conditions of moderate rather than low or high arousal. Moderate arousal often serves us best, but there are times when low or high arousal is linked with better performance.

Hunger and Sex

Stomach signals are one factor in hunger. Glucose (blood sugar) and insulin both play an important role in hunger. Glucose is needed for the brain to function, and low levels of glucose increase hunger. Insulin can cause a rise in hunger.

Leptin, a protein secreted by fat cells, decreases food intake and increases energy expenditure. The hypothalamus plays an important role in regulating hunger. The lateral hypothalamus is involved in stimulating eating; the ventromedial hypothalamus, in restricting eating.

Obesity is a serious problem in the United States. Heredity, basal metabolism, set point, and fat cells are biological factors involved in obesity. Time and place affect eating. Our early ancestors ate fruits to satisfy nutritional needs, but today we fill up on the empty calories in sweets.

Motivation for sexual behavior involves the hypothalamus. Masters and Johnson mapped out the human sexual response pattern, which consists of four physiological phases: excitement, plateau, orgasm, and resolution.

Thoughts and images are central in the sexual lives of humans. Sexual scripts influence sexual behavior, as do sensory/perceptual factors.

Sexual values vary across cultures. These values influence sexual behavior.

Describing sexual practices in the United States has been challenging due to the difficulty of surveying a representative sample of the population. Research suggests that Americans have sex less often than they did in the past. Sex education has sometimes been a controversial issue, but research shows that nations with comprehensive sex education have far lower rates of teen pregnancy and sexually transmitted infections than the United States.

Sexual orientation refers to the direction of a person's erotic attraction. Sexual orientation—heterosexual, gay, lesbian, or bisexual—is most likely determined by a combination of genetic, hormonal, cognitive, and environmental factors. Based on scientific evidence, the APA supports marriage equality for gay people. Research shows that children reared by gay parents are not more likely to be gay and are similar in adjustment to their peers.

3 Beyond Hunger and Sex: Motivation in Everyday Life

According to Maslow's hierarchy of needs, our main needs are satisfied in this sequence: physiological needs, safety, love and belongingness, esteem, and self-actualization. Maslow gave the most attention to self-actualization: the motivation to develop to one's full potential.

Self-determination theory states that intrinsic motivation occurs when individuals are engaged in the pursuit of organismic needs that are innate and universal. These needs include competence, relatedness, and autonomy. Intrinsic motivation is based on internal factors. Extrinsic motivation is based on external factors, such as rewards and punishments.

Self-regulation involves setting goals, monitoring progress, and making adjustments in behavior to attain desired outcomes. Research suggests that setting short-term goals is a good strategy for reaching a long-term goal.

4 Emotion

Emotion is feeling, or affect, that has three components: physiological arousal, conscious experience, and behavioral expression. The biology of emotion focuses on physiological arousal involving the autonomic nervous system and its two subsystems. Skin conductance level and the polygraph have been used to measure emotional arousal.

The James-Lange theory states that emotion results from physiological states triggered by environmental stimuli: Emotion follows physiological reactions. The Cannon-Bard theory states that emotion and physiological reactions occur simultaneously. Contemporary biological views of emotion increasingly highlight neural circuitry and neurotransmitters. LeDoux has charted the neural circuitry of fear, which focuses on the amygdala and consists of two pathways, one direct and the other indirect. It is likely that positive and negative emotions use different neural circuitry and neurotransmitters.

Schachter and Singer's two-factor theory states that emotion is the result of both physiological arousal and cognitive labeling. Lazarus believed that cognition always directs emotion, but Zajonc argued that emotion directs cognition. Both probably were right.

Research on the behavioral component of emotion focuses on facial expressions. The facial feedback hypothesis states that facial expressions can influence emotions, as well as reflect them.

Most psychologists believe that facial expressions of basic emotions are the same across cultures. However, display rules, which involve nonverbal signals of body movement, posture, and gesture, vary across cultures.

Emotions can be classified on the basis of valence (pleasant or unpleasant) and arousal (high or low). Using the dimensions of valence and arousal, emotions can be arranged in a circle, or circumplex model.

Positive emotions likely play an important role in well-being by broadening our focus and allowing us to build resources. Resilience is an individual's capacity to thrive even during difficult times. Research has shown that one way resilient individuals thrive is by experiencing positive emotions.

5 Motivation and Emotion: The Pursuit of Happiness

Happiness is heritable, and there is reason to consider each person as having a happiness set point. Still, many people would like to increase their level of happiness. One obstacle to changing happiness is the hedonic treadmill: the idea that we quickly adapt to changes that might enhance happiness. Another obstacle is that pursuing happiness for its own sake often backfires.

Ways to enhance happiness include engaging in physical activity, helping others, positively self-reflecting, and experiencing meaning (such as by keeping a gratitude journal). Another way to enhance happiness is to pursue personally valued goals passionately.

KEY TERMS

androgens	extrinsic motivation	motivation	self-regulation
asexual	facial feedback hypothesis	need	set point
broaden-and-build model	hierarchy of needs	negative affect	sexual orientation
Cannon-Bard theory	homeostasis	pansexual	two-factor theory of emotion
display rules	human sexual response pattern	polygraph	Yerkes-Dodson law
drive	instinct	positive affect	
emotion	intrinsic motivation	self-actualization	
estrogens	James-Lange theory	self-determination theory	

ANSWERS TO SELF-QUIZZES

Section 1:   1. D; 2. B; 3. D; 4. B
Section 2:   1. C; 2. A; 3. D; 4. D

Section 3:   1. C; 2. C; 3. C; 4. C
Section 4:   1. B; 2. B; 3. A; 4. D

Section 5:   1. A; 2. C; 3. A; 4. A

Design elements: (Preview icon): Jiang Hongyan/Shutterstock; (Marginal notes): Shutterstock/Vadarshop

10 Personality

Symbols of Who We Really Are

Looking in the mirror each day, you have seen yourself. Throughout your life, you have seen your reflected image change from the face of a small child to how you look now. You will see it change further as you age. Our faces are a symbol of ourselves, a basis for personal identity. Yet, faces can change—sometimes drastically. Adults who experience facial disfigurement often experience distress about the fact that people can no longer see who they "really are" (Rifkin & others, 2018). In fact, it was the sense that the face is a vital symbol of personal identity that led to reluctance to attempt face transplants (Alberti, 2020). However, such transplants have transformed the lives of people who receive them. In 2006, Andy Sandness was 21 years old when he attempted to die by suicide, sending a bullet through his chin and destroying most of his face (Bever, 2017). A decade later, another 21-year-old, Rudy Ross, died by suicide. His 19-year-old widow, Lily, who was 8 months pregnant, donated Rudy's face, along with other organs. Andy received Rudy's face in a 60-hour surgery. A year later, Andy was able to meet Lily and her son. It was an intense experience for Lily—seeing her husband's face on another man. For Andy it was an opportunity to express his undying gratitude and to assure Lily that her gift would not be wasted (Bever, 2017).

Andy's appearance has changed forever. He now has Rudy's mole and rosy cheeks, and bears a resemblance to Rudy's young son. But behind that face—that mix of Rudy and Andy—the person is still, and always will be, Andy. As much as faces may define us superficially, there is something else about us that continues to provide a sense of identity. Age and life experiences can change us, but something about us endures throughout life. That "something" is personality, the focus of this chapter.

Image Source/Getty Images

1 Psychodynamic Perspectives

personality
A pattern of enduring, distinctive thoughts, emotions, and behaviors that characterize the way an individual adapts to the world.

psychodynamic perspectives
Theoretical views emphasizing that personality is primarily unconscious (beyond awareness).

 If we knew the dark truth of our existence, we might do something as desperate as the tragic Greek hero Oedipus, who unwittingly murdered his father and married his mother–and then gouged out his own eyes.

Personality is a pattern of enduring, distinctive thoughts, emotions, and behaviors that characterize the way a person adapts to the world. Psychologists have approached these enduring characteristics in a variety of ways, focusing on different aspects of the person.

Psychodynamic perspectives on personality emphasize that personality is primarily unconscious (that is, beyond awareness). According to this viewpoint, those enduring patterns that make up personality are largely unavailable to our awareness, and they powerfully shape our behaviors in ways that we cannot consciously comprehend. Psychodynamic theorists use the word *unconscious* differently from how other psychologists might use the term. From the psychodynamic perspective, aspects of our personality are unconscious because they must be; this lack of awareness is motivated. These mysterious, unconscious forces are simply too frightening to be part of our conscious awareness. Compared to other approaches to personality, psychodynamic approaches put the greatest emphasis on unconscious processes.

Psychodynamic theorists believe that behavior is only a surface characteristic and that to truly understand someone's personality, we have to explore the symbolic meanings of that behavior and the deep inner workings of the mind. Psychodynamic theorists also stress the role of early childhood experience in adult personality. From this vantage point, the adult is a reflection of those childhood experiences that shape our earliest conceptions of ourselves and others. These characteristics were sketched by the architect of psychoanalytic theory, Sigmund Freud.

Freud's Psychoanalytic Theory

Sigmund Freud, one of the most influential thinkers of the twentieth century, was born in Freiberg, Moravia (today part of the Czech Republic) in 1856 and died in London at the age of 83. Freud spent most of his life in Vienna, but he left the city near the end of his career to escape the Holocaust. In his lifetime Freud was a celebrated figure–charming and filled with charisma (Anderson, 2017).

For Freud, the sexual drive was the most important motivator of all human activity. However, Freud did not define the word *sex* as you might think. He felt that sex was anything that provided organ pleasure. Anything pleasurable was sex, to Freud. As we will see, Freud thought that the human sexual drive was the main determinant of personality development, and he felt that psychological disorders, dreams, and all human behavior represent the conflict between this unconscious sexual drive and the demands of civilized human society.

 So, if you hear someone describe the joys of eating a decadent dessert like double-chocolate fudge cake as "better than sex," remember that in Freud's view, eating that cake is sex.

Freud developed *psychoanalysis,* his approach to personality, through his work with patients suffering from hysteria. *Hysteria* refers to physical symptoms that have no physical cause. For instance, a person might be unable to see, even with perfectly healthy eyes, or unable to walk, despite having no physical injury.

Sigmund Freud (1856–1939)
Freud's theories have strongly influenced how people in Western cultures view themselves and the world.
Universal History Archive/UIG/REX/Shutterstock

In Freud's day (the Victorian era, a time marked by strict rules regarding sexual relations and very rigid and limited roles for women), many young women suffered from a condition a called hysteria, which means physical problems that could not be explained by actual physical illness. In his practice, Freud spent many long hours listening to these women talk about their symptoms. Freud came to understand that hysterical symptoms stemmed from unconscious psychological conflicts. These conflicts centered on experiences in which the person's drive for pleasure was thwarted by the social pressures of Victorian society. Furthermore, the particular symptoms were symbolically related to these underlying conflicts. One of Freud's patients, Fraulein Elisabeth Von R., suffered from horrible leg pains that prevented her from standing or walking. The fact that Fraulein Elisabeth could not walk was no accident. Through analysis, Freud discovered that Fraulein Elisabeth had had a number of experiences in which she wanted nothing more than to take a walk but had been prevented from doing so by her duty to her ill father.

Freud came to use hysterical symptoms as his metaphor for understanding dreams, slips of the tongue, and all human behavior. Everything we do, he said, has a multitude of unconscious causes.

Drawing from his work analyzing patients, as well as himself, Freud developed his model of the human personality. He described personality as like an iceberg, existing mostly below the level of awareness, just as the massive part of an iceberg lies beneath the surface of the water. Figure 1 illustrates this analogy and depicts the extensiveness of the unconscious part of our mind, in Freud's view.

STRUCTURES OF PERSONALITY

The three parts of the iceberg in Figure 1 reflect the three structures of personality that Freud described. Freud (1917) called these structures the id, the ego, and the superego. You can get a better feel for these Latin labels by considering their English translations: The id is literally the "it," the ego is the "I," and the superego is the "above-I."

The **id** consists of unconscious drives and is the person's reservoir of sexual energy. This "it" is a pool of amoral and often vile urges pressing for expression. In Freud's view, the id has no contact with reality. The id works according to the *pleasure principle,* the Freudian concept that the id always seeks pleasure and immediate gratification.

The world would be dangerous and scary if personalities were all id. As young children mature, they learn that they cannot slug other children in the face, that they have to use the toilet instead of diapers, and that they must negotiate with others to get the things they want. As children experience the constraints of reality, a new element of personality is formed—the **ego,** the Freudian structure of personality that deals with the demands of reality. Indeed, according to Freud, the ego abides by the *reality principle.* That is, it tries to bring the person pleasure within the norms and constraints of society. The ego helps us to test reality, to see how far we can go without getting into trouble and hurting ourselves. Whereas the id is completely unconscious, the ego is partly conscious. It houses our higher mental functions—reasoning, problem solving, and decision making, for example.

The id and ego do not consider whether something is right or wrong. Rather, the **superego** is the

*We call it a **Freudian** slip when someone makes a mistake in speech or action that seems to express unconscious wishes—such as a typo spelling 'Freud' as 'Fraud.'*

id
The part of the person that Freud called the "it," consisting of unconscious drives; the individual's reservoir of sexual energy.

ego
The Freudian structure of personality that deals with the demands of reality.

superego
The Freudian structure of personality that serves as the harsh internal judge of the individual's behavior; what is often referred to as *conscience.*

One of Freud's most famous essays, 'The Ego and the Id,' was titled 'Das Ich und Das Es' in German, meaning 'The I and the It.'

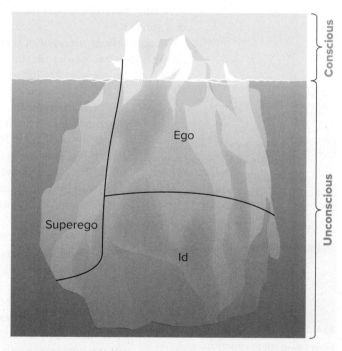

FIGURE 1 **The Conscious and Unconscious Mind: The Iceberg Analogy** The iceberg analogy illustrates how much of the mind is unconscious in Freud's theory. The conscious mind is the part of the iceberg above water; the unconscious mind, the part below water. Notice that the id is totally unconscious, whereas the ego and the superego can operate at either the conscious or the unconscious level.

harsh internal judge of our behavior. The superego is reflected in what we often call *conscience* and evaluates the morality of our behavior. Like the id, the superego does not consider reality; it considers only whether the id's impulses can be satisfied in acceptable moral terms.

The ego acts as a mediator between the conflicting demands of the id and the superego, as well as the real world. Your ego might say, for example, "I will have sex only in a committed relationship and always practice safe sex." Your id, however, screams, "Sex! Now!" and your superego commands, "Sex? Don't even think about it."

DEFENSE MECHANISMS

The conflicts that erupt among the demands of the id, the superego, and reality create a great deal of anxiety for the ego. The ego has strategies for dealing with this anxiety, called defense mechanisms. **Defense mechanisms** are tactics the ego uses to reduce anxiety by unconsciously distorting reality. Freud's daughter Anna introduced and developed many different kinds of defense mechanisms (Wachtel, 2017).

The most primitive defense mechanism is *denial,* in which the ego simply refuses to acknowledge anxiety-producing realities. In denial, for instance, someone might refuse to accept a diagnosis of cancer or the existence of COVID-19. Other defense mechanisms are more complex. For example, imagine that Jason's id is demanding to express an unconscious desire to have sex with his mother. Clearly, acting on this impulse would not please the superego or society at large. If he became aware of this impulse, Jason might recoil in horror. Instead, Jason's ego might use the defense mechanism of displacement, and he might develop a relationship with a girlfriend who looks and acts like his mother. *Displacement* means directing unacceptable impulses at a less threatening target. Through displacement, the ego allows Jason to express his id impulse in a way that will not land him in trouble. Of course, Jason's friends might chuckle at the resemblance between his mother and his girlfriend, but you can bet that Jason will never notice.

Displacement provides the foundation for another defense mechanism, sublimation. *Sublimation* is a special form of displacement in which the person expresses an unconscious wish in a socially valued way, such as a boxer who sublimates his aggressive drive in the ring. Similarly, a person might sublimate their desires to cut into people by becoming a surgeon.

Another defense mechanism is *projection.* In projection, we see in others those impulses that we most fear or despise in ourselves. For instance, our negative attitudes toward people who are different from us may express our unconscious beliefs about ourselves. Projection has been used to explain prejudice.

Reaction formation is a defense mechanism in which a person's conscious experience is exactly the opposite of true unconscious desires. Reaction formation helps to explain hypocrisy and why it is that so many people who rail against the immorality of contemporary society are themselves revealed to be engaged in the very activities they publicly condemned.

Repression is the most powerful and pervasive defense mechanism. Repression pushes unacceptable impulses into the unconscious mind. Repression is the foundation for all of the psychological defense mechanisms, whose goal is to repress threatening impulses, that is, to push them out of awareness. Freud said, for example, that our early childhood experiences, many of which he believed were sexually laden, are too threatening for us to deal with consciously, so we reduce the anxiety of childhood conflict through repression.

Two final points about defense mechanisms are important. First, defense mechanisms are unconscious; we are not aware that we are using them. Second, when used in moderation or on a temporary basis, defense mechanisms are not necessarily unhealthy. Indeed, defense mechanisms that help to relieve anxiety over particular experiences may help people cope (Cramer, 2015; Paradiso & others, 2020). For example, the defense mechanism of *denial* can help a person cope upon first getting the news that death is impending, and the defense mechanism of *sublimation* involves transforming unconscious impulses into activities that actually benefit society. Note that the defense

Anna Freud (1895–1982) *The youngest of Sigmund Freud's six children, Anna Freud not only did influential work on defense mechanisms but also pioneered in the theory and practice of child psychoanalysis.*
Chronicle/Alamy Stock Photo

mechanism of sublimation means that even the very best things that human beings accomplish—a beautiful work of art, an amazing act of kindness—are still explained by unconscious sexual drives and defenses.

PSYCHOSEXUAL STAGES OF PERSONALITY DEVELOPMENT

Freud believed that human beings go through universal stages of personality development and that at each developmental stage we experience sexual pleasure in one part of the body more than in others. Each stage is named for the location of sexual pleasure at that stage. *Erogenous zones* are parts of the body that have especially strong pleasure-giving qualities at particular stages of development. Freud thought that our adult personality is determined by the way we resolve conflicts between these early sources of pleasure—the mouth, the anus, and then the genitals—and the demands of reality.

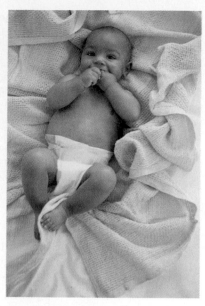

Corbis/VCG/Getty Images

- *Oral stage (first 18 months):* The infant's pleasure centers on the mouth. Chewing, sucking, and biting are the chief sources of pleasure that reduce tension in the infant.

- *Anal stage (18 to 36 months):* During a time when most children are experiencing toilet training, the child's greatest pleasure involves the anus and urethra and their functions. Freud recognized that there is pleasure in "going" and "holding it" as well as in the experience of control over one's parents in deciding when to do either.

- *Phallic stage (3 to 6 years):* The name of Freud's third stage comes from the Latin word *phallus,* which means "penis." Pleasure focuses on the genitals as the child discovers that self-stimulation is enjoyable.

In Freud's view, the phallic stage has special importance in personality development because it triggers the Oedipus complex. This name comes from the Greek tragedy, mentioned earlier, in which Oedipus unknowingly kills his father and marries his mother. The **Oedipus complex** is the boy's intense desire to replace his father and enjoy the affections of his mother. Eventually, the boy recognizes that his father might punish him for these incestuous wishes, specifically by cutting off the boy's penis. *Castration anxiety* refers to the boy's intense fear of being mutilated by his father. To reduce this conflict, the boy identifies with his father, adopting the male gender role. The intense castration anxiety is repressed into the unconscious and serves as the foundation for the development of the superego.

Oedipus complex
According to Freud, a boy's intense desire to replace his father and enjoy the affections of his mother.

> The superego wields a lot of power—it is essentially the internalized castrating father.

Working within the gender binary of his historic time period, Freud recognized that there were differences between boys and girls in the phallic stage. Because a girl does not have a penis, she cannot experience castration anxiety, Freud reasoned. Instead, she compares herself to boys and realizes that she is missing something—a penis. Without experiencing the powerful force of castration anxiety, a girl cannot develop a superego in the same sense that boys do. Given this inability, Freud concluded, women were morally inferior to men, and this inferiority explained their place as second-class citizens in Victorian society. Freud believed that girls experience "castration completed," resulting in *penis envy*—the intense desire to obtain a penis by eventually marrying and bearing a son.

> In Freud's view, **anatomy is destiny.** By this he meant that anatomy (whether a person has a penis or not) determines whether the person will develop a superego.

While noting that his views ran counter to the early feminist thinkers of his time, Freud stood firm that the sexes are not equal in every way. He considered women to be somewhat childlike in their development and thought it was good that fathers, and eventually husbands, should guide them through life. He asserted that the only hope for women's moral development was education.

- *Latency period (6 years to puberty):* This phase is not a developmental stage but rather a kind of psychic time-out. After the drama of the phallic stage, the child sets aside all interest in sexuality. Although we now consider these years extremely important to development, Freud felt that this was a time in which no psychosexual development occurred.

- *Genital stage (adolescence and adulthood):* The genital stage is the time of sexual reawakening, a point when the source of sexual pleasure shifts to someone outside the family.

Stage	Adult Extensions (Fixations)	Sublimations	Reaction Formations
Oral	Smoking, eating, kissing, oral hygiene, drinking, chewing gum	Seeking knowledge, humor, wit, sarcasm, being a food or wine expert	Speech purist, food faddist, prohibitionist, dislike of milk
Anal	Notable interest in one's bowel movements, love of bathroom humor, extreme messiness	Interest in painting or sculpture, being overly giving, great interest in statistics	Extreme disgust with feces, fear of dirt, prudishness, irritability
Phallic	Heavy reliance on masturbation, flirtatiousness, expressions of virility	Interest in poetry, love of love, interest in acting, striving for success	Puritanical attitude toward sex, excessive modesty

Freud believed that in adulthood the individual becomes capable of the two hallmarks of maturity: love and work. However, Freud felt that people are inevitably subject to intense conflict, reasoning that everyone, no matter how healthy or well adjusted, still has an id pressing for expression. Adulthood, even in the best of circumstances, still involves reliving the unconscious conflicts of childhood.

Freud argued that a person may become stuck in any of these developmental stages if they are underindulged or overindulged at a given stage. For example, a parent might stop breastfeeding or bottle-feeding the child too early, be too strict in toilet training, punish the child for masturbating, or smother the child with too much attention. When underindulged or overindulged at a stage, a child may experience fixation. *Fixation* occurs when a particular psychosexual stage colors an individual's adult personality. For instance, an *anal retentive* person (someone who is obsessively neat and organized) is fixated at the anal stage. The idea of fixation explains how, according to Freud's view, childhood experiences can have an enormous impact on adult personality. Figure 2 illustrates possible links between adult personality characteristics and fixation at the oral, anal, and phallic stages.

Psychodynamic Critics and Revisionists

Many psychodynamic thinkers who came after Freud revised the theory in various ways. In particular, Freud's critics have said that his ideas about sexuality, early experience, social factors, and the unconscious mind were misguided. They stress the following points:

■ Sexuality is not the pervasive force behind personality that Freud believed it to be (Knight, 2014). Furthermore, the Oedipus complex is not universal as Freud maintained. Freud's concepts were heavily influenced by the setting in which he lived and worked—turn-of-the-century Vienna, a society that was sexually repressed and male dominated.

■ The first five years of life are not as powerful in shaping adult personality as Freud thought. Later experiences deserve more attention.

■ The ego and conscious processes play a more dominant role in our personality than Freud believed; he claimed that we are forever captive to the unconscious clutches of the id. In addition, the ego has a separate line of development from the id, so achievement, thinking, and reasoning are not always tied to sexual impulses.

■ Sociocultural factors are much more important than Freud believed. In stressing the id's dominance, Freud placed more emphasis on the biological basis of personality. More contemporary psychodynamic scholars have especially emphasized the interpersonal setting of the family and the role of early social relationships in personality development (Barratt, 2017; de Almeida Silva & others, 2016; Tal & Tal, 2017).

A number of dissenters and revisionists to Freud's theory have been influential in the development of psychodynamic theories (Adler, 1927; Erikson, 1968; Fromm, 1947; Horney, 1945; Jung, 1917; Kohut, 1977; Rapaport, 1967; Sullivan, 1953). Erik Erikson, whose psychosocial stages we examined in relation to human development, is among these. Here we consider three other thinkers—Karen Horney, Carl Jung, and Alfred Adler—who made notable revisions to Freud's approach.

Karen Horney (1885–1952) *Horney developed the first feminist criticism of Freud's theory. Horney's view emphasizes women's positive qualities and self-evaluation.*
Bettmann/Getty Images

HORNEY'S SOCIOCULTURAL APPROACH

Karen Horney (1885–1952) rejected the classical psychoanalytic concept that anatomy is destiny and cautioned that some of Freud's most popular ideas were only hypotheses. She insisted that these hypotheses be supported with observable data before being accepted as fact. She also argued that sociocultural influences on personality development should be considered (Mitchell, 2014; Vena, 2015).

Consider Freud's concept of penis envy, which attributed some of the behavior of his female patients to their repressed desire to have a penis. Horney pointed out that women might envy the penis not because of neurotic tendencies but because of the status that society bestows on those who have one. Further, she suggested that both genders envy the attributes of the other, with men coveting women's reproductive capacities (Horney, 1967).

Horney also believed that the need for security, not for sex, is the prime motive in human existence. Horney reasoned that people whose needs for security are met should be able to develop their capacities to the fullest extent. Horney (1964) felt that the unhappiness of modern people comes not so much from conflicts over sexual drives as from hating ourselves because we cannot live up to impossible standards. She viewed psychological health as allowing people to express their talents and abilities freely and spontaneously.

JUNG'S ANALYTICAL THEORY

collective unconscious
Jung's term for the impersonal, deepest layer of the unconscious mind, shared by all human beings because of their common ancestral past.

archetypes
Jung's term for emotionally laden ideas and images in the collective unconscious that have rich and symbolic meaning for all people.

individual psychology
Adler's view that people are motivated by purposes and goals and that perfection, not pleasure, is thus the key motivator in human life.

Freud's contemporary Carl Jung (1875–1961) shared Freud's interest in the unconscious, but he believed that Freud underplayed the unconscious mind's role in personality. In fact, Jung believed that the roots of personality go back to the dawn of human existence. The **collective unconscious** is Jung's name for the transpersonal (meaning it is the same across all people), deepest layer of the unconscious mind, shared by all human beings because of their common ancestral past. In Jung's theory, the experiences of a common past have made a deep, permanent impression on the human mind (Christian, 2017; Greco & Deutsch, 2017).

Jung posited that the collective unconscious contains **archetypes,** emotionally laden ideas and images that have rich and symbolic meaning for all people. Jung said these archetypes emerge in art, literature, religion, and dreams (Bassil-Morozow, 2015; Cwik, 2017; Moreman, 2014). Exploring archetypes was an important step in self-development. Archetypes are essentially predispositions to respond to the environment in particular ways.

Jung used the terms *anima* and *animus* to identify two common archetypes. He believed each of us has a passive feminine side—the anima—and an assertive masculine side—the animus. Another archetype, the *persona,* represents the public mask that we all wear during social interactions. Jung believed that the persona is an essential archetype because it allows us always to keep some secret part of ourselves hidden from others.

Many students find Jung's approach to personality to be odd and unexpected. Yet, many have also encountered a contemporary measure of personality that was inspired by Jung's approach. To read about the Myers-Briggs Type Indicator, see the Psychology in Our World.

Carl Jung (1875–1961) *Swiss psychoanalytic theorist who developed the concepts of the collective unconscious and archetypes.*
ANL/REX/Shutterstock

ADLER'S INDIVIDUAL PSYCHOLOGY

Alfred Adler (1870–1937) was one of Freud's earliest followers, but his approach to personality was drastically different from Freud's. In Adler's **individual psychology,** people are motivated

What's Your Type?

Personality assessment is useful in business settings. Strangely enough, one of the most popular assessment tools for personnel decisions is the Myers-Briggs Type Indicator (MBTI) developed in the 1940s by the mother–daughter team of Katherine Briggs and Isabel Briggs Myers (Briggs & Myers, 1998), neither of whom was trained in psychology or assessment (Saunders, 1991).

Based on a book by Carl Jung (1977), the Myers-Briggs questionnaire provides people with feedback on their personality "type" based on four dimensions:

1. Extraversion (basing one's actions on outward conditions) versus introversion (being more introspective)

2. Sensing (relying on what can be sensed about reality) versus intuiting (relying on gut feelings and unconscious processes)

3. Thinking (relying on logic) versus feeling (relying on emotion)

4. Judgment (using thinking and feeling) versus perception (using sensing and intuiting)

These MBTI dimensions are used to create categories that are labeled with letters; for example, an extraverted person who relies on sensation, thinking, and judgment would be called an ESTJ.

The MBTI has become so popular that people in some organizations introduce themselves as an INTJ or an INSP in the same way that people might exchange their astrological signs. Unfortunately, as in the case of astrology, strong evidence for the actual value of the MBTI types for personnel selection and job performance is weak at best (Hunsley & others, 2015; Pittenger, 2005). In fact, the MBTI is neither reliable (people get different scores with repeated testing) nor valid (it does not predict what it should). For example, there is no evidence that particular MBTI types are better suited to particular occupations (Bjork & Druckman, 1991; Gardner & Martinko, 1996).

Given this lack of empirical support, why does MBTI remain popular? Some practitioners have found it to be useful in their own work (Fornacieri & Lund Dean, 2013; McCaulley, 2000), whether or not they are aware of the lack of scientific evidence for the types. The MBTI has been well marketed, and those who pay for the scale and its training may be motivated to find evidence in their own lives to support the notion that it works. Many students are stunned to learn that this measure originates in the theory of Carl Jung.

It can be fun to learn about our personalities and to be given these letter labels. When we read that an INTJ is someone who is "introspective and likely to sometimes argue a point just for the sake of argument," we might think, "They really figured me out." However, this description could be true of virtually anyone, as is the case for most astrological profiles.

Chantal H/Alamy Stock Photo

The tendency to see ourselves in such vague descriptions is called the Barnum effect, after P. T. Barnum, the famous showman. The wily Barnum—simply by dispensing vague, general descriptions that would likely be true of anyone—convinced people that he could read minds.

The popularity of the Myers-Briggs letter typology attests to the power of marketing, the persistence of confirmation bias (the tendency to use information that supports our ideas rather than refutes them), and the Barnum effect. Most troubling to personality psychologists is that although other psychological measures may not be as exciting and mysterious as those four letters, they are more likely to provide reliable and valid information about job performance. Moreover, valid personality feedback can be obtained for considerably less than the MBTI price tag of $49.95!

by purposes and goals—thus, perfection, not pleasure, is their key motivator. Adler argued that people have the ability to take their genetic inheritance and their environmental experiences and act upon them creatively to become the person they want to be.

Adler thought that everyone strives for superiority by seeking to adapt, improve, and master the environment. Striving for superiority is our response to the uncomfortable feelings of inferiority that we experience as infants and young children when we interact with bigger, more powerful people. *Compensation* is Adler's term for the individual's attempt to overcome imagined or real inferiorities or weaknesses by developing one's own abilities. Adler believed that compensation is normal, and he said that we often make up for a weakness in one ability by excelling in a different ability. For example, a person of small stature and limited physical abilities (like Adler himself) might compensate by excelling in academics. In addition to striving for superiority, Adler believed that each person possesses a natural impulse toward warm relationships with other people. Adler called this the *social interest* (Watts & Bluvshtein, 2020).

Adler believed that birth order could influence how successfully a person would strive for superiority. He viewed firstborn children to be particularly vulnerable given that they begin life as the center of attention but then are knocked off their pedestal by their siblings. Adler in fact believed that the firstborn are more likely to suffer from psychological disorders and to engage in criminal behavior. Youngest children, however, also are potentially in trouble because they are most likely to be spoiled. The healthiest birth order? According to Adler, those (including Adler himself) who are middle-born are in an especially advantageous situation because they have older siblings as built-in inspiration for superiority striving. Importantly, though, Adler did not believe that anyone was doomed by birth order. Rather, sensitive parents could help children in any position in the family to negotiate their needs for superiority.

Many students (but especially middle children) find Adler's approach to the effects of birth order on personality to be fascinating. Adler only theorized about birth order; he did not conduct empirical research to investigate whether birth order affects personality—either as he proposed or in any other way. Importantly, contemporary research using very large samples shows that birth order does not relate in a systematic way to personality (Damian & Roberts, 2015a, 2015b; Rohrer & others, 2015).

Yes, you read that right: BIRTH ORDER IS NOT A STRONG CONTRIBUTOR TO PERSONALITY!

Evaluating the Psychodynamic Perspectives

Although psychodynamic theories have diverged from Freud's original psychoanalytic version, they share some core principles:

- Personality is determined both by current experiences and, as the original psychoanalytic theory proposed, by early life experiences.

- Personality can be better understood by examining it developmentally—as a series of stages that unfold with the individual's physical, cognitive, and socioemotional development.

- We mentally transform our experiences, giving them meaning that shapes our personality.

- The mind is not all consciousness; unconscious motives lie behind some of our puzzling behavior.

- The individual's inner world often conflicts with the outer demands of reality, creating anxiety that is not easy to resolve.

- Personality and adjustment—not just the experimental laboratory topics of sensation, perception, and learning—are rightful and important topics of psychological inquiry.

Thomas Barwick/Getty Images

Psychodynamic perspectives have come under fire for a variety of reasons. Some critics say that psychodynamic theorists overemphasize the influence of early family experiences on personality and do not acknowledge that people retain the capacity for change and adaptation throughout life. Many psychologists believe that Freud and Jung put too much faith in the unconscious mind's ability to control behavior. Others object that Freud placed too much importance on sexuality in explaining personality.

Some have argued, too, that psychoanalysis is not a theory that researchers can test through empirical studies. However, numerous empirical studies on concepts such as defense mechanisms and the unconscious have proved this criticism to be unfounded (Cohen & others, 2017; Di Giuseppe & others, 2020; Porcerelli & others, 2017). At the same time, another version of this argument may be accurate. Although it is certainly possible to test hypotheses derived from psychoanalytic theory through research, the question remains whether psychoanalytically oriented individuals who believe strongly in Freud's ideas would be open to research results that call for serious changes in the theory.

In light of these criticisms, it may be hard to appreciate why Freud continues to have an impact on psychology. Keep in mind that Freud made a number of important contributions, including being the first to propose that childhood is crucial to later functioning, that development might be understood in terms of stages, and that unconscious processes might play a significant role in human life.

self-quiz

1. According to Freud, our conscience is a reflection of the
   A. ego.
   B. collective unconscious.
   C. id.
   D. superego.

2. All of the following are examples of defense mechanisms *except*
   A. sublimation.
   B. repression.
   C. latency.
   D. displacement.

3. A theorist who focused on archetypes is
   A. Karen Horney.
   B. Sigmund Freud.
   C. Alfred Adler.
   D. Carl Jung.

APPLY IT! 4. Simone and her older sister have an intense sibling rivalry. Simone has tried to best her sister in schoolwork, fashion sense, and sporting achievements. Simone's sister complains that Simone needs to get a life of her own. What would Alfred Adler say about Simone's behavior?
   A. Simone is engaging in the defense mechanism of displacement, striving to conquer her sister when it is really her mother she wishes to defeat.
   B. Simone is expressing her animus archetype by engaging in masculine-style competition.
   C. Simone is expressing superiority striving by trying to overcome her sister—a healthy way for middle children to pursue superiority.
   D. Simone lacks a sense of basic trust in the world, and her parents must have been neglectful of her.

2 Humanistic Perspectives

humanistic perspectives
Theoretical views stressing a person's capacity for personal growth and positive human qualities.

Humanistic perspectives stress conscious awareness and a person's capacity for personal growth and positive human qualities. Humanistic psychologists believe that we all have the ability to control our lives and to achieve what we desire (Murphy & others, 2020).

Humanistic perspectives contrast with both psychodynamic perspectives and behaviorism. Humanistic theorists sought to move beyond Freudian psychoanalysis and behaviorism to a theory that might capture the rich and potentially positive aspects of human nature.

Maslow's Approach

A leading architect of the humanistic movement was Abraham Maslow (1908–1970), who is also known for developing a hierarchy of needs to explain motivation. Maslow believed that we can learn the most about human personality by focusing on the very best examples of human beings—self-actualizers.

Recall that at the top of Maslow's (1954, 1971) hierarchy was the need for self-actualization. Self-actualization is the motivation to develop one's full potential as a human being. Maslow described self-actualizers as spontaneous, creative, and possessive of a childlike capacity for awe. According to Maslow, a person at this optimal level of existence would be tolerant of others, have a gentle sense of humor, and be likely to pursue the greater good. Self-actualizers also maintain a capacity for "peak experiences," or breathtaking moments of spiritual insight. As examples of self-actualized individuals, Maslow included Pablo Casals (cellist), Albert Einstein (physicist), Ralph Waldo Emerson (writer), William James (psychologist), Thomas Jefferson (politician), Eleanor Roosevelt (humanitarian, diplomat), and Albert Schweitzer (humanitarian).

Created nearly 50 years ago, Maslow's list of self-actualizers is clearly biased. Maslow focused on highly successful individuals he thought represented the best of the human species. Because Maslow concentrated on people who were successful in a particular historical context, his self-actualizers were limited to those who had opportunities for success in that context. Maslow thus named considerably more men than women, and most of the individuals were from Western cultures and of European ancestry. They were not people of color. They were not members of sexual minority groups or those with diverse gender identities. They were not people with disabilities. Today, we might add to Maslow's list individuals such as Nobel Peace Prize winners the Dalai Lama (Tenzin Gyatso), Tibetan spiritual and political leader; and Malala Yousafzai, the Pakistani girl who survived an assassination attempt in 2012 and at age 17 became the youngest person ever to receive the coveted prize for her activism for access to education.

Rogers's Approach

Yes, another Carl. But Rogers and Jung are quite different in their approaches.

The other key figure in the development of humanistic psychology, Carl Rogers (1902–1987), began his career as a psychotherapist struggling to understand the unhappiness of the people he encountered in therapy. Rogers's groundbreaking work established the foundations for more contemporary studies of personal growth and self-determination.

In the knotted, anxious verbal stream of his clients, Rogers (1961) noted the things that seemed to be keeping them from reaching their full potential. Based on his clinical observations, Rogers devised his own approach to personality. Rogers believed that we are all born with the raw ingredients of a fulfilling life. We simply need the right conditions to thrive. Just as a sunflower seed, once planted in rich soil and given water and sunshine, will grow into a strong and healthy flower, all humans will flourish in the appropriate environment.

This analogy is particularly apt and reveals the differences between Rogers's view of human nature and Freud's. A sunflower seed does not have to be shaped away from its dark natural tendencies by social constraints, nor does it have to reach a difficult compromise between its vile true impulses and reality. Instead, given the appropriate environment, it will grow into a beautiful flower. Rogers believed that, similarly, each person is born with natural capacities for growth and fulfillment. We are also endowed with an innate sense—a gut feeling—that allows us to evaluate whether an experience is good or bad for us. Finally, we are all born with a need for positive regard from others. We need to be loved, liked, or accepted by people around us. As

Carl Rogers (1902–1987) *Rogers was a pioneer in the development of the humanistic perspective.*
Michael Rougier/Time & Life Pictures/Getty Images

children interacting with our parents, we learn early on to value the feeling that they value us, and we gain a sense of valuing ourselves.

EXPLAINING UNHAPPINESS
If we have innate tendencies toward growth and fulfillment, why are so many people so unhappy? The problem arises when our need for positive regard from others is not met *unconditionally*. **Unconditional positive regard** is Rogers's term for being accepted, valued, and treated positively regardless of one's behavior. Unfortunately, others often value us only when we behave in particular ways that meet what Rogers called conditions of worth. **Conditions of worth** are the standards we must live up to in order to receive positive regard from others. For instance, parents might give their child positive regard only when the child achieves in school, succeeds on the soccer field, or chooses a profession that they themselves value.

According to Rogers, as we grow up, people who are central to our lives condition us to move away from our genuine feelings, to earn their love by pursuing those goals that they value, even if those goals do not reflect our deepest wishes.

Rogers's theory includes the idea that we develop a *self-concept,* our conscious representation of who we are and who we wish to become, during childhood. Optimally, this self-concept reflects our genuine, innate desires, but it also can be influenced by conditions of worth. Conditions of worth can become part of who we think we ought to be. As a result, we can become alienated from our genuine feelings and strive to actualize a self that is not who we were meant to be. People who dedicate themselves to such goals might be very successful by outward appearances but might feel utterly unfulfilled. Such individuals might be able to check off all the important boxes in life's to-do lists, and to do all that they are "supposed to do," but never feel truly happy.

unconditional positive regard Rogers's construct referring to the individual's need to be accepted, valued, and treated positively regardless of their behavior.

conditions of worth The standards that the individual must live up to in order to receive positive regard from others.

PROMOTING OPTIMAL FUNCTIONING
To remedy this situation, Rogers believed that the person must reconnect with one's own true feelings and desires. He proposed that to achieve this reconnection, the person must experience a relationship that includes three essential qualities: unconditional positive regard (as defined earlier), empathy, and genuineness.

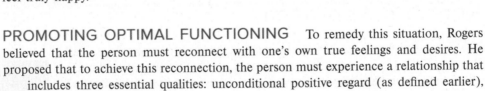
According to Rogers, anyone can help another person thrive, by offering that person unconditional positive regard and empathy, and by being yourself—who you really are—with that person.

First, Rogers said that regardless of what they do, people need unconditional positive regard. Although a person might lack unconditional positive regard in childhood, they can experience this unconditional acceptance from others later, in friendships and/or romantic relationships or during sessions with a therapist. Even when a person's behavior is inappropriate, obnoxious, or unacceptable, that person still needs the respect, comfort, and love of others.

Second, Rogers said that people can become more fulfilled by interacting with people who are empathic toward them. Empathy involves being a sensitive listener and understanding another's true feelings.

Finally, genuineness is a requirement in the individual's path to becoming fully functioning. Being genuine means being open with one's feelings and dropping all pretenses and facades. The importance that Rogers placed on the therapist's acting genuinely in the therapeutic relationship demonstrates his strong belief in the positive character of human nature. For Rogers, we can help others simply by being present for them as the authentic individuals we are.

In sum, Rogers's approach to therapy involved allowing the person to value their authentic self, to overcome conditions of worth, and to live spontaneously and authentically. This optimal outcome is provided by interacting with the person with unconditional positive regard (instead of conditions of worth), accurate empathy, and genuineness.

Thorsten Indra/Alamy Stock Photo

Evaluating the Humanistic Perspectives

The humanistic perspectives emphasize that the way we perceive ourselves and the world is an essential element of personality. Humanistic psychologists also stress that we need to consider the whole person and the positive bent of human nature. Their emphasis on conscious experience has given us the view that personality contains a well of potential that can be developed to its fullest.

Some critics believe that humanistic psychologists are too optimistic about human nature and that they overestimate people's freedom and rationality. Others say that the humanists may promote excessive self-love and narcissism by encouraging people to think so positively about themselves. Still others argue that humanistic approaches do not hold accountable people for their behaviors, if all negative human behavior is seen as emerging out of negative situations.

Self-determination theory, as described in Chapter 9, demonstrates the way that psychologists have tested humanistic ideas that might appear too abstract and difficult to test (Buzinde, 2020; Donald & others, 2020; Lynch & Sheldon, 2020; Vasconcellos & others, 2020). That research, on the ways that following our innate needs for competence, autonomy, and relatedness is crucial for well-being, bears witness to the enduring impact of humanistic perspectives on contemporary personality psychology. Rogers's approach to psychotherapy continues to have influence on practice as well (Behr & others, 2020; Levy, 2020).

Image Source

self-quiz

1. In Maslow's theory, the motivation to develop one's full potential as a human being is called
   A. self-satisfaction.
   B. self-actualization.
   C. self-sufficiency.
   D. self-determination.

2. Rogers proposed that to become fulfilled, the individual requires all of the following *except*
   A. unconditional positive regard.
   B. genuineness.
   C. self-actualization.
   D. empathy.

3. A child who consistently strives for an *A* in math and science to secure the affection of her parents is trying to establish

   A. unconditional positive regard.
   B. conditions of worth.
   C. self-actualization.
   D. empathy.

APPLY IT! 4. Phoebe and Joey, parents to little Jennifer, believe that because so many things in life involve hard work, it is important for Jennifer to earn the good things that happen to her. They make it clear to Jennifer that one of the things she must earn is their approval, and they tell her that they love her only when she does well in school and behaves according to their standards. They are certain that this training will instill in Jennifer the importance of working hard and valuing the good things she gets in

life. What would Carl Rogers say about Phoebe and Joey's parenting style?
   A. They are on the right track, as all children need strict limits and must learn discipline.
   B. What they are doing is fine but will have little influence on Jennifer, because genes matter most to personality.
   C. They are likely to be creating a fixation in Jennifer, and she will spend a lifetime working out her unconscious conflicts.
   D. They are setting Jennifer up to value herself only when she meets certain standards and would be better advised to love her unconditionally.

3 Trait Perspectives

If you are setting up a friend on a blind date, you are likely to describe the person in terms of traits, or stable personality characteristics. Trait perspectives on personality have been the dominant approach for nearly four decades.

Trait Theories

trait theories
Theoretical views stressing that personality consists of broad, enduring dispositions (traits) that tend to lead to characteristic responses.

According to **trait theories,** personality consists of broad, enduring dispositions (traits) that tend to lead to characteristic responses. In other words, we can describe people in terms of the ways they behave, such as whether they are outgoing, friendly, private, or hostile. People who have a strong tendency to behave in certain ways are referred to as "high" on the traits; those with a weak tendency to behave in these ways are "low" on the traits. Although trait theorists differ about which traits make up personality, they agree that traits are the fundamental building blocks of personality.

Gordon Allport (1897–1967), sometimes referred to as the founder of American personality psychology, was particularly bothered by the negative view of humanity that psychoanalysis portrayed. He rejected the notion that the unconscious was central to an understanding of personality. He further believed that to understand healthy people, we must focus on their lives in the present, not on their childhood experiences.

Allport believed that personality psychology should be concerned with understanding healthy, well-adjusted individuals. He described such persons as showing a positive but objective sense of self and others, interest in issues beyond their own experience, a sense of humor, common sense, and a unifying philosophy of life, typically but not always provided by religious faith (Allport, 1961). Allport dedicated himself to the idea that psychology should have relevance to social issues facing contemporary society, and his scholarship has influenced not only personality psychology but also the psychology of religion and prejudice.

In defining personality, Allport (1961) stressed each person's uniqueness and capacity to adapt to the environment. For Allport, the crucial unit for understanding personality is the trait. He defined traits as mental structures that make different situations the same for the person. For Allport, traits are structures inside a person that cause behavior to be similar even in different situations. For instance, if Eli is kind and helpful, they are likely to behave in a kind and helpful fashion whether a person in need is a stranger or a friend. Allport's definition implies that behavior should be consistent across different situations.

We get a sense of the down-to-earth quality of Allport's approach to personality by looking at his study of traits. In the late 1930s, Allport and his colleague H. S. Odbert (1936) sat down with two big unabridged dictionaries and pulled out all the words that could be used to describe a person—a method called the *lexical approach*. This approach reflects the idea that if a trait is important to people in real life, it ought to be represented in the natural language people use to talk about one another. Furthermore, the more important a trait is, the more likely it is that it should be represented by a single word. Allport and Odbert started with 18,000 words and gradually pared down that list to 4,500.

It's called "lexical" because a lexicon is a dictionary or vocabulary. These researchers are generally starting with the words we use to describe other people.

As you can appreciate, 4,500 traits make for a very long questionnaire. Imagine that you are asked to rate a person, Ignacio, on some traits. You use a scale from 1 to 5, with 1 meaning "not at all" and 5 meaning "very much." If you give Ignacio a 5 on "outgoing," what do you think you might give him on "shy"? Clearly, we may not need 4,500 traits to summarize the way we describe personality. Still, how might we whittle down these descriptors further without losing something important?

With advances in statistical methods and the advent of computers, the lexical approach became considerably less unwieldy. Researchers could analyze responses to the words to look for underlying structures that might account for their overlap. Specifically, a statistical procedure called *factor analysis* allowed researchers to identify which traits go together in terms of how they are rated. Factor analysis tells us what items on a scale people are responding to as if they mean the same thing. For example, if Ignacio got a 5 on "outgoing," he probably would get a 5 on "talkative" and a 1 or 2 on "shy." If Sarah is high on "hardworking," she is probably also going to be high on "reliable." Factor analysis helps a scientist identify the dimensions, or factors, that explain this kind of overlap among item ratings. One important characteristic of factor analysis is that it relies on the scientist to interpret the meaning of the factors, and the researcher must make some decisions about how many factors are enough to explain the data.

Fuse/Getty Images

In 1963, W. T. Norman reanalyzed the Allport and Odbert traits and concluded that only five factors were needed to summarize these traits. Norman's research set the stage for the dominant approach in personality psychology today: the five-factor model (Digman, 1990).

The Five-Factor Model of Personality

Pick a friend and jot down 10 of that person's most notable personality traits. Did you perhaps list "reserved" or "a good leader"? "Responsible" or "unreliable"? "Sweet," "kind," or "friendly"? Maybe even "creative"? Personality psychologists have uncovered five broad personality dimensions that are represented in the natural language; these dimensions also summarize the various ways psychologists have studied traits (Allik & others, 2017; Costa & McCrae, 2010, 2013; De Roover & others, 2017; Gorbaniuk & others, 2017).

big five factors of personality The five broad traits that are thought to describe the main dimensions of personality: openness to experience, conscientiousness, extraversion, agreeableness, and neuroticism (emotional instability).

The **big five factors of personality**—the broad traits that are thought to describe the main dimensions of personality—are neuroticism (which refers to the tendency to worry and experience negative emotions), extraversion, openness to experience, agreeableness, and conscientiousness. Although personality psychologists typically refer to the traits as N, E, O, A, and C on the basis of the order in which they emerge in a factor analysis, if you create an anagram from these first letters of the trait names, you get the word *OCEAN*. Figure 3 more fully defines the big five traits.

Take a good look at Figure 3 so that you understand each of the traits. Openness to experience is often the trickiest.

Each of the big five traits has been the topic of extensive research. Here is just a sampling of research findings for each trait that sheds light on the interesting work that the five-factor model has inspired:

- *Neuroticism* is related to feeling negative emotion more often than positive emotion in one's daily life and to experiencing more lingering negative states (Yang & others, 2020). Elevated neuroticism is also a risk factor for a range of psychological disorders (Mineka & others, 2020; Tonarely & others, 2020; Williams & others, 2020). Neuroticism predicts more health complaints

Neuroticism is sometimes identified by its opposite, ***emotional stability***.

Openness to Experience
Imaginative and interested in cognitively engaging with abstract ideas as well as perceptions, nature, and the arts. Someone high in openness enjoys thinking about issues from all sides and is not interested in conventional ways of doing things.

Conscientiousness
Reliable, hard working, and dependable. Someone high in conscientiousness is disciplined, goal directed, and organized.

Extraversion
Outgoing, sociable, and lively. A person high in extraversion is not shy and tends to be enthusiastic with others.

Agreeableness
Kind, nice, and trusting. A person high in agreeableness is likely to be gentle and helpful to others.

Neuroticism
A worrier, anxious, and insecure. Someone who is high on neuroticism is likely to be stressed out by negative events and is prone to experiencing distress. Neuroticism is sometimes labeled by its opposite end. Those low in neuroticism are high in emotional stability.

FIGURE 3 **The Big Five Factors of Personality** Each of the broad supertraits encompasses more narrow traits and characteristics. Use the acronym *OCEAN* to remember the big five personality factors (openness, conscientiousness, extraversion, agreeableness, and neuroticism).

(Kööts Ausmees & others, 2016), risk for cognitive decline with age (Terracciano & others, 2017), and higher incidence of nightmares (Randler & others, 2017). A recent study showed that neuroticism predicted higher levels of FOMO—the fear of missing out (Rozgonjuk & others, 2020). The relationship between neuroticism and mortality is complex, with some studies suggesting that neurotic people live shorter lives but others showing no relationship (Turiano & others, 2020). Researchers have been probing whether some aspects of neuroticism might be related to better rather than worse life outcomes. For example, that part of neuroticism that is about feeling worried and vulnerable is associated with higher income and longer life (Hill & others, 2020; Weiss & Deary, 2020). In addition, personality psychologists continue to debate whether it is the average level of negative affect or the variability in one's moods that is most definitive of neuroticism (Wenzel & Kubiak, 2020).

- People high in *extraversion* are more likely than others to engage in social activities (Emmons & Diener, 1986; Wilt & Revelle, 2017). Extraversion predicts spending more time texting (Schroeder & Sims, 2017). Extraverts also spend more time in bars, cafes, and friends' houses than they do at home (Matz & Harari, 2020). Extraverted CEOs are better able to get their managers to "buy in" to a team vision and, in turn, their firms are more productive (Araujo-Cabrera & others, 2017). Extraverted computer programmers are more creative (Amin & others, 2020). In a recent study of over 100,000 people in 55 countries, among the big five, extraversion uniquely predicted going out (against lockdown orders) during the COVID-19 pandemic (Götz & others, 2020).

- *Openness to experience* is related to liberal values, open-mindedness, tolerance (Brandt & Crawford, 2020; Sutin, 2017), creativity (Madrid & Patterson, 2016), and the capacity to experience awe (Silvia & others, 2015). Openness to experience is associated with being more likely to run for political office but also to be slightly less likely to win (Scott & Medeiros, 2020). Those high on openness to experience are willing to pay more for organic food (Gustavsen & Hegnes, 2020). Openness to experience is the one trait of the big five that predict the ability to produce actually funny jokes (Sutu & others, 2020). Openness to experience is also associated with superior cognitive functioning, achievement, and IQ across the life span (Briley & others, 2014). Those high in openness to experience are more likely to daydream (Eldredge, 2016) and to dress distinctively (Naumann & others, 2009). Openness to experience is viewed as having two components: intellect and openness. Intellect refers to engaging cognitively with abstract ideas and reasoning (DeYoung, 2015). Openness refers to engaging cognitively with perceptions, fantasies, and emotions (Blain & others, 2020). Each of these facets relates to different kinds of creativity: Intellect predicts greater scientific creativity; openness predicts greater artistic creativity (Kaufman & others, 2016) and better humor production (Sutu & others, 2020).

- *Agreeableness* is related to generosity and altruism (Thielmann & others, 2020), to reports of religious faith (Haber & others, 2011), and to more satisfying romantic relationships (Donnellan & others, 2005). Agreeable people have more positive attitudes toward organ donation and are more likely to register to donate (Hill, 2016). There are also links between agreeableness and viewing other people positively (Wood & others, 2010). In online dating profiles, agreeable individuals are less likely than people who score low on this trait to lie about themselves (Hall & others, 2010). Agreeableness is negatively related to traditional and cyberbullying (Pronk & others, 2021). Agreeableness is associated with lower experience of stress while coping with difficult times (Morgan & others, 2017).

- *Conscientiousness* is a key predictor of positive outcomes in a variety of life domains (Jackson & Roberts, 2017); Savelyev, 2020). A meta-analysis found that conscientiousness, but not the other big five factors, was linked to college students' grade point averages (McAbee & Oswald, 2013); conscientiousness has also been linked to better work performance (Brown & others, 2011). Conscientiousness is the trait that captures people's strong desire to achieve. One study showed that after doing poorly on important standardized tests (such as the GRE, MCAT, or LSAT), conscientiousness predicted taking the test again (Barron & others, 2017). Conscientiousness can compensate for low interest in task performance (Song & others, 2020). This means that even when they are not super motivated by a task, conscientious people will double down and see it through. Conscientiousness predicts higher lifetime earnings (Shaffer, 2020) and having a spouse who is high in conscientiousness predicts higher earnings (Averett & others, 2020). Conscientiousness predicts better-quality friendships (Jensen-Campbell & Malcolm, 2007), higher levels of religious faith (Saroglou, 2010), and a forgiving attitude (Balliet, 2010). Low levels of conscientiousness are linked to higher levels of criminal behavior (Wiebe, 2004), substance abuse (Walton & Roberts, 2004), and pathological gambling (Hwang & others, 2012). Conscientiousness is also negatively related to signs of aging in the brain (Yoon & others 2020). Conscientiousness shares a stronger relationship to longevity than do the other four factors (Hill & others, 2011). Why is conscientiousness such a strong predictor of health and longevity? Quite simply, conscientious people engage in healthy behaviors (Turiano & others, 2015).

In many ways, the role of personality traits in our life depends on the situations in which we find ourselves (Gebauer & others, 2020). Personality traits can be strengths or weaknesses, depending on the types of situations we encounter and the kinds of situations we seek out for ourselves (Block, 2010; King & Trent, 2012). Even a trait like agreeableness may be a liability when the situation calls for confrontational behavior. For instance, a person whose marriage is breaking up might wish for a divorce lawyer who treats them kindly but might prefer one who is less than agreeable at the bargaining table. Eminent psychologist Lee Cronbach (1957, p. 679) once said, "If for each environment there is a best organism, for every organism there must be a best environment." If our personalities are not particularly well suited to a situation, we can change that situation or create one that fits better (King & Trent, 2012).

Keep in mind that because the five factors are theoretically independent of one another, a person can be any combination of them. Do you know a neurotic extravert or an agreeable introvert?

Seeking out opportunities to turn our traits into strengths is one of the challenges of life.

CROSS-CULTURAL STUDIES ON THE BIG FIVE
Do the big five show up in the assessment of personality in cultures and languages around the world? Some research suggests that they do: A version of the five factors appears in diverse languages, including Chinese, English, Japanese, Finnish, and Polish (Lovik & others, 2017; Paunonen & others, 1992; Zhou & others, 2009). Still, in some contexts the five factors do not appear to be represented in the natural language or these traits do not fully explain the ways people talk about others (Thalmayer & others, 2020). Among the big five, the factors most likely to emerge across cultures and languages are extraversion, agreeableness, and conscientiousness, with neuroticism and openness to experience being more likely to emerge only in English-speaking samples (De Raad & others, 2010).

THE STUDY OF TRAITS IN ANIMALS
A growing area of research in trait psychology concerns identifying and studying traits in animals (Weiss, 2017). For example, researchers have found evidence for at least some of the big five personality traits in animals, including domestic dogs (Gosling, 2008; Gosling & others, 2003) and hyenas (Gosling & John, 1999). In addition, studies have turned up evidence for general personality traits (such as overall boldness) in a range of animals, including guinea pigs (Brust & Guenther, 2017), sea lions (Ciardelli & others, 2017), possums (Wat & others, 2020), mink (Lehmkuhl Noer & others, 2016), cows (Neave & others, 2020), and reptiles (Siviter & others, 2017). Research has shown consistent individual differences in the behavior of *hoverflies*

Ammit Jack/Shutterstock

(called *flower flies* in the United States) (Odermatt & others, 2017) and even ants (Kolay & others, 2020).

Research on animal personality is important because it allows researchers to study factors like early socialization and genetic heritage, in participants for whom such issues are much easier to control and measure (Fritz & others, 2020; Siviter & others, 2017; White & others, 2020), compared to humans. In addition, comparing the personalities of different groups of nonhuman primates (for instance monkeys vs. Great Apes, our closest cousins) provides a window into the evolution of human personality characteristics (Weiss, 2017). Finally, such studies provide strong evidence for the existence of stable personality characteristics, which was called question by social cognitive approaches, an issue we will address later. If stable behavioral patterns can be observed in nonhuman animals, surely personality might exist in human beings.

The big five are not the only set of traits in personality psychology. An alternative trait approach is called the HEXACO model (Ashton & Lee, 2020). This perspective is similar to the five factors but it breaks up agreeableness into two different traits. One of these traits is agreeableness (including being gentle and kind) and the other is honesty/humility. This last factor is thought to do a better job than any of the five factors in explaining moral (and immoral) behavior (Howard & Van Zandt, 2020; Klein & others, 2020).

As we have reviewed, personality traits are related to a number of consequential life outcomes. A key challenge facing our increasingly diverse society is the existence of prejudice. Prejudice refers to negative attitudes toward someone based on the person's membership in a particular group—whether the group is based on gender, gender identity, age, ethnicity, religion, sexual orientation, disability status, immigrant status, and so on. One clue that prejudice might involve personality is that people who are prejudiced against one group are often prejudiced against many groups (Bierly, 1985; Jones & Splan, 2020). To read about how traits relate to prejudice, see the Intersection.

NEUROTICISM, EXTRAVERSION, AND WELL-BEING

You have probably noticed that some people seem to go through life having fun, while others appear to feel distress at even the slightest problem. You might think that most happiness can be explained by the events that happen to us—of course, you reason, a person is going to be happy if they are doing well in school and has a loving romantic partner, but unhappy if they are doing poorly and have just experienced a painful breakup. In fact, perhaps surprisingly, research has shown that life events explain relatively little about a person's overall well-being.

On average, some people are happier than others. Among the most consistent findings in personality research is the strong relationship between personality traits and well-being (Cosentino & Solano, 2017; Womick & King, 2018). Specifically, extraversion is related to higher levels of well-being, and neuroticism is strongly related to lower levels of well-being (Anglim & others, 2020; Oravecz & others, 2020). These links between extraversion and higher levels of well-being, and between neuroticism and lower levels of well-being, are consistent and have even been found in orangutans (Weiss & others, 2006). What explains these connections?

Traits, Mood, and Subjective Well-Being

To begin, let's define subjective well-being as psychologists do. **Subjective well-being** is a person's assessment of their own level of positive affect relative to negative affect, and the person's evaluation of their life in general (Diener, 2000; Diener & Lucas, 2014; Potter & others, 2020). When psychologists measure subjective well-being, they often focus on a person's positive and negative moods and life satisfaction.

This definition of subjective well-being provides a clue as to why the traits of neuroticism and extraversion might be so strongly related to one's level of well-being. Neuroticism is the tendency to experience negative emotion. Those high in neuroticism experience more negative mood than others, and their moods are more changeable. In contrast, extraversion is more strongly tied to positive mood. In fact, positive mood is recognized as the core of extraversion (Blain, Sassenberg, & others, 2020; Hermes & others, 2011). To the extent that neurotic individuals are more prone

subjective well-being
A person's assessment of their own level of positive affect relative to negative affect, and an evaluation of their life in general.

W Remember you completed the Satisfaction with Life Scale way back in Chapter 1.

Personality and Social Psychology: Do Personality Traits Predict Prejudice?

You likely know prejudice as so many "-isms" (like racism, anti-Semitism, sexism, ageism, ableism) or "phobias" (like homophobia, islamophobia, xenophobia). Prejudice is bias against people who are different from us. Do personality traits have anything to do with holding prejudicial attitudes?

Some have argued that prejudice is more about political ideology than personality (Duckitt, 2001). Specifically, political conservatism and right-wing views are positively related to prejudice (Banton & others, 2020; Marsden & Barnett, 2020; Stern & Crawford, 2020). Generally speaking, personality traits are not political. We can certainly think of liberal and conservative extraverts, for example. So the question is, do traits matter to prejudice even taking political views into account?

Three traits have been most implicated in prejudice: openness to experience, agreeableness, and conscientiousness. Openness and agreeableness are negatively related to prejudice (Esses, 2020; Federico & Aguilera, 2019; Kocaturk & Bozdag, 2020; Metin-Orta & Metin-Camgöz, 2020; Talay & De Coninck, 2020; Tobin & Graziano, 2020). In contrast, conscientiousness is positively related to prejudice (Carlson & others, 2018; Federico & Aguilera, 2019; Ziller & Berning, 2019). Each of these traits is also related to political views. Openness is strongly related to being liberal (DeNeve, 2015; Osborne & Sibley, 2020). Conscientiousness is related to being conservative (De Neve, 2015). One aspect of agreeableness, compassion toward others, is related to liberalism while another aspect, being polite, is related to conservatism (Osborne & others, 2013). Do these traits predict prejudice if we rule out the influence of political ideology?

Researchers addressed this question using two samples (including over 5,000 Americans) (Lin & Alvarez, 2020). Participants completed measures of personality traits, political ideology, and anti-Black prejudice. In both samples, without considering political views, openness and agreeableness related negatively to prejudice and conscientiousness related positively to prejudice. However, the negative relationship between openness and racism was explained by liberalism,

SDI Productions/Getty Images

meaning that liberal political views, not the trait of openness, predicts being low in prejudice.

Both agreeableness and conscientiousness remained related to racism even when taking political ideology into account (Lin & Alvarez, 2020). This means that nice people, regardless of their political views, tend to be lower in anti-Black prejudice. In addition, conscientious people, regardless of their political views, were more likely to endorse anti-Black prejudice.

These results challenge us to think about the meaning of these traits. People who are kind are likely to extend their kindness to others, regardless of group membership. Indeed research shows that agreeable people are more likely to enjoy interacting with diverse others (Diehl, 2020). Results for conscientiousness, a trait associated with many positive life outcomes, are more puzzling. It may be that highly conscientious people, who prefer an orderly and highly structured world, are challenged by diversity.

\\ *Why do you think conscientiousness predicts prejudice?*

to negative emotion, it would seem that this trait might take a toll on overall well-being. Interestingly, however, research has shown that those high on neuroticism can be happy—especially if they are also extraverted (Hotard & others, 1989). Extraversion is strongly related to well-being, even for those high on neuroticism. Why might this be the case?

An early theory about the relationship between extraversion and high levels of well-being was that extraverts engage in behaviors that are related to higher well-being and positive mood, such as socializing with others. Thus, the thinking went, maybe extraverts are happier because they choose to spend more time with other people. Despite the logic of this explanation, research has shown that extraverts are happier than introverts even when they are alone (Lucas & Baird, 2004).

Ingram Publishing/SuperStock

In fact, research supports the conclusion that extraverts are simply happier regardless of what they are doing or with whom they are doing it. Understanding this association between extraversion and happiness continues to motivate research in personality psychology (Smillie, 2013; Wacker, 2017; Weiss & others, 2016). A recent set of studies suggested that the well-being benefits of extraversion appear to be linked most to the high energy level of extraverts, rather than their tendency to experience more positive emotion (Margolis & others, 2020).

Traits and States If you are neurotic or an introvert—or even a neurotic introvert—you may be feeling your mood deflating like a helium-filled balloon in a heat wave. If personality is stable, what good is it to find out that your personality might make you unhappy?

One way to think about these issues is to focus on the difference between *traits* and *states* (Asendorpf & Rauthmann, 2020). Traits are enduring characteristics—they represent the way you generally are. In contrast, states (such as positive and negative moods) are briefer experiences. Having a trait, such as neuroticism, that predisposes you to feelings of worry (a state) does not mean that your overall well-being must suffer. Instead, recognizing that you tend to be neurotic may be an important step in noting when your negative moods are potentially being fed by traits and are not necessarily the result of objective events. Finding out that you have a personality style associated with lowered levels of happiness should not lead you to conclude that you are doomed. Rather, this information can allow you to take steps to improve your life, to foster good habits, and to make the most of your unique qualities. Indeed, increasingly, personality research has begun to examine the ways that behaving out of character—that is, doing things that may not match with one's traits—may lead to boosts in mood and well-being (Sun & others, 2017).

Evaluating the Trait Perspective

Studying people in terms of their personality traits has practical value. Identifying a person's traits allows us to know that individual better. Using traits, psychologists have learned a great deal about the connections between personality and health, ways of thinking, career success, and relations with others.

Strong reliance on traits risks missing the importance of situational factors in personality and behavior. For example, a person might say that she is introverted when meeting new people but very outgoing when with family and friends. Further, some have criticized the trait perspective for painting an individual's personality with very broad strokes. Personality researchers have focused on more specific characteristics (Anglim & others, 2020) and even specific items (Mõttus & others, 2019), to predict behavior and understand personality.

For many modern personality scientists, the five factors are a much better 'sound bite' than, say, the id, ego, and superego.

self-quiz

1. All of the following are among the big five factors of personality *except*
   A. openness to experience.
   B. altruism.
   C. conscientiousness.
   D. extraversion.

2. A researcher is doing a study from the lexical approach to identify the dimensions of personality in the natural language of people whose tribe was only recently discovered. Which of the following traits are most likely to be uncovered?
   A. Extraversion
   B. Agreeableness
   C. Conscientiousness
   D. All of the above

3. The personality factor that is most linked with a higher IQ is
   A. neuroticism.
   B. conscientiousness.
   C. agreeableness.
   D. openness to experience.

APPLY IT! 4. Sigmund, a high-achieving psychoanalyst, always sees his patients in a timely manner and completes his written work ahead of the deadline. A brilliant public speaker, he is often surrounded by enthusiastic admirers, and he enjoys being the center of attention. He has some pretty wild, abstract ideas and has developed a complex theory to explain all of human behavior. He is an

unconventional thinker, to say the least. He does not respond well to criticism and reacts poorly to even slight disapproval. Which of the following best describes Sigmund's personality?
   A. Low conscientiousness, high extraversion, low neuroticism, high openness to experience
   B. High conscientiousness, high extraversion, high openness to experience, high neuroticism
   C. High conscientiousness, low extraversion, low openness to experience, low neuroticism
   D. Low conscientiousness, high extraversion, low openness to experience, high neuroticism

4 Social Cognitive Perspectives

Social cognitive perspectives on personality emphasize conscious awareness, beliefs, expectations, and goals. While incorporating principles from behaviorism, social cognitive psychologists explore the person's ability to reason; to think about the past, present, and future; and to reflect on the self. They emphasize the person's individual interpretation of situations and thus focus on the uniqueness of each person by examining how behavior is tailored to the diversity of situations in which people find themselves.

social cognitive perspectives
Theoretical views emphasizing conscious awareness, beliefs, expectations, and goals.

Social cognitive theorists are not interested in broad traits. Rather, they investigate how more specific factors, such as beliefs and expectations, relate to behavior and performance. In this section we consider the two social cognitive approaches, developed, respectively, by Albert Bandura and Walter Mischel.

Bandura's Social Cognitive Theory

B. F. Skinner, whose work we examined in Chapter 5, believed that there is no such thing as "personality;" rather, he emphasized behavior and felt that internal mental states were irrelevant to psychology. Albert Bandura (1986, 2001, 2007) found Skinner's approach to be too simplistic for understanding human functioning. Bandura took the basic tenets of behaviorism and added recognition of the role of mental processes in determining behavior. While Skinner saw behavior as caused by the situation, Bandura pointed out that the person can cause situations, and sometimes the very definition of the situation itself depends on the person's beliefs about it. For example, is that upcoming exam an opportunity to show your stuff or a threat to your ability to achieve your goals? The test is the same either way, but your unique take on it can influence a host of behaviors (studying hard, worrying, and so on).

Bandura's social cognitive theory states that behavior, environment, and person/cognitive factors are all important in understanding personality. Bandura coined the term *reciprocal determinism* to describe the way behavior, environment, and person/cognitive factors interact to create personality (Figure 4). The environment can determine a person's behavior, and the person can act to change the environment. Similarly, person/cognitive factors can both influence behavior and be influenced by behavior. From Bandura's perspective, then, behavior is a product of a variety of forces, some of which come from the situation and some of

Albert Bandura (b. 1925) *Bandura's practical, problem-solving social cognitive approach has made a lasting mark on personality theory and psychotherapy.*
Cheriss May/NurPhoto/Getty Images

PSYCHOLOGICAL INQUIRY

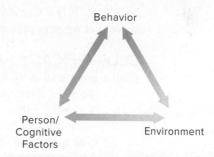

FIGURE 4 **Bandura's Social Cognitive Theory** Bandura's social cognitive theory emphasizes the reciprocal influences of behavior, environment, and person/cognitive factors. Notice that from Bandura's perspective, the three arrows are double-headed, meaning that causation goes in both directions. Consider the following events and experiences. Describe how factors from each of the three points on the triangle spur changes in the other two and how those influences might come back again. >**A Behavior: You study hard and get an** *A* **on your Intro Psych exam.** >**An Environment: Your Intro Psych instructor provides lecture notes online and does a thorough review for each exam.** >**A Person/ Cognitive Factor: You feel very confident about your ability to do well in Intro Psych.**

Vitalii Nesterchuk/
Shutterstock

which the person brings to the situation. In turn, how we behave can influence both the situation and ourselves. We now review the important processes and variables Bandura used to understand personality.

OBSERVATIONAL LEARNING Recall Bandura's belief that observational learning is a key aspect of how we learn. Through observational learning, we form ideas about the behavior of others and then possibly adopt this behavior ourselves. For example, a young child might observe their parent's aggressive outbursts and hostile exchanges with other people; when the child is with their peers, they might interact in a highly aggressive way, showing the same characteristics as the parent. Social cognitive theorists believe that we acquire a wide range of behaviors, thoughts, and feelings through observing others' behavior and that these observations strongly shape our personalities (Bandura, 2010).

In addition to emphasizing the important role of observation and learning, Bandura's social cognitive approach emphasizes characteristics that can influence behavior. These are not broad traits but rather "smaller," more specific beliefs that a person might have about a specific task or situation. Two of the most important beliefs are personal control and self-efficacy.

PERSONAL CONTROL Personal control refers to the degree to which a person believes to be in charge of the outcomes of behaviors. Psychologists commonly describe a sense of behavioral control as coming from inside the person (an *internal locus of control*) or outside the person (an *external locus of control*). When we feel that we ourselves are controlling our choices and behaviors, the locus of control is internal, but when other influences are controlling them, the locus of control is external.

Consider the question of whether you will perform well on your next test. With an internal locus of control, you believe that you are in command of your choices and behaviors, and your answer will depend on what you can realistically do (for example, study hard or attend a special review session). From the perspective of an external locus of control, however, you might say that you cannot predict how things will go because so many factors influence performance, such as whether the test is difficult and if the exam room is too hot or too cold. If a person thinks outcomes in their life depend primarily on luck, that is an external locus of control.

Feeling a strong sense of personal control is vital to many aspects of performance, well-being, and physical health (Lachman & others, 2015; Sheppes & others, 2015; Turiano & others, 2014). Certainly, we do not control everything in our lives. But maintaining sense that we have personal control can be important to positive functioning and success.

SELF-EFFICACY **Self-efficacy** is the belief that one has the competence to accomplish a given goal or task. Self-efficacy is related to a number of positive developments in people's lives, including solving problems, becoming more sociable, adopting healthy habits, undertaking difficult coursework, and successfully searching for a job (Martinez-Calderon & others, 2020; McBride & others, 2020; Petruzziello & others, 2020; Selzler & others, 2020; Thornberry & others, 2020). Self-efficacy provides a starting point to even try to make life changes and can help people persist in the face of obstacles. We will return to the topics of personal control and self-efficacy in the context of making healthy changes in behavior in Chapter 14.

Self-efficacy means having a can-do attitude.

self-efficacy
The belief that one can master a situation and produce positive change.

Mischel's Contributions

Like Albert Bandura, Walter Mischel (1930-2018) was a social cognitive psychologist who explored how personality influences behavior. Mischel left his mark on the field of personality in two ways. First, his critique of the idea of consistency in behavior ignited a flurry of controversy that nearly ended the study of personality. Second, he proposed his own model of personality.

MISCHEL'S CRITIQUE OF CONSISTENCY

Whether we are talking about unconscious sexual conflicts, conscious motives and goals, or traits all of the approaches we have considered so far maintain that these various personality characteristics are an enduring influence on behavior. This shared assumption was attacked in 1968 with the publication of Walter Mischel's *Personality and Assessment,* a book that nearly ended the science of personality psychology.

To understand Mischel's argument, recall Gordon Allport's definition of a trait as a characteristic that ought to make different situations the same for a given person. This quality of traits suggests that a person should behave consistently in different situations—in other words, the person should exhibit *cross-situational consistency.* For example, an outgoing person should act highly sociably whether they are at a party or in a library. Mischel looked at the research compiled on trait prediction of behavior and found it to be lacking. He concluded that there was no evidence for cross-situational consistency in behavior—and thus no evidence for the existence of personality as it had been assumed to exist.

Rather than understanding personality as consisting of broad, internal traits that make for consistent behavior across situations, Mischel said that behavior often changes according to a given situation. He described behavior as discriminative—that is, a person looks at each situation and responds accordingly. Mischel's view is called *situationism,* the idea that behavior changes considerably from one context to another.

It may be hard to grasp how devastating Mischel's critique of personality was. Keep in mind that it occurred in a particular time. In the late 1960s, the United States was dominated by social movements that emphasized the role of situational factors, such as societal limitations, on many groups, including Black Americans and women. The Civil Rights Movement and the Women's Movement both stressed that by changing the circumstances in which people lived, including unjust laws and unfair treatment, we could change people's lives for the better. Mischel's idea that situations matter more to behavior than personal characteristics fit very well with the spirit of the times. Personality psychology, in contrast, focused on the person as the source of the behavior and seemed pretty "old timey" in comparison. The notion that personality itself might not exist, that there was no stable pattern of thoughts, feelings, and behavior underlying a person's actions, was met with a great deal of zeal by many psychologists.

Personality psychologists responded to Mischel's situationist attack in a variety of ways (Donnellan & others, 2009; Funder, 2009; Hogan, 2009). Researchers showed that it is not a matter of *whether* personality predicts behavior, but *when and how* it does so, often in combination with situational factors. The research findings were are follows:

- The narrower and more limited a trait is, the more likely it will predict behavior.

- Some people are consistent on some traits, and other people are consistent on other traits.

- Personality traits exert a stronger influence on an individual's behavior when situational influences are less powerful. A very strong situation is one that contains many clear cues about how a person is supposed to behave. For example, even a highly talkative person typically sits quietly during a class lecture. In weaker situations, however, such as during leisure time, the person may spend most of the time talking.

Keep in mind that people are not randomly assigned to situations in real life as they are in experiments. We select the situations we are in, just as Bandura pointed out. This means that even if situations determine behavior, traits play a role in determining which situations people choose—such as going to a party or staying home to study (Emmons & Diener, 1986).

Let's pause and consider what it means to be consistent. You might believe that being consistent is part of being a genuine, honest person and that tailoring behavior to different situations means being fake. On

Mischel's critique nearly ended the field of personality psychology! Remember—he said there was no such thing as cross-situational consistency in behavior.

For many psychologists, getting rid of theories of personality also meant finally ridding the field of Freud!

David Jakle/Getty Images

the other hand, consider that someone who never changes their behavior to fit a situation might be unpleasant—a "drag" to have around. Think, for example, about someone who cannot put aside their competitive drive even when playing checkers with a four-year-old. Clearly, adaptive behavior might involve sometimes being consistent and sometimes tailoring behavior to the situation.

Over time, Mischel developed an approach to personality that he felt was better suited to capturing the nuances of the relationship between the person and situations in producing behavior. Imagine trying to study personality without using traits or broad motives. What would you focus on? Mischel's answer to this dilemma was his CAPS theory.

CAPS THEORY   Mischel's theory of personality came from his work on *delay of gratification*—putting off a pleasurable experience in the interest of some larger but later reward. Mischel and his colleagues examined how children managed to delay gratification (Mischel & others, 1996; Mischel & Moore, 1980; Mischel & Ayduk, 2011). They placed children in a specific difficult situation—being alone in a room with a tempting cookie within reach. The children were told that if they wanted to at any time, they could ring a bell and eat the cookie. Otherwise, they could wait until the experimenter returned, and then they would get two cookies. The children were then left alone to face this self-control dilemma. In truth, the experimenter was not coming back. The researchers were interested in measuring how long the children could wait before giving in to temptation and eating the cookie.

The children responded in various ways. Some kids sat dead still, focused on the tempting cookie. Some smelled the cookie. Others turned away, sang songs, picked their nose, or did anything but pay attention to the cookie. How did the children who were able to resist temptation do it? The kids who distracted themselves from the cookie by focusing on "cool thoughts" (non-cookie-related things) were better able to delay gratification. In contrast, children who remained focused on the cookie and all its delightful qualities— what Mischel called "hot thoughts"—ate the cookie sooner (Metcalfe & Mischel, 1999). This work demonstrates that avoiding these hot issues might be a good way to see a long-term plan through to completion.

Comstock/Getty Images

These studies have become known as the marshmallow studies. But the reward was often a cookie or a pretzel. Just FYI!

How does this work relate to personality? Over the years, Mischel and his colleagues reported that the amount of time the children were able to delay gratification predicted their academic performance in high school and college (Mischel, 2004) and even their self-regulation skills in their 40s (Casey & others, 2011). In one study, longer delay of gratification at four years of age was linked to a lower body mass index (BMI) three decades later (Schlam & others, 2013). Such results indicate remarkable coherence in personality over time. Note that self-regulation in a child is very different from what it involves for an adult. Mischel's idea is that even though the behaviors are quite different, they make sense as an expression of a personality characteristic.

Now, the results of these studies are striking, but recent research has called into question the links between this single act of delay of gratification and life outcomes. To read about that work, see the Challenge Your Thinking.

To define personality in a way that makes sense of long-term coherence in self-regulation, Mischel conceptualized personality as a set of interconnected **cognitive affective processing systems (CAPS)** (Kross & others, 2010; Mischel, 2004, 2009; Mischel & Ayduk, 2011). According to CAPS theory, our thoughts and emotions about ourselves and the world affect our behavior and become linked in ways that matter to behavior. Personal control and self-efficacy are connections of sorts that a person has made among situations, beliefs, and behaviors. Imagine someone— let's call him Raoul—who is excited by the challenge of a new assignment given by his boss. Raoul may think about all the possible strategies to complete the project and get down to work immediately. Yet this go-getter may respond differently to other challenges, depending on who gives the assignment, what it is, or whether he feels he can do a good job.

Mischel focused on coherence, or whether behaviors make sense across different situations—not whether they are the very same behavior.

cognitive affective processing systems (CAPS) Mischel's theoretical model for describing that individuals' thoughts and emotions about themselves and the world affect their behavior and become linked in ways that matter to that behavior.

Does It Really Matter How Long a Child Waits for That Second Treat?

The famous "marshmallow studies" have become the stuff of psychological legend. You may have even seen the videos on *YouTube*. Despite their popularity, the studies have important limitations. First, compared to today's standards, the number of children in the studies falls far short. In fact, the first study demonstrating the link between the time a child delayed gratification and later outcomes included only 95 children (Mischel & others, 1988). Even later investigations rarely included more than 100 kids in a study (Carlson & others, 2018). In addition, all of the children from the first study came from the same preschool at Stanford University. So these kids were likely to be the offspring of faculty, graduate students, and students associated with an elite university. Subsequent studies have included mostly white and upper-middle-class children from large cities (New York, San Francisco, Seattle) in the United States (Carlson & others, 2018). These features of the studies call into question the degree to which this classic effect can be replicated and generalized. Would a larger, more diverse sample produce similar results?

A team of investigators tested whether the marshmallow study results would replicate in a diverse sample of over 900 children (Watts & others, 2018). The children were four years old when they participated in a delay of gratification task modeled on the original task. The treats were selected among those known to be favored by the kids. Each child was given the standard instructions: They could ring a bell and eat the treat or wait for the experimenter to return and get two treats. Children were left in the room alone with the treat for 7 minutes maximum. The time they waited was recorded. Then, the researchers followed up on the children in the first grade and again at age 15. They collected information about academic achievement and behavioral problems (provided by mothers). In addition, the researchers obtained information about family social class, such as parental education and income.

The results showed that the amount of time children waited was only very weakly related to outcomes in first grade and was not related to outcomes at age 15. More importantly, the differences that emerged among the children in their capacity to delay gratification were explained almost entirely by social class. The researchers reasoned that children from lower income families may take what they can get when they can get it (Watts & others, 2018); after all, who knows if that second treat is really coming? Children who do not live in a world filled with ready access to treats might be more likely to jump the gun, ring the bell, and eat the treat.

Aside from the failure to replicate a classic finding, the study showed two additional interesting results. First, *most kids* were able to wait the full 7 minutes, regardless of social class. The second interesting finding concerned the timing of eating the treat and outcomes at age 15. Children unable to wait for 20 seconds showed some problems at age 15. This result suggests that rather than having to struggle to delay, a child might only have to wait 20 seconds to demonstrate self-control.

Some scientists defended the original findings suggesting that the failure to replicate the previous findings was due to differences in the study design (Falk & others, 2019). Others noted that social class is not a "third variable" but an important support for self-control development (Doebel & others, 2020). In any case, although failing to replicate a classic study may be disappointing, for many this result has been greeted with relief. Even among those of us who would have eaten that marshmallow, the crime of doing so does not lead to a life sentence of low levels of self-control.

Josie Garner/Shutterstock

What Do You Think?

- How long do you think you would have waited for the treat, as a child?
- What factors, other than social class, might affect a person's capacity to delay gratification?

From the CAPS perspective, it makes no sense to ask a person, "How extraverted are you?" because the answer is always, "It depends." A person may be outgoing in one situation (on the first day of class) and not another (right before an exam), and this unique pattern of flexibility is what personality is all about.

Not surprisingly, CAPS theory focuses on how people behave in different situations and how they uniquely interpret situational features. From this perspective, knowing that Crystal is an extravert tells us little about how she will behave in a group discussion in her psychology class. We need to know about Crystal's beliefs and goals in the discussion. For example, does she want to impress the instructor?

Is your own behavior mostly consistent across different situations?

Is she a psychology major? Are several members of the class good friends of hers? We also need to know about her personal understanding of the situation itself: Is this an opportunity to shine, or is she thinking about her test for the next class? Research using the CAPS approach generally involves observing individuals behaving in a variety of contexts in order to identify the patterns of associations that exist among beliefs, emotions, and behavior for each individual person across different situations.

Evaluating the Social Cognitive Perspectives

Social cognitive theory focuses on the interactions of individuals with their environments. The social cognitive approach has fostered a scientific climate for understanding personality that highlights the observation of behavior. Social cognitive theory emphasizes the influence of cognitive processes in explaining personality and suggests that people have the ability to control their environment.

Critics of the social cognitive perspective have suggested a number of weaknesses in this approach, including:

- The social cognitive approach is too concerned with change and situational influences on personality. It does not fully account for the enduring qualities of personality.

- Social cognitive theory ignores the role biology plays in personality.

- In its attempt to incorporate both the situation and the person into its view of personality, social cognitive psychology tends to lead to very specific predictions for each person in any given situation, making generalizations impossible.

self-quiz

1. The following are components of Bandura's social cognitive theory *except*
   A. self-efficacy.
   B. unconscious motivations.
   C. personal control.
   D. observational learning.

2. According to Mischel's 1968 book, behavior is determined by
   A. traits.
   B. biology.
   C. situations and the person's perceptions of them.
   D. unconscious motives.

3. The cognitive affective processing systems (CAPS) approach is centrally concerned with
   A. how personality works in different situations.
   B. how genetic inheritance affects personality.
   C. what biological factors influence personality.
   D. what personality is.

APPLY IT! 4. Omri thinks of himself as an extravert, but he rarely speaks up in his classes. He is especially quiet when he meets new people, especially authority figures such as a new boss. What would Mischel say about Omri's behavior?
   A. Omri is not really an extravert at all; he just does not understand that.
   B. Omri is being discriminative in his behavior. He is probably extraverted in some situations and introverted in others, and that should not be surprising.
   C. Omri is probably fixated at the phallic stage of development.
   D. Omri was not given enough unconditional positive regard in his childhood.

5 Biological Perspectives

The notion that physiological processes influence personality has been around since ancient times. Around 400 B.C.E., Hippocrates, called the "father of medicine," described human beings as having one of four basic personalities based on levels of particular bodily fluids (called *humours*). For Hippocrates, a "sanguine" personality was a happy, optimistic individual who happened to have an abundance of blood. A "choleric" person was quick-tempered with too much yellow bile. A "phlegmatic" personality referred to a placid, sluggish individual with too much phlegm (mucus), and a "melancholic" pessimist had too much black bile.

Although Hippocrates's ideas about bodily fluids have fallen by the wayside, personality psychologists have long acknowledged that personality involves the brain and biological processes, although they often have assumed the processes to exist rather than actually studying them. Freud's psychosexual stages bear witness to his strong belief in the connection between the mind (personality) and the body. Allport defined traits as "neuropsychic" structures and personality as a "psychophysical" system. More recently, advances in method and theory have led to fascinating research on the role of biological processes in personality.

Theories Linking Personality and the Brain

The brain is clearly important in personality, as in other psychological phenomena. Recall the case of Phineas Gage. One of the key effects of Gage's horrific accident was that it changed his personality. He went from being gentle, kind, and reliable to being angry, hostile, and untrustworthy. More generally, any consistent pattern of behavior, thought, and feeling must be reflected in patterns in the brain (Allen & DeYoung, 2017). Personality neuroscience is still a relatively new field but it is gaining steam (Brooks & others, 2020; Montag & others, 2020). Before describing specific findings from research studies, let's review two theoretical approaches to the biology of personality.

EYSENCK'S AROUSAL REGULATION THEORY
British psychologist Hans Eysenck (1967) was among the first to describe the role of a particular brain system in personality. He developed an approach to extraversion/introversion based on the *reticular activation system* (RAS).

Recall that the reticular formation is located in the brainstem and plays a role in wakefulness or arousal. The RAS is the name given to the reticular formation and its connections. Eysenck posited that all of us share an optimal arousal level, a level at which we feel comfortably engaged with the world. However, Eysenck proposed, the RAS of extraverts and introverts differs with respect to the baseline level of arousal. You know that an extravert tends to be outgoing, sociable, and dominant and that an introvert is quieter and more reserved and passive. According to Eysenck, these outward differences in behavior reflect different arousal regulation strategies (Figure 5). Extraverts wake up in the morning underaroused, *below* the optimal level, whereas introverts start out *above* the optimal level.

A warning: Eysenck's approach to introversion/extraversion is important in personality science BUT he was also an advocate for genetic explanations for group differences in IQ.

Introversion

Quiet, reserved, passive

Above optimal level

Keeping distractions to a minimum
Being alone
Reading quietly

Personality Characteristics

Level of Arousal

Typical Activities

Extraversion

Outgoing, social, dominant

Below optimal level

Seeking out distractions
Spending time with friends
Listening to loud music

FIGURE 5 **Eysenck's Reticular Activation System Theory** Eysenck viewed introversion and extraversion as characteristic behavioral patterns that aim to regulate arousal around the individual's baseline level.
(Left): drbimages/E+/Getty Images; (Right): Stockbyte/Getty Images

If *you* were feeling underengaged with life, what might you do? You might listen to loud music or hang out with friends—in other words, behave like an extravert. If, on the other hand, you were feeling overaroused or too stimulated, what would you do? You might spend time alone, keep distractions to a minimum, maybe sit quietly and read a book—in other words, you might act like an introvert.

Research has not shown that extraverts and introverts differ in terms of baseline arousal. Instead, researchers have found that a process similar to what Eysenck proposes for arousal—not involving the activation of RAS, but rather blood flow in the striatum, a part of the basal ganglia—plays a role in dopamine levels (Hermes & others, 2011; Wacker, 2018; Wacker & Smillie, 2015). Recall that dopamine is the neurotransmitter linked with the experience of reward. From this approach, introverts have higher baseline blood flow, and extraverts have lower baseline blood flow to this region of the brain. Because extraverts are motivated to bring their dopamine levels up, they are more likely to seek out arousing and rewarding experiences and thus behave in extraverted ways.

GRAY'S REINFORCEMENT SENSITIVITY THEORY

Building from Eysenck's work, Jeffrey Gray proposed a neuropsychology of personality, called **reinforcement sensitivity theory,** that has been the subject of much research (Bacon & others, 2020; Gray, 1987; Gray & McNaughton, 2000; Krupić & Corr, 2020; Pulver & others, 2020; Smillie & others, 2012). On the basis of animal learning principles, Gray posited that two neurological systems—the *behavioral activation system (BAS)* and the *behavioral inhibition system (BIS)*—could be viewed as underlying personality, as Figure 6 shows.

According to Gray, these systems explain differences in an organism's attention to rewards and punishers in the environment. An organism sensitive to rewards is more likely to learn associations between behaviors and rewards and therefore to show a characteristic pattern of seeking out rewarding opportunities. In contrast, an organism with a heightened sensitivity to punishers in the environment is more likely to learn associations between behaviors and negative consequences. Such an organism shows a characteristic pattern of avoiding such consequences.

In Gray's theory, the BAS is sensitive to rewards in the environment, predisposes one to feelings of positive emotion, and underlies the trait of extraversion (Quilty & others, 2014). In contrast, the BIS is sensitive to punishments and is involved in avoidance learning; it predisposes the individual to feelings of fear and underlies the trait of neuroticism (Thake & Zelenski, 2013). Psychologists often measure the BAS and BIS in humans by using questionnaires that assess a person's attention to rewarding or punishing outcomes (Gomez & others, 2020).

To summarize, Gray's approach described two systems, one for activation and one for inhibition. Those high in behavioral activation should learn associations between behaviors and rewards more quickly. Such people should also experience more positive emotion. In contrast, those high in behavioral inhibition should learn associations between behavior and punisher more rapidly and experience more negative mood.

Gray's conceptual model of reinforcement sensitivity proposed interacting brain systems as primarily responsible for the behavioral manifestations of the BAS and BIS. Research has provided some evidence for the biological underpinnings of these systems, as we will see next.

reinforcement sensitivity theory A theory proposed by Jeffrey Gray identifying two biological systems linked to learning associations between behaviors and rewards or punishers. The behavioral activation system is sensitive to learning about rewards. The behavioral inhibition system is sensitive to learning about punishers.

Behavioral Activation System

Sensitive to
Environmental reward

Behavior
Seek positive consequences/rewards

Character of Emotion
Positive

Personality Trait
Extraversion

Behavioral Inhibition System

Sensitive to
Environmental punishment

Behavior
Avoid negative consequences/punishments

Character of Emotion
Negative

Personality Trait
Neuroticism

FIGURE 6

Gray's Reinforcement Sensitivity Theory Gray theorized that two neurological systems, the BAS and the BIS, explain differences in an organism's attention to environmental rewards and punishments, and in this way shape personality.

(Top): Matthias G. Ziegler/Shutterstock; (Bottom): xubingruo/iStock/Getty Images

THE BRAIN AND NEUROTRANSMITTERS

Research has demonstrated links between brain areas and neurotransmitters, on the one hand, and personality traits, on the other, which fit Gray's model. Here we review evidence for the two most studied traits, extraversion and neuroticism, and then consider the evidence for the other five factors.

Extraversion and the Neuroscience of Reward Gray viewed extraversion as being especially sensitive to rewards in the environment. As such, we might expect that volume and activity in the brain's reward centers (for instance, the ventral tegmental area and the nucleus accumbens) would be associated with extraversion, a hypothesis that has received some support (Allen & DeYoung, 2017; DeYoung & Allen, 2019). For example, one study showed that extraversion predicted greater activity in the nucleus accumbens while participants waited to receive a $5 reward (Wu & others, 2014).

A key neurotransmitter in the process of reward learning is dopamine. Recall that dopamine is a "feel good" neurotransmitter vital to learning that certain behaviors are rewarding and to sending the message "Do it again!" Research has shown that dopamine is also a factor in BAS or extraversion (Depue & Fu, 2013; DeYoung, 2013; Munafo & others, 2008; Wacker & others, 2012). Scholars have suggested that early encounters with warm caregivers and positive life experiences can promote the growth of dopamine-producing cells and receptors. These early experiences can make the brain especially sensitive to rewards, setting the neurochemical stage for extraversion (Depue & Collins, 1999). Although some research supports the idea that extraversion is linked to dopamine and the anticipation of rewards, a recent study found no evidence that extraverts' brains react more intensely after receiving rewards (Hyatt & others, 2020). There is still much to be learned about the role of dopamine in extraversion.

Neuroticism and Avoiding Punishment

Research has also found associations between neuroticism and particular brain structures and neurotransmitters. From Gray's perspective, neuroticism is concerned with avoiding negative outcomes and feeling anxiety. Recall that the amygdalae are parts of the limbic system that are especially active in response to emotionally evocative or survival-relevant stimuli. Studies have shown that neuroticism is related to more persistent activation in the amygdalae in response to aversive or disturbing stimuli (Schuyler & others, 2014; Tabak & others, 2020). In addition, neuroticism is associated with fewer connections between these emotion-related brain structures and those areas of the brain (such as the prefrontal cortex) that help in regulating emotion (Allen & DeYoung, 2017), and thinner cortex (Song, Zhou, & others, 2020).

Even stronger than the link between dopamine and extraversion is the relationship between the neurotransmitter serotonin and neuroticism (Aluja & others, 2018; Xu & others, 2020). In this case, neuroticism is associated with low levels of circulating serotonin. As well, neuroticism may be related to a certain serotonin transporter gene and to the binding of serotonin in the thalamus (Cao & others, 2018; Kruschwitz & others, 2015; Wannemüller & others, 2018). Interestingly, the influence of this gene on personality may depend on experience. Whether individuals who have this genetic characteristic actually develop into worriers depends on their social experiences (Laceulle & others, 2015; Pluess & others, 2010).

A final link between biological processes and neuroticism is the hypothalamic-pituitary-adrenal (HPA) axis. The HPA axis represents the body's stress regulation system. Neuroticism is associated with poorer regulation of the HPA axis and more chronic stress (Montoliu & others, 2020).

Other Traits

Although most research has concerned the more emotional traits of extraversion and neuroticism, some personality neuroscience studies have examined the other three traits of the big five. Findings are generally mixed, but some recent work is suggestive of how personality and behavior may be shaped by (and shape) neural processes. These studies also call our attention to the fact that traits may overlap in their neural correlates. For example, openness to experience, the trait that is most about curiosity and interest, has been associated with dopamine and serotonin (Allen & DeYoung, 2017; DeYoung & Allen, 2019). Agreeableness has been linked to greater neural connections in areas responsible for thinking about other people and to neurotransmitters including oxytocin and serotonin (Perry & others, 2015). Finally, conscientiousness has been found to relate to greater brain volume in areas that are thought to play a role in controlling attention and impulses, and planning (DeYoung & Allen, 2019).

The Brain and Traits

Finding associations between brain activity or neurotransmitters and personality does not tell us, however, about the potential *causal pathways* between these variables. Consider that behavior and experience can influence brain processes and therefore determine brain activity. The link between neuroticism and serotonin provides a telling example. Although neuroticism has been related to the serotonin transporter gene, research demonstrates that individuals with this genetic marker are not inevitably likely to be worriers. For such individuals, the levels of well-being or distress they experience may depend mostly on their environment. For example, if a person has this gene and experiences a warm, supportive environment, they are at *lower* risk for depression and distress (Eley & others, 2004; Vinberg & others, 2010; Way & Gurbaxani, 2008). In short, biological processes take place within a larger social context, and how these processes express themselves may depend on that social world.

Personality and Behavioral Genetics

Behavioral genetics is the study of the inherited underpinnings of behavioral characteristics. A great deal of research in behavioral genetics has involved twin studies, and the hub of this work is, appropriately, the University of Minnesota, Twin Cities.

Twin study findings demonstrate that genetic factors explain a substantial amount of the observed differences in each of the big five traits. Heritability estimates for the five factors are about 50 percent (Bouchard & Loehlin, 2001; Han & Adolphs, 2020 Mõttus & others, 2017; South & Krueger, 2008). Remember that to do these studies, researchers compare identical twins, who share 100 percent of their genes, with fraternal twins, who share just 50 percent. All of the participants complete questionnaires measuring their traits. Then the researchers see if the identical twins are more similar to each other than the fraternal twins. One potential explanation for the strong relationship between personality characteristics and well-being is that the same genetic factors may play a role in traits such as extraversion, neuroticism, and well-being (Caprara & others, 2009; Liu & others, 2017; Weiss & others, 2016).

The heritability statistic describes a group, not an individual, and heritability does not mean that traits are set in stone. Understanding the role of genetic factors in personality is enormously complex. Research on non-twin samples often suggests much lower heritability, for reasons that are not well understood. Furthermore, because genes and environment are often intertwined, it is very difficult to tease apart whether, and how, genes or experience explains enduring patterns of behavior. For instance, a child who is genetically predisposed to disruptive behavior may often end up in a time-out or involved in arguments with parents or teachers. When that child emerges as an adult with a "fighting spirit" or lots of "spunk," are those adult traits the product of genes, experiences, or both?

It might surprise you that well-being, like many psychological characteristics—including intelligence, religiosity, and political attitudes—is influenced by genes.

Finally, most traits are probably influenced by multiple genes. Genome-wide association studies, seeking to identify the specific combinations of genetic characteristics that are linked to individual traits, have produced results that often have failed to replicate (Bull & Varghese, 2020; Genetics of Personality Consortium, 2015; van den Berg & others, 2016). The lack of evidence for specific gene locations to explain genetic factors in personality is sometimes called "missing heritability." Importantly, as always, bear in mind that heritability refers to a group, not a person, and can change over time (Kandler & others, 2020).

Evaluating the Biological Perspectives

Research that explores the biological aspects of personality is likely to remain a key avenue of research. This work ties the field of personality to animal learning models, advances in brain imaging, developmental psychology, and evolutionary theory (Walker & others, 2017; Weiss, 2017). However, a few cautions are necessary in thinking about biological variables and their place in personality.

As we considered earlier, biology can be the effect, not the cause, of personality. To be sure that you grasp this idea, first remember that personality is the individual's characteristic pattern of behavior, thoughts, and feelings. Then recall from previous chapters that behavior, thoughts, and feelings are physical events in the body and brain. If traits predispose individuals to particular, consistent behaviors, thoughts, and emotional responses, traits may play a role in forging particular habitually used pathways in the brain. To the extent that personality represents a person's characteristic pattern of thought, feeling, and behavior, personality may not only be influenced by the brain—it may also play a role in the brain's very structure and functions.

6 Personality Assessment

One of the great contributions of personality psychology to the science of psychology is its development of rigorous methods for measuring mental processes. Psychologists use a number of scientifically developed methods to evaluate personality. They assess personality for different reasons, from clinical evaluation to career counseling and job selection.

Self-Report Tests

The most commonly used method of measuring personality characteristics is the **self-report test,** which directly asks people whether specific items describe their personality traits. Self-report personality tests include items such as:

> I am easily embarrassed.
>
> I love to go to parties.
>
> I like to watch cartoons on TV.

Respondents choose from a limited number of answers (yes or no, true or false, agree or disagree).

One problem with self-report tests is a factor called *social desirability*. To grasp the idea of social desirability, imagine answering the item "I am lazy at times." This statement is probably true for everyone, but would you feel comfortable admitting it? When motivated by social desirability, people say what they think the researcher wants to hear or what

self-report test
Also called an objective test or an inventory, a method of measuring personality characteristics that directly asks people whether specific items describe their personality traits.

they think will make them look better. One way to measure the influence of social desirability is to give individuals a questionnaire that is designed to tap into this tendency. Such a scale typically contains many universally true but threatening items ("I like to gossip at times," "I have never said anything intentionally to hurt someone's feelings"). If scores on a trait measure correlate with this measure of social desirability, we know that the test takers were probably not being straightforward with respect to the trait measure. That is, if a person answers one questionnaire in a socially desirable fashion, then they are probably answering all the questionnaires that way.

Another way to get around social desirability is to design scales so that it is virtually impossible for the respondent to know what the researcher is trying to measure. One way to accomplish this goal is to use an **empirically keyed test,** a type of self-report test that is created by first identifying two groups that are known to be different on the variable one wants to measure. The researcher would give these two groups a large number of questionnaire items and then see which items show the biggest differences between the groups. Those items would then become part of the scale to measure the group difference. For instance, a researcher might want to develop a test that distinguishes between people with a history of substance abuse and those with no such history. The researcher might generate a long list of true/false items that ask about a variety of different topics but that do not even mention substance abuse. These numerous questions are presented to the members of the two groups, and on the basis of the responses, the researcher can then select the items that best differentiate between the members of the differing groups.

Note that an empirically keyed test avoids the issue of social desirability because the items that distinguish between the two groups are not related in any obvious way to the actual purpose of the test. For instance, those without a substance abuse history might typically respond "true" to the item "I enjoy taking long walks," while those with a history of substance abuse might be more likely to respond "false," but this item does not mention substance use, and there is no clear reason why it should distinguish between these groups.

Indeed, an important consideration with respect to empirically keyed tests is that researchers often do not know why a given test item distinguishes between two groups. Imagine, for example, that an empirically keyed test of achievement motivation includes an item such as "I prefer to watch sports on TV instead of romantic movies." A researcher might find that this item does a good job of distinguishing between higher-paid versus lower-paid managers in a work setting. However, does an item such as this example measure achievement motivation or, instead, simply the respondents' gender?

empirically keyed test
A type of self-report test that presents many questionnaire items to two groups that are known to be different in some central way.

An empirically keyed test can have items that seem to have nothing to do with the variable of interest.

MMPI

The **Minnesota Multiphasic Personality Inventory (MMPI)** is the most widely used and researched empirically keyed self-report personality test. The MMPI was initially constructed in the 1940s to assess "abnormal" personality tendencies. The inventory is now in its third iteration, the MMPI-3 (Ben-Porath & Tellegen, 2020). The scale features 335 items and provides information on a variety of personality characteristics. The MMPI also includes a variety of items meant to assess whether the respondent is lying or trying to make a good impression.

Minnesota Multiphasic Personality Inventory (MMPI)
The most widely used and researched empirically keyed self-report personality test.

Some of the MMPI-3 scales measure characteristics associated with psychological disorders, such as depression and schizophrenia. Other scales include masculinity/femininity and introversion.

The MMPI is not only used by clinical psychologists to assess mental health (Whitman & others, 2020). It is also a tool for employment decisions (Marshall & others, 2020), and in forensic settings it is used for assessing criminal risk (Anderson & others, 2020).

ASSESSMENT OF THE BIG FIVE FACTORS

Paul Costa and Robert McCrae (2010) constructed the *Neuroticism Extraversion Openness Personality Inventory—3* (or *NEO-PI-3,* for short), a self-report test geared to assessing the five-factor model: openness, conscientiousness, extraversion, agreeableness, and neuroticism (emotional instability). The test also evaluates six subdimensions that make up the five main factors (Costa & others, 2014; McCrae & others, 2011). Other measures of the big five traits have relied on the lexical approach and offer the advantage of being available for free (Maples-Keller & others, 2019).

Unlike empirically keyed tests, measures of the big five generally contain items that are straightforward; for instance, the trait "talkative" might show up on an extraversion scale. These items have what psychologists call **face validity.** A test item has face validity if it seems on the surface to fit the trait in question. Measures of personality traits include items that are both positive and negative for each trait. For example, to measure agreeableness, we might ask people to rate themselves on traits such as "nice" and "kind" but also "hostile" and "selfish." Such items allow scientists to measure the full range of a trait and avoid a response bias called *yea saying* (that is, having a person simply saying yes to whatever is asked).

face validity

The extent to which a test item appears to fit the particular trait it is measuring.

A key challenge for researchers who use self-report tests is ensuring the accuracy of participants' responses. One way that researchers gauge the truthfulness or accuracy of responses in self-reports is by having someone who knows the test taker well provide ratings of that person (McCredie & Kurtz, 2020; Mõttus & others, 2020). If the test taker's and the acquaintance's ratings agree, the researchers might put more faith in the self-reports than they otherwise would.

An item measuring neuroticism that is very face valid would be "I am a worrier."

It is likely you would be able to give a reasonably good assessment of your own levels of traits such as neuroticism and extraversion. What about the more mysterious aspects of yourself and others? If you are like most people, you view psychological assessments as tools to find out things you do not already know about yourself. For that objective, psychologists might turn to projective tests.

Projective Tests

A **projective test** presents people with an ambiguous stimulus and asks them to describe it or to tell a story about it—to project their own meaning onto the stimulus. This method assumes that the ambiguity of the stimulus allows individuals to interpret it based on their feelings, desires, needs, and attitudes. The test is designed to elicit the person's unconscious feelings and conflicts, providing an assessment that goes deeper than the surface of personality. Projective tests are theoretically aligned with the psychodynamic perspectives on personality, which give more weight to the unconscious than do other perspectives. Projective techniques also require content analysis. The examiner must code the responses for the underlying motivations revealed in the story.

projective test

A personality assessment test that presents individuals with an ambiguous stimulus and asks them to describe it or tell a story about it—to project their own meaning onto the stimulus.

Rorschach Inkblot test

A famous projective test that uses an individual's perception of inkblots to determine their personality.

THE RORSCHACH INKBLOT TEST A famous projective test is the **Rorschach inkblot test,** developed in 1921 by the Swiss psychiatrist Hermann Rorschach. The test consists of 10 cards, half in black/white and half in color, which the individual views one at a time (Figure 7). Test takers are asked to describe what they see in each of the inkblots. A person may say, for example, "I see a crowd of people" or "This is a picture of the female reproductive organs." These responses are scored on the basis of indications of various underlying psychological characteristics (Bender, 2020).

Rorschach scores may predict some psychological outcomes but not others (Mihura & others, 2013). Many researchers are skeptical about the Rorschach (Garb & others, 2001; Hunsley & Bailey, 2001; Weiner, 2004). They have criticized the test's reliability and validity. If the Rorschach were reliable, two different scorers would agree on the personality characteristics of the person being tested. If the Rorschach were valid, it would predict behavior outside of the testing situation; it would predict, for example, whether an individual will become severely depressed, cope successfully with stress, or get along well with others. Conclusions based on research evidence suggest that the Rorschach does not meet these criteria of reliability and validity (Lilienfeld & others, 2000). Thus, many psychologists have serious reservations about the Rorschach's use in diagnosis and clinical practice.

FIGURE 7

Type of Stimulus Used in the Rorschach Inkblot Test What do you see in this figure? Do you see two red fairies playing checkers? Or a red and pink butterfly? A psychologist who relies on the Rorschach test would examine your responses to find out who you are.

MadamSaffa/Shutterstock

Although still administered in clinical circles (Campos & others, 2014), the Rorschach is not commonly used in personality research. However, the projective method itself remains a tool for studying personality, especially in the form of the Thematic Apperception Test (TAT).

THE THEMATIC APPERCEPTION TEST The **Thematic Apperception Test (TAT),** developed by Henry Murray and Christiana Morgan in the 1930s, is designed to elicit stories that reveal a person's unconscious motivations.

Henry Murray (1893–1988) was a personality psychologist who put great stock in the the role of the unconscious in human behavior. His approach to personality was colored by his interest in the theories of Freud and Jung (Anderson, 2017). Murray believed that people are unaware of their motives. Because our motives are outside awareness, we cannot simply ask people what they want. To access unconscious motivations, along with Christiana Morgan, Murray developed the TAT.

The TAT consists of a series of pictures like the one in Figure 8, each on an individual card or slide. The TAT test taker is asked to tell a story about each of the pictures, including events leading up to the situation described, the characters' thoughts and feelings, and the way the situation turns out. The tester assumes that the person projects their own unconscious feelings and thoughts into the story. The TAT (often called "The Picture Story Exercise" or PSE in research) is used to measure unconscious motives for achievement, affiliation, power, intimacy, and a variety of other needs, defense mechanisms, and other characteristics (O'Gorman & others, 2020; Schönbrodt & others, 2020).

Murray posited 22 different unconscious needs to explain behavior. The three needs that have been the focus of most current research are the need for achievement (an enduring concern for attaining excellence and overcoming obstacles), for affiliation (an enduring concern for establishing and maintaining interpersonal connections), and for power (an enduring concern for having impact on the social world).

The scoring that is applied to TAT stories has been applied to other written texts as well. For example, David Winter (2005) analyzed the motives revealed in inaugural addresses of U.S. presidents. He found that certain needs evidenced in these speeches corresponded to later events during the person's presidency. For instance, presidents who scored high on need for achievement, such as Jimmy Carter, were less successful during their terms. Note that the need for achievement is about striving for personal excellence and may have little to do with playing politics, negotiating interpersonal relationships, or delegating responsibility. Presidents who scored high on need for power tended to be judged as more successful (John F. Kennedy, Ronald Reagan), and presidents whose addresses included a great deal of warm, interpersonal imagery (suggesting a high need for affiliation) tended to experience scandal during their presidencies (Richard M. Nixon). An analysis of President Trump's 2017 inaugural address found that it expressed extraordinarily high levels of power motivation as well as high levels of achievement and affiliation motivation and was also high on extraversion and low in agreeableness (Winter, 2018).

FIGURE 8 **Picture from the Thematic Apperception Test (TAT)** What are this man and woman thinking and feeling? How did they come to this situation, and what will happen next? A psychologist who uses the TAT would analyze your story to find out your unconscious motives.

Science History Images/Alamy Stock Photo

Other Assessment Methods

Self-report questionnaires and projective techniques are just two of the multitude of assessment methods developed and used by personality psychologists. Personality psychologists might also measure behavior directly, by observing a person either live or in a video. In addition, cognitive assessments have become more common in personality psychology, as researchers probe such topics as

PSYCHOLOGICAL INQUIRY

Approach	Summary	Assumptions	Typical Methods	Sample Research Question
Psychodynamic	Personality is characterized by unconscious processes. Childhood experiences are of great importance to adult personality.	The most important aspects of personality are unconscious.	Case studies, projective techniques.	How do unconscious conflicts lead to dysfunctional behavior?
Humanistic	Personality evolves out of the person's innate, organismic motives to grow and actualize the self. These healthy tendencies can be undermined by social pressure.	Human nature is basically good. By getting in touch with who we are and what we really want, we can lead happier, healthier lives.	Questionnaires, interviews, observation.	Can situations be changed to support individuals' organismic values and enhance their well-being?
Trait	Personality is characterized by five general traits that are represented in the natural language that people use to describe themselves and others.	Traits are relatively stable over time. Traits predict behavior.	Questionnaires, observer reports.	Are the five factors universal across cultures?
Social Cognitive	Personality is the pattern of coherence that characterizes a person's interactions with the situations they encounter in life. The individual's beliefs and expectations, rather than global traits, are the central variables of interest.	Behavior is best understood as changing across situations. To understand personality, we must understand what each situation means for a given person.	Multiple observations over different situations; video-recorded behaviors rated by coders; questionnaires.	When and why do individuals respond to challenging tasks with fear versus excitement?
Biological	Personality characteristics reflect underlying biological processes such as those carried out by the brain, neurotransmitters, and genes. Differences in behaviors, thoughts, and feelings depend on these processes.	Biological differences among individuals can explain differences in their personalities.	Brain imagining, twin studies, molecular genetic studies.	Do genes explain individual differences in extraversion?

FIGURE 9 **Approaches to Personality Psychology** This figure summarizes the broad approaches to personality described in this chapter. Many researchers in personality do not stick with just one approach but apply the various theories and methods that are most relevant to their research questions.
> **What is one question you have that is relevant to personality psychology? How would each approach address that question?** > **Which approach do you think is most likely to have an impact on the future of personality psychology?** > **How might the popularity of social media and access to online studies influence the various approaches to personality?**

the relationship between personality and processes of attention and memory. Many personality psychologists incorporate friend or peer ratings of individuals' traits or other characteristics. Personality psychologists also employ a host of psychophysiological measures, such as heart rate and skin conductance. Increasingly, personality psychologists are incorporating brain imaging as well.

Whether personality assessments are being used by clinical psychologists, psychological researchers, or other practitioners, the choice of assessment instrument depends on the researcher's theoretical perspective. Figure 9 lists which methods are associated with each perspective, summarizes each approach and its major assumptions, and gives a sample research question for each. Personality psychology is a diverse field, unified by a shared interest in understanding people—all of us.

1. An empirically keyed test is one that
   A. has right and wrong answers.
   B. discriminates between different groups.
   C. has face validity.
   D. has both easy and difficult questions.

2. A problem with self-report tests, and one that researchers try to overcome, is the issue of
   A. social desirability.
   B. memory lapse.
   C. participant bias.
   D. scorer bias.

3. The assessment technique that asks participants to tell a story about the stimuli they see is the

A. Rorschach inkblot test.
B. Minnesota Multiphasic Personality Inventory (MMPI).
C. NEO-P-I.
D. Thematic Apperception Test (TAT).

APPLY IT! 4. Hank applies for a job as a ticket taker at a movie theater. After his interview, he is asked to complete a set of questionnaires. That night, he brags to friends, "They had me answer all these questions about whether I would ever steal from work, gossip about people, or sneak into a movie without paying. I just lied about everything! They'll never know. The job is mine." One of Hank's friends has taken Intro Psych and has news for Hank.

Which of the following best captures what he will say?
A. Good job, Hank. They have no way of knowing you lied. Good luck with the job!
B. Hank, if you lied on all the questions to make yourself look good, they will be able to detect it. It's called social desirability, and you fell for it.
C. Hank, unless the measures involved telling stories, your lies will never be revealed.
D. Hank, you will probably get the job, but your future employer sounds naïve. Didn't they know you could just lie on those tests?

SUMMARY

① Psychodynamic Perspectives

Freud developed psychoanalysis through his work with patients suffering from hysterical symptoms (physical symptoms with no physical cause). Freud viewed these symptoms as representing conflicts between sexual drive and duty. Freud believed that most personality—which, in his theory, includes components he called the id, ego, and superego—is unconscious. The ego uses various defense mechanisms, Freud said, to reduce anxiety.

A number of theorists criticized and revised Freud's approach. Horney said that the need for security, not sex or aggression, is our most important need. Jung developed the concept of the collective unconscious, a storehouse of archetypes. Adler's individual psychology stresses that people are striving toward perfection.

Weaknesses of the psychodynamic perspectives include overreliance on reports from the past and overemphasis of the unconscious mind. Strengths of psychodynamic approaches include recognizing the importance of childhood, conceptualizing development through stages, and calling attention to the role of unconscious processes in behavior.

② Humanistic Perspectives

Humanistic perspectives stress a person's capacity for personal growth and positive human qualities. Maslow developed the concept of a hierarchy of needs, with self-actualization being the highest human need. In Rogers's approach, each of us is born with a tendency toward growth, a sense of what is good and bad for us, and a need for unconditional positive regard. Because we are often denied unconditional positive regard, we may become alienated from our innate growth tendencies. To reconnect with these innate tendencies, Rogers felt, a person required a relationship that included unconditional positive regard, empathy, and genuineness.

The humanistic perspectives promote the positive capacities of human beings. The weaknesses of the approach are a tendency to be too optimistic and an inclination to downplay personal responsibility.

③ Trait Perspectives

Trait theories emphasize that personality consists of traits—broad, enduring dispositions that lead to characteristic responses. Allport stated that traits should produce consistent behavior in different situations, and he used the lexical approach to personality traits, which involves using all the words in the natural language that could describe a person as a basis for understanding the traits of personality.

The current dominant perspective in personality psychology is the five-factor model. The big five traits include openness, conscientiousness, extraversion, agreeableness, and neuroticism. Extraversion is related to enhanced well-being, and neuroticism is linked to lowered well-being.

Studying people in terms of their traits has value. However, trait approaches are criticized for focusing on broad dimensions and not attending to each person's uniqueness.

④ Social Cognitive Perspectives

Social cognitive theory states that behavior, environment, and person/cognitive factors are important in understanding personality. In Bandura's view, these factors reciprocally interact.

Two key concepts in social cognitive theory are self-efficacy and personal control. Self-efficacy is the belief that one can master a situation and produce positive outcomes. Personal control refers to individuals' beliefs about whether the outcomes of their actions depend on their own internal acts or on external events.

In 1968, Mischel's controversial book *Personality and Assessment* stressed that people do not behave consistently across different situations but rather tailor their behavior to suit particular situations. Personality psychologists countered that personality does predict behavior for some people some of the time. Very specific personality characteristics predict behavior better than very general ones, and personality characteristics are more likely to predict behavior in weak versus strong situations.

Mischel developed a revised approach to personality centered on a cognitive affective processing system (CAPS). According to CAPS, personality is best understood as a person's habitual emotional and cognitive reactions to specific situations.

A particular strength of social cognitive theory is its focus on cognitive processes. However, social cognitive approaches have not given adequate attention to enduring individual differences, to biological factors, and to personality as a whole.

Biological Perspectives

Eysenck suggested that the brain's reticular activation system (RAS) plays a role in introversion/extraversion. He thought of these traits as the outward manifestations of arousal regulation. Gray developed a reinforcement sensitivity theory of personality, suggesting that extraversion and neuroticism can be understood as two neurological systems that respond to rewards (the behavioral activation system, or BAS) and punishments (the behavioral inhibition system, or BIS) in the environment.

Dopamine is associated with behavioral approach (extraversion), and serotonin with behavioral avoidance (neuroticism). Behavioral genetics studies have shown that the heritability of personality traits is about 50 percent. Studies of biological processes in personality are valuable but can overestimate the causal role of biological factors.

Personality Assessment

Self-report tests assess personality by asking participants about their preferences and behaviors. One problem in self-report research is the tendency for individuals to respond in socially desirable ways. Empirically keyed tests avoid social desirability problems by using items that distinguish between groups even if we do not know why the items do so.

The most popular test for assessing the big five traits is the NEO-PI-R, which uses self-report items to measure each of the traits. The Minnesota Multiphasic Personality Inventory (MMPI) is the most widely used empirically keyed personality test.

Projective tests, designed to assess unconscious aspects of personality, present individuals with an ambiguous stimulus, such as an inkblot or a picture, and ask them to tell a story about it. Projective tests are based on the assumption that individuals will project their personalities onto these stimuli. The Thematic Apperception Test (TAT) is a projective test that has been used in personality research. Other assessment methods include behavioral observation, obtaining peer reports, and psychophysiological and neuropsychological measures.

KEY TERMS

archetypes	ego	Oedipus complex	social cognitive perspectives
behavioral genetics	empirically keyed test	personality	subjective well-being
big five factors of personality	face validity	projective test	superego
cognitive affective processing systems (CAPS)	humanistic perspectives	psychodynamic perspectives	Thematic Apperception Test (TAT)
collective unconscious	id	reinforcement sensitivity theory	trait theories
conditions of worth	individual psychology	Rorschach inkblot test	unconditional positive regard
defense mechanisms	Minnesota Multiphasic Personality Inventory (MMPI)	self-efficacy	
		self-report test	

ANSWERS TO SELF-QUIZZES

Section 1:   1. D; 2. C; 3. D; 4. C
Section 2:   1. B; 2. C; 3. B; 4. D

Section 3:   1. B; 2. D; 3. D; 4. B
Section 4:   1. B; 2. C; 3. A; 4. B

Section 5:   1. B; 2. C; 3. D; 4. C
Section 6:   1. B; 2. A; 3. D; 4. B

11 Social Psychology

Coming together while socially distanced

In the spring of 2020, people lived under the shadow of COVID-19. Many faced long lockdowns and stay-at-home orders. Human beings need each other, even when safety required them to be apart. All over the world, people found ways to join together while staying separate. Millions of Italians joined together in song, from their balconies, singing their national anthem in a nationwide flash mob (D'Angelo, 2020). In the United States, in New York City, each evening at 7 P.M., New Yorkers leaned out their windows or stood on their fire escapes and applauded the healthcare workers. Separated by necessity, people heeded the call to come together even if at a distance.

Without question, 2020 was a tough year for the most social species on the earth. Human beings are distinct in the extent to which our lives are intertwined with each other. We are remarkable in the ways we reach out to help others and the trust we extend even to strangers. We gain a sense of ourselves from our relationships and the groups to which we belong. As much as we need other people, our lives are also complicated by our social connections. Understanding humans as social creatures is the focus of social psychology.

SelectStock/Getty Images

1 Defining Social Psychology

Human beings are an extremely social species. We need one another to survive. Our thoughts and emotions are often about the people we care about. Our goals often include interpersonal relationships, our families, friends, and romantic partners. The social groups we belong to provide a sense of identity. Our actions are often directed toward (or in response to) another person.

The fundamental social nature of human existence is the focus of social psychology. **Social psychology** is the scientific study of how people think about, influence, and relate to other people. Social psychologists study how a person's thoughts, feelings, and behaviors are influenced by the actual (or imagined) presence of others. Social psychology places many topics from other areas of psychology—including neuroscience, perception, cognition, learning, motivation, and emotion—into a social context.

Social psychology is similar to *sociology,* the study of human societies, organizations, and institutions. Although both sociology and social psychology examine human social behavior, sociologists focus on the group level. In contrast, social psychologists are interested in how individuals influence groups and how groups influence individuals. Unlike sociologists, social psychologists often concentrate on the immediate social situation to understand what causes people to behave as they do. For example, a social psychologist might be interested in studying how the presence of just one other person might influence your behavior.

social psychology
The study of how people think about, influence, and relate to other people.

Think about it. Even if you are the world's biggest loner . . . Did you make your own clothes? Shoes? Toothbrush? People need people.

Features of Social Psychology

Social psychology shares many characteristics with other areas of psychology but it is distinctive in at least two central ways: its connection to real life and its reliance on experimental methods. Let's review each of these features next.

SOCIAL PSYCHOLOGY IS CONNECTED TO REAL LIFE
Since its earliest days, social psychology has been inspired by real-life events. For example, the emergence of social psychology, as a field, can be traced back to the years after the U.S. Civil War (Morawski & Bayer, 2013). Imagine what the country was like then—with the nation recovering from a bloody conflict that nearly ended the union and with people who were once enslaved living among former slave owners. How might these individuals come to live together in harmony and equality? There were many problems to be solved, and people began to consider that the new science of psychology might help solve them. Similarly, after World War II, social psychology experienced extraordinary growth as scientists dedicated themselves to understanding the events that led to the war, the rise of the Nazis, and the Holocaust. During the civil rights movement in the United States, social psychologists were interested in studying all of the ways that changing situational factors might alter a person's life for the better (Stangor, 2016). Today, racism, prejudice, and social justice remain central research topics in social psychology (Easterbrook & Hadden, 2020; Jetten & others, 2020).

Social psychologists not only take inspiration from real-life events, but their research has important implications for many aspects of everyday life. Social psychological research includes topics such as leadership, organizational behavior, marketing, and persuasion.

Chris Schmidt/E+/Getty Images

SOCIAL PSYCHOLOGICAL RESEARCH IS (OFTEN) EXPERIMENTAL

Although some social psychological research is correlational in nature, more often than not social psychologists use experimental methods. This means that social psychologists are likely to manipulate an independent variable, for example, an aspect of the social context, to draw causal conclusions about its effects on some outcome (the dependent variable).

AN EXAMPLE: THE BYSTANDER EFFECT

To put these characteristics together, let's review a classic study to give you a taste of social psychology. In 1964, a young woman named Kitty Genovese was brutally murdered in New York City. She was attacked at about 3 A.M. in a courtyard surrounded by apartment buildings. It took the slayer approximately 30 minutes to kill Genovese. Thirty-eight neighbors watched the gory scene from their windows and heard Genovese's screams. Media reports declared that no one helped or called the police. Those reports turned out to be erroneous (Manning & others, 2007). Nevertheless, inspired by this case, social psychologists John Darley and Bibb Latané (1968) surmised that people might have been less likely to help simply because there were other people present. These scientists devised a number of studies to examine when the presence of others would lead individuals to be less likely to help a person in distress.

The studies involved staging various emergencies. For example, there might be a person in the hallway in distress, or smoke might begin to slowly fill the room from a vent. The independent variable was whether the participant was alone or with other people. The dependent measure was whether people acted to respond to the emergency (or how long it took them to do so).

The studies showed that when alone, a person was likely to take action about 75 percent of the time, but when another bystander was present, the figure dropped to 50 percent (Darley & Latané, 1968). These results demonstrate the **bystander effect,** the tendency of a person to be less likely to help in an emergency when other people are present. What explains the bystander effect? One factor might be that people looked to others for cues about how to behave. Another is deindividuation, a process in which anonymous individuals lose a sense of personal responsibility, a topic we will return to later.

The bystander effect is still in evidence. You have probably seen photos of a person in distress or danger, tweeted and re-tweeted without any indication that someone helped.

bystander effect The tendency of an individual who observes an emergency to be less likely to help when other people are present than when the observer is alone.

The discovery of the bystander effect demonstrates how social psychology often works. Darley and Latané noticed an important event in the social world. They developed a hypothesis about the variables that might have led to the event. Then, they sought to test their hypothesis experimentally.

self-quiz

1. What is one difference between social psychology and sociology?
   A. Social psychology is interested in human social groups, and sociology is not.
   B. Compared with sociology, social psychology is more interested in the behavior of individuals within groups.
   C. Sociology focuses on the person as the main cause of behavior.
   D. Sociology is more likely to rely on experimental lab studies than is social psychology.

2. To explain and predict behavior, _____ psychologists typically look to situational factors.
   A. social
   B. personality

   C. health
   D. cognitive

3. The bystander effect means that a person is
   A. more likely to help if there are other people around.
   B. less likely to help if there are other people around.
   C. likely to be unaffected by the presence or absence of bystanders.
   D. likely to be more aggressive when there are more bystanders around.

APPLY IT! 4. Tracie, a young woman in the armed forces, notices that when she is the only woman in a group, she feels nervous and does not perform as well as she can. When there is at least one other woman in the group, she performs very well. Tracie is interested in studying her experiences as a social psychologist would. Which of the following fits best with how a social psychologist would tackle this problem?
   A. Start a support group of women in the armed forces.
   B. Interview several members of the armed forces and study how institutional factors in the military contribute to troop performance.
   C. Design an experiment in which the gender composition of groups is manipulated and the performance of individuals is measured.
   D. Have a large group of women in the armed forces fill out questionnaires measuring their traits, motives, and beliefs.

2 Social Cognition

social cognition
The area of social psychology exploring how people select, interpret, remember, and use social information.

Social cognition explores how people select, interpret, remember, and use social information (Leblanc & Ramirez, 2020; Stolier & others, 2020). Essentially, it is the way that we think about social situations and other people. Increasingly, social cognitive psychologists have begun to include brain imaging techniques in their research to help identify where and how social cognitive processes occur (Schurz & others, 2020).

Some experts have argued that social cognition–thinking about other people–is more fundamental than thinking about anything else.

Person Perception

person perception
The processes by which an individual uses social stimuli to form impressions of others.

Person perception refers to the processes by which we form impressions of others (Cassidy & others, 2020; Osborne-Crowley, 2020). When we first meet someone, typically the new acquaintance quickly makes an impression on us. First impressions happen very quickly. In one study, just a 100-millisecond exposure to unfamiliar faces was sufficient for individuals to form an impression (Todorov, 2017). Another study showed that impressions that students formed of instructors after very brief exposure at the beginning of a course predicted eventual teaching evaluations and student performance (although quality of instruction still mattered to student learning) (Samudra & others, 2016). A positive first impression predicts more romantic interest in a prospective partner (Kerr & others, 2020).

Are first impressions accurate? Research suggests that they can be. Based on photographs or very brief interactions or video clips, people are able to accurately discern a person's romantic interest in them (Place & others, 2012), propensity for violence (Stillman & others, 2010), leadership (Tskhay & others, 2017), and sexual orientation (Stern & others, 2013). In addition, a large body of evidence suggests that even in cases where two people are only slightly acquainted, ratings of personality traits can be surprisingly accurate, in adults (Carney & others, 2007) and children (Tackett & others, 2016). The first 90 seconds may be all that is needed to predict what a person is likely to do in the rest of a social interaction (Murphy & others, 2015). Social judgments based on very little information can be surprisingly accurate. A set of studies showed that judgments based on very brief clips of contestants in dart-throwing championships (before they even threw the darts) accurately predicted final scores (Furley & Memmert, 2020).

How does the brain make a first impression? Research shows that the dorsomedial prefrontal cortex (or dmPFC, the area of the brain just above your forehead) is involved in first impression formation (Ferrari & others, 2016; Gangopadhyay & others, 2020; Suzuki & O'Doherty, 2020). For example, studies using transcranial magnetic stimulation show that when processing in the dmPFC is disrupted, people are slower to render judgments of a person's traits (Ferrari & others, 2016b). The dmPFC is thought to play a role in our sense of self and in our judgments of others.

What is it about people that gives away so much information so quickly? People reveal themselves in many ways, including body language, emotional expressions, styles of dress, and hairstyle (Martikainen, 2020; Olivola & Todorov, 2017; Zwebner & others, 2017).

The face is a very important social cue. Think about it. We might say a person has a nice face or kind eyes. Can faces really communicate those characteristics? Studies have shown that judgments of political candidates based only on photos of their faces can predict election results as well as success in office (Giacomin & Rule, 2020; Mileva & others, 2020). The faces of candidates are especially likely to influence voters if the voters have little information about the election (Lev-On & Waismel-Manor, 2016).

The face is a powerful social cue.
Hybrid Images/Cultura/Getty Images

Some places have begun to include a photograph of candidates on the election ballots. What do you think of this idea?

The trustworthiness of a face can have profound implications (Zebrowitz, 2018). In a series of studies, participants were shown pictures of unfamiliar people who held political office and rated how corruptible the person seemed to be. The ratings predicted whether the politicians were convicted of political corruption and violation of campaign finance laws (Lin & others, 2018). Other aspects of faces can have additional important implications for social perception, as we now consider.

PHYSICAL ATTRACTIVENESS   Physical attractiveness is a powerful social cue (Schein & others, 2017). Judith Langlois and her colleagues found that even three- to six-month-olds showed a preference for looking at attractive faces versus unattractive faces, as rated by adults (Hoss & Langlois, 2003; Ramsey & others, 2004). This attractiveness bias is visited upon infants as well. Compared to when they view attractive babies, adults show subtle signs of negative emotion in their facial expressions when they look at pictures of unattractive babies (Schein & Langlois, 2015).

You might be asking, what makes a face attractive? Social psychologists have examined this question as well. *People* magazine's "50 Most Beautiful People" issue might lead you to conclude that attractiveness is about being exceptional in some physical way. Consider Beyoncé's radiant smile or Ryan Gosling's icy blue eyes. It turns out, though, that very attractive faces are actually *average*. Using computer technology, researchers have averaged faces together to see how blending faces might affect their attractiveness. Generally, averaged faces are more attractive than individual faces (Jones & Jaeger, 2019). "Average-ness" (along with symmetry and youthfulness) is a key aspect of facial attractiveness. In fact, research suggests that average faces are, literally, "easy on the eyes." A study using EEG (in which electrical activity in the brain is measured through the scalp) showed that average and attractive faces are more quickly processed and identified as human faces by the brain (Trujillo & others, 2014). The ease with which we process attractive faces, in turn, leads to more pleasant mood (Schein & others, 2017).

Do It!

Check out this website to see how the averaging of faces works: www.faceresearch.org/demos/average. Pick some faces you consider unattractive. What happens when you average them together? If you have a digital photograph of yourself and some friends, see what happens when you average those faces. Do you agree that average faces are more attractive than any single face?

Attractive people are generally assumed to have a variety of positive characteristics, including being better adjusted, socially skilled, friendly, likable, extraverted, and apt to achieve superior job performance (Langlois & others, 2000). These positive expectations for physically attractive individuals have been referred to as the "beautiful is good" stereotype.

A **stereotype** is a generalization about a group's characteristics that does not consider any variations from one individual to another. Stereotypes are a natural extension of the limits on human cognitive processing and our reliance on concepts in cognitive processing. We save time and effort by lumping people into groups. Unfortunately, in doing so, we miss the unique characteristics of each member of those groups.

stereotype
A generalization about a group's characteristics that does not consider any variations from one individual to another.

Is there any truth to the "beautiful is good" stereotype? Attractive people may indeed possess a number of positive characteristics (Langlois & others, 2000). Two longitudinal studies are suggestive. One showed that greater facial attractiveness at age nine years predicted higher psychological well-being and lowered depression at age sixty (Datta Gupta & others, 2016). Another showed that, in combination with cognitive ability and self-control, physical attractiveness in adolescence predicted greater education level, which in turn predicted higher income in adulthood (Converse & others, 2016). Does that mean attractiveness is naturally related to, for example, being happier or enjoying career success? Not necessarily.

One way that stereotypes can influence people is **self-fulfilling prophecy**. In a self-fulfilling prophecy, social expectations cause individuals to act in ways that make their expectations come true. Robert Rosenthal and Lenore Jacobsen conducted a classic self-fulfilling prophecy study in 1968. They told grade-school teachers that five students were "late bloomers"—that these students had high levels of ability that would likely shine forth over time. In reality, the students had been selected randomly. Nonetheless, a year later, teachers' expectations for the "late bloomers" were reflected in student performance—the academic performance of these five was beyond that of other students. Self-fulfilling

self-fulfilling prophecy
Social expecta-tions that cause an individual to act in such a way that the expectations are realized.

prophecy effects show the potential power of stereotypes and other sources of expectations on human behavior.

More recent research continues to support the idea that expectations can serve as self-fulfilling prophecies (Gentrup & others, 2020; Willard & Madon, 2016). For example, among older adults, expectations about whether growing old involves loneliness (Pickhartova & others, 2016) predict actual experiences of loneliness. Another study of nearly 5,000 adults showed that negative attitudes about aging predicted slower walking speed two years later (Robertson & others, 2015). Finally, when people meet someone online, expectations can have a large effect on the quality of their interactions (Clark & Green, 2018).

Let's apply self-fulfilling prophecy to physically attractive individuals. If we expect attractive people to have positive characteristics, we treat them differently from others. Remember, even attractive babies are more likely to be met with smiling faces compared with unattractive babies (Schein & Langlois, 2015). Studies have shown that social perceivers are likely to give attractive people better treatment. For example, in one study participants were able to punish or reward others whom they had observed playing a game with a partner (Putz & others, 2016). The results showed that when they cheated, attractive players were punished less than unattractive players. When they behaved prosocially (cooperating with the other player) attractive players were rewarded more than unattractive players. Over the course of a lifetime, positive treatment from others may help to produce the confirmation of positive expectations for attractive people.

The "what is beautiful is good" stereotype suggests that positive expectations can influence subsequent behavior. Could stereotypes that carry negative expectations combine with self-fulfilling prophecy effects as well? Research shows that the answer to that question is, unfortunately, yes. In fact, stereotypes not only influence our views of others but also sometimes influence the feelings and performance of people in stereotyped groups.

stereotype threat
An individual's fast-acting, self-fulfilling fear of being judged based on a negative stereotype about their group.

Stereotype threat is an individual's fast-acting, self-fulfilling fear of being judged based on a negative stereotype about their group (Griffin & Hu, 2015; Schmader & others, 2014). The concept of stereotype threat was introduced by Claude Steele (1997; 2012). He said that people who experience stereotype threat are well aware of stereotypical expectations for them as members of the group. In stereotype-relevant situations, the person experiences anxiety about living "down" to expectations and consequently underperforms. Steele and his colleague, Eliot Aronson (1995, 2004), showed that when a test is presented to Black and non-Latino white students who have first simply checked a box indicating their ethnicity, Black students do not perform as well. In situations where ethnicity was not made salient, no differences in performance emerged.

Stereotype threat is experienced by members of many groups, including those in ethnic minority groups, girls and women, members of sexual minority groups, the elderly, people with low socioeconomic status, individuals with a disability, people with overweight or obesity, Native Americans, and people in recovery from addiction (Cadaret & others, 2017; Gonzalez & others, 2020; Jordano & Touron, 2017; Lewis & Sekaquaptewa 2016; Robinson & others, 2020; Weiss & Perry, 2020). A study of female chess players (who make up a small minority in competitive chess) showed reduced performance when they played against men versus women (Smerdon & others, 2020). Even children have an awareness that when someone tries to do something that does not fit with stereotypes, the person is likely to be anxious (Wegmann, 2017).

Stereotype threat can be activated when people realize they are different in a central way from those around them, like being the only woman in a calculus class or the only man in a nursing program. Stereotype threat has been implicated in the shortage of men in stereotypically feminine roles such as grade school teachers (Kalokerinos & others, 2017) or African Americans in the medical profession (Bullock & others, 2020).

How do you think the teachers influenced the late bloomers? What kind of behaviors led to the kids' enhanced performance?

Stereotype threat affects performance on math tests by women compared to men and on tests of "natural athletic ability" by non-Latino white men. If there is a negative stereotype about a group, members of that group will likely worry about "living down" to those expectations.

Eminent social psychologist Claude Steele introduced the concept of stereotype threat.
Linda A. Cicero/Stanford News Service

What can be done about stereotype threat? A number of interventions have been developed that target aspects of the stereotype threat process, including changing stereotype beliefs, enhancing group belonging, presenting role models, reminding people of other identities they possess, and helping to build confidence. These interventions can be effective (Liu & others, 2020; Vallée & others, 2020).

Attribution

Navigating the social world, we must gather information about the people we encounter. Is a person trustworthy or friendly? We often answer these questions by observing people's actions and drawing inferences about the type of people they are likely to be. Making inferences means taking the information we have and coming up with a good guess about who someone is and what that person is likely to do in the future. The results of those inferences are our attributions. Attributions are explanations of the causes of behavior. The way we explain the behaviors of other people can affect our beliefs about them and how we feel about them (Graham, 2020; Muschetto & Siegel, 2019; Ruybal & Siegel, 2019).

Attribution theory identifies the important dimensions at work in attributions. It states that people are motivated to discover the underlying causes of behavior in order to make sense of that behavior (Heider, 1958; Kelley, 1973; Weiner, 2006). Attributions vary along three dimensions (Jones, 1998):

attribution theory The view that people are motivated to discover the underlying causes of behavior as part of their effort to make sense of the behavior.

1. *Internal and external causes:* Internal attributions include causes inside and specific to the person, such as their traits and abilities. External attributions include causes outside the person, such as situational factors. Did Beth get an *A* on the test because she is smart or because the test was easy?

2. *Stable and unstable causes:* Is the cause relatively enduring and permanent, or is it temporary? Did Aaron blow up at his romantic partner because he is hostile or because he was in a bad mood that day?

3. *Controllable and uncontrollable causes:* We perceive that people have power over some causes (for instance, by preparing delicious food for a picnic) but not others (rain on picnic day). We are more likely to make attributions for behaviors people can control.

ATTRIBUTIONAL ERRORS AND BIASES In attribution theory, the person who produces the behavior is called the *actor*. The person who offers a causal explanation of the actor's behavior is called the *observer*. Actors often explain their own behavior in terms of external (or situational) causes. In contrast, observers frequently explain the actor's behavior in terms of internal causes (such as personality traits). When observers make attributions about behaviors, they often overestimate the importance of internal traits and underestimate the importance of external situations when they seek explanations of another person's behavior (Gilbert & Malone, 1995; Jones & Harris, 1967; Westra, 2020) (Figure 1). This mistake is called the **fundamental attribution error.** Ann might explain that she honked her car horn at someone who was slow to move when the light turned green because she was in a hurry to get to the hospital to see her ill parent, but the driver she honked at might think she was just rude.

fundamental attribution error Observers' overestimation of the importance of internal traits and underestimation of the importance of external situations when they seek explanations of another person's behavior.

Observer Tends to give internal, trait explanations of actor's behavior

"She's late with her report because she can't concentrate on her own responsibilities."

Actor Tends to give external, situational explanations of own behavior

"I'm late with my report because other people keep asking me to help them with their projects."

FIGURE 1 **The Fundamental Attribution Error** In this situation, the supervisor is the observer, and the employee is the actor.

(OBSERVER): Kris Timken/Blend Images LLC; (Actor): John Dowland/Getty Images

Just how fundamental is the fundamental attribution error? Cross-cultural studies show that this error is more common among Westerners. Those from more collectivist cultures are more likely to look to the situation to explain the behavior of others (Imada, 2012; Morris & Peng, 1994; Rips, 2011).

A brain imaging study showed that activity in the brain in an area associated with thinking about other people's mental states (the medial prefrontal cortex or PFC) was related to a higher tendency to commit the fundamental attribution error (Moran & others, 2014). Such results support the idea that this error occurs because people focus on trying to understand the actor's mental states. In this process, the potential influence of the situation may be missed. Indeed, the fundamental attribution error can be reduced by taking the perspective of others as we try to explain their behavior (Hooper & others, 2015). When we put ourselves in the other person's shoes, we might be able to see through that person's eyes, reducing the fundamental attribution error.

Just as we make attributions about the behavior of others, we also make attributions about our own actions. **Self-serving bias** refers to the tendency to take credit for one's own successes and to deny responsibility for one's own failures (Allen & others, 2020; Lee-Bates & others, 2017). Think about taking a psychology exam. If you do well, you are likely to take credit for that success ("I knew that stuff")—that is, to make internal attributions. If you do poorly, however, you are more likely to blame situational factors ("The test was too hard")—that is, to make external attributions.

HEURISTICS IN SOCIAL INFORMATION PROCESSING Just as heuristics are useful in general information processing, they can play a role in *social* information processing. Heuristics can be helpful tools for navigating the complex social landscape, but they can lead to mistakes.

Stereotypes are a type of heuristic in that they allow us to make quick judgments using very little information. Relying on stereotypes can lead to serious errors in social information processing. For example, African American doctors encounter people who assume they are service workers (Goldberg, 2020). The insult of being asked if one is there to "take out the trash" stings. Even former President Barack Obama shared that he had been mistaken for a waiter and asked to bring coffee at an event (Moodley, 2014). Such errors are unfortunately common when people rely on stereotypes rather than attend to people as individuals.

One common heuristic is the false consensus effect. Ask yourself: "How many students at my school support the death penalty?" Your answer is likely to depend on whether you support the death penalty. The **false consensus effect** is the overestimation of the degree to which everybody else thinks or acts the way we do. Note that this means we are using ourselves and our own views as heuristics. Interestingly, research has shown that activity in the brain's reward areas is associated with greater levels of false consensus bias (Welborn & others, 2017), suggesting that it feels good to conclude that others share our views. The false consensus effect shows that we sometimes use ourselves as our best guess about the social world, suggesting the special place of the self in social information processing.

false consensus effect
Observers' overestimation of the degree to which everybody else thinks or acts the way they do.

It's challenging to remember how much the situation can influence behavior. Try avoiding this attributional error for just one day: When you see someone do something, think about the situational factors that might cause that person's behavior.

Interactions like mistaking a doctor for a janitor are called "microaggressions." We will review more about these experiences later.

The Self as a Social Object and Social Comparison

We process social information not only to understand other people but also to understand ourselves. A great deal of research in social cognition revolves around "the self," our mental representation of who we are and what we are like. The self is different from other social objects because we know so much more about ourselves than we do about others. The self is special as well because we value ourselves. One of the most important self-related variables is *self-esteem,* the degree to which we have positive or negative feelings about ourselves (Kernis, 2013).

Blue Jean Images/Getty Images

People with high self-esteem often possess a variety of **positive illusions**—rosy views of themselves that are often a bit better than reality (Molouki & Pronin, 2014). Many of us think of ourselves as "above average" on many valued characteristics, including how trustworthy, objective, and capable we are (Howell & Ratliff, 2017; Sedikides & others, 2014). Some research suggests that holding positive illusions about oneself is linked to psychological health and academic achievement (Chung & others, 2016; Taylor, 2011, 2013, 2014). Other results suggest that positive illusions are not associated with well-being at all (Schimmack & Kim, 2020). Certainly, self-esteem can become problematic if a person's self-views change a great deal from one moment to the next and if it is blown out of proportion (Geukes & others, 2017).

positive illusions
Favorable views of the self that are not necessarily rooted in reality.

Have you ever felt a sense of accomplishment about getting a *B* on a test, only to feel deflated when you found out that your friend in the class got an *A*? Comparing ourselves to other people is one way we come to understand our own behavior. **Social comparison** is the process by which we evaluate our thoughts, feelings, behaviors, and abilities in relation to others. Social comparison tells us what our distinctive characteristics are and aids us in building an identity.

social comparison
The process by which individuals evaluate their thoughts, feelings, behaviors, and abilities in relation to others.

⚐ *Some have argued that self-esteem is a bad thing. Typically such arguments focus on unrealistically high or unstable self-esteem.*

Over 60 years ago, Leon Festinger (1954) proposed a theory of social comparison positing that when people lack an objective way to evaluate their opinions and abilities, they compare themselves with others. Furthermore, to get an accurate view, people are most likely to compare themselves with others who are similar to themselves. Extended and modified over the years, Festinger's social comparison theory continues to provide an important rationale for how people come to know themselves (Putnam-Farr & Morewedge, 2020; Wolff & others, 2020).

Attitudes

attitudes
An individual's opinions and beliefs about people, objects, and ideas—how the person feels about the world.

Attitudes are our opinions and beliefs about people, objects, and ideas—how we feel about the world. Social psychologists are interested in many questions about attitudes, including how attitudes relate to behavior and whether and how attitudes can change (Petty & Briñol, 2015).

ATTITUDES AND BEHAVIOR
It might seem obvious that attitudes should lead to behavior. Attitudes do predict behavior, but only *sometimes*. Specifically, attitudes are likely to guide behavior under the following conditions (Schröder & Wolf, 2017):

- *When the person's attitudes are strong* (Ajzen, 2001): The stronger an attitude toward a particular topic, the more likely a person will act on it. Attitudes tend to be especially strong about issues we believe are morally important. Just labeling a topic moral can increase the strength of attitudes about it (Luttrell & others, 2016).

- *When the person shows a strong awareness of their attitudes and when the person rehearses and practices them* (Fazio & Olsen, 2007; Fazio & others, 1982): For example, a person who has been asked to give a speech about the benefits of recycling is more likely to recycle than is an individual with the same attitude about recycling who has not put the idea into words or defined it in public.

Beyond being places to connect and keep up with others, social media sites like Facebook and Twitter are forums for comparing how we measure up to other people.
santypan/Shutterstock

- **When the person has a vested interest:** People are more likely to act on attitudes when the issue at stake will affect them personally. For example, a classic study examined whether students would show up for a rally protesting a change that would raise the legal drinking age from 18 to 21 (Sivacek & Crano, 1982). Although students in general were against the change, those in the critical age group (from 18 to 20) were more likely to turn out to protest.

What are your attitudes about social issues such as the death penalty, gun ownership, and climate change? How do these views influence your behavior?

BEHAVIOR AND ATTITUDES   Not only do attitudes influence behavior, but behavior can influence attitudes. And changes in behavior can precede changes in attitudes. Social psychologists offer two main explanations of why behavior influences attitudes: cognitive dissonance theory and self-perception theory.

cognitive dissonance
An individual's psychological discomfort (dissonance) caused by two inconsistent thoughts.

Cognitive Dissonance Theory   **Cognitive dissonance**, another concept introduced by Festinger (1957), is the psychological discomfort (dissonance) caused by two inconsistent thoughts. According to the theory, we feel uneasy when we notice an inconsistency between what we believe and what we do. In a classic study, Festinger and J. Merrill Carlsmith (1959) asked college students to engage in a series of very boring tasks, such as sorting spools into trays. These participants were later asked to persuade another student (who was in fact a member of the research team, sometimes called a confederate) to participate in the study by telling them that the task was actually interesting and enjoyable. Half of the participants were randomly assigned to be paid $1 for telling this small fib, and the other half received $20. Afterward, all of the participants rated how interesting and enjoyable the task really was.

Remember, random assignment is used to make sure that groups are equal in every way except for the independent variable—which in this study was how much the participants were paid for the lie. The rating of enjoyment was the dependent variable.

Those who were paid only $1 to tell the lie rated the task as more enjoyable than those who were paid $20. Festinger and Carlsmith reasoned that those paid $20 to tell the lie could attribute their behavior to the high value of the money they received. On the other hand, those who were paid $1 experienced cognitive dissonance: "How could I *lie* for just $1? If I said I liked the task, I must have really liked it." The inconsistency between what they *did* (tell a lie) and what they *were paid for it* (just $1) moved these individuals to change their attitudes about the task. Those who were paid $20 were unlikely to experience dissonance. Only those underpaid for lying would be expected to change their attitudes about the task.

When attitudes and behavior conflict, we can reduce cognitive dissonance in one of two ways: change our behavior to fit our attitudes or change our attitudes to fit our behavior. In the classic study just described, participants changed their attitudes about the task to match their behavior. If you have a positive attitude about recycling but pitched a soda can in the trash, for example, you might feel dissonance ("Wait, I believe in recycling, yet I just pitched that can") and relieve that dissonance by rationalizing ("Recycling is not really *that* important"). Through cognitive dissonance, your behavior changed your attitude.

Effort justification is one type of dissonance reduction. It refers to rationalizing the amount of effort we put into getting something by increasing its value. Effort justification explains strong feelings of loyalty toward a group based on the effort it takes to gain admission into that group. Working hard to get into an organization (such as a Greek society or the Marines) or a profession (such as medicine or law) can change our attitudes about it. According to cognitive dissonance theory, individuals in these situations are likely to think, "If it's this tough to get into, it must be worth it."

What happens in the brain as people experience cognitive dissonance? Research using *f*MRI (which tracks blood oxygen in the brain during information processing) shows that when decisions and behaviors conflict with our values, dissonance is associated with activity in areas of the brain involved in integrating emotion and cognition, and empathy (anterior cingulate cortex and the insula) (de Vries & others, 2015; Izuma & Murayama, 2019).

Social psychologists continue to develop Festinger's approach to cognitive dissonance (Harmon-Jones & Harmon Jones, 2020). Cognitive dissonance can help explain diverse phenomena, such as how people "double down" on conspiracy beliefs even after they have been proven wrong or why people stay in unhealthy relationships. Having invested so much in a movement or a relationship, it can feel impossible to move on.

self-perception theory
Bem's theory on how behaviors influence attitudes, stating that individuals make inferences about their attitudes by perceiving their behavior.

Self-Perception Theory **Self-perception theory** is Daryl Bem's (1967) take on how behavior influences attitudes. According to self-perception theory, individuals make inferences about their attitudes by observing their behavior. That is, behaviors can cause attitudes, because when we are questioned about our attitudes, we think back on our behaviors for information. If someone asked about your attitude toward sushi, you might think, "Well, I rarely eat it, so I must not like it." Your behavior has led you to recognize something about yourself that you had not noticed before.

According to Bem, we are especially likely to look to our behavior to determine our attitudes when those attitudes are unclear.

Figure 2 compares cognitive dissonance theory and self-perception theory. Both theories have merit in explaining the connection between attitudes and behavior, and these opposing views bring to light the complexity that may exist in this connection. Another route to attitude change is persuasion.

✍ *If you have ever played devil's advocate in an argument (arguing a point just for the sake of argument), you might have found yourself realizing that maybe you do hold the views you have pretended to advocate. That's self-perception theory at work.*

Festinger Cognitive Dissonance Theory

We are motivated toward consistency between attitudes and behavior and away from inconsistency.

Example: "I hate my job. I need to develop a better attitude toward it or else quit."

Bem Self-Perception Theory

We make inferences about our attitudes by perceiving and examining our behavior and the context in which it occurs, which might involve inducements to behave in certain ways.

Example: "I am spending all of my time thinking about how much I hate my job. I really must not like it."

FIGURE 2   Two Theories of the Connections between Attitudes and Behavior Although we often think of attitudes as causing behavior, behavior can change attitudes, through either dissonance reduction or self-perception.
Photodisc/Getty Images

Persuasion

Persuasion involves trying to change someone's attitude—and often behavior as well (Briñol & others, 2017). Teachers, lawyers, and sales representatives study techniques that will help them sway their audiences (children, juries, and buyers). Perhaps the most skilled persuaders of all are advertisers, who combine a full array of techniques to sell everything from cornflakes to carpets to cars.

Think about the last advertisement you saw. Who presented the message? What did the message say? How was the message presented? Every act of persuasion is multifaceted. How do these many different aspects relate to attitude change? You might think that if a persuasive argument is based on sound logic, it should win the day. But the quality of an argument is only one aspect of persuasion.

✍ *Advertisers spend huge amounts of money for celebrity endorsements. If an argument is good, why would it matter if LeBron James delivers it?*

elaboration likelihood model
Theory identifying two ways to persuade: a central route and a peripheral route.

TECHNIQUES OF PERSUASION   The **elaboration likelihood model** identifies two pathways of persuasion: a central route and a peripheral route (Petty & Cacioppo, 1986; Teeny & others, 2017). The *central route* works by engaging the audience thoughtfully with a sound, logical argument. The *peripheral route* involves factors such as the attractiveness of the person giving the message or the emotional power of an appeal. The

Making the Sale!

Sooner or later, nearly everyone will be in a position of selling someone something. Social psychologists have studied ways in which social psychological principles influence whether a salesperson makes that sale (Cialdini, 1993, 2016). The effectiveness of these strategies continues to be borne out in research (Cantarero & others, 2017; Guéguen & others, 2016).

Ariel Skelley/Blend Images LLC

One strategy is called the *foot-in-the-door* technique (Freedman & Fraser, 1966). The foot-in-the-door strategy involves making a smaller request ("Would you be interested in a three-month trial subscription to a magazine?") at the beginning, saving the biggest demand ("How about a full year?") for last. The foot-in-the-door strategy works because complying with the smaller request sets the stage for the person to maintain consistency between a prior behavior and the next one (and avoid dissonance).

A different strategy is called the *door-in-the-face* technique (Cialdini & others, 1975). The door-in-the-face technique involves making the biggest pitch first ("Would you be interested in a full-year subscription?"), which the customer probably will reject, and then making a smaller "concessionary" demand ("Okay, then, how about a three-month trial?"). This technique relies on the fact that the customer feels a sense of obligation: You let them off the hook with that big request; maybe they should be nice and take the smaller offer.

Interestingly, even as many classic findings in psychology have not replicated, the door-in-the-face technique still works, some 50 years later (Genschow & others, 2020). In addition, the foot-in-the-door effect appears to work, even with pigeons (Bartonicek & Colombo, 2020).

peripheral route is effective when people are not paying close attention or lack the time or energy to think about the message. As you might guess, television advertisers often use the peripheral route to persuasion on the assumption that during the commercials you are probably not paying full attention to the screen. However, the central route is more persuasive when people have the ability and the motivation to pay attention.

> Attitudes that are changed using the central route are more likely to persist than attitudes that are changed using the peripheral route.

RESISTING PERSUASION

One of the ways people resist persuasion is through counterarguing. This means that as they are exposed to a message, they defend against the information by mentally generating arguments against it. Students who binge

Apple CEO Tim Cook calls on his powers of persuasion whenever Apple introduces a new product.
Karl Mondon/Tribune Content Agency LLC/Alamy Stock Photo

drink, for instance, have been shown to be most likely to counterargue against public service announcements on the dangers of binge drinking (Zhou & Shapiro, 2017). Making messages self-relevant can help to short-circuit the tendency to counterargue.

Of course, sometimes, we are interested in helping people avoid persuasion. Advertisers, salespeople, and politicians work hard to persuade us, sometimes with little concern for whether their products and policies might actually help us. Social psychologists are interested in identifying ways people can effectively resist persuasion (Hornsey & Fielding, 2017). One way to resist persuasion is through *inoculation* (Banas & Richards, 2017; McGuire, 2003; McGuire & Papageorgis, 1961). Just as administering a vaccine inoculates individuals from a virus by introducing a weakened or dead version of that virus to the immune system, giving people a weak version of a persuasive message and allowing them time to argue against it can help individuals avoid persuasion. "Inoculation" helps college students resist plagiarism (Compton & Pfau, 2008) as well as credit card marketing appeals (Compton & Pfau, 2004). When individuals are warned that they are going to be hit with persuasive appeals and are given arguments to help them resist these pitches, they are able to do so.

YOU ARE HEREBY INOCULATED! College students are preyed upon by credit card companies!

self-quiz

1. Stereotype threat refers to
   A. the damage potentially caused by stereotyping others.
   B. the strategy of changing someone's behavior by threatening to use a stereotype.
   C. humans' tendency to categorize people using broad generalizations.
   D. an individual's self-fulfilling fear of being judged based on a negative stereotype about their group.

2. In committing the fundamental attribution error, we overemphasize _____ and underemphasize _____ when making attributions about others' behavior.

   A. internal factors; external factors
   B. external factors; internal factors
   C. controllability; stability
   D. stability; controllability

3. The _____ route to persuasion involves reflecting carefully about the merits of an argument.
   A. peripheral
   B. false consensus
   C. central
   D. emotional

APPLY IT! 4. Thomas has spent long hours working to get his candidate elected president of the student body. When he talks to his mother on election night, Thomas is overjoyed to report that his candidate won by a landslide. His mom points out that Thomas never cared about campus politics before, and she asks him about his sudden interest. Thomas admits that she is right, but notes that he now cares deeply about campus issues and is likely to continue to be involved in politics. What theory best explains Thomas's change?
   A. social comparison theory
   B. self-perception theory
   C. stereotype threat
   D. elaboration likelihood model

3 Social Behavior

We do not just think socially; we also behave in social ways. Two particular behaviors that have interested psychologists represent the extremes of human social activity: altruism and aggression.

Altruism

Over 100,000 adults and children await lifesaving kidney or liver transplants each year. A staggering 8,000 people die each year awaiting a transplant (Donate Life, 2020). Recently, people have begun to offer their own kidneys or parts of their livers to save the lives of others in need. In 2019, while deceased donors accounted for over 32,000 organ donations, an additional 7,300 transplants were made possible by living donors (UNOS, 2020). Recipients of organ donations are sometimes family members, but they can also be strangers. Imagine offering one of your organs to someone you do not know and may never meet. The generosity is staggering. Such extraordinary acts of selfless giving are inspiring but also sometimes puzzling (Carney, 2017).

altruism
Giving to another person with the ultimate goal of benefiting that person, even if it incurs a cost to oneself.

egoism
Giving to another person to ensure reciprocity; to gain self-esteem; to present oneself as powerful, competent, or caring; or to avoid social and self-censure for failing to live up to society's expectations.

In everyday life, we witness and perform "random acts of kindness"—maybe adding a quarter to someone's expired parking meter or giving up our seat on a bus to someone in need. All of these acts are *prosocial behaviors*—they all involve helping another person. Such acts of kindness bear the markings of altruism. **Altruism** means giving aid to another person with the ultimate goal of benefiting that person, even if it incurs a cost to oneself. Are acts of kindness truly altruistic?

Psychologists debate whether human behavior is ever truly altruistic. Altruistic motives contrast with selfish or egoistic motives (Cialdini, 1991; Hawley, 2014; Hirschberger, 2015). **Egoism** means helping another person for personal gain, such as to feel good, or avoid guilt. Kindness might also serve selfish purposes by ensuring *reciprocity*, meaning that we help another person to increase the chances that the person will return the favor.

An example of animal altruism—a baboon plucking bugs from another baboon. Most acts of animal altruism involve kin.
karenfoleyphotography/Shutterstock

Altruism presents a puzzle for evolutionary psychologists. How can a behavior that rewards others, and not oneself, be adaptive? Evolutionary theorists note that helping is especially likely to occur among family members, because helping a relative also means promoting the survival of the family's genes (Buss, 2015). Evolutionary theorists believe that reciprocity in relationships with nonfamily members is essentially the mistaken application of a heuristic that made sense in human evolutionary history—to engage in selfless acts of kindness to one's own family (Chuang & Wu, 2017; Kafashan & others, 2017; Nowak & others, 2000).

Another way evolutionary psychologists explain altruism is called "costly signaling theory" (Zahavi, 1977). Costly signaling theory says that sometimes we engage in a behavior with high personal costs to communicate honest information about ourselves to others. Altruistic behavior, then, could be a costly signal of our traits. Because it comes at a cost, it is unlikely that someone would fake it. The idea is that other people (potential mates or allies) will know that we are genuinely kind if we engage in altruistic behavior (McAndrews, 2002). It may also be that altruistic acts are not only a costly signal of kindness but of other characteristics, including intelligence (Hur, 2020).

Acts of kindness can have a powerful payoff for those who do them. Helping others strongly and consistently leads to increased positive mood (Stukas & others, 2015; Toumbourou, 2016; Snippe & others, 2017). A recent meta-analysis showed that prosocial behavior was linked to greater well-being, especially when the behavior was informal (Hui & others, 2020). Prosocial spending, or spending money on others rather than oneself, is linked to greater well-being, potentially universally (Dunn & others, 2020; Whillans & others, 2020).

Positive mood and prosocial acts are linked, even in early childhood (Eisenberg, 2020; Hammond & Drummond, 2019). One study showed that adolescents experienced well-being benefits for acts of kindness only if those acts were altruistic (Tashjian & others, 2020). Prosocial behavior increases with age, perhaps because altruistic motivation takes center stage in the later years of life (Mayr & Freund, 2020). A meta-analysis showed that older adults who engaged in volunteering had a lower mortality risk than those who did not (Okun & others, 2013).

Does the fact that behaving prosocially leads to feelings of pleasure mean that such behavior is somehow always selfish? Feelings of pleasure are linked with adaptive behaviors—those things we need to do to survive and reproduce. Might it

What was the last altruistic act you committed? What led to your behavior? How did you feel afterward?

INTERSECTION

Social Psychology and Motivation: What Happens When We Don't Get a Chance to Help?

Imagine you are in a class in which you are doing very well. You notice another student who has missed some classes and is struggling. You get the idea to help the person by offering your notes. However, just as you are about to offer those notes, someone else in the class beats you to it. They help the person instead of you. How would you feel?

A recent set of studies (Titova & Sheldon, 2020) posed this and similar scenarios to students. The researchers randomly assigned participants to consider different situations. Some participants imagined they wanted to help and were able to do so. Others imagined wanting to help but having someone else do it, thwarting their prosocial intentions. Still others imagined not wanting to help and seeing someone else help. After imagining the situations, participants rated their mood. The results across studies showed a consistent pattern: those who wanted to help but were prevented from doing so experienced less positive mood compared to the other groups.

These results are interesting because it is a defining feature of a need that we should feel bad when its satisfaction is thwarted. The researchers showed that needs for competence, relatedness, and

SolStock/Getty Images

autonomy did not fully explain the way that thwarted prosocial intentions affected mood. Instead, these results were explained by a fourth need, beneficence motivation. *Beneficence motivation* refers to the pleasure we experience in helping another person (Martela & Ryan, 2020). The idea that we need to help other people fits with the strong link we have seen between prosocial behavior and well-being. Indeed, beneficence motivation explained the way thwarted helping deflated positive mood (Titova & Sheldon, 2020).

The existence of beneficence motivation highlights the deeply social nature of human life. It is interesting to consider that, of course, if you only cared about that other student's well-being, shouldn't you feel happy that the student was helped, regardless of who did the helping? Yet, there is something unsatisfying about getting "scooped" when we want to help. Perhaps one reason for that feeling is when we lose a chance to help another person we experience the thwarting of a basic human need.

\\ **Have your prosocial intentions ever been thwarted?**

---

be that the strong link between pleasure and kindness demonstrates that engaging in prosocial behavior is an important adaptation for humans who are so dependent on one another for survival?

When we help another person, we experience ourselves as competent, interpersonally connected, and, if we helped of our own initiative, autonomous. These are the three basic needs recognized in the self-determination theory (as described in Chapter 9). If helping others meets these important needs, then helping is a great way to boost well-being. What happens when we want to help but are prevented from doing so? To read about research on this topic, and how it suggests yet another human need, see the Intersection.

Setting aside the question of whether such acts are altruistic, in this section we review biological, psychological, and sociocultural factors that predict prosocial behavior. As you read, consider whether you think altruism is a problem to be solved or a natural aspect of human life.

BIOLOGICAL FACTORS IN PROSOCIAL BEHAVIOR

Genetic factors explain between 30 and 53 percent of the differences we see in the tendency to engage in kind acts (Fortuna & Knafo, 2014; Hur, 2020). These genetic factors, in turn, are related to neurotransmitters in the brain (Walsh & others, 2020). High levels of serotonin are associated with prosocial behavior (Contreras-Huerta & others, 2020). Dopamine receptors in the brain are also associated with prosocial behavior (Walsh & others, 2020). Other research has shown that prosocial behavior is related to the balance of GABA (the brain's break pedal) and glutamate (an excitatory neurotransmitter) in the brain. Specifically higher levels of GABA (vs. glutamate) are negatively related to prosocial behavior (Okada & others, 2020). Finally, oxytocin, the neurohormone that plays a role in social bonding, has been linked to enhanced prosocial behavior in experiments in which it is administered nasally (Marsh & others, 2020).

In terms of brain structures, research suggests that when we feel compassion for another person, areas of the midbrain associated with the perception of pain are likely to be active (Simon-Thomas & others, 2012). These same areas are associated with nurturing parental behaviors, suggesting that neural factors associated with the parent-child relationship are involved in kindness toward others. In addition, prosocial behavior is linked to thinning of the cortex in some areas of the left hemisphere (Tamnes & others, 2018).

Oxytocin is not always linked to nice things. It also predicts the tendency to experience envy and to gloat!

PSYCHOLOGICAL FACTORS IN PROSOCIAL BEHAVIOR

Among the psychological factors that play a role in prosocial behavior are empathy, personality, and mood.

empathy
A feeling of oneness with the emotional state of another person.

Empathy

Empathy is a person's feeling of oneness with the emotional state of another. When we feel empathy for someone, we feel what that person is feeling. Empathy allows us to put ourselves in another person's shoes. We can feel empathy even for those we do not particularly like, as was demonstrated when Yankee fans delighted in the playing of "Sweet Caroline" (a tradition of their arch rivals, the Boston Red Sox), following the bombing of the Boston Marathon in 2013. When we are feeling empathy for someone else's plight, we are moved to action—not to make ourselves feel better but out of genuine concern for the other person (Batson & others, 2015).

Personality

Agreeableness is the personality trait most strongly associated with prosocial behaviors (Thielmann & others, 2020; Tobin & Graziano, 2020). The association between agreeableness and brain structures helps to illuminate its role in acts of kindness. Agreeableness is related to greater volume in the posterior cingulate cortex (DeYoung & others, 2010), a brain area associated with understanding other people's beliefs and with empathy (Hubbard & others, 2016; Saxe & Powell, 2006).

Mood

Mood can determine whether or not we engage in kind behaviors. Happy people are more likely than unhappy people to help others (Snippe & others, 2017; Snyder & Lopez, 2007). Indeed, a study in which participants reported their mood and acts of kindness multiple times during a day over several days showed that positive mood predicted more kind acts, which, in turn, led to even more positive mood (Snippe & others, 2017). Does it then follow that being in a bad mood makes people less helpful? Not necessarily, because adults (especially) generally understand that doing good for another person can be a mood booster. When people are in a bad mood, they might be likely to help if they think that doing so will improve their mood.

SOCIOCULTURAL FACTORS IN PROSOCIAL BEHAVIOR

Three sociocultural factors that predict prosocial behavior are socioeconomic status, gender, and the media.

Socioeconomic Status

Some research supports the idea that poorer people are more generous than the wealthy (Piff & others, 2010; Piff & Robinson, 2017). Laboratory

The Samburu people of Kenya are very spiritual, believing in and praying daily to the god Ngai. In addition to the factors discussed here, the presence of established religions within a culture is related to prosocial behaviors.
Eric Lafforgue/Gamma-Rapho/Getty Images

experiments show that compared to more wealthy individuals, those from poorer backgrounds tend be more attuned to the suffering of others (Stellar & others, 2012). It may be that wealth promotes a focus on maintaining one's standing in the world to the detriment of reaching out to help those in need (Kraus & others, 2012). Wealthier people are more likely to engage in public giving while less well-off people are more likely to give privately (Kraus & Callaghan, 2016). Still, some research has suggested limits on these findings. One large-scale study showed that the negative relationship between wealth and prosocial behavior was unique to the United States, a place with very high income inequality (Côté & others, 2015). An even more recently, very large-scale study showed no evidence for the link between income and prosocial behavior (Schmukle & others, 2019), suggesting the differences that can occur when comparing lab studies to large "real world" samples.

Gender   Research on the link between gender and helping has relied on the gender binary and compared those who identify as women to those who identify as men. Given the role of empathy in helping, we might think that women should be more likely to help than men. After all, stereotypes tell us that women are more empathic than men. However, as in most domains, it is useful to think about gender in context. Compared to men, women are more likely to experience an immediate impulse to behave prosocially toward a person in need (Rand & others, 2016). Women are more likely than men to act on these impulses to help when the context involves existing interpersonal relationships or nurturing, such as volunteering time to help a child with a personal problem (Diekman & Clark, 2015; Eagly, 2009). Men, on the other hand, are more likely to help in situations in which a perceived danger is present (for instance, picking up a hitchhiker) and in which they feel competent to help (as in assisting someone with a flat tire) (Eagly & Crowley, 1986). For women, concerns about safety may affect their willingness to help.

Media   Media—including music, TV, film, and video games—can influence prosocial behavior. Listening to music with prosocial lyrics can promote kindness (Greitemeyer, 2009). Watching television shows with positive content predicts prosocial behavior (Hearold, 1986). Playing prosocial video games enhances prosocial thoughts (Greitemeyer & Osswald, 2011) and acts of kindness in children and adolescents (Harrington & O'Connell, 2016). When video games include prosocial content, the players are more likely to behave prosocially during the game (Jin & Li, 2017). A meta-analysis showed that prosocial media does promote prosocial behavior in general and especially when it comes to helping strangers (Coyne & others, 2018).

Some evolutionary scientists have suggested that altruism, especially when it is directed at the members of one's own group, may coexist with hostile actions toward others outside one's group (Arrow, 2007). A soldier may perform selfless acts of altruism for their country, but for a person on the other side of the conflict, that behavior is harmful. Thus, altruism within a group may be linked to aggression. Perhaps the greatest puzzle is that a species capable of incredible acts of kindness can also perpetrate horrifying acts of violence.

Aggression

aggression
Social behavior whose objective is to harm someone, either physically or verbally.

Aggression refers to social behavior whose objective is to harm someone, either physically or verbally. Aggression is common in humans and nonhuman animals. From an evolutionary standpoint, it would appear that it was adaptive for early humans to have the ability to fight. What factors explain aggressive behavior? Psychologists have investigated biological, psychological, and sociocultural factors.

BIOLOGICAL INFLUENCES IN AGGRESSION   Researchers who approach aggression from a biological viewpoint examine the influence of genetics and neurobiological factors.

Genes   The role of genes in aggression is most easily seen in nonhuman animals and in the process of selective breeding. After a number of breedings among only aggressive animals and among only docile animals, vicious and timid strains of animals emerge. The vicious strains attack nearly anything in sight; the timid strains rarely fight, even when attacked. The genetic basis for aggression is less clear in humans. The heritability estimate for aggression is about 50 percent; however, as is often the case, probing the precise genetic locations of aggression has been challenging (Odintsova & others, 2019). The relation of genes to aggression may depend on the type of aggression studied (Veroude & others, 2016). Unprovoked physical aggression and reactive aggression (that is, aggression in response to someone else's provoking behavior) may have different genetic underpinnings (Ilchibaeva & others, 2020).

Eric Isselee/Shutterstock

Neurobiological Factors   Although humans do not have a specific aggression center in the brain, aggressive behavior often results when areas such as the limbic system are stimulated by electric currents (Bartholow, 2018; Lupton & others, 2020). The frontal lobes of the brain—the areas most involved in executive functions such as planning and self-control—have also been implicated in aggression. Research has examined the brains of individuals who have committed the ultimate act of violence: murder (Nordstrom & others, 2011). The results indicate that those who murder may differ from others in deficits in the functioning of these areas of the brain.

Neurotransmitters—particularly, lower levels of serotonin—have been linked to aggressive behavior (Bartholow, 2018; Ilchibaeva & others, 2020), although the association is quite small (Duke & others, 2013).

Hormones are another biological factor that may play a role in aggression. The hormone that is typically implicated in aggressive behavior is testosterone (Zitzmann, 2020). Research on nonhuman animals links testosterone to aggression (Piña-Andrade & others, 2020; Zhao & others, 2020), but results with humans have been less consistent (Carré & Robinson, 2020).

Behavior can influence hormone levels. So, higher testosterone may be an effect of aggressive behavior, not its cause.

PSYCHOLOGICAL INFLUENCES ON AGGRESSION   Psychological influences on aggression include personality characteristics, frustrating circumstances, and cognitive and learning factors.

Personality   Low levels of agreeableness are associated with more aggressive behavior (Tobin & Graziano, 2020). People whose personalities predispose them to feeling callous and cold toward others are more likely to engage in aggression (Plouffe & others, 2020). A constellation of traits, including low agreeableness, low conscientiousness, and high levels of neuroticism, is associated with aggression (Settles & others, 2012). A meta-analysis showed that individuals who are high on hostility and irritability are more likely to behave aggressively, whether provoked or not (Bettencourt & others, 2006).

Frustrating and Aversive Circumstances   Many years ago, John Dollard and his colleagues (1939) proposed that *frustration,* the blocking of an individual's attempts to reach a goal, triggers aggression. The *frustration-aggression hypothesis* states that frustration always leads to aggression. When people feel frustrated and feel like their goals are being thwarted, they are more likely to behave aggressively (Przybylski & others, 2014).

Psychologists later recognized that, besides frustration, a broad range of aversive experiences can cause aggression. They include physical pain, personal insults, crowding, and unpleasant events. Aversive circumstances also include factors in the physical environment, such as the weather. Murder, rape, and assault increase when temperatures are the highest, as well as in the hottest years and the hottest cities (Anderson & Bushman, 2002; Lynott & others, 2017).

Cognitive Determinants   Aspects of the environment may prime us to behave aggressively. Priming involves making something salient to a person, even without the person's awareness. So, anything that makes us think aggressive thoughts could be associated with heightened aggression. Research by Leonard Berkowitz and others has shown that the mere presence of a weapon (such as a gun) may prime hostile thoughts and produce aggression (Anderson & others, 1998; Berkowitz, 1993; Berkowitz & LePage, 1996). The tendency for the presence of firearms to enhance aggression is known as the *weapons effect*. In support of Berkowitz's ideas, a well-known study found that people who lived in a household with a gun were 2.7 times more likely to be murdered than those dwelling in a household without a gun (Kellerman & others, 1993).

The link between a gun in the household and murder was shown in a correlational study. What are some third variables that might explain the association between gun ownership and murder?

Observational Learning   Social cognitive theorists believe that individuals learn aggression through reinforcement and observational learning (Bandura, 2011). Observing others engage in aggressive actions can evoke aggression. One of the strongest predictors of aggression is witnessing aggression in one's own family (Ferguson & others, 2008). Watching television provides a ready opportunity to observe aggression in our culture, which we consider further in the discussion on media violence later.

SOCIOCULTURAL INFLUENCES ON AGGRESSION   Aggression and violence are more common in some cultures than others. In this section, we review sociocultural influences on aggression, including the "culture of honor," gender, and media influences.

The Culture of Honor   Dov Cohen has examined how cultural norms about masculine pride and family honor may foster aggressive behavior (Cohen, 2001; Gul & others, 2020; Nawata, 2020). In *cultures of honor*, a man's reputation is thought to be an essential aspect of his economic survival. Such cultures see insults to a man's honor as diminishing his reputation and view violence as a way to compensate for that loss. In these cultures, family pride might lead to so-called honor killings in which, for example, a woman who has been raped is slain by her male family members so that they, in turn, are not "contaminated" by the rape (Hadi, 2020). Honor killings of gay men have also been reported in countries where being gay is viewed as shameful or stigmatizing.

Cognitive dissonance could lead to even greater valuing of ones personal honor. That honor would have to be really important for a person to harm a family member.

Cohen has examined how, in the United States, white, male southerners are more likely than white, male northerners to be aggressive when honor is at stake. In one study, Cohen and his colleagues (1996) had white men who were from either the North or the South take part in an experiment that required them to walk down a hallway. A member of the study passed all the men, bumping against them and quietly calling them a derogatory name. The southerners were more likely than the northerners to think their masculine reputation was threatened, to become physiologically aroused by the insult, and to engage in actual aggressive or dominant acts. In contrast, the northerners were less likely to perceive a random insult as "fightin' words."

Metinkiyak/Getty Images

Gender   Like research on gender and helping, research on gender and aggression has been limited to the binary conception of gender. The link between gender and aggression depends on the particular *type* of aggression we are talking about. **Overt aggression** refers to physically or verbally harming another person directly. Boys and men tend to be higher in overt aggression than girls and women, throughout the lifespan (Bukowski & others, 2007; Dodge & others, 2006; White & Frabutt, 2006). Of the murders committed in 2019, within the group in which gender of perpetrator could be identified, 64 percent were committed by men (FBI, 2020).

Women's smaller physical size may be one reason they are less likely to engage in overt aggression. To understand aggressive tendencies in girls and women, researchers have focused instead on **relational aggression**,

overt aggression
Physical or verbal behavior that directly harms another person.

relational aggression
Behavior that is meant to harm the social standing of another person.

Men are the victims of murder 70 percent of the time.

behavior that is meant to harm the social standing of another person through activities such as gossiping and spreading rumors (Casper & others, 2020; Kawabata & others, 2014; Wright, 2020). Relational aggression differs from overt aggression in that it requires that the aggressor have a considerable level of social and cognitive skill (Lehman, 2020). To be relationally aggressive, a person must have a good understanding of social circumstances and be motivated to plant rumors that are likely to damage the intended party. Relational aggression is more subtle than overt aggression, and the relationally aggressive individual may not seem to be aggressive to others, as the aggressive acts typically are committed secretly.

Bananastock/age fotostock

It is not clear that girls are more relationally aggressive than boys, but research consistently shows that relational aggression constitutes a greater percentage of girls' overall aggression than it does for boys (Underwood, 2011). Relational aggression is complex, and research suggests that girls who show relational aggression toward others are themselves likely to be the victims of relational aggression particularly if they are unpopular (Ferguson & others, 2016).

Although relational aggression does not lead to the physical injury that might result from overt aggression, it can be extremely painful nevertheless. Phoebe Prince, a 15-year-old who had recently moved from Ireland to the United States with her family, became the target of unrelenting rumors and harassment from a group of popular girls at her high school after she had a brief relationship with a popular senior boy. Prince died by suicide. Even after her death, the girls who had harassed her posted rumors about her on the Facebook page that was set up as a memorial (Cullen, 2010).

Media   Images of violence pervade the U.S. popular media: newscasts, television shows, sports broadcasts, movies, video games, Internet videos, and song lyrics. Do portrayals of violence lead to aggression?

Although some critics reject the conclusion that TV violence causes aggression (Savage & Yancey, 2008), many scholars assert that TV violence can prompt aggressive or antisocial behavior in children (Brown & Tierney, 2011; Bushman & Huesmann, 2012; Comstock, 2012). Of course, TV violence is not the only cause of aggression in children or adults. Like all social behaviors, aggression has multiple determinants (Matos & others, 2012).

As we discussed earlier, research shows that prosocial video games foster prosocial behavior. Do violent video games foster aggression? Correlational studies demonstrate an association between playing violent video games and a number of negative outcomes. A meta-analysis concluded that children and adolescents who play violent video games extensively are more aggressive, less sensitive to real-life violence, and more likely to engage in delinquent acts than their counterparts who spend less time playing the games or do not play them at all (Anderson & others, 2010). However, the effects of exposure to video game violence on aggressive behavior is likely to be small (Hilgard & others, 2017). A recent study of high school students, tested twice one year apart, showed that violent video game play did predict aggressive behavior, but the relationship was quite small and dwarfed by the influence of low agreeableness and deviant peers (López-Fernández & others, 2020).

PeopleImages/Getty Images

As always, correlation does not imply causation. Critics of research showing a link between violent video games and aggression stress that many studies have not consistently measured important and possibly influential third variables, such as family violence, in predicting both video game use and aggression (Ferguson & Kilburn, 2010; Ferguson, 2015; Hilgard & others, 2015). To conclude that violent video games lead to aggression, experiments are required. Experimental evidence shows that playing violent video games, especially those that are highly realistic, can lead to more aggressive thoughts and behaviors in children and adults (DeWall & others, 2013; Gentile & others, 2014; Hollingdale & Greitemeyer, 2014). Importantly, the optimal safeguards against experimental bias have not always been applied in such research. That means researchers have not necessarily examined potential confounds in the studies. The violence present in a video game may not be the "active ingredient" that leads to aggression. For example, a series of studies showed that games that were frustrating and difficult were more likely to lead to aggression, even if the games had no violent content (Przybylski & others, 2014).

Critics of the conclusion that violent video game exposure leads to aggression point out that acts of aggression studied in the laboratory are not generalizable to real-world criminal violence (Ritter & Elsea, 2005; Savage, 2008; Savage & Yancey, 2008). Operationalizing aggression in the laboratory is challenging. Researchers might provide participants the opportunity to "aggress" against another, for instance, by subjecting the individual to a blast of loud noise, dispensing a mild electrical shock, or administering a large dose of hot sauce to swallow. Whether these operational definitions of aggression are applicable to real-life violence is a matter of much debate (Savage & Yancey, 2008).

There is no question that the potential link between video game violence and aggressive behavior is contentious. Scholars can look at the same data and interpret it in very different ways (Boxer & others, 2015; Ferguson, 2015). When you encounter research on this controversial topic, be sure to examine the specific aspects of the studies described.

Increasingly, researchers who study the effects of video games are gamers themselves. How do you think this fact might change the way the research is designed and the conclusions drawn?

self-quiz

1. Egoism is in evidence when
   A. a mother physically protects her children from the ravages of a tornado at risk to her own life.
   B. a person donates bone marrow to a complete stranger.
   C. a coffee shop customer pays for the latte of the stranger in line behind him.
   D. a speaker compliments his staff's achievements in order to make himself look good.

2. Laboratory research on the relationship between socioeconomic status and prosocial behavior shows that on average
   A. the poor give more than the rich.
   B. the rich give more than the poor.

   C. there is no association between socioeconomic status and prosocial behavior.
   D. the extremely rich are most likely to give.

3. With respect to gender and aggression,
   A. boys engage in more overt aggression than girls do.
   B. girls engage in more overt aggression than boys do.
   C. boys engage in more relational aggression than girls do.
   D. boys and girls engage about equally in overt and relational aggression.

APPLY IT! 4. While driving, Nate sees an elderly man struggling to change a flat tire. Nate stops and helps

the man and then continues to his girlfriend's house. When Nate tells his girlfriend about his act of kindness, she says, "I would never have done that." Nate suggests that he just must be nicer than she is. Considering the social psychology of helping, is Nate right?
   A. Nate is right because men are generally more helpful than women.
   B. Nate is right because he engaged in a selfless act of altruism.
   C. Nate is not right because his girlfriend may not have felt safe stopping to help someone on a country road at night.
   D. Nate is not right because he probably got a lot of praise from his girlfriend, rendering his act selfish.

4 Close Relationships

Along with good health and happiness, close relationships figure prominently in many people's notions of a good life. Every day we see commercials lauding the ability of various online matchmaking sites to link us up with the love of our life. In the United States, online dating was a $3 billion industry in 2018 (Lin, 2018). In Europe, spending on dating apps doubled in 2019, reaching $2.2 billion (Freer, 2020).

Because close romantic relationships are important for many people, it is no wonder that social psychologists are interested in studying this vital part of human existence. A vast literature has accumulated in social psychology, examining attraction, love, and intimacy. Research methods in this area have moved beyond having people rate the attractiveness of static pictures to setting up speed dating events (Kerr & others, 2020) and having couples track their feelings using an app on their smartphones (Sels & others, 2020).

Before reviewing this area, let's consider some limitations that characterize this research. First, consider what attraction means. When we find ourselves attracted to someone we are also learning about ourselves and even our identities. For instance, attraction to a person of the same gender or someone who possesses a nonconforming gender identity may provide information about who we are (Stewart & others, 2019; Timmins & others, 2020). Unfortunately, the research on attraction and relationship functioning has not fully explored attraction and relationship processes in people with diverse sexual orientations and gender identities (Thorne & others, 2019). Nor has research generally focused on attraction and couple functioning in minoritized groups, such as people of color. As a result, most of the research we will review is limited to predominantly white and middle- or upper middle-class samples. Although some findings have been demonstrated in more diverse groups (Williamson & Lavner, 2020), it is important to bear this limitation in mind.

If you ever considered being a social psychologist, think of all the great work left for you to do!

Attraction

At the beginning of this chapter, we considered one key factor in interpersonal attraction: physical attractiveness. Research on interpersonal attraction has illuminated a variety of other factors that play a role in the process of becoming attracted to someone (Kerr & others, 2020; Montoya & Horton, 2020).

PROXIMITY, ACQUAINTANCE, AND SIMILARITY   Even in the age of dating apps and message boards, it remains unlikely that you are going to become attracted to someone without meeting the person. Proximity, or physical closeness, is a strong predictor of attraction. You are more likely to become attracted to a person you pass in the hall every day than someone you rarely see. One potential mechanism for the role of proximity in attraction is the mere exposure effect (Zajonc, 1968, 2001). The **mere exposure effect** means that the more we encounter someone or something (a person, a word, an image), the more likely we are to start liking the person or thing even if we do not realize we have seen it before.

mere exposure effect
The phenomenon that the more individuals encounter someone or something, the more probable it is that they will start liking the person or thing even if they do not realize they have seen it before.

Similarity also plays an important role in attraction (Sprecher, 2019). We have all heard that opposites attract, but what is true of magnets is not typically true of human beings. We like to associate with people who are similar to us (Berscheid, 2000). Our friends and lovers are much more like us than unlike us. We share with them similar attitudes, behavior patterns, taste in clothes, intelligence, personality, other friends, values, lifestyle, and physical attractiveness.

Maybe you binged 'Love Is Blind' or 'Dash & Lily' on Netflix. Do you think it's possible to fall in love with someone you've never seen or met?

Similarity also plays a role once we are in a relationship. Actual similarity is associated with relationship satisfaction, but *perceived* similarity (that is, feeling like you are similar to your partner even if you are not) also matters (Sels & others, 2020). Many studies of couples show that what we think our partner is doing may be just as important as what they are actually doing, for our satisfaction in the relationship.

APPROACHES TO ATTRACTION AND RELATIONSHIPS   In addition to identifying the kinds of factors that might influence attraction, psychologists have developed larger frameworks for understanding the attraction process (Montoya & Horton, 2020). One dominant approach comes from evolutionary psychology. Evolutionary psychologists have long focused on gender differences in the variables that account for attraction (Buss, 2015). From this perspective, the goal of romantic relationships is to produce children. For men, this evolutionary task is complicated by the fact that in the human species, paternity is somewhat more mysterious than motherhood. To be sure that a woman is not already pregnant with another man's child, evolutionary psychologists say, men should be more strongly attracted to younger women. Youth might also indicate a woman's fertility. For women, the task of producing offspring is an innately difficult one. From the evolutionary perspective, women search for a mate who will invest his resources in her and her offspring. Evidence for these difference comes from things like personal ads. Heterosexual men tend to look for youth and beauty and offer tangible resources—for example, by describing themselves as a "professional who owns a home" (Buss, 2015). Heterosexual women tend to place ads offering youth and beauty and seek partners with resources. Recent research showed similar gender differences in mate preferences among older heterosexual adults in a retirement community (Vance & others, 2020). Evolutionary psychologists have sought to replicate these patterns in gays, lesbians, and transgender people because they believe that evidence for similar patterns would represent strong evidence for biological pressures (Aristegui & others, 2018). However, regardless of a person's sexual orientation or gender identity, people grow up in cultures and are exposed to cultural values throughout their lives. It cannot be surprising that their preferences might reflect those values.

In fact, cross-cultural variation in gender behavior and mate preference challenges these ideas, suggesting the importance of gender roles in these patterns. For example, Alice Eagly's (2010, 2012) *social role view of gender* asserts that social, not evolutionary, experiences have led to differences in gender behavior. The social role approach acknowledges the biological differences but stresses the ways these differences are played out in a range of cultures and societal contexts. Indeed, in cultures characterized by gender equality women are less likely to prefer mates with economic resources (Kasser & Sharma, 1999).

Contemporary research shows that people share an interest in finding partners who are warm and trustworthy (Valentine & others, 2020). This interest, in which there is no gender difference, continues to characterize not only attraction but satisfaction in relationships. In fact, even evolutionary psychologists have begun to recognize that attraction and the ways we feel at the very beginning of a relationship may be less important overall than our enduring motives for warm, loving relationships and a family (Ko & others, 2020). As we become more committed to a person we may make the transition from casual dating to thinking of ourselves as a part of a relationship (Hadden & others, 2019). These relationships can become the foundation for a sense of home.

The idea that we seek warm, close relationships that provide a kind of home base is reflected in attachment perspectives on close relationships (Mikulincer & Shaver, 2019). Remember from Chapter 8 that the early relationship between an infant and caregivers allows the child to form a sense of the self and world. This internal model is thought to operate throughout life, including in our close relationships.

Love

Some relationships deepen to friendship and perhaps even to love. Social psychologists have long puzzled over exactly what love is (Berscheid, 2006, 2010; Sternberg, 2013). One way to think about love is to consider the types of love that characterize different human relationships—for instance, friendships versus romantic relationships (Hendrick & Hendrick, 2009; Rawlins & Russell, 2013). Here we consider two types of love: romantic love and affectionate love.

Poets, playwrights, and musicians through the ages have celebrated the fiery passion of romantic love—and lamented the searing pain when it fails. Think about songs and books that hit the top of the charts. Chances are that they are about romantic love. **Romantic love**, or **passionate love**, is love with strong components of sexuality and infatuation, and

romantic love or passionate love
Love with strong components of sexuality and infatuation, often predominant in the early part of a love relationship.

it often predominates in the early part of a love relationship (Hendrick & Hendrick, 2006). This is the kind of sexually charged feeling we usually mean when we talk about being "in love" (Berscheid, 1988).

Love is more than just passion, however. **Affectionate love,** or **companionate love**, is the type of love that occurs when a person has a deep, caring affection for another person and desires to have that person near. There is a growing belief that the early stages of love have more romantic ingredients and that, as love matures, passion tends to give way to affection (Berscheid & Regan, 2005).

The way that love changes during a relationship does not mean that the relationship becomes less satisfying. Research using diverse samples has shown that newlyweds report high levels of relationship satisfaction that generally remain stable during the early years of marriage (Williamson & Lavner, 2020). The best predictor of drops in relationship satisfaction is being low on satisfaction from the beginning. Especially among women, being relatively low in relationship satisfaction at the beginning of marriage predicts declines in satisfaction and likelihood of divorce (Williamson & Lavner, 2020).

affectionate love or companionate love
Love that occurs when an individual has a deep, caring affection for another person and desires to have that person near.

As love matures, romantic love tends to evolve into affectionate love.
Ronnie Kaufman/Blend Images

Models of Close Relationships

A key question that is posed in the social psychology of close relationships is what it takes for couples for stay together for the long term. You have probably heard that the U.S. divorce rate is 50 percent, meaning half of marriages will end in divorce. This statistic is misleading, however, because people who have multiple marriages are more likely than others to get divorced. The actual percentage of first marriages in the United States that end in divorce is closer to 30 or 40 percent (Abrams, 2016). Divorce rates have been on the decline in recent decades. The number of marriages ending in divorce hit an all-time low in 2019, with less than 15 marriages per 1,000 ending in divorce—a rate more similar to that of the 1970s (Wang, 2020). It's important to keep in mind that this may or may not be good news. It might be that people are building healthy relationships and staying together. However, it is also possible that people feel less free to divorce because of economic or other pressures. Here we review two approaches to close relationships: social exchange theory and the investment model.

At the beginning of the COVID-19 pandemic, divorce rates spiked as couples faced the prospect of locking down together. Even so, divorce rates leveled off, and it seems the decline in divorce will continue.

SOCIAL EXCHANGE THEORY   The social exchange approach to close relationships focuses on the costs and benefits of one's romantic partner. **Social exchange theory** conceptualizes social relationships as involving an exchange of goods, the objective of which is to minimize costs and maximize benefits. From this perspective, the most important predictor of relationship success is *equity*—that is, having both partners feel that each is doing their "fair share." Essentially, social exchange theory asserts that we keep a mental balance sheet, tallying the pluses and minuses associated with our romantic partner—what we put in ("I paid for our last date") and what we get out ("He brought me flowers").

social exchange theory
The view of social relationships as involving an exchange of goods, the objective of which is to minimize costs and maximize benefits.

As relationships progress, however, equity may no longer apply. Happily committed couples are less likely to keep track of "what I get versus what I give," and they avoid thinking about the costs and benefits of their relationships (Clark & Chrisman, 1994). Indeed, a recent study examined how an exchange orientation—that is, tending to keep track of costs and benefits—was negatively associated with relationship functioning (Jarvis & others, 2019). The research showed that people who were monitoring how they were doing in terms of the exchange factors in their relationships tended to overreact to simple everyday conflicts.

Think of long-term couples you know in which one partner remains committed even when the benefits are hard to see, as when a person's romantic partner is gravely ill for a long time.

THE INVESTMENT MODEL   Another way to think about long-term romantic relationships is to focus on the underlying factors that characterize stable, happy relationships compared to others. Developed by the late Caryl Rusbult, the **investment model** examines the ways that commitment, investment, and the availability of attractive alternative partners predict satisfaction and stability in relationships (Rusbult & others, 2012; Segal & Fraley, 2016). From this perspective, long-term relationships are likely to continue when both partners are committed to the relationship and have invested a great deal in it—and when there are few tempting alternatives (other attractive partners) around.

Commitment to the relationship is especially important and predicts a willingness to sacrifice for a romantic partner. In one study, individuals were given a chance to climb up and down a short staircase, over and over, so that their partner would not have to do so. Those who were more committed to their partner worked harder to climb up and down repeatedly, to spare their loved one the burden (Van Lange & others, 1997).

investment model
A model of long-term relationships that examines the ways that commitment, investment, and the availability of attractive alternative partners predict satisfaction and stability in relationships.

self-quiz

1. With regard to happy relationships, social exchange theory would predict that
   A. we are happiest when we are giving in a relationship.
   B. we are happiest when we are receiving in a relationship.
   C. we are happiest when there is a balance between giving and receiving in a relationship.
   D. equity is important to happiness only in long-lasting relationships.

2. The role of similarity in attraction would predict
   A. that opposites attract.

   B. that relationships with more give and take are best.
   C. that we are attracted to people who are similar to us.
   D. that romantic love is more important than affectionate love.

3. Affectionate love is more common _____, whereas romantic love is more common _____.
   A. in men; in women
   B. in women; in men
   C. early in a relationship; later in a relationship
   D. later in a relationship; early in a relationship

APPLY IT! 4. Daniel and Alexa have been dating for two years. Alexa meets a new guy during spring break and cheats on Daniel. When she tells Daniel about it, he is crushed but forgives her. He reasons that because the two of them have worked together on their relationship for two years, it makes no sense to throw it all away based on one mistake. Which theory would predict that Daniel is likely to cheat on Alexa in the future?
   A. the investment model
   B. social exchange theory
   C. evolutionary theory
   D. theory of affectionate love

5 Social Influence and Group Processes

Social psychologists are interested in understanding how the presence of other people pushes us toward some actions (and away from others). Here we consider two types of social influence—conformity and obedience—and then how the mere presence of others can influence a person's behavior. Finally, we consider how the groups to which we belong influence how we treat others.

Conformity and Obedience

Research on conformity and obedience started in earnest after World War II. Psychologists sought answers to the disturbing question of how ordinary people could be influenced to commit the sort of atrocities inflicted on Jewish people and other minorities during the Holocaust. A central question is, how much will people change their behavior to coincide with what others are doing or demanding?

Think you are a nonconformist? To feel the pressure of conformity, the next time you get on an elevator with other people, DO NOT TURN AROUND to face the door.

CONFORMITY   **Conformity** is a change in a person's behavior to coincide more closely with a group standard. Conformity takes many forms and affects many aspects of people's lives, in negative and positive ways. Conformity is at

conformity
A change in a person's behavior to coincide more closely with a group standard.

work, for example, when a person comes to college and starts to drink alcohol heavily at parties, despite never being a drinker before. Conformity is also involved when we obey the rules and regulations that allow society to run smoothly. Consider how chaotic it would be if people did not conform to social norms such as stopping at a red light and not punching others in the face. Conformity can be a powerful way to increase group cohesion. When behavior is coordinated in a group, people feel connected (Gordon & others, 2020). Even something as simple as marching in step together or singing a song with a group can lead to enhanced cooperation among group members (Wiltermuth & Heath, 2009). Interestingly, research suggests that reciprocity (that is, feeling that people are bonded by mutual helping) is a stronger source of group cooperation than conformity (Romano & Bailliet, 2017).

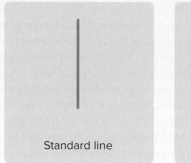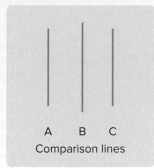

Standard line

A B C
Comparison lines

FIGURE 3 **Asch's Conformity Experiment** The figures show the stimulus materials for the Asch conformity experiment on group influence.

Asch's Experiment

Put yourself in this situation: You are taken into a room where you see five other people seated along a table. A person in a white lab coat enters the room and announces that you are about to participate in an experiment on perceptual accuracy. The group is shown two cards—the first having only a single vertical line on it and the second having three vertical lines of varying length. You are told that the task is to determine which of the three lines on the second card is the same length as the line on the first card. You look at the cards and think, "What a snap. It's so obvious which is the same" (Figure 3).

What you do not know is that the other people in the room are working with the experimenter. On the first several trials, everyone agrees about which line matches the standard. Then on the fourth trial, each of the others picks the same incorrect line. As the last person to make a choice, you have the dilemma of responding as your eyes tell you or conforming to what the others before you said. How would you answer? Solomon Asch conducted this classic experiment on conformity in 1951. Asch instructed the other members of the panel to give incorrect responses on 12 of 18 trials. To his surprise, Asch (1951) found that the volunteer participants conformed to the incorrect answers 35 percent of the time. Subsequent research has supported the notion that the pressure to conform is strong (Fein & others, 1993; Malakh-Pines & Maslach, 2002)—but why do people go along with the group even when faced with clear-cut information such as the lines in the Asch experiment?

Two main factors contribute to conformity: informational social influence and normative social influence. **Informational social influence** refers to the influence other people have on us because we want to be right. Members of the groups may know something we don't. So, we follow the group to be right. In contrast, **normative social influence** is the influence others have on us because we want them to like us. Whether the group is a gang or members of a profession such as medicine, if a particular group is important to us, we might adopt a clothing style that people in the group wear or use the same slang words, and we might assume a certain set of attitudes that characterizes the group's members.

informational social influence
The influence other people have on us because we want to be right.

normative social influence
The influence other people have on us because we want them to like us.

Conformity and the Brain

Conformity is a powerful social force, but why is it so important to us to fit in with a group? Research in social psychology and neuroscience has provided an interesting answer (Ellemers & van Nunspeet, 2020). Research suggests that our brains may actually "feel better" when we fit in.

An fMRI study examined what happens in the brain when people find out that their opinions conflict with those of others (Klucharev & others, 2009). Women were asked to rate a variety of female faces for attractiveness, and their brains were scanned while they received feedback about whether their ratings agreed with those of the other group members. When participants were told that their ratings differed from the group's ratings, they showed enhanced activation in the brain area typically associated with monitoring for errors. In other words, the brain responded

Do you think the brain could learn to like being different?

to judgments that differed from the group judgments as if they were *mistakes*. Furthermore, when the women's ratings were different from the group's ratings, women experienced less activation in the brain's reward centers. The more women's brains responded to being different as an error (and not rewarding), the more they conformed when given a chance to re-rate the faces. These results suggest that, for humans, conformity is rewarding. Interestingly, research has shown that those with more neural connections in the brain's punishment network (the areas of the brain involved in responses to being wrong) are more likely to conform (Du & others, 2020). Other research has shown that activity in the ventromedial prefrontal cortex, an area of the brain implicated in decision making, increases prior to people changing their decisions toward conformity (Li & others, 2020).

Conformity and Culture Individualistic cultures value individuality, individual accomplishments, differences, and uniqueness. Collectivistic cultures value the group and group harmony. It is not surprising, then, that collectivism has been associated with greater levels of conformity. A research review, summarizing 133 experiments following Asch's design, found that individualism within cultures was negatively correlated with conformity (Bond & Smith, 1996).

OBEDIENCE **Obedience** is behavior that complies with the explicit demands of an individual in authority. We are obedient when an authority figure demands that we do something, and we do it. Obedient behavior sometimes can be distressingly cruel. Millions of people throughout history have obeyed commands to commit terrible acts all in the name of "just following orders." How can we come to understand the tendency to obey?

obedience
Behavior that complies with the explicit demands of the individual in authority.

Milgram's studies became the subject of a 1970s TV movie called **The Tenth Level**. *The film starred William Shatner as Stephen Turner—a character based on Stanley Milgram.*

Milgram's Experiment A classic series of experiments by Stanley Milgram (1965, 1974) demonstrated the profound power of obedience. Imagine that, as part of a psychology experiment on the effects of punishment on memory, you are asked to deliver a series of electric shocks to another person. Your role is to be the "teacher" and to punish the mistakes made by the "learner." Each time the learner makes a mistake, you are to increase the intensity of the shock.

You are introduced to the learner, a nice 50-year-old man who mumbles something about having a heart condition. Strapped to a chair in the next room, he communicates with you through an intercom. The apparatus in front of you has 30 switches, ranging from 15 volts (slight) to 450 volts (marked as beyond dangerous, "XXX").

As the trials proceed, the learner quickly runs into trouble and is unable to give the correct answers. As you increase the intensity of the shock, the learner says that he is in pain. At 150 volts, he demands to have the experiment stopped. At 180 volts, he cries out that he cannot stand it anymore. At 300 volts, he yells about his heart condition and pleads to be released. If you hesitate in shocking the learner, the experimenter tells you, "You must go on; the experiment requires that you continue."

Eventually the learner stops responding altogether, and the experimenter tells you that not responding is the same as a wrong answer. The learner is unresponsive. He might be injured or even dead. Would you keep going? Do you think most people would? As shown in Figure 4,

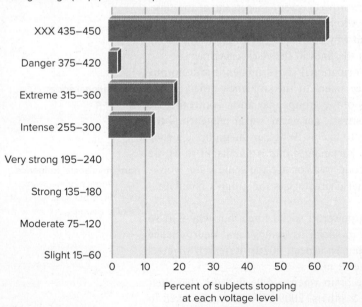

Voltage range (amps) and description

Percent of subjects stopping at each voltage level

FIGURE 4 **Milgram's Obedience Study** In Milgram's experiments, the "learner" was strapped into a chair. The experimenter made it look as if a shock generator was being connected to the learner's body through several electrodes. The chart shows the percentage of "teachers" who stopped shocking the learner at each voltage level.

when Milgram conducted this study, the majority of the teachers obeyed the experimenter: Almost two-thirds delivered the full 450 volts. By the way, the 50-year-old man was a working with Milgram and was not being shocked at all. Of course, the teachers were unaware that the learner was only pretending to be shocked.

"You must go on. The experiment requires that you continue." With those simple statements the experimenter was able to calmly command people (as far as they knew) to shock a man to unconsciousness and possibly death. Such is the power of obedience to authority.

Milgram's studies have long been a subject of controversy. Under today's ethical guidelines, it is unlikely that these experiments would have been approved (Griggs & Whitehead, 2015). Yet we continue to learn from Milgram's data. A meta-analysis of his experiments suggested that the critical decision was at the 150-volt level, when the learner first requested that the experiment be halted. At that point, 80 percent of those who were going to stop did so (Packer, 2008). People who were going to disobey were those who responded to the learner's first request to be set free.

Do Milgram's results apply in more contemporary contexts? To examine this question, Jerry Burger (2009) set up a study very similar to Milgram's, with a key exception: Burger's participants were never allowed to go higher than 150 volts. At 150 volts, the learner, who was working with Burger, asked to end the study, and immediately after participants decided whether to continue, the experiment was ended. The results? They are shown in Figure 5. Surprisingly, Burger's participants were only slightly less likely to obey than Milgram's had been.

Burger had to exclude participants who had heard of Milgram's studies. In effect, people who had learned the lessons of Milgram's work were not given a chance to show their stuff.

Burger (2017) suggested that four key factors likely explain the tendencies of participants to shock the learner. These include:

1. The novelty of the situation. Participants had little idea of how they should or could respond.

2. The normative information provided. This means that participants might have interpreted the experimenter's authority as suggesting how people should behave.

What would you do if you were in Milgram's study?

3. The opportunity to deny responsibility. Participants did not have to feel personally responsible for what happened. They could blame the experimenter.

4. The limited time to think about the decision. Participants were not given very much time to think over their response.

Now is the point when most psychology students would be introduced to the famous Stanford prison experiment (Resnick, 2018). Philip Zimbardo and his students created a simulated prison in the basement of a Stanford University building (Haney & others, 1973; Zimbardo, 1972, 1973, 2007). Newspaper ads recruited men for a two-week study of prison life for which participants would be paid $15 per day (about $97 per day today). After undergoing screening to ensure they were psychologically healthy, 24 men began the study. Each was assigned to the role of either prisoner or guard. To read about what happened next (and why we will not be reviewing the study in detail here), see the Challenge Your Thinking.

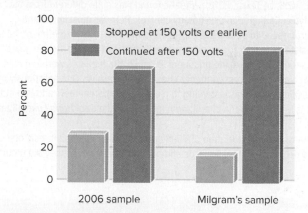

PSYCHOLOGICAL INQUIRY

FIGURE 5 **Obedience Now and Then** This figure shows, side by side, the results of Burger's (2009) study and the results of one of Milgram's studies. The vertical, or Y, axis shows the percent of individuals who stopped or continued after the learner's first expression of a desire to end the study.

SOURCE: Burger, J. "Replicating Milgram: Would people still obey today?" *American Psychology, 64*, 1–11, 2009. American Psychological Association

> Comparing the two sets of results, does the similarity surprise you? Why or why not? > If you had been a "teacher" in either of these studies, what would you have done? How does learning about Milgram's study influence your response?

What Happened in the Stanford Prison Experiment?

The Stanford Prison Experiment is a landmark in social psychology and popular culture. It was even made into a movie (with Billy Crudup playing Zimbardo). But what really happened during the few days those young men were assigned to play guards and prisoners in a mock jail? According to Zimbardo's account, their roles strongly dictated participant behavior. Guards behaved sadistically toward the prisoners, and prisoners became docile and lost their sense of self. One prisoner, said Zimbardo, experienced a psychological breakdown. So horrific were the events that unfolded that the experiment, planned to last for two weeks, was stopped after just six days. Zimbardo claimed the study showed the strong power of roles in determining behavior. He said the prisoners did not ask to leave the study because they had so internalized their roles. Further, he suggested that the guards' actions revealed an important truth about the human capacity for evil: given the opportunity, human beings will engage in vile behavior toward others (Zimbardo, 2007).

A French journalist uncovered audiotapes that were made during the study and interviewed some of the men who had participated all those years ago (Blum, 2018; Le Texier, 2018). What he found was quite different from the story typically shared in psychology courses. Perhaps most importantly, Zimbardo failed to emphasize that prior to the start of the experiment, he and his assistants had strongly encouraged the guards to play a "tough guard" and use fear to control the prisoners (Blum, 2018). If guards were too mild, a research assistant pressed them to be more aggressive. One notoriously mean guard said that he was just trying to please the experimenters and had modeled his behavior after a character in the classic film *Cool Hand Luke* (Blum, 2018).

The typical interpretation of the Stanford Prison Experiment rests on Zimbardo's assertion that guards and prisoners behaved as they did naturally,

Zimbardo recruited college students for the prison study in an ad. Notice it mentioned "prison life." Might that have influenced who applied?
PrisonExp.org

with no interference from him or the students helping with the study. We now know that this is not true. In fact, an attempt to replicate the study showed that without the encouragement given to guards to be tough, it is unlikely any would have behaved aggressively (Haslam & Reicher, 2003).

A serious ethical issue with the study was that numerous prisoners *did* ask to leave the study but were told they could not. In fact, the prisoner who had the mental breakdown said he had faked his distress to get out of the study. He had assumed that he would be able to study while making money in the mock jail (Blum, 2018).

The enduring impact of the Stanford Prison Experiment in psychology is a bit of a puzzle. It was not, after all, much of an experiment and the results were not published in a peer-reviewed journal. Leon Festinger himself called it not science but rather "a happening" (Blum, 2018). A study purporting to show the strong effect of roles on social behavior has transformed into a difficult happening indeed, the meaning of which is now entwined in complex issues of memory, truth, and notoriety.

What Do You Think?

- Why do you think people placed so much stock in the report of the Stanford Prison Experiment?
- Zimbardo and the former participants disagree strongly about what took place during the study. Who do you believe and why?

Group Influence

Scenes of mobs rioting are all too common in the media. A team wins a championship, and fans who would never otherwise break the law can be seen setting fire to cars and looting businesses. People at a political rally who might otherwise never even raise their voices to someone else are captured on video punching people who favor a different candidate. Teammates, fraternity and sorority members, bandmates, and members of various clubs harm and even kill new members during hazing rituals. Make no mistake, the perpetrators in these instances are often people with no history of violence who genuinely have no hostility toward the person harmed. None of them, acting alone, would have performed destructive, even murderous, acts. Why does being in a group lead to such behavior? This central question has driven research in the social psychology of group influence.

The Ku Klux Klan demonstrates a variety of ways that human beings can deindividuate: turning out in groups, acting under cover of darkness, and wearing white hoods to conceal identity.
Greg Mathieson/REX/Shutterstock

DEINDIVIDUATION

deindividuation
The reduction in personal identity and erosion of the sense of personal responsibility when one is part of a group.

One process that sheds light on the behavior of individuals in groups is **deindividuation,** which occurs when being part of a group reduces personal identity and erodes the sense of personal responsibility (Levine & others, 2010). The effects of deindividuation are visible in the wild street revelry that erupts after a team's victory in the World Series or Super Bowl, as well as in mass civic observances such as big-city Fourth of July celebrations. Deindividuation is apparent not just in the behavior of mobs but anytime people feel like they cannot be identified. One explanation for the effects of deindividuation is that groups give us anonymity. When we are part of a group, we may act in an uninhibited way because we believe that no one will be able to identify us. Of course the anonymity provided by social media, online message boards, and comment sections provides a prime context to observe the effects of deindividuation on human behavior (Coles & West, 2016; Mikal & others, 2016).

One particularly horrific example of group influence is when multiple perpetrators rape a woman. Psychologists have probed whether deindividuation plays a role in these violent acts. A recent analysis of 71 victim reports (involving 189 suspects) examined the factors that contribute to the level of violence perpetrated by these groups (Woodhams & others, 2020). The results provided little evidence that deindividuation plays a role in such incidents. Instead it appears that leaders in the groups have a strong influence in pressing for aggression and that group members may mutually reinforce aggressive behavior. Consider that, even if you are in a group and feel anonymous, the actions of one person can make a difference, for better or worse.

> In modern life, we have the strange irony of people acting in mobs and clearly feeling anonymous but then posting about their exploits on social media.

SOCIAL CONTAGION

social contagion
Imitative behavior involving the spread of behavior, emotions, and ideas.

Have you ever noticed that a movie you watched in a crowded theater seemed funnier than it did when you streamed it alone at home? People laugh more when others are laughing. Babies cry when other babies are crying. The effects of others on our behavior can take the form of **social contagion**, imitative behavior involving the spread of behavior, emotions, and ideas (Choi & others, 2017; Cooper & others, 2014; Hodas & Lerman, 2014). Social contagion effects can be observed in varied phenomena such as social fads, the popularity of dog breeds (Herzog, 2006), the spread of bullying (Stubbs-Richardson & May, 2020), and academic engagement (Mendoza & King, 2020) and dropping out of school (Dupéré & others, 2020) among adolescents.

Want to observe social contagion? Sit in a quiet but crowded library and start coughing. You will soon notice others coughing. Similarly, imagine that you are walking down the sidewalk, and you come upon a group of people who are all looking up. How likely is it that you can avoid the temptation of looking up to see what is so interesting to them?

GROUP PERFORMANCE

Are two or three heads better than one? Some studies reveal that we do better in groups; others show that we are more productive when we work alone (Paulus, 1989). We can make sense out of these contradictory findings by looking closely at the circumstances in which performance is being analyzed.

Social Facilitation

If you have ever given a presentation in a class, you might have noticed that you did a much better job standing in front of your classmates than during

social facilitation
Improvement in an individual's performance because of the presence of others.

any of your practice runs. **Social facilitation** occurs when an individual's performance improves because of the presence of others (Herman, 2017; Mendes, 2007). Robert Zajonc (1965) argued that the presence of other individuals arouses us. The arousal produces energy and facilitates our performance in groups. If our arousal is too high, however, we are unable to learn new or difficult tasks efficiently. Social facilitation, then, improves our performance on well-learned tasks. For new or difficult tasks, we might be best advised to work things out on our own before trying them in a group.

Social Loafing Another factor in group performance is the degree to which one's behavior is monitored. **Social loafing** refers to each person's tendency to exert less effort in a group because of reduced accountability for individual effort. The effect of social loafing is lowered group performance (Latané, 1981). The larger the group, the more likely it is that an individual can loaf without detection. Social loafing is less common when group leaders are fair (De Backer & others, 2015).

social loafing
Each person's tendency to exert less effort in a group because of reduced accountability for individual effort.

GROUP DECISION MAKING Many social decisions take place in groups—juries, teams, families, clubs, school boards, and the U.S. Senate, for example. What happens when people put their minds to the task of making a group decision? How do they decide whether a criminal is guilty, whether one country should attack another country, whether a family should stay home or go on vacation, or whether sex education should be part of a school curriculum? Three aspects of group decision making bear special mention: risky shift and group polarization, groupthink, and majority and minority influence.

Risky Shift and Group Polarization Imagine that you have a friend, Lisa, who works as an accountant. All her life Lisa has longed to be a writer. In fact, she believes that she has the next great American novel in her head. She just needs time and energy to devote to writing it. Would you advise Lisa to quit her job and go for it? What if you knew beforehand that her chances of success were 50-50? How about 60-40? How much risk would you advise her to take?

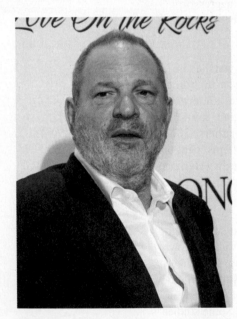

Researchers asked participants to consider fictitious dilemmas like this one and asked them how much risk the characters in the dilemmas should take (Stoner, 1961). When the participants discussed the dilemmas as a group, they were more willing to endorse riskier decisions than when they were queried alone. The so-called **risky shift** is the tendency for a group decision to be riskier than the average decision made by the individual group members (Goethals & Demorest, 1995).

risky shift
The tendency for a group decision to be riskier than the average decision made by the individual group members.

We do not always make riskier decisions in a group than when alone, however; hundreds of research studies show that being in a group moves us more strongly in the direction of the position we initially held (Moscovici, 1985). The **group polarization effect** is the solidification and further strengthening of an individual's position as a consequence of a group discussion or interaction. Initially held views often become more polarized because of group discussion. Group polarization may occur because, during the discussion, people hear new, more persuasive arguments that strengthen their original position. Group polarization also might arise because of social comparison. We may find that our opinion is not as extreme as others' opinions, and we might be influenced to take a stand at least as strong as the most extreme advocate's position.

group polarization effect
The solidification and further strengthening of an individual's position as a consequence of a group discussion or interaction.

Many people were aware that Harvey Weinstein was sexually harassing actresses, but, demonstrating groupthink, those people maintained their silence as part of his organization or the larger Hollywood community.
Arnold Jerocki/EPA/REX/Shutterstock

Groupthink **Groupthink** refers to the impaired group decision making that occurs when making the right decision is less important than maintaining group harmony. Instead of engaging in an open discussion of all the available information, in groupthink, members of a group place the highest value on conformity and unanimity. Members are encouraged to "get with the program," and dissent meets with very strong disapproval.

groupthink
The impaired group decision making that occurs when making the right decision is less important than maintaining group harmony.

Groupthink can result in disastrous decisions. Irving Janis (1972) introduced the concept of groupthink to explain a number of enormous decision-making errors throughout history. Such errors include the lack of U.S. preparation for the Japanese bombing of Pearl Harbor during World War II, the escalation of the Vietnam War in the 1960s, the Watergate coverup in 1974, and the *Challenger* space shuttle disaster in 1986. After the terrorist attacks of September 11, 2001, some suggested that groupthink had interfered with the proper implementation of intelligence. Whistleblower Colleen Rowley, a special agent in the Federal Bureau of Investigation (FBI), revealed that the FBI power hierarchy had been unresponsive to information that might have helped prevent the attacks. Unfortunately, whenever human groups work together to make decisions, groupthink can rear its head (Jaeger, 2020).

Symptoms of groupthink include overestimating the power and morality of one's group, close-mindedness and unwillingness to hear all sides of an argument, and pressure for uniformity (Post & Panis, 2011). Groupthink can occur whenever groups value conformity over accuracy.

Groupthink can be prevented if groups avoid isolation, allow the airing of all sides of an argument, have an impartial leader, include outside experts in the debate, and encourage members who are strongly identified with the group to speak out in dissent (Packer, 2009). When group norms welcome dissent, groupthink can be undermined (Van Bavel & others, 2020).

The decision in 2011 for the Navy SEALS to conduct a raid on Osama bin Laden's compound and to assassinate him involved an open discussion. Although most of President Obama's advisors hedged their bets, then Vice President Biden advised against the raid, while then CIA Director Leon Panetta explicitly recommended going in (Landler, 2012).

Majority and Minority Influence Most groups make decisions by voting, and, even in the absence of groupthink, the majority usually wins. The majority impacts group decision making through both informational influence (they have greater opportunity to share their views) and normative influence (they set group norms). Those who do not go along may be ignored or even given the boot.

In crowdsourcing, individuals provide ideas to solve problems, usually through Internet sites. This way of collecting ideas is thought to reduce the kinds of group biases that exist in face-to-face groups. Are online groups more or less likely to exhibit groupthink or minority influence?

Even so, minority opinion holders can make a difference. Because it is outnumbered, the minority cannot win through normative pressure; instead, it must apply informational pressure. If the minority presents its views consistently and confidently, then the majority is more likely to listen to the minority's perspectives. A powerful way that minority opinion holders can exert influence is by winning over former majority members to their points of view.

Social Identity

Think about the groups of which you are a member—religious and social organizations, your ethnic group, your nationality. When someone asks you to identify yourself, how often do you respond by mentioning these group memberships? How much does it matter to you whether the people you associate with are members of the same groups as you?

Social identity refers to the way people define themselves in terms of their group membership. In contrast to personal identity, which can be highly individualized, social identity assumes some commonalities with others. A person's social identity might include identifying with a religious group, a country, a social organization, a workplace, a political party, a sports team, among other groups. These diverse forms of social identity reflect the numerous ways people connect to groups and social categories. Social psychologist Kay Deaux (2001) identified five distinct types of social identity: ethnicity and religion, political affiliation, vocations and avocations, personal relationships, and stigmatized groups (Figure 6).

social identity
The way individuals define themselves in terms of their group membership.

Ethnic identity and religious identity are central to many individuals' social identity (Cooper & others, 2015; Vedder & Phinney, 2014). Ethnic identity can be a source of pride for people. In the United States, special events celebrate the rich cultural contributions of many different groups to society. Such experiences may provide individuals with an important resource in coping with biases they may encounter in life (Crocker & others, 1998). Feeling connected to one's ethnic group may buffer individuals from the stressful effects of injustice (Marks & others, 2015).

Ethnicity & Religion	Relationships	Vocations & Avocations	Political Affiliation	Stigmatized Identities
Jewish Asian American Southern Baptist West Indian	Parent Spouse Sibling Grandparent	Artist Athlete Psychologist Military veteran	Environmentalist Feminist Republican	Person with obesity Person with AIDS Person who is homeless Person with alcoholism

FIGURE 6 **Types of Social Identity** When we identify ourselves, we draw on a host of different characteristics associated with the various social groups to which we belong.
(Ethnicity): Radius Images/360/Getty Images; (Relationships): CHAINFOTO24/Shutterstock; (Vocation): Rob Melnychuk/Digital Vision/Getty Images; (Political): moodboard/Corbis; (Stigmatized): Stockbyte/Getty Images

Social psychologist Henry Tajfel (1978), a Holocaust survivor, wanted to explain the extreme violence and prejudice that his religious group (Jews) experienced. Tajfel's **social identity theory** states that social identity is a crucial part of self-image and a valuable source of positive feelings about oneself. To feel good about ourselves, we need to feel good about the groups to which we belong. For this reason, we invariably think of the group to which we belong as an *in-group*, a group that has special value in comparison with other groups, called *out-groups*. To improve our self-image, we continually compare our in-groups with out-groups (Joyce & Harwood, 2020). In the process, we often focus more on the differences between the two groups than on their similarities.

What groups define you?

social identity theory
The view that social identity is a crucial part of self-image and a valuable source of positive feelings about oneself.

Research by Tajfel (1978), along with that by many others who have used his theory, showed how easy it is to lead people to think in terms of "us" and "them." In one experiment, Tajfel had participants look at a screen featuring a huge number of dots and estimate how many dots were displayed. He then assigned the participants to groups based on whether they overestimated or underestimated the number of dots. Once assigned to one of the two groups, the participants were asked to award money to other participants. Invariably, individuals awarded money only to members of their own group. If we favor the members of a group that was formed on such trivial bases, it is no wonder that we show intense in-group favoritism when differences are not so trivial.

Such groups are referred to as **minimal groups** *because group assignment is arbitrary and meaningless.*

Research in developmental psychology suggests that in-group favoritism may be part of human nature, based on the ways that infants respond to seeing members of groups interact (Pun & others, 2017). For example, in a series of studies, 17-month-old infants expected characters they watched to be nicer to in-group members than to out-group members (Jin & Baillargeon, 2017). The infants looked longer when in-group members did not show favoritism to each other. Another series of studies showed that six- and eight-year-olds also showed in-group favoritism when it came to giving nice things to group members (Böhm & Buttelmann, 2017). In none of these studies did children appear to desire or expect characters to harm out-groups, however.

The tendency to favor one's own ethnic group over other groups is called **ethnocentrism**. Ethnocentrism does not simply mean taking pride in one's group; it also involves asserting the group's superiority over other groups. As such, ethnocentrism encourages in-group/out-group, we/they thinking (Dovidio & others, 2012). Consequently, ethnocentrism implies that ethnic out-groups are not simply different; they are worse than one's group. Hence, ethnocentrism may underlie prejudice.

As noted in the description of altruism, there may be a thin line between positive feelings and behaviors toward one's own group and hostile feelings and behaviors toward out-groups.

ethnocentrism
The tendency to favor one's own ethnic group over other groups.

prejudice
An unjustified negative attitude toward an individual based on the individual's membership in a group.

Prejudice

Prejudice is a negative attitude toward an individual based on the individual's membership in a particular group. The group can be made up of people of a specific ethnicity, gender, age, religion, gender identity—essentially, people who are different in some way from a prejudiced person. Prejudice as a worldwide phenomenon can be seen in many eruptions of hatred in human history. In the Balkan Peninsula of eastern Europe, the Serbs' prejudice against Bosnians prompted the Serb policy of "ethnic cleansing." The prejudice of the Hutus against the Tutsis in Rwanda led them to go on a murderous rampage, attacking the Tutsis with machetes. Certainly, current events in the United States serve to remind us that prejudice is still alive and well, including acts of violence against people of Asian and Asian/Pacific Island descent, during the COVID-19 pandemic.

Stephen Chung/LNP/REX/Shutterstock

A powerful example of destructive prejudice within the United States is anti-Black prejudice. When slave traders kidnapped African people from their homes, transported them to North America, and sold them to white people, those enslaved people were treated as property and denied the most basic considerations of humanity—as were their descendants. The horrific legacy of slavery has continued in different forms to the present day, including denial of basic rights, segregation, and unjust laws. Even today, skin color impacts the ability of many Americans to access fair treatment in the criminal justice system, education, employment, healthcare, and voting.

The legacy of slavery, segregation, and unjust laws is systemic racism. **Systemic racism** refers to systems, structures, and procedures in a society that disadvantage a racial group and privilege another. To get a sense of how systemic racism works, imagine a college with scholarships that are available to high-achieving children of alumni. This means that applicants who have family members who went to that institution have an opportunity that others do not. Now, Black people were denied entry into many predominantly white institutions of higher education for decades. This means that African American applicants are denied the chance for those scholarships because of a legacy of racism. Systemic racism is bias that is "baked in" to decision making by processes that may have been set in motion generations before (Liverpool, 2020).

systemic racism
Systems, structures, and procedures in a society that disadvantage a racial group and privilege another.

Talking openly about race and ethnicity can be awkward. Even coming up with labels to use to describe different groups of people can be difficult. In the United States, typical labels have changed from Black and white to African American and European American. But even these labels are not without problems. Many times labels are based not on information about the person's origins, but solely on the color of the person's skin. Nelson Mandela, for instance, was not African American. He was African. How do we distinguish between Mandela and F. W. de Klerk, the (white) former South African president who shared the Nobel Peace Prize with Mandela in 1993 for abolishing apartheid?

Many times, when we look at a questionnaire where a person has checked off their race, we rely simply on the person's own assessment of group membership. Former President Barack Obama embodies the complexity of race in America. His mother was European American, his father African, and he labels himself African American. In our discussion here, we will use the terms *Black* and *white* when in a particular study the actual ethnic background of targets of social judgment is not specified. In these instances, the only cue to the person's background, or anything else about that person, is skin color.

Research continues to demonstrate the influence of race in U.S. life. In one study, researchers sent out 5,000 résumés in response to 1,200 job ads placed in newspapers in Chicago and Boston. The résumés were identical in qualifications. They differed only in whether the names of the candidates were stereotypically white or Black. "White" names included Meredith, Emily, Brad, and Greg. "Black" names included Tamika, Lakisha, Darnell, and Kareem. Even with identical qualifications, the applicants with "white-sounding" names were 50 percent more likely to be called for an interview (Bertrand &

Award-winning social psychologist, Jennifer Eberhardt, has conducted important research on the subtle and not so subtle ways skin color can affect social judgments.

Nana Kofi Nti, Courtesy of Dr. Jennifer Eberhardt

Do It!

Want to participate in a social psychological experiment? You can do so online. Hundreds of studies are available at www.socialpsychology.org.

Mullainathan, 2004). Research has shown similar effects in teachers' evaluations of students' math problems (Copur-Gencturk & others, 2020), Uber and Lyft drivers canceling rides (Ge & others, 2020), and counselors' responses to prospective clients (Shin & others, 2016). Clearly, even subtle cues of a person's race can have profound impact (Eberhardt, 2020).

Results such as these are troubling. People who make these biased decisions may not even recognize their own racist behavior.

Because openly expressing racial prejudice is socially unacceptable, people may be unlikely to admit to racist or prejudicial views. So, people may deny being prejudiced while holding prejudicial views at a deeper level. In addition, people may not be consciously aware of their own racial (or gender or age) biases.

To confront this problem, social psychologists examine prejudicial attitudes on two levels—explicit racism and implicit racism. *Explicit racism* is a person's conscious and openly shared attitude, which might be measured using a questionnaire. *Implicit racism* refers to attitudes that exist on a deeper, hidden level (Willard & others, 2015). Explicit prejudice is thought to relate to behaviors that people control. Implicit prejudice is thought to predict more spontaneous, less self-conscious actions, such as facial expressions and body gestures. Implicit attitudes must be measured with a method that does not require awareness. For example, implicit racism is sometimes measured using the Implicit Associations Test (IAT), a computerized survey that assesses the ease with which a person can associate a Black or white person with good things (for example, flowers) or bad things (for example, misery) (Greenwald & others, 2009; Nosek & Banaji, 2007; Nosek & others, 2014). This test is based on the idea that preexisting biases may make it easier to associate some social stimuli with positive rather than negative items. Although the IAT is widely used, scholars have raised concerns about its validity (Blanton & others, 2015; Schimmack, 2019). Another concern is that if bias is labeled implicit, people are more likely to believe that it is out of their control and therefore they are not responsible for their prejudicial actions (Daumeyer & others, 2019). Implicit prejudice is a matter of personal responsibility and can be changed.

FACTORS IN PREJUDICE Why do people develop prejudice? Social psychologists have considered a number of answers to this question. Competition between groups, especially when resources are scarce, can contribute to prejudice. For example, competition between immigrants and established low-income members of a society for jobs can lead to prejudice. People may develop negative attitudes about other groups when they feel bad about themselves or their success in life. Or people may feel that their standing in society is threatened when other groups are given equal opportunities.

How does prejudice develop? Social and developmental psychologists have examined this issue (Costello & Hodson, 2014). Infants and children seem to have a sense of in-group bias. Such bias is likely one factor in prejudice. Another factor is parental attitudes. Parents who hold prejudicial attitudes are more likely to have children who express bias as well, especially if the children identify strongly with their parents (Sinclair & others, 2005). In addition, children who are exposed to information about genetic differences between races, for instance in biology textbooks, are more likely to believe that race differences in intelligence are genetic and are less likely to desire to be friends with children from other groups (Donovan, 2017). Research has also shown that children who encounter many different kinds of people in their lives are less likely to hold prejudicial attitudes (Raabe & Beelman, 2011).

If you want to try the IAT, go to: https://implicit.harvard.edu/implicit/takeatest.html. See what you think!

What are your attitudes toward individuals of an ethnic background different from your own? How do these attitudes compare with those of your parents?

A final factor that might underlie prejudice comes from the limits on our information-processing abilities. Human beings are limited in their capacity for effortful thought, but they face a complex social environment. To simplify the challenge of understanding others' behavior, people use categories or stereotypes. Stereotypes can be a powerful force in developing and maintaining prejudicial attitudes.

STEREOTYPES AND PREJUDICE

Recall that stereotypes are generalizations about a group that deny variations within the group. Researchers have found that we are less likely to detect variations among individuals who belong to "other" groups than among individuals who belong to "our" group. So, we might see the people in our in-group as varied, unique individuals while viewing the members of out-groups as all the same. Thinking that "they all look alike" can be a particular concern in the context of eyewitness identification (Brigham, 1986). At the root of prejudice is a particular kind of stereotype: a generalization about a group that contains negative information that is applied to all members of that group (Wetherell, 2012).

MICROAGGRESSIONS

Whether conscious and intentional or not, bias against people who are different from oneself can be expressed in behaviors called microaggressions. **Microaggression** refers to everyday, subtle, and potentially unintentional acts that communicate bias to members of marginalized groups (Sue 2010; Williams, 2020). Earlier in this chapter, we reviewed the example of African American doctors being mistaken for janitors. Other examples of microaggressions include giving compliments that are embedded in stereotypes. For example, expressing surprise that a woman is good at math, that a gay man likes sports, that a Black person is articulate, or that an Asian person speaks English well are all ways that people from diverse backgrounds can be reminded of the stereotypes for their groups. Despite the prefix *micro-*, these acts can be hurtful and create environments in which people of diverse backgrounds feel uncomfortable at work or in school (Williams, 2020).

microaggression
Everyday, subtle, and potentially unintentional acts that communicate bias to members of marginalized groups.

A key challenge in combating microaggressions is that perpetrators of these acts may be oblivious to the harm they cause and to the attributions that are made about them as a result of the actions (Sue & others, 2019). At the same time, it should not fall on the shoulders of people who are already experiencing the stress of prejudice to call attention to these experiences and explain why they are offensive. For this reason, it is important for people to take responsibility for their own education and become more sensitive to the ways that prejudice affects our everyday interactions.

DISCRIMINATION

When prejudicial attitudes are put into action by people in power, prejudicial behaviors are discriminatory. **Discrimination** refers to a negative or harmful action toward a member of a group simply because the person belongs to that group. Discrimination results when negative emotional reactions combine with prejudicial beliefs and are translated into behavior (Foels & Pratto, 2014). Many forms of discrimination are illegal in the U.S. workplace, including discrimination based on disability status. Since the Civil Rights Act of 1964 (revised in 1991), it has been unlawful to deny someone employment on the basis of gender or race/ethnicity (Parker, 2006).

discrimination
An unjustified negative or harmful action toward a member of a group simply because the person belongs to that group.

Noted social psychologist Jennifer Richeson studies ways to improve intergroup understanding. She has also done important work on the area of wealth-disparities between Black and white people in the United States.
Courtesy of Dr. Jennifer Richeson

Improving Intergroup Relations

The groups we belong to give us a sense of identity and self-esteem. Yet, at the same time, these feelings may lead to negative views of other groups. Certainly, conflicts among various groups are rampant in the modern world. Human beings have innumerable ways to draw lines dividing themselves based on national borders, religious faith, race, ethnicity, age, disability status, gender identity, and sexual orientation. How might we attain a world of peace, harmony and true equality? Social psychologists have strived to identity the factors that might lead to better relationships between different groups (Onyeador & others, 2020; Shelton & Richeson, 2014).

One way might be for people to come to know one another better so that they can get along. However, in daily life many people interact with individuals from other ethnic groups, and this contact does not necessarily lead to warm relations. Indeed, researchers have consistently found that contact by itself—attending the same school or working in the same company—does not necessarily improve relations among people of different ethnic backgrounds. So, rather than focusing on contact per se, researchers have examined how *various features* of a contact situation may be optimal for reducing prejudice and promoting intergroup harmony (Hässler & others, 2020; Stott & others, 2012; Turner & others, 2020). Intergroup contact can reduce prejudice even when it occurs online (White & others, 2020).

Studies have shown that contact is more effective if the people involved think that they are of equal status, feel that an authority figure sanctions their positive relationships, and believe that friendship might emerge from the interaction (Marinucci & others, 2020; Pettigrew & Tropp, 2006). Another important feature of optimal intergroup contact is *task-oriented cooperation*—working together on a shared goal. In Iraq, intergroup conflict has been an issue with Christian refugees coming to the country to escape persecution in their home countries. In a recent study, Christian and Islamic soccer players were assigned to play on teams together. The results showed that playing on the same soccer team reduced prejudice toward teammates of different religious faith (Mousa, 2020).

An example of the power of task-oriented cooperation is Muzafer Sherif's Robbers Cave Study. Even before Jeff Probst started handing out color-coded "buffs" on the TV show *Survivor,* Sherif and his colleagues (1961) had the idea of exploring group processes by assigning 11-year-old boys to two competitive groups (the "Rattlers" and the "Eagles") in a summer camp called Robbers Cave.

Sherif, disguised as a janitor so that he could unobtrusively observe the Rattlers and the Eagles, arranged for the two groups to compete in baseball, touch football, and tug-of-war. If you have watched reality television, you have some idea how this experiment went. In short order, relations between the groups got downright ugly. Members of each group expressed negative opinions of members of the other group, and the Rattlers and Eagles became battling factions. What would bring these clashing groups together? Sherif created tasks that required the joint efforts of both groups, such as working together to repair the camp's only water supply and pooling their money to rent a movie. When the groups were required to work cooperatively to solve problems, the Rattlers and Eagles developed more positive relationships. Figure 7 shows how competitive and cooperative activities changed perceptions of the out-group.

*Although legal segregation is no longer permitted, **voluntary** segregation continues. Look around your classrooms and dining halls. Do you see students separating themselves by ethnic categories? How does this convention impede intergroup relations, and how might it be broken?*

Yes, the Robbers Cave is yet another CLASSIC in social psychology!

PSYCHOLOGICAL INQUIRY

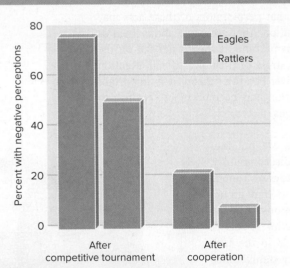

FIGURE 7 **Attitudes toward the Out-Group Following Competitive and Cooperative Activities** This graph shows the negative feelings expressed by members of the Eagles and Rattlers toward the other group after a competitive tournament and cooperative activity.

SOURCE: M. Sherif et al., The Robber's Cave Experiment: Intergroup Conflict and Cooperation (1988).

> When did hostility between the groups peak? When did it drop? > In your own life, what examples are there of holding particular attitudes toward a group different from your own? How have your attitudes changed, and what events preceded those changes? > What is up with the Eagles? What are some reasons why they might be more negative about the Rattlers?

BREAKING THE PREJUDICE HABIT   A variety of studies have been conducted testing different ways of reducing prejudice. Although many studies show positive results, most do not use long-term follow-ups to see if they continue to be effective (Paluck & others, 2021). There is no question that breaking the habit of prejudice is likely to be hard work. Given that much of prejudice in contemporary life is likely to be implicit—meaning that it happens automatically and can occur unconsciously—it is important to ask whether we can change not only explicit attitudes but implicit ones as well. Implicit prejudice rests on automatic associations: You see a person from a different ethnic, racial, gender identity, or religious group, and stereotypical thoughts pop into your head. In this sense, prejudice is a habitual way of thinking.

Can we short-circuit these automatic connections and reduce implicit prejudice? A study by Patricia Devine and her colleagues (2012) suggests that we can. White American college students were randomly assigned to an experimental group that received an intervention to reduce implicit prejudice or a control group that simply completed the dependent measures. Based on the idea that to change racial attitudes people must be aware of them, the first step was to measure implicit racial prejudice using the IAT and give the participants their scores. These scores can be surprising to many people who feel they are not prejudiced against people of color. Next, participants in the experimental group were trained in various ways to break the prejudice habit. This training to introduce new habits of the mind included noticing stereotypic thoughts about Black people and replacing them with more individualized information about the person. Participants were also trained to think about people who do not fit the stereotype of various outgroups, and they were encouraged to put themselves in the other's position and consider things from that person's perspective. Finally, participants were urged to seek out contact with members of the outgroup. The results showed that, six weeks later, participants who had experienced this intervention did show reductions in implicit bias against Black people. In addition, these individuals were more likely to express concerns about the problems of discrimination in U.S. society. These results, and those of contact interventions, suggest that even though the problem of racial tension in the United States and across the world may seem insurmountable, it is possible to improve relations among groups.

self-quiz

1. All of the following are related to deindividuation *except*
   A. doing something as part of a large group.
   B. hearing someone explicitly call your name and express recognition of you.
   C. losing your sense of personal responsibility while taking part in a group activity.
   D. wearing a disguise while taking part in a group activity.

2. Which of the following is the true statement about the difference between conformity and obedience?
   A. Conformity has a stronger influence on behavior than obedience.
   B. Conformity does not involve an explicit command from others.
   C. Conformity happens in small groups, whereas obedience happens in large groups.

   D. Conformity is based on wanting to be right; obedience is based on wanting to be liked.

3. A white woman says she is not racist but avoids sitting near Black individuals. She has
   A. high explicit racism, low implicit racism.
   B. high explicit racism, high implicit racism.
   C. low explicit racism, low implicit racism.
   D. low explicit racism, high implicit racism.

APPLY IT!   4. Will is bothered by the way students segregate themselves in the campus commons, with all the Black students generally eating in one area, the Latinos in another, the Asians in another, and the non-Latino whites in another. Bill starts the Students Together

program to get students to interact more and to promote interethnic harmony. Which of the following strategies is most likely to work, based on the social psychological evidence?
   A. Assign seats in the lunchroom so that students have to sit with members of other ethnic groups.
   B. Identify problems of common interest to students of all groups and initiate ethnically diverse discussion groups aimed at solving those problems.
   C. Post signs in the commons that encourage interethnic contact.
   D. None of the above, because nothing will work.

Defining Social Psychology

Social psychology is the scientific study of how people think about, influence, and relate to other people. This subfield of psychology is relevant to everyday life, it is broad, and it typically relies on experimental methods. An example of social psychology in action is provided by classic research on the bystander effect: the tendency for an individual who observes an emergency to help less when other people are present than when the observer is alone.

Social Cognition

Social cognition refers to the ways we process social information. The face conveys information to social perceivers, including attractiveness. Self-fulfilling prophecy means that our expectations of others can have a powerful impact on their behavior. Attributions are our thoughts about why people behave as they do and about who or what is responsible for the outcome of events. Attribution theory views people as motivated to discover the causes of behavior as part of their effort to make sense of it. The dimensions used to make sense of the causes of human behavior include internal/external, stable/unstable, and controllable/uncontrollable.

The fundamental attribution error is observers' tendency to overestimate traits and to underestimate situations when they explain an actor's behavior. Self-serving bias means attributing our successes to internal causes and blaming our failures on external causes. Heuristics are used as shortcuts in social information processing. One such heuristic is a stereotype—a generalization about a group's characteristics that does not consider any variations among individuals in the group.

The self is our mental representation of our own characteristics. Self-esteem is important and is related to holding unrealistically positive views of ourselves. Stereotype threat is an individual's fast-acting, self-fulfilling fear of being judged based on a negative stereotype about their group. To understand ourselves better, we might engage in social comparison, evaluating ourselves by comparison with others.

Attitudes are our feelings about people, objects, and ideas. We are better able to predict behavior on the basis of attitudes when an individual's attitudes are strong, when the person is aware of their attitudes and expresses them often, and when the attitudes are specifically relevant to the behavior. Sometimes changes in behavior precede changes in attitude.

According to cognitive dissonance theory, our strong need for cognitive consistency causes us to change our behavior to fit our attitudes or to change our attitudes to fit our behavior. Self-perception theory stresses the importance of making inferences about attitudes by observing our own behavior, especially when our attitudes are not clear.

Social Behavior

Altruism is an unselfish interest in helping someone else. Reciprocity often is involved in altruism. Individuals who are in a good mood are more helpful. Empathy is also linked to helping.

Women are more likely to help in situations that are not dangerous and involve caregiving. Men are more likely to help in situations that involve danger or in which they feel competent.

One view of the biological basis of aggression is that early in human evolution, the most aggressive individuals were likely to be the survivors.

Neurobiological factors involved in aggressive behavior include the neurotransmitter serotonin and the hormone testosterone. Psychological factors in aggression include frustrating and aversive circumstances. Sociocultural factors include cross-cultural variations, the culture of honor, and violence in the media. The question of whether men or women are more aggressive depends on the particular *type* of aggression that is involved.

Close Relationships

We tend to be attracted to people whom we see often, whom we are likely to meet, and who are similar to us. Romantic love (passionate love) includes feelings of infatuation and sexual attraction. Affectionate love (companionate love) is more akin to friendship and includes deep, caring feelings for another.

Social exchange theory states that a relationship is likely to be successful if individuals feel that they get out of the relationship what they put in. The investment model focuses on commitment, investment, and the availability of attractive alternatives in predicting relationship success.

Social Influence and Group Processes

Conformity involves a change in behavior to coincide with a group standard. Factors that influence conformity include informational social influence (going along to be right) and normative social influence (going along to be liked).

Obedience is behavior that complies with the explicit demands of an authority. Stanley Milgram conducted landmark research on obedience, demonstrating its power. Another such demonstration is the Stanford Prison Experiment, illustrating the potential effects of obedience not only on individuals who obey but also on those who exercise authority.

People often change their behaviors when they are in a group. Deindividuation refers to the lack of inhibition and diffusion of responsibility that can occur in groups. Social contagion refers to imitative behaviors involving the spread of behavior, emotions, and ideas. Our performance in groups can be improved through social facilitation and lowered because of social loafing.

Risky shift refers to the tendency for a group decision to be riskier than the average decision made by the individual group members. The group polarization effect is the solidification and further strengthening of a position as a consequence of group discussion or interaction. Groupthink involves impaired decision making and avoidance of realistic appraisal to maintain harmony in the group.

Social identity is the way individuals define themselves in terms of their group membership. Social identity theory states that when individuals are assigned to a group, they invariably think of it as the in-group. Identifying with the group allows the person to have a positive self-image. Ethnocentrism is the tendency to favor one's own ethnic group over others.

Prejudice is an unjustified negative attitude toward an individual based on membership in a group. Prejudice can be based on ethnicity, sex, age, religion, sexual orientation, or any characteristic that makes people different from one another.

Prejudice is also based on stereotypes. The cognitive process of stereotyping can lead to discrimination, an unjustified negative or harmful action toward a member of a group simply because they belong to that group. Microaggressions are every day, subtle acts of bias. Discrimination results when negative emotional reactions combine with prejudicial beliefs and are translated into behavior. A key outcome of prejudice and discrimination is health disparities.

An effective strategy for enhancing the effects of intergroup contact is to set up task-oriented cooperation among individuals from different groups. Implicit biases can be targeted by bringing mindful attention to one's thoughts when interacting with people from different backgrounds.

KEY TERMS

affectionate love or companionate love
aggression
altruism
attitudes
attribution theory
bystander effect
cognitive dissonance
conformity
deindividuation
discrimination
egoism

elaboration likelihood model
empathy
ethnocentrism
false consensus effect
fundamental attribution error
group polarization effect
groupthink
informational social influence
investment model
mere exposure effect
microaggression
normative social influence

obedience
overt aggression
person perception
positive illusions
prejudice
relational aggression
risky shift
romantic love or passionate love
self-fulfilling prophecy
self-perception theory
self-serving bias
social cognition

social comparison
social contagion
social exchange theory
social facilitation
social identity
social identity theory
social loafing
social psychology
stereotype
stereotype threat
systemic racism

ANSWERS TO SELF-QUIZZES

Section 1:   1. B; 2. A; 3. B; 4. C
Section 2:   1. D; 2. A; 3. C; 4. B

Section 3:   1. D; 2. A; 3. A; 4. C
Section 4:   1. C; 2. C; 3. D; 4. A

Section 5:   1. B; 2. B; 3. D; 4. B

Design elements: (Preview icon): Jiang Hongyan/Shutterstock; (Marginal notes): Shutterstock/Vadarshop

12 Psychological Disorders

"You are not alone."

At 3 A.M. one morning in 2017, Zachary Burton, a geology graduate student at Stanford University, found himself at the top of a campus parking garage. He had run there, believing that he was being chased by evil people. He called his mother who was able to calm him and contact his girlfriend and roommate who took him to a hospital. He spent 11 days there and was diagnosed with bipolar disorder (Porter, 2019). Zachary had experienced a psychotic break, meaning that his thoughts were divorced from reality. After he was released, Zachary was overwhelmed with the experience of stigma—of feeling ashamed to have others know what he was going through. Eventually he went public with his struggle in a big way. He and his girlfriend collected stories from people, including a Stanford professor, about the experience of struggling with mental illness and turned them into a play. For Zachary, the play was his way of telling those suffering with psychological problems, "You are not alone."

Experiencing a psychological disorder in college is not unusual. Students who experience psychological problems often must juggle the many challenges of university life with the added stress of such disorders (Duffy & others, 2019).

People with psychological disorders are not so different from everyone else. They are siblings, grandparents, parents, aunts, uncles, classmates, coworkers, and friends: People with stories to share. Psychological disorders may make their lives more difficult and our relationships with them more challenging, but psychological disorders cannot take away the humanity that binds us all to each other.

John Giustina/Getty Images

This chapter explores the meaning of the word *abnormal* as it relates to psychology. We examine various theoretical approaches to understanding abnormal behavior and survey the main psychological disorders. We then briefly review suicide. Finally, we delve into how stigma affects the lives of people with psychological disorders, and we consider how even difficult, troubled lives remain valuable and meaningful.

1 Defining and Explaining Abnormal Behavior

What makes behavior "abnormal"? The American Psychiatric Association (2001, 2006, 2013) defines abnormal behavior in medical terms: a mental illness that affects or is manifested in a person's brain and can affect the way the person thinks, behaves, and interacts with others. Abnormal behavior may also be defined by three criteria that distinguish it from normal or typical behavior: **Abnormal behavior** is deviant, maladaptive, or personally distressful over a relatively long period of time. Only one of these criteria needs to be present for a behavior to be labeled "abnormal," but typically two or all three are present.

abnormal behavior
Behavior that is deviant, maladaptive, or personally distressful over a relatively long period of time.

Three Criteria of Abnormal Behavior

Let's take a close look at what each of the three characteristics of abnormal behavior entails:

1. Abnormal behavior is *deviant.* Abnormal behavior is atypical or statistically unusual. However, Shonda Rhimes, Patrick Mahomes, and Bill Gates are atypical in many of their behaviors, and yet we do not categorize them as abnormal. We do often consider atypical behavior abnormal, though, when it deviates from what is acceptable in a culture. A woman who washes her hands three or four times an hour and takes seven showers a day is abnormal because her behavior deviates from cultural norms.

Context matters! If the woman who washes her hands three or four times an hour and takes repeated showers works in a sterile lab with toxic chemicals or live viruses, her behavior might be quite adaptive.

Accomplished individuals such as writer/director/actor Lin Manuel Miranda, tennis champion Naomi Osaka, and NFL quarterback Patrick Mahomes are atypical but not abnormal. However, when atypical behavior deviates from cultural norms, it often is considered abnormal.
(Left): Eduardo Munoz/Reuters/Alamy Stock Photo; (Middle): Thomas Peter/Reuters/Alamy Stock Photo; (Right): Shannon Stapleton/Reuters/Alamy Stock Photo

2. Abnormal behavior is *maladaptive.* In psychology, the term *maladaptive* means that a behavior or characteristic disrupts the person's ability to lead a satisfying life, interfering with their ability to function effectively in the world. Someone who believes that they can endanger others through their breathing may go to great lengths to isolate themselves from people for what they believe is the other person's own good. This belief negatively affects a person's everyday functioning; thus, the behavior is maladaptive. Behavior that presents a danger to the person or those around that person is also considered maladaptive (and abnormal).

3. Abnormal behavior is *personally distressful* over a long period of time. The person engaging in the behavior finds it troubling. A person who secretly makes themselves vomit after every meal may never be seen by others as deviant (because they do not know about it), but this pattern of behavior may cause intense shame, guilt, and despair. People may not realize they have a psychological disorder but still feel extremely distressed by the effects of the disorder on their life.

*Which of these three qualities—deviation from what is acceptable, maladaptiveness, and personal distress—do you think is **most important** to calling a behavior abnormal? Why?*

Culture, Context, and the Meaning of Abnormal Behavior

Consider, for instance, that a symptom of one of Sigmund Freud's most famous patients, Anna O., was that she was not interested in getting married.

Because culture establishes the norms by which people evaluate their own and others' behaviors, culture is at the core of what it means to be normal or abnormal (Stevenson 2020). In evaluating whether behavior is abnormal, culture matters in complex ways. Cultural norms provide guidance about how people should behave and what behavior is healthy or unhealthy. Importantly, however, cultural norms can be mistaken. Consider that smoking cigarettes was once not only judged to be an acceptable habit but also promoted as a healthy way to relax. The point is, definitions of *normal* change as society changes.

Significant, too, is the fact that cultural norms can be limiting, oppressive, and prejudicial (Potter, 2012). People who fight to change the established social order sometimes face the possibility of being labeled deviant—and even mentally ill. In the late nineteenth and early twentieth centuries, for instance, women in Britain who demonstrated for women's right to vote were widely viewed to be mentally ill. When a person's or a group's behavior challenges social expectations, we must open our minds to the possibility that such actions are in fact an adaptive response to injustice. People may justifiably challenge what everyone thinks is true and may express ideas that seem strange. They should be able to make others feel uncomfortable without being labeled abnormal.

Do It!

Spend 15 to 20 minutes observing an area with a large number of people, such as a mall, a cafeteria, or a stadium during a game. Identify and make a list of behaviors you would classify as abnormal. How does your list of behaviors compare with the definition of abnormal provided earlier? What would you change in the list if you were in a different setting, such as a church, a bar, or a library? What does this exercise tell you about the meaning of abnormal?

Further, as people move from one culture to another, interpretations and evaluations of their behavior must take into account the norms in their culture of origin. Historically, people entering the United States from other countries were examined at Ellis Island, and many were judged to be mentally or intellectually impaired simply because of differences in their language and customs.

Cultural variation in what it means to be normal or abnormal makes it very difficult to compare different psychological disorders across different cultures (Alallawi & others, 2020; Carney, 2020). Many of the diagnostic categories we trace in this chapter primarily reflect Western (and often U.S.) notions of normality, and applying these to other cultures can be misleading and even inappropriate. Throughout this chapter, we will see how culture influences the experience of psychological disorders.

Theoretical Approaches to Psychological Disorders

What causes people to develop a psychological disorder, that is, to behave in deviant, maladaptive, and personally distressful ways? Theorists have suggested various approaches to this question.

THE BIOLOGICAL APPROACH   The *biological approach* attributes psychological disorders to organic, internal causes. This perspective primarily focuses on the brain, genetic factors, and neurotransmitter functioning as the sources of abnormality.

The biological approach is evident in the **medical model**, which describes psychological disorders as medical diseases with a biological origin. From the perspective of the medical model, abnormalities are called "mental illnesses," the afflicted individuals are "patients," and they are treated by medical doctors. This perspective generally neglects contextual factors, such as the person's social relationships, that may play a role in psychological disorders.

medical model
The view that psychological disorders are medical diseases with a biological origin.

THE PSYCHOLOGICAL APPROACH   The *psychological approach* emphasizes the contributions of experiences, thoughts, emotions, and personality characteristics in explaining psychological disorders. Psychologists might focus, for example, on the influence of childhood experiences, personality traits, learning experiences, or cognitions in the development and course of psychological disorders.

THE SOCIOCULTURAL APPROACH   The *sociocultural approach* emphasizes the social contexts in which a person lives and charactertistics, including gender, ethnicity, socioeconomic status, family relationships, and culture. For instance, poverty is related to rates of psychological disorders (Byrne & James, 2020; Nurius & others, 2020). The sociocultural perspective places less emphasis on biological processes such as genetics or brain structures and functions, and more emphasis on how a person with these biological characteristics exists in the larger context and adapts to the world.

The sociocultural perspective stresses the ways that culture influences the understanding and treatment of psychological disorders. The frequency and intensity of psychological disorders vary and depend on social, economic, technological, and religious aspects of cultures. Some disorders are culture related, such as *windigo,* a disorder recognized by northern Algonquian Native American groups that involves fear of being bewitched and turned into a cannibal.

Importantly, different cultures may interpret the same pattern of behaviors in different ways. When psychologists look for evidence of the occurrence of a particular disorder in different cultures, they must keep in mind that behaviors considered symptoms of a disorder might not be labeled as illness or dysfunction within a different cultural context. Cultures might have their own interpretations of these behaviors, so researchers must probe whether locals ever observe these patterns of behavior, even if they are not considered illness (Draguns & Tanaka-Matsumi, 2003). For example, in one study researchers interviewed a variety of individuals in Uganda to see whether dissociative disorders, including dissociative identity disorder (which you might know as multiple personality disorder), existed in that culture (van Duijl & others, 2005, 2014). They found that while most dissociative disorders were recognizable to Ugandans, the local healers labeled what Westerners consider dissociative identity disorder as spirit possession.

THE BIOPSYCHOSOCIAL MODEL   Abnormal behavior can be influenced by biological factors (such as genes or neurotransmitters), psychological factors (such as childhood experiences), and sociocultural factors (such as poverty). These factors can operate alone, but they often act in combination with one another.

The perspective that takes the combination of factors into account is called *biopsychosocial.* From the biopsychosocial perspective, biological, psychological, and social factors are all significant ingredients in producing both normal and abnormal behavior. Furthermore, these ingredients may combine in unique ways, so that two people with depression might differ from one another in terms of the key factors associated with the development of the disorder.

Why do we need to consider these multiple factors? As you will see throughout this chapter, generally speaking, there is no one gene or experience that leads inevitably to the development of a psychological disorder. Two people can share the same gene—one develops a disorder but another does not. Similarly, two people might have the same experience, such as childhood neglect, and one might develop a disorder but the other does not. Thus, to understand the development of psychological disorders, we must consider a variety of *interacting* factors from each of the domains of experience.

An important concept that has helped psychologists understand the ways different factors influence the development of psychological disorders is the **vulnerability-stress hypothesis** *(also called the* **diathesis-stress model***)*. The vulnerability-stress hypothesis suggests that preexisting conditions (such as genetic characteristics, personality dispositions, or experiences) may put a person at risk of developing a psychological disorder. This vulnerability in combination with stressful experiences can lead to a psychological disorder. The important bottom line is that the vulnerability-stress hypothesis acknowledges that neither predispositions nor experiences lead directly to the development of a psychological disorder. Rather, they may combine to enhance a person's likelihood of developing a disorder.

Classifying Abnormal Behavior

To understand, treat, and prevent psychological disorders, psychiatrists and psychologists have devised systems classifying abnormal behaviors into specific diagnoses. Classifying psychological disorders provides a common basis for communicating. If one psychologist says that a client is experiencing depression, another psychologist understands that a particular pattern of abnormal behavior has led to this diagnosis. A classification system can also help clinicians predict how likely it is that a particular disorder will occur, who will be most susceptible to it, how the disorder progresses, and what the prognosis (or outcome) for treatment is (Barroilhet & others, 2020; de Pablo & others, 2020).

Further, a classification system may benefit the person suffering from psychological symptoms. Having a name for a problem can be a comfort and a signal that treatments are available. On the other hand, officially naming a problem can also have serious negative implications for the person because of the potential for creating *stigma,* a mark of shame that may cause others to avoid or to act negatively toward an individual. We discuss stigma further at the end of this chapter.

THE *DSM* CLASSIFICATION SYSTEM   In 1952, the American Psychiatric Association (APA) published the first major classification of psychological disorders in the United States, the *Diagnostic and Statistical Manual of Mental Disorders.* Its current version, *DSM-5*, was published in 2013. That edition of the *DSM* was the product of a 14-year revision process. *DSM-5* differs in many ways from its predecessors. Throughout the history of the *DSM,* the number of disorders listed has increased dramatically. For example, *DSM-5* includes new diagnoses such as gambling addiction.

The *DSM* is not the only diagnostic system. The World Health Organization devised the *International Classification of Diseases and Related Problems (ICD-10),* which includes a chapter on mental and behavioral disorders. A goal of the authors of *DSM-5* was to bring diagnoses closer to the *ICD-10,* although the two manuals remain different in important ways.

CRITIQUES OF THE *DSM*   *DSM-5* has been criticized on a number of grounds. A central criticism that applies to all versions of the *DSM* is that it treats psychological disorders as if they are medical illnesses, taking an overly biological view of disorders that may have their roots in social experience (Blashfield & others, 2014). Even as research has shed light on the complex interaction of genetic, neurobiological, cognitive, social, and environmental factors in psychological disorders, *DSM-5* continues to reflect the medical model, neglecting factors such as poverty, unemployment, and trauma. Another general criticism of the *DSM* is that it focuses strictly on problems. Critics argue that emphasizing *strengths* as well as weaknesses might help to destigmatize psychological disorders. Identifying a person's strengths can be an important step toward maximizing their ability to contribute to society (Robbins & others, 2017; Roten, 2007).

Other criticisms of *DSM-5* include the following:

- It relies too much on social norms and subjective judgments.

- Too many new categories of disorders have been added, some of which do not yet have consistent research support and whose inclusion will lead to a significant increase in the number of people being labeled as having a psychological disorder.

- Loosened standards for some existing diagnoses will add to the already very high rates of these.

Studio Works/Alamy Stock Photo

somatic symptom and related disorders
Psychological disorders characterized by bodily symptoms that either are very distressing or interfere with a person's functioning along with excessive thoughts, feeling, and behaviors about the symptoms.

An example of a controversial change in *DSM-5* is the diagnosis of somatic symptom and related disorders (Fabiano & Haslam, 2020). **Somatic symptom and related disorders** are defined as bodily symptoms that either are very distressing or interfere with a person's functioning along with excessive thoughts, feelings, and behaviors about those symptoms (Rief & Martin, 2014). In previous versions of the *DSM*, for these symptoms to be classified as a psychological disorder, it was required that the physical symptoms lack a medical explanation. Now, it is possible for a person to be diagnosed with somatic symptom disorder even if the person has an actual physical illness.

Previously, these symptoms were more like hysterical symptoms: physical problems with no physical cause.

Final Terms and Cautions

Before we begin reviewing various psychological diagnoses, let's clarify common terms and issues. First, keep in mind that many people with psychological disorders have more than one diagnosis, simultaneously. For example, an adult with a diagnosis of depression may also have an anxiety disorder. The term **comorbidity** means that person has two or more disorders at the same time.

Throughout our review, we will be considering the etiology of the disorders. *Etiology* refers to the causes of disorders. For most disorders there is not one known cause. Instead, researchers have identified risk factors that contribute to a person developing a disorder. **Risk factors** are characteristics, experiences, or exposures that increase the likelihood of a person developing a disorder. Risk factors are correlated with the development of a disorder. Of course, correlation does not imply causation. Many people with a particular risk factor may never develop a disorder. One risk factor that we will consider is gender. Be aware that much of the research in this area has focused on gender as a binary construct and comparisons between women and men (and girls and boys) reflect that limited approach to gender.

As we review the disorders, we will also briefly touch on common treatments. These treatments, our focus in Chapter 13, are generally divided into two types, psychotherapy and medication. **Psychotherapy** is a nonmedical process that helps individuals with psychological disorders recognize and overcome their problems. Psychotherapies can involve a variety of activities, including talking, providing support, coaching, and specific procedures to relieve maladaptive behavioral patterns. Medications are drugs that are used to treat the symptoms of psychological disorders. These often involve addressing dysregulation of neurotransmitters in the brain.

A final word of caution: It is very common for people who are learning about psychological disorders to recognize the symptoms and behaviors of disorders in themselves or in people around them. Keep in mind that only trained professionals can diagnose a psychological disorder.

comorbidity
The simultaneous presence of two or more disorders in one person. The conditions are referred to as *"comorbid."*

risk factor
Characteristics, experiences, or exposures that increase the likelihood that a person will develop a psychological disorder.

psychotherapy
A nonmedical process that helps individuals with psychological disorders recognize and overcome their problems.

2 Neurodevelopmental Disorders

Neurodevelopmental disorders are diagnosed in childhood (American Psychiatric Association, 2013) and are traced to genetic differences, atypical brain development, or prenatal exposure to substances that adversely affect development. This class of disorders includes communication disorders (such as language delays), learning disabilities or differences, and intellectual disability (for example, Down syndrome). In this section, we review two neurodevelopmental disorders in depth—autism spectrum disorder and attention-deficit/hyperactivity disorder.

Not all disorders diagnosed in childhood are considered "neurodevelopmental" disorders. For instance, children can be diagnosed with depression or anxiety disorders and these are not generally considered neurodevelopmental.

Autism Spectrum Disorder

Autism spectrum disorder (ASD) is characterized by two key features:

1. People with ASD show persistent deficits in social communication and social interaction across a variety of settings.

2. People with ASD show restrictive repetitive behaviors, interests, and activities.

UrsaHoogle/Getty Images

Typically diagnosed in childhood, the symptoms of ASD are likely to color the person's behavior throughout life (Shulman & others, 2020). For instance, infants may prefer not to be held or cuddled. Children may not develop the ability to speak. Adults may have persistent problems communicating with others. Repetitive behaviors in childhood may include hand flapping and head banging. Children with ASD might become preoccupied by particular objects or activities, like watching a ceiling fan or repeatedly opening and closing a door. Later in life, restrictive interests might involve an enthusiastic or obsessive interest in a topic, such as trains or American history.

Estimates of the number of children with ASD in the United States have increased over the years. The most recent estimate is that 1 in 54 eight-year-olds are affected by the disorder (Maenner & others, 2020). There are likely many reasons for this increase, including earlier diagnoses with increased attention to early warning signs (Winter & others, 2020). ASD is two to five times more common in children assigned male gender at birth than in children assigned female gender at birth (Maenner & others, 2020). ASD occurs in all ethnic groups and at all levels of socioeconomic status.

ASD is called a *spectrum* because the level of impairment experienced by people with ASD is extremely variable (Lord & others, 2020; Påhlman & others, 2020; Su & others, 2020). At one end of the spectrum, some people with ASD may never develop the capacity to speak and have profound difficulties caring for themselves and living independently. At the other end, people with ASD may do well in school, experience successful careers, and have enduring relationships with others. For example, Temple Grandin (2006), an accomplished scientist with ASD, described in her memoir *Thinking in Pictures* how she has had to commit to memory the fact that non-autistic people think in words, not images, and that their facial expressions reveal important information about their feelings.

Some people with ASD are effective advocates for themselves and contrast their ways of thinking, feeling, and existing in the world to "neurotypicals." For these individuals, ASD is a way of existing in a world of others who differ from them in essential ways. These individuals assert that ASD should be understood as a type of valuable *neurodiversity* (Bertilsdotter & others, 2019; Chapman, 2020; Ortiz, 2020). Some members of the neurodiversity movement prefer the term "autistic" rather than "person with autism."

The process of receiving a diagnosis of ASD often begins when parents and pediatricians notice potential signs of ASD, such as language delays. A diagnosis can be made by age 2 (CDC, 2020b) but most children with ASD are diagnosed after age 3 and many are not diagnosed until after age 6 (Harrop & others, 2020). One concern is that children of diverse ethnic, racial, and socioeconomic backgrounds may not receive the evaluations they need to have access to treatments (Winter & others, 2020).

Because early intervention strongly predicts better outcomes (Dimian & others, 2020), researchers have searched for ways to identify children with ASD prior to the age of 3 (Qiu & others, 2020). Parents of children with ASD often report having suspected a problem much earlier in their lives (Viljoen & others, 2020). Still, infants vary greatly in their development and it is likely common for parents to sometimes wonder if their child is developing typically.

In a landmark study, Julie Osterling and Geraldine Dawson (1994) searched for objective evidence of early signs of ASD in an interesting place: home videos. The researchers collected videos of children's first birthday parties from 22 families. For half, the birthday child would later be diagnosed with ASD; the others would develop typically. A team of coders, unaware of whether a child would eventually be diagnosed with ASD or not, carefully watched the videos and coded aspects of the child's behavior.

The results? Even at just one year, compared to their typically developing peers, children who would later be diagnosed with ASD were less likely to look when someone called their name; to point or show an object to another person; and to look at the face of a person interacting with them (Osterling & Dawson, 1994). This study led to additional research on early warning signs of ASD in infants and toddlers. Results indicate that, compared to typically developing children, very young children who will develop ASD show low levels of joint attention (looking at objects with another person), low levels of attention to other children, less interest in the gaze of another person, and lower sophistication in pretend play (Clifford & Dissanayake, 2008; Qiu & others, 2020; Wilson & others, 2017).

To identify ASD risk as early as possible, scientists have tried to identify *biomarkers,* measurable physical qualities that signal the presence of a disorder. Researchers have identified potential markers in blood, saliva, genes, and the digestive tract (El-Ansary & others, 2020; Troisi & others, 2020) that appear to distinguish children with ASD from the typically developing and those with other developmental disorders. Biomarkers for ASD may someday allow interventions to begin even before symptoms appear (Jaffee, 2018). Such tests may be especially valuable for children at risk for ASD, such as those with a family history of the disorder.

Previously, high functioning individuals on the autism spectrum might be diagnosed with 'Asperger's syndrome,' but because ASD now acknowledges the whole spectrum of functioning this diagnosis is no longer used.

There are a number of influencers on Instagram and TikTok who are on the autism spectrum.

To explore the neurodiversity movement, check out http://autisticadvocacy.org/

Focusing on first birthday parties is important. Can you think of two reasons why this step is considered a strength of this work?

For a child with ASD, the world may be an overwhelming cacophony of stimulation.
Stockbyte/Getty Images

UNDERSTANDING AUTISM SPECTRUM DISORDER AND ITS ORIGINS

Theoretical approaches to understanding autism seek to explain the symptoms of ASD as responses to central problems in cognition, sensation, or social motivation. Knowing which processes "come first" might help to develop diagnostic evaluations and to specify directions for treatment.

For example, the *cognitive approach* to ASD emphasizes deficits in theory of mind. Recall from Chapter 4 that theory of mind refers to our understanding that other people have subjective experiences that may differ from our own. From the cognitive perspective, the child with ASD navigates the social world without this key source of information. Simon Baron-Cohen (1995, 2011) proposed that people with ASD are missing this important feature of social development. Support for this approach comes from the fact that people with ASD perform less well on theory of mind tasks compared to typically developing children (Farrar & others, 2017; Rasga & others, 2017).

The *social motivation approach* to ASD suggests that people with ASD are profoundly less interested in social stimuli compared to typically developing people. This difference affects what an infant attends to, the experiences they seek out, and what they learn through experience (Chevallier & others, 2012; Su & others, 2020). Research shows that children with ASD differ from typically developing children in their lack of attention to faces and to the gaze of others during social interactions (Qiu & others, 2020).

Finally, the *sensory processing approach* to ASD focuses on the fact that individuals with ASD process sensory experiences differently from others (Ida-Eto & others, 2017; Vlaeminck & others, 2020). Visually, they are more likely to focus on fine details rather than the "big picture." They may be highly sensitive to sounds and tactile sensations. In addition, those with ASD may have problems organizing and integrating their sensory experiences (Robertson & Baron-Cohen, 2017). Imagine that the world was an overwhelming cacophony of stimulation: Painfully bright lights and colors, loud noises, and scratchy clothes. How might you behave? From the sensory processing perspective, overwhelming sensory experiences are the heart of ASD (Bogdashina, 2016).

What causes ASD? This question is quite complex. Currently, there is strong scientific consensus about two things that *do not* cause ASD. First, particular parenting styles do not lead children to develop ASD (Mandy & Lai, 2016). Second, vaccines do not cause ASD (Anjali & others, 2015). It is worth repeating: Vaccines do not cause ASD.

Risk factors for ASD include male sex, preterm birth (Cogley & others, 2020), older parental age, and prenatal exposure to some chemicals (Christensen & others, 2013). Maternal infections, such as rubella or "German measles" (Hutton, 2016), during pregnancy are also risk factors for ASD (Spann & others, 2017). Finally, complications before, during, and immediately after birth are associated with higher risk for ASD for the infants later in life (Getahun & others, 2017). By far, the two most important factors in ASD are genes and the brain. We review each of these next.

Illusory correlation appears to explain the subjective but mistaken link between vaccines and ASD. Vaccines are often administered at the same time communication delays are noticed. VACCINES DO NOT CAUSE ASD.

Genetic and Brain Differences

Genes play an important role in the development and persistence of ASD (Matoba & others, 2020; Thapar & Rutter, 2020). Children whose siblings have ASD are at increased risk for the disorder (Thapar & Rutter, 2020). Research in molecular genetics has compared the genomes (the entire genetic array) of people with ASD to neurotypical individuals. This research has linked many genetic mutations (that is, alterations or changes in the molecules that make up our genes) to the development of ASD (Hammerschlag & others, 2020; Yousaf & others, 2020). However, within a family, siblings can possess the same genetic mutations and one may develop ASD and the other not (Woodbury-Smith & others, 2017), indicating that genes are not a sole cause of ASD.

How are brain structure and function related to ASD? Research addressing this question has focused on brain areas linked to language and social information processing. For example, a study probed the ways that adolescents with ASD process speech compared to typically developing controls. The results showed that while typically developing adolescents showed activity in both the left and right hemispheres as they processed speech, those with ASD processed speech primarily on the left hemisphere only, without involvement of the right hemisphere (Galilee & others, 2017). Recall that the right hemisphere, though not the center of language processing, contributes to the understanding of word meanings, connotation, and metaphor.

A large-scale postmortem (that is, after death) study compared the donated brains of 418 individuals with ASD to brains of 509 neurotypical adults. The study found that those with ASD had less white matter connections to areas of the brain involved in facial expressions and processing theory of mind (Cheng & others, 2015). These results fit with the deficits associated with ASD, but it is important to keep in mind that behavior affects the brain. So, it is impossible to know if these differences emerge from genetic differences or from a lifetime of behavior. The idea that the brain continues to change with experience is supported by research showing that brain differences in those with ASD increase with age (Lee & others, 2017).

Etiological Diversity   Clearly there is no single genetic or brain characteristic that serves as the key factor in the development of ASD. ASD appears to be *etiologically diverse*, meaning that the underlying causes of this disorder might vary from person to person (Betancur, 2011). One way to think about this complexity is to stay mindful of the fact there are only so many ways to be human. No matter their genes, children will not sprout wings and begin to fly. They will stay, always, human. People on the autism spectrum may share symptoms, such as hand flapping, restricted interests, or difficulty in social interactions, for different underlying reasons. These many reasons may produce similar symptoms because those symptoms are within the realm of what human beings do.

Think about the symptoms of ASD. You likely know someone you might describe as having some levels of these characteristics. In fact, research shows that parents and siblings of children with ASD often show "subclinical" levels of traits (meaning the levels of the traits are not extreme or atypical and do not qualify as abnormal) that are associated with autism, like having restricted interests or being less socially engaged (Bora & others, 2017; Ruzich & others, 2017). Considering ASD in this way helps us recognize that people with the disorder are not qualitatively different from others, they are not only on the autism spectrum but on the *human* spectrum, too.

ktaylorg/Getty Images

TREATMENT   There is no cure for ASD, but treatments can greatly improve the lives of many people with the disorder. Intense behavioral approaches, incorporating principles of operant conditioning, are the standard treatment for ASD (Woo & Leon, 2013). Specifically, **applied behavior analysis**, as reviewed in Chapter 5, has proven effective in treating ASD (Thompson, 2013). How intense is the treatment? Typically, it involves over 20 hours of behavioral intervention per week for one to three years, using reinforcement, shaping, and discrimination (Thompson, 2013). Researchers continue to integrate innovative methods to applied behavior analysis to treat ASD (Koegel & others, 2017).

applied behavior analysis or behavior modification
The use of operant conditioning principles to change human behavior.

In addition to therapy, medications may be used to help alleviate especially troubling symptoms, including tantrums, aggression, or self-harm. The drugs used to treat these symptoms include strong psychiatric medications that are used to treat schizophrenia, an adult disorder we will review later (Alsayouf & others, 2020; Lamy & others, 2020).

People with ASD often have other diagnoses as well. It is not uncommon for children with ASD to receive a dual-diagnosis of ASD and attention-deficit/hyperactivity disorder, our next topic.

Attention-Deficit/Hyperactivity Disorder

attention-deficit/ hyperactivity disorder (ADHD)
One of the most common psychological disorders of childhood, in which individuals show one or more of the following: inattention, hyperactivity, and impulsivity.

Attention-deficit/hyperactivity disorder (or ADHD) is a neurodevelopmental disorder characterized by persistent problems in sustaining attention and difficulty engaging in quiet activities for a prolonged period. The disorder is characterized by three main symptoms (APA, 2013):

1. *Inattention*: The tendency to wander off tasks, lacking persistence, and difficulty sustaining focus.

2. *Hyperactivity*: Excessive activity when it is inappropriate, including excessive fidgeting and restlessness.

3. *Impulsivity*: Taking actions without planning or thinking, including inability to delay gratification.

For a diagnosis of ADHD, symptoms must be observed before the age of 12 and must occur across a variety of settings. So, a fifth grader who is fidgety at school but able to sit for periods of time studying or constructing a complex Lego set at home is unlikely to have ADHD. A child with ADHD might have particular problems in regulating excitement or exuberance and show a lack of caution in behavior.

ADHD occurs in most cultures in about 5 percent of children (APA, 2013), although there are wide differences across countries in the prevalence of ADHD (Taylor, 2017). As with ASD, children assigned male gender at birth are more likely to be diagnosed with ADHD than children assigned female gender at birth (Fairman & others, 2020). Experts previously thought that most children "grow out" of ADHD; however, around 70 percent of adolescents (Sibley & others, 2012) and adults (Asherson & others, 2010) diagnosed as children continue to experience ADHD symptoms (Zhu & others, 2017).

In the United States, ADHD is one of the most common psychological disorders of childhood. In 2018, about 9 percent of children had been diagnosed at some point in their lives. Rates of diagnosis are higher for white/European American children than other ethnic/racial groups (Fairman & others, 2020).

Children with ADHD can be a danger to themselves because they may climb or run into dangerous situations without thinking.

High rates of ADHD have prompted some to suggest that psychiatrists, parents, and teachers are labeling typical childhood behavior a psychological disorder (Timimi, 2017; Wakefield, 2016). These scholars argue that many ADHD diagnoses are the product of medicalizing what is simply annoying but typical childhood behavior (Armstrong 2017; Koutsoklenis & Gaitanidis, 2017). Those who endorse this view argue that in the absence of a specific genetic or brain-related explanation, ADHD should not be considered a psychological diagnosis. However, as we will see, most psychological disorders lack such simple explanations. Moreover, left untreated, ADHD is related to serious negative outcomes.

ADHD can interfere with schooling and social and academic development (Arnold & others, 2020; Colomer & others, 2017). Youths with ADHD are likely to have problems in peer relations, especially because of their inability to regulate excitement (Thorell & others, 2017). Adolescents with ADHD more likely to be bullied (Islam & others, 2020). Compared to typically developing peers, youths with ADHD are more likely to engage in delinquent behavior (Egan & others, 2020). In college students, ADHD symptoms predict lower quality relationships (McKee, 2017). Adults who were diagnosed with ADHD as children show poorer educational outcomes, relationship difficulties, incarceration, and lower socioeconomic status (SES) (Faraone & others, 2015; Owens & others, 2017).

Challenge YOUR THINKING

Does Birth Month Predict ADHD Diagnosis?

Recently researchers identified one fact about a child that predicts ADHD diagnosis: being born in August (Jena & others, 2018). Why August? Many states in the United States have rules for when a child can start kindergarten. Typically, a child who turns five before September 1 may enroll in public school. This means that August-born children start school almost immediately after their fifth birthday. Compare those children to kids born in September. Entering kindergarten, those with September birthdays are almost six years old. So, even though all of the children are in the same grade, kids born in August are up to a year younger than everyone else.

Compared with their (older) classmates, August-born children might show signs of immaturity. A teacher may wonder whether a child's inability to sit still or pay attention indicates ADHD. Might immaturity be mistaken for a disorder?

To find out, researchers (Layton & others, 2018) used the medical records of over 400,000 children born in the United States from 2007 to 2009. In 2015, they collected data about ADHD diagnosis and treatment and compared

FatCamera/Getty Images

children born in August with those born in September. The results? In states with a September 1 cutoff for kindergarten, the rate of ADHD in August-born children was up to 34 percent greater than the rate for those born in September. No other two months differed in this way. In states without a September 1 cutoff, no difference between August- and September-born children was found. Clearly, teachers and physicians must be mindful of children's age (not just their grade in school) when evaluating their behaviors.

What Do You Think?

- Why do you think parents decide to send children born in August to kindergarten?
- If a teacher suggested your child might have ADHD, what would you do?

Importantly, these many negative outcomes can be prevented by effective treatments (Thapar & Cooper, 2016). Indeed, the use of medications to treat ADHD predicts a range of positive outcomes, including lower risk of depression, criminal activity, car accidents, injuries, and better academic performance (Boland & others, 2020).

ADHD diagnosis is based on interviewing the child and parents, combined with standard assessments completed by parents and teachers. ADHD symptoms may occur in other disorders (Cuffe & others, 2020; Mersin Kilic & others, 2020) and so it is important for children to be carefully assessed. Although overdiagnosis is a concern, some groups, such as children from disadvantaged backgrounds, may have less access to evaluation and high-quality treatment (Fairman & others, 2020). Research has shown that an unexpected characteristic—birth month—can predict risk for ADHD diagnosis. To read about this provocative research, see the Challenge Your Thinking.

UNDERSTANDING ADHD AND ITS ORIGINS Risk factors for ADHD include very premature birth (James & others, 2020), low birth weight, and prenatal exposure to alcohol (Eilertsen & others, 2017) and smoking (He & others, 2020). Childhood lead exposure (Kim & others, 2018) in combination with factors such as poverty (Perera & others, 2018) is also associated with ADHD. Exposure to some pesticides can, in conjunction with genetic risk, heighten the chances that a child will develop ADHD (Nilsen & Tulve, 2020). Large-scale longitudinal studies show that children who experience more adverse events, such as physical, sexual, or verbal abuse and neglect, are more likely to be diagnosed with ADHD (Lugo–Candelas & others, 2020).

Genetic differences appear to play a role in ADHD. Twin studies indicate that up to 70 percent of the variance in ADHD is accounted for by genetic variability (Brikell & others, 2020). Genetic studies suggest that ADHD is polygenic, meaning that it is related to the interaction of two or more genes (Doyle & others, 2017). Molecular genetic studies have identified eight to ten specific genetic locations associated with ADHD (Bidwell & others, 2017), including the genes responsible for transporting dopamine and serotonin (Klein & others, 2017). Genetic differences associated with ADHD may not be specific to the disorder but may represent genetic risk for other psychological disorders (Brikell & others, 2020).

These genetic differences are thought to affect brain development in ways that contribute to ADHD. For example, one hypothesis about the origins of ADHD is the *delayed maturation hypothesis* (Klein & others, 2017). This hypothesis says that people with ADHD have brains that develop more slowly than typically developing individuals, especially with regard to the prefrontal cortex (Beare & others, 2017), the area of the brain most associated with planning, thinking, and self-control. Research supporting this hypothesis shows that ADHD is associated with reduced cortical thickness (Boedhoe & others, 2020; Cherkasova & others, 2017). In addition, people with ADHD may have fewer connections between the prefrontal cortex and other brain regions (Ha & others, 2020; Wang & others, 2020).

This research fits well with cognitive approaches to ADHD that emphasize problems in executive function as being the root cause of problems in the lives of those with ADHD (Barkley, 1997; Graham, 2017). Preschoolers who score low on measures of working memory function and capacity to regulate their extreme positive emotions are more likely to show ADHD symptoms in adolescence (Sjöwall & others, 2017). Moreover, it may be that differences in working memory, rather than ADHD per se, explain lower academic performance among children with the disorder (Simone & others, 2017).

Another neurobiological focus has been the areas of the brain that are responsible for reward learning and seeking out rewarding stimuli. People with ADHD have less brain volume in the brain's reward centers (Hoogman & others, 2017; Hulst & others, 2017). ADHD is also associated with dysregulation in the neurotransmitter dopamine and pathways that carry this important neurotransmitter (Volkow & others, 2011). In addition to dopamine, dysregulation of acetylcholine and melatonin in infancy predicts the development of ADHD (Hellmer & Nyström, 2017). Such findings fit with problems that individuals with ADHD have delaying gratification and regulating reward motivation (Martinelli & others, 2017).

TREATMENT   Stimulant medication, such as Ritalin or Adderall, is often the first line of treatment for ADHD because these medications often work (Pelham & others, 2017). It may seem surprising that people who are too active would be treated with stimulants. These medications work because they increase the amount of circulating dopamine (Schrantree & others, 2016). This increase in dopamine means that the child does not have to engage in activity to experience the pleasure that dopamine evokes.

Giving children medication is always best done with extreme care. Although some have been concerned that children who are given stimulant drugs might be at risk for developing substance abuse problems, research results have been mixed. Indeed, it may be that symptoms of ADHD rather than medical treatment of the disorder predict substance use (Molina & Pelham, 2014). Untreated, ADHD has the capacity to cause a great deal of harm to children and their futures.

Children and adults with ADHD can also benefit from nonmedical interventions. Often combined with medication, these interventions include many different

VGstockstudio/Shutterstock

activities, such as physical exercise (Den Heijer & others, 2017), sports training (Altszuler & others, 2017), meditation (Evans & others, 2017), academic tutoring and training (Tamm & others, 2017), and driving safety (Bruce & others, 2017).

Psychotherapy can be effective in treating ADHD in children whose symptoms do not respond to medication or who are very young (Daley & Dupaul, 2016; Sprich & others, 2016). Psychotherapy has also been found to be effective in treating ADHD in adults (Knouse & others, 2017) as well as parents of children with ADHD (Sprich & others, 2016).

3 Anxiety and Anxiety-Related Disorders

Think about how you felt before a make-or-break exam or a big presentation—or perhaps as you noticed police lights flashing behind your speeding car. Did you feel jittery and nervous? These are the feelings of normal anxiety—an unpleasant feeling of fear and dread.

In contrast, **anxiety disorders** involve fears that are uncontrollable, disproportionate to the actual danger the person might be in, and disruptive of ordinary life. They feature motor tension (jumpiness, trembling), hyperactivity (dizziness, a racing heart), and apprehensive expectations and thoughts. *DSM-5* recognizes 12 types of anxiety disorders. In this section, we survey four common anxiety disorders:

anxiety disorders
Disabling (uncontrollable and disruptive) psychological disorders that feature motor tension, hyperactivity, and apprehensive expectations and thoughts.

■ Generalized anxiety disorder

■ Panic disorder

■ Specific phobia

■ Social anxiety disorder

We will also review a disorder that is not classified as an anxiety disorder but is related to the experience of anxiety:

■ Obsessive-compulsive disorder (categorized under "Obsessive-Compulsive and Related Disorders")

Generalized Anxiety Disorder

When you are worrying about getting a speeding ticket, you know why you are anxious; there is a specific cause. **Generalized anxiety disorder** (GAD) is different from such everyday feelings of anxiety in that sufferers experience persistent anxiety for at least six months

generalized anxiety disorder
Psychological disorder marked by persistent anxiety for at least six months, and in which the individual is unable to specify the reasons for the anxiety.

and are unable to specify the reasons for the anxiety (Porta-Casteràs & others, 2020). People with GAD are nervous most of the time. They may worry about their work, relationships, or health. That worry can also take a physical toll and cause fatigue, muscle tension, stomach problems, and difficulty sleeping.

What is the etiology of generalized anxiety disorder? Among the biological factors are genetic predisposition, deficiency in the neurotransmitter GABA, and brain differences (particularly in the amygdala) (Makovac & others, 2016; Zhu & others, 2019). Brain imaging studies show that for people with GAD, the brain responds more strongly to everyday worries (Blair & others, 2017; Li & others, 2020).

> GABA is the neurotransmitter that inhibits neurons from firing—it's like the brain's brake pedal. Problems with GABA are often implicated in anxiety disorders.

Another contributor to GAD are abnormalities in the *respiratory* system (Maric & others, 2020). Interestingly, researchers sometimes use respiratory challenge to study GAD. In such studies, healthy volunteers breathe air that is 7.5 percent carbon dioxide (or CO_2). This level of CO_2 induces feelings of threat and anxiety, suggesting the role of the respiratory system in many anxiety disorders (Huneke & others, 2020), including GAD.

Psychological and sociocultural risk factors for GAD include having harsh (or even impossible) self-standards, overly strict and critical parents, automatic negative thoughts when feeling stressed, and a history of uncontrollable traumas or stressors (such as an abusive parent) (Lee & others, 2020). Women are more likely than men to suffer from GAD (Boehlen & others, 2020). In addition, people who tend to be worriers in general or who are always on the lookout for threats in the environment may be at risk for GAD (Cabrera & others, 2020).

Treatment for GAD often includes medications that mimic the action of GABA (the brain's brake pedal), such as pregabalin (Lyrica) (Baldwin, 2020). Psychotherapy can also be effective in reducing GAD, especially if the person with GAD experiences a strong empathic bond to the therapist (Watts & others, 2020).

Panic Disorder

Much like everyone else, you might sometimes have a specific experience that sends you into a panic. For example, you work all night on a paper, only to have your computer crash before you saved your last changes, or you are about to dash across a street just when you see a large truck coming right at you. Your heart races, your hands shake, and you might break into a sweat.

> A panic attack can be a one-time occurrence. People with **panic disorder** have recurrent attacks that sometimes cause them to be afraid to even leave their homes, a condition called **agoraphobia**.

In **panic disorder**, a person experiences recurrent, sudden onsets of intense terror, often without warning and with no specific cause. Panic attacks can produce severe palpitations, extreme shortness of breath, chest pains, trembling, sweating, dizziness, and a feeling of helplessness (Barber & others, 2020). People with panic disorder may feel that they are having a heart attack or going to die.

panic disorder
Anxiety disorder in which the individual experiences recurrent, sudden onsets of intense apprehension or terror, often without warning and with no specific cause.

During a panic attack, the brain registers fear in the network of the cortico-limbic regions (amygdala, hippocampus, and prefrontal) (Kim & others, 2020; Ohi & others, 2020). Charles Darwin, the scientist who proposed the theory of evolution, suffered from intense panic disorder (Barloon & Noyes, 1997).

What is the etiology of panic disorder? Theories of its origins take into account biological, psychological, and sociocultural factors. In terms of biological factors, people may have a genetic predisposition to the disorder and those genes may be shared by people who are prone to worry (Ohi & others, 2020; Tretiakov & others, 2020). Of particular interest to researchers are genes that direct the action of neurotransmitters such as norepinephrine, GABA, and serotonin (Tanahashi & others, 2020; Tretiakov & others, 2020). Another brain chemical, *lactate,* which plays a role in brain metabolism, has been found to be elevated in people with panic disorder (Kim & others, 2020; Riske & others, 2017). Experimental research has shown that increasing lactate levels can produce panic attacks (Reiman & others, 1989), suggesting that problems in regulating lactate may play a role in panic disorder (Wiese & Boutros, 2019). Other research points to the involvement of a wider range of genes and bodily systems, implicating genes involved in hormone regulation (Demiralay & others, 2017) and responses to stress (Javelot & others, 2020). Compared to other anxiety disorders, panic disorder shares a strong relationship with medical illnesses and it may be that these illnesses affect the body's fight or flight system (Öksüz & others, 2020).

Many experts interpret Edvard Munch's painting The Scream *as an expression of the terror brought on by a panic attack.*
Mariano Garcia/Alamy Stock Photo

With respect to psychological influences, learning processes are one factor that has been considered in panic disorder. Classical conditioning research has shown that learned associations between bodily cues of respiration and fear can play a role in panic attacks (De Houwer, 2020). Interestingly, CO_2 has been found to be a very strong conditioned stimulus for fear, suggesting that humans may be *biologically prepared* to learn an association between high concentrations of CO_2 and fear (Javelot & others, 2020; Savulich & others, 2019). Thus, some researchers have suggested that at the heart of panic attacks are the learned associations between CO_2 and fear (Van Diest, 2019).

An earlier explanation of panic attack was called the **suffocation false alarm theory**. Can you see why it was initially proposed?

In addition, the learning concept of *generalization* may apply to panic disorder. Recall that in classical conditioning, generalization means showing a conditioned response (in this case, fear) to conditioned stimuli other than the particular one used in learning. Research shows that people who suffer from panic attacks are more likely to display overgeneralization of fear learning (Neueder & others, 2019; Wong & Pittig, 2020). Why might those who suffer from panic attacks be more likely to show stronger and more generalized fear associations? Biological predispositions and early experiences with traumatic life events may play a role in setting the stage for such learning (De Venter & others, 2017; Pilecki & others, 2011), suggesting a vulnerability-stress explanation.

In terms of sociocultural factors, in the United States, women are twice as likely as men to have panic attacks (Li & Graham, 2017; Kim & others, 2019). Possible reasons include biological differences in hormones and neurotransmitters (Altemus, 2006; Masdrakis & others, 2017). Compared to men, women are more likely to complain of distressing respiratory experiences during panic attacks (Sheikh & others, 2002). Interestingly, one study showed that healthy women are more likely to experience panic-related emotions when exposed to air enriched with CO_2 (Nillni & others, 2012).

Whenever you encounter gender differences in this discussion, ask yourself whether men or women might be more likely to report having problems or to seek treatment. Research on psychological disorders is often based on individuals who have reported symptoms or sought help. If men are less likely to report symptoms or seek treatment, the data may underestimate the occurrence of psychological disorders in men.

Women may also cope with anxiety-provoking situations differently than men do, and these differences may explain the gender difference in panic disorder (Lovick, 2014; Viswanath & others, 2012). Panic disorder has been observed in a variety of cultures, though there are some cultural differences in the experience of these attacks (Agorastos & others, 2012).

There are a number of effective treatments for panic disorder, including individual and group psychotherapy (Barkowski & others, 2020; Brettschneider & others, 2020). Medications that have a tranquilizing effect can also help those suffering with panic disorder (Quagliato & others, 2019).

Specific Phobia

Many people are afraid of spiders and snakes; indeed, thinking about letting a tarantula crawl over one's face is likely to give anyone the willies. It is not uncommon to be afraid of particular objects or specific environments such as extreme heights. For most of us,

Acrophobia	Fear of high places	Arachnophobia	Fear of spiders	Mysophobia	Fear of dirt
Aerophobia	Fear of flying	Astrapophobia	Fear of lightning	Nyctophobia	Fear of darkness
Ailurophobia	Fear of cats	Cynophobia	Fear of dogs	Ophidiophobia	Fear of nonpoisonous snakes
Algophobia	Fear of pain	Gamophobia	Fear of marriage		
Amaxophobia	Fear of vehicles, driving	Hydrophobia	Fear of water	Thanatophobia	Fear of death
		Melissophobia	Fear of bees	Xenophobia	Fear of strangers

FIGURE 1 **Specific Phobias** This figure features examples of specific phobias—anxiety disorders characterized by irrational and overwhelming fear of a particular object or situation.

(FIRST): Corbis/VCG/Getty Images; (Second): Flying Colours Ltd./Getty Images; (Third): Digital Archive Japan/Alamy Stock Photo; (Fourth): Comstock Images/Alamy Stock Photo; (Fifth): Tom Warner/ WeatherVideoHD.TV; (Sixth): Sigi Kolbe/Getty Images; (Seventh): Arthur Tilley/Getty Images; (Eighth): Morley Read/iStock/Getty Images; (Ninth): G.K. & Vikki Hart/Getty Images

specific phobia
Psychological disorder in which an individual has an irrational, overwhelming, persistent fear of a particular object or situation.

these fears do not interfere with daily life. A fear becomes a phobia when a situation is so dreaded that a person goes to almost any length to avoid it. A snake phobia that keeps a city-dweller from leaving their apartment is clearly disproportionate to the actual chances of encountering a snake. **Specific phobia** is a psychological disorder in which a person has an irrational, overwhelming, persistent fear of a particular object or situation. Specific phobias come in many forms, as shown in Figure 1.

Where do specific phobias originate? Approaches to answering this question typically first acknowledge that fear plays an important role in adaptive behavior. Fear tells us when we are in danger and need to take to action. The importance of this function suggests that fears should be relatively quickly learned, because learning to fear things that will harm us keeps us out of harm's way. Specific phobias, then, might be considered an extreme and unfortunate variant on this adaptive process (Haesen & others, 2017; Muris & Merckelbach, 2012).

Many explanations of specific phobias view these disorders as based on experiences, memories, and learned associations (Veale & others, 2013). For example, the person with a fear of heights perhaps experienced a fall from a high place earlier in life and therefore associates heights with pain (a classical conditioning explanation). Alternatively, they may have heard about or watched others who demonstrated terror of high places (an observational learning explanation), as when a child develops a fear of heights after sitting next to their terrified mother and observing her clutch the handrails, white-knuckled, as the roller coaster creeps steeply uphill. Not all people who have a specific phobia can easily identify experiences that explain these, so other factors may also be at play. Each specific phobia may have its own neural correlates (Klahn & others, 2017; Peñate & others, 2017) and some people may be especially prone to develop phobias (Winder & others, 2020).

Just as the theories of the origins of specific phobia rely largely on associative learning, treatment for specific phobias often rely on learning principles. These psychotherapies often use the concept of extinction to remove the link between a feared stimulus (such as a spider) and feelings of intense anxiety. In exposure therapy, people with specific phobias are exposed to the thing they fear often gradually and sometimes even virtually (Anderson & Molloy, 2020). Interestingly, specific phobia is associated with the development of other psychological disorders, suggesting it may be an early sign of vulnerability to a range of disorders (Wardenaar & others, 2017) and an opportunity for early intervention.

Social Anxiety Disorder

Imagine how you might feel just before you first meet the parents of the person you hope to marry. You might feel fearful of committing some awful gaffe, ruining the impression you hope to make on them. Or imagine getting ready to give a big speech before a crowd and suddenly realizing you have forgotten your notes. **Social anxiety disorder** (also called **social phobia**) is an intense fear of being humiliated or embarrassed in social situations like these (Mohammadi & others, 2020). Not surprisingly, individuals with social anxiety disorder often avoid social interactions. Singers Carly Simon and Barbra Streisand have dealt with social anxiety disorder.

social anxiety disorder or social phobia
An intense fear of being humiliated or embarrassed in social situations.

Where does social anxiety disorder originate? Research supports the role of genetics in the disorder. For example, a large-scale genome-wide association study found significant contribution of specific genetic locations to the disorder (Stein & others, 2017). Interestingly, those same genetic locations were associated with lower levels of the personality trait extraversion. In addition, neural circuitry involving the thalamus, amygdala, and cerebral cortex has been linked to social anxiety disorder (Wang & others, 2018; Yang & others, 2018). A number of neurotransmitters also may be involved, especially serotonin (Frick & others, 2020) and oxytocin (the neurotransmitters that is most involved in social bonds) (Williams & others, 2020; Yoon & Kim, 2020). Social anxiety disorder may involve vulnerabilities, such as genetic characteristics or parenting styles that lay a foundation of risk, combined with learning experiences in a social context (Chubar & others, 2020).

In terms of psychological factors, a longitudinal study showed that children who were prone to blushing were more likely to develop social phobia, suggesting that early experiences with social embarrassment may predispose people to the disorder (Nikolić & others, 2020). Cognitive explanations focus on the person's beliefs about the possibility of social failure and automatic negative thoughts that follow from those beliefs (Nordahl & others, 2017). An additional cognitive factor is that attentional bias toward negative information may play a role in the persistence of the disorder (Evans & others, 2020). As we have seen many times, expectations can color perception and experience. In social anxiety disorder, the person may be constantly looking for indications of social failure and consequently interpret social cues in a negative light. These perceptions then might lead them to avoid social interactions all the more.

Blushing isn't controllable. So blushers may find that being in social situations is uncomfortable!

Treatment for social anxiety disorder often involves psychotherapy that aims at exposing the person to social stimuli while instilling a calm mindset. Such therapy might even incorporate encountering others in virtual reality rather than in person (Niles & others, 2020). Psychotherapy can combat the beliefs, thoughts, and expectations that feed into the disorder (Butler & others, 2020). Medications used to treat social anxiety disorder include drugs that regulate serotonin (typically used for depression) and drugs that serve to reduce anxiety (Rappaport & others, 2020). There might also be a role for oxytocin as a treatment to relieve social anxiety (Jones & others, 2017). Finally, it may be that people with social anxiety disorder will benefit from direct stimulation to the brain that temporarily disrupts the area of the brain involved in attention to threat (Di Lorenzo & others, 2020).

In *DSM-5,* generalized anxiety disorder, panic disorder, specific phobia, and social anxiety disorder are all classified as anxiety disorders. Our next topic, obsessive-compulsive disorder, is not included under the umbrella of anxiety disorders. Instead, this disorder has its own separate category. Nonetheless, anxiety is relevant to obsessive-compulsive disorder, as we shall see.

Obsessive-Compulsive Disorder

Just before leaving on a long road trip, you find yourself checking to be sure you locked the front door. As you pull away in your car, you are stricken with the thought that you forgot to turn off the coffeemaker. Going to bed the night before an early flight, you check your alarm clock a few times to be sure you will wake up on time. These are examples of normal checking behavior.

In contrast, the disorder known as **obsessive-compulsive disorder (OCD)** features anxiety-provoking thoughts that will not go away and/or urges to perform repetitive, ritualistic behaviors to prevent or produce some future situation. *Obsessions* are recurrent thoughts. *Compulsions* are recurrent behaviors. People with OCD dwell on their doubts and repeat their routines sometimes hundreds of times a day. The most common compulsions are excessive checking, cleansing, and counting. Gameshow host Howie Mandel has coped with OCD, as have retired soccer star David Beckham, singer-actor Justin Timberlake, and actor Jessica Alba. Obsessive-compulsive symptoms have been found in many cultures, and culture plays a role in the content of obsessive thoughts or compulsive behaviors (Cordell & Holaway, 2020; Pascual-Vera & others, 2019).

A person with OCD might believe that they must touch the doorway with their left hand whenever they enter a room and count their steps as they walk. If they do not complete this ritual, they may be overcome with a sense of fear that something terrible will happen (Myers & others, 2017).

What is the etiology of obsessive-compulsive disorder? In terms of biological factors, genes are certainly involved (Smit & others, 2020). With regard to the brain, one neuroscientific analysis is that the frontal cortex or basal ganglia are so active in OCD that numerous impulses reach the thalamus, generating obsessive thoughts or compulsive actions (Hauser & others, 2017; Haber & others, 2020). Low levels of the neurotransmitters serotonin, dopamine, and glutamate likely are involved in the brain pathways linked with OCD (Derksen & others, 2020; Karthik & others 2020; Murray & others, 2019). In addition to genetic and neurological issues, there is some evidence that OCD is related to wider dysfunction in the body, including digestive and "gut" issues (Turna & others, 2019) as well as immune functioning (Gerentes & others, 2019; Maia & others, 2019).

In one study, fMRI was used to examine the brain activity of people with OCD before and after treatment (Nakao & others, 2005). Following effective treatment, a number of areas in the frontal cortex showed decreased activation. The limbic system appears to be hyperactivated in those with OCD, even when situations have no uncertainty (Stern & others, 2013). Research with rats has shown that dysfunction in the circuitry connecting the thalamus to the amygdalae (the brain structures that activate in response to survival relevant stimuli) can produce OCD-like symptoms (Ullrich & others, 2017).

In terms of psychological factors, OCD sometimes occurs during a period of life stress such as the birth of a child or a change in occupational or marital status (Brander & others, 2016). A combination of genetic vulnerability and life stress may account for the development of OCD (Alemany-Navarro & others, 2019). In addition, the features of a person's symptoms (such as the type of ritual the person feels compelled to enact) may be predicted by aspects of the person's personality and experience (Vickers & others, 2017). According to the cognitive perspective, what differentiates people with OCD from those who do not have it is the ability to turn off negative, intrusive thoughts by ignoring or effectively dismissing them (Bouvard & others, 2017; Leahy & others, 2012).

Treatments for OCD may rely on behavioral principles that treat OCD symptoms as learned avoidance. This treatment involves having persons *not* engage in their ritual and allowing them to experience the fact that the dreaded outcome never happens.

As long as the person performs the ritual, they never find out that the terrible outcome doesn't happen. The easing of the anxiety exemplifies negative reinforcement (having something bad taken away after performing a behavior).

Repeated experience of this sort is thought to help extinguish the fear response (Hamatani & others, 2020; Pagliaccio & others, 2019). Psychotherapies also target the nonrational beliefs and automatic thoughts that occur in the context of OCD, helping the person to gain control of their mental life. Finally, medications that help in regulating serotonin can be effective in treating OCD (Walkup & Mohatt, 2017). The COVID-19 pandemic presented a unique challenge to treatment of OCD. Consider that hand washing (a common compulsion) was required by the pandemic (Sheu & others, 2020).

OCD-Related Disorders

DSM-5 includes other disorders that are thought to be related to OCD (Abramowitz & Jacoby, 2015). All of these involve repetitive behavior and often anxiety. Here are some examples:

- *Hoarding disorder* involves compulsive collecting, poor organization skills, and difficulty discarding things, along with cognitive deficits in information processing speed, problems with decision making, and procrastination (Levy & others, 2017). People with hoarding disorder find it difficult to throw things away and are troubled by the feeling that they might need items like old newspapers at a later time.

- *Excoriation* (or skin picking) refers to the particular compulsion of picking at one's skin, sometimes to the point of injury. Skin picking is more common among women than men and sometimes occurs as a symptom of autism spectrum disorder.

- *Trichotillomania* (hair pulling) entails compulsively pulling at the hair from the scalp, eyebrows, and other body areas. Hair pulling from the scalp can lead to bald patches that the person can go to great lengths to disguise.

- *Body dysmorphic disorder* involves a distressing preoccupation with imagined or slight flaws in one's physical appearance. People with this disorder cannot stop thinking about how they look and repeatedly compare their appearance to others, check themselves in the mirror, and so forth. Body dysmorphic disorder may include maladaptive behaviors such as compulsive exercise and body building and repeated cosmetic surgery. There is no gender difference in this disorder.

self-quiz

1. Sudden episodes of extreme anxiety or terror that involve symptoms such as heart palpitations, trembling, sweating, and fear of losing control are characteristic of
   A. generalized anxiety disorder.
   B. specific phobia.
   C. obsessive-compulsive disorder.
   D. panic disorder.

2. A person feels the need to tap the door frame 23 times prior to entering a room and if they do not complete this ritual something terrible will happen. This ritual has come to dominate the person's life so they cannot even leave the house in the morning. Which psychological disorder involves these symptoms?
   A. Panic disorder
   B. Generalized anxiety disorder
   C. Obsessive-compulsive disorder
   D. Specific phobia

3. An irrational, overwhelming, persistent fear of a particular object or situation is a defining characteristic of
   A. obsessive-compulsive disorder.
   B. specific phobia.
   C. panic disorder.
   D. generalized anxiety disorder.

APPLY IT! 4. Lately Tina has noticed that her mother appears to be overwhelmed with worry about everything. Her mother has told Tina that she is having trouble sleeping and is experiencing racing thoughts of all the terrible things that might happen at any given moment. Tina's mother is showing signs of
   A. panic disorder.
   B. obsessive-compulsive disorder.
   C. generalized anxiety disorder.
   D. specific phobia.

4 Trauma and Stress-Related Disorders

As we have seen, many psychological disorders involve a combination of factors, including genetic characteristics and experiences. Some psychological disorders, though, are more intimately attached to extremely negative experiences. Traumatic experiences are severe and extremely disturbing. These experiences may involve the threat of death or serious injury or sexual violence, in which a person feels intense fear and helplessness. Traumas can involve a single horrific incident or a pattern of negative experiences. Sometimes such experiences are so extreme that they splinter a person's very sense of self. What traumatic experiences have in common is their capacity to shatter the basic assumptions of life, including beliefs about the fairness of existence and one's place in the world. Traumatic life experiences challenge a person's capacities to cope. In this section, we review two types of disorders that involve trauma: post-traumatic stress disorder and dissociative disorders.

Post-Traumatic Stress Disorder

If you are ever in even a minor car accident, you may have a nightmare or two about it. You might even find yourself reliving the experience for some time. This normal recovery process takes on a devastating character in **post-traumatic stress disorder (PTSD)**, a disorder that develops through exposure to a traumatic event that overwhelms the person's abilities to cope. *DSM-5* expanded the kinds of experiences that might foster PTSD, recognizing that the disorder can occur not only in people who directly experience a trauma but also in those who witness it and those who only *hear* about it (APA, 2013).

PTSD can occur in response to a wide variety of extremely disturbing events. Such events can include natural disasters (such as hurricanes or earthquakes), combat and war-related trauma, unnatural disasters (such as car accidents, plane crashes, terrorist attacks, or mass shootings), childhood abuse and victimization, and sexual violence. Health crises are also linked to the development of PTSD (Sager & others, 2020). One study showed that as many as one in five adults diagnosed with cancer developed PTSD symptoms within four years (Chan & others, 2017). Different traumas can have different consequences for symptoms, such as survivor guilt, lingering feelings of betrayal and powerlessness, or enduring worries about safety (Badour & others, 2017).

PTSD symptoms can follow a trauma immediately or after months or even years (Hiller & others, 2016; Steenkamp & others, 2017). Most individuals who are exposed to a traumatic event experience some of the symptoms in the days and weeks after exposure (NIMH, 2019a). The symptoms of PTSD include the following:

- Flashbacks of the event. A flashback can make the person lose touch with reality and reenact the event for seconds, hours, or, very rarely, days. A person having a flashback—which can come in the form of images, sounds, smells, and/or feelings—usually believes that the traumatic event is happening all over again.

post-traumatic stress disorder (PTSD) Anxiety disorder that develops through exposure to a traumatic event, a severely oppressive situation, cruel abuse, or a natural or unnatural disaster.

Prior to deployment, U.S. troops receive stress-management training aimed at helping to prevent PTSD and other disorders that might be triggered by the high-stress conditions of war.
Brian L Wickliffe/REX/Shutterstock

- A person avoids emotional experiences and avoids talking about emotions with others.
- Emotional numbness and a reduced ability to feel emotions.
- Excessive arousal results in an exaggerated startle response or an inability to sleep.
- Difficulties with memory and concentration.
- Impulsive behavior.

One type of traumatic event linked to PTSD that is unfortunately common for American college students is sexual victimization (Coulter & Rankin, 2020; Hannan & others, 2020). Such traumas can include rape, sexual assault, or sexual harassment. All three of these behaviors can be perpetrated by people of any gender identity, and in opposite-gender and same-gender situations. Rape refers to forced sexual intercourse, including both verbal and physical force. Sexual assault refers to a wider range of behaviors, including unwanted sexual contacts with a victim by a perpetrator. Sexual assault may or may not include force and includes actions such as grabbing or fondling. Sexual harassment is unwelcome conduct of a sexual nature that offends, humiliates, or intimidates another person. In the workplace and at school, sexual harassment includes unwanted sexual advances, requests for sexual favors, and other verbal or physical conduct of a sexual nature. In recent years, the commonplace nature of these inappropriate and often illegal behaviors has come to light in high-profile cases of people using positions of power to force others to engage in sexual acts. Compared to women who have not been victims of a violent crime, women who had experienced sexual victimization are approximately six times more likely to develop PTSD (Gilmore & others, 2018; Kilpatrick, 2000).

How can survivors of sexual victimization be helped? In addition to treatment for their psychological disorders, sexual assault survivors benefit from strong social support (Dworkin & others, 2017). People who have access to information and medical and legal help are better able to cope with this trauma. Unfortunately, too often, victims of sexual assault are unaware of the resources available to them (Holland & others, 2017). For this reason, many colleges and universities have taken steps to better educate their communities about such services and also about the laws surrounding sexual assault and harassment, and their expectations for appropriate behavior. Indeed, an important way to prevent PTSD is to prevent sexual victimization.

What causes PTSD? Clearly, one factor is the traumatic event itself. However, not every person exposed to the same trauma develops PTSD. Therefore, other factors must influence a person's vulnerability to the disorder. These include a history of previous traumatic events and conditions, such as abuse and psychological disorders (Nichter & others, 2020; Wooldridge & others, 2020), cultural background (Alford, 2016), and genetic predisposition (Maul & others, 2020). These preexisting conditions make people more vulnerable to PTSD when combined with stressful events. Thus, PTSD may be best explained by a vulnerability-stress model.

Traumatic life events can profoundly affect the way a person copes with other events. Traumas can alter the delicate balance of neurotransmitters, hormones, and other biological systems, such that PTSD can color the way the body and brain react to stress (Fonkoue & others, 2020; Rauch & others, 2020). PTSD is associated with decreased volume in the hippocampus, the brain structure most associated with integrating memories (Nelson & Tumpap, 2017; Morey & others, 2020).

Treatments for PTSD include psychotherapies using classical conditioning paradigms to break the links between current experiences and deep feelings of trauma (Eftekhari & others, 2020), as well as therapies aimed at changing beliefs and behavior (Márquez & others, 2020). In addition, group therapy with other survivors of similar trauma can be helpful (Levi & others, 2017). Among survivors of sexual violence, self-defense training can be an effective way to reinstate a sense of personal control and efficacy (Geraets & van der Velden, 2020). Finally, medications are also used to help those with PTSD (Miller, 2020), often in conjunction with intense psychotherapy.

Combat veteran and former U.S. Senate candidate Jason Kander has tweeted about his recovery from PTSD @JasonKander.

Sexual Victimization on Campus

Sadly, careful estimates suggest that one in five women will experience sexual assault during their college years (Muehlenhard & others, 2017). A large survey of college campuses, including over 180,000 students, found that 25 percent of women, 7 percent of men, and 23 percent of transgender/genderqueer/nonbinary people reported experiencing sexual violence in college (Cantor & others, 2019). Of course, sexual victimization of any kind is unacceptable in any community, but sexual violence experienced by college students is shocking because it occurs in the context of the college experience itself. Coming to college can be a life-changing experience that prepares a person for a bright future. Unfortunately, for many people college can involve life-changing experiences of another sort entirely.

tommaso79/Shutterstock

In the United States, Title IX, a law passed in 1972 to guarantee gender equality in education and opportunities, is the law that motivates efforts to combat sexual misconduct on college and university campuses. Title IX states, in part, "No person in the United States shall, on the basis of sex, be excluded from participation in, be denied the benefits of, or be subjected to discrimination under any education program or activity receiving Federal financial assistance" (U.S. Department of Justice, 2015).

You may be wondering how Title IX, with its focus on discrimination, is relevant to the issue of sexual victimization on campus. Sexual victimization is associated with lower grades, decreased academic motivation, and higher rates of dropping out of college (Baker & others, 2016; Banyard & others, 2018). Equal access to education means access to a safe learning environment where everyone is supported to achieve. This means that institutions of higher learning must do all that they can proactively to prevent sexual victimization, to assist victims of sexual misconduct, and make every effort to ensure that the victimization is not repeated.

Many universities have implemented educational campaigns to raise awareness about the serious problem of sexual misconduct on campus. Involving all members of a learning community, including faculty, staff, and administrators, is crucial to the effectiveness of such interventions (Sales & Krause, 2017). They are not often victims, but college men have a vital role to play in transforming cultures and intervening in situations that appear ripe for misconduct (Hoxmeier & others, 2017).

Everyone has a right to obtain an education without concern for sexual victimization. Do you know about sexual assault or harassment prevention and victim resources at your school? If not, right now would be a great time to seek out this information. If you or someone you know has been the victim of rape or sexual assault, call 800.656.HOPE (4673) for the Rape, Abuse, Incest National Network (RAINN), for support, information, and to be connected with resources in your area.

Dissociative Disorders

Have you ever been on a long car ride and completely lost track of time, so that you could not even remember a stretch of miles along the road? Have you been so caught up in a daydream that you were unaware of the passage of time? These are examples of normal dissociation. *Dissociation* refers to psychological states in which the person feels disconnected from immediate experience.

At the extreme of dissociation are people who persistently feel a sense of disconnection. **Dissociative disorders** are psychological disorders that involve a sudden loss of memory or change in identity. Under extreme stress or shock, the person's conscious awareness becomes dissociated (separated or split) from previous memories and thoughts (Swart & others, 2020a). Dissociation can be a way to deal with extreme stress (de Thierry, 2020; Thiel & Dekel, 2020). Through dissociation, a person mentally protects their conscious self from the traumatic event.

In *DSM-5*, dissociative disorders are not included as a trauma-related disorder but are placed in close proximity to these to reflect the level of overlap between disorders like PTSD and dissociative disorders (APA, 2013). Indeed, some people with PTSD also experience dissociation (Swart & others, 2020a; Swart & others, 2020b). Both disorders are thought to be rooted, in part, in extremely traumatic life events.

People who develop dissociative disorders may have problems putting together different aspects of consciousness, so that experiences at different levels of awareness might be felt as if they are happening to someone else. The notion that dissociative disorders are related to problems in pulling together memories is supported by findings showing lower volume in the hippocampus and amygdala in individuals with dissociative disorders (McDonald & White, 2013; Lotfinia & others 2020; Vermetten & others, 2006). The hippocampus is especially involved in consolidating memory and organizing life experience into a coherent whole (Xu & others, 2014).

Dissociative disorders are perhaps the most controversial of all diagnostic categories, with some psychologists believing that they are often mistakenly diagnosed and others arguing they are underdiagnosed (Freeland & others, 1993; Rydberg, 2017; Sar & others, 2007). Two kinds of dissociative disorders are dissociative amnesia and dissociative identity disorder.

DISSOCIATIVE AMNESIA

Amnesia is the inability to recall important events. Amnesia can result from a blow to the head that produces trauma in the brain. **Dissociative amnesia** is a type of amnesia characterized by extreme memory loss that stems from extensive psychological stress (Bailey & Brand, 2017; Granacher, 2014). People experiencing dissociative amnesia remember everyday tasks like how to hail a cab and use a phone. They forget only aspects of their own identity and autobiographical experiences.

Sometimes people suffering from dissociative amnesia will also unexpectedly travel away from home, occasionally even assuming a new identity (Mamarde & others, 2013). For instance, on August 28, 2008, Hannah Upp, a 23-year-old middle school teacher in New York City, disappeared while out for a run (Marx & Didziulis, 2009). She had no wallet, no identification, no cell phone, and no money. Her family, friends, and roommates posted flyers around the city and messages on the Internet. As days went by, they became increasingly concerned that something terrible had happened. Finally, Hannah was found floating face down in the New York harbor on September 16, sunburned and dehydrated but alive. She remembered nothing of her experiences. To her, it felt like she had gone out for a run and 10 minutes later was being pulled from the harbor. No one knows how she survived during her two-week disappearance. Strangely enough, after recovering from her first experience, Hannah disappeared again years later and has not been found (Aviv, 2018).

Dissociative amnesia is believed to be caused by extremely stressful events. Treatment for this disorder is likely to involve psychotherapy that provides the person an opportunity to process the traumatic memories in safe context.

dissociative disorders
Psychological disorders that involve a sudden loss of memory or change in identity due to the dissociation (separation) of the individual's conscious awareness from previous memories and thoughts.

dissociative amnesia
Dissociative disorder characterized by extreme memory loss that is caused by extensive psychological stress.

At one point during her dissociative amnesia, Hannah Upp was approached by someone who asked if she was the Hannah everyone was looking for, and she answered no.

DISSOCIATIVE IDENTITY DISORDER

dissociative identity disorder (DID) Formerly called multiple personality disorder, a dissociative disorder in which the individual has two or more distinct personalities or selves, each with its own memories, behaviors, and relationships.

Dissociative identity disorder (DID), formerly called *multiple personality disorder,* is the most dramatic, least common, and most controversial dissociative disorder. People with this disorder have two or more distinct personalities or identities (Brand & others, 2014). Each identity has its own memories, behaviors, and relationships. One identity dominates at one time, another takes over at another time. People sometimes report that a wall of amnesia separates their different identities (Morton, 2017). However, research suggests that memory does transfer across these identities, even if the person believes it does not (Huntjens & others, 2014; Kong & others, 2008). The shift between identities usually occurs under distress (Sar & others, 2007), but sometimes it can also be controlled by the person (Kong & others, 2008).

FIGURE 2 **The Three Faces of Eve** Chris Sizemore, the subject of the 1950s book and film *The Three Faces of Eve,* is shown here with a work she painted, titled *Three Faces in One.*
Gerald Martineau/Getty Images

A famous real-life example of dissociative identity disorder is the "three faces of Eve" case, based on the life of a woman named Chris Sizemore, who died in 2016 (Thigpen & Cleckley, 1957) (Figure 2). Eve White was the original dominant personality. She had no knowledge of her second personality, Eve Black, although Eve Black had alternated with Eve White for a number of years. Eve White was reserved, quiet, and serious. By contrast, Eve Black was carefree, mischievous, and uninhibited. Eve Black would emerge at the most inappropriate times, leaving Eve White with hangovers and bills that she could not explain. During treatment, a third personality emerged: Jane. More mature than the other two, Jane seemed to have developed as a result of therapy.

The factors that contribute to DID remain something of a mystery. People with the disorder often report extraordinarily severe sexual or physical abuse during early childhood (Cook & others, 2017; Ross & Ness, 2010). Some psychologists believe that a child can cope with intense trauma by dissociating from the experience and developing other alternate selves as protectors. One study found that sexual abuse had occurred in as many as 70 percent or more of DID cases (Foote & others, 2006); however, the majority of people who have been sexually abused do not develop DID. The vast majority of people with DID are women. A genetic predisposition might also exist, as the disorder tends to run in families (Dell & Eisenhower, 1990).

Until the 1980s, only about 300 cases of dissociative identity disorder had ever been reported (Suinn, 1984). In the past 40 years, hundreds more cases have been diagnosed (Dorahy & others, 2014). Social cognitive psychologists point out that diagnoses have tended to increase whenever the popular media present a case. From this perspective, people develop multiple identities through social contagion. After exposure to these examples, people may be more likely to view multiple identities as a real condition. Some experts believe, in fact, that dissociative identity disorder is a *social construction*—that it represents a category some people adopt to make sense of their experiences (Spanos, 1996). Rather than being a single person with many conflicting feelings, wishes, and potentially awful experiences, the person compartmentalizes different aspects of the self into independent identities. In some cases, therapists have been accused of creating alternate personalities. Encountering a person who appears to have a fragmented sense of self, the therapist may begin to treat each fragment as its own "personality" (Spiegel, 2006).

Cross-cultural comparisons can shed light on whether dissociative identity disorder is primarily a response to traumatic events or the result of a social cognitive factor like social

The character Charlotte Wells in the Netflix show *Ratched* is an example of someone with DID.

contagion. If dissociation is a response to trauma, individuals with similar levels of traumatic experience should show similar degrees of dissociation, regardless of their exposure to cultural messages about dissociation. In China, the popular media *do not* commonly portray individuals with dissociative disorder, and professional knowledge of the disorder is rare. One study comparing people from China and Canada (where dissociative identity disorder is a widely publicized condition) found reports of traumatic experience to be similar across groups and to relate to dissociative experiences similarly as well (Ross & others, 2008), casting some doubt on the notion that dissociative experiences are entirely a product of social contagion.

Because they are rare, research documenting the effectiveness of treatments for dissociative disorders is limited. One study showed that long-term, in-depth, intensive psychotherapy led to improvement in functioning (Myrick & others, 2017). A key concern is that therapists remain objective in treatment and do not evoke multiple identities in the patient.

*Therapists and patients are making **attributions** to understand abnormal behavior.*

self-quiz

1. Dissociative identity disorder is associated with unusually high rates of
   A. anxiety.
   B. abuse during early childhood.
   C. depression.
   D. divorce.

2. Which of the following is true for post-traumatic stress disorder?
   A. It always occurs immediately after a trauma.
   B. It typically affects only people who have been in combat.
   C. It is associated with intrusive memories and flashbacks.
   D. It is not possible to treat PTSD.

3. The vulnerability-stress approach helps to account for which of the following?
   A. Among survivors of a mass shooting, only some will develop PTSD.
   B. Individuals with dissociative identity disorder have often heard of others with the disorder.
   C. People in a support group all share symptoms of PTSD.
   D. Dissociative identity disorder is quite uncommon.

APPLY IT! 4. Eddie often loses track of time. He is sometimes late for appointments because he is so engrossed in whatever he is doing. While working on a term paper in the library, he gets so caught up in what he is reading that he is shocked when he looks up and sees that the sun has set and it is night. Which of the following best describes Eddie?
   A. Eddie is showing signs of dissociative identity disorder.
   B. Eddie is showing signs of dissociative memory disorder.
   C. Eddie is showing normal dissociative states.
   D. Eddie is at risk for dissociative amnesia.

5 Disorders Involving Emotion and Mood

Our emotions and moods tell us how we are doing in life. We feel good or bad depending on our progress on important goals, the quality of our relationships, and so on. For some people, however, the link between life experiences and emotions is off-kilter. They may feel sad for no reason or a sense of elation in the absence of any great event. Several psychological disorders involve this kind of dysregulation in a person's emotional life. In this section we examine two such disorders—depressive disorders and bipolar disorders.

Depressive Disorders

Everyone feels blue sometimes. A romantic breakup, the death of a loved one, or a personal failure can cast a dark cloud over life. Sometimes, however, a person might feel deeply unhappy for a long time and not know why. **Depressive disorders** are disorders in which the person suffers from *depression,* an unrelenting lack of pleasure in life. Depressive disorders are extremely common. According to a 2017 survey, 7.1 percent of the U.S. population has suffered from depression at some point in their lives, with women being more likely to be affected than men (NIMH, 2019b). Depression often occurs in the

depressive disorders
Mood disorders in which the individual suffers from depression—an unrelenting lack of pleasure in life.

This painting by Vincent Van Gogh, Portrait of Dr. Gachet, reflects the extreme melancholy that characterizes the depressive disorders.
Art Reserve/Alamy Stock Photo

context of chronic physical illnesses, such as diabetes, heart disease, or chronic pain (Khandaker & others, 2020; Liu & others, 2020; Wolock & others, 2020).

A variety of cultures recognize depression, and studies have shown that across cultures depression is characterized as involving an absence of joy, low energy, and high levels of sadness (Chang & others, 2017; Chen & others, 2017). Culture may influence the ways people describe their experiences. For instance, people from Eastern cultures may be less likely to talk about their emotional states, and more likely to describe depressive symptoms in terms of bodily feelings and symptoms, than those from Western cultures (Draguns & Tanaka-Matsumi, 2003). Many successful individuals have been diagnosed with depression. They include musician Sheryl Crow, actors Drew Barrymore, Halle Berry, and Jim Carrey, artist Pablo Picasso, and the late astronaut Buzz Aldrin (the second moon walker).

Major depressive disorder (MDD) involves a significant depressive episode and depressed characteristics, such as lethargy and hopelessness, for at least two weeks. MDD impairs daily functioning, and the National Institute of Mental Health (NIMH) has called it the leading cause of disability in the United States (NIMH, 2019b). The symptoms of major depressive disorder may include

- Depressed mood most of the day
- Reduced interest or pleasure in activities that were once enjoyable
- Significant weight loss or gain or significant decrease or interest in appetite
- Trouble sleeping or sleeping too much
- Fatigue or loss of energy
- Feeling worthless or guilty in an excessive or inappropriate manner
- Problems in thinking, concentrating, or making decisions
- Recurrent thoughts of death and suicide
- No history of manic episodes (periods of euphoric mood)

People who experience less extreme depressive mood for more than two months may be diagnosed with *persistent depressive disorder.* This disorder includes symptoms such as hopelessness, lack of energy, poor concentration, and sleep problems.

What are the causes of depressive disorders? A variety of biological, psychological, and sociocultural factors have been implicated in their development.

BIOLOGICAL FACTORS
The biological factors implicated in depressive disorders include genes, brain structure and function, and neurotransmitters.

Genes explain about 40 percent of the variability we see in depression (Shadrina & others, 2018). Research seeking to identify precise genetic locations associated with depression has produced mixed results (Yu & others, 2018). For instance, depression has been linked to features of the serotonin transporter gene called 5-HTTLPR (Warnke & others, 2020). A classic and controversial study showed that this genetic characteristic predicted depression but only in concert with a stressful social environment, suggesting a stress × gene interaction (Caspi & others, 2003). More recently, over 70 scientists collaborated on a very large meta-analysis combining data from nearly 40,000 people seeking to replicate this pattern (Culverhouse & others, 2018). The results showed no evidence for a stress × gene interaction. The only two predictors of depression were experiencing life stress and being a woman (Culverhouse & others, 2018). Whether the serotonin transporter gene plays a role in depression, with or without stressful events, remains a controversial topic of research.

Keep in mind that depression, sometimes called the "common cold" of psychological disorders, is so commonplace that any biological characteristics associated with the disorder cannot be rare. Indeed, it is estimated that all humans carry some level of genetic risk for depression (Wray & others, 2018).

With regard to brain structures, people with depression show lower levels of brain activity in a section of the prefrontal cortex that is involved in generating actions (Duman & others, 2012; Ghosal & others, 2017) and in regions of the brain associated with the perception of rewards in the environment (Gaffrey & others, 2018; Heshmati & others,

major depressive disorder (MDD)
Psychological disorder involving a major depressive episode and depressed characteristics, such as lethargy and hopelessness, for at least two weeks.

2020). In depression, a person's brain may not recognize opportunities for pleasurable experiences. In addition, brain imaging research shows that when faced with negative stimuli, people with depression are less likely to spontaneously activate those areas of the brain that serve to regulate emotion (Radke & others, 2018). Depression also likely involves problems in neurotransmitter regulation. People with depressive disorder appear to have too few receptors for the neurotransmitters serotonin and norepinephrine (Meng & others, 2020; Wei & others 2020).

PSYCHOLOGICAL FACTORS   Psychological explanations of depression draw on behavioral learning theories and cognitive theories. One behavioral view of depression focuses on *learned helplessness,* which involves an individual's feelings of powerlessness after exposure to aversive circumstances over which the person has no control. One theory is that learned helplessness can lead people to become depressed (Seligman, 1975). Learned helplessness interferes with learning and disrupts motivation (Reznik & others, 2017). When individuals cannot control important outcomes in their lives and have no way to avoid negative experiences, they eventually feel helpless and stop trying to change their situations. This helplessness spirals into hopelessness.

Cognitive explanations of depression focus on the thoughts and beliefs that contribute to this sense of hopelessness (Alloy & others, 2017). Psychiatrist Aaron Beck (1967) proposed that negative thoughts reflect self-defeating beliefs that shape depressed individuals' experiences. These habitual negative thoughts magnify and expand depressed persons' negative experiences (Lam, 2012). For example, a person with depression might overgeneralize about a minor occurrence—say, turning in a work assignment late—and consider oneself to be worthless. A person with depression might view a minor setback such as getting a *D* on a paper as the end of the world. The accumulation of such cognitive distortions can lead to depression (Fazakas-DeHoog & others, 2017; Kiosses & others, 2017).

The way people think can also influence the course of depression. People with depression may ruminate on negative experiences and negative feelings, playing them over and over again in their minds (Zhou & others, 2020). This tendency to ruminate is associated with the development of depression as well as other psychological problems such as binge eating and substance abuse (McLaughlin & others, 2014; Tait & others, 2014). In addition, the relationship between rumination and depression may run both ways, as depression can spur a person to adopt a ruminative mindset (Krause & others, 2017).

Another cognitive view of depression focuses on people's attributions—their attempts to explain what caused something to happen (Moore & others, 2016; Rueger & George, 2017). Depression is thought to be related to a *pessimistic* attributional style. In this style, people regularly explain negative events as having internal causes ("It is my fault I failed the exam"), stable causes ("I'm going to fail again and again"), and global causes ("Failing this exam shows that I won't do well in any of my courses"). Pessimistic attributional style means blaming oneself for negative events and expecting the negative events to recur (Abramson & others, 1978). This pessimistic style can be contrasted with an *optimistic* attributional style, which is essentially its opposite. Optimists make external attributions for bad things that happen ("I did badly on the test because it's hard to know what a professor wants on the first exam"). They also recognize that these causes can change ("I'll do better on the next one") and that they are specific ("It was only one test"). Optimistic attributional style has been related to lowered depression and decreased suicide risk in a variety of samples (Horwitz & others, 2017).

SOCIOCULTURAL FACTORS   People with low socioeconomic status (SES), especially people living in poverty, are more likely to develop depression than their higher-SES counterparts (Joshi & others, 2017; Linder & others, 2020). A longitudinal study of adults revealed that depression increased as one's standard of living and employment circumstances worsened (Lorant & others, 2007). Studies have found very high rates of depression in Native American groups, among whom poverty, hopelessness, and alcoholism are widespread (Kisely & others, 2017).

Lifetime Rate per 100 People

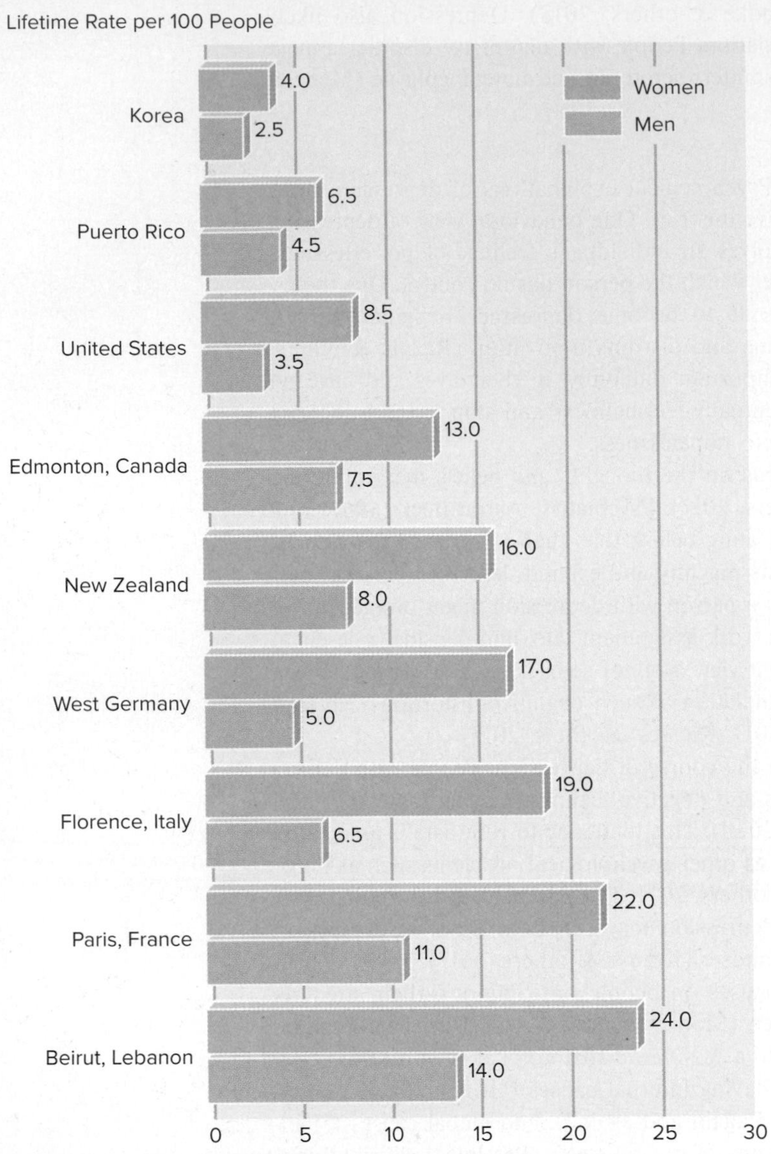

FIGURE 3 **Gender Differences in Depression Across Cultures** This graph shows the rates of depression for men and women in nine cultures (Weissman & Olfson, 1995).

SOURCE: Weissman and Olfson, Science 269(779), 1995. *American Association for the Advancement of Science.*

> **Which cultures have the highest and lowest rates of depression? What might account for these differences?** > **Which cultures have the largest gender difference in depression? What might account for these differences?** > **To be diagnosed with depression, a person has to seek treatment for the disorder. How might gender and culture influence a person's willingness to get treatment?**

Women are nearly twice as likely as men to develop depression (Hodes & others, 2017; Salk & others, 2017). As Figure 3 shows, this gender difference occurs in many countries (Inaba & others, 2005). It also occurs across many different age groups (Salk & others, 2017). Potential explanations of this difference range from differences in biological systems to social structural differences in the treatment of men and women.

There are numerous effective treatments for depression. These include a variety of psychotherapies that target different aspects of the disorder, including reducing self-defeating beliefs and automatic thoughts, reducing distress, and helping to improve emotion regulation. In addition, medications, especially those aimed at regulating serotonin and norepinephrine, are commonly used to treat depression.

Another gender difference to consider: Why might men show lower levels of depression than women?

Bipolar Disorder

Just as we all have down times, we also have times when things seem to be going phenomenally well. For people with bipolar disorder, the ups and downs of life take on an extreme and often harmful tone. **Bipolar disorder** is a disorder characterized by extreme mood swings that include one or more episodes of *mania,* an overexcited, unrealistically optimistic state. A manic episode is like the flipside of a depressive episode. The person who experiences mania feels on top of the world. Imagine all the things you might do if you were filled with excitement and confidence, even if those feelings were not rooted in reality. You may have tremendous energy and might sleep very little. You might do many things that you should not. A manic state features an impulsivity that can get the person in trouble. For example, the person might spend their life savings on a foolish business venture.

Most people with bipolar disorder experience multiple cycles of depression interspersed with mania, usually separated by six months to a year. Unlike depressive disorders, bipolar disorder shows no gender

bipolar disorder
Mood disorder characterized by extreme mood swings that include one or more episodes of mania, an overexcited, unrealistically optimistic state.

There are two types of bipolar disorder. In bipolar I disorder, people have very extreme manic episodes and may even hallucinate, seeing or hearing things that are not there. Bipolar II disorder is less severe, involving a less extreme level of euphoria.

difference. Bipolar disorder does not prevent a person from being successful. Award-winning actor Catherine Zeta-Jones, dancer and choreographer Alvin Ailey, and the late actor-writer Carrie Fisher (Princess Leia) were diagnosed with bipolar disorder.

What factors play a role in the development of bipolar disorder? Genetic influences are stronger predictors of bipolar disorder than of depressive disorders (Gordovez & McMahon, 2020). A person with an identical twin who has bipolar disorder has a 70 percent probability of also having the disorder, and a fraternal twin has a more than 10 percent probability (Baselmans & others, 2020) (Figure 4). Researchers continue to explore the specific genetic location of bipolar disorder (Gordovez & McMahon, 2020).

Other biological processes are also a factor. Like depression, bipolar disorder is associated with differences in brain activity. Figure 5 shows the metabolic activity in the cerebral cortex of an individual cycling through depressive and manic phases. Notice the decrease in metabolic activity in the brain during depression and the increase in metabolic activity during mania (Baxter & others, 1985). In addition to high levels of norepinephrine and low levels of serotonin, studies link high levels of the neurotransmitter glutamate to bipolar disorder (Corcoran & others, 2020; Giridharan & others, 2020; Sonmez & others, 2020).

Bipolar disorder is typically treated with the medication lithium. Lithium is thought to stabilize mood, but the precise way that it does so is not fully understood (Miller & others, 2017). People with bipolar can benefit from psychotherapy (Sylvia & others, 2017). Research continues to examine the optimal match between persons and treatments to most effectively relieve the suffering of those with bipolar disorder.

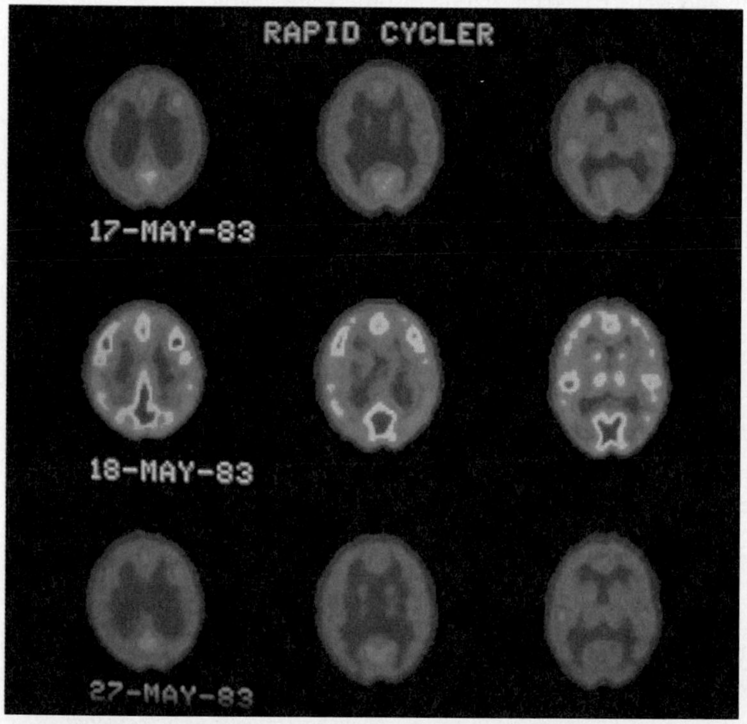

Psychologists and psychiatrists have noted cases of children who appear to suffer from bipolar disorder (Wiggins & others, 2020). A key dilemma in such cases is that lithium is not approved for children's use. The side effects of these drugs could put children's health and development at risk. To address this issue, *DSM-5* includes the diagnosis, *disruptive mood dysregulation disorder,* which is considered a depressive disorder in children who show persistent irritability and recurrent episodes of out-of-control behavior (APA, 2013). This decision is not without controversy. As we saw with ADHD, some children who are perceived to be prone to wild mood swings may be simply behaving like children. In addition, it is notable that cases of suspected bipolar disorder in children are far more common in the United States than in Europe or Canada (Post & others, 2017).

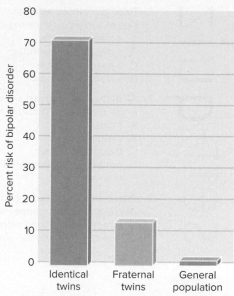

FIGURE 4 **Risk of Bipolar Disorder in Identical and Fraternal Twins If One Twin Has the Disorder, and in the General Population** Notice how much stronger the similarity of bipolar disorder is in identical twins as compared with fraternal twins and the general population. These statistics suggest a strong genetic role in the disorder.

SOURCE: *Annual Review of Neuroscience*, 20, 1997. Annual Reviews. www.annualreviews.org

FIGURE 5 **Brain Metabolism in Mania and Depression** PET scans of an individual with bipolar disorder, who is described as a rapid cycler because of how quickly severe mood changes occurred. (*top and bottom*) The person's brain in a depressed state. (*middle*) A manic state. The PET scans reveal how the brain's energy consumption falls in depression and rises in mania. The red areas in the middle row reflect rapid consumption of glucose.
Courtesy of Dr. Michael Phelps, UCLA School of Medicine

1. To be diagnosed with bipolar disorder, an individual must experience
   A. a manic episode.
   B. a depressive episode.
   C. a psychotic break.
   D. a suicidal episode.

2. All of the following are a symptom of major depressive disorder except
   A. fatigue.
   B. weight change.
   C. thoughts of death.
   D. periods of euphoria and very high energy.

3. A true statement about genes and depression is that genes
   A. directly cause depression. If you have a particular genetic characteristic you will certainly develop depression.
   B. may produce a vulnerability to depression, meaning that if you have a particular genetic characteristic you might be at risk for depression, depending on your experience.
   C. are unrelated to depression.
   D. are the most important factor in determining whether someone develops depression.

APPLY IT! 4. During his first two college years, Barry has felt "down" most of the time. He has had trouble concentrating and difficulty making decisions. Sometimes he is so overwhelmed with deciding on his major and struggling to focus that he feels hopeless. He has problems with loss of appetite and sleeps a great deal of the time, and in general his energy level is low. Barry has found that things he used to love, like watching sports and playing video games, are just no fun anymore. Which of the following is most likely to be true of Barry?
   A. Barry has an anxiety disorder.
   B. Barry is entering the depressive phase of bipolar disorder.
   C. Barry is likely experiencing depression.
   D. Barry is experiencing the everyday blues that everyone gets from time to time.

6 Eating Disorders

For some people, concerns about weight and body image become a serious, debilitating disorder. For such individuals, the very act of eating is an arena where a variety of complex biological, psychological, and cultural issues are played out, often with tragic consequences.

A number of famous people have coped with eating disorders, including Taylor Swift, Paula Abdul, and Kelly Clarkson. Eating disorders are characterized by extreme disturbances in eating behavior—from eating very, very little to eating a great deal. In this section we examine three eating disorders—anorexia nervosa, bulimia nervosa, and binge-eating disorder.

〰 *Disorders of eating can vary across cultures. In Fiji, a disorder known as* **macake** *involves poor appetite and refusing to eat. Very high levels of social concern meet this refusal, and individuals with macake are strongly motivated to start eating and enjoying food again.*

Anorexia Nervosa

anorexia nervosa
Eating disorder that involves the relentless pursuit of thinness through starvation.

Anorexia nervosa is an eating disorder that involves the relentless pursuit of thinness through starvation. Anorexia nervosa is far more common in girls and women than in boys and men, affecting about 0.9 percent of American women in their lifetime (Caceres, 2020). *DSM-5* (APA, 2013) lists these main characteristics of anorexia nervosa:

- Significantly low body weight compared to what is considered normal for age and height, and refusal to maintain weight at a healthy level.

- An intense fear of gaining weight that does not decrease with weight loss.

- A distorted body image. Even when people with anorexia nervosa are extremely thin, they never think they are thin enough.

Over time, anorexia nervosa can lead to physical changes, such as the growth of fine hair all over the body, thinning of bones and hair, severe constipation, and low blood pressure. Dangerous and even life-threatening complications include damage to the heart and thyroid. Anorexia nervosa is said to have the highest mortality rate (about 5.6 percent of people with anorexia nervosa die within 10 years of diagnosis) of any psychological disorder (Hoek, 2006).

Pop singer Demi Lovato has been open about her struggles with disordered eating. She has shared that many women in her family struggled with bulimia nervosa and anorexia. Lovato has been public about her own experiences in the hopes of raising awareness that recovery from eating disorders is possible.
Kristina Bumphrey/StarPix/REX/Shutterstock

Anorexia nervosa typically begins in the teenage years, often following an episode of dieting and some type of life stress (Solmi & others, 2015). Most people with anorexia nervosa are non-Latina white female adolescents or young adults from middle- and upper-income families (Darcy, 2012; Dodge, 2012). They are often high-achieving perfectionists (Forbush & others, 2007). Obsessive thinking about weight and compulsive exercise are also related to anorexia nervosa (Hildebrandt & others, 2012). People with anorexia nervosa seem to be very controlled in what they are doing. Yet, the disorder itself is a type of impulsivity: It involves engaging in very risky behaviors (Lavender & others, 2017).

People with anorexia nervosa may lack personal distress over their symptoms. Recall that personal distress over one's behavior is just one aspect of the definition of abnormal.

Bulimia Nervosa

bulimia nervosa
Eating disorder in which an individual (typically female) consistently follows a binge-and-purge eating pattern.

Bulimia nervosa is an eating disorder in which a person consistently follows a binge-and-purge eating pattern. They go on an eating binge and then purge by self-induced vomiting or the use of laxatives. Bulimia nervosa affects about 1.5 percent of American women in their lifetimes (Caceres, 2020). Most people with bulimia nervosa are preoccupied with food, have a strong fear of becoming overweight, and are depressed or anxious. Because bulimia nervosa occurs within a normal weight range, the disorder is often difficult to detect. A person with bulimia nervosa usually keeps the disorder a secret and experiences a great deal of self-disgust and shame. Bulimia nervosa may be on the rise among people who have diabetes and may engage in purging after indulging in unhealthy food (Coleman & Caswell, 2020).

Corbis/VCG/Getty Images

Bulimia nervosa can lead to complications such as a chronic sore throat, kidney problems, dehydration, and gastrointestinal disorders. The disorder is also related to dental problems, as persistent exposure to the stomach acids in vomit can wear away tooth enamel.

Bulimia nervosa typically begins in late adolescence or early adulthood (Levine, 2002). Like those with anorexia nervosa, many young women who develop bulimia nervosa are highly perfectionistic (Lampard & others, 2012). At the same time, they tend to have low levels of self-efficacy (Sysko & others, 2017). In other words, these are young women with very high standards but very low confidence that they can achieve their goals. Impulsivity, negative emotion, and obsessive-compulsive disorder are also related to bulimia (Roncero & others, 2011). Bulimia nervosa is associated, too, with a high incidence of sexual and physical abuse in childhood (Lo Sauro & others, 2008). Finally, brain imaging research suggests that women with bulimia nervosa may use binge eating as a way to distract themselves from negative feelings (Collins & others, 2017).

Dentists and dental hygienists are sometimes the first to recognize the signs of bulimia nervosa.

Although much more common in women, bulimia can also affect men. Elton John has described his struggles with this eating disorder.

Anorexia Nervosa and Bulimia Nervosa: Causes and Treatments

What is the etiology of anorexia nervosa and bulimia nervosa? For many years researchers thought that sociocultural factors, such as media images of very thin women, were the central determinants of these disorders (Le Grange & others, 2010). Media images that glorify extreme thinness can indeed influence women's body image and eating concerns (White & others, 2016). In addition, emphasis on the thin ideal is related to anorexia nervosa and bulimia nervosa (Benowitz-Fredericks & others, 2012). However, as powerful as these media messages might be, countless girls and women are exposed to media images of unrealistically thin women, but relatively few develop eating disorders. Many young women embark on diets, but comparatively few of them develop eating disorders.

So, it cannot just be media images that lead to these disorders.

In addition, eating disorders occur in cultures that do not emphasize the ideal of thinness, although the disorders may differ from Western descriptions. For instance, in Eastern cultures, people can show the symptoms of anorexia nervosa, but they lack the fear of getting fat that is common in North Americans with the disorder (Pike & others, 2011).

Contemporary researchers have increasingly probed the potential biological underpinnings of eating disorders, examining in particular the interplay of social and biological factors. Genes play a substantial role in both anorexia nervosa and bulimia nervosa (Yilmaz & others, 2020). Indeed, genes influence many psychological characteristics (for example, perfectionism, impulsivity, obsessive-compulsive tendencies, and thinness drive) and behaviors (restrained eating, binge eating, self-induced vomiting) that are associated with anorexia nervosa and bulimia nervosa (Laplana & others, 2014; Tang & others, 2014). These genes are also factors in the regulation of serotonin, and problems in regulating serotonin are related to both anorexia nervosa and bulimia nervosa (Rozenblat & others, 2017).

Even as biological factors play a role in the emergence of eating disorders, eating disorders themselves affect the body, including the brain. While social factors and experiences may play a role in triggering dieting, the physical effects of dieting, bingeing, and purging may change neural networks that then sustain the disordered pattern, in a kind of vicious cycle (Fuglset & others, 2014; Kaye & others, 2013).

Although anorexia and bulimia nervosa are serious disorders, recovery is possible (Fitzpatrick, 2012; Kuipers & others, 2017). Anorexia nervosa may require hospitalization. The first target of intervention is promoting weight gain, in extreme cases using a feeding tube. A common obstacle in the treatment of anorexia nervosa is that people with the disorder deny that anything is wrong. They maintain their belief that thinness and restrictive dieting are correct and not a sign of mental illness (Wilson & others, 2007). Still, drug therapies and psychotherapy (especially cognitive therapy and family therapy) are effective in treating anorexia nervosa, as well as bulimia nervosa (Andries & others, 2014; Bailey & others, 2014; Ciao & others, 2015).

Binge-Eating Disorder

binge-eating disorder (BED)
Eating disorder characterized by recurrent episodes of consuming large amounts of food during which the person feels a lack of control over eating.

Binge-eating disorder (BED) is characterized by recurrent episodes of consuming large amounts of food during which the person feels a lack of control over eating (APA, 2013). Unlike a person with bulimia nervosa, someone with BED does not try to purge. Most people with BED have overweight or obesity (Carrard & others, 2012).

People with BED often eat quickly, eat a great deal when they are not hungry, and eat until they are uncomfortably full. They frequently eat alone because of embarrassment or guilt, and they feel ashamed and disgusted with themselves after overeating. BED is the most common of all eating disorders—affecting people of different gender identities and ethnic groups within the United States more similarly than anorexia nervosa or bulimia nervosa. BED is thought to affect 3.5 percent of women and 2 percent of men in their lifetime (Caceres, 2020). BED is present in 8 percent of people with obesity (Grilo & others, 2010). The complications of BED are those of obesity more generally, including diabetes, hypertension, and cardiovascular disease.

Binge-Eating Disorder: Causes and Treatments

Researchers are examining the role of biological and psychological factors in BED. Genes play a role (Cuesto & others, 2017) as does dopamine, the neurotransmitter related to reward pathways in the brain (Heal & others, 2017). The fact that binge eating often occurs after stressful events suggest that binge eaters use food to regulate their emotions (Weinbach & others, 2018). The areas of the brain and endocrine system that respond to stress are overactive in individuals with BED (Cuesto & others, 2017; Lo Sauro & others, 2008), and this overactivity leads to high levels of circulating cortisol, the hormone most associated with stress. People with BED may be more likely to perceive events as stressful and then seek to manage that stress by binge eating.

Just as treatment for anorexia nervosa first focuses on weight gain, some believe that treatment for BED should first target weight loss (DeAngelis, 2002). Others argue that people with BED must be treated for disordered eating per se, and they insist that if the underlying psychological issues are not addressed, weight loss will not be successful or permanent (Hay & others, 2009). Treatments for BED include psychotherapies that address beliefs and cognitions and emotion regulation as well as medications that regulate mood and assist in weight loss (Brownley & others, 2017).

Unlike individuals with anorexia nervosa or bulimia nervosa, most people with binge-eating disorder are affected by excess weight or obesity.
Digital Vision/Photodisc/Getty Images

self-quiz

1. The main characteristics of anorexia nervosa include all of the following *except*
   A. refusal to maintain healthy body weight.
   B. distorted image of one's body.
   C. strong fears of weight gain even as weight loss occurs.
   D. intense and persistent tremors.

2. A person with bulimia nervosa typically
   A. thinks a lot about food.
   B. is considerably underweight.
   C. is a male.
   D. is not overly concerned about gaining weight.

3. The most common of all eating disorders is
   A. bulimia nervosa.
   B. anorexia nervosa.
   C. binge-eating disorder.
   D. gastrointestinal disease.

APPLY IT! 4. Nancy is a first-year straight-*A* premed major. Nancy's roommate Luci notices that Nancy has lost a great deal of weight and is extremely thin. Luci observes that Nancy works out a lot, rarely finishes meals, and wears bulky sweaters all the time. Luci also notices that Nancy's arms have fine hairs growing on them, and Nancy has mentioned never getting her period

anymore. When Luci asks Nancy about her weight loss, Nancy replies that she is very concerned that she not gain the "freshman 15" and is feeling good about her ability to keep up with her work and keep off those extra pounds. Which of the following is the most likely explanation for what is going on with Nancy?
   A. Nancy likely has bulimia nervosa.
   B. Despite her lack of personal distress about her symptoms, Nancy may have anorexia nervosa.
   C. Nancy has binge-eating disorder.
   D. Given Nancy's overall success, it seems unlikely that she is suffering from a psychological disorder.

7 Schizophrenia

psychosis
A state in which a person's perceptions and thoughts are fundamentally removed from reality.

Schizophrenia is a severe psychological disorder that is characterized by highly disordered thought processes. These disordered thoughts are referred to as *psychotic*. **Psychosis** refers to a severely impaired psychological condition in which a person's thoughts and emotions are divorced from reality. The world of the person with schizophrenia is deeply frightening and chaotic.

Schizophrenia is usually diagnosed in early adulthood, around age 18 for men and 25 for women. People with schizophrenia may see things that are not there, hear voices inside their head, and live in a strange world of twisted logic. They may say odd things, show

schizophrenia
Severe psychological disorder characterized by highly disordered thought processes; individuals suffering from schizophrenia may be referred to as psychotic because they are so far removed from reality.

inappropriate emotion, and move their bodies in peculiar ways. Often, they are socially withdrawn and isolated.

It is difficult to imagine the ordeal of people living with schizophrenia, who constitute about half of the patients in psychiatric hospitals. Although schizophrenia is less common than disorders like depression, those with schizophrenia are more likely to require hospitalization. The suicide risk for people with schizophrenia is eight times that for the general population (Aslan & others, 2020; Roy & Pompili, 2016; Pompili & others, 2007). For many with the disorder, controlling it means using powerful medications to combat symptoms. The most common cause of relapse is that individuals stop taking their medication. They might do so because they feel better and believe they no longer need the drugs, they do not realize that their thoughts are disordered, or the side effects of the medications are too unpleasant.

Symptoms of Schizophrenia

Psychologists generally classify the symptoms of schizophrenia as positive symptoms, negative symptoms, and cognitive deficits.

POSITIVE SYMPTOMS   The positive symptoms of schizophrenia are marked by a distortion or an excess of typical function. They are "positive" because they reflect something added above and beyond normal behavior. Positive symptoms of schizophrenia include hallucinations, delusions, thought disorders, and disorders of movement.

hallucinations
Sensory experiences that occur in the absence of real stimuli.

Hallucinations are sensory experiences that occur in the absence of real stimuli. Hallucinations are usually auditory—the person might complain of hearing voices—or visual, and much less commonly smells or tastes (Galderisi & others, 2014; Thakkar & others, 2020). Culture affects the form hallucinations take, as well as their content and sensory modality (Bauer & others, 2011). Visual hallucinations involve seeing things that are not there, as in the case of Moe Armstrong. At the age of 21, while serving in Vietnam as a Marine medical corpsman, Armstrong experienced a psychotic break. Dead Vietcong soldiers appeared to talk to him and beg him for help and did not seem to realize that they were dead. Armstrong, now a successful businessman and a sought-after public speaker who holds two master's degrees, relies on medication to keep such experiences at bay (Bonfatti, 2005).

delusions
False, unusual, and sometimes magical beliefs that are not part of an individual's culture.

Delusions are false, unusual, and sometimes magical beliefs that are not part of an individual's culture. A delusional person might think that he is Jesus Christ or Muhammad; another might imagine that their thoughts are being broadcast over the radio. It is crucial to distinguish delusions from cultural ideas such as the religious belief that a person can have divine visions or communicate personally with a deity. Generally, psychology and psychiatry do not treat these ideas as delusional.

For people with schizophrenia, delusional beliefs that might seem completely illogical to the outsider are experienced as all too real. At one point in his life, Bill Garrett, a college student with schizophrenia, was convinced that a blister on his hand was a sign of gangrene. So strong was his belief that he tried to cut off his hand with a knife, before being stopped by his family (Park, 2009).

Thought disorder refers to the unusual, sometimes bizarre thought processes that are characteristic positive symptoms of schizophrenia (Morgan & others, 2017). The thoughts of persons with schizophrenia can be disorganized and confused. Often individuals with schizophrenia do not make sense when they talk or write. For example, someone with schizophrenia might say, "Well, Rocky, babe, happening, but where, when, up, top, side, over, you know, out of the way, that's it. Sign off." These incoherent, loose word associations, called *word salad,* have no meaning for the listener. The person might also make up new words (called *neologisms*) (Orlov & others, 2017). In addition, a person with schizophrenia can show **referential thinking**, which means giving personal meaning to completely random events (So & others, 2017). For instance, they might believe that a stranger who happens to make eye contact with them is sending them a message.

referential thinking
Ascribing personal meaning to completely random events.

catatonia
State of immobility and unresponsiveness, lasting for long periods of time.

A final type of positive symptom is *disorders of movement*. A person with schizophrenia may show unusual mannerisms, body movements, and facial expressions. The person may repeat certain motions over and over or, in extreme cases, may become catatonic. **Catatonia** is a state of immobility and unresponsiveness that lasts for long periods of time. Although these symptoms of schizophrenia may lead others to perceive these people as strange, individuals with this disorder can lead creative lives (Figure 6).

NEGATIVE SYMPTOMS
Schizophrenia's positive symptoms are characterized by a distortion or an excess of typical functions. In contrast, schizophrenia's negative symptoms reflect the loss or decrease of typical functions. These negative symptoms can include social withdrawal, behavioral deficits, and apparent lack of emotion. One negative symptom is

flat affect
The display of little or no emotion—a common negative symptom of schizophrenia.

FIGURE 6 People with schizophrenia can be productive and creative. Artist Adam Raven was diagnosed with schizophrenia and thought he would never paint again. However, with treatment he has become an advocate for the importance of creativity in the lives of those with severe psychological disorders.
Alex Lentati/Evening Standard/REX/Shutterstock

flat affect, which means the display of little or no emotion (Cho & others, 2017; Parola & others, 2020). People with schizophrenia also may lack the ability to read the emotions of others (Wearne & McDonald, 2020). They may experience a lack of positive emotional experience in daily life and show a deficient ability to plan, initiate, and engage in goal-directed behavior. Importantly, the appearance of the absence of emotion may mask emotional experience, as individuals with schizophrenia report more negative emotion and less positive emotion than those without the disorder in their daily lives (Cho & others, 2017).

Because negative symptoms are not as obviously part of a psychiatric illness, people with schizophrenia may be perceived as lazy and unwilling to better their lives.

COGNITIVE SYMPTOMS
Cognitive symptoms of schizophrenia include difficulty sustaining attention, problems holding information in memory, deficits in the ability to self-reflect, and inability to interpret information and make decisions (García-Mieres & others, 2020; Nestor & others, 2020; Teigset & others, 2020). These symptoms may be subtle and are often detected only through neuropsychological tests. Researchers now recognize that to understand schizophrenia's cognitive symptoms fully, measures of these symptoms must be tailored to particular cultural contexts (Ishikawa & others, 2017; Mehta & others, 2011).

Causes of Schizophrenia

A great deal of research has focused on schizophrenia's causes, including biological, psychological, and sociocultural factors involved in the disorder.

BIOLOGICAL FACTORS
Research provides strong support for biological explanations of schizophrenia. Especially compelling is the evidence for a genetic predisposition. However, structural brain abnormalities and neurotransmitters also are linked to this severe psychological disorder.

Genes Research supports the notion that schizophrenia is at least partially due to genetic factors (Birnbaum & Weinberger, 2020). As genetic similarity to a person with schizophrenia increases, so does a person's risk of developing schizophrenia, as Figure 7 shows (Cardno & Gottesman, 2000). Researchers are seeking to pinpoint the chromosomal location of

FIGURE 7 Lifetime Risk of Developing Schizophrenia According to Genetic Relatedness As genetic relatedness to an individual with schizophrenia increases, so does the risk of developing schizophrenia.

SOURCE: Irving Gottesman, 2004.

> Which familial relationships have the lowest and highest level of genetic overlap?
> What is the difference in genetic overlap between identical twins and nontwin siblings? > What is the difference in risk of schizophrenia between identical twins and nontwin siblings?

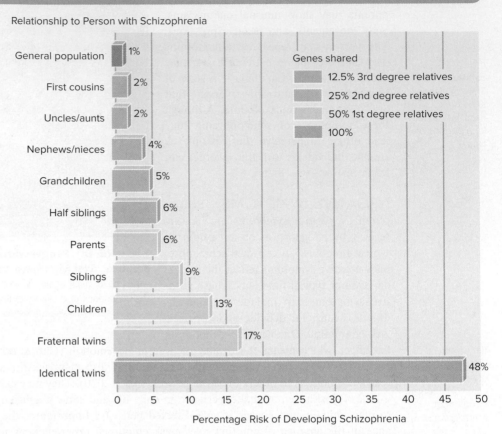

Relationship to Person with Schizophrenia

Genes shared
- 12.5% 3rd degree relatives
- 25% 2nd degree relatives
- 50% 1st degree relatives
- 100%

General population — 1%
First cousins — 2%
Uncles/aunts — 2%
Nephews/nieces — 4%
Grandchildren — 5%
Half siblings — 6%
Parents — 6%
Siblings — 9%
Children — 13%
Fraternal twins — 17%
Identical twins — 48%

Percentage Risk of Developing Schizophrenia

genes involved in susceptibility to schizophrenia (Marshall & others, 2018). Identifying the specific genetic characteristics that enhance vulnerability to a psychotic episode may help in devising early interventions and treatments (Matsumoto & others, 2017).

Structural Brain Differences Studies have found structural brain differences in people with schizophrenia. Imaging techniques such as MRI show enlarged ventricles in the brain (Akudjedu & others, 2020; Tronchin & others, 2020). Ventricles are fluid-filled spaces, and enlargement of the ventricles indicates the deterioration in other brain tissue (Eom & others, 2020). People with schizophrenia also have a smaller frontal cortex (the area in which thinking, planning, and decision making take place) and show less activity in this area than people who do not have schizophrenia (Hassan & others, 2017; Kindler & others, 2020; Nestor & others, 2020).

Still, the differences between the brains of healthy individuals and those with schizophrenia are small. Microscopic studies of brain tissue after death reveal small changes in the distribution or characteristics of brain cells in persons with schizophrenia. It appears that many of these changes occurred prenatally, because they are not accompanied by glial cells, which are always present when a brain injury occurs after birth. It may be that problems in prenatal development such as infections (De Picker & others, 2017) predispose a brain to developing schizophrenic symptoms during puberty and young adulthood (Fatemi & Folsom, 2009). One study sought to trace genetic differences to brain structure and function in schizophrenia (Sekar & others, 2016). Recall that part of brain development involves the process of synaptic pruning. This means that connections that are weak or rarely used are pruned away as the brain matures. This study showed that a specific genetic feature that guides the pruning process was faulty in those with schizophrenia. For these individuals the pruning process was faster or more extensive than in typical development (Sekar

& others, 2016). Such results suggest that during late adolescence and early adulthood (when psychotic symptoms often first appear), brain development may be going awry.

Problems in Neurotransmitter Regulation Whether it is the amount of dopamine, its production, or its uptake, there is good evidence that problems regulating the neurotransmitter dopamine play a role in schizophrenia (Duan & others, 2017; Sonnenschein & others, 2020; Stark & others, 2020). A link between dopamine and psychotic symptoms was first noticed when the drug L-dopa (which increases dopamine levels) was given to people as a treatment for Parkinson disease. In addition to relieving Parkinson symptoms, L-dopa caused some to experience psychosis (Janowsky & others, 1987; Madras, 2013). Further, drugs that reduce psychotic symptoms often block dopamine (van Os & Kapur, 2009).

Previously we have encountered dopamine as the "feel good" neurotransmitter that helps us recognize rewarding stimuli in the environment. How can a neurotransmitter that is associated with good things play a crucial role in schizophrenia?

One way to think about the role of dopamine in schizophrenia is to view dopamine as a neurochemical messenger that in effect shouts out, "Hey! This is important!" whenever we encounter opportunities for reward. Imagine what it might be like to be bombarded with such messages about even the smallest details of life (Kapur, 2003). The person's own thoughts might take on such dramatic proportions that they sound like someone else's voice talking inside the person's head. Fleeting ideas such as "It's raining today because I didn't bring my umbrella to work" suddenly seem not silly but true. Shitij Kapur (2003) has suggested that hallucinations, delusions, and referential thinking may be expressions of the person's attempts to make sense of such extraordinary feelings.

*Poorly regulated dopamine basically tells the person that **everything** is important.*

In addition to dopamine, glutamate is implicated in schizophrenia and in psychotic states more generally (Coyle & others, 2020).

PSYCHOLOGICAL FACTORS

Psychologists used to explain schizophrenia as rooted in childhood experiences with unresponsive parents. Such explanations have fallen by the wayside. Contemporary theorists do recognize that stress may contribute to the development of this disorder. Experiences are now viewed through the lens of the vulnerability-stress hypothesis, described earlier, suggesting that people with schizophrenia may have biological risk factors that interact with experience to produce the disorder.

Vladimir Godnik/fstop/Corbis/Getty Images

SOCIOCULTURAL FACTORS

Sociocultural background is not considered a *cause* of schizophrenia, but sociocultural factors do appear to affect the *course* of the disorder, or how it progresses. Across cultures, people with schizophrenia in developing, nonindustrialized nations tend to have better outcomes than those in developed, industrialized nations (Jablensky, 2000; Myers, 2010). This difference may be due to the fact that in nonindustrialized nations, family and friends are more accepting and supportive of people with schizophrenia. In Western samples, marriage, warm supportive friends (Jablensky & others, 1992; Wiersma & others, 1998), and employment are related to better outcomes for individuals diagnosed with schizophrenia (Rosen & Garety, 2005).

Some experts have called for a focus on preventing schizophrenia by identifying those at risk and then intervening early to reduce the level of disability they experience (Addington & others, 2011; Walder & others, 2012). Early intervention can be important because a strong predictor of relapse (that is, having symptoms again after treatment) is the amount of time a person spends in a psychotic state without treatment (Bora & others, 2017; ten Velden Hegelstad & others, 2012). A large-scale study demonstrated that young adults who received intensive team-based treatment, including medications, therapy, and occupational, educational, and family support rapidly

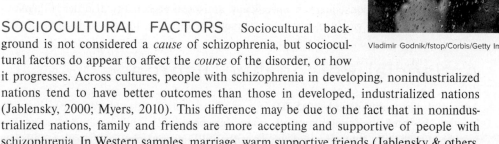

Do It!

If you have never met anyone with schizophrenia, why not get to know Moe Armstrong online? Search for clips of one of Moe's many speeches on YouTube.

following a first psychotic episode, experienced better quality of life two years later, compared to those receiving only medication and community care (Kane & others, 2015). This rapid, comprehensive treatment was especially effective if the person began treatment very early.

self-quiz

1. A negative symptom of schizophrenia is
   A. hallucinations.
   B. flat affect.
   C. delusions.
   D. neologisms.

2. Joel firmly believes that he has superhuman powers. He is likely suffering from
   A. hallucinations.
   B. delusions.
   C. negative symptoms.
   D. referential thinking.

3. The biological factors in schizophrenia include
   A. problems with the body's regulation of dopamine.
   B. abnormalities in brain structure such as enlarged ventricles and a small frontal cortex.
   C. genetic predisposition.
   D. all of the above.

APPLY IT! 4. During a psychiatric hospital internship, Tara approaches a young man sitting alone in a corner, and they have a short conversation. He asks her if she is with the government, and she tells him that she is not. She asks him a few questions and walks away. She tells her advisor later that what disturbed her about the conversation was not so much what the young man said, but that she had this feeling that he just was not really there. Tara was noticing the _____ symptoms of schizophrenia.
   A. positive
   B. negative
   C. cognitive
   D. genetic

8 Personality Disorders

personality disorders
Chronic, maladaptive cognitive-behavioral patterns that are thoroughly integrated into an individual's personality.

Imagine that your personality—the very thing about you that makes you *you*—is the core of your life difficulties. That is what happens with **personality disorders**, which are chronic, maladaptive cognitive-behavioral patterns that are thoroughly integrated into an individual's personality. Because they are fully integrated into the person's personality, these disorders can be difficult to treat. Personality disorders are relatively common. It has been estimated that as many as 15 percent of Americans have symptoms of personality disorders (Grant & others, 2004). Figure 8 reviews the 10 personality disorders included in *DSM-5*.

In this section we survey the two personality disorders that have been the object of most study: antisocial personality disorder and borderline personality disorder. Both are associated with dire consequences. Specifically, antisocial personality disorder is linked to criminal activity and violence; borderline personality disorder, to self-harm and suicide.

Antisocial Personality Disorder

antisocial personality disorder (ASPD)
Psychological disorder characterized by guiltlessness, law-breaking, exploitation of others, irresponsibility, and deceit.

Antisocial personality disorder (ASPD) is a psychological disorder characterized by guiltlessness, law-breaking, exploitation of others, irresponsibility, and deceit. Although they may be superficially charming, individuals with ASPD do not play by the rules, and they often lead a life of crime and violence. ASPD is far more common in men than in women (a 3:1 ratio). ASPD is related to criminal behavior, vandalism, substance abuse, and alcoholism (Chun & others, 2017). Those with ASPD are impulsive, meaning they may act without thinking through their decisions, and they typically do not experience remorse or guilt for their harmful actions.

ASPD is characterized by the following:

- Failure to conform to social norms or obey the law
- Deceitfulness, lying, using aliases, or conning others for personal profit or pleasure
- Impulsivity
- Irritability and aggressiveness; getting into physical fights or perpetrating assaults

Personality Disorder	Description
Paranoid Personality Disorder	Paranoia, suspiciousness, and deep distrust of others. People with this disorder are always on the lookout for danger and the slightest social mistreatment. They may be socially isolated.
Schizoid Personality Disorder	Extreme lack of interest in interpersonal relationships. People with this disorder are emotionally cold and apathetic, and they are generally detached from interpersonal life.
Schizotypal Personality Disorder	Socially isolated and prone to odd thinking. People with this disorder often have elaborate and strange belief systems and attribute unusual meanings to life events and experiences.
Antisocial Personality Disorder	Manipulative, deceitful, and amoral. People with this disorder lack empathy for others, are egocentric, and are willing to use others for their own personal gain.
Borderline Personality Disorder	Emotionally volatile and unstable sense of self. These individuals are prone to mood swings, excessive self-criticism, extreme judgments of others, and are preoccupied with being abandoned.
Histrionic Personality Disorder	Attention-seeking, dramatic, lively, and flirtatious. These individuals are inappropriately seductive in their interactions with others.
Narcissistic Personality Disorder	Self-aggrandizing yet overly dependent on the evaluations of others. People with this disorder view themselves as entitled and better than others. They show deficits in empathy and in understanding the feelings of others.
Avoidant Personality Disorder	Socially inhibited and prone to feelings of inadequacy, anxiety, and shame. These individuals feel inadequate and hold back in social situations. They have unrealistic standards for their own behavior and avoid setting goals, taking personal risks, or pursuing new activities.
Dependent Personality Disorder	Dependent on others for emotional and physical needs. People with this disorder perceive others as powerful and competent and themselves as childlike and helpless.
Obsessive-Compulsive Personality Disorder	Conforming rigidly to rules. These individuals show an excessive attachment to moral codes and are excessively orderly in daily life.

FIGURE 8 **The 10 Personality Disorders Included in *DSM-5*** Diagnoses of these disorders require that the person be over the age of 18, and all involve pervasive aspects of the person that color cognition, emotion, and behavior. Note that some of the labels are potentially confusing. Schizoid and schizotypal personality disorders are not the same thing as schizophrenia (though schizotypal personality disorder may proceed to schizophrenia). Further, obsessive-compulsive personality disorder is not the same thing as obsessive-compulsive disorder.

SOURCE: American Psychiatric Association. (2003). *Diagnostic and Statistical Manual of Mental Disorders* (5th ed.). APA.

- Reckless disregard for the safety of self or others
- Consistent irresponsibility, inconsistent work behavior; not paying bills
- Lack of remorse, indifference to the pain of others, or rationalizing; hurting or mistreating another person

Although ASPD is associated with criminal behavior, not all individuals with ASPD engage in crime, and not all criminals suffer from ASPD. People with ASPD can have successful careers. There are antisocial physicians, clergy members, lawyers, and just about any other occupation. Still, such individuals tend to be exploitative of others, and they break the rules, even if they are never caught.

A number of factors intertwine to produce ASPD. Research has linked ASPD to low levels of activation in the prefrontal cortex and has related these brain differences to poor decision making and problems in learning (Jiang & others, 2016; Yang & Raine, 2009). Both genetic factors and childhood abuse are associated with ASPD (Wilson & others, 2020), but there is evidence that genetic differences may distinguish abused children who go on to commit violent acts from those who do not (Caspi & others, 2002; Lynam & others, 2007).

John Wayne Gacy (top) and Ted Bundy (bottom) exemplify the subgroup of people with ASPD who are also psychopathic.
(Both): Bettmann/Getty Images

People with ASPD show lower levels of autonomic nervous system arousal and are less stressed than others by aversive circumstances, including punishment (Koenig & others, 2020; Raine & Venables, 2017). They have the ability to keep their cool while engaging in deception (Verschuere & others, 2005), suggesting that those with ASPD might be able to fool a polygraph. The underaroused autonomic nervous system may be a key difference between adolescents who become antisocial adults and those whose behavior improves during adulthood (Fagan & others, 2017; Raine & others, 1990). People with ASPD may have difficulty empathizing with other people and show deficits in theory of mind (Newbury-Helps & others, 2017).

Lack of autonomic nervous system activity suggests why individuals with ASPD might be able to fool a polygraph (lie detector).

The term *psychopath* is sometimes used to refer to a subgroup of individuals with ASPD (Crego & Widiger, 2015). Psychopaths are remorseless predators who engage in violence to get what they want. Examples of psychopaths include serial killers John Wayne Gacy, who murdered 33 boys and young men, and Ted Bundy, who confessed to murdering at least 30 young women. Psychopaths tend to show less prefrontal activation than normal individuals and to have structural abnormalities in the amygdala, as well as the hippocampus, the brain structure most closely associated with memory (Kaya & others, 2020; Thijssen & Kiehl, 2017). Importantly, brain differences are most pronounced in "unsuccessful psychopaths"—individuals who have been arrested for their behaviors (Yang & others, 2005). In contrast, "successful psychopaths"—individuals who have engaged in antisocial behavior but have not gotten caught—are more similar to healthy controls in terms of brain structure and function. However, in their behavior, successful psychopaths demonstrate a lack of empathy and a willingness to act immorally; they victimize others to enrich their own lives. Psychopaths show deficiencies in learning about fear and have difficulty processing information related to the distress of others, such as sad or fearful faces (Dolan & Fullam, 2006).

Their functioning frontal lobes might help successful psychopaths avoid getting caught.

A key challenge in treating individuals with ASPD, including psychopaths, is their ability to con even sophisticated mental health professionals (Dunbar & others, 2020). They may encounter mental health professionals in the context of substance abuse treatment but many never seek therapy, and others end up in prison, where treatment is rarely an option. Nevertheless, treatment can help those with ASPD especially if they get help early.

Borderline Personality Disorder

borderline personality disorder (BPD)
Psychological disorder characterized by a pervasive pattern of instability in interpersonal relationships, self-image, and emotions, and of marked impulsivity beginning by early adulthood and present in a variety of contexts.

Borderline personality disorder (BPD) is a pervasive pattern of instability in interpersonal relationships, self-image, and emotions, and of marked impulsivity beginning by early adulthood and present in various contexts. People with BPD are insecure, impulsive, and emotional (Spong & others, 2020). BPD is related to self-harming behaviors such as *cutting* (injuring oneself with a sharp object but without suicidal intent) and to suicide (Goodman & others, 2017; Homan & others, 2017).

At the very core of BPD is profound instability in mood, in sense of self, and in relationships. BPD is characterized by four essential features (Trull & Brown, 2013):

1. Unstable affect
2. Unstable sense of self and identity, including self-destructive impulsive behavior and chronic feelings of emptiness

3. Negative interpersonal relationships that are unstable, intense, and characterized by extreme shifts between idealization and devaluation

4. Self-harm, including recurrent suicidal behavior, gestures, or threats or self-mutilating behavior

People with BPD are prone to wild mood swings and are very sensitive to how others treat them. They often feel as if they are riding a nonstop emotional rollercoaster, and their loved ones may have to work hard to avoid upsetting them. Individuals with BPD tend to see things as either all good or all bad, a thinking style called *splitting*. For example, they typically view other people as either hated enemies with no positive qualities or as beloved, idealized friends who can do no wrong.

Borderline personality disorder is far more common in women than in men. Women make up 75 percent of those with the disorder (Chun & others, 2017; Oltmanns & Powers, 2012). A renowned expert on BPD, Marsha Linehan revealed that she herself has struggled with the disorder (Carey, 2011).

The potential causes of BPD are likely complex and include biological factors and childhood experiences. The role of genes in BPD has been demonstrated in a variety of studies and across cultures (McCarthy & others, 2014). The heritability of BPD is between 30 and 70 percent (Conde & others, 2017; Distel & others, 2008).

Many people with borderline personality disorder report experiences of childhood sexual abuse, as well as physical abuse and neglect (Kulacaogula & others, 2017). It is not clear, however, whether abuse is a primary cause of the disorder (Trull & Brown, 2013). Childhood abuse experiences may combine with genetic factors in promoting BPD.

Cognitive factors associated with BPD include a tendency to hold a set of irrational beliefs (Leahy & others, 2012). These include thinking that one is powerless and innately unacceptable and that other people are dangerous and hostile (Arntz, 2005). People with BPD also display *hypervigilance:* the tendency to be constantly on the alert, looking for threatening information in the environment (Sieswerda & others, 2007).

Decades ago, experts thought that BPD was untreatable. However, many people with BPD do show improvement over time—as many as 50 percent within two years of starting treatment (Gunderson, 2008). One key aspect of successful treatment appears to be a reduction in social stress, such as that due to leaving an abusive romantic partner or establishing a sense of trust in a therapist (Gunderson & others, 2003). Effective treatment for BPD includes a program of psychotherapies that are brought together to help people with the disorder on many different levels, including behavior, cognition, emotion, and sense of self (Choi-Kain & others, 2017).

> Movie depictions of BPD include **Single White Female**, and **Obsessed**. Where these films get it wrong is that they show BPD as leading to more harm to others than to the self. The TV show **Crazy Ex-Girlfriend** captured the self-defeating nature of the disorder.

> This would be a vulnerability-stress hypothesis explanation for BPD.

> To recognize the severe toll of BPD on those suffering from it (and on their families and friends), the U.S. House of Representatives declared May to be National Borderline Personality Disorder Awareness Month.

self-quiz

1. Individuals with ASPD
   A. are incapable of having successful careers.
   B. are typically women.
   C. are typically men.
   D. rarely engage in criminal behavior.

2. People with BPD
   A. pay little attention to how others treat them.
   B. rarely have problems with anger or strong emotion.
   C. tend to have suicidal thoughts or engage in self-harming actions.
   D. tend to have a balanced viewpoint of people and things rather than to see them as all good or all bad.

3. All of the following are true of BPD except

   A. BPD can be caused by a combination of nature and nurture—genetic inheritance and childhood experience.
   B. recent research has shown that people with BPD respond positively to treatment.
   C. a common symptom of BPD is impulsive behavior such as binge eating and reckless driving.
   D. BPD is far more common in men than in women.

APPLY IT! 4. Your new friend Maureen tells you that she was diagnosed with borderline personality disorder at the age of 23. She feels hopeless when she considers that her mood swings and unstable self-esteem are part of her very personality. Despairing, she asks, "How will I ever change?" Which of the following statements about Maureen's condition is accurate?

   A. Maureen should seek therapy and strive to improve her relationships with others, as BPD is treatable.
   B. Maureen's concerns are realistic, because a personality disorder like BPD is unlikely to change.
   C. Maureen should seek treatment for BPD because there is a high likelihood that she will end up committing a criminal act.
   D. Maureen is right to be concerned, because BPD is most often caused by genetic factors.

9 Suicide

Approximately 90 percent of people who die by suicide are estimated to have a diagnosable psychological disorder (National Alliance on Mental Illness, 2013). Many people who, to the outside eye, seem to be leading successful and fulfilling lives have died by suicide. Examples include comedian-actor Robin Williams, actor Marilyn Monroe, novelist Ernest Hemingway, and grunge music icon Kurt Cobain (who died by suicide after lifelong battles with ADHD and bipolar disorder).

Sadly, suicide has been on the rise in the United States in recent years. Consider that in 2000 the number of men who died by suicide in the United States was 17.7/100,000. For women, it was 4/100,000. In 2018, for men, the suicide rate was 22.8/100,000 and for women it was 6.2/100,000 (NIMH, 2020). Clearly there is a large gender difference in suicide risk with men being more likely than women to die by suicide. With age, these differences become even more pronounced. Among men over the age of 75, the suicide rate is remarkably high, 39.9/100,000 (NIMH, 2020). Given these grim statistics, psychologists work to reduce the frequency and intensity of suicidal impulses. Figure 9 provides good advice on what to do and what not to do if you encounter someone you believe is considering suicide.

Why would people seek to end their own lives? Biological, psychological, and sociocultural circumstances can be contributing factors.

BIOLOGICAL FACTORS   Genetic factors appear to play a role in suicide (Ruderfer & others, 2020). The Hemingways are one famous family that has been plagued by suicide. Five Hemingways, spread across generations, died by suicide, including the writer Ernest Hemingway and his granddaughter Margaux, a model and actor. Similarly, in 2009, Nicholas Hughes—a successful marine biologist and the son of Sylvia Plath, a poet who died by suicide—tragically hanged himself. Of course, people whose parents died by suicide may be more likely to consider suicide an option, so these links are not simply genetics.

Suicide is correlated with low levels of the neurotransmitter serotonin (Di Narzo & others, 2014). People who attempt suicide and who have low serotonin levels are 10 times more likely to attempt suicide again than are attempters who have high serotonin levels (Courtet & others, 2004). Poor physical health, especially when it is chronic, is another risk factor for suicide (Costanza & others, 2020; Webb & others, 2012).

PSYCHOLOGICAL FACTORS   Psychological factors that can contribute to suicide risk include psychological disorders and traumatic experiences (Miller & others, 2013; Ramberg & others, 2014; Scheer & others, 2020). An immediate and highly stressful circumstance—such as the loss of a loved one, losing one's job, work stress, failing in school, or an unwanted pregnancy—can lead people to wish to end their lives (Davidson & others, 2020; Harris & others, 2020;

What to Do

1. Ask direct, straightforward questions in a calm manner. For example, "Are you thinking about hurting yourself?"

2. Be a good listener and be supportive. Emphasize that unbearable pain can be survived.

3. Take the suicide threat very seriously. Ask questions about the person's feelings, relationships, and thoughts about the type of method to be used. If a gun, pills, rope, or other means is mentioned and a specific plan has been developed, the situation is dangerous. Stay with the person until help arrives.

4. Encourage the person to get professional help and assist them in getting help. If the person is willing, take them to a mental health facility or hospital.

What Not to Do

1. Don't ignore the warning signs.

2. Don't refuse to talk about suicide if the person wants to talk about it.

3. Don't react with horror, disapproval, or repulsion.

4. Don't offer false reassurances ("Everything will be all right") or make judgments ("You should be thankful for . . .").

5. Don't abandon the person after the crisis seems to have passed or after professional counseling has begun.

FIGURE 9   When Someone Is Threatening Suicide Do not ignore the warning signs if you think someone you know is considering suicide. Talk to a counselor if you are reluctant to say anything to the person yourself.

Nathan Lau/Design Pics

Howarth & others, 2020). In addition, substance abuse is linked with suicide (Kaplan & others, 2014). Finally, a longitudinal study showed that people victimized by bullies in adolescence were at higher risk for suicide than those who were not bullied (Foss Sigurdson & others, 2018).

An expert on suicide, Thomas Joiner (2005) has proposed a comprehensive theory of the phenomenon. His *interpersonal theory of suicide* states that suicide involves two factors (Hames & others, 2013; Joiner, 2005; Joiner & Ribeiro, 2011; Smith & others, 2013):

- A desire to die
- The acquired capability for suicide

The desire to die, from this perspective, emerges when a person's social needs are not met. People who feel they do not belong, are chronically lonely, and who perceive themselves to be a burden on others are more likely to experience a desire to die (Hames & others, 2013; Lamis & others, 2014).

Even among those with a desire to die, however, a key variable is the person's acquired capability to complete a suicide attempt. For most people, a fear of death and a strong desire to avoid feeling pain prevent us from suicide. Research suggests that one way people might overcome these natural motivations is to develop a tolerance for pain through previous experiences of injury (Joiner & others, 2012). Through past experiences, then, the interpersonal theory asserts that individuals are able to extinguish the fear response.

SOCIOCULTURAL FACTORS Chronic economic hardship can be a factor in suicide (Kerr & others, 2017). Poverty is a strong predictor of deaths by suicide among children (those under the age of 19 for whom suicide is the second leading cause of death) (Hoffmann & others, 2020). In the United States, adolescents' suicide attempts vary across ethnic groups. As Figure 10 illustrates, more than 20 percent of Native American/Alaska Native (NA/AN) female adolescents reported that they had attempted suicide in the previous year, and suicide accounts for almost 20 percent of NA/AN deaths in 15- to 19-year-olds (Goldston & others, 2008). A major risk factor in the high rate of suicide attempts by NA/AN adolescents is their elevated rate of alcohol abuse.

Suicide rates vary worldwide; the lowest rates occur in countries with strong cultural and religious norms against suicide. Nations with the highest suicide rates in 2018 were Lithuania, Russia, Guyana, and South Korea (World Population Review, 2018). Among the nations with the lowest rates are the Bahamas, Jamaica, Grenada, and Barbados. Of 176 nations, the United States ranks 25th (World Population Review, 2018).

Research has linked suicide to the culture of honor. Recall that in honor cultures, individuals are more likely to defend their personal honor with aggression. A set of studies examined suicide and depression in the United States, comparing states in the South (a region with a culture of honor) with other areas. Even accounting for a host of other factors, suicide rates were higher in southern states (Osterman & Brown, 2011). Honor-related suicides are more likely to occur in the context of public, reputation-damaging events (Roberts & others, 2018). Compared with other suicide deaths, people who die by honor-related suicides are more likely to be suffering from depression but not seeking treatment (Roberts & others, 2018).

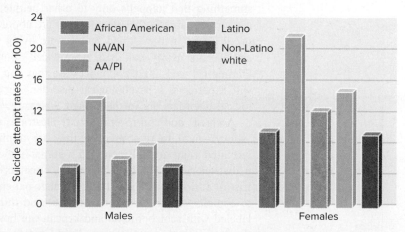

FIGURE 10 **Suicide Attempts by U.S. Adolescents from Different Ethnic Groups** Note that the data shown are for one-year rates of self-reported suicide attempts. NA/AN = Native Americans/Alaska Native; AA/PI = Asian American/Pacific Islander.

As we have already seen, men are far more likely than women to die by suicide, even if women are more likely to make suicide attempts. You might note that, aside from anti-social personality disorder, this is the only time we have seen the gender difference in the direction of men being higher than women, for what is, of course, a permanent solution to potentially temporary problems. This difference suggests important unmet needs of men in terms of undiagnosed psychological disorders.

As a final note as we leave this difficult topic, if you are thinking right now about ending your own life for whatever reason, please know that you are not alone. Reach out to a loved one or call the National Suicide Prevention hotline at 800-273-8255. The world cannot bear to lose you.

self-quiz

1. The neurotransmitter that has been linked to suicide is
   A. GABA.
   B. norepinephrine.
   C. serotonin.
   D. oxytocin.

2. The interpersonal theory of suicide suggests that suicide involves two things. First, the desire to die, and second,
   A. a genetic predisposition.
   B. a family history of suicide.
   C. the acquired capability for suicide.
   D. a severe life stressor.

3. In terms of sociocultural factors and suicide, research supports which of the following?

   A. Within culture of honor states, suicide is less likely because people are more religious.
   B. Individuals of high income levels are particularly at risk for suicide.
   C. Women are more likely than men to die by suicide.
   D. In the United States, older white men have a high rate of suicide.

APPLY IT! 4. Jerome is writing a paper on the influence of culture on suicide. He tells you that he is going to focus on how different countries have different suicide rates because culture basically means differences in country of origin. What is your advice to Jerome?
   A. Jerome is on the right track.
   B. Jerome should consider that within a country there may be cultural differences that play a role in suicide, such as the culture of honor in the southern United States.
   C. Jerome should note that gender does not have a role to play in suicide in most cultures.
   D. Jerome should note that the suicide rate in the United States is the highest in the world.

10 Combatting Stigma

Putting a label on a person with a psychological disorder can make the disorder seem like something that happens only to other people. The truth is that psychological disorders are not just about other people; they are about people, period. In 2019, nearly one in five of all adults in the United States were living with a psychological disorder, and among adolescents nearly half had experienced some kind of psychological disorder in their lifetime (NIMH, 2020). Chances are that you, someone you know, or someone you love will experience a psychological disorder. Figure 11 shows how common many psychological disorders are in the United States.

A classic and controversial study illustrates that labels of psychological disorder can be "sticky"—hard to remove once they are applied to a person. David Rosenhan (1973) recruited eight adults (including a stay-at-home mother, a psychology graduate student, a pediatrician, and some psychiatrists), none with a psychological disorder, to see a psychiatrist at various hospitals. These "pseudo-patients" were instructed to act normally except to complain about hearing voices that said things like "empty" and "thud." All eight were labeled with schizophrenia and kept in the hospital from 3 to 52 days. Once admitted to the hospital, they stopped describing symptoms, expressed an interest in leaving, and behaved cooperatively. None of the mental health professionals they encountered ever questioned the diagnosis that was given to these people, and all were discharged with the label "schizophrenia in remission." The label "schizophrenia" had stuck to the

Only other patients questioned whether pseudo-patients were really legit.

pseudo-patients and caused the professionals around them to interpret their quite normal behavior as abnormal. Clearly, once someone has been labeled with a psychological disorder, that label colors how others perceive everything else the person does.

Labels of psychological disorder carry with them a wide array of implications for the person. Is the person still able to be a good friend? A good parent? A competent worker? A significant concern for people with psychological disorders is the negative attitudes that others might have about people struggling with a psychological disorder (Angermeyer & Schomerus, 2017; Boysen & others, 2020; Chen & others, 2020). Stigma can be a barrier for individuals coping with a psychological disorder, as well as for their families and loved ones (Corrigan & Penn, 2015; Yin & others, 2020). Negative attitudes about people with psychological disorders are common in many cultures, and cultural norms and values influence these attitudes (Abdullah & Brown, 2020). Negative attitudes about people with psychological disorders are more common in rural (vs. urban) areas (Schroeder & others, 2020). Members of the military and police may have higher levels of stigmatizing psychological problems and might therefore be less likely to seek treatment (Jetelina & others, 2020; Nichter & others, 2020). Fear of stigma can prevent people from seeking treatment and from talking about their problems with family and friends (Beers & Joshi, 2020; Ferrie & others, 2020).

	Number of U.S. Adults in a Given Year (Millions)	Percentage of U.S. Adults
Anxiety Disorders		
General anxiety disorder	6.8	3.1%
Panic disorder	6.0	2.7%
Specific phobia	19.2	8.7
PTSD	7.7	3.5%
Major Depressive Disorder	14.8	6.7%
Bipolar Disorder	5.7	2.6%
Schizophrenia	2.4	1.1%

FIGURE 11   The 12-Month Prevalence of the Most Common Psychological Disorders If you add up the numbers in this figure, you will see that the totals are higher than the numbers given in the text. The explanation is that people are frequently diagnosed with more than one psychological disorder. An individual who has both a depressive and an anxiety disorder would be counted in both of those categories.
Ryan McVay/Getty Images

Consequences of Stigma

The stigma attached to psychological disorders can provoke prejudice and discrimination toward individuals who are struggling with these problems, complicating an already difficult situation. Having a disorder and experiencing the stigma associated with it can also negatively affect one's physical health.

PREJUDICE AND DISCRIMINATION   Labels of psychological disorders can be damaging because they may lead to negative stereotypes, which play a role in prejudice. For example, the label "schizophrenic" often has negative connotations such as "frightening" and "dangerous."

Vivid cases of extremely harmful behavior by people with psychological disorders can perpetuate the stereotype that people with such disorders are violent. You have probably heard of such cases. In 2007, Cho Seung-Hui, a college student, murdered 32 students and faculty at Virginia Tech University before dying by suicide. James Eagan Holmes committed a mass shooting in an Aurora, Colorado, movie theater in July 2012, leaving 12 dead and 58 injured. The following December, Adam Lanza killed his mother and then killed 20 children and 6 adults at Sandy Hook Elementary School in Newtown, Connecticut, before dying by suicide. In 2018, Nikolas Cruz allegedly murdered 17 people in a Florida high school.

Aside from having committed these notorious acts of violence, these four men have something else in common—they were all described as having a history of psychological disorders. For Cho, it was depression and anxiety; for Lanza, it was a type of autism spectrum disorder as well as anxiety, depression, and anorexia. Holmes was reported to suffer from schizophrenia or a personality disorder. Although a specific diagnosis has not been identified, Cruz had a history of mental illness and treatments. Such vivid cases may give the erroneous impression that individuals who suffer from psychological disorders are prone to violence. That impression is called an *illusory correlation*, because it is an illusion. Consider that these young men share other qualities as well. For instance, they were all young men. They are no more representative of people with psychological disorders than they are representative of young men.

People with psychological disorders are often aware of the negative stigma attached to these conditions and may themselves have previously held such negative attitudes. Negative views of people with psychological disorders are apparent even in children and adolescents (DuPont-Reyes & others, 2020; Link & others, 2020). Seeking the assistance they need may involve accepting a stigmatized identity. Even mental health and medical professionals can fall prey to prejudicial attitudes toward those who are coping with psychological disorders (Lien & others, 2020; Nordt & others, 2006). Improved knowledge about the neurobiological and genetic processes involved in many psychological disorders appears to be a promising direction for interventions to reduce such prejudice. Research shows that information about the role of genes in these disorders reduces prejudicial attitudes (Won-Pat-Borja & others, 2012).

Among the most feared aspects of stigma is discrimination, or acting prejudicially toward a person who is a member of a stigmatized group. In the workplace, discrimination against a person with a psychological disorder violates the law. The Americans with Disabilities Act (ADA) of 1990 made it illegal to refuse employment or a promotion to someone with a psychological disorder when the person's condition does not prevent performance of the job's essential functions (Cleveland & others, 1997). A person's appearance or behavior may be unusual or irritating, but as long as that individual is able to complete the duties required of a position, employment or promotion cannot be denied.

People with psychological disorders may hesitate to reveal their diagnoses because they fear being stigmatized by others. To read about research on social reactions to disclosures of diagnoses, see the Intersection.

PHYSICAL HEALTH People with psychological disorders are more likely to be physically ill and two times more likely to die than their psychologically healthy counterparts (Gittelman, 2008). They are also more likely to have obesity, to smoke, to drink excessively, and to lead sedentary lives (Geoffroy & others, 2014; Kagabo & others, 2020; McCoy & Morgan, 2020; Smith & others, 2014; Tranter & Robertson, 2020)

You might be thinking that these physical health issues are the least of their worries. If people struggling with schizophrenia want to smoke, why not let them? This type of thinking sells short the capacity of psychological and psychiatric treatments to help those with psychological disorders. Research has shown that health-promotion programs can work well for individuals with a severe psychological disorder (Koenig & Butters, 2014; Mairs & Bradshaw, 2013). When we disregard the potential of physical health interventions for people with psychological disorders to make positive life changes, we reveal our biases.

Overcoming Stigma

How can we combat the stigma of psychological disorders? One obstacle to changing people's attitudes toward individuals with psychological disorders is that mental illness is often "invisible." A person we know can have a disorder without us ever knowing. We may be unaware of *many* good lives around us that are being lived under a cloud of psychological disorder, because worries about being stigmatized keep the affected individuals from "coming out." Thus, stigma

If a friend discloses to you that they have a psychological disorder, you might wonder how best to behave. One tip is to think about how you would act if the friend told you that they have arthritis or diabetes. You can't "cure" those conditions and you can't "cure" your friend. You do not have to have all the answers, but being there to listen and support may be the best way to be a friend.

A recent study showed that even having video game characters with psychological disorders can help inform people and improve attitudes (Ferrie & others, 2020)

Clinical Psychology and Social Psychology: How Does the Stigma of Mental Illness Affect Social Interactions?

Imagine participating in the following study. You fill out a form that asks some basic questions and whether you have been hospitalized in the last 12 months. You are told that, as part of the study, you will be interacting with a partner a week later. To help you get to know the person (who is in the next room), you are provided with her information form. On the form, the person has indicated that she was hospitalized in the last year because she was diagnosed with schizophrenia. Then the experimenter tells you that you can sign up for one of three topics for next week. On the list, the person in the next room has already indicated her preferred topic. You can avoid meeting the person if you pick a different topic. Would you?

College students in an experiment faced this situation (Lucas & Phelan, 2019). The partner (who did not actually exist) was identified either as not having been hospitalized (the control condition) or as having one of three disorders:

Eric Audras/Onoky/SuperStock

panic disorder, depression, or schizophrenia (the experimental conditions). In the control and panic disorder conditions, 90 percent of the participants selected the same topic as the partner. When the partner mentioned having depression only 60 percent signed up for that interaction. When the partner mentioned having schizophrenia, less than half of the participants selected that topic.

A small entry on a form mentioning a psychological disorder was all that it took to move people to avoid contact with a person. This powerful stigma can have profound implications. Certainly, throughout our review of psychological disorders we have not encountered evidence that being socially isolated from others is optimal for anyone.

A great deal of research is aimed at understanding the stigma attached to psychological disorders and reducing prejudice against those who are already suffering (Chow & others, 2020; DuPont-Reyes & others 2020; Ferchaud & others, 2020; Lien & others 2020). In addition to education, contact with a person who has a disorder can improve attitudes (Morgan & others, 2018). If your first response to someone with a disorder is to avoid contact, much is lost, including the opportunity to reduce your own prejudices and potentially to make a new friend.

\\ **How is having a psychological disorder different from other stigmatized identities?**

leads to a catch-22: Positive examples of people coping with psychological disorders are often missing from our experience because those who are doing well shun public disclosure of their disorders (Corrigan & Penn, 2015).

A critical step toward eliminating stigma is to resist thinking of people with disorders as completely defined by their condition. Instead, it is vital to recognize that they are still people, who demonstrate considerable strengths both in confronting their disorder and in carrying on despite their problems—and their achievements. By creating a positive environment for people with disorders, we encourage more of them to become confidently "visible" and empower them to be positive role models.

After reading this chapter, you know that many admired individuals have dealt with psychological disorders. Their diagnoses do not detract from their accomplishments. To the contrary, their accomplishments are all the more remarkable in the context of the challenges they have faced.

Do It!

Visit the Schizophrenia Oral History Website, at http://www.schizophreniaoralhistories.com/.

self-quiz

1. The percentage of Americans 18 years of age and older who suffer from a diagnosable psychological disorder in a given year is closest to
   A. 15 percent.
   B. 26 percent.
   C. 40 percent.
   D. 46 percent.

2. The stigma attached to psychological disorders can have implications for
   A. the physical health of an individual with such a disorder.
   B. the psychological well-being of an individual with such a disorder.

C. other people's attitudes and behaviors toward the individual with such a disorder.
D. all of the above.

3. Labeling psychological disorders can lead to damaging
   A. stereotyping.
   B. discrimination.
   C. prejudice.
   D. all of the above.

APPLY IT! 4. Liliana has applied for a job after graduation doing data entry for a polling firm. During her second interview, Liliana asks the human resources manager whether the job's health benefits include prescription drug coverage, as she

is on antianxiety medication for generalized anxiety disorder. Which of the following statements is most applicable, legally and otherwise, in light of Liliana's request?
   A. The human resources manager should tell the hiring committee to avoid hiring Liliana because she has a psychological disorder.
   B. It is illegal for the firm to deny Liliana employment simply because she has a psychological disorder.
   C. Liliana should not have asked that question, because she will not be hired.
   D. Liliana must be given the job, or the firm could face a lawsuit.

SUMMARY

① Defining and Explaining Abnormal Behavior

Abnormal behavior is deviant, maladaptive, or personally distressful. Theoretical perspectives on the causes of psychological disorders include biological, psychological, sociocultural, and biopsychosocial approaches.

Biological approaches to disorders describe psychological disorders as diseases with origins in structural, biochemical, and genetic factors. Psychological approaches include the behavioral, social cognitive, and trait perspectives. Sociocultural approaches place emphasis on the larger social context in which a person lives, including marriage, socioeconomic status, ethnicity, gender, and culture. Biopsychosocial approaches view the interactions among biological, psychological, and social factors as significant forces in producing both normal and abnormal behavior.

The classification of disorders provides a shorthand for communication, allows clinicians to make predictions about disorders, and helps them to decide on appropriate treatment. The *Diagnostic and Statistical Manual of Mental Disorders (DSM)* is the classification system clinicians use to diagnose psychological disorders. Some psychologists contend that the *DSM* perpetuates the medical model of psychological disorders, labels everyday problems as psychological disorders, and fails to address strengths.

② Neurodevelopmental Disorders

Neurodevelopmental disorders are a class of psychological disorders that are diagnosed in childhood and often traced to atypical brain development, genetic differences, or prenatal exposure to substances that

adversely affect development. Autism spectrum disorder (ASD) is typically diagnosed in early childhood and involves deficits in communication and social relationships. The level of impairment experienced by people with ASD is variable. ASD is linked to genetics and brain development but no single cause has been identified. Treatments include intense behavioral-oriented therapies as well as medication. Attention-deficit/hyperactivity disorder (ADHD) is characterized by problems maintaining attention, excessive, inappropriate activity and high levels of impulsivity. ADHD is typically diagnosed in childhood and, left untreated, can have numerous negative consequences for people. Genetic and brain differences, especially in the prefrontal cortex and reward centers, are linked to ADHD. Treatment for ADHD typically involves stimulant medications. Nonmedical treatments such as psychotherapy, often offered in conjunction with medication, can also be effective in treating ADHD.

③ Anxiety and Anxiety-Related Disorders

Anxiety disorders are characterized by unrealistic and debilitatingly high levels of anxiety. Generalized anxiety disorder involves a high level of anxiety with no specific reason for the anxiety. Panic disorder involves attacks marked by the sudden onset of intense terror.

Specific phobias entail an irrational, overwhelming fear of a particular object, such as snakes, or a situation, such as flying. Social anxiety disorder refers to the intense fear that one will do something embarrassing or humiliating in public. Obsessive-compulsive disorder involves anxiety-provoking thoughts that will not go away (obsession) and/or urges to perform repetitive, ritualistic behaviors to prevent or produce some future situation (compulsion). A variety of experiential, psychological, and genetic factors have been shown to relate to these disorders.

④ Trauma and Stress-Related Disorders

Some psychological disorders are strongly tied to specific traumatic events. One such disorder, post-traumatic stress disorder (PTSD), includes flashbacks, emotional avoidance, emotional numbing, and excessive arousal. A variety of experiential, psychological, and genetic factors have been shown to relate to PTSD. Dissociative disorders are also linked to traumatic experience. Dissociative amnesia involves memory loss caused by extensive psychological stress. In dissociative identity disorder, formerly called multiple personality disorder, two or more distinct personalities are present in the same individual; this disorder is rare.

⑤ Disorders Involving Emotion and Mood

In depressive disorders, the individual experiences a serious depressive episode and depressed characteristics such as lethargy and hopelessness. Biological explanations of depressive disorders focus on heredity, neurophysiological abnormalities, and neurotransmitter deregulation. Psychological explanations include behavioral and cognitive perspectives. Sociocultural explanations emphasize socioeconomic and ethnic factors, as well as gender.

Bipolar disorder is characterized by extreme mood swings that include one or more episodes of mania (an overexcited, unrealistic, optimistic state). Most individuals with bipolar disorder go through multiple cycles of depression interspersed with mania. Genetic influences are stronger predictors of bipolar disorder than depressive disorder, and biological processes are also a factor in bipolar disorder.

⑥ Eating Disorders

Anorexia nervosa is characterized by extreme underweight and starvation. The disorder is related to perfectionism and obsessive-compulsive tendencies. Bulimia nervosa involves a pattern of binge eating followed by purging through self-induced vomiting. In contrast, binge-eating disorder involves binge eating without purging.

Anorexia nervosa and bulimia nervosa are much more common in women than in men, but there is no gender difference in binge-eating disorder. Although sociocultural factors were once primary in explaining eating disorders, newer evidence points to the role of biological factors.

⑦ Schizophrenia

Schizophrenia is a severe psychological disorder characterized by highly disordered thought processes. Positive symptoms of schizophrenia are behaviors and experiences that are present in individuals with schizophrenia but absent in healthy people; they include hallucinations, delusions, thought disorder, and disorders of movement. Negative symptoms of schizophrenia are behaviors and experiences that are part of healthy human life that are absent for those with this disorder; they include flat affect and an inability to plan or engage in goal-directed behavior.

Biological factors (heredity, structural brain abnormalities, and problems in neurotransmitter regulation, especially dopamine), psychological factors (vulnerability-stress hypothesis or diathesis-stress model), and sociocultural factors may be involved in schizophrenia. Psychological and sociocultural factors are not viewed as stand-alone causes of schizophrenia, but they are related to the course of the disorder.

⑧ Personality Disorders

Personality disorders are chronic, maladaptive cognitive-behavioral patterns that are thoroughly integrated into an individual's personality. Two common types are antisocial personality disorder (ASPD) and borderline personality disorder (BPD).

Antisocial personality disorder is characterized by guiltlessness, law-breaking, exploitation of others, irresponsibility, and deceit. Individuals with this disorder often lead a life of crime and violence. Psychopaths—remorseless predators who engage in violence to get what they want—are a subgroup of individuals with ASPD.

Borderline personality disorder is a pervasive pattern of instability in interpersonal relationships, self-image, and emotions. This disorder is related to self-harming behaviors such as cutting and suicide.

Biological factors for ASPD include genetic, brain, and autonomic nervous system differences. The potential causes of BPD are complex and include biological and cognitive factors as well as childhood experiences.

⑨ Suicide

Severe depression and other psychological disorders can lead individuals to want to end their lives. Theorists have proposed biological, psychological, and sociocultural explanations of suicide. The interpersonal theory of suicide suggests that dying by suicide requires the desire to die and the acquired capability to kill oneself.

⑩ Combatting Stigma

Stigma can create a significant barrier for people coping with a psychological disorder, as well as for their loved ones. Fear of being labeled can prevent individuals with a disorder from getting treatment and from talking about their problems with family and friends. In addition, the stigma attached to psychological disorders can lead to prejudice and discrimination toward individuals who are struggling with these problems. Having a disorder and experiencing the stigma associated with it can also negatively affect the physical health of such individuals. We can help to combat stigma by acknowledging the strengths and the achievements of individuals coping with psychological disorders.

KEY TERMS

abnormal behavior

anorexia nervosa

antisocial personality disorder (ASPD)

anxiety disorders

applied behavior analysis

attention-deficit/hyperactivity disorder (ADHD)

binge-eating disorder (BED)

bipolar disorder

borderline personality disorder (BPD)

bulimia nervosa

catatonia

comorbidity

delusions

depressive disorders

dissociative amnesia

dissociative disorders

dissociative identity disorder (DID)

DSM-5

flat affect

generalized anxiety disorder

hallucinations

major depressive disorder (MDD)

medical model

obsessive-compulsive disorder (OCD)

panic disorder

personality disorders

post-traumatic stress disorder (PTSD)

psychosis

psychotherapy

referential thinking

risk factors

schizophrenia

social anxiety disorder or social phobia

somatic symptom and related disorders

specific phobia

vulnerability-stress hypothesis or diathesis-stress model

ANSWERS TO SELF-QUIZZES

Section 1: 1. A; 2. C; 3. C; 4. A
Section 2: 1. A; 2. D; 3. C; 4. A
Section 3: 1. D; 2. C; 3. B; 4. C
Section 4: 1. B; 2. C; 3. A; 4. C

Section 5: 1. A; 2. D; 3. B; 4. C
Section 6: 1. D; 2. A; 3. C; 4. B
Section 7: 1. B; 2. B; 3. D; 4. B
Section 8: 1. C; 2. C; 3. D; 4. A

Section 9:   1. C; 2. C; 3. D; 4. B
Section 10: 1. B; 2. D; 3. D; 4. B

Design elements: (Preview icon): Jiang Hongyan/Shutterstock; (Marginal notes): Shutterstock/Vadarshop

13 Therapies

Using Play to Express What Words Cannot

Far too many children experience traumatic life events (Merrick & others, 2019). These events can disrupt the development of a sense of themselves as someone who matters and a sense that the world is a safe, predictable place (Murphy & others, 2014). Traumatic life events in childhood are linked to problems in adolescence and adulthood (Lehto & others, 2019). Helping children who have experienced trauma is challenging because they may not be able to put their experiences into words. Children may not be able to talk about their traumatic events but they are able to do one thing: play. Play therapy is one way that psychologists work with children to help them express their experiences. Although play therapists may come from different perspectives, they share the belief that play can be therapeutic, that through play a child can bring structure to their traumatic experiences and learn to cope with them. Through play, children can "speak the unspeakable" (McClain, 2020). In play therapy, children who are accustomed to adults telling them "no" or "stop" are given a chance to work through their experiences and find a way to safely exist in the world.

Sometimes, we all need to process events that happen in our lives. Talking about our experiences can help us benefit from the support of others, from sharing who we really are with others, and from getting a reality check. For all of us, there is help and hope in unburdening ourselves to people who are trained to help.

FatCamera/Getty Images

1 Approaches to Treating Psychological Disorders

To begin our survey of treatments for psychological disorders, we review three broad approaches to the topic—psychological, biological, and sociocultural. All therapies share the goal of relieving the suffering of people with psychological disorders, but they differ in their focus and in the types of professionals who are able to deliver them. Let's take a look at each of these broad perspectives on therapy.

The Psychological Approach to Therapy

clinical psychology
An area of psychology that integrates science and theory to prevent and treat psychological disorders.

Clinical psychology is the area of psychology that integrates science and theory to prevent and treat psychological disorders. To treat psychological disorders, clinical psychologists use psychotherapy. **Psychotherapy** is a nonmedical process that helps individuals with psychological disorders recognize and overcome their problems. Psychotherapists employ a number of strategies, including talking, interpreting, listening, rewarding, challenging, and modeling.

psychotherapy
A nonmedical process that helps individuals with psychological disorders recognize and overcome their problems.

Psychotherapy is practiced by a variety of professionals, including clinical psychologists, counselors, and social workers. Figure 1 lists the main types of mental health professionals, their degrees, the years of education required, and the nature of their training. Licensing and certification are two ways in which society retains control over psychotherapy practitioners (Callahan & others, 2020; MacKain & Noel, 2020). Laws at the state level are used to license or certify such professionals. These laws specify the training individuals must have and provide for some assessment of an applicant's skill through formal examination.

There is a shortage of people trained to help with many widespread psychological disorders, such as substance abuse disorder.

During graduate school, those seeking a PhD in clinical psychology begin to see clients under the supervision of a licensed clinical psychologist. In addition to completing their graduate work, these individuals complete a clinical internship, one year spent providing supervised therapy in an accredited site, to hone their therapeutic skills. Psychotherapy may be given alone or in conjunction with medication. In most U.S. states, this privilege is reserved for psychiatrists or other medical doctors.

The Biological Approach to Therapy

biological therapies
Also called *biomedical* therapies, treatments that reduce or eliminate the symptoms of psychological disorders by altering aspects of body functioning.

The biological approach to therapy adopts the medical model, which views psychological disorders as akin to diseases requiring specific treatments, typically medications. Such **biological therapies,** also called *biomedical therapies,* are treatments that reduce or eliminate the symptoms of psychological disorders by altering aspects of body functioning.

Professional Type	Degree	Education Beyond Bachelor's Degree	Nature of Training
Clinical Psychologist	PhD or PsyD	5–7 years	Requires both clinical and research training. Includes a 1-year internship in a psychiatric hospital or mental health facility. Some universities have developed PsyD programs, which have a stronger clinical than research emphasis. The PsyD training program takes as long as the clinical psychology PhD program and also requires the equivalent of a 1-year internship.
Psychiatrist	MD	7–9 years	Four years of medical school, plus an internship and residency in psychiatry, is required. A psychiatry residency involves supervision in therapies, including psychotherapy and biomedical therapy.
Counseling Psychologist	MA, PhD, PsyD, or EdD	3–7 years	Similar to clinical psychologist but with emphasis on counseling and therapy. Some counseling psychologists specialize in vocational counseling. Some counselors complete master's degree training, others PhD or EdD training, in graduate schools of psychology or education.
School Psychologist	MA, PhD, PsyD, or EdD	3–7 years	Training in graduate programs of education or psychology. Emphasis on psychological assessment and counseling practices involving students' school-related problems. Training is at the master's or doctoral level.
Social Worker	MS W/DSW or PhD	2–5 years	Graduate work in a school of social work that includes specialized clinical training in mental health facilities.
Psychiatric Nurse	RN, MA, or PhD	0–5 years	Graduate work in a school of nursing with special emphasis on care of people with psychological disorders in hospital settings and mental health facilities.
Occupational Therapist	BS, MA, or PhD	0–5 years	Emphasis on occupational training with focus on people with psychological or physical disability. Stresses getting individuals back into the mainstream of work.
Pastoral Counselor	None to PhD or DD (Doctor of Divinity)	0–5 years	Requires ministerial background and training in psychology. An internship in a mental health facility as a chaplain is recommended.
Counselor	MA or MEd	2 years	Graduate work in a department of psychology or department of education with specialized training in counseling techniques.

FIGURE 1 Main Types of Mental Health Professionals A wide range of professionals with varying levels of training have taken on the challenge of helping people with psychological disorders.

Generally, those who administer biological therapies are required to have completed the training to become a medical doctor (that is, an MD). *Psychiatrists* are medical doctors who specialize in treating psychological disorders. In the United States, psychiatrists complete medical school and then spend an additional four years in a psychiatric residency program. During their residency, these people continue training in diagnosing disorders, understanding the effects of drug therapies on disorders, and practicing psychotherapy (Yager, 2020; Yager & Kay, 2020). For licensing in the United States, these individuals are required to demonstrate proficiency in different approaches to psychotherapy that we will review later in this chapter. As noted earlier, although psychologists and psychiatrists can both administer psychotherapy, for the most part, only psychiatrists can prescribe medications. When a psychotherapist believes that a client would benefit from medication, they might refer that client to a trusted psychiatrist to obtain a prescription.

For a number of years, psychologists have sought to obtain the right to prescribe drug treatments to their clients. As of this writing, psychologists can prescribe medication in the military (anywhere) and in the states of Illinois, Iowa, Idaho, Louisiana, and New Mexico. In these states, clinical psychologists complete additional training in medicine to qualify for prescription privileges. A key motivator for states granting psychologists these privileges is the lack of psychiatrists to treat those with psychological disorders.

Those who support the idea that psychologists ought to have prescription privileges argue that this change in regulations would make treatment more efficient for those with psychological disorders. In addition, they note that a psychologist's likely first impulse is to treat with psychotherapy, rather than medication, reducing the potential overuse of strong drugs when they are not needed. Those who oppose such a change note that psychoactive drugs are powerful and affect many body systems. They feel that additional training simply cannot replace the level of expertise required to prescribe and monitor these medicines. Further, those with psychological disorders may also have physical illnesses that psychologists would not be trained to diagnose or treat.

What do you think? Should psychologists have prescription privileges? Do they in your state?

The Sociocultural Approach to Therapy

In the treatment of psychological disorders, psychological therapies seek to modify emotions, thinking, and behavior, and biological therapies change the body. In contrast, **sociocultural therapies** view the person as part of a system of relationships that are influenced by social and cultural factors. This approach to therapy involves acknowledging the relationships, roles, identities, and cultural contexts that characterize a person's life and often bring them into the therapeutic context (Chakawa & others, 2020; Thompson, Namusoke, & others, 2020; Wampler & McWey, 2020).

From the sociocultural approach, to help a person suffering, we must recognize the many layers of social connections that exist in the person's life and how these may contribute to not only the development of psychological problems but their treatment as well. The sociocultural approach to therapy recognizes and seeks to intervene within these social connections, recognizing the person as a member of a couple, a family, and a group. Sociocultural approaches acknowledge as well that individuals' lives may be shaped in important ways by societal norms, roles, and cultural expectations and beliefs.

Sociocultural approaches to therapy may be practiced by trained psychotherapists, social workers, psychiatric social workers (those especially trained to intervene with those with psychological disorders), as well as paraprofessionals. A *paraprofessional* is a person who has been taught by a professional to provide some mental health services but who does not have formal mental health training (Niec & others, 2020; Reddy & others, 2020). The paraprofessional may have personally had the disorder; for example, a chemical dependency counselor may also be a person recovering from addiction.

sociocultural therapies
Treatments that acknowledge the relationships, roles, and cultural contexts that characterize an individual's life, often bringing them into the therapeutic context.

2 Psychotherapy

Although there are different types of psychotherapy, they all involve a trained professional engaging in an interpersonal relationship with someone who is suffering. Later in this section, we will review the specific types of psychotherapies that have been developed from broad theoretical approaches in psychology. Before we do so, let's consider some general issues that are common to all forms of psychotherapy.

Central Issues in Psychotherapy

Psychotherapy can be a very desirable type of treatment. Many people would prefer to talk to someone than to take medicines for their problems (Duncan & others, 2010). Still, a person seeking psychotherapy might wonder whether it works, whether some therapies work better than others, what factors influence a therapy's effectiveness, and how forms of therapy differ. In this section, we address these questions.

DOES PSYCHOTHERAPY WORK? There are many debates in psychology, but the effectiveness of psychotherapy for psychological disorders is not one of them. Many studies have been conducted to answer the question, "Does psychotherapy work?" and the answer is a resounding yes (American Psychological Association, 2012). A large body of research and multiple meta-analyses support this conclusion (Barkowski & others, 2020; Butler & others, 2006; Clemens, 2010; Duncan & Reese, 2013; Lambert, 2001; Lipsey & Wilson, 1993; Luborsky & others, 2002; Slotema & others, 2020; Wampold, 2001). People who experience 12 to 14 sessions of psychotherapy are more likely to improve compared to those who receive a placebo treatment or no treatment at all and are likely to maintain these improvements for two to three years, provided that the treatment they receive is based on sound psychological theory (Lambert, 2013). Figure 2 shows the results of an analysis of treatment outcomes.

So, psychotherapy works, but does one type work better than the others? In some ways, this question pits the different brands of psychotherapy against each other in a kind of horse race (Gaines & others, 2020). It is important to bear in mind that it is human nature for psychologists to have allegiances or loyalties to the brand of psychotherapy they use and in which they were trained. The results of many studies addressing the "which therapy is the best" question show that, although there is strong research supporting the notion that therapy is effective, no one therapy has been shown to be significantly more effective than the others, overall (Baldwin & Imel, 2020; Wachtel & others, 2020).

Indeed, the answer to the question which type of therapy works best may be, "It depends

Psychotherapy works! And without a lot of side effects! Are you surprised by these conclusions?

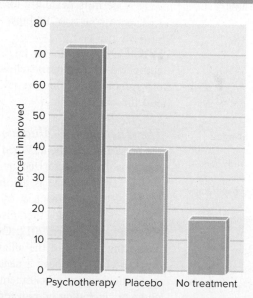

PSYCHOLOGICAL INQUIRY

FIGURE 2 **The Effects of Psychotherapy** In a review of studies, more than 70 percent of individuals who saw a therapist improved, whereas less than 40 percent who received a placebo and less than 20 percent who received no treatment improved (Lambert, 2001). >**Why do you think participants in the "no treatment" group improved?** >**Do these results allow us to make a causal claim about the effectiveness of therapy? Why or why not? Does one therapy work better than others?**

on the person's diagnosis." That is, some therapies might be more effective than others for particular psychological disorders (Raeder & others, 2020; Palpacuer & others, 2017). Some psychologists believe that therapies should be tailored to particular disorders, and therapists should rely on research evidence when they decide on a course of treatment (Chambless, 2002; Duncan & Reese, 2013). From this perspective, ideally, each disorder would be treated using the particular type of therapy that has been shown by research to work best for that disorder (Christopherson & VanScoyoc, 2013; Chu & Leino, 2017).

Other psychologists argue that targeting particular disorders with specific treatments amounts to taking an overly medical approach (Berg & Slaattelid, 2017; Shedler, 2020). In medicine, drugs (like antibiotics) are specifically prescribed to treat particular illnesses (like strep throat), but, these critics argue, psychotherapies are not like medicines that target a particular disease. They stress that psychotherapies are best understood as promoting better functioning through common characteristics that lead to benefits. These psychologists are concerned that closely dictating what therapists should do takes away the flexibility that could be vital for improvement.

Relying on empirical research in making treatment decisions is challenging because the evidence is not always clear-cut. For instance, research is not always conducted on appropriate samples (that is, people who are actually diagnosed with particular disorders), or the samples may not be representative of the people clinicians see in practice (Shedler, 2020; Wampold, 2013). A serious problem in relying on empirical evidence to guide treatment decisions is that such evidence is limited or nonexistent for many disorders. Moreover, relying on research to direct decisions leaves the clinician in the dark when research-recommended treatments do not work for a particular client.

A special task force appointed by the president of the American Psychological Association reached a compromise between these two positions. The task force endorsed evidence-based practice. **Evidence-based practice** means that decisions about treatment are made using the best available research and considering the therapist's clinical judgment and client characteristics, culture, and preferences (APA Presidential Task Force on Evidence-Based Practice, 2006). Although evidence-based practice recognizes the importance of scientific evidence, it also permits flexibility for clinicians to consider a broad range of factors in deciding on a course of therapy.

Some people think this compromise is a good one, and others think it is a cop-out: After all, as scientists we should be putting our money where our mouth is. What do you think?

evidence-based practice
Integration of the best available research with clinical expertise in the context of client characteristics, culture, and preferences.

FACTORS IN EFFECTIVE PSYCHOTHERAPY Research has also addressed the common factors that play an important role in determining the effectiveness of psychotherapy. Here we review three such factors: the therapeutic alliance, the therapist, and the client.

The Therapeutic Alliance   The **therapeutic alliance** is the relationship between the therapist and the client. When therapists and clients feel they are engaged in a real working relationship characterized by trust, respect, and cooperation, the therapeutic alliance is strong. This alliance is an important element of successful psychotherapy (Albaum & others, 2020; Leibovich & others, 2019).

It is important for therapists to monitor the quality of the relationship with each client (Brattland & others, 2019; Goldberg & others, 2019). Clients of therapists who did not assess the quality of the alliance were two times more likely to drop out of therapy and far more likely to have a negative outcome (Hubble & Miller, 2004).

Contemporary psychotherapists work in a context that makes establishing an alliance complex, including administering therapy remotely (Wehmann & others, 2020). Sociocultural factors are also important considerations, including client gender identity, sexual orientation, race, disability, and immigrant status (Budge & others, 2020; Carretier & others, 2020; Drescher & Fadus, 2020; Manyam & Davis, 2020). Moreover, clients and therapists can disagree about the quality of their alliance (Kivity & others, 2020; Penix & others, 2020). Of course, such assessments are subjective. It is possible that more objective, physiological measures might provide information about the therapeutic alliance (Deits-Lebehn & others, 2020). To read about one possible biological indicator of trust, see the Intersection.

therapeutic alliance
The relationship between the therapist and client—an important element of successful psychotherapy.

Neuroscience and Psychotherapy: Does Oxytocin Reflect the Therapeutic Alliance?

The hormone and neurotransmitter oxytocin is associated with human bonding, from parent-child attachment, to the feeling of love for a romantic partner (Young, 2015). The bond between client and therapist is different from these, but it is a social bond, and the quality of that bond matters to the effectiveness of therapy (Albaum & others, 2020). Could oxytocin provide a window into the social process of therapy? Recent research suggests it might.

We might think of oxytocin as a kind of love potion, but it does not always increase with feelings of calm, warmth, or intimacy. Rather, it can increase whenever social relationships are important, including when we are in competition with another person or in the midst of conflict. To understand what oxytocin means in the context of therapy, it is important to consider what is called the *social salience hypothesis* (Shamay-Tsoory & Abu-Akel, 2016). According to this hypothesis, the job of oxytocin is to regulate the importance of social cues as we traverse the environment. Oxytocin is deeply connected to our interpersonal lives, whether things are going well or poorly.

Research has tracked oxytocin levels in clients and therapists across the course of therapy. Results indicate that oxytocin increases above average levels when the therapeutic alliance is threatened—when the client felt in conflict with or alienated from the therapist (Jobst & others, 2018; Zilcha-Mano & others, 2018).

Interestingly, not just changes in oxytocin but average levels can provide important information about a person's interpersonal capacities. Those who are generally high in oxytocin are more likely to easily bond with others. So, it may be that therapists who experience higher levels of oxytocin are more effective (Zilcha-Mano & others, 2020). Indeed, combining information about oxytocin in therapists and clients during the course of therapy may provide a way to see the process of therapy, beyond the self-reports of clients and therapists.

A number of logistic hurdles prevent the widespread use of oxytocin as a measure of the interpersonal process of therapy. But these kinds of studies show us how our bodies and brains are involved in all of our interactions, even those with a trusted therapist.

FilippoBacci/Getty Images

\\ **Do you think your average level of oxytocin is high? Why or why not?**

The Therapist   Whether psychotherapy works or not depends more on the particular therapist than on the type of therapy used. Therapists differ in their level of expertise, their deep knowledge about psychological problems, and their treatment (Bernhardt & others, 2020; Owen & others, 2019). Therapists with high expertise are those who not only possess a great deal of knowledge but continue to learn, monitor client progress, and make changes when necessary (Heinonen & Nissen-Lie, 2020).

The match between a therapist's style and client personality can influence whether therapy is effective. For example, if a client has a strong resistance to changing, a therapist who is more laid back and less directive may have greater success than one who engages directly with the client (Beutler & others, 2012).

The Client   Another major factor in therapeutic outcomes is the person seeking treatment. Indeed, meta-analyses suggest that the quality of the client's participation is the most important determinant of whether therapy is successful (Bohart & Tallman, 2010; McKay & others, 2006). Even though people often seek therapy due to difficulties and problems in their life, it is their strengths, abilities, skills, and motivation that account for therapeutic success (Fuertes & Nutt Williams, 2017; Larsen & others, 2020; Wampold & Brown, 2005). In one study, clients' failure to develop increased self-esteem over the course of therapy was the main reason they dropped out of therapy (Kegel & Flückiger, 2014). A review of the extensive evidence on therapeutic efficacy concluded: "The data make abundantly clear that therapy does not make clients work, but rather clients make therapy work" (Hubble & Miller, 2004, p. 347).

The remainder of this section focuses on four main approaches to psychotherapy: psychodynamic, humanistic, behavioral, and cognitive therapy, as well as integrative approaches to therapy that combine many different techniques.

Psychodynamic Therapies

psychodynamic therapies
Treatments that stress the importance of the unconscious mind, extensive interpretation by the therapist, and the role of early childhood experiences in the development of an individual's problems.

The **psychodynamic therapies** stress the importance of the unconscious mind, extensive interpretation by the therapist, and the role of early childhood experiences in the development of an individual's problems. The goal of psychodynamic therapies is to help individuals gain insight into the unconscious conflicts that are the source of their problems. Many psychodynamic approaches grew out of Freud's psychoanalytic theory of personality. Today some therapists with a psychodynamic perspective practice Freudian techniques, but others do not.

psychoanalysis
Freud's therapeutic technique for analyzing an individual's unconscious thoughts.

PSYCHOANALYSIS
Psychoanalysis is Freud's therapeutic technique for analyzing a person's unconscious conflicts. Freud believed that a person's current problems could be traced to childhood experiences, many of which involved unconscious sexual conflicts. Only through extensive questioning, probing, and analyzing was Freud able to put together the pieces of the client's personality and help the person become aware of how these early experiences were affecting present behavior. The psychoanalyst's goal is to bring unconscious conflicts into conscious awareness, thus giving the client insight into core problems and freeing the individual from unconscious influences. To reach the shadowy world of the unconscious, psychoanalytic therapists use the therapeutic techniques of free association, interpretation, dream analysis, analysis of transference, and analysis of resistance.

free association
A psychoanalytic technique that involves encouraging individuals to say aloud whatever comes to mind, no matter how trivial or embarrassing.

Free association involves encouraging individuals to say aloud whatever comes to mind, no matter how trivial or embarrassing. When Freud detected a person resisting the spontaneous flow of thoughts, he probed further. He believed that the crux of the person's problem probably lurked below this point of resistance. Encouraging people to talk freely, Freud reasoned, would allow their deepest thoughts and feelings to emerge. *Catharsis* is the release of emotional tension a person experiences when reliving an emotionally charged and conflicting experience.

interpretation
A psychoanalyst's search for symbolic, hidden meanings in what the client says and does during therapy.

Interpretation plays an important role in psychoanalysis. Interpretation means that the analyst does not take the patient's statements and behavior at face value; rather, to understand the source of the person's conflicts, the therapist searches for symbolic, hidden meanings in the individual's words and deeds. From time to time, the therapist suggests possible meanings of the person's statements and behavior.

dream analysis
A psychoanalytic technique for interpreting a person's dreams.

Dream analysis is a psychoanalytic technique for interpreting a person's dreams. Psychoanalysts believe that dreams contain information about unconscious thoughts, wishes, and conflicts (Freud, 1899/1911). From this perspective, dreams provide our unconscious with an outlet to express our unconscious wishes, a mental theater in which our deepest and most secret desires can be played out. According to Freud, every dream, even our worst nightmare, contains a hidden, disguised wish. Nightmares might express a wish for punishment, or the sheer horror we feel during the nightmare might itself be the disguise.

Freud distinguished between the dream's manifest content and latent content. *Manifest content* refers to the conscious, remembered aspects of a dream. If you wake up remembering a dream about being back in sixth grade with your teacher scolding

you for not turning in your homework, that is the dream's manifest content. *Latent content* refers to the unconscious, hidden aspects that are symbolized by the manifest content. To understand your dream, a psychoanalyst might ask you to free-associate to each of the elements of the manifest content. What comes to your mind when you think of being in sixth grade or of your teacher? According to Freud, the latent meaning of a dream is locked inside the dreamer's unconscious mind. The psychoanalyst's goal is to unlock that secret meaning by having the individual free-associate about the manifest dream elements. The analyst interprets the dream by examining the manifest content for disguised unconscious wishes and needs, especially those that are sexual and aggressive. Dream symbols can mean different things to different dreamers. Freud (1899/1911) recognized that the true meaning of any dream symbol depends on the individual.

The dreamer "knows" what the dream means, but its meaning is locked in the unconscious mind.

Freud believed that transference was an essential aspect of the analyst–patient relationship. **Transference** is the psychoanalytic term for the client's relating to the analyst in ways that reproduce or relive important relationships in the client's life. A client might interact with an analyst as if the analyst were a parent or lover, for example. Transference can be used therapeutically as a model of how individuals relate to important people in their lives (Herron, 2015).

Resistance is the psychoanalytic term for the client's unconscious defense strategies that prevent the analyst from understanding the person's problems. Resistance occurs because it is painful for the client to bring conflicts into conscious awareness. By resisting analysis, the person does not have to face the threatening truths that underlie their problems. Showing up late or missing sessions, arguing with the psychoanalyst, and faking free associations are examples of resistance. A major goal of the analyst is to break through this resistance.

Freud would say that the appearance of resistance means the analyst is getting very close to the truth.

transference
A client's relating to the psychoanalyst in ways that reproduce or relive important relationships in the client's life.

resistance
A client's unconscious defense strategies that prevent the person from gaining insight into their psychological problems.

CONTEMPORARY PSYCHODYNAMIC THERAPIES
Psychodynamic therapy has changed extensively since its beginnings many years ago. Nonetheless, many contemporary psychodynamic therapists still probe unconscious thoughts about early childhood experiences to get clues to their clients' current problems, and they try to help individuals gain insight into their emotionally laden, repressed conflicts (Cohen & Kaplan, 2020). However, contemporary psychoanalysts accord more power to the conscious mind and to a person's current relationships, and they generally place less emphasis on sex.

In addition, clients today rarely lie on a couch or see their therapist several times a week, as was the norm in early psychodynamic therapy. Instead, they sit in a comfortable chair facing the therapist, and weekly appointments are typical.

Some contemporary psychodynamic therapists focus on the self in social contexts, as Heinz Kohut (1977) recommended. In Kohut's view, early social relationships with attachment figures such as parents are critical. As we develop, we internalize those relationships, and they serve as the basis for our sense of self. Kohut (1977) believed that the therapist's job is to replace unhealthy childhood relationships with the healthy relationship the therapist provides. From Kohut's perspective, the therapist needs to interact with the client in empathic and understanding ways. Empathy and understanding are also cornerstones for humanistic therapies, as we next consider.

Humanistic Therapies

The underlying philosophy of humanistic therapies is captured by the metaphor of how an acorn, if provided with appropriate conditions, will grow—pushing naturally toward its actualization as an oak tree (Schneider, 2002). In **humanistic therapies**, people are encouraged toward self-understanding and personal growth. The humanistic therapies are unique in their emphasis on the person's self-healing capacities. In contrast to psychodynamic therapies, humanistic therapies emphasize conscious rather than unconscious thoughts, the present rather than the past, and self-fulfillment rather than illness (House & others, 2017).

humanistic therapies
Treatments that uniquely emphasize people's self-healing capacities and that encourage clients to understand themselves and to grow personally.

Brand X Pictures/PunchStock

Client-centered therapy (also called *Rogerian therapy* or *nondirective therapy*) is a form of humanistic therapy, developed by Carl Rogers, in which the therapist provides a warm, supportive interpersonal context to improve the client's self-concept and to encourage the client to gain insight into problems (Rogers, 1961, 1980). The goal of client-centered therapy is to help clients identify and understand their own genuine feelings and become more *congruent*, bringing their actual self closer to their genuine self.

client-centered therapy
Also called *Rogerian therapy* or *nondirective therapy*, a form of humanistic therapy, developed by Rogers, in which the therapist provides a warm, supportive atmosphere to improve the client's self-concept and to encourage the client to gain insight into problems.

Do It!

To experience Rogerian therapy firsthand, a quick YouTube search for Carl Rogers and therapy will give you a chance to see his approach in action.

One way to achieve this goal is through active listening and **reflective speech**, a technique in which the therapist mirrors the client's own feelings back to the client. For example, as a client is describing their grief over the traumatic loss of a spouse in a car accident, the therapist might suggest, "You sound angry," to help them identify their true feelings.

Rogers believed that humans require three essential elements to grow: unconditional positive regard, empathy, and genuineness. These three elements are reflected in his approach to therapy:

- *Unconditional positive regard:* The therapist constantly recognizes the inherent value of the client, providing a context for personal growth and self-acceptance.

- *Empathy:* The therapist strives to put themselves in the client's shoes—to feel the emotions the client is feeling.

- *Genuineness:* The therapist is a real person in the relationship with the client, sharing feelings and not hiding behind a façade of objectivity.

reflective speech
A technique in which the therapist mirrors the client's own feelings back to the client.

If you were a therapist, would you be able to maintain positive regard for a client while also being genuine? Imagine that your client admitted to harming children. Could you successfully offer the person positive regard?

For genuineness to coexist with unconditional positive regard, that regard must be a sincere expression of the therapist's true feelings. The therapist may distinguish between the person's behavior and the person's self. Although the person is acknowledged as a valuable human being, the client's behavior can be evaluated negatively: "You are a good person but your actions are not." Rogers's positive view of humanity extended to his view of therapists. He believed that by being genuine with the client, the therapist could help the client improve.

Behavior Therapies

Psychodynamic and humanistic methods are sometimes called *insight therapies* because they encourage self-awareness as the path to psychological health. Behavior therapies take a different approach. Insight and self-awareness are not the keys to helping individuals develop more adaptive behavior patterns, behavior therapists say. Rather, behavior therapists offer action-oriented strategies to help people change their behavior. Specifically, **behavior therapies** use principles of learning to reduce or eliminate maladaptive behavior. Behavior therapists say that people can become aware of why they are depressed and yet still be depressed. They strive to eliminate the *symptoms* or *behaviors* rather than trying to get individuals to gain insight into, or awareness of, why they have those symptoms (McGuire & others, 2020; Vinograd & Craske, 2020).

Although initially based almost exclusively on the learning principles of classical and operant conditioning, behavior therapies have become more diverse (Spiegler & Guevremont, 2016). As social cognitive theory grew in popularity, behavior therapists increasingly included observational learning, cognitive factors, and self-instruction in their treatments. Through self-instruction, therapists try to get people to change what they say to themselves as we will review below. Both classical and operant conditioning have led to psychotherapeutic techniques.

behavior therapies
Treatments, based on behavioral and social cognitive theories, that use principles of learning to reduce or eliminate maladaptive behavior.

CLASSICAL CONDITIONING TECHNIQUES Various techniques of classical conditioning have been used in treating psychological disorders. For example, in the treatment of specific phobias, clinicians have used systematic desensitization. **Systematic desensitization** is a behavior therapy that treats anxiety by teaching the client to associate deep relaxation with increasingly intense anxiety-producing situations.

Figure 3 shows a desensitization hierarchy. *Desensitization* involves exposing someone to a feared situation in a real or an imagined way (McMullen & others, 2017). Desensitization is based on the classical conditioning process of extinction. During extinction, the conditioned stimulus is presented without the unconditioned stimulus, leading to a decreased conditioned response. During systematic desensitization, the person continually maintains a state of relaxation in the face of increasingly anxiety-producing stimuli.

systematic desensitization
A behavior therapy that treats anxiety by teaching the client to associate deep relaxation with increasingly intense anxiety-producing situations.

Another application of classical conditioning to changing behavior is aversive conditioning. Aversive conditioning entails repeated pairings of a problematic unconditioned stimulus (such as alcohol or nicotine) with aversive conditioned stimuli to decrease their positive associations. Through aversive conditioning, people can learn to avoid behaviors such as smoking, overeating, and drinking alcohol. Electric shocks, nausea-inducing substances, and verbal insults are some of the noxious stimuli used in aversive conditioning (Sommer & others, 2006). Figure 4 illustrates how classical conditioning is the backbone of aversive conditioning.

OPERANT CONDITIONING TECHNIQUES The idea behind using operant conditioning as a therapeutic approach is that just as maladaptive behavior patterns are learned, they can be unlearned. Therapy involves conducting a careful analysis of the person's environment to determine the consequences that are associated with problematic behaviors. Specifically, the question to be answered is, "How is the problematic behavior being reinforced?" Especially important is changing the consequences of the person's behavior to ensure that healthy, adaptive replacement behaviors are followed by positive reinforcement.

1 A month before an examination
2 Two weeks before an examination
3 A week before an examination
4 Five days before an examination
5 Four days before an examination
6 Three days before an examination
7 Two days before an examination
8 One day before an examination
9 The night before an examination
10 On the way to the university on the day of an examination
11 Before the unopened doors of the examination room
12 Awaiting distribution of examination papers
13 The examination paper lies facedown before her
14 In the process of answering an examination paper

FIGURE 3   A Desensitization Hierarchy Involving Test Anxiety In this hierarchy, the individual begins with the least feared circumstance (a month before the exam) and moves through each of the circumstances until reaching the most feared circumstance (answering the exam questions on test day). At each step, the person replaces fear with deep relaxation and successful visualization.
Hero Images/Getty Images

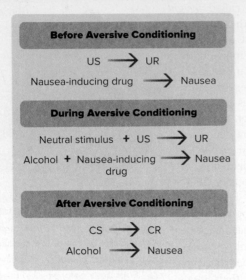

FIGURE 4 **Classical Conditioning: The Backbone of Aversive Conditioning** This figure shows how classical conditioning can produce a conditioned aversion to alcohol. Recall the abbreviations US (unconditioned stimulus), UR (unconditioned response), CS (conditioned stimulus), and CR (conditioned response).

> What is the conditioned stimulus? > What is the likely effect of alcohol prior to aversion therapy? Is this effect learned or not? > What role, if any, does a person's motivation play in aversive conditioning?

These ideas are used in applied behavior analysis, which involves establishing positive reinforcement connections between behaviors and rewards so that individuals engage in appropriate behavior and extinguish inappropriate behavior. Consider, for example, a person with obsessive-compulsive disorder (OCD) who engages in a compulsive ritual such as checking that they have locked the door of their house 10 times every time they the house. If they do not complete this ritual, they are overcome with anxiety that something dreadful will happen. Note that whenever they complete the ritual and nothing dreadful does happen, and their anxiety is relieved. This compulsion is a behavior that is reinforced by the relief of anxiety and the fact that nothing dreadful happens. Of course, the dreadful outcome never would have happened but the person never gets to see that this is the case. Such a ritual, then, could be viewed as avoidance learning. An operant conditioning–based therapy would involve stopping the behavior to extinguish this avoidance. Specifically, allowing the person to experience the lack of catastrophic consequences in the absence of repeatedly checking the lock, as well as training them to relax, might help to eliminate the compulsive rituals. Indeed, behavior therapy is effective in treating OCD (Thompson, Twohig, & others, 2020; Wu & others, 2020).

BECAUSE it does not rely on cognitive ability, therapy directed at changing behaviors can be remarkably useful.

It may strike you as odd that behavioral approaches do not emphasize gaining insight and self-awareness. However, for the very reason that they do not stress these goals, such treatments may be particularly useful in people whose cognitive abilities are limited, such as adults with developmental disabilities or children. Applied behavior analysis can be used, for instance, to treat people on the autism spectrum who engage in self-injurious behaviors such as head banging (Odom & others, 2020; Wilder & others, 2020).

As reviewed in Chapter 12, applied behavior analysis has proven effective in treating people on the autism spectrum (Thompson, 2013). Research reviews and meta-analyses support the efficacy of early intervention using high-intensity behavior analysis over extended time frames for improving cognitive and language functioning in children with autism spectrum disorders (Weitlauf & others, 2014; Yu & others, 2020).

Applied behavior analysis is the established treatment for autism spectrum disorder, but it is controversial as adults on the spectrum have spoken out about their experiences of the intense training.

Cognitive Therapies

cognitive therapies
Treatments emphasizing that cognitions (thoughts) are the main source of psychological problems; therapies that attempt to change the individual's feelings and behaviors by changing cognitions.

Cognitive therapies emphasize that *cognitions,* or thoughts, are the main source of psychological problems, and they attempt to change the person's feelings and behaviors by changing cognitions. *Cognitive restructuring,* a general concept for changing a pattern of thought that is presumed to be causing maladaptive behavior or emotion, is central to cognitive therapies. Cognitive therapies differ from psychoanalytic therapies by focusing more on overt symptoms than on deep-seated unconscious thoughts, by providing more structure to the individual's thoughts, and by being less concerned about the origin of the problem.

Compared with humanistic therapies, cognitive therapies are more directive and they are based on specific cognitive techniques.

FOUNDATIONS OF COGNITIVE THERAPIES

All cognitive therapies involve two basic assumptions: Human beings have control over their feelings, and how people feel about something depends on how they think about it (Beck, 2006; Ellis, 2005). This approach to therapy was introduced by Aaron Beck, whose research and practice laid the foundation for applying cognitive principles to treating psychological disorders. Cognitive therapy involves getting people to recognize the connections between their thoughts and their feelings and helping them use thinking to change their feelings (Beck & Haigh, 2014; Müller-Engelmann & Steil, 2017). *Cognitive restructuring,* a general concept for changing a pattern of thought that is presumed to be causing maladaptive behavior or emotion, is central to cognitive therapies.

Unfortunately, thoughts that lead to emotions can happen so rapidly that we are not even aware of them. Thus, the first goal of therapy is to bring these automatic thoughts into awareness so that they can be changed. The therapist helps clients to identify their own automatic thoughts and to keep records of their thought content and emotional reactions.

With the therapist's assistance, clients learn to recognize logical errors in their thinking and to challenge the accuracy of these automatic thoughts. Logical errors in thinking can lead people to the following erroneous beliefs (Butcher & others, 2013):

Aaron Beck (b. 1921) *Beck's method stresses that the goal of therapy should be to help people recognize and eliminate illogical and self-defeating thinking.*
Clem Murray/MCT/Newscom

- Perceiving the world as harmful while ignoring evidence to the contrary—for example, when a young person still feels worthless after a friend has just told them how much they are loved.

- Overgeneralizing on the basis of limited examples—such as feeling hopeless because of one romantic break-up.

- Magnifying the importance of undesirable events—such as seeing one failure as the end of the world.

- Engaging in absolutist thinking—such as exaggerating the importance of someone's mildly critical comment and perceiving it as proof of one's total inadequacy.

When was the last time you were upset? What kinds of thoughts did you have about the situation?

Cognitive therapists guide individuals in identifying their irrational and self-defeating thoughts. Then they get clients to challenge these thoughts and to consider more positive ways of thinking. Again, cognitive therapies all involve the basic assumption that human beings have control over their feelings and that how individuals feel about something depends on how they think about it.

Figure 5 describes some of the most widely used cognitive therapy techniques.

Do It!

The Beck Institute for Cognitive Behavior Therapy maintains a website chronicling the latest developments in cognitive therapy and including a variety of videos of Beck himself. Check out the site at www.beckinstitute.org/

cognitive-behavior therapy (CBT)
A therapy that combines cognitive therapy and behavior therapy with the goal of developing the client's self-efficacy.

COGNITIVE-BEHAVIOR THERAPY

Cognitive-behavior therapy (CBT) is a combination of cognitive therapy, with its emphasis on reducing self-defeating thoughts, and behavior therapy, with its emphasis on changing behavior. In CBT, the therapist takes a directive role, engaging in a dialogue to help the client identify automatic thoughts and the feelings they produce and working with the client to change those thoughts while also focusing on changing behavior (Beck, 2020). In CBT, the client may be given homework assignments directed at these behavioral changes.

Self-instructional methods are cognitive-behavior techniques aimed at teaching people to modify their own behavior. Using self-instructional techniques, cognitive-behavior therapists prompt clients to change what they say to themselves. The therapist gives the client examples of constructive statements, known as *reinforcing self-statements,* which the client

Cognitive Therapy Technique	Description	Example
Challenge Idiosyncratic Meanings	Explore personal meaning attached to the client's words and ask the client to consider alternatives.	When a client says they will be "devastated" by their spouse leaving, ask exactly how they would be devastated and ways they could avoid being devastated.
Question the Evidence	Systematically examine the evidence for the client's beliefs or assertions.	When a client says they can't live without their spouse, explore how they lived without the spouse before they married.
Reattribution	Help the client distribute responsibility for events appropriately.	When a client says that their child's failure in school must be the parent's fault, explore other possible explanations.
Examine Options and Alternatives	Help the client generate alternative actions to maladaptive ones.	If a client considers leaving school, explore whether tutoring or going part-time to school are good alternatives.
Decatastrophize	Help the client evaluate whether the client is overestimating the nature of a situation.	If a client states that failure in a course means giving up the dream of medical school, question whether this is a necessary conclusion.
Fantasize Consequences	Explore fantasies of a feared situation: if unrealistic, the client may recognize this; if realistic, work on effective coping strategies.	Help a client who fantasizes "falling apart" when asking the boss for a raise to role-play the situation and develop effective skills for making the request.
Examine Advantages and Disadvantages	Examine advantages and disadvantages of an issue, to instill a broader perspective.	If a client says they were "just born depressed and will always be that way," explore the advantages and disadvantages of holding that perspective versus other perspectives.
Turn Adversity to Advantage	Explore ways that difficult situations can be transformed into opportunities.	If a client has just been laid off, explore whether this is an opportunity for them to return to school.
Guided Association	Help the client see connections between different thoughts or ideas.	Draw the connections between a client's anger at their spouse for going on a business trip and their fear of being alone.
Scaling	Ask the client to rate her emotions or thoughts on scales to help gain perspective.	If a client says they were overwhelmed by an emotion, ask them to rate their feelings on a scale from 0 (not at all) to 100 (I fell down in a faint).
Thought Stopping	Provide the client with ways of stopping a cascade of negative thoughts.	Teach an anxious client to picture a stop sign or hear a bell when anxious thoughts begin to snowball.
Distraction	Help the client find benign or positive distractions to take attention away from negative thoughts or emotions temporarily.	Have a client count to 200 by 13s when they feel themselves becoming anxious.
Labeling of Distortions	Provide labels for specific types of distorted thinking to help the client gain more distance and perspective.	Have a client keep a record of the number of times a day they engage in all-or-nothing thinking—seeing things as all bad or all good.

FIGURE 5 **Cognitive Therapy Techniques** Cognitive therapists develop strategies to help change the way people think.

	Cause of Problem	Therapy Emphasis	Nature of Therapy and Techniques
Psychodynamic Therapies	Client's problems are symptoms of deep-seated, unresolved unconscious conflicts.	Discover underlying unconscious conflicts and work with client to develop insight.	Psychoanalysis, including free association, dream analysis, resistance, and transference: therapist interprets heavily, operant conditioning.
Humanistic Therapies	Client is not functioning at an optimal level of development.	Provide a context for inherent potential for growth.	Person-centered therapy, including unconditional positive regard, genuineness, accurate empathy, and active listening; self-appreciation emphasized.
Behavior Therapies	Client has learned maladaptive behavior patterns.	Learn adaptive behavior patterns through changes in the environment or cognitive processes.	Observation of behavior and its controlling conditions; specific advice given about what should be done; therapies based on classical conditioning, operant conditioning.
Cognitive Therapies	Client has developed inappropriate thoughts.	Change feelings and behaviors by changing cognitions.	Conversation with client designed to get them to change irrational and self-defeating beliefs.

FIGURE 6   Therapy Comparisons Different therapies address the same problems in very different ways. Many therapists use the tools that seem right for any given client and their problems.

SOURCE: A. Freeman and M. A. Reinecke, 2003. Cognitive therapy. In A. S. Gurman, ed., *Essential Psychotherapies* (pp. 224–271). American Psychological Association.

can repeat to take positive steps to cope with stress or meet a goal. The therapist also encourages the client to practice the statements through role playing and strengthens the client's newly acquired skills through reinforcement.

An important aspect of cognitive-behavior therapy is self-efficacy: the belief that one can master a situation and produce positive outcomes (Bandura, 2001, 2010, 2012). At each step of the therapy process, clients need to bolster their confidence by telling themselves messages such as "I'm going to master my problem," "I can do it," "I'm improving," and "I'm getting better." As they gain confidence and engage in adaptive behavior, the successes become intrinsically motivating. Before too long, individuals persist (with considerable effort) in their attempts to solve personal problems because of the positive outcomes that were set in motion by self-efficacy.

CBT is the most common form of therapy used today, and it has proved effective in treating a host of disorders. It has been integrated into treatment of anxiety disorders, depression, personality disorders, eating disorders, and, in conjunction with medication, bipolar disorder and schizophrenia. The four psychotherapies—psychodynamic, humanistic, behavior, and cognitive—are compared in Figure 6.

Therapy Integrations

integrative therapy
Using a combination of techniques from different therapies based on the therapist's judgment of which particular methods will provide the greatest benefit for the client.

As many as 50 percent of therapists identify themselves as not adhering to one particular method. Rather, they refer to themselves as "integrative" or "eclectic." **Integrative therapy** is a combination of techniques from different therapies based on the therapist's judgment of which particular methods will provide the greatest benefit for the client. Integrative therapy is characterized by openness to various ways of applying diverse therapies. For example, a therapist might use a behavioral approach to treat an individual with panic disorder and a cognitive approach to treat a client with major depressive disorder. The notion that common factors such as therapeutic alliance contribute to therapy outcomes suggest the wisdom of combining diverse techniques (Wachtel & others, 2020).

What might the popularity of integrative approaches mean for the training of new therapists?

Seeking Therapy? There Is Probably an App for That

one photo/Shutterstock

You have probably seen ads for smartphone apps that are meant to help improve psychological functioning. You may have even used one for yourself. While those apps target things like relaxation and mindfulness (Gál & others, 2021), psychotherapy apps, sometimes called "mHealth," meaning *mobile health-related interventions,* are more interactive—providing immediate feedback aimed at modifying thoughts and behaviors. Therapy apps are meant to be mobile extensions of evidence-based therapies. The idea of using apps to administer psychotherapy is exciting because it might meet an enormous need for psychological help to many people (Schueller & Torous, 2020). However, demonstrating the effectiveness of therapy apps, in general and across specific disorders, remains a goal for the future, rather than a current reality (Wright & Mishkind, 2020; Porras-Segovia & others, 2020). Still, apps may be used by psychotherapists as a complement to their in-person (or remote) interventions.

Although possibly less portable, *cybertherapy* (or *e-therapy*) involves getting therapeutic help online. The efficacy of most forms of therapy when administered online has yet to be demonstrated definitively. In comparisons of Internet vs. face-to-face CBT for depression, equivalent outcomes were found with long-term positive effects still present three years after treatment (Andersson & Hedman, 2013; Andersson & others, 2013).

However, e-therapy websites and mHealth apps are controversial among mental health professionals (Emmelkamp, 2011; Lui & others, 2017). For one thing, many of these sites do not include the most basic information about the therapists' qualifications (Norcross & others, 2013). It is notable that, because cybertherapy occurs at a distance, these sites typically exclude people who are having thoughts of suicide. Further, confidentiality, a crucial aspect of the therapeutic relationship, cannot always be guaranteed on a website. On the plus side, though, people who are unwilling or unable to seek out face-to-face therapy may be more disposed to get help online, and studies indicate some success of Internet-based therapies (Andersson & Titov, 2014; Norcross & others, 2013). Certainly the COVID-19 pandemic nudged many psychotherapists and their clients into online rather than in-person formats (Feijt & others, 2020), which will likely increase what we know about their effectiveness.

Because clients present a wide range of problems, it makes sense for therapists to use the best tools in each case rather than to adopt a "one size fits all" program. Sometimes a given psychological disorder is so difficult to treat that it requires therapists to bring all of their tools to bear (Navarro-Haro & others, 2020; Zeitler & others, 2020). For example, borderline personality disorder involves emotional instability, impulsivity, and self-harming behaviors. This disorder responds to a therapy called *dialectical behavior therapy* (DBT) (Lungu & Linehan, 2017). Like psychodynamic approaches, DBT assumes that early childhood experiences are important to the development of borderline personality disorder. However, DBT employs a variety of techniques, including homework assignments, cognitive interventions, intensive individual therapy, and group sessions involving others with the disorder. Group sessions focus on mindfulness training as well as emotional and interpersonal skills training.

DBT was developed by Marsha Linehan, a renowned psychologist, who "came out" as suffering from borderline personality disorder herself.

Therapy integrations are conceptually compatible with the biopsychosocial model of abnormal behavior. That is, many therapists believe that abnormal behavior involves biological, psychological, and social factors. Many single-therapy approaches concentrate on one aspect of the person more than others; for example, drug therapies focus on biological factors, and cognitive therapies probe psychological factors. One integrative method is to combine psychotherapy with drug therapy. Combined cognitive therapy and drug therapy have been effective in treating a variety of disorders. A mental health team that includes a psychiatrist and a clinical psychologist might conduct this integrative therapy. The use of medication to treat psychological disorders is a type of biological approach to therapy, our next topic.

3 Biological Therapies

As mentioned at the beginning of the chapter, *biological therapies* or *biomedical therapies* are treatments that reduce or eliminate the symptoms of psychological disorders by altering aspects of body functioning. Drug therapy is the most common form of biomedical therapy. Electroconvulsive therapy and psychosurgery are much less widely used biomedical therapies.

Drug Therapy

Although people have long used medicines and herbs to alleviate symptoms of emotional distress, it was not until the twentieth century that drug treatments revolutionized mental healthcare. Psychotherapeutic drugs are used mainly in three diagnostic categories: anxiety disorders, mood disorders, and schizophrenia. In this section you will read about the effectiveness of drugs for these various disorders—antianxiety drugs, antidepressant drugs, and antipsychotic drugs.

 Have you ever taken a prescription drug for a psychological problem? As you read, see if that medication is described. Does the description fit with your experience?

antianxiety drugs
Commonly known as tranquilizers, drugs that reduce anxiety by making individuals calmer and less excitable.

ANTIANXIETY DRUGS **Antianxiety drugs** are commonly known as *tranquilizers*. These drugs reduce anxiety by making people calmer and less excitable. Benzodiazepines are the antianxiety drugs that generally offer the greatest relief for anxiety symptoms, though these drugs are potentially addictive. They work by binding to the receptor sites of neurotransmitters that become overactive during anxiety. The most frequently prescribed benzodiazepines include Xanax, Valium, and Librium (Balon & Starcevic, 2020). A nonbenzodiazepine—buspirone or BuSpar—is commonly used to treat generalized anxiety disorder.

Benzodiazepines are relatively fast-acting, taking effect within hours. Side effects of benzodiazepines include drowsiness, loss of coordination, fatigue, and mental slowing. These effects can be hazardous when a person is driving or operating machinery, especially when the individual first starts taking the medication. Benzodiazepines also have been linked to complications in pregnancy and problems in babies exposed prenatally (Grigoriadis & others, 2020). Further, the combination of benzodiazepines with alcohol and with other medications can lead to problems, including overdose and death (Dai & others, 2020). Antianxiety medications are best used only temporarily for symptomatic relief. Too often, they are overused and can become addictive.

The death of actor Heath Ledger was due to a fatal combination of antianxiety drugs, alcohol, pain killers, and sleeping pills.

antidepressant drugs
Drugs that regulate mood.

ANTIDEPRESSANT DRUGS   **Antidepressant drugs** regulate mood. The four main classes of antidepressant drugs are tricyclics, such as Elavil; tetracyclics such as Avanza; monoamine oxidase (MAO) inhibitors, such as Nardil; and selective serotonin reuptake inhibitors, such as Prozac, Paxil, and Zoloft (de Sousa Tomaz & others, 2020; Sadek, 2020). These antidepressants are all thought to help alleviate depressed mood through their effects on neurotransmitters in the brain. In different ways, they all allow the person's brain to increase or maintain its level of important neurotransmitters. You will notice that many antidepressant drugs influence serotonin. Low levels of this neurotransmitter are associated with negative mood and aggression. Let's take a close look at each of these types of antidepressants.

- *Tricyclics*, so-called because of their three-ringed molecular structure, are believed to work by increasing the level of some neurotransmitters, especially norepinephrine and serotonin (Sadek, 2020). Tricyclics reduce the symptoms of depression in approximately 60 to 70 percent of cases. Tricyclics usually take two to four weeks to improve mood. Adverse side effects may include restlessness, faintness, trembling, sleepiness, and memory difficulties. Older antidepressant drugs, such as the tricyclics, reduced depression as effectively or more effectively than the newer antidepressant drugs (Undurraga & Baldessarini, 2012; Ulrich & others, 2020).

- *Tetracyclic antidepressants,* named for their four-ringed structure, are also called *noradrenergic and specific serotonergic antidepressants* (NaSSAs). These drugs have effects on both norepinephrine and serotonin, enhancing brain levels of these neurotransmitters.

- *MAO inhibitors* are thought to work by blocking the enzyme monoamine oxidase, which breaks down the neurotransmitters serotonin and norepinephrine in the brain (Undurraga & Baldessarini, 2017). The idea behind these medications is that the blocking action of MAO inhibitors allows these neurotransmitters to stick around and help regulate mood. MAO inhibitors are not as widely used as tricyclics because of potentially severe side effects. MAO inhibitors may be especially risky because of their potential interactions with certain foods and drugs. Cheese and other fermented foods—including alcoholic beverages, such as red wine—can interact with the inhibitors to raise blood pressure and, over time, cause a stroke.

- *Selective serotonin reuptake inhibitors* (SSRIs) target serotonin and work mainly by interfering only with the reabsorption of serotonin in the brain. Figure 7 shows how this process is thought to work. The positive effects of serotonin in the synapse may be related to the role of this neurotransmitter in maintaining the brain's adaptability and capacity to change with experience (Kraus & others, 2017). SSRIs are the most commonly prescribed antidepressants due to their effectiveness in reducing the symptoms of depression with fewer side effects than other antidepressants (Sadek, 2020). Nonetheless, they can have side effects, including insomnia, anxiety, headache, and impaired sexual functioning, and produce severe withdrawal symptoms if the individual abruptly stops taking them (Cosci, 2017).

Effexor is also commonly prescribed for depression. It inhibits the reuptake of serotonin and norepinephrine and is thus called a serotonin-norepinephrine reuptake inhibitor (SNRI) (Jasiak & Bostwick, 2014).

Beyond their usefulness in treating mood disorders, antidepressant drugs are often effective for a number of anxiety disorders, as well as some eating and sleep disorders. Increasingly, antidepressants are prescribed for other common problems, such as chronic pain. Such prescriptions are called "off label" because they involve using a drug for reasons other than those recommended.

MEDICATION FOR BIPOLAR DISORDER

lithium
The lightest of the solid elements in the periodic table of elements, widely used to treat bipolar disorder.

Lithium is widely used to treat bipolar disorder. Lithium is the lightest of the solid elements in the periodic table of elements. A recent research review concluded that there is good evidence for lithium's efficacy in treating the mania side of bipolar disorder and some, but less robust, evidence for its effectiveness in treating bipolar depression and mixed episodes (Smilowitz & others, 2020). Continuing research seeks to identify genetic and brain characteristics of people most likely to benefit from lithium (Moreira & others, 2017; Spuhler & others, 2017).

The amount of lithium that circulates in the bloodstream must be carefully monitored because the effective dosage is precariously close to toxic levels. Alternative mood stabilizers have been developed to treat bipolar disorder but lithium continues to outperform these in terms of effectiveness (Hafeman & others, 2020).

The effectiveness of lithium depends on the person's staying on the medication. Some consumers may be troubled by the association between lithium and weight gain, and others may go off the drug when they are feeling well (Smilowitz & others, 2020). People with bipolar disorder may also be treated with other medications as needed for depression or mania. In addition, sometimes those with bipolar are treated with antipsychotic drugs, our next topic.

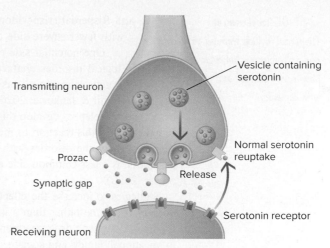

FIGURE 7 How the Antidepressant Prozac Works
Secreted by a transmitting neuron, serotonin moves across the synaptic gap and binds to receptors in a receiving neuron. Excess serotonin in the synaptic gap is normally reabsorbed by the transmitting neuron. The antidepressant Prozac blocks this reuptake of serotonin by the transmitting neuron, however, leaving excess serotonin in the synaptic gap. The excess serotonin is transmitted to the receiving neuron and circulated through the brain. The result is hypothesized to be a reduction of the serotonin deficit found in depressed individuals.

Antipsychotic Drugs

Antipsychotic drugs are powerful drugs that diminish agitated behavior, reduce tension, decrease hallucinations, improve social behavior, and produce better sleep patterns in people with severe psychological disorders, especially schizophrenia. Before antipsychotic drugs were developed in the 1950s, few, if any, interventions brought relief from the torment of psychotic symptoms.

Neuroleptics are the most extensively used class of antipsychotic drugs. When taken in sufficient doses, neuroleptics reduce symptoms of schizophrenia (Leucht & others, 2020). The most widely accepted explanation for the effectiveness of neuroleptics is their effect on dopamine and dopamine pathways in the brain. One study showed that, indeed, antipsychotic medications affect blood flow to areas of the brain that are rich in dopamine and that these medications appear to increase the brain's capacity to process this dopamine, leading to reduced symptoms (Eisenberg & others, 2017).

Medications do not cure schizophrenia; they treat its symptoms, not its causes. If an individual with schizophrenia stops taking the drug, the symptoms return. Neuroleptic drugs have substantially reduced the length of hospital stays for people with schizophrenia. However, when people are able to return to the community (because the drug therapy reduces their symptoms), many have difficulty coping with the demands of society. In the absence of symptoms, people may have trouble justifying taking these powerful drugs (the very drugs that reduced their symptoms). Neuroleptics do have potentially severe side effects, including stroke. Drugs that treat disturbed thought by reducing dopamine can also induce a lack of pleasure (Kapur, 2003).

Newer antipsychotic drugs—called *atypical antipsychotic medications*—that were introduced in the 1990s appear to influence dopamine and serotonin. The two most widely used medications in this group, Clozaril (clozapine)

antipsychotic drugs
Powerful drugs that diminish agitated behavior, reduce tension, decrease hallucinations, improve social behavior, and produce better sleep patterns in individuals with a severe psychological disorder, especially schizophrenia.

Recall that people with schizophrenia have difficulty regulating the neurotransmitter dopamine, which is associated with the experience of reward.

and Risperdal (risperidone), show promise for reducing schizophrenia symptoms with fewer severe side effects than neuroleptics (Laffont & others, 2015).

One potential side effect of antipsychotic drugs is *tardive dyskinesia*, a neurological disorder characterized by involuntary movements of the facial muscles, tongue, and mouth, as well as extensive twitching of the neck, arms, and legs (Bashir & Jankovic, 2020). Up to 20 percent of individuals with schizophrenia who take neuroleptics develop this disorder, and researchers have sought to identify potential genetic risk for this reaction to medications (Choi & others, 2020). Various treatments have been tested for the disorder with mixed effectiveness (Artukoglu & others, 2020). Tardive dyskinesia is a less common side effect of newer medications but can still occur (Stämpfli & others, 2020).

Strategies to increase the effectiveness of antipsychotic drugs involve administering small dosages over time rather than a large initial dose, and combining drug therapy with psychotherapy. Along with drug treatment, individuals with schizophrenia may benefit from training in vocational, family, and social skills (Csillag & others, 2016).

Figure 8 lists the drugs used to treat various psychological disorders, their effectiveness, and their side effects. Note that for some anxiety disorders, such as agoraphobia, MAO inhibitors (antidepressant drugs) might be used instead of antianxiety drugs.

Psychological Disorder	Drug	Effectiveness	Side Effects
Everyday Anxiety and Anxiety Disorders			
Everyday Anxiety	Antianxiety drugs; antidepressant drugs	Substantial improvement short term	Antianxiety drugs: less powerful the longer people take them; may be addictive. Antidepressant drugs: see Depressive disorders
Generalized Anxiety Disorder	Antianxiety drugs	Not very effective	Less powerful the longer people take them; may be addictive
Panic Disorder	Antianxiety drugs	About half show improvement	Less powerful the longer people take them; may be addictive
Agoraphobia	Tricyclic drugs and MAO inhibitors	Majority show improvement	Tricyclics: restlessness, fainting, and trembling MAO inhibitors: toxicity
Specific Phobias	Antianxiety drugs	Not very effective	Less powerful the longer people take them; may be addictive
Mood Disorders			
Depressive Disorders	Tricyclic drugs, MAO inhibitors, and SSRI drugs	Majority show moderate improvement	Tricyclics: cardiac problems, mania, confusion, memory loss, fatigue MAO inhibitors: toxicity SSRI drugs: nausea, nervousness, insomnia, and, in a few cases, suicidal thoughts
Bipolar Disorder	Lithium	Large majority show substantial improvement	Toxicity
Schizophrenia Spectrum			
Schizophrenia	Neuroleptics; atypical antipsychotic medications	Majority show partial improvement	Neuroleptics: irregular heartbeat, low blood pressure, uncontrolled fidgeting, tardive dyskinesia, and immobility of face Atypical antipsychotic medications: less extensive side effects than with neuroleptics, but can have a toxic effect on white blood cells

FIGURE 8 Drug Therapy for Psychological Disorders This figure summarizes the types of drugs used to treat various psychological disorders.

Electroconvulsive Therapy and Transcranial Magnetic Stimulation

The goal of **electroconvulsive therapy (ECT)** is to set off a seizure in the brain, much like what happens spontaneously in some seizure disorders. The idea of causing a seizure to help cure a psychological disorder has been around since ancient times. Hippocrates, the ancient Greek founder of medicine, first noticed that malaria-induced convulsions would sometimes cure individuals who were thought to be insane (Endler, 1988). Following Hippocrates, many other medical doctors noted that head traumas, seizures, and convulsions brought on by fever would sometimes lead to the apparent cure of psychological problems.

electroconvulsive therapy (ECT)
Also called *shock therapy*, a treatment, sometimes used for depression, that sets off a seizure in the brain.

In the early twentieth century, doctors induced seizures by insulin overdose and other means and used this procedure primarily to treat schizophrenia. In 1937, Ugo Cerletti, an Italian neurologist specializing in seizure disorders, developed the procedure by which seizures could be induced using electrical shock. With colleagues, he developed a fast, efficient means of causing seizures in humans, and ECT gained wide use in psychiatric hospitals (Endler, 1988). Unfortunately, in earlier years, ECT was used indiscriminately, sometimes even to punish patients, as in the book and film *One Flew Over the Cuckoo's Nest*.

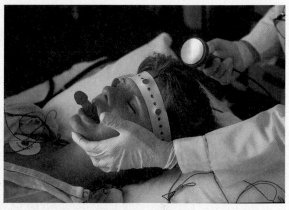

Electroconvulsive therapy (ECT), commonly called shock therapy, causes a seizure in the brain. ECT is given mainly to treat major depressive disorder.
Will McIntyre/Science Source

Today, doctors use ECT primarily to treat severe depression that has not responded to any other treatment (Gbyl & others, 2020; Nuninga & others, 2020). In contemporary usage, a small electric current lasting for one second or less passes through electrodes placed on the person's head. The current stimulates a seizure that lasts for approximately a minute, involving little discomfort. The patient receives anesthesia and muscle relaxants before the current is applied; this medication allows the person to sleep through the procedure, minimizes convulsions, and reduces the risk of physical injury (Stripp & others, 2017). Increasingly, ECT is applied only to the brain's right side. The person awakens shortly afterward with no conscious memory of the treatment.

ECT is considered a key way to prevent suicide in the extremely depressed (Fink & others, 2014; Ryan & Oquendo, 2020). However, it does have potential negative side effects, including memory loss and other cognitive impairments that are generally more severe than drug side effects (Basso & others, 2020; Caverzasi & others, 2008). Some people treated with ECT report prolonged and profound memory loss (Choi & others, 2011). As such, any benefits of ECT must be weighed against potential cognitive impairment (Blanken & others, 2020).

Recent research suggests that, when compared to other treatments, ECT may not differ in terms of preventing suicide in the extremely depressed (Peltzman & others, 2020). In addition, other experimental treatments, such as the use of repeated doses of ketamine (a medication used as an anesthesia), perform as well as ECT at preventing suicide with fewer side effects (Basso & others, 2020). Finally, the body of research testing ECT against appropriate control conditions has been criticized (Read & others, 2020).

Ketamine is used in veterinary medicine and was once considered only a drug of abuse for humans. Now it is emerging as a viable treatment for severe depression.

ECT involves setting off an electrical storm in the brain. More recently, ways to affect the brain in a more nuanced way have been used to treat psychological disorders. In particular, transcranial magnetic stimulation (TMS) has become a more common way to treat depression and other disorders (Fitzgerald, 2020). TMS uses a magnet rather than electrical current to affect the brain, and it has been approved by the Food and Drug Administration as a treatment for people who have not responded to at least one antidepressant medication (Brunelin & others, 2014). TMS appears to have far less wide-reaching impact on the brain than ECT (or medications) (Chau & others, 2017). Currently, TMS is administered multiple times, sometimes in the same day (Chen & others, 2020). Repetitive TMS (rTMS) appears to be safe and effective, with only minor side effects, including headache, fatigue, and dizziness (Hett & others, 2020). Commonly, the area of the brain that is targeted by rTMS is the left dorsolateral prefrontal cortex (DLPFC), an area of the brain involved in planning, attention, and other executive functions (Chen & others, 2020). Other brain areas may be targeted, and in the future it may be preferable to combine rTMS with brain imaging (Cash & others, 2020) or electroencephalography (EEG) (Metin & others, 2020) to pinpoint the precise brains areas for treatment for each person (and each disorder).

TMS can also be used to target brain areas inside the brain, rather than on the surface. Deep TMS is effective in treating OCD (Roth & others, 2020). It has been found to be effective in treating not only depression but also the negative symptoms of schizophrenia (Hasan & others, 2017). For each disorder, the precise area of the brain to be stimulated is chosen to reflect the problems associated with the disorder.

Psychosurgery

psychosurgery
A biological therapy, with irreversible effects, that involves removal or destruction of brain tissue to improve the individual's adjustment.

Psychosurgery is a biological intervention that involves the removal or permanent destruction of brain tissue to improve the individual's adjustment. The effects of psychosurgery cannot be reversed.

In the 1930s, Portuguese physician Antonio Egas Moniz developed a surgical procedure in which an instrument was inserted into the brain and rotated, severing fibers that connect the frontal lobe and the thalamus. Moniz theorized that by severing the connections between these structures, the surgeon could alleviate the symptoms of severe mental disorders (Soares & others, 2013). In 1949, Moniz received the Nobel Prize for developing the procedure, which he felt should be used with extreme caution and only as a last resort.

After hearing about Moniz's procedure, American neurologist Walter Freeman became the champion of *prefrontal lobotomies* (a term Freeman coined). Freeman developed his own technique, performed using a device similar to an ice pick, in surgeries that lasted mere minutes. A dynamic and charismatic advocate of this procedure, in the 1950s and 1960s Freeman traveled the country in a van he called the "lobotomobile," demonstrating the surgery in state-run mental institutions. In his career, Freeman performed over 3,000 lobotomies (Caruso & Sheehan, 2017; El-Hai, 2005).

In 1939 a former patient shot Moniz, leaving him a paraplegic. Moniz subsequently retired.

Prefrontal lobotomies were conducted on tens of thousands of patients from the 1930s through the 1960s. These numbers speak not only to Freeman's persuasive charm but also to the desperation many physicians felt in treating institutionalized patients with severe psychological disorders (Lerner, 2005).

Subsequent research called the lobotomy procedure into question (Landis & Erlick, 1950; Mettler, 1952). Many people who received lobotomies suffered permanent and profound brain damage (Soares & others, 2013). Ethical concerns arose because, in many instances, giving consent for the lobotomy was a requirement for release from an institution. Like ECT, lobotomies were being used as a form of punishment and control. In the late 1970s new regulations classified the procedure as experimental and established safeguards for patients.

It makes more sense to target the amygdala rather than the frontal lobes, considering the functions of these brain areas.

Fortunately, crude lobotomies are no longer performed, and Freeman's technique is certainly not typical of contemporary psychosurgery. Modern psychosurgery is quite precise and involves making just a small lesion in the amygdala or another part of the limbic system (Fountas & Smith, 2007; Rzesnitzek & others, 2020; Neumaier & others, 2017). Today, only several hundred patients who have severely debilitating conditions undergo psychosurgery each year. Psychosurgery may be performed for OCD, major depression, or bipolar disorder rather than for schizophrenia (Shelton & others, 2010; van Vliet & others, 2013). Just as Moniz originally suggested, the procedure is now used only as a last resort—and with the utmost caution.

Another technique that involves at least some surgery is called deep brain stimulation. In *deep brain stimulation*, doctors surgically implant electrodes in the brain that emit signals to alter the brain's electrical circuitry. Deep brain stimulation was initially developed as a treatment for Parkinson disease but now also is being used to treat individuals with

Who Should Decide What Treatment Is Best for a Person?

When someone has a physical illness, that person has a great deal of autonomy in deciding on a course of treatment. Should the same be true for a person with a psychological disorder? Sometimes a person with a disorder may not have insight into their own illness. A person with severe depression may not recognize that the feelings of hopelessness are a symptom of the disorder. Someone experiencing psychosis may not recognize that their perceptions are not real. People with a substance abuse disorder may not realize their risk of overdose. In such cases, professionals or family members may believe that it is in the person's best interest to take action (Goulet & others, 2020). Additionally, when people engage in behaviors that break the law, they may be forced to undergo treatment as part of a court order. If behavior presents a danger to a person, their health, or to other people, they may be forced into treatment. This type of treatment can range from assisted outpatient treatment to forced hospitalization. Requiring a person to undergo treatment poses legal and ethical challenges (Hui & others, 2020) and potentially violates the person's autonomy (Szmukler, 2020). Forced hospitalization can be extremely stressful, and people might feel as if they have been imprisoned. We know that when people have a say in their treatment, better outcomes are likely (Swift & others, 2018). Does treatment work if it is forced on a person? Unfortunately, very little research has addressed this question (Burns & others, 2017; Jain & others, 2018).

Recently, people with psychological disorders have begun to create *psychiatric advance directives* or PADs (NAMI, 2020). These directives are similar to those that many people have for medical treatments. Medical advance directives specify who should make decisions if one is unable to do so and whether or not a person wishes to receive extreme life-supportive measures if they lose consciousness. PADs are legal documents that persons with a serious psychological disorder can create to specify the types of

shutter_o/Shutterstock

treatment they prefer and those they do not wish to receive (NAMI, 2020). They also allow the person to provide names of people who should participate in their treatment decisions. PADs allow people to make decisions about psychiatric treatment before they are too ill to play an active role in decision making.

Although not all states in the United States treat PADs as legally binding, these documents can be an important source of information for medical professionals from the persons who know the most about their illnesses—the patients themselves. Patients can specify what treatments have worked in the past and what side effects they have experienced. For instance, one man with bipolar and borderline personality disorder suggested that rather than being restrained, isolated, or medicated, it would be very helpful to hear someone say, "Everything is going to be okay" (Belluck, 2018). Although PADs are controversial, many psychiatrists have positive attitudes toward them (Hotzy & others, 2020). It remains unclear how psychiatrists should resolve conflicts that might erupt between their own opinion on optimal treatment and patient preferences. These directives have the potential to help people with psychological disorders maintain a sense of personal control and autonomy during challenging times (Easter & others, 2020).

What Do You Think?

- Would a doctor be justified in ignoring a psychiatric advance directive if they felt it was not the right course of action?
- How does a psychiatric advance directive differ from a medical advance directive?

---

treatment-resistant depression and obsessive-compulsive disorder (Li & others, 2020; McIntyre & others, 2015). For instance, deep brain stimulation of the nucleus accumbens (part of the brain's reward pathways) has been effective in treating severe depression (Bais & others, 2014). As with TMS, researchers continue to hone in on specific areas for stimulation and target only those areas of the brain (Li & others, 2020).

Certainly as we have reviewed biological approaches to treating psychological disorders we have frequently bumped up against the complex decisions that such treatments involve. Medications work for some people and not others. Some people suffer debilitating side effects and other do not. Moreover, ECT and psychosurgery have a shameful history in which they were used as punitive measures rather than treatments. Think about this: The very people whose lives are most affected by these treatments and their risks are also suffering from disorders that may affect their capacity to make decisions for themselves. Recently, people with psychological disorders have started to find a way to have their voices heard in their treatment decisions. To read about this topic, see the Challenge Your Thinking.

4 Sociocultural Approaches and Issues in Treatment

This section focuses on sociocultural approaches to the treatment of psychological disorders. These methods view the individual as part of a system of relationships that are influenced by various social and cultural factors. We first review common sociocultural approaches and then survey various cultural perspectives on therapy.

Group Therapy

People who share a psychological problem may benefit from observing others cope with a similar problem. In turn, helping others can improve individuals' feelings of competence and efficacy. The sociocultural approach known as **group therapy** brings together people who share a psychological disorder in sessions that are typically led by a mental health professional.

group therapy
A sociocultural approach to the treatment of psychological disorders that brings together individuals who share a particular psychological disorder in sessions that are typically led by a mental health professional.

Advocates of group therapy point out that individual therapy puts the client outside the normal context of the relationships—family, marital, or peer-group relationships, for example—where many psychological problems develop. Yet such relationships may hold the key to successful therapy, these advocates say. By taking the context of important groups into account, group therapy may be more successful than individual therapy.

Group therapy takes many diverse forms—including psychodynamic, humanistic, behavior, and cognitive therapy—plus approaches that do not reflect the major psychotherapeutic perspectives (Hare & Graziano, 2020). Group therapy can be useful not only for psychological disorders but for people coping with a common physical illness such as breast cancer (Ye & others, 2020). An effective therapy provided to people in a group context is more cost-effective than having a single individual in treatment alone (Dunlap & others, 2020; Hare & Graziano, 2020). Six additional features make group therapy an attractive treatment format (Yalom & Leszcz, 2005):

1. *Information:* Individuals receive information about their problems from either the group leader or other group members.

2. *Universality:* Many people develop the sense that no one else has frightening and unacceptable impulses. In the group, individuals observe that others suffer and feel anguish as well.

3. *Altruism:* Group members support one another with advice and sympathy and learn that they have something to offer others.

digitalskillet/Getty Images

4. *Experience of a positive family group:* A therapy group often resembles a family (in family therapy, the group *is* a family), with the leaders representing parents and the other members siblings. In this new family, old wounds may be healed, and new, more positive family ties may be made.

5. *Development of social skills:* Feedback from peers may correct flaws in the individual's interpersonal skills. Self-centered individuals may see that they are self-centered if five other group members inform them about this quality; in one-on-one therapy, the individual might not believe the therapist.

6. *Interpersonal learning:* The group can serve as a training ground for practicing new behaviors and relationships. A hostile person may learn that they can get along better with others by behaving less aggressively, for example.

Family and Couples Therapy

Relationships with family members and significant others are an important part of human life. Sometimes these vital relationships can benefit from a helpful outsider. **Family therapy** is group therapy among family members (Szapocznik & Hervis, 2020). **Couples therapy** is group therapy involving married or unmarried couples whose major problem lies within their relationship. These approaches stress that although one person may have psychological symptoms, these symptoms are a function of the family or couple relationship (Hewison & others, 2014).

In family therapy, four of the most widely used techniques are:

1. *Validation:* The therapist expresses an understanding and acceptance of each family member's feelings and beliefs and thus validates the person. The therapist finds something positive to say to each family member.

2. *Reframing:* The therapist helps families reframe problems as family problems, not an individual's problems. The problems of an adolescent engaged in delinquency are reframed in terms of how each family member contributed to the situation. The parents' lack of attention or marital conflict may be involved, for example.

3. *Structural change:* The family therapist tries to restructure the coalitions in a family. So, if a parent feels they are going it alone in caring and disciplining the children, a therapist might suggest that the other parent take a stronger disciplinarian role to relieve the other's burden. Restructuring might be as simple as suggesting that the parents explore satisfying ways of being together, such as going out once a week for a quiet dinner.

4. *Detriangulation:* In some families, one member is the scapegoat for two other members who are in conflict but pretend not to be. For example, parents of a child struggling with an eating disorder might insist that their marriage is fine but find themselves in subtle conflict over parenting. The therapist tries to disentangle, or detriangulate, this situation by shifting attention away from the child to the conflict between the parents.

Couples therapy proceeds similarly to family therapy. Relationship conflict frequently involves poor communication. The therapist tries to improve the communication between the partners and help them understand and solve their problems. Couples therapy addresses diverse problems such as alcohol abuse, jealousy, sexual issues, infidelity, gender roles, two-career families, divorce, remarriage, and the special concerns of blended families (Jarnecke & others, 2020).

Self-Help Support Groups

Self-help support groups are voluntary organizations of people who get together on a regular basis to discuss topics of common interest. The self-help support group is conducted by a paraprofessional or a member of the common-interest group. Self-help

family therapy
Group therapy with family members.

couples therapy
Group therapy involving married or unmarried couples whose major problem lies within their relationship.

support groups play a key and valuable role in the U.S. mental health system (Norcross & others, 2013). There are thousands of self-help support groups for mental health support in the United States, with more than 1 million members (Goldstrom & others, 2006). In addition to reaching so many people in need of help, these groups are important because they use community resources and are relatively inexpensive. They also serve people who are otherwise less likely to receive help, such as those with a limited education or low income.

Self-help support groups provide members with a sympathetic audience for social sharing and emotional release. The social support, role modeling, and sharing of concrete strategies for solving problems that unfold in self-help groups add to their effectiveness. A woman who has been raped might not believe a male therapist who tells her that, with time, she will put the pieces of her life back together. The same message from another rape survivor—someone who has had to work through the same feelings of rage, fear, and violation—might be more believable.

There are myriad self-help groups, including groups for cocaine addiction, dieters, victims of child abuse, and people with various medical conditions (heart disease, cancer, diabetes, and so on). Alcoholics Anonymous (AA) is one of the best-known self-help groups. Mental health professionals often recommend AA for clients struggling with alcoholism (Kelly, 2017). Some studies show a positive effect for AA, but others do not (Zemore, 2017). One study found that AA reduced drinking by improving self-efficacy related to not drinking in social contexts, fostering positive changes in social networks, increasing spirituality/religiousness, and reducing negative affect (Kelly & others, 2012). AA paired with individual therapy for alcohol use disorder may be optimal (Breuninger & others, 2020).

For people who tend to cope by seeking information and affiliation with similar peers, self-help support groups can reduce stress and promote adjustment. However, as with any group therapy, there is a possibility that negative emotions will spread through the group, especially if the members face circumstances that deteriorate over time, as terminal cancer patients do. Group leaders who are sensitive to the spread of negative emotions can minimize such effects.

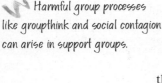
Harmful group processes like groupthink and social contagion can arise in support groups.

In addition to face-to-face groups, a multitude of online support groups has also emerged. Online support groups have promise, but they can have downsides. In the absence of guidance from a trained professional, members may lack the expertise and knowledge to provide optimal advice. The emergence of pro-anorexia (or "pro-ana") sites on Instagram, TikTok, and the web, which *promote* anorexia, exemplifies the potentially negative side of social media "support" phenomena (Branley & Covey, 2017).

Community mental health includes services such as medical care, one-on-one counseling, self-help support groups, workshops, and supported residences like halfway houses.
SolStock/Getty Images

Community Mental Health

The community mental health movement began in the 1960s, when society recognized that locking away people with psychological disorders and disabilities was inhumane and inappropriate. The deplorable conditions inside some psychiatric facilities spurred the movement as well. The central idea behind the community mental health movement was that people with disorders ought to remain within society and with their families and should receive treatment in community mental health centers. This movement also reflected economic concerns, as it was thought that institutionalizing people was more expensive than treating them in the community at large. Thus, with the passage of the Community Mental Health Act of

1963, large numbers of people with psychological disorders were transferred from mental institutions to community-based facilities, a process called *deinstitutionalization*. Although at least partially motivated by a desire to help people with psychological disorders more effectively, deinstitutionalization has been implicated in rising rates of homelessness. The success of community mental health services depends on the resources and commitment of the communities in which they occur.

Community mental health involves training teachers, ministers, family physicians, nurses, and others who directly interact with community members to offer lay counseling and workshops on topics such as coping with stress, reducing drug use, and assertiveness training (Lim & others, 2012). Advocates and providers of community mental health believe that the best way to treat a psychological disorder is through prevention (Feinstein & others, 2012; Kopelovich & others, 2020).

Community psychologists move beyond focus on the individual, embedding the person with a disorder in a larger familial, local, societal, and cultural context. A key goal of community psychology is to empower individuals and groups to enact change in their lives and communities (Society for Community Research and Action, 2017). Community psychologists are dedicated to helping people who are disenfranchised from society, such as those living in poverty, to lead happier, more productive lives (Easter & others, 2020). All community mental health programs may rely on financial support from local, state, and federal governments.

How much would you be willing to pay in taxes to support community mental health programs?

Cultural Perspectives

The psychotherapies discussed earlier in this chapter—psychodynamic, humanistic, behavior, and cognitive—center on the individual. This focus is generally compatible with the cultural values of the United States and Western Europe, where the emphasis is on the individual rather than the group (family, community, or ethnic group). However, these psychotherapies may not be as effective with people from collectivist cultures that place more importance on the group (Sue & others, 2016). Some psychologists argue that family therapy is likely to be more effective with people in cultures that place a high value on the family, such as Latino and Asian cultures (Guo, 2005). Adapting or tailoring the therapist relationship to cultural background and to religious/spiritual orientation improves therapy effectiveness (Norcross, 2011).

If you think about psychotherapy as a conversation among people, you can appreciate the profound and intricate ways culture can influence the psychotherapeutic process. Throughout our exploration of psychology, we have noted the ways in which culture is expressed in language, in our modes of talking to one another, and in the things we talk about. Placing the conversation that is psychotherapy within a cultural framework is enormously complex (Qureshi, 2020). Cultures may differ, for instance, in terms of how they view the appropriateness of talking with an elder about personal problems or of talking about one's feelings at all. Cultural concerns in therapy include factors such as socioeconomic status, immigrant status, ethnicity, gender, country of origin, refugee status, current culture, and religious beliefs and traditions (Schouler-Ocak & Kastrup, 2020).

When therapists engage with clients in a culturally sensitive way, a key goal is to not immediately assume too much about how culture influences the person. Each person must be treated as an individual.

Cross-cultural competence refers both to how skilled a therapist feels about being able to manage cultural issues that might arise in therapy and to how the client perceives the therapist's ability (Tehee, 2020). Dominant features of cross-cultural competence are demonstrating respect for cultural beliefs and practices and balancing the goals of a particular therapeutic approach with the goals and values of a culture. It may be most important that therapists demonstrate humility as they negotiate relationships with clients (Jones & Branco, 2020). Cultural humility means that even if a person is well versed in another culture they cannot know precisely what it is like to belong to that culture (Davis & others, 2018).

cross-cultural competence
A therapist's assessment of their abilities to manage cultural issues in therapy and the client's perception of those abilities.

ETHNICITY   Many ethnic minority people prefer discussing problems with parents, friends, and relatives rather than mental health professionals (Diller, 2015; Sue & others, 2016). Might therapy progress best, then, when the therapist and the client are from the same ethnic background? Researchers have found that when there is an ethnic match between the therapist and the client and when ethnic-specific services are provided, clients are less likely to drop out of therapy early, and in many cases these clients have better treatment outcomes (Jackson & Greene, 2000). Ethnic-specific services include culturally appropriate greetings and arrangements (for example, serving tea rather than coffee to Chinese American clients), providing flexible hours for treatment, and employing a bicultural/bilingual staff (Nystul, 1999). Increasingly, psychotherapists are themselves a diverse group and the importance of race, ethnicity, and culture in training clinical psychologists has become a high priority (Ramos & others, 2020).

How would you feel receiving treatment from a therapist who differed from you in ethnic background? In gender? In religious faith?

Therapy can be effective when the therapist and the client are from different ethnic backgrounds if the therapist has excellent clinical skills and is culturally sensitive (Akhtar, 2006). Culturally skilled, humble psychotherapists have good knowledge of their clients' cultural groups, understand sociopolitical influences on clients, and have competence in working with culturally diverse groups (Austad, 2009).

GENDER   A feminist informed approach to therapy requires that we place people in the context of social expectations and power (Zerbe Enns & others, 2020). Traditionally, the goal of therapy has been autonomy or self-determination for the client. However, even these goals prioritize stereotypically masculine concerns and may neglect the importance of relatedness and connection with others. Thus, some psychologists argue that therapy goals should involve increased attention to relatedness and connection with others, especially for women, or should emphasize both autonomy/self-determination and relatedness/connection to others (Notman & Nadelson, 2002).

Feminist therapists believe that traditional psychotherapy continues to carry considerable gender bias and has not adequately addressed the specific concerns of women in an unjust society. Thus, several alternative, nontraditional therapies have arisen that aim to help clients break free from traditional gender roles and stereotypes. In terms of improving clients' lives, the goals of feminist therapists are no different from those of other therapists. However, feminist therapists believe that women must become alert to the possibility of bias and discrimination in their own lives to achieve their mental health goals (Zerbe Enns & others, 2020).

Similar issues are implicated in the treatment of people with minoritized sexual orientations and gender identities (Cronin & others, 2020; Huffman & others, 2020). Members of these communities may have difficulty seeking treatment because of concerns about encountering warm, welcoming treatment. In addition, it is important for therapists to bear in mind the experience of stress that may occur from encountering prejudice and discrimination.

self-quiz

1. A family therapist who attempts to change the alliances among members of a family is using the technique of
   A. reflective speech.
   B. negative reinforcement.
   C. detriangulation.
   D. dream analysis.

2. Many homeless individuals suffer from psychological disorders. Programs to help such individuals are typically part of
   A. cognitive-behavior therapy.
   B. community mental health.
   C. psychoanalysis.
   D. biomedical treatments.

3. Deinstitutionalization is
   A. the release of a convict from the prison system.

B. the transfer of mental health clients from institutions to community agencies.
C. the process of having someone admitted to a treatment center against their will.
D. discharging someone with a psychological disorder from treatment.

APPLY IT! 4. Frank, an Asian American, is struggling with depression. When he tells a friend that he plans to get therapy, Frank mentions that he hopes the therapist is Asian American. His friend responds that Frank is biased and that he should be open to a therapist from any background. Based on research findings, what is Frank's wisest course of action if he genuinely wants his therapy to succeed?
   A. Frank should insist on an Asian American therapist because he can benefit from counseling only if his therapist shares his ethnicity.
   B. Frank is being close-minded and should take whatever therapist he gets.
   C. Frank's concern is valid but hopefully he will be open to a therapist who understands the cultural issues that might affect Frank's life.
   D. Frank should seek out a therapist who does not share his ethnicity because that way he will be forced to think outside the box.

1 Approaches to Treating Psychological Disorders

Three approaches to treating psychological disorders are psychotherapy, biological therapies, and sociocultural approaches. Psychotherapy is the process that mental health professionals use to help individuals recognize, define, and overcome their disorders and improve their adjustment. Biological treatments involve drugs and other procedures that change the functioning of the body. Sociocultural approaches to treating psychological disorders emphasize that helping a person requires acknowledging the relationships, roles, and cultural contexts that characterize a person's life; these characteristics and features are often brought into the therapeutic context.

2 Psychotherapy

Psychotherapy is the process that mental health professionals use to help individuals recognize, define, and overcome their disorders and improve their adjustment.

In Freudian psychoanalysis, psychological disorders stem from unresolved unconscious conflicts, believed to originate in early family experiences. A therapist's interpretation of free association, dreams, transference, and resistance provides paths for understanding the client's unconscious conflicts. Although psychodynamic therapy has changed, many contemporary psychodynamic therapists still probe the unconscious mind for early family experiences that might provide clues to clients' current problems.

In humanistic therapies, the analyst encourages clients to understand themselves and to grow personally. Client-centered therapy, developed by Rogers, is a type of humanistic therapy that includes active listening, reflective speech, unconditional positive regard, empathy, and genuineness.

Behavior therapies use learning principles to reduce or eliminate maladaptive behavior. They are based on the behavioral and social cognitive theories of personality. Behavior therapies seek to eliminate symptoms or behaviors rather than to help individuals gain insight into their problems.

The two main behavior therapy techniques based on classical conditioning are systematic desensitization and aversive conditioning. In systematic desensitization, individuals with anxiety are treated by helping them to associate deep relaxation with increasingly intense anxiety-producing situations. In aversive conditioning, repeated pairings of the undesirable behavior with aversive stimuli decrease the behavior's pleasant associations.

In operant conditioning approaches to behavior therapy, an analysis of the person's environment determines which factors need modification. Applied behavior analysis is the application of operant conditioning to change behavior. Its main goal is to replace maladaptive behaviors with adaptive ones.

Cognitive therapies emphasize that the person's cognitions (thoughts) are the root of abnormal behavior. Cognitive therapies attempt to change the person's feelings and behaviors by changing cognitions. The most popular form of therapy is cognitive-behavior therapy, which combines cognitive and behavior therapy techniques and emphasizes self-efficacy and self-instructional methods.

As many as 50 percent of practicing therapists refer to themselves as "integrative" or "eclectic." Integrative therapy combines techniques from different therapies based on the therapist's judgment of which methods will provide the greatest benefit.

3 Biological Therapies

Biological approaches to therapy include drugs, electroconvulsive therapy (ECT), and psychosurgery. Psychotherapeutic drugs that treat psychological disorders fall into three main categories: antianxiety drugs, antidepressant drugs, and antipsychotic drugs.

Benzodiazepines are the most commonly used antianxiety drugs. Antidepressant drugs regulate mood; the four main classes are tricyclics, tetracyclics, MAO inhibitors, and SSRI drugs. Lithium is used to treat bipolar disorder. Antipsychotic drugs are administered to treat severe psychological disorders, especially schizophrenia.

Practitioners use electroconvulsive therapy, which triggers a brain seizure, to alleviate severe depression when other interventions have failed. Transcranial magnetic stimulation involves a less invasive way to disrupt brain activity in targeted areas. Psychosurgery is an irreversible procedure in which brain tissue is destroyed. Though rarely used today, psychosurgery is much more precise now than in the days of prefrontal lobotomies. Deep brain stimulation, which applies electrical stimulation in precise brain locations, is also used to treat some forms of depression.

4 Sociocultural Approaches and Issues in Treatment

Group therapies emphasize that relationships can hold the key to successful therapy. Family therapy is group therapy involving family members. Four widely used family therapy techniques are validation, reframing, structural change, and detriangulation. Couples therapy is group therapy with couples whose major problem is within their relationship.

Self-help support groups are voluntary organizations of individuals who get together on a regular basis to discuss topics of common interest. They are conducted by a paraprofessional or a member of the common-interest group.

The community mental health movement emerged out of the belief that the mental healthcare system was not adequately reaching people in poverty and deinstitutionalized individuals. Empowerment is a common goal of community mental health.

Psychotherapies' traditional focus on the individual may be successful in individualistic cultures, but individual-centered psychotherapies may not work as well in collectivistic cultures. Therapy is often more effective when there is an ethnic match between the therapist and the client, although culturally sensitive therapy can be provided by a therapist from a different ethnic background.

The psychotherapeutic emphasis on autonomy may pose a problem for the many women who value connectedness in relationships. Feminist-based therapies emphasize the importance of raising women's awareness of the influence of traditional gender roles and stereotypes in their lives.

KEY TERMS

antianxiety drugs
antidepressant drugs
antipsychotic drugs
behavior therapies
biological or biomedical therapies
client-centered therapy
clinical psychology
cognitive-behavior therapy (CBT)

cognitive therapies
couples therapy
cross-cultural competence
dream analysis
electroconvulsive therapy (ECT)
evidence-based practice
family therapy
free association

group therapy
humanistic therapies
integrative therapy
interpretation
lithium
psychoanalysis
psychodynamic therapies
psychosurgery

psychotherapy
reflective speech
resistance
sociocultural therapies
systematic desensitization
therapeutic alliance
transference

ANSWERS TO SELF-QUIZZES

Section 1:   1. C; 2. D; 3. B; 4. A
Section 2:   1. C; 2. B; 3. D; 4. D

Section 3:   1. B; 2. B; 3. C; 4. C
Section 4:   1. C; 2. B; 3. B; 4. C

14 Health Psychology

The COVID-19 Revelation

Perhaps no event in modern times has revealed the link between the ways we think, feel, and behave, and our physical health more than the COVID-19 pandemic. Without question, other infectious diseases have profoundly affected us, from flu epidemics to HIV. But aspects of the virus itself, its virulence and unpredictability, and aspects of the contemporary world, including increasing globalization and the rapid spread of (mis)information, rendered this crisis extraordinary in many ways. The pandemic revealed a great deal about human psychology. It showed how we use information to guide our behavior and the dangers of bias in information processing. It showed us, as well, the tension that can occur among conflicting motivations, the incredible heroism of many, the capacity to withstand superhuman challenges, and long-standing inequities that exist in our societies. We learned the terrible lessons of loneliness, worry, and grief. We experienced the importance of clear communication and leadership and the potential of scientific innovation to solve complex problems. The pandemic showed how these many processes ultimately affect our physical health.

Certainly, our response to the COVID-19 pandemic depended on contexts, behaviors, motivations, thoughts, and feelings—in other words, on factors at the very heart of the science of psychology.

The focus of this chapter is health psychology, the field devoted to promoting healthy practices and understanding the psychological processes that underlie health and illness. After first defining the field, we examine the ways psychologists explain the process of making healthy life changes and the resources on which individuals can draw to achieve positive change. Next we survey the psychology of stress and coping and consider psychological perspectives on making wise choices in four vital areas: physical activity, diet, the decision to smoke or not to smoke, and safer sex. Fittingly, the chapter closes with a look at psychology's role in shaping a good life.

1 Health Psychology and Behavioral Medicine

health psychology
A subfield of psychology that emphasizes psychology's role in establishing and maintaining health and preventing and treating illness.

behavioral medicine
An interdisciplinary field that focuses on developing and integrating behavioral and biomedical knowledge to promote health and reduce illness; overlaps with and is sometimes indistinguishable from health psychology.

Health psychology emphasizes psychology's role in establishing and maintaining health and preventing and treating illness. Health psychology reflects the belief that choices, behaviors, and psychological characteristics can play important roles in health (Hennessy & others, 2020; O'Connor & others, 2021; Suls & others, 2020). A related discipline, **behavioral medicine**, is an interdisciplinary field that focuses on developing and integrating behavioral and biomedical knowledge to promote health and reduce illness (Ruiz & Revenson, 2020; Weiss & Schwartz, 2019). The concerns of health psychology and behavioral medicine overlap. Health psychology primarily focuses on behavioral, social, emotional, motivational, and cognitive factors, whereas behavioral medicine centers on behavioral, social, and biomedical factors. Note that what sets health psychology apart from behavioral medicine is its interest in what goes on inside the person's head (emotion, cognition, and motivation).

Related to health psychology and behavioral medicine are the fields of health promotion and public health. *Health promotion* involves helping people change their lifestyle to optimize their health and assisting them in achieving balance in physical, emotional, social, spiritual, and intellectual health and wellness (Fernandez & others, 2019; Oliffe & others, 2020). Health promotion can be a goal of a company's human resources department, as well as state and city health departments, and it is sometimes a specialty for social workers and other members of the helping professions. *Public health* is concerned with studying health and disease in large populations to guide policymakers (Melvin & others, 2020). Public health experts identify public health concerns, set priorities, and design interventions for health promotion. An important goal of public health is to ensure that all populations have access to cost-effective healthcare and health promotion services. For many people, the COVID-19 pandemic was their first exposure to the field of public health, as experts appeared on television and social media to convey information about the appropriate steps to take to prevent infection.

A job in health promotion or public health can involve creating attention-grabbing public service advertisements and brochures to alert the public to health-related issues. If you have seen one of thetruth.com's antismoking ads on TV (or their hashtag #FinishIT on twitter) or noticed a "Click It or Ticket" sign on a highway, you have a good feel for what health promotion and public health are all about. All over the world, public health officials devised ways to give people a sense of the 6 feet of social distancing required by the pandemic, from an adult kangaroo (in Australia) to a moose's antlers (U.S. Park Service).

During the COVID-19 pandemic, public health officials tailored their social distancing messages to their particular locales.

(Left): Margarita Young/Alamy Stock Photo; (Middle): Kenneth Martin/Alamy Stock Photo; (Right): Elena Berd/Shutterstock

The Biopsychosocial Model

The interests of health psychologists and behavioral medicine researchers are broad. The biopsychosocial model in the context of psychological disorders applies to health psychology as well, because health psychology integrates biological, psychological, and social factors (Suls & others, 2019; Woods, 2019).

For example, stress is a focal point of study across the broad field of psychology. Study of the brain and behavior acknowledges the impact of stress on the autonomic nervous system (O'Connor & others, 2021). Furthermore, a person's state of consciousness, as well as their process of thinking about events in particular ways, can influence the experience of stress (Sala & others, 2020). Stressful events affect our emotions, which are themselves psychological and physical events (Wickrama & others, 2020). Aspects of our personalities, too, may be associated with stress and can influence our health (Weston & others, 2020). Finally, social contexts can shape both an individual's experience of stress and their ability to cope with it (Karatekin & Ahluwalia, 2020).

Connections between Mind and Body

From the biopsychosocial perspective, the many diverse aspects of each human being are tightly intertwined. The body and mind are deeply connected. Although the mind is responsible for much of what happens in the body, it is not the only factor. Even as we consider the many ways that psychological processes contribute to health and disease, we must understand that sometimes illness happens for other reasons—affecting even those who have led healthy lives. After suffering a heart attack, one health psychologist ruefully noted that none of his colleagues in the field had thought to ask him whether heart disease was part of his family history, ignoring the obvious question that a medical doctor would ask first.

While it might be fascinating to think about how the mind may influence bodily health, it is also important to appreciate that the body may influence the mind as well. Health psychology and behavioral medicine are concerned not only with how psychological states influence health, but also with how health and illness may influence the person's psychological experience, including cognitive abilities, stress, and coping (Thongseiratch & Chandeying, 2020; Weaver & Szigethy, 2020). A person who is feeling psychologically run down may not realize that the level of fatigue is the beginning stage of an illness. In turn, being physically healthy can be a source of psychological wellness (Fekete & others, 2020).

1. Health psychologists believe that a key factor in health involves an individual's
   A. psychological characteristics.
   B. lifestyle.
   C. behaviors.
   D. All of the above

2. Health psychology overlaps in significant ways with
   A. philosophy.
   B. behavioral medicine.
   C. neuroscience.
   D. behaviorism.

3. The experience of stress can depend on
   A. one's state of consciousness.
   B. one's personality.
   C. one's social situation.
   D. all of the above.

APPLY IT! 4. Anastasia is committed to getting all *A*s this semester. In her pursuit of academic excellence, she decides to sleep only three hours a night, to drink a lot of coffee, and to stop wasting time at the gym. She studies nearly 12 hours every night. During finals week, Anastasia is so exhausted that she sleeps through one of her exams and fails another because she cannot concentrate. Which of the following best explains what happened?
   A. Anastasia probably didn't study as hard as she claimed.
   B. Anastasia forgot that the body can affect the functioning of the mind.
   C. Anastasia took too many hard classes this semester.
   D. Anastasia set her goals too high.

2 Making Positive Life Changes

health behaviors
Practices that have an impact on physical well-being.

One of health psychology's missions is to help people identify and implement ways they can effectively change their behaviors for the better. **Health behaviors**—practices that affect our physical well-being—include adopting a healthy approach to stress, exercising, eating right, brushing one's teeth, performing breast and testicular exams, not smoking, drinking in moderation (or not at all), and practicing safe sex. Before exploring what health psychologists have learned about the best ways to make healthy behavioral changes, we first focus on the process of change itself.

Theoretical Models of Change

In many instances, changing behaviors begins by changing attitudes. Psychologists have sought to understand how changing attitudes can lead to behavioral changes.

theory of reasoned action
Theoretical model stating that effective change requires individuals to have specific intentions about their behaviors, as well as positive attitudes about a new behavior, and to perceive that their social group looks positively on the new behavior as well.

A number of theoretical models have addressed the factors that likely play roles in effective health behavior changes. The **theory of reasoned action** is one powerful approach to this issue. The theory suggests that effective change requires three important things (Ajzen, 2012a, 2012b; Ajzen & Fishbein, 1980, 2005):

1. Specific intentions about the behavioral change

2. Positive attitude about the new behavior

3. Belief that one's social group looks upon the new behavior favorably

If, for example, you smoke and want to quit smoking, you will be more successful if you devise an explicit intention of quitting, feel good about it, and believe that your friends support you. Icek Ajzen (pronounced "I-zen") modified the theory of reasoned action to include the fact that not all of our behaviors are under our control. The **theory of planned behavior** includes the three basic ideas of the theory of reasoned action but adds a fourth: the person's perceptions of control over the outcome (Ajzen, 2002, 2012a, 2012b, 2015, 2020).

theory of planned behavior
Theoretical model that includes the basic ideas of the theory of reasoned action but adds the person's perceptions of control over the outcome.

The theory of reasoned action and its extension, the theory of planned behavior, have accurately predicted whether individuals successfully enact healthy behaviors (Ajzen & Schmidt, 2020). Just a few of the types of behaviors that have been predicted by the theory include cancer screening (Roncancio & others, 2013), HIV prevention (Jozaghi & Carleton, 2015; Kalichman, 2007), preventing

W *Remember the key difference between the theory of reasoned action and the theory of planned behavior is a sense of control!*

the use and abuse of drugs (Lorenzo-Blanco & others, 2016; Roberto & others, 2014), preventing binge drinking in college students (Case & others, 2016; Elliott & Ainsworth, 2012), increasing exercise (Plotnikoff & others, 2011), changing eating habits (Hackman & Knowlden, 2014; Menozzi & others, 2017), increasing willingness to intervene in dating violence (Casey & others, 2017; Lemay & others, 2017), reducing cyberbullying in adolescents (Heirman & Walrave, 2012), pro-environmental behaviors (Yuriev & others, 2020), bicycle sharing (Si & others, 2020), and how parents model healthy behaviors for their children (Hamilton & others, 2020).

The Stages of Change Model

stages of change model
Theoretical model describing a five-step process by which individuals give up bad habits and adopt healthier lifestyles.

The **stages of change model** describes the process by which individuals give up bad habits and adopt healthier lifestyles. The model breaks down behavioral changes into five steps, recognizing that real change does not occur overnight with one monumental decision, even if that night is New Year's Eve (Norcross & others, 2011; Prochaska & others, 2013). Rather, change occurs in progressive stages, each characterized by particular issues and challenges (Figure 1). Those stages are

Have you made a healthy life change recently? As we go over these stages, ask yourself whether they apply to your experience.

- Precontemplation
- Contemplation
- Preparation/determination
- Action/willpower
- Maintenance

Stage	Description	Example
Precontemplation **1**	Individuals are not yet ready to think about changing and may not be aware that they have a problem that needs to be changed.	People with obesity are not aware that they have a weight problem.
Contemplation **2**	Individuals acknowledge that they have a problem but may not yet be ready to change.	People with obesity know they have a weight problem but aren't yet sure they want to commit to losing weight.
Preparation/ determination **3**	Individuals are preparing to take action.	People with obesity explore options they can pursue in losing weight.
Action/willpower **4**	Individuals commit to making a behavioral change and enact a plan.	People with obesity start an exercise program and make dietary changes.
Maintenance **5**	Individuals are successful in continuing their behavior change over time.	People with obesity are able to stick with their exercise and diet regimens for 6 months.

FIGURE 1 **Stages of Change Model Applied to Losing Weight** The stages of change model has been applied to many different health behaviors, including losing weight.

PRECONTEMPLATION
The *precontemplation stage* occurs when individuals are not yet genuinely thinking about change. They may even be unaware that they have a problem behavior. Individuals who drink to excess but are not aware that their drinking is affecting their work may be in the precontemplation phase. At this stage, raising one's consciousness about the problem is crucial.

A woman who smokes may find her consciousness raised by the experience of becoming pregnant. Someone who is stopped for drunk driving may be forced to take a good look at their drinking. Similarly, people affected by obesity may not recognize their problem until they see photos of themselves taken at a family reunion or until they learn that an order of a McDonald's Big Mac, large fries, and large chocolate shake amounts to over 2,000 calories, the recommended caloric intake for an adult for an entire day.

People in the precontemplation phase commonly deny that their behavior is a problem and may defend it, claiming, "I don't drink/smoke/spend/eat that much." But people may discover they actually do "drink/smoke/spend/eat" that much when they start keeping track.

CONTEMPLATION
In the *contemplation stage,* people acknowledge the problem but may not be ready to commit to change. As the name of the stage suggests, at this point people are actively thinking about change. They might reevaluate themselves and the place of a behavior in their life. They understandably may have mixed feelings about giving up a bad habit. For example, how will they deal with missing their friends on a smoke break? Or going out drinking? Or packing a healthy lunch instead of heading to the drive-thru? They may weigh the short-term gains of the harmful behavior against the long-term benefits of changing. Future rewards can be difficult to pursue when immediate pleasures beckon. Sure, it would be nice to be more fit but exercising will take awhile to have an effect and there are all those shows to watch in one's Netflix queue.

PREPARATION/DETERMINATION
In the *preparation/determination stage,* people are getting ready to take action. At this point, self-belief and especially beliefs about one's ability to "see it through" are very important. A key consideration is the feeling that one is truly ready to change.

During the preparation/determination stage, people start thinking concretely about how they might take on their new challenge. For example, they explore options for the best ways to quit smoking or drinking or to start an exercise program. Some people who smoke might consider trying a nicotine patch or participating in a support group for people wanting to quit. People seeking to get fit might think about joining a gym or setting the alarm clock for a 6:00 A.M. run.

ACTION/WILLPOWER
At the *action/willpower stage,* people commit to making a real behavioral change and enact an effective plan. An important challenge at this stage is to find ways to support the new, healthy behavior pattern. One approach is to establish reinforcements or rewards for the new behavior. People who have quit smoking might focus on how much better food tastes after they have given up cigarettes. People who have successfully improved their diet and exercise habits might treat themselves to new workout clothes. Acknowledging, enjoying, and celebrating accomplishments can motivate consistent behavior.

Another source of support for new behaviors is the person's social network. Friends, family, and members of a support group can help through their encouraging words and behaviors. Members of a family might all quit smoking at the same time or join the person in physical activities or healthier eating.

Finally, people may focus on alternative behaviors that replace the unhealthy ones. Instead of bar hopping, they might join a group dedicated to activities not associated with drinking alcohol, such as a dance club or community theater group. In other words, effective change also involves avoiding temptations.

MAINTENANCE
In the *maintenance stage,* people successfully avoid temptation and consistently pursue healthy behaviors. They may become skilled at anticipating

Can you quit smoking if you are spending time with smokers? Can you avoid binge drinking if you regularly go to keg parties?

tempting situations and avoid them or actively prepare for them. If people who are trying to quit smoking know that they always enjoy a cigarette after a big meal out with friends, they might mentally prepare themselves for that temptation before going out. People with new exercise routines might post a consciousness-raising photograph on the refrigerator or leave their walking shoes by the door. People who feel like they have control over their decision to make a life change are more likely to make it to the maintenance stage (Vancampfort & others, 2016).

At some point, people in maintenance may find that actively fighting the urge to indulge in unhealthy behaviors is no longer necessary. *Transcendence* means that they are no longer consciously engaged in maintaining their healthy lifestyle; rather, the lifestyle has become a part of who they are. They are now nonsmokers, healthy eaters, or committed runners.

kali9/Getty Images

RELAPSE   One challenge during the maintenance stage is to avoid **relapse**, a return to former unhealthy patterns. Relapse is a common aspect of change, and it can be discouraging. However, the majority of people who eventually do change do not succeed on the first try. Rather, they try and fail and try again, cycling through the five stages several times before achieving a stable, healthy lifestyle. Consequently, experts in changing health behavior consider relapse to be normal (Prochaska & Norcross, 2013).

Relapse is a normal part of change. What does this principle suggest about recovery from drug addiction?

If you have ever tried to adopt a healthier lifestyle by eating healthier foods, starting an exercise program, or quitting smoking, you might know how bad you feel when you experience relapse. One slip, however, does not mean that you are a failure and will never reach your goal. Rather, when a slipup occurs, you have an opportunity to learn, to think about what led to the relapse, and to devise a strategy for preventing it in the future.

EVALUATION OF THE STAGES OF CHANGE MODEL   The stages of change model has been applied successfully to a broad range of behaviors. These include cigarette smoking (Hsu & others, 2020), exercise (Kleis & others, 2020), healthy food choices (Nor & others, 2020), cancer prevention behaviors (Choi & others, 2013), safe-sex practices (Mogro-Wilson & others, 2020), substance use and abuse (Field & others, 2020), and improving academic behaviors (Moreira & others, 2020). The model has also been used to predict adherence to safe practices at work (Rothmore & others, 2015) and to behavior change in criminal offenders (Yong & others, 2015).

Despite its relevance to a variety of behaviors, the stages of change model is controversial (Brug & others, 2005; Joseph & others, 1999). Some critics have questioned whether the stages are mutually exclusive and whether individuals move from one stage to another in the order proposed (Littrell & Girvin, 2002). A meta-analysis showed that the sequence of stages may vary depending on the specific domain of health change (Rosen, 2000). Critics of the model also point out that it refers more to attitudes that change than to behaviors (West, 2005). On the more positive side, some evidence suggests that the stages of change model does a good job of capturing the ways that individuals make positive life changes (Gantiva & others, 2015; Lippke & others, 2009, 2010). A meta-analysis of 39 studies that encompassed more than 8,000 psychotherapy clients found that the stages of change model was effective in predicting psychotherapy outcomes (Norcross & others, 2011).

The stages of change model can be a tool for therapists and medical professionals who are trying to help clients institute healthy behavior patterns. Sometimes, sharing the model with individuals who are trying to change provides them with a useful language for understanding the change process, for reducing uncertainty, and for developing realistic expectations for the difficult journey.

1. The theoretical model that breaks down behavioral change into five distinct steps is the
   A. theory of planned behavior.
   B. theory of reasoned action.
   C. cognitive theory of change.
   D. stages of change model.

2. When people who are trying to change a behavior return to unhealthy patterns, we say that they are in a state of
   A. denial.
   B. relapse.
   C. plateau.
   D. maintenance.

3. The stages of change model
   A. is not at all controversial.
   B. applies to a wide variety of behaviors.
   C. does not apply to cigarette smoking.
   D. does not apply to safe-sex practices.

APPLY IT! 4. Malcolm has been trying to quit smoking for two years. During his last attempt, he went three full months without smoking but then had a cigarette after a big fight with his romantic partner. He is feeling hopeless about his chances of quitting. What does

the stages of change model say about Malcolm's situation?
   A. Relapse is a normal part of change. Malcolm might think about why he relapsed and try to move on from there with a new strategy.
   B. Malcolm has blown it and will probably never quit smoking.
   C. Malcolm is stuck in the contemplation phase of change.
   D. Malcolm is unusual in that he had a relapse after three full months. He probably has a particularly strong addiction to cigarettes.

3 Resources for Effective Life Change

Making positive changes to promote health can be challenging. Fortunately, we all have various psychological, social, and cultural resources at our disposal to help us in the journey to a healthier lifestyle. In this section we consider some of these tools that can help us achieve effective change and, ultimately, a healthier life.

Motivation

Recall that motivation refers to the "why" of behavior. Motivational tools for self-change involve changing for the right reasons. Motivation is important at every stage in the change process (Serafini & others, 2016). Moreover, change is most effective when you are doing it for you. When people feel that they freely choose their new behaviors, change is more likely to persist.

Self-determination theory distinguishes between *intrinsic motivation*, doing something because you want to and enjoy it, and *extrinsic motivation*, doing something for external rewards. Creating a context in which people feel more in control, more autonomous, and more competent is associated with enhanced outcomes for a broad array of health behaviors (Ntoumanis & others, 2020; Sheeran & others, 2020).

Westend61/Getty Images

Planning and goal setting are crucial to making effective change. People who come up with specific strategies, or **implementation intentions**, for dealing with the challenges of making a life change are more successful than others at navigating change (Mutter & others, 2020). "Implement" means put into action. So, implementation intentions are strategies for actually putting our health goals to work. Setting short-term, achievable goals also allows individuals to experience the emotional payoff of small successes along the way to self-change. These feelings of satisfaction can help to motivate continued effort toward achieving health goals (Harris & others, 2014). Implementation intentions are especially effective if they are tied to a person's identity and sense of self (Kendzierski & others, 2015).

implementation intentions
Specific strategies for dealing with the challenges of making a life change.

Enjoying the payoffs of our efforts to change also means that we must monitor our goal progress. It is important to get feedback on our progress in the pursuit of any goal. If we find out that we are falling short, then we can try to identify areas that need work. On the other hand, discovering that we are doing well can be a potent motivator for future progress.

Social Relationships

Research has shown, again and again, that social ties are an important, if not the most important, variable in predicting health. In a landmark study, social isolation had six times the effect on mortality rates that cigarette smoking had (House & others, 1988). Loneliness has wide-ranging negative impact on the body and mind (Smith & others, 2020). Longitudinal studies show that chronic loneliness predicts dying from a range of causes, over time (Luo & others, 2012; Patterson & Veenstra, 2010). Social isolation is a predictor of mortality among those who have cardiovascular disease (Kraav & others, 2020; Yu & others, 2020) and among the very old (Wang & others, 2020). Loneliness is thought to play such a large role in health because it is linked to higher stress, poorer sleep, poorer health behaviors, and less physical activity (Christiansen & others, 2016).

One way that social connections make a difference in our lives is what psychologists call social support. **Social support** is information and feedback from others indicating that one is loved and cared for, esteemed and valued, and included in a network of communication and mutual obligation. Social support has three types of benefits (Taylor, 2018):

social support
Information and feedback from others indicating that one is loved and cared for, esteemed and valued, and included in a network of communication and mutual obligation.

1. *Tangible assistance:* Family and friends can provide goods and services in stressful circumstances, as when gifts of food are given after the death of a loved one.

2. *Information:* Individuals who extend support can also recommend specific strategies to help the person under stress cope. Friends may notice that a coworker is overloaded with work and suggest ways of better managing time or delegating tasks.

3. *Emotional support:* People under stress often suffer emotionally and may develop depression, anxiety, or loss of self-esteem. Friends and family can reassure the stressed person that they are valuable and loved. Knowing that others care allows a person to manage stress with greater assurance.

Social support can help as a person prepares for a stressful experience, allowing the person to see the event in a less negative light. It can also help after the stressful event, helping the person recover. Social support can act as a shield, protecting people from negative psychological and physical consequences of difficult or traumatic events (De Maria & others, 2020; Forbes & others, 2020). For example, it is not uncommon for symptoms of PTSD to be associated with increases in physical illness symptoms but social support can disrupt this relationship. A study of military personnel who had been deployed to Iran, Iraq, and Afghanistan found that among whose who felt their unit was "like a family" and those with high level of support from family and friends after returning from combat, PTSD symptoms were less likely to be associated with physical symptoms (Luciano & McDevitt-Murphy, 2017).

Social support is important for children as well. A relevant study examined the ways that social support buttressed the well-being of 430 children who were orphaned in Rwanda. These children were left orphaned by war and illness and lived in a variety of sometimes very difficult circumstances, including on the street, in orphanages, in foster homes, or in child-headed households. Among these children, social support especially from adults was associated with higher psychological well-being and less distress (Caserta & others, 2017). In fact, the negative association between adult social support and distress was similar to that for having access to three meals a day.

Find a quiet place to write.

Pick just one topic to explore through writing.

Dedicate yourself to at least 20 minutes of writing about that topic.

While writing, do not be concerned with grammar or spelling; just let yourself go and write about all of the emotions, thoughts, and feelings associated with the experience you are writing about.

If you feel that writing about something negative is not for you, try writing about your most positive life experiences, about the people you care about, or all the things you feel grateful for in life.

FIGURE 2 **Harnessing the Power of Writing**
Try this simple exercise to explore the health benefits of writing.
Ryan McVay/Getty Images

One way that people gain support during difficult times is through *social sharing*—turning to others who act as a sounding board or a willing ear. Sometimes sharing our thoughts and feelings does not have to be very social to be helpful. James Pennebaker and his colleagues (Pennebaker, 1997a, 1997b, 2004) have demonstrated that writing about traumatic life events for 20 minutes a day over two or three days is associated with improved health, fewer illnesses, greater immune system function, and superior reactions to vaccines. Although writing about trauma is usually linked to increased distress in the short term, over the long run it brings physical and psychological health benefits (Baddeley & Pennebaker, 2011; Frattaroli, 2006; Pennebaker & Chung, 2007, 2011; Smyth, 1998). Studies have shown that expressive writing can lead to benefits for many different groups (Bryan & Lu, 2016; Greenbaum & Javdani, 2017; Travagin & others, 2015). Although, typically, expressive writing studies have focused on sharing about negative life events and expressing negative feelings (Lee & others, 2016), studies have found health benefits for writing about one's best possible future self (Loveday & others, 2017) and intensely positive life experiences (Burton & King, 2004, 2008; King, 2002). If you would like to give this simple intervention a try, see Figure 2.

Getting support from others is important, but *giving* support can also have benefits (Crocker & others, 2017). A study of 423 older adult couples who were followed for five years revealed how helping others benefits physical health (Brown & others, 2003). At the beginning of the study, the couples were asked about the extent to which they had given or received emotional or practical help in the past year. Five years later, those who said they had helped others were half as likely to have died. Indeed, in general prosocial behavior is related to health and longer life (Fritz & others, 2020), although the exact mechanism of this effect is not known.

Religious Faith

Religious faith is strongly related to maintaining a healthy lifestyle and to good health (Fastame & others, 2017; Shattuck & Muehlenbein, 2020). Religious faith can be thought of as the ultimate "biopsychosocial" variable in predicting enhanced health and longevity (Page & others, 2020), as it impinges on behaviors, social support, and a sense of meaning in life. Many religions frown on excess and promote moderation. Weekly religious attendance relates to a host of healthy behaviors, including not smoking, taking vitamins, walking regularly, wearing seat belts, exercising strenuously, sleeping soundly, and drinking moderately or not at all (Haber & others, 2011; Hill & others, 2006; Mellor & Freeborn, 2011). A number of studies have linked religious participation to a longer and healthier life (Campbell & others, 2009; Krause, 2006; McCullough & Willoughby, 2009).

In addition to promoting healthier behaviors, religious participation may benefit health through its relationship to social support (Le & others, 2016; Hovey & others, 2014). Belonging to a faith community may give people access to a warm group of others who are available during times of need. This community is there to provide transportation to the doctor, to check in with the person during hard times, and simply to stand next to the individual during a worship service, as a member of the community. The social connections promoted by religious activity can forestall anxiety and depression and help to prevent isolation and loneliness (Page & others, 2020). For example, a study of Black Americans found that church-based social support was protective against depressive symptoms (Chatters & others, 2014).

Religious faith and spirituality more generally may also be important factors in good health because they provide a sense of life meaning and a buffer against the effects of stressful life events (Appel & others, 2020; Reynolds & others, 2020; Williams, 2020). Religious thoughts can play a role in maintaining hope and stimulating motivation for positive life changes. One study showed that among people coping with diagnoses of heart failure or cancer, those who were higher in religious faith increased in the sense of meaning in life over time (George & Park, 2017). Belief in the enduring meaningfulness of one's life can help one keep perspective and see life's hassles in the context of the big picture.

How might these results apply to a person who is not religious?

Personality Characteristics

Personality traits are powerful instruments in the self-change toolbox. Here we survey some of the personality characteristics related to health.

CONSCIENTIOUSNESS   Conscientious people are responsible and reliable; they like structure and seeing a task to its completion. Conscientiousness is not the sexiest trait, but it is the most important of the big five traits when it comes to health, healthy living, and longevity (Steptoe & Jackson, 2020; Turiano & others, 2015). Various studies show that conscientious people tend to do all the things that they are told are good for their health, such as getting regular exercise, eating healthy foods, avoiding drinking and smoking, wearing seat belts, monitoring their blood pressure, and checking smoke detectors (O'Connor & others, 2009; Turiano & others, 2012; Wilson & others, 2016). Conscientious people are less likely to die than their counterparts who are less conscientious (Hill & others, 2011; Roberts & others, 2013).

PERSONAL CONTROL   Another personality characteristic associated with taking the right steps toward a long, healthy life is a sense of personal control, what is referred to as an *internal locus of control*. Feeling in control can reduce stress during difficult times (Lachman & others, 2015) and can lead to the development of problem-solving strategies to deal with life's challenges. A person with a good sense of personal control might reason, "If I stop smoking now, I will not develop lung cancer."

A sense of personal control has been linked to a lower risk for common chronic diseases such as cancer and cardiovascular disease (Climie & others 2020; Sturmer & others, 2006). Further, like conscientiousness, a sense of personal control might also help people avoid a risky lifestyle that involves health-compromising behaviors. A recent study showed that during the Great Recession (from 2007 to 2009), economic hardship was associated with declines in personal control and increases in distress. However, a high sense of personal control buffered the effects of economic problems on distress (Koltai & Stuckler, 2020). Overall, across a wide range of studies, a sense of personal control has been related to emotional well-being, successful coping with stressful events, healthy behavior change, and good health (Hughes & others, 2012; Sheppes & others, 2015; Turiano & others, 2014; Zhang & Jang, 2017).

A person with a low level of personal control may feel that whatever happens happens–it is meant to be or a matter of (good or bad) luck.

How might culture influence the roles of self-efficacy and personal control in health?

SELF-EFFICACY   Self-efficacy is the belief that one can master a situation and produce positive outcomes. Self-efficacy affects behavior in many situations, ranging from solving personal problems to starting an exercise routine. Self-efficacy influences whether individuals try to develop healthy habits, how much effort they expend in coping with stress, how long they persist in the face of obstacles, and how much stress they experience (Thompson & others, 2017).

Self-efficacy is related to success in a wide variety of positive life changes. These include achieving weight loss (Nezami & others, 2016), exercising (Strachan & others, 2016), quitting smoking (Berndt & others, 2012; Shadel & others, 2017), reducing substance abuse (Goldsmith & others, 2012), and leading a healthier lifestyle (Axelsson & others, 2012). Self-efficacy is linked to cardiovascular functioning following heart failure; individuals high in self-efficacy not only are less likely to suffer a second hospitalization due to heart failure but also are likely to live longer (Maeda & others, 2012; Sarkar & others, 2009). If there is a problem to be fixed, self-efficacy—having a can-do attitude—is related to finding a solution.

Throughout this book, we have examined the placebo effect as a positive response to a treatment that has no medicinal power. The placebo effect results from the individual's belief in the effectiveness of the treatment. Can you really adopt a healthier exercise routine? Maybe or maybe not, but believing that you can allows you to harness the placebo effect. Self-efficacy is the power of belief in yourself.

OPTIMISM   One factor that is often linked to positive functioning and adjustment is optimism (Steptoe & Jackson, 2020). Optimism refers to looking on the bright side and having positive expectations for the future. Optimism is associated with taking proactive steps to protect one's health, while pessimism is linked to engaging in health-compromising behaviors (Carver & Scheier, 2014). Numerous studies reveal that optimists generally function more effectively and are physically and mentally healthier than pessimists (Avvenuti & others, 2016). For instance, in a sample of men with coronary heart disease, optimists were less likely to experience negative health events over the course of a year (Hevey & others, 2014). One study showed that being optimistic and having an optimistic spouse were both associated with better health (Kim & others, 2014). Optimism has been linked to more effective immune system functioning and better health (Engberg & others, 2013). And a study of centenarians (people who live to be 100 or more years of age) revealed that those who were in better health had a higher level of optimism than their pessimistic counterparts (Tigani & others, 2012). Optimists are also more likely to seek out potentially threatening health risk information (Taber & others, 2015).

Optimism has also been defined as a way of explaining events that happen. An optimistic attributional style means seeing the causes of bad events as external, temporary, and specific (Seligman, 1990). In contrast, pessimists identify bad events as caused by internal, stable, and global causes. Explaining life events optimistically is associated with positive outcomes, including a better quality of life (Carver & Scheier, 2014).

As you think about the traits we have examined—conscientiousness, personal control, self-efficacy, and optimism—and their relationship to good health, an important practical tip to keep in mind is that you can *cultivate* these qualities. Studies show that even conscientiousness, the most stable of these characteristics, can increase, especially in young adulthood. Keep in mind, too, that these various resources relate to each other in important ways. A person with a high level of self-efficacy may be more optimistic about health decisions and challenges (Hajek & Helmut-König, 2017).

Even as we think about the value of these various characteristics, it is important to always bear in mind that not everything is under a person's control, and that includes health outcomes. Research pointing to the links between various personality characteristics and better health outcomes can sometimes create a burden on people who are struggling with illness (Jones & Ruthig, 2015; Ruthig & Holfeld, 2016). Not everything that happens to us is in our control, and psychological factors can only go so far in predicting health. To read about this issue, see Challenge Your Thinking.

How Powerful Is the Power of Positive Thinking?

Research demonstrating the role of psychological variables in health, disease, and mortality is extremely appealing because it gives us a sense we have some control over our physical health. Yet as reassuring as such findings might be, these factors are not a psychological recipe for immortality. The plain truth is that eventually everyone dies. When scientists find a link between some psychological factor and an important health outcome, the popular media often latch on to the results as if they mean that such factors play a causal role in disease. The popularity of such ideas can lead to victim blaming: thinking that a person is ill or has died because of a deficit of self-efficacy or optimism.

A compelling case in point is provided by research on "fighting spirit" in combatting breast cancer. In a study published over four decades ago, 69 women were interviewed three months after undergoing surgery for breast cancer (Greer & others, 1979). Based on the interviews, the researchers categorized the women's responses to breast cancer as denial, fighting spirit, quiet resignation, or helplessness. The researchers then followed up on the women five years later to see whether they had experienced a recurrence. The results of the follow-up study showed that women whose responses were characterized by either denial or fighting spirit were less likely to have had a recurrence of cancer. This study led to the conclusion that women with breast cancer should be encouraged to adopt a fighting attitude toward their cancer. The idea that a fighting spirit is important to cancer survival continues to hold sway in interventions for people coping with the disease (Coyne & Tennen, 2010).

Crucially, this finding, based on a single study with a small sample, has not withstood the test of time. Subsequent research, especially studies employing much larger samples, has failed to show any link between adopting a fighting spirit and breast cancer outcomes (Petticrew & others, 2002; Phillips & others, 2008; Watson & others, 2005).

Although the reality that a fighting spirit does not improve a woman's chances of beating cancer might seem disappointing, many have welcomed this news. As one expert commented, such findings "may help to remove any continuing feelings of guilt or sense of blame for breast cancer relapse from those women who worry because they cannot always maintain a fighting spirit or a positive attitude" (Dobson, 2005). The widespread belief that adopting a fighting spirit was key to cancer survival imposed a burden on people already dealing with a difficult life experience.

King Lawrence/Blend Images

Does this conclusion mean that psychosocial variables have no role to play in disease? Certainly not. One study that found no effect of fighting spirit did show that initial helplessness in response to diagnoses was a predictor of poorer outcomes among women with breast cancer (Watson & others, 2005). Knowing that a person feels helpless early on may prompt professionals to provide much-needed information about treatment and the potential for long-term recovery. Indeed, among the factors that (happily) complicate this type of research are that many cancers have effective treatments and that, especially with early detection, relatively few individuals die or experience a recurrence. A study of people with Stage IV colorectal cancer showed that, although hope and optimism were unrelated to disease progression, depression predicted poorer outcomes (Schofield & others, 2016). Family and friends can provide support that helps a person maintain quality of life without pressuring the person the "be positive" (Hill, 2016). Professionals can also use information about psychological characteristics to build in behavioral supports that might be needed to help a person stick with treatment and optimize her outcomes.

People deal with potentially life-threatening diagnoses in different ways. Dutch swimmer Maarten van der Weijden was diagnosed with leukemia in 2001 at the age of 20 but went on to win Olympic gold in 2008. With respect to his diagnosis, he remarked, "I . . . simply surrendered to the doctors. You always hear those stories that you have to think positively, that you have to fight to survive. This can be a great burden for patients. It has never been proven that you can cure cancer by thinking positively or by fighting" (quoted in Coyne & others, 2010, p. 40).

What Do You Think?

- In the 1979 study, fighting spirit and denial both were associated with better outcomes. Why do you think people latched on to fighting spirit rather than denial as a key intervention?

- If someone you love were diagnosed with cancer, how would the research reported here influence the support you would provide to that person?

1. All of the following are powerful tools for self-change *except*
   A. loneliness.
   B. religious faith.
   C. personality traits.
   D. motivation.

2. The benefits of social support include all of the following *except*
   A. information.
   B. tangible assistance.
   C. emotional support.
   D. victim blaming.

3. Optimists explain the causes of bad events as
   A. external.
   B. internal.
   C. global.
   D. stable.

APPLY IT! 4. Daniel was recently diagnosed with diabetes. His doctor gave him a new diet to control his condition. Which of the following situations offers the *best* chances that Daniel will stick with the diet?
   A. Daniel loves junk food and does not want to follow the diet, but his mother and aunt, both living with diabetes, are pressuring him to follow the doctor's recommendations.
   B. Daniel has always had trouble following through on doing what is good for him, though he says he wants more structure in his life.
   C. Clark, Daniel's roommate, has a brother with diabetes, and Clark tells Daniel about how his brother has coped and what diet he follows; Clark offers to introduce them.
   D. Daniel has a pessimistic personality and expects things to work out badly.

4 Toward a Healthier Mind (and Body): Controlling Stress

If you could change one thing about your behavior, what would you choose? Would the change perhaps have to do with feeling stressed out much of the time? Maybe you wish you could stop facing every daily challenge with tension. Let's look at the problems that can arise when you feel chronically stressed and the ways you can better manage your stress.

Stress and Its Stages

 Getting married, having a baby, graduating from college, getting a promotion at work—although positive life events, these achievements can be stressful because they are also major life changes.

Stress is the response to environmental stressors, the circumstances and events that threaten people and tax their coping abilities. Hans Selye (1974, 1983), the founder of stress research, focused on the body's response to stressors, especially the wear and tear due to the demands placed on the body. After observing patients with different problems—the death of someone close, loss of income—Selye concluded that any number of environmental events or stimuli would produce the same stress symptoms: loss of appetite, muscular weakness, and decreased interest in the world. Selye realized that whenever people experienced stressors, they experienced a common phenomenon, stress.

General adaptation syndrome (GAS) is Selye's term for the common effects on the body when demands are placed on it (Figure 3). The GAS consists of three stages: alarm, resistance, and exhaustion. Selye's model is especially useful in helping us understand the link between stress and health.

The body's first reaction to a stressor, in the *alarm stage,* is a temporary state of shock during which resistance to illness and stress falls below normal limits. In trying to cope with the initial effects of stress, the body releases hormones that, in a short time, adversely affect the functioning of the immune system, the body's network of natural defenses. During this time the person is vulnerable to infections from illness and injury.

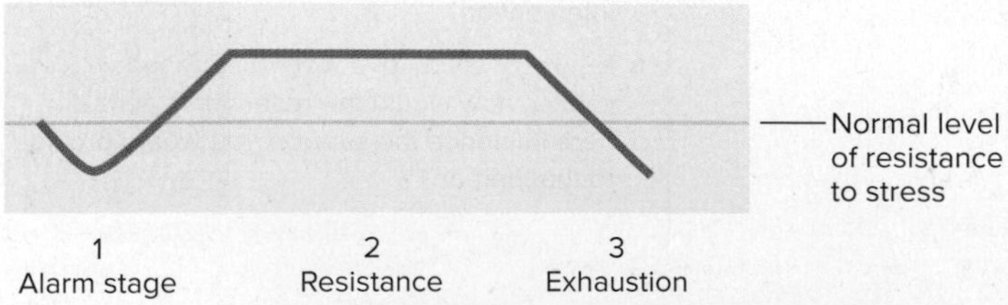

general adaptation syndrome (GAS) Selye's term for the common effects of stressful demands on the body, consisting of three stages: alarm, resistance, and exhaustion.

FIGURE 3   Selye's General Adaptation Syndrome The general adaptation syndrome (GAS) describes an individual's response to stress in terms of three stages: (1) alarm, in which the body mobilizes its resources; (2) resistance, in which the body strives mightily to endure the stressor; and (3) exhaustion, in which resistance becomes depleted.

— Normal level of resistance to stress

1
Alarm stage

2
Resistance

3
Exhaustion

In the *resistance stage* of Selye's general adaptation syndrome, glands throughout the body manufacture different hormones that protect the individual. Endocrine and sympathetic nervous system activity are not as high as in the alarm stage, although they still are elevated. During the resistance stage, the body's immune system can fight off infection with remarkable efficiency. Similarly, hormones that reduce the inflammation normally associated with injury circulate at high levels.

If the body's all-out effort to combat stress fails and the stress persists, the person moves into the *exhaustion stage*. At this point, wear and tear takes its toll—the person might collapse in exhaustion, and vulnerability to disease increases. Serious, possibly irreversible damage to the body—such as a heart attack or even death—may occur.

The body system that plays the greatest role in Selye's GAS model is the **hypothalamic-pituitary-adrenal axis (HPA axis)**. The HPA axis is a complex set of interactions among the hypothalamus (part of the brain's limbic system), the pituitary gland (the master gland of the endocrine system), and the adrenal glands (endocrine system glands located on top of each kidney) (Spencer & Deak, 2017). The HPA axis regulates various body processes, including digestion, immune system responses, emotion, and energy expenditure. The axis also controls reactions to stressful events, and these responses will be our focus here.

When the brain detects a threat in the environment, it signals the hypothalamus to release corticotropin-releasing hormone (CRH). In turn, CRH stimulates the pituitary gland to produce another hormone that causes the adrenal glands to release cortisol. Cortisol is itself the "stress hormone" that directs cells to make sugar, fat, and protein available so the body can take quick action. Cortisol also suppresses the immune system.

Stress can be acute or chronic. Acute stress can sometimes be adaptive, and in acute stress cortisol plays an important role in helping us to take the necessary action to avoid dire consequences. Typically, once the body has dealt with a given stressor, cortisol level returns to normal. However, under chronic stress, the HPA axis can remain activated over the long haul.

The activity of the HPA axis varies from one person to the next. These differences may be explained by genes as well as by stressful experiences (Starr & others, 2020). Research with nonhuman animals and humans has shown that prenatal and postnatal stress can influence the development of the HPA axis (Laurent & others, 2014; Mustoe & others, 2014; Reilly & Gunnar, 2019). When the HPA is chronically active, various systems in the body suffer (O'Connor & others, 2021).

hypothalamic-pituitary-adrenal axis (HPA axis)
The complex set of interactions among the hypothalamus, the pituitary gland, and the adrenal glands that regulate various body processes and control reactions to stressful events.

Stress and the Immune System

Chronic stress can have serious implications for the body, in particular the immune system. Interest in links between stress and the immune system spawned the field of **psychoneuroimmunology**, which explores connections among psychological factors (such as attitudes and emotions), the nervous system, and the immune system (Acabchuk & others, 2017; Debnath & others, 2020).

The immune system and the central nervous system are similar in their modes of receiving, recognizing, and integrating signals from the external environment (Arisi, 2014). The central nervous system and the immune system both possess "sensory" elements that receive information from the environment and other parts of the body and "motor" elements that carry out an appropriate response. Both systems also rely on chemical mediators for communication. CRH, a key hormone shared by the central nervous system and the immune system, is produced in the hypothalamus, as we saw earlier, and unites the stress and immune responses.

Stress can profoundly influence the immune system. Acute stressors (sudden, stressful, one-time life events) can produce immunological changes, but because the stressors are temporary, these changes are not likely to be problematic in healthy people (Schakel & others, 2019). However, chronic stressors (long-lasting agents of stress) are associated with an increasing downturn in immune system responsiveness

psychoneuroimmunology
A new field of scientific inquiry that explores connections among psychological factors (such as attitudes and emotions), the nervous system, and the immune system.

Recall that the immune system can learn through classical conditioning.

(Acabchuk & others, 2017; Dai & others, 2020; Schakel & others, 2019; Thames & others, 2019). This effect has been linked to lowered immune system functioning in a number of contexts, including worries about living next to a damaged nuclear reactor, close relationship problems, depression, loneliness, and burdensome caregiving for a family member with progressive illness, and coping with prejudice (Fagundes & others, 2013; Jaremka & others, 2013a, 2013b, 2014). The basic idea is that chronic stress wears down the body and depresses the immune system, leaving the person susceptible to disease (O'Connor & others, 2021).

Sheldon Cohen and his colleagues have carried out a number of studies on the effects of stress, emotion, and social support on immunity and susceptibility to infectious disease (Cohen, 2020; Cohen & Lemay, 2007; Cohen & others, 2009, 2012). In one such study, Cohen and his colleagues (1998) focused on 276 adults who were exposed to viruses and then quarantined for five days. Figure 4 shows the dramatic results. The longer the participants had experienced major stress in their lives before the study, the more likely they were to catch a cold. Cohen concluded that stress-triggered changes in the immune system and hormones might create greater vulnerability to infection. These findings suggest that when we are under stress, we need to take better care of ourselves than usual (Cohen, 2020).

Meditation is a great way to cope with stress and has positive benefits for the immune system.

Stress and Cardiovascular Disease

People who have experienced stressful life changes, like the death of a spouse, the loss of a job, or a natural disaster are at increase risk of heart attack (Leor & others, 1996; Mostofsky & others, 2012; Musey & others, 2020). You may have heard of someone "dying of a broken heart," and it is the case that people, especially women, who have experienced grief are at an increased risk for acute coronary issues (Peters & others, 2015). Even more than acute stress, though, chronic stress predicts risk for cardiovascular disease (Osborne & others, 2020). One likely reason for this link is changes in health behaviors related to stress, like overeating, smoking, or avoiding exercise. However, it is also clear that chronic stress affects the cardiovascular system itself (Osborne & others, 2020).

Just as personality characteristics such as a sense of control or self-efficacy can help buffer an individual against stress, other personality characteristics have been shown to worsen stress, with special significance for cardiovascular illness. In particular, people who are impatient or quick to anger or who display frequent hostility have an increased risk for cardiovascular disease (Espnes & Byrne, 2016).

In the late 1950s, a secretary for two California cardiologists, Meyer Friedman and Ray Rosenman, observed that the chairs in their waiting rooms were tattered and worn, but only on the front edges. The cardiologists had also noticed the impatience of their cardiac patients, who often arrived exactly on time and were in a great hurry to leave. Intrigued by this consistency, they conducted a study of 3,000 healthy men between the ages of 35 and 59 over eight years to find out whether people with certain behavioral characteristics might be prone to heart problems (Friedman & Rosenman, 1974). During the eight years, one group of men had twice as many heart attacks or other forms of heart disease as the other men. Further, autopsies of the men who died revealed that this same group had coronary arteries that were more obstructed than those of the other men.

Friedman and Rosenman described the common personality characteristics of the men who developed coronary disease as the **Type A behavior pattern.** They theorized that a cluster of characteristics—being excessively competitive, hard-driven, impatient, and hostile—is related to the incidence of heart disease.

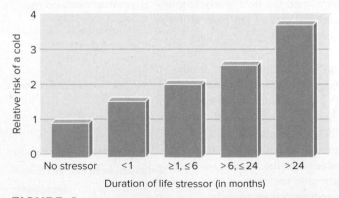

FIGURE 4 Stress and the Risk of Developing a Cold
In a study by Cohen and others (1998), the longer individuals had a life stressor, the more likely they were to develop a cold. The four-point scale is based on the odds (0 = lower; 4 = higher) of getting a cold.

Type A behavior pattern
A cluster of characteristics—including being excessively competitive, hard-driven, impatient, and hostile—that is related to a higher incidence of heart disease.

Rosenman and Friedman labeled the behavior of the healthier group, who were typically relaxed and easygoing, the **Type B behavior pattern.**

Further research on the link between Type A behavior and coronary disease indicates that the association is not as strong as Friedman and Rosenman believed (Benjamin, 2020). The Type A behavior component most consistently associated with coronary problems is hostility (Appleton & others, 2016; Ohira & others, 2012). People who are hostile outwardly or who turn anger inward are more likely to develop heart disease than their less angry counterparts (Eng & others, 2003; Matthews & others, 2004). Such people have intense physiological reactions to stress: Their hearts race, their breathing quickens, and their muscles tense up. One study found that hostility was a better predictor of coronary heart disease in older men than smoking, drinking, high caloric intake, or high levels of LDL (bad) cholesterol (Niaura & others, 2002). It is important to bear in mind that hard-driving successful professionals, the people who first inspired research on Type A behavior pattern, are not a high-risk group for heart disease. Type A probably did not pertain to people most likely to be affected by coronary heart disease who have low levels of education and have less access to healthcare (Rosengren & others, 2019).

Another pattern of behavior, the **Type D behavior pattern**, describes people who are generally distressed, frequently experience negative emotions, and are socially inhibited (Lodder, 2020). Even after adjustment for depression, Type D individuals face a threefold increased risk of adverse cardiovascular outcomes (Denollet & Conraads, 2011). A meta-analysis also found that Type D persons with cardiovascular disease are at a higher risk for major adverse cardiac events and have a lower health-related quality of life (O'Dell & others, 2011).

It is important to bear in mind that the body's internal reactions to stress do not magically lead to poor health. The way people behave when they are stressed can contribute to disease risk. In addition, as we will discuss later, people do have ways to cope with stress that can help alleviate its negative effects.

Type B behavior pattern
A cluster of characteristics—including being relaxed and easygoing—that is related to a lower incidence of heart disease.

Type D behavior pattern
A cluster of characteristics—including being generally distressed, having negative emotions, and being socially inhibited—that is related to adverse cardiovascular outcomes.

Students who set high standards sometimes think they are Type A but remember that the key feature of "Type A" is **hostility!** Type A folks often have explosive tempers.

Back in the day, research on Type A behavior pattern was funded (and favored) by the tobacco industry—because it seemed to provide a way to explain (away) the serious health consequences of smoking.

Stress and Cancer

Given the association of stress with poor health behaviors such as smoking, it is not surprising that stress has also been related to cancer risk (Nezu & others, 2013). Stress sets in motion biological changes involving the autonomic, endocrine, and immune systems. If the immune system is not compromised, it appears to help provide resistance to cancer and slow its progression. This is the idea behind contemporary immunotherapy treatment—boosting our bodies' natural defenses against cancer. However, the physiological effects of stress inhibit a number of cellular immune responses (Dai & others, 2020; Hayakawa, 2012). People with cancer show diminished natural killer cell (NK-cell) activity in the blood (Rosental & others, 2012) (Figure 5). Low NK-cell activity is linked with the development of further malignancies, and the length of survival for the cancer patient is related to NK-cell activity (Hegmans & Aerts, 2014; Leung, 2014). The activation of the HPA by chronic stress affects the endocrine and immune systems in ways that damage the body's capacity to resist cancer and may promote tumor growth (Dai & others, 2020; Kruk & others, 2019).

Given the damaging effects of especially chronic stress, psychologists have been interested in identifying sources of stress and ways to deal with stress effectively. Unfortunately, an important source of stress for many people is the experience of prejudice, as we now consider.

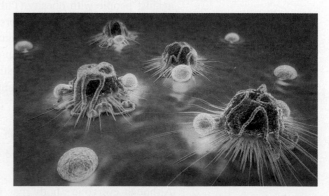

FIGURE 5 NK Cells and Cancer Two natural killer (NK) cells (*yellow*) are shown attacking a leukemia cell (*red*). Notice the blisters that the leukemia cell has developed to defend itself. Nonetheless, the NK cells are surrounding the leukemia cell and are about to destroy it. YAY!

Christoph Burgstedt/Alamy Stock Photo

Stress and Prejudice

It is not hard to see how prejudice and discrimination are stressful. Given how important social relationships are to human beings, feeling excluded and treated unfairly because of your identity is stressful. The negative effects of prejudice and discrimination have been documented in a variety of samples. For example, prejudice and discrimination directed at members of sexual minority groups as well as transgender people predicts stress as well as physical illness (Feinstein & others 2020; Kassing & others 2020). The stigma that is attached to various identities, including sexual minority, immigrant status, and veteran status, leads to stress (Albuja & others 2019; Hipes & Gemoets, 2019; Xia & Ma, 2020). Hate hurts when it is encountered, and that is true of people in general. Anti-Black racism may play a role in the continuing health disparities that exist in the United States.

Health disparities refer to often preventable differences in physical functioning (including disease, injury, violence) and psychological functioning (including depression and anxiety) that are experienced by socially disadvantaged groups. In the United States, Black people have a lower life expectancy (75 years) than white people (78.9 years) (Xu & others, 2020). Although these numbers are improvements over previous years, there remain important differences between Black and white Americans. The differential effect of the COVID-19 pandemic on people of color has widened the differences (Wrigley-Field, 2020).

When people try to explain health disparities, they often note differences in access to resources such as health information, healthcare, nutrition, safe neighborhoods, opportunities for physical activity, and so forth. But why do people of different races experience differences in access to resources? One answer is systemic racism (Dover & others, 2020; Egede & Walker, 2020). Evidence also points to the idea that racism plays a role in differences in the quality of healthcare people receive (Beck & others, 2020; Chaney & others, 2019; Dover & others, 2020; Williams & others, 2019). Events like the police killing of George Floyd in 2020 led many people to recognize that the COVID-19 pandemic was not the only public health emergency facing the United States. Many public health and medical experts declared racism a public health crisis (APHA, 2020; O'Reilly, 2020; Serchen & others, 2020).

There is simply no question that, at the level of the individual, experiencing discrimination is stressful (Dover & others, 2020). Research with Black participants has demonstrated that experiences with prejudice take a toll on the body, affecting the stress response and immune system function (Currie & others, 2020; Thames & others, 2019). The chronic stress of injustice, unfair treatment, and daily experiences of prejudice and discrimination might certainly contribute to health disparities.

Recently, psychologists have been interested in how prejudice and discrimination against people with obesity might affect their health. Stereotypes of people with obesity include that they are lazy or incompetent, and even health professionals have shown bias against people with obesity (Puhl & others, 2020). People with obesity are well aware of the negative judgments of others and may find these judgments inescapable, as if they are trapped inside a body that stigmatizes them (Williams & Annandale, 2020). The effect of chronic stress on the body promotes weight gain, and stress predicts obesity over time (Tomiyama, 2019). Could the stigma attached to obesity contribute to the negative health effects of obesity? To read about this possibility, see the Intersection.

Cognitive Appraisal and Coping with Stress

Although everyone's body may respond similarly to stressors, not everyone perceives the same events as stressful. Whether an experience "stresses you out" depends, at least to some extent, on how you think about that experience. For example, you may perceive an upcoming job interview as a threatening obligation, whereas your roommate may perceive it as a challenging opportunity—a chance to shine. You might view a *D* on a paper as a crushing blow; a friend may view the same

Let's face it. Just reading about the negative effects of stress can be stressful.

Health Psychology and Social Psychology: Can Weight-Based Bias Affect the Health Consequences of Obesity?

Negative stereotypes of people with obesity harm these people in important ways. As with other stigmatized groups, being treated poorly due to one's weight is associated with stress (Puhl & others, 2020). Now, obesity is recognized as a risk factor for numerous diseases. We might think that these links are simply a matter of additional adipose (fat) tissue directly leading to physical problems. However, it is also possible that experiencing discrimination because of one's weight leads to stress that takes a toll on the body, over and above the effects of obesity itself.

To examine this possibility, researchers studied a sample of over 3,000 adults in a longitudinal study spanning four years (Daly & others, 2019). A variety of medical measures were taken, including measures of cardiovascular function, cholesterol, blood glucose, and so on, that served as biomarkers of health as well as weight and height to provide a measure of obesity. In all, there were 1,100

Prasit photo/Getty Images

people in the sample with obesity. In interviews, participants rated questions about their recent experiences, such as, "You were treated with less respect or courtesy" and were asked why they believed it happened. Answers mentioning weight were used as measures of weight discrimination. The researchers then used obesity as well as perceived weight discrimination to predict declines in physical health over time.

The results of the study were stunning. Yes, obesity was related to declines in indicators of health but so was perceived weight discrimination. In fact, the experience of discrimination explained more than a quarter of the effects of obesity itself on physical health (Daly & others, 2019). Think about that: Part of the negative effect of obesity on health is the stress of being stigmatized for one's weight.

These results show how the mind and body are linked and embedded in a wider social group. When people are treated poorly because of their health status, it only adds to their suffering and physical risk. Indeed, many health conditions are the source of stigma (Bagcchi, 2020; Noronha, 2020; Yuvaraj & others, 2020). No matter a person's health status, that person is still a person. As human beings we need each other, and kindness may be a vital medicine for those dealing with health conditions.

\\ **What are some differences in the stigmatization of obesity versus racial, gender, or other identities?**

---

grade as an incentive to work harder. Some people responded to COVID-19 restrictions with a great deal of stress, while others took the many changes that occurred in stride.

STEPS IN COGNITIVE APPRAISAL

Cognitive appraisal refers to a person's interpretation of an event as harmful, threatening, or challenging, and the person's determination of whether they have the resources to cope effectively with the event. **Coping** is essentially a kind of problem solving. It involves managing taxing circumstances, expending effort to solve life's problems, and seeking to master or reduce stress.

Richard Lazarus articulated the importance of cognitive appraisal to stress and coping (1993, 2000). In Lazarus's view, people appraise events in two steps: primary appraisal and secondary appraisal. In *primary appraisal,* individuals interpret whether an event involves *harm or loss* that has already occurred, a *threat* of some future danger, or a *challenge* to be overcome. Lazarus believed that perceiving a stressor as a challenge to be

coping
Managing taxing circumstances, expending effort to solve life's problems, and seeking to master or reduce stress.

cognitive appraisal
Individuals' interpretation of the events in their life as harmful, threatening, or challenging and their determination of whether they have the resources to cope effectively with the events.

Corbis/VCG/Getty Images

overcome rather than as a threat is a great strategy for reducing stress. To understand Lazarus's concept of primary appraisal, consider two students, each with a failing grade in a psychology class at midterm. Sam is almost frozen by the stress of the low grade and looks at the rest of the term as a threatening prospect. In contrast, Alex does not become overwhelmed by the harm already done and the threat of future failures. They see the low grade as a challenge that they can address and overcome.

In *secondary appraisal,* people evaluate their resources and determine how effectively they can be marshaled to cope with the event. This appraisal is secondary because it both comes after primary appraisal and depends on the degree to which the event is appraised as harmful, threatening, or challenging. Sam might have some helpful resources for coping with his low midterm grade, but he views the stressful circumstance as so harmful and threatening that he does not take stock of and use his resources. Alex, in contrast, evaluates the resources they can call on to improve their grade. These include asking the instructor for suggestions about how to improve their studying for the tests, managing time to include more study hours, and consulting with high-achieving classmates.

Keep in mind that not everything is in a person's head! You're allowed to be unhappy about life events. There are limits to positive thinking and sometimes taking action to change a situation is best.

TYPES OF COPING
Research has identified two types of coping. **Problem-focused coping** is the cognitive strategy of squarely facing one's troubles and trying to solve them. For example, if you are having trouble with a class, you might go to the campus study skills center and sign up for a program to learn how to study more effectively. Having done so, you have faced your problem and attempted to do something about it. Problem-focused coping might involve coming up with goals and implementation intentions, the problem-solving steps we examined earlier.

Emotion-focused coping entails responding to the stress that one is feeling—trying to manage the emotional reaction—rather than confronting the root problem. If you use emotion-focused coping, you might avoid going to a class that is a problem for you. Instead, you might say the class does not matter, deny that you are having difficulty with it, joke about it with your friends, or pray that you will do better.

Emotion-focused coping can be adaptive in situations in which there is no solution to a problem, such as grieving over a loved one's death, when in fact it makes sense to focus on feeling better and accepting the present circumstances.

In some circumstances, emotion-focused coping can be beneficial in dealing with life's problems. Denial is one of the main protective psychological mechanisms for navigating the flood of feelings that occurs when the reality of death or dying becomes too great. For example, one study found that following the death of a loved one, bereaved people who directed their attention away from their negative feelings had fewer health problems and were rated as better adjusted by their friends, compared to bereaved individuals who did not use this coping strategy (Coifman & others, 2007). In other circumstances, however, emotion-focused coping can be a problem. Denying that the person you dated does not love you anymore when they become engaged to someone else keeps you from getting on with life.

Many people successfully use both problem-focused and emotion-focused coping when adjusting to a stressful circumstance. For example, in one study, individuals said they employed both problem-focused and emotion-focused coping strategies in 98 percent of the stressful situations they encounter (Folkman & Lazarus, 1980). Over the long term, though, problem-focused coping usually works best (Pavani & others, 2016; Tuncay & Musabak, 2015).

problem-focused coping
The coping strategy of squarely facing one's troubles and trying to solve them.

emotion-focused coping
The coping strategy that involves responding to the stress that one is feeling—trying to manage one's emotional reaction—rather than focusing on the root problem itself.

Strategies for Successful Coping

A stressful circumstance becomes considerably less stressful when a person successfully copes with it. Effective coping is associated with a sense of personal control, a healthy immune system, personal resources, and positive emotions.

Research supports the idea that positive reappraisal is a type of coping that is associated with positive outcomes (Bailey & others, 2020; Finkelstein-Fox & others, 2020). **Positive reappraisal** means reinterpreting a potentially stressful experience as less threatening, more valuable, or even beneficial. In positive reappraisal, we look at a situation differently, so that it is not as threatening. Negative events can, for instance, be learning experiences. They can also help us recognize our own strengths. Positive reappraisal can happen at the primary appraisal stage—if a person is able to change their view of a stressor in the moment. A recent study measured distress in a sample of young adults before and during the COVID-19 pandemic (and subsequent lockdowns) (Shanahan & others, 2020). Results showed that positive reappraisal predicted less increase in distress over time. Positive reappraisal does not only mean changing things inside one's head. It can also mean finding purpose in negative events and seeking to make changes in the world. A recent study linked such a pattern to civic engagement among Black Americans (Riley & others, 2020).

Multiple coping strategies often work better than a single strategy, as is true with any problem-solving challenge. People who have experienced a stressful life event or a cluster of difficulties might actively embrace problem solving and consistently take advantage of opportunities for positive experiences, even in the context of the bad times they are going through. Positive emotion can give them a sense of the big picture, help them devise possible solutions, and allow them to make creative connections.

Optimism can play a strong role in effective coping (Taylor & others, 2012). Lisa Aspinwall has found, for example, that optimistic people are more likely to attend to and remember potentially threatening health-related information than are pessimists (Aspinwall, 1998, 2011; Aspinwall & others, 2009; Aspinwall & Pengchit, 2013). Aspinwall views optimism as a resource that allows individuals to engage constructively with potentially frightening information. Optimists engage with life from a place of strength, so when an optimist finds out, for instance, that a favorite pastime, tanning, is related to an elevated risk of skin cancer, the information is important but not overwhelming. In contrast, pessimists are already living in a bleak world and prefer not to hear more bad news.

Optimists are not just denying that anything bad can happen. They are actively engaged with reality, even when it contains threatening news.

Another personality trait proposed to promote superior coping during difficult times is hardiness. **Hardiness** is characterized by a sense of commitment rather than alienation, and of control rather than powerlessness, as well as a perception of problems as challenges rather than threats (Maddi & others, 2017). Hardiness is the trait displayed by the basketball player whose team is down by two points with seconds remaining on the clock when the player shouts, "Coach! Give me the ball!" Many of us would shrink from such a high-pressure moment.

The links among hardiness, stress, and illness were the focus of the Chicago Stress Project, which studied male business executives 32 to 65 years of age over a five-year period (Kobasa & others, 1982; Maddi, 1998). During the five years, most of the executives experienced stressful events such as divorce, job transfers, a close friend's death, inferior work-performance evaluations, and reporting to an unpleasant boss. Figure 6 shows how hardiness buffered these individuals from stress-related illness (Kobasa & others, 1986).

Other researchers also have found support for the role of hardiness in illness and health and especially coping with difficult times (Abdollahi & others, 2014; Pitts & others, 2016). For instance, one study found that mothers of children with cancer were less likely to suffer from traumatic stress symptoms if they were high in hardiness (Stoppelbein & others, 2017).

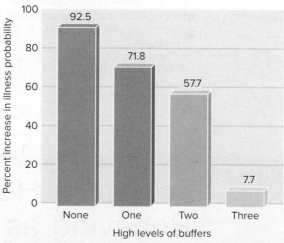

FIGURE 6 **Illness in High-Stress Business Executives** In one study of high-stress business executives, a low level of all three buffers (hardiness, exercise, and social support) involved a high probability of at least one serious illness in that year. High levels of one, two, and all three buffers decreased the likelihood of at least one serious illness occurring in the year of the study.

Stress Management Programs

"Avoid stress" may be good advice, but life is full of potentially stressful experiences. Sometimes just checking email or answering a phone can be an invitation for stress.

Because many people have difficulty regulating stress, psychologists have developed techniques that individuals can learn (Benjet, 2020). **Stress management programs** teach individuals how to appraise stressful events, develop coping skills, and put these skills to practical use. Some stress management programs teach a range of techniques to handle stress; others focus on a specific technique, such as relaxation or assertiveness training.

Stress management programs are often taught through workshops, which are becoming more common in the workplace. Aware of the high cost in lost productivity due to stress-related disorders, many organizations have become increasingly motivated to help their workers identify and cope with stressful circumstances. Colleges and universities similarly run stress management programs for students.

Do stress management programs work? Studies show that they can. Programs including physical exercise, relaxation, mindfulness, and so on can reduce psychological stress and physiological markers of stress (Orosz & others, 2020; Stier-Jarmer & others, 2016).

Coping effectively with stress is essential for physical and mental health. Still, there is a lot more we can do to promote our health. Healthful living—establishing healthy habits and evaluating and changing behaviors that interfere with good health—helps us avoid the damaging effects of stress. Just as the biopsychosocial perspective predicts, healthy changes in one area of life can have benefits that flow to other areas.

self-quiz

1. Selye's term for the pattern of common effects on the body when demands are placed on it is
   A. exhaustion syndrome.
   B. the Type A behavior pattern.
   C. the Type B behavior pattern.
   D. general adaptation syndrome.

2. A personality trait that is characterized by a sense of commitment and control, as well as by a perception of problems as challenges rather than threats, is
   A. self-efficacy.
   B. self-determination.
   C. hardiness.
   D. self-confidence.

3. Dealing with difficult circumstances, expending effort to solve life's problems, and seeking to control or reduce stress are key aspects of
   A. coping.
   B. cognitive appraisal.
   C. primary appraisal.
   D. secondary appraisal.

APPLY IT! 4. In addition to taking a full load of classes, Bonnie works at two part-time jobs and helps her sister care for two toddlers. Bonnie is achievement oriented and strives to get As in all of her courses. Because of her many commitments, she is often in a hurry and regularly does more than one thing at a time, but she tells people that she enjoys her busy routine. Which answer best assesses whether Bonnie is Type A and at risk for cardiovascular disease?
   A. Bonnie's hurriedness and achievement orientation indicate that she is Type A and probably at risk for cardiovascular disease.
   B. Although Bonnie may experience stress, the lack of hostility mentioned in this description suggests that she is not Type A or at risk for cardiovascular disease.
   C. Bonnie is a "hot reactor" and thus at risk for cardiovascular disease.
   D. Bonnie is Type D, but her enjoyment of life means that she is not at risk for cardiovascular disease.

5 Toward a Healthier Body (and Mind): Behaving as If Your Life Depends upon It

There is no escaping it: Getting stress under control is crucial for a healthy mind and body. It is also important to make wise behavioral choices in four additional life domains where healthy habits can benefit both body and mind. In this section we examine the advantages of becoming physically active, eating right, quitting smoking, and practicing safe sex.

Becoming Physically Active

Imagine that there was a time when, to change a TV channel, people had to get up and walk a few feet to turn a knob. Consider the time when people physically had to go to the library and hunt through card catalogs and shelves to find information rather than going online and googling. As our daily tasks have gotten increasingly easy, we have become less active, and inactivity is a serious health problem (Wuest & Fisette, 2015). A report of three nationally representative samples of U.S. adults showed that those who reported themselves as less active than their peers were 71 percent more likely to die during the study period (Zahrt & Crum, 2017).

Any activity that expends physical energy can be part of a healthy lifestyle. It can be as simple as taking the stairs instead of an elevator, walking or biking to class instead of driving, or getting up and dancing instead of sitting at the bar. In addition to its link to longevity, physical activity correlates with other positive outcomes, including a lower probability of developing cardiovascular disease and cancer, coping with treatment for illness and better response to treatment, healthier body weight, better mood, better cognitive function, healthier aging, lowered anxiety, and lowered depression (Ashdown-Franks & others, 2020; Lee, 2020; Ludyga & others, 2020). Physical activity is a strong predictor of better functioning.

Even pigs benefit from exercise; Figure 7 shows the positive effects of physical activity in hogs. Being physically active is like investing energy in a wellness bank account: Activity enhances physical well-being and gives us the ability to face life's potential stressors energetically (Donatelle, 2015).

exercise
Structured activities whose goal is to improve health.

Exercise is one special type of physical activity. **Exercise** formally refers to structured activities whose goal is to improve health. Although exercise designed to strengthen muscles and bones or to improve flexibility is important to fitness, many health experts stress the benefits of **aerobic exercise**, which is sustained activity—jogging, swimming, or cycling, for example—that stimulates heart and lung functioning.

aerobic exercise
Sustained activity—jogging, swimming, or cycling, for example—that stimulates heart and lung functioning.

In one study, exercise meant the difference between life and death for middle-aged and older adults (Blair & others, 1989). More than 10,000 people were divided into categories of low fitness, medium fitness, and high fitness. Then they were studied over eight years. As shown in Figure 8, sedentary participants (low fitness) were more than twice as likely to die during the study's eight-year time span than those who were moderately fit, and more than three times as likely to die as those who were highly fit. The positive effects of physical fitness occurred for both men and women. Further, another study revealed that adults aged 60 and over who were in the lowest fifth in terms of physical fitness as determined by a treadmill test were four times more likely to die over a 12-year period than their counterparts who were in the top fifth of physical fitness (Sui & others, 2007). This study also showed that older adults who were overweight but physically fit had a lower mortality risk over the 12 years than their typical-weight counterparts who were low in fitness (Sui & others, 2007). In addition, a longitudinal study found that men who exercised regularly at 72 years of age had a 30 percent higher probability of being alive at 90 years of age than their sedentary counterparts (Yates & others, 2008).

Health experts recommend that adults engage in at least 30 minutes of moderate physical activity on most, preferably all, days of the week and that children exercise for 60 minutes daily. Most specialists advise that exercisers raise their heart rate to at least 60 percent of their maximum rate. Only about one-fifth of adults are active at these

FIGURE 7   The Jogging Hog Experiment Jogging hogs reveal the dramatic effects of exercise on health. In one investigation, a group of hogs was trained to run approximately 100 miles per week (Bloor & White, 1983). Then the researchers narrowed the arteries that supplied blood to the hogs' hearts. The hearts of the jogging hogs developed extensive alternate pathways for blood supply, and 42 percent of the threatened heart tissue was salvaged, compared with only 17 percent in a control group of non-jogging hogs.
Courtesy of Colin M. Bloor/UC San Diego Health Sciences

Fitness here refers to the body's ability to supply fuel during sustained physical activity.

Make time in your day to exercise. It might mean waking up earlier or forgoing watching TV. The benefits are well worth it.

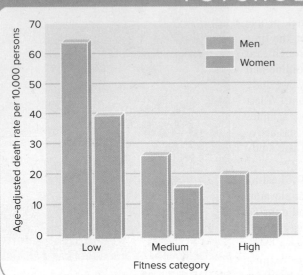

FIGURE 8 **Physical Fitness and Mortality** This graph presents the results of an eight-year longitudinal study of more than 10,000 men and women (Blair & others, 1989). The horizontal, or X, axis shows the participants divided by their levels of fitness as well as their sex. The vertical, or Y, axis shows the death rates.

> Which groups had the highest and lowest death rates? > Comparing the results for men and women, what role does biological sex play in mortality? > This is a correlational study, so causation cannot be assumed. What third variables might explain the results?

recommended levels. Figure 9 lists physical activities that qualify as moderate and, for comparison, vigorous. Both moderate and intense activities can produce important physical and psychological gains and improve quality of life (Focht, 2012).

Eating Right

Healthy eating habits are an important part of a healthy life. Eating right means selecting sensible, nutritious foods that maximize health and wellness. Despite our seemingly boundless food options, many of us are unhealthy eaters. We take in too much sugar and not enough foods high in vitamins, minerals, and fiber, such as fruits, vegetables, and grains. We eat too much fast food and too few well-balanced meals—choices that increase our fat and cholesterol intake, both of which are implicated in long-term health problems (Bertoia & others, 2016; Lynch & others, 2015).

Moderate	Vigorous
walking briskly (3–4 mph)	walking briskly uphill or with a load
swimming, moderate effort	swimming, fast treading crawl
cycling for pleasure or transportation (≤10 mph)	cycling, fast or racing (>10mph)
racket sports, table tennis	racket sports, singles tennis, racketball
conditioning exercise, general calisthenics	conditioning exercise, stair ergometer, ski machine
golf, pulling cart or carrying clubs	golf, practice at driving range
canoeing, leisurely (2.0–3.9 mph)	canoeing, rapidly(≥4 mph)
home care, general cleaning	moving furniture
mowing lawn, power mower	mowing lawn, hand mower
home repair, painting	fix-up projects, including weight bearing work

FIGURE 9 **Moderate and Vigorous Physical Activities** At minimum, adults should strive for 30 minutes of moderate activity each day. That activity can become even more beneficial if we "pump it up" to vigorous.

(Leaves): Reed Kaestner/Getty Images; (Dog): Stockbyte/PunchStock

Healthy eating is not just about weight loss but about committing to lifelong healthy food habits. Several health goals can be accomplished through a sound nutritional plan. Not only does a well-balanced diet provide more energy, but it also can lower blood pressure and lessen the risk for cancer (Eguchi & others, 2012). Two studies totaling more than 110,000 U.S. adults found that a high level of red meat consumption was linked to an increased risk of earlier death due to cardiovascular disease and cancer (Pan & others, 2012).

The biggest health risk facing modern North Americans is being affected by excess body fat (Ogden & others, 2015; Schiff, 2015). "Overweight" and "obese" refer to ranges of weight that are greater than what experts consider healthy for a person's height. In recent years, the percentage of individuals who are classified as overweight or obese has been increasing. As Figure 10 shows, the prevalence of overweight or obesity in the United States changed little from 1960 to 1980 (Flegal & others, 2012; Ogden & Carroll, 2010; Ogden & others, 2012). However, from 2009 to 2016, the vast majority of U.S. adults were classified as overweight or obese (CDC, 2018a). In 2018, obesity affected over 42 percent of Americans (Hales & others, 2020).

Losing weight and opting for healthier foods can be difficult, especially when you are just starting out. Many weight-loss fads promise weight loss with no effort, no hunger, and no real change in one's food consumption. These promises are unrealistic. Making genuine, enduring changes in eating behavior is hard work—but this does not mean that pessimism is required. Rather, positive expectations and self-efficacy are important because the task at hand is challenging.

The National Weight Control Registry is an ongoing study of people who have lost at least 40 pounds and kept it off for at least two years. Research on those who have successfully maintained weight loss gives us important tips on how people who keep the weight off achieve this goal (Raynor & others, 2005). They show consistency in what they eat, plan their meals, weigh themselves, and stay active, especially limiting time spent watching TV or other screens (Thomas & others, 2014).

One key practice is eating breakfast, especially whole-grain cereals.

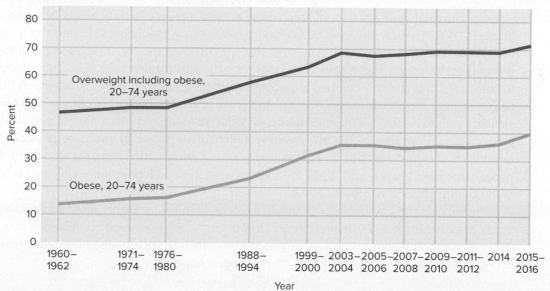

PSYCHOLOGICAL INQUIRY

FIGURE 10 **Changes in the Percentage of U.S. Adults 20 to 74 Years of Age Classified as Overweight or Obese, 1960–2018** Having overweight or obesity poses the greatest overall health risk for Americans today. In this graph, the vertical, or Y, axis shows the percentage of people classified as overweight or obese, and the horizontal, or X, axis shows the years for these values.
> Thinking of these lines as data points, is time positively or negatively correlated with overweight/obesity? Why? > Find the year of your birth on the X axis. How has the body weight of Americans changed during your lifetime? > What years show the steepest rise in weight gain? What factors might explain this increase?

Environments That Support Active Lifestyles

French food—featuring delicious cheeses, heavy sauces, and buttery pastries—is one of the richest cuisines on the planet. Yet the obesity rate in France is less than half that of the United States (OECD, 2017). Why are Americans so much more likely to be affected by obesity than the French? Part of the gap may be due to differences in food portions (Rozin & others, 2003), but another factor is how people get to their food. In the United States, grocery shopping typically involves a trip in a car, whereas in France individuals are more likely to walk or ride a bicycle (Ferrières, 2004).

Naufal MQ/Getty Images

A major obstacle to promoting exercise in the United States is that many cities are not designed in ways that promote walking or cycling. Advocates for change say that by making life too easy and too accommodating to cars and drivers, urban designers have created an *obesogenic* (obesity-promoting) environment—a context where it is challenging for people to engage in healthy activities (Henderson, 2008; Lydon & others, 2011). Such environments are more common in places where people of color live, another sign of systemic racism (Bell & others, 2019). Countries such as the Netherlands and Denmark have adopted urban planning strategies that promote walking and biking and discourage car use. In the Netherlands, 60 percent of all journeys taken by people over age 60 are by bicycle (Henderson, 2008).

Environmental contexts that invite physical activity increase activity levels (Mackenbach & others, 2016). For example, one quasi-experimental study examined the effects of changes to the physical environment on activity. The study focused on an urban neighborhood in which a greenway (a biking and walking trail) was retrofitted to connect with pedestrian sidewalks. Researchers counted the number of people outside engaging in physical activity in that neighborhood for a two-hour period at various times over two years. Compared to two other similar neighborhoods, the neighborhood with the trail featured more people walking and biking (Fitzhugh & others, 2010). Environmental characteristics that welcome physical activity are also associated with health and wellness. In one study, elderly people who lived near parks, tree-lined streets, and areas for taking walks showed higher longevity over a five-year study period (Takano & others, 2002).

Other aspects of city design can influence obesity rates, including access to nutritious foods (Lydon & others, 2011) and the perceived safety of neighborhoods (Eisenstein & others, 2011). By shedding light on how environmental factors influence healthy lifestyles, research on human behavior can meaningfully impact public policy.

So, just don't buy the junk food! If it is not around, you won't eat it.

The truth is that keeping weight off is an ongoing process. Moreover, the longer a person keeps the weight off, the less likely they are to gain it back (McGuire & others, 1999). Further, research suggests that making small changes in the availability of junk food can have an impact on eating and weight (Rozin & others, 2011).

Quitting Tobacco Use

Another health-related goal is giving up tobacco, including smoking. Evidence from a number of studies underscores the dangers of exposure to tobacco and being around people who smoke (American Cancer Society, 2020). Smoking negatively affects nearly all bodily organ systems and it is the leading preventable cause of death. Secondhand smoke is implicated in as many as 9,000 lung cancer deaths a year. Children of people who smoke are at special risk for respiratory and middle-ear diseases (CDC, 2017).

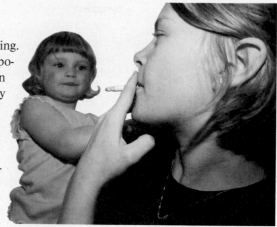

Hannah Maule-Finch/Image Source/Getty Images

Fewer people smoke today than in the past, and almost half of the living adults who ever smoked have quit. Still, many American adults continue to smoke. Exposure to tobacco through chewing or dipping also predicts higher risk of cardiovascular disease, a variety of cancers, and dental problems (American Cancer Society, 2015).

Quitting tobacco use has enormous health benefits. When individuals stop smoking, their risk of fatal lung cancer declines over time. Nicotine addiction makes quitting a challenge. Nicotine, the active drug in tobacco, is a stimulant that increases a person's energy and alertness, a pleasurable and reinforcing experience (Grando, 2014). In addition, nicotine stimulates neurotransmitters that have a calming or pain-reducing effect (Malin & others, 2017).

Research confirms that giving up smoking (and other forms of tobacco use) can be difficult, especially in the early days of quitting. There are various ways to quit (Bradizza & others, 2017; Piper & others, 2017), including:

- *Going cold turkey:* Some individuals succeed when they stop using tobacco without making any major lifestyle changes. They decide they are going to quit, and they do.

- *Using a substitute source of nicotine:* Nicotine gum, the nicotine patch, the nicotine inhaler, and nicotine spray work on the principle of supplying small amounts of nicotine to diminish the intensity of withdrawal.

- *Seeking therapeutic help:* Some people get professional help to kick the tobacco habit.

No one method for quitting smoking or other tobacco use is foolproof (Piper & others, 2017). Often a combination of approaches is the best strategy. Furthermore, quitting for good typically requires more than one try, as the stages of change model suggests.

Practicing Safe Sex

Satisfying sexual experiences are part of a happy life, for many people. Sexual behavior also has important implications for physical health. We have previously examined research findings on unplanned pregnancy and on educational approaches that can help prevent teen pregnancy. Here we look at another aspect of sexuality: protecting oneself from sexually transmitted infections (STIs). Naturally, by not having sex, people can avoid both unplanned pregnancy and STIs. However, even for those whose goal is abstinence, knowledge about preventing unwanted pregnancy and STIs is important, because as the stages of change model suggests, we sometimes fall short of our goals.

PROTECTING AGAINST SEXUALLY TRANSMITTED INFECTIONS

A **sexually transmitted infection (STI)** is an infection that is contracted primarily through sexual activity—vaginal intercourse as well as oral and anal sex. Figure 11 shows the symptoms of STIs and their long-term consequences. Clearly, preventing STIs is important.

sexually transmitted infection (STI)
An infection that is contracted primarily through sexual activity—vaginal intercourse as well as oral and anal sex.

	Symptoms	Long-Term Consequences
Bacterial Infections Gonorrhea	Burning when urinating and, in men, white, yellow, or green discharge from the penis. Gonorrhea can also infect the throat when passed via oral sex.	In women, untreated gonorrhea can lead to pelvic inflammatory disease, increasing the risk of pregnancy complications and infertility. In men, it can cause an inflammation in the genital tract, which can lead to scarring, blocked sperm ducts, and infertility issues.
Syphilis	The first sign of syphilis is a small painless sore, called a chancre. Within a few weeks of the chancre healing, the person may develop a rash that begins on the trunk, eventually covering the body.	Complications of late-stage syphilis include damage to the brain, nerves, eyes, heart, blood vessels, liver, bones and joints. Babies born to women who have syphilis can become infected and may suffer a number of complications as a result.
Chlamydia	Most people who have chlamydia have no symptoms. In women, symptoms may include abnormal vaginal discharge and/or a burning sensation when urinating. Men may experience discharge from the penis, a burning sensation when urinating, and pain/swelling in the testicles.	If left untreated, chlamydia can cause serious, permanent damage to a woman's reproductive system.
Viral Infections Genital herpes	Symptoms can be extremely mild and include pain/itching in the genitals, small red bumps or tiny white blisters that become scabs.	Herpes sores increase the risk of transmitting other STIs and may lead to bladder problems in some cases. Babies born to infected mothers can experience serious problems, including brain damage, blindness, and death.
Human papilloma virus (HPV)	HPV infections appear as warts on or around the genital area. Other varieties of HPV can cause warts on hands, fingers, elbows and feet.	Certain strains of HPV can cause cervical cancer and may also contribute to cancers of the genitals, anus, mouth, and upper respiratory tract.
Human immunodeficiency virus (HIV)	A person can carry HIV for a long time without experiencing symptoms. However, some people may develop a set of flu-like symptoms.	Left untreated, HIV may develop into AIDS.

Note. All sexually transmitted infections are preventable. You can greatly reduce your chances of being infected by taking appropriate measures, including using latex condoms. Regular STI testing is important for people who are sexually active. Some of these infections can be spread by means other than sexual contact, such as by needle sharing among those who use injectable drugs of abuse. Information is from the CDC (2018a).

FIGURE 11 **Sexually Transmitted Infections**

SOURCE: CDC (2018a).

acquired immune deficiency syndrome (AIDS)

A sexually transmitted infection, caused by the human immunodeficiency virus (HIV), that destroys the body's immune system.

Acquired immune deficiency syndrome (AIDS) is caused by the sexually transmitted human immunodeficiency virus (HIV), which destroys the body's immune system. HIV is not curable, but early treatment can prevent the development of AIDS. Without treatment, most people with HIV are vulnerable to germs that a healthy immune system can destroy. People behaviorally vulnerable to HIV may be eligible for medications called "Pre-Exposure Prophylaxis," or PrEP (CDC, 2018b; Raifman & Sherman, 2018). PrEP involves taking medication daily and is effective in reducing the risk of HIV if used consistently (CDC, 2018b).

Although new HIV diagnoses have dropped, the numbers of diagnoses of other STIs have continued to rise, reaching record highs in 2017 (Howard, 2018).

Many sensual activities such as kissing, French kissing, cuddling, massage, and mutual masturbation (that does not involve the exchange of bodily fluids) involve no risk of an STI. Sexual activities that involve penetration, including vaginal, anal, and oral sex, are riskier behaviors that can be made less risky with the use of proper protection.

Latex condoms are key in the prevention of STIs. In your own sexual experience, the wisest course of action is always to protect yourself by using a latex condom. When correctly used, latex condoms help to prevent or significantly reduce the risk of transmission of many STIs (Hodges & Holland, 2018; Johnson & others, 2018). Unfortunately, in young adult relationships, experiences of commitment and trust tend to be negatively related to consistent condom use (Agnew & others, 2017). That is, couples may believe that once they are committed to each other, condoms are no longer necessary. Such a belief may leave both partners vulnerable to STIs.

Programs to promote safe sex are especially effective if they include the eroticization of condom use—that is, making condoms a part of the sensual experience of foreplay (Hood & others, 2017; Scott-Sheldon & Johnson, 2006). Fear tactics are relatively less effective, and programs emphasizing active skill building (for example, role-playing the use of condoms), self-efficacy, and positive attitudes about condom use are effective with most groups (Albarracín & others, 2006, 2016; Durantini & others, 2006). Making safe sex sexy is a great way to practice safe sex (Ellis & others, 2018).

self-quiz

1. Regular physical activity and, in particular, exercise are associated with all of the following *except*
   A. weight loss.
   B. increased self-esteem.
   C. less incidence of depression.
   D. premature death in middle-age and older adults.

2. The biggest health risk facing most Americans today is
   A. heart disease.
   B. cancer.
   C. overweight and obesity.
   D. stress.

3. Typically, the *best* approach to quitting smoking is to
   A. go cold turkey.
   B. use a nicotine patch.
   C. use a combination of methods.
   D. get help from a therapist.

APPLY IT! 4. J. C. and Veronica promote student health causes on their college campus. This year, they are targeting wise sexual choices. Which of the following is the most promising strategy for their campaign?
   A. They should focus on fear of disease as a motivator for condom use.
   B. They should focus on promoting nonrisky sexual activities, eroticizing condom use, and teaching students skills for effective condom use, reminding students that even if they do not intend to have sex, it is best to be safe.
   C. They should focus only on students who are already engaging in sexual behavior.
   D. They should encourage students, before having sex, simply to ask their partners how many sexual partners they have had.

6 Psychology and Your Good Life

In this discussion of health psychology, we have examined how the mental and physical aspects of your existence intertwine and influence each other. The field of health psychology illustrates how all of the various areas of psychology converge to reveal that interplay.

As a human being, you are both a physical entity and a system of mental processes that are themselves reflected in that most complex of physical organs, the brain. At every moment, both body and mind are present and affecting each other. Caring for your brain and mind—the resources that make it possible for you to read this book, study for tests, listen to lectures, fall in love, share with friends, help others, and make a difference in the world—is worthy of being a life mission.

Many pages ago, we defined psychology as the scientific study of behavior and mental processes, broadly meaning the things we do, think, and feel. Reflect for a moment on the psychological dimensions of vision. When we studied the human visual system, we examined the processes by which those amazing sense organs, our eyes, detect color, light, dark, shape, and depth. We probed the ways that the brain takes that information and turns it into perception—how a pattern of colors, shapes, and light comes to be perceived as a flower, a fall day, a sunset. Visual systems, we discovered, are generally the same from one person to the next. Thus, you can memorize the different parts of the human eye and know that your understanding is true for just about all the human eyes you will encounter in life.

An experience as deceptively simple as taking in and perceiving a sunset becomes stunningly complex in the context of a human life.
Dirima/Getty Images

Do It!

Think back to all the topics of this book. Which did you find most interesting? Go to your school's library and locate the journals that are devoted to that subject (you can ask a librarian for help). Browse a recent issue. What specific topics are scientists studying? If a particular study described in this book sounded interesting, you can probably obtain it online. Do a Google "Scholar" search on the authors and take a look at the original article. What did the authors conclude? What did you learn?

However, even something as deceptively simple as perceiving a sunset through the sense of vision becomes amazingly complex when we put it in the context of a human life. Is that sunset the first you see while on your honeymoon, or right after a painful romantic breakup, or as a new parent? Placing even the most ordinary moment in the context of a human life renders it extraordinary and fascinating.

This fascination is a primary motivation for the science of psychology itself. Humans have always pondered the mysteries of behavior, thought, and emotion. Why do we do the things we do? How do we think and feel? In this book, we have explored the broad range of topics that have interested psychologists throughout the history of this young science.

Coming to the close of this introduction to psychology allows you to take stock of what psychology has come to mean to you now, as well as to consider what it might mean to you in the future. Whether or not you continue coursework in psychology, this book has highlighted opportunities for your future exploration about yourself and your world. In each of the real-life examples of human experience described in these pages—moments of heroism, weakness, joy, pain, and more—psychology has had a lesson to share with respect to the person that is you. Making the most of what you have learned about psychology means making the most of yourself and your life.

SUMMARY

① Health Psychology and Behavioral Medicine

Health psychology emphasizes biological, psychological, and social factors in human health. Closely aligned with health psychology is behavioral medicine, which combines medical and behavioral knowledge to reduce illness and promote health. These approaches demonstrate the biopsychosocial model by examining the interaction of biological, psychological, and social variables as they relate to health and illness. Stress is an example of a biological, psychological, and social construct.

Health psychology and behavioral medicine bring the relationship of the mind and body to the forefront. These approaches examine the reciprocal mind–body relationship: how the body is influenced by psychological states and how mental life is influenced by physical health.

② Making Positive Life Changes

The theory of reasoned action suggests that we can make changes by devising specific intentions for behavioral change. We are more likely to follow through on our intentions if we feel good about the change and if we believe that others around us also support the change. The theory of planned behavior incorporates these factors as well as our perceptions of control over the behavior.

The stages of change model posits that personal change occurs in a series of five steps: precontemplation, contemplation, preparation/determination, action/willpower, and maintenance. Each stage has its own challenges. Relapse is a natural part of the journey toward change.

③ Resources for Effective Life Change

Motivation is an important part of sustaining behavioral change. Change is more effective when people do it for intrinsic reasons (because they want to) rather than extrinsic reasons (to gain rewards). Implementation intentions are the specific ways individuals plan to institute changes.

Social relationships are strongly associated with health and survival. Social support refers to the aid provided by others to a person in need. Support can take the form of tangible assistance, information, or emotional support. Social support is strongly related to functioning and coping with stress.

Religious faith is associated with enhanced health. One reason for this connection is that religions often frown on excess and promote healthy behavior. In addition, religious participation allows individuals to benefit from a social group, and religion provides a meaning system on which to rely in times of difficulty.

Personality characteristics related to positive health behaviors include conscientiousness, personal control, self-efficacy, and optimism. Conscientious individuals are likely to engage in healthy behaviors and live longer. Personal control is associated with better coping with stress. Self-efficacy is the person's belief in their own ability to master a situation and produce positive outcomes. Optimism refers to a particular explanatory style as well as to the inclination to have positive expectations for the future. Studies have shown that both of these types of optimism relate to positive health outcomes.

④ Toward a Healthier Mind (and Body): Controlling Stress

Stress is the response of individuals when life circumstances threaten them and tax their ability to cope. Selye characterized the stress response with his concept of a general adaptation syndrome (GAS), which has three stages: alarm, resistance, and exhaustion.

Chronic stress takes a toll on the body's natural disease-fighting abilities. Stress is also related to cardiovascular disease and cancer.

To kick the stress habit means remembering that stress is a product of how we think about events in our lives. Taking control of our appraisals allows us to see potentially threatening events as challenges. Hardiness is associated with thriving during stressful times.

The Type A behavior pattern, particularly the hostility component, is associated with stressing out angrily when things are going badly. This hostility leads to poor health outcomes. There is growing interest in the Type D behavior pattern, seen in individuals who experience generalized and frequent distress, negative emotions, and social inhibition; research has associated this pattern with an elevated risk of cardiovascular problems. When a person is unable to manage stress alone, stress management programs provide options for relief.

 Toward a Healthier Body (and Mind): Behaving as If Your Life Depends upon It

Exercise has many positive psychological and physical benefits. Tips for increasing one's activity level include starting small by making changes in one's routine to incorporate physical activity and keeping track of progress.

Overweight and obesity pose the greatest health risks to Americans today. They can be largely avoided by eating right, which means selecting nutritious foods and maintaining healthy eating habits for a lifetime, not just while on a diet. A combination of healthy eating and exercise is the best way to achieve weight loss.

Despite widespread knowledge that smoking causes cancer, some people still smoke. Methods of quitting include going cold turkey, using a substitute source of nicotine, and seeking therapy. Quitting for good is difficult and usually takes more than one try. Usually a combination of methods is the best strategy for quitting.

Practicing safe sex is another aspect of health behavior of interest to health psychologists. Condoms help prevent both unwanted pregnancy and the transmission of sexually transmitted infections (STIs). Interventions to promote condom use are most successful when they include making condom use sexy, promoting contraceptive skills and self-efficacy, and encouraging positive attitudes about condoms.

 Psychology and Your Good Life

Psychology is all about you. This book has aimed to show the relevance of psychology to your health and wellness and to help you appreciate the many, and deep, connections between this comparatively new science and your life.

KEY TERMS

acquired immune deficiency syndrome (AIDS)
aerobic exercise
behavioral medicine
cognitive appraisal
coping
emotion-focused coping

exercise
general adaptation syndrome (GAS)
hardiness
health behaviors
health psychology
hypothalamic-pituitary-adrenal axis (HPA axis)

implementation intentions
positive reappraisal
problem-focused coping
psychoneuroimmunology
relapse
sexually transmitted infection (STI)
social support

stages of change model
stress management program
theory of planned behavior
theory of reasoned action
Type A behavior pattern
Type B behavior pattern
Type D behavior pattern

ANSWERS TO SELF-QUIZZES

Section 1:   1. D; 2. B; 3. D; 4. B
Section 2:   1. D; 2. B; 3. B; 4. A

Section 3:   1. A; 2. D; 3. A; 4. C
Section 4:   1. D; 2. C; 3. A; 4. B

Section 5:   1. D; 2. C; 3. C; 4. B

glossary

A

abnormal behavior Behavior that is deviant, maladaptive, or personally distressful over a relatively long period of time.

absolute threshold The minimum amount of stimulus energy that a person can detect.

accommodation An individual's adjustment of their schemas to new information.

acquired immune deficiency syndrome (AIDS) A sexually transmitted infection, caused by the human immunodeficiency virus (HIV), that destroys the body's immune system.

acquisition The initial learning of the connection between the unconditioned stimulus and the conditioned stimulus when these two stimuli are paired.

action potential The brief wave of positive electrical charge that sweeps down the axon.

activation-synthesis theory Theory that dreaming occurs when the cerebral cortex synthesizes neural signals generated from activity in the lower part of the brain and that dreams result from the brain's attempts to find logic in random brain activity that occurs during sleep.

addiction A physical or a psychological dependence, or both, on a drug.

adrenal glands Glands at the top of each kidney that are responsible for regulating moods, energy level, and the ability to cope with stress.

aerobic exercise Sustained activity—jogging, swimming, or cycling, for example—that stimulates heart and lung functioning.

affectionate love or companionate love Love that occurs when an individual has a deep, caring affection for another person and desires to have that person near.

afferent nerves Also called sensory nerves; nerves that carry information about the external environment to the brain and spinal cord via sensory receptors.

aggression Social behavior whose objective is to harm someone, either physically or verbally.

agonist A drug that mimics or increases a neurotransmitter's effects.

alcoholism Disorder that involves long-term, repeated, uncontrolled, compulsive, and excessive use of alcoholic beverages and that impairs the drinker's health and social relationships.

algorithms Strategies—including formulas, instructions, and the testing of all possible solutions—that guarantee a solution to a problem.

all-or-nothing principle The principle that once the electrical impulse reaches a certain level of intensity (its threshold), it fires and moves all the way down the axon without losing any intensity.

altruism Giving to another person with the ultimate goal of benefiting that person, even if it incurs a cost to oneself.

amnesia The loss of memory.

amygdala An almond-shaped structure within the base of the temporal lobe that is involved in the discrimination of objects that are necessary for the organism's survival, such as appropriate food, mates, and social rivals.

androgens The class of sex hormones that predominates in males, produced by the testes in males and by the adrenal glands in all people.

anorexia nervosa Eating disorder that involves the relentless pursuit of thinness through starvation.

antagonist A drug that blocks a neurotransmitter's effects.

anterograde amnesia A memory disorder that affects the retention of new information and events.

antianxiety drugs Commonly known as tranquilizers, drugs that reduce anxiety by making individuals calmer and less excitable.

antidepressant drugs Drugs that regulate mood.

antipsychotic drugs Powerful drugs that diminish agitated behavior, reduce tension, decrease hallucinations, improve social behavior, and produce better sleep patterns in individuals with a severe psychological disorder, especially schizophrenia.

antisocial personality disorder (ASPD) Psychological disorder characterized by guiltlessness, law-breaking, exploitation of others, irresponsibility, and deceit.

anxiety disorders Disabling (uncontrollable and disruptive) psychological disorders that feature motor tension, hyperactivity, and apprehensive expectations and thoughts.

apparent movement The perception that a stationary object is moving.

applied behavior analysis or behavior modification The use of operant conditioning principles to change human behavior.

archetypes Jung's term for emotionally laden ideas and images in the collective unconscious that have rich and symbolic meaning for all people.

artificial intelligence (AI) A scientific field that focuses on creating machines capable of performing activities that require intelligence when they are done by people.

asexual A person experiences a lack of sexual attraction to others and may feel no sexual orientation.

assimilation An individual's incorporation of new information into existing knowledge.

association cortex Sometimes called association areas, the region of the cerebral cortex that is the site of the highest intellectual functions, such as thinking and problem solving.

associative learning Learning that occurs when an organism makes a connection, or an association, between two events.

Atkinson-Shiffrin theory Theory stating that memory storage involves three separate systems: sensory memory, short-term memory, and long-term memory.

attention-deficit/hyperactivity disorder (ADHD) One of the most common psychological disorders of childhood, in which individuals show one or more of the following: inattention, hyperactivity, and impulsivity.

attitudes An individual's opinions and beliefs about people, objects, and ideas—how the person feels about the world.

attribution theory The view that people are motivated to discover the underlying causes of behavior as part of their effort to make sense of the behavior.

auditory nerve The nerve structure that receives information about sound from the hair cells of the inner ear and carries these neural impulses to the brain's auditory areas.

authoritarian parenting A restrictive, punitive parenting style in which the parent exhorts the child to follow the parent's directions and to value hard work and effort.

authoritative parenting A parenting style that encourages the child to be independent but that still places limits and controls on behavior.

autism spectrum disorder A neurodevelopmental disorder characterized by deficits in social communication and social interaction across a variety of settings as well as restrictive repetitive behaviors, interests, and activities.

autobiographical memory A special form of episodic memory, consisting of a person's recollections of their life experiences.

automatic processes States of consciousness that require little attention and do not interfere with other ongoing activities.

autonomic nervous system The body system that takes messages to and from the body's internal organs, monitoring such processes as breathing, heart rate, and digestion.

availability heuristic A prediction about the probability of an event based on the ease of recalling or imagining similar events.

aversive conditioning A form of treatment that consists of repeated pairings of a stimulus with a very unpleasant stimulus.

avoidance learning An organism's learning that it can altogether avoid a negative stimulus by making a particular response.

axon The part of the neuron that carries information away from the cell body toward other cells.

B

barbiturates Depressant drugs, such as Nembutal and Seconal, that decrease central nervous system activity.

basal ganglia Large neuron clusters located above the thalamus and under the cerebral cortex that work with the cerebellum and the cerebral cortex to control and coordinate voluntary movements.

base rate neglect The tendency to ignore information about general principles in favor of very specific but vivid information.

behavior Everything we do that can be directly observed.

behavior therapies Treatments, based on behavioral and social cognitive theories, that use principles of learning to reduce or eliminate maladaptive behavior.

behavioral approach An approach to psychology emphasizing the scientific study of observable behavioral responses and their environmental determinants.

behavioral genetics The study of the inherited underpinnings of behavioral characteristics.

behavioral medicine An interdisciplinary field that focuses on developing and integrating behavioral and biomedical knowledge to promote health and reduce illness; overlaps with and is sometimes indistinguishable from health psychology.

behaviorism A theory of learning that focuses solely on observable behaviors, discounting the importance of such mental activity as thinking, wishing, and hoping.

big five factors of personality The five broad traits that are thought to describe the main dimensions of personality: openness to experience, conscientiousness, extraversion, agreeableness, and neuroticism (emotional instability).

binding In the sense of vision, the bringing together and integration of what is processed by different neural pathways or cells.

binge-eating disorder (BED) Eating disorder characterized by recurrent episodes of consuming large amounts of food during which the person feels a lack of control over eating.

binocular cues Depth cues that depend on the combination of the images in the left and right eyes and on the way the two eyes work together.

biological approach An approach to psychology focusing on the body, especially the brain and nervous system.

biological rhythms Periodic physiological fluctuations in the body, such as the rise and fall of hormones and accelerated/decelerated cycles of brain activity, that can influence behavior.

biological therapies Also called *biomedical therapies*, treatments that reduce or eliminate the symptoms of psychological disorders by altering aspects of body functioning.

biopsychosocial approach A perspective on human behavior that asserts that biological, psychological, and social factors are all significant ingredients in producing behavior. All of these levels are important to understanding human behavior.

bipolar disorder Mood disorder characterized by extreme mood swings that include one or more episodes of mania, an overexcited, unrealistically optimistic state.

borderline personality disorder (BPD) Psychological disorder characterized by a pervasive pattern of instability in interpersonal relationships, self-image, and emotions, and of marked impulsivity beginning by early adulthood and present in a variety of contexts.

bottom-up processing The operation in sensation and perception in which sensory receptors register information about the external environment and send it up to the brain for interpretation.

brain stem The stemlike brain area that includes much of the hindbrain (excluding the cerebellum) and the midbrain; connects with the spinal cord at its lower end and then extends upward to encase the reticular formation in the midbrain.

broaden-and-build model Fredrickson's model of positive emotion, stating that the function of positive emotions lies in their effects on an individual's attention and ability to build resources.

bulimia nervosa Eating disorder in which an individual (typically female) consistently follows a binge-and-purge eating pattern.

bystander effect The tendency of an individual who observes an emergency to be less likely to help when other people are present than when the observer is alone.

C

Cannon-Bard theory The proposition that emotion and physiological reactions occur simultaneously.

case study or case history An in-depth look at a single individual.

catatonia State of immobility and unresponsiveness, lasting for long periods of time.

cell body The part of the neuron that contains the nucleus, which directs the manufacture of substances that the neuron needs for growth and maintenance.

central nervous system (CNS) The brain and spinal cord.

cerebral cortex Part of the forebrain, the outer layer of the brain, responsible for the most complex mental functions, such as thinking and planning.

chromosomes In the human cell, threadlike structures that come in 23 pairs, one member of each pair originating from each parent, and that contain DNA.

circadian rhythms Daily behavioral or physiological cycles that involve the sleep/wake cycle, body temperature, blood pressure, and blood sugar level.

classical conditioning Learning process in which a neutral stimulus becomes associated with an innately meaningful stimulus and acquires the capacity to elicit a similar response.

client-centered therapy Also called *Rogerian therapy or nondirective therapy,* a form of humanistic therapy, developed by Rogers, in which the therapist provides a warm, supportive atmosphere to improve the client's self-concept and to encourage the client to gain insight into problems.

clinical psychology An area of psychology that integrates science and theory to prevent and treat psychological disorders.

cognition The way in which information is processed and manipulated in remembering, thinking, and knowing.

cognitive affective processing systems (CAPS) Mischel's theoretical model for describing that individuals' thoughts and emotions about themselves and the world affect their behavior and become linked in ways that matter to that behavior.

cognitive appraisal Individuals' interpretation of the events in their life as harmful, threatening, or challenging and their determination of whether they have the resources to cope effectively with the events.

cognitive approach An approach to psychology emphasizing the mental processes involved in knowing: how we direct our attention, perceive, remember, think, and solve problems.

cognitive dissonance An individual's psychological discomfort (dissonance) caused by two inconsistent thoughts.

cognitive theory of dreaming Theory proposing that one can understand dreaming by applying the same cognitive concepts used in studying the waking mind.

cognitive therapies Treatments emphasizing that cognitions (thoughts) are the main source of psychological problems; therapies that attempt to change the individual's feelings and behaviors by changing cognitions.

cognitive-behavior therapy (CBT) A therapy that combines cognitive therapy and behavior therapy with the goal of developing the client's self-efficacy.

collective unconscious Jung's term for the impersonal, deepest layer of the unconscious mind, shared by all human beings because of their common ancestral past.

comorbidity The simultaneous presence of two or more disorders in one person. The conditions are referred to as "*comorbid.*"

concept A mental category that is used to group objects, events, and characteristics.

concrete operational stage Piaget's third stage of cognitive development, lasting from about 7 to 11 years of age, during which the individual uses operations and replaces intuitive reasoning with logical reasoning in concrete situations.

conditioned response (CR) The learned response to the conditioned stimulus that occurs after a conditioned stimulus–unconditioned stimulus pairing.

conditioned stimulus (CS) A previously neutral stimulus that eventually elicits a conditioned response after being paired with the unconditioned stimulus.

conditions of worth The standards that the individual must live up to in order to receive positive regard from others.

cones The receptor cells in the retina that allow for color perception.

confederate A person who is given a role to play in a study so that the social context can be manipulated.

confirmation bias The tendency to search for and use information that supports one's ideas rather than refutes them.

conformity A change in a person's behavior to coincide more closely with a group standard.

connectionism (parallel distributed processing [PDP]) The theory that memory is stored throughout the brain in connections among neurons, several of which may work together to process a single memory.

consciousness An individual's awareness of external events and internal sensations under a condition of arousal, including awareness of the self and thoughts about one's experiences.

control group The participants in an experiment who are as much like the experimental group as possible and who are treated in every way like the experimental group except for a manipulated factor, the independent variable.

controlled processes The most alert states of human consciousness, during which individuals actively focus their efforts toward a goal.

convergence A binocular cue to depth and distance in which the muscle movements in an individual's two eyes provide information about how deep and/or far away something is.

convergent thinking Thinking that produces the single best solution to a problem.

coping Managing taxing circumstances, expending effort to solve life's problems, and seeking to master or reduce stress.

core knowledge approach A perspective on infant cognitive development that holds that babies are born with domain-specific knowledge systems.

corpus callosum The large bundle of axons that connects the brain's two hemispheres, responsible for relaying information between the two sides.

correlational research Research that examines the relationships between variables, whose purpose is to examine whether and how two variables change together.

counterconditioning A classical conditioning procedure for changing the relationship between a conditioned stimulus and its conditioned response.

couples therapy Group therapy involving married or unmarried couples whose major problem lies within their relationship.

creativity The ability to think about something in novel and unusual ways and to devise unconventional solutions to problems.

critical thinking The process of thinking deeply and actively, asking questions, and evaluating the evidence.

cross-cultural competence A therapist's assessment of their abilities to manage cultural issues in therapy and the client's perception of those abilities.

cross-sectional design A research design in which a group of people is assessed on a psychological variable at one point in time.

culture-fair tests Intelligence tests that are intended to be culturally unbiased.

D

decay theory Theory stating that when an individual learns something new, a neurochemical memory trace forms, but over time this trace disintegrates; suggests that the passage of time always increases forgetting.

decision making The mental activity of evaluating alternatives and choosing among them.

deductive reasoning Reasoning from a general case that is known to be true to a specific instance.

defense mechanisms Tactics the ego uses to reduce anxiety by unconsciously distorting reality.

deindividuation The reduction in personal identity and erosion of the sense of personal responsibility when one is part of a group.

delay of gratification Putting off the pleasure of an immediate reward in order to gain a larger, later reward.

delusions False, unusual, and sometimes magical beliefs that are not part of an individual's culture.

demand characteristics Any aspects of a study that communicate to the participants how the experimenter wants them to behave.

dendrites Treelike fibers projecting from a neuron, which receive information and orient it toward the neuron's cell body.

deoxyribonucleic acid (DNA) A complex molecule in the cell's chromosomes that carries genetic information.

dependent variable The outcome; the factor that can change in an experiment in response to changes in the independent variable.

depressants Psychoactive drugs that slow down mental and physical activity.

depressive disorders Mood disorders in which the individual suffers from depression—an unrelenting lack of pleasure in life.

depth perception The ability to perceive objects three dimensionally.

development The pattern of continuity and change in human capabilities that occurs throughout life, involving both growth and decline.

difference threshold The degree of difference that must exist between two stimuli before the difference is detected.

discrimination An unjustified negative or harmful action toward a member of a group simply because the person belongs to that group.

discrimination (classical conditioning) The process of learning to respond to certain stimuli and not others.

discrimination (operant conditioning) Responding appropriately to stimuli that signal that a behavior will or will not be reinforced.

display rules Sociocultural standards that determine when, where, and how emotions should be expressed.

dissociative amnesia Dissociative disorder characterized by extreme memory loss that is caused by extensive psychological stress.

dissociative disorders Psychological disorders that involve a sudden loss of memory or change in identity due to the dissociation (separation) of the individual's conscious awareness from previous memories and thoughts.

dissociative identity disorder (DID) Formerly called multiple personality disorder, a dissociative disorder in which the individual has two or more distinct personalities or selves, each with its own memories, behaviors, and relationships.

divergent thinking Thinking that produces many solutions to the same problem.

divided attention Concentrating on more than one activity at the same time.

divided consciousness view of hypnosis Hilgard's view that hypnosis involves a splitting of consciousness into two separate components, one following the hypnotist's commands and the other acting as a "hidden observer."

dominant-recessive genes principle The principle that if one gene of a pair is dominant and one is recessive, the dominant gene overrides the recessive gene. A recessive gene exerts its influence only if both genes of a pair are recessive.

double-blind experiment An experimental design in which neither the experimenter nor the participants are aware of which participants are in the experimental group and which are in the control group until the results are calculated.

dream analysis A psychoanalytic technique for interpreting a person's dreams.

drive An aroused state that occurs because of a physiological need.

DSM-5 The *Diagnostic and Statistical Manual of Mental Disorders (DSM-5)*, 5th ed.; the major classification of psychological disorders in the United States.

dyslexia A learning disability characterized by difficulty with learning to read fluently and with accurate comprehension, despite normal intelligence.

E

efferent nerves Also called motor nerves; nerves that carry information out of the brain and spinal cord to other areas of the body.

ego The Freudian structure of personality that deals with the demands of reality.

egoism Giving to another person to ensure reciprocity; to gain self-esteem; to present oneself as powerful, competent, or caring; or to avoid social and self-censure for failing to live up to society's expectations.

elaboration The formation of a number of different connections around a stimulus at a given level of memory encoding.

elaboration likelihood model Theory identifying two ways to persuade: a central route and a peripheral route.

electroconvulsive therapy (ECT) Also called *shock therapy*, a treatment, sometimes used for depression, that sets off a seizure in the brain.

emerging adulthood The transitional period from adolescence to adulthood, spanning approximately 18 to 25 years of age.

emotion Feeling, or affect, that can involve physiological arousal (such as a fast heartbeat), conscious experience (thinking about being in love with someone), and behavioral expression (a smile or grimace).

emotion-focused coping The coping strategy that involves responding to the stress that one is feeling—trying to manage one's emotional reaction—rather than focusing on the root problem itself.

empathy A feeling of oneness with the emotional state of another person.

empirical method Gaining knowledge through the observation of events, the collection of data, and logical reasoning.

empirically keyed test A type of self-report test that presents many questionnaire items to two groups that are known to be different in some central way.

encoding The first step in memory; the process by which information gets into memory storage.

endocrine system The body system consisting of a set of glands that regulate the activities of certain organs by releasing their chemical products into the bloodstream.

episodic memory The retention of information about the where, when, and what of life's happenings—that is, how individuals remember life's episodes.

estrogens The class of sex hormones that predominates in females, produced mainly by the ovaries.

ethnocentrism The tendency to favor one's own ethnic group over other groups.

eugenics The belief in the possibility of improving the human species by discouraging reproduction among those with less desirable characteristics and enhancing reproduction among those with more desirable characteristics, such as high intelligence.

evidence-based practice Integration of the best available research with clinical expertise in the context of client characteristics, culture, and preferences.

evolutionary approach An approach to psychology centered on evolutionary ideas such as adaptation, reproduction, and natural selection as the basis for explaining specific human behaviors.

executive attention The ability to plan action, allocate attention to goals, detect errors and compensate for them, monitor progress on tasks, and deal with novel or difficult circumstances.

executive function Higher-order, complex cognitive processes, including thinking, planning, and problem solving.

exercise Structured activities whose goal is to improve health.

experiment A carefully regulated procedure in which the researcher manipulates one or more variables that are believed to influence some other variable.

experimental group The participants in an experiment who receive the drug or other treatment under study—that is, those who are exposed to the change that the independent variable represents.

experimenter bias The influence of the experimenter's expectations on the outcome of research.

explicit memory or declarative memory The conscious recollection of information, such as specific facts or events and, at least in humans, information that can be verbally communicated.

external validity The degree to which an experimental design actually reflects the real-world issues it is supposed to address.

extinction (classical conditioning) The weakening of the conditioned response when the unconditioned stimulus is absent.

extinction (operant conditioning) Decreases in the frequency of a behavior when the behavior is no longer reinforced.

extrinsic motivation Motivation that involves external incentives such as rewards and punishments.

F

face validity The extent to which a test item appears to fit the particular trait it is measuring.

facial feedback hypothesis The idea that facial expressions can influence emotions and reflect them.

false consensus effect Observers' overestimation of the degree to which everybody else thinks or acts the way they do.

family therapy Group therapy with family members.

feature detectors Neurons in the brain's visual system that respond to particular features of a stimulus.

figure-ground relationship The principle by which we organize the perceptual field into stimuli that stand out (figure) and those that are left over (ground).

fixation Using a prior strategy and failing to look at a problem from a fresh new perspective.

flashbulb memory The memory of emotionally significant events that people often recall with more accuracy and vivid imagery than everyday events.

flat affect The display of little or no emotion—a common negative symptom of schizophrenia.

forebrain The brain's largest division and its most forward part.

formal operational stage Piaget's fourth stage of cognitive development, which begins at 11 to 15 years of age and continues through the adult years; it features thinking about things that are not concrete, making predictions, and using logic to come up with hypotheses about the future.

free association A psychoanalytic technique that involves encouraging individuals to say aloud whatever comes to mind, no matter how trivial or embarrassing.

frequency theory Theory on how the inner ear registers the frequency of sound, stating that the perception of a sound's frequency depends on how often the auditory nerve fires.

frontal lobes The portion of the cerebral cortex behind the forehead, involved in personality, intelligence, and the control of voluntary muscles.

functional fixedness Failing to solve a problem as a result of fixation on a thing's usual functions.

functionalism James's approach to mental processes, emphasizing the functions and purposes of the mind and behavior in the individual's adaptation to the environment.

fundamental attribution error Observers' overestimation of the importance of internal traits and underestimation of the importance of external situations when they seek explanations of another person's behavior.

G

gender The social and psychological aspects of being male, female, both or neither.

gender identity A person's inner concept of themselves in relation to the ideas of being male, female, both, or neither.

gender roles Roles that reflect the society's expectations for how people of different genders should think, act, and feel.

gender similarities hypothesis Hyde's proposition that people of different genders are much more similar than they are different.

gene × environment (g × e) interaction The interaction of a specific measured variation in DNA and a specific measured aspect of the environment.

general adaptation syndrome (GAS) Selye's term for the common effects of stressful demands on the body, consisting of three stages: alarm, resistance, and exhaustion.

generalization (classical conditioning) The tendency of a new stimulus that is similar to the original conditioned stimulus to elicit a response that is similar to the conditioned response.

generalization (operant conditioning) Performing a reinforced behavior in a different situation.

generalized anxiety disorder Psychological disorder marked by persistent anxiety for at least six months, and in which the individual is unable to specify the reasons for the anxiety.

genes The units of hereditary information, consisting of short segments of chromosomes composed of DNA.

genotype An individual's genetic heritage; their actual genetic material.

gestalt psychology A school of thought interested in how people naturally organize their perceptions according to certain patterns.

gifted Possessing high intelligence (an IQ of 130 or higher) and/or superior talent in a particular area.

glands Organs or tissues in the body that create chemicals that control many bodily functions.

glial cells The second of two types of cells in the nervous system; glial cells (also called glia) provide support, nutritional benefits, and other functions and keep neurons running smoothly.

group polarization effect The solidification and further strengthening of an individual's position as a consequence of a group discussion or interaction.

group therapy A sociocultural approach to the treatment of psychological disorders that brings together individuals who share a particular psychological disorder in sessions that are typically led by a mental health professional.

groupthink The impaired group decision making that occurs when making the right decision is less important than maintaining group harmony.

H

habituation Decreased responsiveness to a stimulus after repeated presentations.

hallucinations Sensory experiences that occur in the absence of real stimuli.

hallucinogens Psychoactive drugs that modify a person's perceptual experiences and produce visual images that are not real.

hardiness A personality trait characterized by a sense of commitment rather than alienation and of control rather than powerlessness; a perception of problems as challenges rather than threats.

health behaviors Practices that have an impact on physical well-being.

health psychology A subfield of psychology that emphasizes psychology's role in establishing and maintaining health and preventing and treating illness.

heritability The proportion of observable differences in a group that can be explained by differences in the genes of the group's members.

heuristics Shortcut strategies or guidelines that suggest a solution to a problem but do not guarantee an answer.

hierarchy of needs Maslow's theory that human needs must be satisfied in the following sequence: physiological needs, safety, love and belongingness, esteem, and self-actualization.

hindbrain Located at the skull's rear, the lowest portion of the brain, consisting of the medulla, cerebellum, and pons.

hindsight bias The tendency to report falsely, after the fact, that one has accurately predicted an outcome.

hippocampus The structure in the limbic system that has a special role in the storage of memories.

homeostasis The body's tendency to maintain an equilibrium, or steady state.

hormones Chemical messengers that are produced by the endocrine glands and carried by the bloodstream to all parts of the body.

human sexual response pattern According to Masters and Johnson, the characteristic sequence of physiological changes that humans experience during sexual activity, consisting of four phases: excitement, plateau, orgasm, and resolution.

humanistic approach An approach to psychology emphasizing a person's positive qualities, the capacity for positive growth, and the freedom to choose any destiny.

humanistic perspectives Theoretical views stressing a person's capacity for personal growth and positive human qualities.

humanistic therapies Treatments that uniquely emphasize people's self-healing capacities and that encourage clients to understand themselves and to grow personally.

hypnosis An altered state of consciousness or a psychological state of altered attention and expectation in which the individual is unusually receptive to suggestions.

hypothalamic-pituitary-adrenal axis (HPA axis) The complex set of interactions among the hypothalamus, the pituitary gland, and the adrenal glands that regulate various body processes and control reactions to stressful events.

hypothalamus A small forebrain structure, located just below the thalamus, that monitors three pleasurable activities—eating, drinking, and sexual behavior—as well as emotion, stress, and reward.

hypothesis A testable prediction that derives logically from a theory.

I

id The part of the person that Freud called the "it," consisting of unconscious drives; the individual's reservoir of sexual energy.

implementation intentions Specific strategies for dealing with the challenges of making a life change.

implicit memory or nondeclarative memory Memory in which behavior is affected by prior experience without a conscious recollection of that experience.

independent variable A manipulated experimental factor; the variable that the experimenter changes to see what its effects are.

individual psychology Adler's view that people are motivated by purposes and goals and that perfection, not pleasure, is thus the key motivator in human life.

inductive reasoning Reasoning from specific observations to make generalizations.

infant attachment The close emotional bond between an infant and its caregiver.

infinite generativity The ability of language to produce an endless number of meaningful sentences.

informational social influence The influence other people have on us because we want to be right.

inner ear The part of the ear that includes the oval window, cochlea, and basilar membrane and whose function is to convert sound waves into neural impulses and send them to the brain.

insight learning A form of problem solving in which the organism develops a sudden insight into or understanding of a problem's solution.

instinct An innate (unlearned) biological pattern of behavior that is assumed to be universal throughout a species.

instinctive drift The tendency of animals to revert to instinctive behavior that interferes with learning.

integrative therapy Using a combination of techniques from different therapies based on the therapist's judgment of which particular methods will provide the greatest benefit for the client.

intellectual disability A condition of limited mental ability in which an individual has a low IQ, usually below 70 on a traditional intelligence test, and has difficulty adapting to everyday life.

intelligence All-purpose ability to do well on cognitive tasks, to solve problems, and to learn from experience.

intelligence quotient (IQ) An individual's mental age divided by chronological age multiplied by 100.

interference theory The theory that people forget not because memories are lost from storage but because other information gets in the way of what they want to remember.

internal validity The degree to which changes in the dependent variable are due to the manipulation of the independent variable.

interpretation A psychoanalyst's search for symbolic, hidden meanings in what the client says and does during therapy.

intrinsic motivation Motivation based on internal factors such as organismic needs (competence, relatedness, and autonomy), as well as curiosity, challenge, and fun.

investment model A model of long-term relationships that examines the ways that commitment, investment, and the availability of attractive alternative partners predict satisfaction and stability in relationships.

J

James-Lange theory The theory that emotion results from physiological states triggered by stimuli in the environment.

K

kinesthetic senses Senses that provide information about movement, posture, and orientation.

L

language A form of communication—whether spoken, written, or signed—that is based on a system of symbols.

latent content According to Freud, a dream's hidden content; its unconscious and true meaning.

latent learning or implicit learning Unreinforced learning that is not immediately reflected in behavior.

law of effect Thorndike's law stating that behaviors followed by positive outcomes are strengthened and that behaviors followed by negative outcomes are weakened.

learned helplessness Through experience with unavoidable aversive stimuli, an organism learns that it has no control over negative outcomes.

learning A systematic, relatively permanent change in behavior that occurs through experience.

levels of processing A continuum of memory processing from shallow to intermediate to deep, with deeper processing producing better memory.

limbic system A set of subcortical brain structures central to emotion, memory, and reward processing.

lithium The lightest of the solid elements in the periodic table of elements, widely used to treat bipolar disorder.

long-term memory A relatively permanent type of memory that stores huge amounts of information for a long time.

longitudinal design A special kind of systematic observation, used by correlational researchers, that involves obtaining measures of the variables of interest in multiple waves over time.

loss aversion The tendency to strongly prefer to avoid losses compared to attempting to acquire gains.

M

major depressive disorder (MDD) Psychological disorder involving a major depressive episode and depressed characteristics, such as lethargy and hopelessness, for at least two weeks.

manifest content According to Freud, the surface content of a dream, containing dream symbols that disguise the dream's true meaning.

medical model The view that psychological disorders are medical diseases with a biological origin.

meditation The attainment of a peaceful state of mind in which thoughts are not occupied by worry; the meditator is mindfully present to their thoughts and feelings but is not consumed by them.

memory The retention of information or experience over time as the result of three key processes: encoding, storage, and retrieval.

mental age (MA) An individual's level of mental development relative to that of others.

mental processes The thoughts, feelings, and motives that people experience privately but that cannot be observed directly.

mere exposure effect The phenomenon that the more individuals encounter someone or something, the more probable it is that they will start liking the person or thing even if they do not realize they have seen it before.

microaggression Everyday, subtle, and potentially unintentional acts that communicate bias to members of marginalized groups.

midbrain Located between the hindbrain and forebrain, an area in which many nerve-fiber systems ascend and descend to connect the higher and lower portions of the brain; in particular, the midbrain relays information between the brain and the eyes and ears.

middle ear The part of the ear that channels and amplifies sound through the eardrum, hammer, anvil, and stirrup to the inner ear.

mindfulness The state of being alert and mentally present for one's everyday activities.

Minnesota Multiphasic Personality Inventory (MMPI) The most widely used and researched empirically keyed self-report personality test.

monocular cues Powerful depth cues available from the image in one eye, either the right or the left.

morphology A language's rules for word formation.

motivated forgetting Forgetting that occurs when something is so painful or anxiety-laden that remembering it is intolerable.

motivation The force that moves people to behave, think, and feel the way they do.

motor cortex A region in the cerebral cortex, located just behind the frontal lobes, that processes information about voluntary movement.

myelin sheath A layer of fat cells that encases and insulates most axons.

N

natural selection Darwin's principle of an evolutionary process in which organisms that are best adapted to their environment will survive and produce offspring.

naturalistic observation The observation of behavior in a real-world setting.

nature An individual's biological inheritance, especially genes.

need A deprivation that energizes the drive to eliminate or reduce the deprivation.

negative affect Negative emotions such as anger, guilt, and sadness.

negative punishment The removal of a stimulus following a given behavior in order to decrease the frequency of that behavior.

negative reinforcement The removal of a stimulus following a given behavior in order to increase the frequency of that behavior.

neglectful parenting A parenting style characterized by a lack of parental involvement in the child's life.

neocortex The outermost part of the cerebral cortex, making up 80 percent of the human brain's cortex.

nervous system The body's electrochemical communication circuitry.

neural networks Networks of nerve cells that integrate sensory input and motor output.

neurons One of two types of cells in the nervous system; neurons are the nerve cells that handle the information-processing function.

neuroscience The scientific study of the structure, function, development, genetics, and biochemistry of the nervous system, emphasizing that the brain and nervous system are central to understanding behavior, thought, and emotion.

neurotransmitters Chemical substances that are stored in very tiny sacs within the neuron's terminal buttons and involved in transmitting information across a synaptic gap to the next neuron.

noise Irrelevant and competing stimuli—not only sounds but also any distracting stimuli for the senses.

normal distribution A symmetrical, bell-shaped curve, with a majority of test scores (or other data) falling in the middle of the possible range and few scores (or other data points) appearing toward the extremes.

normative social influence The influence other people have on us because we want them to like us.

nurture An individual's environmental and social experiences.

O

obedience Behavior that complies with the explicit demands of the individual in authority.

object permanence Piaget's term for the crucial accomplishment of understanding that objects and events continue to exist even when they cannot directly be seen, heard, or touched.

observational learning Learning that occurs through observing and imitating another's behavior.

obsessive-compulsive disorder (OCD) Disorder in which the individual has anxiety-provoking thoughts that will not go away and/or urges to perform repetitive, ritualistic behaviors to prevent or produce some future situation.

occipital lobes Structures located at the back of the head that respond to visual stimuli.

Oedipus complex According to Freud, a boy's intense desire to replace his father and enjoy the affections of his mother.

olfactory epithelium The lining of the roof of the nasal cavity, containing a sheet of receptor cells for smell.

open-mindedness The state of being receptive to other ways of looking at things.

operant conditioning or instrumental conditioning A form of associative learning in which the consequences of a behavior change the probability of the behavior's occurrence.

operational definition A definition that provides an objective description of how a variable is going to be measured and observed in a particular study.

opioids A class of drugs that act on the brain's endorphin receptors. These include opium and its natural derivatives (sometimes called opiates) as well as chemicals that do not occur naturally but that have been created to mimic the activity of opium. These drugs (also called narcotics) depress activity in the central nervous system and eliminate pain.

opponent-process theory Theory stating that cells in the visual system respond to complementary pairs of red-green and blue-yellow colors; a given cell might be excited by red and inhibited by green, whereas another cell might be excited by yellow and inhibited by blue.

optic nerve The structure at the back of the eye, made up of axons of the ganglion cells, that carries visual information to the brain for further processing.

outer ear The outermost part of the ear, consisting of the pinna and the external auditory canal.

ovaries Sex-related endocrine glands that produce hormones involved in sexual development and reproduction.

overt aggression Physical or verbal behavior that directly harms another person.

P

pain The sensation that warns an individual of damage to the body.

pancreas A dual-purpose gland under the stomach that performs both digestive and endocrine functions.

panic disorder Anxiety disorder in which the individual experiences recurrent, sudden onsets of intense apprehension or terror, often without warning and with no specific cause.

pansexual A person's sexual attractions do not depend on the biological sex, gender, or gender identity of others.

papillae Rounded bumps above the tongue's surface that contain the taste buds, the receptors for taste.

parallel processing The simultaneous distribution of information across different neural pathways.

parasympathetic nervous system The part of the autonomic nervous system that calms the body.

parietal lobes Structures at the top and toward the rear of the head that are involved in registering spatial location, attention, and motor control.

perception The process of organizing and interpreting sensory information so that it makes sense.

perceptual constancy The recognition that objects are constant and unchanging even though sensory input about them is changing.

perceptual set A predisposition or readiness to perceive something in a particular way.

peripheral nervous system (PNS) The network of nerves that connects the brain and spinal cord to other parts of the body.

permissive parenting A parenting style characterized by the placement of few limits on the child's behavior.

person perception The processes by which an individual uses social stimuli to form impressions of others.

personality A pattern of enduring, distinctive thoughts, emotions, and behaviors that characterize the way an individual adapts to the world.

personality disorders Chronic, maladaptive cognitive-behavioral patterns that are thoroughly integrated into an individual's personality.

phenotype An individual's observable characteristics.

phonology A language's sound system.

physical dependence The physiological need for a drug that causes unpleasant withdrawal symptoms such as physical pain and a craving for the drug when it is discontinued.

pituitary gland A pea-sized gland just beneath the hypothalamus that controls growth and regulates other glands.

place theory Theory on how the inner ear registers the frequency of sound, stating that each frequency produces vibrations at a particular spot on the basilar membrane.

placebo In a drug study, a harmless substance that has no physiological effect, given to participants in a control group so that they are treated identically to the experimental group except for the active agent.

placebo effect The situation where participants' expectations, rather than the experimental treatment, produce an experimental outcome.

plasticity The brain's special capacity for change.

polygraph A machine, commonly called a lie detector, that monitors changes in the body, and is used to try to determine whether someone is lying.

population The entire group about which the researcher wants to draw conclusions.

positive affect Pleasant emotions such as joy, happiness, and interest.

positive illusions Favorable views of the self that are not necessarily rooted in reality.

positive punishment The presentation of a stimulus following a given behavior in order to decrease the frequency of that behavior.

positive reappraisal Reinterpreting a potentially stressful experience as positive, valuable, or even beneficial

positive reinforcement The presentation of a stimulus following a given behavior in order to increase the frequency of that behavior.

post-traumatic stress disorder (PTSD) Anxiety disorder that develops through exposure to a traumatic event, a severely oppressive situation, cruel abuse, or a natural or unnatural disaster.

pragmatics The useful character of language and the ability of language to communicate even more meaning than is verbalized.

preferential looking A research technique that involves giving an infant a choice of what object to look at.

prefrontal cortex An important part of the frontal lobes that is involved in higher cognitive functions such as planning, reasoning, and self-control.

prejudice An unjustified negative attitude toward an individual based on the individual's membership in a group.

preoperational stage Piaget's second stage of cognitive development, lasting from about two to seven years of age, during which thought is more symbolic than sensorimotor thought.

preparedness The species-specific biological predisposition to learn in certain ways but not others.

primary reinforcer A reinforcer that is innately satisfying; one that does not take any learning on the organism's part to make it pleasurable.

priming The activation of information that people already have in storage to help them remember new information better and faster.

proactive interference Situation in which material that was learned earlier disrupts the recall of material that was learned later.

problem solving The mental process of finding an appropriate way to attain a goal when the goal is not readily available.

problem-focused coping The coping strategy of squarely facing one's troubles and trying to solve them.

procedural memory Memory for skills.

projective test A personality assessment test that presents individuals with an ambiguous stimulus and asks them to describe it or tell a story about it—to project their own meaning onto the stimulus.

prosocial behavior Behavior that is intended to benefit other people.

prospective memory Remembering information about doing something in the future; includes memory for intentions.

prototype model A model emphasizing that when people evaluate whether a given item reflects a certain concept, they compare the item with the most typical item(s) in that category and look for a "family resemblance" with that item's properties.

psychoactive drugs Drugs that act on the nervous system to alter consciousness, modify perception, and change moods.

psychoanalysis Freud's therapeutic technique for analyzing an individual's unconscious thoughts.

psychodynamic approach An approach to psychology emphasizing unconscious thought, the conflict between biological drives (such as the drive for sex) and society's demands, and early childhood family experiences.

psychodynamic perspectives Theoretical views emphasizing that personality is primarily unconscious (beyond awareness).

psychodynamic therapies Treatments that stress the importance of the unconscious mind, extensive interpretation by the therapist, and the role of early childhood experiences in the development of an individual's problems.

psychological dependence The strong desire to repeat the use of a drug for emotional reasons, such as a feeling of well-being and reduction of stress.

psychology The scientific study of behavior and mental processes.

psychoneuroimmunology A new field of scientific inquiry that explores connections among psychological factors (such as attitudes and emotions), the nervous system, and the immune system.

psychosis A state in which a person's perceptions and thoughts are fundamentally removed from reality.

psychosurgery A biological therapy, with irreversible effects, that involves removal or destruction of brain tissue to improve the individual's adjustment.

psychotherapy A nonmedical process that helps individuals with psychological disorders recognize and overcome their problems.

puberty A period of rapid skeletal and sexual maturation that occurs mainly in early adolescence.

punishment A consequence that decreases the likelihood that a behavior will occur.

R

random assignment Researchers' assignment of participants to groups by chance, to reduce the likelihood that an experiment's results will be due to preexisting differences between groups.

random sample A sample that gives every member of the population an equal chance of being selected.

reasoning The mental activity of transforming information to reach conclusions.

referential thinking Ascribing personal meaning to completely random events.

reflective speech A technique in which the therapist mirrors the client's own feelings back to the client.

reinforcement The process by which a stimulus or an event (a reinforcer) following a particular behavior increases the probability that the behavior will happen again.

reinforcement sensitivity theory A theory proposed by Jeffrey Gray identifying two biological systems linked to learning associations between behaviors and rewards or punishers. The behavioral activation system is sensitive to learning about rewards. The behavioral inhibition system is sensitive to learning about punishers.

relapse A return to former unhealthy patterns.

relational aggression Behavior that is meant to harm the social standing of another person.

reliability The extent to which a test yields a consistent, reproducible measure of performance.

REM sleep A stage of sleep characterized by rapid eye movement when most vivid dreams occur.

replication The process in which a scientist attempts to reproduce a study to see if the same results emerge.

representativeness heuristic The tendency to make judgments about group membership based on physical appearances or the match between a person and one's stereotype of a group rather than on available base rate information.

research participant bias In an experiment, the influence of participants' expectations, and of their thoughts on how they should behave, on their behavior.

resilience A person's ability to recover from or adapt to difficult times.

resistance A client's unconscious defense strategies that prevent the person from gaining insight into their psychological problems.

resting potential The stable, negative charge of an inactive neuron.

reticular activation system A network of structures including the brain stem, medulla, and thalamus that determine arousal, one aspect of consciousness.

reticular formation A system in the midbrain comprising a diffuse collection of neurons involved in stereotyped patterns of behavior such as walking, sleeping, and turning to attend to a sudden noise.

retina The multilayered light-sensitive surface in the eye that records electromagnetic energy and converts it to neural impulses for processing in the brain.

retrieval The memory process that occurs when information that was retained in memory comes out of storage.

retroactive interference Situation in which material that was learned later disrupts the retrieval of information that was learned earlier.

retrograde amnesia Memory loss for a segment of the past but not for new events.

retrospective memory Remembering information from the past.

risk factor Characteristics, experiences, or exposures that increase the likelihood that a person will develop a psychological disorder.

risky shift The tendency for a group decision to be riskier than the average decision made by the individual group members.

rods The receptor cells in the retina that are sensitive to light but not very useful for color vision.

romantic love or passionate love Love with strong components of sexuality and infatuation, often predominant in the early part of a love relationship.

Rorschach inkblot test A famous projective test that uses an individual's perception of inkblots to determine their personality.

S

sample The subset of the population chosen by the investigator for study.

schedules of reinforcement Specific patterns that determine when a behavior will be reinforced.

schema A preexisting mental concept or framework that helps people to organize and interpret information. Schemas from prior encounters with the environment influence the way individuals encode, make inferences about, and retrieve information.

schizophrenia Severe psychological disorder characterized by highly disordered thought processes; individuals suffering from schizophrenia may be referred to as psychotic because they are so far removed from reality.

science The use of systematic methods to observe the natural world, including human behavior, and to draw conclusions.

script A schema for an event, often containing information about physical features, people, and typical occurrences.

secondary reinforcer A reinforcer that acquires its positive value through an organism's experience; a secondary reinforcer is a learned or conditioned reinforcer.

secure attachment The ways that infants use their caregiver, usually their mother, as a secure base from which to explore the environment.

selective attention The act of focusing on a specific aspect of experience while ignoring others.

self-actualization The motivation to develop one's full potential as a human being—the highest and most elusive of Maslow's proposed needs.

self-determination theory Deci and Ryan's theory asserting that all humans have three basic, innate organismic needs: competence, relatedness, and autonomy.

self-efficacy The belief that one can master a situation and produce positive change.

self-fulfilling prophecy Social expectations that cause an individual to act in such a way that the expectations are realized.

self-perception theory Bem's theory on how behaviors influence attitudes, stating that individuals make inferences about their attitudes by perceiving their behavior.

self-regulation The process by which an organism effortfully controls its behavior in order to pursue important objectives.

self-report test Also called an objective test or an inventory, a method of measuring personality characteristics that directly asks people whether specific items describe their personality traits.

self-serving bias The tendency to take credit for one's successes and to deny responsibility for one's failures.

semantic memory A person's knowledge about the world.

semantics The meaning of words and sentences in a particular language.

semicircular canals Three fluid-filled circular tubes in the inner ear containing the sensory receptors that detect head motion caused when an individual tilts or moves the head and/or the body.

sensation The process of receiving stimulus energies from the external environment and transforming those energies into neural energy.

sensorimotor stage Piaget's first stage of cognitive development, lasting from birth to about two years of age, during which infants construct an understanding of the world by coordinating sensory experiences with motor (physical) actions.

sensory adaptation A change in the responsiveness of the sensory system based on the average level of surrounding stimulation.

sensory memory Memory system that involves holding information from the world in its original sensory form for only an instant, not much longer than the brief time it is exposed to the visual, auditory, and other senses.

sensory receptors Specialized cells that detect stimulus information and transmit it to sensory (afferent) nerves and the brain.

serial position effect The tendency to recall the items at the beginning and end of a list more readily than those in the middle.

set point The weight maintained when the individual makes no effort to gain or lose weight.

sexual orientation The direction of an individual's erotic interests, today viewed as a continuum from exclusive male–female relations to exclusive same-gender relations.

sexually transmitted infection (STI) An infection that is contracted primarily through sexual activity—vaginal intercourse as well as oral and anal sex.

shaping Rewarding successive approximations of a desired behavior.

short-term memory Limited-capacity memory system in which information is usually retained for only as long as 30 seconds unless strategies are used to retain it longer.

signal detection theory An approach to perception that focuses on decision making about stimuli in the presence of uncertainty.

sleep A natural state of rest for the body and mind that involves the reversible loss of consciousness.

social anxiety disorder or social phobia An intense fear of being humiliated or embarrassed in social situations.

social cognition The area of social psychology exploring how people select, interpret, remember, and use social information.

social cognitive behavior view of hypnosis The perspective that hypnosis is a normal state in which the hypnotized person behaves the way they believe that a hypnotized person should behave.

social cognitive perspectives Theoretical views emphasizing conscious awareness, beliefs, expectations, and goals.

social comparison The process by which individuals evaluate their thoughts, feelings, behaviors, and abilities in relation to others.

social contagion Imitative behavior involving the spread of behavior, emotions, and ideas.

social exchange theory The view of social relationships as involving an exchange of goods, the objective of which is to minimize costs and maximize benefits.

social facilitation Improvement in an individual's performance because of the presence of others.

social identity The way individuals define themselves in terms of their group membership.

social identity theory The view that social identity is a crucial part of self-image and a valuable source of positive feelings about oneself.

social loafing Each person's tendency to exert less effort in a group because of reduced accountability for individual effort.

social psychology The study of how people think about, influence, and relate to other people.

social support Information and feedback from others indicating that one is loved and cared for, esteemed and valued, and included in a network of communication and mutual obligation.

sociocultural approach An approach to psychology that examines the ways in which social and cultural environments influence behavior.

sociocultural therapies Treatments that acknowledge the relationships, roles, and cultural contexts that characterize an individual's life, often bringing them into the therapeutic context.

somatic nervous system The body system consisting of the sensory nerves, whose function is to convey information from the skin and muscles to the central nervous system about conditions such as pain and temperature, and the motor nerves, whose function is to tell muscles what to do.

somatic symptom and related disorders Psychological disorders characterized by bodily symptoms that either are very distressing or interfere with a person's functioning along with excessive thoughts, feeling, and behaviors about the symptoms.

somatosensory cortex A region in the cerebral cortex that processes information about body sensations, located at the front of the parietal lobes.

specific phobia Psychological disorder in which an individual has an irrational, overwhelming, persistent fear of a particular object or situation.

spontaneous recovery The process in classical conditioning by which a conditioned response can recur after a time delay, without further conditioning.

stages of change model Theoretical model describing a five-step process by which individuals give up bad habits and adopt healthier lifestyles.

standardization The development of uniform procedures for administering and scoring a test, and the creation of norms (performance standards) for the test.

stem cells Unique primitive cells that have the capacity to develop into most types of human cells.

stereotype A generalization about a group's characteristics that does not consider any variations from one individual to another.

stereotype threat An individual's fast-acting, self-fulfilling fear of being judged based on a negative stereotype about their group.

stimulants Psychoactive drugs, including caffeine, nicotine, amphetamines, and cocaine, that increase the central nervous system's activity.

storage The retention of information over time and how this information is represented in memory.

stream of consciousness Term used by William James to describe the mind as a continuous flow of changing sensations, images, thoughts, and feelings.

stress The responses of individuals to environmental stressors.

stress management program A regimen that teaches individuals how to appraise stressful events, how to develop skills for coping with stress, and how to put these skills into use in everyday life.

stressors Circumstances and events that threaten individuals and tax their coping abilities and that cause physiological changes to ready the body to handle the assault of stress.

structuralism Wundt's approach to discovering the basic elements, or structures, of mental processes.

subgoals Intermediate goals or problems to solve that put one in a better position for reaching a final goal or solution.

subjective well-being A person's assessment of their own level of positive affect relative to negative affect, and an evaluation of their life in general.

subliminal perception The detection of information below the level of conscious awareness.

substance use disorder A psychological disorder in which a person's use of psychoactive drugs (such as alcohol or opiates) affects their health, ability to work, and engage in social relationships.

superego The Freudian structure of personality that serves as the harsh internal judge of the individual's behavior; what is often referred to as *conscience*.

suprachiasmatic nucleus (SCN) A small brain structure that uses input from the retina to synchronize its own rhythm with the daily cycle of light and dark; the body's way of monitoring the change from day to night.

sustained attention The ability to maintain attention to a selected stimulus for a prolonged period of time.

sympathetic nervous system The part of the autonomic nervous system that arouses the body to mobilize it for action and thus is involved in the experience of stress.

synapses Tiny spaces between neurons; the gaps between neurons are referred to as synaptic gaps.

syntax A language's rules for combining words to form acceptable phrases and sentences.

systematic desensitization A behavior therapy that treats anxiety by teaching the client to associate deep relaxation with increasingly intense anxiety-producing situations.

systemic racism Systems, structures, and procedures in a society that disadvantage a racial group and privilege another.

T

temperament An individual's behavioral style and characteristic way of responding.

temporal lobes Structures in the cerebral cortex that are located just above the ears and are involved in hearing, language processing, and memory.

testes Sex-related endocrine glands in the scrotum that produce hormones involved in sexual development and reproduction.

thalamus The forebrain structure that sits at the top of the brain stem in the brain's central core and serves as an important relay station.

Thematic Apperception Test (TAT) A projective test that is designed to elicit stories that reveal something about an individual's personality.

theory A broad idea or set of closely related ideas that attempts to explain observations and to make predictions about future observations.

theory of mind Individuals' understanding that they and others think, feel, perceive, and have private experiences.

theory of planned behavior Theoretical model that includes the basic ideas of the theory of reasoned action but adds the person's perceptions of control over the outcome.

theory of reasoned action Theoretical model stating that effective change requires individuals to have specific intentions about their behaviors, as well as positive attitudes about a new behavior, and to perceive that their social group looks positively on the new behavior as well.

therapeutic alliance The relationship between the therapist and client—an important element of successful psychotherapy.

thermoreceptors Sensory nerve endings under the skin that respond to changes in temperature at or near the skin and provide input to keep the body's temperature at 98.6 degrees Fahrenheit.

thinking The process of manipulating information mentally by forming concepts, solving problems, making decisions, and reflecting critically or creatively.

third variable problem The circumstance where a variable that has not been measured accounts for the relationship between two other variables. Third variables are also known as *confounds*.

tip-of-the-tongue (TOT) phenomenon A type of effortful retrieval associated with a person's feeling that they know something (say, a word or a name) but cannot quite pull it out of memory.

tolerance The need to take increasing amounts of a drug to get the same effect.

top-down processing The operation in sensation and perception, launched by cognitive processing at the brain's higher levels, that allows the organism to sense what is happening and to apply that framework to information from the world.

trait theories Theoretical views stressing that personality consists of broad, enduring dispositions (traits) that tend to lead to characteristic responses.

tranquilizers Depressant drugs, such as Valium and Xanax, that reduce anxiety and induce relaxation.

transference A client's relating to the psychoanalyst in ways that reproduce or relive important relationships in the client's life.

triarchic theory of intelligence Sternberg's theory that intelligence comes in three forms: analytical, creative, and practical.

trichromatic theory Theory stating that color perception is produced by three types of cone receptors in the retina that are particularly sensitive to different, but overlapping, ranges of wavelengths.

two-factor theory of emotion Schachter and Singer's theory that emotion is determined by two factors: physiological arousal and cognitive labeling.

Type A behavior pattern A cluster of characteristics—including being excessively competitive, hard-driven, impatient, and hostile—that is related to a higher incidence of heart disease.

Type B behavior pattern A cluster of characteristics—including being relaxed and easygoing—that is related to a lower incidence of heart disease.

Type D behavior pattern A cluster of characteristics—including being generally distressed, having negative emotions, and being socially inhibited—that is related to adverse cardiovascular outcomes.

U

unconditional positive regard Rogers's construct referring to the individual's need to be accepted, valued, and treated positively regardless of their behavior.

unconditioned response (UR) An unlearned reaction that is automatically elicited by the unconditioned stimulus.

unconditioned stimulus (US) A stimulus that produces a response without prior learning.

unconscious thought According to Freud, a reservoir of unacceptable wishes, feelings, and thoughts that are beyond conscious awareness.

V

validity The soundness of the conclusions that a researcher draws from an experiment. In the realm of testing, the extent to which a test measures what it is intended to measure.

variable Anything that can change.

vestibular sense Sense that provides information about balance and movement.

visual cortex Located in the occipital lobe, the part of the cerebral cortex involved in vision.

volley principle Principle addressing limitations of the frequency theory of hearing, stating that a cluster of nerve cells can fire neural impulses in rapid succession, producing a volley of impulses.

vulnerability-stress hypothesis or diathesis-stress model A theory holding that preexisting conditions (genetic characteristics, personality dispositions, experiences, and so on) put an individual at risk of developing a psychological disorder.

W

Weber's law The principle that two stimuli must differ by a constant minimum percentage (rather than a constant amount) to be perceived as different.

wisdom Expert knowledge about the practical aspects of life.

working memory A combination of components, including short-term memory and attention, that allow individuals to hold information temporarily as they perform cognitive tasks; a kind of mental workbench on which the brain manipulates and assembles information to guide understanding, decision making, and problem solving.

Y

Yerkes-Dodson law The psychological principle stating that performance is best under conditions of moderate arousal rather than either low or high arousal.

references

A

Aalsma, M. C., Lapsley, D. K., & Flannery, D. J. (2006). Personal fables, narcissism, and adolescent adjustment. *Psychology in the Schools, 43*(4), 481–491.

Abati, E., Citterio, G., Bresolin, N., Comi, G. P., & Corti, S. (2020). Glial cells involvement in spinal muscular atrophy: Could SMA be a neuroinflammatory disease? *Neurobiology of Disease, 140,* 104870.

ABC7 News. (2020, November 11). 92-year-old woman with dementia performs 'Moonlight Sonata' on piano. https://abc7ny.com/society/92-year-old-woman-with-dementia-performs-moonlight-sonata/7869709/

Abdollahi, A., Abu Talib, M., Yaacob, S. N., & Ismail, Z. (2014). Hardiness as a mediator between perceived stress and happiness in nurses. *Journal of Psychiatric and Mental Health Nursing, 21*(9), 789–796.

Abdullah, T., & Brown, T. L. (2020). Diagnostic labeling and mental illness stigma among Black Americans: An experimental vignette study. *Stigma and Health, 5*(1), 11–21.

Abel, E. L. (2020). *Fetal alcohol syndrome: From mechanism to prevention.* CRC Press.

Abi-Dargham, A. (2017). A dual hit model for dopamine in schizophrenia. *Biological Psychiatry, 81*(1), 2–4.

Abma, J. C., & Martinez, G. M. (2017). Sexual activity and contraceptive use among teenagers in the United States, 2011–2015. *National Health Statistics Reports, 104.* National Center for Health Statistics, CDC.

Abney, D. H., Suanda, S. H., Smith, L. B., & Yu, C. (2020). What are the building blocks of parent–infant coordinated attention in free-flowing interaction? *Infancy, 25*(6), 871–887.

Abramowitz, J. S., & Jacoby, R. J. (2015). Obsessive-compulsive and related disorders: A critical review of the new diagnostic class. *Annual Review of Clinical Psychology, 11,* 165–186.

Abrams, A. (2016, December 4). Divorce rate in U.S. drops to nearly 40-year low. *Time* http://time.com/4575495/divorce-rate-nearly-40-year-low/

Abrams, L., & Rodriguez, E. L. (2005). Syntactic class influences phonological priming of tip-of-the-tongue resolution. *Psychonomic Bulletin and Review, 12*(6), 1018–1023.

Abramson, C. I. (2009). A study in inspiration: Charles Henry Turner (1867-1923) and the investigation of insect behavior. *Annual Review of Entomology, 54,* 343–359.

Abramson, L., Uzefovsky, F., Toccaceli, V., & Knafo-Noam, A. (2020). The genetic and environmental origins of emotional and cognitive empathy: Review and meta-analyses of twin studies. *Neuroscience & Biobehavioral Reviews, 114,* 113–133.

Abramson, L. Y., Seligman, M. E. P., & Teasdale, J. (1978). Learned helplessness in humans: Critique and reformulation. *Journal of Abnormal Psychology, 87*(1), 49–74.

Acabchuk, R. L., Kamath, J., Salamone, J. D., & Johnson, B. T. (2017). Stress and chronic illness: The inflammatory pathway. *Social Science & Medicine, 185,* 166–170. https://doi.org/10.1016/j.socscimed.2017.04.039

Accardi, M., Cleere, C., Lynn, S. J., & Kirsch, I. (2013). Placebo versus "standard" hypnosis rationale: Attitudes, expectancies, hypnotic responses, and experiences. *American Journal of Clinical Hypnosis, 56*(2), 103–114.

Adamantidis, A. R., Schmidt, M. H., Carter, M. E., Burda-kov, D., Peyron, C., & Scammell, T. E. (2020). A circuit perspective on narcolepsy. *Sleep, 43*(5), zsz296.

Adams, Z., & Browning, J. (2020). How colour qualia became a problem. *Journal of Consciousness Studies, 27*(5-6), 14–25.

Addington, J., & others. (2011). At clinical high risk for psychosis: Outcome for nonconverters. *American Journal of Psychiatry, 168*(8), 800–805.

Ader, R. (1974). Letter to the editor: Behaviorally conditioned immunosuppression. *Psychosomatic Medicine, 36,* 183–184.

Ader, R. (2000). On the development of psychoneuroimmunology. *European Journal of Pharmacology, 405*(1–3), 167–176.

Ader, R., & Cohen, N. (1975). Behaviorally conditioned immunosuppression. *Psychosomatic Medicine, 37*(4), 333–340.

Ader, R., & Cohen, N. (2000). Conditioning and immunity. In R. Ader, D. L. Felton, & N. Cohen (Eds.), *Psychoneuroimmunology* (3rd ed.). Academic.

Adler, A. (1927). *The theory and practice of individual psychology.* Harcourt Brace.

Adolph, K. E., & Hoch, J. E. (2020). The importance of motor skills for development. *Building Future Health and Well-Being of Thriving Toddlers and Young Children, 95,* 136–144.

Agnew, C. R., Harvey, S. M., VanderDrift, L. E., & Warren, J. (2017). Relational underpinnings of condom use: Findings from the project on partner dynamics. *Health Psychology, 36*(7), 713–720. https://doi.org/10.1037/hea0000488

Agorastos, A., Haasen, C., & Huber, C. G. (2012). Anxiety disorders through a transcultural perspective: Implications for migrants. *Psychopathology, 45*(2), 67–77.

Ahmann, E. (2014). Encouraging positive behavior in 'challenging' children: The nurtured heart approach. *Pediatric Nursing, 40*(1), 38–42.

Aida, S., Matsuda, Y., & Shimono, K. (2020). Interaction of disparity size and depth structure on perceived numerosity in a three-dimensional space. *PlosOne, 15*(4), e0230847.

Ainsworth, M. D. S. (1979). Infant–mother attachment. *American Psychologist, 34*(10), 932–937.

Ainsworth, M. S., Blehar, M. C., Waters, E., & Wall, S. (1978). *Patterns of attachment: A psychological study of the strange situation.* Erlbaum.

Ajzen, I. (2001). Nature and operation of attitudes. *Annual Review of Psychology, 52,* 27–58.

Ajzen, I. (2002). Perceived behavioral control, self-efficacy, locus of control, and the theory of planned behavior. *Journal of Applied Social Psychology, 32*(4), 665–683.

Ajzen, I. (2012a). Attitudes and persuasion. In K. Deaux & M. Snyder (Eds.), *The Oxford handbook of personality and social psychology.* Oxford University Press.

Ajzen, I. (2012b). The theory of planned behavior. In P. A. M. Lange, A. W. Kruglanski, & E. T. Higgins (Eds.), *Handbook of theories of social psychology.* Sage.

Ajzen, I. (2015). The theory of planned behaviour is alive and well, and not ready to retire: A commentary on Sniehotta, Presseau, and Araújo-Soares. *Health Psychology Review, 9*(2), 131–137. https://doi.org/10.1080/17437199.2014.883474

Ajzen, I. (2020). The theory of planned behavior: Frequently asked questions. *Human Behavior and Emerging Technologies, 2*(4), 314–324.

Ajzen, I., & Fishbein, M. (1980). *Understanding attitudes and predicting social behavior.* Prentice-Hall.

Ajzen, I., & Fishbein, M. (2005). The influence of attitudes on behavior. In D. Albarracin, B. T. Johnson, & M. P. Zanna (Eds.), *The handbook of attitudes* (pp. 173–221). Erlbaum.

Ajzen, I., & Schmidt, P. (2020). Changing behaviour using the theory of planned behavior. In M. S. Hagger, L. D. Cameron, K. Hamilton, N. Hankonen, & T. Lintunen (Eds.), *The handbook of behavior change* (pp. 17–31). Cambridge University Press.

Akhtar, S. (2006). Technical challenges faced by the immigrant psychoanalyst. *Psychoanalytic Quarterly, 75*(1), 21–43.

Akhtar, S., Justice, L. V., Morrison, C. M., & Conway, M. A. (2018). Fictional first memories. *Psychological science, 29*(10), 1612–1619.

Aknin, L. B., Dunn, E. W., Proulx, J., Lok, I., & Norton, M. I. (2020). Does spending money on others promote happiness? A registered replication report. *Journal of Personality and Social Psychology 119*(2), e15–e26.

Aknin, L. B., Wiwad, D., & Hanniball, K. B. (2018). Buying well-being: Spending behavior and happiness. *Social and Personality Psychology Compass, 12* (5), e12386.

Akudjedu, T. N., Tronchin, G., McInerney, S., Scanlon, C., Kenney, J. P., McFarland, J., . . . Hallahan, B. (2020). Progression of neuroanatomical abnormalities after first-episode of psychosis: A 3-year longitudinal sMRI study. *Journal of Psychiatric Research, 130,* 137–151.

Alallawi, B., Hastings, R. P., & Gray, G. (2020). A systematic scoping review of social, educational, and psychological research on individuals with autism spectrum disorder and their family members in Arab countries and cultures. *Review Journal of Autism and Developmental Disorders, 7,* 364–382.

Albarracin, D., Durantini, M. R., & Earl, A. (2006). Empirical and theoretical conclusions of an analysis of outcomes of HIV-prevention interventions. *Current Directions in Psychological Science, 15*(2), 73–78.

Albarracin, D., Wilson, K., Durantini, M. R., Sunderrajan, A., & Livingood, W. (2016). A meta-intervention to increase completion of an HIV-prevention intervention: Results from a randomized controlled trial in the state of Florida. *Journal of Consulting and Clinical Psychology, 84*(12), 1052–1065.

Albaum, C., Tablon, P., Roudbarani, F., & Weiss, J. A. (2020). Predictors and outcomes associated with therapeutic alliance in cognitive behaviour therapy for children with autism. *Autism, 24*(1), 211–220.

Alberti, F. B. (2020). Face transplants as surgical acts and psychosocial processes. *The Lancet, 395,* 1106–1107.

Albuja, A. F., Gaither, S. E., Sanchez, D. T., Straka, B., & Cipollina, R. (2019). Psychophysiological stress responses to bicultural and biracial identity denial. *Journal of Social Issues, 75*(4), 1165–1191.

Alcock, K., Watts, S., & Horst, J. (2020). What am I supposed to be looking at? Controls and measures in intermodal preferential looking. *Infant Behavior and Development, 60,* 101449.

Alemany-Navarro, M., Costas, J., Real, E., Segalàs, C., Bertolin, S., Domènech, L., . . . Alonso, P. (2019). Do polygenic risk and stressful life events predict pharmacological treatment response in obsessive compulsive disorder? A gene-environment interaction approach. *Translational Psychiatry, 9*(1), 1–10.

Alessio, H., Marron, K. H., Cramer, I. M., Hughes, M., Betz, K., Stephenson, S., . . . Bunger, A. L. (2020). Effects of cardiovascular health factors and personal listening behaviors on hearing sensitivity in college-aged students. *Annals of Otology, Rhinology & Laryngology,* 0003489420909403.

Alexander, M. G., & Fisher, T. D. (2003). Truth and consequences: Using the bogus pipeline to examine sex differences in self-reported sexuality. *Journal of Sex Research, 40*(1), 27–35.

Alfonsi, V., Scarpelli, S., D'Atri, A., Stella, G., & De Gennaro, L. (2020). Later school start time: The impact of sleep on academic performance and health in the adolescent population. *International Journal of Environmental Research and Public Health, 17*(7), 2574. https://doi.org/10.3390/ijerph17072574

Alford, C. F. (2016). *Trauma, Culture, and PTSD.* Springer.

Alia-Klein, N., Gan, G., Gilam, G., Bezek, J., Bruno, A., Denson, T. F., . . . Palumbo, S. (2020). The feeling of anger: From brain networks to linguistic expressions. *Neuroscience & Biobehavioral Reviews, 108,* 480–497.

Allen, J. P., Narr, R. K., Kansky, J., & Szwedo, D. E. (2020). Adolescent peer relationship qualities as predictors of long-term romantic life satisfaction. *Child Development, 91*(1), 327–340.

Allen, K., & Diamond, L. M. (2012). Same-sex relationships. In M. Fine & F. Fincham (Eds.), *Family theories* (pp. 123–144). Routledge.

Allen, M. S., Robson, D. A., Martin, L. J., & Laborde, S. (2020). Systematic review and meta-analysis of self-serving attribution biases in the competitive context of organized sport. *Personality and Social Psychology Bulletin, 46*(7), 1027–1043.

Allen, T.A., & DeYoung, C. G. (2017). Personality neuroscience and the five factor model. In T.A. Widiger (Ed.), *Oxford Handbook of the Five Factor Model* (pp. 319–352). Oxford University Press.

Alley, J. C., & Diamond, L. M. (2020). Oxytocin and human sexuality: Recent developments. *Current Sexual Health Reports, 12,* 182–185. https://link.springer.com/article/10.1007/s11930-020-00274-4

Allik, J., Church, A. T., Ortiz, F. A., Rossier, J., Hřebíčková, M., de Fruyt, F., et al. (2017). Mean profiles of the NEO personality inventory. *Journal of Cross-Cultural Psychology, 48*(3), 402–420. https://doi.org/10.1177/0022022117692100

Alloy, L. B., Salk, R., Stange, J. P., & Abramson, L. Y. (2017). Cognitive vulnerability and unipolar depression. In R. J. DeRubeis & D. R. Strunk (Eds.), *The Oxford Handbook of Mood Disorders* (pp. 142–153). Oxford University Press.

Allport, G. W. (1961). *Pattern and growth in personality.* Holt, Rinehart & Winston.

Allport, G. W., & Odbert, H. (1936). *Trait-names: A psycho-lexical study* (no. 211). Psychological Review Monographs.

Alluri, V., Toiviainen, P., Burunat, I., Kliuchko, M., Vuust, P., & Brattico, E. (2017). Connectivity patterns during music listening: Evidence for action-based processing in musicians. *Human Brain Mapping, 38,* 2955–2970.

Alsayouf, H. A., Talo, H., Biddappa, M. L., Qasaymeh, M., Qasem, S., & De Los Reyes, E. (2020). Pharmacological intervention in children with autism spectrum disorder with standard supportive therapies significantly improves core signs and symptoms: A single-center, retrospective case series. *Neuropsychiatric Disease and Treatment, 16,* 2779.

Altemus, M. (2006). Sex differences in depression and anxiety disorders: Potential biological determinants. *Hormones and Behavior, 50*(4), 534–538.

Altena, E., Baglioni, C., Espie, C. A., Ellis, J., Gavriloff, D., Holzinger, B., . . . Riemann, D. (2020). Dealing with sleep problems during home confinement due to the COVID-19 outbreak: Practical recommendations from a task force of the European CBT-I Academy. *Journal of Sleep Research,* e13052.

Altenor, A., Volpicelli, J. R., & Seligman, M. E. P. (1979). Debilitating shock escape is produced by both short- and long-duration inescapable shock: Learned helplessness vs. learned inactivity. *Bulletin of the Psychonomic Society, 14*(5), 337–339.

Altszuler, A. R., Morrow, A. S., Merrill, B. M., Bressler, S., Macphee, F. L., Gnagy, E. M., et al. (2017). The effects of stimulant medication and training on sports competence among children with ADHD. *Journal of Clinical Child & Adolescent Psychology.* https://doi.org/10.1080/15374416.2016.1270829

Aluja, A., Balada, F., Blanco, E., Fibla, J., & Blanch, A. (2018). Twenty candidate genes predicting neuroticism and sensation seeking personality traits: A multivariate analysis association approach. *Personality and Individual Differences.* https://doi.org/10.1016/j.paid.2018.03.041

Alvarez, M. J., & Garcia-Marques, L. (2011). Cognitive and contextual variables in sexual partner and relationship perception. *Archives of Sexual Behavior, 40*(2), 407–417

Alves, F. J., De Carvalho, E. A., Aguilar, J., De Brito, L. L., & Bastos, G. S. (2020). Applied behavior analysis for the treatment of autism: A systematic review of assistive technologies. *IEEE Access, 8,* 118664–118672.

American Association on Intellectual and Developmental Disabilities (AAIDD). (2010). *Intellectual disability: Definition, classification, and systems of supports (11th edition).* Author.

American Cancer Society. (2015). Cigarette smoking. *American Cancer Society.* www.cancer.org/cancer/cancercauses/tobaccocancer/cigarettesmoking/cigarette-smoking-toc

American Cancer Society. (2020). Tobacco and cancer. https://www.cancer.org/cancer/cancer-causes/tobacco-and-cancer.html

American Osteopathic Association (AOA). (2020). Headphones and hearing loss. https://osteopathic.org/what-is-osteopathic-medicine/headphones-hearing-loss/

American Psychiatric Association (APA). (2001). *Mental illness.* Author.

American Psychiatric Association (APA). (2006). *American Psychiatric Association practice guidelines for the treatment of psychiatric disorders.* Author.

American Psychiatric Association (APA). (2013). *Diagnostic and statistical manual of mental disorders (DSM-5).* Author.

American Psychological Association. (2004, July 28). APA supports legalization of APA of same-sex civil marriages and opposes discrimination against lesbian and gay parents. *American Psychological Association.* www.apa.org/news/press/releases/2004/07/gay-marriage.aspx (accessed December 22, 2014)

American Psychological Association. (2011). Mary Whiton Calkins, 1905 APA president. https://www.apa.org/about/governance/president/bio-mary-whiton-calkins

American Psychological Association (APA). (2012). Research shows psychotherapy is effective but underutilized. *APA.org.* www.apa.org/news/press/releases/2012/08/psychotherapy-effective.aspx

American Public Health Association (APHA). (2020). Racism and health. https://www.apha.org/topics-and-issues/health-equity/racism-and-health

Amin, A., Basri, S., Rahman, M., Capretz, L. F., Akbar, R., Gilal, A. R., & Shabbir, M. F. (2020). The impact of personality traits and knowledge collection behavior on programmer creativity. *Information and Software Technology, 128,* 106405.

Amunts, K., Schlaug, G., Jancke, L., Steinmetz, H., Schleicher, A., Dabringhaus, A., & Zilles, K. (1997). Motor cortex and hand motor skills: Structural compliance in the human brain. *Human Brain Mapping, 5*(3), 206–215.

An, D., & Carr, M. (2017). Learning styles theory fails to explain learning and achievement: Recommendations for alternative approaches. *Personality and Individual Differences, 116,* 410–416. https://doi.org/10.1016/j.paid.2017.04.050

Anastas, T. M., Miller, M. M., Hollingshead, N. A., Stewart, J. C., Rand, K. L., & Hirsh, A. T. (2020). The unique and interactive effects of patient race, patient socioeconomic status, and provider attitudes on chronic pain care decisions. *Annals of Behavioral Medicine,* kaaa016.

Anastasi, A., & Urbina, S. (1996). *Psychological testing* (7th ed.). Prentice-Hall.

Andel, R., Crowe, M., Pedersen, N. L., Mortimer, J., Crimmins, E., Johansson, B., & Gatz, M. (2005). Complexity of work and risk of Alzheimer's disease: A population-based study of Swedish twins. *Journals of Gerontology: Series B: Psychological Sciences and Social Sciences, 60*(5), 251–258.

Andersen, J. V., Jakobsen, E., Westi, E. W., Lie, M. E., Voss, C. M., Aldana, B. I., . . . Waagepetersen, H. S. (2020). Extensive astrocyte metabolism of γ-aminobutyric acid (GABA) sustains glutamine synthesis in the mammalian cerebral cortex. Glia. https://doi.org/10.1002/glia.23872

Anderson, C. A., Anderson, K. B., Dorr, N., DeNeve, K. M., & Flanagan, M. (2000). Temperature and aggression. In M. P. Zanna (Ed.), *Advances in experimental social psychology* (vol. 32, pp. 63–133). Academic.

Anderson, C. A., Benjamin, A. J., Jr., & Bartholow, B. D. (1998). Does the gun pull the trigger? Automatic priming effects of weapon pictures and weapon names. *Psychological Science, 9*(4), 308–314.

Anderson, C. A., & Bushman, B. J. (2002). Human aggression. *Annual Review of Psychology, 53,* 27–51.

Anderson, C. A., & others. (2010). Violent video game effects on aggression, empathy, and prosocial behavior in Eastern and Western countries: A meta-analytic review. *Psychological Bulletin, 136*(2), 151–173.

Anderson, J. L., Brockhaus, R., Kloefer, J., & Sellbom, M. (2020). Utility of the MMPI-2-RF in sexual violence risk assessment. *International Journal of Forensic Mental Health, 19*(4), 403–415.

Anderson, J. W. (2017). An interview with Henry A. Murray on his meeting with Sigmund Freud. *Psychoanalytic Psychology, 34*(3), 322–331. https://doi.org/10.1037/pap0000073

Anderson, M. C., & Hanslmayr, S. (2014). Neural mechanisms of motivated forgetting. *Trends in Cognitive Sciences, 18*(6), 279–292.

Anderson, N. D., & Craik, F. I. (2017). 50 years of cognitive aging theory. *The Journals of Gerontology: Series B, 72*(1), 1–6.

Anderson, P. L., & Molloy, A. (2020). Maximizing the impact of virtual reality exposure therapy for anxiety disorders. *Current Opinion in Psychology.* https://doi.org/10.1016/j.copsyc.2020.10.001

Anderson, V., Catroppa, C., Morse, S., Haritou, F., & Rosenfeld, J. (2005). Functional plasticity or vulnerability after early brain injury? *Pediatrics, 116*(6), 1374–1382.

Anderson, V., & others. (2009). Childhood brain insult: Can age at insult help us predict outcome? *Brain, 132*(Pt. 1), 45–56.

Andersson, G., & Hedmen, E. (2013). Effectiveness of guided Internet-based cognitive behavior therapy in regular clinical settings. *Verhaltenstherapie, 23,* 140–148.

Andersson, G., & others. (2013). Randomised controlled non-inferiority trial with 3-year follow-up of internet-delivered versus face-to-face group cognitive behavioural therapy for depression. *Journal of Affective Disorders, 151*(3), 986–994.

Andersson, G., & Titov, N. (2014). Advantages and limitations of Internet-based interventions for common mental disorders. *World Psychiatry, 13*(1), 4–11.

Andreatta, M., Genheimer, H., Wieser, M. J., & Pauli, P. (2020). Context-dependent generalization of conditioned responses to threat and safety signals. *International Journal of Psychophysiology, 155,* 140–151.

Andreou, M., & Skrimpa, V. (2020). Theory of mind deficits and neurophysiological operations in autism spectrum disorders: A review. *Brain Sciences, 10*(6), 393.

Andrew, S. (2020, November 17). Dolly Parton helped fund Moderna's Covid-19 vaccine research. CNN.com. https://www.cnn.com/2020/11/17/entertainment/dolly-parton-covid-moderna-vaccine-trnd/index.html.

Andrews, S. J., Fulton-Howard, B., & Goate, A. (2020). Interpretation of risk loci from genome-wide association studies of Alzheimer's disease. *The Lancet Neurology, 19*(4), 326–335.

Andries, A., Frystyk, J., Flyvbjerg, A., & Støving, R. K. (2014). Dronabinol in severe, enduring anorexia nervosa: A randomized controlled trial. *International Journal of Eating Disorders, 47*(1), 18–23.

Angermeyer, M. C., & Schomerus, G. (2017). State of the art of population-based attitude research on mental health: A systematic review. *Epidemiology and Psychiatric Sciences, 26*(3), 252–264.

Anglim, J., Horwood, S., Smillie, L. D., Marrero, R. J., & Wood, J. K. (2020). Predicting psychological and subjective well-being from personality: A meta-analysis. *Psychological Bulletin, 146*(4), 279–323.

Anjali J., Marshall, J., Buikema, A., Bancroft, T., Kelly, J.P., & Newschaffer, C.J. (2015). Autism occurrence by MMR vaccine status among US children with older siblings with and without autism. *JAMA, 313*(15): 1534–1540

Aoyama, Y., Uchida, H., Sugi, Y., Kawakami, A., Fujii, M., Kiso, K., et al. (2017). Immediate effect of subliminal priming with positive reward stimuli on standing balance in healthy individuals: A randomized controlled trial. *Medicine, 96*(28), e7494.

APA Presidential Task Force on Evidence-Based Practice. (2006). Evidence-based practice in psychology. *American Psychologist, 61*(4), 271–285.

Appel, J. E., Park, C. L., Wortmann, J. H., & van Schie, H. T. (2020). Meaning violations, religious/spiritual struggles, and meaning in life in the face of stressful life events. *The International Journal for the Psychology of Religion, 30*(1), 1–17.

Appleton, K. M., Woodside, J. V., Arveiler, D., Haas, B., Amouyel, P., Montaye, M., et al. (2016). A role for behavior in the relationships between depression and hostility and cardiovascular disease incidence, mortality, and all-cause mortality: The prime study. *Annals of Behavioral Medicine, 50*(4), 582–591. https://doi.org/10.1007/s12160-016-9784-x

Araujo-Cabrera, Y., Suarez-Acosta, M. A., & Aguiar-Quintana, T. (2017). Exploring the influence of CEO extraversion and openness to experience on firm performance: The mediating role of top management team behavioral integration. *Journal of Leadership & Organizational Studies, 24*(2), 201–215. https://doi.org/10.1177/1548051816655991

Aravena, V., García, F. E., Téllez, A., & Arias, P. R. (2020). Hypnotic intervention in people with fibromyalgia: A randomized controlled trial. *American Journal of Clinical Hypnosis, 63*(1), 49–61.

Arcaro, M. J., Thaler, L., Quinlan, D. J., Monaco, S., Khan, S., Valyear, K. F., . . . Culham, J. C. (2019). Psychophysical and neuroimaging responses to moving stimuli in a patient with the Riddoch phenomenon due to bilateral visual cortex lesions. *Neuropsychologia, 128*, 150–165.

Archibald, J., MacMillan, E. L., Enzler, A., Jutzeler, C. R., Schweinhardt, P., & Kramer, J. L. (2020). Excitatory and inhibitory responses in the brain to experimental pain: A systematic review of MR spectroscopy studies. *NeuroImage*, 116794.

Areh, I. (2011). Gender-related differences in eyewitness testimony. *Personality and Individual Differences, 50*(5), 559–563.

Arend, I., Yuen, K., Ashkenazi, S., & Henik, A. (2020). Space counts! Brain correlates of spatial and numerical representations in synaesthesia. *Cortex, 122*, 300–310.

Arisi, G. M. (2014). Nervous and immune systems signals and connections: Cytokines in hippocampus physiology and pathology. *Epilepsy & Behavior, 38*, 43–47.

Aristegui, I., Castro Solano, A., & Buunk, A. P. (2018). Mate preferences in Argentinean transgender people: An evolutionary perspective. *Personal Relationships, 25*(3), 330–350.

Arkadir, D., Bergman, H., & Fahn, S. (2014). Redundant dopaminergic activity may enable compensatory axonal sprouting in Parkinson disease. *Neurology, 82*(12), 1093–1098.

Armstrong, T. (2017). *The Myth of the ADHD Child, Revised Edition: 101 Ways to Improve Your Child's Behavior and Attention Span Without Drugs, Labels, or Coercion*. Penguin.

Armstrong, T., Rockloff, M., & Browne, M. (2020). Gamble with your head and not your heart: A conceptual model for how thinking-style promotes irrational gambling beliefs. *Journal of Gambling Studies, 36*, 183–206.

Arndt, J., & Goldenberg, J. L. (2017). Where health and death intersect: Insights from a terror management health model. *Current Directions in Psychological Science, 26*(2), 126–131.

Arnett, J. J. (2006). Emerging adulthood: Understanding the new way of coming of age. In J. J. Arnett & J. L. Tanner (Eds.), *Emerging adults in America* (pp. 3–19). American Psychological Association.

Arnett, J. J. (2007). Socialization in emerging adulthood. In J. E. Grusec & P. D. Hastings (Eds.), *Handbook of socialization* (pp. 208–239). Oxford University Press.

Arnett, J. J. (2010). Oh, grow up! Generational grumbling and the new life stage of emerging adulthood. *Perspectives on Psychological Science, 5*(1), 89–92.

Arnett, J. J. (Ed.). (2012). *Adolescent psychology around the world*. Psychology Press.

Arnett, J. J. (2014). *Emerging adulthood: The winding road from the late teens through the twenties* (2nd ed.). Oxford University Press.

Arnett, J. J., & Fischel, E. (2013). *When will my grown-up kid grow up?* Workman.

Arnold, J. F., & Sade, R. M. (2020). Regulating marijuana use in the United States: Moving past the gateway hypothesis of drug use. *The Journal of Law, Medicine & Ethics, 48*(2), 275–278.

Arnold, L. E., Hodgkins, P., Kahle, J., Madhoo, M., & Kewley, G. (2020). Long-term outcomes of ADHD: Academic achievement and performance. *Journal of Attention Disorders, 24*(1), 73–85.

Arntz, A. (2005). Cognition and emotion in borderline personality disorder. *Journal of Behavior Therapy and Experimental Psychiatry, 36*(3), 167–172.

Aronowitz, S. V., Mcdonald, C. C., Stevens, R. C., & Richmond, T. S. (2020). Mixed studies review of factors influencing receipt of pain treatment by injured black patients. *Journal of Advanced Nursing, 76*(1), 34–46.

Arrow, H. (2007). The sharp end of altruism. *Science, 318*(5850), 581–582.

Arrowood, R. B., Cox, C. R., Kersten, M., Routledge, C., Shelton, J. T., & Hood Jr, R. W. (2017). Ebola salience, death-thought accessibility, and worldview defense: A terror management theory perspective. *Death Studies, 41*, 585–591.

Artukoglu, B. B., Li, F., Szejko, N., & Bloch, M. H. (2020). Pharmacologic treatment of tardive dyskinesia: A meta-analysis and systematic review. *The Journal of Clinical Psychiatry, 81*(4). https://www.psychiatrist.com/JCP/article/Pages/pharmacologic-treatment-of-tardive-dyskinesia.aspx

Asch, S. E. (1951). Effects of group pressure on the modification and distortion of judgments. In H. S. Guetzkow (Ed.), *Groups, leadership, and men* (pp. 177–190). Carnegie University Press.

Asendorpf, J. B., & Rauthmann, J. F. (2020). States and situations, traits and environments. In P. J. Corr & G. Matthews (Eds.), *The Cambridge handbook of personality psychology* (2nd ed., pp. 56–58). Cambridge University Press.

Ashdown-Franks, G., Firth, J., Carney, R., Carvalho, A. F., Hallgren, M., Koyanagi, A., . . . Vancampfort, D. (2020). Exercise as medicine for mental and substance use disorders: A meta-review of the benefits for neuropsychiatric and cognitive outcomes. *Sports Medicine, 50*, 151–170.

Asherson, P., Adamou, M., Bolea, B., Muller, U., Dunn, S., Pitts, M., Thome, J., & Young, S. (2010). Is ADHD a valid diagnosis in adults? Yes. *British Medical Journal, 340*, 736–737.

Ashok, A. H., Mizuno, Y., Volkow, N. D., & Howes, O. D. (2017). Association of stimulant use with dopaminergic alterations in users of cocaine, amphetamine, or methamphetamine: A systematic review and meta-analysis. *JAMA Psychiatry, 74*(5), 511–519.

Ashton, M. C., & Lee, K. (2020). Objections to the HEXACO model of personality structure—And why those objections fail. *European Journal of Personality, 34*, 492–510.

Aslan, M., Radhakrishnan, K., Rajeevan, N., Sueiro, M., Goulet, J. L., Li, Y., . . . Harvey, P. D. (2020). Suicidal ideation, behavior, and mortality in male and female US veterans with severe mental illness. *Journal of Affective Disorders, 267*, 144–152.

Aspinwall, L. G. (1998). Rethinking the role of positive affect in self-regulation. *Motivation and Emotion, 22*(1), 1–32.

Aspinwall, L. G. (2011). Future-oriented thinking, proactive coping, and the management of potential threats to health and well-being. In S. Folkman (Ed.), *The Oxford handbook of stress, health, and coping*. Oxford University Press.

Aspinwall, L. G., Leaf, S. L., & Leachman, S. A. (2009). Meaning and agency in the context of genetic testing for familial cancer. In P. T. P. Wong (Ed.), *The human quest for meaning* (2nd ed.). Erlbaum.

Aspinwall, L. G., & Pengchit, W. (2013). Positive psychology. In M. D. Gellman & J. R. Turner (Eds.). *Encyclopedia of behavioral medicine*. Springer.

Asselmann, E., & Specht, J. (2020). Taking the ups and downs at the rollercoaster of love: Associations between major life events in the domain of romantic relationships and the Big Five personality traits. *Developmental Psychology, 56*(9), 1803–1816

Atkinson, R. C., & Shiffrin, R. M. (1968). Human memory: A proposed system and its control processes. *Psychology of Learning and Motivation, 2*, 89–195.

Augustine, R. A., Ladyman, S. R., Bouwer, G. T., Alyousif, Y., Sapsford, T. J., Scott, V., et al. (2017). Prolactin regulation of oxytocin neurone activity in pregnancy and lactation. *The Journal of Physiology, 595*(11), 3591–3605.

Austad, C. S. (2009). *Counseling and psychotherapy today*. McGraw-Hill.

Austen, J. M., & Sanderson, D. J. (2020). Cue duration determines response rate but not rate of acquisition of Pavlovian conditioning in mice. *Quarterly Journal of Experimental Psychology*. https://doi.org/10.1177/1747021820937696

Averett, S. L., Bansak, C., & Smith, J. K. (2020). Behind every high earning man is a conscientious woman: The impact of spousal personality on earnings and marriage. *Journal of Family and Economic Issues*. https://link.springer.com/article/10.1007/s10834-020-09692-x

Aviv, R. (2018). How a young woman lost her identity. *The New Yorker*. https://www.newyorker.com/magazine/2018/04/02/how-a-young-woman-lost-her-identity

Avvenuti, G., Baiardini, I., & Giardini, A. (2016). Optimism's explicative role for chronic diseases. *Frontiers In Psychology, 7*.

Axelsson, M., Lötvall, J., Cliffordson, C., Lundgren, J., & Brink, E. (2012). Self-efficacy and adherence as mediating factors between personality traits and health-related quality of life. *Quality of Life Research, 22*(3), 567–575.

Azhari, A., Truzzi, A., Neoh, M. J. Y., Balagtas, J. P. M., Tan, H. H., Goh, P. P., . . . Esposito, G. (2020). A decade of infant neuroimaging research: What have we learned and where are we going? *Infant Behavior and Development, 58*, 101389.

B

Bach, D. (April 15, 2015). Man with restored sight provides new insight into how vision develops. *UW News*. http://www.washington.edu/news/2015/04/15/man-with-restored-sight-provides-new-insight-into-how-vision-develops/

Bachhuber, M. A., Saloner, B., Cunningham, C. O., & Barry, C. L. (2014). Medical cannabis laws and opioid analgesic overdose mortality in the United States, 1999–2010. *JAMA Internal Medicine, 174*(10), 1668–1673.

Bacon, A. M., McDaid, C., Williams, N., & Corr, P. J. (2020). What motivates academic dishonesty in students? A reinforcement sensitivity theory explanation. *British Journal of Educational Psychology, 90*(1), 152–166.

Baddeley, A. (2013). On applying cognitive psychology. *British Journal of Psychology, 104*(4), 443–456.

Baddeley, A. D. (1993). Working memory and conscious awareness. In A. F. Collins, S. E. Gatherhole, M. A. Conway, & P. E. Morris (Eds.), *Theories of memory*. Erlbaum.

Baddeley, A. D. (1998). *Human memory* (rev. ed.). Allyn & Bacon.

Baddeley, A. D. (2003). Working memory and language: An overview. *Journal of Communication Disorders, 36*(3), 189–208.

Baddeley, A. D. (2008). What's new in working memory? *Psychological Review, 13*(3), 2–5.

Baddeley, A. D. (2012). Working memory: Theories, models, and controversies. *Annual Review of Psychology, 63*, 1–29.

Baddeley, A. D., & Hitch, G. J. (2019). The phonological loop as a buffer store: An update. *Cortex, 112,* 91–106.

Baddeley, A. D., Hitch, G. J., & Allen, R. J. (2019). From short-term store to multicomponent working memory: The role of the modal model. *Memory & Cognition, 47*(4), 575–588.

Baddeley, J. L., & Pennebaker, J. W. (2011). A postdeployment expressive writing intervention for military couples: A randomized controlled trial. *Journal of Traumatic Stress, 24*(5), 581–585.

Baddeley, J. L., & Singer, J. A. (2010). A loss in the family: Silence, memory, and narrative identity after bereavement. *Memory, 189*(2), 198–207.

Badin, A. S., Fermani, F., & Greenfield, S. A. (2017). The features and functions of neuronal assemblies: Possible dependency on mechanisms beyond synaptic transmission. *Frontiers in Neural Circuits, 10,* 114.

Badour, C. L., Resnick, H. S., & Kilpatrick, D. G. (2017). Associations between specific negative emotions and DSM-5 PTSD among a national sample of interpersonal trauma survivors. *Journal of Interpersonal Violence, 32*(11), 1620–1641.

Badura, K. L., Grijalva, E., Galvin, B. M., Owens, B. P., & Joseph, D. L. (2020). Motivation to lead: A meta-analysis and distal-proximal model of motivation and leadership. *Journal of Applied Psychology, 105*(4), 331–354.

Baer, M., Dane, E., & Madrid, H. (2020). Zoning out or breaking through? Linking daydreaming to creativity in the workplace. *Academy of Management Journal.* https://doi.org/10.5465/amj.2017.1283

Bagcchi, S. (2020). Stigma during the COVID-19 pandemic. *The Lancet. Infectious Diseases, 20*(7), 782.

Bagemihl, B. (1999). *Biological exuberance: Animal homosexuality and natural diversity.* St. Martin's Press.

Baglioni, C., Bostanova, Z., Bacaro, V., Benz, F., Hertenstein, E., Spiegelhalder, K., . . . Feige, B. (2020). A systematic review and network meta-analysis of randomized controlled trials evaluating the evidence base of melatonin, light exposure, exercise, and complementary and alternative medicine for patients with insomnia disorder. *Journal of Clinical Medicine, 9*(6), 1949.

Bahrick, H. P. (1984). Semantic memory content in permastore: Fifty years of memory for Spanish learned in school. *Journal of Experimental Psychology, 113*(1), 1–29.

Bahrick, H. P. (2000). Long-term maintenance of knowledge. In E. Tulving & F. I. M. Craik (Eds.), *The Oxford handbook of memory* (pp. 347–362). Oxford University Press.

Bahrick, H. P., Bahrick, P. O., & Wittlinger, R. P. (1974). Long-term memory: Those unforgettable high-school days. *Psychology Today, 8*(7), 50–56.

Bahtiyar, S., Karaca, K. G., Henckens, M. J., & Roozendaal, B. (2020). Norepinephrine and glucocorticoid effects on the brain mechanisms underlying memory accuracy and generalization. *Molecular and Cellular Neuroscience, 103537.*

Bai, Y., Maruskin, L. A., Chen, S., Gordon, A. M., Stellar, J. E., McNeil, G. D., & . . . Keltner, D. (2017). Awe, the diminished self, and collective engagement: Universals and cultural variations in the small self. *Journal of Personality and Social Psychology, 113*(2), 185–209. https://doi.org/10.1037/pspa0000087

Bailey, A. P., Parker, A. G., Colautti, L. A., Hart, L. M., Liu, P., & Hetrick, S. E. (2014). Mapping the evidence for the prevention and treatment of eating disorders in young people. *Journal of Eating Disorders, 2,* 5.

Bailey, C., Venta, A., & Langley, H. (2020). The bilingual [dis]advantage. *Language and Cognition, 12*(2), 225–281.

Bailey, J. M. (2003). Biological perspectives on sexual orientation. In L. D. Garnets & D. C. Kimmel (Eds.), *Psychological perspectives on lesbian, gay, and bisexual experiences* (2nd ed., pp. 50–85). Columbia University Press.

Bailey, S., Boivin, J., Cheong, Y., Bailey, C., Kitson-Reynolds, E., & Macklon, N. (2020). A feasibility and acceptability study of the positive reappraisal coping intervention: A supportive intervention for women with recurrent pregnancy loss. *Reproductive BioMedicine Online.* https://doi.org/10.1016/j.rbmo.2020.01.022

Bailey, T. D., & Brand, B. L. (2017). Traumatic dissociation: Theory, research, and treatment. *Clinical Psychology: Science and Practice, 24,* 170–185.

Baillargeon, R., Scott, R. M., & Bian, L. (2016). Psychological reasoning in infancy. *Annual Review of Psychology, 67,* 159–186. https://doi.org/10.1146/annurev-psych-010213-115033

Bais, M., Figee, M., & Denys, D. (2014). Neuromodulation in obsessive-compulsive disorder. *Psychiatric Clinics of North America, 37*(3), 393–413.

Baker, M. R., Frazier, P. A., Greer, C., Paulsen, J. A., Howard, K., Meredith, L. N., et al. (2016). Sexual victimization history predicts academic performance in college women. *Journal of Counseling Psychology, 63*(6), 685–692.

Baker, N., Lu, H., Erlikhman, G., & Kellman, P. J. (2020). Local features and global shape information in object classification by deep convolutional neural networks. *Vision Research, 172,* 46–61.

Bakermans-Kranenburg, M. J., & van IJzendoorn, M. H. (2011). Differential susceptibility to rearing environment depending on dopamine-related genes: New evidence and a meta-analysis. *Development and Psychopathology, 23*(1), 39–52.

Balcetis, E., & Dunning, D. (2010). Wishful seeing: More desired objects are seen as closer. *Psychological Science, 21*(1), 147–152.

Baldwin, D. S. (2020). Pharmacological treatment of generalized anxiety disorder (GAD). In A. L. Gerlach & A. T. Gloster (Eds.), *Generalized anxiety disorder and worrying: A comprehensive handbook for clinicians and researchers* (pp.**297**–318). John Wiley & Sons.

Baldwin, M. W., & Ko, M. C. (2020). Functional evolution of vertebrate sensory receptors. *Hormones and Behavior, 124,* 104771.

Baldwin, S. A., & Imel, Z. E. (2020). Studying specificity in psychotherapy with meta-analysis is hard. *Psychotherapy Research, 30*(3), 294–296.

Ball, C. L., Smetana, J. G., & Sturge Apple, M. L. (2017). Following my head and my heart: Integrating preschoolers' empathy, theory of mind, and moral judgments. *Child Development, 88*(2), 597–611.

Balliet, D. (2010). Conscientiousness and forgivingness: A meta-analysis. *Personality and Individual Differences, 48*(3), 259–263.

Balliet, D., Li, N. P., & Joireman, J. (2011). Relating trait self-control and forgiveness within prosocials and proselfs: Compensatory versus synergistic models. *Journal of Personality and Social Psychology, 101*(5), 1090–1105.

Balon, R., & Starcevic, V. (2020). Role of benzodiazepines in anxiety disorders. In Y. K. Kim (Ed.), *Anxiety disorders* (pp. 367–388). Springer.

Balter, M. (2010). Did working memory spark creative culture? *Science, 328*(5975), 160–163.

Balthazart, J. & Court, L. (2017). Human sexual orientation: The importance of evidentiary convergence. *Archives of Sexual Behavior, 46,* 1595–1600.

Banas, J. A., & Richards, A. S. (2017). Apprehension or motivation to defend attitudes? Exploring the underlying threat mechanism in inoculation-induced resistance to persuasion. *Communication Monographs, 84*(2), 164–178. https://doi.org/10.1080/03637751.2017.1307999

Bandura, A. (1986). *Social foundations of thought and action.* Prentice-Hall.

Bandura, A. (2001). Social cognitive theory. *Annual Review of Psychology 52,* 1–26.

Bandura, A. (2007). Social cognitive theory. In W. Donsbach (Ed.), *International encyclopedia of communication.* Sage.

Bandura, A. (2010). Self efficacy. In D. Matsumoto (Ed.), *Cambridge dictionary of psychology.* Cambridge University Press.

Bandura, A. (2011). Social cognitive theory. In P. A. M. van Lange, A. W. Kruglanski, & E. T. Higgins (Eds.), *Handbook of social psychological theories.* Sage.

Bandura, A. (2012). Social cognitive theory. *Annual Review of Clinical Psychology* (vol. 8). Annual Reviews.

Bandura, A. (2016a). *Moral disengagement: How people do harm and live with themselves.* Worth Publishers.

Bandura, A. (2016b). The power of observational learning through social modeling. In R. J. Sternberg, S. T. Fiske, & D. J. Foss (Eds.), *Scientists making a difference: One hundred eminent behavioral and brain scientists talk about their most important contributions* (pp. 235–239). Cambridge University Press.

Bandura, A., & Bussey, K. (2004). On broadening the cognitive, motivational, and sociostructural scope of theorizing about gender development and functioning: Comment on Martin, Ruble, and Szkrybalo (2002). *Psychological Bulletin, 130*(5), 691–701.

Bandura, A., & Rosenthal, T. L. (1966). Vicarious classical conditional as a function of arousal level. *Journal of Personality and Social Psychology, 3,* 54–62.

Bandura, A., Ross, D., & Ross, S. A. (1961). Transmission of aggression through imitation of aggressive models. *Journal of Abnormal and Social Psychology, 63*(3), 575–582.

Banks, A. P., Gamblin, D. M., & Hutchinson, H. (2020). Training fast and frugal heuristics in military decision making. *Applied Cognitive Psychology, 34*(3), 699–719.

Banton, O., West, K., & Kinney, E. (2020). The surprising politics of anti-immigrant prejudice: How political conservatism moderates the effect of immigrant race and religion on infrahumanization judgements. *British Journal of Social Psychology, 59*(1), 157–170.

Banyard, V. L., Demers, J. M., Cohn, E. S., Edwards, K. M., Moynihan, M. M., Walsh, W. A., & Ward, S. K. (2020). Academic correlates of unwanted sexual contact, intercourse, stalking, and intimate partner violence: An understudied but important consequence for college students. *Journal of Interpersonal Violence 35*(21–22), 4375–4392.

Banyard, V. L., & Williams, L. M. (2007). Women's voices on recovery: A multi-method study of the complexity of recovery from child sexual abuse. *Child Abuse and Neglect, 31*(3), 275–290.

Barber, J. P., Milrod, B., Gallop, R., Solomonov, N., Rudden, M. G., McCarthy, K. S., & Chambless, D. L. (2020). Processes of therapeutic change: Results from the Cornell-Penn Study of Psychotherapies for Panic Disorder. *Journal of Counseling Psychology, 67*(2), 222.

Barber, S. J., Kireeva, D., Seliger, J., & Jayawickreme, E. (2020). Wisdom once gained is not easily lost: Implicit theories about wisdom and age-related cognitive declines. *Innovation in Aging, 4*(2), igaa010.

Bard, P. (1934). Emotion. In C. Murchison (Ed.), *Handbook of general psychology.* Clark University Press.

Barker, M. D., & Hicks, C. B. (2020). Treating deficits in auditory processing abilities. *Language, Speech, and Hearing Services in Schools, 51*(2), 416–427.

Barkley, R. A. (1997). Behavioral inhibition, sustained attention, and executive functions: Constructing a unifying theory of ADHD. *Psychological Bulletin, 121*(1), 65–95.

Barkowski, S., Schwartze, D., Strauss, B., Burlingame, G. M., & Rosendahl, J. (2020). Efficacy of group psychotherapy for anxiety disorders: A systematic review and meta-analysis. *Psychotherapy Research, 30,* 965–982. https://doi.org/10.1080/10503307.2020.1729440

Barloon, T., & Noyes, R., Jr. (1997). Charles Darwin and panic disorder. *Journal of the American Medical Association, 277*(2), 138–141.

Baron-Cohen, S. (1995). *Mindblindness: An essay on autism and theory of mind.* MIT Press.

Baron-Cohen, S. (2011). The empathizing-systemizing (E-S) theory of autism: A cognitive developmental account. In U. Goswami (Ed.), *Wiley-Blackwell Handbook of Childhood Cognitive Development (2nd edition).* (pp. 626–640). Wiley-Blackwell.

Barratt, B.B. (2017). Seven questions for Kleinian psychology. *Psychoanalytic Psychology, 34,* 332–345.

Barrera, M. E., & Maurer, D. (1981). Recognition of mother's photographed face by the three-month-old infant. *Child Development, 52*(2), 714–716.

Barrett, K. C., Barrett, F. S., Jiradejvong, P., Rankin, S. K., Landau, A. T., & Limb, C. J. (2020). Classical creativity: A functional magnetic resonance imaging (fMRI) investigation of pianist and improviser Gabriela Montero. *NeuroImage, 209,* 116496.

Barroilhet, S. A., Pellegrini, A. M., McCoy, T. H., & Perlis, R. H. (2020). Characterizing DSM-5 and ICD-11 personality disorder features in psychiatric inpatients at scale using electronic health records. *Psychological Medicine, 50*(13), 2221-2229.

Barron, L. G., Randall, J. G., Trent, J. D., Johnson, J. F., & Villado, A. J. (2017). Big Five traits: Predictors of retesting propensity and score improvement. *International Journal of Selection and Assessment, 25*(2), 138-148. https://doi.org/10.1111/ijsa.12166

Barrouillet, P. (2015). Theories of cognitive development: From Piaget to today. *Developmental Review, 38*, 1-12.

Bartholow, B. D. (2018). The aggressive brain: Insights from neuroscience. *Current Opinion in Psychology, 19.*

Bartonicek, A., & Colombo, M. (2020). Claw-in-the-door: pigeons, like humans, display the foot-in-the-door effect. *Animal Cognition, 23*, 893-900.

Barušs, I., & Rabier, V. (2014). Failure to replicate retrocausal recall. *Psychology of Consciousness: Theory, Research, and Practice, 1*(1), 82-91.

Baselmans, B. M., Yengo, L., van Rheenen, W., & Wray, N. R. (2020). Risk in relatives, heritability, SNP-based heritability and genetic correlations in psychiatric disorders: A review. *Biological Psychiatry, 89*, 11-19.

Bashir, H. H., & Jankovic, J. (2020). Treatment of tardive dyskinesia. *Neurologic Clinics, 38*(2), 379-396.

Bassil Morozow, H. (2015). Analytical psychology and cinema. *The Journal of Analytical Psychology, 60*(1), 132-136. https://doi.org/10.1111/1468-5922.12135

Basso, L., Bönke, L., Aust, S., Gärtner, M., Heuser-Collier, I., Otte, C., . . . Grimm, S. (2020). Antidepressant and neurocognitive effects of serial ketamine administration versus ECT in depressed patients. *Journal of Psychiatric Research, 123*, 1-8.

Batson, D., Lishner, D., & Stocks, E. (2015). The empathy-altruism hypothesis. In D. A. Schroeder & W. G. Graziano (Eds.). *The Oxford handbook of prosocial behavior.* Oxford University Press.

Batterink, L. J., & Paller, K. A. (2017). Sleep-based memory processing facilitates grammatical generalization: evidence from targeted memory reactivation. *Brain and language, 167*, 83-93.

Baucal, A., Gillespie, A., Krstić, K., & Zittoun, T. (2020). Reproducibility in psychology: Theoretical distinction of different types of replications. *Integrative Psychological and Behavioral Science, 54*(1), 152-157.

Bauer, J. J., McAdams, D. P., & Sakaeda, A. R. (2005). The crystallization of desire and the crystallization of discontent in narratives of life-changing decisions. *Journal of Personality, 73*(5), 1181-1213.

Bauer, M., Hämmerli, S., & Leeners, B. (2020). Unmet needs in sex education—What adolescents aim to understand about sexuality of the other sex. *Journal of Adolescent Health, 67*, 245-252.

Bauer, P. J. (2013). Memory. In P. D. Zelazo (Ed.), *The Oxford handbook of developmental psychology* (vol. 1, pp. 505-541). Oxford University Press.

Bauer, S. M., & others. (2011). Culture and the prevalence of hallucinations in schizophrenia. *Comprehensive Psychiatry, 52*(3), 319-325.

Baum, G. L., Cui, Z., Roalf, D. R., Ciric, R., Betzel, R. F., Larsen, B., . . . Ruparel, K. (2020). Development of structure-function coupling in human brain networks during youth. *Proceedings of the National Academy of Sciences, 117*(1), 771-778.

Baumeister, R. F. (2000). Gender differences in erotic plasticity: The female sex drive as socially flexible and responsive. *Psychological Bulletin, 126*(3), 347-374.

Baumrind, D. (1991). Parenting styles and adolescent development. In J. Brooks-Gunn, R. Lerner, & A. C. Petersen (Eds.), *The encyclopedia of adolescence* (vol. 2). Garland.

Baumrind, D. (1993). The average expectable environment is not good enough: A response to Scarr. *Child Development, 64*(5), 1299-1307.

Baxter, L. R., Jr., Phelps, M. E., Mazziotta, J. C., Schwartz, J. M., Gerner, R. H., Selin, C. E., & Sumida, R. M. (1985). Cerebral metabolic rates for glucose in mood disorders: Studies with positron emission tomography and fluorodeoxyglucose F 18. *Archives of General Psychiatry, 42*(5), 441-447.

Beare, R., Adamson, C., Bellgrove, M. A., Vilgis, V., Vance, A., Seal, M. L., & Silk, T. J. (2017). Altered structural connectivity in ADHD: a network based analysis. *Brain Imaging and Behavior, 11*(3), 846-858.

Beaujean, A. A., & Woodhouse, N. (2020). Wechsler Intelligence Scale for Children (WISC). In B. J. Carducci (Editor-in-Chief) & C. Nave (Vol. Ed.), *The Wiley-Blackwell encyclopedia of personality and individual differences: Vol. II. Measurement and Assessment* (pp. 465-471). John Wiley & Sons.

Becht, A. I., & Mills, K. L. (2020). Modeling individual differences in brain development. *Biological Psychiatry, 88*(1), 63-69.

Beck, A. (1967). *Depression.* Harper & Row.

Beck, A. F., Edwards, E. M., Horbar, J. D., Howell, E. A., McCormick, M. C., & Pursley, D. M. (2020). The color of health: How racism, segregation, and inequality affect the health and well-being of preterm infants and their families. *Pediatric Research, 87*(2), 227-234.

Beck, A. T. (2006). How an anomalous finding led to a new system of psychotherapy. *Nature Medicine, 12*(10), 1139-1141.

Beck, A. T., & Haigh, E. A. (2014). Advances in cognitive theory and therapy: The generic cognitive model. *Annual Review of Clinical Psychology, 10*, 1-24.

Beck, J. S. (2020). *Cognitive behavior therapy: Basics and beyond.* Guilford Publications.

Becker, E. (1971). *The birth and death of meaning.* Free Press.

Beebe, K. R. (2014). Hypnotherapy for labor and birth. *Nursing and Women's Health, 18*(1), 48-58.

Beeli, G., Esslen, M., & Jancke, L. (2005). Synaesthesia: When coloured sounds taste sweet. *Nature, 434*(7029), 38.

Beer, T. (2020, September 2). Majority of Republicans believe the QAnon conspiracy theory is partly or mostly true, survey finds. Forbes.com. https://www.forbes.com/sites/tommybeer/2020/09/02/majority-of-republicans-believe-the-qanon-conspiracy-theory-is-partly-or-mostly-true-survey-finds/?sh=1683f7a35231

Beers, N., & Joshi, S. V. (2020). Increasing access to mental health services through reduction of stigma. *Pediatrics, 145*(6), e20200127.

Beh, H. G. (2015). The right to comprehensive sex education. In S. M. Coupet & E. Marrus (Eds.), *Children, sexuality, and the law* (pp. 163-185). New York University Press. https://doi.org/10.18574/nyu/9780814723852.003.0007

Behr, M., Aich, G., & Scheurenbrand, C. (2020). Person-centered and experiential psychotherapy and transactional analysis—contributions of two humanistic approaches to challenging or confounded counselling situations. *Person-Centered & Experiential Psychotherapies, 19.* https://doi.org/10.1080/14779757.2020.1748694

Behrman, B. W., & Davey, S. L. (2001). Eyewitness identification in actual criminal cases: An archival analysis. *Law and Human Behavior, 25*(5), 475-491.

Beijers, R., Hartman, S., Shalev, I., Hastings, W., Mattern, B. C., de Weerth, C., & Belsky, J. (2020). Testing three hypotheses about effects of sensitive–insensitive parenting on telomeres. *Developmental Psychology, 56*(2), 237-250. https://doi.org/10.1037/dev0000879

Bek, J., Arakaki, A. I., Lawrence, A., Sullivan, M., Ganapathy, G., & Poliakoff, E. (2020). Dance and Parkinson's: A review and exploration of the role of cognitive representations of action. *Neuroscience & Biobehavioral Reviews, 109*, 16-28.

Bell, A. P., Weinberg, M. S., & Hammersmith, S. K. (1981). *Sexual preference: Its development in men and women.* Indiana University Press.

Bell, C. N., Kerr, J., & Young, J. L. (2019). Associations between obesity, obesogenic environments, and structural racism vary by county-level racial composition. *International Journal of Environmental Research and Public Health, 16*(5), 861.

Belluck, P. (2018, December 3). Specifying mental health care before they're too ill to choose. *New York Times,* A1.

Bem, D. (1967). Self-perception: An alternative explanation of cognitive dissonance phenomena. *Psychological Review, 74*(3), 183-200.

Bender, S. (2020). The Rorschach test. In B. J. Carducci, C. S. Nave, J. S. Mio, & R. E. Riggio (Eds.), *The Wiley encyclopedia of personality and individual differences: Measurement and assessment* (vol. 2, pp. 367-376). John Wiley & Sons.

Benedetti, F., Pollo, A., Lopiano, L., Lanotte, M., Vighetti, S., & Rainero, I. (2003). Conscious expectation and unconscious conditioning in analgesic; Motor and hormonal placebo/nocebo responses. *Journal of Neuroscience, 23*(10), 4315-4323.

Benjamin Jr, A. J. (2020). Type A/B personalities. In J. S. Mio & R. E. Riggio (Eds.), *The Wiley encyclopedia of personality and individual differences: Clinical, applied, and cross-cultural research* (pp. 383-386). John Wiley & Sons. https://doi.org/10.1002/9781119547181.ch328

Benjamin Jr., L. T., Henry, K. D., & McMahon, L. R. (2005). Inez Beverly Prosser and the education of African Americans. *Journal of the History of the Behavioral Sciences, 41*, 43-62.

Benjet, C. (2020). Stress management interventions for college students in the context of the COVID-19 pandemic. *Clinical Psychology.* https://dx.doi.org/10.1111%2Fcpsp.12353

Benowitz-Fredericks, C. A., Garcia, K., Massey, M., Vasagar, B., & Borzekowski, D. L. (2012). Body image, eating disorders, and the relationship to adolescent media use. *Pediatric Clinics of North America, 59*(3), 693-704.

Ben-Porath, Y. S., & Tellegen, A. (2020). *Minnesota Multiphasic Personality Inventory-3 (MMPI-3).* Pearson.

Berg, H., & Slaattelid, R. (2017). Facts and values in psychotherapy—A critique of the empirical reduction of psychotherapy within evidence based practice. *Journal of Evaluation in Clinical Practice.* http://onlinelibrary.wiley.com/doi/10.1111/jep.12739/full

Berger, T., Lee, H., & Thuret, S. (2020a). Neurogenesis right under your nose. *Nature Neuroscience, 23*(3), 297-298.

Berger, T., Lee, H., & Thuret, S. (2020). Neurogenesis right under your nose. *Nature Neuroscience, 23*(3), 297-298.

Berger, T., Lee, H., Young, A. H., Aarsland, D., & Thuret, S. (2020b). Adult hippocampal neurogenesis in major depressive disorder and Alzheimer's disease. *Trends in Molecular Medicine.* https://doi.org/10.1016/j.molmed.2020.03.010

Bergman, L. R., Corovic, J., Ferrer-Wreder, L., & Modig, K. (2014). High IQ in early adolescence and career success in adulthood: Findings from a Swedish longitudinal study. *Research In Human Development, 11*(3), 165-185. https://doi.org/10.1080/15427609.2014.936261

Bergmann, O., Liebel, J., Bernard, S., Alkass, K., Yeung, M.S.Y., Steier, P., Kutschera, W., Johnson, L., Landen, M., Druid, H., Spalding, K.L., & Frisen, J. (2012). The age of olfactory bulb neurons in humans. *Neuron, 74*, 634-639. https://doi.org/10.1016/j.neuron.2012.03.0.

Bergstrom, H. C. (2020). Assaying fear memory discrimination and generalization: Methods and concepts. *Current Protocols in Neuroscience, 91*(1), e89.

Berke, D. S., Reidy, D. E., Miller, J. D., & Zeichner, A. (2017). Take it like a man: Gender-threatened men's experience of gender role discrepancy, emotion activation, and pain tolerance. *Psychology of Men & Masculinity, 18*(1), 62-69.

Berkowitz, L. (1993). *Aggression.* McGraw-Hill.

Berkowitz, L., & LePage, A. (1996). Weapons as aggression-eliciting stimuli. In S. Fein & S. Spencer (Eds.), *Readings in social psychology: The art and science of research* (pp. 67-73). Houghton Mifflin.

Berndt, N. C., Hayes, A. F., Verboon, P., Lechner, L., Bolman, C., & De Vries, H. (2012). Self-efficacy mediates the impact of craving on smoking abstinence in low to moderately anxious patients: Results of a moderated mediation approach. *Psychology of Addictive Behaviors, 27*(1), 113-124.

Bernhardt, I. S., Nissen-Lie, H. A., & Råbu, M. (2020). The embodied listener: a dyadic case study of how therapist and patient reflect on the significance of therapist's personal presence for the therapeutic change process. *Psychotherapy Research.* https://doi.org/10.1080/1050330 7.2020.1808728

Berscheid, E. (1988). Some comments on love's anatomy. Or, whatever happened to an old-fashioned lust? In R. J. Sternberg & M. L. Barnes (Eds.), *Anatomy of love* (pp. 359-374). Yale University Press.

Berscheid, E. (2000). Attraction. In A. Kazdin (Ed.), *Encyclopedia of psychology.* American Psychological Association and Oxford University Press.

Berscheid, E. (2006). Searching for the meaning of "love." In R. J. Sternberg & K. Weis (Eds.), *The new psychology of love* (pp. 171-183). Yale University Press.

Berscheid, E. (2010). Love in the fourth dimension. *Annual Review of Psychology, 61,* 1-25.

Berscheid, E., & Regan, P. C. (2005). *The psychology of interpersonal relationships.* Prentice-Hall.

Bertilsdotter Rosqvist, H., Kourti, M., Jackson-Perry, D., Brownlow, C., Fletcher, K., Bendelman, D., & O'Dell, L. (2019). Doing it differently: Emancipatory autism studies within a neurodiverse academic space. *Disability & Society, 34*(7-8), 1082-1101.

Bertoia, M. L., Rimm, E. B., Mukamal, K. J., Hu, F. B., Willett, W. C., & Cassidy, A. (2016). Dietary flavonoid intake and weight maintenance: Three prospective cohorts of 124,086 US men and women followed for up to 24 years. *BMJ: British Medical Journal,* 352.

Bertrand, M., & Mullainathan, S. (2004). Are Emily and Greg more employable than Lakisha and Jamal? A field experiment on labor market discrimination. *American Economic Review, 94*(4), 991-1013.

Besedovsky, L., Lange, T., & Born, J. (2012). Sleep and immune function. *European Journal of Physiology, 463*(1), 121-137.

Best, J. R., Rosano, C., Aizenstein, H. J., Tian, Q., Boudreau, R. M., Ayonayon, H. N., et al. (2017). Long-term changes in time spent walking and subsequent cognitive and structural brain changes in older adults. *Neurobiology of Aging,* 57, 153-161

Betancur, C. (2011). Etiological heterogeneity in autism spectrum disorders: more than 100 genetic and genomic disorders and still counting. *Brain Research, 1380,* 42-77.

Bettencourt, B., Talley, A., Benjamin, A. J., & Valentine, J. (2006). Personality and aggressive behavior under provoking and neutral conditions: a meta-analytic review. *Psychological Bulletin, 132*(5), 751-777.

Beuriat, P. A., Cristofori, I., Richard, N., Bardi, L., Loriette, C., Szathmari, A., . . . Claude, L. (2020). Cerebellar lesions at a young age predict poorer long-term functional recovery. *Brain Communications, 2*(1), fcaa027.

Beutler, L. E., Forrester, B., Gallagher-Thompson, D., Thompson, L., & Tomlins, J. B. (2012). Common, specific, and treatment fit variables in psychotherapy outcome. *Journal of Psychotherapy Integration, 22*(3), 255-281.

Bever, L. (2017, November 10). The unforgettable moment a widow touched the face that once belonged to her husband. *Washington Post.* https://www.washingtonpost.com/news/to-your-health/wp/2017/11/10/the-unforgettable-moment-a-widow-touched-the-face-that-once-belonged-to-her-husband/?noredirect=on&utmterm=.0d2 f23e260cf

Biddlestone, M., Green, R., & Douglas, K. (2020). Cultural orientation, powerlessness, belief in conspiracy theories, and intentions to reduce the spread of COVID-19. *British Journal of Social Psychology.* https://doi.org/10.1111/bjso.12397

Bidwell, L., Gray, J. C., Weafer, J., Palmer, A. A., de Wit, H., & MacKillop, J. (2017). Genetic influences on ADHD symptom dimensions: Examination of a priori candidates, gene-based tests, genome-wide variation, and SNP heritability. *American Journal of Medical Genetics Part B: Neuropsychiatric Genetics, 174*(4), 458-466.

Bierly, M. M. (1985). Prejudice toward contemporary outgroups as a generalized attitude. *Journal of Applied Social Psychology, 15*(2), 189-199.

Birch, L. L., Savage, J. S., & Fisher, J. O. (2015). Right sizing prevention. Food portion size effects on children's

eating and weight. *Appetite, 88,* 11-16. https://doi.org/ 10.1016/j.appet.2014.11.021

Birnbaum, R., & Weinberger, D. R. (2020). Translational science update. Pharmacological implications of emerging schizophrenia genetics: Can the bridge from 'genomics' to 'therapeutics' be defined and traversed? *Journal of Clinical Psychopharmacology, 40*(4), 323-329.

Biswas-Diener, R., Vitterso, J., & Diener, E. (2005). Most people are pretty happy, but there is cultural variation: The Inughuit, the Amish, and the Maasai. *Journal of Happiness Studies,* 6, 205-226.

Biton, A., Traut, N., Poline, J. B., Aribisala, B. S., Bastin, M. E., Bülow, R., . . . Hagenaars, S. (2020). Polygenic architecture of human neuroanatomical diversity. *Cerebral Cortex, 30*(4), 2307-2320.

Bjork, R. A., & Druckman, D. (1991). *In the mind's eye: Enhancing human performance.* National Academy Press.

Bjork, R. A., Dunlosky, J., & Kornell, N. (2013). Self-regulated learning: Beliefs, techniques, and illusions. *Annual Review of Psychology* 64, 417-444.

Björklund, A., & Lindvall, O. (2017). Replacing dopamine neurons in Parkinson's disease: How did it happen? *Journal of Parkinson's Disease, 7*(s1), S23-S33.

Black, D. (2020). The global workspace theory, the phenomenal concept strategy, and the distribution of consciousness. *Consciousness and Cognition, 84,* 102992.

Blackie, L. E., Jayawickreme, E., Tsukayama, E., Forgeard, M. J., Roepke, A. M., & Fleeson, W. (2017). Post-traumatic growth as positive personality change: Developing a measure to assess within-person variability. *Journal of Research in Personality, 69,* 22-32.

Blaess, S., Stott, S. R. W., & Ang, S. L. (2020). The generation of midbrain dopaminergic neurons. In J. Rubenstein, P. Rakic, B. Chen, & K.Y. Kwan (Eds.), *Patterning and cell type specification in the developing CNS and PNS* (pp. 369-398). Academic Press.

Blagrove, M., & Akehurst, L. (2000). Personality and dream recall frequency: Further negative findings. *Dreaming, 10*(3), 139-148.

Blain, S. D., Longenecker, J. M., Grazioplene, R. G., Klimes-Dougan, B., & DeYoung, C. G. (2020). Apophenia as the disposition to false positives: A unifying framework for openness and psychoticism. *Journal of Abnormal Psychology, 129*(3), 279-292.

Blain, S. D., Sassenberg, T. A., Xi, M., Zhao, D., & DeYoung, C. G. (2020). Extraversion but not depression predicts reward sensitivity: Revisiting the measurement of anhedonic phenotypes. *Journal of Personality and Social Psychology.* https://doi.org/10.1037/pspp0000371

Blaine, B.E., & McClure, K.J. (2018). *Understanding the Psychology of Diversity* (3rd ed.). Sage.

Blair, K. S., Otero, M., Teng, C., Geraci, M., Ernst, M., Blair, R. J. R., et al. (2017). Reduced optimism and a heightened neural response to everyday worries are specific to generalized anxiety disorder, and not seen in social anxiety. *Psychological Medicine, 47*(10), 1806-1815.

Blair, S. N., Kohl, H. W., Paffenbarger, R. S., Clark, D. G., Cooper, K. H., & Gibbons, L. W. (1989). Physical fitness and all-cause mortality: A prospective study of healthy men and women. *Journal of the American Medical Association, 262*(17), 2395-2401.

Blakemore, J. E. O., Berenbaum, S. E., & Liben, L. S. (2009). *Gender development.* Psychology Press.

Blanco, M., & others. (2009). *Investigating critical incidents, driver restart period, sleep quantity, and crash countermeasures in commercial operations using naturalistic data collection: A final report* (Contract No. DTFH61-01-00049, Task Order #23). Federal Motor Carrier Safety Administration.

Blanken, M. A., Oudega, M. L., Schouws, S. N., van Zanten, J. S., Gatchel, J. R., Regenold, W. T., & Dols, A. (2020). Is ECT a viable option to treat depression in older adults with bipolar disorder who are vulnerable to cognitive side effects? *Bipolar Disorders.* https://doi.org/10.1111/bdi.13005

Blankenship, T. L., Slough, M. A., Calkins, S. D., Deater-Deckard, K., Kim-Spoon, J., & Bell, M. A. (2019).

Attention and executive functioning in infancy: Links to childhood executive function and reading achievement. *Developmental Science, 22*(6), e12824.

Blanton, H., Jaccard, J., Strauts, E., Mitchell, G., & Tetlock, P. E. (2015). Toward a meaningful metric of implicit prejudice. *Journal of Applied Psychology, 100*(5), 1468-1481. https://doi.org/10.1037/a0038379

Blashfield, R. K., Keeley, J. W., Flanagan, E. H., & Miles, S. R. (2014). The cycle of classification: DSM-1 through DSM-5. *Annual Review of Clinical Psychology* 10, 25-21.

Blass, R. (2017). Understanding Freud's conflicted view of the object-relatedness of sexuality and its implications for contemporary psychoanalysis: A re-examination of Three Essays on the Theory of Sexuality. In P. Van Haute, & H. Westerink (Eds.), *Deconstructing normativity? Re-reading Freud's 1905 Three Essays* (pp. 6-27). Routledge/Taylor & Francis Group.

Bleidorn, W. (2015). What account for personality maturation in early adulthood? *Current Directions in Psychological Science, 24,* 245-252.

Bleidorn, W., Klimstra, T. A., Denissen, J. A., Rentfrow, P. J., Potter, J., & Gosling, S. D. (2013). Personality maturation around the world: A cross-cultural examination of social-investment theory. *Psychological Science, 24,* 2530-2540.

Block, J. (1982). Assimilation, accommodation, and the dynamics of personality development. *Child Development, 53*(2), 281-295.

Block, J. (2010). The five-factor framing of personality and beyond: Some ruminations. *Psychological Inquiry, 21*(1), 2-25.

Block, J., & Kremen, A. M. (1996). IQ and ego-resiliency: Conceptual and empirical connections and separateness. *Journal of Personality and Social Psychology, 70*(2), 349-361.

Bloom, B. (1985). *Developing talent in young people.* Ballantine.

Bloom, P. (2004). Myths of word learning. In D. G. Hall & S. R. Waxman (Eds.), *Weaving a lexicon* (pp. 205-224). MIT Press.

Bloor, C., & White, F. (1983). Unpublished manuscript. University of California, San Diego.

Blum, B. (2018, June 7). The lifespan of a lie. *The Medium.* https://medium.com/s/trustissues/the-lifespan-of-a-lie-d869212b1f62

Blum, D. J., & Zeitzer, J. M. (2020). Irregular sleep-wake, non-24-hour sleep-wake, jet lag, and shift work sleep disorders. In E. H. During & C. A. Kushida (Eds.), *Clinical sleep medicine: A comprehensive guide for mental health and other medical professionals* (pp. 245-271). American Psychiatric Association.

Blume, C., Del Giudice, R., Lechinger, J., Wislowska, M., Heib, D. P., Hoedlmoser, K., & Schabus, M. (2017). Preferential processing of emotionally and self-relevant stimuli persists in unconscious N2 sleep. *Brain and Language, 167,* 72-82.

Boag, R. J., Strickland, L., Heathcote, A., Neal, A., & Loft, S. (2019). Cognitive control and capacity for prospective memory in complex dynamic environments. *Journal of Experimental Psychology: General, 148*(12), 2181-2206.

Boddice, R. (2017). *Pain: A very short introduction.* Oxford University Press.

Boedhoe, P. S., Van Rooij, D., Hoogman, M., Twisk, J. W., Schmaal, L., Abe, Y., . . . Arango, C. (2020). Subcortical brain volume, regional cortical thickness, and cortical surface area across disorders: Findings from the ENIGMA ADHD, ASD, and OCD Working Groups. *American Journal of Psychiatry, 177*(9), 834-843.

Boehlen, F. H., Herzog, W., Schellberg, D., Maatouk, I., Schoettker, B., Brenner, H., & Wild, B. (2020). Gender-specific predictors of generalized anxiety disorder symptoms in older adults: Results of a large population-based study. *Journal of Affective Disorders, 262,* 174-181.

Bogaert, A. F. (2000). Birth order and sexual orientation in a national probability sample. *Journal of Sex Research, 37*(4), 361-368.

Bogaert, A. F., Ashton, M. C., & Lee, K. (2018). Personality and sexual orientation: Extension to asexuality and the HEXACO model. *The Journal of Sex Research, 55* (8), 951-961.

Bogdashina, O. (2016). *Sensory perceptual issues in autism and asperger syndrome: Different sensory experiences-different perceptual worlds.* Jessica Kingsley Publishers.

Bohart, A. C., & Tallman, K. (2010). Clients: The neglected common factor in psychotherapy. In B. L. Duncan, S. D. Miller, B. E. Wampold, & M. A. Hubble (Eds.), *The heart and soul of change: Delivering what works in therapy* (2nd ed., pp. 83–111). American Psychological Association.

Bohbot, V. D., Anderson, N., Trowbridge, A., Sham, R., Konishi, K., Kurdi, V., & Behrer, L. (2011, November 16). *A spatial memory intervention program aimed at stimulating the hippocampus while controlling for caudate nucleus-dependent response strategies led to significant improvements on independent virtual navigation tasks in healthy elderly.* Society for Neuroscience Convention.

Böhm, R., & Buttelmann, D. (2017). The impact of resource valence on children's other-regarding preferences. *Developmental Psychology, 53*(9), 1656–1665. https://doi.org/10.1037/dev0000365

Bohn, M., Kordt, C., Braun, M., Call, J., & Tomasello, M. (2020). Learning novel skills from iconic gestures: A developmental and evolutionary perspective. *Psychological Science.* https://doi.org/10.1177/0956797620921519

Boland, H., DiSalvo, M., Fried, R., Woodworth, K. Y., Wilens, T., Faraone, S. V., & Biederman, J. (2020). A literature review and meta-analysis on the effects of ADHD medications on functional outcomes. *Journal of Psychiatric Research, 123,* 21–30.

Boldt, L. J., Goffin, K. C., & Kochanska, G. (2020). The significance of early parent-child attachment for emerging regulation: A longitudinal investigation of processes and mechanisms from toddler age to preadolescence. *Developmental Psychology, 56*(3), 431–443.

Boleda, G. (2020). Distributional semantics and linguistic theory. *Annual Review of Linguistics,* 6, 213–234.

Bonanno, G. A. (2004). Loss, trauma, and human resilience: Have we underestimated the human capacity to thrive after extremely aversive events? *American Psychologist, 59*(1), 20–28.

Bonanno, G. A. (2005). Resilience in the face of potential trauma. *Current Directions in Psychological Science, 14*(3), 135–138.

Bond, R., & Smith, P. B. (1996). Culture and conformity: A meta-analysis of studies using Asch's (1952, 1956) line judgment task. *Psychological Bulletin, 119*(1), 111–137.

Bonfatti, J. F. (2005). Finding inspiration: Hope holds the key. *NAMI.* www.nami.org/Template.cfm?Section=Spotlight_1&Template=/ContentManagement/ContentDisplay.cfm&ContentID=25281&lstid=537

Bonus, J. A., Wulf, T., & Matthews, N. L. (2020, in press). The cost of clairvoyance: Enjoyment and appreciation of a popular movie as a function of affective forecasting errors. *Journal of Media Psychology: Theories, Methods, and Applications.*

Bora, E., Aydın, A., Saraç, T., Kadak, M. T., & Köse, S. (2017). Heterogeneity of subclinical autistic traits among parents of children with autism spectrum disorder: Identifying the broader autism phenotype with a data driven method. *Autism Research, 10*(2), 321–326.

Bosker, H. R., Sjerps, M. J., & Reinisch, E. (2020). Spectral contrast effects are modulated by selective attention in "cocktail party" settings. *Attention, Perception, & Psychophysics, 82,* 1318–1332.

Bosman, J. (2020, December 2). As hospitals fill, travel nurses race to virus hot spots. *New York Times.* https://www.nytimes.com/2020/12/02/us/covid-travel-nurses.html?searchResultPosition=1

Bouchard, T. J., Jr., & Loehlin, J. C. (2001). Genes, evolution, and personality. *Behavior Genetics, 31*(3), 243–273.

Bouchard, T. J., Lykken, D. T., Tellegen, A., & McGue, M. (1996). Genes, drives, environment, and experience. In D. Lubinski & C. Benbow (Eds.), *Psychometrics and social issues concerning intellectual talent.* Johns Hopkins University Press.

Bouloukaki, I., Grote, L., McNicholas, W. T., Hedner, J., Verbraecken, J., Parati, G., . . . Steiropoulos, P. (2020). Mild obstructive sleep apnea increases hypertension risk, challenging traditional severity classification. *Journal of Clinical Sleep Medicine, 16*(6), 889–898.

Bouris, A., & Hill, B. J. (2017). Exploring the mother-adolescent relationship as a promotive resource for sexual and gender minority youth. *Journal of Social Issues, 73*(3), 618–636.

Bouton, M. E., Broomer, M. C., Rey, C. N., & Thrailkill, E. A. (2020). Unexpected food outcomes can return a habit to goal-directed action. *Neurobiology of Learning and Memory, 169,* 107163.

Bouvard, M., Fournet, N., Denis, A., Sixdenier, A., & Clark, D. (2017). Intrusive thoughts in patients with obsessive compulsive disorder and non-clinical participants: a comparison using the International Intrusive Thought Interview Schedule. *Cognitive Behaviour Therapy, 46*(4), 287–299.

Bova, A. (2019). Parental strategies in argumentative dialogues with their children at mealtimes. *Language and Dialogue, 9*(3), 379–401.

Bowen, H. J. (2020). Examining memory in the context of emotion and motivation. *Current Behavioral Neuroscience Reports, 7,* 193–202.

Bowlby, J. (1969). *Attachment and loss* (vol. 1). Hogarth.

Bowlby, J. (1989). *Secure and insecure attachment.* Basic.

Boxer, P., Groves, C. L., & Docherty, M. (2015). Video games do indeed influence children and adolescents' aggression, prosocial behavior, and academic performance: A clearer reading of Ferguson (2015). *Perspectives on Psychological Science, 10*(5), 671–673. https://doi.org/10.1177/1745691615592239

Boyd, R. L., Pasca, P., & Lanning, K. (2020). The personality panorama: Conceptualizing personality through big behavioural data. *European Journal of Personality.* https://doi.org/10.1002/per.2254

Boysen, G. A., Isaacs, R. A., Tretter, L., & Markowski, S. (2020). Evidence for blatant dehumanization of mental illness and its relation to stigma. *The Journal of Social Psychology, 160*(3), 346–356.

Bradford, A. C., Bradford, W. D., Abraham, A., & Adams, G. B. (2018). Association between US state medical cannabis laws and opioid prescribing in the Medicare Part D population. *JAMA Internal Medicine, 178*(5), 667–672.

Bradizza, C. M., Stasiewicz, P. R., Zhuo, Y., Ruszczyk, M., Maisto, S. A., Lucke, J. F., et al. (2017). Smoking cessation for pregnant smokers: Development and pilot test of an emotion regulation treatment supplement to standard smoking cessation for negative affect smokers. *Nicotine & Tobacco Research, 19*(5), 578–584. https://doi.org/10.1093/ntr/ntw398

Brand, B. L., Loewenstein, R. J., & Spiegel, D. (2014). Dispelling myths about dissociative identity disorder treatment: An empirically based approach. *Psychiatry, 77*(2), 169–189.

Brander, G., Pérez-Vigil, A., Larsson, H., & Mataix-Cols, D. (2016). Systematic review of environmental risk factors for obsessive-compulsive disorder: a proposed roadmap from association to causation. *Neuroscience & Biobehavioral Reviews, 65,* 36–62.

Brandt, M. J., & Crawford, J. T. (2020). Worldview conflict and prejudice. *Advances in Experimental Social Psychology, 61,* 1–66.

Branley, D. B., & Covey, J. (2017). Pro-ana versus pro-recovery: A content analytic comparison of social media users' communication about eating disorders on Twitter and Tumblr. *Frontiers in Psychology, 8,* 1356.

Brannon, L. (1999). *Gender: Psychological perspectives* (2nd ed.). Allyn & Bacon.

Brashier, N. M., & Marsh, E. J. (2020). Judging truth. *Annual Review of Psychology, 71,* 499–515.

Bratman, G. N., Hamilton, J. P., Hahn, K. S., Daily, G. C., & Gross, J. J. (2015). Nature experience reduces rumination and subgenual prefrontal cortex activation. *Proceedings of the National Academy of Sciences, 112*(28), 8567–8572.

Brattland, H., Koksvik, J. M., Burkeland, O., Klöckner, C. A., Lara-Cabrera, M. L., Miller, S. D., . . . Iversen, V. C. (2019). Does the working alliance mediate the effect of routine outcome monitoring (ROM) and alliance feedback on psychotherapy outcomes? A secondary analysis from a randomized clinical trial. *Journal of Counseling Psychology, 66*(2), 234–246.

Braun, M. H., Lukowiak, K., Karnik, V., & Lukowiak, K. (2012). Differences in neuronal activity explain differences in memory forming abilities of different populations of Lymnaea stagnalis. *Neurobiology of Learning and Memory, 97*(1), 173–182.

Braunstein, G. D. (2007). Management of female sexual dysfunction in postmenopausal women by testosterone administration: Safety issues and controversies. *Journal of Sexual Medicine, 4*(4 Pt. 1), 859–866.

Breedlove, S. M. (2017). Prenatal influences on human sexual orientation: Expectations versus data. *Archives of Sexual Behavior, 46,* 1583–1592.

Breland, K., & Breland, M. (1961). The misbehavior of animals. *American Psychologist, 16,* 681–684.

Brennan, J. F., & Houde, K. A. (2017). *History and systems of psychology.* Cambridge University Press.

Brescoll, V. L. (2016). Leading with their hearts? How gender stereotypes of emotion lead to biased evaluations of female leaders. *The Leadership Quarterly, 27*(3), 415–428. https://doi.org/10.1016/j.leaqua.2016.02.005

Brettschneider, C., Gensichen, J., Hiller, T. S., Breitbart, J., Schumacher, U., Lukaschek, K., . . . König, H. H. (2020). Cost-effectiveness of practice team-supported exposure training for panic disorder and agoraphobia in primary care: A cluster-randomized trial. *Journal of General Internal Medicine,* 35, 1120–1126.

Breuninger, M. M., Grosso, J. A., Hunter, W., & Dolan, S. L. (2020). Treatment of alcohol use disorder: Integration of Alcoholics Anonymous and cognitive behavioral therapy. *Training and Education in Professional Psychology, 14*(1), 19–26.

Brewer, C. (2020). Partial ALDH inhibition to facilitate controlled drinking rather than abstinence has already been tried and it works. *Proceedings of the National Academy of Sciences, 117*(14), 7572.

Brewer, J. B., Zuo, Z., Desmond, J. E., Glover, G. H., & Gabrieli, J. D. E. (1998). Making memories: Brain activity that predicts how well visual experience will be remembered. *Science, 281*(5380), 1185–1187.

Brewer, N., Weber, N., & Guerin, N. (2020). Police lineups of the future? *American Psychologist, 75*(1), 76–91. https://doi.org/10.1037/amp0000465.

Brewer, N., & Wells, G. L. (2011). Eyewitness identification. *Current Directions in Psychological Science, 20*(1), 24–27.

Brewin, C. R. (2020, in press). Tilting at Windmills: Why Attacks on Repression Are Misguided. *Perspectives on Psychological Science.*

Brickman, P., & Campbell, D. T. (1971). Hedonic relativism and planning the good society. In M. H. Appley (Ed.), *Adaptation-level theory* (pp. 287–302). Academic.

Bridges, J.S., & Etaugh, C.A. (2017). *Women's Lives: A Psychological Exploration* (4th ed.). Routledge.

Briggs, K. C., & Myers, I. B. (1998). *Myers-Briggs Type Indicator.* Consulting Psychologists Press.

Brigham, J. C. (1986). Race and eyewitness identifications. In S. Worschel & W. G. Austin (Eds.), *Psychology of intergroup relations.* Nelson-Hall.

Brigham, J. C., Bennett, L. B., Meissner, C. A., & Mitchell, T. L. (2007). The influence of race on eyewitness memory. In R.C.L. Lindsay, D. F. Ross, J. D. Read, & M. P. Toglia (Eds.), *The handbook of eyewitness memory* (vol. 2). Erlbaum.

Brikell, I., Larsson, H., Lu, Y., Pettersson, E., Chen, Q., Kuja-Halkola, R., . . . Martin, J. (2020). The contribution of common genetic risk variants for ADHD to a general factor of childhood psychopathology. *Molecular Psychiatry, 25*(8), 1809–1821.

Briley, D. A., Domiteaux, M., & Tucker-Drob, E. M. (2014). Achievement-relevant personality: Relations with the big five and validation of an efficient instrument. *Learning and Individual Differences, 32,* 26–39.

Brinkmann, S. (2020). Psychology as a science of life. *Theory & Psychology, 30*(1), 3–17.

Briñol, P., Petty, R. E., Durso, G. O., & Rucker, D. D. (2017). Power and persuasion: Processes by which perceived power can influence evaluative judgments. *Review of General Psychology, 21*(3), 223–241. https://doi.org/10.1037/gpr0000119

Brito, R. G., Rasmussen, L. A., & Sluka, K. A. (2017). Regular physical activity prevents development of chronic muscle pain through modulation of supraspinal opioid and serotonergic mechanisms. *Pain Reports, 2*(5), e618.

Broberg, D. J., & Bernstein, I. L. (1987). Candy as a scapegoat in the prevention of food aversions in children receiving chemotherapy. *Cancer, 60*(9), 2344–2347.

Brody, G.H., Yu, T., & Shalev, I. (2017). Risky family processes prospectively forecast shorter telomere length. *Health Psychology, 36,* 438–444.

Brody, L. R. (1999). *Gender, emotion, and the family.* Harvard University Press.

Brody, N. (2007). Does education influence intelligence? In P. C. Kyllonen, R. D. Roberts, & L. Stankov (Eds.), *Extending intelligence.* Erlbaum.

Brooker, R. J., Widmaier, E. P., Graham, L., & Stiling, P. (2015). *Principles of Biology.* McGraw-Hill.

Brooks, D., Hulst, H. E., de Bruin, L., Glas, G., Geurts, J. J., & Douw, L. (2020). The multilayer network approach in the study of personality neuroscience. *Brain Sciences, 10*(12), 915.

Brooks, J. A., & Freeman, J. B. (2019). Neuroimaging of person perception: A social-visual interface. *Neuroscience Letters, 693,* 40–43.

Brooks, J. R., Hong, J. H., Madubata, I. J., Odafe, M. O., Cheref, S., & Walker, R. L. (2020, in press). The moderating effect of dispositional forgiveness on perceived racial discrimination and depression for African American adults. *Cultural Diversity and Ethnic Minority Psychology.*

Brown, A. (2017, June 13). Five key findings about LGBT Americans. Pew Research. http://www.pewresearch.org/fact-tank/2017/06/13/5-key-findings-about-lgbt-americans/

Brown, A. S. (2004). *The déjà vu experience.* Psychology Press.

Brown, A. S., & Marsh, E. J. (2010). Digging into déjà vu: Recent research findings on the possible mechanisms. In B. H. Ross (Ed.,) *The psychology of learning and motivation* (vol. 53, pp. 33–62). Academic Press.

Brown, D. (2007). Evidence-based hypnotherapy for asthma: A critical review. *International Journal of Clinical and Experimental Hypnosis, 55*(2), 220–249.

Brown, J. P., Martin, D., Nagaria, Z., Verceles, A. C., Jobe, S. L., & Wickwire, E. M. (2020). Mental health consequences of shift work: An updated review. *Current Psychiatry Reports, 22*(2), 7.

Brown, P., & Tierney, C. (2011). Media role in violence and the dynamics of bullying. *Pediatric Reviews, 32*(10), 453–454.

Brown, R. (1973). *A first language: The early stages.* Harvard University Press.

Brown, R., & LeDoux, J. (2020). Higher-order memory schema and consciousness experience. *Cognitive Neuropsychology, 37,* 213–215.

Brown, S. D., Lent, R. W., Telander, K., & Tramayne, S. (2011). Social cognitive career theory, conscientiousness, and work performance: A meta-analytic path analysis. *Journal of Vocational Behavior, 79*(1), 81–90.

Brown, S. L., Nesse, R. N., Vinokur, A. D., & Smith, D. M. (2003). Providing social support may be more beneficial than receiving it: Results from a prospective study of mortality. *Psychological Science, 14*(4), 320–327.

Brownell, C. A., Lemerise, E. A., Pelphrey, K. A., & Roisman, G. I. (2015). Measuring socioemotional behavior and development. In R. M. Lerner & M. E. Lamb (Ed.), *Handbook of child psychology and developmental science* (7th ed., vol. 3). Wiley.

Brownley, K. A., Berkman, N. D., Peat, C. M., Lohr, K. N., & Bulik, C. M. (2017). Binge-Eating disorder in adults. *Annals of Internal Medicine, 166*(3), 231–232.

Broz, F., & others. (2014). The ITALK Project: A developmental robotics approach to the study of individual, social, and linguistic learning. *Topics in Cognitive Science, 6*(3), 534–544.

Bruce, C. R., Unsworth, C. A., Dillon, M. P., Tay, R., Falkmer, T., Bird, P., & Carey, L. M. (2017). Hazard perception skills of young drivers with attention deficit hyperactivity disorder (ADHD) can be improved with computer based driver training: An exploratory randomised controlled trial. *Accident Analysis & Prevention, 109,* 70–77.

Bruck, M., & Ceci, S. J. (2012). Forensic developmental psychology in the courtroom. In D. Faust & M. Ziskin (Eds.), *Coping with psychiatric and psychological testimony.* Cambridge University Press.

Bruck, M., & Ceci, S. J. (2013). Expert testimony in a child sex abuse case: Translating memory development research. *Memory, 21*(5), 556–565.

Bruehl, S., Burns, J. W., Gupta, R., Buvanendran, A., Chont, M., Orlowska, D., . . . & France, C. R. (2017). Do resting plasma β-endorphin levels predict responses to opioid analgesics? *The Clinical Journal of Pain, 33*(1), 12–20.

Brug, J., Conner, M., Harré, N., Kremers, S., McKellar, S., & Whitelaw, S. (2005). The transtheoretical model and stages of change: A critique. Observations by five commentators on the paper by Adams, J. and White, M. (2004) Why don't stage-based activity promotion interventions work? *Health Education Research, 20*(2), 244–258.

Brumberg-Kraus, J. (2020). The role of ritual in eating. *Handbook of Eating and Drinking: Interdisciplinary Perspectives,* 333–348.

Brunelin, J., & others. (2014). The efficacy and safety of low frequency repetitive transcranial magnetic stimulation for treatment-resistant depression: The results from a large multicentre French RCT. *Brain Stimulation, 7*(6), 855–863.

Brunstein, J. C. (1993). Personal goals and subjective well-being: A longitudinal study. *Journal of Personality and Social Psychology, 65*(5), 1061–1070.

Brus, M., Keller, M., & Lévy, F. (2013). Temporal features of adult neurogenesis: Differences and similarities across mammalian species. *Frontiers in Neuroscience, 7,* 135.

Brust, V., & Guenther, A. (2017). Stability of the guinea pigs personality—Cognition—Linkage over time. *Behavioural Processes, 134,* 4–11. https://doi.org/10.1016/j.beproc.2016.06.009

Bryan, J. L., & Lu, Q. (2016). Vision for improvement: Expressive writing as an intervention for people with Stargardt's disease, a rare eye disease. *Journal of Health Psychology, 21*(5), 709–719. https://doi.org/10.1177/1359105314536453

Bryan, S. M. (2012, January 28). Wolves' senses tapped to keep them clear of cattle. Associated Press. http://bangordailynews.com/2012/01/28/outdoors/wolves-senses-tapped-to-keep-them-clear-of-cattle/

Budge, S. L., Sinnard, M. T., & Hoyt, W. T. (2020). Longitudinal effects of psychotherapy with transgender and nonbinary clients: A randomized controlled pilot trial. *Psychotherapy.* https://doi.org/10.1037/pst0000310

Buhidma, Y., Rukavina, K., Chaudhuri, K. R., & Duty, S. (2020). Potential of animal models for advancing the understanding and treatment of pain in Parkinson's disease. *NPJ Parkinson's Disease, 6*(1), 1–7.

Bukowski, W. M., Brendgen, M., & Vitaro, F. (2007). Peers and socialization: Effects on externalizing and internalizing problems. In J. E. Grusec & P. D. Hastings (Eds.), *Handbook of socialization: Theory and research* (pp. 355–381). Guilford.

Bull, C., & Varghese, F. P. (2020). Human Genome Project and personality. In B. J. Carducci, C. S. Nave, J. S. Mio, & R. E. Riggio (Eds.), *The Wiley encyclopedia of personality and individual differences: Measurement and assessment* (pp. 403–407). John Wiley & Sons.

Bullock, J. L., Lockspeiser, T., del Pino-Jones, A., Richards, R., Teherani, A., & Hauer, K. E. (2020). They don't see a lot of people my color: A mixed methods study of racial/ethnic stereotype threat among medical students on core clerkships. *Academic Medicine, 95*(11), S58–S66.

Burger, J. (2009). Replicating Milgram: Would people still obey today? *American Psychologist, 64*(1), 1–11.

Burger, J. M. (2017). *Obedience.* Oxford University Press.

Burgos, J. E. (2018). Is a nervous system necessary for learning? *Perspectives on Behavior Science, 41*(2), 343–368.

Burke, B. L., Martens, A., & Faucher, E. H. (2010). Two decades of terror management theory: A meta-analysis of mortality salience research. *Personality and Social Psychology Review, 14*(2), 155–195.

Burman, A., Garcia-Milian, R., & Whirledge, S. (2020). Gene X environment: The cellular environment governs the transcriptional response to environmental chemicals. *Human Genomics, 14,* 1–14.

Burnette, J. L., Hoyt, C. L., Russell, V. M., Lawson, B., Dweck, C. S., & Finkel, E. (2020). A growth mind-set intervention improves interest but not academic performance in the field of computer science. *Social Psychological and Personality Science, 11*(1), 107–116.

Burns, T., Kisely, S., & Rugkåsa, J. (2017). Randomised controlled trials and outpatient commitment. *The Lancet Psychiatry, 4*(12), 843–845.

Burris, M., Miller, E., Romero-Daza, N., & Himmelgreen, D. (2020). Food insecurity and age at menarche in Tampa Bay, Florida. *Ecology of Food and Nutrition, 59,* 346–366.

Burton, C. M., & King, L. A. (2004). The health benefits of writing about peak experiences. *Journal of Research in Personality, 38*(2), 150–163.

Burton, C. M., & King, L. A. (2008). The effects of (very) brief writing on health: The 2-minute miracle. *British Journal of Health Psychology, 13*(1), 9–14.

Bushman, B. J., Wang, M. C., & Anderson, C. A. (2005). Is the curve relating temperature to aggression linear or curvilinear? Assaults and temperature in Minneapolis reexamined. *Journal of Personality and Social Psychology, 89*(1), 62–66.

Bushman, L., & Huesmann, L. R. (2012). Effects of media violence on aggression. In D. G. Singer & J. L. Singer (Eds.), *Handbook of children and the media* (2nd ed.). Sage.

Buss, D. M. (2015). *Evolutionary psychology* (5th ed.). Pearson.

Buss, D. M. (2016). *The Handbook of Evolutionary Psychology* (2nd Ed.). Wiley.

Buss, D. M. (2020). Evolutionary psychology is a scientific revolution. *Evolutionary Behavioral Sciences, 14*(4), 316–323.

Buss, D. M., Durkee, P. K., Shackelford, T. K., Bowdle, B. F., Schmitt, D. P., Brase, G. L., . . . Trofimova, I. (2020, in press). Human status criteria: Sex differences and similarities across 14 nations. *Journal of Personality and Social Psychology.*

Buss, D. M., & Foley, P. (2020). Mating and marketing. *Journal of Business Research, 120,* 492–497.

Butcher, J. N., Hooley, J. M., & Mineka, S. M. (2013). *Abnormal psychology* (16th ed.). Pearson.

Butler, A. C., Chapman, J. E., Forman, E. M., & Beck, A. T. (2006). The empirical status of cognitive-behavioral therapy: A review of meta-analyses. *Clinical Psychology Review, 26*(1), 17–31.

Butler, R. M., O'Day, E. B., Swee, M. B., Horenstein, A., & Heimberg, R. G. (2020). Cognitive behavioral therapy for social anxiety disorder: Predictors of treatment outcome in a quasi-naturalistic setting. *Behavior Therapy.* https://doi.org/10.1016/j.beth.2020.06.002

Butter, M., & Knight, P. (Eds.). (2020). *Routledge handbook of conspiracy theories.* Routledge.

Butt, Y. M., Smith, M. L., Tazelaar, H. D., Vaszar, L. T., Swanson, K. L., Cecchini, M. J., . . . Khoor, A. (2019). Pathology of vaping-associated lung injury. *New England Journal of Medicine, 381*(18), 1780–1781.

Buzinde, C. N. (2020). Theoretical linkages between well-being and tourism: The case of self-determination theory and spiritual tourism. *Annals of Tourism Research, 83,* 102920.

Byrne, A., & Preston, C. A. (2019). Mr Fantastic meets the Invisible Man: An illusion of invisible finger stretching. *Perception, 48*(2), 185–188.

Byrne, P., & James, A. (2020). Placing poverty-inequality at the centre of psychiatry. *BJPsych Bulletin, 44*(5), 187–190.

Bzdok, D., & Dunbar, R. I. M. (2020). The neurobiology of social distance. Trends in Cognitive Sciences. https://www.cell.com/trends/cognitive-sciences/pdf/S1364-6613(20)30140-6.pdf

C

Cabrera, I., Brugos, D., & Montorio, I. (2020). Attentional biases in older adults with generalized anxiety disorder. *Journal of Anxiety Disorders*. https://doi.org/10.1016/j.janxdis.2020.102207

Caceres, V. (2020, February 14). Eating disorders statistics. *U.S. News & World Report*. https://health.usnews.com/conditions/eating-disorder/articles/eating-disorder-statistics

Cadaret, M. C., Hartung, P. J., Subich, L. M., & Weigold, I. K. (2017). Stereotype threat as a barrier to women entering engineering careers. *Journal of Vocational Behavior, 99*, 40–51. https://doi.org/10.1016/j.jvb.2016.12.002

Cadete, D., & Longo, M. R. (2020). A continuous illusion of having a sixth finger. *Perception, 49*(8), 807–821.

Cadoni, C., Pisanu, A., Simola, N., Frau, L., Porceddu, P. F., Corongiu, S., et al. (2017). Widespread reduction of dopamine cell bodies and terminals in adult rats exposed to a low dose regimen of MDMA during adolescence. *Neuropharmacology, 123*, 385–394.

Cahn, L. (2020, May 11). Service dogs who saved the life of veterans. RD.com. https://www.rd.com/list/service-dogs-who-saved-the-life-of-veterans/

Callaghan, T., & Corbit, J. (2018). Early prosocial development across cultures. *Current Opinion in Psychology, 20*, 102–106.

Callahan, J. L., Bell, D. J., Davila, J., Johnson, S. L., Strauman, T. J., & Yee, C. M. (2020). The enhanced examination for professional practice in psychology: A viable approach? *American Psychologist, 75*(1), 52–65.

Callaway, E. (2020, November 30). "It will change everything": DeepMind's AI makes gigantic leap in solving protein structures. *Nature*. https://www.nature.com/articles/d41586-020-03348-4

Calma, A., & Davies, M. (2020). Critical thinking in business education: Current outlook and future prospects. *Studies in Higher Education*. https://doi.org/10.1080/03075079.2020.1716324

Calmes, J. (2012, May 23). When a boy found a familiar feel in a pat of the head of state. *The New York Times*. www.nytimes.com/2012/05/24/us/politics/indelible-image-of-a-boys-pat-on-obamas-head-hangs-in-white-house.html

Cameron, C. A., McKay, S., Susman, E. J., Wynne-Edwards, K., Wright, J. M., & Weinberg, J. (2017). Cortisol stress response variability in early adolescence: Attachment, affect and sex. *Journal of Youth and Adolescence, 46*(1), 104–120.

Cameron, R. (2016). The paranormal as an unhelpful concept in psychotherapy and counselling research. *European Journal of Psychotherapy and Counselling, 18*(2), 142–155. https://doi.org/10.1080/13642537.2016.1170060

Campbell, J. D., Yoon, D. P., & Johnstone, B. (2009). Determining relationships between physical health and spiritual experience, religious practice, and congregational support in a heterogeneous sample. *Journal of Religion and Health, 21*(2), 127–136.

Campbell, L., Campbell, B., & Dickinson, D. (2004). *Teaching and learning through multiple intelligences*. Allyn & Bacon.

Campbell, L., Overall, N., Rubin, H., & Lackenhauer, S. D. (2013). Inferring a partner's ideal discrepancies: Accuracy, projection, and the communicative role of interpersonal behavior. *Journal of Personality and Social Psychology, 105*(2), 217–233.

Campione-Barr, N., Lindell, A. K., & Giron, S. E. (2020). Developmental changes in parental authority legitimacy and over-time associations with adjustment: Differences in parent, first-born, and second-born perspectives. *Developmental Psychology, 56*(5), 951–969.

Campos, L., Bernardes S., & Godinho C. (2020). Food as a way to convey masculinities: How conformity to hegemonic masculinity norms influences men's and women's food consumption. *Journal of Health Psychology, 25*(12):1–15.

Campos, R. C., Mesquita, I., Besser, A., & Blatt, S. J. (2014). Neediness and depression in women. *Bulletin of the Menninger Clinic, 78*(1), 16–33.

Cannet, C., Pilotto, A., Rocha, J. C., Schäfer, H., Spraul, M., Berg, D., . . . Piel, D. (2020). Lower plasma cholesterol, LDL-cholesterol and LDL-lipoprotein subclasses in adult phenylketonuria (PKU) patients compared to healthy controls: Results of NMR metabolomics investigation. *Orphanet Journal of Rare Diseases, 15*(1), 1–7.

Cannon, W. B. (1927). The James-Lange theory of emotions: A critical examination and an alternative theory. *American Journal of Psychology, 39*, 106–124.

Cans, A. S. U., Wang, Y., Fathali, H. M., Mishra, D., Olsson, T., Keighron, J., & Skibicka, K. (2020). Counting the number of glutamate molecules in single synaptic vesicles. *Biophysical Journal, 118*(3), 454a.

Cansino, S., Torres-Trejo, F., Estrada-Manilla, C., Pérez-Loyda, M., Vargas-Martínez, C., Tapia-Jaimes, G., & Ruiz-Velasco, S. (2020). Contributions of cognitive aging models to the explanation of source memory decline across the adult lifespan. *Experimental Aging Research, 46*(3), 194–213.

Cantarero, K., Gamian-Wilk, M., & Dolinski, D. (2017). Being inconsistent and compliant: The moderating role of the preference for consistency in the door-in-the-face technique. *Personality and Individual Differences, 11*, 554–557. https://doi.org/10.1016/j.paid.2016.07.005

Cantor, D., Fisher, B., Chibnall, S., Harps, S., Townsend, R., Thomas, G., . . . Madden, K. (2019). *Report on the AAU campus climate survey on sexual assault and misconduct*. The Association of American Universities.

Cantor, N., & Sanderson, C. A. (1999). Life task participation and well-being: The importance of taking part in daily life. In D. Kahneman, E. Diener, & N. Schwarz (Eds.), *Well-being: The foundations of hedonic psychology* (pp. 230–243). Russell Sage Foundation.

Cao, H., Harneit, A., Walter, H., Erk, S., Braun, U., Moessnang, C., . . . & Romanczuk-Seiferth, N. (2018). The 5-HTTLPR polymorphism affects network-based functional connectivity in the visual-limbic system in healthy adults. *Neuropsychopharmacology, 43* (2), 406–414.

Caprara, G. V., Fagnani, C., Alessandri, G., Steca, P., Gigantesco, A., Cavalli Sforza, L. L., & Stazi, M. A. (2009). Human optimal functioning: The genetics of positive orientation towards self, life, and the future. *Behavior Genetics, 39*(3), 277–284.

Cardno, A. G., & Gottesman, I. I. (2000). Twin studies of schizophrenia: From bow-and-arrow concordances to Star Wars Mx and functional genomics. *American Journal of Medical Genetics, 97*(1), 12–17.

Cardozo, M. J., Almuedo-Castillo, M., & Bovolenta, P. (2020). Patterning the vertebrate retina with morphogenetic signaling pathways. *The Neuroscientist, 26*(2), 185–196.

Carey, B. (2009, August 9). After injury, fighting to regain a sense of self. *New York Times*. www.nytimes.com/2009/08/09/health/research/09brain.html

Carey, B. (2011). Finding purpose after living with delusion. *The New York Times*. www.nytimes.com/2011/11/26/health/man-uses-his-schizophrenia-to-gather-clues-for-daily-living.html?pagewanted=all

Carlier, M. E. M., & Harmony, T. (2020). Development of auditory sensory memory in preterm infants. *Early Human Development, 145*, 105045.

Carlson, K. A., Winquist, J. R. (2018). *An Introduction to Statistics: An Active Learning Approach (*2nd ed.). Sage.

Carlson, S. M., Shoda, Y., Ayduk, O., Aber, L., Schaefer, C., Sethi, A., . . . & Mischel, W. (2018). Cohort effects in children's delay of gratification. *Developmental Psychology, 54* (8), 1395–1407.

Carman, C. A. (2011). Stereotypes of giftedness in current and future educators. *Journal for the Education of the Gifted, 34*(5), 790–812.

Carney, D. R., Colvin, C. R., & Hall, J. A. (2007). A thin slice perspective on the accuracy of first impressions. *Journal of Research in Personality, 41*(5), 1054–1072.

Carney, E. F. (2017). The psychology of extraordinary altruism. *Nature Reviews Nephrology, 13*, 383.

Carney, J. (2020). Culture and mood disorders: The effect of abstraction in image, narrative and film on depression and anxiety. *Medical Humanities, 46*(4), 430–443.

Carothers, B. J., & Reis, H. T. (2013). Men and women are from earth: Examining the latent structure of gender. *Journal of Personality and Social Psychology, 104*(2), 385–407.

Carr, A., Cullen, K., Keeney, C., Canning, C., Mooney, O., Chinseallaigh, E., & O'Dowd, A. (2020). Effectiveness of positive psychology interventions: A systematic review and meta-analysis. *The Journal of Positive Psychology, 25*(12), 1–21.

Carrard, I., Van der Linden, M., & Golay, A. (2012). Comparison of obese and nonobese individuals with binge eating disorder: Delicate boundary between binge eating disorder and non-purging bulimia nervosa. *European Eating Disorders Review, 20*(5), 350–354.

Carré, J. M., & Robinson, B. A. (2020). Testosterone administration in human social neuroendocrinology: Past, present, and future. *Hormones and Behavior, 122*, 104754.

Carretier, E., Grau, L., Mansouri, M., Moro, M. R., & Lachal, J. (2020). Qualitative assessment of transcultural psychotherapy by adolescents and their migrant families: Subjective experience and perceived effectiveness. *PloS One, 15*(8), e0237113.

Carroll, J. L. (2013). *Sexuality now* (4th ed.). Cengage.

Carroll, K., Pottinger, A. M., Wynter, S., & DaCosta, V. (2020). Marijuana use and its influence on sperm morphology and motility: identified risk for fertility among Jamaican men. *Andrology, 8*(1), 136–142.

Carskadon, M. A. (2020). The time has come to expand our studies of school timing for adolescents. *Journal of Biological Rhythms*. https://doi.org/10.1177/0748730420940080

Carskadon, M. A., & Barker, D. H. (2020). Editorial perspective: Adolescents' fragile sleep—shining light on a time of risk to mental health. *Journal of Child Psychology and Psychiatry*. https://doi.org/10.1111/jcpp.13275

Carstensen, L. L. (2006). The influence of a sense of time on human development. *Science, 312*(5782), 1913–1915.

Carstensen, L. L., & others. (2011). Emotional experience improves with age: Evidence based on over 10 years of experience sampling. *Psychology and Aging, 26*(1), 21–33.

Carter, A. R., McAvoy, M. P., Siegel, J. S., Hong, X., Astafiev, S. V., Rengachary, J., et al. (2017). Differential white matter involvement associated with distinct visuospatial deficits after right hemisphere stroke. *Cortex, 88*, 81–97.

Carter, O., van Swinderen, B., Leopold, D., Collin, S., & Maier, A. (2020). Perceptual rivalry across animal species. *Journal of Comparative Neurology*. https://doi.org/10.1002/cne.24939

Carter, R. (1998). *Mapping the mind*. Berkeley: University of California Press.

Caruso, J. P., & Sheehan, J. P. (2017). Psychosurgery, ethics, and media: a history of Walter Freeman and the lobotomy. *Neurosurgical Focus, 43*(3), E6.

Caruso, S., Mauro, D., Scalia, G., Palermo, C. I., Rapisarda, A. M. C., & Cianci, A. (2018). Oxytocin plasma levels in orgasmic and anorgasmic women. *Gynecological Endocrinology, 34*(1), 69–72.

Carver, C. S., & Scheier, M. F. (2014). Dispositional optimism. *Trends in Cognitive Science, 18*(6), 293–299.

Case, P., Sparks, P., & Pavey, L. (2016). Identity appropriateness and the structure of the theory of planned behaviour. *British Journal of Social Psychology, 55*(1), 109–125. https://doi.org/10.1111/bjso.12115

Cásedas, L., Pirruccio, V., Vadillo, M. A., & Lupiáñez, J. (2020). Does mindfulness meditation training enhance executive control? A systematic review and meta-analysis of randomized controlled trials in adults. *Mindfulness, 11*, 411–424.

Caserta, T. A., Punamäki, R., & Pirttilä Backman, A. (2017). The buffering role of social support on the psychosocial wellbeing of orphans in Rwanda. *Social Development, 26*(2), 402–422. https://doi.org/10.1111/sode.12188

Casey, B. J., & others. (2011). Behavioral and neural correlates of delay of gratification 40 years later. *Proceedings of the National Academy of Sciences USA, 108*(36), 14998–15003.

Casey, E. A., Lindhorst, T., & Storer, H. L. (2017). The situational-cognitive model of adolescent bystander behavior: Modeling bystander decision-making in the context of bullying and teen dating violence. *Psychology of Violence, 7*(1), 33–44. https://doi.org/10.1037/vio0000033

Cash, R. F., Cocchi, L., Lv, J., Fitzgerald, P. B., & Zalesky, A. (2020). Functional magnetic resonance imaging-guided personalization of transcranial magnetic stimulation treatment for depression. *JAMA Psychiatry*. https://doi.org/10.1001/jamapsychiatry.2020.3794

Cash, R., Udupa, K., Gunraj, C., Mazzella, F., Daskalakis, Z. J., Wong, A. H., et al. (2017). Influence of the BDNF Val66Met polymorphism on the balance of excitatory and inhibitory neurotransmission and relationship to plasticity in human cortex. *Brain Stimulation: Basic, Translational, and Clinical Research in Neuromodulation, 10*(2), 502.

Casper, D. M., Card, N. A., & Barlow, C. (2020). Relational aggression and victimization during adolescence: a meta-analytic review of unique associations with popularity, peer acceptance, rejection, and friendship characteristics. *Journal of Adolescence, 80*, 41–52.

Caspi, A., & others. (2002). Role of genotype in the cycle of violence in maltreated children. *Science, 297*(5582), 851–854.

Caspi, A., Sugden, K., Moffitt, T. E., Taylor, A., Craig, I. W., Harrington, H., . . . Poulton, R. (2003). Influence of life stress on depression: Moderation by a polymorphism in the 5-HTT gene. *Science, 301*(5631), 386–389. https://doi.org/10.1126/science.1083968

Cassidy, B. S., Hughes, C., & Krendl, A. C. (2020). Age differences in neural activity related to mentalizing during person perception. *Aging, Neuropsychology, and Cognition.* https://doi.org/10.1080/13825585.2020.1718060

Cathers-Schiffman, T. A., & Thompson, M. S. (2007). Assessment of English- and Spanish-speaking students with the WISC-III and Leiter-R. *Journal of Psychoeducational Assessment, 25*(1), 41–52.

Catricalà, E., Conca, F., Fertonani, A., Miniussi, C., & Cappa, S. F. (2020). State-dependent TMS reveals the differential contribution of ATL and IPS to the representation of abstract concepts related to social and quantity knowledge. *Cortex, 123*, 30–41.

Cavadel, E. W., & Frye, D. A. (2017). Not just numeracy and literacy: Theory of mind development and school readiness among low-income children. *Developmental Psychology, 53*(12), 2290-2303. https://doi.org/10.1037/dev0000409

Cavazza, N., Guidetti, M., & Butera, F. (2017). Portion size tells who I am, food type tells who you are: Specific functions of amount and type of food in same- and opposite-sex dyadic eating contexts. *Appetite, 112,* 96–101. https://doi.org/10.1016/j.appet.2017.01.019

Caverzasi, E., & others. (2008). Complications in major depressive disorder therapy: A review of magnetic resonance spectroscopy studies. *Functional Neurology, 23*(3), 129–132.

CDC. (2020a). America's drug overdose epidemic: Putting data to action. https://www.cdc.gov/injury/features/prescription-drug-overdose/index.html

CDC. (2020b). Cigarette and tobacco use. https://www.cdc.gov/tobacco/

Centers for Disease Control and Prevention (CDC). (2017). *Health Effects of Cigarette Smoking.* https://www.cdc.gov/tobacco/data_statistics/fact_sheets/health_effects/effects_cig_smoking/index.htm

Centers for Disease Control and Prevention (CDC). (2018) How does loud noise cause hearing loss? https://www.cdc.gov/nceh/hearing_loss/how_does_loud_noise_cause_hearing_loss.html

Centers for Disease Control and Prevention (CDC). (2018a). Table 21. Selected health conditions and risk factors, by age: United States, selected years 1988–1994 through 2015–2016. https://www.cdc.gov/nchs/data/hus/2018/021.pdf

Centers for Disease Control and Prevention (CDC). (2018b). Pre-exposureprophylaxis. https://www.hiv.gov/hiv-basics/hiv-prevention/using-hiv-medication-to-reduce-risk/pre-exposure-prophylaxis

Centers for Disease Control and Prevention (CDC). (2020b, March 13). Screening and diagnosis of autism spectrum disorder. https://www.cdc.gov/ncbddd/autism/screening.html

Cha, J., Guffanti, G., Gingrich, J., Talati, A., Wickramaratne, P., Weissman, M., & Posner, J. (2018). Effects of serotonin transporter gene variation on impulsivity mediated by default mode network: A family study of depression. *Cerebral Cortex, 28*(6), 1911–1921. https://doi.org/10.1093/cercor/bhx097

Chae, D. H., Clouston, S., Hatzenbuehler, M. L., Kramer, M. R., Cooper, H. L. F., Wilson, S. M., . . . Link, B. (2015). Association between an internet-based measure of area racism and black mortality. *PLoS ONE, 10*(4), e0122963. https://doi.org/10.1371/journal.pone.0122963

Chakawa, A., Frye, W., Travis, J., & Brestan-Knight, E. (2020). Parent-child interaction therapy: Tailoring treatment to meet the sociocultural needs of an adoptive foster child and family. *Journal of Family Social Work, 23*(1), 53–70.

Chaladze, G. (2016). Heterosexual male carriers could explain persistence of homosexuality in men: Individual-based simulations of an X-linked inheritance model. *Archives of Sexual Behavior, 45*(7), 1705–1711. https://doi.org/10.1007/s10508-016-0742-2

Chambless, D. L. (2002). Beware the dodo bird: The dangers of overgeneralization. *Clinical Psychology: Science and Practice, 9*(1), 13–16.

Chan, C. M. H., Ng, C. G., Taib, N. A., Wee, L. H., Krupat, E., & Meyer, F. (2018). Course and predictors of post traumatic stress disorder in a cohort of psychologically distressed patients with cancer: A 4 year follow up study. *Cancer 124*(2), 406–416.

Chandra, A., Mosher, W. D., Copen, C., & Sionean, C. (2011). Sexual behavior, sexual attraction, and sexual identity in the United States: Data from the 2006–2008 National Survey of Family Growth. *National Health Statistics Reports,* No. 36. www.cdc.gov/nchs/data/nhsr/nhsr036.pdf (accessed December 21, 2014)

Chaney, C., Lopez, M., Wiley, K. S., Meyer, C., & Valeggia, C. (2019). Systematic review of chronic discrimination and changes in biology during pregnancy among African American women. *Journal of Racial and Ethnic Health Disparities, 6*(6), 1208–1217.

Chang, A. K., Bijur, P. E., & Esses, D., Barnaby, D. F., & Baer, J. (2017). Effect of a single dose of oral opioid and nonopioid analgesics on acute extremity pain in the emergency department. A randomized clinical trial. *Journal of the American Medical Association, 318,* 1661–1667. https://doi.org/10.1001/jama.2017.16190

Chang, M. X. L., Jetten, J., Cruwys, T., & Haslam, C. (2017). Cultural identity and the expression of depression: A social identity perspective. *Journal of Community & Applied Social Psychology, 27*(1), 16–34.

Chapman, R. (2020). Neurodiversity, disability, wellbeing. In H. Bertilsdotter Rosqvist, N. Chown, & A. Stenning (Eds.), *Neurodiversity studies: A new critical paradigm.* Routledge.

Chatters, L. M., Taylor, R. J., Woodward, A. T., & Nicklett, E. J. (2014). Social support from church and family members and depressive symptoms among older African Americans. *American Journal of Geriatric Psychiatry.* https://doi.org/10.1016/j.jagp.2014.04.008

Chaudhry, S. (2020). The modification of aging process. *Annals of Geriatric Education and Medical Sciences, 7*(1), 1–8.

Chaves, J. F. (2000). Hypnosis. In A. Kazdin (Ed.), *Encyclopedia of psychology.* American Psychological Association and Oxford University Press.

Chen, C., Hewitt, P. L., & Flett, G. L. (2017). Ethnic variations in other-oriented perfectionism's associations with depression and suicide behaviour. *Personality and individual differences, 104*, 504–509.

Chen, F. R., Rothman, E. F., & Jaffee, S. R. (2017). Early puberty, friendship group characteristics, and dating abuse in US girls. *Pediatrics*, e20162847.

Chen, L., Hudaib, A. R., Hoy, K. E., & Fitzgerald, P. B. (2020). Efficacy, efficiency and safety of high-frequency repetitive transcranial magnetic stimulation applied more than once a day in depression: A systematic review. *Journal of Affective Disorders, 277*, 986–996

Chen, L., Mishra, A., Newton, A. T., Morgan, V. L., Stringer, E. A., Rogers, B. P., & Gore, J. C. (2011). Fine-scale functional connectivity in somatosensory cortex revealed by high-resolution fMRI. *Magnetic Resonance Imaging, 29*(10), 1330–1337.

Chen, P., & Jacobson, K. C. (2012). Developmental trajectories of substance use from early adolescence to young adulthood: Gender and racial/ethnic differences. *Journal of Adolescent Health, 50*(2), 154–163.

Chen, S., Bagrodia, R., Pfeffer, C. C., Meli, L., & Bonanno, G. A. (2020). Anxiety and resilience in the face of natural disasters associated with climate change: *A review and methodological critique. Journal of Anxiety Disorders, 76*, 102297.

Chen, S. X., Benet-Martinez, V., & Ng, J. C. (2014). Does language affect personality perception? A functional approach to testing the Whorfian hypothesis. *Journal of Personality, 82*(2), 130–143.

Chen, S. X., Mak, W. W., & Lam, B. C. (2020). Is it cultural context or cultural value? Unpackaging cultural influences on stigma toward mental illness and barrier to help-seeking. *Social Psychological and Personality Science.* https://doi.org/10.1177%2F1948550619897482

Chen, X., Latham, G.P., Piccolo, R.F., & Itzchakov, G. (2020). An enumerative review and a meta-analysis of primed goal effects on organizational behavior. *Applied Psychology: An International Review.* https://doi.org/10.1111/apps.12239.

Chen, Y., Shen, J., Ke, K., & Gu, X. (2020). Clinical potential and current progress of mesenchymal stem cells for Parkinson's disease: A systematic review. *Neurological Sciences, 41*, 1051–1061.

Cheng, C. H., Niddam, D. M., Hsu, S. C., Liu, C. Y., & Tsai, S. Y. (2017). Resting GABA concentration predicts inhibitory control during an auditory Go-Nogo task. *Experimental Brain Research, 12,* 3833–3841.

Cheng, E. R., & Carroll, A. E. (2020). Delaying school start times to improve population health. *JAMA Pediatrics, 174*(7):641–643. https://doi.org/10.1001/jamapediatrics.2020.0351

Cheng, M. F. (2017). Adult neurogenesis in injury-induced self-repair: Use it or lose it. *Brain Plasticity, 2*(2), 115–126.

Cheng, W., Rolls, E. T., Gu, H., Zhang, J., & Feng, J. (2015). Autism: reduced connectivity between cortical areas involved in face expression, theory of mind, and the sense of self. *Brain, 138*(5), 1382–1393.

Cheong, J. L., Burnett, A. C., Treyvaud, K., & Spittle, A. J. (2020). Early environment and long-term outcomes of preterm infants. *Journal of Neural Transmission, 127*(1), 1–8.

Chéreau, R., Bawa, T., Fodoulian, L., Carleton, A., Pagès, S., & Holtmaat, A. (2020). Dynamic perceptual feature selectivity in primary somatosensory cortex upon reversal learning. *Nature Communications, 11*(1), 1–14.

Cherkasova, M. V., Faridi, N., Casey, K. F., Larcher, K., O'Driscoll, G. A., Hechtman, L., et al. (2017). Differential associations between cortical thickness and striatal dopamine in treatment-naïve adults with ADHD vs. healthy controls. *Frontiers in Human Neuroscience, 11*, 421.

Chess, S., & Thomas, A. (1977). Temperamental individuality from childhood to adolescence. *Journal of Child Psychiatry, 16*(2), 218–226.

Chevallier, C., Kohls, G., Troiani, V., Brodkin, E. S., & Schultz, R. T. (2012). The social motivation theory of autism. *Trends in Cognitive Sciences, 16*, 231–239.

Chivers, M. L. (2017). The specificity of women's sexual response and its relationship with sexual orientations: A review and ten hypotheses. *Archives of Sexual Behavior, 46*(5), 1161–1179.

Chivers, M. L., & Brotto, L. A. (2017). Controversies of women's sexual arousal and desire. *European Psychologist, 22*(1), 5–26. https://doi.org/10.1027/1016-9040/a000274

Chivers, M. L., Seto, M. C., & Blanchard, R. (2007). Gender and sexual orientation differences in sexual response to sexual activities versus gender of actors in sexual films. *Journal of Personality and Social Psychology, 93*(6), 1108–1121.

Cho, H., Gonzalez, R., Lavaysse, L. M., Pence, S., Fulford, D., & Gard, D. E. (2017). Do people with schizophrenia experience more negative emotion and less positive emotion in their daily lives? A meta-analysis of experience sampling studies. *Schizophrenia Research, 183,* 49–55.

Choi, H., Kensinger, E. A., & Rajaram, S. (2017). Mnemonic transmission, social contagion, and emergence of collective memory: Influence of emotional valence, group structure, and information distribution. *Journal of Experimental Psychology: General, 146*(9), 1247–1265. https://doi.org/10.1037/xge0000327

Choi, J. H., Chung, K., & Park, K. (2013). Psychosocial predictors of four health-promoting behaviors for cancer prevention using the stage of change of Transtheoretical Model. *Psycho-Oncology, 22*(10), 2253–2261.

Choi, J., Lisanby, S. H., Medalia, A., & Prudic, J. (2011). A conceptual introduction to cognitive remediation for memory deficits associated with right unilateral electroconvulsive therapy. *Journal of Electroconvulsive Therapy, 27*(4), 286–291.

Choi, K. Y., Choo, J. M., Lee, Y. J., Lee, Y., Cho, C. H., Kim, S. H., & Lee, H. J. (2020). Association between the IL10 rs1800896 polymorphism and tardive dyskinesia in schizophrenia. *Psychiatry Investigation, 17*(10), 1031–1036.

Choi-Kain, L. W., Finch, E. F., Masland, S. R., Jenkins, J. A., & Unruh, B. T. (2017). What works in the treatment of borderline personality disorder. *Current Behavioral Neuroscience Reports, 4*(1), 21–30.

Chomsky, N. (1975). *Reflections on language.* Pantheon.

Chorney, Saryn. (2017, October 19). Lulu the CIA's Bomb-Sniffing Dog Fails Out of School; Just Wasn't That Into It. *People Magazine.* http://people.com/pets/bomb-sniffing-cia-dog-fails-school/

Chow, G., Bird, M., Gabana, N., Cooper, B., & Swanbrow Becker, M. (2020). A program to reduce stigma toward mental illness and promote mental health literacy and help-seeking in NCAA Division I student-athletes. *Journal of Clinical Sport Psychology.* https://doi.org/10.1123/jcsp.2019-0104

Chowdhury, T., Bindu, B., Singh, G. P., & Schaller, B. (2017). Sleep disorders: Is the trigemino-cardiac reflex a missing link? *Frontiers in Neurology, 8.*

Christensen, J., Grønborg, T.K., Sørensen, M.J., Schendel, D., Parner, E.T., Pedersen, L.H., & Vestergaard, M. (2013). Prenatal valproate exposure and risk of autism spectrum disorders and childhood autism. *JAMA, 309*(16):1696–1703. https://doi.org/10.1001/jama.2013.2270.

Christian, T. M. (2017). An appearance of the collective unconscious in a military mental health clinic. *Psychological Perspectives, 60*(2), 162–169.

Christiansen, J., Larsen, F. B., & Lasgaard, M. (2016). Do stress, health behavior, and sleep mediate the association between loneliness and adverse health conditions among older people? *Social Science & Medicine, 15,* 280–286. https://doi.org/10.1016/j.socscimed.2016.01.020

Christopherson, E. R., & VanScoyoc, S. M. (2013). *Treatments that work with children: Empirically supported strategies for managing childhood problems* (2nd ed.). American Psychological Association.

Chu, J., & Leino, A. (2017). Advancement in the maturing science of cultural adaptations of evidence-based interventions. *Journal of Consulting and Clinical Psychology, 85*(1), 45–57.

Chuang, Y., & Wu, P. (2017). Kin altruism: Testing the predictions of evolutionary theories in Taiwan. *Evolutionary Behavioral Sciences.* https://doi.org/10.1037/ebs0000094

Chubar, V., Van Leeuwen, K., Bijttebier, P., Van Assche, E., Bosmans, G., Van den Noortgate, W., . . . Claes, S. (2020). Gene–environment interaction: New insights into perceived parenting and social anxiety among adolescents. *European Psychiatry, 63*(1) e64.

Chun, S., Harris, A., Carrion, M., Rojas, E., Stark, S., Lejuez, C., et al. (2017). A psychometric investigation of gender differences and common processes across borderline and antisocial personality disorders. *Journal of Abnormal Psychology, 126*(1), 76–88.

Chung, G. H., Lee, J., & Gonzales-Backen, M. (2020). Context of biethnic acceptance, biethnic affirmation, and life satisfaction among Korean biethnic adolescents. *Journal of Adolescence, 80,* 242–253.

Chung, J., Schriber, R. A., & Robins, R. W. (2016). Positive illusions in the academic context: A longitudinal study of academic self-enhancement in college. *Personality and Social Psychology Bulletin, 42*(10), 1384–1401. https://doi.org/10.1177/0146167216662866

Cialdini, R. (2016). *Pre-suasion: A revolutionary way to influence and persuade.* Simon & Schuster.

Cialdini, R. B. (1991). Altruism or egoism? That is (still) the question. *Psychological Inquiry, 2*(2), 124–126.

Cialdini, R. B. (1993). *Influence: Science and practice.* HarperCollins.

Cialdini, R. B., Vincent, J. E., Lewis, S. K., Catalan, J., Wheeler, D., & Darby, B. L. (1975). Reciprocal concessions procedure for inducing compliance: The door-in-the-face technique. *Journal of Personality and Social Psychology, 31*(2), 206–215.

Ciao, A. C., Accurso, E. C., Fitzsimmons-Craft, E. E., Lock, J., & Le Grange, D. (2015). Family functioning in two treatments for adolescent anorexia nervosa. *International Journal of Eating Disorders, 48*(1), 81–90.

Ciardelli, L. E., Weiss, A., Powell, D. M., & Reiss, D. (2017). Personality dimensions of the captive California sea lion (*Zalophus californianus*). *Journal of Comparative Psychology, 131*(1), 50–58. https://doi.org/10.1037/com0000054

Cioffi, C.E., Levitsky, D.A., Pacanowski, C.R., & Bertz, F. (2015). A nudge in a healthy direction. The effect of nutrition labels on food purchasing behaviors in university dining facilities. *Appetite, 92,* 7–14. https://doi.org/10.1016/j.appet.2015.04.053

Ciotu, C. I., Tsantoulas, C., Meents, J., Lampert, A., McMahon, S. B., Ludwig, A., & Fischer, M. J. (2019). Noncanonical ion channel behaviour in pain. *International Journal of Molecular Sciences, 20*(18), 4572.

Cirstea, C.M., Choi, I., Lee, P., Peng, H., Kaufman, C.L., & Frey, S.H. (2017). Magnetic resonance spectroscopy of current hand amputees reveals evidence for neuronal-level changes in former sensorimotor cortex. *Journal of Neurophysiology, 117,* 1821–1830. https://doi.org/10.1152/jn.00329.2016

Cissé, Y., Ishibashi, M., Jost, J., Toossi, H., Mainville, L., Adamantidis, A., . . . Jones, B. E. (2020). Discharge and role of GABA pontomesencephalic neurons in cortical activity and sleep-wake states examined by optogenetics and juxtacellular recordings in mice. *Journal of Neuroscience, 40*(31), 5970–5989.

Clark, J. L., & Green, M. C. (2018). Self-fulfilling prophecies: Perceived reality of online interaction drives expected outcomes of online communication. *Personality and Individual Differences 133,* 73–76. https://doi.org/10.1016/j.paid.2017.08.031

Clark, M. S., & Chrisman, K. (1994). Resource allocation in intimate relationships: Trying to make sense of a confusing literature. In M. J. Lerner & G. Mikula (Eds.), *Entitlement and the affectional bond: Justice in close relationships* (pp. 65–88). Plenum.

Clark, R. D., & Hatfield, E. (1989). Gender differences in receptivity to sexual offers. *Journal of Psychology and Human Sexuality, 2*(1), 39–55.

Cleary, A. M. (2019). The biasing nature of the tip-of-the-tongue experience: When decisions bask in the glow of the tip-of-the-tongue state. *Journal of Experimental Psychology: General, 148*(7), 1178–1191.

Cleary, A. M., & Claxton, A. B. (2018). Déjà Vu: An illusion of prediction. *Psychological Science, 29*(4), 635–644.

Cleary, A. M., Huebert, A. M., McNeely-White, K. L., & Spahr, K. S. (2019). A postdictive bias associated with déjà vu. *Psychonomic bulletin & review, 26*(4), 1433–1439.

Clemens, N. A. (2010). Evidence base for psychotherapy: Two perspectives. *Journal of Psychiatric Practice, 16*(3), 183–186.

Cleveland, J. N., Barnes-Farrell, J. L., & Ratz, J. M. (1997). Accommodation in the workplace. *Human Resource Management Review, 7*(1), 77–107.

Clifford, S. M., & Dissanayake, C. (2008). The early development of joint attention in infants with autistic disorder using home video observations and parental interview. *Journal of Autism and Developmental Disorders, 38*(5), 791–805.

Climie, R., Fuster, V., & Empana, J. P. (2020). Health literacy and primordial prevention in childhood—an opportunity to reduce the burden of cardiovascular disease. *JAMA Cardiology, 5*(12), 1323–1324.

Clocksin, H. E., Hawks, Z. W., White, D. A., & Christ, S. E. (2020). Inter- and intra-tract analysis of white matter abnormalities in individuals with early-treated phenylketonuria (PKU). *Molecular Genetics and Metabolism.* https://doi.org/10.1016/j.ymgme.2020.12.001

Clouston, S. A., Smith, D. M., Mukherjee, S., Zhang, Y., Hou, W., Link, B. G., & Richards, M. (2020). Education and cognitive decline: An integrative analysis of global longitudinal studies of cognitive aging. *The Journals of Gerontology: Series B, 75*(7), e151–e160.

Cobb, S., & Battin, B. (2004). Second-impact syndrome. *Journal of School Nursing, 20*(5), 262–267.

Cobianchi, S., de Cruz, J., & Navarro, X. (2014). Assessment of sensory thresholds and nociceptive fiber growth after sciatic nerve injury reveals the differential contribution of collateral reinnervation and nerve regeneration to neuropathic pain. *Experimental Neurology, 255,* 1–11.

Coburn, A., Vartanian, O., & Chatterjee, A. (2017). Buildings, beauty, and the brain: A neuroscience of architectural experience. *Journal of Cognitive Neuroscience, 29,* 1521–1531.

Cogley, C., O'Reilly, H., Bramham, J., & Downes, M. (2020). A systematic review of the risk factors for autism spectrum disorder in children born preterm. *Child Psychiatry & Human Development.* https://doi.org/10.1007/s10578-020-01071-9

Cohen, D. (2001). Cultural variation: Considerations and implications. *Psychological Bulletin, 127*(4), 451–471.

Cohen, D., Kim, E., & Hudson, N. W. (2017). Religion, repulsion, and reaction formation: Transforming repellent attractions and repulsions. *Journal of Personality and Social Psychology.* https://doi.org/10.1037/pspp0000151

Cohen, D., Nisbett, R. E., Bowdle, B. F., & Schwarz, N. (1996). Insult, aggression, and the southern culture of honor: An "experimental ethnography." *Journal of Personality and Social Psychology, 70*(5), 945–960.

Cohen, J. N., & Kaplan, S. C. (2020). Understanding and treating anxiety disorders: A psychodynamic approach. In E. Bui, M. E. Charney, & A. W. Baker (Eds.), *Clinical handbook of anxiety disorders* (pp. 315–332). Humana.

Cohen, S. (2020). Psychosocial vulnerabilities to upper respiratory infectious illness: Implications for susceptibility to coronavirus disease 2019 (COVID-19). *Perspectives on Psychological Science.* https://doi.org/10.1177/1745691620942516

Cohen, S., & Lemay, E. (2007). Why would social networks be linked to affect and health practices? *Health Psychology, 26*(4), 410–417.

Cohen, S., Doyle, W. J., Alper, C. M., Janicki-Deverts, D., & Turner, R. B. (2009). Sleep habits and susceptibility to the common cold. *Archives of Internal Medicine, 169*(1), 62–67.

Cohen, S., Frank, E., Doyle, W. J., Skoner, D. P., Rabin, B. S., & Gwaltney Jr, J. M. (1998). Types of stressors that increase susceptibility to the common cold in healthy adults. *Health Psychology, 17*(3), 214–223

Cohen, S., Janicki-Deverts, D., Doyle, W. J., Miller, G. E., Frank, E., Rabin, B. S., & Turner, R. B. (2012). Chronic stress, glucocorticoid receptor resistance, inflammation, and disease risk. *Proceedings of the National Academy of Sciences USA, 109*(16), 5995–5999.

Cohn, N. (2020). Visual narrative comprehension: Universal or not? *Psychonomic Bulletin & Review, 27*(2), 266–285.

Coifman, K. G., Bonanno, G. A., Ray, R. D., & Gross, J. J. (2007). Does repressing coping promote resilience? Affective-autonomic response discrepancy during bereavement. *Journal of Personality and Social Psychology, 92*(4), 745–758.

Colapinto, J. (2000). *As nature made him.* HarperAcademic.

Coleman, S. E., & Caswell, N. (2020). Diabetes and eating disorders: An exploration of 'diabulimia.' *BMC Psychology, 8*(1), 1–7.

Coles, B. A., & West, M. (2016). Weaving the internet together: Imagined communities in newspaper comment threads. *Computers In Human Behavior, 60,* 44–53. https://doi.org/10.1016/j.chb.2016.02.049

Collins, B., Breithaupt, L., McDowell, J. E., Miller, L. S., Thompson, J., & Fischer, S. (2017). The impact of acute stress on the neural processing of food cues in bulimia nervosa: Replication in two samples. *Journal of Abnormal Psychology, 126*(5), 540–551.

Colomer, C., Berenguer, C., Roselló, B., Baixauli, I., & Miranda, A. (2017). The impact of inattention, hyperactivity/impulsivity symptoms, and executive functions on learning behaviors of children with ADHD. *Frontiers in Psychology, 8.*

Commonwealth of Massachusetts v. Porter 31285–330 (1993, Massachusetts).

Compton, J., & Pfau, M. (2004). Use of inoculation to foster resistance to credit card marketing targeting college students. *Journal of Applied Communication Research, 32*(4), 343–364.

Compton, J., & Pfau, M. (2008). Inoculating against pro-plagiarism justifications: Rational and affective strategies. *Journal of Applied Communication Research, 36*(1), 98–119.

Comstock, G. (2012). The use of television and other film-related media. In D. G. Singer & J. L. Singer (Eds.), *Handbook of children and the media* (2nd ed.). Sage.

Conde, L. C., Amin, N., Hottenga, J. J., Gizer, I., Trull, T., van Duijn, C., et al. (2017). The first genome-wide association meta-analysis of borderline personality disorder features. *European Neuropsychopharmacology, 27,* S504–S505.

Conley, T. D. (2011). Perceived proposer personality characteristics and gender differences in acceptance of casual sex offers. *Journal of Personality and Social Psychology, 100*(2), 309–329.

Conley, T. D., Moors, A. C., Matsick, J. L., Ziegler, A., & Valentine, B. A. (2011). Women, men and the bedroom: Methodological and conceptual insights that narrow, reframe, and eliminate gender differences in sexuality. *Current Directions in Psychological Science, 20*(5), 296–300.

Contreras-Huerta, L. S., Lockwood, P. L., Bird, G., Apps, M. A., & Crockett, M. J. (2020). Prosocial behavior is associated with transdiagnostic markers of affective sensitivity in multiple domains. *Emotion.* http://dx.doi.org/10.1037/emo0000813

Converse, P. D., Thackray, M., Piccone, K., Sudduth, M. M., Tocci, M. C., & Miloslavic, S. A. (2016). Integrating self-control with physical attractiveness and cognitive ability to examine pathways to career success. *Journal of Occupational and Organizational Psychology, 89*(1), 73–91. https://doi.org/10.1111/joop.12107

Conway, M., & Rubin, D. (1993). The structure of autobiographical memory. In A. F. Collins, S. E. Gathercole, M. A. Conway, & P. E. Morris (Eds.), *Theories of memory.* Erlbaum.

Cook, A., Spinazzola, J., Ford, J., Lanktree, C., Blaustein, M., Cloitre, M., et al. (2017). Complex trauma in children and adolescents. *Psychiatric Annals, 35*(5), 390–398.

Cooper, C. R., Gonzalez, E., & Wilson, A. R. (2015). Identities, cultures, and schooling: How students navigate racial-ethnic, indigenous, immigrant, social class, and gender identities on their pathways through school. In K. C. McLean & M. Syed (Eds.), *The Oxford handbook of identity development.* Oxford University Press.

Cooper, E. A., Garlick, J., Featherstone, E., Voon, V., Singer, T., Critchley, H. D., & Harrison, N. A. (2014). You turn me cold: Evidence for temperature contagion. *PloS One, 9*(12), e116126.

Cooper, R. M., & Zubek, J. P. (1958). Effects of enriched and restricted early environments on the learning ability of bright and dull rats. *Canadian Journal of Psychology, 12*(3), 159–164.

Cooper, S., & Bidaisee, S. (2017). Evaluation of risks and factors linked to precocious puberty. *Journal of Translational Diagnostic Technology, 2*(1), 27–29.

Copur-Gencturk, Y., Cimpian, J. R., Lubienski, S. T., & Thacker, I. (2020). Teachers' bias against the mathematical ability of female, black, and Hispanic students. *Educational Researcher, 49*(1), 30–43.

Corbett, B., & Duarte, A. (2020). How proactive interference during new associative learning impacts general and specific memory in young and old. *Journal of Cognitive Neuroscience.* https://doi.org/10.1162/jocn_a_01582

Corcoran, M., Hawkins, E. L., O'Hora, D., Whalley, H. C., Hall, J., Lawrie, S. M., & Dauvermann, M. R. (2020). Are working memory and glutamate concentrations involved in early-life stress and severity of psychosis? *Brain and Health,* e01616.

Cordell, E., & Holaway, R. (2020). Obsessive-compulsive and related disorders. In L. T. Benuto, F. R. Gonzalez, & J. Singer (Eds.), *Handbook of cultural factors in behavioral health* (pp. 293–307). Springer.

Cornelius, T., & Kershaw, T. (2017). Perception of partner sexual history: Effects on safe-sex intentions. *Health Psychology, 36*(7), 704–712. https://doi.org/10.1037/hea0000474

Corrigan, P. W., & Penn, D. L. (2015). Lessons from social psychology on discrediting psychiatric stigma. *Stigma And Health, 1*(S), 2–17. https://doi.org/10.1037/2376-6972.1.S.2

Corum, J., & Zimmer, C. (2020, December 8). How Moderna's vaccine works. *The New York Times.* https://www.nytimes.com/interactive/2020/health/moderna-covid-19-vaccine.html.

Cosci, F. (2017). Withdrawal symptoms after discontinuation of a noradrenergic and specific serotonergic antidepressant: A case report and review of the literature. *Personalized Medicine in Psychiatry, 1,* 81–84.

Cosentino, A. C., & Solano, A. C. (2017). The high five: Associations of the five positive factors with the big five and well-being. *Frontiers in Physiology, 8,* art. no. 1250.

Cosmides, L. (2011). Evolutionary psychology. *Annual Review of Psychology, 62.* Annual Reviews.

Cossins, D. (2018, April 12). Discriminating algorithms: 5 times AI showed prejudice. *New Scientist.* https://www.newscientist.com/article/2166207-discriminating-algorithms-5-times-ai-showed-prejudice

Costa, P. J., & McCrae, R. R. (2006). Age changes in personality and their origins: Comment on Roberts, Walton, and Viechtbauer (2006). *Psychological Bulletin, 132*(1), 26–28. https://doi.org/10.1037/0033-2909.132.1.26

Costa, P. T., & McCrae, R. R. (2010). *NEO Inventories, professional manual.* Psychological Assessment Resources.

Costa, P. T., & McCrae, R. R. (2013). A theoretical context for adult temperament. In T. D. Wachs, R. R. McCrae, & G. A. Kohnstamm (Eds.), *Temperament in context.* Psychology Press.

Costa, P. T., Weiss, A., Duberstein, P. R., Friedman, B., & Siegler, I. C. (2014). Personality facets and all-cause mortality among Medicare patients aged 66 to 102 years: A follow-on study of Weiss and Costa (2005). *Psychosomatic Medicine, 76*(5), 370–378.

Costa, S. P., & Neves, P. (2017). Forgiving is good for health and performance: How forgiveness helps individuals cope with the psychological contract breach. *Journal of Vocational Behavior, 100,* 124–136. https://doi.org/10.1016/j.jvb.2017.03.005

Costanza, A., Amerio, A., Radomska, M., Ambrosetti, J., Di Marco, S., Prelati, M., . . . Michaud, L. (2020). Suicidality assessment of the elderly with physical illness in the emergency department. *Frontiers in Psychiatry, 11.*

Costello, K., & Hodson, G. (2014). Explaining dehumanization among children: The interspecies model of prejudice. *British Journal of Social Psychology, 53*(1), 175–197. https://doi.org/10.1111/bjso.12016

Côté S., House J., & Willer R., (2015). High economic inequality leads higher-income individuals to be less generous. *Proceedings of the National Academy of Science, 112,* 15838–15843.

Coubart, A., Izard, V., Spelke, E. S., Marie, J., & Streri, A. (2014). Dissociation between small and larger numerosities in newborn infants. *Developmental Science, 17*(1), 11–22.

Coughlin, J. M., Rubin, L. H., Du, Y., Rowe, S. P., Crawford, J. L., Rosenthal, H. B., . . . Speck, C. L. (2020). High availability of the α7-nicotinic acetylcholine receptor in brains of individuals with mild cognitive impairment: A pilot study using 18F-ASEM PET. *Journal of Nuclear Medicine, 61*(3), 423–426.

Coulter, R. W., & Rankin, S. R. (2020). College sexual assault and campus climate for sexual-and gender-minority undergraduate students. *Journal of Interpersonal Violence, 35*(5-6), 1351–1366.

Courtet, P., & others. (2004). Serotonin transporter gene may be involved in short-term risk of subsequent suicide attempts. *Biological Psychiatry, 55*(1), 46–51.

Cowan, N. (2008). What are the differences between long-term, short-term, and working memory? In W. S. Sossin, L. C. Lacaille, V. F. Castellucci, & S. Belleville (Eds.), *Essence of memory: Progress in brain research* (vol. 169, pp. 323–338). Elsevier.

Cowan, N. (2010). The magical mystery four: How is working memory capacity limited, and why? *Current Directions in Psychological Science, 19*(1), 51–57.

Cowan, N. (2017). The many faces of working memory and short-term storage. *Psychonomic Bulletin & Review, 24*(4), 1158–1170.

Cowan, N., Li, D., Moffitt, A., Becker, T. M., Martin, E. A., Saults, J. S., & Christ, S. E. (2011a). A neural region of abstract working memory. *Journal of Cognitive Neuroscience, 23*(10), 2852–2863.

Cowan, N., Morey, C. C., Aubuchon, A. M., Zwilling, C. E., Gilchrist, A. L., & Saults, J. S. (2011b). New insights into an old problem: Distinguishing storage from processing in the development of working memory. In P. Barrouillet & V. Gaillard (Eds.), *Cognitive development and working memory: A dialogue between neo-Piagetian theories and cognitive approaches* (pp. 137–150). Psychology Press.

Cowan, N., Rouder, J. N., Blume, C. L., & Saults, J. S. (2012). Models of verbal working memory capacity: What does it take to make them work? *Psychological Review, 119*(3), 480–499.

Cowan, R. L., Roberts, D. M., & Joers, J. M. (2008). Neuroimaging in human MDMA (Ecstasy) users. *Annals of the New York Academy of Sciences, 1139,* 291–298.

Cox, R. E., & Bryant, R. A. (2008). Advances in hypnosis research: Methods, designs, and contributions of intrinsic and instrumental hypnosis. In M. R. Nash & A. J. Barnier (Eds.), *The Oxford handbook of hypnosis: Research theory and practice* (pp. 311–336). Oxford University Press.

Coyle, J. T., Ruzicka, W. B., & Balu, D. T. (2020). Fifty years of research on schizophrenia: the ascendance of the glutamatergic synapse. *American Journal of Psychiatry, 177*(12), 1119–1128.

Coyne, J. C., & Tennen, H. (2010). Positive psychology in cancer care: Bad science, exaggerated claims, and unproven medicine. *Annals of Behavioral Medicine, 39*(1), 16–26.

Coyne, J. C., Tennen, H., & Ranchor, A. V. (2010). Positive psychology and cancer care: A story line resistant to evidence. *Annals of Behavioral Medicine, 39*(1), 35–42.

Coyne, S. M., Padilla-Walker, L. M., Holmgren, H. G., Davis, E. J., Collier, K. M., Memmott-Elison, M. K., & Hawkins, A. J. (2018). A meta-analysis of prosocial media on prosocial behavior, aggression, and empathic concern: A multidimensional approach. *Developmental Psychology, 54*(2), 331–347.

Craik, F. I. M., & Lockhart, R. S. (1972). Levels of processing: A framework for memory research. *Journal of Verbal Learning and Verbal Behavior, 11*(6), 671–684.

Craik, F. I. M., & Tulving, E. (1975). Depth of processing and retention of words in episodic memory. *Journal of Experimental Psychology, 104*(3), 268–294.

Cramer, P. (2015). Defense mechanisms: 40 years of empirical research. *Journal of Personality Assessment, 97*(2), 114–122. https://doi.org/10.1080/00223891.2014.947997

Crane, R. S., Brewer, J., Feldman, C., Kabat-Zinn, J., Santorelli, S., Williams, J. M. G., & Kuyken, W. (2017). What defines mindfulness-based programs? The warp and the weft. *Psychological Medicine, 47*(6), 990–999.

Crans, R. A., Wouters, E., Valle-León, M., Taura, J., Massari, C. M., Fernández-Dueñas, V., . . . Ciruela, F. (2020). Striatal dopamine D2-muscarinic acetylcholine M1 receptor-receptor interaction in a model of movement disorders. *Frontiers in Pharmacology, 11*, 194.

Crego, C., & Widiger, T. A. (2015). Psychopathy and the *DSM. Journal of Personality, 83*(6), 665–677. https://doi.org/10.1111/jopy.12115

Crocker, J., Canevello, A., & Brown, A. A. (2017). Social motivation: Costs and benefits of selfishness and otherishness. *Annual Review of Psychology, 68*, 299–325. https://doi.org/10.1146/annurev-psych-010416-044145

Crocker, J., Major, B., & Steele, C. (1998). Social stigma. In D. T. Gilbert, S. T. Fiske, & G. Lindzey (Eds.), *The handbook of social psychology* (4th ed., vol. 2, pp. 504–553). McGraw-Hill.

Cronbach, L. J. (1957). The two disciplines of scientific psychology. *American Psychologist, 12*, 671–684.

Cronin, T. J., Pepping, C. A., Halford, W. K., & Lyons, A. (2020). Minority stress and psychological outcomes in sexual minorities: The role of barriers to accessing services. *Journal of Homosexuality.* https://doi.org/10.1080/00918369.2020.1804264

Crönlein, T., Wetter, T. C., Rupprecht, R., & Spiegelhalder, K. (2020). Cognitive behavioral treatment for insomnia is equally effective in insomnia patients with objective short and normal sleep duration. *Sleep Medicine, 66*, 271–275.

Cross, M. P., & Pressman, S. D. (2020). Say cheese? The connections between positive facial expressions in student identification photographs and health care seeking behavior. *Journal of Health Psychology, 25*(13–14), 2511–2519.

Crowell, C. A., Davis, S. W., Beynel, L., Deng, L., Lakhlani, D., Hilbig, S. A., . . . Lisanby, S. H. (2020). Older adults benefit from more widespread brain network integration during working memory. *NeuroImage, 218*, 116959.

Csikszentmihalyi, M. (1990). Flow: *The psychology of optimal experience.* Harper Perennial.

Csillag, C., Nordentoft, M., Mizuno, M., Jones, P. B., Killackey, E., Taylor, M., . . . & McDaid, D. (2016). Early intervention services in psychosis: from evidence to wide implementation. *Early Intervention in Psychiatry, 10*(6), 540–546.

Cuesto, G., Everaerts, C., León, L. G., & Acebes, A. (2017). Molecular bases of anorexia nervosa, bulimia nervosa and binge eating disorder: shedding light on the darkness. *Journal of Neurogenetics, 31*(4), 266–287.

Cuffe, S. P., Visser, S. N., Holbrook, J. R., Danielson, M. L., Geryk, L. L., Wolraich, M. L., & McKeown, R. E. (2020). ADHD and psychiatric comorbidity: Functional outcomes in a school-based sample of children. *Journal of Attention Disorders, 24*(9), 1345–1354.

Cullen, K. (2010, January 24). The untouchable mean girls. *Boston Globe.* www.boston.com/news/local/massachusetts/articles/2010/01/24/the_untouchable_mean_girls/

Culverhouse, R. C., Saccone, N. L., Horton, A. C., Ma, Y., Anstey, K. J., Banaschewski, T., . . . Goldman, N. (2018). Collaborative meta-analysis finds no evidence of a strong interaction between stress and 5-HTTLPR genotype contributing to the development of depression. *Molecular Psychiatry, 23*(1), 133–142.

Cummings, M. (2016). *Human heredity* (11th ed.). Cengage.

Curot, J., Pariente, J., Hupé, J. M., Lotterie, J. A., Mirabel, H., & Barbeau, E. J. (2019). Déjà vu and prescience in a case of severe episodic amnesia following bilateral hippocampal lesions. *Memory.* https://doi.org/10.1080/09658211.2019.1673426

Currie, C. L., Copeland, J. L., Metz, G. A., Moon-Riley, K. C., & Davies, C. M. (2020). Past-year racial discrimination and allostatic load among Indigenous adults in Canada: The role of cultural continuity. *Psychosomatic Medicine, 82*(1), 99.

Curtiss, S. (2014). The case of Chelsea: The effects of late age at exposure to language on language performance and evidence for the modularity of language and mind. In C. T. Schutze & L. Stockall (Eds.), Connections: Papers by and for Sarah Van Wagenen. *UCLA Working Papers in Linguistics, 18*, 115–146.

Cutler, R. R., & Kokovay, E. (2020). Rejuvenating subventricular zone neurogenesis in the aging brain. *Current Opinion in Pharmacology, 50*, 1–8.

Cwik, A. J. (2017). What is a Jungian analyst dreaming when myth comes to mind? Thirdness as an aspect of the anima media natura. *The Journal Of Analytical Psychology, 62*(1), 107–129. https://doi.org/10.1111/1468-5922.12284

D

D'Angelo, B. (2020, March 14). Italians sing from balconies to show unity against coronavirus. Boston25news.com. https://www.boston25news.com/news/trending/italians-sing-balconies-show-unity-against-coronavirus/QAZIPF2LY5EQTNCTTPBUA62FY4/

da Cunha-Bang, S., Hjordt, L. V., Perfalk, E., Beliveau, V., Bock, C., Lehel, S., et al. (2017). Serotonin 1B receptor binding is associated with trait anger and level of psychopathy in violent offenders. *Biological Psychiatry, 82*(4), 267–274. https://doi.org/10.1016/j.biopsych.2016.02.030

da Silva, R. G., Ribeiro, M. H. D. M., Mariani, V. C., & dos Santos Coelho, L. (2020). Forecasting Brazilian and American COVID-19 cases based on artificial intelligence coupled with climatic exogenous variables. *Chaos, Solitons & Fractals, 139*, 110027.

D'Souza, D., Brady, D., Haensel, J. X., & D'Souza, H. (2020). Is mere exposure enough? The effects of bilingual environments on infant cognitive development. *Royal Society Open Science, 7*(2), 180191.

Dabney, W., Kurth-Nelson, Z., Uchida, N., Starkweather, C. K., Hassabis, D., Munos, R., & Botvinick, M. (2020). A distributional code for value in dopamine-based reinforcement learning. *Nature, 577*(7792), 671–675.

Dahlgren, K., Ferris, C., & Hamann, S. (2020). Neural correlates of successful emotional episodic encoding and retrieval: An SDM meta-analysis of neuroimaging studies. *Neuropsychologia, 107495.*

Dai, S., Mo, Y., Wang, Y., Xiang, B., Liao, Q., Zhou, M., . . . Guo, C. (2020). Chronic stress promotes cancer development. *Frontiers in Oncology, 10*, 1492.

Dai, Z., Abate, M. A., Long, D. L., Smith, G. S., Halki, T. M., Kraner, J. C., & Mock, A. R. (2020). Quantifying enhanced risk from alcohol and other factors in polysubstance-related deaths. *Forensic Science International, 313*, 110352.

Dakin, C. J., Kumar, P., Forbes, P. A., Peters, A., & Day, B. L. (2020). Variance based weighting of multisensory head rotation signals for verticality perception. *PlosOne, 15*(1), e0227040.

Daley, D., & DuPaul, G. (2016). Nonpharmacological interventions for preschool children with attention-deficit/hyperactivity disorder. *Journal of the American Academy of Child & Adolescent Psychiatry, 55*(10), S329–S330.

Daley, R. T., Bowen, H. J., Fields, E. C., Parisi, K. R., Gutchess, A., & Kensinger, E. A. (2020). Neural mechanisms supporting emotional and self-referential information processing and encoding in older and younger adults. *Social Cognitive and Affective Neuroscience, 15*, 405–421.

Daly, M., Sutin, A. R., & Robinson, E. (2019). Perceived weight discrimination mediates the prospective association between obesity and physiological dysregulation: Evidence from a population-based cohort. *Psychological Science, 30*(7), 1030–1039.

Damian, R. I., & Roberts, B. W. (2015a). The associations of birth order with personality and intelligence in a representative sample of U.S. high school students. *Journal of Research In Personality, 58*, 96–105. https://doi.org/10.1016/j.jrp.2015.05.005

Damian, R. I., & Roberts, B. W. (2015b). Settling the debate on birth order and personality. *PNAS: Proceedings of the National Academy of Sciences, 112*, 14119–14120. https://doi.org/10.1073/pnas.1519064112

Danner, D. D., Snowdon, D. A., & Friesen, W. V. (2001). Positive emotions in early life and longevity: Findings from the Nun Study. *Journal of Personality and Social Psychology, 80*(5), 804–813.

Darcy, E. (2012). Gender issues in child and adolescent eating disorders. In J. Lock (Ed.), *The Oxford handbook of child and adolescent eating disorders: Developmental perspectives.* Oxford University Press.

Darley, J. M., & Latané, B. (1968). Bystander intervention in emergencies: Diffusion of responsibility. *Journal of Personality and Social Psychology, 8*(4), 377–383.

Darwin, C. (1965). *The expression of emotions in man and animals.* University of Chicago Press. (original work published in 1872).

Datta Gupta, N., Etcoff, N. L., & Jaeger, M. M. (2016). Beauty in mind: The effects of physical attractiveness on psychological well-being and distress. *Journal of Happiness Studies, 17*(3), 1313–1325. https://doi.org/10.1007/s10902-015-9644-6

Daumeyer, N. M., Onyeador, I. N., Brown, X., & Richeson, J. A. (2019). Consequences of attributing discrimination to implicit vs. explicit bias. *Journal of Experimental Social Psychology, 84*, 103812.

Davidson, J. E., Proudfoot, J., Lee, K., Terterian, G., & Zisook, S. (2020). A longitudinal analysis of nurse suicide in the United States (2005–2016) with recommendations for action. *Worldviews on Evidence-Based Nursing, 17*(1), 6–15.

Davidson, R. J., & others. (2003). Alterations in brain and immune function produced by mindfulness meditation. *Psychosomatic Medicine, 65*(4), 564–570.

Davies, G., & others. (2011). Genome-wide association studies establish that human intelligence is highly heritable and polygenic. *Molecular Psychiatry, 16*, 996–1005.

Davies, P. T., & Cicchetti, D. (2014). How and why does the 5-HTTLPR gene moderate associations between maternal unresponsiveness and children's disruptive problems? *Child Development, 85*(2), 484–500.

Davis, D. E., DeBlaere, C., Owen, J., Hook, J. N., Rivera, D. P., Choe, E., . . . Placeres, V. (2018). The multicultural orientation framework: A narrative review. *Psychotherapy, 55*(1), 89–100.

Davis, J. T., & Hines, M. (2020). How large are gender differences in toy preferences? A systematic review and meta-analysis of toy preference research. *Archives of Sexual Behavior, 49*(2), 373–394.

Davis, K. C., & Raizen, D. M. (2017). A mechanism for sickness sleep: lessons from invertebrates. *The Journal of Physiology.* https://doi.org/10.1113/JP273009

Davis, M. A., Anthony, D. L., & Pauls, S. D. (2015). Seeking and receiving social support on Facebook for surgery. *Social Science & Medicine, 131*, 40–47.

Davis, S. R. (2013). Androgen therapy in women, beyond libido. *Climacteric, 16*, Suppl. 1, 18–24.

Davydenko, M., & Peetz, J. (2017). Time grows on trees: The effect of nature settings on time perception. *Journal of Environmental Psychology, 54*, 20–26.

Day, M. A., Ward, L. C., Grover, M. P., Ehde, D. M., Illingworth, O. R., & Jensen, M. P. (2020). The roles of race, sex and cognitions in response to experimental pain. *European Journal of Pain.* https://doi.org/10.1002/ejp.1552

de Almeida Silva, V., Louzã, M. R., da Silva, M. A., & Nakano, E. Y. (2016). Ego defense mechanisms and types of object relations in adults with ADHD. *Journal of Attention Disorders, 20*(11), 979–987. https://doi.org/10.1177/1087054712459559

De Backer, M., Boen, F., De Cuyper, B., Høigaard, R., & Broek, G. V. (2015). A team fares well with a fair coach: Predictors of social loafing in interactive female sport teams. *Scandinavian Journal of Medicine & Science in Sports, 25*(6), 897–908. https://doi.org/10.1111/sms.12303

de Guzman, N. S., & Nishina, A. (2014). A longitudinal study of body dissatisfaction and pubertal timing in an ethnically diverse adolescent sample. *Body Image, 11*(1), 68–71.

De Houwer, J. (2020). Revisiting classical conditioning as a model for anxiety disorders: A conceptual analysis and brief review. *Behaviour Research and Therapy, 127,* 103558.

De Maria, M., Tagliabue, S., Ausili, D., Vellone, E., & Matarese, M. (2020). Perceived social support and health-related quality of life in older adults who have multiple chronic conditions and their caregivers: A dyadic analysis. *Social Science & Medicine, 262,* 113193.

De Neve, J. (2015). Personality, childhood experience, and political ideology. *Political Psychology, 36*(1), 55–73.

de Pablo, G. S., Catalan, A., & Fusar-Poli, P. (2020). Clinical validity of DSM-5 attenuated psychosis syndrome: advances in diagnosis, prognosis, and treatment. *JAMA Psychiatry, 77*(3), 311–320.

De Picker, L. J., Morrens, M., Chance, S. A., & Boche, D. (2017). Microglia and brain plasticity in acute psychosis and schizophrenia illness course: A meta-review. *Frontiers in Psychiatry, 8,* 238.

De Venter, M., Van Den Eede, F., Pattyn, T., Wouters, K., Veltman, D. J., Penninx, B. W., & Sabbe, B. G. (2017). Impact of childhood trauma on course of panic disorder: Contribution of clinical and personality characteristics. *Acta Psychiatrica Scandinavica, 135*(6), 554–563.

de Vries, J., Byrne, M., & Kehoe, E. (2015). Cognitive dissonance induction in everyday life: An fMRI study. *Social Neuroscience, 10*(3), 268–281. https://doi.org/10.1080/17470919.2014.990990

Deane, C., Vijayakumar, N., Allen, N. B., Schwartz, O., Simmons, J. G., Bousman, C. A., . . . Whittle, S. (2020). Parenting × brain development interactions as predictors of adolescent depressive symptoms and well-being: Differential susceptibility or diathesis-stress? *Development and Psychopathology, 32*(1), 139–150.

DeAngelis, T. (2002). Binge-eating disorder: What's the best treatment? *Monitor on Psychology.* www.apa.org/monitor/mar02/binge.html

Deaux, K. (2001). Social identity. In J. Worell (Ed.), *Encyclopedia of gender and women.* Academic.

Debnath, M., Berk, M., & Maes, M. (2020). Changing dynamics of psychoneuroimmunology during the COVID-19 pandemic. *Brain, Behavior, & Immunity-Health, 5,* 100096.

DeCarlo, L. T. (2020). An item response model for true-false exams based on signal detection theory. *Applied Psychological Measurement, 44*(3), 234–248.

DeCasien, A. R., & Higham, J. P. (2019). Primate mosaic brain evolution reflects selection on sensory and cognitive specialization. *Nature Ecology & Evolution, 3*(10), 1483–1493.

Decety, J., Meidenbauer, K. L., & Cowell, J. M. (2017). The development of cognitive empathy and concern in preschool children: A behavioral neuroscience investigation. *Developmental Science.* http://onlinelibrary.wiley.com/doi/10.1111/desc.12570/full

Deits-Lebehn, C., Baucom, K. J., Crenshaw, A. O., Smith, T. W., & Baucom, B. R. (2020). Incorporating physiology into the study of psychotherapy process. *Journal of Counseling Psychology, 67*(4), 488–499.

DeJager, B., Houlihan, D., Filter, K. J., Mackie, P. F., & Klein, L. (2020). Comparing the effectiveness and ease of implementation of token economy, response cost, and a combination condition in rural elementary school classrooms. *Journal of Rural Mental Health, 44*(1), 39–50.

Dell, P. F., & Eisenhower, J. W. (1990). Adolescent multiple personality disorder: A preliminary study of eleven cases. *Journal of the American Academy of Child & Adolescent Psychiatry, 29*(3), 359–366.

Demiralay, C., Agorastos, A., Yassouridis, A., Jahn, H., Wiedemann, K., & Kellner, M. (2017). Copeptin–A potential endocrine surrogate marker of CCK-4-induced panic symptoms? *Psychoneuroendocrinology, 76,* 14–18.

Demiray, B., & Freund, A. M. (2014). Michael Jackson, Bin Laden, and I: Functions of positive and negative, public and private flashbulb memories. *Memory.* https://doi.org/10.1080/09658211.2014.907428

Den Heijer, A. E., Groen, Y., Tucha, L., Fuermaier, A. B., Koerts, J., Lange, K. W., et al. (2017). Sweat it out? The effects of physical exercise on cognition and behavior in children and adults with ADHD: A systematic literature review. *Journal of Neural Transmission, 124*(1), 3–26.

Dennett, D. C. (2017). *Consciousness explained.* Little, Brown.

Dennett, D. C. (2020). A history of qualia. *Topoi, 39*(1), 5–12.

Denollet, J., & Conraads, V. M. (2011). Type D personality and vulnerability to adverse outcomes of heart disease. *Cleveland Clinic Journal of Medicine, 78,* Suppl. 1, S13–S19.

Depue, R. A., & Collins, P. F. (1999). Neurobiology of the structure of personality: Dopamine, facilitation of incentive motivation, and extraversion. *Behavioral and Brain Sciences, 22*(3), 491–569.

Depue, R. A., & Fu, Y. (2013). On the nature of extraversion: Variation in conditioned contextual activation of dopamine-facilitated affective, cognitive, and motor processes. *Frontiers in Human Neuroscience, 7,* 288.

De Raad, B., & others. (2010). Only three factors of personality description are fully replicable across languages: A comparison of 14 trait taxonomies. *Journal of Personality and Social Psychology, 98*(1), 160–173.

Derksen, M., Feenstra, M., Willuhn, I., & Denys, D. (2020). The serotonergic system in obsessive-compulsive disorder. In C. P. Müller & K. A. Cunningham (Eds.), *Handbook of behavioral neuroscience* (vol. 31, pp. 865–891). Elsevier.

Derntl, B., Habel, U., Robinson, S., Windischberger, C., Kryspin-Exner, I., Gur, R. C., & Moser, E. (2012). Culture but not gender modulates amygdala activation during explicit emotion recognition. *BMC Neuroscience, 13,* 54.

De Roover, K., Vermunt, J.K., Timmerman, M.E., Ceulemans, E. (2017). Mixture simultaneous factor analysis for capturing differences in latent variables between higher level units of multilevel data. *Structural Equation Modeling, 24*(4), 506–523.

de Sousa Tomaz, V., Chaves Filho, A. J. M., Cordeiro, R. C., Jucá, P. M., Soares, M. V. R., Barroso, P. N., . . . Macedo, D. S. (2020). Antidepressants of different classes cause distinct behavioral and brain pro- and anti-inflammatory changes in mice submitted to an inflammatory model of depression. *Journal of Affective Disorders, 268,* 188–200.

de Souto Barreto, P., Andrieu, S., Rolland, Y., Vellas, B., & DSA MAPT Study Group. (2018). Physical activity domains and cognitive function over three years in older adults with subjective memory complaints: Secondary analysis from the MAPT trial. *Journal of Science and Medicine in Sport, 21*(1), 52–57.

DeSouza, D. D., Stimpson, K. H., Baltusis, L., Sacchet, M. D., Gu, M., Hurd, R., . . . Spiegel, D. (2020). Association between anterior cingulate neurochemical concentration and individual differences in hypnotizability. *Cerebral Cortex, 30*(6), 3644–3654.

de Thierry, B. (2020). *The simple guide to complex trauma and dissociation: What it is and how to help.* Jessica Kingsley Publishers.

Detloff, A. M., Hariri, A. R., & Strauman, T. J. (2020). Neural signatures of promotion versus prevention goal priming: fMRI evidence for distinct cognitive-motivational systems. *Personality Neuroscience, 3,* e1.

Devine, P. G., Forscher, P. S., Austin, A. J., & Cox, W. T. (2012). Long-term reduction in implicit race bias: A prejudice habit-breaking intervention. *Journal of Experimental Social Psychology, 48*(6), 1267–1278.

Devine, R. T., White, N., Ensor, R., & Hughes, C. (2016). Theory of mind in middle childhood: Longitudinal associations with executive function and social competence. *Developmental Psychology, 52*(5), 758–771

Devita, M., Montemurro, S., Zangrossi, A., Ramponi, S., Marvisi, M., Villani, D., . . . & Mondini, S. (2017). Cognitive and motor reaction times in obstructive sleep apnea syndrome: a study based on computerized measures. *Brain and Cognition, 117,* 26–32.

DeWall, C. N., Anderson, C. A., & Bushman, B. J. (2013). Aggression. In H. A. Tennen, J. M. Suls, & I. B. Weiner (Eds.), *Handbook of psychology* (2nd ed., vol. 5, pp. 449–466). Wiley.

DeYoung, C. G. (2013). The neuromodulator of exploration: A unifying theory of the role of dopamine in personality. *Frontiers in Human Neuroscience, 7,* 762.

DeYoung, C. G. (2015). Cybernetic Big Five Theory. *Journal of Research in Personality, 56,* 33–58.

DeYoung, C. G., & Allen, T. A. (2019). Personality neuroscience: A developmental perspective. In D. P. McAdams, R. L. Shiner, & J. L. Tackett (Eds.), *Handbook of personality development* (pp. 79–105). The Guilford Press.

DeYoung, C. G., Hirsh, J. B., Shane, M. S., Papademetris, X., Rajeevan, N., & Gray, J. R. (2010). Testing predictions from personality neuroscience: Brain structure and the big five. *Psychological Science, 21*(6), 820–828.

di Giannantonio, M., Northoff, G., & Salone, A. (2020). The interface between psychoanalysis and neuroscience: The state of the art. *Frontiers in Human Neuroscience.* https://doi.org/10.3389/fnhum.2020.00199

Di Giuseppe, M., Perry, J. C., Conversano, C., Gelo, O. C. G., & Gennaro, A. (2020). Defense mechanisms, gender, and adaptiveness in emerging personality disorders in adolescent outpatients. *The Journal of Nervous and Mental Disease, 208*(12), 933–941.

Di Lorenzo, G., Jannini, T. B., Longo, L., Rossi, R., Siracusano, A., & Dell'Osso, B. (2020). Transcranial magnetic stimulation in the treatment of anxiety disorders. In B. Dell'Osso & G. Di Lorenzo (Eds.), *Non-invasive brain stimulation in psychiatry and clinical neurosciences* (pp. 175–190). Springer.

Di Narzo, A. F., & others. (2014). A unique gene expression signature associated with serotonin 2C receptor RNA editing in the prefrontal cortex and altered in suicide. *Human Molecular Genetics, 23*(18), 4801–4813.

Diamond, L. M. (2013a). Gender and same-sex sexuality. In D. T. Tolman & L. M. Diamond (Eds.), *APA handbook on sexuality and psychology.* American Psychological Association.

Diamond, L. M. (2013b). Sexuality and same-sex sexuality in relationships. In J. Simpson & J. Davidio (Eds.), *Handbook of personality and social psychology.* American Psychological Association.

Diamond, L. M. (2020). Gender fluidity and nonbinary gender identities among children and adolescents. *Child Development Perspectives, 14*(2), 110–115.

Diamond, L. M., Dickenson, J. A., & Blair, K. L. (2017). Stability of sexual attractions across different timescales: The roles of bisexuality and gender. *Archives of Sexual Behavior, 46*(1), 193–204. https://doi.org/10.1007/s10508-016-0860-x

Diamond, L. M., & Savin-Williams, R. C. (2015). Same-sex activity in adolescence: Multiple meanings and implications. In R. F. Fassinger & S. L. Morrow (Eds.), *Sex in the margins.* American Psychological Association.

Diamond, M., & Sigmundson, H. K. (1997). Sex reassignment at birth: Long-term review and clinical implications. *Archives of Pediatric and Adolescent Medicine, 151*(3), 298–304.

Diederich, A., & Trueblood, J. S. (2018). A dynamic dual process model of risky decision making. *Psychological Review, 125*(2), 270–292.

Diehl, D. K. (2020). The relationship between personality traits and interracial contact on campus. *SAGE Open, 10*(4), 2158244020965251.

Diekman, A. B., & Clark, E. K. (2015). Beyond the damsel in distress: Gender differences and similarities in enacting prosocial behavior. In D. A. Schroeder & W. G. Graziano (Eds.), *The Oxford handbook of prosocial behavior* (pp. 376–391). Oxford University Press. https://doi.org/10.1093/oxfordhb/9780195399813.013.028

Diener, E. (1999). Introduction to the special section on the structure of emotion. *Journal of Personality and Social Psychology, 76*(5), 803–804.

Diener, E. (2000). Subjective well-being: The science of happiness and a proposal for a national index. *American Psychologist, 55*(1), 34–43.

Diener, E., & Biswas-Diener, R. (2002). Will money increase subjective well-being? *Social Indicators Research, 57*(2), 119–169. https://doi.org/0.1023/A:1014411319119

Diener, E., & Chan, M. Y. (2011). Happy people live longer: Subjective well-being contributes to health and well-being. *Applied Psychology: Health and Well-Being, 3*(1), 1–43.

Diener, E., & Diener, C. (1996). Most people are happy. *Psychological Science, 7*(3), 181–185.

Diener, E., Emmons, R. A., Larsen, R. J., & Griffin, S. (1985). The Satisfaction with Life Scale. *Journal of Personality Assessment, 49*(1), 71–75.

Diener, E., & Lucas, R. (2014). New findings on personality and well-being. In M. Mikulincer & P. Shaver (Eds.), *APA handbook of personality and social psychology.* American Psychological Association.

Diener, E., Ng, W., Harter, J., & Arora, R. (2010). Wealth and happiness across the world: Material prosperity predicts life evaluation, whereas psychosocial prosperity predicts positive feeling. *Journal of Personality and Social Psychology, 99*(1), 52–61.

Diener, E., Tay, L., & Oishi, S. (2013). Rising income and the subjective well-being of nations. *Journal of Personality and Social Psychology, 104*(2), 267–276.

Digman, J. M. (1990). Personality structure: Emergence of the five-factor model. *Annual Review of Psychology 41*, 417–440.

Dijkhuizen, J., Gorgievski, M., Veldhoven, M., & Schalk, R. (2017). Well-being, personal success and business performance among entrepreneurs: A two-wave study. *Journal of Happiness Studies.* https://doi.org/10.1007/s10902-017-9914-6

Dillen, Y., Kemps, H., Gervois, P., Wolfs, E., & Bronckaers, A. (2020). Adult neurogenesis in the subventricular zone and its regulation after ischemic stroke: implications for therapeutic approaches. *Translational Stroke Research, 11*(1), 60–79.

Diller, J. (2015). *Cultural diversity* (5th ed.). Cengage.

Dimian, A. F., Symons, F. J., & Wolff, J. J. (2020). Delay to early intensive behavioral intervention and educational outcomes for a Medicaid-enrolled cohort of children with autism. *Journal of Autism and Developmental Disorders.* https://doi.org/10.1007/s10803-020-04586-1

Dinkel, D., & Snyder, K. (2020). Exploring gender differences in infant motor development related to parent's promotion of play. *Infant Behavior and Development, 59*, 101440.

Dirkx, M. F., den Ouden, H. E., Aarts, E., Timmer, M. H., Bloem, B. R., Toni, I., & Helmich, R. C. (2017). Dopamine controls Parkinson's tremor by inhibiting the cerebellar thalamus. *Brain, 140*(3), 721–734.

Dissel, S. (2020). Drosophila as a model to study the relationship between sleep, plasticity, and memory. *Frontiers in Physiology, 11*, 533.

Distel, M. A., & others. (2008). Heritability of borderline personality disorder features is similar across three countries. *Psychological Medicine, 38*(9), 1219–1229.

Dittmar, H., Bond, R., Hurst, M., & Kasser, T. (2014). The relationship between materialism and personal well-being: A meta-analysis. *Journal of Personality and Social Psychology, 107*(5), 879–924.

Dixon, R. A., McFall, G. P., Whitehead, B. P., & Dolcos, S. (2013). Cognitive development in adulthood and aging. In R. M. Lerner, M. A. Easterbrooks, J. Mistry, & I. B. Weiner (Eds.), *Handbook of psychology* (2nd ed., vol. 6, pp. 451–474). Wiley.

Dobson, R. (2005). "Fighting spirit" after cancer diagnosis does not improve outcome. *British Medical Journal, 330*(7456), 865.

Dodge, E. (2012). Family evolution and process during the child and adolescent years in eating disorders. In J. Lock (Ed.), *The Oxford handbook of child and adolescent eating disorders: Developmental perspectives.* Oxford University Press.

Dodge, K. A., Coie, J. D., & Lynam, D. (2006). Aggression and antisocial behavior in youth. In W. Damon & R. Lerner (Eds.), *Handbook of child psychology* (6th ed.). Wiley.

Doebel, S., Michaelson, L. E., & Munakata, Y. (2020). Good things come to those who wait: Delaying gratification likely does matter for later achievement (a commentary on Watts, Duncan, & Quan, 2018). *Psychological Science, 31*(1), 97–99.

Doi, D., Magotani, H., Kikuchi, T., Ikeda, M., Hiramatsu, S., Yoshida, K., . . . Takahashi, J. (2020). Pre-clinical study of induced pluripotent stem cell-derived dopaminergic progenitor cells for Parkinson's disease. *Nature Communications, 11*(1), 1–14.

Dolan, I. J., Strauss, P., Winter, S., & Lin, A. (2020). Misgendering and experiences of stigma in health care settings for transgender people. *The Medical Journal of Australia, 212*(4), 150–151.

Dolan, M., & Fullam, R. (2006). Face affect recognition deficits in personality-disordered offenders: Association with psychopathy. *Psychological Medicine, 36*(11), 1563–1569.

Dolean, D., & Călugăr, A. (2020). How reliably can we measure a child's true IQ? Socio-economic status can explain most of the inter-ethnic differences in general non-verbal abilities. *Frontiers in Psychology, 11*, 2000.

Dollard, J., Doob, L. W., Miller, N. E., Mowrer, O. H., & Sears, R. R. (1939). *Frustration and aggression.* Yale University Press.

Domhoff, G. W. (2017). *The emergence of dreaming: Mind-wandering, embodied simulation, and the default network.* Oxford University Press.

Domhoff, G. W., & Schneider, A. (2020). From adolescence to young adulthood in two dream series: The consistency and continuity of characters and major personal interests. *Dreaming, 30*(2), 140–161. https://doi.org/10.1037/drm0000133

Domjan, M. (2015). *Principles of learning and behavior* (7th ed.). Cengage.

Domjan, M. (2016). Elicited versus emitted behavior: Time to abandon the distinction. *Journal of the Experimental Analysis of Behavior, 105*(2), 231–245.

Donald, J. N., Bradshaw, E. L., Ryan, R. M., Basarkod, G., Ciarrochi, J., Duineveld, J. J., . . . Sahdra, B. K. (2020). Mindfulness and its association with varied types of motivation: A systematic review and meta-analysis using self-determination theory. *Personality and Social Psychology Bulletin, 46*(7), 1121–1138.

Donaldson, S. I., Csikszentmihalyi, M., & Nakamura, J. (Eds.). (2020). *Positive psychological science: Improving everyday life, health, work, education, and societies across the globe.* Routledge.

Donate Life. (2020). Organ, eye and tissue donation statistics. https://www.donatelife.net/statistics/?gclid=CjwKCAiA8ov_BRAoEiwAOZogwVXGgwahzg6CSVgavwIWPRWbzJOcOH3LCYj2RyzgQc-C45A0hTrcRyhoCJeEQAvD_BwE

Donatelle, R. J. (2015). *Access to health* (14th ed.). Pearson.

Dong, L., Li, G., Gao, Y., Lin, L., Zheng, Y., & Cao, X. B. (2020). Exploring the form-And time-dependent effect of low-frequency electromagnetic fields on maintenance of hippocampal long-term potentiation. *European Journal of Neuroscience.* https://doi.org/10.1111/ejn.14705

Donnellan, M. B., Larsen-Rife, D., & Conger, R. D. (2005). Personality, family history, and competence in early adult romantic relationships. *Journal of Personality and Social Psychology, 88*(3), 562–576.

Donnellan, M. B., Lucas, R. E., & Fleeson, W. (2009). Introduction to personality and assessment at age 40: Reflections on the legacy of the person–situation debate and the future of person–situation integration. *Journal of Research in Personality, 43*(2), 117–119.

Donnelly, G. E., Zheng, T., Haisley, E., & Norton, M. I. (2018). The amount and source of millionaires' wealth (moderately) predict their happiness. *Personality and Social Psychology Bulletin, 44* (5), 684–699.

Donovan, B. M. (2017). Learned inequality: Racial labels in the biology curriculum can affect the development of racial prejudice. *Journal of Research in Science Teaching, 54*(3), 379–411. https://doi.org/10.1002/tea.21370

Dorahy, M. J., & van der Hart, O. (2014). DSM-5's "PTSD with dissociative symptoms": Challenges and future directions. *Journal of Trauma and Dissociation.* PMID: 24983300

Douglas, K. M., Sutton, R. M., & Cichocka, A. (2017). The psychology of conspiracy theories. *Current Directions in Psychological Science, 26*(6), 538–542.

Douven, I. (2019). Putting prototypes in place. *Cognition, 193*, 104007.

Dover, T. L., Hunger, J. M., & Major, B. (2020). Health consequences of prejudice and discrimination. In K. Sweeny, M. L. Robbins, & L. M. Cohen (Eds.), *The Wiley Encyclopedia of Health Psychology* (pp. 231–238). John Wiley & Sons.

Dovidio, J. F., Piliavin, J. A., Schroeder, D. A., & Penner, L. A. (2017). *The social psychology of prosocial behavior.* Psychology Press.

Dovidio, J. F., Saguy, T., Gaertner, S. L., & Thomas, E. L. (2012). From attitudes to (in)action: The darker side of "we." In J. Dixon, M. Keynes, & M. Levine (Eds.), *Beyond prejudice.* Cambridge University Press.

Dowd, E. W., & Golomb, J. D. (2019). The binding problem after an eye movement. *Attention, Perception, & Psychophysics, 82*(1), 168–180

Doyle, A. E., Martin, J., Vuijk, P. J., Capawana, M. R., O'Keefe, S. M., Lee, B. A., et al. (2017). Translating discoveries in attention-deficit/hyperactivity disorder (ADHD) genomics to the clinic. *Journal of the American Academy of Child & Adolescent Psychiatry, 56*(10), S229.

Draguns, J. G., & Tanaka-Matsumi, J. (2003). Assessment of psychopathology across and within cultures: Issues and findings. *Behavior Research and Therapy, 41*(7), 755–776.

Drescher, J., & Fadus, M. (2020). Issues arising in psychotherapy with lesbian, gay, bisexual, and transgender patients. *Focus, 18*(3), 262–267.

Dreyer, B. P. (2020). Racial/ethnic bias in pediatric care and the criminalization of poverty and race/ethnicity—seek and ye shall find. *JAMA Pediatrics, 174*(8), 751–752.

Du, Y., Wang, Y., Yu, M., Tian, X., & Liu, J. (2020). Resting-state functional connectivity of the punishment network associated with conformity. *Frontiers in Behavioral Neuroscience.* https://doi.org/10.3389/fnbeh.2020.617402

Duan, J., Sanders, A., Drigalenko, E., Göring, H. H., & Gejman, P. (2017). Transcriptomic signatures of schizophrenia revealed by dopamine perturbation of lymphoblastoid cell lines. *European Neuropsychopharmacology, 27*, S400.

Duckitt, J. (2001). A dual-process cognitive-motivational theory of ideology and prejudice. *Advances in Experimental Social Psychology, 33*, 41–113. https://doi.org/10.1016/S0065-2601(01)80004-6

Dudukovic, N. M., & Kuhl, B. A. (2017). Cognitive control in memory encoding and retrieval. In T. Egner (Ed.), *The Wiley handbook of cognitive control* (pp. 355–375). Wiley.

Duffy, A., Saunders, K. E., Malhi, G. S., Patten, S., Cipriani, A., McNevin, S. H., . . . Geddes, J. (2019). Mental health care for university students: A way forward? *The Lancet Psychiatry, 6*(11), 885–887.

Duke, A. A., Bègue, L., Bell, R., & Eisenlohr-Moul, T. (2013). Revisiting the serotonin-aggression relationship in humans: A meta-analysis. *Psychological Bulletin, 139*(5), 1148–1172.

Duman, R. S., Li, N., Liu, R. J., Duric, V., & Aghajanian, G. (2012). Signaling pathways underlying the rapid antidepressant actions of ketamine. *Neuropharmacology, 62*(1), 35–41.

Dunbar Jr, E. T., Koltz, R. L., Elliott, A., & Hurt-Avila, K. M. (2020). The role of clinical supervision in treating clients with antisocial personality disorder. *The Journal of Counselor Preparation and Supervision, 13*(3), 4.

Dunbar, R. I. M., & Shultz, S. (2007). Evolution in the social brain. *Science, 317*(5843), 1344–1347.

Duncan, A. E., & others. (2006). Exposure to paternal alcoholism does not predict development of alcohol-use disorders in offspring: Evidence from an offspring-of-twins study. *Journal of Studies on Alcohol, 67*(5), 649–656.

Duncan, B. L., Miller, S. D., Wampold, B. E., & Hubble, M. A. (Eds.). (2010). *The heart and soul of change: Delivering what works in therapy* (2nd ed.). American Psychological Association.

Duncan, G. J., Magnuson, K., & Votruba-Drzal, E. (2015). Children and socioeconomic status. In M. H. Bornstein & T. Leventhal (Eds.), *Handbook of child psychology and developmental science* (7th ed., vol. 4). Wiley.

Duncan, B. L., & Reese, R. J. (2013). Empirically supported treatments, evidence-based treatments, and evidence-based practice. In G. Stricker, T. A. Widiger, & I. B. Weiner (Eds.), *Handbook of Psychology* (2nd ed., vol. 8, pp. 489–514). Wiley.

Dunlap, L. J., O'Farrell, T. J., Schumm, J. A., Orme, S. S., Murphy, M., & Murchowski, P. M. (2020). Group versus standard behavioral couples' therapy for alcohol use disorder patients: Cost-effectiveness. *Journal of Studies on Alcohol and Drugs, 81*(2), 152–163.

Dunn, A. R., Stout, K. A., Ozawa, M., Lohr, K. M., Hoffman, C. A., Bernstein, A. I., et al. (2017). Synaptic vesicle glycoprotein 2C (SV2C) modulates dopamine release and is disrupted in Parkinson disease. *Proceedings of the National Academy of Sciences, 114*(11), E2253–E2262.

Dunn, E. W., Aknin, L. B., & Norton, M. I. (2008). Spending money on others promotes happiness. *Science, 319*(5870), 1687–1688.

Dunn, E. W., Whillans, A. V., Norton, M. I., & Aknin, L. B. (2020). Prosocial spending and buying time: Money as a tool for increasing subjective well-being. In *Advances in Experimental Social Psychology* (vol. 61, pp. 67–126). Academic Press.

Dunne, L., & Opitz, B. (2020). Attention control processes that prioritise task execution may come at the expense of incidental memory encoding. *Brain and Cognition, 144*, 105602.

Dupéré, V., Dion, E., Cantin, S., Archambault, I., & Lacourse, E. (2020). Social contagion and high school dropout: The role of friends, romantic partners, and siblings. *Journal of Educational Psychology.* https://psycnet.apa.org/doi/10.1037/edu0000484

DuPont-Reyes, M. J., Villatoro, A. P., Phelan, J. C., Painter, K., & Link, B. G. (2020). Adolescent views of mental illness stigma: An intersectional lens. *American Journal of Orthopsychiatry, 90*(2), 201–211.

Durankus, F., Ciftdemir, N. A., Ozbek, U. V., Duran, R., & Acunas, B. (2020). Comparison of sleep problems between term and preterm born preschool children. *Sleep Medicine, 75*, 484–490.

Durante, M. A., Kurtenbach, S., Sargi, Z. B., Harbour, J. W., Choi, R., Kurtenbach, S., . . . Goldstein, B. J. (2020). Single-cell analysis of olfactory neurogenesis and differentiation in adult humans. *Nature Neuroscience, 23*(3), 323–326.

Durantini, M. R., Albarracin, D., Mitchell, A. L., Earl, A. N., & Gillette, J. C. (2006). Conceptualizing the influence of social agents of behavior change: A meta-analysis of the effectiveness of HIV-prevention interventionists for different groups. *Psychological Bulletin, 132*, 212–248.

Duverne, S., & Koechlin, E. (2017). Hierarchical control of behaviour in human prefrontal cortex. In T. Egner (Ed.), *The Wiley Handbook of Cognitive Control* (pp. 207–220). Wiley-Blackwell.

Dweck, C. S. (2012). Mindsets and human nature: Promoting change in the Middle East, the school yard, the racial divide, and willpower. *American Psychologist, 67*(8), 614–622.

Dweck, C. S. (2017). *Mindset: Changing the way you think to fulfill your potential.* Random House.

Dworkin, E. R., Ullman, S. E., Stappenbeck, C., Brill, C. D., & Kaysen, D. (2017). Proximal relationships between social support and PTSD symptom severity: A daily diary study of sexual assault survivors. *Depression and Anxiety.*

Dys, A. (2009, February 4). Rock Hill man apologizes on TV for 1961 attack on congressman. *The Rock Hill Herald.* https://www.heraldonline.com/news/local/article12249260.html

Dzokoto, V., Wallace, D. S., Peters, L., & Gentsi-Enchill, E. (2014). Attention to emotion and non-Western faces: Revisiting the facial feedback hypothesis. *Journal of General Psychology, 141*(2), 151–168.

E

Eagle, M. N. (2020). Toward an integrated psychoanalytic theory: Foundation in a revitalized ego psychology. *Clinical Social Work Journal.* https://doi.org/10.1007/s10615-020-00768-7

Eagly, A. H. (2009). The his and hers of prosocial behavior: An examination of the social psychology of gender. *American Psychologist, 64*(8), 644–658.

Eagly, A. H. (2010). Gender roles. In J. Levine & M. Hogg (Eds.), *Encyclopedia of group processes and intergroup relations.* Sage.

Eagly, A. H. (2012). Science, feminism, and the investigation of gender. In R. W. Proctor & E. J. Capaldi (Eds.), *Psychology of science: Implicit and explicit reasoning.* Oxford University Press.

Eagly, A. H. (2020). Do the social roles that women and men occupy in science allow equal access to publication? *Proceedings of the National Academy of Sciences, 117*(11), 5553–5555.

Eagly, A. H., & Crowley, M. (1986). Gender and helping behavior: A meta-analytic review of the social psychological literature. *Psychological Bulletin, 100*(3), 283–308.

Eagly, A. H., & Wood, W. (2011). Feminism and the evolution of sex differences and similarities. *Sex Roles, 64*(9–10), 758–767.

Eagly, A. H., & Wood, W. (2012). Social role theory. In P. A. M. Van Lange, A. W. Kruglanski, & E. T. Higgins (Eds.), *Handbook of theories of social psychology* (pp. 458–476). Sage.

Eagly, A. H., & Wood, W. (2013). The nature–nurture debates: 25 years of challenges in understanding the psychology of gender. *Perspectives on Psychological Science, 8*(3), 340–357.

Earnshaw, V. A., Reisner, S. L., Juvonen, J., Hatzenbuehler, M. L., Perrotti, J., & Schuster, M. A. (2017). LGBTQ bullying: Translating research to action in pediatrics. *Pediatrics*, e20170432.

Easter, M. M., Swanson, J. W., Robertson, A. G., Moser, L. L., & Swartz, M. S. (2020). Impact of psychiatric advance directive facilitation on mental health consumers: empowerment, treatment attitudes and the role of peer support specialists. *Journal of Mental Health.* https://doi.org/10.1080/09638237.2020.1714008

Easterbrook, M. J., & Hadden, I. R. (2020). Tackling educational inequalities with social psychology: Identities, contexts, and interventions. *Social Issues and Policy Review.* https://doi.org/10.1111/sipr.12070

Ebaid, D., & Crewther, S. G. (2020). Time for a systems biological approach to cognitive aging?–A critical review. *Frontiers in Aging Neuroscience, 12*, 114.

Eberhardt, J. L. (2020). *Biased: Uncovering the hidden prejudice that shapes what we see, think, and do.* Penguin Books.

Ebert, S. (2020). Theory of mind, language, and reading: Developmental relations from early childhood to early adolescence. *Journal of Experimental Child Psychology, 191*, 104739.

Eckstein, M. K., Starr, A., & Bunge, S. A. (2019). How the inference of hierarchical rules unfolds over time. *Cognition, 185*, 151–162.

Eckstein, M.P., Koehler, K., Welbourne, L.E., & Akbas, E. (2017). Humans, but not deep neural networks, often miss giant targets in scenes. *Current Biology, 27*, 2827–2832.

Edenberg, H. J., Gelernter, J., & Agrawal, A. (2019). Genetics of alcoholism. *Current Psychiatry Reports, 21*(4), 26.

Edenbrandt, A. K., Gamborg, C., & Jellesmark Thorsen, B. (2020). Observational learning in food choices: The effect of product familiarity and closeness of peers. *Agribusiness.* https://doi.org/10.1002/agr.21638

Eeles, E., Pinsker, D., Burianova, H., & Ray, J. (2020). Dreams and the daydream retrieval hypothesis. *Dreaming. 30*(1), 68–78. https://doi.org/10.1037/drm0000123

Eftekhari, A., Crowley, J. J., Mackintosh, M. A., & Rosen, C. S. (2020). Predicting treatment dropout among veterans receiving prolonged exposure therapy. *Psychological Trauma: Theory, Research, Practice, and Policy, 12*(4), 405–412.

Egan, V., Bull, E., & Trundle, G. (2020). Individual differences, ADHD, adult pathological demand avoidance, and delinquency. *Research in Developmental Disabilities, 105*, 103733.

Ege, R., Van Opstal, A. J., & Van Wanrooij, M. M. (2019). Perceived target range shapes human sound-localization behavior. *Eneuro, 6*(2). https://doi.org/10.1523/ENEURO.0111-18.2019

Egede, L. E., & Walker, R. J. (2020). Structural racism, social risk factors, and Covid-19–a dangerous convergence for black Americans. *New England Journal of Medicine, 383*(12), e77.

Eguchi, E., & others. (2012). Healthy lifestyle behaviors and cardiovascular mortality among Japanese men and women: The Japan collaborative cohort study. *European Heart Journal, 33*(4), 467–477.

Eilertsen, E. M., Gjerde, L. C., Reichborn-Kjennerud, T., Ørstavik, R. E., Knudsen, G. P., Stoltenberg, C., et al. (2017). Maternal alcohol use during pregnancy and offspring attention-deficit hyperactivity disorder (ADHD): A prospective sibling control study. *International Journal of Epidemiology, 46*, 633–1640.

Eisenberg, D. P., Yankowitz, L., Ianni, A. M., Rubinstein, D. Y., Kohn, P. D., Hegarty, C. E., et al. (2017). Presynaptic dopamine synthesis capacity in schizophrenia and striatal blood flow change during antipsychotic treatment and medication-free conditions. *Neuropsychopharmacology, 42*, 2232–2241.

Eisenberg, M., Gower, A., Brown, C., Wood, B., & Porta, C. (2017). "They want to put a label on it:" Patterns and interpretations of sexual orientation and gender identity labels among adolescents. *Journal of Adolescent Health, 60*(2), S27–S28.

Eisenberg, N. (2020). Considering the role of positive emotion in the early emergence of prosocial behavior: Commentary on Hammond and Drummond (2019). *Developmental Psychology, 56*(4), 843–845.

Eisenberg, N., & Spinrad, T. R. (2014). Multidimensionality of prosocial behavior: Re-thinking the conceptualization and development of prosocial behavior. In L. Padilla-Walker & G. Gustavo (Eds.), *Prosocial behavior.* Oxford University Press.

Eisenstein, A. R., & others. (2011). Environmental correlates of overweight and obesity in community residing older adults. *Journal of Aging and Health, 23*(6), 994–1009.

Eissa, N., Mujawar, Q., Abdul-Salam, T., Zohni, S., & El-Matary, W. (2020). The immune-sleep crosstalk in inflammatory bowel disease. *Sleep Medicine, 73*, 38–46.

Ejelöv, E., & Luke, T. J. (2020). "Rarely safe to assume": Evaluating the use and interpretation of manipulation checks in experimental social psychology. *Journal of Experimental Social Psychology, 87*, 103937.

Ekman, P. (1980). *The face of man.* Garland.

Ekman, P. (1996). Lying and deception. In N. L. Stein, C. Brainerd, P. A. Ornstein, & B. Tversky (Eds.), *Memory for everyday emotional events.* Erlbaum.

Ekman, P. (2003). Emotions inside out: 130 years after Darwin's "The expression of emotions in man and animal." *Annals of the New York Academy of Science, 1000*, 1–6.

Ekman, P., & Friesen, W. V. (1969). The repertoire of non-verbal behavior: Categories, origins, usage, and coding. *Semiotica, 1*(1) 49–98.

Ekman, P., & Friesen, W. V. (1971). Constants across cultures in the face and emotion. *Journal of Personality and Social Psychology, 17*(2), 124–129.

Ekman, P., Davidson, R. J., & Friesen, W. V. (1990). The Duchenne smile: Emotional expression and brain physiology: II. *Journal of Personality and Social Psychology, 58*(2), 342–353.

Ekman, P., Levenson, R. W., & Friesen, W. V. (1983). Autonomic nervous system activity distinguishes among emotions. *Science, 221*(4616), 1208–1210.

Ekman, P., & O'Sullivan, M. (1991). Facial expression: Methods, means, and moues. In R. S. Feldman and B. Rime (Eds.), *Fundamentals of nonverbal behavior*. Cambridge University Press.

El-Ansary, A., Hassan, W. M., Daghestani, M., Al-Ayadhi, L., & Ben Bacha, A. (2020). Preliminary evaluation of a novel nine-biomarker profile for the prediction of autism spectrum disorder. *Plos One, 15*(1), e0227626.

Eldred, M. (2017). Love hormones and mental health. *Science, 355*(6330), 1170–1171.

Eldredge, J. H. (2016). Imagined interaction frequency and the five-factor model of personality. *Imagination, Cognition and Personality, 35*(4), 351–358. https://doi.org/10.1177/0276236616636215

Eley, T. C., & others. (2004). Gene-environment interaction analysis of serotonin system markers with adolescent depression. *Molecular Psychiatry, 9*(10), 908–915.

El-Hai, J. (2005). *The lobotomist: A maverick medical genius and his tragic quest to rid the world of mental illness*. Wiley.

Ellemers, N., & van Nunspeet, F. (2020). Neuroscience and the social origins of moral behavior: How neural underpinnings of social categorization and conformity affect everyday moral and immoral behavior. *Current Directions in Psychological Science, 29*(5), 513–520.

Ellermann, C., Coenen, A., Niehues, P., Leitz, P., Kochhäuser, S., Dechering, D. G., . . . Frommeyer, G. (2020). Proarrhythmic effect of acetylcholine-esterase inhibitors used in the treatment of Alzheimer's disease: Benefit of rivastigmine in an experimental whole-heart model. *Cardiovascular Toxicology, 20*(2), 168–175.

Elliott, M. A., & Ainsworth, K. (2012). Predicting university undergraduates' binge-drinking behavior: A comparative test of the one- and two-component theories of planned behavior. *Addictive Behaviors, 37*(1), 92–101.

Ellis, A. (2005). Why I (really) became a therapist. *Journal of Clinical Psychology, 61*(8), 945–948.

Ellis, C. T., Skalaban, L. J., Yates, T. S., Bejjanki, V. R., Córdova, N. I., & Turk-Browne, N. B. (2020). Reimagining fMRI for awake behaving infants. *Nature Communications, 11*(1), 1–12.

Ellis, E. M., Rajagopal, R., & Kiviniemi, M. T. (2018). The interplay between feelings and beliefs about condoms as predictors of their use. *Psychology & Health, 33* (2), 176–192.

Elman, J. (2020). Jahvid Best became an Olympic sprinter after concussions ended his NFL career. *Sportscasting*. https://www.sportscasting.com/jahvid-best-became-an-olympic-sprinter-after-concussions-ended-his-nfl-career/

El-Sheikh, M. (2013). *Auburn university child sleep, health, and development center*. Auburn University.

Emens, J. (2020). Non-24-hour sleep-wake rhythm disorder. In R.R. Auger (Ed.), *Circadian rhythm sleep-wake disorders* (pp. 123–136). Springer.

Emmelkamp, P. M. (2011). Effectiveness of cybertherapy in mental health: A critical appraisal. *Studies in Health Technology and Informatics, 167*, 3–8.

Emmons, R. A., & Diener, E. (1986). Situation selection as a moderator of response consistency and stability. *Journal of Personality and Social Psychology, 51*(5), 1013–1019.

Emmons, R. A., & King, L. A. (1988). Conflict among personal strivings: Immediate and long-term implications for psychological and physical well-being. *Journal of Personality and Social Psychology, 54*(6), 1040–1048.

Emmons, R. A., & McCullough, M. E. (2003). Counting blessings versus burdens: An experimental investigation of gratitude and subjective well-being in daily life. *Journal of Personality and Social Psychology, 84*(2), 377–389.

Emmons, R. A., & McCullough, M. E. (Eds.). (2004). *The psychology of gratitude*. Oxford University Press.

Endler, N. S. (1988). The origins of electroconvulsive therapy (ECT). *Convulsive Therapy, 4*(1), 5–23.

Eng, P. M., Fitzmaurice, G., Kubzansky, L. D., Rimm, E. B., & Kawachi, I. (2003). Anger expression and risk of stroke and coronary heart disease among male health professionals. *Psychosomatic Medicine, 65*(1), 100–110.

Engberg, H., Jeune, B., Andersen-Ranberg, K., Martinussen, T., Vaupel, J. V., & Christensen, K. (2013). Optimism and survival: Does an optimistic outlook predict better survival at advanced ages? A twelve-year follow-up of Danish nonagenarians. *Aging: Clinical and Experimental Research, 25*(5), 517–525.

Engel, A. K., & Singer, W. (2001). Temporal binding and the neural correlates of sensory awareness. *Trends in Cognitive Science, 5*(1), 16–25.

English, T., & Carstensen, L. (2014). Selective narrowing of social networks across adulthood is associated with improved emotional experience in daily life. *International Journal of Behavioral Development, 38*(2), 195–202.

Ensembl Human. (2010). *Explore the homo sapiens genome*. www.ensembl.org/Homo_sapiens/index.html (accessed November 9, 2014)

Ensor, T. M., Bancroft, T. D., & Hockley, W. E. (2019). Listening to the picture-superiority effect: Evidence for the conceptual-distinctiveness account of picture superiority in recognition. *Experimental Psychology, 66*(2), 134–153.

Eom, T. Y., Han, S. B., Kim, J., Blundon, J. A., Wang, Y. D., Yu, J., . . . Horner, L. (2020). Schizophrenia-related microdeletion causes defective ciliary motility and brain ventricle enlargement via microRNA-dependent mechanisms in mice. *Nature Communications, 11*(1), 1–17.

Erickson, J. T. (2020). Central serotonin and autoresuscitation capability in mammalian neonates. *Experimental Neurology, 326*, 113162.

Erickson, K. I., Miller, D. L., & Roecklein, K. A. (2012). The aging hippocampus: Interactions between exercise, depression, and BDNF. *Neuroscientist, 18*(1), 82–97.

Erikson, E. (1969). *Gandhi's truth*. Norton.

Erikson, E. H. (1968). *Identity: Youth and crisis*. Norton.

Eriksson, P. L., Wängqvist, M., Carlsson, J., & Frisén, A. (2020). Identity development in early adulthood. *Developmental Psychology, 56*(10), 1968–1983.

Ernst, J., Johnson, M., & Burcak, F. (2019). The nature and nurture of resilience: Exploring the impact of nature preschools on young children's protective factors. *International Journal of Early Childhood Environmental Education, 6*(2), 7–18.

Espnes, G. A., & Byrne, D. (2016). Type A behavior and cardiovascular disease. In M. E. Alvarenga & D. Byrne (Eds.), *Handbook of psychocardiology* (pp. 645–664). Springer Science + Business Media. https://doi.org/10.1007/978-981-287-206-7_30

Esses, V. M. (2020). Prejudice and discrimination toward immigrants. *Annual Review of Psychology, 72*. https://doi.org/10.1146/annurev-psych-080520-102803

Esteves, M., Ganz, E., Sousa, N., & Leite-Almeida, H. (2020). Asymmetrical brain plasticity: Physiology and pathology. *Neuroscience*.

Etzel, L. C., & Shalev, I. (2021). Effects of psychological stress on telomeres as genome regulators. In G. Fink (Ed.), *Stress: Genetics, epigenetics and genomics*. (pp. 109–117). Academic Press.

Euesden, J., Gowrisankar, S., Qu, A. X., Jean, P. S., Hughes, A. R., & Pulford, D. J. (2020). Cognitive decline in Alzheimer's disease: Limited clinical utility for GWAS or polygenic risk scores in a clinical trial setting. *Genes, 11*(5), 501.

Evans, B. G., & Iverson, P. (2007). Plasticity in vowel perception and production: A study of accent change in young adults. *Journal of the Acoustical Society of America, 121*(6), 3814–3826.

Evans, S., Ling, M., Hill, B., Rinehart, N., Austin, D., & Sciberras, E. (2017). Systematic review of meditation-based interventions for children with ADHD. *European Child & Adolescent Psychiatry*.

Evans, T. C., Bar-Haim, Y., Fox, N. A., Pine, D. S., & Britton, J. C. (2020). Neural mechanisms underlying heterogeneous expression of threat-related attention in social anxiety. *Behaviour Research and Therapy, 132*, 103657.

Eysenck, H. J. (1967). *The biological basis of personality*. Thomas.

F

Fabiano, F., & Haslam, N. (2020). Diagnostic inflation in the DSM: A meta-analysis of changes in the stringency of psychiatric diagnosis from DSM-III to DSM-5. *Clinical Psychology Review, 80*, 101889.

Fagan, S. E., Zhang, W., & Gao, Y. (2017). Social adversity and antisocial behavior: Mediating effects of autonomic nervous system activity. *Journal of Abnormal Child Psychology, 45*, 1553–1564.

Fagundes, C. P., Glaser, R., Hwang, B. S., Malarkey, W. B., & Kiecolt-Glaser, J. K. (2013). Depressive symptoms enhance stress-induced inflammatory responses. *Brain, Behavior, and Immunity, 31*, 172–176.

Fairman, K. A., Peckham, A. M., & Sclar, D. A. (2020). Diagnosis and treatment of ADHD in the United States: Update by gender and race. *Journal of Attention Disorders, 24*(1), 10–19.

Falk, A., Kosse, F., & Pinger, P. (2019). Re-revisiting the marshmallow test: A direct comparison of studies by Shoda, Mischel, and Peake (1990) and Watts, Duncan, and Quan (2018). *Psychological Science, 31*, 100–104.

Fan, J. E., Wammes, J. D., Gunn, J. B., Yamins, D. L., Norman, K. A., & Turk-Browne, N. B. (2020). Relating visual production and recognition of objects in human visual cortex. *Journal of Neuroscience, 40*(8), 1710–1721.

Fan, M., Sun, D., Zhou, T., Heianza, Y., Lv, J., Li, L., & Qi, L. (2020). Sleep patterns, genetic susceptibility, and incident cardiovascular disease: A prospective study of 385 292 UK biobank participants. *European Heart Journal, 41*(11), 1182–1189.

Faraone, S.V., Asherson, P. A., Banaschewski, T., Biederman, J., Buitelaar, J.K. et al. (2015). Attention-deficit hyperactivity disorder. *Nature Reviews Disease Primers, 1*, 15020. https://doi.org/10.1038/nrdp.2015.20

Farr, R. H., & Patterson, C. J. (2013). Coparenting among lesbian, gay, and heterosexual couples: Associations with adopted children's outcomes. *Child Development, 84*(4), 1226–1240.

Farr, R. H., Bruun, S. T., & Patterson, C. J. (2019). Longitudinal associations between coparenting and child adjustment among lesbian, gay, and heterosexual adoptive parent families. *Developmental Psychology, 55*(12), 2547–2560. https://doi.org/10.1037/dev0000828

Farr, R. H., Vázquez, C. P., & Patterson, C. J. (2020). LGBTQ adoptive parents and their children. In A. E. Goldberg & K. R. Allen (Eds.), *LGBTQ-parent families* (pp. 45–64). Springer.

Farrar, M. J., Seung, H. K., & Lee, H. (2017). Language and false-belief task performance in children with autism spectrum disorder. *Journal of Speech, Language, and Hearing Research, 60*(7), 1999–2013.

Fasko Jr, D., & Fair, F. (Eds). (2020). *Critical thinking and reasoning*. Brill Sense.

Fastame, M. C., Hitchcott, P. K., & Penna, M. P. (2017). Does social desirability influence psychological well-being: Perceived physical health and religiosity of Italian elders? A developmental approach. *Aging & Mental Health, 21*(4), 348–353. https://doi.org/10.1080/13607 863.2015.1074162

Fatemi, S. H., & Folsom, T. D. (2009). The neurodevelopmental hypothesis of schizophrenia, revisited. *Schizophrenia Bulletin, 35*(3), 528–548.

Fattinger, S., de Beukelaar, T. T., Ruddy, K. L., Volk, C., Heyse, N. C., Herbst, J. A., . . . & Huber, R. (2017). Deep sleep maintains learning efficiency of the human brain. *Nature Communications, 8*(1), 1–14.

Favaretto, M., De Clercq, E., Briel, M., & Elger, B. S. (2020). Working through ethics review of big data research projects: An investigation into the experiences of Swiss and American researchers. *Journal of Empirical Research on Human Research Ethics*. https://doi.org/10.1177%2F1556264620935223

Fawcett, J. M., & Hulbert, J. C. (2020). The many faces of forgetting: Toward a constructive view of forgetting in everyday life. *Journal of Applied Research in Memory and Cognition, 9*, 1–18.

Faymonville, M. E., Boly, M., & Laureys, S. (2006). Functional neuroanatomy of the hypnotic state. *Journal of Physiology, Paris, 99*, 463–469.

Fazakas-DeHoog, L. L., Rnic, K., & Dozois, D. J. (2017). A cognitive distortions and deficits model of suicide ideation. *Europe's Journal of Psychology, 13*, 178–193.

Fazio, R. H., & Olsen, A. (2007). Attitudes. In M. A. Hogg & J. Cooper (Eds.), *The Sage handbook of social psychology* (concise 2nd ed.). Sage.

Fazio, R. H., Chen, J.-M., McDonel, E. C., & Sherman, S. J. (1982). Attitude accessibility, attitude-behavior consistency, and the strength of the object-evaluation association. *Journal of Experimental Social Psychology, 18*(4), 339–357.

Federal Bureau of Investigation (FBI). (2020). FBI releases 2019 crime statistics. FBI.gov. https://www.fbi.gov/news/pressrel/press-releases/fbi-releases-2019-crime-statistics

Federal Communications Commission (FCC) (2017). The dangers of distracted driving. https://www.fcc.gov/consumers/guides/dangers-texting-while-driving

Federico, C. M., & Aguilera, R. (2019). The distinct pattern of relationships between the big five and racial resentment among white Americans. *Social Psychological and Personality Science, 10*(2), 274–284.

Feijt, M., de Kort, Y., Bongers, I., Bierbooms, J., Westerink, J., & IJsselsteijn, W. (2020). Mental health care goes online: Practitioners' experiences of providing mental health care during the COVID-19 pandemic. *Cyberpsychology, Behavior, and Social Networking, 23*(12), 860–864.

Fein, S., Goethals, G. R., Kassin, S. M., & Cross, J. (1993, August). *Social influence and presidential debates.* Paper presented at the meeting of the American Psychological Association, Toronto.

Feinstein, B. A., Dyar, C., Milstone, J. S., Jabbour, J., & Davila, J. (2020). Use of different strategies to make one's bisexual + identity visible: Associations with dimensions of identity, minority stress, and health. *Stigma and Health.* https://doi.org/10.1037/sah0000225

Feinstein, E. C., Richter, L., & Foster, S. E. (2012). Addressing the critical health problem of adolescent substance use through health care, research, and public policy. *Journal of Adolescent Health, 50*(5), 431–436.

Feitosa-Santana, C. (2017). Understanding how the mind works: The neuroscience of perception, behavior, and creativity. In R. Zuanon (Ed.), *Projective processes and neuroscience in art and design* (pp. 239–252). Information Science Reference/IGI Global. https://doi.org/10.4018/978-1-5225-0510-5.ch014

Fekete, C., Siegrist, J., Post, M. W., Tough, H., Brinkhof, M. W., & SwiSCI Study Group. (2020). Does engagement in productive activities affect mental health and well-being in older adults with a chronic physical disability? Observational evidence from a Swiss cohort study. *Aging & Mental Health, 24*(5), 732–739.

Feldman, R., & Bakermans-Kranenburg, M. J. (2017). Oxytocin: a parenting hormone. *Current Opinion in Psychology, 15*, 13–18.

Feng, Y., Herdman, M., van Nooten, F., Cleeland, C., Parkin, D., Ikeda, S., . . . Devlin, N. J. (2020). An exploration of differences between Japan and two European countries in the self-reporting and valuation of pain and discomfort on the EQ-5D. *Quality of Life Research, 29*, 755–763.

Ferchaud, A., Seibert, J., Sellers, N., & Escobar Salazar, N. (2020). Reducing mental health stigma through identification with video game avatars with mental illness. *Frontiers in Psychology, 11*, 2240.

Ferguson, C. J. (2015). Do angry birds make for angry children? A meta-analysis of video game influences on children's and adolescents' aggression, mental health, prosocial behavior, and academic performance. *Perspectives on Psychological Science, 10*(5), 646–666. https://doi.org/10.1177/1745691615592234

Ferguson, C. J., & Kilburn, J. (2010). Much ado about nothing: The misestimation and overinterpretation of violent video game effects in Eastern and Western nations: Comment on Anderson et al. (2010). *Psychological Bulletin, 136*(2), 174–178.

Ferguson, C. J., Rueda, S. M., Cruz, A. M., Ferguson, D. E., Fritz, S., & Smith, S. M. (2008). Violent video games and aggression: Causal relationship or byproduct of family violence and intrinsic violence motivation? *Criminal Justice and Behavior, 35*(3), 311–332.

Ferguson, G. M., Iturbide, M. I., & Raffaelli, M. (2020). Proximal and remote acculturation: Adolescents' perspectives of biculturalism in two contexts. *Journal of Adolescent Research, 35*(4), 431–460.

Ferguson, S., Zimmer-Gembeck, M. J., & Duffy, A. L. (2016). A longitudinal study of relational aggression and victimisation in early adolescence: Gender differences in the moderating effects of social status. *Journal of Relationships Research, 7.* https://doi.org/10.1017/jrr.2016.9

Fernandez, E. J. (2020). Training petting zoo sheep to act like petting zoo sheep: An empirical evaluation of response-independent schedules and shaping with negative reinforcement. *Animals, 10*(7), 1122.

Fernandez, M. E., Ruiter, R. A., Markham, C. M., & Kok, G. (2019). Theory- and evidence-based health promotion program planning: Intervention mapping. *Frontiers in Public Health, 7*, 209.

Ferrannini, E., & Mari, A. (2020). Physiology of insulin secretion. In S. Melmed, R. J. Auchus, A. B. Goldfine, R. J. Koenig, & C. J. Rosen (Eds.), *William's textbook of endocrinology* (14th ed., pp. 1338–148e3). Elsevier.

Ferrarelli, F., & others. (2013). Experienced mindfulness meditators exhibit higher parietal-occipital EEG gamma activity during NREM sleep. *PLoS One, 8*(8), e73417.

Ferrari, C., Lega, C., Vernice, M., Tamietto, M., Mende-Siedlecki, P., Vecchi, T., et al. (2016b). The dorsomedial prefrontal cortex plays a causal role in integrating social impressions from faces and verbal descriptions. *Cerebral Cortex, 26*(1), 156–165. https://doi.org/10.1093/cercor/bhu186

Ferrari, C., Vecchi, T., Todorov, A., & Cattaneo, Z. (2016). Interfering with activity in the dorsomedial prefrontal cortex via TMS affects social impressions updating. *Cognitive, Affective & Behavioral Neuroscience, 16*(4), 626–634. https://doi.org/10.3758/s13415-016-0419-2

Ferrari, M., & Westrate, N. (Eds.). (2013). *Personal wisdom.* Springer.

Ferrie, J. E., Shipley, M. J., Akbaraly, T. N., Marmot, M. G., Kivimäki, M., & Singh-Manoux, A. (2011). Change in sleep duration and cognitive function: Findings from the Whitehall II Study. *Sleep, 34*(5), 565–573.

Ferrie, J., Miller, H., & Hunter, S. C. (2020). Psychosocial outcomes of mental illness stigma in children and adolescents: A mixed-methods systematic review. *Children and Youth Services Review*, 104961.

Ferrières, J. (2004). The French paradox: Lessons for other countries. *Heart, 90*(1), 107–111.

Fesce, R. (2020). Subjectivity as an emergent property of information processing by neuronal networks. *Frontiers in Neuroscience, 14*.

Festa, M., Durm, T., Lünebach, M., & Gauterin, F. (2020). Difference thresholds for the perception of sinusoidal vertical stimuli of whole-body vibrations in ranges of amplitude and frequency relevant to ride comfort. *Vibration, 3*(2), 116–131.

Festinger, L. (1954). A theory of social comparison processes. *Human Relations, 7*(2), 117–140.

Festinger, L. (1957). *A theory of cognitive dissonance.* Row Peterson.

Festinger, L., & Carlsmith, J. M. (1959). Cognitive consequences of forced compliance. *Journal of Abnormal and Social Psychology, 58*(2), 203–211.

Field, C. A., Richards, D. K., Castro, Y., Alonso Cabriales, J., Wagler, A., & von Sternberg, K. (2020). The effects of a brief motivational intervention for alcohol use through stages of change among nontreatment seeking injured patients. *Alcoholism: Clinical and Experimental Research, 44*(11), 2361–2372.

Fielding, N.G., Lee, R.M., & Blank, G. (Eds.) (2017). *The SAGE Handbook of Online Research Methods* (2nd ed.). Sage.

Filley, C. M. (2020). Social cognition and white matter: Connectivity and cooperation. *Cognitive and Behavioral Neurology, 33*(1), 67–75.

Finch, C. E. (2009). The neurobiology of middle-age has arrived. *Neurobiology and Aging, 30*(4), 515–520.

Fincham, F. D. (2020, in press). Towards a psychology of divine forgiveness. *Psychology of Religion and Spirituality.*

Fink, M., Kellner, C. H., & McCall, W. V. (2014). The role of ECT in suicide prevention. *The Journal of ECT, 30*(1), 5–9.

Finkel, A. G., Ivins, B. J., Yerry, J. A., Klaric, J. S., Scher, A., & Sammy Choi, Y. (2017). Which matters more? A retrospective cohort study of headache characteristics and diagnosis type in soldiers with MTBI/concussion. *Headache: The Journal of Head and Face Pain, 57*(5), 719–728.

Finkelstein-Fox, L., Park, C. L., & Kalichman, S. C. (2020). Health benefits of positive reappraisal coping among people living with HIV/AIDS: A systematic review. *Health Psychology Review, 14*(3), 394–426.

Fischer, B. (Ed.). (2019). *The Routledge handbook of animal ethics.* Routledge.

Fischer, R. (2006). Congruence and functions of personal and cultural values: Do my values reflect my culture's values? *Personality and Social Psychology Bulletin, 32*(11), 1419–1431.

Fisher, O., & Oyserman, D. (2017). Assessing interpretations of experienced ease and difficulty as motivational constructs. *Motivation Science, 3*, 133–163. https://doi.org/10.1037/mot0000055

Fisher, T. D., Moore, Z. T., & Pittenger, M. J. (2012). Sex on the brain? An examination of frequency of sexual cognitions as a function of gender, erotophilia, and social desirability. *Journal of Sex Research, 49*(1), 69–77.

Fitzgerald, P. B. (2020). An update on the clinical use of repetitive transcranial magnetic stimulation in the treatment of depression. *Journal of Affective Disorders, 276*, 90–103.

Fitzhugh, E. C., Bassett, D. R., & Evans, M. F. (2010). Urban trails and physical activity: A natural experiment. *American Journal of Preventive Medicine, 39*(3), 259–262.

Fitzpatrick, C., Archambault, I., Janosz, M., & Pagani, L. S. (2015). Early childhood working memory forecasts high school dropout risk. *Intelligence, 53*, 160–165.

Fitzpatrick, K. K. (2012). Developmental considerations when treating anorexia nervosa in adolescents and young adults. In J. Lock (Ed.), *The Oxford handbook of child and adolescent eating disorders: Developmental perspectives.* Oxford University Press.

Fivush, R. (2011). The development of autobiographical memory. *Annual Review of Psychology, 62*, 559–582.

Flaccus, G. (2017, May 28). Mayor: 'Heroes' Died Protecting Women From Anti-Muslim Rant. Associated Press. http://hosted2.ap.org/APDEFAULT/386c25518f46418 6bf7a2ac026580ce7/Article_2017-05-28-US–Fatal%20 Stabbing-Portland/id-cc9d904c773345e792141b-5737b3ef15

Flegal, K. M., Carroll, M. D., Kit, B. K., & Ogden, C. L. (2012). Prevalence of obesity and trends in the distribution of body mass index among U.S. adults, 1999–2010. *Journal of the American Medical Association, 307*, 491–497.

Fleming, K. A., Heintzelman, S. J., & Bartholow, B. D. (2016). Specifying associations between conscientiousness and executive functioning: Mental set shifting, not prepotent response inhibition or working memory updating. *Journal of Personality, 84*(3), 348–360.

Flor, H. (2014). Psychological pain interventions and neurophysiology: Implications for a mechanisms-based approach. *American Psychologist, 69*(2), 188–196.

Floyd, K., York, C., & Ray, C. D. (2020). Heritability of affectionate communication: A twins study. *Communication Monographs.* https://doi.org/10.1080/03637751.2020.1760327

Flynn, J. R. (1999). Searching for justice: The discovery of IQ gains over time. *American Psychologist, 54*(1), 5–20.

Flynn, J. R. (2006). The history of the American mind in the 20th century: A scenario to explain gains over time and a case for the irrelevance of g. In P. C. Kyllonen, R. D. Roberts, & L. Stankov (Eds.), *Extending intelligence.* Erlbaum.

Flynn, J. R. (2011). Secular changes in intelligence. In R. J. Sternberg & S. B. Kaufman (Eds.), *The Cambridge handbook of intelligence* (pp. 647–665). Cambridge University Press.

Flynn, J. R. (2013). *Are we getting smarter?* Cambridge University Press.

Flynn, James R. (2009). *What is intelligence: Beyond the Flynn effect* (expanded paperback ed.). Cambridge University Press.

Focht, B. C. (2012). Exercise and health-related quality of life. In E. O. Acevedo (Ed.), *The Oxford handbook of exercise psychology.* Oxford University Press.

Foels, R., & Pratto, F. (2014). The hidden dynamics of discrimination: How ideologies organize power and influence intergroup relations. In M. Mikulincer, P. R. Shaver, J. F. Dovidio, & J. A. Simpson (Eds.), *APA Handbook of Personality and Social Psychology* (vol. 2). American Psychological Association.

Fok, H., Hui, C., Bond, M. H., Matsumoto, D., & Yoo, S. H. (2008). Integrating personality, context, relationship, and emotion type into a model of display rules. *Journal of Research in Personality, 42*(1), 133-150.

Folkman, S., & Lazarus, R. S. (1980). An analysis of coping in a middle-aged community sample. *Journal of Health and Social Behavior, 21*(3), 219-239.

Fonkoue, I. T., Marvar, P. J., Norrholm, S., Li, Y., Kankam, M. L., Jones, T. N., . . . Park, J. (2020). Symptom severity impacts sympathetic dysregulation and inflammation in post-traumatic stress disorder (PTSD). *Brain, Behavior, and Immunity, 83*, 260-269.

Foote, B., Smolin, Y., Kaplan, M., Legatt, M. E., & Lipschitz, D. (2006). Prevalence of dissociative disorders in psychiatric outpatients. *American Journal of Psychiatry, 163*(4), 566-568.

Foote, M. (2012). Freedom from high-stakes testing: A formula for small school success. In M. Hantzopoulos & A. R. Tyner-Mullings (Eds.), *Critical small schools: Beyond privatization in New York City urban educational reform* (pp. 121-133). IAP Information Age Publishing.

Forbes, C. N., Tull, M. T., Xie, H., Christ, N. M., Brickman, K., Mattin, M., & Wang, X. (2020). Emotional avoidance and social support interact to predict depression symptom severity one year after traumatic exposure. *Psychiatry Research, 284*, 112746.

Forbes, M. K., Eaton, N. R., & Krueger, R. F. (2017). Sexual quality of life and aging: A prospective study of a nationally representative sample. *The Journal of Sex Research, 54*(2), 137-148.

Forbush, K., Heatherton, T. F., & Keel, P. K. (2007). Relationships between perfectionism and specific disordered eating behaviors. *International Journal of Eating Disorders, 40*(1), 37-41.

Ford, T. C., Crewther, D. P., & Abu-Akel, A. (2020). Psychosocial deficits across autism and schizotypal spectra are interactively modulated by excitatory and inhibitory neurotransmission. *Autism, 24*(2), 364-373.

Fornaciari, C. J., & Lund Dean, K. (2013). I, S, T, and J Grading Techniques for Es, Ns, Fs, and Ps: Insights from the MBTI on managing the grading process. *Journal of Management Education, 37*(6), 828-853. https://doi.org/10.1177/1052562912461737

Forsberg, A., Johnson, W., & Logie, R. H. (2020). Cognitive aging and verbal labeling in continuous visual memory. *Memory & Cognition, 48*, 1196-1213.

Fortuna, K., & Knafo, A. (2014). Parental and genetic contributions to prosocial behavior during childhood. In L. M. Padilla-Walker & G. Carlo (Eds.), *Prosocial development: A multidimensional approach* (pp. 70-89). Oxford University Press. https://doi.org/10.1093/acprof:oso/9780199964772.003.0004

Foss Sigurdson, J., Undheim, A. M., Lance Wallander, J., Lydersen, S., & Sund, A. M. (2018). The longitudinal association of being bullied and gender with suicide ideations, self-harm, and suicide attempts from adolescence to young adulthood: A cohort study. *Suicide and Life-Threatening Behavior, 48*(2), 169-182.

Fountas, K. N., & Smith, J. R. (2007). Historical evolution of stereotactic amygdalotomy for the management of severe aggression. *Journal of Neurosurgery, 106*(4), 710-713.

Francis, R., Estlin, T., Doran, G., Johnstone, S., Gaines, D., Verma, V., . . . Bornstein, B. (2017). AEGIS autonomous targeting for ChemCam on Mars Science Laboratory: Deployment and results of initial science team use. *Science Robotics, 2*, eaan4582. https://doi.org/10.1126/scirobotics.aan4582

Frankenhuis, W. E., & Nettle, D. (2020). The strengths of people in poverty. *Current Directions in Psychological Science, 29*(1), 16-21.

Frankl, V. E. (1946/1984). *Man's search for meaning* (3rd ed.). First Washington Square Press. (Original work published 1946.)

Frankl, V. E. (1946/1984). *Man's search for meaning* (3rd ed.). First Washington Square Press. (original work published 1946)

Frattaroli, J. (2006). Experimental disclosure and its moderators: A meta-analysis. *Psychological Bulletin, 132*(6), 823-865.

Fredrick, S., & Loewenstein, G. (1999). Hedonic adaptation. In D. Kahneman, E. Diener, & N. Schwarz (Eds.), *Well-being: The foundations of hedonic psychology* (pp. 302-329). Russell Sage Foundation.

Fredrickson, B. L. (1998). What good are positive emotions? *Review of General Psychology, 2*(3), 300-319.

Fredrickson, B. L. (2013). Positive emotions broaden and build. In P. Devine & A. Plant (Eds.), *Advances in experimental social psychology* (vol. 47, pp. 1-53). Elsevier.

Fredrickson, B. L., & Siegel, D. J. (2017). Broaden-and-build theory meets interpersonal neurobiology as a lens on compassion and positivity resonance. In P. Gilbert (Eds.), *Compassion: Concepts, research and applications* (pp. 203-217). Routledge/Taylor & Francis Group.

Fredrickson, B. L., Cohn, M. A., Coffey, K. A., Pek, J., & Finkel, S. M. (2008). Open hearts build lives: Positive emotions, induced through loving-kindness meditation, build consequential personal resources. *Journal of Personality and Social Psychology, 95*(5), 1045-1062.

Fredrickson, B. L., Tugade, M. M., Waugh, C. E., & Larkin, G. R. (2003). What good are positive emotions in crisis? A prospective study of resilience and emotions following the terrorist attacks on the United States on September 11th, 2001. *Journal of Personality and Social Psychology, 84*(2), 365-376.

Freedman, J. L., & Fraser, S. C. (1966). Compliance without pressure: The foot-in-the-door technique. *Journal of Personality and Social Psychology, 4*(2), 195-202.

Freeland, A., Manchanda, R., Chiu, S., Sharma, V., & Merskey, H. (1993). Four cases of supposed multiple personality disorder: Evidence of unjustified diagnoses. *Canadian Journal of Psychiatry, 38*(4), 245-247.

Freeman, L., Wu, O. C., Sweet, J., Cohen, M., Smith, G. A., & Miller, J. P. (2020). Facial sensory restoration after trigeminal sensory rhizotomy by collateral sprouting from the occipital nerves. *Neurosurgery, 86*(5), e436-e441.

Freer, A. (2020, February 17). In-app spending on dating apps doubles in 2019. App Business. https://www.businessofapps.com/news/in-app-spending-on-dating-apps-doubles-in-2019/

Freiberg, A. S. (2020). Why we sleep: A hypothesis for an ultimate or evolutionary origin for sleep and other physiological rhythms. *Journal of Circadian Rhythms, 18.*

Freidin, R. (2020). *Adventures in English syntax.* Cambridge University Press.

French, B. H., & Neville, H. A. (2017). What is nonconsensual sex? Young women identify sources of coerced sex. *Violence Against Women, 23*(3), 368-394. https://doi.org/10.1177/1077801216641517

Frenda, S. J., Nichols, R. M., & Loftus, E. F. (2011). Current issues and advances in misinformation research. *Current Directions in Psychological Science, 20*(1), 20-23.

Frere, P. B., Vetter, N. C., Artiges, E., Filippi, I., Miranda, R., Vulser, H., . . . Walter, H. (2020). Sex effects on structural maturation of the limbic system and outcomes on emotional regulation during adolescence. *NeuroImage, 210*, 116441.

Freud, S. (1911). *The interpretation of dreams* (3rd ed.). A. A. Brill (Trans.). Macmillan. (original work published 1899).

Freud, S. (1917). *A general introduction to psychoanalysis.* Washington Square Press.

Freud, S. (1996). Number 23091. In R. Andrews, M. Seidel, & M. Biggs (Eds.), *Columbia World of Quotations.* Columbia University Press. (Original work published in 1918.)

Frick, A., Engman, J., Alaie, I., Björkstrand, J., Gingnell, M., Larsson, E. M., . . . Furmark, T. (2020). Neuro-imaging, genetic, clinical, and demographic predictors of treatment response in patients with social anxiety disorder. *Journal of Affective Disorders, 261*, 230-237.

Friedman, L. M., Rapport, M. D., Raiker, J. S., Orban, S. A., & Eckrich, S. J. (2017). Reading comprehension in boys with ADHD: The mediating roles of working memory and orthographic conversion. *Journal of Abnormal Child Psychology, 45*(2), 273-287.

Friedman, M., & Rosenman, R. (1974). *Type A behavior and your heart.* Knopf.

Friedman, R., Myers, P., & Benson, H. (1998). Meditation and the relaxation response. In H. S. Friedman (Ed.), *Encyclopedia of mental health* (vol. 2). Academic.

Friedrich, M., Mölle, M., Friederici, A. D., & Born, J. (2020). Sleep-dependent memory consolidation in infants protects new episodic memories from existing semantic memories. *Nature Communications, 11*(1), 1-9.

Frijda, N. H. (2007). *The laws of emotion.* Erlbaum.

Fritz, M. M., Walsh, L. C., Cole, S. W., Epel, E., & Lyubomirsky, S. (2020). Kindness and cellular aging: A pre-registered experiment testing the effects of prosocial behavior on telomere length and well-being. *Brain, Behavior, & Immunity-Health, 11*, 100187.

Fritz, R. G., Zimmermann, E., Meier, M., Mestre-Francés, N., Radespiel, U., & Schmidtke, D. (2020). Neurobiological substrates of animal personality and cognition in a nonhuman primate (Microcebus murinus). *Brain and Behavior, 10*(9), e01752.

Fromm, E. (1947). *Man for himself.* Holt, Rinehart & Winston.

Froum, S., & Neymark, A. (2019). Vaping and oral health: It's worse than you think. *Perio-Implant Advisory.* https://www.perioimplantadvisory.com/clinical-tips/article/16412201/vaping-and-oral-health-its-worse-than-you-think

Frumkin, K. (2020). Behavioral conditioning, the placebo effect, and emergency department pain management. *The Journal of Emergency Medicine, 59*, 303-310.

Fu, W., Wang, C., Zou, L., Guo, Y., Lu, Z., Yan, S., & Mao, J. (2020). Psychological health, sleep quality, and coping styles to stress facing the COVID-19 in Wuhan, China. *Translational Psychiatry, 10*(1), 1-9.

Fuensalida-Novo, S., Parás-Bravo, P., Jiménez-Antona, C., Castaldo, M., Wang, K., Benito-González, E., . . . Fernández-De-Las-Peñas, C. (2020). Gender differences in clinical and psychological variables associated with the burden of headache in tension-type headache. *Women & Health, 60*(6), 652-663.

Fuentes, J. J., Fonseca, F., Elices, M., Farre, M., & Torrens, M. (2019). Therapeutic use of LSD in psychiatry: A systematic review of randomized-controlled clinical trials. *Frontiers in Psychiatry, 10*, 943.

Fuertes, J. N., & Nutt Williams, E. (2017). Client-focused psychotherapy research. *Journal of Counseling Psychology, 64*(4), 369-375.

Fuglset, T. S., Endestad, T., Landrø, N. I., & Rø, Ø. (2014). Brain structure alterations associated with weight changes in young females with anorexia nervosa: A case series. *Neurocase, 21*(2), 169-177

Funder, D. C. (2009). Persons, behaviors, and situations: An agenda for personality psychology in the postwar era. *Journal of Research in Personality, 43*(2), 120-126.

Furley, P., & Memmert, D. (2020). The expression of success: Are thin-slices of pre-performance nonverbal behavior prior to throwing darts predictive of performance in professional darts? *Journal of Nonverbal Behavior.* https://doi.org/10.1007/s10919-020-00342-2

G

Gaffrey, M. S., Barch, D. M., Bogdan, R., Farris, K., Petersen, S. E., & Luby, J. L. (2018). Amygdala reward reactivity mediates the association between preschool stress response and depression severity. *Biological Psychiatry, 83*(2), 128-136.

Gaines, A. N., Goldfried, M. R., & Constantino, M. J. (2020). Revived call for consensus in the future of psychotherapy. *Evidence-Based Mental Health.* http://dx.doi.org/10.1136/ebmental-2020-300208

Gal, D., & Wilkie, J. (2010). Real men don't eat quiche: Regulation of gender-expressive choices by men. *Social Psychological and Personality Science, 1,* 291–301.

Gál, É., Ştefan, S., & Cristea, I. A. (2021). The efficacy of mindfulness meditation apps in enhancing users' well-being and mental health related outcomes: A meta-analysis of randomized controlled trials. *Journal of Affective Disorders, 279,* 131–142.

Galderisi, S., Vignapiano, A., Mucci, A., & Boutros, N. N. (2014). Physiological correlates of positive symptoms in schizophrenia. *Current Topics in Behavioral Neuroscience, 21,* 103–128.

Galesloot, T., Grotenhuis, A. J., Fleshner, N. E., James, N. D., Bryan, R. T., Cheng, K. K., . . . Vermeulen, S. H. (2020). Meta-analysis of six genome-wide association studies identifies novel loci associated with recurrence and progression of non-muscle invasive bladder cancer. *European Urology Open Science, 19,* e1369.

Galilee, A., Stefanidou, C., & McCleery, J. P. (2017). Atypical speech versus non-speech detection and discrimination in 4 to 6-yr old children with autism spectrum disorder: An ERP study. *PloS one, 12*(7), e0181354.

Gallistel, C. R., & Papachristos, E. B. (2020). Number and time in acquisition, extinction and recovery. *Journal of the Experimental Analysis of Behavior, 113*(1), 15–36.

Gammon, S., & Ramshaw, G. (2020). Distancing from the Present: Nostalgia and Leisure in Lockdown. *Leisure Sciences.* https://doi.org/10.1080/01490400.2020.1773993

Gamond, L., Vecchi, T., Ferrari, C., Merabet, L. B., & Cattaneo, Z. (2017). Emotion processing in early blind and sighted individuals. *Neuropsychology, 31*(5), 516–524.

Gangopadhyay, P., Chawla, M., Dal Monte, O., & Chang, S. W. (2020). Prefrontal-amygdala circuits in social decision-making. *Nature Neuroscience.* https://www.nature.com/articles/s41593-020-00738-9

Gannon, T. A., Wood, J. L., Pina, A., Tyler, N., Barnoux, M. F., & Vasquez, E. A. (2014). An evaluation of mandatory polygraph testing for sexual offenders in the United Kingdom. *Sexual Abuse, 26*(2), 178–203.

Gantiva, C., Ballén, Y., Casas, M., Camacho, K., Guerra, P., & Vila, J. (2015). Influence of motivation to quit smoking on the startle reflex: Differences between smokers in different stages of change. *Motivation & Emotion, 39*(2), 293–298. https://doi.org/10.1007/s11031-014-9449-7

Gao, Y., Gan, T., Jiang, L., Yu, L., Tang, D., Wang, Y., . . . Ding, G. (2020). Association between shift work and risk of type 2 diabetes mellitus: A systematic review and dose-response meta-analysis of observational studies. *Chronobiology International, 37*(1), 29–46.

Garb, H. N., Wood, J. M., Nezworski, M. T., Grove, W. M., & Stejskal, W. J. (2001). Toward a resolution of the Rorschach controversy. *Psychological Assessment, 13*(4), 433–448.

Garcia, J. (1989). Food for Tolman: Cognition and cathexis in concert. In T. Archer & L. Nilsson (Eds.), *Aversion, avoidance, and anxiety.* Erlbaum.

Garcia, J., & Koelling, R. A. (1966). Relation of cue to consequence in avoidance learning. *Psychonomic Science, 4*(1), 123–124.

Garcia, J., & Koelling, R. A. (2009). Specific hungers and poison avoidance as adaptive specializations of learning. In D. Shanks (Ed.), *Psychology of learning.* Sage.

Garcia, R., Bouleti, C., Li, A., Frasca, D., El Harrouchi, S., Marechal, J., . . . Degand, B. (2020). Hypnosis versus placebo during atrial flutter ablation: The PAINLESS study: A randomized controlled trial. *JACC: Clinical Electrophysiology.* https://doi.org/10.1016/j.jacep.2020.05.028

García-Mieres, H., Lundin, N. B., Minor, K. S., Dimaggio, G., Popolo, R., Cheli, S., & Lysaker, P. H. (2020). A cognitive model of diminished expression in schizophrenia: The interface of metacognition, cognitive symptoms and language disturbances. *Journal of Psychiatric Research, 131,* 169–176.

Gardani, F. (2020). Borrowing matter and pattern in morphology. An overview. *Morphology, 30,* 263–282.

Gardner, H. (1983). *Frames of mind.* Basic.

Gardner, H. (1993). *Multiple intelligences.* Basic.

Gardner, H. (2002). The pursuit of excellence through education. In M. Ferrari (Ed.), *Learning from extraordinary minds.* Erlbaum.

Gardner, M. K. (2014). Theories of intelligence. In M. A. Bray & T. J. Kehle (Eds.), *Oxford handbook of school psychology.* Oxford University Press.

Gardner, W. L., & Martinko, M. J. (1996). Using the Myers-Briggs Type Indicator to study managers: A literature review and research agenda. *Journal of Management, 22,* 45–83.

Garrett, B. L. (2011). *Convicting the innocent: Where criminal prosecutions go wrong.* Harvard University Press.

Gašparović, A. Č. (2020). Free radical research in cancer. *Antioxidants, 9*(2), 157.

Gates, G. (2017). *In US, More Adults Identifying as LGBT.* Gallup. http://news.gallup.com/poll/201731/lgbt-identification-rises.aspx

Gbyl, K., Rostrup, E., Raghava, J. M., Andersen, C., Rosenberg, R., Larsson, H. B. W., & Videbech, P. (2020). Volume of hippocampal subregions and clinical improvement following electroconvulsive therapy in patients with depression. *Progress in Neuro-Psychopharmacology and Biological Psychiatry, 104,* 110048.

Ge, Y., Knittel, C. R., MacKenzie, D., & Zoepf, S. (2020). Racial discrimination in transportation network companies. *Journal of Public Economics, 190,* 104205.

Geary, N. (2020). Control-theory models of body-weight regulation and body-weight-regulatory appetite. *Appetite, 144,* 104440.

Gebauer, J. E., Eck, J., Entringer, T. M., Bleidorn, W., Rentfrow, P. J., Potter, J., & Gosling, S. D. (2020). The well-being benefits of person-culture match are contingent on basic personality traits. *Psychological Science, 31*(10), 1283–1293.

Gedam, S. R., Patil, P. S., & Shivji, I. A. (2018). Childhood night terrors and sleepwalking: diagnosis and treatment. *Open Journal of Psychiatry & Allied Sciences, 9*(1), 73–74.

Geller, E. S. (2002). The challenge of increasing proenvironmental behavior. In R. B. Bechtel & A. Churchman (Eds.), *Handbook of environmental psychology* (pp. 525–540). Wiley.

Geller, E. S. (2006). Occupational injury prevention and applied behavior analysis. In A. C. Gielen, D. A. Sleet, & R. J. DiClemente (Eds.), *Injury and violence prevention: Behavioral science theories, methods, and applications* (pp. 297–322). Jossey-Bass.

Genetics of Personality Consortium. (2015). Meta-analysis of genome-wide association studies for neuroticism, and the polygenic association with major depressive disorder. *JAMA Psychiatry, 72*(7), 642–650. https://doi.org/10.1001/jamapsychiatry.2015.0554

Gennetian, L. A., Tamis-LeMonda, C. S., & Frank, M. C. (2020). Advancing transparency and openness in child development research: Opportunities. *Child Development Perspectives, 14*(1), 3–8.

Genschow, O., Westfal, M., Crusius, J., Bartosch, L., Feikes, K. I., Pallasch, N., & Wozniak, M. (2020). Does social psychology persist over half a century? A direct replication of Cialdini et al.'s (1975) classic door-in-the-face technique. *Journal of Personality and Social Psychology.* In press.

Gentile, D. A., Li, D., Khoo, A., Prot, S., & Anderson, C. A. (2014). Mediators and moderators of long-term effects of violent video games on aggressive behavior: Practice, thinking, and action. *JAMA Pediatrics, 168*(5), 450–457.

Gentrup, S., Lorenz, G., Kristen, C., & Kogan, I. (2020). Self-fulfilling prophecies in the classroom: Teacher expectations, teacher feedback and student achievement. *Learning and Instruction, 66,* 101296.

Geoffroy, M. C., Li, L., & Power, C. (2014). Depressive symptoms and body mass index: Co-morbidity and direction of association in a British birth cohort followed over 50 years. *Psychological Medicine, 44*(12), 2641–2652.

George, L. S., & Park, C. L. (2017). Does spirituality confer meaning in life among heart failure patients and cancer survivors? *Psychology of Religion and Spirituality, 9*(1), 131–136. https://doi.org/10.1037/rel0000103

Geraerts, E., Lindsay, D. S., Merckelbach, H., Jelicic, M., Raymaekers, L., Arnold, M. M., & Schooler, J. W. (2009). Cognitive mechanisms underlying recovered-memory experiences of childhood sexual abuse. *Psychological Science, 20*(1), 92–98.

Geraets, A. F., & van der Velden, P. G. (2020). Low-cost non-professional interventions for victims of sexual violence: A systematic review. *Aggression and Violent Behavior, 53,* 101425.

Gerchen, M. F., Rentsch, A., Kirsch, M., Kiefer, F., & Kirsch, P. (2019). Shifts in the functional topography of frontal cortex-striatum connectivity in alcohol use disorder. *Addiction Biology, 24*(6), 1245–1253.

Gerentes, M., Pelissolo, A., Rajagopal, K., Tamouza, R., & Hamdani, N. (2019). Obsessive-compulsive disorder: Autoimmunity and neuroinflammation. *Current Psychiatry Reports, 21*(8), 78.

Gerrits, R., Verhelst, H., & Vingerhoets, G. (2020). Mirrored brain organization: Statistical anomaly or reversal of hemispheric functional segregation bias? *Proceedings of the National Academy of Sciences, 117*(25), 14057–14065.

Gertel, V. H., Zhang, H., & Diaz, M. T. (2020). Stronger right hemisphere functional connectivity supports executive aspects of language in older adults. *Brain and Language, 206,* 104771.

Getahun, D., Fassett, M. J., Peltier, M. R., Wing, D. A., Xiang, A. H., Chiu, V., & Jacobsen, S. J. (2017). Association of perinatal risk factors with autism spectrum disorder. *American Journal of Perinatology, 34,* 295–304.

Geukes, K., Nestler, S., Hutteman, R., Dufner, M., Küfner, A. P., Egloff, B., et al. (2017). Puffed-up but shaky selves: State self-esteem level and variability in narcissists. *Journal of Personality and Social Psychology, 112*(5), 769–786. https://doi.org/10.1037/pspp0000093

Geuter, S., Koban, L., & Wager, T. D. (2017). The cognitive neuroscience of placebo effects: Concepts, predictions, and physiology. *Annual Review of Neuroscience, 40,* 167–188.

Ghosal, S., Hare, B. D., & Duman, R. S. (2017). Prefrontal cortex GABAergic deficits and circuit dysfunction in the pathophysiology and treatment of chronic stress and depression. *Current Opinion in Behavioral Sciences, 14,* 1–8.

Giacomin, M., & Rule, N. O. (2020). How static facial cues relate to real-world leaders' success: A review and meta-analysis. *European Review of Social Psychology, 31*(1), 120–148.

Giang, D. W., Goodman, A. D., Schiffer, R. B., Mattson, D. H., Petrie, M., Cohen, N., & Ader, R. (1996). Conditioning of cyclophosphamide-induced leukopenia in humans. *Journal of Neuropsychiatry and Clinical Neuroscience, 8*(2), 194–201.

Gibbs, J. C. (2014). *Moral development and reality: Beyond the theories of Kohlberg and Hoffman* (3rd ed.). Oxford University Press.

Giehl, K., Ophey, A., Reker, P., Rehberg, S., Hammes, J., Barbe, M. T., . . . & van Eimeren, T. (2020). Effects of Home-Based Working Memory Training on Visuo-Spatial Working Memory in Parkinson's Disease: A Randomized Controlled Trial. *Journal of Central Nervous System Disease, 12.* https://doi.org/10.1177/1179573519899469.

Giesbers, S. A. H., Hendriks, A. H. C., Hastings, R. P., Jahoda, A., Tournier, T., & Embregts, P. J. C. M. (2020). Family-based social capital of emerging adults with and without mild intellectual disability. *Journal of Intellectual Disability Research, 64*(10), 757–769.

Gilbert, D. T., & Malone, P. S. (1995). The correspondence bias. *Psychological Bulletin, 117*(1), 21–38.

Gilligan, C. (1982). *In a different voice.* Harvard University Press.

Gilmore, A. K., Walsh, K., Badour, C. L., Ruggiero, K. J., Kilpatrick, D. G., & Resnick, H. S. (2018). Suicidal ideation, posttraumatic stress, and substance abuse based on forcible and drug- or alcohol-facilitated/incapacitated rape histories in a national sample of women. *Suicide and Life-Threatening Behavior, 48*(2), 183–192. https://doi.org/10.1111/sltb.12337

Giordano, S., & Holm, S. (2020). Is puberty delaying treatment 'experimental treatment'? *International Journal of*

Transgender Health, 21. https://doi.org/10.1080/26895269.2020.1747768

Girault, J. A. (2020). Epigenetic tinkering with neurotransmitters. *Science, 368*(6487), 134–135.

Giridharan, V. V., Sayana, P., Pinjari, O. F., Ahmad, N., da Rosa, M. I., Quevedo, J., & Barichello, T. (2020). Postmortem evidence of brain inflammatory markers in bipolar disorder: A systematic review. *Molecular Psychiatry, 25*(1), 94–113.

Girme, Y. U., Jones, R. E., Fleck, C., Simpson, J. A., & Overall, N. C. (2020). Infants' attachment insecurity predicts attachment-relevant emotion regulation strategies in adulthood. *Emotion.* https://doi.org/10.1037/emo0000721

Gittelman, M. (2008). Editor's introduction: Why are the mentally ill dying? *International Journal of Mental Health, 37*(1), 3–12.

Giusino, D., Fraboni, F., De Angelis, M., & Pietrantoni, L. (2019). Commentary: Principles, approaches and challenges of applying big data in safety psychology research. *Frontiers in Psychology, 10*, 2801.

Giustino, T. F., Ramanathan, K. R., Totty, M. S., Miles, O. W., & Maren, S. (2020). Locus coeruleus norepinephrine drives stress-induced increases in basolateral amygdala firing and impairs extinction learning. *Journal of Neuroscience, 40*(4), 907–916.

Glaw, X. M., Garrick, T. M., Terwee, P. J., Patching, J. R., Blake, H., & Harper, C. (2009). Brain donation: Who and why? *Cell and Tissue Banking, 10*(3), 241–246.

Glogger-Frey, I., Gaus, K., & Renkl, A. (2017). Learning from direct instruction: Best prepared by several self-regulated or guided invention activities? *Learning and Instruction, 51*, 26–35.

Godden, D. R., & Baddeley, A. D. (1975). Context-dependent memory in two natural environments: On land and under water. *British Journal of Psychology, 66*(3), 325–331.

Goethals, G. R., & Demorest, A. P. (1995). The risky shift is a sure bet. In M. E. Ware & D. E. Johnson (Eds.), *Demonstrations and activities in teaching of psychology* (vol. 3). Erlbaum.

Goffin, K. C., Kochanska, G., & Yoon, J. E. (2020). Children's theory of mind as a mechanism linking parents' mind-mindedness in infancy with children's conscience. *Journal of Experimental Child Psychology, 193*, 104784.

Goldberg, E. (2020, August 11). For doctors of color, microaggressions are all too familiar. *New York Times,* D4. https://www.nytimes.com/2020/08/11/health/microaggression-medicine-doctors.html

Goldberg, S. B., Rowe, G., Malte, C. A., Ruan, H., Owen, J. J., & Miller, S. D. (2019). Routine monitoring of therapeutic alliance to predict treatment engagement in a veterans affairs substance use disorders clinic. *Psychological Services,17*(3), 291–299.

Golden, A., & Kessler, C. (2020). Obesity and genetics. *Journal of the American Association of Nurse Practitioners, 32*(7), 493–496.

Goldfarb, A. H., Kraemer, R. R., & Baiamonte, B. A. (2020). Endogenous opiates and exercise-related hypoalgesia. In A. Hackney & N. Constantini (Eds.), *Endocrinology of physical activity and sport. Contemporary endocrinology* (pp. 19–39). Humana.

Goldfarb, E. S., & Lieberman, L. D. (2020). Three decades of research: The case for comprehensive sex education. *Journal of Adolescent Health.* https://doi.org/10.1016/j.jadohealth.2020.07.036

Goldin-Meadow, S. (2014a). Language and the manual modality: How our hands help us talk and think. In N. J. Enfield & others (Eds.), *The Cambridge handbook of linguistic anthropology.* Cambridge University Press.

Goldin-Meadow, S. (2014b). Widening the lens: What the manual modality reveals about language, learning, and cognition. *Philosophical Transactions of the Royal Society, Series B, 369*(1651), 20130295.

Golds, L., de Kruiff, K., & MacBeth, A. (2020). Disentangling genes, attachment, and environment: A systematic review of the developmental psychopathology literature on gene–environment interactions and attachment. *Development and Psychopathology, 32*(1), 357–381.

Goldschmidt, A. B., Crosby, R. D., Cao, L., Pearson, C. M., Utzinger, L. M., Pacanowski, C. R., et al. (2017). Contextual factors associated with eating in the absence of hunger among adults with obesity. *Eating Behaviors, 26*, 33–39.

Goldsmith, A. A., Thompson, R. D., Black, J. J., Tran, G. Q., & Smith, J. P. (2012). Drinking refusal self-efficacy and tension-reduction alcohol expectancies moderating the relationship between generalized anxiety and drinking behaviors in young adult drinkers. *Psychology of Addictive Behaviors, 26*(1), 59–67.

Goldstein, M. H., King, A. P., & West, M. J. (2003). Social interaction shapes babbling: Testing parallels between birdsong and speech. *Proceedings of the National Academy of Sciences USA, 100*(13), 8030–8035.

Goldston, D. B., Molock, S. D., Whibeck, L. B., Murakami, J. L., Zayas, L. H., & Hall, G. C. (2008). Cultural considerations in adolescent suicide prevention and psychosocial treatment. *American Psychologist, 63*(1), 14–31.

Goldstrom, I. D., Campbell, J., Rogers, J. A., Lambert, D. B., Blacklow, B., Henderson, M. J., & Manderscheid, R. W. (2006). National estimates for mental health mutual support groups, self-help organizations, and consumer-operated services. *Administration and Policy in Mental Health, 33*(1), 92–103.

Goleman, D., Kaufman, P., & Ray, M. (1993). *The creative mind.* Plume.

Golombok, S., Mellish, L., Jennings, S., Casey, P., Tasker, F., & Lamb, M. E. (2014). Adoptive gay father families: Parent–child relationships and children's psychological adjustment. *Child Development, 85*(2), 456–468.

Gomez, R., Watson, S., Wynen, J. V., Trawley, S., Stavropoulos, V., & Corr, P. J. (2020). Reinforcement sensitivity theory of personality questionnaire: Factor structure based on CFA and ESEM, and associations with ADHD. *Journal of Personality Assessment.* https://doi.org/10.1080/00223891.2020.1769113

Gomez-Nicola, D., & Perry, V. H. (2014). Microglial dynamics and role in the healthy and diseased brain: A paradigm of functional plasticity. *Neuroscientist.* https://doi.org/10.1177/1073858414530512

Gonthier, C., & Roulin, J. L. (2020). Intraindividual strategy shifts in Raven's matrices, and their dependence on working memory capacity and need for cognition. *Journal of Experimental Psychology: General, 149*(3), 564–579.

Gonzalez, V. M., Burroughs, A., & Skewes, M. C. (2020). Belief in the American Indian/Alaska Native biological vulnerability myth and drinking to cope: Does stereotype threat play a role? *Cultural Diversity and Ethnic Minority Psychology.* https://doi.org/10.1037/cdp0000366

González-Roldán, A. M., Terrasa, J. L., Sitges, C., van der Meulen, M., Anton, F., & Montoya, P. (2020). Age-related changes in pain perception are associated with altered functional connectivity during resting state. *Frontiers in Aging Neuroscience, 12*, 116. https://doi.org/10.3389/fnagi.2020.00116

Gonzalez-Vallejo, C., Lassiter, G. D., Bellezza, F. S., & Lindberg, M. J. (2008). "Save angels perhaps": A critical examination of unconscious thought theory and the deliberation-without-attention effect. *Review of General Psychology, 12*(3), 282–296.

Goode, E., & Schwartz, J. (2011, August 28). Police lineups start to face fact: Eyes can lie. *New York Times,* A1.

Goodman, M., Tomas, I. A., Temes, C. M., Fitzmaurice, G. M., Aguirre, B. A., & Zanarini, M. C. (2017). Suicide attempts and self-injurious behaviours in adolescent and adult patients with borderline personality disorder. *Personality and Mental Health, 11*, 157–163.

Gorbaniuk, O., Razmus, W., Slobodianyk, A., Mykhailych, O., Troyanowskyj, O., Kashchuk, M., et al. (2017). Searching for a common methodological ground for the study of politicians' perceived personality traits: A multilevel psycholexical approach. *Journal of Research in Personality, 70*, 27–44.

Gordon, I., Gilboa, A., Cohen, S., Milstein, N., Haimovich, N., Pinhasi, S., & Siegman, S. (2020). Physiological and behavioral synchrony predict group cohesion and performance. *Scientific Reports, 10*(1), 1–12.

Gordon, I., Zagoory-Sharon, O., Leckman, J. F., & Feldman, R. (2010). Oxytocin and the development of parenting in humans. *Biological Psychiatry, 68*(4), 377–382.

Gordon, N., Tsuchiya, N., Koenig-Robert, R., & Hohwy, J. (2019). Expectation and attention increase the integration of top-down and bottom-up signals in perception through different pathways. *PLoS Biology, 17*(4), e3000233.

Gordon, R., Smith-Spark, J. H., Newton, E. J., & Henry, L. A. (2020). Working memory and high-level cognition in children: An analysis of timing and accuracy in complex span tasks. *Journal of Experimental Child Psychology, 191*, 104736.

Gordovez, F. J. A., & McMahon, F. J. (2020). The genetics of bipolar disorder. *Molecular Psychiatry, 25*, 544–559.

Goritz, C., & Frisén, J. (2012). Neural stem cells and neurogenesis in the adult. *Cell: Stem Cell, 10*(6), 657–659.

Gorman, I., Belser, A. B., Jerome, L., Hennigan, C., Shechet, B., Hamilton, S., . . . Feduccia, A. A. (2020). Posttraumatic growth after MDMA-assisted psychotherapy for posttraumatic stress disorder. *Journal of Traumatic Stress, 33*(2), 161–170.

Gosling, S. D. (2008). Personality in nonhuman animals. *Social and Personality Psychology Compass, 2*(2), 985–1001.

Gosling, S. D., & John, O. P. (1999). Personality dimensions in nonhuman animals: A cross-species review. *Current Directions in Psychological Science, 8*(3), 69–75.

Gosling, S. D., Kwan, V. S., & John, O. (2003). A dog's got personality: A cross-species comparison of personality judgments in dogs and humans. *Journal of Personality and Social Psychology, 85*(6), 1161–1169.

Gothe, N. P. (2020). Examining the effects of light versus moderate to vigorous physical activity on cognitive function in African American adults. *Aging & Mental Health.* https://doi.org/10.1080/13607863.2020.1768216

Gott, J., Rak, M., Bovy, L., Peters, E., van Hooijdonk, C. F., Mangiaruga, A., . . . Weber, F. (2020). Sleep fragmentation and lucid dreaming. *Consciousness and Cognition, 84*, 102988.

Gottlieb, G. (2007). Probabilistic epigenesis. *Developmental Science, 10*(1), 1–11.

Gottman, J. M. (2006, April 29). Love special: Secrets of long-term love. *New Scientist, 2549*, 40.

Gottman, J. M., Swanson, C., & Swanson, K. (2002). A general systems theory of marriage: Nonlinear difference equation modeling of marital interaction. *Personality and Social Psychology Review, 6*(4), 326–340.

Götz, F. M., Gvirtz, A., Galinsky, A. D., & Jachimowicz, J. M. (2020). How personality and policy predict pandemic behavior: Understanding sheltering-in-place in 55 countries at the onset of COVID-19. *American Psychologist.* Advance online publication. https://doi.org/10.1037/amp0000740

Gould, S. J. (1981). *The mismeasure of man.* W.W. Norton.

Goulet, M. H., Pariseau-Legault, P., Côté, C., Klein, A., & Crocker, A. G. (2020). Multiple stakeholders' perspectives of involuntary treatment orders: a meta-synthesis of the qualitative evidence toward an exploratory model. *International Journal of Forensic Mental Health, 19*(1), 18–32.

Goyal, M. K., Kuppermann, N., Cleary, S. D., Teach, S. J., & Chamberlain, J. M. (2015). Racial disparities in pain management of children with appendicitis in emergency departments. *JAMA Pediatrics, 169*(11), 996–1002.

Graber, J. A. (2013). Pubertal timing and the development of psychopathology in adolescence and beyond. *Hormones and Behavior, 64*(2), 262–269.

Graham, J., Nosek, B. A., Haidt, J., Iyer, R., Koleva, S., & Ditto, P. H. (2011). Mapping the moral domain. *Journal of Personality and Social Psychology, 101*(2), 366–385.

Graham, S. (2017). Attention-deficit hyperactivity disorder (ADHD), learning disabilities (LD), and executive functioning: Recommendations for future research. *Contemporary Educational Psychology, 50*, 97–101. https://doi.org/10.1016/j.cedpsych.2017.01.001

Graham, S. (2020). An attributional theory of motivation. *Contemporary Educational Psychology, 61*, 101861.

Granacher, R. P. (2014). Commentary: Dissociative amnesia and the future of forensic psychiatry assessment.

Journal of the American Academy of Psychiatry and Law, 42(2), 214–218.

Grandin, T. (2006). *Thinking in pictures: My life with autism.* Random House.

Grando, S. A. (2014). Connections of nicotine to cancer. *Nature Reviews. Cancer, 14*(6), 419–429.

Grant, B. F., & others. (2004). Prevalence and co-occurrence of substance use disorders and independent mood and anxiety disorders: Results from the National Epidemiologic Survey on Alcohol and Related Conditions. *Archives of General Psychiatry, 61*(8), 807–816.

Grant, L. K., Cain, S. W., Chang, A. M., Saxena, R., Czeisler, C. A., & Anderson, C. (2018). Impaired cognitive flexibility during sleep deprivation among carriers of the Brain Derived Neurotrophic Factor (BDNF) Val66Met allele. *Behavioural Brain Research, 338,* 51–55.

Gratz, K. L., Tull, M. T., Richmond, J. R., Edmonds, K. A., Scamaldo, K., & Rose, J. P. (2020). Thwarted belongingness and perceived burdensomeness explain the associations of COVID-19 social and economic consequences to suicide risk. *Suicide and Life-Threatening Behavior.* https://doi.org/10.1111/sltb.12654

Gray, J. A. (1987). *The psychology of fear and stress.* Cambridge University Press.

Gray, J. A., & McNaughton, N. (2000). *The neuropsychology of anxiety: An enquiry into the functions of the septo-hippocampal system.* Oxford University Press.

Gray, J. M., & Diller, J. W. (2017). Evaluating the work of applied animal behaviorists as applied behavior analysis. *Behavior Analysis: Research And Practice, 17*(1), 33–41. https://doi.org/10.1037/bar0000041

Grayot, J. D. (2020). Dual process theories in behavioral economics and neuroeconomics: A critical review. *Review of Philosophy and Psychology, 11*(1), 105–136.

Grayson, D. S., Bliss-Moreau, E., Bennett, J., Lavenex, P., & Amaral, D. G. (2017). Neural Reorganization Due to Neonatal Amygdala Lesions in the Rhesus Monkey: Changes in Morphology and Network Structure. *Cerebral Cortex, 27*(6), 3240–3253.

Graziani, A. R., Guidetti, M., & Cavazza, N. (2020). Food for boys and food for girls: Do preschool children hold gender stereotypes about food? *Sex Roles.* https://doi.org/10.1007/s11199-020-01182-6

Graziano, M. S., Guterstam, A., Bio, B. J., & Wilterson, A. I. (2020). Toward a standard model of consciousness: Reconciling the attention schema, global workspace, higher-order thought, and illusionist theories. *Cognitive Neuropsychology, 37*(3-4), 155–172.

Greco, F. A., & Deutsch, C. K. (2017). Carl Gustav Jung and the psychobiology of schizophrenia. *Brain: A Journal of Neurology, 140*(1).

Green, J. P., Page, R. A., Handley, G. W., & Rasekhy, R. (2005). The "hidden observer" and ideomotor responding: A real-simulator comparison. *Contemporary Hypnosis, 22*(3), 123–137.

Green, T. A., Baracz, S. J., Everett, N. A., Robinson, K. J., & Cornish, J. L. (2020). Differential effects of GABA A receptor activation in the prelimbic and orbitofrontal cortices on anxiety. *Psychopharmacology.* https://doi.org/10.1007/s00213-020-05606-9.

Greenbaum, C. A., & Javdani, S. (2017). Expressive writing intervention promotes resilience among juvenile justice-involved youth. *Children and Youth Services Review, 73,* 220–229. https://doi.org/10.1016/j.childyouth.2016.11.034

Greene, C. M., Maloney-Derham, R., & Mulligan, K. (2020). Effects of perceptual load on eyewitness memory are moderated by individual differences in cognitive ability. *Memory, 28*(4), 450–460.

Greene, N. R., & Naveh-Benjamin, M. (2020). A specificity principle of memory: Evidence from aging and associative memory. *Psychological Science, 31*(3), 316–331.

Greenhill, C. (2020). Three more genes linked with childhood obesity. *Nature Reviews Endocrinology, 16*(8), 402–403.

Greenwald, A. G., & Lai, C. K. (2020). Implicit social cognition. *Annual Review of Psychology, 71,* 419–445.

Greenwald, A. G., Poehlman, T. A., Uhlmann, E., & Banaji, M. R. (2009). Understanding and using the Implicit Association Test: III. Meta-analysis of predictive validity. *Journal of Personality and Social Psychology, 97*(1), 17–41.

Greer, S., Morris, T., & Pettingale, K. W. (1979). Psychological response to breast cancer: Effect on outcome. *Lancet, 2*(8146) 785–787.

Greitemeyer, T. (2009). Effects of songs with prosocial lyrics on prosocial thoughts, affect, and behavior. *Journal of Experimental Social Psychology, 45*(1), 186–190.

Greitemeyer, T., & Osswald, S. (2011). Playing prosocial video games increases the accessibility of prosocial thoughts. *Journal of Social Psychology, 151*(2), 121–128.

Griffin, B., & Hu, W. (2015). The interaction of socio-economic status and gender in widening participation in medicine. *Medical Education, 49*(1), 103–113.

Griggs, R. A., & Whitehead, G. I. (2015). Coverage of Milgram's obedience experiments in social psychology textbooks: Where have all the criticisms gone? *Teaching of Psychology, 42*(4), 315–322. https://doi.org/10.1177/0098628315603065

Grigorenko, E. L., Jarvin, L., Tan, M., & Sternberg, R. J. (2009). Something new in the garden: Assessing creativity in academic domains. *Psychology Science Quarterly, 50*(2), 295–307.

Grigoriadis, S., Graves, L., Peer, M., Mamisashvili, L., Ruthirakuhan, M., Chan, P., . . . Steiner, M. (2020). Pregnancy and delivery outcomes following benzodiazepine exposure: A systematic review and meta-analysis. *The Canadian Journal of Psychiatry.* https://doi.org/10.1177/0706743720904860

Grilo, C. M., Masheb, R. M., & White, M. A. (2010). Significance of overvaluation of shape/weight in binge-eating disorder: Comparative study with overweight and bulimia nervosa. *Obesity, 18*(3), 499–504.

Grimm, K. J., Davoudzadeh, P., & Ram, N. (2017). Developmental methodology: IV. Developments in the analysis of longitudinal data. *Monographs of the Society for Research in Child Development, 82*(2), 46–66. https://doi.org/10.1111/mono.12298

Grodd, W., Kumar, V. J., Schüz, A., Lindig, T., & Scheffler, K. (2020). The anterior and medial thalamic nuclei and the human limbic system: Tracing the structural connectivity using diffusion-weighted imaging. *Scientific Reports, 10*(1), 1–25.

Groh, A. M., & Narayan, A. J. (2019). Infant attachment insecurity and baseline physiological activity and physiological reactivity to interpersonal stress: A meta-analytic review. *Child Development, 90*(3), 679–693.

Groh, A. M., Narayan, A. J., Bakermans-Kranenburg, M. J., Roisman, G. I., Vaughn, B. E., Fearon, R. M., & IJzendoorn, M. H. (2017). Attachment and temperament in the early life course: A meta-analytic review. *Child Development, 88*(3), 770–795.

Grønning Hansen, M., Laterza, C., Palma-Tortosa, S., Kvist, G., Monni, E., Tsupykov, O., . . . Martino, G. (2020). Grafted human pluripotent stem cell-derived cortical neurons integrate into adult human cortical neural circuitry. *Stem Cells Translational Medicine.* https://doi.org/10.1002/sctm.20-0134

Gross, J. T., Stern, J. A., Brett, B. E., & Cassidy, J. (2017). The multifaceted nature of prosocial behavior in children: Links with attachment theory and research. *Social Development, 26,* 661–678.

Grossman, I., Na, J., Varnum, M. E. W., Park, D. C., Kitayama, S., & Nisbett, R. E. (2010). Reasoning about social conflicts improves into old age. *Proceedings of the National Academy of Sciences USA, 107*(16), 7246–7250.

Grosz, M. P., Rohrer, J. M., & Thoemmes, F. (2020, in press). The taboo against explicit causal inference in nonexperimental psychology. *Perspectives on Psychological Science.*

Gualtieri, S., & Denison, S. (2018). The development of the representativeness heuristic in young children. *Journal of Experimental Child Psychology, 174,* 60–76.

Guardino, C. M., Schetter, C. D., Hobel, C. J., Lanzi, R. G., Schafer, P., Thorp, J. M., & Shalowitz, M. U. (2017). Chronic stress and C-reactive protein in mothers during the first postpartum year. *Psychosomatic Medicine, 79*(4), 450–460.

Guéguen, N., Martin, A., Silone, F., & David, M. (2016). Foot-in-the-door technique and reduction of driver's aggressiveness: A field study. *Transportation Research Part F: Traffic Psychology and Behaviour,* 361–365. https://doi.org/10.1016/j.trf.2015.10.006

Guitard, D., & Cowan, N. (2020). Do we use visual codes when information is not presented visually? *Memory & Cognition.* https://doi.org/10.3758/s13421-020-01054-0

Gul, P., Cross, S. E., & Uskul, A. K. (2020). Implications of culture of honor theory and research for practitioners and prevention researchers. *American Psychologist.* https://doi.org/10.1037/amp0000653

Gülgöz, S., & Sahin-Acar, B. (Eds.). (2020). *Autobiographical Memory Development: Theoretical and Methodological Approaches.* Routledge.

Gunderson, J. (2008). Borderline personality disorder: An overview. *Social Work in Mental Health, 6*(1–2), 5–12.

Gunderson, J. G., & others. (2003). Plausibility and possible determinants of sudden "remissions" in borderline patients. *Psychiatry: Interpersonal and Biological Processes, 66*(2), 111–119.

Gunnerud, H. L., ten Braak, D., Reikerås, E. K. L., Donolato, E., & Melby-Lervåg, M. (2020). Is bilingualism related to a cognitive advantage in children? A systematic review and meta-analysis. *Psychological Bulletin.* Advance online publication. https://doi.org/10.1037/bul0000301

Guo, D., & Yang, J. (2020). Interplay of the long axis of the hippocampus and ventromedial prefrontal cortex in schema-related memory retrieval. *Hippocampus, 30*(3), 263–277.

Guo, Y. (2005). Filial therapy for children's behavioral and emotional problems in mainland China. *Journal of Child and Adolescent Psychiatric Nursing, 18*(4), 171–180.

Gupta, M. A. (2020). Spontaneous reporting of onset of disturbing dreams and nightmares related to early life traumatic experiences during the COVID-19 pandemic by patients with posttraumatic stress disorder in remission. *Journal of Clinical Sleep Medicine,* jcsm-8562.

Gurrentz, B. (2019). *Cohabitation over the last 20 years: Measuring and understanding the changing demographics of unmarried partners, 1996-2017.* Working Paper. U.S. Census Bureau. https://www.census.gov/library/working-papers/2019/demo/SEHSD-WP2019-10.html

Gust, K., Caccese, C., Larosa, A., & Nguyen, T. V. (2020). Neuroendocrine effects of lactation and hormone-gene-environment interactions. *Molecular Neurobiology, 57,* 2074–2084. https://doi.org/10.1007/s12035-019-01855-8

Gustavsen, G. W., & Hegnes, A. W. (2020). Individuals' personality and consumption of organic food. *Journal of Cleaner Production, 245,* 118772.

Guttman, N., & Kalish, H. I. (1956). Discriminability and stimulus generalization. *Journal of Experimental Psychology, 51*(1), 79–88.

Guye, S., & von Bastian, C. C. (2017). Working memory training in older adults: Bayesian evidence supporting the absence of transfer. *Psychology And Aging, 32*(8), 732–746. https://doi.org/10.1037/pag0000206

H

Ha, S., Lee, H., Choi, Y., Kang, H., Jeon, S. J., Ryu, J. H., . . . Lee, D. S. (2020). Maturational delay and asymmetric information flow of brain connectivity in SHR model of ADHD revealed by topological analysis of metabolic networks. *Scientific Reports, 10*(1), 1–13.

Haaker, J., Yi, J., Petrovic, P., & Olsson, A. (2017). Endogenous opioids regulate social threat learning in humans. *Nature Communications, 8,* 15495. https://doi.org/10.1038/ncomms15495

Haber, J. R., Koenig, L. B., & Jacob, T. (2011). Alcoholism, personality, religion/spirituality: An integrative review. *Current Drug Abuse Reviews, 4*(4), 250–260.

Haber, S. N., Yendiki, A., & Jbabdi, S. (2020). Four deep brain stimulation targets for obsessive-compulsive disorder: Are they different? *Biological Psychiatry.* https://doi.org/10.1016/j.biopsych.2020.06.031

Hackman, C. L., & Knowlden, A. P. (2014). Theory of reasoned action and theory of planned behavior—based dietary interventions in adolescents and young adults: A systematic review. *Adolescent Health, Medicine, and Therapeutics, 5,* 101–114.

Hadamitzky, M., Lückemann, L., Pacheco-López, G., & Schedlowski, M. (2020). Pavlovian conditioning of immunological and neuroendocrine functions. *Physiological Reviews, 100*(1), 357–405.

Hadden, B. W., Harvey, S. M., Settersten Jr, R. A., & Agnew, C. R. (2019). What do I call us? The investment model of commitment processes and changes in relationship categorization. *Social Psychological and Personality Science, 10*(2), 235–243.

Hadi, A. (2020). 'Honor' killings in misogynistic society: A feminist perspective. *Academic Journal of Interdisciplinary Studies, 9*(3), 29–29.

Haesen, K., Beckers, T., Baeyens, F., & Vervliet, B. (2017). One-trial overshadowing: Evidence for fast specific fear learning in humans. *Behaviour Research and Therapy, 90,* 16–24.

Hafeman, D. M., Rooks, B., Merranko, J., Liao, F., Gill, M. K., Goldstein, T. R., . . . Strober, M. (2020). Lithium versus other mood-stabilizing medications in a longitudinal study of youth diagnosed with bipolar disorder. *Journal of the American Academy of Child & Adolescent Psychiatry, 59*(10), 1146–1155.

Hag, L., Abrahamsen, E. R., & Hougaard, D. D. (2020). Normative vestibular-ocular reflex gain values for the vertical semicircular canals. *Research in Vestibular Science, 19*(2), 62–70.

Hagedorn, J. M., Moeschler, S., Goree, J., Weisbein, J., & Deer, T. R. (2020). Diversity and inclusion in pain medicine. *Regional Anesthesia & Pain Medicine.*

Haghighi, S. S., Ghorbani, M., Dehnavi, F., Safaie, M., & Moghimi, S. (2020). Motivated forgetting increases the recall time of learnt items: Behavioral and event related potential evidence. *Brain Research, 1729,* 146624.

Haidle, M. N. (2010). Working memory capacity and the evolution of modern cognitive capacities—Implications from animal and early human tool use. *Current Anthropology, 51,* Suppl. 1 Wenner-Gren Symposium, S149–S166.

Hajek, A. & Helmut-König, H. (2017). The role of self-efficacy, self-esteem and optimism for using routine health check-ups in a population-based sample. A longitudinal perspective. *Preventive Medicine, 105,* 47–51.

Halberstadt, A. L. (2017). Hallucinogenic drugs: A new study answers old questions about LSD. *Current Biology, 27*(4), R156–R158.

Halberstadt, J., & Catty, S. (2008). Analytic thought disrupts familiarity-based decision making. *Social Cognition, 26*(6), 755–765.

Hales, C. M., Carroll, M. D., Fryar, C. D., & Ogden, C. L. (2020). Prevalence of obesity and severe obesity among adults: United States, 2017–2018. NCHS Data Brief, No 360. National Center for Health Statistics.

Halfon N., & Forrest C.B. (2018) The emerging theoretical framework of life course health development. In N. Halfon, C. Forrest, R. Lerner, & E. Faustman (Eds.), *Handbook of life course health development.* Springer.

Hall, J. A., Park, N., Song, H., & Cody, M. J. (2010). Strategic misrepresentation in online dating: The effects of gender, self-monitoring, and personality traits. *Journal of Social and Personal Relationships, 27*(1), 117–135.

Hall, W. J. (2017). Psychosocial risk and protective factors for depression among lesbian, gay, bisexual, and queer youth: A systematic review. *Journal of Homosexuality, 65,* 263–316.

Hall, W., West, R., Marsden, J., Humphreys, K., Neale, J., & Petry, N. (2018). It is premature to expand access to medicinal cannabis in hopes of solving the US opioid crisis. *Addiction, 113*(6), 987–988.

Halldorsdottir, T., &. Binder, E.B. (2017). Gene × Environment Interactions: From Molecular Mechanisms to Behavior. *Annual Review of Psychology, 68,* 215–241.

Haller, C. S., Bosma, C. M., Kapur, K., Zafonte, R., & Langer, E. J. (2017). Mindful creativity matters: trajectories of reported functioning after severe traumatic brain injury as a function of mindful creativity in

patients' relatives: a multilevel analysis. *Quality of life research, 26*(4), 893–902.

Hamano, Y. H., Sugawara, S. K., Yoshimoto, T., & Sadato, N. (2020). The motor engram as a dynamic change of the cortical network during early sequence learning: An fMRI study. *Neuroscience Research, 153,* 27–39.

Hamatani, S., Tsuchiyagaito, A., Nihei, M., Hayashi, Y., Yoshida, T., Takahashi, J., . . . Shimizu, E. (2020). Predictors of response to exposure and response prevention-based cognitive behavioral therapy for obsessive-compulsive disorder. *BMC Psychiatry, 20*(1), 1–8.

Hames, J. L., Hagan, C. R., & Joiner, T. E. (2013). Interpersonal processes in depression. *Annual Review of Clinical Psychology, 9,* 355–377.

Hamilton, J. L., Hamlat, E. J., Stange, J. P., Abramson, L. Y., & Alloy, L. B. (2014). Pubertal timing and vulnerabilities to depression in early adolescence: Differential pathways to depressive symptoms by sex. *Journal of Adolescence, 37*(2), 165–174.

Hamilton, K., van Dongen, A., & Hagger, M. S. (2020). An extended theory of planned behavior for parent-for-child health behaviors: A meta-analysis. *Health Psychology, 39*(10), 863–878.

Hammerschlag, A. R., Byrne, E. M., Bartels, M., Wray, N. R., Middeldorp, C. M., Agbessi, M., . . . Awadalla, P. (2020). Refining attention-deficit/hyperactivity disorder and autism spectrum disorder genetic loci by integrating summary data from genome-wide association, gene expression and DNA methylation studies. *Biological Psychiatry.* https://doi.org/10.1016/j.biopsych.2020.05.002

Hammond, D., Lillico, H. G., Vanderlee, L., White, C. M., & Reid, J. L. (2015). The impact of nutrition labeling on menus: A naturalistic cohort study. *American Journal of Health Behavior, 39*(4), 540–548.

Hammond, S. I., & Drummond, J. K. (2019). Rethinking emotions in the context of infants' prosocial behavior: The role of interest and positive emotions. *Developmental Psychology, 55*(9), 1882–1888.

Hammond, S. I., Al-Jbouri, E., Edwards, V., & Feltham, L. E. (2017). Infant helping in the first year of life: Parents' recollection of infants' earliest prosocial behaviors. *Infant Behavior and Development, 47,* 54–57.

Han, S. (2016). Culture, self, and brain: Sociocultural influences on neurocognitive processes of the self. In M. J. Gelfand, C. Chiu, & Y. Hong (Eds.), *Handbook of advances in culture and psychology.* (pp. 77–112). Oxford University Press.

Han, W., Shin, J. O., Ma, J. H., Min, H., Jung, J., Lee, J., . . . Bok, J. (2020). Distinct roles of stereociliary links in the nonlinear sound processing and noise resistance of cochlear outer hair cells. *Proceedings of the National Academy of Sciences, 117*(20), 11109–11117.

Han, Y., & Adolphs, R. (2020). Estimating the heritability of psychological measures in the Human Connectome Project dataset. *PloS one, 15*(7), e0235860.

Haney, C., Banks, C., & Zimbardo, P. (1973). Interpersonal dynamics in a simulated prison. *International Journal of Criminology and Penology, 1*(1), 69–97.

Hanna, E., Ward, L. M., Seabrook, R. C., Jerald, M., Reed, L., Giaccardi, S., & Lippman, J. R. (2017). Contributions of social comparison and self-objectification in mediating associations between Facebook use and emergent adults' psychological well-being. *Cyberpsychology, Behavior, and Social Networking, 20*(3), 172–179. https://doi.org/10.1089/cyber.2016.0247

Hannan, S. M., Zimnick, J., & Park, C. (2020). Consequences of sexual violence among college students: Investigating the role of PTSD symptoms, rumination, and institutional betrayal. *Journal of Aggression, Maltreatment & Trauma.* https://doi.org/10.1080/10926771.2020.1796871

Hannum, R. D., Rosellini, R. A., & Seligman, M. E. P. (1976). Learned helplessness in the rat: Retention and immunization. *Developmental Psychology, 12*(5), 449–454.

Hanowski, R. J., Olson, R. L., Hickman, J. S., & Bocanegra, J. (2009, September). *Driver distraction in commercial vehicle operations.* Paper presented at the First International Conference on Driver Distraction and Inattention, Gothenburg, Sweden.

Hansson, A. C., Gründer, G., Hirth, N., Noori, H. R., Spanagel, R., & Sommer, W. H. (2019). Dopamine and opioid systems adaptation in alcoholism revisited: Convergent evidence from positron emission tomography and postmortem studies. *Neuroscience & Biobehavioral Reviews, 106,* 141–164.

Hardin, J. S., Jones, N. A., Mize, K. D., & Platt, M. (2020). Parent-training with kangaroo care impacts infant neurophysiological development & mother-infant neuroendocrine activity. *Infant Behavior and Development, 58,* 101416.

Hare, M. M., & Graziano, P. A. (2020). The cost-effectiveness of parent–child interaction therapy: Examining standard, intensive, and group adaptations. *Administration and Policy in Mental Health and Mental Health Services Research.* https://doi.org/10.1007/s10488-020-01083-6

Harker, L. A., & Keltner, D. (2001). Expressions of positive emotion in women's college yearbook pictures and their relationship to personality and life outcomes across adulthood. *Journal of Personality and Social Psychology, 80*(1), 112–124.

Harkin, B., Webb, T. L., Chang, B. I., Prestwich, A., Conner, M., Kellar, I., et al. (2016). Does monitoring goal progress promote goal attainment? A meta-analysis of the experimental evidence. *Psychological Bulletin, 142*(2), 198–229. https://doi.org/10.1037/bul0000025

Harlow, H. F. (1958). The nature of love. *American Psychologist, 13*(12), 673–685.

Harmon-Jones, E., & Harmon-Jones, C. (2020). Cognitive dissonance processes serve an action-oriented adaptive function. *Behavioral and Brain Sciences,* 43, e38.

Harper, Q., Worthington, E. L., Jr., Griffin, B. J., Lavelock, C. R., Hook, J. N., Vrana, S. R., & Greer, C. L. (2014). Efficacy of a workbook to promote forgiveness: A randomized controlled trial with university students. *Journal of Clinical Psychology.* https://doi.org/10.1002/jclp.22079

Harrington, B., & O'Connell, M. (2016). Video games as virtual teachers: Prosocial video game use by children and adolescents from different socioeconomic groups is associated with increased empathy and prosocial behaviour. *Computers In Human Behavior, 63,* 650–658. https://doi.org/10.1016/j.chb.2016.05.062

Harris, D. M., & Kay, J. (1995). I recognize your face but I can't remember your name: Is it because names are unique? *British Journal of Psychology, 86*(Pt. 3), 345–358.

Harris, L. M., Huang, X., Linthicum, K. P., Bryen, C. P., & Ribeiro, J. D. (2020). Sleep disturbances as risk factors for suicidal thoughts and behaviours: A meta-analysis of longitudinal studies. *Scientific Reports, 10*(1), 1–11.

Harris, P. R., & others. (2014). Combining self-affirmation with implementation intentions to promote fruit and vegetable consumption. *Health Psychology, 33*(7), 729–736.

Harrop, C., Libsack, E., Bernier, R., Dapretto, M., Jack, A., McPartland, J. C., . . . GENDAAR Consortium. (2020). Do biological sex and early developmental milestones predict the age of first concerns and eventual diagnosis in autism spectrum disorder? *Autism Research.* https://doi.org/10.1002/aur.2446

Hart, B., & Risley, T.R. (1995). *Meaningful differences in the everyday experience of young American children.* Brookes.

Hart, R. (2020). *Positive psychology: The basics.* Routledge.

Hartenbaum, N., & others. (2006). Sleep apnea and commercial motor vehicle operators. *Chest, 130*(3), 902–905.

Hartmann, E. (1993). Nightmares. In M. A. Carskadon (Ed.), *Encyclopedia of sleep and dreams.* Macmillan.

Harvey, J. R., Plante, A. E., & Meredith, A. L. (2020). Ion channels controlling circadian rhythms in suprachiasmatic nucleus excitability. *Physiological Reviews.* https://doi.org/10.1152/physrev.00027.2019

Hasan, A., Wobrock, T., Guse, B., Langguth, B., Landgrebe, M., Eichhammer, P., . . . & Winterer, G. (2017). Structural brain changes are associated with response of negative symptoms to prefrontal repetitive transcranial magnetic stimulation in patients with schizophrenia. *Molecular Psychiatry, 22*(6), 857–864.

Hasantash, M., & Afraz, A. (2020). Richer color vocabulary is associated with better color memory but not color perception. *Proceedings of the National Academy of Sciences.* https://doi.org/10.1073/pnas.2001946117

Haselton, M. G. (2006, April 29). How to pick a perfect mate. *New Scientist, 2549,* 36.

Haslam, N. (2011). Genetic essentialism, neuroessentialism, and stigma: Commentary on Dar-Nimrod and Heine (2011). *Psychological Bulletin, 137*(5), 819–824. https://doi.org/10.1037/a0022386

Haslam, S. A., & Reicher, S. (2003). Beyond Stanford: Questioning a role-based explanation of tyranny. *SPSP Dialogue, 18,* 22–25.

Hässler, T., Uluğ, Ö. M., Kappmeier, M., & Travaglino, G. A. (2020). Intergroup contact and social change: An integrated contact-collective action model. *Journal of Social Issues.* In press.

Hastings, M. H., Smyllie, N. J., & Patton, A. P. (2020). Molecular-genetic manipulation of the suprachiasmatic nucleus circadian clock. *Journal of Molecular Biology.* https://doi.org/10.1016/j.jmb.2020.01.019

Hauser, T. U., Iannaccone, R., Dolan, R. J., Ball, J., Hättenschwiler, J., Drechsler, R., et al. (2017). Increased fronto-striatal reward prediction errors moderate decision making in obsessive–compulsive disorder. *Psychological Medicine, 47*(7), 1246–1258.

Hausken, S. E., Lie, H. C., Lien, N., Sleddens, E. F., Melbye, E. L., & Bjelland, M. (2019). The reliability of the general functioning scale in Norwegian 13–15-year-old adolescents and association with family dinner frequency. *Nutrition Journal, 18*(1), 20.

Hawley, P. H. (2014). Evolution, prosocial behavior, and altruism: A roadmap for understanding where the proximate meets the ultimate. In L. M. Padilla-Walker & G. Carlo (Eds.), *Prosocial development: A multidimensional approach* (pp. 43–69). Oxford University Press. https://doi.org/10.1093/acprof:oso/9780199964772.003.0003

Haws, K. L., & Liu, P. J. (2016). Half-size me? How calorie and price information influence ordering on restaurant menus with both half and full entrée portion sizes. *Appetite, 97,* 127–137. https://doi.org/10.1016/j.appet.2015.11.031

Hay, P. P., Bacaltchuk, J., Stefano, S., & Kashyap, P. (2009). Psychological treatments for bulimia nervosa and binging. *Cochrane Database of Systematic Reviews,* (4), CD000562.

Hayakawa, Y. (2012). Targeting NKG2D in tumor surveillance. *Expert Opinion on Therapeutic Targets, 16*(6), 587–599.

Hayama, S., Chang, L., Gumus, K., King, G. R., & Ernst, T. (2016). Neural correlates for perception of companion animal photographs. *Neuropsychologia, 85,* 278–286. https://doi.org/10.1016/j.neuropsychologia.2016.03.018

Hayflick, L. (1977). The cellular basis for biological aging. In C. E. Finch & L. Hayflick (Eds.), *Handbook of the biology of aging.* Van Nostrand.

Hazelden Betty Ford Foundation (June 10, 2015). *Survey Finds Risky Opioid Use Among College-Age Youth, With Limited Knowledge of the Danger or Where to Get Help.* http://www.hazeldenbettyford.org/about-us/news-media/press-release/2015-opioid-use-among-college-youth

He, W., Fan, C., & Li, L. (2017). Transcranial magnetic stimulation reveals executive control dissociation in the rostral prefrontal cortex. *Frontiers In Human Neuroscience, 11.* https://doi.org/10.3389/fnhum.2017.00464

He, W., Qian, F., & Cao, J. (2017). Pinning-controlled synchronization of delayed neural networks with distributed-delay coupling via impulsive control. *Neural Networks, 85,* 1–9.

He, Y., Chen, J., Zhu, L. H., Hua, L. L., & Ke, F. F. (2020). Maternal smoking during pregnancy and ADHD: results from a systematic review and meta-analysis of prospective cohort studies. *Journal of Attention Disorders, 24*(12), 1637–1647.

Heal, D. J., Hallam, M., Prow, M., Gosden, J., Cheetham, S., Choi, Y. K., et al. (2017). Dopamine and μ-opioid receptor dysregulation in the brains of binge-eating female rats possible relevance in the psychopathology

and treatment of binge-eating disorder. *Journal of Psychopharmacology, 31*(6), 770–783.

Hearold, S. (1986). A synthesis of 1043 effects of television on social behavior. In G. Comstock (Ed.), *Public communication of behavior* (pp. 65–133). Academic.

Heaven, W.D. (2020, July 17). Predictive policing algorithms are racist. They need to be dismantled. *Technology Review.* https://www.technologyreview.com/2020/07/17/1005396/predictive-policing-algorithms-racist-dismantled-machine-learning-bias-criminal-justice/

Hebb, D. O. (1949). *The organization of behavior: A neuropsychological theory.* Wiley.

Hebb, D. O. (1980). *Essay on mind.* Erlbaum.

Hegarty, P., Ansara, Y. G., & Barker, M. J. (2018). Nonbinary gender identities. In N. K. Dess, J. Marecek, & L. C. Bell (Eds.), *Gender, sex, and sexualities: Psychological perspectives* (pp. 53–76). Oxford University Press.

Hegmans, J. P., & Aerts, J. G. (2014). Immunomodulation in cancer. *Current Opinion in Pharmacology, 17,* 17–21.

Heider, F. (1958). *The psychology of interpersonal relations.* Erlbaum.

Heil, P., & Matysiak, A. (2020). Absolute auditory threshold: testing the absolute. *European Journal of Neuroscience, 51*(5), 1224–1233.

Heineman, K. R., Schendelaar, P., van den Heuvel, E. R., & Hadders-Algra, M. (2017). Investigate associations between motor development measured with the Infant Motor Profile (IMP) in infancy and IQ at the age of 4 years. *European Journal of Paediatric Neurology, 21,* e156.

Heinonen, E., & Nissen-Lie, H. A. (2020). The professional and personal characteristics of effective psychotherapists: A systematic review. *Psychotherapy Research, 30*(4), 417–432.

Heirman, W., & Walrave, M. (2012). Predicting adolescent perpetration in cyberbullying: An application of the theory of planned behavior. *Psicothema, 24*(4), 614–620.

Héjja-Brichard, Y., Rima, S., Rapha, E., Durand, J. B., & Cottereau, B. R. (2020). Stereomotion processing in the nonhuman primate brain. *Cerebral Cortex, 30,* 4528–4543.

Hellmer, K., & Nyström, P. (2017). Infant acetylcholine, dopamine, and melatonin dysregulation: Neonatal biomarkers and causal factors for ASD and ADHD phenotypes. *Medical Hypotheses, 100,* 64–66.

Helmore, E. (August 20, 2017). 'He's trying to save lives': The ex-addict judge on the frontline of the opiate crisis. *The Guardian.* https://amp.theguardian.com/us-news/2017/aug/20/opioid-crisis-america-buffalo-new-york-trump-national-emergency

Helms, J. E., Jernigan, M., & Mascher, J. (2005). The meaning of race in psychology and how to change it: A methodological perspective. *American Psychologist, 60,* 27–36.

Henderson, M. (2008, February 18). Welcome to the town that will make you lose weight. *The Times.* www.the-times.co.uk/tto/health/article1881007.ece

Hendrick, C., & Hendrick, S. S. (2006). Styles of romantic love. In R. J. Sternberg & K. Weis (Eds.), *The new psychology of love* (pp. 149–170). Yale University Press.

Hendrick, C., & Hendrick, S. S. (2009). Love. In S. Lopez & C. R. Snyder (Eds.), *The Oxford handbook of positive psychology* (2nd ed., pp. 447–454). Oxford University Press.

Hennessey, B. A. (2011). Intrinsic motivation and creativity: Have we come full circle? In R. A. Beghetto & J. C. Kaufman (Eds.), *Nurturing creativity in the classroom.* Cambridge University Press.

Hennessy, E. A., Johnson, B. T., Acabchuk, R. L., McCloskey, K., & Stewart-James, J. (2020). Self-regulation mechanisms in health behavior change: A systematic meta-review of meta-analyses, 2006–2017. *Health Psychology Review, 14*(1), 6–42.

Henrich, J., Heine, S. J., & Norenzayan, A. (2010). The weirdest people in the world? *Behavioral and Brain Sciences, 33*(2–3), 61–83.

Henriques, G., & Michalski, J. (2020). Defining behavior and its relationship to the science of psychology. *Integrative Psychological and Behavioral Science, 54*(2), 328–353.

Henry, J. D., MacLeod, M. S., Phillips, L. H., & Crawford, J. R. (2004). A meta-analytic review of prospective memory and aging. *Psychology and Aging, 19*(1), 27–39.

Hepting, U., & Solle, R. (1973). Sex-specific differences in color coding. *Archiv fur Psychologie, 125*(2–3), 184–202.

Herman, C. P. (2017). The social facilitation of eating or the facilitation of social eating? *Journal of Eating Disorders, 5.* https://doi.org/10.1186/s40337-017-0146-2

Hermans, E. J., Battaglia, F. P., Atsak, P., de Voogd, L. D., Fernández, G., & Roozendaal, B. (2014). How the amygdala affects emotional memory by altering brain network properties. *Neurobiology of Learning and Memory, 112C,* 2–16.

Hermes, M., Hagemann, D., Naumann, E., & Walter, C. (2011). Extraversion and its positive emotional core—Further evidence from neuroscience. *Emotion, 11*(2), 367–378. https://doi.org/10.1037/a0021550

Hermiz, J., Hossain, L., Arneodo, E. M., Ganji, M., Rogers, N., Vahidi, N., . . . Gilja, V. (2020). Stimulus driven single unit activity from micro-electrocorticography. *Frontiers in Neuroscience, 14,* 55.

Herrnstein, R. J., & Murray, C. (1994). *The bell curve.* Free Press.

Herron, W. G. (2015). Transference, countertransference, and resistance. In D. O. Morris, R. A. Javier, & W. G. Herron, *Specialty competencies in psychoanalysis in psychology* (pp. 101–123). Oxford University Press.

Herting, M. M., & Sowell, E. R. (2017). Puberty and structural brain development in humans. *Frontiers in Neuroendocrinology, 44,* 122–137.

Hertzog, C., Kramer, A. F., Wilson, R. S., & Lindenberger, U. (2009). Enrichment effects on adult cognitive development. *Psychological Perspectives in the Public Interest, 9*(1), 1–65.

Herzog, H. (2006). Forty-two thousand and one Dalmatians: Fads, social contagion, and dog breed popularity. *Society & Animals, 14*(4), 383–397.

Heshmati, M., Christoffel, D. J., LeClair, K., Cathomas, F., Golden, S. A., Aleyasin, H., . . . Russo, S. J. (2020). Depression and social defeat stress are associated with inhibitory synaptic changes in the nucleus accumbens. *Journal of Neuroscience, 40*(32), 6228–6233.

Hett, D., Rogers, J., Humpston, C., & Marwaha, S. (2020). Repetitive transcranial magnetic stimulation (rTMS) for the treatment of depression in adolescence: A systematic review. *Journal of Affective Disorders, 12460.*

Hevey, D., McGee, H. M., & Horgan, J. H. (2014). Comparative optimism among patients with coronary heart disease (CHD) is associated with fewer adverse clinical events 12 months later. *Journal of Behavioral Medicine, 37*(2), 300–307. https://doi.org/10.1007/s10865-012-9487-0

Heward, W. L. (2013). *Exceptional children* (10th ed.). Pearson.

Hewison, D., Clulow, C., & Drake, H. (2014). *Couple therapy for depression: A clinician's guide to integrative practice.* Oxford University Press.

Heyselaar, E., Segaert, K., Walvoort, S. J., Kessels, R. P., & Hagoort, P. (2017). The role of nondeclarative memory in the skill for language: Evidence from syntactic priming in patients with amnesia. *Neuropsychologia, 101,* 97–105.

Hicks, J. A., Trent, J., Davis, W., & King, L. A. (2012). Positive affect, meaning in life, and future time perspective: An application of Socioemotional Selectivity Theory. *Psychology and Aging, 27*(1), 181–189.

Highfield, R. (2008, April 30). Harvard's baby brain research lab. *The Telegraph.* www.telegraph.co.uk/science/science-news/3341166/Harvards-baby-brain-research-lab.html.

Hildebrandt, T., Bacow, T., Markella, M., & Loeb, K. L. (2012). Anxiety in anorexia nervosa and its management using family-based treatment. *European Eating Disorders Review, 20*(1), e1–e16.

Hilgard, E. R. (1977). *Divided consciousness: Multiple controls in human thought and action.* Wiley.

Hilgard, E. R. (1992). Dissociation and theories of hypnosis. In E. Fromm & M. R. Nash (Eds.), *Contemporary hypnosis research.* Guilford.

Hilgard, J., Engelhardt, C. R., Bartholow, B. D., & Rouder, J. N. (2015). How much evidence is p > .05? Stimulus pre-testing and null primary outcomes in violent video games research. *Psychology of Popular Media Culture.* https://doi.org/10.1037/ppm0000102

Hilgard, J., Engelhardt, C. R., & Rouder, J. N. (2017). Overstated evidence for short-term effects of violent games on affect and behavior: A reanalysis of Anderson et al. (2010). *Psychological Bulletin, 143*(7), 757–774. https://doi.org/10.1037/bul0000074

Hill, E. M. (2016). Posthumous organ donation attitudes, intentions to donate, and organ donor status: Examining the role of the big five personality dimensions and altruism. *Personality and Individual Differences, 88,* 182–186. https://doi.org/10.1016/j.paid.2015.09.021

Hill, P. L., Jackson, J. J., Roberts, B. W., Lapsley, D. K., & Brandenberger, J. W. (2011). Change you can believe in: Changes in goal setting during emerging and young adulthood predict later adult well-being. *Social Psychological and Personality Science, 2*(2), 123–131.

Hill, R. Z., & Bautista, D. M. (2020). Getting in touch with mechanical pain mechanisms. *Trends in Neurosciences, 43,* 311–325.

Hill, T. D., Burdette, A. M., Ellison, C. G., & Musick, M. A. (2006). Religious attendance and the health behaviors of Texas adults. *Preventive Medicine, 42*(4), 309–312.

Hill, W. D., Weiss, A., Liewald, D. C., Davies, G., Porteous, D. J., Hayward, C., . . . Deary, I. J. (2020). Genetic contributions to two special factors of neuroticism are associated with affluence, higher intelligence, better health, and longer life. *Molecular Psychiatry, 25*(11), 3034–3052.

Hiller, R. M., Meiser-Stedman, R., Fearon, P., Lobo, S., MacKinnon, A., Fraser, A., & Halligan, S. L. (2016). Changes in the prevalence and symptom severity of child PTSD in the year following trauma: A meta-analytic study. *Journal of Child Psychology and Psychiatry, 57*(8), 884–898.

Himmer, L., Schönauer, M., Heib, D. P. J., Schabus, M., & Gais, S. (2019). Rehearsal initiates systems memory consolidation, sleep makes it last. *Science Advances, 5*(4), eaav1695.

Hindman, A. H., Wasik, B. A., & Snell, E. K. (2016). Closing the 30 million word gap: Next steps in designing research to inform practice. *Child Development Perspectives, 10,* 134–139. https://doi.org/10.1111/cdep.12177

Hingson, R. W., Heeren, T., & Winter, M. R. (2006). Age at drinking onset and alcohol dependence: Age of onset, duration, and severity. *Archives of Pediatric and Adolescent Medicine, 160*(7), 739–746.

Hinterberger, T., Schmidt, S., Kamei, T., & Walach, H. (2014). Decreased electrophysiological activity represents the conscious state of emptiness in meditation. *Frontiers in Psychology, 5,* 99.

Hinton, E. C., Brunstrom, J. M., Fay, S. H., Wilkinson, L. L., Ferriday, D., Rogers, P. J., & de Wijk, R. (2013). Using photography in 'The Restaurant of the Future'. A useful way to assess portion selection and plate cleaning? *Appetite, 63,* 31–35. https://doi.org/10.1016/j.appet.2012.12.008

Hipes, C., & Gemoets, D. (2019). Stigmatization of war veterans with posttraumatic stress disorder (PTSD): Stereotyping and social distance findings. *Society and Mental Health, 9*(2), 243–258.

Hirai, M., & Senju, A. (2020). The two-process theory of biological motion processing. *Neuroscience & Biobehavioral Reviews, 111,* 114–124.

Hirbec, H., Déglon, N., Foo, L. C., Goshen, I., Grutzendler, J., Hangen, E., . . . Rion, S. (2020). Emerging technologies to study glial cells. *Glia.* https://doi.org/10.1002/glia.23780

Hirschberger, G. (2015). Terror management and prosocial behavior: A theory of self-protective altruism. In D. A. Schroeder & W. G. Graziano (Eds.), *The Oxford handbook of prosocial behavior* (pp. 166–187). Oxford University Press.

Hirsh-Pasek, K., & Golinkoff, R. M. (2014). Early language and literacy: Six principles. In S. Gilford (Ed.), *Head Start teacher's guide.* Teacher's College Press.

Hitch, G. J., Allen, R. J., & Baddeley, A. D. (2020). Attention and binding in visual working memory: Two forms of attention and two kinds of buffer storage. *Attention, Perception, & Psychophysics, 82*(1), 280–293.

Ho, M. Y., Van Tongeren, D. R., & You, J. (2020). The role of self-regulation in forgiveness: A regulatory model of forgiveness. *Frontiers in Psychology, 11,* 1084.

Hobson, A. J., & Maxwell, B. (2017). Supporting and inhibiting the well-being of early career secondary school teachers: Extending self-determination theory. *British Educational Research Journal, 43*(1), 168–191. https://doi.org/10.1002/berj.3261

Hobson, J. A. (1999). Dreams. In R. Conlan (Ed.), *States of mind.* Wiley.

Hobson, J. A. (2000). Dreams: Physiology. In A. Kazdin (Ed.), *Encyclopedia of psychology.* American Psychological Association and Oxford University Press.

Hobson, J. A. (2002). *Dreaming.* Oxford University Press.

Hobson, J. A. (2004). Freud returns? Like a bad dream. *Scientific American, 290*(5), 89.

Hobson, J. A., & Voss, U. (2011). A mind to go out of: Reflections on primary and secondary consciousness. *Consciousness and Cognition, 20*(4), 993–997.

Hobson, J. A., Pace-Schott, E. F., & Stickgold, R. (2000). Dreaming and the brain: Toward a cognitive neuroscience of conscious states. *Behavior and Brain Sciences, 23*(6), 793–842.

Hockett, C. F. (1960). The origin of speech. *Scientific American, 203,* 88–96.

Hodapp, R. M., Griffin, M. M., Burke, M. M., & Fisher, M. H. (2011). Intellectual disabilities. In R. J. Sternberg & S. B. Kaufman (Eds.), *The Cambridge handbook of intelligence.* Cambridge University Press.

Hodas, N. O., & Lerman, K. (2014). The simple rules of social contagion. *Scientific Reports, 4,* 4343.

Hodes, G. E., Walker, D. M., Labonté, B., Nestler, E. J., & Russo, S. J. (2017). Understanding the epigenetic basis of sex differences in depression. *Journal of Neuroscience Research, 95*(1–2), 692–702.

Hodges, A. L., & Holland, A. C. (2018). Common sexually transmitted infections in women. *Nursing Clinics of North America, 53* (2), 189–202.

Hoek, H. W. (2006). Incidence, prevalence and mortality of anorexia nervosa and other eating disorders. *Current Opinion in Psychiatry, 19*(4), 389–394.

Hoekzema, E., Barba-Müller, E., Pozzobon, C., Picado, M., Lucco, F., García-García, D., et al. (2017). Pregnancy leads to long-lasting changes in human brain structure. *Nature neuroscience, 20*(2), 287–296.

Hoffman, K. M., Trawalter, S., Axt, J. R., & Oliver, M. N. (2016). Racial bias in pain assessment and treatment recommendations, and false beliefs about biological differences between blacks and whites. *Proceedings of the National Academy of Sciences, 113*(16), 4296–4301.

Hoffmann, J. A., Farrell, C. A., Monuteaux, M. C., Fleegler, E. W., & Lee, L. K. (2020). Association of pediatric suicide with county-level poverty in the United States, 2007–2016. *JAMA Pediatrics, 174*(3), 287–294.

Hogan, R. (2009). Much ado about nothing: The person–situation debate. *Journal of Research in Personality, 43*(2), 249.

Hogan, R., & Sherman, R. A. (2020). Personality theory and the nature of human nature. *Personality and Individual Differences, 152,* 109561.

Holdsworth, E. A., & Appleton, A. A. (2020). Adverse childhood experiences and reproductive strategies in a contemporary US population. *American Journal of Physical Anthropology, 171*(1), 37–49.

Holla, B., Bharath, R. D., Venkatasubramanian, G., & Benegal, V. (2019). Altered brain cortical maturation is found in adolescents with a family history of alcoholism. *Addiction Biology, 24*(4), 835–845.

Holland, K. J., Rabelo, V. C., & Cortina, L. M. (2017). (Missing) knowledge about sexual assault resources: Undermining military mental health. *Violence and Victims, 32*(1), 60–77.

Hollingdale, J., & Greitemeyer, T. (2014). The effect of online violent video games on levels of aggression. *PLoS One, 9*(11), e111790.

Hollis, S., Heller, G., Stevenson, M., & Schofield, P. (2014). Recurrent mild traumatic brain injury among a cohort of rugby union players. *British Journal of Sports Medicine, 48*(7), 609.

Holmes, S. (1993). Food avoidance in patients undergoing cancer chemotherapy. *Support Care Cancer, 1*(6), 326–330.

Homan, K. J., Greenberg, J. S., & Mailick, M. R. (2020). Generativity and well-being of midlife and aging parents with children with developmental or mental health problems. *Research on Aging, 42*(3–4), 95–104.

Homan, K. J., Sim, L. A., Fargo, J. D., & Twohig, M. P. (2017). Five-year prospective investigation of self-harm/suicide-related behaviors in the development of borderline personality disorder. *Personality Disorders: Theory, Research, and Treatment, 8*(2), 183–188.

Honig, M., Ma, W. J., & Fougnie, D. (2020). Humans incorporate trial-to-trial working memory uncertainty into rewarded decisions. *Proceedings of the National Academy of Sciences, 117*(15), 8391–8397.

Hood, K. B., Shook, N. J., & Belgrave, F. Z. (2017). "Jimmy cap before you tap": Developing condom use messages for African American women. *The Journal of Sex Research, 54*(4–5), 651–664.

Hoogman, M., Bralten, J., Hibar, D. P., Mennes, M., Zwiers, M. P., Schweren, L. S., et al. (2017). Subcortical brain volume differences in participants with attention deficit hyperactivity disorder in children and adults: A cross-sectional mega-analysis. *The Lancet Psychiatry, 4*(4), 310–319.

Hooper, J., & Teresi, D. (1993). *The 3-pound universe.* Tarcher/Putnam.

Hooper, N., Erdogan, A., Keen, G., Lawton, K., & McHugh, L. (2015). Perspective taking reduces the fundamental attribution error. *Journal of Contextual Behavioral Science, 4*(2), 69–72. https://doi.org/10.1016/j.jcbs.2015.02.002

Hope, T. M., Leff, A. P., Prejawa, S., Bruce, R., Haigh, Z., Lim, L., . . . & Seghier, M. L. (2017). Right hemisphere structural adaptation and changing language skills years after left hemisphere stroke. *Brain, 140*(6), 1718–1728.

Horney, K. (1945). *Our inner conflicts.* Norton.

Horney, K. (1964). *The neurotic personality of our time.* W W Norton & Co.

Horney, K. (1967). *Feminine psychology: collected essays, 1922–1937.* H. Kelman (Ed.). Norton.

Hornsey, M. J., & Fielding, K. S. (2017). Attitude roots and Jiu Jitsu persuasion: Understanding and overcoming the motivated rejection of science. *American Psychologist, 72*(5), 459–473. https://doi.org/10.1037/a0040437

Horowitz, A., & Franks, B. (2020). What smells? Gauging attention to olfaction in canine cognition research. *Animal Cognition, 23*(1), 11–18.

Horwitz, A. G., Berona, J., Czyz, E. K., Yeguez, C. E., & King, C. A. (2017). Positive and negative expectations of hopelessness as longitudinal predictors of depression, suicidal Ideation, and suicidal behavior in high-risk adolescents. *Suicide and Life-threatening Behavior, 47*(2), 168–176.

Hoss, R. A., & Langlois, J. H. (2003). Infants prefer attractive faces. In O. Pascalis & A. Slater (Eds.), *The development of face processing in infancy and early childhood: Current perspectives* (pp. 27–38). Nova Science.

Hotard, S. R., McFatter, R. M., McWhirter, R. M., & Stegall, M. E. (1989). Interactive effects of extraversion, neuroticism, and social relationships on subjective well-being. *Journal of Personality and Social Psychology, 57*(2), 321–331.

Hotzy, F., Cattapan, K., Orosz, A., Dietrich, B., Steinegger, B., Jaeger, M., . . . Bridler, R. (2020). Psychiatric advance directives in Switzerland: Knowledge and attitudes in patients compared to professionals and usage in clinical practice. *International Journal of Law and Psychiatry, 68,* 101514.

Hou, Y., Kim, S. Y., Hazen, N., & Benner, A. D. (2017). Parents' perceived discrimination and adolescent adjustment in Chinese American families: Mediating family processes. *Child Development, 88*(1), 317–331. https://doi.org/10.1111/cdev.12603

House, J. S., Landis, K. R., & Umberson, D. (1988). Social relationships and health. *Science, 241*(4865), 540–545.

House, R., Kalisch, D., & Maidman, J. (Eds.). (2017). *Humanistic psychology: Current trends and future prospects.* Routledge.

Hovey, J. D., Morales, L. R., Hurtado, G., & Seligman, L. D. (2014). Religion-based emotional social support mediates the relationship between intrinsic religiosity and mental health. *Archives of Suicide Research, 18*(4), 376–391.

Howard, J. (2018, August 28). Rates of three STDs in US reach record high, CDC says. *CNN.* https://www.cnn.com/2018/08/28/health/std-rates-united-states-2018-bn/index.html

Howard, M. C., & Van Zandt, E. C. (2020). The discriminant validity of honesty-humility: A meta-analysis of the HEXACO, big five, and dark triad. *Journal of Research in Personality, 103982.*

Howarth, E. J., O'Connor, D. B., Panagioti, M., Hodkinson, A., Wilding, S., & Johnson, J. (2020). Are stressful life events prospectively associated with increased suicidal ideation and behaviour? A systematic review and meta-analysis. *Journal of Affective Disorders, 266,* 731–742.

Howe, M. J. A., Davidson, J. W., Moore, D. G., & Sloboda, J. A. (1995). Are there early childhood signs of musical ability? *Psychology of Music, 23*(2), 162–176.

Howell, J. L., & Ratliff, K. A. (2017). Not your average bigot: The better-than-average effect and defensive responding to implicit association test feedback. *British Journal of Social Psychology, 56*(1), 125–145. https://doi.org/10.1111/bjso.12168

Howes, M. B. (2006). *Human memory.* Sage.

Howes, M. J. R., Fang, R., & Houghton, P. J. (2017). Effect of Chinese herbal medicine on Alzheimer's disease. *International Review of Neurobiology.* https://doi.org/10.1016/bs.irn.2017.02.003

Hoxmeier, J. C., Acock, A. C., & Flay, B. R. (2017). Students as prosocial bystanders to sexual assault: Demographic correlates of intervention norms, intentions, and missed opportunities. *Journal of Interpersonal Violence.*

Hoyer, D., Hannon, J. P., & Martin, G. R. (2002). Molecular, pharmacological, and functional diversity of 5-HT receptors. *Pharmacology, Biochemistry, and Behavior, 71*(4), 533–554.

Hoyt, L. T., Niu, L., Pachucki, M. C., & Chaku, N. (2020). Timing of puberty in boys and girls: Implications for population health. *SSM-Population Health, 10,* 100549.

Hsieh, L. T., Hung, D. L., Tzeng, O. J., Lee, J. R., & Cheng, S. K. (2009). An event-related potential investigation of the processing of remember/forget cues and item encoding in item-method directed forgetting. *Brain Research, 1250,* 190–201.

Hsu, C. Y., Liao, H. E., & Huang, L. C. (2020). Exploring smoking cessation behaviors of outpatients in outpatient clinics: Application of the transtheoretical model. *Medicine, 99*(27), e20971.

Hu, Y., Hitch, G. J., Baddeley, A. D., Zhang, M., & Allen, R. J. (2014). Executive and perceptual attention play different roles in visual working memory: Evidence from suffix and strategy effects. *Journal of Experimental Psychology: Human Perception and Performance, 40*(4), 1665–1678.

Huang, C. C., & Chang, Y. C. (2009). The long-term effects of febrile seizures on the hippocampal neuronal plasticity—clinical and experimental evidence. *Brain Development, 31*(5), 383–387.

Huang, Y., & Spelke, E. (2014). Core knowledge and the emergence of symbols: The case of maps. *Journal of Cognition and Development.* https://doi.org/10.1080/15248372.2013.784975

Huart, C., Rombaux, P., & Hummel, T. (2013). Plasticity of the human olfactory system: The olfactory bulb. *Molecules, 18*(9), 11586–11600.

Hubbard, J., Harbaugh, W. T., Srivastava, S., Degras, D., & Mayr, U. (2016). A general benevolence dimension that links neural, psychological, economic, and life-span data on altruistic tendencies. *Journal of Experimental Psychology: General, 145*(10), 1351–1358. https://doi.org/10.1037/xge0000209

Hubble, M. A., & Miller, S. D. (2004). The client: Psychotherapy's missing link for promoting a positive psychology. In A. Linley & S. Joseph (Eds.), *Positive psychology in practice* (pp. 335–353). Wiley.

Hubel, D. H., & Wiesel, T. N. (1963). Receptive fields of cells in striate cortex of very young, visually inexperienced kittens. *Journal of Neurophysiology, 26,* 994–1002.

Hudson, A., & Jacques, S. (2014). Put on a happy face! Inhibitory control and socioemotional knowledge predict emotion regulation in 5- to 7-year-olds. *Journal of Experimental Child Psychology, 123,* 36–52.

Hudson, A. N., Van Dongen, H. P., & Honn, K. A. (2020). Sleep deprivation, vigilant attention, and brain function: A review. *Neuropsychopharmacology, 45*(1), 21–30.

Hudson, N. W., Lucas, R. E., & Donnellan, M. B. (2020). Are we happier with others? An investigation of the links between spending time with others and subjective well-being. *Journal of Personality and Social Psychology.* https://doi.org/10.1037/pspp0000290

Huff, C. R. (2002). What can we learn from other nations about the problem of wrongful conviction? *Judicature, 86*(2), 91–97.

Huffman, J. M., Warlick, C., Frey, B., & Kerr, B. (2020). Religiosity, spirituality, gender identity, and sexual orientation of sexual minorities. *Translational Issues in Psychological Science, 6*(4), 356–371. https://doi.org/10.1037/tps0000262

Hughes, A. E., Berg, C. A., & Wiebe, D. J. (2012). Emotional processing and self-control in adolescents with type 1 diabetes. *Journal of Pediatric Psychology, 37*(8), 925–934.

Hui, B. P., Ng, J. C., Berzaghi, E., Cunningham-Amos, L. A., & Kogan, A. (2020). Rewards of kindness? A meta-analysis of the link between prosociality and well-being. *Psychological Bulletin, 146*(12), 1084–1116.

Hui, K., Cooper, R. B., & Zaheer, J. (2020). Engaging patients and families in the ethics of involuntary psychiatric care. *The American Journal of Bioethics, 20*(6), 82–84.

Hull, E. M., & Dominguez, J. M. (2006). Getting his act together: Roles of glutamate, nitricoxide, and dopamine in the medial preoptic area. *Brain Research, 1126*(1), 66–75.

Hulst, B. M., Zeeuw, P., Bos, D. J., Rijks, Y., Neggers, S. F., & Durston, S. (2017). Children with ADHD symptoms show decreased activity in ventral striatum during the anticipation of reward, irrespective of ADHD diagnosis. *Journal of Child Psychology and Psychiatry, 58*(2), 206–214.

Humphrey, N. (2008). *Seeing red: A study in consciousness.* Harvard University Press.

Humphreys, K. L., Camacho, M. C., Roth, M. C., & Estes, E. C. (2020). Prenatal stress exposure and multimodal assessment of amygdala–medial prefrontal cortex connectivity in infants. *Developmental Cognitive Neuroscience, 46,* 100877.

Huneke, N. T., Broulidakis, M. J., Darekar, A., Baldwin, D. S., & Garner, M. (2020). Brain functional connectivity correlates of response in the 7.5% CO₂ inhalational model of generalized anxiety disorder: A pilot study. *International Journal of Neuropsychopharmacology, 23*(4), 268–273.

Hunsley, J., & Bailey, J. M. (2001). Whither the Rorschach? An analysis of the evidence. *Psychological Assessment, 13*(4), 472–485.

Hunsley, J., Lee, C. M., Wood, J. M., & Taylor, W. (2015). Controversial and questionable assessment techniques. In S. O. Lilienfeld, S. J. Lynn, & J. M. Lohr (Eds.), *Science and pseudoscience in clinical psychology, 2nd ed* (pp. 42–82). Guilford Press.

Hunt, E., & Carlson, J. (2007). Considerations relating to the study of group differences in intelligence. *Perspectives on Psychological Science, 2,* 194–213.

Hunter, K. (2020). Bedtime and sleep problems in children. In M. Knox (Ed.), *Clinician's toolkit for children's behavioral health* (pp. 219–234). Academic Press.

Huntjens, R. J. C., Wessel, I., Hermans, D., & van Minnen, A. (2014). Autobiographical memory specificity in dissociative identity disorder. *Journal of Abnormal Psychology, 123*(2), 419–428.

Hur, Y. M. (2020). Relationships between cognitive abilities and prosocial behavior are entirely explained by shared genetic influences: A Nigerian twin study. *Intelligence, 82,* 101483.

Hurme, M., Koivisto, M., Revonsuo, A., & Railo, H. (2017). Early processing in primary visual cortex is necessary for conscious and unconscious vision while late processing is necessary only for conscious vision in neurologically healthy humans. *Neuroimage, 150,* 230–238. https://doi.org/10.1016/j.neuroimage.2017.02.060

Huttenlocher, P. R. (1999). Dendritic synaptic development in human cerebral cortex: Time course and critical periods. *Developmental Neuropsychology, 16*(3), 347–349.

Hutton, J. (2016). Does rubella cause autism: A 2015 reappraisal? *Frontiers in Human Neuroscience, 10,* 25. https://doi.org/10.3389/fnhum.2016.00025

Hwang, J. Y., & others. (2012). Multidimensional comparison of personality characteristics of the Big Five model, impulsiveness, and affect in pathological gambling and obsessive-compulsive disorder. *Journal of Gambling Studies, 28*(3), 351–362.

Hyatt, C., Hallowell, E., Owens, M., Weiss, B., Sweet, L., & Miller, J. (2020). An fMRI investigation of the relations between extraversion, internalizing psychopathology, and neural activation following reward receipt in the Human Connectome Project sample. *Personality Neuroscience, 3,* E13. https://doi.org/10.1017/pen.2020.11

Hyde, J. S. (2005). The gender similarities hypothesis. *American Psychologist, 60*(6), 581–592.

Hyde, J. S. (2007). New directions in the study of gender similarities and differences. *Current Directions in Psychological Science, 16*(5), 259–263.

Hyde, J. S. (2014). Gender similarities and differences. *Annual Review of Psychology 65,* 373–398.

Hyde, J. S., Bigler, R. S., Joel, D., Tate, C. C., & van Anders, S. M. (2018). The future of sex and gender in psychology: Five challenges to the gender binary. *American Psychologist.* https://doi.org/10.1037/amp0000307

Hyde, L. W., Shaw, D. S., & Moilanen, K. L. (2010). Developmental precursors of moral disengagement and the role of moral disengagement in the development of antisocial behavior. *Journal of Abnormal Child Psychology, 38*(2), 197–209.

Hyman, S. (2001, October 23). *Basic and clinical neuroscience in the post-genomic era.* Paper presented at the centennial symposium on the Celebration of Excellence in Neuroscience, Rockefeller University, New York City.

I

Iacono, W. G., & Lykken, D. T. (1997). The validity of the lie detector: Two surveys of scientific opinion. *Journal of Applied Psychology, 82*(3), 426–433.

Iannilli, E., Singh, P. B., Schuster, B., Gerber, J., & Hummel, T. (2012). Taste laterality studied by means of umani and salt stimuli: An fMRI study. *NeuroImage, 60*(1), 426–435.

Ida-Eto, M., Hara, N., Ohkawara, T., & Narita, M. (2017). Mechanism of auditory hypersensitivity in human autism using autism model rats. *Pediatrics International, 59*(4), 404–407.

Iemmola, F., & Ciani, A. C. (2009). New evidence of genetic factors influencing sexual orientation in men: Female fecundity increase in the maternal line. *Archives of Sexual Behavior, 38*(3), 393–399.

Ikeda, B. E., Collins, C. E., Alvaro, F., Marshall, G., & Garg, M. L. (2006). Well-being and nutrition-related side effects in children undergoing chemotherapy. *Nutrition and Dietetics, 63*(4), 227–239.

Ilchibaeva, T. V., Tsybko, A. S., Kondaurova, E. M., Kovetskaya, A. I., Kozhemyakina, R. V., & Naumenko, V. S. (2020). Expression patterns of serotonin receptors 1H and 7 in the brain of rats with genetically determined fear-induced aggressive behavior or the lack of aggression. *Neurochemical Journal, 14*(2), 180–186.

Imada, T. (2012). Cultural narratives of individualism and collectivism: A content analysis of textbook stories in the United States and Japan. *Journal of Cross-Cultural Psychology, 43*(4), 576–591.

Imeri, L., & Opp, M. R. (2009). How (and why) the immune system makes us sleep. *Nature Reviews: Neuroscience, 10*(3), 199–210.

Inaba, A., Thoits, P. A., Ueno, K., Gove, W. R., Evenson, R. J., & Sloan, M. (2005). Depression in the United States and Japan: Gender, marital status, and SES patterns. *Social Science & Medicine, 61*(11), 2280–2292.

Inayat, S., Nazariahangarkolaee, M., Singh, S., McNaughton, B. L., Whishaw, I. Q., & Mohajerani, M. H. (2020). Low acetylcholine during early sleep is important for motor memory consolidation. *Sleep, 43*(6), zsz297.

Infurna, F. J. (2020). What does resilience signify? An evaluation of concepts and directions for future research. *Gerontology, 66*, 323-331.

Irwing, P., Booth, T., Nyborg, H., & Rushton, J. P. (2012). Are *g* and the general factor of personality (GFP) correlated? *Intelligence, 40*(3), 296-305.

Isbilen, E. S., McCauley, S. M., Kidd, E., & Christiansen, M. H. (2020). Statistically Induced Chunking Recall: A Memory-Based Approach to Statistical Learning. *Cognitive Science, 44*(7), e12848.

Isenschmid, D. S. (2020). Cocaine. In B. S. Levine & S. Kerrigan (Eds.), *Principles of forensic toxicology* (pp. 371-387). Springer.

Ishihara, T., Sugasawa, S., Matsuda, Y., & Mizuno, M. (2017). The beneficial effects of game-based exercise using age-appropriate tennis lessons on the executive functions of 6-12-year-old children. *Neuroscience Letters, 642*, 97-101.

Ishikawa, R., Ishigaki, T., Kikuchi, A., Matsumoto, K., Kobayashi, S., Morishige, S., et al. (2017). Cross-cultural validation of the cognitive biases questionnaire for psychosis in Japan and examination of the relationships between cognitive biases and schizophrenia symptoms. *Cognitive Therapy and Research, 41*(2), 313-323.

Ishkhanyan, B., Michel Lange, V., Boye, K., Mogensen, J., Karabanov, A., Hartwigsen, G., & Siebner, H. R. (2020). Anterior and posterior left inferior frontal gyrus contribute to the implementation of grammatical determiners during language production. *Frontiers in Psychology, 11*, 685.

Islam, M. I., Khanam, R., & Kabir, E. (2020). Bullying victimization, mental disorders, suicidality and self-harm among Australian high schoolchildren: Evidence from nationwide data. *Psychiatry Research, 292*, 113364.

Ives, A. M., & Bertke, A. S. (2017). Stress Hormones Epinephrine and Corticosterone Selectively Modulate Herpes Simplex Virus 1 (HSV-1) and HSV-2 Productive Infections in Adult Sympathetic, but Not Sensory, Neurons. *Journal of Virology, 91*(13), e00582-17.

Iwaniuk, A. N., & Wylie, D. R. (2020). Sensory systems in birds: What we have learned from studying sensory specialists. *Journal of Comparative Neurology.* https://doi.org/10.1002/cne.24896

Iyer, V. J., Finch, D. G., & Kalu, C. O. (2018). Unlocking the locked-in syndrome: capacity evaluation and a multidisciplinary approach to care. *Psychiatric Annals, 48*(9), 448-451.

Izuma, K., & Murayama, K. (2019). Neural basis of cognitive dissonance. In E. Harmon-Jones (Ed.), *Cognitive dissonance: Reexamining a pivotal theory in psychology* (pp. 227-245). American Psychological Association.

Izzetoglu, M., Shewokis, P. A., Tsai, K., Dantoin, P., Sparango, K., & Min, K. (2020). Short-term effects of meditation on sustained attention as measured by fNIRS. *Brain Sciences, 10*(9), 608.

J

Jablensky, A. (2000). Epidemiology of schizophrenia: The global burden of disease and disability. *European Archives of Psychiatry and Clinical Neuroscience, 250*(6), 274-285.

Jablensky, A., & others. (1992). Schizophrenia: Manifestations, incidence and course in different cultures: A World Health Organization ten-country study. *Psychological Medicine, Monograph Suppl., 20*, 1-97.

Jackson, J. C., Castelo, N., & Gray, K. (2020). Could a rising robot workforce make humans less prejudiced? *American Psychologist.* Advance online publication. http://dx.doi.org/10.1037/amp0000582

Jackson, J. J., & Roberts, B. W. (2017). Conscientiousness. In T.A. Widiger (Ed.), *Oxford Handbook of the Five Factor Model.* Oxford University Press.

Jackson, L. C., & Greene, B. (2000). *Psychotherapy with African-American women.* Guilford.

Jackson, T. L., Braun, J. M., Mello, M., Triche, E. W., & Buka, S. L. (2017). The relationship between early childhood head injury and later life criminal behaviour: a longitudinal cohort study. *Journal of Epidemiology and Community Health, 71*(8), 800-805.

Jacobs, T. L., Epel, E. S., Lin, J., Blackburn, E. H., Wolkowitz, O. M., Bridwell, D. A., et al. (2011). Intensive meditation training, immune cell telomerase activity, and psychological mediators. *Psychoneuroendocrinology, 36*(5), 664-681.

Jacobsen, P. B., & others. (1993). Formation of food aversions in cancer patients receiving repeated infusions of chemotherapy. *Behavior Research Therapy, 31*(8), 739-748.

Jaeger, E. L. (2020). Not the desired outcome: Groupthink undermines the work of a literacy council. *Small Group Research, 51*(4), 517-541.

Jafari, Z., Kolb, B. E., & Mohajerani, M. H. (2017). Effect of acute stress on auditory processing: a systematic review of human studies. *Reviews in the Neurosciences, 28*(1), 1-13.

Jaffee, S. (2018). Promises and pitfalls in the development of biomarkers that can promote early intervention in children at risk. *Journal of Child Psychology and Psychiatry, 59*(2), 97-98.

Jain, A., Christopher, P., & Appelbaum, P. S. (2018). Civil commitment for opioid and other substance use disorders: Does it work? *Psychiatric Services, 69*(4), 374-376.

James, G. M., Baldinger-Melich, P., Philippe, C., Kranz, G. S., Vanicek, T., Hahn, A., . . . & Mitterhauser, M. (2017). Effects of selective serotonin reuptake inhibitors on inter-regional relation of serotonin transporter availability in major depression. *Frontiers in human neuroscience, 11.*

James, S. N., Rommel, A. S., Rijsdijk, F., Michelini, G., McLoughlin, G., Brandeis, D., . . . Kuntsi, J. (2020). Is association of preterm birth with cognitive-neurophysiological impairments and ADHD symptoms consistent with a causal inference or due to familial confounds? *Psychological Medicine, 50*(8), 1278-1284.

James, W. (1950 [1890]). *Principles of psychology.* Dover.

Jamet, E., Gonthier, C., Cojean, S., Colliot, T., & Erhel, S. (2020). Does multitasking in the classroom affect learning outcomes? A naturalistic study. *Computers in Human Behavior, 106*, 106264.

Jamieson, D., Beaudequin, D. A., McLoughlin, L. T., Parker, M. J., Lagopoulos, J., & Hermens, D. F. (2020). Associations between sleep quality and psychological distress in early adolescence. *Journal of Child & Adolescent Mental Health.* https://doi.org/10.2989/1728 0583.2020.1811288.

Jäncke, L., Liem, F., & Merillat, S. (2020). Are language skills related to structural features in Broca's and Wernicke's area? *European Journal of Neuroscience.* https://doi.org/10.1111/ejn.15038

Janecka, I. P. (2019). Cognitive robotics': A convergent evolution. *EC Neurology, 11*, 62-82.

Janis, I. (1972). *Victims of groupthink: A psychological study of foreign-policy decisions and fiascos.* Houghton Mifflin.

Janowsky, D. S., Addario, D., & Risch, S. C. (1987). *Psychopharmacology case studies* (2nd ed.). Guilford.

Jaramillo, N., Buhi, E. R., Elder, J. P., & Corliss, H. L. (2017). Associations between sex education and contraceptive use among heterosexually active, adolescent males in the United States. *Journal of Adolescent Health, 60*(5), 534-540. https://doi.org/10.1016/j.jadohealth.2016.11.025

Jaremka, L. M., Fagundes, C. P., Peng, J., Bennett, J. M., Glaser, R., Malarkey, W. B., & Kiecolt-Glaser, J. K. (2013a). Loneliness promotes inflammation during acute stress. *Psychological Science, 24*(7), 1089-1097.

Jaremka, L. M., Glaser, R., Malarkey, W. B., & Kiecolt-Glaser, J. K. (2013b). Marital distress prospectively predicts poorer cellular immune function. *Psychoneuroendocrinology, 38*(11), 2713-2719.

Jaremka, L. M., & others. (2014). Pain, depression, and fatigue: Loneliness as a longitudinal risk factor. *Health Psychology, 33*(9), 948-957.

Jarnecke, A. M., Ridings, L. E., Teves, J. B., Petty, K., Bhatia, V., & Libet, J. (2020). The path to couples therapy: A descriptive analysis on a veteran sample. *Couple and Family Psychology: Research and Practice, 9*(2), 73.

Jarvis, S. N., McClure, M. J., & Bolger, N. (2019). Exploring how exchange orientation affects conflict and intimacy in the daily life of romantic couples. *Journal of Social and Personal Relationships, 36*(11-12), 3575-3587.

Jasaui, Y., Kirton, A., Wilkes, T. C., McLellan, Q., Kahl, C., Swansburg, R., & MacMaster, F. (2017). 647-glutamate and response to repetitive transcranial magnetic stimulation in youth with treatment resistant depression. *Biological Psychiatry, 81*(10), S262.

Jasiak, N. M., & Bostwick, J. R. (2014). Risk of QT/QTc prolongation among newer non-SSRI antidepressants. *Annals of Pharmacotherapy, 48*(12), 1620-1628.

Javaheipour, N., Shahdipour, N., Noori, K., Zarei, M., Camilleri, J. A., Laird, A. R., . . . Khazaie, H. (2019). Functional brain alterations in acute sleep deprivation: An activation likelihood estimation meta-analysis. *Sleep Medicine Reviews, 46*, 64-73.

Javelot, H., Weiner, L., Hingray, C., Freire, R. C., & Nardi, A. E. (2020). COVID-19 and its psychological consequences: Beware of the respiratory subtype of panic disorder. *Respiratory Physiology & Neurobiology, 282,* **103530.**

Jayakumar, V., & Simpson, T. L. (2020). Multiple-criterion signal detection theory analysis of corneal pneumatic stimuli. *Investigative Ophthalmology & Visual Science, 61*(7), 393-393.

Jayawickreme, E. & Blackie, L.E.R. (2016). *Exploring the psychological benefits of hardship: A critical reassessment of posttraumatic growth.* Springer.

Jayawickreme, E., Brocato, N. W., & Blackie, L. E. (2017). Wisdom gained? Assessing relationships between adversity, personality and well-being among a late adolescent sample. *Journal of Youth and Adolescence, 46*(6), 1179-1199.

Jebb, A. T., Morrison, M., Tay, L., & Diener, E. (2020). Subjective well-being around the world: Trends and predictors across the life span. *Psychological Science, 31*(3), 293-305.

Jeffers, S. L., Hill, R., Krumholz, M. F., & Winston-Proctor, C. (2020). Themes of gerotranscendence in narrative identity within structured life review. *GeroPsych, 33,* 77-84.

Jena, A. B., Barnett, M., & Layton, T. J. (2018). The link between August birthdays and ADHD. *The New York Times*, Opinion. https://www.nytimes.com/2018/11/28/opinion/august-birthdays-adhd.html

Jennings, K. J., & de Lecea, L. (2020). Neural and hormonal control of sexual behavior. *Endocrinology, 161*(10), bqaa150.

Jensen, M. P., & Patterson, D. R. (2014). Hypnotic approaches for chronic pain management: Clinical implications of recent research findings. *American Psychologist, 69*(2), 167-177.

Jensen-Campbell, L. A., & Malcolm, K. T. (2007). The importance of conscientiousness in adolescent interpersonal relationships. *Personality and Social Psychology Bulletin, 33*(3), 368-383.

Jess, R. L., & Dozier, C. L. (2020). Increasing handwashing in young children: A brief review. *Journal of Applied Behavior Analysis, 53*(3), 1219-1224.

Jessen, L., Smith, E. P., Ulrich-Lai, Y., Herman, J. P., Seeley, R. J., Sandoval, D., & D'alessio, D. (2017). Central nervous system GLP-1 receptors regulate islet hormone secretion and glucose homeostasis in male rats. *Endocrinology, 158*(7), 2124-2133.

Jetelina, K. K., Molsberry, R. J., Gonzalez, J. R., Beauchamp, A. M., & Hall, T. (2020). Prevalence of mental illness and mental health care use among police officers. *JAMA Network Open, 3*(10), e2019658-e2019658.

Jetten, J., Mols, F., & Selvanathan, H. P. (2020). How economic inequality fuels the rise and persistence of the yellow vest movement. *International Review of Social Psychology, 33*(1), 2. http://doi.org/10.5334/irsp.356

Ji, X., Saha, S., Kolpakova, J., Guildford, M., Tapper, A. R., & Martin, G. E. (2017). Dopamine receptors differentially control binge alcohol drinking-mediated synaptic plasticity of the core nucleus accumbens direct and indirect pathways. *Journal of Neuroscience, 37*(22), 5463–5474.

Jia, Y. F., Wininger, K., Ho, A. M. C., Peyton, L., Baker, M., & Choi, D. S. (2020). Astrocytic glutamate transporter 1 (GLT1) deficiency reduces anxiety- and depression-like behaviors in mice. *Frontiers in Behavioral Neuroscience, 14*, 57.

Jiang, T., Chen, Z., & Sedikides, C. (2020). Self-concept clarity lays the foundation for self-continuity: The restorative function of autobiographical memory. *Journal of personality and social psychology, 119*(4), 945–959.

Jiang, W., Li, G., Liu, H., Shi, F., Wang, T., Shen, C., et al. (2016). Reduced cortical thickness and increased surface area in antisocial personality disorder. *Neuroscience, 337*, 143–152.

Jin, K., & Baillargeon, R. (2017). Infants possess an abstract expectation of ingroup support. *PNAS Proceedings of the National Academy of Sciences of the United States of America, 114*(31), 8199–8204. https://doi.org/10.1073/pnas.1706286114

Jin, Y., Hamiez, J., & Hao, J. (2017). Algorithms for the minimum sum coloring problem: A review. *Artificial Intelligence Review, 47*(3), 367–394. https://doi.org/10.1007/s10462-016-9485-7

Jin, Y., & Li, J. (2017). When newbies and veterans play together: The effect of video game content, context and experience on cooperation. *Computers in Human Behavior, 68*, 556–563. https://doi.org/10.1016/j.chb.2016.11.059

Jobst, A., Sabaß, L., Hall, D., Brücklmeier, B., Buchheim, A., Hall, J., & Padberg, F. (2018). Oxytocin plasma levels predict the outcome of psychotherapy: A pilot study in chronic depression. *Journal of Affective Disorders, 227*, 206–213. http://dx.doi.org/10.1016/j.jad.2017.10.037

Joel, D. (2011). Male or female? Brains are intersex. *Frontiers in Integrative Neuroscience, 5*, 57.

Joel, D., Berman, Z., Tavor, I., Wexler, N., Gaber, O., Stein, Y., . . . Assaf, Y. (2015). Sex beyond the genitalia: The human brain mosaic. *Proceedings of the National Academy of Sciences of the United States of America, 112*, 15468–15473. https://doi.org/10.1073/pnas.1509654112

Johannsson, M., Snaedal, J., Johannesson, G. H., Gudmundsson, T. E., & Johnsen, K. (2015). The acetylcholine index: An electroencephalographic marker of cholinergic activity in the living human brain applied to Alzheimer's disease and other dementias. *Dementia and Geriatric Cognitive Disorders, 39*, 132–142. https://doi.org/10.1159/000367889

Johansson, J., Salami, A., Lundquist, A., Wåhlin, A., Andersson, M., & Nyberg, L. (2020). Longitudinal evidence that reduced hemispheric encoding/retrieval asymmetry predicts episodic-memory impairment in aging. *Neuropsychologia, 137*, 107329.

Johnson, G. B. (2015). *The living world* (8th ed.). McGraw-Hill.

Johnson, J. S., & Newport, E. L. (1991). Critical period effects on universal properties of language: The status of subjacency in the acquisition of a second language. *Cognition, 39*(3), 215–258.

Johnson, J., & Wood, A. M. (2017). Integrating positive and clinical psychology: Viewing human functioning as continua from positive to negative can benefit clinical assessment, interventions and understandings of resilience. *Cognitive Therapy And Research, 41*(3), 335–349. https://doi.org/10.1007/s10608-015-9728-y

Johnson, S. P., & Hannon, E. H. (2015). Perceptual development. In R. M. Lerner, L. S. Liben, & U. Mueller (Eds.), *Handbook of child psychology and developmental science* (7th ed., vol. 2). Wiley.

Johnson, W. D., O'Leary, A., & Flores, S. A. (2018). Perpartner condom effectiveness against HIV for men who have sex with men. *AIDS, 32*(11), 1499–1505.

Johnston, L. D., Miech, R. A., O'Malley, P. M., Bachman, J. G., Schulenberg, J. E., & Patrick, M. E. (2020). Monitoring the future national survey results on drug use 1975–2019: *Overview, key findings on adolescent drug use*. Institute for Social Research, University of Michigan.

Joiner, T. E. (2005). *Why people die by suicide.* Harvard University Press.

Joiner, T. E., & Ribeiro, J. D. (2011). Assessment and management of suicidal behavior in children and adolescents. *Pediatric Annals, 40*(6), 319–324.

Joiner, T. E., Ribeiro, J. D., & Silva, C. (2012). Nonsuicidal self-injury, suicidal behavior, and their co-occurrence as viewed through the lens of the interpersonal theory of suicide. *Current Directions in Psychological Science, 21*(5), 342–347.

Jonasdottir, S. S., Minor, K., & Lehmann, S. (2020). Gender differences in nighttime sleep patterns and variability across the adult lifespan: A global-scale wearables study. *Sleep*, zsaa169. https://doi.org/10.1093/sleep/zsaa169

Jones, A. L., & Jaeger, B. (2019). Biological bases of beauty revisited: The effect of symmetry, averageness, and sexual dimorphism on female facial attractiveness. *Symmetry, 11*(2), 279.

Jones, B. E. (2020). Arousal and sleep circuits. *Neuropsychopharmacology, 45*(1), 6–20.

Jones, C., Barrera, I., Brothers, S., Ring, R., & Wahlestedt, C. (2017). Oxytocin and social functioning. *Dialogues in Clinical Neuroscience, 19*(2), 193–201.

Jones, C. T., & Branco, S. F. (2020). The interconnectedness between cultural humility and broaching in clinical supervision: working from the multicultural orientation framework. *The Clinical Supervisor.* https://doi.org/10.1080/07325223.2020.1830327

Jones, E. E. (1998). Major developments in five decades of social psychology. In D. T. Gilbert, S. T. Fiske, & G. Lindzey (Eds.), *The Handbook of Social Psychology* (4th ed., vol. 1). McGraw-Hill.

Jones, E. E., & Harris, V. A. (1967). The attribution of attitudes. *Journal of Experimental Social Psychology, 3*(1), 1–24.

Jones, J. M. & Splan, E. D. (2020). Personality and prejudice. In B. J. Carducci, C. S. Nave, J. S. Mio, & R. E. Riggio (Eds.), *The Wiley encyclopedia of personality and individual differences: Clinical, applied, and cross-cultural research* (vol. IV, pp. 275–279). John Wiley & Sons.

Jones, K. M., & Ruthig, J. C. (2015). The impact of positive thinking, gender, and empathy on social attributions for cancer outcomes. *Current Psychology: A Journal for Diverse Perspectives on Diverse Psychological Issues, 34*(4), 762–771. https://doi.org/10.1007/s12144-014-9288-4

Jones, N. P., Chase, H. W., & Fournier, J. C. (2016). Brain mechanisms of anxiety's effects on cognitive control in major depressive disorder. *Psychological Medicine, 46*(11), 2397–2409. https://doi.org/10.1017/S0033291716001185

Jooyoung, P., Lu, F., & Hedgcock, W. (2017). Forward and backward planning and goal pursuit. *Psychological Science.* https://doi.org/10.1177/0956797617715510

Jordano, M. L., & Touron, D. R. (2017). Stereotype threat as a trigger of mind-wandering in older adults. *Psychology And Aging, 32*(3), 307–313. https://doi.org/10.1037/pag0000167

Joseph, J. (2006). *The missing gene.* Algora.

Joseph, J., Breslin, C., & Skinner, H. (1999). Critical perspectives on the transtheoretical model and stages of change. In J. A. Tucker, D. M. Donovan, & G. A. Marlatt (Eds.), *Changing addictive behavior: Bridging clinical and public health strategies* (pp. 160–190). Guilford.

Joshi, S., Mooney, S. J., Rundle, A. G., Quinn, J. W., Beard, J. R., & Cerdá, M. (2017). Pathways from neighborhood poverty to depression among older adults. *Health & Place, 43*, 138–143.

Josselyn, S. A., & Tonegawa, S. (2020). Memory engrams: Recalling the past and imagining the future. *Science, 367*(6473).

Joyce, J., & Harwood, J. (2020). Social identity motivations and intergroup media attractiveness. *Group Processes & Intergroup Relations, 23*(1), 71–90.

Jozaghi, E., & Carleton, R. (2015). The identification of subtypes among injection drug users: HIV and hepatitis C differences as indicated with the theory of planned behaviour. *Journal of Substance Use, 20*(2), 119–127. https://doi.org/10.3109/14659891.2013.866987

Jung, C. (1917). *Analytic psychology.* Moffat, Yard.

Jung, C. G. (1921/1977). *The collected works of C. G. Jung. vol. 6. Psychological types.* Princeton University Press.

K

Kabat-Zinn, J. (1990). *Full catastrophe living: Using the wisdom of your body and mind to face stress, pain, and illness.* Delta.

Kabat-Zinn, J. (2006). *Coming to our senses: Healing ourselves and the world through mindfulness.* Hyperion.

Kabat-Zinn, J. (2009, March 18). *This analog life: Reconnecting with what is important in an always uncertain world.* Presentation at the 7th Annual Conference at the Center for Mindful Meditation.

Kabat-Zinn, J. (2019). *Mindfulness for all: The wisdom to transform the world.* Hachette Books.

Kabat-Zinn, J., & Davidson, R. (Eds.). (2012). *The mind's own physician.* New Harbinger.

Kabat-Zinn, J., Lipworth, L., & Burney, R. (1985). The clinical use of mindfulness meditation for the self-regulation of chronic pain. *Journal of Behavioral Medicine, 8*(2), 163–190.

Kabat-Zinn, J., & others. (1998). Influence of a mindfulness meditation-based stress reduction intervention on rates of skin clearing in patients with moderate to severe psoriasis undergoing phototherapy (UVB) and photochemotherapy (PUVA). *Psychosomatic Medicine, 60*(5), 625–632.

Kaczmarek, L. D., Behnke, M., Kashdan, T. B., Kusiak, A., Marzec, K., Mistrzak, M., & Włodarczyk, M. (2018). Smile intensity in social networking profile photographs is related to greater scientific achievements. *The Journal of Positive Psychology, 13*(5), 435–439.

Kafashan, S., Sparks, A., Rotella, A., & Barclay, P. (2017). Why heroism exists: Evolutionary perspectives on extreme helping. In S. T. Allison, G. R. Goethals, & R. M. Kramer,(Eds.), *Handbook of heroism and heroic leadership* (pp. 36–57). Routledge/Taylor & Francis Group.

Kagabo, R., Gordon, A. J., & Okuyemi, K. (2020). Smoking cessation in inpatient psychiatry treatment facilities: A review. *Addictive Behaviors Reports, 11*, 100255.

Kagan, J. (2013). Temperamental contributions to inhibited and uninhibited profiles. In P. D. Zelazo (Ed.), *The Oxford handbook of developmental psychology.* Oxford University Press.

Kahn, A., & others. (1992). Sleep and cardiorespiratory characteristics of infant victim of sudden death: A prospective case-control study. *Sleep, 15*(4), 287–292.

Kahneman, D., Knetsch, J. L., & Thaler, R. H. (1990). Experimental tests of the endowment effect and the coase theorem. *Journal of Political Economy, 98*(6), 1325–1348.

Kahneman, D., & Tversky, A. (1984). Choices, values and frames. *American Psychologist, 39*(4), 341–350.

Kalichman, S. C. (2007). The theory of reasoned action and advances in HIV/AIDS prevention. In I. Ajzen, D. Albarracin, & R. Hornik (Eds.), *Prediction and change of health behavior.* Erlbaum.

Kalokerinos, E. K., Kjelsaas, K., Bennetts, S., & Hippel, C. (2017). Men in pink collars: Stereotype threat and disengagement among male teachers and child protection workers. *European Journal of Social Psychology.* https://doi.org/10.1002/ejsp.2246

Kamali, A., Karbasian, N., Sherbaf, F. G., Wilken, L. A., Aein, A., Sair, H. I., . . . Riascos, R. F. (2020). Uncovering the dorsal thalamo-hypothalamic tract of the human limbic system. *Neuroscience, 432*, 55–62.

Kamin, L. J. (1968). Attention-like processes in classical conditioning. In M. R. Jones (Ed.), *Miami symposium on the prediction of behavior: Aversive stimuli.* University of Miami Press.

Kammel, L. G., & Correa, S. M. (2020). Selective sexual differentiation of neurone populations may contribute to sex-specific outputs of the ventromedial nucleus of the hypothalamus. *Journal of Neuroendocrinology, 32*(1), e12801.

Kanazawa, S. (2017). Possible evolutionary origins of human female sexual fluidity. *Biological Reviews, 92*(3), 1251-1274.

Kandel, E. R., & Schwartz, J. H. (1982). Molecular biology of learning: Modulation of transmitter release. *Science, 218*(4571), 433-443.

Kandler, C., Bratko, D., Butković, A., Hlupić, T. V., Tybur, J. M., Wesseldijk, L. W., . . . Lewis, G. J. (2020). How genetic and environmental variance in personality traits shift across the life span: Evidence from a cross-national twin study. *Journal of Personality and Social Psychology.* In press.

Kane, J. M., Robinson, D. G., Schooler, N. R., Mueser, K. T., Penn, D. L., Rosenheck, R. A., et al. (2015). Comprehensive versus usual community care for first-episode psychosis: 2-year outcomes from the NIMH RAISE early treatment program. *American Journal of Psychiatry, 173*(4), 362-372.

Kang, Y., Gray, J. R., & Dovidio, J. F. (2014). The nondiscriminating heart: Lovingkindness meditation training decreases implicit intergroup bias. *Journal of Experimental Psychology: General, 143*(3), 1306-1313.

Kanwisher, N. (2006). Neuroscience: What's in a face? *Science, 311*(5761), 617-618.

Kaplan, M. S., Huguet, N., McFarland, B. H., Caetano, R., Conner K. R., Giesbrecht, N., & Nolte, K. B. (2014). Use of alcohol before suicide in the United States. *Annals of Epidemiology, 24*(8), 588-592.

Kapoor, E., Faubion, S., Hurt, R. T., Fischer, K., Schroeder, D., Fokken, S., & Croghan, I. T. (2020). A selective serotonin receptor agonist for weight loss and management of menopausal vasomotor symptoms in overweight midlife women: A pilot study. *Menopause, 27*(11), 1228-1235.

Kaptchuk, T. J., Hemond, C. C., & Miller, F. G. (2020). Placebos in chronic pain: evidence, theory, ethics, and use in clinical practice. *BMJ, 370.* https://doi.org/10.1136/bmj.m1668

Kapur, S. (2003). Psychosis as a state of aberrant salience: A framework linking biology, phenomenology, and pharmacology. *American Journal of Psychiatry, 160*(1), 13-23.

Karatekin, C., & Ahluwalia, R. (2020). Effects of adverse childhood experiences, stress, and social support on the health of college students. *Journal of Interpersonal Violence, 35*(1-2), 150-172.

Karl, S., Boch, M., Virányi, Z., Lamm, C., & Huber, L. (2020). Training pet dogs for eye-tracking and awake fMRI. *Behavior Research Methods, 52,* 838-856. https://doi.org/10.3758/s13428-019-01281-7

Karnath, H. O., & Rennig, J. (2017). Investigating structure and function in the healthy human brain: Validity of acute versus chronic lesion-symptom mapping. *Brain Structure and Function, 222*(5), 2059-2070.

Karpati, F.J., Giacosa, C., Foster, N.E.V., Penhune, V.B., Hyde, K.L. (2017). Dance and music share gray matter structural correlates. *Brain Research, 1657,* 62-3.

Karthik, S., Sharma, L. P., & Narayanaswamy, J. C. (2020). Investigating the role of glutamate in obsessive-compulsive disorder: Current perspectives. *Neuropsychiatric Disease and Treatment, 16,* 1003.

Kasemsap, K. (2017). Mastering cognitive neuroscience and social neuroscience perspectives in the information age. In M. A. Dos Santos (Ed.). *Applying neuroscience to business practice* (pp. 82-113). Hershey, PA: Business Science Reference/IGI Global. https://doi.org/10.4018/978-1-5225-1028-4.ch005

Kashiwagi, M., Kanuka, M., Tatsuzawa, C., Suzuki, H., Morita, M., Tanaka, K., . . . Yanagisawa, M. (2020). Widely distributed neurotensinergic neurons in the brainstem regulate NREM sleep in mice. *Current Biology.*

Kasser, T. (2011). Cultural values and the well-being of future generations: A cross-national study. *Journal of Cross-Cultural Psychology, 42*(2), 206-215.

Kasser, T., & Ryan, R. M. (1993). A dark side of the American dream: Correlates of financial success as a central life aspiration. *Journal of Personality and Social Psychology, 65*(2), 410-422.

Kasser, T., & Ryan, R. M. (1996). Further examining the American dream: Differential correlates of intrinsic and extrinsic goals. *Personality and Social Psychology Bulletin, 22*(3), 280-287.

Kasser, T., Ryan, R. M., Couchman, C. E., & Sheldon, K. M. (2004). Materialistic values: Their causes and consequences. In T. Kasser & A. D. Kanner (Eds.), *Psychology and consumer culture: The struggle for a good life in a materialistic world* (pp. 11-28). American Psychological Association.

Kasser, T., & Sharma, Y. S. (1999). Reproductive freedom, educational equality, and females' preference for resource-acquisition characteristics in mates. *Psychological Science, 10*(4), 374-377.

Kassing, F., Casanova, T., Griffin, J. A., Wood, E., & Stepleman, L. M. (2020). The effects of polyvictimization on mental and physical health outcomes in an LGBTQ sample. *Journal of Traumatic Stress.* https://doi.org/10.1002/jts.22579

Katthagen, T., Kaminski, J., Heinz, A., Buchert, R., & Schlagenhauf, F. (2020). Striatal dopamine and reward prediction error signaling in unmedicated schizophrenia patients. *Schizophrenia Bulletin.*

Kaufman, S. B., Quilty, L. C., Grazioplene, R. G., Hirsh, J. B., Gray, J. R., Peterson, J. B., & DeYoung, C. G. (2016). Openness to experience and intellect differentially predict creative achievement in the arts and sciences. *Journal of Personality, 84*(2), 248-258. https://doi.org/10.1111/jopy.12156

Kaufman, T. M., Baams, L., & Veenstra, R. (2020). Disparities in persistent victimization and associated internalizing symptoms for heterosexual versus sexual minority youth. *Journal of Research on Adolescence, 30,* 516-531.

Kawabata, Y., Tseng, W. L., & Crick, N. R. (2014). Adaptive, maladaptive, mediational, and bidirectional processes of relational and physical aggression, relational and physical victimization, and peer liking. *Aggressive Behavior, 40*(3), 273-287.

Kawasaki, M., Uno, Y., Mori, J., Kobata, K., & Kitago, K. (2014). Transcranial magnetic stimulation-induced global propagation of transient phase resetting associated with directional information flow. *Frontiers in Human Neuroscience, 8,* 173.

Kaya, S., Yildirim, H., & Atmaca, M. (2020). Reduced hippocampus and amygdala volumes in antisocial personality disorder. *Journal of Clinical Neuroscience.*

Kaye, W. H., Wierenga, C. E., Bailer, U. F., Simmons, A. N., & Bischoff-Grethe, A. (2013). Nothing tastes as good as skinny feels: The neurobiology of anorexia nervosa. *Trends in Neuroscience, 36*(2), 110-120.

Kayser, D. (2017). Using facial expressions of emotion as a means for studying music-induced emotions. *Psychomusicology: Music, Mind, and Brain, 27*(3), 219-222. https://doi.org/10.1037/pmu0000187

Kazour, F., Richa, S., Char, C. A., Atanasova, B., & El-Hage, W. (2020). Olfactory memory in depression: State and trait differences between bipolar and unipolar disorders. *Brain Sciences, 10*(3), 189.

Kegel, A. F., & Flückiger, C. (2014). Predicting psychotherapy dropouts: A multilevel approach. *Clinical Psychology and Psychotherapy.* https://doi.org/10.1002/cpp.1899

Kell, H. J., & Lubinski, D. (2014). The study of mathematically precocious youth at maturity: Insights into elements of genius. In D. K. Simonton (Ed.), *The Wiley handbook of genius* (pp. 397-421). Wiley-Blackwell.

Keller, N. E., & Dunsmoor, J. E. (2020). The effects of aversive-to-appetitive counterconditioning on implicit and explicit fear memory. *Learning & Memory, 27*(1), 12-19.

Keller, N. E., Hennings, A. C., & Dunsmoor, J. E. (2020). Behavioral and neural processes in counterconditioning: Past and future directions. *Behaviour Research and Therapy, 125,* 103532.

Kellerman, A. L., & others. (1993). Gun ownership as a risk factor for homicide in the home. *New England Journal of Medicine, 329*(15), 1084-1091.

Kelley, H. H. (1973). The processes of causal attribution. *American Psychologist, 28*(2), 107-128.

Kelly, J. F. (2017). Is Alcoholics Anonymous religious, spiritual, neither? Findings from 25 years of mechanisms of behavior change research. *Addiction, 112*(6), 929-936.

Kelly, J. F., Hoeppner, B., Sout, R. L., & Pagano, M. (2012). Determining the relative importance of the mechanisms of behavior change within Alcoholics Anonymous: A multiplier mediator analysis. *Addiction, 107*(2), 289-299.

Keltner, D., Kogan, A., Piff, P. K., & Saturn, S. R. (2014). The sociocultural appraisals, values, and emotions (SAVE) framework of prosociality: Core processes from gene to meme. *Annual Review of Psychology 65,* 425-460.

Kempuraj, D., Ahmed, M. E., Selvakumar, G. P., Thangavel, R., Raikwar, S. P., Zaheer, S. A., . . . Zaheer, A. (2020). Psychological stress-induced immune response and risk of Alzheimer's disease in veterans from Operation Enduring Freedom and Operation Iraqi Freedom. *Clinical Therapeutics, 42*(6), 974-982.

Kendzierski, D., Ritter, R. L., Stump, T. K., & Anglin, C. L. (2015). The effectiveness of an implementation intentions intervention for fruit and vegetable consumption as moderated by self-schema status. *Appetite, 95228-238.* https://doi.org/10.1016/j.appet.2015.07.007

Kenny, L., Cribb, S. J., & Pellicano, E. (2019). Childhood executive function predicts later autistic features and adaptive behavior in young autistic people: a 12-year prospective study. *Journal of Abnormal Child Psychology, 47*(6), 1089-1099.

Kensinger, E. A., & Ford, J. H. (2020). Retrieval of emotional events from memory. *Annual Review of Psychology, 71,* 251-272.

Kernis, M. H. (Ed.). (2013). *Self-esteem issues and answers: A sourcebook of current perspectives.* Psychology Press.

Kerr, L. G., Tissera, H., McClure, M. J., Lydon, J. E., Back, M. D., & Human, L. J. (2020). Blind at first sight: The role of distinctively accurate and positive first impressions in romantic interest. *Psychological Science.* https://doi.org/10.1177/0956797620919674

Kerr, W. C., Kaplan, M. S., Huguet, N., Caetano, R., Giesbrecht, N., & McFarland, B. H. (2017). Economic recession, alcohol, and suicide rates: comparative effects of poverty, foreclosure, and job loss. *American Journal of Preventive Medicine, 52*(4), 469-475.

Kervezee, L., Kosmadopoulos, A., & Boivin, D. B. (2020). Metabolic and cardiovascular consequences of shift work: The role of circadian disruption and sleep disturbances. *European Journal of Neuroscience, 51*(1), 396-412.

Kerzel, D., & Andres, M. K. S. (2020). Object features reinstated from episodic memory guide attentional selection. *Cognition, 197,* 104158.

Kettenmann, H., & Verkhratsky, A. (2016). Glial cells: neuroglia. In D.W. Pfaff, N.D. Volkov (Eds.), *Neuroscience in the 21st century: From basic to clinical* (2nd ed., pp. 547-578). Springer.

Khan, A., Schmidt, L. A., & Chen, X. (2017). Cultural neuroscience of emotion: Toward a developmental framework. *Psychology & Neuroscience, 10*(1), 11-40. https://doi.org/10.1037/pne0000078

Khandaker, G. M., Zuber, V., Rees, J. M., Carvalho, L., Mason, A. M., Foley, C. N., . . . Burgess, S. (2020). Shared mechanisms between coronary heart disease and depression: Findings from a large UK general population-based cohort. *Molecular Psychiatry, 25*(7), 1477-1486.

Khurana, V. G., & Kaye, A. H. (2012). An overview of concussion in sport. *Journal of Clinical Neuroscience, 19*(1), 1-11.

Kihlstrom, J. F. (2020). Hypnosis. In V. Ziegler-Hill & T. K. Shackelford (Eds.), *Encyclopedia of personality and individual differences* (pp. 2103-2107). Springer International Publishing.

Killen, M., & Smetana, J. G. (Eds.). (2014). *Handbook of moral development* (2nd ed.). Psychology Press.

Kilpatrick, D. G. (2000). *The Mental Health Impact of Rape.* Medical University of South Carolina: National Violence Against Women Prevention Research Center. https://mainweb-v.musc.edu/vawprevention/research/mentalimpact.shtml

Kim, E. S., Chopik, W. J., & Smith, J. (2014). Are people healthier if their partners are more optimistic? The dyadic effect of optimism on health among older adults. *Journal of Psychosomatic Research, 76*(6), 447–453.

Kim, H. (2011). Differential neural activity in the recognition of old and new events: An activation likelihood estimation meta-analysis. *Human Brain Mapping, 34*(4), 814–836.

Kim, H. J., Greenspan, J. D., Ohrbach, R., Fillingim, R. B., Maixner, W., Renn, C. L., . . . Dorsey, S. G. (2019). Racial/ethnic differences in experimental pain sensitivity and associated factors—cardiovascular responsiveness and psychological status. *PLoS ONE, 14*(4), e0215534. https://doi.org/10.1371/journal.pone.0215534

Kim, J. I., Kim, J. W., Lee, J. M., Yun, H. J., Sohn, C. H., Shin, M. S., et al. (2018). Interaction between DRD2 and lead exposure on the cortical thickness of the frontal lobe in youth with attention-deficit/hyperactivity disorder. *Progress in Neuro-Psychopharmacology and Biological Psychiatry, 82*, 169–176.

Kim, K., Jang, E. H., Kim, A. Y., Fava, M., Mischoulon, D., Papakostas, G. I., . . . Jeon, H. J. (2019). Pre-treatment peripheral biomarkers associated with treatment response in panic symptoms in patients with major depressive disorder and panic disorder: A 12-week follow-up study. *Comprehensive Psychiatry, 95*, 152140.

Kim, S. W., Kim, M. K., Kim, B., Choi, T. K., & Lee, S. H. (2020). White matter connectivity differences between treatment responders and non-responders in patients with panic disorder. *Journal of Affective Disorders, 260*, 527–535.

Kindler, J., Lim, C. K., Weickert, C. S., Boerrigter, D., Galletly, C., Liu, D., . . . Lenroot, R. (2020). Dysregulation of kynurenine metabolism is related to proinflammatory cytokines, attention, and prefrontal cortex volume in schizophrenia. *Molecular Psychiatry, 25*(11), 2860-2872.

King, L. A. (2002). Gain without pain: Expressive writing and self-regulation. In S. J. Lepore & J. Smyth (Eds.), *The writing cure* (pp. 119–134). American Psychological Association.

King, L. A. (2008). Personal goals and life dreams: Positive psychology and motivation in daily life. In W. Gardner & J. Shah (Eds.), *Handbook of motivation science* (pp. 518–532). Guilford.

King, L. A., & Hicks, J. A. (2006). Narrating the self in the past and the future: Implications for maturity. *Research in Human Development, 3*(2–3), 121–138.

King, L. A., & Hicks, J. A. (2007). Whatever happened to "what might have been"? Regret, happiness, and maturity. *American Psychologist, 62*(7), 625–636.

King, L. A., & Hicks, J. A. (2012). Positive affect and meaning in life: The intersection of hedonism and eudaimonia. In P. T. Wong (Ed.), *The human quest for meaning* (2nd ed., pp. 125–142). Routledge.

King, L. A., & Hicks, J. A. (2020, in press). The science of meaning in life. *Annual Review of Psychology, 72.*

King, L. A., Hicks, J. A., Krull, J., & Del Gaiso, A. K. (2006). Positive affect and the experience of meaning in life. *Journal of Personality and Social Psychology, 90*(1), 179–196.

King, L. A., & Trent, J. (2012). Personality strengths. In H. Tennen & J. M. Suls (Eds.), *Handbook of psychology: Personality and social psychology* (vol. 5). Wiley.

King, L. S., Camacho, M. C., Montez, D. F., Humphreys, K. L., & Gotlib, I. H. (2020). Naturalistic language input is associated with resting-state functional connectivity in infancy. *Journal of Neuroscience.* JN-RM-0779-20. https://doi.org/10.1523/JNEUROSCI.0779-20.2020

King, R. B., & Mendoza, N. B. (2020). Achievement goal contagion: mastery and performance goals spread among classmates. *Social Psychology of Education, 23*(3), 795–814.

Kinsella, G. J., Pike, K. E., & Wright, B. J. (2020). Who benefits from cognitive intervention in older age? The role of executive function. *The Clinical Neuropsychologist, 34*(4), 826–844.

Kinsey, A. C., Martin, C. E., & Pomeroy, W. B. (1953). *Sexual behavior in the human female.* Saunders.

Kinsey, A. C., Pomeroy, W. B., & Martin, C. E. (1948). *Sexual behavior in the human male.* Saunders.

Kiosses, D. N., Gross, J. J., Banerjee, S., Duberstein, P. R., Putrino, D., & Alexopoulos, G. S. (2017). Negative emotions and suicidal ideation during psychosocial treatments in older adults with major depression and cognitive impairment. *The American Journal of Geriatric Psychiatry, 25*(6), 620–629.

Kirby, D. B. (2008). The impact of abstinence and comprehensive sex and STD/HIV education programs on adolescent sexual behavior. *Sexuality Research & Social Policy: A Journal of the NSRC, 5*(3), 18–27. https://doi.org/10.1525/srsp.2008.5.3.18

Kirk, H., Gray, K., Ellis, K., Taffe, J., & Cornish, K. (2017). Impact of attention training on academic achievement, executive functioning, and behavior: A randomized controlled trial. *American Journal on Intellectual and Developmental Disabilities, 122*(2), 97–117.

Kisely, S., Alichniewicz, K. K., Black, E. B., Siskind, D., Spurling, G., & Toombs, M. (2017). The prevalence of depression and anxiety disorders in indigenous people of the Americas: A systematic review and meta-analysis. *Journal of Psychiatric Research, 84*, 137–152.

Kivikangas, J. M., Fernández-Castilla, B., Järvelä, S., Ravaja, N., & Lönnqvist, J. E. (2020). Moral foundations and political orientation: Systematic review and meta-analysis. *Psychological Bulletin.* https://doi.org/10.1037/bul0000308

Kivity, Y., Levy, K. N., Kolly, S., & Kramer, U. (2020). The therapeutic alliance over 10 sessions of therapy for borderline personality disorder: Agreement and congruence analysis and relation to outcome. *Journal of Personality Disorders, 34*(1), 1–21.

Klahn, A. L., Klinkenberg, I. A., Lueken, U., Notzon, S., Arolt, V., Pantev, C., et al. (2017). Commonalities and differences in the neural substrates of threat predictability in panic disorder and specific phobia. *NeuroImage: Clinical, 14*, 530–537.

Klein, M., Onnink, M., van Donkelaar, M., Wolfers, T., Harich, B., Shi, Y., et al. (2017). Brain imaging genetics in ADHD and beyond—mapping pathways from gene to disorder at different levels of complexity. *Neuroscience & Biobehavioral Reviews, 80*, 115–155.

Klein, R. A., Ratliff, K. A., Vianello, M., Adams, R. J., Bahník, Š., Bernstein, M. J., & . . . Nosek, B. A. (2014). Investigating variation in replicability: A 'many labs' replication project. *Social Psychology, 45*(3), 142–152. https://doi.org/10.1027/1864-9335/a000178

Klein, S. A., Thielmann, I., Hilbig, B. E., & Heck, D. W. (2020). On the robustness of the association between honesty-humility and dishonest behavior for varying incentives. *Journal of Research in Personality, 88*, 104006.

Kleis, R. R., Hoch, M. C., Hogg-Graham, R., & Hoch, J. M. (2020). The effectiveness of the transtheoretical model to improve physical activity in healthy adults: A systematic review. *Journal of Physical Activity and Health.* https://doi.org/10.1123/jpah.2020-0334.

Klocek, J. (2017). History of medical and psychological hypnosis. In G. R. Elkins (Ed.), *Handbook of medical and psychological hypnosis: Foundations, applications, and professional issues* (pp. 3–7). Springer.

Klonoff, E. A. (2009). Disparities in the provision of medical care: An outcome in search of an explanation. *Journal of Behavioral Medicine, 32*(1), 48–63.

Klosterhalfen, S., & others. (2000). Pavlovian conditioning of taste aversion using a motion sickness paradigm. *Psychosomatic Medicine, 62*(5), 671–677.

Klucharev, V., Hytönen, K., Rijpkema, M., Smidts, A., & Fernández, G. (2009). Reinforcement learning signal predicts social conformity. *Neuron, 61*(1), 140–151.

Knapp, S., & VandeCreek, L. (2000). Recovered memories of childhood abuse: Is there an underlying consensus? *Professional Psychology: Research and Practice, 31*(4), 365–371.

Knight, R. (2014). A hundred years of latency: From Freudian psychoanalytic theory to dynamic systems nonlinear development in middle childhood. *Journal of the American Psychoanalytic Association, 62*(2), 203–235.

Knoblach, R. A., Schwartz, J. A., McBride, M., & Beaver, K. M. (2020). The association between genetic predisposition and parental socialization: An examination of gene–environment correlations using an adoption-based design. *International Journal of Offender Therapy and Comparative Criminology, 64*(2-3), 187–209.

Knoll, A. R., Otani, H., Skeel, R. L., & Van Horn, K. R. (2017). Learning style, judgements of learning, and learning of verbal and visual information. *British Journal of Psychology, 108*(3), 544–563. https://doi.org/10.1111/bjop.12214

Knouse, L. E., Teller, J., & Brooks, M. A. (2017). Meta-analysis of cognitive–behavioral treatments for adult ADHD. *Journal of Consulting and Clinical Psychology, 85*(7), 737–750.

Knudsen, E. I. (2020). Evolution of neural processing for visual perception in vertebrates. *Journal of Comparative Neurology.* https://doi.org/10.1002/cne.24871

Ko, A., Pick, C. M., Kwon, J. Y., Barlev, M., Krems, J. A., Varnum, M. E., . . . Crispim, A. C. (2020). Family matters: Rethinking the psychology of human social motivation. *Perspectives on Psychological Science, 15*(1), 173–201.

Ko, Y. W., Liao, C. P., Clark, R. W., Hsu, J. Y., Tseng, H. Y., & Huang, W. S. (2020). Aposematic coloration of prey enhances memory retention in an agamid lizard. *Animal Behaviour, 161*, 1–13.

Kobasa, S. C., Maddi, S. R., Puccetti, M. C., & Zola, M. (1986). Relative effectiveness of hardiness, exercise, and social support as resources against illness. *Journal of Psychosomatic Research, 29*(5), 525–533.

Kobasa, S., Maddi, S., & Kahn, S. (1982). Hardiness and health: A prospective study. *Journal of Personality and Social Psychology, 42*(1), 168–177.

Kocaturk, M., & Bozdag, F. (2020). Xenophobia among university students: Its relationship with five factor model and dark triad personality traits. *International Journal of Educational Methodology, 6*(3), 545–554.

Koć-Januchta, M., Höffler, T., Thoma, G., Prechtl, H., & Leutner, D. (2017). Visualizers versus verbalizers: Effects of cognitive style on learning with texts and pictures—An eye-tracking study. *Computers In Human Behavior, 68*, 170–179. https://doi.org/10.1016/j.chb.2016.11.028

Koegel, R. L., Koegel, L. K., Vernon, T. W., & Brookman-Frazee, L. I. (2017). Pivotal response treatment for individuals with autism spectrum disorder. In J.R. Weisz & A.E. Kazdin (Eds.), *Evidence-Based Psychotherapies for Children and Adolescents, 3rd ed.* (pp 290–307). Guilford.

Koenig, A. M., & Butters, M. A. (2014). Cognition in late-life depression: Treatment considerations. *Current Treatment Options in Psychiatry, 1*(1), 1–14.

Koenig, A. M., Eagly, A. H., Mitchell, A. A., & Ristikari, T. (2011). Are leader stereotypes masculine? A meta-analysis of three research paradigms. *Psychological Bulletin, 137*(4), 616–642.

Koenig, J., Thayer, J. F., & Kaess, M. (2020). Psychophysiological concomitants of personality pathology in development. *Current Opinion in Psychology.* https://doi.org/10.1016/j.copsyc.2020.12.004

Koenis, M. M., Romeijn, N., Piantoni, G., Verweij, I., Van der Werf, Y. D., Van Someren, E. J., & Stam, C. J. (2011). Does sleep restore the topology of functional brain networks. *Human Brain Mapping, 34*(2), 487–500.

Köffer, J., Scheiper-Welling, S., Verhoff, M. A., Bajanowski, T., & Kauferstein, S. (2020). Post-mortem genetic investigation of cardiac disease–associated genes in sudden infant death syndrome (SIDS) cases. *International Journal of Legal Medicine*, 1–6.

Kohlberg, L. (1958). *The development on modes of moral thinking and choice in the years 10 to 16.* Unpublished doctoral dissertation, University of Chicago, Chicago, Illinois.

Kohlberg, L. (1986). A current statement on some theoretical issues. In S. Modgil & C. Modgil (Eds.), *Lawrence Kohlberg.* Falmer.

Kohler, P. K., Manhart, L. E., & Lafferty, W. E. (2008). Abstinence-only and comprehensive sex education and the initiation of sexual activity and teen pregnancy. *Journal Of Adolescent Health, 42*(4), 344–351. https://doi.org/10.1016/j.jadohealth.2007.08.026

Köhler, W. (1925). *The mentality of apes.* Harcourt Brace Jovanovich.

Kohut, H. (1977). *Restoration of the self.* International Universities Press.

Koizumi, H., Mohammad, S., Ozaki, T., Muto, K., Matsuba, N., Kim, J., . . . Ikeda, M. (2020). Intracellular interplay between cholecystokinin and leptin signalling for satiety control in rats. *Scientific Reports, 10*(1), 1–15.

Kok, B. E., Catalino, L. I., & Fredrickson, B. L. (2008). The broadening, building, buffering effects of positive emotion. In S. J. Lopez (Ed.), *Positive psychology: Exploring the best of people* (vol. 3). Greenwood.

Kok, B. E., Waugh, C. E., & Fredrickson, B. L. (2013). Meditation and health: The search for mechanisms of action. *Social and Personality Psychology Compass, 7*(1), 27–39.

Kokou-Kpolou, C. K., Megalakaki, O., Laimou, D., & Kousouri, M. (2020). Insomnia during COVID-19 pandemic and lockdown: Prevalence, severity, and associated risk factors in France population. *Psychiatry Research, 290,* 113128.

Kolay, S., Boulay, R., & d'Ettorre, P. (2020). Regulation of ant foraging: A review of the role of information use and personality. *Frontiers in Psychology, 11,* 734.

Koletzko, S. H., Herrmann, M., & Brandstätter, V. (2015). Unconflicted goal striving: Goal ambivalence as a mediator between goal self-concordance and well-being. *Personality and Social Psychology Bulletin, 41*(1), 140–156.

Koltai, J., & Stuckler, D. (2020). Recession hardships, personal control, and the amplification of psychological distress: Differential responses to cumulative stress exposure during the US Great Recession. *SSM-Population Health, 10,* 100521.

Kong, F., Heller, A. S., van Reekum, C. M., & Sato, W. (2020). Positive neuroscience: The neuroscience of human flourishing. *Frontiers in Human Neuroscience, 14,* 47.

Kong, L. L., Allen, J. J. B., & Glisky, E. L. (2008). Interidentity memory transfer in dissociative identity disorder. *Journal of Abnormal Psychology, 117*(3), 686–692.

Königs, M., Pouwels, P. J., Ernest van Heurn, L., Bakx, R., Jeroen Vermeulen, R., Carel Goslings, J., & . . . Oosterlaan, J. (2017). Relevance of neuroimaging for neurocognitive and behavioral outcome after pediatric traumatic brain injury. *Brain Imaging And Behavior.* https://doi.org/10.1007/s11682-017-9673-3

Konrad, C., Adolph, D., Herbert, J. S., Neuhoff, L., Mohr, C., Jagusch-Poirier, J., . . . Schneider, S. (2020). A new 3-day standardized eyeblink conditioning protocol to assess extinction learning from infancy to adulthood. *Frontiers in Behavioral Neuroscience, 14,* 135.

Koopmann-Holm, B., Sze, J., Jinpa, T., & Tsai, J. L. (2020). Compassion meditation increases optimism towards a transgressor. *Cognition and Emotion, 34*(5), 1028–1035.

Kööts-Ausmees, L., Schmidt, M., Esko, T., Metspalu, A., Allik, J., & Realo, A. (2016). The role of the five-factor personality traits in general self-rated health. *European Journal of Personality, 30*(5), 492–504. https://doi.org/10.1002/per.2058

Kopelovich, S. L., Monroe-DeVita, M., Buck, B. E., Brenner, C., Moser, L., Jarskog, L. F., . . . Chwastiak, L. A. (2020). Community mental health care delivery during the COVID-19 pandemic: Practical strategies for improving care for people with serious mental illness. *Community Mental Health Journal.* https://doi.org/10.1007/s10597-020-00662-z

Kosslyn, S. M., Thompson, W. L., Kim, I. J., Rauch, S. L., & Alpert, N. M. (1996). Individual differences in cerebral blood flow in Area 17 predict the time to evaluate visualized letters. *Journal of Cognitive Neuroscience, 8*(1), 78–82.

Köster, M., Kayhan, E., Langeloh, M., & Hoehl, S. (2020). Making sense of the world: Infant learning from a predictive processing perspective. *Perspectives on Psychological Science.* https://doi.org/10.1177/1745691619895071

Kotter-Grühn, D., & Smith, J. (2011). When time is running out: Changes in positive future perception and their relationships to changes in well-being in old age. *Psychology and Aging, 26*(2), 381–387.

Koutsoklenis, A., & Gaitanidis, A. (2017). Interrogating the effectiveness of educational practices: A critique of evidence-based psychosocial treatments for children diagnosed with attention-deficit/hyperactivity disorder. *Frontiers in Education, 2,* 11.

Kovacs, B., & Kleinbaum, A. M. (2020). Language-style similarity and social networks. *Psychological Science, 31*(2), 202–213.

Kovarski, K., Latinus, M., Charpentier, J., Cléry, H., Roux, S., Houy-Durand, E., et al. (2017). Facial expression related vMMN: Disentangling emotional from neutral change detection. *Frontiers in Human Neuroscience, 11.*

Kraav, S. L., Awoyemi, O., Junttila, N., Vornanen, R., Kauhanen, J., Toikko, T., . . . Tolmunen, T. (2020). The effects of loneliness and social isolation on all-cause, injury, cancer, and CVD mortality in a cohort of middle-aged Finnish men. A prospective study. *Aging & Mental Health.* https://doi.org/10.1080/13607863.2020.1830945.

Kraus, C., Castrén, E., Kasper, S., & Lanzenberger, R. (2017). Serotonin and neuroplasticity–Links between molecular, functional and structural pathophysiology in depression. *Neuroscience & Biobehavioral Reviews, 77,* 317–326

Kraus, M. W., & Callaghan, B. (2016). Social class and prosocial behavior: The moderating role of public versus private contexts. *Social Psychological and Personality Science, 7*(8), 769–777. https://doi.org/10.1177/1948550616659120

Kraus, M. W., Onyeador, I. N., Daumeyer, N. M., Rucker, J. M., & Richeson, J. A. (2019). The misperception of racial economic inequality. *Perspectives on Psychological Science, 14*(6), 899–921.

Kraus, M. W., Piff, P. K., Mendoza-Denton, R., Rheinschmidt, M. L., & Keltner, D. (2012). Social class, solipsism, and contextualism: How the rich are different from the poor. *Psychological Review, 119*(3), 546–572. https://doi.org/10.1037/a0028756

Krause, A. J., Simon, E. B., Mander, B. A., Greer, S. M., Saletin, J. M., Goldstein-Piekarski, A. N., & Walker, M. P. (2017). The sleep-deprived human brain. *Nature Reviews Neuroscience, 18,* 404–418.

Krause, E. D., Vélez, C. E., Woo, R., Hoffmann, B., Freres, D. R., Abenavoli, R. M., & Gillham, J. E. (2017). Rumination, depression, and gender in early adolescence: A longitudinal study of a bidirectional model. *The Journal of Early Adolescence.*

Krause, N. (2006). Religion and health in late life. In J. E. Birren & K. W. Schaie (Eds.), *Handbook of the psychology of aging* (6th ed., pp. 500–515). Elsevier.

Krieger, I. (2017). *Counseling transgender and non-binary youth: The essential guide.* Jessica Kingsley Publishers.

Kristeller, J. L., & Epel, E. (2014). Mindful eating and mindless eating: The science and the practice. In A. Ie, C. T. Ngnoumen, & E. J. Langer (Eds.), *The Wiley Blackwell handbook of mindfulness, Vols. I and II* (pp. 913–933). Wiley-Blackwell.

Kroger, J., Martinussen, M., & Marcia, J. E. (2010). Identity change in adolescence and young adulthood: A meta-analysis. *Journal of Adolescence, 33*(5), 683–698.

Kropp Lopez, A. K., Nichols, S. D., Chung, D. Y., Kaufman, D. E., McCall, K. L., & Piper, B. J. (2020). Prescription opioid distribution after the legalization of recreational marijuana in Colorado. *International Journal of Environmental Research and Public Health, 17*(9), 3251.

Kross, E., Mischel, W., & Shoda, Y. (2010). Enabling self-control: A cognitive affective processing system (CAPS) approach to problematic behavior. In J. Maddux & J. Tangney (Eds.), *Social psychological foundations of clinical psychology.* Guilford.

Krueger, J. M. (2020). Sleep and circadian rhythms: Evolutionary entanglement and local regulation. *Neurobiology of Sleep and Circadian Rhythms.* https://doi.org/10.1016/j.nbscr.2020.100052

Krüger, T. H., Giraldi, A., & Tenbergen, G. (2020). The neurobiology of sexual responses and its clinical relevance. In M. Lew-Starowicz, A. Giraldi, & T. H. C.

Krüger (Eds.), *Psychiatry and sexual medicine* (pp. 71–84). Springer.

Kruk, J., Aboul-Enein, B. H., Bernstein, J., & Gronostaj, M. (2019). Psychological stress and cellular aging in cancer: A meta-analysis. *Oxidative Medicine and Cellular Longevity, 1270397.* https://doi.org/10.1155/2019/1270397.

Krupić, D., & Corr, P. J. (2020). How reinforcement sensitivity theory relates to self-determination theory. *Personality and Individual Differences, 155,* 109705.

Kruschwitz, J. D., Walter, M., Varikuti, D., Jensen, J., Plichta, M. N., Haddad, L., . . . & Walter, H. (2015). 5-HT-TLPR/rs25531 polymorphism and neuroticism are linked by resting state functional connectivity of amygdala and fusiform gyrus. *Brain Structure and Function, 220*(4), 2373–2385.

Kruse, E., Chancellor, J., Ruberton, P. M., & Lyubomirsky, S. (2014). An upward spiral between gratitude and humility. *Social Psychological and Personality Science, 5*(7), 805–814.

Kruse, O., Klein, S., Tapia León, I., Stark, R., & Klucken, T. (2020). Amygdala and nucleus accumbens involvement in appetitive extinction. *Human Brain Mapping, 41*(7), 1833–1841.

Kübler-Ross, E. (1969). *On Death and Dying.* MacMillan.

Kucera, K. L., Yau, R. K., Register-Mihalik, Marshall, S. W., Thomas, L. C., et al. (2017). Traumatic brain and spinal cord fatalities among high school and college football players—United States, 2005–2014. *MMWR. Morbidity and Mortality Weekly Report, 65,* 1465–1469

Kuhn, D. (2008). Formal operations from a twenty-first century perspective. *Human Development, 51*(1), 48–55.

Kuipers, G. S., den Hollander, S., van der Ark, L. A., & Bekker, M. J. (2017). Recovery from eating disorder 1 year after start of treatment is related to better mentalization and strong reduction of sensitivity to others. *Eating And Weight Disorders, 22*(3), 535–547. https://doi.org/10.1007/s40519-017-0405-x

Kulacaoglu, F., Solmaz, M., Ardic, F. C., Akin, E., & Kose, S. (2017). The relationship between childhood traumas, dissociation, and impulsivity in patients with borderline personality disorder comorbid with ADHD. *Psychiatry and Clinical Psychopharmacology, 27*(4), 393–402.

Kumar, A. J., Martins, D. O., Arruda, B. P., Lee, V. Y., Chacur, M., & Nogueira, M. I. (2020). Impairment of nociceptive responses after neonatal anoxia correlates with somatosensory thalamic damage: A study in rats. *Behavioural Brain Research, 390,* 112690.

Kumar, A., Killingsworth, M. A., & Gilovich, T. (2020). Spending on doing promotes more moment-to-moment happiness than spending on having. *Journal of Experimental Social Psychology, 88,* 103971.

Kuo, M. C., Kiu, I. P., Ting, K. H., & Chan, C. C. (2014). Age-related effects on perceptual and semantic encoding in memory. *Neuroscience, 261,* 95–106.

Kurson, R. (2007). *Crashing through: A true story of risk, adventure, and the man who dared to see.* Random House.

Kuvaas, B., Buch, R., Weibel, A., Dysvik, A., & Nerstad, C. L. (2017). Do intrinsic and extrinsic motivation relate differently to employee outcomes? *Journal of Economic Psychology, 61,* 244–258. https://doi.org/10.1016/j.joep.2017.05.004

Kuyper, P. (1972). The cocktail party effect. *Audiology, 11,* 277–282.

Kwak, S., & Chang, M. C. (2020). Impaired consciousness due to injury of the ascending reticular activating system in a patient with bilateral pontine infarction: A case report. *Translational Neuroscience, 11*(1), 264–268.

Kwon, D., Pfefferbaum, A., Sullivan, E. V., & Pohl, K. M. (2020). Regional growth trajectories of cortical myelination in adolescents and young adults: Longitudinal validation and functional correlates. *Brain Imaging and Behavior, 14*(1), 242–266.

L

Labbe, D., Rytz, A., Godinot, N., Ferrage, A., & Martin, N. (2017). Is portion size selection associated with expected satiation, perceived healthfulness or expected tastiness? A case study on pizza using a photograph-based computer task. *Appetite, 108,* 311–316.

Labouvie-Vief, G. (1986, August). *Modes of knowing and life-span cognition.* Paper presented at the meeting of the American Psychological Association, Washington, DC.

Labouvie-Vief, G. (2006). Emerging structures of adult thought. In J. J. Arnett & J. L. Tanner (Eds.), *Emerging adults in America* (pp. 60–84). American Psychological Association.

Laceulle, O. M., van Aken, M. A., Ormel, J., & Nederhof, E. (2015). Stress-sensitivity and reciprocal associations between stressful events and adolescent temperament. *Personality and Individual Differences, 81,* 76–83.

Lachman, M. E., Teshale, S., & Agrigoroaei, S. (2015). Midlife as a pivotal point in the life course: Balancing growth and decline at the crossroads of youth and old age. *International Journal of Behavioral Development, 39*(1), 20–31.

Lack, C.W., & Rousseau, J. (2016). *Critical Thinking, Science, and Pseudoscience: Why We Can't Trust Our Brains.* Springer.

Laffont, C. M., Gomeni, R., Zheng, B., Heidbreder, C., Fudala, P. J., & Nasser, A. F. (2015). Population pharmacokinetic modeling and simulation to guide dose selection for RBP-7000, a new sustained-release formulation of risperidone. *Journal of Clinical Pharmacology, 55*(1), 93–103.

Lafraire, J., Rioux, C., Hamaoui, J., Girgis, H., Nguyen, S., & Thibaut, J. P. (2020). Food as a borderline domain of knowledge: The development of domain-specific inductive reasoning strategies in young children. *Cognitive Development, 56,* 100946.

LaFrance, A. (2020, June). The prophecies of Q: American conspiracy theories are entering a dangerous new phase. *The Atlantic.* https://www.theatlantic.com/magazine/archive/2020/06/qanon-nothing-can-stop-what-is-coming/610567/

Lafuse, W. P., Wu, Q., Saljoughian, N., Turner, J., & Rajaram, M. (2020). Psychological stress creates an immune-suppressive environment that increases the susceptibility of aged mice to *Mycobacterium tuberculosis* infection. *Journal of Immunology, 204*(1), 16.

Lahti, A., White, D., Kraguljac, N., & Reid, M. (2017). Hippocampal glutamate and functional connectivity as biomarkers of treatment response to antipsychotic medication. *Schizophrenia Bulletin, 43*(suppl_1), S5–S5.

Laible, D. J., & Karahuta, E. (2014). Prosocial behaviors in early childhood: Helping others, responding to the distress of others, and working with others. In L. Padilla-Walker & G. Carlo (Eds.), *Prosocial behavior.* Oxford University Press.

Laird, S. (2016, July 21). Concussions ended his promising NFL career, so he became an Olympian instead. *Mashable.* http://mashable.com/2016/07/21/jahvid-best-nfl-olympics/#zBKcgSdyw5qx

Lake, S., Socias, M. E., & Milloy, M. J. (2020). Evidence shows that cannabis has fewer relative harms than opioids. *CMAJ, 192*(7), e166–e167.

Lamb, M. E., Malloy, L. C., Hershkowitz, I., & La Rooy, D. (2015). Children and the law. In R. E. Lerner (Ed.), *Handbook of child psychology and developmental science* (7th ed., vol. 3). Wiley.

Lam, R. (2012). *Depression* (2nd ed.). Oxford University Press.

Lambert, K., Hyer, M., Bardi, M., Rzucidlo, A., Scott, S., Terhune-Cotter, B., . . . & Kinsley, C. (2016). Natural-enriched environments lead to enhanced environmental engagement and altered neurobiological resilience. *Neuroscience, 330,* 386–394.

Lambert, M. J. (2001). The effectiveness of psychotherapy: What has a century of research taught us about the effects of treatment. *Georgia Psychologist, 55*(2), 9–11.

Lambert, M. J. (2013). Outcome in psychotherapy: The past and important advances. *Psychotherapy, 50*(1), 42–51.

Laming, D. (2010). Serial position curves in free recall. *Psychological Review, 117*(1), 93–133.

Lamis, D. A., Ballard, E. D., & Patel, A. B. (2014). Loneliness and suicide ideation in drug-using college students. *Suicide and Life-Threatening Behavior, 44*(6), 629–640.

Lampard, A. M., Byrne, S. M., McLean, N., & Fursland, A. (2012). The Eating Disorder Inventory-2 perfectionism scale: Factor structure and associations with dietary restraint and weight and shape concern in eating disorders. *Eating Behaviors, 13*(1), 49–53.

Lamy, M., Pedapati, E. V., Dominick, K. L., Wink, L. K., & Erickson, C. A. (2020). Recent advances in the pharmacological management of behavioral disturbances associated with autism spectrum disorder in children and adolescents. *Pediatric Drugs, 22,* 473–483.

Lander, I. A. (2016). Exploring the place of forgiveness therapy in social work practice. *Journal of Social Work Practice, 30,* 69–80. https://doi.org/10.1080/02650533.2015.1081879

Landis, C., & Erlick, D. (1950). An analysis of the Porteus Maze Test as affected by psychosurgery. *American Journal of Psychology, 63*(4), 557–566.

Landler, M. (2012, January 30). From Biden, a vivid account of Bin Laden decision. The Caucus, *New York Times* blog. http://thecaucus.blogs.nytimes.com/2012/01/30/from-biden-a-vivid-account-of-bin-laden-decision/?_r=0

Lane, A., Luminet, O., Nave, G., & Mikolajczak, M. (2016). Is there a publication bias in behavioural intranasal oxytocin research on humans? Opening the file drawer of one laboratory. *Journal of Neuroendocrinology, 28*(4).

Lane, A., Mikolajczak, M., Treinen, E., Samson, D., Corneille, O., de Timary, P., & Luminet, O. (2015). Failed replication of oxytocin effects on trust: The envelope task case. *Plos ONE, 10*(9),

Lane, R. D. (2020). Promoting the integration of psychodynamic and emotion-focused psychotherapies through advances in affective science and neuroscience. *Clinical Social Work Journal.* https://doi.org/10.1007/s10615-020-00759-8

Lange, C. G. (1922). *The emotions.* Williams & Wilkins.

Langer, E., Blank, A., & Chanowitz, B. (1978). The mindlessness of ostensibly thoughtful action: The role of "placebic" information in interpersonal interaction. *Journal of Personality and Social Psychology, 36*(6), 635–642.

Langer, E. J. (2000). Mindful learning. *Current Directions in Psychological Science, 9*(6), 220–223.

Langer, J. J. (1991). *Holocaust testimonies: The ruins of memory.* Yale University Press.

Langer, K., Hagedorn, B., Stock, L., Otto, T., Wolf, O., & Jentsch, V. L. (2020). Acute stress improves the effectivity of cognitive emotion regulation in men. *Scientific Reports, 10*(1). https://doi.org/10.1038/s41598-020-68137-5

Langeslag, S. E., & Surti, K. (2017). The effect of arousal on regulation of negative emotions using cognitive reappraisal: An ERP study. *International Journal of Psychophysiology, 118,* 18–26. https://doi.org/10.1016/j.ijpsycho.2017.05.012

Langlois, J. H., Kalakanis, L., Rubenstein, A. J., Larson, A., Hallam, M., & Smoot, M. (2000). Maxims or myths of beauty? A meta-analytic and theoretical review. *Psychological Bulletin, 126*(3), 390–423.

Langstrom, N., Rahman, Q., Carlstrom, E., & Lichtenstein, P. (2010). Genetic and environmental effects on same-sex sexual behaviour: A population study of twins in Sweden. *Archives of Sexual Behavior, 39*(1), 75–80.

Lanter, A., & Singer-Dudek, J. (2020). The effects of an observational conditioning-by-denial intervention on the establishment of three observational learning cusps. *European Journal of Behavior Analysis.* https://doi.org/10.1080/15021149.2020.1724001

Laplana, M., Royo, J. L., Garcia, L. F., Aluja, A., Gomez-Skarmeta, J. L., & Fibla, J. (2014). SIRPB1 copy-number polymorphism as candidate quantitative trait locus for impulsive-disinhibited personality. *Genes, Brain, and Behavior, 13*(7), 653–662.

Larsen, D. J., Whelton, W. J., Rogers, T., McElheran, J., Herth, K., Tremblay, J., . . . Domene, J. (2020). Multidimensional hope in counseling and psychotherapy scale. *Journal of Psychotherapy Integration, 30*(3), 407–422. https://doi.org/10.1037/int0000198

Lashley, K. (1950). In search of the engram. In *Symposium of the Society for Experimental Biology* (vol. 4, pp. 454–482). Cambridge University Press.

Latané, B. (1981). The psychology of social impact. *American Psychologist, 36*(4), 343–356.

Latchoumane, C. F. V., Ngo, H. V. V., Born, J., & Shin, H. S. (2017). Thalamic spindles promote memory formation during sleep through triple phase-locking of cortical, thalamic, and hippocampal rhythms. *Neuron, 95*(2), 424–435.

Laub, R., & Frings, C. (2020). Why star retrieves scar: Binding and retrieval of perceptual distractor features. *Journal of Experimental Psychology: Learning, Memory, and Cognition, 46*(2), 350–363.

Laurent, H. K., Neiderhiser, J. M., Natsuaki, M. N., Shaw, D. S., Fisher, P. A., Reiss, D., & Leve, L. D. (2014). Stress system development from age 4.5 to 6: Family environment predictors and adjustment implications of HPA activity stability versus change. *Developmental Psychobiology, 56,* 340–354.

Lavender, J. M., Goodman, E. L., Culbert, K. M., Wonderlich, S. A., Crosby, R. D., Engel, S. G., & et al. (2017). Facets of impulsivity and compulsivity in women with anorexia nervosa. *European Eating Disorders Review, 25*(4), 309–313. https://doi.org/10.1002/erv.2516

Layton, T. J., Barnett, M. L., Hicks, T. R., & Jena, A. B. (2018). Attention-deficit/hyperactivity disorder and month of school enrollment. *New England Journal of Medicine, 379,* 2122–2130.

Lazarus, R. S. (1991). On the primacy of cognition. *American Psychologist, 39*(2), 124–129.

Lazarus, R. S. (1993). Coping theory and research: Past, present, and future. *Psychosomatic Medicine, 55*(3), 234–247.

Lazarus, R. S. (2000). Toward better research on stress and coping. *American Psychologist, 55*(6), 665–673.

Lazarus, R. S. (2003). Does the positive psychology movement have legs? *Psychological Inquiry, 14*(2), 93–109.

Le, D., Holt, C. L., Hosack, D. P., Huang, J., & Clark, E. M. (2016). Religious participation is associated with increases in religious social support in a national longitudinal study of African Americans. *Journal of Religion and Health, 55*(4), 1449–1460. https://doi.org/10.1007/s10943-015-0143-1

Le Grange, D., Lock, J., Loeb, K., & Nicholls, D. (2010). Academy for Eating Disorders position paper: The role of the family in eating disorders. *International Journal of Eating Disorders, 43*(1), 1–5.

Le Nguyen, K. D., Lin, J., Algoe, S. B., Brantley, M. M., Kim, S. L., Brantley, J., . . . Fredrickson, B. L. (2019). Loving-kindness meditation slows biological aging in novices: Evidence from a 12-week randomized controlled trial. *Psychoneuroendocrinology, 108,* 20–27.

Le Texier, T. (2018). *Histoire d'un mensonge. Enquête sur l'expérience de Stanford.* Zones (published in French).

Leahy, R. L., Holland, S. J. F., & McGinn, L. K. (2012). *Treatment plans and interventions for depression and anxiety disorders* (2nd ed.). Guilford.

Leary, M. R. (2020). The need to belong, the sociometer, and the pursuit of relational value: Unfinished business. *Self and Identity.* https://doi.org/10.1080/15298868.2020.1779120

Leary, M. R., & Guadagno, J. (2011). The sociometer, self-esteem, and the regulation of interpersonal behavior. In K. D. Vohs & R. F. Baumeister (Eds.), *Handbook of self-regulation: Research, theory, and applications* (2nd ed., pp. 339–354). Guilford.

Leavy, P. (2017). *Research design: Quantitative, qualitative, mixed methods, arts-based, and community-based participatory research.* Guilford.

Leblanc, H., & Ramirez, S. (2020). Linking social cognition to learning and memory. *Journal of Neuroscience, 40*(46), 8782–8798.

Lecce, S., Caputi, M., Pagnin, A., & Banerjee, R. (2017). Theory of mind and school achievement: The mediating role of social competence. *Cognitive Development, 44,* 85–97.

Lecompte, V., Robins, S., King, L., Solomonova, E., Khan, N., Moss, E., . . . Turecki, G. (2020). Examining the role of mother-child interactions and DNA methylation of the oxytocin receptor gene in understanding child controlling attachment behaviors. *Attachment & Human*

Development, January 2020, 1–19. https://doi.org/10.10 80/14616734.2019.1708422

LeDoux, J. E. (1996). *The emotional brain: The mysterious underpinnings of emotional life.* Simon & Schuster.

LeDoux, J. E. (2001). Emotion, memory, and the brain. www.cns.nyu.edu/ledoux/overview.htm

LeDoux, J. E. (2012). Evolution of human emotion: A view through fear. *Progress in Brain Research, 195,* 431–442.

LeDoux, J. E. (2013). The slippery slope of fear. *Trends in Cognitive Science, 17*(4), 155–156.

LeDoux, J. E. (2014). Coming to terms with fear. *Proceedings of the National Academy of Sciences USA, 111*(8), 2871–2878.

LeDoux, J. E., & Pine, D. S. (2016). Using neuroscience to help understand fear and anxiety: A two-system framework. *The American Journal of Psychiatry, 173*(11), 1083–1093. https://doi.org/10.1176/appi.ajp.2016.16030353

Lee, A. C., Thavabalasingam, S., Alushaj, D., Çavdaroğlu, B., & Ito, R. (2020). The hippocampus contributes to temporal duration memory in the context of event sequences: A cross-species perspective. *Neuropsychologia, 137,* 107300.

Lee, A. S., Azmitia, E. C., & Whitaker-Azmitia, P. M. (2017). Developmental microglial priming in postmortem autism spectrum disorder temporal cortex. *Brain, Behavior, and Immunity, 62,* 193–202

Lee-Bates, B., Billing, D. C., Caputi, P., Carstairs, G. L., Linnane, D., & Middleton, K. (2017). The application of subjective job task analysis techniques in physically demanding occupations: Evidence for the presence of self-serving bias. *Ergonomics, 60*(9), 1240–1249. https://doi.org/10.1080/00140139.2016.1262063

Lee, H. (2016). Which feedback is more effective for pursuing multiple goals of differing importance? The interaction effects of goal importance and performance feedback type on self-regulation and task achievement. *Educational Psychology, 36*(2), 297–322. https://doi.org/10.1080/01443410.2014.995596

Lee, H., Kim, Y., & Terry, J. (2020). Adverse childhood experiences (ACEs) on mental disorders in young adulthood: Latent classes and community violence exposure. *Preventive Medicine, 134,* 106039.

Lee, I. S., Yoon, S. S., Lee, S. H., Lee, H., Park, H. J., Wallraven, C., & Chae, Y. (2013). An amplification of feedback from facial muscles strengthened sympathetic activations to emotional facial cues. *Autonomic Neuroscience, 179*(1–2), 37–42.

Lee, J. C., Hall, D. L., & Wood, W. (2018). Experiential or material purchases? Social class determines purchase happiness. *Psychological Science, 29.* https://doi.org/10.1177/0956797617736386

Lee, J. Y., Kang, B. C., Park, J. W., & Park, H. J. (2020). Changes in cortical auditory evoked potentials by ipsilateral, contralateral and binaural speech stimulation in normal-hearing adults. *Clinical and Experimental Otorhinolaryngology, 13*(2), 133–140.

Lee, K., Quinn, P. C., Pascalis, O., & Slater, A. (2013). Development of face-processing ability in children. In P. D. Zelazo (Ed.), *The Oxford handbook of developmental psychology.* Oxford University Press.

Lee, N. (2020). The benefits of exercise effect on cancer: A review. *Exercise Science, 29*(1), 4–9.

Lee, S. W., Kim, I., Yoo, J., Park, S., Jeong, B., & Cha, M. (2016). Insights from an expressive writing intervention on Facebook to help alleviate depressive symptoms. *Computers in Human Behavior, 62,* 613–619. https://doi.org/10.1016/j.chb.2016.04.034

Leerkes, E. M., Su, J., Calkins, S. D., O'Brien, M., & Supple, A. J. (2017). Maternal physiological dysregulation while parenting poses risk for infant attachment disorganization and behavior problems. *Development and Psychopathology, 29*(1), 245–257.

Lefevre, E. M., Pisansky, M. T., Toddes, C., Baruffaldi, F., Pravetoni, M., Tian, L., . . . Rothwell, P. E. (2020). Interruption of continuous opioid exposure exacerbates drug-evoked adaptations in the mesolimbic dopamine system. *Neuropsychopharmacology,* 1–12.

Légal, J. B., Chekroun, P., Coiffard, V., & Gabarrot, F. (2017). Beware of the gorilla: Effect of goal priming on

inattentional blindness. *Consciousness and Cognition, 55,* 165–171.

Lehman, B. (2020). Good grades and a bad reputation: Gender, academics, and relational aggression. *Youth & Society, 52*(3), 490–509.

Lehmkuhl Noer, C., Kjær Needham, E., Wiese, A., Johannes Skovbjerg Balsby, T., & Dabelsteen, T. (2016). Personality matters: Consistency of inter-individual variation in shyness-boldness across non-breeding and pre-breeding season despite a fall in general shyness levels in farmed American mink (Neovison vison). *Applied Animal Behaviour Science, 181,* 191–199. https://doi.org/10.1016/j.applanim.2016.05.003

Lehto, K., Karlsson, I., Lundholm, C., & Pedersen, N. L. (2019). Genetic risk for neuroticism predicts emotional health depending on childhood adversity. *Psychological Medicine, 49*(2), 260–267.

Leibovich, L., McCarthy, K. S., & Zilcha-Mano, S. (2019). How do supportive techniques bring about therapeutic change: The role of therapeutic alliance as a potential mediator. *Psychotherapy.* 57(2), 151–159.

Leippe, M. R., Bergold, A. N., & Eisenstadt, D. (2017). Prejudice and terror management at trial: Effects of defendant race/ethnicity and mortality salience on mock-jurors' verdict judgments. *The Journal of Social Psychology, 157*(3), 279–294.

Leist, A., Muniz-Terrera, G., & Solomon, A. (2020). Using cohort data to emulate lifestyle interventions: Long-term beneficial effects of initiating physical activity on cognitive decline and dementia. *Alzheimer's and Dementia: The Journal of the Alzheimer's Association.* http://hdl.handle.net/10993/44917

Lemaitre, D., Hurtado, M. L., De Gregorio, C., Oñate, M., Martínez, G., Catenaccio, A., & Wishart, T. M. (2020). Collateral sprouting of peripheral sensory neurons exhibits a unique transcriptomic profile. *Molecular Neurobiology,* 1–18.

Lemay, E. J., O'Brien, K. M., Kearney, M. S., Sauber, E. W., & Venaglia, R. B. (2017). Using conformity to enhance willingness to intervene in dating violence: A theory of planned behavior analysis. *Psychology of Violence.* https://doi.org/10.1037/vio0000114

Lenneberg, E. H., Rebelsky, F. G., & Nichols, I. A. (1965). The vocalization of infants born to deaf and hearing parents. *Human Development, 8,* 23–37.

Leonard, H. (2020). Elaine Lebar. https://www.halleonard.com/biography/205/elaine-lebar

Leone, T., & Brown, L. (2020). Timing and determinants of age at menarche in low-income and middle-income countries. *BMJ Global Health.* In press.

Leor, J., Poole, W. K., & Kloner, R. A. (1996). Sudden cardiac death triggered by an earthquake. *New England Journal of Medicine, 334,* 413–419. https://doi.org/10.1056/NEJM199602153340701

Lerman, A. (2020). *The non-disclosing patient.* Springer.

Lerner, B. H. (2005). Last-ditch medical therapy—Revisiting lobotomy. *New England Journal of Medicine, 353*(2), 119–121.

Lerner, J. E., & Hawkins, R. L. (2016). Welfare, liberty, and security for all? U.S. sex education policy and the 1996 Title V Section 510 of the Social Security Act. *Archives of Sexual Behavior, 45*(5), 1027–1038. https://doi.org/10.1007/s10508-016-0731-5

Leshinskaya, A., & Caramazza, A. (2016). For a cognitive neuroscience of concepts: Moving beyond the grounding issue. *Psychonomic Bulletin & Review, 23,* 991–1001. https://doi.org/10.3758/s13423-015-0870-z

Lesicko, A. M., & Llano, D. A. (2020). Circuit mechanisms underlying the segregation and integration of parallel processing streams in the inferior colliculus. *Journal of Neuroscience.* https://doi.org/10.1523/JNEURO-SCI.0646-20.2020

Letzen, J. E., Remeniuk, B., Smith, M. T., Irwin, M. R., Finan, P. H., & Seminowicz, D. A. (2020). Individual differences in pain sensitivity are associated with cognitive network functional connectivity following one night of experimental sleep disruption. *Human Brain Mapping, 41*(3), 581–593.

Leucht, S., Crippa, A., Siafis, S., Patel, M. X., Orsini, N., & Davis, J. M. (2020). Dose-response meta-analysis of

antipsychotic drugs for acute schizophrenia. *American Journal of Psychiatry, 177*(4), 342–353.

Leung, A. K., Maddux, W. W., Galinsky, A. D., & Chiu, C. (2008). Multicultural experience enhances creativity: The when and how. *American Psychologist, 63*(3), 169–181.

Leung, W. (2014). Infusions of allogeneic natural killer cells as cancer therapy. *Clinical Cancer Research, 20*(13), 3390–3400.

Leventhal, H., & Tomarken, A. J. (1986). Emotion: Today's problems. *Annual Review of Psychology 37*(1), 565–610.

Levi, O., Shoval-Zuckerman, Y., Fruchter, E., Bibi, A., Bar-Haim, Y. and Wald, I. (2017), Benefits of a psychodynamic group therapy (PGT) model for treating veterans with PTSD. *Journal of Clinical Psychology, 73,* 1247–1258.

Levine, B., Svoboda, E., Turner, G. R., Mandic, M., & Mackey, A. (2009). Behavioral and functional neuroanatomical correlates of anterograde autobiographical memory in isolated retrograde amnesic patient M.L. *Neuropsychologia, 47*(11), 2188–2196.

Levine, M., Cassidy, C., & Jentzsch, I. (2010). The implicit identity effect: Identity primes, group size, and helping. *British Journal of Social Psychology, 49*(Pt 4), 785–802.

Levine, R. L. (2002). Endocrine aspects of eating disorders in adolescents. *Adolescent Medicine, 13*(1), 129–144.

Lev-On, A., & Waismel-Manor, I. (2016). Looks that matter: The effect of physical attractiveness in low- and high-information elections. *American Behavioral Scientist, 60*(14), 1756–1771. https://doi.org/10.1177/0002764216676249

Levy, B. R., Slade, M. D., Kunkel, S. R., & Kasl, S. V. (2002). Longevity increased by positive self-perceptions of aging. *Journal of Personality and Social Psychology, 83*(2), 261–270.

Levy, H. C., Worden, B. L., Gilliam, C. M., D'Urso, C., Steketee, G., Frost, R. O., & Tolin, D. F. (2017). Changes in saving cognitions mediate hoarding symptom change in cognitive-behavioral therapy for hoarding disorder. *Journal of Obsessive-Compulsive and Related Disorders, 14,* 112–118.

Levy, I. P. (2020). "Real recognize real": Hip-hop spoken word therapy and humanistic practice. *The Journal of Humanistic Counseling, 59*(1), 38–53.

Lewis, D. G., Al-Shawaf, L., Conroy-Beam, D., Asao, K., & Buss, D. M. (2017). Evolutionary psychology: A how-to guide. *American Psychologist, 72*(4), 353–373. https://doi.org/10.1037/a0040409

Lewis Jr., N. A., & Earl, A. (2018). Seeing more and eating less: Effects of portion size granularity on the perception and regulation of food consumption. *Journal of Personality and Social Psychology, 114* (5), 786–803. https://doi.org/10.1037/pspp0000183.supp

Lewis Jr, N. A., & Sekaquaptewa, D. (2016). Beyond test performance: A broader view of stereotype threat. *Current Opinion in Psychology, 11,* 40–43.

Lewis Jr, N., & Wai, J. (2020, in press). Communicating what we know, and what isn't so: Science communication in psychology. *Perspectives on Psychological Science.*

Lewkowicz, D. J. (2010). Infant perception of audiovisual speech synchrony. *Developmental Psychology, 46*(1), 66–77.

Lewkowicz, D. J., & Hansen-Tift, A. M. (2012). Infants deploy selective attention to the mouth of a talking face when learning speech. *Proceedings of the National Academy of Sciences USA, 109*(5), 1431–1436.

Li, A., MacNeill, B., Curiel, H., & Poling, A. (2017). Risperidone in combination with other drugs: Experimental research in individuals with autism spectrum disorder. *Experimental and Clinical Psychopharmacology, 25*(5), 434–439.

Li, G., Shao, C., Chen, Q., Wang, Q., & Yang, K. (2017). Accumulated GABA activates presynaptic GABAB receptors and inhibits both excitatory and inhibitory synaptic transmission in rat midbrain periaqueductal gray. *NeuroReport, 28*(6), 313–318.

Li, J., Zhong, Y., Ma, Z., Wu, Y., Pang, M., Wang, C., . . . Zhang, N. (2020). Emotion reactivity-related brain network analysis in generalized anxiety disorder: A task fMRI study. *BMC Psychiatry, 20*(1), 1–13.

Li, K., Kadohisa, M., Kusunoki, M., Duncan, J., Bundesen, C., & Ditlevsen, S. (2020). Distinguishing between parallel and serial processing in visual attention from neurobiological data. *Royal Society Open Science, 7*(1), 191553.

Li, L., Qin, L., Xu, Z., Yin, Y., Wang, X., Kong, B., . . . Cao, K. (2020). Artificial intelligence distinguishes COVID-19 from community acquired pneumonia on chest T. *Radiology*. https://doi.org/10.1148/radiol.2020200905

Li, N., Baldermann, J. C., Kibleur, A., Treu, S., Akram, H., Elias, G. J., . . . Barcia, J. A. (2020). A unified connectomic target for deep brain stimulation in obsessive-compulsive disorder. *Nature Communications, 11*(1), 1–12.

Li, S. H., & Graham, B. M. (2017). Why are women so vulnerable to anxiety, trauma-related and stress-related disorders? The potential role of sex hormones. *The Lancet Psychiatry, 4*(1), 73–82.

Li, X., Chen, W., & Popiel, P. (2015). What happens on Facebook stays on Facebook? The implications of Facebook interaction for perceived, receiving, and giving social support. *Computers in Human Behavior, 51*, 106–113.

Li, Y., Cho, H., Wang, F., Canela-Xandri, O., Luo, C., Rawlik, K., . . . Wang, Q. K. (2020). Statistical and functional studies identify epistasis of cardiovascular risk genomic variants from genome-wide association studies. *Journal of the American Heart Association, 9*(7), e014146.

Li, Y., & Epley, N. (2009). When the best appears to be saved for last: Serial position effects on choice. *Journal of Behavioral Decision Making, 22*(4), 378–389.

Li, Y., Wang, J., Ye, H., & Luo, J. (2020). Modulating the activity of vmPFC regulates informational social conformity: A tDCS study. *Frontiers in Psychology, 11*, 566977.

Liang, B., Williams, L. M., & Siegel, J. A. (2006). Relational outcomes of childhood sexual trauma in female survivors: A longitudinal study. *Journal of Interpersonal Violence, 21*(1), 42–57.

Liben, L. S., Bigler, R. S., & Hilliard, L. J. (2014). Gender development: From universality to individuality. In E. T. Gershoff, R. S. Mistry, & D. A. Crosby (Eds.), *Societal contexts of child development.* Oxford University Press.

Lieberz, J., Scheele, D., Spengler, F. B., Matheisen, T., Schneider, L., Stoffel-Wagner, B., . . . Hurlemann, R. (2020). Kinetics of oxytocin effects on amygdala and striatal reactivity vary between women and men. *Neuropsychopharmacology, 45*(7), 1134–1140.

Lien, Y. Y., Lin, H. S., Lien, Y. J., Tsai, C. H., Wu, T. T., Li, H., & Tu, Y. K. (2020). Challenging mental illness stigma in healthcare professionals and students: a systematic review and network meta-analysis. *Psychology & Health.* https://doi.org/10.1080/08870446.2020.1828413

Liew, S. C., & Aung, T. (2020). Sleep deprivation and its association with diseases—A review. *Sleep Medicine.* https://doi.org/10.1016/j.sleep.2020.07.048

Lilienfeld, S. O., Wood, J. M., & Garb, H. N. (2000, November). The scientific status of projective techniques. *Psychological Science in the Public Interest, 1*(2), 27–66.

Lim, A., Nakamura, B. J., Higa-McMillan, C. K., Shimabukuro, S., & Slavin, L. (2012). Effects of workshop trainings on evidence-based practice knowledge and attitudes among youth community mental health providers. *Behavior Research and Therapy, 50*(6), 397–406.

Lin, C., Adolphs, R., & Alvarez, R. M. (2018). Inferring whether officials are corruptible from looking at their faces. *Psychological Science.* https://doi.org/10.1177/0956797618788882

Lin, C., & Alvarez, R. M. (2020). Personality traits are directly associated with anti-black prejudice in the United States. *PloS one, 15*(7), e0235436.

Lin, J. Y., Kuo, W. W., Baskaran, R., Kuo, C. H., Chen, Y. A., Chen, W. S. T., . . . Huang, C. Y. (2020). Swimming exercise stimulates IGF1/PI3K/Akt and AMPK/SIRT1/PGC1α survival signaling to suppress apoptosis and inflammation in aging hippocampus. *Aging (Albany NY), 12*(8), 6852.

Lin, M. (2018). Online dating industry: The business of love. https://www.toptal.com/finance/business-model-consultants/online-dating-industry#:~:text=Online%20Dating%20Industry%20Market%20Size,growing%20since%20the%20previous%20year

Lin, Y., & others. (2014). Efficacy of REACH forgiveness across cultures. *Journal of Clinical Psychology, 70*(9), 781–793.

Lin, Y., Stavans, M., & Baillargeon, R. (2020). Infants' physical reasoning and the cognitive architecture that supports it. To appear in O. Houdé & G. Borst (Eds.-in-chief), *Cambridge handbook of cognitive development.* Cambridge University Press.

Lind, C., Walsh, C., Mccaffrey, G., Wardle, M. L., Johansson, B., & Juby, B. (2019). Youth strengths arise from the ashes of adversity. *International Journal of Adolescence and Youth, 24*(3), 274–281.

Lindberg, L.D., Santelli, J.S., & Desai, S. (2016). Understanding the decline in adolescent fertility in the United States, 2007–2012. *Journal of Adolescent Health, 59*, 577–583. https://doi.org/10.1016/j.jadohealth.2016.06.024

Linder, A., Gerdtham, U. G., Trygg, N., Fritzell, S., & Saha, S. (2020). Inequalities in the economic consequences of depression and anxiety in Europe: A systematic scoping review. *European Journal of Public Health, 30*(4), 767–777.

Link, B. G., DuPont-Reyes, M. J., Barkin, K., Villatoro, A. P., Phelan, J. C., & Painter, K. (2020). A school-based intervention for mental illness stigma: A cluster randomized trial. *Pediatrics, 145*(6), e20190780.

Linkovski, O., Katzin, N., & Salti, M. (2017). Mirror neurons and mirror-touch synesthesia. *The Neuroscientist, 23*(2), 103–108. https://doi.org/10.1177/1073858416652079

Lipp, O. V., Waters, A. M., Luck, C. C., Ryan, K. M., & Craske, M. G. (2020). Novel approaches for strengthening human fear extinction: The roles of novelty, additional USs, and additional GSs. *Behaviour Research and Therapy, 124*, 103529.

Lippke, S., Schwarzer, R., Ziegelmann, J. P., Scholz, U., & Schüz, B. (2010). Testing stage-specific effects of a stage-matched intervention: A randomized controlled trial targeting physical exercise and its predictors. *Health Education & Behavior, 37*(4), 533–546. https://doi.org/10.1177/1090198109359386

Lippke, S., Ziegelmann, J. P., Schwarzer, R., & Velicer, W. F. (2009). Validity of stage assessment in the adoption and maintenance of physical activity and fruit and vegetable consumption. *Health Psychology, 28*(2), 183–193.

Lipsey, M. W., & Wilson, D. B. (1993). The efficacy of psychological, educational, and behavioral treatment: Confirmation from meta-analysis. *American Psychologist, 48*(12), 1181–1209.

Liszkowski, U., Schaffer, M., Carpenter, M., & Tomasello, M. (2009). Prelinguistic infants, but not chimpanzees, communicate about absent entities. *Psychological Science, 20*(5), 654–660.

Litim, N., Morissette, M., & Di Paolo, T. (2017). Metabotropic glutamate receptors as therapeutic targets in Parkinson's disease: An update from the last 5 years of research. *Neuropharmacology, 115*, 166–179.

Littleton, H. (2014). Interpersonal violence on college campuses: Understanding risk factors and working to find solutions. *Trauma Violence Abuse, 15*(4), 297–303.

Littleton, S. H., Berkowitz, R. I., & Grant, S. F. (2020). Genetic determinants of childhood obesity. *Molecular Diagnosis & Therapy, 24*, 653–663.

Littrell, J. H., & Girvin, H. (2002). Stages of change: A critique. *Behavior Modification, 26*(2), 223–273.

Liu, C. Y., Huang, W. L., Kao, W. C., & Gau, S. S. F. (2017). Influence of disruptive behavior disorders on academic performance and school functions of youths with attention-deficit/hyperactivity disorder. *Child Psychiatry & Human Development, 48*, 870–880.

Liu, L., Escudero, P., Quattropani, C., & Robbins, R. A. (2020). Factors affecting infant toy preferences: Age, gender, experience, motor development, and parental attitude. *Infancy, 25*(5), 593–617.

Liu, S., Brooks, N. B., & Spelke, E. S. (2019). Origins of the concepts cause, cost, and goal in prereaching infants. *Proceedings of the National Academy of Sciences, 116*(36), 17747–17752.

Liu, S., Liu, P., Wang, M., & Zhang, B. (2020). Effectiveness of stereotype threat interventions: A meta-analytic review. *Journal of Applied Psychology.* https://doi.org/10.1037/apl0000770

Liu, S., & Spelke, E. S. (2017). Six-month-old infants expect agents to minimize the cost of their actions. *Cognition, 160*, 35–42.

Liu, X., & Engel, S. A. (2020). Higher-level meta-adaptation mitigates visual distortions produced by lower-level adaptation. *Psychological Science, 31*(6), 654–662.

Liu, X., Haagsma, J., Sijbrands, E., Buijks, H., Boogaard, L., Mackenbach, J. P., . . . Polinder, S. (2020). Anxiety and depression in diabetes care: Longitudinal associations with health-related quality of life. *Scientific Reports, 10*(1), 1–9.

Liu, Y., & Gu, X. (2020). Media multitasking, attention, and comprehension: a deep investigation into fragmented reading. *Educational Technology Research and Development, 68*(1), 67–87.

Liverpool, L. (2020). Racism in close up. *New Scientist, 248*(3309), 41–45.

Liyanarachchi, S., Gudmundsson, J., Ferkingstad, E., He, H., Jonasson, J. G., Tragante, V., . . . Mayordomo, J. I. (2020). Assessing thyroid cancer risk using polygenic risk scores. *Proceedings of the National Academy of Sciences, 117*(11), 5997–6002.

Lo Sauro, C., Ravaldi, C., Cabras, P. L., Faravelli, C., & Ricca, V. (2008). Stress, hypothalamic-pituitary-adrenal axis, and eating disorders. *Neuropsychobiology, 57*(3), 95–115.

Lodder, P. (2020). A re-evaluation of the type D personality effect. *Personality and Individual Differences, 167*, 110254.

Loftus, E. F. (1975). Leading questions and the eye-witness report. *Cognitive Psychology, 7*(4), 560–572.

Loftus, E. F. (1993). Psychologists in the eyewitness world. *American Psychologist, 48*(5), 550–552.

Logan, A. M., Mammel, A. E., Robinson, D. C., Chin, A. L., Condon, A. F., & Robinson, F. L. (2017). Schwann cell-specific deletion of the endosomal PI 3-kinase Vps34 leads to delayed radial sorting of axons, arrested myelination, and abnormal ErbB2-ErbB3 tyrosine kinase signaling. *Glia.* https://doi.org/10.1002/glia.23173

Lok, I., Dunn, E. W., Vazire, S., & Chopik, W. (2020). Under what conditions does prosocial spending promote happiness? *Collabra: Psychology, 6*, 5.

Lokensgard, K. H. (2014). Blackfoot nation. In D. A. Leeming (Ed.), *Encyclopedia of psychology and religion.* Springer.

Londerée, A. M., & Wagner, D. D. (2020). The orbitofrontal cortex spontaneously encodes food health and contains more distinct representations for foods highest in tastiness. *Social Cognitive and Affective Neuroscience*, nsaa083. https://doi.org/10.1093/scan/nsaa083

London News. (2020, July 22). Former aspiring singer writes book with his left eyelid after suffering massive stroke which left him with lock-in syndrome, which destroys body's nerves. https://londonnewsonline.co.uk/former-aspiring-singer-writes-book-with-his-left-eyelid-after-suffering-massive-stroke-which-left-him-with-lock-in-syndrome-which-destroys-bodys-nerves/

Longobardi, C., Settanni, M., Lin, S., & Fabris, M. A. (2020). Student–teacher relationship quality and prosocial behaviour: The mediating role of academic achievement and a positive attitude towards school. *British Journal of Educational Psychology.* https://doi.org/10.1111/bjep.12378

Lonsdorf, E. V. (2017). Sex differences in nonhuman primate behavioral development. *Journal of Neuroscience Research, 95*(1–2), 213–221.

Lopez, J., Coll, J., Haimel, M., Kandasamy, S., Tarraga, J., Furio-Tari, P., et al. (2017). HGVA: the Human Genome Variation Archive. *Nucleic Acids Research, 45*(W1), W189–W194.

López-Fernández, F. J., Mezquita, L., Etkin, P., Griffiths, M. D., Ortet, G., & Ibáñez, M. I. (2020). The role of violent video game exposure, personality, and deviant peers in aggressive behaviors among adolescents: A two-wave longitudinal study. *Cyberpsychology, Behavior, and Social Networking.* https://doi.org/10.1089/cyber.2020.0030

López-López, A., Matías-Pompa, B., Fernández-Carnero, J., Gil-Martínez, A., Alonso-Fernández, M., Alonso Pérez, J. L., & González Gutierrez, J. L. (2020). Blunted pain modulation response to induced stress in women with fibromyalgia with and without posttraumatic stress disorder comorbidity: New evidence of hypo-reactivity to stress in fibromyalgia. *Behavioral Medicine.* https://doi.org/10.1080/08964289.2020.1758611

Lorant, V., Croux, C., Weich, S., Deliège, D., Mackenbach, J., & Ansseau, M. (2007). Depression and socioeconomic risk factors: 7-year longitudinal population study. *British Journal of Psychiatry, 190,* 293-298.

Lord, C., McCauley, J. B., Pepa, L. A., Huerta, M., & Pickles, A. (2020). Work, living, and the pursuit of happiness: Vocational and psychosocial outcomes for young adults with autism. *Autism, 24*(7), 1691-1703.

Lorenzo-Blanco, E. I., Schwartz, S. J., Unger, J. B., Zambo-anga, B. L., Des Rosiers, S. E., Baezconde-Garbanati, L., et al. (2016). Alcohol use among recent immigrant Latino/a youth: Acculturation, gender, and the Theory of Reasoned Action. *Ethnicity & Health, 21*(6), 609-627. https://doi.org/10.1080/13557858.2016.1179723

Lorenz, T. (2020). Cake, no matter how you slice it. *The New York Times, July 19,* ST3.

LoSavio, S. T., Cohen, L. H., Laurenceau, J.-P., Dasch, K. B., Parrish, B. P., & Park, C. L. (2011). Reports of stress-related growth from daily negative events. *Journal of Social and Clinical Psychology, 30*(7), 760-785.

Losin, E. A. R., Woo, C. W., Medina, N. A., Andrews-Hanna, J. R., Eisenbarth, H., & Wager, T. D. (2020). Neural and sociocultural mediators of ethnic differences in pain. *Nature Human Behaviour, 4,* 517-530.

Lotfinia, S., Soorgi, Z., Mertens, Y., & Daniels, J. (2020). Structural and functional brain alterations in psychiatric patients with dissociative experiences: A systematic review of magnetic resonance imaging studies. *Journal of Psychiatric Research, 128,* 5-15.

Loveday, P. M., Lovell, G. P., & Jones, C. M. (2017). The best possible selves intervention: A review of the literature to evaluate efficacy and guide future research. *Journal of Happiness Studies.* https://doi.org/10.1007/s10902-016-9824-z

Lovick, T. A. (2014). Sex determinants of experimental panic attacks. *Neuroscience and Biobehavioral Reviews, 46*(Pt. 3), 465-471.

Lovik, A., Nassiri, V., Verbeke, G., Molenberghs, G., & Sodermans, A. K. (2017). Psychometric properties and comparison of different techniques for factor analysis on the Big Five Inventory from a Flemish sample. *Personality And Individual Differences, 117,* 122-129. https://doi.org/10.1016/j.paid.2017.05.048

Lu, J., Sherman, D., Devor, M., & Saper, C. B. (2006). A putative flip-flop switch for control of REM sleep. *Nature, 441,* 589-594.

Lu, L. C., Lan, S. H., Hsieh, Y. P., Lin, L. Y., Chen, J. C., & Lan, S. J. (2020). Massage therapy for weight gain in preterm neonates: A systematic review and meta-analysis of randomized control trials. *Complementary Therapies in Clinical Practice.* https://doi.org/10.1016/j.ctcp.2020.101168

Luan, S., Reb, J., & Gigerenzer, G. (2019). Ecological rationality: Fast-and-frugal heuristics for managerial decision making under uncertainty. *Academy of Management Journal, 62*(6), 1735-1759.

Lubinski, D., Benbow, C. P., Webb, R. M., & Bleske-Rechek, A. (2006). Tracking exceptional human capital over two decades. *Psychological Science, 17*(3), 194-199.

Lubinski, D., Webb, R. M., Morelock, M. J., & Benbow, C. P. (2001). Top 1 in 10,000: A 10-year follow-up of the profoundly gifted. *Journal of Applied Psychology, 86*(4), 718-729.

Luborsky, L., & others. (2002). The dodo bird verdict is alive and well—mostly. *Clinical Psychology: Science and Practice, 9*(1), 2-12.

Lucas, J. W., & Phelan, J. C. (2019). Influence and social distance consequences across categories of race and mental illness. *Society and Mental Health, 9*(2), 143-157.

Lucas, R. E., & Baird, B. M. (2004). Extraversion and emotional reactivity. *Journal of Personality and Social Psychology, 86*(3), 473-485.

Luciano, M. T., & McDevitt-Murphy, M. E. (2017). Posttraumatic stress and physical health functioning: Moderating effects of deployment and postdeployment social support in OEF/OIF/OND veterans. *Journal of Nervous and Mental Disease, 205*(2), 93-98.

Ludyga, S., Gerber, M., Pühse, U., Looser, V. N., & Kamijo, K. (2020). Systematic review and meta-analysis investigating moderators of long-term effects of exercise on cognition in healthy individuals. *Nature Human Behaviour, 4,* 603-612.

Lugaresi, E., Provini, F., & Montagna, P. (2020). The neuroanatomy of sleep. Considerations on the role of the thalamus in sleep and a proposal for a caudorostral organization. *European Journal of Anatomy, 8*(2), 85-93.

Lugo-Candelas, C., Corbeil, T., Wall, M., Posner, J., Bird, H., Canino, G., . . . Duarte, C. S. (2020). ADHD and risk for subsequent adverse childhood experiences: Understanding the cycle of adversity. *Journal of Child Psychology and Psychiatry.*

Lui, J. H., Marcus, D. K., & Barry, C. T. (2017). Evidence-based apps? A review of mental health mobile applications in a psychotherapy context. *Professional Psychology: Research and Practice, 48*(3), 199-210.

Lund, H. G., Reider, B. D., Whiting, A. B., & Prichard, J. R. (2010). Sleep patterns and predictors of disturbed sleep in a large population of college students. *Journal of Adolescent Health, 46*(2), 124-132.

Lunde, C. E., & Sieberg, C. B. (2020). Walking the tightrope: A proposed model of chronic pain and stress. *Frontiers in Neuroscience, 14,* 270.

Lungu, A., & Linehan, M. M. (2017). Dialectical behavior therapy: Overview, characteristics, and future directions. In S. G. Hofman & G. J. G. Asmundsen (Eds.), *The science of cognitive behavioral therapy* (pp. 429-459). Academic Press.

Luo, W., & Zhou, R. (2020). Can working memory task-related EEG biomarkers measure fluid intelligence and predict academic achievement in healthy children? *Frontiers in Behavioral Neuroscience, 14,* 2.

Luo, Y., Cao, Z., Wang, D., Wu, L., Li, Y., Sun, W., & Zhu, Y. (2014). Dynamic study of the hippocampal volume by structural MRI in a rat model of depression. *Neurological Sciences, 35*(11), 1777-1783.

Luo, Y., Hawkley, L. C., Waite, L. J., & Cacioppo, J. T. (2012). Loneliness, health, and mortality in old age: A national longitudinal study. *Social Science and Medicine, 74*(6), 907-914.

Lupton, A., Abu-Suwa, H., Bolton, G. C., & Golden, C. (2020). The implications of brain tumors on aggressive behavior and suicidality: A review. *Aggression and Violent Behavior.* https://doi.org/10.1016/j.avb.2020.101416

Luria, A. R. (1973). *The working brain.* Penguin.

Lushington, K., Pamula, Y., Martin, J., & Kennedy, J. D. (2014). Developmental changes in sleep: Infancy and preschool years. In A. R. Wolfson & E. Montgomery-Downs (Eds.), *Oxford handbook of infant, child, and adolescent sleep and behavior.* Oxford University Press.

Luttrell, A., Petty, R. E., Briñol, P., & Wagner, B. C. (2016). Making it moral: Merely labeling an attitude as moral increases its strength. *Journal of Experimental Social Psychology, 65,* 82-93. https://doi.org/10.1016/j.jesp.2016.04.003

Lydon, C. A., Rohmeier, K. D., Yi, S. C., Mattaini, M. A., & Williams, W. L. (2011). How far do you have to go to get a cheeseburger around here? The realities of an environmental design approach to curbing the consumption of fast-food. *Behavior and Social Issues, 20,* 6-23.

Lykken, D. (1999). *Happiness: What studies on twins show us about nature, nurture, and the happiness set-point.* Golden Books.

Lykken, D. T. (1987). The probity of the polygraph. In S. M. Kassin & L. S. Wrightsman (Eds.), *The psychology of evidence and trial procedures.* Sage.

Lykken, D. T. (1998). *A tremor in the blood: Uses and abuses of the lie detector.* Plenum.

Lykken, D. T. (2001). Lie detection. In W. E. Craighead & C. B. Nemeroff (Eds.), *The Corsini encyclopedia of psychology and behavioral science* (3rd ed.). Wiley.

Lynam, D. R., Caspi, A., Moffitt, T. E., Loeber, R., & Stouthamer-Loeber, M. (2007). Longitudinal evidence that psychopathy scores in early adolescence predict adult psychopathy. *Journal of Abnormal Psychology, 116*(1), 155-165.

Lynch, A., Elmore, B., & Kotecki, J. (2015). *Choosing health* (2nd ed.). Pearson.

Lynch, K. E., Morandini, J. S., Dar-Nimrod, I., & Griffiths, P. E. (2019). Causal reasoning about human behavior genetics: Synthesis and future directions. *Behavior Genetics, 49*(2), 221-234.

Lynch, M. F., & Sheldon, K. M. (2020). Conditional regard, self-concept, and relational authenticity: Revisiting some key Rogerian concepts cross-culturally, through multilevel modeling. *Journal of Humanistic Psychology, 60*(2), 168-186.

Lynn, S. J., & Green, J. P. (2011). The sociocognitive and dissociation theories of hypnosis: Toward a rapprochement. *International Journal of Clinical and Experimental Hypnosis, 59*(3), 277-293.

Lynn, S. J., Maxwell, R., & Green, J. P. (2017). The hypnotic induction in the broad scheme of hypnosis: A sociocognitive perspective. *American Journal of Clinical Hypnosis, 59*(4), 363-384. https://doi.org/10.1080/00029157.2016.1233093

Lynott, D., Corker, K. S., Connell, L., & O'Brien, K. S. (2017). The effect of haptic and ambient temperature experience on prosocial behavior. *Archives of Scientific Psychology, 5*(1), 10-18. https://doi.org/10.1037/arc0000031

Lyubomirsky, S. (2008). *The how of happiness: A scientific approach to getting the life you want.* Penguin.

Lyubomirsky, S. (2011). *The way to happiness: Action plan for a happy life.* Kinneret.

Lyubomirsky, S. (2013). *The myth of happiness.* Penguin.

Lyubomirsky, S., Boehm, J. K., Kasri, F., & Zehm, K. (2011a). The cognitive and hedonic costs of dwelling on achievement-related negative experiences: Implications for enduring happiness and unhappiness. *Emotion, 11*(5), 1152-1167.

Lyubomirsky, S., Dickerhoof, R., Boehm, J. K., & Sheldon, K. M. (2011b). Becoming happier takes both a will and a proper way: An experimental longitudinal intervention to boost well-being. *Emotion, 11*(2), 391-402.

M

Maccallum, F., Malgaroli, M., & Bonanno, G. A. (2017). Networks of loss: Relationships among symptoms of prolonged grief following spousal and parental loss. *Journal of Abnormal Psychology, 126*(5), 652-662.

Maccoby, E. E. (2002). Gender and group processes: A developmental perspective. *Current Directions in Psychological Science, 11*(2), 54-58.

MacGillavry, D. W., & Ullrich, D. (2020). A novel theory on the predictive value of variation in the β-endorphin system on the risk and severity of PTSD. *Military Psychology, 32*(3), 247-260.

MacKain, S. J., & Noel, N. E. (2020). Master's-level psychology training in substance use disorder treatment: One model for expanding the workforce. *Training and Education in Professional Psychology, 14*(1), 27-33.

Mackenbach, J. D., Lakerveld, J., Van Lenthe, F. J., Teixeira, P. J., Compernolle, S., De Bourdeaudhuij, I., et al. (2016) Interactions of individual perceived barriers and neighbourhood destinations with obesity-related behaviours in Europe. *Obesity Reviews, 17,* 68-80. https://doi.org/10.1111/obr.12374.

MacMullin, L. N., Bokeloh, L. M., Nabbijohn, A. N., Santarossa, A., van der Miesen, A. I., Peragine, D. E., & VanderLaan, D. P. (2020). Examining the relation between gender nonconformity and psychological well-being in children: The roles of peers and parents. *Archives of Sexual Behavior.* https://link.springer.com/article/10.1007/s10508-020-01832-6

Maddi, S. (1998). Hardiness. In H. S. Friedman (Ed.), *Encyclopedia of mental health* (vol. 3). Academic.

Maddi, S. R., Matthews, M. D., Kelly, D. R., Villarreal, B. J., Gundersen, K. K., & Savino, S. C. (2017). The continuing role of hardiness and grit on performance and retention in West Point cadets. *Military Psychology, 29*(5), 355–358. https://doi.org/10.1037/mil0000145

Maddux, W. W., & Galinsky, A. D. (2007, September). *Cultural borders and mental barriers: Living in and adapting to foreign countries facilitates creativity.* Working Paper No. 2007/51/B. INSEAD.

Madhok, D. Y., Yue, J. K., Sun, X., Suen, C. G., Coss, N. A., Jain, S., . . . Track-Tbi Investigators. (2020). Clinical predictors of 3- and 6-month outcome for mild traumatic brain injury patients with a negative head CT scan in the emergency department: A TRACK-TBI pilot study. *Brain Sciences, 10*(5), 269.

Madhyastha, T. M., Hamaker, E. L., & Gottman, J. M. (2011). Investigating spousal influence using moment-to-moment affect data from marital conflict. *Journal of Family Psychology, 25*(2), 292–300.

Madras, B. K. (2013). History of the discovery of the antipsychotic dopamine D2 receptor: A basis for the dopamine hypothesis of schizophrenia. *Journal of the History of the Neurosciences, 22*(1), 62–78.

Madrid, H. P., & Patterson, M. G. (2016). Creativity at work as a joint function between openness to experience, need for cognition and organizational fairness. *Learning and Individual Differences, 51,* 409–416. https://doi.org/10.1016/j.lindif.2015.07.010

Maeda, U., Shen, B. J., Schwarz, E. R., Farrell, K. A., & Mallon, S. (2012). Self-efficacy mediates the association of social support and depression with treatment adherence in heart failure patients. *International Journal of Behavioral Medicine, 20*(1), 88–96.

Maenner, M. J., Shaw, K. A., Baio, J., Washington, A., Patrick, M., & Dietz, P. M. (2020). Prevalence of autism spectrum disorder among children aged 8 years—Autism and Developmental Disabilities Monitoring Network, 11 sites, United States, 2016. *MMWR Surveillance Summary, 69,* 1–12. http://dx.doi.org/10.15585/mmwr.ss6904a1external icon

Maguire, E. A., Gadian, G. D., Johnsrude, I. S., Good, C. D., Ashburner, J., Frackowiak, R. S. J., & Frith, C. D. (2000). Navigation-related structural change in the hippocampi of taxi drivers. *Proceedings of the National Academy of Sciences USA, 97*(9), 4398–4403.

Mahmoud, F. A., Aktas, A., Walsh, D., & Hullihen, B. (2011). A pilot study of taste changes among hospice inpatients with advanced cancer. *American Journal of Hospice and Palliative Care, 28*(7), 487–492.

Mahmut, M. K., & Croy, I. (2019). The role of body odors and olfactory ability in the initiation, maintenance and breakdown of romantic relationships–a review. *Physiology & Behavior, 207,* 179–184.

Maia, A., Oliveira, J., Lajnef, M., Mallet, L., Tamouza, R., Leboyer, M., & Oliveira-Maia, A. J. (2019). Oxidative and nitrosative stress markers in obsessive–compulsive disorder: A systematic review and meta-analysis. *Acta Psychiatrica Scandinavica, 139*(5), 420–433.

Maier, M., Rosenbaum, D., Haeussinger, F., Brüne, M., Enzi, B., Plewnia, C., . . . Ehlis, A. (2019). Forgiveness and cognitive control–provoking revenge via theta-burst-stimulation of the DLPFC. *Brain Stimulation: Basic, Translational, and Clinical Research in Neuromodulation, 12*(2), 484.

Maier, N. R. F. (1931). Reasoning in humans. II. The solution of a problem and its appearance in consciousness. *Journal of Comparative Psychology, 12*(2), 181–194.

Mairs, H. J., & Bradshaw, T. (2013). Promoting mental and physical health in adults with psychosis. *Nursing Standard, 27*(47), 50–56.

Maitre, N. L., Key, A. P., Chorna, O. D., Slaughter, J. C., Matusz, P. J., Wallace, M. T., & Murray, M. M. (2017). The dual nature of early-life experience on somatosensory processing in the human infant brain. *Current Biology, 27*(7), 1048–1054.

Mak, C., Whittingham, K., Cunnington, R., & Boyd, R. N. (2017). Efficacy of mindfulness-based interventions for

attention and executive function in children and adolescents–a systematic review. *Mindfulness.*

Makovac, E., Watson, D. R., Meeten, F., Garfinkel, S. N., Cercignani, M., Critchley, H. D., & Ottaviani, C. (2016). Amygdala functional connectivity as a longitudinal biomarker of symptom changes in generalized anxiety. *Social Cognitive and Affective Neuroscience, 11*(11), 1719–1728.

Malakh-Pines, A., & Maslach, C. (2002). *Experiencing social psychology* (4th ed.). McGraw-Hill.

Malcolm-Smith, S., Solms, M., Turnbull, O., & Tredoux, C. (2008). Threat in dreams: An adaptation? *Consciousness and Cognition, 17*(4), 1281–1291.

Maldei, T., Baumann, N., & Koole, S. L. (2020). The language of intuition: a thematic integration model of intuitive coherence judgments. *Cognition and Emotion, 34,* 1183–1198.

Malin, D. H., Anderson, A. N., & Goyarzu, P. (2017). Transmitters and receptors in nicotine withdrawal syndrome. In F. S. Hall, J. W. Young, & A. Der-Avakian (Eds.), *Negative affective states and cognitive impairments in nicotine dependence* (pp. 133–151). Elsevier Academic Press. https://doi.org/10.1016/B978-0-12-802574-1.00008-9

Malnic, B., & Glezer, I. (2020). Olfactory loss of function as a possible symptom of COVID-19. *JAMA Otolaryngology-Head & Neck Surgery.* https://doi.org/10.1001/jamaoto.2020.1588

Mamarde, A., Navkhare, P., Singam, A., & Kanoje, A. (2013). Recurrent dissociative fugue. *Indian Journal of Psychological Medicine, 35*(4), 400–401.

Mandler, G. (1980). Recognizing: The judgment of previous occurrence. *Psychological Review, 87*(3), 252–271.

Mandy, W., & Lai, M. (2016). Annual research review: The role of the environment in the developmental psychopathology of autism spectrum condition. *Journal of Child Psychology and Psychiatry, 57*(3), 271–292. https://doi.org/10.1111/jcpp.12501

Manger, P. R. (2020). Consistencies and variances in the anatomical organization of aspects of the mammalian brain stem. In J. H. Kaas (Ed.), *Evolutionary neuroscience* (pp. 377–396). Academic Press.

Mangino, M., Roederer, M., Beddall, M. H., Nestle, F. O., & Spector, T. D. (2017). Innate and adaptive immune traits are differentially affected by genetic and environmental factors. *Nature communications, 8.*

Manning, R., Levine, M., & Collins, A. (2007). The Kitty Genovese murder and the social psychology of helping: The parable of the 38 witnesses. *American Psychologist, 62*(6), 555–562.

Mantonakis, A., Rodero, P., Lesschaeve, I., & Hastie, R. (2009). Order in choice: Effects of serial position on preferences. *Psychological Science, 20*(11), 1309–1312.

Mantzios, M., Skillett, K., & Egan, H. (2019). Examining the effects of two mindful eating exercises on chocolate consumption: An experimental study. *European Journal of Health Psychology, 26*(4), 120–128.

Manyam, S. B., & Davis, T. L. (2020). Trauma group therapy with African American children and adolescents: A 30-plus year content analysis. *The Journal for Specialists in Group Work, 45*(1), 56–75.

Manzi, F., Ishikawa, M., Di Dio, C., Itakura, S., Kanda, T., Ishiguro, H., . . . Marchetti, A. (2020). The understanding of congruent and incongruent referential gaze in 17-month-old infants: An eye-tracking study comparing human and robot. *Scientific Reports, 10*(1), 1–10.

Maples-Keller, J. L., Williamson, R. L., Sleep, C. E., Carter, N. T., Campbell, W. K., & Miller, J. D. (2019). Using item response theory to develop a 60-item representation of the NEO PI-R using the International Personality Item Pool: Development of the IPIP–NEO-60. *Journal of Personality Assessment, 101*(1), 4–15.

Marchetti, B., Tirolo, C., L'Episcopo, F., Caniglia, S., Testa, N., Smith, J. A., . . . Serapide, M. F. (2020). Parkinson's disease, aging and adult neurogenesis: Wnt/β-catenin signalling as the key to unlock the mystery of endogenous brain repair. *Aging Cell, 19*(3), e13101.

Marcia, J. E. (1980). Ego identity development. In J. Adelson (Ed.), *Handbook of adolescent psychology.* Wiley.

Marcia, J. E. (2002). Identity and psychosocial development in adulthood. *Identity: An International Journal of Theory and Research, 2*(1), 7–28.

Marcondes, L. A., Nachtigall, E. G., Zanluchi, A., de Carvalho Myskiw, J., Izquierdo, I., & Furini, C. R. G. (2020). Involvement of medial prefrontal cortex NMDA and AMPA/kainate glutamate receptors in social recognition memory consolidation. *Neurobiology of Learning and Memory, 168,* 107153.

Marcus, G. F. (2001). *The algebraic mind.* MIT Press.

Mares, D. (2013). Climate change and levels of violence in socially disadvantaged neighborhoods. *Journal of Urban Health, 90*(4), 768–783.

Mares, S., & McMahon, C. (2020). Attachment security: Influences on social and emotional competence. In R. Midford, G. Nutton, B. Hyndman, & S. Silburn, (Eds.), *Health and education interdependence: Thriving from birth to adulthood* (pp. 55–74). Springer.

Margolis, S., Stapley, A. L., & Lyubomirsky, S. (2020). The association between extraversion and well-being is limited to one facet. *Journal of Personality, 88*(3), 478–484.

Maric, V., Ramanathan, D., & Mishra, J. (2020). Respiratory regulation & interactions with neuro-cognitive circuitry. *Neuroscience & Biobehavioral Reviews, 112,* 95–106.

Marinkovic, K., & others. (2011). Right hemisphere has the last laugh: Neural dynamics of joke appreciation. *Cognitive and Affective Behavioral Neuroscience, 11*(1), 113–130.

Marinucci, M., Maunder, R., Sanchez, K., Thai, M., McKeown, S., Turner, R. N., & Stevenson, C. (2020). Intimate intergroup contact across the lifespan. *Journal of Social Issues.* https://doi.org/10.1111/josi.12399

Markram, H. (2012). The human brain project. *Scientific American, 306*(6), 50–55.

Marks, A. K., Ejesi, K., McCullough, M. B., & Garcia Coll, C. (2015). The implications of discrimination for child and adolescent development. In R. M. Lerner & M. E. Lamb (Eds.), *Handbook of child psychology and developmental science* (7th ed., vol. 3). Wiley.

Márquez, Y. I., Deblinger, E., & Dovi, A. T. (2020). The value of trauma-focused cognitive behavioral therapy (tf-cbt) in addressing the therapeutic needs of trafficked youth: A case study. *Cognitive and Behavioral Practice, 27*(3), 253–269.

Marsden, A. D., & Barnett, M. D. (2020). The role of empathy in the relationship between social political ideology and sexual prejudice in heterosexual college students in the US. *Archives of Sexual Behavior 49,* 1853–1861.

Marsh, E. J., & Roediger, H. L. (2013). Episodic and autobiographical memory. In A. F. Healy, R. W. Proctor, & I. B. Weiner (Eds.), *Handbook of psychology* (2nd ed., vol. 4, pp. 472–494). Wiley.

Marsh, N., Marsh, A. A., Lee, M. R., & Hurlemann, R. (2020). Oxytocin and the neurobiology of prosocial behavior. *The Neuroscientist.* https://doi.org/10.1177/1073858420960111

Marshall, C. R., Howrigan, D. P., Merico, D., Thiruvahindrapuram, B., Wu, W., Greer, D. S., et al. (2018). Contribution of copy number variants to schizophrenia from a genome-wide study of 41, 321 subjects. *Nature Genetics.*

Marshall, D. S. (1971). Sexual behavior in Mangaia. In D. S. Marshall & R. C. Suggs (Eds.), *Human sexual behavior: Variations in the ethnographic spectrum* (pp. 103–162). Basic.

Marshall, P. J., & Meltzoff, A. N. (2014). Neural mirroring mechanisms and imitation in human infants. *Philosophical Transactions of the Royal Society of London. Series B, Biological Sciences, 369*(1644), 20130620.

Marshall, R. E., Milligan-Saville, J. S., Steel, Z., Bryant, R. A., Mitchell, P. B., & Harvey, S. B. (2020). Pre-employment MMPI-2 measures and later psychological injury-related absenteeism among police officers. *Policing: A Journal of Policy and Practice.* https://doi.org/10.1093/police/paaa017

Marteau, T. (2015). Downsizing: policy options to reduce portion sizes to help tackle obesity. *British Medical Journal, 351,* h5863

Martela, F., & Ryan, R. M. (2020). Distinguishing between basic psychological needs and basic wellness enhancers:

the case of beneficence as a candidate psychological need. *Motivation and Emotion, 44*(1), 116–133.

Martikainen, J. (2020). How students categorize teachers based on visual cues: Implications of nonverbal communication for classroom management. *Scandinavian Journal of Educational Research, 64*(4), 569–588.

Martin, G. L., & Pear, J. (2014). *Behavior modification* (10th ed.). Pearson.

Martin, J. A., Hamilton, B. E., Osterman, M. J. K., & Driscoll, A. K. (2019). *Births: Final data for 2018 National Vital Statistics Reports, 68*. National Center for Health Statistics, CDC.

Martin, V., Mathieu, L., Diaz, J., Salman, H., Alterio, J., Chevarin, C., . . . Stockmeier, C. A. (2020). Key role of the 5-HT1A receptor addressing protein Yif1B in serotonin neurotransmission and SSRI treatment. *Depression and Anxiety, 13*, 14. https://doi.org/10.1503/jpn.190134

Martinelli, M. K., Mostofsky, S. H., & Rosch, K. S. (2017). Investigating the impact of cognitive load and motivation on response control in relation to delay discounting in children with ADHD. *Journal of Abnormal Child Psychology, 45*(7), 1339–1353.

Martinez-Calderon, J., Meeus, M., Struyf, F., & Luque-Suarez, A. (2020). The role of self-efficacy in pain intensity, function, psychological factors, health behaviors, and quality of life in people with rheumatoid arthritis: a systematic review. *Physiotherapy Theory and Practice, 36*(1), 21–37.

Marx, R. F., & Didziulis, V. (2009, February 27). A life, interrupted. *New York Times*. www.nytimes.com/2009/03/01/nyregion/thecity/01miss.html?pagewanted=all (accessed January 5, 2015)

Masdrakis, V. G., Papageorgiou, C., & Markianos, M. (2017). Associations of plasma leptin to clinical manifestations in reproductive aged female patients with panic disorder. *Psychiatry Research, 255*, 161–166.

Mashour, G. A., & Hudetz, A. G. (2017). Bottom-up and top-down mechanisms of general anesthetics modulate different dimensions of consciousness. *Frontiers in Neural Circuits, 11*. https://doi.org/10.3389/fncir.2017.0004444.

Mashour, G. A., Roelfsema, P., Changeux, J. P., & Dehaene, S. (2020). Conscious processing and the global neuronal workspace hypothesis. *Neuron, 105*(5), 776–798.

Maslow, A. H. (1954). *Motivation and personality*. Harper & Row.

Maslow, A. H. (1971). *The farther reaches of human nature*. Viking.

Massa, L. J., & Mayer, R. E. (2006). Testing the ATI hypothesis: Should multimedia instruction accommodate verbalizer-visualizer cognitive style? *Learning and Individual Differences, 16*(4), 321–336.

Masten, A. S., & Motti-Stefanidi, F. (2020). Multisystem resilience for children and youth in disaster: Reflections in the context of COVID-19. *Adversity and Resilience Science, 1*(2), 95–106.

Masters, W. H., & Johnson, V. E. (1966). *Human sexual response*. Little, Brown.

Masuda, T., & Nisbett, R. E. (2001). Attending holistically vs. analytically: Comparing the context sensitivity of Japanese and Americans. *Journal of Personality and Social Psychology, 81*(5), 922–934.

Masuda, T., & Nisbett, R. E. (2006). Culture and change blindness. *Trends in Cognitive Science, 30*(2), 381–399.

Mate, I., Madrid, J. A., & De la Fuente, M. D. (2014). Chronobiology of the neuroimmunoendocrine system and aging. *Current Pharmaceutical Design, 20*(29), 4642–4655.

Mathur, M. B., Reichling, D. B., Lunardini, F., Geminiani, A., Antonietti, A., Ruijten, P. A., . . . Szuts, J. (2020). Uncanny but not confusing: Multisite study of perceptual category confusion in the uncanny valley. *Computers in Human Behavior, 103*, 21–30.

Matias-Garcia, J. A., & Cubero-Pérez, R. (2020). Heterogeneity in the conceptions of intelligence of university teaching staff. *Culture & Psychology*. https://doi.org/10.1177/1354067X20936926

Matkarimov, B. T., & Saparbaev, M. K. (2017). Aberrant DNA glycosylase-initiated repair pathway of free radicals induced DNA damage: implications for age-related diseases and natural aging. *Biopolymers and Cell, 33*(1), 3–23.

Matoba, N., Liang, D., Sun, H., Aygün, N., McAfee, J. C., Davis, J. E., . . . Kosuri, S. (2020). Common genetic risk variants identified in the SPARK cohort support DDHD2 as a candidate risk gene for autism. *Translational Psychiatry, 10*(1), 1–14.

Matos, A. P., Ferreira, J. A., & Haase, R. F. (2012). Television and aggression: A test of a mediated model with a sample of Portuguese students. *Journal of Social Psychology, 152*(1), 75–91.

Matott, M. P., Kline, D. D., & Hasser, E. M. (2017). Glial EAAT2 regulation of extracellular nTS glutamate critically controls neuronal activity and cardiorespiratory reflexes. *The Journal of Physiology, 595*, 6045–6063.

Matsuda, T., Hiyama, T. Y., Kobayashi, K., Kobayashi, K., & Noda, M. (2020). Distinct CCK-positive SFO neurons are involved in persistent or transient suppression of water intake. *Nature Communications, 11*(1), 1–15.

Matsumoto, D., & Hwang, H. C. (2014). Judgments of subtle facial expressions of emotion. *Emotion, 14*(2), 349–357.

Matsumoto, M., Walton, N. M., Yamada, H., Kondo, Y., Marek, G. J., & Tajinda, K. (2017). The impact of genetics on future drug discovery in schizophrenia. *Expert Opinion on Drug Discovery, 12*, 673–686.

Matthews, K. A., Gump, B. B., Harris, K. F., Haney, T. L., & Barefoot, J. C. (2004). Hostile behaviors predict cardiovascular mortality among men enrolled in the multiple risk factor intervention trial. *Circulation, 109*(1), 66–70.

Matthews, N., Welch, L., Festa, E. K., Bruno, A. A., & Schafer, K. (2020). Global depth perception alters local timing sensitivity. *Plos One, 15*(1), e0228080.

Matz, S. C., & Harari, G. M. (2020). Personality–place transactions: Mapping the relationships between big five personality traits, states, and daily places. *Journal of Personality and Social Psychology*. https://doi.org/10.1037/pspp0000297

Mauceri, D., Buchthal, B., Hemstedt, T. J., Weiss, U., Klein, C. D., & Bading, H. (2020). Nasally delivered VEGFD mimetics mitigate stroke-induced dendrite loss and brain damage. *Proceedings of the National Academy of Sciences, 117*(15), 8616–8623.

Maul, S., Giegling, I., Fabbri, C., Corponi, F., Serretti, A., & Rujescu, D. (2020). Genetics of resilience: Implications from genome-wide association studies and candidate genes of the stress response system in posttraumatic stress disorder and depression. *American Journal of Medical Genetics Part B: Neuropsychiatric Genetics, 183*(2), 77–94.

Maxcey, A. M., Janakiefski, L., Megla, E., Smerdell, M., & Stallkamp, S. (2019). Modality-specific forgetting. *Psychonomic Bulletin & Review, 26*(2), 622–633.

Maxfield, M., Pyszczynski, T., Greenberg, J., Pepin, R., & Davis, H. P. (2012). The moderating role of executive functioning in older adults' responses to a reminder of mortality. *Psychology and Aging, 27*(1), 256–263.

Maxwell, A. L., Loxton, N. J., & Hennegan, J. M. (2017). Exposure to food cues moderates the indirect effect of reward sensitivity and external eating via implicit eating expectancies. *Appetite, 111*, 135–141.

Maxwell, W. L. (2012). Traumatic brain injury in the neonate, child, and adolescent human: An overview of pathology. *International Journal of Developmental Neuroscience, 30*(3), 167–183.

May, M. (2003, May 25). *Vision diary*. www.guardian.co.uk/g2/story/0,3604,1029268,00.html

May, R. W., Bauer, K. N., Seibert, G. S., Jaurequi, M. E., & Fincham, F. D. (2020). School burnout is related to sleep quality and perseverative cognition regulation at bedtime in young adults. *Learning and Individual Differences, 78*, 101821.

Mayer, A. R., Quinn, D. K., & Master, C. L. (2017). The spectrum of mild traumatic brain injury: A review. *Neurology, 89*(6), 623–632.

Mayr, U., & Freund, A. M. (2020). Do we become more prosocial as we age, and if so, why? *Current Directions in Psychological Science*. https://doi.org/10.1177/0963721420910811

Mazur, K., Machaj, D., Mazur, D., Asztabska, A., & Płaczek, A. (2020). The use of melatonin in the treatment of jet lag–clinical review. *Journal of Education, Health and Sport, 10*(5), 175–179.

Mazursky-Horowitz, H., Thomas, S. R., Woods, K. E., Chrabaszcz, J. S., Deater-Deckard, K., & Chronis-Tuscano, A. (2017). Maternal executive functioning and scaffolding in families of children with and without parent-reported ADHD. *Journal of Abnormal Child Psychology*. https://doi.org/10.1007/s10802-017-0289-2

McAbee, S. T., & Oswald, F. L. (2013). The criterion-related validity of personality measures for predicting GPA: A meta-analytic validity comparison. *Psychological Assessment, 25*(2), 532–544.

McAdams, D. P. (2001). The psychology of life stories. *Review of General Psychology, 5*(2), 100–122.

McAdams, D. P. (2006). *The redemptive self: Stories Americans live by*. Oxford University Press.

McAdams, D. P. (2009). *The person* (5th ed.). Wiley.

McAdams, D. P. (2011a). Life narratives. In K. L. Fingerman, C. A. Berg, J. Smith, & T. C. Antonucci (Eds.), *Handbook of lifespan development*. Springer.

McAdams, D. P. (2011b). Narrative identity. In S. J. Schwartz, K. Luyckx, & V. L. Vignoles (Eds.), *Handbook of identity theory and research*. Springer.

McAdams, D. P. (2018). Narrative identity: What is it? What does it do? How do you measure it? *Imagination, Cognition and Personality, 37*(3), 359–372.

McAndrew, F. T. (2002). New evolutionary perspectives on altruism: Multilevel-selection and costly-signaling theories. *Current Directions in Psychological Science, 11*(2), 79–82.

McBride, E., Oswald, W. W., Beck, L. A., & Vashlishan Murray, A. (2020). "I'm just not that great at science": Science self-efficacy in arts and communication students. *Journal of Research in Science Teaching, 57*(4), 597–622.

McBride, K. R., Sanders, S. A., Hill, B. J., & Reinisch, J. M. (2017). Heterosexual women's and men's labeling of anal behaviors as having "had sex". *The Journal of Sex Research*.

McCarthy, M. J., Liang, S., Spadoni, A. D., Kelsoe, J. R., & Simmons, A. N. (2014). Whole brain expression of bipolar disorder associated genes: Structural and genetic analyses. *PLoS One, 9*(6), e100204.

McCaulley, M. H. (2000). Myers-Briggs Type Indicator: A bridge between counseling and consulting. *Consulting Psychology Journal: Practice and Research, 52*, 117–132.

McClain, D. (2020, July 21). Play therapy can help kids speak the unspeakable. *The New York Times*. https://nyti.ms/2CSdWD

McClelland, J. L. (2011). Memory as a constructive process: The parallel-distributed processing approach. In S. Nalbantian, P. Matthews, & J. L. McClelland (Eds.), *The memory process*. MIT Press.

McClelland, J. L., Botvinick, M. M., Noelle, D. C., Plaut, D. C., Rogers, T. T., Seidenberg, M. S., & Smith, L. B. (2010). Letting structure emerge: Connectionist and dynamical systems approaches to cognition. *Trends in Cognitive Sciences, 14*(8), 348–356.

McClelland, J. L., & Rumelhart, D. E. (2009). Why there are complementary learning systems in the hippocampus and neocortex: Insights from the successes and failures of connectionist models of learning and memory. In D. Shanks (Ed.), *Psychology of learning*. Sage.

McCoy, S. M., & Morgan, K. (2020). Obesity, physical activity, and sedentary behaviors in adolescents with autism spectrum disorder compared with typically developing peers. *Autism, 24*(2), 387–399.

McCrae, R. R., Harwood, T. M., & Kelly, S. L. (2011). The NEO inventories. In T. M. Harwood, L. E. Beutler, & G. Groth-Marnat (Eds.), *Integrative assessment of adult personality* (3rd ed.). Guilford.

McCredie, M. N., & Kurtz, J. E. (2020). Prospective prediction of academic performance in college using self-and informant-rated personality traits. *Journal of Research in Personality, 85*, 103911.

McCullough, M. E., Emmons, R. A., & Tsang, J. (2002). The grateful disposition: A conceptual and empirical topography. *Journal of Personality and Social Psychology, 82*(1), 112-127.

McCullough, M. E., Kurzban, R., & Tabak, B. A. (2011). Evolved mechanisms for revenge and forgiveness. In M. Mikulincer & P. R. Shaver (Eds.), *Human aggression and violence: Causes, manifestations, and consequences* (pp. 221-239). American Psychological Association.

McCullough, M. E., Luna, L. R., Berry, J. W., Tabak, B. A., & Bono, G. (2010). On the form and function of forgiving: Modeling the time-forgiveness relationship and testing the valuable relationships hypothesis. *Emotion, 10*(3), 358-376.

McCullough, M. E., & Willoughby, B. L. (2009). Religion, self-regulation, and self-control: Associations, explanations, and implications. *Psychological Bulletin, 135*(1), 69-93.

McCullough, M., Kurzban, R., & Tabak, B. A. (2013). Putting revenge and forgiveness in an evolutionary context. *Behavioral and Brain Sciences, 36*(1), 41-58.

McCutcheon, R. A., Krystal, J. H., & Howes, O. D. (2020). Dopamine and glutamate in schizophrenia: Biology, symptoms and treatment. *World Psychiatry, 19*(1), 15-33.

McDaniel, M. A., & Einstein, G. O. (2020). Training learning strategies to promote self-regulation and transfer: The knowledge, belief, commitment, and planning framework. *Perspectives on Psychological Science, 15*(6), 1363-1381.

McDonald, R. J., & White, N. M. (2013). A triple dissociation of memory systems: Hippocampus, amygdala, & dorsal striatum. *Behavioral Neuroscience, 127*(6), 835-853.

McElvaney, R., Greene, S., & Hogan, D. (2014). To tell or not to tell? Factors influencing young people's informal disclosures of child sexual abuse. *Journal of Interpersonal Violence, 29*(5), 928-947.

McEwen, B. S. (2017). Neurobiological and systemic effects of chronic stress. *Chronic Stress, 1*, 1-11 http://journals.sagepub.com/doi/abs/10.1177/2470547017692328

McGrath, M., Low, M. A., Power, E., McCluskey, A., & Lever, S. (2020). Addressing sexuality among people living with chronic disease and disability: A systematic mixed methods review of knowledge, attitudes and practices of healthcare professionals. *Archives of Physical Medicine and Rehabilitation.* https://doi.org/10.1016/j.apmr.2020.09.379

McGue, M., Rustichini, A., & Iacono, W. G. (2017). Cognitive, noncognitive, and family background contributions to college attainment: A behavioral genetic perspective. *Journal of Personality, 85*(1), 65-78.

McGuire, J. F., Ricketts, E. J., Scahill, L., Wilhelm, S., Woods, D. W., Piacentini, J., . . . Peterson, A. L. (2020). Effect of behavior therapy for Tourette's disorder on psychiatric symptoms and functioning in adults. *Psychological Medicine, 50*(12), 2046-2056.

McGuire, M. T., Wing, R. R., Klem, M. L., Lang, W., & Hill, J. O. (1999). What predicts weight regain in a group of successful weight losers? *Journal of Consulting and Clinical Psychology, 67*(2), 177-185.

McGuire, W. J. (2003). Doing psychology my way. In R. J. Sternberg (Ed.), *Psychologists defying the crowd: Stories of those who battled the establishment and won* (pp. 119-137). American Psychological Association.

McGuire, W. J., & Papageorgis, D. (1961). The relative efficacy of various types of prior belief-defense in producing immunity against persuasion. *Journal of Abnormal and Social Psychology, 62*(2), 327-337.

McHugh, R. K., Kneeland, E. T., Edwards, R. R., Jamison, R., & Weiss, R. D. (2020). Pain catastrophizing and distress intolerance: prediction of pain and emotional stress reactivity. *Journal of Behavioral Medicine, 43*(4), 623-629.

McIlvain, G., Clements, R. G., Magoon, E. M., Spielberg, J. M., Telzer, E. H., & Johnson, C. L. (2020). Viscoelasticity of reward and control systems in adolescent risk taking. *NeuroImage*, 116850.

McIntosh, W. D., Harlow, T. F., & Martin, L. L. (1995). Linkers and non-linkers: Goal beliefs as a moderator of the effects of everyday hassles on rumination, depression, and physical complaints. *Journal of Applied Social Psychology, 25*(14), 1231-1244.

McIntyre, C. C., Chaturvedi, A., Shamir, R. R., & Lempka, S. F. (2015). Engineering the next generation of clinical deep brain stimulation. *Brain Stimulation, 8*(1), 21-26.

McKay, K. M., Imel, Z. E., & Wampold, B. E. (2006). Psychiatrist effects in the psychopharmacological treatment of depression. *Journal of Affective Disorders, 92*(2-3), 287-290.

McKee, T. E. (2017). Peer relationships in undergraduates with ADHD symptomatology: Selection and quality of friendships. *Journal of Attention Disorders, 21*(12), 1020-1029.

McLaughlin, K. A., Aldao, A., Wisco, B. E., & Hilt, L. M. (2014). Rumination as a transdiagnostic factor underlying transitions between internalizing symptoms and aggressive behavior in early adolescents. *Journal of Abnormal Psychology, 123*(1), 13-23.

McMichael, B. J., Van Horn, R. L., & Viscusi, W. K. (2020). The impact of cannabis access laws on opioid prescribing. *Journal of Health Economics, 69*, 102273.

McMullen, V., Mahfood, S. L., Francis, G. L., & Bubenik, J. (2017). Using prediction and desensitization techniques to treat dental anxiety: A case example. *Behavioral Interventions, 32*(1), 91-100.

McNicholas, F., Healy, E., White, M., Sherdian-Pereira, M., O'Connor, N., Coakley, S., & Dooley, B. (2014). Medical, cognitive, and academic outcomes of very low birth weight infants at 10-14 years in Ireland. *Irish Journal of Medical Science, 183*(4), 525-532.

McNicholas, P. J., Floyd, R. G., Woods, I. J., Singh, L. J., Manguno, M. S., & Maki, K. E. (2017). State special education criteria for identifying intellectual disability: A review following revised diagnostic criteria and Rosa's Law. *School Psychology Quarterly*. https://doi.org/10.1037/spq0000208

McNulty, J. K. (2011). The dark side of forgiveness: The tendency to forgive predicts continued psychological and physical aggression in marriage. *Personality and Social Psychology Bulletin, 37*(6), 770-783.

McNulty, J. K. (2020). Highlighting the dark side of forgiveness and the need for a contextual approach. In E. L. Worthington Jr., & N. G. Wade (Eds.), *The handbook of forgiveness* (2nd ed., pp. 33-42). Routledge.

McNulty, J. K., & Russell, V. M. (2016). Forgive and forget, or forgive and regret? Whether forgiveness leads to less or more offending depends on offender agreeableness. *Personality and Social Psychology Bulletin, 42*, 616-631. https://doi.org/10.1177/0146167216637841

McNulty, J. K., Wenner, C. A., & Fisher, T. D. (2016). Longitudinal associations among relationship satisfaction, sexual satisfaction, and frequency of sex in early marriage. *Archives of Sexual Behavior, 45*(1), 85-97.

Meaney M. J. (2017) Epigenetics and the biology of gene × environment interactions. In P. Tolan & B. Leventhal (Eds.), *Gene-environment transactions in developmental psychopathology. Advances in Development and Psychopathology: Brain Research Foundation Symposium Series, Vol 2.* (pp. 59-94). Springer.

Medina, A., Torres, J., Kazama, A. M., Bachevalier, J., & Raper, J. (2020). Emotional responses in monkeys differ depending on the stimulus type, sex, and neonatal amygdala lesion status. *Behavioral Neuroscience, 134*(2), 153-165.

MedlinePlus. (2014). *Alcoholism.* www.nlm.nih.gov/medlineplus/alcoholism.html

Mehta, U. M., Thirthalli, J., Gangadhar, B. N., & Keshaven, M. S. (2011). Need for culture specific tools to assess social cognition in schizophrenia. *Schizophrenia Research, 133*(1-3), 255-256.

Mellor, J. M., & Freeborn, B. A. (2011). Religious participation and risky health behaviors among adolescents. *Health Economics, 20*(10), 1226-1240. https://doi.org/10.1002/hec.1666

Melton, L. (2005). How brain power can help you cheat old age. *New Scientist, 2530*, 32

Meltzer, A. L., Makhanova, A., Hicks, L. L., French, J. E., McNulty, J. K., & Bradbury, T. N. (2017). Quantifying the sexual afterglow: The lingering benefits of sex and their implications for pair-bonded relationships. *Psychological Science, 28*(5), 587-598. https://doi.org/10.1177/0956797617691361

Meltzer, A. L., & McNulty, J. K. (2016). Who is having more and better sex? The Big Five as predictors of sex in marriage. *Journal of Research in Personality, 63*, 62-66.

Melvin, S. C., Wiggins C., Burse. N., Thompson. E., & Monger, M. (2020). The role of public health in COVID-19 emergency response efforts from a rural health perspective. *Preventing Chronic Disease, 17*, 200256. http://dx.doi.org/10.5888/pcd17.200256

Melzack, R. (1973). *The puzzle of pain.* Basic Books.

Memmott-Elison, M. K., Holmgren, H. G., Padilla-Walker, L. M., & Hawkins, A. J. (2020). Associations between prosocial behavior, externalizing behaviors, and internalizing symptoms during adolescence: a meta-analysis. *Journal of Adolescence, 80*, 98-114.

Mendes, N., Hanus, D., & Call, J. (2007). Raising the level: Orangutans use water as a tool. *Biology Letters, 3*(5), 453-455.

Mendes, W. B. (2007). Social facilitation. In R. Baumeister & K. Vohs (Eds.), *Encyclopedia of social psychology.* Sage.

Mendoza, N. B., & King, R. B. (2020). The social contagion of student engagement in school. *School Psychology International, 41*(5), 454-474.

Meng, L., Bai, X., Zheng, Y., Chen, D., & Zheng, Y. (2020). Altered expression of norepinephrine transporter participate in hypertension and depression through regulated TNF-α and IL-6. *Clinical and Experimental Hypertension, 42*(2), 181-189.

Menozzi, D., Sogari, G., Veneziani, M., Simoni, E., & Mora, C. (2017). Eating novel foods: An application of the theory of planned behaviour to predict the consumption of an insect-based product. *Food Quality and Preference, 59*, 27-34. https://doi.org/10.1016/j.foodqual.2017.02.001

Mercier, H., & Sperber, D. (2011). Why do humans reason? Arguments for an argumentative theory. *Behavioral and Brain Sciences, 34*(2), 57-111.

Mercure, E., Evans, S., Pirazzoli, L., Goldberg, L., Bowden-Howl, H., Coulson-Thaker, K., . . . MacSweeney, M. (2020). Language experience impacts brain activation for spoken and signed language in infancy: Insights from unimodal and bimodal bilinguals. *Neurobiology of Language, 1*(1), 9-32.

Merrick, M. T., Ford, D., C., Ports, K., A., Guinn, A. S., Chen, J., & Mercy, J. A. (2019). Vital signs: Estimated proportion of adult health problems attributable to adverse childhood experiences and implications for prevention—25 states, 2015-2017. *MMWR Morbidity and Mortality Weekly Report, 68*, 999-1005. http://dx.doi.org/10.15585/mmwr.mm6844e1

Mersin Kilic, S., Dondu, A., Memis, C. O., Ozdemiroglu, F., & Sevincok, L. (2020). The clinical characteristics of ADHD and obsessive-compulsive disorder comorbidity. *Journal of Attention Disorders, 24*(12), 1757-1763.

Mertens, G., & Engelhard, I. M. (2020). A systematic review and meta-analysis of the evidence for unaware fear conditioning. *Neuroscience & Biobehavioral Reviews, 108*, 254-268.

Messenger, J. C. (1971). Sex and repression in an Irish folk community. In D. S. Marshall & R. C. Suggs (Eds.), *Human sexual behavior.* Basic.

Metcalfe, J., & Mischel, W. (1999). A hot/cool system analysis of delay of gratification: Dynamics of will power. *Psychological Review, 106*(1), 3-19.

Metin-Orta, I., & Metin-Camgöz, S. (2020). Attachment style, openness to experience, and social contact as predictors of attitudes toward homosexuality. *Journal of Homosexuality, 67*(4), 528-553.

Metin, S. Z., Balli Altuglu, T., Metin, B., Erguzel, T. T., Yigit, S., Arıkan, M. K., & Tarhan, K. N. (2020). Use of EEG for predicting treatment response to transcranial magnetic stimulation in obsessive compulsive disorder. *Clinical EEG and Neuroscience, 51*(3), 139-145.

Mettler, F. A. (Ed.). (1952). *Psychosurgical problems.* Blakiston.

Mez, J., Daneshvar, D. H., Kiernan, P. T., Abdolmohammadi, B., Alvarez, V. E., et al. (2017). Clinicopathological evaluation of chronic traumatic encephalopathy in players of American football. *Journal of the American Medical Association, 318*, 360–370.

Michalik, N. M., Eisenberg, N., Spinrad, T. L., Ladd, B., Thompson, M., & Valiente, C. (2007). Longitudinal relations among parental emotional expressivity and sympathy and prosocial behavior in adolescence. *Social Development, 16*(2), 286–309.

Michalowski, V. I., Gerstorf, D., Hülür, G., Drewelies, J., Ashe, M. C., Madden, K. M., & Hoppmann, C. A. (2020). Intraindividual variability and empathic accuracy for happiness in older couples. *GeroPsych.* https://doi.org/10.1024/1662-9647/a000233

Miech, R. A., Patrick, M. E., O'Malley, P. M., Johnston, L. D., & Bachman, J. G. (2020). Trends in reported marijuana vaping among US adolescents, 2017–2019. *JAMA 323*(5), 475–476. https://doi.org/10.1001/jama.2019.20185

Mierop, A., Mikolajczak, M., Stahl, C., Béna, J., Luminet, O., Lane, A., & Corneille, O. (2020, in press). How can intranasal oxytocin research be trusted? A systematic review of the interactive effects of intranasal oxytocin on psychosocial outcomes. *Perspectives on Psychological Science.*

Mihura, J. L., Meyer, G. J., Dumitrascu, N., & Bombel, G. (2013). The validity of individual Rorschach variables: Systematic reviews and meta-analyses of the comprehensive system. *Psychological Bulletin, 139*(3), 548–605. https://doi.org/10.1037/a0029406

Mikal, J. P., Rice, R. E., Kent, R. G., & Uchino, B. N. (2016). 100 million strong: A case study of group identification and deindividuation on Imgur.Com. *New Media & Society, 18*(11), 2485–2506.

Mikkelsen, E. N., Gray, B., & Petersen, A. (2020). Unconscious processes of organizing: Intergroup conflict in mental health care. *Journal of Management Studies.* https://doi.org/10.1111/joms.12611

Mikolajczak, M., Pinon, N., Lane, A., de Timary, P., Luminet O. (2010). Oxytocin not only increases trust when money is at stake, but also when confidential information is in the balance. *Biological Psychology, 85*, 182–184

Mikulincer, M., & Shaver, P. R. (2019). A behavioral systems approach to romantic love relationships: Attachment, caregiving, and sex. In R. J. Sternberg & K. Sternberg (Eds.), *The new psychology of love* (pp. 259–279). Cambridge University Press.

Miles-Novelo, A., & Anderson, C. A. (2019). Climate change and psychology: Effects of rapid global warming on violence and aggression. *Current Climate Change Reports, 5*(1), 36–46.

Mileva, M., Tompkinson, J., Watt, D., & Burton, A. M. (2020). The role of face and voice cues in predicting the outcome of student representative elections. *Personality and Social Psychology Bulletin, 46*(4), 617–625.

Milgram, S. (1965). Some conditions of obedience and disobedience to authority. *Human Relations, 18*, 56–76.

Milgram, S. (1974). *Obedience to authority.* Harper & Row.

Miller, A. B., Esposito-Smythers, C., Weismoore, J. T., & Renshaw, K. D. (2013). The relation between child maltreatment and adolescent suicidal behavior: A systematic review and critical examination of the literature. *Clinical Child and Family Psychology Review, 16*(2), 146–172.

Miller, B. L., & Cummings, J. L. (Eds.). (2017). *The human frontal lobes: Functions and disorders.* Guilford Publications.

Miller, D. B., & O'Callaghan, J. P. (2006). The pharmacology of wakefulness. *Metabolism, 55,* Suppl. 2, S13–S19.

Miller, G. A. (1956). The magical number seven, plus or minus two: Some limits on our capacity for processing information. *Psychological Review, 63*(2), 81–97.

Miller, J. J., Fletcher, K., & Kabat-Zinn, J. (1995). Three-year follow-up and clinical implications of a mindfulness meditation-based stress reduction intervention in the treatment of anxiety disorders. *General Hospital Psychiatry, 17*(3), 192–200.

Miller, M. W. (2020). Leveraging genetics to enhance the efficacy of PTSD pharmacotherapies. *Neuroscience Letters, 726*, 133562.

Miller, N., Ament, S. A., Shekhtman, T., Roach, J. C., Study, T. B. G., & Kelsoe, J. R. (2017). Search for risk variants in TrkB and BDNF that predispose to lithium responsiveness in bipolar disorder. *Biological Psychiatry, 81*(10), S96.

Miller, N. E. (1985). The value of behavioral research on animals. *American Psychologist, 40*(4), 432–440.

Milner, A. D., & Goodale, M. A. (1995). *The visual brain in action.* Oxford University Press.

Milojev, P., & Sibley, C. G. (2017). Normative personality trait development in adulthood: A 6-year cohort-sequential growth model. *Journal of Personality And Social Psychology, 112*(3), 510–526. https://doi.org/10.1037/pspp0000121

Mineka, S., & Ohman, A. (2002). Phobias and preparedness: The selective, automatic, and encapsulated nature of fear. *Biological Psychiatry, 52*(10), 927–937.

Mineka, S., Williams, A. L., Wolitzky-Taylor, K., Vrshek-Schallhorn, S., Craske, M. G., Hammen, C., & Zinbarg, R. E. (2020). Five-year prospective neuroticism–stress effects on major depressive episodes: Primarily additive effects of the general neuroticism factor and stress. *Journal of Abnormal Psychology, 129*(6), 646–657.

Mischel, W. (2004). Toward an integrative science of the person. *Annual Review of Psychology 55*, 1–22.

Mischel, W. (2009). From *Personality and Assessment* (1968) to personality science, 2009. *Journal of Research in Personality, 43*(2), 282–290.

Mischel, W., & Ayduk, O. (2011). Willpower in a cognitive-affective processing system: The dynamics of delay of gratification. In K. D. Vohs & R. F. Baumeister (Eds.), *Handbook of self-regulation* (2nd ed.). Guilford.

Mischel, W., Cantor, N., & Feldman, S. (1996). Principles of self-regulation: The nature of willpower and self-control. In E. T. Higgins & A. W. Kruglanski (Eds.), *Social psychology: Handbook of basic principles.* Guilford.

Mischel, W., & Moore, B. S. (1980). The role of ideation in voluntary delay for symbolically presented rewards. *Cognitive Therapy and Research, 4*(2), 211–221.

Mischel, W., Shoda, Y., & Peake, P. K. (1988). The nature of adolescent competencies predicted by preschool delay of gratification. *Journal of Personality and Social Psychology, 54*(4), 687–696. https://doi.org/10.1037/0022-3514.54.4.687

Misra, U. K., Kalita, J., Tripathi, G., & Bhoi, S. K. (2017). Role of β endorphin in pain relief following high rate repetitive transcranial magnetic stimulation in migraine. *Brain Stimulation, 10*(3), 618–623.

Mistry, J., Contreras, M., & Dutta, R. (2013). Culture and child development. In R. M. Lerner, M. A. Easterbrooks, J. Mistry, & I. B. Weiner (Eds.), *Handbook of psychology* (2nd ed., vol. 6, pp. 265–286). Wiley.

Mitchell, J. (2014). *Individualism and moral character: Karen Horney's depth psychology.* Transaction Publishers.

Miyamoto, Y., Nisbett, R. E., & Masuda, T. (2006). Culture and the physical environment. *Psychological Science, 17*(2), 113–119.

Modreanu, R. M., Buhmann, C., & Hauptmann, B. (2020). Nine-years follow-up of cavernoma located in basal ganglia mimicking Parkinson's disease. *Clinical Neurology and Neurosurgery, 190*, 105664.

Mogil, J. S. (2020). Qualitative sex differences in pain processing: Emerging evidence of a biased literature. *Nature Reviews Neuroscience, 21*, 353–365.

Mogro-Wilson, C., Drake, A., Coman, E., Sanghavi, T., Martin-Peele, M., & Fifield, J. (2020). Increasing condom usage for African-American and Hispanic young fathers in a community based intervention. *Ethnicity & Health, 25*(3), 408–419.

Mohammadi, M. R., Salehi, M., Khaleghi, A., Hooshyari, Z., Mostafavi, S. A., Ahmadi, N., . . . Amanat, M. (2020). Social anxiety disorder among children and adolescents: A nationwide survey of prevalence, socio-demographic characteristics, risk factors and co-morbidities. *Journal of Affective Disorders, 263*, 450–457.

Moieni, M., Irwin, M. R., Seeman, T. E., Robles, T. F., Lieberman, M. D., Breen, E. C., . . . Cole, S. W. (2020). Feeling needed: Effects of a randomized generativity intervention on well-being and inflammation in older women. *Brain, Behavior, and Immunity, 84*, 97–105.

Molina, B., & Pelham, W.E. (2014). The attention deficit/hyperactivity disorder (ADHD) substance use connection. *Annual Review of Clinical Psychology, 10*, 607–639.

Molouki, S., & Pronin, E. (2014). Self and other. In M. Mikulincer, P. R. Shaver, E. Borgida, & J. A. Bargh (Eds.), *APA handbook of personality and social psychology* (vol. 1). American Psychological Association.

Money, J., & Tucker, P. (1975). *Sexual signatures: On being a man or woman.* Little Brown.

Monnereau, C., Jansen, P. W., Tiemeier, H., Jaddoe, V. W., & Felix, J. F. (2017). Influence of genetic variants associated with body mass index on eating behavior in childhood. *Obesity, 25*(4), 765–772.

Monni, E., Cusulin, C., Cavallaro, M., Lindvall, O., & Kokaia, Z. (2014). Human fetal striatum-derived neural stem (NS) cells differentiate to mature neurons in vitro and in vivo. *Current Stem Cell Research and Therapy.* https://doi.org/10.2174/1574888X09666140321115803

Montag, C., Ebstein, R. P., Jawinski, P., & Markett, S. (2020). Molecular genetics in psychology and personality neuroscience: On candidate genes, genome wide scans, and new research strategies. *Neuroscience & Biobehavioral Reviews, 118*, 163–174.

Montenegro, G., Alves, L., Zaninotto, A. L., Falcão, D. P., & de Amorim, R. B. (2017). Hypnosis as a valuable tool for surgical procedures in the oral and maxillofacial area. *American Journal of Clinical Hypnosis, 59*(4), 414–421. https://doi.org/10.1080/00029157.2016.1172057

Montoliu, T., Hidalgo, V., & Salvador, A. (2020). Personality and hypothalamic–pituitary–adrenal axis in older men and women. *Frontiers in Psychology, 11.*

Montoya, R. M., & Horton, R. S. (2020). Understanding the attraction process. *Social and Personality Psychology Compass, 14*(4), e12526.

Moodley, K. (2014). Barack Obama: I've been mistaken for a valet and a waiter. *The Independent.* https://www.independent.co.uk/news/world/americas/barack-obama-i-ve-been-mistaken-valet-and-waiter-9931054.html?amp

Moojen, S. M. P., Gonçalves, H. A., Bassôa, A., Navas, A. L., de Jou, G., & Miguel, E. S. (2020). Adults with dyslexia: How can they achieve academic success despite impairments in basic reading and writing abilities? The role of text structure sensitivity as a compensatory skill. *Annals of Dyslexia, 70*, 115–140.

Mooneyham, B. W., Mrazek, M. D., Mrazek, A. J., Mrazek, K. L., Ihm, E. D., & Schooler, J. W. (2017). An integrated assessment of changes in brain structure and function of the insula resulting from an intensive mindfulness-based intervention. *Journal of Cognitive Enhancement: 1*(3), 327–336. https://doi.org/10.1007/s41465-017-0034-3

Moore, D. F., Jaffee, M., Ling, G., & Radovitzky, R. (2020). Overview of traumatic brain injury (TBI). In J. W. Tsao (Ed.), *Traumatic Brain Injury* (2nd ed., pp. 1–13). Springer.

Moore, D. S. (2013). Behavioral genetics, genetics, and epigenetics. In P. D. Zelazo (Ed.), *Handbook of Developmental Psychology.* Oxford University Press.

Moore, M. T., Dawkins Jr, M. R., Fisher, J. W., & Fresco, D. M. (2016). Depressive realism and attributional style: Replication and Extension. *International Journal of Cognitive Therapy, 9*(1), 1–12.

Moradi, Z., Najlerahim, A., Macrae, C. N., & Humphreys, G. W. (2020). Attentional saliency and ingroup biases: From society to the brain. *Social Neuroscience, 15*(3), 324–333.

Moran, J. M., Jolly, E., & Mitchell, J. P. (2014). Spontaneous mentalizing predicts the fundamental attribution error. *Journal of Cognitive Neuroscience, 26*(3), 569–576. https://doi.org/10.1162/jocn_a_00513

Moravčík, M., Schmid, M., Burch, N., Lisý, V., Morrill, D., Bard, N., et al. (2017). DeepStack: Expert-level artificial intelligence in heads-up no-limit poker. *Science, 356*(6337), 508–513. https://doi.org/10.1126/science.aam6960

Morawetz, C., Alexandrowicz, R. W., & Heekeren, H. R. (2017). Successful emotion regulation is predicted by amygdala activity and aspects of personality: A latent variable approach. *Emotion, 17*(3), 42–441.

Morawski, J. G., & Bayer, B. M. (2013). Social psychology. In D. K. Freedheim & I. B. Weiner (Eds.), *Handbook of psychology* (2nd ed., vol. 1, pp. 248–278). Wiley.

Moreira, J., Courtin, C., Geoffroy, P. A., Curis, E., Bellivier, F., & Cynthia, M. C. (2017). Lithium response in bipolar disorder: No difference in GADL1 gene expression between cell lines from excellent-responders and non-responders. *Psychiatry Research, 251*, 217–220.

Moreira, P. A., Faria, V., Cunha, D., Inman, R. A., & Rocha, M. (2020). Applying the transtheoretical model to adolescent academic performance using a person-centered approach: A latent cluster analysis. *Learning and Individual Differences, 78*, 101818.

Moreman, C. M. (2014). On the relationship between birds and spirits of the dead. *Society & Animals: Journal of Human-Animal Studies, 22*(5), 481–502. https://doi.org/10.1163/15685306-12341328

Moreno-Stokoe, C. M., & Damian, M. F. (2020). Employing natural control for confounding factors in the hunt for the bilingual advantage in attention: Evidence from school children in Gibraltar. *Journal of Cognition, 3*(1), 5.

Morey, R. A., Garrett, M. E., Stevens, J. S., Clarke, E. K., Haswell, C. C., van Rooij, S. J., . . . Dennis, M. F. (2020). Genetic predictors of hippocampal subfield volume in PTSD cases and trauma-exposed controls. *European Journal of Psychotraumatology, 11*(1), 1785994.

Morgan, A. J., Reavley, N. J., Ross, A., San Too, L., & Jorm, A. F. (2018). Interventions to reduce stigma towards people with severe mental illness: Systematic review and meta-analysis. *Journal of Psychiatric Research, 103*, 120–133.

Morgan, C. A., & Southwick, S. (2014). Perspective: I believe what I remember, but it may not be true. *Neurobiology of Learning and Memory, 112*, 101–103.

Morgan, C. J., Coleman, M. J., Ulgen, A., Boling, L., Cole, J. O., Johnson, F. V., et al. (2017). Thought disorder in schizophrenia and bipolar disorder probands, their relatives, and nonpsychiatric controls. *Schizophrenia Bulletin, 43*(3), 523–535.

Morris, M., Mankowska, A., & Heilman, K. M. (2020). Upper vertical spatial neglect with a right temporal lobe stroke. *Cognitive and Behavioral Neurology, 33*(1), 63–66.

Morris, M. W., & Peng, K. (1994). Culture and cause: American and Chinese attributions for social and physical events. *Journal of Personality and Social Psychology, 67*(6), 949–971.

Morrison, J. D., & Mayer, L. (2017). Physical activity and cognitive function in adults with multiple sclerosis: An integrative review. *Disability and Rehabilitation, 39*(19), 1909–1920.

Mortimer, J. A., Snowdon, D. A., & Markesbery, W. R. (2009). The effect of APOE-epsilon4 on dementia is mediated by Alzheimer neuropathology. *Alzheimer Disease and Associated Disorders, 23*(2), 152–157.

Morton, J. (2017). Interidentity amnesia in dissociative identity disorder. *Cognitive Neuropsychiatry, 22*, 315–330.

Moscovici, S. (1985). Social influence and conformity. In G. Lindzey & E. Aronson (Eds.), *Handbook of social psychology* (3rd ed., vol. 2). Random House.

Mostofsky, E., Maclure, M., Sherwood, J. B., Tofler, G. H., Muller, J. E., & Mittleman, M. A. (2012). Risk of acute myocardial infarction after the death of a significant person in one's life: The Determinants of Myocardial Infarction Onset Study. *Circulation, 125*(3), 491–496.

Mõttus, R., Allik, J., & Realo, A. (2020). Do self-reports and informant-ratings measure the same personality constructs? *European Journal of Psychological Assessment, 36*(2), 289.

Mõttus, R., Kandler, C., Bleidorn, W., Riemann, R., & McCrae, R. R. (2017). Personality traits below facets: The consensual validity, longitudinal stability, heritability, and utility of personality nuances. *Journal of Personality and Social Psychology, 112*(3), 474–490. https://doi.org/10.1037/pspp0000100

Mõttus, R., Sinick, J., Terracciano, A., Hřebíčková, M., Kandler, C., Ando, J., . . . Jang, K. L. (2019). Personality characteristics below facets: A replication and meta-analysis of cross-rater agreement, rank-order stability, heritability, and utility of personality nuances. *Journal of Personality and Social Psychology, 117*(4), e35–e50.

Mousa, S. (2020). Building social cohesion between Christians and Muslims through soccer in post-ISIS Iraq. *Science, 369*, 866–870.

Mrkva, K., Johnson, E. J., Gächter, S., & Herrmann, A. (2020). Moderating loss aversion: Loss aversion moderators, but reports of its death are greatly exaggerated. *Journal of Consumer Psychology, 30*(3), 407–428.

Muehlenhard, C. L., Peterson, Z. D., Humphreys, T. P., & Jozkowski, K. N. (2017). Evaluating the one-in-five statistic: women's risk of sexual assault while in college. *The Journal of Sex Research, 54*(4–5), 549–576.

Muehlroth, B. E., Rasch, B., & Werkle-Bergner, M. (2020). Episodic memory consolidation during sleep in healthy aging. *Sleep Medicine Reviews, 52*, 101304.

Mueller, D. L. (2010). Mechanisms maintaining peripheral tolerance. *Nature Immunology, 11*(1), 21–27.

Muller, C. P., & Cunningham, K. A. (2020). *Handbook of the behavioral neurobiology of serotonin.* Academic Press.

Müller-Engelmann, M., & Steil, R. (2017). Cognitive restructuring and imagery modification for posttraumatic stress disorder (CRIM-PTSD): A pilot study. *Journal of Behavior Therapy and Experimental Psychiatry, 54*, 44–50.

Mullol, J., Mariño-Sánchez, F., Valls, M., Alobid, I., & Marin, C. (2020). The sense of smell in chronic rhinosinusitis. *Journal of Allergy and Clinical Immunology, 145*(3), 773–776.

Munafo, M. R., Yalcin, B., Willis-Owen, S. A., & Flint, J. (2008). Association of the dopamine D4 receptor (DRD4) gene and approach-related personality traits: Meta-analysis and new data. *Biological Psychiatry, 63*(2), 197–206.

Munn, B., Zeater, N., Pietersen, A. N., Solomon, S. G., Cheong, S. K., Martin, P. R., & Gong, P. (2020). Fractal spike dynamics and neuronal coupling in the primate visual system. *The Journal of Physiology, 598*(8), 1551–1571.

Murgas, K. A., Wilson, A. M., Michael, V., & Glickfeld, L. L. (2020). Unique spatial integration in mouse primary visual cortex and higher visual areas. *Journal of Neuroscience, 40*(9), 1862–1873.

Muris, P., & Merckelbach, H. (2012). Specific phobia: Phenomenology, epidemiology, and etiology. In T. E. Davis, T. H. Ollendick, & L.-G. Öst (Eds.), *Intensive one-session treatment of specific phobias.* Springer.

Murphy, A. C., Bertolero, M. A., Papadopoulos, L., Lydon-Staley, D. M., & Bassett, D. S. (2020). Multimodal network dynamics underpinning working memory. *Nature Communications, 11*(1), 1–13.

Murphy, A., Steele, M., Dube, S. R., Bate, J., Bonuck, K., Meissner, P., . . . Steele, H. (2014). Adverse childhood experiences (ACEs) questionnaire and adult attachment interview (AAI): Implications for parent child relationships. *Child Abuse & Neglect, 38*(2), 224–233.

Murphy, D., Joseph, S., Demetriou, E., & Karimi-Mofrad, P. (2020). Unconditional positive self-regard, intrinsic aspirations, and authenticity: Pathways to psychological well-being. *Journal of Humanistic Psychology, 60*(2), 258–279.

Murphy, N. A., Hall, J. A., Mast, M. S., Ruben, M. A., Frauendorfer, D., Blanch-Hartigan, D., et al. (2015). Reliability and validity of nonverbal thin slices in social interactions. *Personality and Social Psychology Bulletin, 41*(2), 199–213. https://doi.org/10.1177/0146167214559902

Murphy, S. E., De Cates, A. N., Gillespie, A. L., Godlewska, B. R., Scaife, J. C., Wright, L. C., . . . Harmer, C. J. (2020). Translating the promise of 5HT 4 receptor agonists for the treatment of depression. *Psychological Medicine.* https://doi.org/10.1017/S0033291720000604

Murray, G. K., Knolle, F., Ersche, K. D., Craig, K. J., Abbott, S., Shabbir, S. S., . . . Robbins, T. W. (2019). Dopaminergic drug treatment remediates exaggerated cingulate prediction error responses in obsessive-compulsive disorder. *Psychopharmacology, 236*(8), 2325–2336.

Muschetto, T., & Siegel, J. T. (2019). Attribution theory and support for individuals with depression: The impact of controllability, stability, and interpersonal relationship. *Stigma and Health, 4*(2), 126–135.

Musey Jr, P. I., Schultebraucks, K., & Chang, B. P. (2020). Stressing out about the heart: A narrative review of the role of psychological stress in acute cardiovascular events. *Academic Emergency Medicine, 27*(1), 71–79.

Mustoe, A. C., Taylor, J. H., Birnie, A. K., Huffman, M. C., & French, J. A. (2014). Gestational cortisol and social play shape development of marmosets' HPA functioning and behavioral responses to stressors. *Developmental Psychobiology, 56*, 1229–1243.

Mutter, E. R., Oettingen, G., & Gollwitzer, P. M. (2020). An online randomised controlled trial of mental contrasting with implementation intentions as a smoking behaviour change intervention. *Psychology & Health, 35*(3), 318–345.

Muzzulini, B., Tinti, C., Conway, M. A., Testa, S., & Schmidt, S. (2020). Flashbulb memory: referring back to Brown and Kulik's definition. *Memory, 28*(6), 766–782.

Myers, N. L. (2010). Culture, stress, and recovery form schizophrenia: Lessons from the field for global mental health. *Culture, Medicine, and Psychiatry, 34*(3), 500–528.

Myers, S. G., Grøtte, T., Haseth, S., Guzey, I. C., Hansen, B., Vogel, P. A., & Solem, S. (2017). The role of metacognitive beliefs about thoughts and rituals: A test of the metacognitive model of obsessive-compulsive disorder in a clinical sample. *Journal of Obsessive-Compulsive and Related Disorders, 13*, 1–6.

Myrick, A. C., Webermann, A. R., Loewenstein, R. J., Lanius, R., Putnam, F. W., & Brand, B. L. (2017). Six-year follow-up of the treatment of patients with dissociative disorders study. *European Journal of Psychotraumatology, 8*(1), Article 1344080.

N

Nagele, D. A., Hooper, S. R., Hildebrant, K., McCart, M., Dettmer, J., & Glang, A. (2019). Under-identification of students with long term disability from moderate to severe TBI. *Physical Disabilities: Education and Related Services, 38*(1), 10–25.

Nakajima, S. (2020). Effect of pretrial running on running-based taste aversion learning in rats. *Journal of Experimental Psychology: Animal Learning and Cognition, 46*(3), 273–285.

Nakao, T., & others. (2005). A functional MRI comparison of patients with obsessive-compulsive disorder and normal controls during a Chinese character Stroop task. *Psychiatry Research: Neuroimaging, 139*(2), 101–114.

Nalepka, P., Kallen, R.W., Chemero, A., Saltzman, E., & Richardson, M.J. (2017). Herd those sheep: Emergent multiagent coordination and behavioral-mode switching. *Psychological Science 28*, 630–650.

Nalipay, M. J. N., King, R. B., & Cai, Y. (2020). Autonomy is equally important across East and West: Testing the cross-cultural universality of self-determination theory. *Journal of Adolescence, 78*, 67–72.

Nascimento, M. G., Kosminsky, M., & Chi, M. (2020). Gender role in pain perception and expression: An integrative review. *Brazilian Journal of Pain, 3*(1), 58–62.

Nash, M. R. (2001). The truth and the hype about hypnosis. *Scientific American, 285*(1), 46–49, 52–55.

National Alliance for the Mentally Ill (NAMI). (2020). Psychiatric advance directives (PAD). https://www.nami.org/Advocacy/Policy-Priorities/Improve-Care/Psychiatric-Advance-Directives-(PAD)

National Alliance on Mental Illness. (2013). Mental illness: Facts and numbers. *NAMI.* www.nami.org/factsheets/mentalillness_factsheet.pdf

National Highway Traffic Safety Administration. (2020). What is distracted driving? https://www.nhtsa.gov/risky-driving/distracted-driving#:~:text=Overview,by%20preventing%20this%20dangerous%20behavior

National Human Genome Research Institute. (2012). Genome-wide association studies. *Genome.* www.genome.gov/12011238

National Institute of Child Health and Human Development (NICHD). (2019). Noise induced hearing loss. https://www.nidcd.nih.gov/health/noise-induced-hearing-loss/

National Institute of Drug Addiction (NIDA). (2020a). Opioid overdose crisis. https://www.drugabuse.gov/drug-topics/opioids/opioid-overdose-crisis

National Institute of Drug Addiction (NIDA). (2020b). MDMA (ecstasy/molly). DrugFacts. https://www.drugabuse.gov/publications/drugfacts/mdma-ecstasymolly

National Institute of Mental Health (NIMH). (2019a). Posttraumatic stress disorder. https://www.nimh.nih.gov/health/topics/post-traumatic-stress-disorder-ptsd/index.shtml

National Institute of Mental Health (NIMH). (2019b). Major depression. https://www.nimh.nih.gov/health/statistics/major-depression.shtml

National Institute of Mental Health (NIMH). (2020a). Mental illness. https://www.nimh.nih.gov/health/statistics/mental-illness.shtml#:~:text=Mental%20illnesses%20are%20common%20in,mild%20to%20moderate%20to%20severe.

National Institute of Mental Health (NIMH). (2020b). Suicide. https://www.nimh.nih.gov/health/statistics/suicide.shtml

National Institute on Deafness and Other Communication Disorders (NIDCD). (2017). Cochlear implants. U.S. Department of Health and Human Services, National Institutes of Health. www.nidcd.nih.gov/health/hearing/pages/coch.aspx

National Institutes of Health (NIH). (2020). What was the human genome project and why has it been important? https://ghr.nlm.nih.gov/primer/hgp/description.

National Sleep Foundation. (2011, March 7). *Annual Sleep in America Poll exploring connections with communications technology use and sleep.* Author.

National Sleep Foundation. (2020a). How much sleep do we really need? https://www.sleepfoundation.org/articles/how-much-sleep-do-we-really-need

National Sleep Foundation (2020b). Sleep guidelines during the COVID-19 pandemic. https://www.sleepfoundation.org/sleep-guidelines-covid-19-isolation#guidelines

Naumann, L. P., Vazire, S., Rentfrow, P. J., & Gosling, S. D. (2009). Personality judgments based on physical appearance. *Personality and Social Psychology Bulletin, 35*(12), 1661–1671.

Navarro-Haro, M. V., Botella, V. G., Badenes-Ribera, L., Borao, L., & García-Palacios, A. (2020). Dialectical behavior therapy in the treatment of comorbid borderline personality disorder and eating disorder in a naturalistic setting: A six-year follow-up study. *Cognitive Therapy and Research.* https://doi.org/10.1007/s10608-020-10170-9

Nawata, K. (2020). A glorious warrior in war: Cross-cultural evidence of honor culture, social rewards for warriors, and intergroup conflict. *Group Processes & Intergroup Relations, 23*(4), 598–611.

Neave, H. W., Costa, J. H., Weary, D. M., & Von Keyserlingk, M. A. (2020). Long-term consistency of personality traits of cattle. *Royal Society Open Science, 7*(2), 191849.

Nechifor, R. E., Ciobanu, D., Vonica, C. L., Popita, C., Roman, G., Bala, C., ... Rusu, A. (2020). Social jetlag and sleep deprivation are associated with altered activity in the reward-related brain areas: An exploratory resting-state fMRI study. *Sleep Medicine, 72*, 12–19.

Needham, A., Barrett, T., & Peterman, K. (2002). A pick-me-up for infants' exploratory skills: Early simulated experiences reaching for objects using "sticky mittens" enhances young infants' object exploration skills. *Infant Behavior and Development, 25*(3), 279–295.

Needham, A. W., Wiesen, S. E., Hejazi, J. N., Libertus, K., & Christopher, C. (2017). Characteristics of brief sticky mittens training that lead to increases in object exploration. *Journal of Experimental Child Psychology, 164*, 209–224

Neikrug, A. B., & Ancoli-Israel, S. (2010). Sleep disorders in the older adult: A mini-review. *Gerontology, 56*(2), 181–189.

Nelson, C. A. (2013). Brain development and behavior. In A. M. Rudolph, C. Rudolph, L. First, G. Lister, & A. A. Gershon (Eds.), *Rudolph's pediatrics* (22nd ed.). McGraw-Hill.

Nelson, M. D., & Tumpap, A. M. (2017). Posttraumatic stress disorder symptom severity is associated with left hippocampal volume reduction: a meta-analytic study. *CNS Spectrums, 22*(4), 363–372.

Nelson, N. L., & Russell, J. A. (2014). Dynamic facial expressions allow differentiation of displays intended to convey positive and hubristic pride. *Emotion, 14*(5), 857–864.

Nelson, T. D. (Ed.). (2018). *Getting Grounded in Social Psychology: The Essential Literature for Beginning Researchers.* Psychology Press Ltd.

Neniskyte, U., & Gross, C. T. (2017). Errant gardeners: glial-cell-dependent synaptic pruning and neurodevelopmental disorders. *Nature Reviews Neuroscience, 18*(11), 658–670.

Nesti, A., Beykirch, K. A., MacNeilage, P. R., Barnett-Cowan, M., & Bülthoff, H. H. (2014). The importance of stimulus noise analysis for self-motion studies. *PLoS One, 9*(4), e94570.

Nestler, S., Blank, H., & Egloff, B. (2010). Hindsight ≠ hindsight: Experimentally induced dissociations between hindsight components. *Journal of Experimental Psychology: Learning, Memory, and Cognition, 36*(6), 1399–1413.

Nestor, P. G., Forte, M., Ohtani, T., Levitt, J. J., Newell, D. T., Shenton, M. E., ... McCarley, R. W. (2020). Faulty executive attention and memory interactions in schizophrenia: Prefrontal gray matter volume and neuropsychological impairment. *Clinical EEG and Neuroscience, 51*(4), 267–274.

Nettis, M. A., Veronese, M., Nikkheslat, N., Mariani, N., Lombardo, G., Sforzini, L., ... Pariante, C. M. (2020). PET imaging shows no changes in TSPO brain density after IFN-α immune challenge in healthy human volunteers. *Translational Psychiatry, 10*(1), 1–11.

Neueder, D., Pauli, P., & Andreatta, M. (2019). Contextual fear conditioning and fear generalization in individuals with panic attacks. *Frontiers in Behavioral Neuroscience, 13*, 152.

Neufeld, J., Sinke, C., Dillo, W., Emrich, H. M., Szycik, G. R., Dima, D., Bleich, S., & Zedler, M. (2012). The neural correlates of coloured music: A functional MRI investigation of auditory-visual synesthesia. *Neuropsychologia, 50*(1), 85–89.

Neumaier, F., Paterno, M., Alpdogan, S., Tevoufouet, E. E., Schneider, T., Hescheler, J., & Albanna, W. (2017). Surgical approaches in psychiatry: a survey of the world literature on psychosurgery. *World Neurosurgery, 97*, 603–634.

Neville, H. J. (2006). Different profiles of plasticity within human cognition. In Y. Munakata & M. H. Johnson (Eds.), *Attention and performance.* Oxford University Press.

Newbury-Helps, J., Feigenbaum, J., & Fonagy, P. (2017). Offenders with antisocial personality disorder display more impairments in mentalizing. *Journal of Personality Disorders, 31*(2), 232–255.

Newcombe, S., & O'Brien-Kop, K. (Eds.). (2020). *Routledge handbook of yoga and meditation studies.* Routledge.

Newell, K. M. (2020). What are fundamental motor skills and what is fundamental about them? *Journal of Motor Learning and Development, 8*(2), 280–314.

Newton, E. K., Thompson, R. A., & Goodman, M. (2016). Individual differences in toddlers' prosociality: Experiences in early relationships explain variability in prosocial behavior. *Child Development, 87*(6), 1715–1726. https://doi.org/10.1111/cdev.12631

Nezami, B. T., Lang, W., Jakicic, J. M., Davis, K. K., Polzien, K., Rickman, A. D., et al. (2016). The effect of self-efficacy on behavior and weight in a behavioral weight-loss intervention. *Health Psychology, 35*(7), 714–722. https://doi.org/10.1037/hea0000378

Nezu, A. M., Nezu, C. M., Felgoise, S. H., & Greenberg, L. M. (2013). Psychosocial oncology. In A. M. Nezu, C. M. Nezu, P. A. Geller, & I. B. Weiner (Eds.), *Handbook of psychology* (2nd ed., vol. 9, pp. 271–291). Wiley.

Ng, R., Allore, H. G., & Levy, B. R. (2020). Self-acceptance and interdependence promote longevity: Evidence from a 20-year prospective cohort study. *International Journal of Environmental Research and Public Health, 17*(16), 5980.

Ng, W., & Diener, E. (2018). Affluence and subjective well-being: Does income inequality moderate their associations? *Applied Research in Quality of Life, 1–16.*

Niarchou, M., Byrne, E. M., Trzaskowski, M., Sidorenko, J., Kemper, K. E., McGrath, J. J., ... Wray, N. R. (2020). Genome-wide association study of dietary intake in the UK biobank study and its associations with schizophrenia and other traits. *Translational Psychiatry, 10*(1), 1–11.

Niaura, R., Todaro, J. F., Strood, L., Spiro, A., Ward, K. D., & Weiss, S. (2002). Hostility, the metabolic syndrome, and incident coronary heart disease. *Health Psychology, 21*(6), 588–593.

Nicholas, C. L., Jordan, A. S., & Trinder, J. (2017). The functions of sleep. In D. Mansfield, N. Antic, S. Rajaratnam, & M. Naughton (Eds.), *Sleep Medicine.* IP Communications.

Nicholson, A. A., Rabellino, D., Densmore, M., Frewen, P. A., Paret, C., Kluetsch, R., et al. (2017). The neurobiology of emotion regulation in posttraumatic stress disorder: Amygdala downregulation via real-time fMRI neurofeedback. *Human Brain Mapping, 38*, 541–560.

Nichter, B., Hill, M., Norman, S., Haller, M., & Pietrzak, R. H. (2020). Mental health treatment utilization among US military veterans with suicidal ideation: Results from the National Health and Resilience in Veterans Study. *Journal of Psychiatric Research, 130*, 61–67.

Nickels, N., Kubicki, K., & Maestripieri, D. (2017). Sex differences in the effects of psychosocial stress on cooperative and prosocial behavior: Evidence for 'flight or fight' in males and 'tend and befriend' in females. *Adaptive Human Behavior and Physiology, 3*(2), 171–183.

Nickerson, R. S., & Adams, M. J. (1979). Long-term memory for a common object. *Cognitive Psychology, 11*(3), 287–307.

Niec, L. N., Egan, R., Schoonover, C., & Brodd, I. (2020). Selection and training of paraprofessionals in core parent-child interaction therapy skills. *Children and Youth Services Review, 111*, 104818.

Niego, A., & Benítez-Burraco, A. (2019). Williams syndrome, human self-domestication, and language evolution. *Frontiers in Psychology, 10*, 521.

Nie, J., & Hashino, E. (2020). Generation of inner ear organoids from human pluripotent stem cells. In J. R. Spence (Ed.), *Methods in cell biology* (vol. 159, pp. 303–321). Academic Press.

Nie, Y., Chua, B. L., Yeung, A. S., Ryan, R. M., & Chan, W. Y. (2015). The importance of autonomy support and the mediating role of work motivation for well-being: Testing self-determination theory in a Chinese work organisation. *International Journal of Psychology, 50*(4), 245–255. https://doi.org/10.1002/ijop.12110

Niemiec, C. P., & Coulson, J. C. (2017). Need-supportive parenting and its role in the wellbeing and recovery of individuals: A self-determination theory perspective. In M. Slade, L. Oades, & A. Jarden (Eds.), *Wellbeing, recovery and mental health* (pp. 300–310). Cambridge University Press. https://doi.org/10.1017/9781316339275.025

Nikolić, M., Majdandžić, M., Colonnesi, C., de Vente, W., Möller, E., & Bögels, S. (2020). The unique contribution of blushing to the development of social anxiety disorder symptoms: results from a longitudinal study. *Journal of Child Psychology and Psychiatry.* https://doi.org/10.1111/jcpp.13221

Niles, A. N., Axelsson, E., Andersson, E., Hedman-Lagerlöf, E., Carlbring, P., Andersson, G., ... Ljótsson, B. (2020). Internet-based cognitive behavior therapy for depression, social anxiety disorder, and panic disorder: Effectiveness and predictors of response in a teaching clinic. *Behaviour Research and Therapy, 136*, 103767.

Nillni, Y. I., Berenz, E. C., Rohan, K. J., & Zvolensky, M. J. (2012). Sex differences in panic-relevant responding to a 10% carbon dioxide–enriched air biological challenge. *Journal of Anxiety Disorders, 26*(1), 165–172.

Nilsen, F. M., & Tulve, N. S. (2020). A systematic review and meta-analysis examining the interrelationships between chemical and non-chemical stressors and inherent characteristics in children with ADHD. *Environmental Research, 180*, 108884.

Nilsson, T. (2020). What came out of visual memory: Inferences from decay of difference-thresholds. *Attention, Perception, & Psychophysics, 82*, 2963-2984.

Nimrod, G. (2020). Aging well in the digital age: Technology in processes of selective optimization with compensation. *The Journals of Gerontology: Series B, 75*(9), 2008-2017.

Nisbett, R. E. (2005). Heredity, environment, and race differences in IQ: A commentary on Rushton and Jensen. *Psychology, Public Policy, and Law, 11*(2), 302-310.

Nisbett, R. E. (2009). *Intelligence and how to get it: Why schools and cultures count*. WW Norton & Company.

Noble, K. G., & Giebler, M. A. (2020). The neuroscience of socioeconomic inequality. *Current Opinion in Behavioral Sciences, 36*, 23-28.

Nor, N. M., Sidek, S., Saad, N., Jaafar, N. H., & Shukri, N. A. M. (2020). Adhering to lifestyle change recommendations via the trans-theoretical model: A mixed-methods study among type 2 diabetes patients. *Nutrition & Food Science*. https://www.emerald.com/insight/content/doi/10.1108/NFS-02-2020-0043/full/html

Norcross, J. C., Campbell, L. F., Grohol, J. M., Santrock, J. W., Selagea, F., & Sommer, R. (2013). *Self-help that works: Resources to improve emotional health and strengthen relationships* (4th ed.). Oxford University Press.

Norcross, J. C. (Ed.). (2011). *Psychotherapy relationships that work: Evidence-based responsiveness*. Oxford University Press.

Norcross, J. C., Krebs, P. M., & Prochaska, J. O. (2011). Stages of change. *Journal of Clinical Psychology, 67*(2), 143-154.

Nordahl, H., Nordahl, H. M., Hjemdal, O., & Wells, A. (2017). Cognitive and metacognitive predictors of symptom improvement following treatment for social anxiety disorder: A secondary analysis from a randomized controlled trial. *Clinical Psychology & Psychotherapy*. https://doi.org/10.1002/cpp.2083

Nordberg, A. (2020, December 4). Family dinners and so much baking: How the pandemic has changed the way we eat. *San Francisco Chronicle*. https://www.sfchronicle.com/culture/article/Family-dinners-and-so-much-baking-How-the-15755194.php

Nordstrom, B. R., & others. (2011). Neurocriminology. *Advances in Genetics, 75*, 255-283.

Nordt, C., Rossler, W., & Lauber, C. (2006). Attitudes of mental health professionals toward people with schizophrenia and major depression. *Schizophrenia Bulletin, 32*(4), 709-714.

Norman, E., Pfuhl, G., Sæle, R. G., Svartdal, F., Låg, T., & Dahl, T. I. (2019). Metacognition in psychology. *Review of General Psychology, 23*(4), 403-424.

Noronha, J. L. (2020). Cancer stigma–why don't we sit down and talk about it? *Cancer Research, Statistics, and Treatment, 3*(2), 167.

Norris, D., Kalm, K., & Hall, J. (2020). Chunking and redintegration in verbal short-term memory. *Journal of Experimental Psychology: Learning, Memory, and Cognition, 46*(5), 872-893. http://dx.doi.org/10.1037/xlm0000762

Nosek, B. A., & Banaji, M. R. (2007). Implicit attitude. In P. Wilken, T. Bayne, & A. Cleeremans (Eds.), *The Oxford companion to consciousness*. Oxford: Oxford University Press.

Nosek, B. A., Bar-Anan, Y., Sriram, N., Axt, J., & Greenwald, A. G. (2014). Understanding and using the Brief Implicit Association Test: Recommended scoring procedures. *PLoS ONE, 9*(12), e110938.

Notman, M. T., & Nadelson, C. C. (2002). Women's issues. In M. Hersen & W. H. Sledge (Eds.), *Encyclopedia of psychotherapy*. Academic.

Nouchi, R., & others. (2013). Brain training game boosts executive functions, working memory, and processing speed in the young adults: A randomized controlled trial. *PLoS One, 8*(2), e55518.

Nowak, M. A., Page, K. M., & Sigmund, K. (2000). Fairness versus reason in the ultimatum game. *Science, 289*(5485), 1773-1775.

Ntoumanis, N., Ng, J. Y., Prestwich, A., Quested, E., Hancox, J. E., Thøgersen-Ntoumani, C., . . . Williams, G. C. (2020). A meta-analysis of self-determination theory-informed intervention studies in the health domain: Effects on motivation, health behavior, physical, and psychological health. *Health Psychology Review*. https://doi.org/10.1080/17437199.2020.1718529.

Núñez, R., Allen, M., Gao, R., Rigoli, C. M., Relaford-Doyle, J., & Semenuks, A. (2019). What happened to cognitive science? *Nature Human Behaviour, 3*(8), 782-791.

Nuninga, J. O., Mandl, R. C., Boks, M. P., Bakker, S., Somers, M., Heringa, S. M., . . . Sommer, I. E. (2020). Volume increase in the dentate gyrus after electroconvulsive therapy in depressed patients as measured with 7T. *Molecular Psychiatry, 25*(7), 1559-1568.

Nurius, P., LaValley, K., & Kim, M. H. (2020). Victimization, poverty, and resilience resources: Stress process considerations for adolescent mental health. *School Mental Health, 12*(1), 124-135.

Nystul, M. S. (1999). *Introduction to counseling*. Allyn & Bacon.

O

Oakley, D. A., & Halligan, P. W. (2011). Using hypnosis to gain insights into healthy and pathological cognitive functioning. *Consciousness and Cognition, 20*(2), 328-331.

Oakley, M., Farr, R. H., & Scherer, D. G. (2017). Same-sex parent socialization: Understanding gay and lesbian parenting practices as cultural socialization. *Journal of GLBT Family Studies, 13*(1), 56-75. https://doi.org/10.1080/1550428X.2016.1158685

O'Brien, E., & Roney, E. (2017, June 7). Worth the wait? Leisure can be just as enjoyable with work left undone. *Psychological Science*. https://doi.org/10.1177/0956797617701749

O'Barr, W. M. (2006). Multiculturalism in the marketplace: Targeting Latinas, African American women, and gay consumers. Advertising and Society Review. https://muse.jhu.edu/journals/advertising_and_society_review/v007/7.4unit11.html (accessed December 22, 2014)

O'Connor, D. B., Conner, M., Jones, F., McMillan, B., & Ferguson, E. (2009). Exploring the benefits of conscientiousness: An investigation of the role of daily stressors and health behaviors. *Annals of Behavioral Medicine, 37*(2), 184-196.

O'Connor, D. B., Thayer, J. F., & Vedhara, D. (2021). Stress and health: A review of psychobiological processes. *Annual Review of Psychology, 71*. https://doi.org/10.1146/annurev-psych-062520-122331

O'Connor, D., Potler, N. V., Kovacs, M., Xu, T., Ai, L., Pellman, J., et al. (2017). The healthy brain network serial scanning initiative: a resource for evaluating inter-individual differences and their reliabilities across scan conditions and sessions. *GigaScience, 6*(2), 1-4.

O'Dell, K. R., Masters, K. S., Spielmans, G. I., & Maisto, S. A. (2011). Does type-D personality predict outcomes among patients with cardiovascular disease? A meta-analytic review. *Journal of Psychosomatic Research, 71*(4), 199-206.

O'Gorman, E. T., Cobb, H. R., Galtieri, L. R., & Kurtz, J. E. (2020). Stimulus characteristics in picture story exercise cards and their effects on the social cognition and object relations scale–global rating method. *Journal of Personality Assessment, 102*(2), 250-258.

O'Reilly, K. B. (2020). AMA: Racism is a threat to public health. https://www.ama-assn.org/delivering-care/health-equity/ama-racism-threat-public-health

O'Toole, S. E., Monks, C. P., & Tsermentseli, S. (2017). Executive function and theory of mind as predictors of aggressive and prosocial behavior and peer acceptance in early childhood. *Social Development, 26*, 907-920.

Oberauer, K. (2019). Is rehearsal an effective maintenance strategy for working memory? *Trends in Cognitive Sciences, 23*(9), 798-809.

Odermatt, J., Frommen, J. G., & Menz, M. H. (2017). Consistent behavioural differences between migratory and resident hoverflies. *Animal Behaviour, 127*, 187-195. https://doi.org/10.1016/j.anbehav.2017.03.015

Odintsova, V. V., Roetman, P. J., Ip, H. F., Pool, R., Van der Laan, C. M., Tona, K. D., . . . Boomsma, D. I. (2019). Genomics of human aggression: Current state of genome-wide studies and an automated systematic review tool. *Psychiatric Genetics, 29*(5), 170-190.

Odom, S. L., Hall, L. J., & Suhrheinrich, J. (2020). Implementation science, behavior analysis, and supporting evidence-based practices for individuals with autism. *European Journal of Behavior Analysis, 21*(1), 55-73.

Ogden, C. L., & Carroll, M. D. (2010). Prevalence of overweight, obesity, and extreme obesity among adults: United States, trends 1960-1962 through 2007-2008. *NCHS Health E-Stat*. www.cdc.gov/nchs/data/hestat/obesity_adult_07_08/obesity_adult_07_08.pdf (accessed January 13, 2015)

Ogden, C. L., Carroll, M. D., Fryar, C.D., & Flegal K. M. (2015). *Prevalence of childhood and adult obesity in the United States, 2011-2014*. NCHS Data brief, #219. https://www.cdc.gov/nchs/data/databriefs/db219.pdf

Ogden, C. L., Carroll, M. D., Kit, B. K., & Flegal, K. M. (2012, January). Prevalence of obesity in the United States, 2009-2010. *NCHS Data Brief, (82)*, 1-8.

Ogilvie, R. D., & Wilkinson, R. T. (1988). Behavioral versus EEG-based monitoring of all-night sleep/wake patterns. *Sleep, 11*(2), 139-155.

Ohi, K., Otowa, T., Shimada, M., Sugiyama, S., Muto, Y., Tanahashi, S., . . . Shioiri, T. (2020). Shared transethnic genetic basis of panic disorder and psychiatric and related intermediate phenotypes. *European Neuropsychopharmacology*. https://doi.org/10.1016/j.euroneuro.2020.11.003

Ohira, T., Diez Roux, A. V., Polak, J. F., Homma, S., Iso, H., & Wasserman, B. A. (2012). Associations of anger, anxiety, and depressive symptoms with carotid arterial wall thickness: The multi-ethnic study of atherosclerosis. *Psychosomatic Medicine, 74*(5), 517-525.

Ohman, A., & Mineka, S. (2001). Fears, phobias, and preparedness: Toward an evolved module of fear and fear learning. *Psychological Review, 108*(3), 483-522.

Ohman, A., & Mineka, S. (2003). The malicious serpent: Snakes as a prototypical stimulus for an evolved module of fear. *Current Directions in Psychological Science, 12*(1), 5-9.

Ohman, A., & Soares, J. J. P. (1998). Emotional conditioning to masked stimuli: Expectancies for aversive outcomes following nonrecognized fear-relevant stimuli. *Journal of Experimental Psychology, 127*(1), 69-82.

Oishi, S., & Diener, E. (2014). Residents of poor nations have a greater sense of meaning in life than residents of wealthy nations. *Psychological Science, 25*(2), 422-430.

Okada, N., Yahata, N., Koshiyama, D., Morita, K., Sawada, K., Kanata, S., . . . Koike, S. (2020). Neurometabolic underpinning of the intergenerational transmission of prosociality. *NeuroImage*. https://doi.org/10.1016/j.neuroimage.2020.116965

Okada, N., Yahata, N., Koshiyama, D., Morita, K., Sawada, K., Kanata, S., . . . Koike, S. (2020). Smaller anterior subgenual cingulate volume mediates the effect of girls' early sexual maturation on negative psychobehavioral outcome. *Neuroimage, 209*, 116478.

Öksüz, Ö., Günver, G., Oba, M. Ç., & Arıkan, K. (2020). Psychiatry to dermatology; panic disorder. *Journal of Clinical Neuroscience, 81*, 316-320.

Okun, M. A., Yeung, E. W., & Brown, S. (2013). Volunteering by older adults and risk of mortality: A meta-analysis. *Psychology and Aging, 28*(2), 564-577.

Olajide, O., Gbadamosi, I., Yawson, E., Arogundade, T., Asogwa, N., & Adeniyi, P. (2020). Hippocampal degeneration involves glutamate excitoxicity during the pathogenesis of Alzheimer's disease. *Biological Psychiatry, 87*(9), S294-S295.

Olds, J. M. (1958). Self-stimulation experiments and differential reward systems. In H. H. Jasper, L. D. Proctor, R. S. Knighton, W. C. Noshay, & R. T. Costello (Eds.), *Reticular formation of the brain*. Boston: Little, Brown.

Olds, J. M. (1958). Self-stimulation experiments and differential reward systems. In H. H. Jasper, L.D. Proctor,

R.S. Knighton, W. C. Noshay, & R. T. Costello (Eds.), *Reticular formation of the brain.* Little, Brown.

Olds, J. M., & Milner, P. M. (1954). Positive reinforcement produced by electrical stimulation of the septal area and other areas of the rat brain. *Journal of Comparative and Physiological Psychology, 47*(6), 419–427.

Oliffe, J. L., Rossnagel, E., Bottorff, J. L., Chambers, S. K., Caperchione, C., & Rice, S. M. (2020). Community-based men's health promotion programs: Eight lessons learnt and their caveats. *Health Promotion International, 35*(5), 1230–1240.

Oliver, Z. J., Cristino, F., Roberts, M. V., Pegna, A. J., & Leek, E. C. (2017). Stereo viewing modulates three-dimensional shape processing during object recognition: A high-density ERP study. *Journal of Experimental Psychology: Human Perception and Performance.* https://doi.org/10.1037/xhp0000444

Olivola, C. Y., & Todorov, A. (2017). The biasing effects of appearances go beyond physical attractiveness and mating motives. *Behavioral And Brain Sciences, 40.* https://doi.org/10.1017/S0140525X16000595

Oltmanns, T. F., & Powers, A. D. (2012). Gender and personality disorders. In T. Widiger (Ed.), *The Oxford handbook of personality disorders.* Oxford University Press.

Onishi, E., Kobayashi, T., Dexter, E., Marino, M., Maeno, T., & Deyo, R. A. (2017). Comparison of opioid prescribing patterns in the United States and Japan: primary care physicians' attitudes and perceptions. *The Journal of the American Board of Family Medicine, 30*(2), 248–254.

Onyeador, I. N., Wittlin, N. M., Burke, S. E., Dovidio, J. F., Perry, S. P., Hardeman, R. R., . . . van Ryn, M. (2020). The value of interracial contact for reducing anti-black bias among non-black physicians: A cognitive habits and growth evaluation (CHANGE) study report. *Psychological Science, 31*(1), 18–30.

Open Science Collaboration. (2015). Estimating the reproducibility of psychological science. *Science, 349,* 943.

Oravecz, Z., Dirsmith, J., Heshmati, S., Vandekerckhove, J., & Brick, T. R. (2020). Psychological well-being and personality traits are associated with experiencing love in everyday life. *Personality and Individual Differences, 153,* 109620.

Ordóñez-Carabaño, Á., Prieto-Ursúa, M., & Dushimimana, F. (2020). Reconciling the irreconcilable: The role of forgiveness after the Rwandan genocide. *Peace and Conflict: Journal of Peace Psychology, 26*(2), 213–216. https://doi-org.proxy.mul.missouri.edu/10.1037/pac0000432

Oren, C., & Shamay-Tsoory, S. G. (2019). Women's fertility cues affect cooperative behavior: Evidence for the role of the human putative chemosignal estratetraenol. *Psychoneuroendocrinology, 101,* 50–59.

Organisation for Economic Co-Operation and Development (OECD). (2017). Obesity update. *OECD.* http://www.oecd.org/els/health-systems/Obesity-Update-2017.pdf

Orlov, N. D., Tracy, D. K., Joyce, D., Patel, S., Rodzinka-Pasko, J., Dolan, H., et al. (2017). Stimulating cognition in schizophrenia: A controlled pilot study of the effects of prefrontal transcranial direct current stimulation upon memory and learning. *Brain Stimulation, 10*(3), 560–566.

Orosz, A., Federspiel, A., Eckert, A., Seeher, C., Dierks, T., Tschitsaz, A., & Cattapan, K. (2020). Exploring the effectiveness of a specialized therapy program for burnout using subjective report and biomarkers of stress, *Clinical Psychology & Psychotherapy.* https://doi.org/10.1002/cpp.2539

Ortiz, L. A. (2020). Reframing neurodiversity as competitive advantage: Opportunities, challenges, and resources for business and professional communication educators. *Business and Professional Communication Quarterly, 83*(3), 261–284.

Osborne-Crowley, K. (2020). Social cognition in the real world: Reconnecting the study of social cognition with social reality. *Review of General Psychology, 24*(2), 144–158.

Osborne, D., & Sibley, C. G. (2020). Does openness to experience predict changes in conservatism? A nine-wave longitudinal investigation into the personality roots to ideology. *Journal of Research in Personality, 87,* 103979.

Osborne, D., Wootton, L. W., & Sibley, C. G. (2013). Are liberals agreeable or not? Politeness and compassion differentially predict political conservatism via distinct ideologies. *Social Psychology, 44*(5), 354–360. https://doi-org.proxy.mul.missouri.edu/10.1027/1864-9335/a000132

Osborne, M. T., Shin, L. M., Mehta, N. N., Pitman, R. K., Fayad, Z. A., & Tawakol, A. (2020). Disentangling the links between psychosocial stress and cardiovascular disease. *Circulation: Cardiovascular Imaging, 13*(8), e010931.

Ostendorf, A. L., Schlüter, H., & Hackländer, R. P. (2020). Sounds in the classroom: Auditory context-dependent memory in school-aged children. *Open Psychology, 2*(1), 106–118.

Osterhaus, C., Putnick, D. L., Kristen-Antonow, S., Kloo, D., Bornstein, M. H., & Sodian, B. (2020). Theory of mind and diverse intelligences in 4-year-olds: Modelling associations of false beliefs with children's numerate-spatial, verbal, and social intelligence. *British Journal of Developmental Psychology.* https://doi.org/10.1111/bjdp.12336

Osterling, J., & Dawson, G. (1994). Early recognition of children with autism: A study of first birthday home videotapes. *Journal of Autism and Developmental Disorders, 24*(3), 247–257.

Osterman, L. L., & Brown, R. P. (2011). Culture of honor and violence against the self. *Personality and Social Psychology Bulletin, 37*(12), 1611–1623.

Ostir, G. V., Markides, K. S., Black, S. A., & Goodwin, J. S. (2000). Emotional well-being predicts subsequent functional independence and survival. *Journal of the American Geriatrics Society, 48*(5), 473–478.

Oswald, F. L. (2020). Future research directions for big data in psychology. In S. E. Woo, L. Tay, & R. W. Proctor (Eds.), *Big data in psychological research* (pp. 427–441). American Psychological Association. https://doi.org/10.1037/0000193-020

Otake, K., Shimai, S., Tanaka-Matsumi, J., Otsui, K., & Fredrickson, B. L. (2006). Happy people becoming happier through kindness: A counting kindnesses intervention. *Journal of Happiness Studies, 7*(3), 361–375.

Otgaar, H., Wang, J., Dodier, O., Howe, M. L., Lilienfeld, S. O., Loftus, E. F., . . . & Patihis, L. (2020). Skirting the issue: What does believing in repression mean? *Journal of Experimental Psychology: General, 149*(10), 2005–2006. https://doi.org/10.1037/xge0000982

Ottaviani, G., & Buja, L. M. (2020). Pathology of unexpected sudden cardiac death: Obstructive sleep apnea is part of the challenge. *Cardiovascular Pathology, 47,* 107221.

Owen, J., Drinane, J. M., Kivlighan III, M., Miller, S., Kopta, M., & Imel, Z. (2019). Are high-performing therapists both effective and consistent? A test of therapist expertise. *Journal of Consulting and Clinical Psychology, 87*(12), 1149–1156.

Owens, E. B., Zalecki, C., Gillette, P., & Hinshaw, S. P. (2017). Girls with childhood ADHD as adults: Cross-domain outcomes by diagnostic persistence. *Journal of Consulting and Clinical Psychology, 85*(7), 723.

Owens, R.E. (2015). *Language development: An introduction (9th ed.).* Pearson.

Oxenham, A.J. (2018). How we hear: The perception and neural coding of sound. *Annual Review of Psychology, 69,* 27–50.

Ozana, A., & Ganel, T. (2019). Obeying the law: speed–precision tradeoffs and the adherence to Weber's law in 2D grasping. *Experimental Brain Research, 237*(8), 2011–2021.

P

Pachai, A. A., Acai, A., LoGiudice, A. B., & Kim, J. A. (2016). The mind that wanders: Challenges and potential benefits of mind wandering in education. *Scholarship of Teaching and Learning in Psychology, 2*(2), 134–146. https://doi.org/10.1037/stl0000060

Packer, D. J. (2008). Identifying systematic disobedience in Milgram's obedience experiments: A meta-analytic review. *Perspectives on Psychological Science, 3*(4), 301–304.

Packer, D. J. (2009). Avoiding groupthink: Whereas weakly identified members stay silent, strongly identified members dissent about collective matters. *Psychological Science, 20*(5), 546–548.

Page, R. L., Peltzer, J. N., Burdette, A. M., & Hill, T. D. (2020). Religiosity and health: A holistic biopsychosocial perspective. *Journal of Holistic Nursing, 38*(1), 89–101.

Pagliaccio, D., Middleton, R., Hezel, D., Steinman, S., Snorrason, I., Gershkovich, M., . . . Marsh, R. (2019). Task-based fMRI predicts response and remission to exposure therapy in obsessive-compulsive disorder. *Proceedings of the National Academy of Sciences, 116*(41), 20346–20353.

Paglieri, F. & others. (2014). Nonhuman gamblers: Lessons from robots, primates, and robots. *Frontiers in Behavioral Neuroscience, 8,* 33.

Påhlman, M., Gillberg, C., & Himmelmann, K. (2020). Autism and attention-deficit/hyperactivity disorder in children with cerebral palsy: High prevalence rates in a population-based study. *Developmental Medicine & Child Neurology.* https://doi.org/10.1111/dmcn.14736

Paivio, A. (1971). *Imagery and verbal processes.* Holt, Rinehart & Winston.

Paivio, A. (1986). *Mental representations: A dual coding approach.* Oxford University Press.

Paivio, A. (2007). *Mind and Its evolution: A dual coding theoretical approach.* Erlbaum.

Paivio, A., & Sadoski, M. (2011). Lexicons, contexts, events, and images: Commentary on Elman (2009) from the perspective of dual coding theory. *Cognitive Science, 35*(1), 198–209.

Pakhomov, S. V. S., & Hemmy, L. S. (2014). A computational linguistic measure of clustering behavior on semantic verbal fluency task predicts risk of future dementia in the Nun Study. *Cortex, 55*(1), 97–106.

Paksarian, D., Rudolph, K. E., Stapp, E. K., Dunster, G. P., He, J., Mennitt, D., . . . Merikangas, K. R. (2020). Association of outdoor artificial light at night with mental disorders and sleep patterns among US adolescents. *JAMA Psychiatry.* https://doi.org/10.1001/jamapsychiatry.2020.1935

Palmatier, J. J., & Rovner, L. (2014). Credibility assessment: Preliminary Process Theory, the polygraph process, and construct validity. *International Journal of Psychophysiology.* PMID: 24933412

Palmqvist, L., Danielsson, H., Jönsson, A., & Rönnberg, J. (2020). Cognitive abilities and life experience in everyday planning in adolescents with intellectual disabilities: Support for the difference model. *Journal of Intellectual Disability Research, 64*(3), 209–220.

Palpacuer, C., Gallet, L., Drapier, D., Reymann, J. M., Falissard, B., & Naudet, F. (2017). Specific and non-specific effects of psychotherapeutic interventions for depression: Results from a meta-analysis of 84 studies. *Journal of Psychiatric Research, 87,* 95–104.

Paluck, E. L., Porat, R., Clark, C. S., & Green, D. P. (2021). Prejudice reduction: Progress and challenges. *Annual Review of Psychology, 72.* https://doi.org/10.1146/annurev-psych-071620-030619

Pan, A., & others. (2012). Red meat consumption and mortality: Results from 2 prospective cohort studies. *Archives of Internal Medicine, 172*(7), 555–563.

Panagiotaropoulos, T. I., Wang, L., & Dehaene, S. (2020). Hierarchical architecture of conscious processing and subjective experience. *Cognitive Neuropsychology, 37,* 180–183.

Paradiso, S., Brown, W. S., Porcerelli, J. H., Tranel, D., Adolphs, R., & Paul, L. K. (2020). Integration between cerebral hemispheres contributes to defense mechanisms. *Frontiers in Psychology, 11,* 1534.

Paralympic Movement. (2012). *Believe in yourself.* www.paralympic.org/believe-in-yourself/our-heroes.

Park, D. C., & Bischof, G. N. (2011). The aging mind: Neuroplasticity in response to cognitive training. *Dialogues in Clinical Neuroscience, 15*(1), 109–119.

Park, D. C., Lodi-Smith, J., Drew, L., Haber, S., Hebrank, A., Bischof, G. N., & Aamodt, W. (2014). The impact of sustained engagement on cognitive function in older adults: The Synapse Project. *Psychological Science, 25*(1), 103–112.

Park, J. H., Gorky, J., Ogunnaike, B., Vadigepalli, R., & Schwaber, J. S. (2020). Investigating the effects of brainstem neuronal adaptation on cardiovascular homeostasis. *Frontiers in Neuroscience, 14,* 470.

Park, J. J. (2020). Do we really know what we see? The role of cognitive bias in how we view race in higher education. *Change: The Magazine of Higher Learning, 52*(2), 46–49.

Park, M. (2009). Teen tries to quiet the voices caused by schizophrenia. *CNN.* www.cnn.com/2009/HEALTH/04/24/schizophrenia.soloist.brain/index.html

Park, S., Roh, S. H., & Lee, J. Y. (2019). Body regional heat pain thresholds using the method of limit and level: A comparative study. *European Journal of Applied Physiology, 119*(3), 771–780.

Parker, P. S. (2006). *Race, gender, and leadership.* Erlbaum.

Parola, A., Simonsen, A., Bliksted, V., & Fusaroli, R. (2020). Voice patterns in schizophrenia: A systematic review and Bayesian meta-analysis. *Schizophrenia Research, 216*, 24–40.

Parris, B. A. (2017). The role of frontal executive functions in hypnosis and hypnotic suggestibility. *Psychology of Consciousness: Theory, Research, and Practice, 4*(2), 211–229. https://doi.org/10.1037/cns0000106

Parry, D. A., & le Roux, D. B. (2019). Media multitasking and cognitive control: A systematic review of interventions. *Computers in Human Behavior, 92*, 316–327.

Parsons, T. D., Carlew, A. R., Magtoto, J., & Stonecipher, K. (2017). The potential of function-led virtual environments for ecologically valid measures of executive function in experimental and clinical neuropsychology. *Neuropsychological Rehabilitation, 27*(5), 777–807.

Pascalls, O., & Kelly, D. J. (2008). Face processing. In M. M. Haith & J. B. Benson (Eds.), *Encyclopedia of infant and early childhood development.* Elsevier.

Pascual-Vera, B., Akin, B., Belloch, A., Bottesi, G., Clark, D. A., Doron, G., . . . Jiménez-Ros, A. (2019). The cross-cultural and transdiagnostic nature of unwanted mental intrusions. *International Journal of Clinical and Health Psychology, 19*(2), 85–96.

Pashler, H. & Harris, C. R. (2012). Is the replicability crisis overblown? Three arguments examined. *Perspectives on Psychological Science, 7*, 531–536.

Pashler, H., McDaniel, M., Rohrer, D., & Bjork, R. (2008). Learning styles: Concepts and evidence. *Psychological Science in the Public Interest, 9*(3), 105–119.

Passer, M. (2017). *Research Methods* (2nd ed.). Macmillan Higher Education.

Patterson, A. C., & Veenstra, G. (2010). Loneliness and risk of mortality: A longitudinal investigation in Alameda County, California. *Social Science & Medicine, 71*(1), 181–186. https://doi.org/10.1016/j.socscimed.2010.03.024

Patterson, C. J. (2012). Family lives of lesbian and gay adults. In G. W. Petersen & K. R. Bush (Eds.), *Handbook of marriage and the family.* Springer.

Patterson, C. J. (2013). Family lives of gay and lesbian adults. In G. W. Peterson & K. R. Bush (Eds.), *Handbook of marriage and the family* (3rd ed.). Springer.

Patterson, C. J. (2014). Sexual minority youth and youth with sexual minority parents. In G. B. Melton, A. Ben-Arieh, J. Cashmore, G. S. Goodman, & N. K. Worley (Eds.), *Sage handbook of child research.* Sage.

Patterson, C. J., & Farr, R. H. (2014). Children of lesbian and gay parents: Reflections on the research-policy interface. In H. R. Schaffer & K. Durkin (Eds.), *Blackwell handbook of developmental psychology in action.* Blackwell.

Patterson, C. J., Riskind, R. G., & Tornello, S. L. (2014). Sexual orientation and parenting: A global perspective. In A. Abela & J. Walker (Eds.), *Contemporary issues in family studies.* Wiley.

Patterson, C. J., & Wainwright, J. L. (2010). Adolescents with same-sex parents: Findings from the National Longitudinal Study of Adolescent Health. In D. Brodzinsky, A. Pertman, & D. Kunz (Eds.), *Lesbian and gay adoption: A new American reality.* Oxford University Press.

Pattillo, R. (2010). Are students as good at multi-tasking as they think? *Nurse Educator, 35*(1), 24.

Paul, A., Chaker, Z., & Doetsch, F. (2017). Hypothalamic regulation of regionally distinct adult neural stem cells and neurogenesis. *Science 356*(6345), 1383–1386.

Paulus, P. B. (1989). An overview and evaluation of group influence. In P. B. Paulus (Ed.), *Psychology of group influence.* Erlbaum.

Paunonen, S., Jackson, D., Trzebinski, J., & Forserling, F. (1992). Personality structures across cultures: A multimethod evaluation. *Journal of Personality and Social Psychology, 62*(3), 447–456.

Pavani, J., Le Vigouroux, S., Kop, J., Congard, A., & Dauvier, B. (2016). Affect and affect regulation strategies reciprocally influence each other in daily life: The case of positive reappraisal, problem-focused coping, appreciation and rumination. *Journal of Happiness Studies, 17*(5), 2077–2095. https://doi.org/0.1007/s10902-015-9686-9

Pavlov, I. P. (1927). *Conditioned reflexes.* G. V. Anrep (Trans.). Dover.

Pavot, W., & Diener, E. (2008). The Satisfaction with Life Scale and the emerging construct of life satisfaction. *Journal of Positive Psychology, 3*(2), 137–152.

Pawluski, J. L., Paravatou, R., Even, A., Cobraiville, G., Fillet, M., Kokras, N., . . . Charlier, T. D. (2020). Effect of sertraline on central serotonin and hippocampal plasticity in pregnant and non-pregnant rats. *Neuropharmacology, 166*, 107950.

Payne, B. K., Brown-Iannuzzi, J. L., & Loersch, C. (2016). Replicable effects of primes on human behavior. *Journal of Experimental Psychology: General, 145*(10), 1269–1279. https://doi.org/10.1037/xge0000201

Peachman, R.R. (January 31, 2017). Raising a transgender child. *New York Times*, D4.

Pearlstein, J. G., Staudenmaier, P. J., West, A. E., Geraghty, S., & Cosgrove, V. E. (2020). Immune response to stress induction as a predictor of cognitive-behavioral therapy outcomes in adolescent mood disorders: A pilot study. *Journal of Psychiatric Research, 120*, 56–63.

Pebsworth, P., & Radhakrishna, S. (2020). Using conditioned taste aversion to reduce human-nonhuman primate conflict: A comparison of four potentially illness-inducing drugs. *Applied Animal Behaviour Science, 225*, 104948.

Peers, P. V., Astle, D. E., Duncan, J., Murphy, F. C., Hampshire, A., Das, T., & Manly, T. (2020). Dissociable effects of attention vs working memory training on cognitive performance and everyday functioning following fronto-parietal strokes. *Neuropsychological Rehabilitation, 30*(6), 1092–1114.

Pelham, W. E., Smith, B. H., Evans, S. W., Bukstein, O., Gnagy, E. M., Greiner, A. R., & Sibley, M. H. (2017). The effectiveness of short-and long-acting stimulant medications for adolescents with ADHD in a naturalistic secondary school setting. *Journal of Attention Disorders, 21*(1), 40–45.

Pelleymounter, M. A., Cullen, M. J., Baker, M. B., Hecht, R., Winters, D., Boone, T., & Collins, F. (1995). Effects of the obese gene product on body weight regulation in ob/ob mice. *Science, 269*(5223), 540–543.

Pellicano, E., Kenny, L., Brede, J., Klaric, E., Lichwa, H., & McMillin, R. (2017). Executive function predicts school readiness in autistic and typical preschool children. *Cognitive Development, 43*, 1–13.

Peltzman, T., Shiner, B., & Watts, B. V. (2020). Effects of electroconvulsive therapy on short-term suicide mortality in a risk-matched patient population. *The Journal of ECT, 36*(3), 187–192.

Peñate, W., Fumero, A., Viña, C., Herrero, M., Marrero, R. J., & Rivero, F. (2017). A meta-analytic review of neuroimaging studies of specific phobia to small animals. *The European Journal of Psychiatry, 31*(1), 23–36.

Pena-Orbea, C., Kolla, B. P., & Mansukhani, M. P. (2020). Jet lag sleep disorder. In R. R. Auger (Ed.), *Circadian rhythm sleep-wake disorders* (pp. 199–205). Springer.

Penfield, W. (1947). Some observations in the cerebral cortex of man. *Proceedings of the Royal Society, 134*(876), 349.

Peng, P., Namkung, J., Barnes, M., & Sun, C. (2016). A meta-analysis of mathematics and working memory: Moderating effects of working memory domain, type of mathematics skill, and sample characteristics. *Journal Of Educational Psychology, 108*(4), 455–473. https://doi.org/10.1037/edu0000079

Peng, Z., Dai, C., Ba, Y., Zhang, L., Shao, Y., & Tian, J. (2020). Effect of sleep deprivation on the working memory-related N2-P3 components of the event-related potential waveform. *Frontiers in Neuroscience, 14.*

Penix, E. A., Swift, J. K., Russell, K. A., & Trusty, W. T. (2020). Client and therapist agreement in moment-to-moment helpfulness ratings in psychotherapy: A microprocess approach. *Journal of Clinical Psychology, 77*(1), 36–48.

Pennebaker, J. W. (1997a). *Opening up: The healing power of expressing emotions* (rev. ed.). Guilford.

Pennebaker, J. W. (1997b). Writing about emotional experiences as a therapeutic experience. *Psychological Science, 8*(3), 162–166.

Pennebaker, J. W. (2004). *Writing to heal: A guided journal for recovering from trauma and emotional upheaval.* New Harbinger.

Pennebaker, J. W., & Chung, C. K. (2007). Expressive writing, emotional upheavals, and health. In H. S. Friedman & R. C. Silver (Eds.), *Foundations of health psychology* (pp. 263–284). Oxford University Press.

Pennebaker, J. W., & Chung, C. K. (2011). Expressive writing: Connections to physical and mental health. In H. S. Friedman (Ed.), *The Oxford handbook of health psychology* (pp. 417–437). Oxford University Press.

Pennycook, G., McPhetres, J., Zhang, Y., Lu, J. G., & Rand, D. G. (2020). Fighting COVID-19 misinformation on social media: Experimental evidence for a scalable accuracy-nudge intervention. *Psychological Science*, 31(7), 770–780.

Perera, F. P., Wheelock, K., Wang, Y., Tang, D., Margolis, A. E., Badia, G., et al. (2018). Combined effects of prenatal exposure to polycyclic aromatic hydrocarbons and material hardship on child ADHD behavior problems. *Environmental Research, 160*, 506–513.

Perkins, D. N. (1984). Creativity by design. *Educational Leadership, 42*(1) 18–25.

Perkins, S. C., Finegood, E. D., & Swain, J. E. (2013). Poverty and language development: Roles of parenting and stress. *Innovations In Clinical Neuroscience, 10*, 10–19.

Perogamvros, L., Baird, B., Seibold, M., Riedner, B., Boly, M., & Tononi, G. (2017). The phenomenal contents and neural correlates of spontaneous thoughts acrosswakefulness, NREM sleep, and REM sleep. *Journal of Cognitive Neuroscience, 29*, 1766–1777

Perry, A., Mankuta, D., & Shamay-Tsoory, S. G. (2015). OT promotes closer interpersonal distance among highly empathic individuals. *Social Cognitive and Affective Neuroscience, 10*(1), 3–9.

Pert, C. B. (1999). *Molecules of emotion.* Simon & Schuster.

Pert, C. B., & Snyder, S. H. (1973). Opiate receptor: Demonstration in a nervous tissue. *Science, 179*(4077), 1011.

Pešlová, E., Mareček, R., Shaw, D. J., Kašpárek, T., Pail, M., & Brázdil, M. (2018). Hippocampal involvement in nonpathological déjà vu: Subfield vulnerability rather than temporal lobe epilepsy equivalent. *Brain and Behavior*, e00996. https://doi.org/10.1002/brb3.996.

Pesta, B. J., Kirkegaard, E. O., te Nijenhuis, J., Lasker, J., & Fuerst, J. G. (2020). Racial and ethnic group differences in the heritability of intelligence: A systematic review and meta-analysis. *Intelligence, 78*, 101408.

Peter, S., Urbanus, B. H., Klaassen, R. V., Wu, B., Boele, H. J., Azizi, S., . . . & Spijker, S. (2020). AMPAR Auxiliary Protein SHISA6 Facilitates Purkinje Cell Synaptic Excitability and Procedural Memory Formation. *Cell Reports, 31*(2), 107515.

Peters, M. N., George, P., & Irimpen, A. M. (2015). The broken heart syndrome: Takotsubo cardiomyopathy. *Trends in Cardiovascular Medicine, 25*(4), 351–357.

Petersen, J. L., & Hyde, J. S. (2010). A meta-analytic review of research on gender differences in sexuality, 1973–2007. *Psychological Bulletin, 136*(1), 21–38.

Petersen, J. L., & Hyde, J. S. (2011). Gender differences in sexual attitudes and behaviors: A review of meta-analytic results and large datasets. *Journal of Sex Research, 48*(2–3), 149–165. https://doi.org/10.1080/00224499.2011.551851

Peterson, A. C., Zhang, S., Hu, S., Chao, H. H., & Li, C. S. R. (2017). The effects of age, from young to middle adulthood, and gender on resting state functional connectivity of the dopaminergic midbrain. *Frontiers in Human Neuroscience, 11*.

Peterson, M. H., Griffith, R. L., Isaacson, J. A., O'Connell, M. S., & Mangos, P. M. (2011). Applicant faking, social desirability, and the prediction of counterproductive work behaviors. *Human Performance, 24*(3), 270–290.

Petruzziello, G., Mariani, M. G., Chiesa, R., & Guglielmi, D. (2020). Self-efficacy and job search success for new graduates. *Personnel Review.* https://www.emerald.com/insight/content/doi/10.1108/PR-01-2019-0009/full/html

Petsko, C. D., & Bodenhausen, G. V. (2020). Multifarious person perception: How social perceivers manage the complexity of intersectional targets. *Social and Personality Psychology Compass, 14*(2), e12518.

Petticrew, M., Bell, R., & Hunter, D. (2002). Influence of psychological coping on survival and recurrence in people with cancer: Systematic review. *British Medical Journal, 325*(7372), 1066–1069.

Pettigrew, T. F., & Tropp, L. R. (2006). A meta-analytic test of intergroup contact theory. *Journal of Personality and Social Psychology, 90*(5), 751–783.

Petty, R. E., & Briñol, P. (2015). Processes of social influence through attitude change. In M. Mikulincer, P. R. Shaver, E. Borgida, & J. A. Bargh, M. (Eds.), *APA handbook of personality and social psychology, Volume 1: Attitudes and social cognition* (pp. 509–545). American Psychological Association. https://doi.org/10.1037/14341-016

Petty, R. E., & Cacioppo, J. T. (1986). The elaboration likelihood of persuasion. In L. Berkowitz (Ed.), *Advances in experimental social psychology* (vol. 19). Academic.

Pettygrove, D.M., Hammond, S.I., Karahuta, E.L., Waugh, W.E., & Brownell, C. (2013). From cleaning up to helping out: Parental socialization and children's early prosocial behavior. *Infant Behavior and Development, 36*, 843–846. https://doi.org/10.1016/j.infbeh.2013.09.005se

Peyron, C., & Rampon, C. (2020). Young neurons tickle memory during REM sleep. *Neuron, 107*(3), 397–398.

Pezdek, K. (2003). Event memory and autobiographical memory for the events of September 11, 2001. *Applied Cognitive Psychology, 17*(9), 1033–1045.

Pezzo, M. V. (2011). Hindsight bias: A primer for motivational researchers. *Social and Personality Psychology Compass, 5*(9), 665–678.

Pezzuti, L., Tommasi, M., Saggino, A., Dawe, J., & Lauriola, M. (2020). Gender differences and measurement bias in the assessment of adult intelligence: Evidence from the Italian WAIS-IV and WAIS-R standardizations. *Intelligence, 79*, 101436.

Pfeuffer, C. U., Aufschnaiter, S., Thomaschke, R., & Kiesel, A. (2020). Only time will tell the future: Anticipatory saccades reveal the temporal dynamics of time-based location and task expectancy. *Journal of Experimental Psychology: Human Perception and Performance.* https://doi.org/10.1037/xhp0000850

Philippi, C. L., Bruss, J., Boes, A. D., Albazron, F. M., Deifelt Streese, C., Ciaramelli, E., . . . Tranel, D. (2020). Lesion network mapping demonstrates that mind-wandering is associated with the default mode network. *Journal of Neuroscience Research.* https://doi.org/10.1002/jnr.24648

Phillips, F., & Fleming, R. W. (2020). The veiled virgin illustrates visual segmentation of shape by cause. *Proceedings of the National Academy of Sciences, 117*(21), 11735–11743.

Phillips, K. A., Osborne, R. H., Giles, G. G., Dite, G. S., Apicella, C., Hopper, J. L., & Milne, R. L. (2008). Psychosocial factors and survival of young women with breast cancer: A population-based prospective cohort study. *Journal of Clinical Oncology, 26*(28), 4666–4671.

Phillips, K. J., & Mudford, O. C. (2008). Functional analysis skills training for residential caregivers. *Behavioral Interventions, 23*(1), 1–12.

Piaget, J. (1952). *The origins of intelligence in children.* Oxford University Press

Picchioni, D., Reith, R. M., Nadel, J. L., & Smith, C. B. (2014). Sleep, plasticity, and the pathophysiology of neurodevelopmental disorders: The potential roles of protein synthesis and other cellular processes. *Brain Sciences, 4*(1), 150–201.

Piff, P. K., Kraus, M. W., Côté, S., Cheng, B. H., & Keltner, D. (2010). Having less, giving more: The influence of social class on prosocial behavior. *Journal of Personality and Social Psychology, 99*(5), 771–784.

Piff, P. K., & Robinson, A. R. (2017). Social class and prosocial behavior: Current evidence, caveats, and questions. *Current Opinion in Psychology, 18*, 6–10.

Pike, K. M., Yamamiya, Y., & Konishi, H. (2011). Eating disorders in Japan: Cultural context, clinical features, and future directions. In R. H. Striegel-Moore, S. A. Wonderlich, B. T. Walsh, & J. E. Mitchell (Eds.), *Developing an evidence-based classification of eating disorders: Scientific findings for DSM-5* (pp. 335–349). American Psychiatric Association.

Pikhartova, J., Bowling, A., & Victor, C. (2016). Is loneliness in later life a self-fulfilling prophecy? *Aging & Mental Health, 20*(5), 543–549. https://doi.org/10.1080/13607863.2015.1023767

Pilditch, T. D., Lagator, S., & Lagnado, D. (2020). Strange but true: Corroboration and base rate neglect. *Journal of Experimental Psychology: Learning, Memory, and Cognition.* Advance online publication. https://doi.org/10.1037/xlm0000816

Pilecki, B., Arentoft, A., & McKay, D. (2011). An evidence-based causal model of panic disorder. *Journal of Anxiety Disorders, 25*(3), 381–388.

Pillemer, D. B. (1998). *Momentous events: Vivid memories.* Harvard University Press.

Pina, M. M., & Cunningham, C. L. (2017). Ethanol-seeking behavior is expressed directly through an extended amygdala to midbrain neural circuit. *Neurobiology of Learning And Memory, 13*, 783–791. https://doi.org/10.1016/j.nlm.2016.11.013

Piña-Andrade, S., Ramos, G., Cárdenas-León, M., Martínez, A., Romero-Morales, L., Martínez-Torres, M., . . . Luis, J. (2020). Testosterone dependent territorial aggression is modulated by cohabitation with a female in male Mongolian gerbils *(Meriones unguiculatus). Hormones and Behavior, 117*, 104611.

Pinker, S. (1994). *The Language Instinct.* HarperCollins.

Pinquart, M., & Kauser, R. (2018). Do the associations of parenting styles with behavior problems and academic achievement vary by culture? Results from a meta-analysis. *Cultural Diversity and Ethnic Minority Psychology.* https://doi.org/10.1037/cdp0000149

Piper, M. E., Cook, J. W., Schlam, T. R., Smith, S. S., Bolt, D. M., Collins, L. M., et al. (2017). Toward precision smoking cessation treatment II: Proximal effects of smoking cessation intervention components on putative mechanisms of action. *Drug and Alcohol Dependence, 17*, 150–158. https://doi.org/10.1016/j.drugalcdep.2016.11.027

Pisani, F., Fusco, C., & Spagnoli, C. (2020). Linking acute symptomatic neonatal seizures, brain injury and outcome in preterm infants. *Epilepsy & Behavior, 112*, 107406. https://doi.org/10.1016/j.yebeh.2020.107406

Pittenger, D. J. (2005). Cautionary comments regarding the Myers-Briggs Type Indicator. *Consulting Psychology Journal: Practice and Research, 57*(3), 210–221. https://doi.org/10.1037/1065-9293.57.3.210

Pitts, B. L., Safer, M. A., Russell, D. W., & Castro-Chapman, P. L. (2016). Effects of hardiness and years of military service on posttraumatic stress symptoms in U.S. Army medics. *Military Psychology, 28*(4), 278–284. https://doi.org/10.1037/mil0000106

Place, S. S., Todd, P. M., Zhuang, J., Penke, L., & Asendorpf, J. B. (2012). Judging romantic interest of others from thin slices is a cross-cultural ability. *Evolution and Human Behavior, 33*(5), 547–550.

Pliatsikas, C. (2020). Understanding structural plasticity in the bilingual brain: The dynamic restructuring model. *Bilingualism: Language and Cognition, 23*(2), 459–471.

Plomin, R. (2018). *Blueprint: How DNA makes us who we are.* MIT Press.

Plomin, R., & von Stumm, S. (2018). The new genetics of intelligence. *Nature Reviews Genetics, 19*, 148–159.

Plotnikoff, R. C., Lubans, D. R., Costigan, S. A., Trinh, L., Spence, J. C., Downs, S., & McCargar, L. (2011). A test of the theory of planned behavior to explain physical activity in a large population sample of adolescents from Alberta, Canada. *Journal of Adolescent Health, 49*(5), 547–549.

Plouffe, R. A., Wilson, C. A., & Saklofske, D. H. (2020). Examining the relationships between childhood exposure to intimate partner violence, the dark tetrad of personality, and violence perpetration in adulthood. *Journal of Interpersonal Violence.* https://doi.org/10.1177/0886260520948517

Pluess, M., Belsky, J., Way, B. M., & Taylor, S. E. (2010). 5-HTTLPR moderates effects of current life events on neuroticism: Differential susceptibility to environmental influences. *Progress in Neuro-Psychopharmacology and Biological Psychiatry, 34* (6), 1070–1074.

Poels, E. M., Kamperman, A. M., Vreeker, A., Gilden, J., Boks, M. P., Kahn, R. S., . . . Bergink, V. (2020). Lithium use during pregnancy and the risk of miscarriage. *Journal of Clinical Medicine, 9*(6), 1819.

Poetsch, M. S., Strano, A., & Guan, K. (2020). Role of leptin in cardiovascular diseases. *Frontiers in Endocrinology, 11*, 354.

Poh, J. H., & Chee, M. W. (2017). Degradation of cortical representations during encoding following sleep deprivation. *Neuroimage, 153*, 131–138.

Pohjola, H. (2020). Acquired disability in young women: A challenge for identity? *Journal of Youth Studies, 23*(2), 127–139.

Polverino, A., Grimaldi, M., Sorrentino, P., Jacini, F., D'Ursi, A. M., & Sorrentino, G. (2018). Effects of Acetylcholine on β-Amyloid-Induced cPLA2 Activation in the TB Neuroectodermal Cell Line: Implications for the Pathogenesis of Alzheimer's Disease. *Cellular and Molecular Neurobiology, 38*(4), 817–826. https://doi.org/10.1007/s10571-017-0555-4

Pompili, M., & others. (2007). Suicide risk in schizophrenia: Learning from the past to change the future. *Annals of General Psychiatry, 6*, 10.

Poon, K. T., Chen, Z., & Wong, W. Y. (2020). Beliefs in conspiracy theories following ostracism. *Personality and Social Psychology Bulletin.* https://doi.org/10.1177/0146167219898944

Porcerelli, J. H., Cramer, P., Porcerelli, D. J., & Arterbery, V. E. (2017). Defense mechanisms and utilization in cancer patients undergoing radiation therapy: A pilot study. *Journal of Nervous and Mental Disease, 205*(4), 466–470. https://doi.org/doi:10.1097/NMD.0000000000000674

Porras-Segovia, A., Díaz-Oliván, I., Gutiérrez-Rojas, L., Dunne, H., Moreno, M., & Baca-Garcia, E. (2020). Apps for depression: Are they ready to work? *Current Psychiatry Reports, 22*(3), 11.

Porta-Casteràs, D., Fullana, M. A., Tinoco, D., Martinez-Zalacain, I., Pujol, J., Palao, D. J., . . . Cardoner, N. (2020). Prefrontal-amygdala connectivity in trait anxiety and generalized anxiety disorder: Testing the boundaries between healthy and pathological worries. *Journal of Affective Disorders, 267*, 211–219.

Porter, S.A. (2019). A psychotic break led this Stanford Ph.D. student to create a play on mental illness. *Des Moines Register.* https://www.desmoinesregister.com/story/entertainment/2019/11/15/psychotic-break-led-stanford-ph-d-student-create-play-des-moines-iowa/4192629002/

Posner, J., Russell, J., & Peterson, B. S. (2005). The circumplex model of affect: An integrative approach to affective neuroscience, cognitive development, and psychopathology. *Developmental Psychopathology, 17*(3), 715–734.

Posner, M. I., & Shulman, G. L. (2019). Cognitive science. In E. Hearst (Ed.), *The first century of experimental psychology* (pp. 371–406). Routledge.

Posporelis, S., Coughlin, J. M., Marsman, A., Pradhan, S., Tanaka, T., Wang, H., et al. (2017). Decoupling of brain temperature and glutamate in recent onset of schizophrenia: A 7T proton magnetic resonance spectroscopy study. *Biological Psychiatry: Cognitive Neuroscience and Neuroimaging.* https://doi.org/10.1016/j.bpsc.2017.04.003

Post, J. M., & Panis, L. K. (2011). Crimes of obedience: "Groupthink" at Abu Ghraib. *International Journal of Group Psychotherapy, 61*(1), 48-66.

Post, R. M., Altshuler, L. L., Kupka, R., McElroy, S. L., Frye, M. A., Rowe, M., et al. (2017). More childhood onset bipolar disorder in the United States than Canada or Europe: Implications for treatment and prevention. *Neuroscience & Biobehavioral Reviews, 74*, Part A, 204-213

Potasiewicz, A., Krawczyk, M., Gzielo, K., Popik, P., & Nikiforuk, A. (2020). Positive allosteric modulators of alpha 7 nicotinic acetylcholine receptors enhance procognitive effects of conventional anti-Alzheimer drugs in scopolamine-treated rats. *Behavioural Brain Research, 385*, 112547.

Potter, C., Ferriday, D., Griggs, R. L., Hamilton-Shield, J. P., Rogers, P. J., & Brunstrom, J. M. (2017). Parental beliefs about portion size, not children's own beliefs, predict child BMI. *Pediatric Obesity.*

Potter, N. N. (2012). Mad, bad, or virtuous? The moral, cultural, and pathologizing features of deviance. *Theory & Psychology, 22*(1), 23-45.

Potter, S., Drewelies, J., Wagner, J., Duezel, S., Brose, A., Demuth, I., . . . Gerstorf, D. (2020). Trajectories of multiple subjective well-being facets across old age: The role of health and personality. *Psychology and Aging 35*(6), 894-909.

Pottratz, S. T., Hutchinson, J. C., Karageorghis, C. I., Mullin, E. M., & Zenko, Z. (2020). Prime movers: Effects of subliminal primes, music, and music video on psychological responses to exercise. *Annals of Behavioral Medicine*, kaaa036.

Powell, B., Cooper, G., Hoffman, K., & Marvin, B. (2014). *The circle of security intervention: Enhancing attachment in early parent-child relationships.* Guilford.

Powell, D., Pacula, R. L., & Jacobson, M. (2018). Do medical marijuana laws reduce addictions and deaths related to pain killers? *Journal of Health Economics, 58*, 29-42.

Power, R. A., Parkhill, J., & de Oliveira, T. (2017). Microbial genome-wide association studies: Lessons from human GWAS. *Nature Reviews Genetics, 18*(1), 41-50.

Prather, A. A., Pressman, S. D., Miller, G. E., & Cohen, S. (2020). Temporal links between self-reported sleep and antibody responses to the influenza vaccine. *International Journal of Behavioral Medicine.* https://doi.org/10.1007/s12529-020-09879-4

Prat, Y. (2019). Animals have no language, and humans are animals too. *Perspectives on Psychological Science, 14*(5), 885-893.

Pratt, M. W., Lawford, H. L., Matsuba, M. K., & Villar, F. (2020). The life span development of generativity. In L. A. Jensen (Ed.), *The Oxford handbook of moral development: An interdisciplinary perspective* (pp. 366-379). Oxford University Press.

Preckel, F., Baudson, T. G., Krolak-Schwerdt, S., & Glock, S. (2015). Gifted and maladjusted? Implicit attitudes and automatic associations related to gifted children. *American Educational Research Journal, 52*(6), 1160-1184. https://doi.org/10.3102/0002831215596413

Preiss, D. D., Cosmelli, D., & Kaufman, J. C. (Eds.). (2020). *Creativity and the wandering mind: Spontaneous and controlled cognition.* Elsevier.

Preller, K. H., Herdener, M., Pokorny, T., Planzer, A., Kraehenmann, R., Stämpfli, P., . . . & Vollenweider, F. X. (2017). The fabric of meaning and subjective effects in LSD-induced states depend on serotonin 2A receptor activation. *Current Biology, 27*(3), 451-457.

Pressman, S. D., & Cross, M. P. (2018). Moving beyond a one-size-fits-all view of positive affect in health research. *Current Directions in Psychological Science, 27*(5), 339-344.

Presti, D.E. (2016). *Foundational concepts in neuroscience: A brain-mind odyssey (Norton Series on Interpersonal Neurobiology)* (1st ed.). W.W. Norton.

Price, D. D., Finniss, D. G., & Benedetti, F. (2008). A comprehensive review of the placebo effect. *Annual Review of Psychology, 59*, 565-590.

Price-Blackshear, M. A., Kamble, S. V., Mudhol, V., Sheldon, K. M., & Bettencourt, B. A. (2017). Mindfulness practices moderate the association between intergroup anxiety and outgroup attitudes. *Mindfulness, 8*, 1172-1183.

Prichard, A., Cook, P. G., Spivak, M., Chhibber, R., & Berns, G. S. (2018). Awake fMRI reveals brain regions for novel word detection in dogs. *Frontiers in Neuroscience, 12*, 737.

Prisciandaro, J. J., Schacht, J. P., Prescot, A. P., Brenner, H. M., Renshaw, P. F., Brown, T. R., & Anton, R. F. (2019). Intraindividual changes in brain GABA, glutamate, and glutamine during monitored abstinence from alcohol in treatment-naive individuals with alcohol use disorder. *Addiction Biology*, e12810.

Pritchard, A., Richardson, M., Sheffield, D., & McEwan, K. (2020). The relationship between nature connectedness and eudaimonic well-being: A meta-analysis. *Journal of Happiness Studies, 21*(3), 1145-1167.

Prochaska, J. O., & Norcross, J. C. (2013). *Systems of psychotherapy: A transtheoretical analysis* (8th ed.). Cengage.

Prochaska, J. O., Norcross, J. C., & DiClemente, C. C. (2013). Applying the stages of change. In G. P. Koocher, J. C. Norcross, & B. A. Greene (Eds.), *Psychologists' desk reference* (3rd ed., pp. 176-181). Oxford University Press.

Pronk, J., Olthof, T., de Vries, R. E., & Goossens, F. A. (2021). HEXACO personality correlates of adolescents' involvement in bullying situations. *Aggressive Behavior.* https://doi.org/10.1002/ab.21947

Pronk, T. M., Karremans, J. C., Overbeek, G., Vermulst, A. A., & Wigboldus, D. H. J. (2010). What it takes to forgive: When and why executive functioning facilitates forgiveness. *Journal of Personality and Social Psychology, 98*(1), 119-131.

Protzko, J. (2017). Raising IQ among school-aged children: Five meta-analyses and a review of randomized controlled trials. *Developmental Review, 46*, 81-101.

Protzko, J., Aronson, J., & Blair, C. (2013). How to make a young child smarter: Evidence from the Database of Raising Intelligence. *Perspectives on Psychological Science, 8*(1), 25-40.

Provenzo, E. F. (2002). *Teaching, learning, and schooling in American culture: A critical perspective.* Allyn & Bacon.

Przybylski, A. K., Deci, E. L., Rigby, C. S., & Ryan, R. M. (2014). Competence-impeding electronic games and players' aggressive feelings, thoughts, and behaviors. *Journal of Personality and Social Psychology, 106*(3), 441-457. https://doi.org/10.1037/a0034820

Puhl, R. M., Himmelstein, M. S., & Pearl, R. L. (2020). Weight stigma as a psychosocial contributor to obesity. *American Psychologist, 75*(2), 274-289.

Pulver, A., Kiive, E., & Harro, J. (2020). Reward sensitivity, affective neuroscience personality, symptoms of attention-deficit/hyperactivity disorder, and TPH2-703G/T (rs4570625) genotype. *Acta Neuropsychiatrica, 32*(5), 247-256.

Pun, A., Ferera, M., Diesendruck, G., Hamlin, J.K., & Baron, A.S. (2017). Foundations of infants' social group evaluations. *Developmental Science.* https://doi.org/10.1111/desc.12586

Putnam-Farr, E., & Morewedge, C. K. (2020). Which social comparisons influence happiness with unequal pay? *Journal of Experimental Psychology: General.* https://doi-org.proxy.mul.missouri.edu/10.1037/xge0000965

Putt, S. S. (2020, in press). Evolution of the human body with respect to language evolution. In J. Stanlaw (Ed.), *The international encyclopedia of linguistic anthropology.* https://doi.org/10.1002/9781118786093.iela0125

Putz, Á., Palotai, R., Csertő, I., & Bereczkei, T. (2016). Beauty stereotypes in social norm enforcement: The effect of attractiveness on third-party punishment and reward. *Personality and Individual Differences, 88*, 230-235. https://doi.org/10.1016/j.paid.2015.09.025

Pyszczynski, T., Greenberg, J., Solomon, S., Arndt, J., & Schimel, J. (2004). Why do people need self-esteem? A theoretical and empirical review. *Psychological Bulletin, 130*(3), 435-468.

Pyszczynski, T., Lockett, M., Greenberg, J., & Solomon, S. (2020). Terror management theory and the COVID-19 pandemic. *Journal of Humanistic Psychology.* https://doi.org/10.1177%2F0022167820959488

Q

Qi, Z., Beach, S. D., Finn, A. S., Minas, J., Goetz, C., Chan, B., & Gabrieli, J. D. (2017). Native-language N400 and P600 predict dissociable language-learning abilities in adults. *Neuropsychologia, 98*, 177-191. https://doi.org/10.1016/j.neuropsychologia.2016.10.005

Qian, J., Scheer, F. A., Hu, K., & Shea, S. A. (2020). The circadian system modulates the rate of recovery of systolic blood pressure after exercise in humans. *Sleep, 43*(4), zsz253.

Qiu, A., Mori, S., & Miller, M. I. (2015). Diffusion tensor imaging for understanding brain development: Why and how. *Annual Review of Psychology 66*, 853-876.

Qiu, N., Tang, C., Zhai, M., Huang, W., Weng, J., Li, C., . . . Fang, H. (2020). Application of the still-face paradigm in early screening for high-risk autism spectrum disorder in infants and toddlers. *Frontiers in Pediatrics, 8*, 290.

Quadrelli, E., Geangu, E., & Turati, C. (2019). Human action sounds elicit sensorimotor activation early in life. *Cortex, 117*, 323-335.

Quagliato, L. A., Cosci, F., Shader, R. I., Silberman, E. K., Starcevic, V., Balon, R., . . . Freire, R. C. (2019). Selective serotonin reuptake inhibitors and benzodiazepines in panic disorder: A meta-analysis of common side effects in acute treatment. *Journal of Psychopharmacology, 33*(11), 1340-1351.

Quam, S., VanHook, C., Szoko, N., Passarello, A., Miller, E., & Culyba, A. J. (2020). Racial identity, masculinities, and violence exposure: Perspectives from male adolescents in marginalized neighborhoods. *Journal of Adolescent Health, 67*(5), 638-644.

Quesque, F., & Rossetti, Y. (2020). What do theory-of-mind tasks actually measure? Theory and practice. *Perspectives on Psychological Science, 15*(2), 384-396.

Quilty, L. C., DeYoung, C. G., Oakman, J. M., & Bagby, R. M. (2014). Extraversion and behavioral activation: Integrating the components of approach. *Journal of Personality Assessment, 96*(1), 87-94. https://doi.org/10.1080/00223891.2013.834440

Qureshi, A. (2020). Cultural competence in psychotherapy. In M. Schouler-Ocak & M. C. Kastrup (Eds.), *Intercultural psychotherapy* (pp. 119-130). Springer.

R

Raabe, T., & Beelmann, A. (2011). Development of ethnic, racial, and national prejudice in childhood and adolescence: A multinational meta-analysis of age differences. *Child Development, 82*(6), 1715-1737. https://doi.org/10.1111/j.1467-8624.2011.01668.x

Raasakka, A., & Kursula, P. (2020). Flexible players within the sheaths: The intrinsically disordered proteins of myelin in health and disease. *Cells, 9*(2), 470. https://doi.org/10.3390/cells9020470

Racine, M., Solé, E., Sánchez-Rodríguez, E., Tomé-Pires, C., Roy, R., Jensen, M. P., . . . Cane, D. (2020). An evaluation of sex differences in patients with chronic pain undergoing an interdisciplinary pain treatment program. *Pain Practice, 20*(1), 62-74.

Racsmány, M., Conway, M. A., & Demeter, G. (2010). Consolidation of episodic memory during sleep: Long-term effects of retrieval practice. *Psychological Science, 21*(1), 80-85.

Rada, R., Strohmaier, S., Drucker, A. M., Eliassen, A. H., & Schernhammer, E. S. (2020). Night shift work surrounding pregnancy and offspring risk of atopic disease. *Plos One, 15*(4), e0231784.

Radke, S., Hoffstaedter, F., Löffler, L., Kogler, L., Schneider, F., Blechert, J., & Derntl, B. (2018). Imaging the up's and down's of emotion regulation in lifetime depression. *Brain Imaging and Behavior, 12*(1), 156-167.

Raeder, F., Merz, C. J., Margraf, J., & Zlomuzica, A. (2020). The association between fear extinction, the ability to accomplish exposure and exposure therapy outcome in specific phobia. *Scientific Reports, 10*(1), 1-11.

Raifman, J., Charlton, B. M., Arrington-Sanders, R., Chan, P. A., Rusley, J., Mayer, K. H., . . . McConnell, M. (2020). Sexual orientation and suicide attempt disparities among US adolescents: 2009–2017. *Pediatrics, 145*(3), e20191658.

Raifman, J., & Sherman, S. G. (2018). US guidelines that empower women to prevent HIV with preexposure prophylaxis. *Sexually Transmitted Diseases, 45*(6), e38–e39.

Railo, H., Andersson, E., Kaasinen, V., Laine, T., & Koivisto, M. (2014). Unlike in clinical blindsight patients, unconscious processing of chromatic information depends on early visual cortex in healthy humans. *Brain Stimulation, 7*(3), 415–420.

Raine, A., & Venables, P. H. (2017). Adolescent daytime sleepiness as a risk factor for adult crime. *Journal of Child Psychology and Psychiatry, 58*(6), 728–735.

Raine, A., Venables, P. H., & Williams, M. (1990). Relationships between N1, P300, and contingent negative variation recorded at age 15 and criminal behavior at age 24. *Psychophysiology, 27*(5), 567–575.

Raj, M., Wiltermuth, S. S., & Adams, G. S. (2020). The social costs of forgiving following multiple-victim transgressions. *Journal of Personality and Social Psychology, 119*, 344–366. https://doi.org/10.1037/pspi0000215

Ramírez-Esparza, N., García-Sierra, A., & Kuhl, P. K. (2014). Look who's talking: Speech style and social context in language input to infants are linked to concurrent and future speech development. *Developmental Science, 17*(6), 880–891. https://doi.org/10.1111/desc.12172

Ramaekers, J. G., Hutten, N., Mason, N. L., Dolder, P., Theunissen, E. L., Holze, F., . . . Kuypers, K. P. (2020). A low dose of lysergic acid diethylamide decreases pain perception in healthy volunteers. *Journal of Psychopharmacology.* https://doi.org/10.1177/0269881120940937.

Ramberg, M., Stanley, B., Ystgaard, M., & Mehlum, L. (2014). Depressed suicide attempters with posttraumatic stress disorder. *Archives of Suicide Research.* PMID: 25058681

Ramirez-Esparza, N., Gosling, S. D., Benet-Martinez, V., Potter, J. P., & Pennebaker, J. W. (2006). Do bilinguals have two personalities? A special case of cultural frame switching. *Journal of Research in Personality, 40*(2), 99–120.

Ramos, G., Brookman-Frazee, L., Kodish, T., Rodriguez, A., & Lau, A. S. (2020). Community providers' experiences with evidence-based practices: The role of therapist race/ethnicity. *Cultural Diversity and Ethnic Minority Psychology.* https://doi.org/10.1037/cdp0000357

Ramsey, J. L., Langlois, J. H., Hoss, R. A., Rubenstein, A. J., & Griffin, A. M. (2004). Origins of a stereotype: Categorization of facial attractiveness by 6-month-old infants. *Developmental Science, 7*(2), 201–211.

Rand, D. G., Brescoll, V. L., Everett, J. C., Capraro, V., & Barcelo, H. (2016). Social heuristics and social roles: Intuition favors altruism for women but not for men. *Journal of Experimental Psychology: General, 145*(4), 389–396. https://doi.org/10.1037/xge0000154

Randler, C., Schredl, M., Göritz, A.S. (2017). Chronotype, sleep behavior, and the big five personality factors. *SAGE Open, 7*(3).

Rankin, K., Walsh, L. C., & Sweeny, K. (2019). A better distraction: Exploring the benefits of flow during uncertain waiting periods. *Emotion, 19*(5), 818–828. https://doi.org/10.1037/emo0000479

Rano, S., Manzura, E., & Nargiza, R. (2020). The relation of linguistic and pragmatic factors in the speech realization of grammatical meaning. *Journal of Critical Reviews, 7*(9), 232–239.

Rapaport, D. (1967). On the psychoanalytic theory of thinking. In M. M. Gill (Ed.), *The collected papers of David Rapaport.* Basic.

Rapaport, S. (1994, November 28). Interview. *U.S. News and World Report,* 94.

Rappaport, L. M., Hunter, M. D., Russell, J. J., Pinard, G., Bleau, P., & Moskowitz, D. S. (2020). Emotional and interpersonal mechanisms in community SSRI treatment of social anxiety disorder. *Journal of Psychiatry and Neuroscience, 45*(6), e190164–e190164.

Rasga, C., Quelhas, A. C., & Byrne, R. M. (2017). How children with autism reason about other's intentions: False-belief and counterfactual inferences. *Journal of Autism and Developmental Disorders, 47*(6), 1806–1817.

Rasmussen, K. R., Stackhouse, M., Boon, S. D., Comstock, K., & Ross, R. (2019). Meta-analytic connections between forgiveness and health: The moderating effects of forgiveness-related distinctions. *Psychology & Health, 34*(5), 515–534.

Rasmussen, R., & Yonehara, K. (2020). Contributions of retinal direction selectivity to central visual processing. *Current Biology, 30*(15), R897–R903.

Rasskazova, E., Ivanova, T., & Sheldon, K. (2016). Comparing the effects of low-level and high-level worker need-satisfaction: A synthesis of the self-determination and Maslow need theories. *Motivation And Emotion, 40*(4), 541–555. https://doi.org/10.1007/s11031-016-9557-7

Rathbone, C. J., O'Connor, A. R., & Moulin, C. J. (2017). The tracks of my years: Personal significance contributes to the reminiscence bump. *Memory & Cognition, 45*(1), 137–150.

Rauch, S. A., King, A., Kim, H. M., Powell, C., Rajaram, N., Venners, M., . . . Liberzon, I. (2020). Cortisol awakening response in PTSD treatment: Predictor or mechanism of change. *Psychoneuroendocrinology,* 118, 104714.

Raufer, S., Idoff, C., Zosuls, A., Marino, G., Blanke, N., Bigio, I. J., . . . Nakajima, H. H. (2020). Anatomy of the human osseous spiral lamina and cochlear partition bridge: Relevance for cochlear partition motion. *Journal of the Association for Research in Otolaryngology, 21,* 171–182.

Rawlins, W. K., & Russell, L. D. (2013). Friendship, positive being-with-others, and the edifying practices of storytelling and dialogue. In M. Hojjat & D. Cramer (Eds.), *Positive psychology and love.* Oxford University Press.

Ray, J. V., Frick, P. J., Thornton, L. C., Wall Myers, T. D., Steinberg, L., & Cauffman, E. (2017). Callous-unemotional traits predict self-reported offending in adolescent boys: The mediating role of delinquent peers and the moderating role of parenting practices. *Developmental Psychology, 53*(2), 319–328.

Raynor, H. A., Jeffrey, R. W., Phelan, S., Hill, J. O., & Wing, R. R. (2005). Amount of food groups variety consumed in the diet and long-term weight loss maintenance. *Obesity Research, 13*(5), 883–890.

Raza, K. (2020). Artificial intelligence against COVID-19: A meta-analysis of current research. In A. E. Hassanien, N. Dey, & S. Elghamrawy (Eds.), *Big data analytics and artificial intelligence against COVID-19: Innovation vision and approach. Studies in Big Data* (vol. 78, pp. 165–176). Springer.

Read, J., Kirsch, I., & McGrath, L. (2020). Electroconvulsive therapy for depression: A review of the quality of ECT versus sham ECT trials and meta-analyses. *Ethical Human Psychology and Psychiatry.* https://doi.org/10.1891/EHPP-D-19-00014

Rebello, C. J., Kirwan, J. P., & Greenway, F. L. (2020). Obesity, the most common comorbidity in SARS-CoV-2: Is leptin the link? *International Journal of Obesity, 44,* 1810–1817.

Reber, A. S., & Alcock, J. E. (2020). Searching for the impossible: Parapsychology's elusive quest. *American Psychologist. 75*(3), 391–399.

Rebok, G. W., & others. (2014). Ten-year effects of the advanced cognitive training for independent and vital elderly cognitive training trial on cognition and everyday functioning in older adults. *Journal of the American Geriatrics Society, 62*(1), 16–24.

Reddy, L. A., Alperin, A., & Glover, T. A. (2020). A critical review of the professional development literature for paraprofessionals supporting students with externalizing behavior disorders. *Psychology in the Schools.* https://doi.org/10.1002/pits.22381

Rehman, F., Munawar, A., Iftikhar, A., Qasim, A., Hassan, J., Samiullah, F., . . . Qasim, N. (2020). Design and development of AI-based mirror neurons agent towards emotion and empathy. *International Journal of Advanced Computer Science and Applications, 11*(3), 386–395.

Reichardt, C. S. (2009). Quasi-experimental designs. In R. E. Millsap & A. Maydeu-Olivares (Eds.), *The Sage handbook of quantitative methods* (pp. 46–71). Sage.

Reichenbach, A., & Bringmann, A. (2020). Glia of the human retina. *Glia, 68*(4), 768–796.

Reid, V. M., Kaduk, K., & Lunn, J. (2019). Links between action perception and action production in 10-week-old infants. *Neuropsychologia, 126,* 69–74.

Reilly, E. B., & Gunnar, M. R. (2019). Neglect, HPA axis reactivity, and development. *International Journal of Developmental Neuroscience, 78,* 100–108.

Reiman, E. M., Fusselman, M. J., Fox, P. T., & Raichle, M. E. (1989). Neuroanatomical correlates of anticipatory anxiety. *Science, 243*(4894 Pt. 1), 1071–1074.

Reinders, M. J., Huitink, M., Dijkstra, S. C., Maaskant, A. J., & Heijnen, J. (2017). Menu-engineering in restaurants-adapting portion sizes on plates to enhance vegetable consumption: a real-life experiment. *International Journal of Behavioral Nutrition and Physical Activity, 14*(1), 41.

Reinoso-Suárez, F., De Andrés, I., Rodrigo-Angulo, M. L., De la Roza, C., Nuñez, A., & Garzón, M. (2020). The anatomy of dreaming, and REM sleep. *European Journal of Anatomy, 3*(3), 163–175.

Reis, H. T., & Lee, K. Y. (2016). Promise, peril, and perspective: Addressing concerns about reproducibility in social-personality psychology. *Journal of Experimental Social Psychology, 66,* 148–152.

Reitz, A. K., Motti-Stefanidi, F., & Asendorpf, J. B. (2016). Me, us, and them: Testing sociometer theory in a socially diverse real-life context. *Journal of Personality and Social Psychology, 110*(6), 908–920. https://doi.org/10.1037/pspp0000073

Rendell, P. G., & Craik, F. I. M. (2000). Virtual week and actual week: Age-related differences in prospective memory. *Applied Cognitive Psychology, 14*(7), S43–S62.

Rescorla, R. A. (1966). Predictability and number of pairings in Pavlovian fear conditioning. *Psychonomic Science, 4*(11), 383–384.

Rescorla, R. A. (1988). Pavlovian conditioning: It's not what you think it is. *American Psychologist, 43*(3), 151–160.

Rescorla, R. A. (2003). Contemporary study of Pavlovian conditioning. *Spanish Journal of Psychology, 6*(2), 185–195.

Rescorla, R. A. (2005). Spontaneous recovery of excitation but not inhibition. *Journal of Experimental Psychology: Animal Behavior Processes, 31*(3), 277–288.

Rescorla, R. A. (2009). A theory of Pavlovian conditioning: Variations in the effectiveness of reinforcement and non-reinforcement. In D. Shanks (Ed.), *Psychology of learning.* Sage.

Rescorla, R. A., & Wagner, A. R. (2009). A theory of attention: Variations in the associability of stimuli with reinforcement. In D. Shanks (Ed.), *Psychology of learning.* Sage.

Resnick, B. (2018, June 13). The most famous psychological studies are often wrong. Textbooks need to catch up. Vox.com. https://www.vox.com/2018/6/13/17449118/stanford-prison-experiment-fraud-psychology-replication

Revdal, E., Arntsen, V., Doan, T. P., Kvello-Alme, M., Kvistad, K. A., Bråthen, G., & Brodtkorb, E. (2020). Experiential seizures related to the hippocampal-parahippocampal spatial representation system. *Epilepsy & Behavior Reports,* 14, 100386.

Reyes-Mendez, M. E., Osuna-López, F., Herrera-Zamora, J. M., Navarro-Polanco, R. A., Moreno-Galindo, E. G., & Alamilla, J. (2020). Functional pre- and postsynaptic changes between the retinohypothalamic tract and suprachiasmatic nucleus during rat postnatal development. *Journal of Biological Rhythms, 35*(1), 28–44.

Reynolds, C. J., Smith, S. M., & Conway, P. (2020). Intrinsic religiosity attenuates the negative relationship between social disconnectedness and meaning in life. *Psychology of Religion and Spirituality.* https://doi.org/10.1037/rel0000318

Reynolds, G. D., & Richards, J. E. (2019). Infant visual attention and stimulus repetition effects on object recognition. *Child Development, 90*(4), 1027–1042.

Reznik, S. J., Nusslock, R., Pornpattananangkul, N., Abramson, L. Y., Coan, J. A., & Harmon-Jones, E.

(2017). Laboratory-induced learned helplessness attenuates approach motivation as indexed by posterior versus frontal theta activity. *Cognitive, Affective, & Behavioral Neuroscience, 17,* 904–916.

Richardson, H., & Saxe, R. (2020). Development of predictive responses in theory of mind brain regions. *Developmental Science, 23*(1), e12863.

Riddle, J., Hwang, K., Cellier, D., Dhanani, S., & D'Esposito, M. (2019). Causal evidence for the role of neuronal oscillations in top–down and bottom–up attention. *Journal of Cognitive Neuroscience, 31*(5), 768–779.

Rief, W., & Martin, A. (2014). How to use the new DSM-5 somatic symptom disorder diagnosis in research and practice: a critical evaluation and a proposal for modifications. *Annual Review of Clinical Psychology,* 10, 339–367.

Rifkin, W. J., Kantar, R. S., Ali-Khan, S., Plana, N. M., Diaz-Siso, J. R., Tsakiris, M., & Rodriguez, E. D. (2018). DDS facial disfigurement and identity: A review of the literature and implications for facial transplantation. *AMA Journal of Ethics, 20,* 309–323.

Riggle, E. B., Rostosky, S. S., Black, W. W., & Rosenkrantz, D. E. (2017). Outness, concealment, and authenticity: Associations with LGB individuals' psychological distress and well-being. *Psychology of Sexual Orientation and Gender Diversity, 4*(1), 54–62. https://doi.org/10.1037/sgd0000202

Riley, T. N., DeLaney, E., Brown, D., Lozada, F. T., Williams, C. D., Dick, D. M., & Spit For Science Working Group. (2020). The associations between African American emerging adults' racial discrimination and civic engagement via emotion regulation. *Cultural Diversity and Ethnic Minority Psychology.* https://doi.org/10.1037/cdp0000335

Rinaldi, L., & Marelli, M. (2020). The use of number words in natural language obeys Weber's law. *Journal of Experimental Psychology: General, 149*(7), 1215–1230.

Rindermann, H., Becker, D., & Coyle, T. R. (2020). Survey of expert opinion on intelligence: Intelligence research, experts' background, controversial issues, and the media. *Intelligence, 78,* 101406.

Rindermann, H., & Thompson, J. (2013). Ability rise in NAEP and narrowing ethnic gaps? *Intelligence, 41*(6), 821–831.

Rips, L. J. (2011). Causation from perception. *Perspectives on Psychological Science, 6*(1), 77–97.

Riske, L., Thomas, R. K., Baker, G. B., & Dursun, S. M. (2017). Lactate in the brain: An update on its relevance to brain energy, neurons, glia and panic disorder. *Therapeutic Advances in Psychopharmacology, 7*(2), 85–89.

Risley, T. R., & Hart, B. (2006). Promoting early language development. In N. F. Watt, C. Ayoub, R. H. Bradley, J. E. Puma, & W. A. LeBoeuf (Eds.), *The crisis in youth mental health: Critical issues and effective programs, vol. 4: Early intervention programs and policies* (pp. 83–88). Praeger.

Ritskes, R., Ritskes-Hoitinga, M., Stodkilde-Jorgensen, H., Baerentsen, K., & Hartman, T. (2003). MRI scanning during Zen meditation: The picture of enlightenment? *Constructivism in the Human Sciences, 8*(1), 85–90.

Ritter, D., & Elsea, M. (2005). Hot sauce, toy guns, and graffiti: A critical account of current laboratory aggression paradigms. *Aggressive Behavior, 31*(5), 407–419.

Rivas-Drake, D., Syed, M., Umaña-Taylor, A., Markstrom, C., French, S., Schwartz, S. J., & Lee, R. (2014). Feeling good, happy, and proud: A meta-analysis of positive ethnic–racial affect and adjustment. *Child Development, 85*(1), 77–102.

Robbins, B. D., Kamens, S. R., & Elkins, D. N. (2017). DSM-5 reform efforts by the Society for Humanistic Psychology. *Journal of Humanistic Psychology, 57,* 602–624.

Roberto, A. J., Shafer, M. S., & Marmo, J. (2014). Predicting substance-abuse treatment providers' communication with clients about medication assisted treatment: A test of the theories of reasoned action and planned behavior. *Journal of Substance Abuse Treatment, 47*(5), 307–313. https://doi.org/10.1016/j.jsat.2014.06.002

Roberts, B. W., Donnellan, M. B., & Hill, P. L. (2013). Personality trait development in adulthood: Findings and implications. In H. M. Tennen, J. M. Suls, & I. B. Weiner (Eds.), *Handbook of psychology* (2nd ed., vol. 5, pp. 183–196). Wiley.

Roberts, B. W., Wood, D., & Caspi, A. (2008). The development of personality traits in adulthood. In O. P. John, R. W. Robins, & L. A. Pervin (Eds.), *Handbook of personality: Theory and research, 3rd ed* (pp. 375–398). Guilford Press.

Roberts, K., Miller, M., & Azrael, D. (2018). Honor-related suicide in the United States: A study of national violent death reporting system data. *Archives of Suicide Research.* https://doi.org/10.1080/13811118.2017.1411299

Roberts, S. O., Bareket-Shavit, C., Dollins, F. A., Goldie, P. D., & Mortenson, E. (2020, in press). Racial inequality in psychological research: Trends of the past and recommendations for the future. *Perspectives on Psychological Science.* https://doi.org/10.1177%2F1745691620927709.

Robertson, C. E., & Baron-Cohen, S. (2017). Sensory perception in autism. *Nature Reviews Neuroscience, 18,* 671–684

Robertson, D. A., Savva, G. M., King-Kallimanis, B. L., & Kenny, R. A. (2015). Negative perceptions of aging and decline in walking speed: A self-fulfilling prophecy. *Plos ONE, 10*(4).

Robinson, E., Haynes, A., Sutin, A., & Daly, M. (2020). Self-perception of overweight and obesity: A review of mental and physical health outcomes. *Obesity Science & Practice, 6*(5), 552–561.

Rodrigue, K. M., & Kennedy, K. M. (2011). The cognitive consequences of structural changes to the aging brain. In K. W. Schaie & S. L. Willis (Eds.), *Handbook of the psychology of aging, 7th ed.* (pp. 73–92). Elsevier.

Roelke, A., & Hofmann, M. J. (2020). Functional connectivity of the left inferior frontal gyrus during semantic priming. *Neuroscience Letters, 735,* 135236.

Rogers, C. R. (1961). *On becoming a person.* Houghton Mifflin.

Rogers, C. R. (1980). *A way of being.* Houghton Mifflin.

Rogers, G., & others. (2009). The harmful health effects of recreational ecstasy: A systematic review of observational evidence. *Health Technology Assessment, 13*(6), 1–315.

Rognum, I. J., & others. (2014). Serotonin metabolites in the cerebrospinal fluid in sudden infant death syndrome. *Journal of Neuropathology and Experimental Neurology, 73*(2), 115–122.

Roh, E., Song, D., & Kim, M. S. (2016). Emerging role of the brain in the homeostatic regulation of energy and glucose metabolism. *Experimental Molecular Medicine, 48,* e216. https://doi.org/10.1038/emm.2016.4

Rohrer, J. M., Egloff, B., & Schmukle, S. C. (2015). Examining the effects of birth order on personality. *PNAS Proceedings of The National Academy of Sciences of The United States of America, 112*(46), 14224–14229. https://doi.org/10.1073/pnas.1506451112

Romano, A., & Bailliet, B. (2017). Reciprocity outperforms conformity to promote cooperation. *Psychological Science.* https://doi.org/10.1177/0956797617714828.

Romano, M., Ma, R., Moscovitch, M., & Moscovitch, D. A. (2020). Autobiographical memory bias. In J. S. Abramowitz & S. M. Blakey (Eds.), *Clinical handbook of fear and anxiety: Maintenance processes and treatment mechanisms* (p. 183–202). American Psychological Association. https://doi.org/10.1037/0000150-011

Romeo, A. N., & Običan, S. G. (2020). Teratogen update: Antithyroid medications. *Birth Defects Research, 112*(15), 1150–1170.

Romero, A., Piña-Watson, B., Stevens, A. K., Schwartz, S. J., Unger, J. B., Zamboanga, B. L., . . . Baezconde-Garbanati, L. (2020). Disentangling relationships between bicultural stress and mental well-being among Latinx immigrant adolescents. *Journal of Consulting and Clinical Psychology, 88*(2), 149–159.

Romijnders, K. A., Wilkerson, J. M., Crutzen, R., Kok, G., Bauldry, J., & Lawler, S. M. (2017). Strengthening social ties to increase confidence and self-esteem among sexual and gender minority youth. *Health Promotion Practice, 18*(3), 341–347.

Roncancio, A. M., Ward, K. K., & Fernandez, M. E. (2013). Understanding cervical cancer screening intentions among Latinas using an expanded theory of planned behavior model. *Behavioral Medicine, 39*(3), 66–72. https://doi.org/10.1080/08964289.2013.799452

Roncero, M., Perpina, C., & Garcia-Soriano, G. (2011). Study of obsessive compulsive beliefs: Relationship with eating disorders. *Behavioral and Cognitive Psychotherapy, 39*(4), 457–470.

Roozendaal, B., & Mirone, G. (2020). Opposite effects of noradrenergic and glucocorticoid activation on accuracy of an episodic-like memory. *Psychoneuroendocrinology, 114,* 104588.

Rosano, C., Guralnik, J., Pahor, M., Glynn, N. W., Newman, A. B., Ibrahim, T. S., et al. (2017). Hippocampal response to a 24-month physical activity intervention in sedentary older adults. *The American Journal of Geriatric Psychiatry, 25*(3), 209–217.

Rose, N. S., & Craik, F. I. (2012). A processing approach to the working memory/long-term memory distinction: Evidence from the levels-of-processing span task. *Journal of Experimental Psychology: Learning, Memory, and Cognition, 38*(4), 1019–1029.

Rose, R. J., Koskenvuo, M., Kaprio, J., Sarna, S., & Langinvainio, H. (1988). Shared genes, shared experiences, and similarity of personality: Data from 14,228 adult Finnish co-twins. *Journal of Personality and Social Psychology, 54*(1), 161–171.

Rosen, C. S. (2000). Is the sequencing of change processes by stage consistent across health problems? A meta-analysis. *Health Psychology, 19*(6), 593–604. https://doi.org/10.1037/0278-6133.19.6.593

Rosen, K., & Garety, P. (2005). Predicting recovery from schizophrenia: A retrospective comparison of characteristics at onset of people with single and multiple episodes. *Schizophrenia Bulletin, 31*(3), 735–750.

Rosen, M. L., Sheridan, M. A., Sambrook, K. A., Dennison, M. J., Jenness, J. L., Askren, M. K., et al. (2017). Salience network response to changes in emotional expressions of others is heightened during early adolescence: Relevance for social functioning. *Developmental Science.*

Rosenbaum, R. S., & others. (2005). The case of K.C.: Contributions of a memory-impaired person to memory theory. *Neuropsychologia, 43*(7), 989–1021.

Rosengren, A., Smyth, A., Rangarajan, S., Ramasundarahettige, C., Bangdiwala, S. I., AlHabib, K. F., . . . Gupta, R. (2019). Socioeconomic status and risk of cardiovascular disease in 20 low-income, middle-income, and high-income countries: The Prospective Urban Rural Epidemiologic (PURE) study. *The Lancet Global Health, 7*(6), e748–e760.

Rosenhan, D. L. (1973). On being sane in insane places. *Science, 179*(4070), 250–258.

Rosental, B., Appel, M. Y., Yossef, R., Hadad, U., Brusilovsky, M., & Porgador, A. (2012). The effect of chemotherapy/radiotherapy on cancerous pattern recognition by NK cells. *Current Medicinal Chemistry, 19*(12), 1780–1791.

Rosenthal, R. (1966). *Experimenter effects in behavioral research.* Appleton-Century-Crofts.

Rosenthal, R. (1966). *Experimenter effects in behavioral research.* Appleton-Century-Crofts.

Rosenthal, R., & Jacobson, L. (1968). *Pygmalion in the classroom: Teacher expectation and student intellectual development.* Holt, Rinehart & Winston.

Rosnow, R. L., & Rosenthal, R. (2012). *Beginning behavioral research* (7th ed.). Prentice-Hall.

Ross, C. A., & Ness, L. (2010). Symptom patterns in dissociative identity disorder patients and the general population. *Journal of Trauma and Dissociation, 11*(4), 458–468.

Ross, C. A., & others. (2008). A cross-cultural test of the trauma model of dissociation. *Journal of Trauma & Dissociation, 9*(1), 35–49.

Ross, J., Hutchison, J., & Cunningham, S. J. (2020). The me in memory: The role of the self in autobiographical memory development. *Child Development, 91*(2), e299–e314.

Roten, R. G. (2007). DSM-IV and the taxonomy of roles: How can the taxonomy of roles complement the DSM-IV to create a more holistic diagnostic tool? *The Arts in Psychotherapy, 34*(1), 53–68.

Roth, Y., Tendler, A., Arikan, M. K., Vidrine, R., Kent, D., Muir, O., . . . Tolin, K. (2020). Real-world efficacy of deep TMS for obsessive-compulsive disorder: Post-marketing data collected from twenty-two clinical sites. *Journal of Psychiatric Research*. https://doi.org/10.1016/j.jpsychires.2020.11.009

Rothbaum, F., Kakinuma, M., Nagaoka, R., & Azuma, H. (2007). Attachment and AMAE: Parent–child closeness in the United States and Japan. *Journal of Cross-Cultural Psychology, 38*(4), 465–486.

Rothbaum, F., Weisz, J., Pott, M., Miyake, K., & Morelli, G. (2000). Attachment and culture: Security in the United States and Japan. *American Psychologist, 55*(10), 1093–1104.

Rothmore, P., Aylward, P., & Karnon, J. (2015). The implementation of ergonomics advice and the stage of change approach. *Applied Ergonomics, 51*, 370–376. https://doi.org/10.1016/j.apergo.2015.06.013

Roy, A., & Pompili, M. (2016). Suicide risk in schizophrenia. In D. Wasserman (Ed.), *Suicide: An Unnecessary Death, 2nd edition.* (pp. 105–112). Oxford University Press.

Rozenblat, V., Ong, D., Fuller-Tyszkiewicz, M., Akkermann, K., Collier, D., Engels, R. C., et al. (2017). A systematic review and secondary data analysis of the interactions between the serotonin transporter 5-HTTLPR polymorphism and environmental and psychological factors in eating disorders. *Journal of Psychiatric Research, 84*, 62–72.

Rozgonjuk, D., Sindermann, C., Elhai, J. D., & Montag, C. (2020). Individual differences in fear of missing out (FoMO): Age, gender, and the big five personality trait domains, facets, and items. *Personality and Individual Differences*, 110546.

Rozin, P. (2020). Why we know so little about the psychology of eating in humans. In H.L. Meisleman, (Ed.), *Handbook of Eating and Drinking: Interdisciplinary Perspectives,* (pp. 1557–1576). Springer.

Rozin, P., Kabnick, K., Pete, E., Fischler, C., & Shields, C. (2003). The ecology of eating: Smaller portion sizes in France than in the United States help explain the French paradox. *Psychological Science, 14*(5), 450–454.

Rozin, P., Scott, S., Dingley, M., Urbanek, J. K., Jiang, H., & Kaltenbach, M. (2011). Nudge to nobesity I: Minor changes in accessibility decrease food intake. *Judgment and Decision Making, 6*(4), 323–332.

Ruderfer, D. M., Walsh, C. G., Aguirre, M. W., Tanigawa, Y., Ribeiro, J. D., Franklin, J. C., & Rivas, M. A. (2020). Significant shared heritability underlies suicide attempt and clinically predicted probability of attempting suicide. *Molecular Psychiatry, 25*(10), 2422–2430.

Rudy, D., & Grusec, J. E. (2006). Authoritarian parenting in individualist and collectivist groups: Associations with maternal emotion and cognition and children's self-esteem. *Journal of Family Psychology, 20*(1), 68–78.

Rueger, S. Y., & George, R. (2017). Indirect effects of attributional style for positive events on depressive symptoms through self-esteem during early adolescence. *Journal of Youth and Adolescence, 46*(4), 701–708.

Ruiz, J. M., & Revenson, T. A. (2020). Behavioral medicine in the COVID-19 era: Dawn of the golden age. *Annals of Behavioral Medicine, 54*, 541–543. https://doi.org/10.1093/abm/kaaa057

Rusbult, C. E., Agnew, C. R., & Arriaga, X. B. (2012). The investment model of commitment processes. In P. A. M van Lange, A. W. Kruglanski, & E. T. Higgins (Eds.), *Handbook of theories of social psychology*. Sage.

Ruth, K. S., Day, F., Tyrrell, J., Thompson, D. J., Wood, A. R., Mahajan, A., . . . Beaumont, R. N. (2020). Using human genetics to understand the disease impacts of testosterone in men and women. *Nature Medicine, 26* (2), 252–258. https://doi.org/10.1038/s41591-020-0751-5

Ruthig, J. C., & Holfeld, B. (2016). Positive thinking and social perceptions of a male vs. female peer's cancer experience. *The Journal of Social Psychology, 156*(2),154–167. https://doi.org/10.1080/00224545.2015.1052361

Ruvalcaba, Y., & Eaton, A. A. (2020). Nonconsensual pornography among US adults: A sexual scripts framework on victimization, perpetration, and health correlates for women and men. *Psychology of Violence, 10*(1), 68.

Ruybal, A. L., & Siegel, J. T. (2019). Attribution theory and reducing stigma toward women with postpartum depression: Examining the role of perceptions of stability. *Stigma and Health, 4*(3), 320.

Ruzich, E., Allison, C., Smith, P., Ring, H., Auyeung, B., & Baron-Cohen, S. (2017). The Autism-Spectrum Quotient in siblings of people with Autism. *Autism Research, 10*(2), 289–297.

Ryan, E. P., & Oquendo, M. A. (2020). Suicide risk assessment and prevention: Challenges and opportunities. *Focus, 18*(2), 88–99.

Ryan, R. M., & Deci, E. L. (2011). A self-determination theory perspective on social, institutional, cultural, and economic supports for autonomy and their importance for well-being. In V. I. Chirkov, R. M. Ryan, & K. M. Sheldon (Eds.), *Human autonomy in cross-cultural context: Perspectives on the psychology of agency, freedom, and well-being* (pp. 45–64). Springer Science + Business Media.

Ryan, R. M., & Deci, E. L. (2017). *Self-determination theory: Basic psychological needs in motivation, development, and wellness.* Plenum Press.

Ryan, R. M., & Deci, E. L. (2020). Intrinsic and extrinsic motivation from a self-determination theory perspective: Definitions, theory, practices, and future directions. *Contemporary Educational Psychology, 61*, 101860.

Ryan, R. M., Huta, V., & Deci, E. L. (2008). Living well: A self-determination theory perspective on eudaimonia. *Journal of Happiness Studies, 9*(1), 139–170.

Rydberg, J. A. (2017). Research and clinical issues in trauma and dissociation: Ethical and logical fallacies, myths, misreports, and misrepresentations. *European Journal of Trauma & Dissociation, 1*, 89–99.

Rymer, R. (1993). *Genie.* HarperCollins.

Rzesnitzek, L., Hariz, M., & Krauss, J. K. (2020). Psychosurgery in the history of stereotactic functional neurosurgery. *Stereotactic and Functional Neurosurgery, 98*(4), 241–247.

S

Sacks, O. (2006, June 19). Stereo Sue. *New Yorker*, 64–73.

Sadek, J. (2020) *Clinician's guide to psychopharmacology.* Springer.

Sager, Z. S., Wachen, J. S., Naik, A. D., & Moye, J. (2020). Post-traumatic stress disorder symptoms from multiple stressors predict chronic pain in cancer survivors. *Journal of Palliative Medicine, 23.* https://doi.org/10.1089/jpm.2019.0458

Sailor, K. A., Schinder, A. F., & Lledo, P. M. (2017). Adult neurogenesis beyond the niche: its potential for driving brain plasticity. *Current Opinion in Neurobiology, 42*, 111–117.

Saito, K. (2017). Effects of sound, vocabulary, and grammar learning aptitude on adult second language speech attainment in foreign language classrooms. *Language Learning.* https://doi.org/10.1111/lang.12244

Sakai, J. (2020). Core concept: How synaptic pruning shapes neural wiring during development and, possibly, in disease. *Proceedings of the National Academy of Sciences, 117*(28), 16096–16099.

Sakai, N. (2020). Top-down processing in food perception: Beyond the multisensory processing. *Acoustical Science and Technology, 41*(1), 182–188.

Sakaluk, J. K., Todd, L. M., Milhausen, R., Lachowsky, N. J., & Undergraduate Research Group in Sexuality (URGiS). (2014). Dominant heterosexual scripts in emerging adulthood: Conceptualization and measurement. *Journal of Sexual Research, 51*(5), 516–531.

Sala, G., & Gobet, F. (2020). Working memory training in typically developing children: A multilevel meta-analysis. *Psychonomic Bulletin & Review, 27*, 423–434.

Sala, M., Rochefort, C., Lui, P. P., & Baldwin, A. S. (2020). Trait mindfulness and health behaviours: A meta-analysis. *Health Psychology Review, 14*(3), 345–393.

Sale, M.V., Reid, L.B., Cocchi, L., Pagnozzi, A.M., Rose, S.E., Mattingley, J.B. (2017). Brain changes following four weeks of unimanual motor training: Evidence from behavior, neural stimulation, cortical thickness, and functional MRI. *Human Brain Mapping, 38*, 4773–4787.

Sales, J., & Krause, K. (2017). Schools must include faculty and staff in sexual violence prevention efforts. *Journal of American College Health, 65*(8), 585–587.

Salk, R. H., Hyde, J. S., & Abramson, L. Y. (2017). Gender differences in depression in representative national samples: Meta-analyses of diagnoses and symptoms. *Psychological Bulletin, 143*(8), 783–822.

Salthouse, T. A. (1994). The nature of the influence of speed on adult age differences in cognition. *Developmental Psychology, 30*(2), 240–259.

Salthouse, T. A. (2012). Consequences of age-related cognitive declines. *Annual Review of Psychology* (vol. 63). Annual Reviews.

Salthouse, T. A. (2017). Contributions of the individual differences approach to cognitive aging. *The Journals of Gerontology: Series B, 72*(1), 7–15.

SAMHSA. (2018). National Survey on Drug Use and Health (NSDUH). Table 2.1B—Tobacco product and alcohol use in lifetime, past year, and past month among persons aged 12 or older, by age group: Percentages, 2017 and 2018. https://www.samhsa.gov/data/sites/default/files/cbhsq-reports/NSDUHDetailedTabs2018R2/NSDUHDetTabsSect2pe2018.htm#tab2-1b

Samudra, P. G., Min, I., Cortina, K. S., & Miller, K. F. (2016). No second chance to make a first impression: The 'thin-slice' effect on instructor ratings and learning outcomes in higher education. *Journal of Educational Measurement, 53*(3), 313–331. https://doi.org/10.1111/jedm.12116

San Martin, A. H., Serrano, J. P., Cambriles, T. D., Arias, E. M. A., Méndez, J. M., del Yerro Álvarez, M. J., & Sánchez, M. G. (2020). Sleep characteristics in health workers exposed to the COVID-19 pandemic. *Sleep Medicine.* https://doi.org/10.1016/j.sleep.2020.08.013

Sanchez, D., & Awad, G. H. (2016). Ethnic group differences in racial identity attitudes, perceived discrimination and mental health outcomes in African American, Black Caribbean and Latino Caribbean college students. *International Journal of Culture And Mental Health, 9*(1), 31–43. https://doi.org/10.1080/17542863.2015.1081955

Sand, A., & Nilsson, M. E. (2016). Subliminal or not? Comparing null-hypothesis and Bayesian methods for testing subliminal priming. *Consciousness And Cognition: An International Journal, 44*, 29–40. https://doi.org/10.1016/j.concog.2016.06.012

SantaCruz, K. S., Sonnen, J. A., Pezhouh, M. K., Desrosiers, M. F., Nelson, P. T., & Tyas, S. L. (2011). Alzheimer disease pathology in subjects without dementia in 2 studies of aging: The Nun Study and the Adult Changes in Thought Study. *Journal of Neuropathology and Experimental Neurology, 70*(10), 832–840.

Santelli, J. S., Kantor, L. M., Grilo, S. A., Speizer, I. S., Lindberg, L. D., Heitel, J., et al. (2017). Abstinence-only-until-marriage: An updated review of U.S. policies and programs and their impact. *Journal of Adolescent Health, 61*(3), 273–280. https://doi.org/10.1016/j.jadohealth.2017.05.031

Santtila, P., Sandnabba, N. K., Harlaar, N., Varjonen, M., Alanko, K., & von der Pahlen, B. (2008). Potential for homosexual response is prevalent and genetic. *Biological Psychology, 77*(1), 102–105.

Sapolsky, R. M. (2004). *Why zebras don't get ulcers, 3rd Edition.* Henry Holt.

Sar, V., Akyuz, G., & Dogan, O. (2007). Prevalence of dissociative disorders among women in the general population. *Psychiatry Research, 149*(1–3), 169–176.

Sarkar, U., Ali, S., & Whooley, M. A. (2009). Self-efficacy as a marker of cardiac function and predictor of heart failure hospitalization and mortality in patients with stable coronary heart disease: Findings from the Heart and Soul Study. *Health Psychology, 28*(2), 166–173.

Saroglou, V. (2010). Religiousness as a cultural adaptation of basic traits: A five-factor model perspective. *Personality and Social Psychology Review, 14*(1), 108-125.

Sasser, T. R., Bierman, K. L., Heinrichs, B., & Nix, R. L. (2017). Preschool intervention can promote sustained growth in the executive-function skills of children exhibiting early deficits. *Psychological Science, 28,* 1719-1730.

Sateia, M. J., Buysse, D. J., Krystal, A. D., Neubauer, D. N., & Heald, J. L. (2017). Clinical practice guideline for the pharmacologic treatment of chronic insomnia in adults: An American Academy of Sleep Medicine Clinical Practice Guideline. *Journal of Clinical Sleep Medicine, 13*(2), 307-349.

Saunders, F. W. (1991). *Katherine and Isabel: Mother's light, daughter's journey.* Consulting Psychologists Press.

Savage, J. (2008). The role of exposure to media violence in the etiology of violent behavior: A criminologist weighs in. *American Behavioral Scientist, 51*(8), 1123-1136.

Savage, J., & Yancey, C. (2008). The effects of media violence exposure on criminal aggression: A meta-analysis. *Criminal Justice and Behavior, 35*(6), 772-791.

Savelyev, P. A. (2020). Conscientiousness, extraversion, college education, and longevity of high-ability individuals. *Journal of Human Resources,* 0918-9720R2.

Savin-Williams, R. C. (2015). The new sexual-minority teenager. In J. S. Kaufman & D. A. Powell (Eds.), *The meaning of sexual identity in the 21st century.* Cambridge Scholars.

Savulich, G., Hezemans, F. H., van Ghesel Grothe, S., Dafflon, J., Schulten, N., Brühl, A. B., . . . Robbins, T. W. (2019). Acute anxiety and autonomic arousal induced by CO_2 inhalation impairs prefrontal executive functions in healthy humans. *Translational Psychiatry, 9*(1), 1-10.

Saxe, L., & Ben-Shakhar, G. (1999). Admissibility of polygraph tests: The application of scientific standards post-Daubert. *Psychology, Public Policy and Law, 5*(1), 203-223.

Saxe, R., & Powell, L. J. (2006). It's the thought that counts: Specific brain regions for one component of theory of mind. *Psychological Science, 17*(8), 692-699.

Scammell, T. E., Arrigoni, E., & Lipton, J. O. (2017). Neural circuitry of wakefulness and sleep. *Neuron, 93*(4), 747-765.

Scarr, S. (1984, May). Interview. *Psychology Today, 18*(5), 59-63.

Schachter, S., & Singer, J. E. (1962). Cognitive, social, and physiological determinants of emotional state. *Psychological Review, 69,* 379-399.

Schacter, D. L. (2001). *The seven sins of memory.* Houghton Mifflin.

Schaie, K. W. (1994). The course of adult intellectual development. *American Psychologist, 49*(4), 304-313.

Schaie, K. W. (2007). Generational differences: The age-cohort period model. In J. E. Birren & K. W. Schaie (Eds.), *Encyclopedia of gerontology.* Elsevier.

Schaie, K. W. (2010). Adult intellectual abilities. *Corsini encyclopedia of psychology.* Wiley.

Schaie, K. W. (2012). *Developmental influences on adult intellectual development: The Seattle Longitudinal Study* (2nd ed.). Oxford University Press.

Schakel, L., Veldhuijzen, D. S., Crompvoets, P. I., Bosch, J. A., Cohen, S., van Middendorp, H., . . . Evers, A. W. (2019). Effectiveness of stress-reducing interventions on the response to challenges to the immune system: A meta-analytic review. *Psychotherapy and Psychosomatics, 88*(5), 274-286.

Schalk, G., Kapeller, C., Guger, C., Ogawa, H., Hiroshima, S., Lafer-Sousa, R., et al. (2018). Facephenes and rainbows: Causal evidence for functional and anatomical specificity of face and color processing in the human brain. *Proceedings of the National Academy of Sciences, 114*(46), 12285-12290.

Schank, R., & Abelson, R. (1977). *Scripts, plans, goals, and understanding.* Erlbaum.

Scheer, V., Blanco, C., Olfson, M., Lemogne, C., Airagnes, G., Peyre, H., . . . Hoertel, N. (2020). A comprehensive model of predictors of suicide attempt in individuals with panic disorder: Results from a national 3-year prospective study. *General Hospital Psychiatry, 67,* 127-135.

Schein, S. S., & Langlois, J. H. (2015). Unattractive infant faces elicit negative affect from adults. *Infant Behavior & Development, 38,* 130-134. https://doi.org/10.1016/j.infbeh.2014.12.009

Schein, S. S., Trujillo, L. T., & Langlois, J. H. (2017). Attractiveness bias: A cognitive explanation. *Behavioral and Brain Sciences, 40.* https://doi.org/10.1017/S0140525X16000649

Schepers, J., Gebhardt, C., Bracke, A., Eiffler, I., & und Halbach, O. V. B. (2020). Structural and functional consequences in the amygdala of leptin-deficient mice. *Cell and Tissue Research, 382*(2), 421-426.

Schiff, W. J. (2015). *Nutrition essentials: A personal approach.* McGraw-Hill.

Schiller, M., Ben-Shaanan, T. L., & Rolls, A. (2020). Neuronal regulation of immunity: Why, how and where? *Nature Reviews Immunology.* https://www.nature.com/articles/s41577-020-0387-1

Schimmack, U. (2019). The implicit association test: a method in search of a construct. *Perspectives on Psychological Science.* https://doi.org/10.1177/1745691619863798

Schimmack, U., & Kim, H. (2020). An integrated model of social psychological and personality psychological perspectives on personality and wellbeing. *Journal of Research in Personality, 84,* 103888.

Schirmer, A. (2013). Sex differences in emotion. In J. Armony & P. Vuilleumier (Eds.), *The Cambridge handbook of human affective neuroscience* (pp. 591-610). Cambridge University Press.

Schirmer, A., & Adolphs, R. (2017). Emotion perception from face, voice, and touch: comparisons and convergence. *Trends in Cognitive Sciences, 21,* 216-228. https://doi.org/10.1016/j.tics.2017.01.001

Schkade, D. A., & Kahneman, D. (1998). Does living in California make people happy? A focusing illusion in judgments of life satisfaction. *Psychological Science, 9*(5), 340-346.

Schlam, T. R., Wilson, N. L., Shoda, Y., Mischel, W., & Ayduk, O. (2013). Preschoolers' delay of gratification predicts body mass 30 years later. *Journal of Pediatrics, 162*(1), 90-93.

Schlinger, H. D. (2017). Commentary: Can inner experience be apprehended in high fidelity? Examining brain activation and experience from multiple perspectives. *Frontiers In Psychology, 8.*

Schlinger, H. D. (2017). The importance of analysis in applied behavior analysis. *Behavior Analysis: Research and Practice.* https://doi.org/10.1037/bar0000080

Schmader, T., Hall, W., & Croft, A. (2014). Stereotype threat in intergroup relations. In M. Mikulincer, P. R. Shaver, J. F. Dovidio, & J. A. Simpson (Eds.), *APA handbook of personality and social psychology* (vol. 2). American Psychological Association.

Schmidt, K. T., Schroeder, J. P., Foster, S. L., Squires, K., Smith, B. M., Pitts, E. G., . . . & Weinshenker, D. (2017). Norepinephrine regulates cocaine-primed reinstatement via α1-adrenergic receptors in the medial prefrontal cortex. *Neuropharmacology, 119,* 134-140.

Schmidt, S. N., Sojer, C. A., Hass, J., Kirsch, P., & Mier, D. (2020). fMRI adaptation reveals: The human mirror neuron system discriminates emotional valence. *Cortex, 128,* 270-280.

Schmukle, S. C., Korndörfer, M., & Egloff, B. (2019). No evidence that economic inequality moderates the effect of income on generosity. *Proceedings of the National Academy of Sciences, 116*(20), 9790-9795.

Schneider, D., Zickerick, B., Thönes, S., & Wascher, E. (2020). Encoding, storage, and response preparation—distinct EEG correlates of stimulus and action representations in working memory. *Psychophysiology, 57*(6), e13577

Schneider, K. J. (2002). Humanistic psychotherapy. In M. Hersen & W. H. Sledge. (Eds.), *Encyclopedia of psychotherapy.* Academic.

Schneiderman, I., Kanat-Maymon, Y., Zagoory-Sharon, O., & Feldman, R. (2014). Mutual influences between partners' hormones shape conflict dialog and relationship duration at the initiation of romantic love. *Social Neuroscience, 9*(4), 337-351.

Schoenfeld, E. A., Loving, T. J., Pope, M. T., Huston, T. L., & Štulhofer, A. (2017). Does sex really matter? Examining the connections between spouses' nonsexual behaviors, sexual frequency, sexual satisfaction, and marital satisfaction. *Archives of Sexual Behavior, 46*(2), 489-501.

Schofield, G., Larsson, B., & Ward, E. (2017). Risk, resilience and identity construction in the life narratives of young people leaving residential care. *Child & Family Social Work, 22*(2), 782-791.

Schofield, P. E., Stockler, M.R., Zannino, D., et al. (2016). Hope, optimism and survival in a randomised trial of chemotherapy for metastatic colorectal cancer. *Support Care Cancer, 24,* 401-408. https://doi.org/10.1007/s00520-015-2792-8

Schönbrodt, F. D., Hagemeyer, B., Brandstätter, V., Czikmantori, T., Gröpel, P., Hennecke, M., . . . Kopp, P. M. (2020). Measuring implicit motives with the picture story exercise (PSE): Databases of expert-coded German stories, pictures, and updated picture norms. *Journal of Personality Assessment.* https://doi.org/10.1080/00223891.2020.1726936

Schooler, J. W. (2002). Re-representing consciousness: Dissociations between experience and metaconsciousness. *Trends in Cognitive Sciences, 6*(8), 339-344.

Schooler, J. W., Ambadar, Z., & Bendiksen, M. (1997). A cognitive corroborative case study approach for investigating discovered memories of sexual abuse. In J. D. Read & D. S. Lindsay (Eds.), *Recollections of trauma: Scientific evidence and clinical practice* (pp. 379-387). Plenum Press.

Schooler, J. W., Ariely, D., & Loewenstein, G. (2003). The explicit pursuit and assessment of happiness can be self-defeating. In I. Brocas & J. Carrillo (Eds.), *The psychology of economic decisions.* Oxford University Press.

Schouler-Ocak, M., & Kastrup, M. C. (Eds.). (2020). *Intercultural psychotherapy.* Springer.

Schrantee, A., Tamminga, H. G., Bouziane, C., Bottelier, M. A., Bron, E. E., Mutsaerts, H. J. M., et al. (2016). Age-dependent effects of methylphenidate on the human dopaminergic system in young vs adult patients with attention-deficit/hyperactivity disorder: A randomized clinical trial. *JAMA Psychiatry, 73*(9), 955-962.

Schreiner, M. S., van Schaik, J. E., Sučević, J., Hunnius, S., & Meyer, M. (2020). Let's talk action: Infant-directed speech facilitates infants' action learning. *Developmental Psychology, 56*(9), 1623-1631.

Schröder, T., & Wolf, I. (2017). Modeling multi-level mechanisms of environmental attitudes and behaviours: The example of carsharing in Berlin. *Journal of Environmental Psychology, 52,* 136-148. https://doi.org/10.1016/j.jenvp.2016.03.007

Schroeder, B. L., & Sims, V. K. (2017). Texting as a multidimensional behavior: Individual differences and measurement of texting behaviors. *Psychology of Popular Media Culture.* https://doi.org/10.1037/ppm0000148

Schroeder, S., Tan, C. M., Urlacher, B., & Heitkamp, T. (2020). The role of rural and urban geography and gender in community stigma around mental illness. *Health Education & Behavior,* 1090198120974963.

Schueller, S. M., & Torous, J. (2020). Scaling evidence-based treatments through digital mental health. *American Psychologist, 75*(8), 1093-1104

Schultz, D. P., & Schultz, S. E. (2016). *A history of modern psychology* (11th ed.). Cengage.

Schultz, W., Stauffer, W. R., & Lak, A. (2017). The phasic dopamine signal maturing: from reward via behavioural activation to formal economic utility. *Current Opinion in Neurobiology, 43,* 139-148.

Schulz-Stübner, S., Krings, T., Meister, I. G., Rex, S., Thron, A., & Rossaint, R. (2004). Clinical hypnosis modulates functional magnetic resonance imaging signal intensities and pain perception in a thermal stimulation paradigm. *Regional Anesthesia and Pain Medicine, 29*(6), 549-556.

Schurz, M., Radua, J., Tholen, M. G., Maliske, L., Margulies, D. S., Mars, R. B., . . . Kanske, P. (2020). Toward a hierarchical model of social cognition: A neuroimaging meta-analysis and integrative review of empathy and theory of mind. *Psychological Bulletin.* http://dx.doi.org/10.1037/bul0000303

Schuyler, B. S., Kral, T. R. A., Jacquart, J., Burghy, C. A., Weng, H. Y., Perlman, D. M., . . . Davidson, R. J. (2014). Temporal dynamics of emotional responding: Amygdala recovery predicts emotional traits. *Social Cognitive and Affective Neuroscience, 9*(2), 176–181.

Schwartz, S. (2010). Life goes on in dreams. *Sleep, 33*(1), 15–16.

Scott, C., & Medeiros, M. (2020). Personality and political careers: What personality types are likely to run for office and get elected? *Personality and Individual Differences, 152,* 109600.

Scott, R. M., & Baillargeon, R. (2013). Do infants really expect others to act efficiently? A critical test of the rationality principle. *Psychological Science, 24*(4), 466–474.

Scott-Sheldon, L. A. J., & Johnson, B. T. (2006). Eroticizing creates safer sex: A research synthesis. *Journal of Primary Prevention, 27*(6), 619–640.

Scrimgeour, M. B., Mariotti, E. C., & Blandon, A. Y. (2017). Children's physiological regulation and sbling conflict as correlates of children's conscience development. *Social Development, 26*(2), 329–348.

Scroggs, B., & Vennum, A. (2020). Gender and sexual minority group identification as a process of identity development during emerging adulthood. *Journal of LGBT Youth.* https://doi.org/10.1080/19361653.2020.1722780

Scullin, M. K., Ball, B. H., & Bugg, J. M. (2020). Structural correlates of commission errors in prospective memory. *Cortex, 124,* 44–53.

Scullin, M. K., Gao, C., Fillmore, P., Roberts, R. L., Pruett, N., & Bliwise, D. L. (2019). Rapid eye movement sleep mediates age-related decline in prospective memory consolidation. *Sleep, 42*(6), zsz055.

Seaborn, K., Pennefather, P., & Fels, D. I. (2020). Eudaimonia and hedonia in the design and evaluation of a cooperative game for psychosocial well-being. *Human-Computer Interaction, 35*(4), 289–337.

Seabrook, R. C., Ward, L. M., Cortina, L. M., Giaccardi, S., & Lippman, J. R. (2017). Girl power or powerless girl? Television, sexual scripts, and sexual agency in sexually active young women. *Psychology of Women Quarterly, 41*(2), 240–253. https://doi.org/10.1177/0361684316677028

Seblova, D., Berggren, R., & Lövdén, M. (2020). Education and age-related decline in cognitive performance: Systematic review and meta-analysis of longitudinal cohort studies. *Ageing Research Reviews, 58,* 101005.

Sedgh G, Finer LB, Bankole A, Eilers MA, Singh S. (2015). Adolescent pregnancy, birth, and abortion rates across countries: levels and recent trends. *Journal of Adolescent Health, 56,* 223–30.

Sedikides, C., Meek, R., Alicke, M. D., & Taylor, S. (2014). Behind bars but above the bar: Prisoners' consider themselves more prosocial than non-prisoners. *British Journal of Social Psychology, 53*(2), 396–403.

Seelye, K. Q. (2020, July 19). John Lewis, towering figure of civil rights era, dies at 80. *New York Times,* A1.

Seery, M. D., Holman, E. A., & Silver, R. C. (2010). Whatever does not kill us: Cumulative lifetime adversity, vulnerability, and resilience. *Journal of Personality and Social Psychology, 99*(6), 1025–1041.

Segal, N., & Fraley, R. C. (2016). Broadening the investment model: An intensive longitudinal study on attachment and perceived partner responsiveness in commitment dynamics. *Journal of Social and Personal Relationships, 33*(5), 581–599. https://doi.org/10.1177/0265407515584493

Segall, M. H., Campbell, D. T., & Herskovits, M. J. (1966). *The influence of culture on visual perception.* Bobbs-Merrill.

Sekar, A., Bialas, A. R., de Rivera, H., Davis, A., Hammond, T. R., Kamitaki, N., et al. (2016). Schizophrenia risk from complex variation of complement component 4. *Nature, 530*(7589), 177–183.

Seligman, M. (2016). How positive psychology happened and where it is going. In R. J. Sternberg, S. T. Fiske, D. J. Foss (Eds.). *Scientists making a difference: One hundred eminent behavioral and brain scientists talk about their most important contributions* (pp. 478–480). Cambridge University Press.

Seligman, M. E. P. (1970). On the generality of the laws of learning. *Psychological Review, 77*(5), 406–418.

Seligman, M. E. P. (1975). *Helplessness: On depression, development and death.* Freeman.

Seligman, M. E. P. (1990). *Learned optimism.* Knopf.

Seligman, M. E. P., & Csikszentmihalyi, M. (2000). Positive psychology: An introduction. *American Psychologist, 55*(1), 5–14.

Seligman, M. E. P., & Maier, S. F. (1967). Failure to escape traumatic shock. *Journal of Experimental Psychology, 74*(1), 1–9.

Sellon, J. B., Ghaffari, R., & Freeman, D. M. (2017). Geometric requirements for tectorial membrane traveling waves in the presence of cochlear loads. *Biophysical Journal, 112*(6), 1059–1062.

Sels, L., Ruan, Y., Kuppens, P., Ceulemans, E., & Reis, H. (2020). Actual and perceived emotional similarity in couples' daily lives. *Social Psychological and Personality Science, 11*(2), 266–275.

Selye, H. (1974). *Stress without distress.* Saunders.

Selye, H. (1983). The stress concept: Past, present, and future. In C. I. Cooper (Ed.), *Stress research.* Wiley.

Selzler, A. M., Habash, R., Robson, L., Lenton, E., Goldstein, R., & Brooks, D. (2020). Self-efficacy and health-related quality of life in chronic obstructive pulmonary disease: A meta-analysis. *Patient Education and Counseling, 103*(4), 682–692.

Senan, S., Ali, M. S., Vadivel, R., & Arik, S. (2017). Decentralized event-triggered synchronization of uncertain Markovian jumping neutral-type neural networks with mixed delays. *Neural Networks, 86,* 32–41.

Senaratna, C. V., Perret, J. L., Lodge, C. J., Lowe, A. J., Campbell, B. E., Matheson, M. C., et al. (2017). Prevalence of obstructive sleep apnea in the general population: A systematic review. *Sleep Medicine Reviews, 34,* 70–81.

Senzaki, S., Masuda, T., & Ishii, K. (2014). When is perception top-down and when is it not? Culture, narrative, and attention. *Cognitive Science, 38*(7), 1493–506.

Serafini, K., Shipley, L., & Stewart, D. G. (2016). Motivation and substance use outcomes among adolescents in a school-based intervention. *Addictive Behaviors, 53,* 74–79. https://doi.org/10.1016/j.addbeh.2015.10.004

Serchen, J., Doherty, R., Atiq, O., & Hilden, D. (2020). Racism and health in the United States: A policy statement from the American College of Physicians. *Annals of Internal Medicine, 173*(7), 556–557.

Serdarevic, F., Tiemeier, H., Jansen, P. R., Alemany, S., Xerxa, Y., Neumann, A., . . . Ghassabian, A. (2020). Polygenic risk scores for developmental disorders, neuromotor functioning during infancy, and autistic traits in childhood. *Biological Psychiatry, 87*(2), 132–138.

Sereno, M. I. (2017). Finding and understanding cortical maps using neuroimaging. *The Journal of the Acoustical Society of America, 141*(5), 3558–3558.

Settles, R. E., Fischer, S., Cyders, M. A., Combs, J. L., Gunn, R. L., & Smith, G. T. (2012). Negative urgency: A personality predictor of externalizing behavior characterized by neuroticism, low conscientiousness, and disagreeableness. *Journal of Abnormal Psychology, 121*(1), 160–172.

Sewell, K. K., McGarrity, L. A., & Strassberg, D. S. (2017). Sexual behavior, definitions of sex, and the role of self-partner context among lesbian, gay, and bisexual adults. *The Journal of Sex Research, 54*(7), 825–831.

Seydell-Greenwald, A., Chambers, C. E., Ferrara, K., & Newport, E. L. (2020). What you say versus how you say it: Comparing sentence comprehension and emotional prosody processing using fMRI. *NeuroImage, 209,* 116509.

Seymour, T. L., Seifert, C. M., Shafto, M. G., & Mosmann, A. L. (2000). Using response time measures to assess "guilty knowledge." *Journal of Applied Psychology, 85*(1), 30–37.

Shadel, W. G., Martino, S. C., Setodji, C., Cervone, D., & Witkiewitz, K. (2017). Does self-efficacy causally influence initial smoking cessation? An experimental study. *Addictive Behaviors, 73,* 199–203. https://doi.org/10.1016/j.addbeh.2017.05.018

Shadrina, M., Bondarenko, E. A., & Slominsky, P. A. (2018). Genetics factors in major depression disease. *Frontiers in Psychiatry, 9,* 334.

Shaffer, J. A. (2020). Forethought and intelligence: How conscientiousness, future planning, and general mental ability predict net worth. *Personality and Individual Differences, 159,* 109853.

Shahror, R. A., Linares, G. R., Wang, Y., Hsueh, S. C., Wu, C. C., Chuang, D. M., . . . Chen, K. Y. (2020). Transplantation of mesenchymal stem cells overexpressing fibroblast growth factor 21 facilitates cognitive recovery and enhances neurogenesis in a mouse model of traumatic brain injury. *Journal of Neurotrauma, 37*(1), 14–26.

Shalev, I., Entringer, S., Wadhwa, P. D., Wolkowitz, O. M., Puterman, E., Lin, J., & Epel, E. S. (2013). Stress and telomere biology: A lifespan perspective. *Psychoneuroendocrinology, 38,* 1835–1842. http://dx.doi.org/10.1016/j.psyneuen.2013.03.010

Shalev, N., Brosnan, M. B., & Chechlacz, M. (2020). Right lateralized brain reserve offsets age-related deficits in ignoring distraction. *Cerebral Cortex Communications, 1*(1), tgaa049.

Shamay-Tsoory, S. G., & Abu-Akel, A. (2016). The social salience hypothesis of oxytocin. *Biological Psychiatry, 79,* 194–202. http://dx.doi.org/10.1016/j.biopsych.2015.07.020

Shanahan, L., Steinhoff, A., Bechtiger, L., Murray, A. L., Nivette, A., Hepp, U., . . . Eisner, M. (2020). Emotional distress in young adults during the COVID-19 pandemic: Evidence of risk and resilience from a longitudinal cohort study. *Psychological Medicine.* https://doi.org/10.1017/S003329172000241X

Shankar, A., & Hinds, P. (2017). Perceived Discrimination: Associations With Physical and Cognitive Function in Older Adults. *Health Psychology.* https://doi.org/10.1037/hea0000522

Shapiro-Mendoza, C. K., & others. (2014). Classification system for the Sudden Unexpected Infant Death Case Registry and its application. *Pediatrics, 134*(1), e210–e219.

Shapiro, R. (August 25, 2017). Dwayne 'The Rock' Johnson is wowed by boy who copied his skills to save toddler's life. *Huffington Post.* https://www.huffingtonpost.com/entry/dwayne-johnson-the-rock-boy-save-life-hero_us_599fc670e4b06d67e336ce3b?ncid=tweetlnkush pmg00000041

Sharma, A., Park, S., & Nicolau, J. L. (2020). Testing loss aversion and diminishing sensitivity in review sentiment. *Tourism Management, 77,* 104020.

Shattuck, E. C., & Muehlenbein, M. P. (2020). Religiosity/spirituality and physiological markers of health. *Journal of Religion and Health, 59*(2), 1035–1054.

Shayman, C. S., Peterka, R. J., Gallun, F. J., Oh, Y., Chang, N. Y. N., & Hullar, T. E. (2020). Frequency-dependent integration of auditory and vestibular cues for self-motion perception. *Journal of Neurophysiology, 123*(3), 936–944.

Shea, J. L., Wongt, P. Y., & Chen, Y. (2014). Free testosterone: Clinical utility and important analytical aspects of measurement. *Advances in Clinical Chemistry, 63,* 59–84.

Shedler, J. (2020). Where is the evidence for "evidence-based" therapy? In M. Leuzinger-Bohleber, M. Solms, & S. E. Arnold (Eds.), *Outcome research and the future of psychoanalysis: Clinicians and researchers in dialogue.* Routledge.

Sheeran, P., Wright, C. E., Avishai, A., Villegas, M. E., Lindemans, J. W., Klein, W. M., . . . Ntoumanis, N. (2020). Self-determination theory interventions for health behavior change: Meta-analysis and meta-analytic structural equation modeling of randomized controlled trials. *Journal of Consulting and Clinical Psychology 88*(8), 726–737.

Sheikh, J. I., Leskin, G. A., & Klein, D. F. (2002). Gender differences in panic disorder: Findings from the National Comorbidity Survey. *American Journal of Psychiatry, 159*(1), 55–58.

Sheikh, U. A., Carreiras, M., & Soto, D. (2019). Decoding the meaning of unconsciously processed words using fMRI-based MVPA. *NeuroImage, 191,* 430–440.

Sheldon, K. M. (2002). The self-concordance model of healthy goal-striving: When personal goals correctly represent the person. In E. L. Deci & R. M. Ryan (Eds.), *Handbook of self-determination research* (pp. 65–86). University of Rochester Press.

Sheldon, K. M., & Elliot, A. J. (1998). Not all personal goals are personal: Comparing autonomous and controlled reasons for goals as predictors of effort and attainment. *Personality and Social Psychology Bulletin, 24*(5), 546–557.

Sheldon, K. M., Holliday, G., Titova, L., & Benson, C. (2020). Comparing Holland and self-determination theory measures of career preference as predictors of career choice. *Journal of Career Assessment, 28*(1), 28–42.

Sheldon, K. M., Kasser, T., Houser-Marko, L., Jones, T., & Turban, D. (2005). Doing one's duty: Chronological age, felt autonomy, and subjective well-being. *European Journal of Personality, 19*(2), 97–115.

Sheldon, K. M., & Lyubomirsky, S. (2007). Is it possible to become happier? (And if so, how?). *Social and Personality Psychology Compass, 1*(1), 129–145.

Sheldon, K. M., & Lyubomirsky, S. (2012). The challenge of staying happier: Testing the happiness adaptation model. *Personality and Social Psychology Bulletin, 38*(5), 670–680.

Sheldon, K. M., Osin, E. N., Gordeeva, T. O., Suchkov, D. D., & Sychev, O. A. (2017). Evaluating the dimensionality of self-determination theory's relative autonomy continuum. *Personality and Social Psychology Bulletin, 43*(9), 1215–1238. https://doi.org/10.1177/0146167217711915

Shelton, J. N., & Richeson, J. A. (2014). Interacting across racial lines. In M. Mikulincer, P. R. Shaver, J. F. Dovidio, & J. A. Simpson (Eds.), *APA handbook of personality and social psychology* (vol. 2). American Psychological Association.

Shelton, R. C., Osuntokun, O., Heinloth, A. N., & Corya, S. A. (2010). Therapeutic options for treatment-resistant depression. *CNS Drugs, 24*(2), 131–161.

Shen, W., Zhao, Y., Hommel, B., Yuan, Y., Zhang, Y., Liu, Z., & Gu, H. (2019). The impact of spontaneous and induced mood states on problem solving and memory. *Thinking Skills and Creativity, 32*, 66–74.

Shepard, R. N. (1967). Recognition memory for words, sentences, and pictures. *Journal of Verbal Learning and Verbal Behavior, 6*, 156–163.

Sheppard, S. M., Keator, L. M., Breining, B. L., Wright, A. E., Saxena, S., Tippett, D. C., & Hillis, A. E. (2020). Right hemisphere ventral stream for emotional prosody identification: Evidence from acute stroke. *Neurology, 94*(10), e1013–e1020.

Sheppes, G., Suri, G., & Gross, J. J. (2015). Emotion regulation and psychopathology. *Annual Review of Clinical Psychology, 11*, 379–405.

Sherif, M., Harvey, O. J., White, B. J., Hood, W. R., & Sherif, C. W. (1961). *Intergroup cooperation and competition: The Robbers Cave experiment*. University of Oklahoma Press.

Sherry, S. B., Sherry, D. L., Macneil, M. A., Smith, M. M., Mackinnon, S. P., Stewart, S. H., & Antony, M. M. (2014). Does socially prescribed perfectionism predict daily conflict? A 14-day daily diary study of romantic couples using self- and partner-reports. *Personality and Individual Differences, 61-62*, 24–27.

Sheu, J. C., McKay, D., & Storch, E. A. (2020). COVID-19 and OCD: Potential impact of exposure and response prevention therapy. *Journal of Anxiety Disorders, 76*, 102314.

Shi, B., Cao, X., Chen, Q., Zhuang, K., & Qiu, J. (2017). Different brain structures associated with artistic and scientific creativity: a voxel-based morphometry study. *Scientific Reports, 7*, 42911.

Shields, S. A. (1991). Gender in the psychology of emotion. In K. T. Strongman (Ed.), *International review of studies of emotion* (vol. 1). Wiley.

Shimizu, M., Gillis, B. T., Buckhalt, J. A., & El-Sheikh, M. (2020). Linear and nonlinear associations between sleep and adjustment in adolescence. *Behavioral Sleep Medicine, 18*(5), 690–704.

Shin, J., & Grant, A. M. (2020). When putting work off pays off: The curvilinear relationship between procrastination and creativity. *Academy of Management Journal*. https://doi.org/10.5465/amj.2018.1471

Shin, R. Q., Smith, L. C., Welch, J. C., & Ezeofor, I. (2016). Is Allison more likely than Lakisha to receive a callback from counseling professionals? A racism audit study. *The Counseling Psychologist, 44*(8), 1187–1211. https://doi.org/10.1177/0011000016668814

Shiota, M. N., Campos, B., Oveis, C., Hertenstein, M. J., Simon-Thomas, E., & Keltner, D. (2017). Beyond happiness: Building a science of discrete positive emotions. *American Psychologist, 72*(7), 617–643.

Shover, C. L., Davis, C. S., Gordon, S. C., & Humphreys, K. (2019). Association between medical cannabis laws and opioid overdose mortality has reversed over time. *Proceedings of the National Academy of Sciences, 116*(26), 12624–12626.

Shrout, P. E., & Rodgers, J. L. (2018). Psychology, Science, and Knowledge Construction: Broadening Perspectives from the Replication Crisis. *Annual Review of Psychology, 69*, 487–510.

Shu, L. L., Gino, F., & Bazerman, M. H. (2011). Dishonest deed, clear conscience: When cheating leads to moral disengagement and motivated forgetting. *Personality and Social Psychology Bulletin, 37*(3), 330–349.

Shulman, C., Rice, C. E., Morrier, M. J., & Esler, A. (2020). The role of diagnostic instruments in dual and differential diagnosis in autism spectrum disorder across the lifespan. *Child and Adolescent Psychiatric Clinics, 29*(2), 275–299.

Shumlich, E. J., & Fisher, W. A. (2020). An exploration of factors that influence enactment of affirmative consent behaviors. *The Journal of Sex Research, 57*, 1108–1121.

Si, H., Shi, J. G., Tang, D., Wu, G., & Lan, J. (2020). Understanding intention and behavior toward sustainable usage of bike sharing by extending the theory of planned behavior. *Resources, Conservation and Recycling, 152*, 104513.

Sibley, M. H., Pelham Jr., W. E., Molina, B. S. G., Gnagy, E. M., Waschbusch, D. A., Garefino, A. C., Kuriyan, A. B., Babinski, D. E., & Karch, K. M. (2012). Diagnosing ADHD in adolescence. *Journal of Consulting and Clinical Psychology, 80*, 139–150.

Siebert, R., Taubert, N., Spadacenta, S., Dicke, P. W., Giese, M. A., & Thier, P. (2020). A naturalistic dynamic monkey head avatar elicits species-typical reactions and overcomes the uncanny valley. *Eneuro, 7*(4).

Siebner, H. R., Hartwigsen, G., Kassuba, T., & Rothwell, J. C. (2009). How does transcranial magnetic stimulation modify neuronal activity in the brain? Implications for studies of cognition. *Cortex, 45*(9), 1035–1042.

Siegel, J. M. (2005). Clues to the functions of mammalian sleep. *Nature, 437*(7063), 1264–1271.

Siegel, S. (2016). The heroin overdose mystery. *Current Directions In Psychological Science, 25*(6), 375–379. https://doi.org/10.1177/0963721416664404

Sieswerda, S., Arntz, A., Mertens, I., & Vertommen, S. (2007). Hypervigilance in patients with borderline personality disorder: Specificity, automaticity, and predictors. *Behaviour Research and Therapy, 45*(5), 1011–1024.

Silva, F. R. D., Guerreiro, R. D. C., Andrade, H. D. A., Stieler, E., Silva, A., & de Mello, M. T. (2020). Does the compromised sleep and circadian disruption of night and shiftworkers make them highly vulnerable to 2019 coronavirus disease (COVID-19)? *Chronobiology International*, 1–11.

Silver, N., & Hovick, S. R. (2018). A schema of denial: The influence of rape myth acceptance on beliefs, attitudes, and processing of affirmative consent campaign messages. *Journal of Health Communication, 23*, 505–513.

Silvia, P. J., Fayn, K., Nusbaum, E. C., & Beaty, R. E. (2015). Openness to experience and awe in response to nature and music: Personality and profound aesthetic experiences. *Psychology of Aesthetics, Creativity, and the Arts, 9*(4), 376.

Simcock, G., & Hayne, H. (2002). Breaking the barrier? Children fail to translate their preverbal memories into language. *Psychological Science, 13*(3), 225–231.

Simmons, J. P., & Simonsohn, U. (2017). Power posing: P-curving the evidence. *Psychological Science, 28*, 687–693.

Simmons, W. K., Burrows, K., Avery, J. A., Kerr, K. L., Taylor, A., Bodurka, J., . . . Drevets, W. C. (2020). Appetite changes reveal depression subgroups with distinct endocrine, metabolic, and immune states. *Molecular Psychiatry, 25*(7), 1457–1468.

Simone, A. N., Marks, D. J., Bédard, A. C., & Halperin, J. M. (2018). Low working memory rather than ADHD symptoms predicts poor academic achievement in school-aged children. *Journal of Abnormal Child Psychology, 46*(2), 277–290. https://doi.org/10.1007/s10802-017-0288-3

Simon, E. J. (2015). *Biology*. Pearson.

Simon, H. A. (1969). *The sciences of the artificial*. MIT Press.

Simons, D. J., & Chabris, C. F. (1999). Gorillas in our midst: Sustained inattentional blindness for dynamic events. *Perception, 28*(9), 1059–1074.

Simons, D. J., & Chabris, C. F. (2011). What people believe about how memory works: A representative survey of the U.S. population. *PLoS One, 6*(8), e22757.

Simon-Thomas, E. R., Godzik, J., Castle, E., Antonenko, O., Ponz, A., Kogan, A., & Keltner, D. J. (2012). An fMRI study of caring vs self-focus during induced compassion and pride. *Social Cognitive Affective Neuroscience, 7*(6), 635–648.

Sin, N. L., & Lyubomirsky, S. (2009). Enhancing well-being and alleviating depressive symptoms with positive psychology interventions: A practice-friendly meta-analysis. *Journal of Clinical Psychology, 65*(5), 467–487.

Sinclair, S., Dunn, E., & Lowery, B. S. (2005). The relationship between parental racial attitudes and children's implicit prejudice. *Journal of Experimental Social Psychology, 41*(3), 283–289. https://doi.org/10.1016/j.jesp.2004.06.003

Singer, J. A., & Blagov, P. (2004). The integrative function of narrative processing: Autobiographical memory, self-defining memories, and the life story of identity. In D. R. Beike, J. M. Lampinen, & D. A. Behrend (Eds.), *The self and memory* (pp. 117–138). Psychology Press.

Singer, J. A., & Conway, M. A. (2008). Should we forget about forgetting? *Memory Studies, 1*(3), 279–285.

Singer, J. A., & Conway, M. A. (2011). Reconsidering therapeutic action: Loewald, cognitive neuroscience, and the integration of memory's duality. *International Journal of Psychoanalysis, 92*(5), 1183–1207.

Singh, A. K., & Singh, R. (2020). Efficacy and safety of lorcaserin in obesity: A systematic review and meta-analysis of randomized controlled trials. *Expert Review of Clinical Pharmacology, 13*(2), 183–190.

Singh, U., Upadhya, M., Basu, S., Singh, O., Kumar, S., Kokare, D. M., & Singru, P. S. (2020). Transient receptor potential vanilloid 3 (TRPV3) in the cerebellum of rat and its role in motor coordination. *Neuroscience, 424*, 121–132.

Singhal, J. (2017). Why behavioral approaches to fighting poverty are so controversial. *New York Magazine*. http://nymag.com/scienceofus/2017/02/word-gap-controversy.html

Sivacek, J., & Crano, W. D. (1982). Vested interest as a moderator of attitude–behavior consistency. *Journal of Personality and Social Psychology, 43*(2), 210–221.

Sivandzade, F., Alqahtani, F., Sifat, A., & Cucullo, L. (2020). The cerebrovascular and neurological impact of chronic smoking on post-traumatic brain injury outcome and recovery: an in vivo study. *Journal of Neuroinflammation, 17*, 1–18.

Siviter, H., Charles Deeming, D., Rosenberger, J., Burman, O. P., Moszuti, S. A., & Wilkinson, A. (2017). The impact of egg incubation temperature on the personality of oviparous reptiles. *Animal Cognition, 20*(1), 109–116. https://doi.org/10.1007/s10071-016-1030-1

Sjöwall, D., Bohlin, G., Rydell, A. M., & Thorell, L. B. (2017). Neuropsychological deficits in preschool as predictors of ADHD symptoms and academic achievement in late adolescence. *Child Neuropsychology, 23*(1), 111–128.

Skinner, A. L., & Perry, S. (2020). Are attitudes contagious? Exposure to biased nonverbal signals can create novel social attitudes. *Personality and Social Psychology Bulletin, 46*(4), 514–524.

Skinner, B. F. (1938). *The behavior of organisms: An experimental analysis.* Appleton-Century-Crofts.

Skinner, B. F. (1957). *Verbal behavior.* Appleton-Century-Crofts.

Skinner, E., Graham, J. P., Brule, H., Rickert, N., & Kindermann, T. (2020). "I get knocked down but I get up again": Integrative frameworks for studying the development of motivational resilience in school. *International Journal of Behavioral Development.* https://doi.org/10.1177/0165025420924122

Skolin, I., Wahlin, Y. B., Broman, D. A., Koivisto Hursti, U., Vikstrom, L. M., & Hernell, O. (2006). Altered food intake and taste perception in children with cancer after start of chemotherapy: Perspectives of children, parents, and nurses. *Supportive Care in Cancer, 14*(4), 369–378.

Skvortsova, A., Veldhuijzen, D. S., Pacheco-Lopez, G., Bakermans-Kranenburg, M., van IJzendoorn, M., Smeets, M. A., . . . van der Wee, N. J. (2020). Placebo effects in the neuroendocrine system: Conditioning of the oxytocin responses. *Psychosomatic Medicine, 82*(1), 47–56.

Sladek, M. R., Umaña-Taylor, A. J., Oh, G., Spang, M. B., Tirado, L. M. U., Vega, L. M. T., . . . Wantchekon, K. A. (2020). Ethnic-racial discrimination experiences and ethnic-racial identity predict adolescents' psychosocial adjustment: Evidence for a compensatory risk-resilience model. *International Journal of Behavioral Development.* https://doi.org/10.1177%2F0165025420912013

Sliwa, J., & Freiwald, W. A. (2017). A dedicated network for social interaction processing in the primate brain. *Science, 356*(6339), 745–749.

Slotema, C. W., Wilhelmus, B., Arends, L. R., & Franken, I. H. (2020). Psychotherapy for posttraumatic stress disorder in patients with borderline personality disorder: A systematic review and meta-analysis of its efficacy and safety. *European Journal of Psychotraumatology, 11*(1), 1796188.

Slotnik, S. D. (2017). *Cognitive Neuroscience of Memory, 1st edition.* Cambridge University Press.

Smerdon, D., Hu, H., McLennan, A., von Hippel, W., & Albrecht, S. (2020). Female chess players show typical stereotype-threat effects: Commentary on Stafford (2018). *Psychological Science,* 0956797620924051.

Smillie, L. D. (2013). Why does it feel good to act like an extravert? *Social and Personality Psychology Compass, 7*(12), 878–887.

Smillie, L. D., Cooper, A., Wilt, J., & Revelle, W. (2012). Do extraverts get more bang for the buck? Refining the affective-reactivity hypothesis of extraversion. *Journal of Personality and Social Psychology, 103*(2), 306–326.

Smilowitz, S., Aftab, A., Aebi, M., Levin, J., Tatsuoka, C., & Sajatovic, M. (2020). Age-related differences in medication adherence, symptoms, and stigma in poorly adherent adults with bipolar disorder. *Journal of Geriatric Psychiatry and Neurology, 33*(5), 250–255.

Smit, D. J., Cath, D., Zilhão, N. R., Ip, H. F., Denys, D., den Braber, A., . . . Boomsma, D. I. (2020). Genetic meta-analysis of obsessive–compulsive disorder and self-report compulsive symptoms. *American Journal of Medical Genetics Part B: Neuropsychiatric Genetics, 183*(4), 208–216.

Smith, A. R., Silva, C., Covington, D. W., & Joiner, T. E. (2013). An assessment of suicide-related knowledge and skills among health professionals. *Health Psychology, 33*(2), 110–119.

Smith, J. L., & Bryant, F. B. (2019). Enhancing positive perceptions of aging by savoring life lessons. *Aging & Mental Health, 23*(6), 762–770.

Smith, K. Gavey, S., Riddell, N.E., Kontari, P., & Victor, C. (2020). The association between loneliness, social isolation and inflammation: A systematic review and meta-analysis. *Neuroscience & Biobehavioral Reviews.* https://doi.org/10.1016/j.neubiorev.2020.02.002

Smith, L., Grabovac, I., Yang, L., López-Sánchez, G. F., Firth, J., Pizzol, D., . . . Jackson, S. E. (2020). Sexual activity and cognitive decline in older age: A prospective cohort study. *Aging Clinical and Experimental Research, 32*(1), 85–91.

Smith, M. C., Bibi, U., & Sheard, D. E. (2003). Evidence for the differential impact of time and emotion on personal and event memories for September 11, 2001. *Applied Cognitive Psychology, 17*(9), 1047–1055.

Smith, R. E., Horn, S. S., & Bayen, U. J. (2012). Prospective memory in young and older adults: The effects of ongoing task load. *Neuropsychology, Development, and Cognition: Aging, Neuropsychology, and Cognition, 19*(4), 495–514.

Smith, T. C., & others. (2014). Longitudinal assessment of mental disorders, smoking, and hazardous drinking among a population-based cohort of U.S. service members. *Journal of Addiction Medicine, 8*(4), 271–281.

Smoczek, M., Vital, M., Wedekind, D., Basic, M., Zschemisch, N. H., Pieper, D. H., . . . Buettner, M. (2020). A combination of genetics and microbiota influences the severity of the obesity phenotype in diet-induced obesity. *Scientific Reports, 10*(1), 1–12.

Smyth, J. (1998). Written emotional expression: Effect sizes, outcome types, and moderating variables. *Journal of Consulting and Clinical Psychology, 66*(1), 174–184.

Snippe, E., Jeronimus, B. F., aan het Rot, M., Bos, E. H., de Jonge, P., & Wichers, M. (2017). The reciprocity of prosocial behavior and positive affect in daily life. *Journal of Personality.* https://doi.org/10.1111/jopy.12299

Snow, J. (2017, March 28). Computers learn to cooperate better than humans. *Science Magazine.* http://www.sciencemag.org/news/2017/03/computers-learn-cooperate-better-humans.

Snowdon, D. A. (2003). Healthy aging and dementia: Findings from the Nun Study. *Annals of Internal Medicine, 139*(5 Pt. 2), 450–454.

Snowdon, D. A. (2007, April). *Aging with grace: Findings from the Nun Study.* Paper presented at the 22nd annual Alzheimer's regional conference, Seattle, Washington.

Snowling, M. J., Hulme, C., & Nation, K. (2020). Defining and understanding dyslexia: past, present and future. *Oxford Review of Education, 46*(4), 501–513.

Snyder, C. R., & Lopez, S. J. (Eds.). (2007). *Positive psychology: The scientific and practical explorations of human strengths.* Sage.

So, S. H., Chau, A. K. C., Peters, E., Swendsen, J., Garety, P., & Kapur, S. (2017). Moment-to-moment associations between emotional disturbances, aberrant salience and persecutory delusions. *European Psychiatry, 41,* S838.

Soares, M. S., & others. (2013). Psychosurgery for schizophrenia: History and perspectives. *Neuropsychiatric Disease and Treatment, 9,* 509–515.

Society for Community Research and Action (2017). *What is community psychology?* http://www.scra27.org/

Sokol-Hessner, P., & Rutledge, R. B. (2019). The psychological and neural basis of loss aversion. *Current Directions in Psychological Science, 28*(1), 20–27.

Sokolowski, H. M., Fias, W., Mousa, A., & Ansari, D. (2017). Common and distinct brain regions in both parietal and frontal cortex support symbolic and nonsymbolic number processing in humans: A functional neuroimaging meta-analysis. *NeuroImage, 146,* 376–394.

Solmi, M., Veronese, N., Manzato, E., Sergi, G., Favaro, A., Santonastaso, P., & Correll, C. U. (2015). Oxidative stress and antioxidant levels in patients with anorexia nervosa: A systematic review and exploratory meta-analysis. *International Journal of Eating Disorders, 48*(7), 826–841. https://doi.org/10.1002/eat.22443

Solomon, S., Greenberg, J., & Pyszczynski, T. (1991). Terror management theory of self-esteem. In C. R. Snyder & D. R. Forsyth (Eds.), *Handbook of social and clinical psychology: The health perspective* (pp. 21–40). Pergamon.

Sommer, M., Hajak, G., Dohnel, K., Schwerdtner, J., Meinhardt, J., & Muller, J. L. (2006). Integration of emotion and cognition in patients with psychopathy. *Progress in Brain Research, 156,* 457–466.

Sommer, V., & Vasey, P. L. (Eds.). (2006). *Homosexual behaviour in animals: An evolutionary perspective.* Cambridge University Press.

Sommerville, J. A., Schmidt, M. F., Yun, J. E., & Burns, M. (2013). The development of fairness expectations and prosocial behavior in the second year of life. *Infancy, 18,* 40–66. https://doi.org/10.1111/j.1532-7078.2012.00129.x7.

Song, J., Gaspard, H., Nagengast, B., & Trautwein, U. (2020). The conscientiousness × interest compensation (CONIC) model: Generalizability across domains, outcomes, and predictors. *Journal of Educational Psychology, 112*(2), 271.

Song, L., Zhou, Z., Meng, J., Zhu, X., Wang, K., Wei, D., & Qiu, J. (2020). Rostral middle frontal gyrus thickness mediates the relationship between genetic risk and neuroticism trait. *Psychophysiology,* e13728.

Song, S. (2006, March 27). Mind over medicine. *Time, 167,* 13.

Sonmez, A. I., Lewis, C. P., Port, J. D., Cabello-Arreola, A., Blacker, C. J., Seewoo, B. J., . . . Croarkin, P. E. (2020). Glutamatergic correlates of bipolar symptoms in adolescents. *Journal of Child and Adolescent Psychopharmacology, 30*(10), 599–605.

Sonnenschein, S. F., Gomes, F. V., & Grace, A. A. (2020). Dysregulation of midbrain dopamine system and the pathophysiology of schizophrenia. *Frontiers in Psychiatry, 11,* 613.

Sorensen, N. L., Maloney, S. K., Pillow, J. J., & Mark, P. J. (2020). Endocrine consequences of circadian rhythm disruption in early life. *Current Opinion in Endocrine and Metabolic Research, 11,* 65–71.

Souders, D. J., Boot, W. R., Blocker, K., Vitale, T., Roque, N. A., & Charness, N. (2017). Evidence for narrow transfer after short-term Cognitive training in older adults. *Frontiers in Aging Neuroscience, 9.*

South, S. C., & Krueger, R. F. (2008). An interactionist perspective on genetic and environmental contributions to personality. *Social and Personality Psychology Compass, 2*(2), 929–948.

Southwell, D. G., Nicholas, C. R., Basbaum, A., Stryker, M. P., Kriegstein, A. R., Rubenstein, J. L., & Alvarez-Buylla, A. (2014). Interneurons from embryonic development to cell-based therapy. *Science, 344*(6180), 1240622.

Spann, M. N., Sourander, A., Surcel, H. M., Hinkka-Yli-Salomäki, S., & Brown, A. S. (2017). Prenatal toxoplasmosis antibody and childhood autism. *Autism Research, 10*(5), 769–777.

Spanos, N. P. (1996). *Multiple identities and false memories: A sociocognitive perspective.* American Psychological Association.

Sparrow, B., Liu, J., & Wegner, D. M. (2011). Google effects on memory: Cognitive consequences of having information at our fingertips. *Science, 333*(6043), 776–778.

Spearman, C. (1904). "General intelligence" objectively determined and measured. *American Journal of Psychology, 15*(2), 201–293.

Speer, M. E., & Delgado, M. R. (2020). The social value of positive autobiographical memory retrieval. *Journal of Experimental Psychology: General, 149*(4), 790.

Spelke, E. S., Bernier, E. P., & Snedeker, J. (2013). Core social cognition. In M. R. Banaji & S. A. Gelman (Eds.), *Navigating the social world: What infants, children, and other species can teach us.* Oxford University Press.

Spence, C., & Frings, C. (2020). Multisensory feature integration in (and out) of the focus of spatial attention. *Attention, Perception, & Psychophysics, 82*(1), 363–376.

Spence, K. W. (1938). Gradual versus sudden solution of discrimination problems by chimpanzees. *Journal of Comparative Psychology, 25*(1), 213–224.

Spence, M., Stancu, V., Dean, M., Livingstone, M. E., Gibney, E. R., & Lähteenmäki, L. (2016). Are food-related perceptions associated with meal portion size decisions? A cross-sectional study. *Appetite, 103,* 377–385. https://doi.org/10.1016/j.appet.2016.04.039

Spencer, K. (November 5, 2017). Opioids on the Quad. *The New York Times,* ED22.

Spencer, R. L., & Deak, T. (2017). A users guide to HPA axis research. *Physiology & Behavior, 17,* 843–865. https://doi.org/10.1016/j.physbeh.2016.11.014

Spengler, M., Brunner, M., Damian, R. I., Lüdtke, O., Martin, R., & Roberts, B. W. (2015). Student characteristics and behaviors at age 12 predict occupational success 40 years later over and above childhood IQ and parental socioeconomic status. *Developmental Psychology, 51*(9), 1329-1340. https://doi.org/10.1037/dev0000025

Spenser, K., Bull, R., Betts, L., & Winder, B. (2020). Underpinning prosociality: Age related performance in theory of mind, empathic understanding, and moral reasoning. *Cognitive Development, 56,* 100928.

Sperling, G. (1960). The information available in brief visual presentations. *Psychological Monographs: General and Applied, 74*(11), 1-29.

Sperling, G., Sun, P., Liu, D., & Lin, L. (2020). Theory of the perceived motion direction of equal-spatial-frequency plaid stimuli. *Psychological Review, 127*(3), 305-326.

Sperry, R. W. (1968). Hemisphere deconnection and unity in conscious awareness. *American Psychologist, 23*(10), 723-733.

Sperry, R. W. (1974). Lateral specialization in surgically separated hemispheres. In F. O. Schmitt & F. G. Worden (Eds.), *The neurosciences: Third study program.* MIT Press.

Spiegel, D. (2006). Editorial: Recognizing traumatic dissociation. *American Journal of Psychiatry, 163*(4), 566-568.

Spiegel, D. (2010). Hypnosis testing. In A. F. Barabasz, K. Olness, R. Boland, & S. Kahn (Eds.), *Medical hypnosis primer: Clinical and research evidence* (pp. 11-18). Routledge/Taylor & Francis.

Spiegler, M. D., & Guevremont, D. C. (2016). *Contemporary behavior therapy* (6th ed.). Cengage.

Spong, A. J., Clare, I. C. H., Galante, J., Crawford, M. J., & Jones, P. B. (2020). Brief psychological interventions for borderline personality disorder. A systematic review and meta-analysis of randomised controlled trials. *Clinical Psychology Review,* 101937.

Sprecher, S. (2019). Does (Dis) similarity information about a new acquaintance lead to liking or repulsion? An experimental test of a classic social psychology issue. *Social Psychology Quarterly, 82*(3), 303-318.

Sprich, S. E., Safren, S. A., Finkelstein, D., Remmert, J. E., & Hammerness, P. (2016). A randomized controlled trial of cognitive behavioral therapy for ADHD in medication-treated adolescents. *Journal of Child Psychology and Psychiatry, 57*(11), 1218-1226.

Sprott, R. A., & Benoit Hadcock, B. (2018). Bisexuality, pansexuality, queer identity, and kink identity. *Sexual and Relationship Therapy, 33* (1-2), 214-232.

Spuhler, K., Huang, C., Ananth, M., Bartlett, E., Ding, J., He, X., et al. (2017). Using PET and MRI to assess pretreatment markers of lithium treatment responsiveness in bipolar depression. *Journal of Nuclear Medicine, 58*(supplement 1), 138-138.

Spurny, B., Seiger, R., Moser, P., Vanicek, T., Reed, M. B., Heckova, E., . . . Trattnig, S. (2020). Hippocampal GABA levels correlate with retrieval performance in an associative learning paradigm. *NeuroImage, 204,* 116244.

Squire, L. R. (2007). Memory systems as a biological concept. In H. L. Roediger, Y. Dudai, & S. Fitzpatrick (Eds.), *Science of memory: Concepts.* Oxford University Press.

Sroufe, L. A. (2020). Principles of development: The case of dependency. *Attachment & Human Development.* https://doi.org/10.1080/14616734.2020.1751992

Staff, J., Maggs, J. L., Bucci, R., & Mongilio, J. (2019). Changes in externalizing behaviors after children first have an alcoholic drink and first drink heavily. *Journal of Studies on Alcohol and Drugs, 80*(4), 472-479.

Staffa, M., Rossi, S., Tapus, A., & Khamassi, M. (2020). Special issue on behavior adaptation, interaction, and artificial perception for assistive robotics. *International Journal of Social Robotics.* https://doi.org/10.1007/s12369-020-00655-8

Stafford, B. K., & Huberman, A. D. (2017). Signal Integration in Thalamus: Labeled Lines Go Cross-Eyed and Blurry. *Neuron, 93*(4), 717-720.

Stämpfli, D., Weiler, S., & Burden, A. M. (2020). Movement disorders and use of risperidone and methylphenidate: a review of case reports and an analysis of the WHO database in pharmacovigilance. *European Child & Adolescent Psychiatry.* https://doi.org/10.1007/s00787-020-01589-2

Stange-Marten, A., Nabel, A. L., Sinclair, J. L., Fischl, M., Alexandrova, O., Wohlfrom, H., et al. (2017). Input timing for spatial processing is precisely tuned via constant synaptic delays and myelination patterns in the auditory brainstem. *Proceedings of the National Academy of Sciences, 114*(24), E4851-E4858.

Stangor, C. (2016). The study of stereotyping, prejudice, and discrimination within social psychology: A quick history of theory and research. In T. D. Nelson (Ed.), *Handbook of prejudice, stereotyping, and discrimination, 2nd ed* (pp. 3-27). Psychology Press.

Stanislaw, H., Howard, J., & Martin, C. (2020). Helping parents choose treatments for young children with autism: A comparison of applied behavior analysis and eclectic treatments. *Journal of the American Association of Nurse Practitioners, 32*(8), 571-578.

Stanlaw, J. (2020, in press). Color and color nomenclature. In J. Stanlaw (Ed.), *The international encyclopedia of linguistic anthropology.* https://doi.org/10.1002/9781118786093.iela0062

Stanley, J. T., & Isaacowitz, D. M. (2011). Age-related differences in profiles of mood-change trajectories. *Developmental Psychology, 47*(2), 318-330.

Stanovich, K. E. (2013). *How to think straight about psychology* (10th ed.). Pearson.

Stanton, S. C., Campbell, L., & Loving, T. J. (2014). Energized by love: Thinking about romantic relationships increases positive affect and blood glucose levels. *Psychophysiology, 51*(10), 990-995.

Stark, T., Di Bartolomeo, M., Di Marco, R., Drazanova, E., Platania, C. B. M., Iannotti, F. A., . . . Piscitelli, F. (2020). Altered dopamine D3 receptor gene expression in MAM model of schizophrenia is reversed by peripubertal cannabidiol treatment. *Biochemical Pharmacology,* 114004.

Starr, L. R., Stroud, C. B., Shaw, Z. A., & Vrshek-Schallhorn, S. (2020). Stress sensitization to depression following childhood adversity: Moderation by HPA axis and serotonergic multilocus profile scores. *Development and Psychopathology.* https://doi.org/10.1017/S0954579420000474

Staudinger, U. M., & Gluck, J. (2011). Psychological wisdom research. *Annual Review of Psychology 62,* 215-241.

Stavans, M., Lin, Y., Wu, D., & Baillargeon, R. (2019). Catastrophic individuation failures in infancy: A new model and predictions. *Psychological Review, 126*(2), 196-225.

Steblay, N., Dysart, J., & Wells, G. L. (2011). Seventy-two tests of the sequential lineup superiority effect: A meta-analysis and policy discussion. *Psychology, Public Policy, and Law, 17*(1), 99-139.

Steele C. M. (1997). A threat in the air: How stereotypes shape intellectual identity and performance. *American Psychologist, 52,* 613-629. https://doi.org/10.1037/0003-066X.52.6.613

Steele, C. M. (2012). Conclusion: Extending and applying stereotype threat research: A brief essay. In M. Inzlicht, T. Schmader, M. Inzlicht, & T. Schmader (Eds.), *Stereotype threat: Theory, process, and application* (pp. 297-303). Oxford University Press.

Steele, C. M., & Aronson, J. (1995). Stereotype threat and the intellectual test performance of African Americans. *Journal of Personality and Social Psychology, 69*(5), 797-811.

Steele, C. M., & Aronson, J. A. (2004). Stereotype threat does not live by Steele and Aronson (1995) alone. *American Psychologist, 59*(1), 47-48.

Steenhuis, I., & Poelman, M. (2017). Portion size: Latest developments and interventions. *Current Obesity Reports, 6*(1), 10-17.

Steenkamp, M. M., Schlenger, W. E., Corry, N., Henn-Haase, C., Qian, M., Li, M., et al. (2017). Predictors of PTSD 40 years after combat: Findings from the National Vietnam Veterans Longitudinal Study. *Depression and Anxiety, 34,* 711-722

Steger, M. F., & Frazier, P. (2005). Meaning in life: One link in the chain from religion to well-being. *Journal of Counseling Psychology, 52*(4), 574-582.

Steger, M. F., Frazier, P., Oishi, S., & Kaler, M. (2006). The meaning in life questionnaire: Assessing the presence of and search for meaning in life. *Journal of Counseling Psychology, 53*(1), 80-93.

Stein, M. B., Chen, C. Y., Jain, S., Jensen, K. P., He, F., Heeringa, S. G., . . . & Sun, X. (2017). Genetic risk variants for social anxiety. *American Journal of Medical Genetics Part B: Neuropsychiatric Genetics, 174*(2), 120-131.

Stein, R. (2003, September 2). *Blinded by the light.* www.theage.com.au/articles/2003/09/01/ 1062403448264.html?from=storyrhs

Steinbauer, P., Deindl, P., Fuiko, R., Unterasinger, L., Cardona, F., Wagner, M., . . . Giordano, V. (2020). Long-term impact of systematic pain and sedation management on cognitive, motor, and behavioral outcomes of extremely preterm infants at preschool age. *Pediatric Research.* https://doi.org/10.1038/s41390-020-0979-2

Steinbrook, R. (1992). The polygraph test—A flawed diagnostic method. *New England Journal of Medicine, 327*(2), 122-123.

Steketee, M. B., & others. (2014). Regulation of intrinsic axon growth ability at retinal ganglion cell growth cones. *Investigative Ophthalmology and Visual Science, 55*(7), 4369-4377.

Stellar, J. E., Manzo, V. M., Kraus, M. W., & Keltner, D. (2012). Class and compassion: Socioeconomic factors predict responses to suffering. *Emotion, 12*(3), 449-459.

Stenling, A., Sörman, D. E., Lindwall, M., Hansson, P., Körning Ljungberg, J., & Machado, L. (2020). Physical activity and cognitive function: Between-person and within-person associations and moderators. *Aging, Neuropsychology, and Cognition.* https://doi.org/10.1080/138 25585.2020.1779646

Stephens-Davidowitz, S. (2015, December 8). How many American men are gay? *The New York Times,* SR5.

Stephens, R. G., Dunn, J. C., Hayes, B. K., & Kalish, M. L. (2020). A test of two processes: The effect of training on deductive and inductive reasoning. *Cognition, 199,* 104223.

Steptoe, A., & Jackson, S. E. (2020). Association of noncognitive life skills with mortality at middle and older ages in England. *JAMA Network Open, 3*(5), e204808-e204808.

Steptoe, W., Steed, A., & Slater, M. (2013). Human tails: Ownership and control of extended humanoid avatars. *IEEE Transactions on Visualization and Computer Graphics, 19,* 583-590.

Sterling, J., Jost, J. T., & Bonneau, R. (2020). Political psycholinguistics: A comprehensive analysis of the language habits of liberal and conservative social media users. *Journal of Personality and Social Psychology, 118,* 805-834.

Stern, C., & Crawford, J. T. (2020). Ideological conflict and prejudice: An adversarial collaboration examining correlates and ideological (a) symmetries. *Social Psychological and Personality Science.* https://doi.org/10.1177/1948550620904275

Stern, C., West, T. V., Jost, J. T., & Rule, N. O. (2013). The politics of gaydar: Ideological differences in the use of gendered cues in categorizing sexual orientation. *Journal of Personality and Social Psychology, 104*(3), 520-541.

Stern, E. R., Welsh, R. C., Gonzalez, R., Fitzgerald, K. D., Abelson, J. L., & Taylor, S. F. (2013). Subjective uncertainty and limbic hyperactivation in obsessive-compulsive disorder. *Human Brain Mapping, 34*(8), 1956-1970.

Stern, Y., Alexander, G. E., Prohovnik, I., & Mayeux, R. (1992). Inverse relationship between education and parietotemporal perfusion deficit in Alzheimer's disease. *Annals of Neurology, 32*(3), 371-375.

Stern, Y., Scarmeas, N., & Habeck, C. (2004). Imaging cognitive reserve. *International Journal of Psychology, 39*(1), 18-26.

Sternberg, R. J. (1988). *The triarchic mind: A new theory of human intelligence.* Viking.

Sternberg, R. J. (2013). Searching for love. *The Psychologist, 26*(2), 98-101.

Sternberg, R. J. (2019). A theory of adaptive intelligence and its relation to general intelligence. *Journal of Intelligence, 7*(4), 23.

Sternberg, R. J. (2020). Creativity from start to finish: A "straight-A" model of creative process and its relation to intelligence. *The Journal of Creative Behavior, 54*(2), 229–241.

Sternberg, R. J., & Bridges, S. L. (2014). Varieties of genius. In D. K. Simonton (Ed.), *The Wiley handbook of genius* (pp. 185–197). Wiley-Blackwell.

Sternberg, R. J., & Halpern, D. F. (Eds.). (2020). *Critical thinking in psychology*. Cambridge University Press.

Stevenson, A. (2020). *Cultural issues in psychology* (2nd ed.). Routledge.

Stewart, G. R., Wallace, G. L., Cottam, M., & Charlton, R. A. (2020). Theory of mind performance in younger and older adults with elevated autistic traits. *Autism Research, 13*(5), 751–762.

Stewart, J. L., Spivey, L. A., Widman, L., Choukas-Bradley, S., & Prinstein, M. J. (2019). Developmental patterns of sexual identity, romantic attraction, and sexual behavior among adolescents over three years. *Journal of Adolescence, 77*, 90–97.

Stickgold, R. (2001). Watching the sleeping brain watch us: Sensory processing during sleep. *Trends in Neuroscience, 24*(6), 307–309.

Stier-Jarmer, M., Frisch, D., Oberhauser, C., Berberich, G., & Schuh, A. (2016). The effectiveness of a stress reduction and burnout prevention program: A randomized controlled trial of an outpatient intervention in a health resort setting. *Deutsches Ärzteblatt International, 113*(46), 781–788.

Stillman, T. F., Maner, J. K., & Baumeister, R. F. (2010). A thin slice of violence: Distinguishing violent from nonviolent sex offenders at a glance. *Evolution and Human Behavior, 31*(4), 298–303.

Stoel-Gammon, C., & Menn, L. (2013). Phonological development: Learning sounds and sound patterns. In J. Berko Gleason & N. B. Ratner (Eds.), *The development of language* (8th ed.). Pearson.

Stolier, R. M., Hehman, E., & Freeman, J. B. (2020). Trait knowledge forms a common structure across social cognition. *Nature Human Behaviour, 4*(4), 361–371.

Stoner, J. (1961). *A comparison of individual and group decisions, including risk*. Unpublished master's thesis, School of Industrial Management, MIT.

Stoppelbein, L., McRae, E., & Greening, L. (2017). A longitudinal study of hardiness as a buffer for posttraumatic stress symptoms in mothers of children with cancer. *Clinical Practice In Pediatric Psychology, 5*(2), 149–160. https://doi.org/10.1037/cpp0000168

Stott, C., Drury, J., & Reicher, S. (2012). From prejudice to collective action. In J. Dixon, M. Keynes, & M. Levine (Eds.), *Beyond prejudice*. Cambridge University Press.

Strachan, S. M., Perras, M. M., Brawley, L. R., & Spink, K. S. (2016). Exercise in challenging times: The predictive utility of identity, self-efficacy, and past exercise. *Sport, Exercise, and Performance Psychology, 5*(3), 247–258. https://doi.org/10.1037/spy0000064

Strauss, J. A., & Grand, J. A. (2020). Promoting robust and reliable big data research in psychology. In S. E. Woo, L. Tay, & R. W. Proctor (Eds.), *Big data in psychological research* (pp. 373–391). American Psychological Association. https://doi.org/10.1037/0000193-017

Streff, F. M., & Geller, E. S. (1986). Strategies for motivating safety belt use: The application of applied behavior analysis. *Health Education Research, 1*(1), 47–59.

Strickland, L., Elliott, D., Wilson, M. D., Loft, S., Neal, A., & Heathcote, A. (2019). Prospective memory in the red zone: Cognitive control and capacity sharing in a complex, multistimulus task. *Journal of Experimental Psychology: Applied, 25*(4), 695–715. https://doi.org/10.1037/xap0000224

Stripp, T. K., Jorgensen, M. B., & Olsen, N. V. (2017). Anaesthesia for electroconvulsive therapy-new tricks for old drugs: a systematic review. *Acta Neuropsychiatrica.* https://www.cambridge.org/core/journals/acta-neuro-psychiatrica/article/anaesthesia-for-electroconvulsive-therapy-new-tricks-for-old-drugs-a-systematic-review/319 6A1565F953740A8D49B5C4F95FBDA

Strother, L., Zhou, Z., Coros, A. K., & Vilis, T. (2017). An fMRI study of visual hemifield integration and cerebral lateralization. *Neuropsychologia, 100*, 35–43.

Stubbs-Richardson, M., & May, D. C. (2020). Social contagion in bullying: An examination of strains and types of bullying victimization in peer networks. *American Journal of Criminal Justice*, 1–22.

Stukas, A. A., Snyder, M., & Clary, E. G. (2015). Volunteerism and community involvement: Antecedents, experiences, and consequences for the person and the situation. In D. A. Schroeder & W. G. Graziano (Eds.), *The Oxford handbook of prosocial behavior*. Oxford University Press.

Štukelj, G. (2020). On the simplicity of simple heuristics. *Adaptive Behavior, 28*(4), 261–271.

Sturmer, T., Hasselbach, P., & Amelang, M. (2006). Personality, lifestyle, and risk of cardiovascular disease and cancer: Follow-up of population-based cohort. *British Medical Journal, 332*(7554), 1359.

Su, P. L., Rogers, S. J., Estes, A., & Yoder, P. (2020). The role of early social motivation in explaining variability in functional language in toddlers with autism spectrum disorder. *Autism.* https://doi.org/10.1177%2F1362361320953260

Sue, D., Sue, D. W., Sue, S., & Sue, D. M. (2016). *Understanding abnormal behavior* (11th ed.). Cengage.

Sue, D. W. (2010). *Microaggressions in everyday life: Race, gender, and sexual orientation.* John Wiley & Sons.

Sue, D. W., Alsaidi, S., Awad, M. N., Glaeser, E., Calle, C. Z., & Mendez, N. (2019). Disarming racial microaggressions: Microintervention strategies for targets, white allies, and bystanders. *American Psychologist, 74*(1), 128–142.

Sui, X., LaMonte, M. J., Laditka, J. N., Hardin, J. W., Chase, N., Hooker, S. P., & Blair, S. N. (2007). Cardiorespiratory fitness and adiposity as mortality predictors in older adults. *Journal of the American Medical Association, 298*(21), 2507–2516.

Suinn, R. M. (1984). *Fundamentals of abnormal psychology.* Nelson-Hall.

Sullivan, H. S. (1953). *The interpersonal theory of psychiatry.* Norton.

Suls, J., Green, P. A., & Boyd, C. M. (2019). Multimorbidity: Implications and directions for health psychology and behavioral medicine. *Health Psychology, 38*(9), 772782.

Suls, J., Mogavero, J. N., Falzon, L., Pescatello, L. S., Hennessy, E. A., & Davidson, K. W. (2020). Health behaviour change in cardiovascular disease prevention and management: Meta-review of behaviour change techniques to affect self-regulation. *Health Psychology Review, 14*(1), 43–65.

Sultzer, D. L., Melrose, R. J., Riskin-Jones, H., Narvaez, T. A., Veliz, J., Ando, T. K., . . . & Mandelkern, M. A. (2017). Cholinergic receptor binding in Alzheimer Disease and healthy aging: Assessment in vivo with Positron Emission Tomography imaging. *The American Journal of Geriatric Psychiatry, 25*(4), 342–353.

Sumner, R. L., McMillan, R., Spriggs, M. J., Campbell, D., Malpas, G., Maxwell, E., . . . & Sundram, F. (2020). Ketamine Enhances visual sensory evoked potential long-term potentiation in patients with Major Depressive Disorder. *Biological Psychiatry: Cognitive Neuroscience and Neuroimaging, 5*(1), 45–55.

Sumontha, J., Farr, R. H., & Patterson, C. J. (2017). Children's gender development: Associations with parental sexual orientation, division of labor, and gender Ideology. *Psychology of Sexual Orientation and Gender Diversity.* https://doi.org/10.1037/sgd0000242

Sun, J., Stevenson, K., Kabbani, R., Richardson, B., & Smillie, L. D. (2017). The pleasure of making a difference: Perceived social contribution explains the relation between extraverted behavior and positive affect. *Emotion, 17*(5), 794–810. https://doi.org/10.1037/emo0000273

Sun, L., Zhou, H., Cichon, J., & Yang, G. (2020). Experience and sleep-dependent synaptic plasticity: From structure to activity. *Philosophical Transactions of the Royal Society B, 375*(1799), 20190234.

Suo, C., Gates, N., Singh, M. F., Saigal, N., Wilson, G. C., Meiklejohn, J., . . . & Baune, B. T. (2017). Midlife managerial experience is linked to late life hippocampal morphology and function. *Brain imaging and behavior, 11*(2), 333–345.

Sussman, T. J., Jin, J., & Mohanty, A. (2020). The impact of top-down factors on threat perception biases in health and anxiety. In T. Aue & H. Okon-Singer (Eds.), *Cognitive biases in health and psychiatric disorders* (pp. 215–241). Academic Press.

Sutin, A. R. (2017). Openness. In T. A. Widiger (Ed.), *Oxford Handbook of the Five Factor Model.* Oxford University Press.

Sutton, R. M., & Douglas, K. M. (2020). Agreeing to disagree: Reports of the popularity of Covid-19 conspiracy theories are greatly exaggerated. *Psychological Medicine.* https://doi.org/10.1017/S0033291720002780

Sutu, A., Phetmisy, C. N., & Damian, R. I. (2020). Open to laugh: The role of openness to experience in humor production ability. *Psychology of Aesthetics, Creativity, and the Arts.* https://doi.org/10.1037/aca0000298

Suzuki, H., Savitz, J., Teague, T. K., Gandhapudi, S. K., Tan, C., Misaki, M., et al. (2017). Altered populations of natural killer cells, cytotoxic T lymphocytes, and regulatory T cells in major depressive disorder: Association with sleep disturbance. *Brain, behavior, and immunity, 66*, 193–200.

Suzuki, S., & O'Doherty, J. P. (2020). Breaking human social decision making into multiple components and then putting them together again. *Cortex, 127*, 221–230.

Swami, V., Voracek, M., Stieger, S., Tran, U. S., & Furnham, A. (2014). Analytic thinking reduces belief in conspiracy theories. *Cognition, 133*(3), 572–585.

Swart, S., Wildschut, M., Draijer, N., Langeland, W., Hoogendoorn, A. W., & Smit, J. H. (2020b). The course of (comorbid) trauma-related, dissociative and personality disorders: Two year follow up of the Friesland Study Cohort. *European Journal of Psychotraumatology, 11*(1), 1750171.

Swart, S., Wildschut, M., Draijer, N., Langeland, W., & Smit, J. H. (2020a). Dissociative subtype of posttraumatic stress disorder or PTSD with comorbid dissociative disorders: Comparative evaluation of clinical profiles. *Psychological Trauma: Theory, Research, Practice, and Policy, 12*(1), 38.

Swift, J. K., Callahan, J. L., Cooper, M., & Parkin, S. R. (2018). The impact of accommodating client preference in psychotherapy: A meta-analysis. *Journal of Clinical Psychology, 74*(11), 1924–1937.

Sylvia, L. G., Salcedo, S., Peters, A. T., da Silva Magalhães, P. V., Frank, E., Miklowitz, D. J., et al. (2017). Do sleep disturbances predict or moderate the response to psychotherapy in bipolar disorder? *The Journal of Nervous and Mental Disease, 205*(3), 196–202.

Sylwander, C., Larsson, I., Andersson, M., & Bergman, S. (2020). The impact of chronic widespread pain on health status and long-term health predictors: A general population cohort study. *BMC Musculoskeletal Disorders, 21*(1), 1–11.

Synard, J., & Gazzola, N. (2017). Happiness, eudaimonia, and other holy grails: What can job loss teach us about 'One-size-fits-all' theories of well-being? *The Journal of Positive Psychology, 12*, 246–262. https://doi.org/10.1080/17439760.2016.1225116

Syrjala, K. L., Jensen, M. P., Mendoza, M. E., Yi, J. C., Fisher, H. M., & Keefe, F. J. (2014). Psychological and behavioral approaches to cancer pain management. *Journal of Clinical Oncology, 32*(16), 1703–1711.

Sysko, R., Ojserkis, R., Schebendach, J., Evans, S. M., Hildebrandt, T., & Walsh, B. T. (2017). Impulsivity and test meal intake among women with bulimia nervosa. *Appetite, 11*, 21–28. https://doi.org/10.1016/j.appet.2017.01.005

Szapocznik, J., & Hervis, O. E. (2020). *Brief strategic family therapy.* American Psychological Association.

Szmukler, G. (2020). Involuntary detention and treatment: Are we edging toward a "paradigm shift"? *Schizophrenia Bulletin, 46*(2), 231–235.

Szocs, C., & Lefebvre, S. (2017). Spread or stacked? Vertical versus horizontal food presentation, portion size perceptions, and consumption. *Journal of Business Research, 75*, 249–257. https://doi.org/10.1016/j.jbusres.2016.07.022

Szucs, D., & Ioannidis, J. P. (2020). Sample size evolution in neuroimaging research: An evaluation of highly-cited studies (1990–2012) and of latest practices (2017–2018) in high-impact journals. *Neuroimage*, 117164. https://doi.org/10.1016/j.neuroimage.2020.117164

Tabak, B. A., Young, K. S., Torre, J. B., Way, B. M., Burklund, L. J., Eisenberger, N. I., . . . Craske, M. G. (2020). Preliminary evidence that CD38 moderates the association of neuroticism on amygdala-subgenual cingulate connectivity. *Frontiers in Neuroscience, 14.*

Taber, J. M., Klein, W. P., Ferrer, R. A., Lewis, K. L., Biesecker, L. G., & Biesecker, B. B. (2015). Dispositional optimism and perceived risk interact to predict intentions to learn genome sequencing results. *Health Psychology, 34*(7), 718–728. https://doi.org/10.1037/hea0000159

Tackett, J. L., Herzhoff, K., Kushner, S. C., & Rule, N. (2016). Thin slices of child personality: Perceptual, situational, and behavioral contributions. *Journal of Personality and Social Psychology, 110*(1), 150–166. https://doi.org/10.1037/pspp0000044

Taga, K. A., Markey, C. N., & Friedman, H. S. (2006). A longitudinal investigation of associations between boys'pubertal timing and adult behavioral health and well-being. *Journal of Youth and Adolescence, 35*(3), 380–390.

Taherkhani, A., Belatreche, A., Li, Y., Cosma, G., Maguire, L. P., & McGinnity, T. M. (2020). A review of learning in biologically plausible spiking neural networks. *Neural Networks, 122,* 253–272.

Tait, R. J., Brinker, J., Moller, C. I., & French, D. J. (2014). Rumination, substance use, and self-harm in a representative Australian adult sample. *Journal of Clinical Psychology, 70*(3), 283–293.

Tajfel, H. (1978). The achievement of group differentiation. In H. Tajfel (Ed.), *Differentiation between social groups.* Academic.

Tajiri, N., & others. (2014). Behavioral and histopathological assessment of adult ischemic rat brains after intracerebral transplantation of NSI-566RSC cell lines. *PLoS One, 9*(3), e91408.

Takacs, Z. K., & Kassai, R. (2019). The efficacy of different interventions to foster children's executive function skills: A series of meta-analyses. *Psychological Bulletin, 145*(7), 653–697.

Takano, T., Nakamura, K., & Watanabe, M. (2002). Urban residential environments and senior citizens' longevity in megacity areas: The importance of walkable green space. *Journal of Epidemiology and Community Health, 56*(12), 913–916.

Takmakov, P. A. (2017). Electrochemistry of a Robust Neural Interface. *The Electrochemical Society Interface, 26*(3), 49–51.

Tal, R., & Tal, K. (2017). Child-parent relationship therapy—A dialogue with Winnicott's theory. *International Journal Of Play Therapy, 26*(3), 151–159. https://doi.org/10.1037/pla0000051

Talami, F., Vaudano, A. E., & Meletti, S. (2020). Motor and limbic system contribution to emotional laughter across the lifespan. *Cerebral Cortex, 30*(5), 3381–3391.

Talay, L., & De Coninck, D. (2020). Exploring the link between personality traits and European attitudes towards refugees. *International Journal of Intercultural Relations, 77,* 13–24.

Tamm, L., Denton, C. A., Epstein, J. N., Schatschneider, C., Taylor, H., Arnold, L. E., et al. (2017). Comparing treatments for children with ADHD and word reading difficulties: A randomized clinical trial. *Journal of Consulting and Clinical Psychology, 85*(5), 434–446.

Tamnes, C. K., Overbye, K., Ferschmann, L., Fjell, A. M., Walhovd, K. B., Blakemore, S. J., & Dumontheil, I. (2018). Social perspective taking is associated with self-reported prosocial behavior and regional cortical thickness across adolescence. *Developmental Psychology, 54*(9), 1745–1757.

Tanahashi, S., Tanii, H., Konishi, Y., Otowa, T., Sasaki, T., Tochigi, M., . . . Okada, M. Association of serotonin transporter gene (5-HTTLPR/rs25531) polymorphism with comorbidities of panic disorder. *Neuropsychobiology.* https://doi.org/10.1159/000512699

Tang, A., Crawford, H., Morales, S., Degnan, K. A., Pine, D. S., & Fox, N. A. (2020). Infant behavioral inhibition predicts personality and social outcomes three decades later. *Proceedings of the National Academy of Sciences, 117*(18), 9800–9807.

Tang, C., Hamilton, L. S., & Chang, E. F. (2017). Intonational speech prosody encoding in the human auditory cortex. *Science, 357,* 797–801.

Tang, R., & others. (2014). Candidate genes and functional noncoding variants identified in a canine model of obsessive-compulsive disorder. *Genome Biology, 15*(3), R25.

Tang, W., Kochubey, O., Kintscher, M., & Schneggenburger, R. (2020). A VTA to basal amygdala dopamine projection contributes to signal salient somatosensory events during fear learning. *Journal of Neuroscience, 40*(20), 3969–3980.

Tang, Y., & others. (2014). Neural stem cell protects aged rat brain from ischemia-reperfusion injury through neurogenesis and angiogenesis. *Journal of Cerebral Blood Flow and Metabolism.* https://doi.org/10.1093/brain/awp174

Tarokh, L., & Carskadon, M. A. (2010). Developmental changes in the human sleep EEG during early adolescence. *Sleep, 33*(6), 801–809.

Tashjian, S. M., Rahal, D., Karan, M., Eisenberger, N., Galván, A., Cole, S. W., & Fuligni, A. J. (2020). Evidence from a randomized controlled trial that altruism moderates the effect of prosocial acts on adolescent well-being. *Journal of Youth and Adolescence.* https://doi.org/10.1007/s10964-020-01362-3

Tassone, A., Liu, J. J., Reed, M. J., & Vickers, K. (2020). Multitasking in the classroom: testing an educational intervention as a method of reducing multitasking. *Active Learning in Higher Education, 21*(2), 128–141.

Tay, L., & Diener, E. (2011). Needs and subjective well-being around the world. *Journal of Personality and Social Psychology, 101*(2), 354–365.

Tay, L., Li, M., Diener, E., & Myers, D. (2013). Religiosity and subjective well-being: An international perspective. In C. Kim-Prieto (Ed.), *The positive psychology of religion and spirituality.* Springer.

Tay, L., Li, M., Diener, E., & Myers, D. (2013). Religiosity and subjective well-being: An international perspective. In C. Kim-Prieto (Ed.), *The positive psychology of religion and spirituality.* Springer.

Taylor, E. (2017). Attention deficit hyperactivity disorder: Overdiagnosed or diagnoses missed? *Archives of Disease in Childhood, 102*(4), 376–379.

Taylor, J. P., Ashworth, S. J., Petrovich, S., & Young, C. A. (2017). Inducing an availability heuristic on the Wason selection task overrides the matching bias. *Journal of Cognitive Psychology, 29*(4), 508–519. https://doi.org/10.1080/20445911.2017.1281282

Taylor, S. E. (2001). Toward a biology of social support. In C. R. Snyder & S. J. Lopez (Eds.), *Handbook of positive psychology.* Oxford University Press.

Taylor, S. E. (2007). Social support. In H. S. Friedman & R. C. Silver (Eds.), *Foundations of health psychology.* Oxford University Press.

Taylor, S. E. (2011). Positive illusions: How ordinary people become extraordinary. In M. A. Gernsbacher, R. W. Pew, L. M. Hough, & J. R. Pomerantz (Eds.), *Psychology and the real world: Essays illustrating fundamental contributions to society* (pp. 224–228). Worth.

Taylor, S. E. (2011). Tend and befriend theory. In P. A. M. Van Lange, A. W. Kruglanski, & E. T. Higgins (Eds.), *Handbook of theories of social psychology: Collection* (vol. 1, pp. 32–49). Sage.

Taylor, S. E. (2011a). Tend and befriend theory. In A. M. van Lange, A. W. Kruglanski, & E. T. Higgins (Eds.), *Handbook of theories of social psychology.* Sage.

Taylor, S. E. (2011b). Affiliation and stress. In S. S. Folkman (Ed.), *Oxford handbook of stress, health, and coping.* Oxford University Press.

Taylor, S. E. (2013). Social cognition and health. In D. E. Carlston (Ed.), *The Oxford handbook of social cognition.* Oxford University Press.

Taylor, S. E. (2014). Social cognition and health. In M. Mikulincer, P. R. Shaver, E. Borgida, & J. A. Bargh (Eds.), *APA handbook of personality and social psychology* (vol. 1). American Psychological Association.

Taylor, S. E. (2014a). Social cognition and health. In M. Mikulincer, P. R. Shaver, E. Borgida, & J. A. Bargh (Eds.), *APA handbook of personality and social psychology* (vol. 1). American Psychological Association.

Taylor, S. E. (2018). *Health psychology* (10th ed.). McGraw-Hill.

Taylor, Z. E., Widaman, K. F., Robins, R. W., Jochem, R., Early, D. R., & Conger, R. D. (2012). Dispositional optimism: A psychological resource for Mexican-origin mothers experiencing economic stress. *Journal of Family Psychology, 26*(1), 133–139.

Tedesco, A. M., Bianchini, F., Piccardi, L., Clausi, S., Berthoz, A., Molinari, M., & . . . Leggio, M. (2017). Does the cerebellum contribute to human navigation by processing sequential information? *Neuropsychology, 31*(5), 564–574. https://doi.org/10.1037/neu0000354

Teeny, J., Briñol, P., & Petty, R. E. (2017). The elaboration likelihood model: Understanding consumer attitude change. In C. V. Jansson-Boyd, M. & J. Zawisza (Eds.), *Routledge international handbook of consumer psychology* (pp. 390–410). Routledge/Taylor & Francis Group.

Tehee, M., Isaacs, D., & Rodríguez, M. M. D. (2020). The elusive construct of cultural competence. In L. T. Benuto, F. R. Gonzalez, & J. Singer (Eds.), *Handbook of cultural factors in behavioral health* (pp. 11–24). Springer.

Teigset, C. M., Mohn, C., & Rund, B. R. (2020). Perinatal complications and executive dysfunction in early-onset schizophrenia. *BMC Psychiatry, 20*(1), 1–12.

ten Velden Hegelstad, W., & others. (2012). Long-term follow-up of the TIPS early detection in psychosis study: Effects on 10-year outcome. *American Journal of Psychiatry, 169*(4), 374–380.

Tennie, C., Call, J., & Tomasello, M. (2010). Evidence for emulation in chimpanzees in social settings using the floating peanut task. *PLoS One, 5*(5), e10544.

Teodorescu, M., & others. (2006). Correlates of daytime sleepiness in patients with asthma. *Sleep Medicine, 7*(8), 607–613.

Terman, L. (1925). *Genetic studies of genius: Vol. 1: Mental and physical traits of a thousand gifted children.* Stanford University Press.

Terr, L. C. (1988). What happens to early memories of trauma? A study of twenty children under age five at the time of documented traumatic events. *Journal of the American Academy of Child and Adolescent Psychiatry, 27*(1), 96–104.

Terracciano, A., Stephan, Y., Luchetti, M., Albanese, E., & Sutin, A. R. (2017). Personality traits and risk of cognitive impairment and dementia. *Journal of Psychiatric Research, 89,* 22–27. https://doi.org/10.1016/j.jpsychires.2017.01.011

Thagard, P. (2014). Artistic genius and creative cognition. In D. K. Simonton (Ed.), *The Wiley handbook of genius* (pp. 120–138). Wiley-Blackwell.

Thake, J., & Zelenski, J. M. (2013). Neuroticism, BIS, and reactivity to discrete negative mood inductions. *Personality and Individual Differences, 54*(2), 208–213. https://doi.org/10.1016/j.paid.2012.08.041

Thakkar, K. N., Mathalon, D. H., & Ford, J. M. (2020). Reconciling competing mechanisms posited to underlie auditory verbal hallucinations. *Philosophical Transactions of the Royal Society B, 376*(1817), 20190702.

Thakral, P. P., Madore, K. P., Addis, D. R., & Schacter, D. L. (2020). Reinstatement of event details during episodic simulation in the hippocampus. *Cerebral Cortex, 30*(4), 2321–2337.

Thalmann, M., Souza, A. S., & Oberauer, K. (2019). How does chunking help working memory? *Journal of Experimental Psychology: Learning, Memory, and Cognition, 45*(1), 37–55.

Thalmayer, A. G., Saucier, G., Ole-Kotikash, L., & Payne, D. (2020). Personality structure in east and west Africa: Lexical studies of personality in Maa and Supyire-Senufo. *Journal of Personality and Social Psychology, 119*(5), 1132–1152.

Thames, A. D., Irwin, M. R., Breen, E. C., & Cole, S. W. (2019). Experienced discrimination and racial differences in leukocyte gene expression. *Psychoneuroendocrinology, 106,* 277–283.

Thapar, A., & Cooper, M. (2016). Attention-deficit hyperactivity disorder. *Lancet, 387*, 1240–1250.

Thapar, A., & Rutter, M. (2020). Genetic advances in autism. *Journal of Autism and Developmental Disorders.* https://doi.org/10.1007/s10803-020-04685-z

The Week Staff. (2011, November 4). Is eyewitness testimony too unreliable to trust? *The Week.* http://theweek.com/article/index/221008/is-eyewitness-testimony-too-unreliable-to-trust

Thibodeau, P. H., Blonder, A., & Flusberg, S. J. (2020). A connectionist account of the relational shift and context sensitivity in the development of generalisation. *Connection Science.* https://doi.org/10.1080/09540091.2020.1728519

Thiel, F., & Dekel, S. (2020). Peritraumatic dissociation in childbirth-evoked posttraumatic stress and postpartum mental health. *Archives of Women's Mental Health, 23*(2), 189–197.

Thielmann, I., Spadaro, G., & Balliet, D. (2020). Personality and prosocial behavior: A theoretical framework and meta-analysis. *Psychological Bulletin, 146*(1), 30–90.

Thigpen, C. H., & Cleckley, H. M. (1957). *The three faces of Eve.* McGraw-Hill.

Thijssen, S., & Kiehl, K. A. (2017). Functional connectivity in incarcerated male adolescents with psychopathic traits. *Psychiatry Research: Neuroimaging, 265*, 35–44.

Thimmapuram, J., Yommer, D., Tudor, L., Bell, T., Dumitrescu, C., & Davis, R. (2020). Heartfulness meditation improves sleep in chronic insomnia. *Journal of Community Hospital Internal Medicine Perspectives, 10*(1), 10–15.

Thomas, A., & Daley, A. J. (2020). Women's views about physical activity as a treatment for vasomotor menopausal symptoms: A qualitative study. *BMC Women's Health, 20*(1), 1–11.

Thomas, J. G., Bond, D. S., Phelan, S., Hill, J. O., & Wing, R. R. (2014). Weight-loss maintenance for 10 years in the National Weight Control Registry. *American Journal of Preventive Medicine, 46*, 17–23.

Thomas, J. I. (2019). Current status of consciousness research from the neuroscience perspective. *Acta Scientific Neurology, 2*(2), 38–44.

Thomas, M. S. C., & Johnson, M. H. (2008). New advances in understanding sensitive periods in brain development. *Current Directions in Psychological Science, 17*(1), 1–5.

Thompson, B. L., Twohig, M. P., & Luoma, J. B. (2020). Psychological flexibility as shared process of change in acceptance and commitment therapy and exposure and response prevention for obsessive–compulsive disorder: A single case design study. *Behavior Therapy.* https://doi.org/10.1016/j.beth.2020.04.011

Thompson, C. E., Namusoke, J., & Isaac De Barros, K. (2020). On Pan-Africanism, feminism, and psychotherapy: The perspectives of three Black scholar-practitioners from the US, Uganda, and St. Kitts/US. *Women & Therapy.* https://doi.org/10.1080/02703149.2020.1775980

Thompson, K. J., & Tobin, A. B. (2020). Crosstalk between the M1 muscarinic acetylcholine receptor and the endocannabinoid system: A relevance for Alzheimer's disease? *Cellular Signalling*, 109545.

Thompson, R. A. (2015a). Early attachment and later development: New questions. In J. Cassidy & P. R. Shaver (Eds.), *Handbook of attachment* (3rd ed.). Guilford.

Thompson, R. A. (2015b). Relationships, regulation, and development. In R. M. Lerner & Michael E. Lamb (Eds.), *Handbook of child psychology and developmental science* (7th ed., vol. 3). Wiley.

Thompson, R. A., Meyer, S., & McGinley, M. (2014). Understanding values in relationships: The development of conscience. In M. Killen & J. G. Smetana (Eds.), *Handbook of moral development* (2nd ed., pp. 267–298). Psychology Press.

Thompson, T. (2013). Autism research and services for young children: History, progress, and challenges. *Journal of Applied Research on Intellectual Disabilities, 26*(2), 1–27.

Thompson, T., Mitchell, J. A., Johnson-Lawrence, V., Watkins, D. C., & Modlin, C. J. (2017). Self-rated health and health care access associated with African American men's health self-efficacy. *American Journal of Men's Health, 11*(5), 1385–1387. https://doi.org/10.1177/1557988315598555

Thongseiratch, T., & Chandeying, N. (2020). Chronic illnesses and student academic performance. *Journal of Health Science and Medical Research, 38*(3), 245–253.

Thorell, L. B., Sjöwall, D., Diamantopoulou, S., Rydell, A. M., & Bohlin, G. (2017). Emotional functioning, ADHD symptoms, and peer problems: A longitudinal investigation of children age 6–9.5 years. *Infant and Child Development, 26*(4), E2008.

Thornberry, A., Garcia, T. J., Peck, J., & Sefcik, E. (2020). Occupational health nurses' self-efficacy in smoking cessation interventions: An integrative review of the literature. *Workplace Health & Safety, 68*(11), 533–543.

Thorndike, E. L. (1898). Animal intelligence: An experimental study of the associative processes in animals. *Psychological Review,* monograph supplements, no. 8. Macmillan.

Thorne, S. R., Hegarty, P., & Hepper, E. G. (2019). Equality in theory: From a heteronormative to an inclusive psychology of romantic love. *Theory & Psychology, 29*(2), 240–257.

Thorstad, R., & Wolff, P. (2019). Predicting future mental illness from social media: A big-data approach. *Behavior Research Methods, 51*(4), 1586–1600.

Tigani, X., Artemiadis, A. K., Alexopoulos, E. C., Chrousos, G. P., & Darviri, C. (2012). Self-rated health in centenarians: A nation-wide cross-sectional Greek study. *Archives of Gerontology and Geriatrics, 54*(3), e342–e348.

Tiger, M., Veldman, E. R., Ekman, C. J., Halldin, C., Svenningsson, P., & Lundberg, J. (2020). A randomized placebo-controlled PET study of ketamine's effect on serotonin 1B receptor binding in patients with SSRI-resistant depression. *Translational Psychiatry, 10*(1), 1–8.

Timimi, S. (2017). Non-diagnostic based approaches to helping children who could be labelled ADHD and their families. *International Journal of Qualitative Studies on Health and Well-being, 12*(sup1). https://doi.org/10.1080/17482631.2017.1298270

Timmins, L., Rimes, K. A., & Rahman, Q. (2020). Is being queer gay? Sexual attraction patterns, minority stressors, and psychological distress in non-traditional categories of sexual orientation. *The Journal of Sex Research.* https://doi.org/10.1080/00224499.2020.1849527

Titova, L., & Sheldon, K. M. (2020). Thwarted beneficence: Not getting to help lowers mood. *The Journal of Positive Psychology.* https://doi.org/10.1080/17439760.2020.1858339

Tobin, R. M., & Graziano, W. G. (2020). Agreeableness. In B. J. Carducci, C.S. Nave, J. S. Mio, & R. E. Riggio (Eds.), *The Wiley encyclopedia of personality and individual differences: Models and theories* (vol. I, pp. 105–110). John Wiley & Sons.

Todd, B. K., Fischer, R. A., Di Costa, S., Roestorf, A., Harbour, K., Hardiman, P., & Barry, J. A. (2018). Sex differences in children's toy preferences: A systematic review, meta-regression, and meta-analysis. *Infant and Child Development.*

Todd, K. H., Deaton, C., D'Adamo, A. P., & Goe, L. (2000). Ethnicity and analgesic practice. *Annals of Emergency Medicine, 35*(1), 11–16.

Todorov, A. (2017). *Face value: The irresistible influence of first impressions.* Princeton University Press.

Tolman, E. C. (1932). *Purposive behavior in animals and man.* Appleton-Century-Crofts.

Tolman, E. C., & Honzik, C. H. (1930). Degrees of hunger, reward and non-reward, and maze performance in rats. *University of California Publications in Psychology, 4*, 21–256.

Tomiyama, A. J. (2019). Stress and obesity. *Annual Review of Psychology, 70*, 703–718

Tomljenovic, H., Bubic, A., & Erceg, N. (2020). It just doesn't feel right–the relevance of emotions and intuition for parental vaccine conspiracy beliefs and vaccination uptake. *Psychology & Health, 35*(5), 538–554.

Tonarely, N. A., Sherman, J. A., Grossman, R. A., Shaw, A. M., & Ehrenreich-May, J. (2020). Neuroticism as an underlying construct in youth emotional disorders. *Bulletin of the Menninger Clinic, 84*(3), 214–236.

Tong, E. W. (2015). Differentiation of 13 positive emotions by appraisals. *Cognition And Emotion, 29*(3), 484–503. https://doi.org/10.1080/02699931.2014.922056

Tong, E. W., & Jia, L. (2017). Positive emotion, appraisal, and the role of appraisal overlap in positive emotion co-occurrence. *Emotion, 17*(1), 40–54. https://doi.org/10.1037/emo0000203

Tornero, D., Tsupykov, O., Granmo, M., Rodriguez, C., Grønning-Hansen, M., Thelin, J., et al. (2017). Synaptic inputs from stroke-injured brain to grafted human stem cell-derived neurons activated by sensory stimuli. *Brain, 140*(3), 692–706.

Toumbourou, J. W. (2016). Beneficial action within altruistic and prosocial behavior. *Review of General Psychology, 20*(3), 245–258. https://doi.org/10.1037/gpr0000081

Tran, T. B., & Joormann, J. (2015). The role of Facebook use in mediating the relation between rumination and adjustment after a relationship breakup. *Computers in Human Behavior, 49*, 56–61.

Tranter, S., & Robertson, M. (2020). Improving the physical health of people with a mental illness: Holistic nursing assessments. *Mental Health Practice, 23*(3). https://doi.org/10.7748/mhp.2019.e1334

Travagin, G., Margola, D., & Revenson, T. A. (2015). How effective are expressive writing interventions for adolescents? A meta-analytic review. *Clinical Psychology Review, 36*, 42–55. https://doi.org/10.1016/j.cpr.2015.01.003

Travis, F. (2020). Temporal and spatial characteristics of meditation EEG. *Psychological Trauma: Theory, Research, Practice, and Policy, 12*(2), 111–115.

Trawalter, S., & Hoffman, K. M. (2015). Got pain? Racial bias in perceptions of pain. *Social and Personality Psychology Compass, 9*(3), 146–157.

Trent, J., Lavelock, C., & King, L. A. (2013). Processing fluency, positive affect, and meaning in life. *Journal of Positive Psychology, 8*(2), 135–139.

Tretiakov, A., Malakhova, A., Naumova, E., Rudko, O., & Klimov, E. (2020). Genetic biomarkers of panic disorder: A systematic review. *Genes, 11*(11), 1310.

Trevino, A. E., Sinnott-Armstrong, N., Andersen, J., Yoon, S. J., Huber, N., Pritchard, J. K., . . . Paşca, S. P. (2020). Chromatin accessibility dynamics in a model of human forebrain development. *Science, 367*(6476).

Triandis, H. C. (2000). Cross-cultural psychology: History of the field. In A. Kazdin (Ed.), *Encyclopedia of psychology.* American Psychological Association and Oxford University Press.

Trindade, I. A., Mendes, A. L., & Ferreira, N. B. (2020). The moderating effect of psychological flexibility on the link between learned helplessness and depression symptomatology: A preliminary study. *Journal of Contextual Behavioral Science, 15*, 68–72.

Trivedi, P., & Bhargava, N. (2017). Effect of left and right hemisphere of brain in both eye open and close state on minimum power values and frequency of alpha wave activity. *Brain, 6*, 233–237.

Troisi, J., Autio, R., Beopoulos, T., Bravaccio, C., Carraturo, F., Corrivetti, G., . . . Gea, M. (2020). Genome, environment, microbiome and metabolome in autism (GEMMA) study design: Biomarkers identification for precision treatment and primary prevention of autism spectrum disorders by an integrated multi-omics systems biology approach. *Brain Sciences, 10*(10), 743.

Tromholt, M. (2016). The Facebook experiment: Quitting Facebook leads to higher levels of well-being. *Cyberpsychology, Behavior, And Social Networking, 19*(11), 661–666. https://doi.org/10.1089/cyber.2016.0259

Tronchin, G., Akudjedu, T. N., Ahmed, M., Holleran, L., Hallahan, B., Cannon, D. M., & McDonald, C. (2020). Progressive subcortical volume loss in treatment-resistant schizophrenia patients after commencing clozapine treatment. *Neuropsychopharmacology, 45*, 1353–1361.

Trujillo, L. T., Jankowitsch, J. M., & Langlois, J. H. (2014). Beauty is in the ease of the beholding: A neurophysiological test of the averageness theory of facial attractiveness. *Cognitive, Affective & Behavioral Neuroscience, 14*(3), 1061–1076. https://doi.org/10.3758/s13415-013-0230-2

Trull, T. J., & Brown, W. C. (2013). Borderline personality disorder: A five-factor model perspective. In T. A. Widiger & P. T. Costa (Eds.), *Personality disorders and the five-factor model of personality* (3rd ed., pp. 119–132). American Psychological Association.

Tryon, R. C. (1940). Genetic differences in maze-learning ability in rats. In *39th Yearbook, National Society for the Study of Education*. Chicago: University of Chicago Press.

Trzaskowski, M., Yang, J., Visscher, P. M., & Plomin, R. (2014). DNA evidence for strong genetic stability and increasing heritability of intelligence from age 7 to 12. *Molecular Psychiatry, 19*(3), 380–384.

Tskhay, K. O., Zhu, R., & Rule, N. O. (2017). Perceptions of charisma from thin slices of behavior predict leadership prototypicality judgments. *The Leadership Quarterly, 28*(4), 555–562. https://doi.org/10.1016/j.leaqua.2017.03.003

Tsur, N., Defrin, R., Shahar, G., & Solomon, Z. (2020). Dysfunctional pain perception and modulation among torture survivors: the role of pain personification. *Journal of Affective Disorders, 265*, 10–17.

Tubbs, A. S., Khader, W., Fernandez, F., & Grandner, M. A. (2020). The common denominators of sleep, obesity, and psychopathology. *Current Opinion in Psychology, 34*, 84–88.

Tuck, I., & Anderson, L. (2014). Forgiveness, flourishing, and resilience: The influences of expressions of spirituality on mental health recovery. *Issues in Mental Health Nursing, 35*(4), 277–282.

Tucker-Drob, E. M. (2011). Individual differences methods for randomized experiments. *Psychological Methods, 16*(3), 298–318.

Tugade, M. M., Fredrickson, B. L., & Feldman Barrett, L. (2004). Psychological resilience and positive emotional granularity: Examining the benefits of positive emotions on coping and health. *Journal of Personality, 72*(6), 1161–1190.

Tulving, E. (1972). Episodic and semantic memory. In E. Tulving & W. Donaldson (Eds.), *Origins of memory*. Academic.

Tulving, E. (1983). *Elements of episodic memory*. Oxford University Press.

Tulving, E. (1989). Remembering and knowing the past. *American Scientist, 77*(4), 361–367.

Tulving, E. (2000). Concepts of memory. In E. Tulving & F. I. M. Craik (Eds.), *The Oxford handbook of memory*. Oxford University Press.

Tuncay, T., & Musabak, I. (2015). Problem-focused coping strategies predict posttraumatic growth in veterans with lower-limb amputations. *Journal of Social Service Research, 41*(4), 466–483. https://doi.org/10.1080/01488376.2015.1033584

Turiano, N. A., Chapman, B. P., Agrigoroaei, S., Infurna, F. J., & Lachman, M. (2014). Perceived control reduces mortality risk at low, not high, education levels. *Health Psychology, 33*(8), 883–890.

Turiano, N. A., Chapman, B. P., Gruenewald, T. L., & Mroczek, D. K. (2015). Personality and the leading behavioral contributors of mortality. *Health Psychology, 34*(1), 51–60. https://doi.org/10.1037/hea0000038

Turiano, N. A., Chapman, B. P., Gruenewald, T. L., & Mroczek, D. K. (2015). Personality and the leading behavioral contributors of mortality. *Health Psychology, 34*, 51–60.

Turiano, N. A., Graham, E. K., Weston, S., Booth, T., Harrison, F., James, B. D., . . . Mroczek, D. K. (2020). Is healthy neuroticism associated with longevity? A coordinated integrative data analysis. *Collabra: Psychology, 6*(1), 33. http://doi.org/10.1525/collabra.268

Turiano, N. A., Pitzer, L., Armour, C., Karlamangla, A., Ryff, C. D., & Mroczek, D. K. (2012). Personality trait level and change as predictors of health outcomes: Findings from a national study of Americans (MIDUS). *Journal of Gerontology B: Psychological Sciences and Social Sciences, 67*(1), 4–12.

Turkewitz, J. (2017, September 21). 'The Pills are Everywhere': The opioid crisis and its youngest victims. *New York Times*, A15.

Turkheimer, E. (2011). Genetics and human agency: Comment on Dar-Nimrod and Heine. *Psychological Bulletin, 137*(5), 825–828.

Turna, J., Kaplan, K. G., Patterson, B., Bercik, P., Anglin, R., Soreni, N., & Van Ameringen, M. (2019). Higher prevalence of irritable bowel syndrome and greater gastrointestinal symptoms in obsessive-compulsive disorder. *Journal of Psychiatric Research, 118*, 1–6.

Turner, R. N., Hodson, G., & Dhont, K. (2020). The role of individual differences in understanding and enhancing intergroup contact. *Social and Personality Psychology Compass, 14*(6), e12533.

Tussis, L., Sollmann, N., Boeckh-Behrens, T., Meyer, B., & Krieg, S. M. (2017). Identifying cortical first and second language sites via navigated transcranial magnetic stimulation of the left hemisphere in bilinguals. *Brain and Language, 168*, 106–116.

Twenge, J. M., Sherman, R. A., & Wells, B. E. (2017). Declines in sexual frequency among American adults, 1989–2014. *Archives of Sexual Behavior*.

Tyas, S. L., Salazar, J. C., Snowdon, D. A., Desrosier, M. F., Riley, K. P., Mendiondo, M. S., & Kryscio, R. J. (2007). Transitions to mild cognitive impairments, dementia, and death: Findings from the Nun Study. *American Journal of Epidemiology, 165*(11), 1231–1238.

U

Uddin, M. S., Sufian, M. A., Hossain, M. F., Kabir, M. T., Islam, M. T., Rahman, M. M., & Rafe, M. R. (2017). Neuropsychological effects of caffeine: Is caffeine addictive. *Journal of Psychology and Psychotherapy, 7*(295), 2161–0487.

Ullman, A. D. (1952). Review of "Antabuse" in the treatment of alcoholism. *Psychological Bulletin, 49*(5), 557–558.

Ullrich, M., Weber, M., Post, A. M., Popp, S., Grein, J., Zechner, M., et al. (2017). OCD-like behavior is caused by dysfunction of thalamo-amygdala circuits and upregulated TrkB/ERK-MAPK signaling as a result of SPRED2 deficiency. *Molecular Psychiatry*. https://doi.org/10.1038/mp.2016.232

Ulrich, S., Ricken, R., Buspavanich, P., Schlattmann, P., & Adli, M. (2020). Efficacy and adverse effects of tranylcypromine and tricyclic antidepressants in the treatment of depression: A systematic review and comprehensive meta-analysis. *Journal of Clinical Psychopharmacology, 40*(1), 63–74.

Ünal, E., & Papafragou, A. (2020). Relations between language and cognition: evidentiality and sources of knowledge. *Topics in Cognitive Science, 12*(1), 115–135.

Underhill, S. M., Colt, M. S., & Amara, S. G. (2020). Amphetamine stimulates endocytosis of the norepinephrine and neuronal glutamate transporters in cultured locus coeruleus neurons. *Neurochemical Research, 45*:1410–1419. https://doi.org/10.1007/s11064-019-02939-6

Underwood, M. K. (2011). Aggression. In M. K. Underwood & L. Rosen (Eds.), *Social development*. Guilford.

UNDOC. (2020). UNDOC world drug report 2020; global drug use rising; while COVID-19 has far-reaching impact on global drug markets. https://www.unodc.org/unodc/press/releases/2020/June/media-advisory—global-launch-of-the-2020-world-drug-report.html

Undurraga, J., & Baldessarini, R. J. (2012). Randomized, placebo-controlled trials of antidepressants for acute major depression: Thirty-year meta-analytic review. *Neuropsychopharmacology, 37*(4), 851–864.

Undurraga, J., & Baldessarini, R. J. (2017). Direct comparison of tricyclic and serotonin-reuptake inhibitor antidepressants in randomized head-to-head trials in acute major depression: Systematic review and meta-analysis. *Journal of Psychopharmacology, 31*(9), 1184–1189.

United Nations. (2011, July 19). Happiness should have greater role in development policy—UN Member States. *UN News Centre*. www.un.org/apps/news/story.asp?NewsID=39084 (accessed November 3, 2014)

United Network for Organ Sharing (UNOS). (2020). Living donation. https://unos.org/transplant/living-donation/

University of Western Ontario. (2018). Researchers map brain of blind patient who can see motion. *Science Daily*. www.sciencedaily.com/releases/2018/06/180612105716.htm

Unsworth, N., & Robison, M. K. (2020). Working memory capacity and sustained attention: A cognitive-energetic perspective. *Journal of Experimental Psychology: Learning, Memory, and Cognition, 46*(1), 77–103.

Urbina, S. (2011). Tests of intelligence. In R. J. Sternberg & S. B. Kaufman (Eds.), *The Cambridge handbook of intelligence*. Cambridge University Press.

Urgolites, Z. J., Wixted, J. T., Goldinger, S. D., Papesh, M. H., Treiman, D. M., Squire, L. R., & Steinmetz, P. N. (2020). Spiking activity in the human hippocampus prior to encoding predicts subsequent memory. *Proceedings of the National Academy of Sciences, 117* (24) 13767–13770

Urquhart, J. A., & O'Connor, A. R. (2014). The awareness of novelty for strangely familiar words: A laboratory analogue of the déjà vu experience. *PeerJ, 2*, e666.

Urquhart, J. A., Sivakumaran, M. H., MacFarlane, J. A., & O'Connor, A. R. (2018). fMRI evidence supporting the role of memory conflict in the déjà vu experience. *Memory*, 1–12. https://doi.org/10.1080/09658211.2018.1524496

Urry, H. L., & others. (2004). Making a life worth living: Neural correlates of well-being. *Psychological Science, 15*(6), 367–372.

U.S. Census Bureau. (2020). Table MS-2: Estimated median age at first marriage. https://www2.census.gov/programs-surveys/demo/tables/families/time-series/marital/ms2.xls; https://www.census.gov/data/tables/time-series/demo/families/marital.html

U.S. Department of Justice. (2015). *Overview of Title IX of the Education Amendments of 1972, 20 U.S.C. A§ 1681 Et. Seq.* https://www.justice.gov/crt/overview-title-ix-education-amendments-1972-20-usc-1681-et-seq

Uysal, A., Ascigil, E., & Turunc, G. (2017). Spousal autonomy support, need satisfaction, and well-being in individuals with chronic pain: A longitudinal study. *Journal of Behavioral Medicine, 40*(2), 281–292. https://doi.org/10.1007/s10865-016-9783-1

V

Vaidya, A. R., Pujara, M. S., Petrides, M., Murray, E. A., & Fellows, L. K. (2019). Lesion studies in contemporary neuroscience. *Trends in Cognitive Sciences, 23*(8), 653–671.

Vail, K. E., Juhl, J., Arndt, J., Vess, M., Routledge, C., & Rutjens, B. T. (2012). When death is good for life: Considering the positive trajectories of terror management. *Personality and Social Psychology Review, 16*(4), 303–329.

Vaillant, G. (2003). A 60-year follow-up of alcoholic men. *Addiction, 98*(8), 1043–1051.

Valadi, S., & Gabbard, C. (2020). The effect of affordances in the home environment on children's fine-and gross motor skills. *Early Child Development and Care, 190*(8), 1225–1232.

Valentin, D. (2020). Femininities & masculinities: Sex, gender, and stereotypes in food studies. *Current Opinion in Food Science, 33*, 156–164.

Valentine, K. A., Li, N. P., Meltzer, A. L., & Tsai, M. H. (2020). Mate preferences for warmth-trustworthiness predict romantic attraction in the early stages of mate selection and satisfaction in ongoing relationships. *Personality and Social Psychology Bulletin, 46*(2), 298–311.

Valenza, G., Citi, L., Lanata, A., Scilingo, E. P., & Barbieri, R. (2014, May 21). Revealing real-time emotional responses: A personalized assessment based on heartbeat dynamics. *Scientific Reports, 4*, 4998.

Valenzuela, R., Codina, N., Castillo, I., & Pestana, J. V. (2020). Young university students' academic self-regulation profiles and their associated procrastination: Autonomous functioning requires self-regulated operations. *Frontiers in Psychology, 11*.

Vallée, B., Magoutier, F., Voisin, D., & Montalan, B. (2020). Reducing the effects of the stereotype threat that girls perform less well than boys in mathematics: the efficacy of a mixed debate in a real classroom situation. *Social Psychology of Education, 23*(5), 1327–1341.

Vallée-Tourangeau, F., & Vallée-Tourangeau, G. (2020). Mapping systemic resources in problem solving. *New Ideas in Psychology, 59*, 100812.

Van Bavel, J. J., Reinero, D. A., Harris, E., Robertson, C. E., & Pärnamets, P. (2020). Breaking groupthink: Why scientific identity and norms mitigate ideological episte-mology. *Psychological Inquiry, 31*(1), 66–72.

van den Berg, L., & Gredebäck, G. (2020). The sticky mit-tens paradigm: A critical appraisal of current results and explanations. *Developmental Science,* e13036.

van den Berg, S. M., de Moor, M. M., Verweij, K. H., Krueger, R. F., Luciano, M., Arias Vasquez, A., et al. (2016). Meta-analysis of genome-wide association stud-ies for extraversion: Findings from the genetics of per-sonality consortium. *Behavior Genetics, 46*(2), 170–182. https://doi.org/10.1007/s10519-015-9735-5

van der Kaap-Deeder, J., Audenaert, E., Vandevelde, S., Soe-nens, B., Van Mastrigt, S., Mabbe, E., & Vansteenkiste, M. (2017). Choosing when choices are limited: The role of perceived afforded choice and autonomy in prisoners' well-being. *Law and Human Behavior.* https://doi.org/10.1037/lhb0000259

van Der Ploeg, M. M., Brosschot, J. F., Quirin, M., Lane, R. D., & Verkuil, B. (2020). Inducing unconscious stress: Subliminal anger and relax primes show similar cardio-vascular activity patterns. *Journal of Psychophysiology, 34*(3), 192–201.

van der Wal, R. C., Karremans, J. C., & Cillessen, A. N. (2014). It takes two to forgive: The interactive role of relationship value and executive control. *Personality and Social Psychology Bulletin, 40*(6), 803–815. https://doi.org/10.1177/0146167214525807

Van Diest, I. (2019). Interoception, conditioning, and fear: The panic threesome. *Psychophysiology, 56*(8), e13421.

van Duijl, M., Cardeña, E., & de Jong, J. (2005). The valid-ity of DSM-IV dissociative disorders categories in south-west Uganda. *Transcultural Psychiatry, 42*(2), 219–241.

van Duijl, M., Kleijn, W., & de Jong, J. (2014). Unravelling the spirits' message: a study of help-seeking steps and explanatory models among patients suffering from spirit possession in Uganda. *International Journal of Mental Health Symptoms, 8,* 24.

Van Lange, P. A., Rusbult, C. E., Drigotas, S. M., Arriaga, X. B., Witcher, B. S., & Cox, C. L. (1997). Willingness to sacrifice in close relationships. *Journal of Personality and Social Psychology, 72*(6), 1373–1395.

van Lent, M., & Souverijn, M. (2020). Goal setting and rais-ing the bar: A field experiment. *Journal of Behavioral and Experimental Economics, 87,* 101570.

van Os, J., & Kapur, S. (2009). Schizophrenia. *Lancet, 374*(9690), 635–645.

Van Overwalle, F., Manto, M., Cattaneo, Z., Clausi, S., Ferrari, C., Gabrieli, J. D., . . . Michelutti, M. (2020). Consensus paper: Cerebellum and social cognition. *The Cerebellum.* https://doi.org/10.1007/s12311-020-01155-1

Van Petegem, S., Vansteenkiste, M., Soenens, B., Zimmer-mann, G., Antonietti, J. P., Baudat, S., & Audenaert, E. (2017). When do adolescents accept or defy to maternal prohibitions? The role of social domain and commu-nication style. *Journal of Youth and Adolescence, 46*(5), 1022–1037.

van Prooijen, J. W. (2017). Why education predicts de-creased belief in conspiracy theories. *Applied Cognitive Psychology, 31*(1), 50–58.

van Prooijen, J.-W., & van Vugt, M. (2018). Conspiracy theories: Evolved functions and psychological me-chanisms. *Perspectives on Psychological Science, 13*(6), 770–788.

van Rensburg, D. C. C. J., van Rensburg, A. J., Fowler, P., Fullagar, H., Stevens, D., Halson, S., . . . Roach, G. D. (2020). How to manage travel fatigue and jet lag in ath-letes? A systematic review of interventions. *British Jour-nal of Sports Medicine, 54,* 16.

Van Riper, M. (2007). Families of children with Down syn-drome: Responding to "a change in plans" with resil-ience. *Journal of Pediatric Nursing, 22*(2), 116–128.

van Roekel, E., Verhagen, M., Engels, R. E., & Kuppens, P. (2017). Variation in the serotonin transporter polymor-phism (5-HTTLPR) and inertia of negative and positive emotions in daily life. *Emotion.* https://doi.org/10.1037/emo0000336

van Scheppingen, M. A., Jackson, J. J., Specht, J., Hutte-man, R., Denissen, J. A., & Bleidorn, W. (2016).

Personality trait development during the transition to parenthood: A test of social investment theory. *Social Psychological and Personality Science, 7*(5), 452–462. https://doi.org/10.1177/1948550616630032

van Vliet, I. M., & others. (2013). An evaluation of irrevers-ible psychosurgical treatment of patients with obsessive-compulsive disorder in the Netherlands, 2001–2008. *Journal of Nervous and Mental Disease, 201*(3), 226–228.

Van Zalk, N., & Van Zalk, M. (2020). Early adolescent dis-closure and parental knowledge regarding online activi-ties: Social anxiety and parental rule-setting as moderators. *Current Psychology, 39*(1), 287–298.

Vancampfort, D., Moens, H., Madou, T., De Backer, T., Vallons, V., Bruyninx, P., et al. (2016). Autonomous motivation is associated with the maintenance stage of behaviour change in people with affective disorders. *Psychiatry Research, 240,* 267–271. https://doi.org/10.1016/j.psychres.2016.04.005

Vance, G., Shackelford, T. K., Weekes-Shackelford, V. A., & Abed, M. G. (2020). Later life sex differences in sexual psychology and behavior. *Personality and Individual Dif-ferences, 157,* 109730.

Vanderwal, T., Eilbott, J., Finn, E. S., Craddock, R. C., Turnbull, A., & Castellanos, F. X. (2017). Individual dif-ferences in functional connectivity during naturalistic viewing conditions. *NeuroImage, 157,* 521–530.

Vasconcellos, D., Parker, P. D., Hilland, T., Cinelli, R., Owen, K. B., Kapsal, N., . . . Lonsdale, C. (2020). Self-determination theory applied to physical education: A systematic review and meta-analysis. *Journal of Educa-tional Psychology, 112*(7), 1444–1469.

Veale, D., Murphy, P., Ellison, N., Kanakam, N., & Costa, A. (2013). Autobiographical memories of vomiting in people with a specific phobia of vomiting (emetophobia). *Journal of Behavioral Therapy and Experimental Psychia-try, 44*(1), 14–20.

Vedder, P., & Phinney, J. S. (2014). Identity formation in bicultural youth: A developmental perspective. In V. Benet-Martínez & Y.-Y. Hong (Eds.), *The Oxford hand-book of multicultural identity.* Oxford University Press.

Velik, R. (2012). From simple receptors to complex multi-modal percepts: A first global picture on the mecha-nisms involved in perceptual binding. *Frontiers in Psychology, 3,* 259.

Veltkamp, G. M., Recio, G., Jacobs, A. M., & Conrad, M. (2013). Is personality modulated by language? Interna-tional *Journal of Bilingualism, 17*(4), 496–504.

Vena, J. (2015). Karen Horney. In D. L. Dobbert & T. X. Mackey (Eds.), *Deviance: Theories on behaviors that defy social norms* (pp. 48–57). Praeger/ABC-CLIO.

Verduyn, P., Lee, D. S., Park, J., Shablack, H., Orvell, A., Bayer, J., Ybarra, O., Jonides, J., & Kross, E. (2015). Passive Facebook usage undermines affective well-being: Experimental and longitudinal evidence. *Journal of Ex-perimental Psychology: General, 144*(2), 480–488.

Vermetten, E., Schmahl, C., Lindner, S., Loewenstein, R. J., & Bremner, J. D. (2006). Hippocampal and amygdalar volumes in dissociative identity disorder. *American Jour-nal of Psychiatry, 163*(4), 630–636.

Vernon, D., Hocking, I., & Tyler, T. C. (2016). An evidence-based review of creative problem solving tools: A practitio-ner's resource. *Human Resource Development Review, 15*(2), 230–259. https://doi.org/10.1177/1534484316641512

Veroude, K., Zhang-James, Y., Fernàndez-Castillo, N., Bakker, M. J., Cormand, B., & Faraone, S. V. (2016). Genetics of aggressive behavior: An overview. *Ameri-can Journal of Medical Genetics Part B: Neuropsychiat-ric Genetics, 171*(1), 3–43. https://doi.org/10.1002/ajmg.b.32364

Versace, V., Schwenker, K., Langthaler, P. B., Golasze-wski, S. M., Sebastianelli, L., Brigo, F., . . . Nardone, R. (2019). Facilitation of auditory comprehension after theta burst stimulation of Wernicke's area in stroke patients: A pilot study. *Frontiers in Neurology, 10,* 1319.

Verschuere, B., Crombez, G., De Clercq, A., & Koster, E. H. (2005). Psychopathic traits and autonomic responding to concealed information in a prison sample. *Psycho-physiology, 42*(2), 239–245.

Vickers, K., Ein, N., Koerner, N., Kusec, A., McCabe, R. E., Rowa, K., & Antony, M. M. (2017). Self-reported hy-giene-related behaviors among individuals with contami-nation-related obsessive-compulsive disorder, individuals with anxiety disorders, and nonpsychiatric controls. *Journal of Obsessive-Compulsive and Related Disorders, 14,* 71–83.

Vidal, F. (2018). Phenomenology of the locked-in syndrome: An overview and some suggestions. *Neuroethics.* https://doi.org/10.1007/s12152-018-9388-1

Viljoen, M., Mahdi, S., Shelly, J., & de Vries, P. J. (2020). Parental perspectives of functioning in their children with autism spectrum disorder: A global scoping review. *Autism.* https://doi.org/10.1177/1362361320950055

Vinberg, M., Mellerup, E., Andersen, P. K., Bennike, B., & Kessing, L. V. (2010). Variations in 5-HTTLPR: Relation to familiar risk of affective disorder, life events, neuroti-cism, and cortisol. *Progress in Neuro-Psychopharmacology & Biological Psychiatry, 34*(1), 86–91.

Vinograd, M., & Craske, M. G. (2020). Using neuroscience to augment behavioral interventions for depression. *Har-vard Review of Psychiatry, 28*(1), 14–25.

Viswanath, B., Maroky, A. S., Math, S. B., John, J. P., Benegal, V., Hamza, A., & Chaturvedi, S. K. (2012). Psychological impact of the tsunami on elderly survi-vors. *American Journal of Geriatric Psychiatry, 20*(5), 402–407.

Vlaeminck, F., Vermeirsch, J., Verhaeghe, L., Warreyn, P., & Roeyers, H. (2020). Predicting cognitive development and early symptoms of autism spectrum disorder in preterm children: The value of temperament and sen-sory processing. *Infant Behavior and Development, 59,* 101442.

Vogel, E. H., Ponce, F. P., & Brandon, S. E. (2020). Can the stimulus processing assumptions of the sometimes-opponent-process (SOP) model explain instances of con-textual learning? *Journal of Experimental Psychology: Animal Learning and Cognition, 46*(3), 205–214.

Vohs, K. D., & Luce, M. F. (2010). Judgment and decision making. In R. F. Baumeister & E. J. Finkel (Eds.), *Ad-vanced social psychology: The state of the science* (pp. 733–756). Oxford University Press.

Volenec, V., & Reiss, C. (2020). Formal generative phonol-ogy. *Radical: A Journal of Phonology, 2,* 1–65.

Volkow, N. D., Michaelides, M., & Baler, R. (2019). The neuroscience of drug reward and addiction. *Physiological Reviews, 99*(4), 2115–2140.

Volkow, N. D., Wang, G. J., Newcorn, J. H., Kollins, S. H., Wigal, T. L., Telang, F., et al. (2011). Motivation defi-cit in ADHD is associated with dysfunction of the dopamine reward pathway. *Molecular Psychiatry, 16*(11), 1147–1154.

Volpe, J. J. (2020). Commentary–Marijuana use during preg-nancy and premature birth: A problem likely to worsen. *Journal of Neonatal-Perinatal Medicine, 13*(1), 1–3.

von Békésy, G. (1960). Vibratory patterns of the basilar membrane. In E. G. Wever (Ed.), *Experiments in hear-ing.* McGraw-Hill.

von Dawans, B., Ditzen, B., Trueg, A., Fischbacher, U., & Heinrichs, M. (2019). Effects of acute stress on so-cial behavior in women. *Psychoneuroendocrinology, 99,* 137–144.

Von Neumann, J. (1958). *The computer and the brain.* Yale University Press.

Vosberg, D. E., Syme, C., Parker, N., Richer, L., Pausova, Z., & Paus, T. (2020). Sex continuum in the brain and body during adolescence and psychological traits. *Na-ture Human Behaviour,* 1–8.

Vrticka, P., Lordier, L., Bediou, B., & Sander, D. (2014). Human amygdala response to dynamic facial expres-sions of positive and negative surprise. *Emotion, 14*(1), 161–169. https://doi.org/10.1037/a0034619

Vygotsky, L. S. (1962). *Thought and language.* MIT Press.

W

Wachtel, P. L. (2017). Psychoanalysis and the Moebius strip: Reexamining the relation between the internal world and the world of daily experience. *Psychoanalytic*

Psychology, 34(1), 58–68. https://doi.org/10.1037/pap0000101

Wachtel, P. L., Siegel, J. P., & Baer, J. C. (2020). The scope of psychotherapy integration: Introduction to a special issue. *Clinical Social Work Journal, 48*, 231–235

Wacker, J. (2017). Effects of positive emotion, extraversion, and dopamine on cognitive stability-flexibility and frontal eeg asymmetry. *Psychophysiology, 55*(1), e12727. https://doi.org/10.1111/psyp.12727

Wacker, J. (2018). Effects of positive emotion, extraversion, and dopamine on cognitive stability-flexibility and frontal EEG asymmetry. *Psychophysiology, 55* (1), e12727.

Wacker, J., Mueller, E. M., Hennig, J., & Stemmler, G. (2012). How to consistently link extraversion and intelligence to the catechol-O-methyltransferase (COMT) gene: On defining and measuring psychological phenotypes in neurogenetic research. *Journal of Personality and Social Psychology, 102*(2), 427–444.

Wacker, J., & Smillie, L. D. (2015). Trait extraversion and dopamine function. *Social and Personality Psychology Compass, 9* (6), 225–238.

Wagner, A. D., Schacter, D. L., Rotte, M., Koutstaal, B., Maril, A., Dale, A. M., . . . Buckner, R. L. (1998). Building memories: Remembering and for-getting of verbal experiences as predicted by brain activity. *Science, 281*(5380), 1185–1187.

Wagner, M. T., Mithoefer, M. C., Mithoefer, A. T., MacAulay, R. K., Jerome, L., Yazar-Klosinski, B., & Doblin, R. (2017). Therapeutic effect of increased openness: Investigating mechanism of action in MDMA-assisted psychotherapy. *Journal of Psychopharmacology, 31*(8), 967–974.

Wahlberg, L. U. (2020). Brain implants. In R. Lanza. R. Langer, J. P. Vacant, & A. Atala (Eds.), *Principles of tissue engineering* (5th ed., pp. 1025–1035). Academic Press.

Wai, J., Lubinski, D., & Benbow, C. P. (2005). Creativity and occupational accomplishments among intellectually precocious youths: An age 13 to age 33 longitudinal study. *Journal of Educational Psychology, 97*(3), 484–492.

Wakefield, J. C. (2016). Diagnostic issues and controversies in DSM-5: return of the false positives problem. *Annual Review of Clinical Psychology, 12*, 105–132.

Wakita, M. (2020). Language evolution from a perspective of Broca's area. In N. Masataka (Ed.), *The origins of language revisited* (pp. 97–113). Springer.

Walder, D. J., Ospina, L., Daly, M., Statucka, M., & Raparia, E. (2012). Early neurodevelopment and psychosis risk: Role of neurohormones and biological sex in modulating genetic, prenatal, and sensory processing factors in brain development. In X. Anastassion-Hadjicharalambous (Ed.), *Psychosis: Causes, diagnosis and treatment.* Nova Science.

Walker, B. R., Jackson, C. J., & Frost, R. (2017). A comparison of revised reinforcement sensitivity theory with other contemporary personality models. *Personality and Individual Differences, 109*, 232–236. https://doi.org/10.1016/j.paid.2016.12.053

Walker, L. J. (2014a). Moral personality, motivation, and identity. In M. Killen & J. G. Smetana (Eds.), *Handbook of moral development* (2nd ed.). Psychology Press.

Walker, L. J. (2014b). Prosocial exemplarity in adolescence and adulthood. In L. Padilla-Walker & G. Carlo (Eds.), *Prosocial behavior.* Oxford University Press.

Walkup, J. T., & Mohatt, J. W. (2017). 1.2 Psychopharmacology of obsessive-compulsive disorder: Weighing the evidence versus the risk. *Journal of the American Academy of Child & Adolescent Psychiatry, 56*(10), S134.

Wallace, J., Covassin, T., Nogle, S., Gould, D., & Kovan, J. (2017). Knowledge of concussion and reporting behaviors in high school athletes with or without access to an athletic trainer. *Journal of Athletic Training, 52*(3), 228–235.

Wallace-Wells, D. (2020, December 7). We had the vaccine the whole time. *New York Times Magazine.* https://nymag.com/intelligencer/2020/12/moderna-covid-19-vaccine-design.html.

Wallien, M. S. C., Veenstra, R., Kreukels, B. P. C., & Cohen-Kettenis, P. T. (2010). Peer group status of gender dysphoric children: A sociometric study. *Archives of Sexual Behavior, 39*(2), 553–560.

Walsh, J. J., Christoffel, D. J., Wu, X., Pomrenze, M. B., & Malenka, R. C. (2020). Dissecting neural mechanisms of prosocial behaviors. *Current Opinion in Neurobiology, 68*, 9–14.

Walter, N., Bilandzic, H., Schwarz, N., & Brooks, J. J. (2020). Metacognitive approach to narrative persuasion: The desirable and undesirable consequences of narrative disfluency. *Media Psychology.* https://doi.org/10.1080/15213269.2020.1789477

Walton, K., Breen, A., Gruson-Wood, J., Jewell, K., Haycraft, E., & Haines, J. (2020). Dishing on dinner: A life course approach to understanding the family meal context among families with preschoolers. *Public Health Nutrition*, 1–11. https://doi.org/10.1017/S1368980020001779

Walton, K. E., & Roberts, B. W. (2004). On the relationship between substance use and personality traits: Abstainers are not maladjusted. *Journal of Research in Personality, 38*(6), 515–535.

Wampler, K.S., & McWey, L. M. (Eds.). (2020). *The handbook of systemic family therapy.* Wiley.

Wampold, B. E. (2001). *The great psychotherapy debate: Models, methods, and findings.* Erlbaum.

Wampold, B. E. (2013). The good, the bad, and the ugly: A 50-year perspective on the outcome problem. *Psychotherapy, 50*(1), 16–24.

Wampold, B. E., & Brown, G. S. (2005). Estimating variability in outcomes attributable to therapists: A naturalistic study of outcomes of managed care. *Journal of Consulting and Clinical Psychology, 73*(5), 914–923.

Wang, A. R., Groome, A. M., Taniguchi, L., Eshel, N., & Bentzley, B. S. (2020). The role of dopamine in reward-related behavior: Shining new light on an old debate. *Journal of Neurophysiology.*

Wang, C. C., Shih, H. C., Shyu, B. C., & Huang, A. C. W. (2017). Effects of thalamic hemorrhagic lesions on explicit and implicit learning during the acquisition and retrieval phases in an animal model of central post-stroke pain. *Behavioural brain research, 317*, 251–262.

Wang, H., Leng, Y., Zhao, E., Fleming, J., Brayne, C., & CC75C Study Collaboration. (2020). Mortality risk of loneliness in the oldest old over a 10-year follow-up. *Aging & Mental Health, 24*(1), 35–40.

Wang, W. (2020, November 10). The U.S. divorce rate has hit a 50-year low. IF Studies.org. https://ifstudies.org/blog/the-us-divorce-rate-has-hit-a-50-year-low#:~:text=It%20is%20even%20slightly%20lower,in%20divorce%20per%201%2C000%20marriages.&text=The%20drop%20in%20the%20divorce,because%20of%20lockdown%2Drelated%20stress

Wang, W., Zhou, X., & Liu, Y. (2020). Characterization and evaluation of umami taste: A review. *Trends in Analytical Chemistry, 127*, 115876.

Wang, X., Cheng, B., Luo, Q., Qiu, L., & Wang, S. (2018). Gray matter structural alterations in social anxiety disorder: A voxel-based meta-analysis. *Frontiers in Psychiatry, 9*, 449.

Wang, Y., Di, M., Zhao, J., Hu, S., Yao, Z., & Wang, Y. (2020). Attentional modulation of unconscious inhibitory visuomotor processes: An EEG study. *Psychophysiology*, e13561.

Wang, Y., & Yu, C. (2017). Social interaction-based consumer decision-making model in social commerce: The role of word of mouth and observational learning. *International Journal of Information Management, 37*(3), 179–189.

Wang, Y., Zuo, C., Xu, Q., Hao, L., & Zhang, Y. (2020). Attention-deficit/hyperactivity disorder is characterized by a delay in subcortical maturation. *Progress in Neuro-Psychopharmacology and Biological Psychiatry, 104*, 110044.

Wannemüller, A., Moser, D., Kumsta, R., Jöhren, H. P., & Margraf, J. (2018). The return of fear: variation of the serotonin transporter gene predicts outcome of a highly standardized exposure-based one-session fear treatment. *Psychotherapy and psychosomatics, 87* (2), 95–104.

Ward, J., & Filiz, G. (2020). Synaesthesia is linked to a distinctive and heritable cognitive profile. *Cortex, 126*, 134–140.

Ward, S. J., & King, L. A. (2016). Socrates' dissatisfaction, a happiness arms race, and the trouble with eudaimonic well-being. In J. Vittersø (Ed.), *Handbook of Eudaimonic Wellbeing.* (pp. 523–531). Springer.

Ward, S. J., & King, L. A. (2016). Socrates' dissatisfaction, a happiness arms race, and the trouble with eudaimonic well-being. In J. Vittersø (Ed.), *Handbook of Eudaimonic Wellbeing* (pp. 523–531). Springer.

Ward, S. J., & King, L. A. (2019). Exploring the place of financial status in the good life: Income and meaning in life. *Journal of Positive Psychology, 14*(3), 312–323. https://doi.org/10.1080/17439760.2017.1402075

Wardenaar, K. J., Lim, C. C., Al-Hamzawi, A. O., Alonso, J., Andrade, L. H., Benjet, C., et al. (2017). The cross-national epidemiology of specific phobia in the World Mental Health Surveys. *Psychological Medicine, 47*(10), 1744–1760.

Ware, A. T., Kirkovski, M., & Lum, J. A. (2020). Meta-analysis reveals a bilingual advantage that is dependent on task and age. *Frontiers in Psychology, 11.*

Warneken, F., & Tomasello, M. (2007). Helping and cooperation at 14 months of age. *Infancy, 11*, 271– 294. https://doi.org/10.1080/15250000701310395

Warnke, K., Brandt, J., Jörgens, S., Arolt, V., Beer, K., Domschke, K., . . . Schwarte, K. (2020). Association of 5-HTTLPR/rs25531 with depressive symptoms in patients with coronary heart disease: A prospective study. *Journal of Affective Disorders, 277*, 531–539.

Wat, K. K., Banks, P. B., & McArthur, C. (2020). Linking animal personality to problem-solving performance in urban common brushtail possums. *Animal Behaviour, 162*, 35–45.

Watanabe, H., & Mizunami, M. (2007). Pavlov's cockroach: Classical conditioning of salivation in an insect. *PloS One, 2*(6), e529.

Waters, L. (2020). Using positive psychology interventions to strengthen family happiness: A family systems approach. *The Journal of Positive Psychology, 15*(5), 645–652.

Watson, D. (2001). Positive affectivity: The disposition to experience pleasurable emotional states. In C. R. Snyder & S. J. Lopez (Eds.), *Handbook of positive psychology.* Oxford University Press.

Watson, J. B., & Rayner, R. (1920). Conditioned emotional reactions. *Journal of Experimental Psychology, 3*(1), 1–14.

Watson, M., Homewood, J., Haviland, J., & Bliss, J. M. (2005). Influence of psychological response on breast cancer survival: A 10-year follow-up of a population-based cohort. *European Journal of Cancer, 41*(12), 1710–1714.

Watts, R. E., & Bluvshtein, M. (2020). Adler's theory and therapy as a river: A brief discussion of the profound influence of Alfred Adler. *The Journal of Individual Psychology, 76*(1), 99–109.

Watts, S., Marchand, A., Bouchard, S., Gosselin, P., Langlois, F., Belleville, G., & Dugas, M. J. (2020). Telepsychotherapy for generalized anxiety disorder: Impact on the working alliance. *Journal of Psychotherapy Integration, 30*(2), 208.

Watts, T. W., Duncan, G. J., & Quan, H. (2018). Revisiting the marshmallow test: A conceptual replication investigating links between early delay of gratification and later outcomes. *Psychological Science, 29*, 1159–1177.

Waugh, W. E., & Brownell, C. A. (2017). "Help Yourself!" What can toddlers' helping failures tell us about the development of prosocial behavior? *Infancy, 22*, 665–680.

Way, B. M., & Gurbaxani, B. M. (2008). A genetics primer for social health research. *Social and Personality Psychology Compass, 2*(2), 785–816.

Wearne, T. A., & McDonald, S. (2020). Social cognition v. emotional intelligence in first-episode psychosis: Are they the same? *Psychological Medicine.* https://doi.org/10.1017/S0033291720000185

Weaver, E., & Szigethy, E. (2020). Managing pain and psychosocial care in IBD: A primer for the practicing gastroenterologist. *Current Gastroenterology Reports, 22*(4), 1–12.

Webb, R. T., Kontopantelis, E., Doran, T., Qin, P., Creed, F., & Kapur, N. (2012). Suicide risk in primary care patients with major physical diseases: A case-control study. *Archives of General Psychiatry, 69*(3), 256–264.

Webb, W. B. (2000). Sleep. In A. Kazdin (Ed.), *Encyclopedia of psychology*. American Psychological Association and Oxford University Press.

Webster, G. D., Smith, C. V., Orozco, T., Jonason, P. K., Gesselman, A. N., & Greenspan, R. L. (2020). Missed connections and embarrassing confessions: Using big data to examine sex differences in sexual omission and commission regret. *Evolutionary Behavioral Sciences.* https://doi-org.proxy.mul.missouri.edu/10.1037/ebs0000199

Wechsler, D. (1939). *The measurement of adult intelligence*. Williams & Wilkins.

Wechsler, H., Lee, J. E., Kuo, M., & Lee, H. (2000). College binge drinking in the 1990s—A continuing health problem: Results of the Harvard University School of Public Health 1999 College Alcohol Study. *Journal of American College Health, 48*(5), 199–210.

Wechsler, H., Lee, J. E., Kuo, M., Seibring, M., Nelson, T. F., & Lee, H. (2002). Trends in college binge drinking during a period of increased prevention efforts: Findings from 4 Harvard School of Public Health college alcohol study surveys: 1993–2001. *Journal of American College Health, 50*(5), 203–217.

Wegmann, K. M. (2017). 'His skin doesn't match what he wants to do': Children's perceptions of stereotype threat. *American Journal of Orthopsychiatry.* https://doi.org/10.1037/ort0000238

Wehmann, E., Köhnen, M., Härter, M., & Liebherz, S. (2020). Therapeutic alliance in technology-based interventions for the treatment of depression: Systematic review. *Journal of Medical Internet Research, 22*(6), e17195.

Wei, S., Smits, M. G., Tang, X., Kuang, L., Meng, H., Ni, S., . . . Zhou, X. (2020). Efficacy and safety of melatonin for sleep onset insomnia in children and adolescents: a meta-analysis of randomized controlled trials. *Sleep Medicine, 68*, 1–8.

Wei, Y. B., McCarthy, M., Ren, H., Carrillo-Roa, T., Shekhtman, T., DeModena, A., . . . Henigsberg, N. (2020). A functional variant in the serotonin receptor 7 gene (HTR7), rs7905446, is associated with good response to SSRIs in bipolar and unipolar depression. *Molecular Psychiatry, 25*(6), 1312–1322.

Weinbach, N., Sher, H., & Bohon, C. (2018). Differences in emotion regulation difficulties across types of eating disorders during adolescence. *Journal of Abnormal Child Psychology.*

Weiner, B. (2006). *Social motivation, justice, and the moral emotions: An attributional approach*. Erlbaum.

Weiner, I. B. (2004). Rorschach assessment: Current status. In M. Hersen (Ed.), *Comprehensive handbook of psychological assessment* (vol. 2). Wiley.

Weingarten, E., Chen, Q., McAdams, M., Yi, J., Hepler, J., & Albarracín, D. (2016). From primed concepts to action: A meta-analysis of the behavioral effects of incidentally presented words. *Psychological Bulletin, 142*(5), 472–497.

Weinstein, A., & Weinstein, Y. (2014). Exercise addiction—diagnosis, bio-psychological mechanisms, and treatment issues. *Current Pharmaceutical Design, 20*(25), 4062–4069.

Weinstein, N., Legate, N., Ryan, W. S., & Hemmy, L. (2019). Autonomous orientation predicts longevity: New findings from the Nun Study. *Journal of Personality, 87*(2), 181–193.

Weir, W. (1984, October 15). Another look at subliminal "facts." *Advertising Age*, 46.

Weisman, O., Zagoory-Sharon, O., & Feldman, R. (2014). Oxytocin administration, salivary testosterone, and father-infant social behavior. *Progress in Neuro-Psychopharmacology and Biological Psychiatry, 49*, 47–52.

Weiss, A. (2018). Personality traits: A view from the animal kingdom. *Journal of Personality. 86*(1), 12–22

Weiss, A., Baselmans, B. L., Hofer, E., Yang, J., Okbay, A., Lind, P. A., et al. (2016). Personality polygenes, positive affect, and life satisfaction. *Twin Research and Human Genetics, 19*(5), 407–417. https://doi.org/10.1017/thg.2016.65

Weiss, A., & Deary, I. J. (2020). A new look at neuroticism: Should we worry so much about worrying? *Current Directions in Psychological Science, 29*(1), 92–101.

Weiss, A., King, J. E., & Perkins, L. (2006). Personality and subjective well-being in orangutans (*Pongo pygmaeus and Pongo abelii*). *Journal of Personality and Social Psychology, 90*(3), 501–511.

Weiss, D., & Perry, E. L. (2020). Implications of generational and age metastereotypes for older adults at work: The role of agency, stereotype threat, and job search self-efficacy. *Work, Aging and Retirement, 6*(1), 15–27.

Weiss, L. G., & Saklofske, D. H. (2020). Mediators of IQ test score differences across racial and ethnic groups: The case for environmental and social justice. *Personality and Individual Differences, 161*, 109962.

Weiss, S. M., & Schwartz, G. E. (2019). Behavioral medicine: A retro/prospective view of the field. *Journal of Behavioral Medicine, 42*(1), 5–11.

Weissman, M. M., & Olfson, M. (1995). Depression in women: Implications for health care research. *Science, 269*(5225), 799–801.

Weitlauf, A. S., & others. (2014). *Therapies for children with autism spectrum disorder: Behavioral interventions update.* Comparative Effectiveness Review No. 137. AHRQ Publication No. 14-EHC036-EF. Agency for Healthcare Research and Quality. www.effectivehealthcare.ahrq.gov/reports/final.cfm

Welborn, B. L., Gunter, B. C., Vezich, I. S., & Lieberman, M. D. (2017). Neural correlates of the false consensus effect: Evidence for motivated projection and regulatory restraint. *Journal of Cognitive Neuroscience, 29*(4), 708–717. https://doi.org/10.1162/jocn_a_01084

Welcome, S. E., Paivio, A., McRae, K., & Joanisse, M. F. (2011). An electrophysiological study of task demands on concreteness effects: Evidence for dual coding theory. *Experimental Brain Research, 212*(3), 347–358.

Wellman, H. M., & Woolley, J. D. (1990). From simple desires to ordinary beliefs: The early development of everyday psychology. *Cognition, 35*(3), 245–275.

Wells, G. L. (1993). What do we know about eyewitness identification? *American Psychologist, 48*(5), 553–571.

Wells, G. L., Kovera, M. B., Douglass, A. B., Brewer, N., Meissner, C. A., & Wixted, J. T. (2020). Policy and procedure recommendations for the collection and preservation of eyewitness identification evidence. *Law and Human Behavior, 44*(1), 3–36. https://doi.org/10.1037/lhb0000359

Wells, G. L., Steblay, N. K., & Dysart, J. E. (2011). *A test of the simultaneous vs. sequential lineup methods: An initial report of the AJS National Eyewitness Identification Field Studies.* American Judicature Society.

Wen, H., & Hockenberry, J. M. (2018). Association of medical and adult-use marijuana laws with opioid prescribing for Medicaid enrollees. *JAMA Internal Medicine, 178*(5), 673–679.

Wenzel, M., & Kubiak, T. (2020). Neuroticism may reflect emotional variability when correcting for the confound with the mean. *Proceedings of the National Academy of Sciences.* https://doi.org/10.1073/pnas.2017910117

Werner, A., Wu, C., Zachariae, R., Nohr, E. A., Uldbjerg, N., & Hansen, Å. M. (2020). Effects of antenatal hypnosis on maternal salivary cortisol during childbirth and six weeks postpartum—a randomized controlled trial. *PloS one, 15*(5), e0230704.

Wertheimer, M., & Puente, A. E. (2020). *A brief history of psychology*. Routledge.

West, R. (2005). Time for a change: Putting the transtheoretical (stages of change) model to rest. *Addiction, 100*(8), 1036–1039.

Westerhof, G. J., Bohlmeijer, E. T., & McAdams, D. P. (2017). The relation of ego integrity and despair to personality traits and mental health. *The Journals of Gerontology: Series B, 72*(3), 400–407.

Weston, S. J., Edmonds, G. W., & Hill, P. L. (2020). Personality traits predict dietary habits in middle-to-older adults. *Psychology, Health & Medicine, 25*(3), 379–387.

Westra, E. (2020). Getting to know you: Accuracy and error in judgments of character. *Mind & Language, 35*(5), 583–600.

Wetherell, M. (2012). The prejudice problematic. In J. Dixon, M. Keynes, & M. Levine (Eds.), *Beyond prejudice*. Cambridge University Press.

Whillans, A. V., Aknin, L. B., Ross, C. J., Chen, L., & Chen, F. S. (2020). Common variants of the oxytocin receptor gene do not predict the positive mood benefits of prosocial spending. *Emotion, 20*(5), 734–749.

White, E. K., Warren, C. S., Cao, L., Crosby, R. D., Engel, S. G., Wonderlich, S. A., et al. (2016). Media exposure and associated stress contribute to eating pathology in women with anorexia nervosa: Daily and momentary associations. *International Journal of Eating Disorders, 49*(6), 617–621. https://doi.org/10.1002/eat.22490

White, F. A., Borinca, I., Vezzali, L., Reynolds, K. J., Blomster Lyshol, J. K., Verrelli, S., & Falomir-Pichastor, J. M. (2020). Beyond direct contact: The theoretical and societal relevance of indirect contact for improving intergroup relations. *Journal of Social Issues.* https://doi.org/10.1111/josi.12400

White, J. W., & Frabutt, J. M. (2006). Violence against girls and women: An integrative developmental perspective. In J. Worell & C. D. Goodheart (Eds.), *Handbook of girls' and women's psychological health: Gender and well-being across the lifespan* (pp. 85–93). Oxford University Press.

White, R. C., & Aimola Davies, A. (2008). Attention set for numbers: Expectation and perceptual load in inattentional blindness. *Journal of Experimental Psychology: Human Perception and Performance, 34*(5), 1092–1107.

White, R. E., Prager, E. O., Schaefer, C., Kross, E., Duckworth, A. L., & Carlson, S. M. (2017). The "Batman Effect": Improving perseverance in young children. *Child Development, 88*(5), 1563–1571.

White, S. J., Pascall, D. J., & Wilson, A. J. (2020). Towards a comparative approach to the structure of animal personality variation. *Behavioral Ecology, 31*(2), 340–351.

Whiten, A. (2020). Wild chimpanzees scaffold youngsters' learning in a high-tech community. *Proceedings of the National Academy of Sciences, 117*(2), 802–804.

Whitman, M. R., Tylicki, J. L., Mascioli, R., Pickle, J., & Ben-Porath, Y. S. (2020). Psychometric properties of the Minnesota Multiphasic Personality Inventory-3 (MMPI-3) in a clinical neuropsychology setting. *Psychological Assessment.* https://doi.org/10.1037/pas0000969

WHO. (2020). Information sheet on opioid overdose. https://www.who.int/news-room/fact-sheets/detail/opioid-overdose

Whorf, B. L. (1956). *Language, thought, and creativity*. Wiley.

Wickersham, I. R., & Feinberg, E. H. (2012). New technologies for imaging synaptic partners. *Current Opinion in Neurobiology, 22*(1), 121–127.

Wickrama, K. A., Klopack, E. T., & O'Neal, C. W. (2020). How midlife chronic stress combines with stressful life events to influence later life mental and physical health for husbands and wives in enduring marriages. *Journal of Aging and Health.* https://doi.org/10.1177/0898264320952905

Widdowson, A. O., Ranson, J. A., Siennick, S. E., Rulison, K. L., & Osgood, D. W. (2020). Exposure to persistently delinquent peers and substance use onset: A test of Moffitt's social mimicry hypothesis. *Crime & Delinquency, 66*(3), 420–445.

Widome, R., Berger, A. T., Iber, C., Wahlstrom, K., Laska, M. N., Kilian, G., . . . Erickson, D. J. (2020). Association of delaying school start time with sleep duration, timing, and quality among adolescents. *JAMA Pediatrics, 174*(7):697–704. https://doi.org/10.1001/jamapediatrics.2020.0344

Wiebe, R. P. (2004). Delinquent behavior and the Five Factor model: Hiding in the adaptive landscape? *Individual Differences Research, 2*(1), 38–62.

Wiemer, J., Gerdes, A. B., & Pauli, P. (2012). The effects of an unexpected spider stimulus on skin conductance responses and eye movements: An inattentional blindness study. *Psychological Research, 77*(2), 155–166.

Wiersma, D., Nienhuis, F. J., Slooff, C. J., & Giel, R. (1998). Natural course of schizophrenic disorders: A 15-year followup of a Dutch incidence cohort. *Schizophrenia Bulletin, 24*(1), 75–85.

Wiese, A. D., & Boutros, N. N. (2019). Diagnostic utility of sodium lactate infusion and CO_2-35% inhalation for panic disorder. *Neuropsychobiology, 78*(2), 59–69.

Wiggins, J. L., Briggs-Gowan, M. J., Brotman, M. A., Leibenluft, E., & Wakschlag, L. S. (2020, in press). Don't miss the boat: Toward a developmental nosology for disruptive mood dysregulation disorder in early childhood. *Journal of the American Academy of Child & Adolescent Psychiatry*.

Wilder, D. A., Ertel, H. M., & Cymbal, D. J. (2020). A review of recent research on the manipulation of response effort in applied behavior analysis. *Behavior Modification*. https://doi.org/10.1177%2F0145445520908509

Wilder, D. A., Ertel, H. M., & Cymbal, D. J. (2020). A review of recent research on the manipulation of response effort in applied behavior analysis. *Behavior Modification*. https://doi.org/10.1177/0145445520908509

Willard, G., Isaac, K. J., & Carney, D. R. (2015). Some evidence for the nonverbal contagion of implicit racial bias. *Organizational Behavior and Human Decision Processes, 128*, 96-107.

Willard, J., & Madon, S. (2016). Understanding the connections between self-fulfilling prophecies and social problems. In S. Trusz & P. Bąbel (Eds.), *Interpersonal and intrapersonal expectancies* (pp. 117–124). Routledge/Taylor & Francis Group.

Willcox, B. J., & Willcox, D. C. (2014). Caloric restriction, caloric restriction mimetics, and healthy aging in Okinawa: Controversies and clinical implications. *Current Opinion in Clinical Nutrition and Metabolic Care, 17*(1), 51–58.

Willcox, D. C., Scapagnini, G., & Willcox, B. J. (2014). Healthy aging diets other than Mediterranean: A focus on the Okinawan diet. *Mechanisms of Aging and Development, 136–137*, 148–162.

Williams, A. L., Craske, M. G., Mineka, S., & Zinbarg, R. E. (2020). Reciprocal effects of personality and general distress: Neuroticism vulnerability is stronger than scarring. *Journal of Abnormal Psychology*. https://doi.org/10.1037/abn0000635

Williams, A. V., Duque-Wilckens, N., Ramos-Maciel, S., Campi, K. L., Bhela, S. K., Xu, C. K., . . . Trainor, B. C. (2020). Social approach and social vigilance are differentially regulated by oxytocin receptors in the nucleus accumbens. *Neuropsychopharmacology*, 45, 1423–1430.

Williams, C. (2020). *Religion and the meaning of life: An existential approach*. Cambridge University Press.

Williams, D. R., Lawrence, J. A., & Davis, B. A. (2019). Racism and health: Evidence and needed research. *Annual Review of Public Health, 40*, 105–125.

Williams, I. A., Obeso, I., & Jahanshahi, M. (2020). Dopaminergic medication improves cognitive control under low cognitive demand in Parkinson's disease. *Neuropsychology, 34*(5), 551.

Williams, J. D., & Gruzelier, J. H. (2001). Differentiation of hypnosis and relaxation by analysis of narrow band theta and alpha frequencies. *International Journal of Clinical and Experimental Hypnosis, 49*(3), 185–206.

Williams, J. L., Aiyer, S. M., Durkee, M. I., & Tolan, P. H. (2014). The protective role of ethnic identity for urban adolescent males facing multiple stressors. *Journal of Youth and Adolescence, 43*(10), 1728–1741.

Williams, L. M. (1995). Recovered memories of abuse in women with documented child sexual victimization histories. *Journal of Traumatic Stress, 8*(4), 649–673.

Williams, L. M. (2003). Understanding child abuse and violence against women: A life-course perspective. *Journal of Interpersonal Violence, 18*(4), 441–451.

Williams, L. M. (2004). Researcher-advocate collaborations to end violence against women. *Journal of Interpersonal Violence, 19*(11), 1350–1357.

Williams, M. T. (2020). Microaggressions: Clarification, evidence, and impact. *Perspectives on Psychological Science, 15*(1), 3–26.

Williams, O., & Annandale, E. (2020). Obesity, stigma and reflexive embodiment: Feeling the 'weight' of expectation. *Health, 24*(4), 421–441.

Williams, P. T. (2020). Quantile-specific heritability may account for gene-environment interactions involving coffee consumption. *Behavior Genetics, 50*(2), 119–126.

Williams, S. E., van Zanten, J. J. V., Trotman, G. P., Quinton, M. L., & Ginty, A. T. (2017). Challenge and threat imagery manipulates heart rate and anxiety responses to stress. International *Journal of Psychophysiology, 117*, 111–118.

Williamson, H. C., & Lavner, J. A. (2020). Trajectories of marital satisfaction in diverse newlywed couples. *Social Psychological and Personality Science, 11*(5), 597–604.

Willyard, C. (2018). New human gene tally reignites debate. *Nature, 558*, 354–355.

Wilson, A. E., O'Connor, D. B., Lawton, R., Hill, P. L., & Roberts, B. W. (2016). Conscientiousness and fruit and vegetable consumption: Exploring behavioural intention as a mediator. *Psychology, Health & Medicine, 21*(4), 469–475. https://doi.org/10.1080/13548506.2015.1093644

Wilson, G. T., Grilo, C. M., & Vitousek, K. M. (2007). Psychological treatment of eating disorders. *American Psychologist, 62*(3), 199–216.

Wilson, H. E. (2015). Social and emotional characteristics and early childhood mathematical and literacy giftedness: Observations from parents and childcare providers using the ECLS-B. *Journal for the Education of the Gifted, 38*(4), 377–404. https://doi.org/10.1177/0162353215607323

Wilson, K. P., Carter, M. W., Wiener, H. L., DeRamus, M. L., Bulluck, J. C., Watson, L. R., et al. (2017). Object play in infants with autism spectrum disorder: A longitudinal retrospective video analysis. *Autism & Developmental Language Impairments, 2*, 1–12.

Wilson, N., Robb, E., Gajwani, R., & Minnis, H. (2020). Nature and nurture? A review of the literature on childhood maltreatment and genetic factors in the pathogenesis of borderline personality disorder. *Journal of Psychiatric Research*. https://doi.org/10.1016/j.jpsychires.2020.12.025

Wilson, R. S., Barnes, L. L., Aggarwal, N. T., Boyle, P. A., Hebert, L. E., Mendes de Leon, C. F., & Evans, D. A. (2010). Cognitive activity and the cognitive morbidity of Alzheimer disease. *Neurology, 75*(11), 990–996.

Wilt, J., & Revelle, W. (2017). Extraversion. In T.A. Widiger (Ed.), *Oxford Handbook of the Five Factor Model*. Oxford University Press.

Wiltermuth, S. S., & Heath, C. (2009). Synchrony and cooperation. *Psychological Science, 20*(1), 1–5.

Winder, J. R., Mangen, K. H., Martinez-Snyder, A. E., & Valentiner, D. P. (2020). Anxiety sensitivity, disgust sensitivity and aversive reactions to four stimuli. *Behavioural and Cognitive Psychotherapy*. https://doi.org/10.1017/S1352465820000570

Wingen, T., Berkessel, J. B., & Englich, B. (2020). No replication, no trust? How low replicability influences trust in psychology. *Social Psychological and Personality Science, 11*(4), 454–463.

Winner, E. (2000). The origins and ends of giftedness. *American Psychologist, 55*(1), 159–169.

Winner, E. (2006). Development in the arts. In W. Damon & R. Lerner (Eds.), *Handbook of child psychology* (6th ed.). Wiley.

Winston, A. S. (2020). Why mainstream research will not end scientific racism in psychology. *Theory & Psychology, 30*(3), 425–430.

Winter, A. S., Fountain, C., Cheslack-Postava, K., & Bearman, P. S. (2020). The social patterning of autism diagnoses reversed in California between 1992 and 2018. *Proceedings of the National Academy of Sciences, 117*(48), 30295–30302.

Winter, D. G. (2005). Measuring the motives of political actors at a distance. In J. M. Post (Ed.), *The psychological assessment of political leaders: With profiles of Saddam Hussein and Bill Clinton* (pp. 153–177). University of Michigan Press.

Winter, D. G. (2018). What does Trump really want? *Analyses of Social Issues and Public Policy, 18*(1), 155–171.

Wirth, A., Holst, K., & Ponimaskin, E. (2017). How serotonin receptors regulate morphogenic signalling in neurons. *Progress in Neurobiology, 151*, 35–56.

Wise, R. A., & Robble, M. A. (2020). Dopamine and addiction. *Annual Review of Psychology, 71*, 79–106.

Witelson, S. F., Kigar, D. L., & Harvey, T. (1999). The exceptional brain of Albert Einstein. *Lancet, 353*(9174), 2149–2153.

Witherington, D. C., & Lickliter, R. (2017). Integrating development and evolution in psychological science: Evolutionary developmental psychology, developmental systems, and explanatory pluralism. *Human Development, 59*(4), 200–234. https://doi.org/10.1159/000450715

Wixted, J. T. (2020). The forgotten history of signal detection theory. *Journal of Experimental Psychology: Learning, Memory, and Cognition, 46*(2), 201–233. https://doi.org/10.1037/xlm0000732

Wolff, F., Wigfield, A., Möller, J., Dicke, A.-L., & Eccles, J. S. (2020). Social, dimensional, and temporal comparisons by students and parents: An investigation of the 2I/E model at the transition from elementary to junior high school. *Journal of Educational Psychology, 112*(8), 1644–1660.

Wolf, T., & Zimprich, D. (2020). What characterizes the reminiscence bump in autobiographical memory? New answers to an old question. *Memory & Cognition*, 48, 607–622.

Wolock, E. R., Queen, A. H., Rodríguez, G. M., & Weisz, J. R. (2020). Chronic illness and internalizing symptomatology in a transdiagnostic clinical sample of youth. *Journal of Pediatric Psychology, 45*, 633–642.

Womick, J., & King, L. A. (2018). Personality and well-being. In B. J. Carducci (Editor-in-Chief) & J. S. Mio & R. E. Riggio (Vol. Eds.), *The Wiley-Blackwell encyclopedia of personality and individual differences: Vol. IV. Clinical, applied, and cross-cultural research*. John Wiley & Sons.

Wong, A. H., & Pittig, A. (2020). Costly avoidance triggered by categorical fear generalization. *Behaviour Research and Therapy*, 129, 103606.

Wong, C. K., Marshall, N. S., Grunstein, R. R., Ho, S. S., Fois, R. A., Hibbs, D. E., et al. (2017). Spontaneous adverse event reports associated with Zolpidem in the United States 2003–2012. *Journal of Clinical Sleep Medicine, 13*(2), 223–234.

Wong, D., Kuwabara, H., Roberts, J., Gapasin, L., Mishra, C., Brasic, J., . . . & Kem, W. (2017). Test-retest and DXMB-A occupancy of human alpha 7 nicotinic acetylcholine receptors measured by 18F-ASEM. *Journal of Nuclear Medicine, 58*(suppl. 1), 134–134.

Wong, D. L., & others. (2012). Epinephrine: A short- and long-term regulator of stress and developmental illness. *Cellular and Molecular Neurobiology, 32*(5), 737–748.

WonPat-Borja, A. J., Yang, L. H., Link, B. G., & Phelan, J. C. (2012). Eugenics, genetics, and mental illness stigma in Chinese Americans. *Social Psychiatry and Psychiatric Epidemiology, 47*(1), 145–156.

Woo, C.C., & Leon, M. (2013). Environmental enrichment as an effective treatment for autism: A randomized controlled trial. *Behavioral Neuroscience, 127*, 487–497.

Wood, D., Harms, P., & Vazire, S. (2010). Perceiver effects as projective tests: What your perceptions of others say about you. *Journal of Personality and Social Psychology, 99*(1), 174–190.

Wood, K., & Simons, D. J. (2019). Processing without noticing in inattentional blindness: A replication of Moore and Egeth (1997) and Mack and Rock (1998). *Attention, Perception, & Psychophysics, 81*(1), 1–11.

Wood, S. K., & Valentino, R. J. (2017). The brain norepinephrine system, stress and cardiovascular vulnerability. *Neuroscience & Biobehavioral Reviews, 74*, 393–400.

Woodbury-Smith, M., Nicolson, R., Zarrei, M., Ryan, K., Yuen, C., Walker, S., et al. (2017). Variable phenotype expression in a family segregating microdeletions of the NRXN1 and MBD5 autism spectrum disorder susceptibility genes. *NPJ Genomic Medicine, 2*, 1. https://doi.org/10.1038/s41525-017-0020-9

Woodhams, J., Taylor, P. J., & Cooke, C. (2020). Multiple perpetrator rape: Is perpetrator violence the result of victim resistance, deindividuation, or leader–follower dynamics? *Psychology of Violence, 10*(1), 120–129.

Woods, S. A., Edmonds, G. W., Hampson, S. E., & Lievens, F. (2020). How our work influences who we are: Testing a theory of vocational and personality development over fifty years. *Journal of Research in Personality, 85,* 103930.

Woods, S. B. (2019). Biopsychosocial theories. In B. H. Fiese, M. Celano, K. Deater-Deckard, E. N. Jouriles, & M. A. Whisman (Eds.), *APA handbook of contemporary family psychology: Foundations, methods, and contemporary issues across the lifespan* (pp. 75–92). American Psychological Association.

Woody, W. D. & Viney, W. (2017). *A History of Psychology: The Emergence of Science and Applications,* (6th ed.). Routledge.

Woolcott, G. (2020). *Reconceptualising information processing for education.* Springer.

Wooldridge, J. S., Bosch, J., Crawford, J. N., Morland, L., & Afari, N. (2020). Relationships among adverse childhood experiences, PTSD symptom clusters, and health in women Veterans. *Stress and Health.* https://doi.org/10.1002/smi.2953

Woolfolk, A. (2015). *Educational psychology* (13th ed.). Pearson.

World Health Organization (WHO). (2020, April). *Obesity and overweight.* https://www.who.int/news-room/fact-sheets/detail/obesity-and-overweight

World Population Review. (2018). Suicide rate by country 2018. http://worldpopulationreview.com/countries/suicide-rate-by-country/

Worthington, E. L., Davis, D. E., Hook, J. N., Miller, A. J., Gartner, A. L., & Jennings, D. J. (2011). Promoting forgiveness as a religious or spiritual intervention. In J. D. Aten, M. R. McMinn, & E. L. Worthington (Eds.), *Spiritually oriented interventions for counseling and psychotherapy* (pp. 169–195). American Psychological Association.

Wraw, C., Deary, I. J., Der, G., & Gale, C. R. (2016). Intelligence in youth and mental health at age 50. *Intelligence, 58,* 69–79. https://doi.org/10.1016/j.intell.2016.06.005

Wray, N. R., Ripke, S., Mattheisen, M., Trzaskowski, M., Byrne, E. M., Abdellaoui, A., . . . Bacanu, S. A. (2018). Genome-wide association analyses identify 44 risk variants and refine the genetic architecture of major depression. *Nature Genetics, 50*(5), 668–681.

Wright, A. J. (2020). Equivalence of remote, digital administration and traditional, in-person administration of the Wechsler Intelligence Scale for Children (WISC-V). *Psychological Assessment, 32,* 809–817.

Wright, J. H., & Mishkind, M. (2020). Computer-assisted CBT and mobile apps for depression: Assessment and integration into clinical care. *Focus, 18*(2), 162–168.

Wright, M. F. (2020). The role of technologies, behaviors, gender, and gender stereotype traits in adolescents' cyber aggression. *Journal of Interpersonal Violence, 35*(7–8), 1719–1738.

Wrigley-Field, E. (2020). U.S. racial inequality may be as deadly as COVID-19. *Proceedings of the National Academy of Science, 117,* 21854–21856.

Wruck, W., & Adjaye, J. (2020). Meta-analysis of human prefrontal cortex reveals activation of GFAP and decline of synaptic transmission in the aging brain. *Acta Neuropathologica Communications, 8*(1), 1–18.

Wu, C.C., Samanez-Larkin, G.R., Katovich, K., & Knutson, K. (2014). Affective traits link to reliable neural markers of incentive anticipation. *NeuroImage, 84,* 279–289.

Wu, C., Zhen, Z., Huang, L., Huang, T., & Liu, J. (2020). COMT-polymorphisms modulated functional profile of the fusiform face area contributes to face-specific recognition ability. *Scientific Reports, 10*(1), 1–12.

Wu, M. S., Thamrin, H., & Pérez, J. (2020). Exposure with response prevention for obsessive-compulsive disorder in children and adolescents. In T. S. Peris, E. A. Storch, & J. F. McGuire (Eds.), *Exposure therapy for children with anxiety and OCD* (pp. 245–268). Academic Press.

Wuest, D. A., & Fisette, J. L. (2015). *Foundations of physical education, exercise science, and sport* (18th ed.). McGraw-Hill.

Wynn, T., & Coolidge, F. L. (2010). Beyond symbolism and language: An introduction to Supplement 1, working memory. *Current Anthropology, 51,* Suppl. 1, 5–16.

Wynn, T., Coolidge, F. L., & Bright, M. (2009). Hohlenstein-Stadel and the evolution of human conceptual thought. *Cambridge Archaeological Journal, 19*(1), 73–83.

X

Xia, Y., & Ma, Z. (2020). Social integration, perceived stress, locus of control, and psychological wellbeing among Chinese emerging adult migrants: A conditional process analysis. *Journal of Affective Disorders, 267,* 9–16.

Xiao, X., Dong, Q., Gao, J., Men, W., Poldrack, R. A., & Xue, G. (2017). Transformed neural pattern reinstatement during episodic memory retrieval. *Journal of Neuroscience, 37*(11), 2986–2998.

Xie, W., & Zhang, W. (2017). Familiarity speeds up visual short-term memory consolidation. *Journal of Experimental Psychology: Human Perception and Performance, 43,* 1207–1221. https://doi.org/10.1037/xhp0000355

Xu, H., & others. (2013). The function of BMP4 during neurogenesis in the adult hippocampus in Alzheimer's disease. *Ageing Research and Reviews, 12*(1), 157–164.

Xu, J., & others. (2014). Nondirective meditation activates default mode network and areas associated with memory retrieval and emotional processing. *Frontiers in Human Neuroscience, 8,* 86.

Xu, J. Q., Murphy, S. L., Kochanek, K. D., & Arias E. (2020). Mortality in the United States, 2018. NCHS Data Brief, No 355. National Center for Health Statistics.

Xu, Q., Liu, F., Qin, W., Jiang, T., & Yu, C. (2020). Multiscale neurobiological correlates of human neuroticism. *Human Brain Mapping, 41*(16), 4730–4743.

Xu, R., Wu, J., Lang, L., Hu, J., Tang, H., Xu, J., & Sun, B. (2020). Implantation of glial cell line-derived neurotrophic factor-expressing adipose tissue-derived stromal cells in a rat Parkinson's disease model. *Neurological Research, 42,* 712–720.

Y

Yager, J. (2020). Trends in psychiatric residency education and practice from 1944 to 2019: A loving, informal, and highly personal review served with gently roasted sacred cow. *Journal of Psychiatric Practice, 26*(6), 493–502.

Yager, J., & Kay, J. (2020). Clinical curiosity in psychiatric residency training: Implications for education and practice. *Academic Psychiatry, 44*(1), 90–94.

Yakunchikov, D. Y., Olechowski, C. J., Simmonds, M. K., Verrier, M. J., Rashiq, S., McWilliams, L. A., et al. (2017). The effect of social observational learning, empathy and catastrophizing in chronic pain patients during acute pain Induction. *Pain Medicine, 18*(5), 871–878.

Yalom, I. D., & Leszcz, M. (2005). *The theory and practice of group psychotherapy* (5th ed.). Basic Books.

Yang, J., Mao, Y., Niu, Y., Wei, D., Wang, X., & Qiu, J. (2020). Individual differences in neuroticism personality trait in emotion regulation. *Journal of Affective Disorders, 265,* 468–474.

Yang, S., & Sternberg, R. J. (1997a). Conceptions of intelligence in ancient Chinese philosophy. *Journal of Theoretical and Philosophical Psychology, 17*(2), 101–119.

Yang, S., & Sternberg, R. J. (1997b). Taiwanese Chinese people's conceptions of intelligence. *Intelligence, 25*(1), 21–36.

Yang, X., Liu, J., Meng, Y., Xia, M., Cui, Z., Wu, X., et al. (2018). Network analysis reveals disrupted functional brain circuitry in drug-naive social anxiety disorder. *NeuroImage 190,* 213–223.

Yang, Y., & Raine, A. (2009). Prefrontal structural and functional brain imaging findings in antisocial, violent, and psychopathic individuals: A meta-analysis. *Psychiatry Research, 174*(2), 81–88.

Yang, Y., Raine, A., Lencz, T., Bihrle, S., LaCasse, L., & Colletti, P. (2005). Volume reduction in prefrontal gray matter in unsuccessful criminal psychopaths. *Biological Psychiatry, 57*(10), 1103–1108.

Yates, L. B., Djoussé, L., Kurth, T., Buring, J. E., & Gaziano, J. M. (2008). Exceptional longevity in men:

Modifiable factors associated with survival and function to age 90 years. *Archives of Internal Medicine, 168*(3), 284–290.

Ye, T., Fleming, S. M., & Hamilton, A. F. (2020). Spontaneous attribution of false beliefs in adults examined using a signal detection approach. *Quarterly Journal of Experimental Psychology, 73*(4), 555–567.

Ye, Z. J., Zhang, Z., Zhang, X. Y., Tang, Y., Liang, J., Sun, Z., . . . Yu, Y. L. (2020). Effectiveness of adjuvant supportive-expressive group therapy for breast cancer. *Breast Cancer Research and Treatment, 180*(1), 121–134.

Yegorov, Y. E. (2020). Healthy aging: Antioxidants, uncouplers and/or telomerase? *Molecular Biology, 54*(3), 311–316.

Yeomans, J. S. (2012). Muscarinic receptors in the brain stem and mesopontine cholinergic arousal functions. *Handbook of Experimental Pharmacology, 208,* 243–259.

Yeomans, M. R. (2020). Satiety. In H. L. Meiselman (Ed.), *Handbook of eating and drinking: Interdisciplinary perspectives* (pp. 293–313). Springer.

Yeung, Timothy Yu-Cheong. (2020). Did the COVID-19 Pandemic Trigger Nostalgia? Evidence of Music Consumption on Spotify (August 21, 2020). Available at SSRN: https://ssrn.com/abstract=3678606or http://dx.doi.org/10.2139/ssrn.3678606

Yi, K., & Kim, C. (2020). Dissociable neural correlates of spatial attention and response inhibition in spatially driven interference. *Neuroscience Letters, 135111.*

Yilmaz, Z., Halvorsen, M., Bryois, J., Yu, D., Thornton, L. M., Zerwas, S., . . . Erdman, L. (2020). Examination of the shared genetic basis of anorexia nervosa and obsessive-compulsive disorder. *Molecular Psychiatry, 25*(9), 2036–2046.

Yin, M., Li, Z., & Zhou, C. (2020). Experience of stigma among family members of people with severe mental illness: A qualitative systematic review. *International Journal of Mental Health Nursing, 29*(2), 141–160.

Yin, X., Ma, Y., Xu, X., & Yang, H. (2019). The effect of self-referencing on memory for different kinds of source information. *Memory, 27*(4), 519–527.

Yoder, R. M., Chan, J. M., & Taube, J. S. (2017). Acetylcholine contributes to the integration of self-movement cues in head direction cells. *Behavioral Neuroscience, 131*(4), 312–324. https://doi.org/10.1037/bne0000205

Yong, A. D., Williams, M. M., Provan, H., Clarke, D., & Sinclair, G. (2015). How do offenders move through the stages of change? *Psychology, Crime & Law, 21*(4), 375–397. https://doi.org/10.1080/1068316X.2014.989166

Yong, E. (2019). The human brain project hasn't lived up to its promise. *The Atlantic.* https://www.theatlantic.com/science/archive/2019/07/ten-years-human-brain-project-simulation-markram-ted-talk/594493/.

Yong, Z., Hsieh, P., & Milea, D. (2017). Seeing the sound after visual loss: Functional MRI in acquired auditory-visual synesthesia. *Experimental Brain Research, 235*(2), 415–420. https://doi.org/10.1007/s00221-016-4802-6

Yoo, C. (2020). Cohort effects associated with reduced sleep duration in adolescents. *Sleep Medicine, 67,* 184–190.

Yoon, B., Baker, S. L., Korman, D., Tennant, V. R., Harrison, T. M., Landau, S., & Jagust, W. J. (2020). Conscientiousness is associated with less amyloid deposition in cognitively normal aging. *Psychology and Aging, 35*(7), 993–999.

Yoon, S., & Kim, Y. K. (2020). The role of the oxytocin system in anxiety disorders. In Y. K. Kim (Ed.), *Anxiety disorders* (pp. 103–120). Springer.

Young, L. J. (2015). Oxytocin, social cognition and psychiatry. *Neuropsychopharmacology, 40,* 243–244. http://dx.doi.org/10.1038/npp.2014.186

Yousaf, A., Waltes, R., Haslinger, D., Klauck, S. M., Duketis, E., Sachse, M., . . . Nöthen, M. (2020). Quantitative genome-wide association study of six phenotypic subdomains identifies novel genome-wide significant variants in autism spectrum disorder. *Translational Psychiatry, 10*(1), 1–11.

Yu, B., Steptoe, A., Chen, L. J., Chen, Y. H., Lin, C. H., & Ku, P. W. (2020). Social isolation, loneliness, and all-cause mortality in patients with cardiovascular disease: A 10-year follow-up study. *Psychosomatic Medicine, 82*(2), 208–214.

Yu, C., Arcos-Burgos, M., Baune, B. T., Arolt, V., Dannlowski, U., Wong, M. L., & Licinio, J. (2018). Low-frequency and rare variants may contribute to elucidate the genetics of major depressive disorder. *Translational Psychiatry, 8*(1), 70.

Yu, C. E., Xie, S. Y., & Wen, J. (2020). Coloring the destination: The role of color psychology on Instagram. *Tourism Management, 80*, 104110.

Yu, Q., Li, E., Li, L., & Liang, W. (2020). Efficacy of interventions based on applied behavior analysis for autism spectrum disorder: A meta-analysis. *Psychiatry Investigation, 17*(5), 432.

Yu, Q., Li, E., Li, L., & Liang, W. (2020). Efficacy of interventions based on applied behavior analysis for autism spectrum disorder: A meta-analysis. *Psychiatry Investigation, 17*(5), 432-443.

Yue, C., Zhang, Y., Jian, M., Herold, F., Yu, Q., Mueller, P., . . . Zou, L. (2020, March). Differential effects of tai chi chuan (motor-cognitive training) and walking on brain networks: A resting-state fMRI study in Chinese women aged 60. *Healthcare, 8*(1), 67.

Yu, S., Levesque-Bristol, C., & Maeda, Y. (2018). General need for autonomy and subjective well-being: A meta-analysis of studies in the us and east asia. *Journal of Happiness Studies. 19*(6), 1863-1882. https://doi.org/10.1007/s10902-017-9898-2

Yuriev, A., Dahmen, M., Paillé, P., Boiral, O., & Guillaumie, L. (2020). Pro-environmental behaviors through the lens of the theory of planned behavior: A scoping review. *Resources, Conservation and Recycling, 155*, 104660.

Yuvaraj, A., Mahendra, V. S., Chakrapani, V., Yunihastuti, E., Santella, A. J., Ranauta, A., & Doughty, J. (2020). HIV and stigma in the healthcare setting. *Oral Diseases, 26*, 103-111.

Z

Zahavi, A. (1977). Reliability in communication systems and the evolution of altruism. In B. Stonehouse & C. M. Perrins (Eds.), *Evolutionary ecology* (pp. 253-259). MacMillan Press.

Zahrt, O. H., & Crum, A. J. (2017). Perceived physical activity and mortality: evidence from three nationally representative U.S. samples. *Health Psychology.* https://doi.org/10.1037/hea0000531

Zajonc, R. B. (1965). Social facilitation. *Science, 149*(3681), 269-274.

Zajonc, R. B. (1968). Attitudinal effects of mere exposure. *Journal of Personality and Social Psychology, 9*(2, Pt 2), 1-27.

Zajonc, R. B. (1984). On the primacy of affect. *American Psychologist, 39*(2), 117-123.

Zajonc, R. B. (2001). Mere exposure: A gateway to the subliminal. *Current Directions in Psychological Science, 10*(6), 224-228.

Zald, D. H., & Treadway, M. T. (2017). Reward Processing, Neuroeconomics, and Psychopathology. *Annual Review of Clinical Psychology, 13*, 471-495.

Zarkhin, F. (2017, May 31). Surviving victim of Friday train stabbing a poet and PSU student. The Oregonian/OregonLive. www.oregonlive.com/portland/index.ssf/2017/05/surviving_victim_of_friday_tra.html

Zebrowitz, L. (2018). *Reading faces: Window to the soul?* Routledge.

Zedelius, C. M., Protzko, J., Broadway, J. M., & Schooler, J. W. (2020). What types of daydreaming predict creativity? Laboratory and experience sampling evidence. *Psychology of Aesthetics, Creativity, and the Arts.* https://doi.org/10.1037/aca0000342

Zee, P., Thorpy, M., & Pagel, J. (2020). The diagnosis and management of narcolepsy. *The Medical Roundtable General Medicine Edition, 1*(2), 96-103.

Zeidner, M., & Shani-Zinovich, I. (2011). Do academically gifted and nongifted students differ on the big-five and adaptive status? Some recent data and conclusions. *Personality and Individual Differences, 51*(5), 566-570. https://doi.org/10.1016/j.paid. 2011.05.007

Zeiss, C. J. (2020). Utility of spontaneous animal models of Alzheimer's disease in preclinical efficacy studies. *Cell and Tissue Research, 380*, 273-286.

Zeitler, M. L., Bohus, M., Kleindienst, N., Knies, R., Ostermann, M., Schmahl, C., & Lyssenko, L. (2020). How to assess recovery in borderline personality disorder: Psychosocial functioning and satisfaction with life in a sample of former DBT study patients. *Journal of Personality Disorders, 34*(3), 289-307.

Zelazo, P. D. (2020). Executive function and psychopathology: A neurodevelopmental perspective. *Annual Review of Clinical Psychology, 16*, 431-454.

Zemore, S. E. (2017). Implications for future research on drivers of change and alternatives to Alcoholics Anonymous. *Addiction, 112*(6), 940-942.

Zerbe Enns, C., Bryant-Davis, T., & Díaz, L. C. (2020). Transnational feminist therapy: Recommendations and illustrations. *Women & Therapy*, 1-22.

Zhai, S., Tanimura, A., Graves, S. M., Shen, W., & Surmeier, D. J. (2018). Striatal synapses, circuits, and Parkinson's disease. *Current Opinion in Neurobiology, 48*, 9-16.

Zhang, A., & Jang, Y. (2017). The role of internal health locus of control in relation to self-rated health in older adults. *Journal of Gerontological Social Work, 60*(1), 68-78. https://doi.org/10.1080/01634372.2016.1267672

Zhang, B., Zhou, Z. G., Zhou, Y., & Chen, Y. C. (2020). Increased attention to snake images in cynomolgus monkeys: an eye-tracking study. *Zoological Research, 41*(1), 32.

Zhang, H., & Bramham, C. R. (2020). Bidirectional dysregulation of AMPA receptor-mediated synaptic transmission and plasticity in brain disorders. *Frontiers in Synaptic Neuroscience, 12*, 26.

Zhang, J., Zhou, C., & Yu, R. (2020). Oxytocin amplifies the influence of good intentions on social judgments. *Hormones and Behavior, 117*, 104589.

Zhao, F., Zhang, W., Zhu, D., Wang, X., Qin, W., & Liu, F. (2020). Long-term pingju opera training induces plasticity changes in cerebral blood flow: An arterial spin labeling MRI study. *Neuroscience, 436*, 27-33.

Zhao, J., Deng, B., Qin, Y., Men, C., Wang, J., Wei, X., & Sun, J. (2017). Weak electric fields detectability in a noisy neural network. *Cognitive Neurodynamics, 11*, 81-90. https://doi.org/10.1007/s11571-016-9409-x

Zhao, W. J., Walasek, L., & Bhatia, S. (2020). Psychological mechanisms of loss aversion: A drift-diffusion decomposition. *Cognitive Psychology, 123*, 101331.

Zhao, X., Castelli, F. R., Wang, R., Auger, A. P., & Marler, C. A. (2020). Testosterone-related behavioral and neural mechanisms associated with location preferences: A model for territorial establishment. *Hormones and Behavior, 121*, 104709.

Zhou, H. X., Chen, X., Shen, Y. Q., Li, L., Chen, N. X., Zhu, Z. C., . . . Yan, C. G. (2020). Rumination and the default mode network: Meta-analysis of brain imaging studies and implications for depression. *Neuroimage, 206*, 116287.

Zhou, S., & Shapiro, M. A. (2017). Reducing resistance to narrative persuasion about binge drinking: The role of self-activation and habitual drinking behavior. *Health Communication, 32*(10), 1297-1309. https://doi.org/10.1080/10410236.2016.1219931

Zhou, X., Saucier, G., Gao, D., & Liu, J. (2009). The factor structure of Chinese personality terms. *Journal of Personality, 77*(2), 363-400.

Zhou, Y., Wang, F., Tang, J., Nussinov, R., & Cheng, F. (2020). Artificial intelligence in COVID-19 drug repurposing. *The Lancet Digital Health, 2*, e667-e676.

Zhu, G., Sun, X., Yang, Y., Du, Y., Lin, Y., Xiang, J., & Zhou, N. (2019). Reduction of BDNF results in GABAergic neuroplasticity dysfunction and contributes to late-life anxiety disorder. *Behavioral Neuroscience, 133*(2), 212.

Zhu, L., Brescoll, V. L., Newman G., & Uhlmann E. L. (2015) Macho nachos: The implicit effects of gender stereotypes on preferences for healthy and unhealthy foods. *Social Psychology, 46*, 182-196.

Zhu, W., Wadley, V. G., Howard, V. J., Hutto, B., Blair, S. N., & Hooker, S. P. (2016). Objectively measured physical activity and cognitive function in older adults. *Medicine and Science in Sports and Exercise, 49*(1), 47-53.

Zhu, Y., Liu, W., Li, Y., Wang, X., & Winterstein, A. G. (2017). Prevalence of ADHD in publicly insured adults. *Journal of Attention Disorders.*

Zhu, Z., Ho, S. M., & Bonanno, G. A. (2013). Cultural similarities and differences in the perception of valence and intensity: A comparison of Americans and Hong Kong Chinese. *American Journal of Psychology, 126*(3), 261-273.

Zietsch, B. P., & others. (2008). Genetic factors predisposing to homosexuality may increase mating success in heterosexuals. *Evolution and Human Behavior, 29*(6), 424-433.

Zilcha-Mano, S., Porat, Y., Dolev, T., & Shamay-Tsoory, S. (2018). Oxytocin as a neurobiological marker of ruptures in the working alliance. *Psychotherapy and Psychosomatics, 87*, 126-127. http://dx.doi.org/10.1159/000487190

Zilcha-Mano, S., Shamay-Tsoory, S., Dolev-Amit, T., Zagoory-Sharon, O., & Feldman, R. (2020). Oxytocin as a biomarker of the formation of therapeutic alliance in psychotherapy and counseling psychology. *Journal of Counseling Psychology, 67*(4), 523-535.

Ziller, C., & Berning, C. C. (2019). Personality traits and public support of minority rights. *Journal of Ethnic and Migration Studies.* https://doi.org/10.1080/1369183X.2019.1617123

Zillikens, M. C., Demissie, S., Hsu, Y. H., Yerges-Armstrong, L. M., Chou, W. C., Stolk, L., et al. (2017). Large meta-analysis of genome-wide association studies identifies five loci for lean body mass. *Nature Communications, 8*(1), 80.

Zimbardo, P. G. (1972). Pathology of imprisonment. *Society, 9*(6), 4-8.

Zimbardo, P. G. (1973). On the ethics of intervention in human psychological research: With special reference to the Stanford prison experiment. *Cognition, 2*(2), 243-256.

Zimbardo, P. G. (2007). *The Lucifer effect: Understanding how good people turn evil.* Random House.

Zitzmann, M. (2020). Testosterone, mood, behaviour and quality of life. *Andrology, 8*(6), 1598-1605.

Zollig, J., Mattli, F., Sutter, C., Aurelio, A., & Martin, M. (2012). Plasticity of prospective memory through a familiarization intervention in old adults. *Neuropsychology, Development, and Cognition: Aging, Neuropsychology, and Cognition, 19*(1-2), 168-194.

Zuraikat, F. M., Smethers, A. D., & Rolls, B. J. (2020). The influence of portion size on eating and drinking. *Handbook of eating and drinking: Interdisciplinary perspectives* (pp. 679-714). Springer.

Zwebner, Y., Sellier, A., Rosenfeld, N., Goldenberg, J., & Mayo, R. (2017). We look like our names: The manifestation of name stereotypes in facial appearance. *Journal of Personality and Social Psychology, 112*(4), 527-554. https://doi.org/10.1037/pspa0000076

name index

subject index

Note: Page numbers followed by *f* represent figures or illustrations. Page numbers in **bold** represent defined terms.

conspiracy theories, 258–259
contamination stories, 240
contemplation stage, 531*f*, 532
content analysis, 401
context-dependent memory, 226
contiguity, in conditioning, 174
contingency, in conditioning, 174, 181
continuity hypothesis, 145
continuous positive airway pressure
    (CPAP), 143, 143*f*
continuous reinforcement, 186
contour, 104
contraception, 343
control groups, 26–28
controlled processes, **129**–130, 131*f*
conventional level, in moral
    development, 324
convergence, **105**
convergent thinking, **259**, 260
Convicting the Innocent (Garrett), 231
Cool Hand Luke (film), 434
coping, **545**
    strategies for, 546–547
    types of, 546
core knowledge approach, **304**
cornea, 98, 98*f*
coronavirus (COVID-19) pandemic,
    1, 15, 23, 204, 307, 527–528
    artificial intelligence usage during, 247
    as bioweapon, 258
    and cybertherapy, 512
    differential effect of, 544
    and distress in young adults, 547
    misinformation about, 27
    and nightmares, 143
    and public health, 528, 529*f*
    and sense of smell, 120
    and sleep problems, 136–137
    and social distance, 406
    and Thanksgiving 2020, 331
    vaccines, 245
corpus callosum, **67**–68, 67*f*
correlational coefficient, 21
correlational research, 18, **21**–25,
    22*f*, 30*f*
    applications of, 30–31
    causal relationships and, 21–23, 58
    correlations in, 22*f*
    longitudinal designs, 23–25
correlational studies *vs.* experimental
    studies, 25–26
corticosteroids, 46
corticotropin-releasing hormone
    (CRH), 541
cortisol
    binge eating and, 479
    jet lag and, 134
    stress and, 541
costly signaling theory, 419
counseling psychologist, 499*f*
counseling psychology, 13*f*
counselor, 499*f*
counterconditioning, **176**
counterintuitive, 3
couples therapy, **521**
COVID-19. *See* coronavirus (COVID-19)
    pandemic
CPAP (continuous positive airway
    pressure), 143, 143*f*
crack, 156
creative intelligence, 271
creativity, **259**
    characteristics of, 260
    insight learning and, 197
    intrinsic motivation and, 352
    mood and, 363
    personality and, 383–384
    in problem-solving, 250–251, 252*f*,
    259–260

CRH (corticotropin-releasing
    hormone), 541
criterion, 262
criterion validity, 262
critical thinking, **3**, 258
    reasoning and, 252
cross-cultural competence, **523**
cross-cultural psychology, 13*f*
    conformity and, 432
    dissociative identity disorder and,
    470–471
cross-cultural research, 12
cross-sectional designs, **285**
cross-situational consistency, 391
crystal methamphetamine/meth, 155
CTE (chronic traumatic
    encephalopathy), 74
CT scans, 57
cultural context, cognitive development
    in, 305
cultural-familial intellectual disability, 270
culture. *See also* collectivist cultures;
    ethnicity; individualistic cultures
    big five factors of personality
    and, 385
    conformity and, 432
    culture-fair tests, 264, 264*f*
    death/dying and, 326
    defined, 12
    eating disorders and, 476, 478
    expression of emotions and, 360
    of honor, 424
    identity within, 315–316
    infant attachment and, 311
    influence on learning, 199*f*, 200
    language and, 277
    pain and, 118
    and parenting, 314
    perception and, 94
    perspectives on therapy, 523–524
    psychotherapy and, 502, 523–524
    self-determination theory and,
    351–352
    sexual motivation and, 342–343
    suicide rates and, 489
    taste and, 119–120
    terror management theory and, 326
culture-fair tests, **264**, 264*f*
culture of honor, 424, 489
curiosity, in psychology, 3
cutaneous senses, 116–118
cutting, 486
cyberbullying, 531
cybertherapy, 512
cynophobia, 462*f*
cytokines, 141
Cytoxan, 177

D

The Dark Knight (film), 151
"The Dark Side of the American Dream"
    (Kasser and Ryan), 17
data, 17
data analysis, 17
dating violence, 531
daydreaming, **131**, 145
DBT (dialectical behavior therapy), 512
death
    Bonanno's theory of grieving, 327
    Kübler-Ross's stages of dying, 327
    meaning from reality of, 328
    terror management theory, 326
debriefing, 35
decay theory, **235**
deception, in research, 35
decibels (dB), 109
decision making, **253**–257, 254*f*
    group, 436–437

declarative memory, **216**
deductive reasoning, **253**, 253*f*
deep brain stimulation, 518–519
deep learning, 171
deep sleep, 139, 141
Deepstack, 247
default mode network, 131, 145
defense mechanisms, 372–373
dehumanization, 129
deindividuation, **435**
deinstitutionalization, 523
déjà vu, 236–237
delayed grief/trauma, 327
delayed maturation hypothesis, 458
delayed reinforcement, 189–190
delay of gratification, **189**, 392
delta sleep, 139
delta waves, 138, 138*f*, 141
delusions, **480**
demand characteristics, **28**
dementia, 284
dendrites, **47**–48, 48*f*, 293, 293*f*
denial, 327, 372, 546
deoxyribonucleic acid (DNA),
    76, 76*f*, 81
dependent personality disorder, 485*f*
dependent variables, **26**
depressants, **148**–152, 159*f*. *See also*
    psychoactive drugs
    alcohol, 148–151, 149*f*, 150*f*
    barbiturates, 151
    opioids, 151–152
    tranquilizers, 151
depression, 471, 493. *See also* depressive
    disorders
    artificial light and, 135
    drug treatment for, 514–515, 515*f*
    in dying, 327
    electroconvulsive therapy for,
    517–518, 517*f*
    meditation and, 164
    neurotransmitters and, 53, 472–473
    sleep and, 141
depressive disorders, **471**–474, 516*f*
    biological factors in, 472–473
    bipolar disorder, 474–475, 475*f*
    drug therapies for, 514
    electroconvulsive therapy for,
    517–518, 517*f*
    major depressive disorder (MDD),
    472, 491*f*
    persistent depressive disorder, 472
    psychological factors in, 473
    sociocultural factors in,
    473–474, 474*f*
    suicide and, 472–473
depth of processing, 207, 207*f*, 208
depth perception, **104**–106, 106*f*
descriptive research, 18–21, 30*f*
    applications of, 30–31
    case studies, 19
    observation, 18
    surveys and interviews, 19
    value of, 19–21
desensitization, 507, 507*f*
desensitization hierarchy, 507, 507*f*
desynchronous waves, 137
detriangulation, 521
development, 284–329, **285**
    active, as lifelong process, 328–329
    of brain, 293, 293*f*, 294*f*
    cognitive. *See* cognitive development
    cognitive processes, 288
    domains of, 288
    early experiences and, 286–287
    emerging adulthood, 316–317
    infant attachment in, 310–311, 310*f*
    Marcia's theory of identity status,
    314–315, 315*f*

nature and nurture in, 286–288
    of personality, 373–374, 374*f*
    physical. *See* physical development
    physical processes, 288
    prenatal, 289–291, 290*f*
    sleep and, 140
    socioemotional. *See* socioemotional
    development
    socioemotional processes, 288
    of visual system, 101
developmental psychology, 13*f*
    childhood rationality, 307
    children's stereotypes, 338
    research methods in, 285
    sleep and, 218
deviant behavior, 447
diabetes, sleep and, 141
*Diagnostic and Statistical Manual of Mental
    Disorders* (DSM-5), 450–451
dialectical behavior therapy (DBT), 512
diathesis-stress model, **450**
diet, 338, 550–552
dietary supplements, 266
difference threshold, **91**
difficult child, 308
disabilities, intellectual, 269–270
discovered memories, 229–230
discrimination, 174, **186**, 441
    Civil Rights Act of 1964 and, 441
    in classical conditioning, 174
    in operant conditioning, 186
    psychological disorders and,
    455, 491–492
    sexual orientation and, 347
disease. *See also specific diseases*
    sleep and, 141
dismissive attachment style, 311
disorders of movement, 481
displacement, 372
display rules, **360**
disruptive mood dysregulation
    disorder, 475
dissociation, 469
dissociative disorders, **469**–471
    dissociative amnesia, **469**
    dissociative identity disorder (DID),
    470–471, 470*f*
dissonance, 415
distortion, in eyewitness testimony, 230
divergent thinking, **259**
diversity. *See* culture; ethnicity
divided attention, **206**
divided consciousness view of hypnosis,
    161–162, 161*f*
DNA (deoxyribonucleic acid),
    76, 76*f*, 81
dogs
    brain activity of, 59
    conditioning studies in, 185
    Pavlov's stimulus studies on, 172–175,
    172*f*, 173*f*
    personality in, 385
    rescue, 168, 198
    sense of smell in, 120
    service, 168, 186
dominant-recessive genes principle, **77**
door-in-the-face technique, 417
dopamine, 53
    aggression and, 80
    antipsychotic drugs, 514
    binge eating and, 478
    cocaine and, 156
    in drug addiction, 147
    Ecstasy and, 156
    functions of, 53
    orgasm and, 341
double-blind experiment, **29**, 231
Down syndrome, 269*f*, 270
dream analysis, **504**–505

fMRI (functional magnetic resonance imaging), 57–58, 57f
focal objects, 94
focused attention, 163
folic acid, 289
foot-in-the-door technique, 417
forebrain, **61**, 139, 145–146
forensic psychology, 13f
Fore tribe, 360
forgetting, 233–237
  amnesia and, 236–237
  decay theory and, 235
  encoding failure in, 233–234, 234f
  interference and, 234–235, 235f
  motivated, 229
  prospective memory and, 236
  retrieval failure in, 234–237, 235f
  tip-of-the-tongue phenomenon, 235–236
forgiveness, 5–6
formal operational stage, 301f, **303**–304
fovea, 98f, 99
fraternal twins, 78
free association, **504**
Freedom Riders, 5
free radicals, 298
free-radical theoy, of aging, 298
frequency
  of action potentials, 88
  of sexual behavior, 345
  of sound wave, 109, 110f
frequency theory, **114**
Freud, Sigmund, 10–11, 10f
Freud's psychoanalytic theory
  critics and revisionists of, 374–377, 375f
  defense mechanisms in, 372–373, 374f
  dreams in, 144
  evaluating, 377–378
  hysteria in, 370–371
  personality development stages in, 373–374, 374f
  personality structures in, 371–372
  psychoanalysis, 10–11, 370–371, 504–505
  unconscious in, 131f, 133
friendship, in marriage, 318
frontal lobes, 64f, **65**, 66f. *See also* prefrontal cortex
  aggression and, 423
  association cortex in, 64f, 67
  in different animals, 61, 62f
  memory and, 222, 222f
frustration-aggression hypothesis, 423
functional differences, 270
functional fixedness, **251**
functionalism, 7
functional magnetic resonance imaging (fMRI), 57–58, 57f, 165
fundamental attribution error, **412**, 412f
fusiform face area, 69, 70f

G

GABA (gamma-aminobutyric acid), 53, 139, 149
  anxiety disorders and, 460
Gage, Phineas T., brain injury of, 65, 65f, 67, 395
galant reflex, 291f
gamma-aminobutyric acid (GABA), 53, 139, 149, 460
gamophobia, 462f
ganglion cells, 99, 99f
GAS (general adaptation syndrome), **540**–541, 540f
gastric signals, 335–336, 335f
gays, 344–347

gender, **319**
  and chromosomes, 320
  prosocial behavior and, 422
  social role view of, 428
  therapy and, 524
gender development, 319–323
  biology and, 320
  cognitive aspects of, 320, 322
  identities, 321
  nature and nurture, 322–323
  socioemotional experience and, 322
gender differences
  aggression and, 424–425
  in alcohol effects, 149
  in altruism, 422
  in attraction, 427–428
  in depression, 473–474, 474f
  in emotional experiences, 359, 361
  in Freud's psychoanalytic theory, 373
  in growth, 294, 294f
  pain and, 118
  in panic attacks, 461
  psychotherapy and, 524
  in same-sex partners, 347
  in sexual attitudes and behavior, 345–347
  in suicide, 489f, 490
  and suicide attempts, 490
gender expression, 319–320
gender identity, **319**
gender roles, **320**, 322, 428
gender schemas, 320
gender similarities hypothesis, **322**
gene × environment (G × E) interaction, **80**
gene-gene interaction, 78
general adaptation syndrome (GAS), **540**–541, 540f
general events, 227
generalization, **174**, **185**
  in classical conditioning, 174, 461
  concepts and, 249
  inductive reasoning and, 252
  in operant conditioning, 185–186, 186f
  panic attacks and, 461
generalized anxiety disorder, **459**–460, 516f
generative people, 240
generativity *vs.* stagnation, 312f, 318
genes, 8, **76**–77, 76f, 481–482
  aggression and, 423
  and behavior, 81, 83
  chromosomes/DNA and, 76, 76f
  dominant-recessive genes principle, 77
  and environment, 79–81
  and intelligence quotient, 267–268
  linkage analysis, 78
  number of, in humans, 76
  role in affectionate communication, 78
genetic
  behavior, 78–79, 79f
  behavior genetics, 377, 385, 398
  chromosomes/DNA/genes and, 76–77, 76f
  and DNA testing, 81
  dominant-recessive genes principle, 77
  environment and, 79–81
  evolution and, 8
  genome-wide association method, 78
  genotypes, 80
  intelligence and, 265–267, 266f
  Mendel's early discoveries in, 77
  molecular, 77
  phenotypes, 80
  selective breeding, 77–78
genetic disorders
  Down syndrome, 269f, 270
  in ob mice, 336
  phenylketonuria, 286
  Williams syndrome, 276

genetic essentialism, 81
genetic expression, 80
genetic influences
  on aggression, 423
  on antisocial personality disorder, 485
  on anxiety disorders, 460, 463
  on bipolar disease, 475
  on eating disorders, 477–478
  on happiness, 77, 364
  on intelligence, 265–267, 266f
  on obesity, 337
  on personality, 398
  on schizophrenia, 482, 482f
  on suicide, 488
genetic mutations, 81
genital herpes, 554f
genital stage, 373–374
genome, 77
genome-wide association method, 78, 398
genotype, 286
genotypes, **80**
genuineness, 380, 506
germinal period, 289
gestalt psychology, **104**, 104f
giftedness, **268**–269
girls. *See* women and girls
glands, **71**
glial cells, **47**
global brain workspace, 128
glucose, 336
glutamate, 53
goals
  happiness and, 366
  health, 534
  life, 127
  in life, 534
  self-regulation and, 352–353
  subgoals, 250
gonorrhea, 554f
gratification, delay of, 392
grieving
  Bonanno's theory of, 327
  patterns of, 327
gripping reflex, 291f
group differences, in intelligence, 267–268
group influence, 434–437
  decision making, 436–437
group performance, 435–436
group polarization effect, **436**
group results, individual needs and, 37
group therapies, **520**–521
groupthink, **436**–437
growth hormone, sleep and, 140
growth mindset, 200–201
growth spurt, in puberty, 294, 294f

H

habit breaking, 176
habituation, drug, **178**–180, 179f
hair cells, 112, 112f
hallucinations, **480**
hallucinogens, **157**–158, 159f
hammer, 111, 111f
happiness, 26, 363
  biological factors in, 364
  brain activity and, 56
  explaining unhappiness, 380
  goals and, 366
  hedonic treadmill and, 364
  money and, 365
  neuroticism and extraversion and, 386–388
  in nonindustrialized cultures, 20–21
  obstacles in pursuit of, 364–365
  operational definitions for, 16
  Satisfaction with Life Scale, 16, 20

happiness gene, 77
hardiness, 547
hashish, 157
healing, parasympathetic nervous system and, 354, 355f
health behaviors, **530**
health disparities, 544
health promotion, 528
health psychology, 13f, 27, **528**–556
  biopsychosocial model in, 529
  cognitive appraisal and, 544, 545–546
  coping in. *See* coping
  defined, 528
  diet and, 550–552
  mind and body relationship in, 529, 555–556
  motivation in, 534–535
  personality characteristics and, 537–538
  physical activity and, 549–550, 549f, 550f
  positive thinking and, 539
  religious faith and, 536–537
  safe sex and, 553–555
  smoking cessation and, 530, 553
  social support and, 535–536
  stages of change model in, 531–533, 531f
  stress and, 540–548, 540f, 542f, 543f
  stress management programs, 548
  supportive environments and, 552
  theoretical models of change, 530–531
  tobacco use, 553
  weight-based bias, 545
  weight loss and, 551, 551f, 552
hearing, absolute threshold in, 90, 91f
heart disease, 141, 529, 538
  hostility and, 542
  smoking and, 553
hedonic treadmill, 364
helping, 325, 420
hemispheres
  association areas, 64f, 67, 128
  Broca's area, 67, 67f
  corpus callosum and, 67–68, 68f
  lateralization in, 68–70
  left hemisphere, 68–69, 69f
  lobes in, 64–65, 64f
  location of, 64f
  right hemisphere, 68, 69, 69f
  split-brain research, 67–70, 69f
  visual processing in, 69f, 100, 101f
  Wernicke's area, 67, 67f
heritability, **265**, 398. *See also* genetic influences
heroin, 151, 290
heuristics, **250**–251
  availability, 254f, **256**
  base rate neglect, 254f, 256
  decision making and, 254–257, 254f
  problem solving and, 250–251
  representativeness, 254f, **256**, **307**
  in social information processing, 413
HEXACO model, 386
hierarchy of needs, **349**, 350f
higher-level consciousness, 129–130, 131f
hindbrain, **59**–61, 60f
hindsight bias, 254f, **255**
hippocampus, **62**
  exercise and, 296
  location and function of, 60f
  and memory, 44
  memory and, 222, 222f, 467
  neurogenesis in, 73
  REM sleep in, 138
history of psychology, 6–8, 6f, 7f

N

NAc. *See* nucleus accumbens (NAc)
Naloxone, 152
naps, 218
narcissistic personality disorder, 485*f*
narcolepsy, 143
narcotics (opiates), 151–153, 159*f*
Nardil, 514
NaSSAs (noradrenergic and specific
    serotonergic antidepressants), 514
National Weight Control Registry, 551
Native Americans, 449, 473,
    489, 489*f*
nativist approach, 304–305
natural disasters, 466
naturalistic observation, **33**
naturalist intelligence, 271
natural killer (NK) cells, 543, 543*f*
natural selection, 7–8
nature, in development, **286**–288. *See also*
    genetic influences
    Piaget on, 304
Nazi prisoner experiments, 34
need, **333**
negative affect, 310, **362**
negative correlations, 21, 22*f*
negative emotions, 543
negative punishment, **188**, 189*f*
negative reinforcement, **183**–185,
    184*f*, 189*f*
neglectful parenting, 314
Nembutal, 151
neocortex, **64**
neologisms, **480**
nervous system, **43**–46, 82.
    See also brain; neurons
    central, **45**, 45*f*, 82
    characteristics of, 43–44
    defined, 43
    divisions of, 45–46, 45*f*
    embryological development of, 58, 60*f*
    glial cells, 47
    neural networks, 44–45, 55, 55*f*,
      145–146, 220
    neurogenesis, 73
    neurons. *See* neurons
    pathways in, 44–45
    peripheral, **45**–46, 45*f*, 82
    sympathetic. *See* sympathetic nervous
      system
neural impulse, 48–50, 49*f*, 50*f*
neural networks, **44**–45, 55, 55*f*,
    145–146
    example of, 55*f*
    long-term memory and, 220
neural tube developmen, 289
neurobiological factors, in aggression, 423
neurochemical messengers, 52–54
neurodevelopmental disorders, 452–459
    attention-deficit/hyperactivity disorder,
      456–459
    autism spectrum disorder, 452–456
neurodiversity, 453
neurogenesis, 73, 298
neuroleptics, 515–516, 516*f*
neurology, 85
neurons, 44, **47**–55, 48*f*, 82
    binding by, 101–102
    cell structure of, 47–48
    collateral sprouting, 73
    feature detectors and, 101
    glial cells and, 47
    graded currents from sensory
      receptors, 88
    memory and, 221–222
    mirror, 47
    neural impulse, 48–50, 49*f*, 50*f*
    neurogenesis, 73

sleep and, 141
synapses and neurotransmitters,
    50–54, 51*f*
    in visual processing, 101
neuroscience, **9**–10, 43, 59
neuroscientists, 43
neuroticism
    HPA and, 397
    and mortality, 384
    neurotransmitters and, 396–398
    in reinforcement sensitivity theory,
      396, 396*f*
    serotonin and, 397
    subjective well-being and, 386–388
    in trait theories, 383, 383*f*, 386–388
Neuroticism Extraversion Openness
    Personality Inventory-Revised
    (NEO-PI-R), 400
neurotransmitters, **51**–52. *See also specific*
    neurotransmitters
    cocaine and, 156, 156*f*
    depression and, 472–473
    drugs and, 54, 151
    emotions and, 356–357, 357*f*
    examples of, 52–54
    hunger and, 337
    MAO inhibitors and, 514
    midbrain and, 61
    oxytocin, 503
    panic attacks, 460–461
    personality and, 396–398
    reuptake of, 52
    schizophrenia and, 481
    sexual behavior and, 341
    sleep and, 139, 144
    working of, 51*f*
NFL players, and CTE, 74
nicotine, 152–155, 154*f*, 155*f*, 159*f*, 553.
    See also tobacco
nicotine gum, 553
nicotine inhaler, 553
nicotine patch, 553
nicotine spray, 553
nightmares, 143
night terror, 143
NIHL (noise-induced hearing loss), 113
nine-dot problem, 252*f*, 283*f*
9/11 terrorist attacks, 227–228, 228*f*, 437
NK (natural killer) cells, 543, 543*f*
nodes (locations of neural activity), 220
noise, **91**
noise-induced hearing loss (NIHL), 113
nonbinary gender, 321
nonconscious thought. *See* unconscious
    thought
nondeclarative memory, **217**. *See also*
    implicit memory
nondirective therapy, 506
noradrenergic and specific serotonergic
    antidepressants (NaSSAs), 514
norepinephrine, 53
    adrenal glands and, 72
    bipolar disorder and, 475
    cocaine and, 156
    depression and, 53, 473, 474*f*, 514
    Ecstasy and, 156
    functions of, 53, 72
    panic disorder and, 460
    sleep and, 139
"normalcy," views of, 448
normal distribution, **263**, 264*f*
normative social influence, **431**
nostalgia, 204
novel stimuli, 93
nucleus accumbens (NAc)
    addiction and, 147, 148*f*
    alcohol and, 149
    cocaine and, 156*f*
    in reward pathway, 147, 148*f*, 519

Nun Study, 298–299, 298*f*
nurture, in development, **286**, 287–288.
    See also environment
    Piaget on, 304
nyctophobia, 462*f*

O

obedience, **432**–433, 432*f*, 433*f*
obesity, 337–339, 531*f*, 532. *See also*
    weight loss
    biology of, 337
    diet and, 551–552
    leptin and, 336, 336*f*
    portion size and, 339
    psychological factors in, 337–339
    rates, 552
    sleep and, 141
    and weight-based bias, 545
obesogenic environment, 552
objective evaluation of work, 260
objectivity, 4, 260
object permanence, **300**, 301*f*
ob mice, 336
observation
    in descriptive research, 18
    naturalistic, **33**
    in scientific method, 14–15, 15*f*
observational learning, **170**, 191–194,
    194*f*, 390
    aggression and, 424
    Bandura's model of, 191–192,
      194*f*, 390
    and interpersonal biases, 193
obsessions, 464
obsessive-compulsive disorder (OCD),
    459, 463–**464**, 485*f*
    operant conditioning and, 508
    related disorders, 465
occipital lobes, **64**, 64*f*, 66*f*, 89
    hypnosis and, 161
    visual information in, 89
occupational therapist, 499*f*
Oedipus complex, **373**
Old Order Amish, 20–21
Olds studies, 63
olfactory bulb, 73
olfactory epithelium, **120**, 120*f*
olfactory sense, 120–121, 120*f*
Omega-3 fatty acids, 266
On Death and Dying (Kübler-Ross), 327
One Flew Over the Cuckoo's Nest
    (film), 517
one-third rule for alcoholism, 151
online support groups, 522
On the Origin of Species (Darwin), 7
open-mindedness, **258**
open monitoring, 163
openness, 383*f*, 384, 387
    neurotranmitters and, 397
operant conditioning, 180–191, **181**
    applied behavior analysis, 190–191
    classical conditioning *vs.*, 170, 170*f*
    discrimination in, 186
    extinction in, 186
    generalization in, 185–186, 186*f*
    learning and, 170, 170*f*
    positive and negative reinforcement in,
      183–185, 184*f*, 189*f*
    punishment in, 188, 189*f*
    reinforcement, principles of. *See*
      reinforcement
    shaping in, 183
    Skinner's approach to, 182, 182*f*
    as therapy, 507–508
    Thorndike's law of effect, 181, 182*f*
    timing and consequences of behavior,
      188–190
operational definition, **16**–17

operations, in Piaget's theory, 301*f*, 302
ophidiophobia, 462*f*
ophthalmology, 85
Opiate Crisis Intervention Court, 153
opiates. *See* opioids
opioids, **151**–153, 158, 159*f*
opium, 54
opponent-process theory, **103**, 103*f*
optical illusions, 108, 108*f*
optic chiasm, 100, 101*f*
optic nerve, 98*f*, **99**, 99*f*
optimism, 538, 547
optimistic attributional style, 473
optimum arousal theory, 333–335, 334*f*
oral sex, 344–345
oral stage, 373, 374*f*
organic intellectual disability, 270
orgasm, 54, 341
outer ear, **110**, 111*f*
out-group, 438, 441–442, 442*f*
oval window, 111, 112*f*
ovaries, 71*f*, **72**
overgeneralizing, 37, 509
overlaps, 105
overlearning, 334*f*, 335
overt aggression, **424**–425
oxytocin, 54, 341, 463, 503

P

packaged foods, 340
PADs (psychiatric advance
    directives), 519
pain, **117**–118
    expectation for reduction of, 195
    hypnosis and, 161–162, 161*f*
    meditation and, 164
    neurotransmitters and, 54
    perception of, 117, 162
    treatment of, 151, 159*f*
pancreas, 71*f*, **72**
panic attacks, 164, 460–461
panic disorder, **460**–461, 461*f*, 516*f*
papillae, **118**
parallel distributed processing
    (PDP), **220**
parallel processing, **101**, 132
Paralympics, 351*f*
paranoid personality disorder, 485*f*
paraprofessionals, 500, 521
parasympathetic nervous system,
    45*f*, **46**, 354, 355*f*
parathyroid gland, 71*f*
parents/parenting
    adolescent development and, 314–316
    childhood socioemotional development
      and, 313–314
    cultural context of, 314
    generativity and, 312*f*, 318
    oxytocin and, 54
    same-sex, 348
    styles, 313–314
    teenagers as, 343
parietal lobes, 64*f*, **65**, 66*f*, 89, 161
Parkinson disease, 53, 61, 73
partial reinforcement, 186
passionate love, **428**
pastoral counselor, 499*f*
pattern recognition, 108*f*
Pavlov's studies, 172–175, 172*f*, 173*f*
Paxil, 514
PDP (parallel distributed processing), **220**
peak experiences, 379
peer pressure, 300
peer review, 38
peers, identity and, 316
penis envy, 373, 375
penny exercise, 234, 234*f*
perceived similarity, 427

sensation, **85**–86, 124. *See also* perception
  bottom-up processing, **86**–87, 86*f*
  brain and, 87–90, 88*f*
  hypnosis and, 157–158
  information flow in, 89*f*
  kinesthetic senses, 121–124
  noise in, 91
  purposes of, 87
  sensory adaptation, 95
  and social psychology, 119
  specialization in psychology, 13*f*
  thresholds in, 90–91*f*, 90–93
  top-down processing, **86**–87, 86*f*
  vestibular sense, 121–124, 122*f*
sensorimotor cortex, 161
sensorimotor stage, **300**, 301*f*
sensory adaptation, **95**
sensory memory, **211**–212, 211*f*
  short-term memory *vs.*, 212
  Sperling's experiment, 211–212, 211*f*
sensory nerves, 44–46
sensory processing approach, to autism-
    spectrum disorder, 454
sensory receptors, **87**–90, 88*f*
September 11 attack, 227–228, 228*f*, 437
serial position effect, 224, 224*f*
serotonin
  aggression and, 397, 423
  antipsychotic drugs, 514
  cocaine and, 156
  depression and, 463, 514, 515*f*
  eating disorders and, 478
  Ecstasy and, 156
  functions and pathways of, 53–54, 53*f*
  hunger and, 337
  LSD and, 157
  memory and, 221
  neuroticism and, 397
  sleep and, 139
  social phobia and, 463
  suicide and, 488
service dogs, 168, 186
set point, **337**, 364
sex and sexuality
  alcohol and, 149
  assigned, 294
  attitudes and practices, 344
  biology of, 340–342
  cognitive and sensory/perceptual
      factors in, 342
  cultural factors in, 342–344
  education, 343–344
  Freud on, 370–371
  gender differences in attitudes,
      345–347
  human sexual response pattern,
      341–342
  oral, 344
  preventing unwanted pregnancies, 343
  safe sex, 553–555
  sex hormones, 341
  sexual behavior and orientation, 344–349
  sexually transmitted infections and,
      343, 553–555, 554*f*
sex hormones
  androgens, 341
  estrogens, 295–296, 341
  sex drive and, 341
  testosterone, 295, 341, 423
sexual abuse. *See also* child sexual abuse
  dissociative identity disorder and, 470
sexual arousal, 341, 344, 346–347
sexual behavior
  defining, 344
  frequency of, 345
  research on, 344–347
sexually transmitted infections (STIs),
    343, **553**–555, 554*f*
sexual orientation, **319**, 347–349

sexual scripts, 342
sexual victimization
  on campuses, 468
  PTSD and, 467
shading, 105
shape constancy, 107, 107*f*
shaping, **183**
shiftable attention, 93
shift work, sleep and, 135
shock therapy, 517
short-term memory, **212**–215, 214*f*
SIDS (sudden infant death syndrome),
    143–144
signal detection theory, **92**–93, 93*f*
similarity
  actual, 427
  gestalt principle of, 104*f*
  perceived, 427
single research studies, 38
single-unit recording, 56
situationism, 391
six-matchstick problem, 252*f*, 283*f*
60/60 rule, 113
size constancy, 107, 107*f*
skepticism, in psychology, 3
skin conductance level (SCL), 355
Skinner box, 182, 183*f*
Skinner's approach to operant
    conditioning, 182, 182*f*, 183*f*
skin senses, 116–118, 125
sleep, **133**–144, 166. *See also* dreams
  amount needed by animals, 135*f*
  biological rhythms and, 133–135, 134*f*
  and brain, 139
  COVID-19 pandemic and, 136–137
  cycling, 139, 139*f*
  deprivation of, 136–137
  and developmental psychology, 218
  and disease, 141
  disorders, 141–144
  function of, 135–136
  and memory, 136
  need for, 135–136
  neurotransmitters and, 53
  REM, 137, **138**–139, 138*f*
  stages of, 137–139, 138*f*
  subconscious awareness during, 132
  throughout life span, 140–141, 140*f*
sleep apnea, 143–144
sleep debt, 136
sleep eating, 142
sleep spindles, 138*f*, 139
sleep talking, 138, 142
sleepwalking, 138, 142
slow pathway, for pain perception, 117
slow-to-warm-up child, 308
slow-wave sleep, 138
smell, 120–121, 120*f*
  absolute threshold in, 91*f*
  emotion and, 121
  "smelly T-shirt" paradigm, 121
smiling, 278, 279*f*
smoking, 152–155, 154*f*, 155*f*, 190, 553.
      See also tobacco
snakes, fear of, 462, 462*f*
SNS. *See* sympathetic nervous
    system (SNS)
social anxiety disorder, **463**
social behavior, 418–426
  altruism, 418–422
social bonding
  memory and, 241
  oxytocin and, 54
social category, 267
social cognition, **409**–418. *See also* social
    psychology
  person perception, 409–412
social cognitive behavior view of
    hypnosis, **162**

social cognitive perspectives, **389**–394
  Bandura's social cognitive theory,
      389–390, 389*f*
  CAPS theory, 392–394
  evaluation of, 394
  on hypnosis, 162
  Mischel's critique of consistency,
      391–392
  observational learning, 390
  personal control and, 390
  on personality, 389–394
  self-efficacy and, 390
social comparison, **414**
social construction, dissociative, 470
social contagion, **435**
social desirability, 399–400
social distance, during COVID-19
    pandemic, 406
social exchange theory, **429**
social facilitation, 435–**436**
social identity, **437**–438, 438*f*
social identity theory, **438**
social information processing, autism
    and, 455
social interactions, mental illness
    and, 493
social interest, 377
social learning theory, 192–193
social loafing, **436**
social media
  adolescent egocentrism and, 306
  downsides of "support" on, 522
  language on, 273
  and misinformation about
      COVID-19, 27
social motivation approach, to autism-
    spectrum disorder, 454
social perception, 269
social phobia, **463**
social psychology, 13*f*, 406–443,
    407–408, 493
  aggression and, 422–426
  altruism and, 418–422
  attitudes and, 414–416, 416*f*
  attraction and, 427–428
  attribution theory and,
      412–413, 412*f*
  children's stereotypes, 338
  close relationships and, 426–430
  cognitive dissonance theory and,
      415–416
  conformity and, 430–432, 431*f*
  consciousness and, 130
  group influence and, 434–437
  heuristics in social information
      processing, 413
  intergroup relations and, 441–443
  learning and, 193
  love and, 428–429
  and motivation, 420
  obedience and, 432–433, 432*f*, 433*f*
  person perception and, 409–412
  persuasion and, 416–418
  prejudice and, 438–441, 443
  self as a social object and, 413–414
  sensation/perception and, 119
  social behavior and, 418–422
  social identity and, 437–438, 438*f*
  weight-based bias, 545
social role view of gender, 428
social salience hypothesis, 503
social sharing, 522, 536
social skills, 270, 521
social support, **535**–536
social workers, 498, 499*f*, 500
sociocultural approach, **12**, 375, 375*f*
  binge eating disorder, 478
  to psychological disorders, 449
  in treatment, 520–524

sociocultural factors
  personality and, 374
  in suicide, 489–490, 489*f*
sociocultural therapies, 500
socioeconomic status, 421–422
socioemotional development, 309–319
  Erikson's theory, 311–319, 312*f*
  gender development and, 322
  in infancy, 309–311, 310*f*
  moral, 325
socioemotional processes, in
    development, 288
socioemotional selectivity theory, 318
sociology, 407
sodium ions, 50, 50*f*
somatic nervous system, **45**–46, 45*f*
somatic symptom and related
    disorders, **451**
somnambulism, 142
somniloquy, 142
sound, 109–110, 110*f*. *See also*
    auditory system
  intensity of, 115
  localizing, 114–115, 115*f*
  timing of, 115
sound shadow, 115, 115*f*
spam emails, 171
Spanish language, memory for,
    216, 216*f*
spatial intelligence, 271
specific phobia, 461–**462**, 462*f*
spectrum, 453
speech, 67, 67*f*, 68, 69*f*
Sperling's sensory memory experiment,
    211–212, 211*f*
spina bifida, 289
split-brain research, 67–70, 69*f*.
    See also hemispheres
splitting, 487
spontaneous recovery, in classical
    conditioning, **175**, 175*f*
sport psychology, 13*f*
SSRIs (selective serotonin reuptake
    inhibitors), 514, 516*f*
stages of change model, **531**–533, 531*f*
standardization, of tests, **262**, 262*f*
Stanford-Binet intelligence test,
    262–263, 264*f*
Stanford Prison Experiment, 433–434
A Star in His Own Imagination
    (Allen), 126
startle reflex, 291*f*
Star Wars: The Last Jedi (film), 3
statistics, in data analysis, 17
stem cells, **75**
stereotypes, 87, 256, **410**
  "beautiful is good," 410
  children's, 338
  explicit, 338
  implicit, 338
  negative, 545
  prejudice and, 441
  self-fulfilling prophecies and,
      410–411
  social perception and, 269
stereotype threat, **411**–412
sticky mittens, 292, 292*f*
stigma of psychological disorders,
    490–494, 491*f*
stimulants, **152**–156, 159*f*. *See also*
    psychoactive drugs
  caffeine, 152, 159*f*
  cocaine, 155–156, 156*f*, 159*f*
  MDMA (Ecstasy), 156, 159*f*
  nicotine, 152–155, 154*f*, 155*f*, 159*f*
stimuli, perception of, 93–95
stirrup, 111, 111*f*, 112*f*
STIs (sexually transmitted infections),
    343, **553**–555, 554*f*